COLD SPRING HARBOR SYMPOSIA ON QUANTITATIVE BIOLOGY

VOLUME XLV—PART 1

COLD SPRING HARBOR SYMPOSIA ON QUANTITATIVE BIOLOGY

VOLUME XLV

MOVABLE GENETIC ELEMENTS

COLD SPRING HARBOR LABORATORY
1981

COLD SPRING HARBOR SYMPOSIA ON QUANTITATIVE BIOLOGY
VOLUME XLV

International Standard Book Number 0-87969-044-5
Library of Congress Catalog Card Number 34-8174

Printed in the United States of America

COLD SPRING HARBOR SYMPOSIA ON QUANTITATIVE BIOLOGY

Founded in 1933 by
REGINALD G. HARRIS
Director of the Biological Laboratory 1924 to 1936

Previous Symposia Volumes

I (1933) Surface Phenomena
II (1934) Aspects of Growth
III (1935) Photochemical Reactions
IV (1936) Excitation Phenomena
V (1937) Internal Secretions
VI (1938) Protein Chemistry
VII (1939) Biological Oxidations
VIII (1940) Permeability and the Nature of Cell Membranes
IX (1941) Genes and Chromosomes: Structure and Organization
X (1942) The Relation of Hormones to Development
XI (1946) Heredity and Variation in Microorganisms
XII (1947) Nucleic Acids and Nucleoproteins
XIII (1948) Biological Applications of Tracer Elements
XIV (1949) Amino Acids and Proteins
XV (1950) Origin and Evolution of Man
XVI (1951) Genes and Mutations
XVII (1952) The Neuron
XVIII (1953) Viruses
XIX (1954) The Mammalian Fetus: Physiological Aspects of Development
XX (1955) Population Genetics: The Nature and Causes of Genetic Variability in Population
XXI (1956) Genetic Mechanisms: Structure and Function
XXII (1957) Population Studies: Animal Ecology and Demography
XXIII (1958) Exchange of Genetic Material: Mechanism and Consequences
XXIV (1959) Genetics and Twentieth Century Darwinism
XXV (1960) Biological Clocks
XXVI (1961) Cellular Regulatory Mechanisms
XXVII (1962) Basic Mechanisms in Animal Virus Biology
XXVIII (1963) Synthesis and Structure of Macromolecules
XXIX (1964) Human Genetics
XXX (1965) Sensory Receptors
XXXI (1966) The Genetic Code
XXXII (1967) Antibodies
XXXIII (1968) Replication of DNA in Microorganisms
XXXIV (1969) The Mechanism of Protein Synthesis
XXXV (1970) Transcription of Genetic Material
XXXVI (1971) Structure and Function of Proteins at the Three-dimensional Level
XXXVII (1972) The Mechanism of Muscle Contraction
XXXVIII (1973) Chromosome Structure and Function
XXXIX (1974) Tumor Viruses
XL (1975) The Synapse
XLI (1976) Origins of Lymphocyte Diversity
XLII (1977) Chromatin
XLIII (1978) DNA: Replication and Recombination
XLIV (1979) Viral Oncogenes

The Symposium Volumes are published by The Cold Spring Harbor Laboratory, Box 100, Cold Spring Harbor, New York 11724.

Symposium Participants

ABOU-SABE, MORAD A., Dept. of Microbiology, Rutgers University, New Brunswick, New Jersey
AHMED, ASAD, Dept. of Genetics, University of Alberta, Edmonton, Canada
ANISIMOV, P. I., Antiplague Research Institute, Saratov, USSR
APPLEBY, DAVID, Institute of Toxicology, Albany Medical College, New York
APPLEBY, NORMA, Institute of Toxicology, Albany Medical College, New York
ARBER, WERNER, Biozentrum, University of Basel, Switzerland
ARNHEIM, NORMAN, Dept. of Biochemistry, State University of New York, Stony Brook
ATKINS, JOHN F., Dept. of Biology, University of Utah, Salt Lake City
AUERSWALD, E. A., Dept. of Microbiology, University of Heidelberg, Federal Republic of Germany
AUGUSTINE, ANDREI, Dept. of Microbiology, Columbia University College of Physicians and Surgeons, New York, New York
AUSUBEL, FREDERICK, The Biological Laboratories, Harvard University, Cambridge, Massachusetts
BABCOCK, MARLA, D-2 Carolina Apts., Carrboro, North Carolina
BADE, ERNESTO, Faculty of Biology, University of Konstanz, Federal Republic of Germany
BAKER, ROBERT, Dept. of Molecular Biology, University of Southern California, Los Angeles
BALTIMORE, DAVID, Dept. of Cancer Research, Massachusetts Institute of Technology, Cambridge
BANK, ARTHUR, Hammer Health Sciences Center, Columbia University, New York, New York
BEDBROOK, J. R., Plant Industry Division, CSIRO, Canberra, Australia
BENNETT, P. M., Dept. of Bacteriology, University of Bristol, England
BENSON, SPENCER, Dept. of Cancer Biology, Frederick Cancer Research Center, Frederick, Maryland
BERG, CLAIRE M., Dept. of Biological Sciences, University of Connecticut, Storrs
BERG, DOUGLAS, Dept. of Microbiology, Washington University Medical School, St. Louis, Missouri
BERNARDI, ALBERTO, Genetics Center, CNRS, Gif-sur-Yvette, France
BERNINGER, MARK, Bethesda Research Laboratories, Inc., Rockville, Maryland
BERTANI, GIUSEPPE, Dept. of Microbial Genetics, Karolinska Institute, Stockholm, Sweden
BERTRAND, KEVIN, Dept. of Microbiology, University of California, Irvine
BESEMER, JURGEN, Dept. of Genetics, University of Köln, Federal Republic of Germany
BIEK, DONALD, Dept. of Biology, University of Utah, Salt Lake City
BIEZUNSKI, NAOMI, Dept. of Biology, Case Western Reserve University, Cleveland, Ohio
BINGHAM, PAUL, NIEHS, National Institutes of Health, Research Triangle Park, North Carolina
BIRSHTEN, BARBARA, Dept. of Cell Biology, Albert Einstein College of Medicine, Bronx, New York
BLAKESLEY, ROBERT, Bethesda Research Laboratories, Inc., Rockville, Maryland
BLOCK, KARIN, Institute of Genetics, University of Lund, Sweden
BOCCARA, MARTINE, INSERM, Pasteur Institute, Lille, France
BOISTARD, PIERRE, Dept. of Vegetable Pathology, Institut National de la Recherche Agronomique, Versailles, France
BORST, PIET, Dept. of Biochemistry, University of Amsterdam, Holland
BOTCHAN, MICHAEL, Dept. of Molecular Biology, University of California, Berkeley
BOTSTEIN, DAVID, Dept. of Biology, Massachusetts Institute of Technology, Cambridge
BOURRET, ROBERT, Dept. of Biology, Massachusetts Institute of Technology, Cambridge
BRENNER, SYDNEY, Laboratory of Molecular Biology, Medical Research Council, Cambridge, England
BROREIN, WILLIAM, JR., 1199 Beacon St., Brookline, Massachusetts
BUKHARI, AHMAD I., Cold Spring Harbor Laboratory, New York
BURR, BENJAMIN, Dept. of Biology, Brookhaven National Laboratory, Upton, New York
BURTON, WILLIAM, Bethesda Research Laboratories, Inc., Rockville, Maryland
CABEZON, TERESA, Smith Kline-RIT, Rixensart, Belgium
CAIRNS, JOHN, Imperial Cancer Research Fund Laboratories, London, England
CALOS, MICHELE, Dept. of Molecular Biology, University of Geneva, Switzerland
CAMPBELL, ALLAN M., Dept. of Biological Sciences, Stanford University, California
CANNON, FRANK, ARC Unit of Nitrogen Fixation, University of Sussex, Brighton, England
CARDILLO, THOMAS, Dept. of Radiation Biology and Biophysics, University of Rochester, New York
CARLSON, MARIAN, Dept. of Biology, Massachusetts Institute of Technology, Cambridge

CASADABAN, MALCOLM, Dept. of Biophysics and Theoretical Biology, University of Chicago, Illinois

CASEY, JAMES, Dept. of Chemistry, California Institute of Technology, Pasadena

CHACONAS, GEORGE, Cold Spring Harbor Laboratory, New York

CHANDLER, MICHAEL, Dept. of Molecular Biology, University of Geneva, Switzerland

CHATTORAJ, DHRUBA, NICHHD, National Institutes of Health, Bethesda, Maryland

CLOWES, ROYSTON, Dept. of Biology, University of Texas at Dallas, Richardson

COHEN, S. N., Dept. of Genetics, Stanford University School of Medicine, California

COLLINS, JOHN, Gessellschaft Biotechnologische fur Schungsstehle, Braunschweig, Federal Republic of Germany

COMEAU, ANNE M., Dept. of Biology, Brandeis University, Waltham, Massachusetts

CORNELIS, GUY, Dept. of Microbiology, Catholic University of Louvain, Brussels, Belgium

COZZARELLI, NICHOLAS, Dept. of Biochemistry, University of Chicago, Illinois

CROW, JAMES F., Dept. of Genetics, University of Wisconsin, Madison

DATTA, NAOMI, Hammersmith Hospital, Royal Postgraduate Medical School, London, England

DAVIS, MARK M., California Institute of Technology, Pasadena

DAVISON, JOHN, Institute of Cell Pathology, Brussels, Belgium

DE BRUIJN, FRANS J., The Biological Laboratories, Harvard University, Cambridge, Massachusetts

DECARIS, BERNARD, Dept. of Genetics, University of Paris, Orsay, France

DE CROMBRUGGHE, BENOIT, NCI, National Institutes of Health, Bethesda, Maryland

DEMOPULOS-RODRIGUEZ, JAMES, Dept. of Genetics, University of Wisconsin, Madison

DEONIER, RICHARD, Dept. of Molecular Biology, University of South California, Los Angeles

DOOLITTLE, W. FORD, Dept. of Biochemistry, Dalhousie University, Halifax, Canada

DOONER, HUGO K., Dept. of Genetics, University of Wisconsin, Madison

DOWSETT, ANDREW, Dept. of Genetics, Rockefeller University, New York, New York

DRESSLER, DAVID, The Biological Laboratories, Harvard University, Cambridge, Massachusetts

DUBOW, MICHAEL, Cold Spring Harbor Laboratory, New York

DUJON, BERNARD, The Biological of Laboratories, Harvard University, Cambridge, Massachusetts

DUNSMUIR, PAMELA, Sidney Farber Cancer Institute, Boston, Massachusetts

EDEN, FRANCINE, NCI, National Institutes of Health, Bethesda, Maryland

EGEL, RICHARD, Institute of Genetics, University of Copenhagen, Denmark

ELDER, ROBERT, Stanford University, California

EMERICK, ANNE, Cetus Corporation, Berkeley, California

ENDOW, SHARYN, Dept. of Microbiology and Immunology, Duke University Medical Center, Durham, North Carolina

ENGELS, WILLIAM, Dept. of Genetics, University of Wisconsin, Madison

ERNST, JOACHIM, Dept. of Radiological Biology and Biophysics, University of Rochester, New York

ERREDE, BEVERLY, Dept. of Radiation Biology and Biophysics, University of Rochester, New York

FAELEN, MICHAEL, Dept. of Genetics, Université Libre de Bruxelles, Rhode-St.-Genèse, Belgium

FARABAUGH, PHILIP, Dept. of Biochemistry, Cornell University, Ithaca, New York

FASY, THOMAS M., Dept. of Pathology, Mt. Sinai School of Medicine, New York, New York

FEDOROFF, NINA, Dept. of Embryology, Carnegie Institution of Washington, Baltimore, Maryland

FEINGOLD, JAY, Dept. of Molecular Biology, Albert Einstein College of Medicine, Bronx, New York

FENNEWALD, MICHAEL, Dept. of Biochemistry, University of Chicago, Illinois

FINCHAM, JOHN R. S., Dept. of Genetics, University of Edinburgh, Scotland

FINK, GERALD, Cornell University, Ithaca, New York

FLAVELL, ANDREW, Sidney Farber Cancer Institute, Boston, Massachusetts

FLAVELL, RICHARD B., Plant Breeding Institute, Cambridge, England

FLEISSNER, ERWIN, Sloan-Kettering Institute for Cancer Research, New York, New York

FLYNN, ANN E., New England Biological Laboratories, Inc., Beverly, Massachusetts

FOSTER, T. J., Dept. of Microbiology, Trinity College, Dublin, Ireland

FOX, MAURICE, Dept. of Biology, Massachusetts Institute of Technology, Cambridge

FRANKE, ARTHUR, Dental Research Institute, University of Michigan, Ann Arbor

FRIEDMAN, DAVID, Dept. of Microbiology, University of Michigan, Ann Arbor

FRIEDMAN, LINDA, Dept. of Radiation Biology and Biophysics, University of Rochester, New York

FRITSCH, EDWARD, Dept. of Biology, California Institute of Technology, Pasadena

FURTEK, DOUGLAS, Dept. of Genetics, University of Wisconsin, Madison

GALAS, DAVID, Dept. of Molecular Biology, University of Geneva, Switzerland

GELLERT, MARTIN, NIAMDD, National Institutes of Health, Bethesda, Maryland

GLANSDORFF, NICOLAS, Dept. of Microbiology, University of Brussels, Belgium

GLUZMAN, YAKOV, Cold Spring Harbor Laboratory, New York

GOLDBERG, GREGORY, Dept. of Genetics, University of Wisconsin, Madison

GOLDBERG, S., Columbia University Medical School, New York, New York

GREER, HELEN, The Biological Laboratories, Harvard University, Cambridge, Massachusetts
GRINDLEY, NIGEL D., Dept. of Biological Sciences, University of Pittsburgh, Pennsylvania
GRINSTED, JOHN, Dept. of Bacteriology, University of Bristol, England
GRONKOVA, ROSA, Dept. of Microbiology, University of Pennsylvania, Philadelphia
GUYER, MARK S., NIAMDD, National Institutes of Health, Bethesda, Maryland
HABER, JAMES D., Rosenstiel Center, Brandeis University, Waltham, Massachusetts
HARSHEY, RASIKA, Cold Spring Harbor Laboratory, New York
HARRIMAN, PHILIP D., Genetic Biology Program, National Science Foundation, Washington, DC
HEFFRON, FRED, Dept. of Biochemistry, University of California, San Francisco
HELMS, CYNTHIA, Dept. of Microbiology, Rutgers Medical School, Newark, New Jersey
HERSKOWITZ, IRA, Dept. of Molecular Biology, University of Oregon, Eugene
HICKS, JAMES B., Cold Spring Harbor Laboratory, New York
HILL, CHARLES, Hershey Medical Center, Pennsylvania State University, Hershey
HIRSCHEL, BERNARD, Dept. of Microbiology, Washington University School of Medicine, St. Louis, Missouri
HONJO, TASUKU, Dept. of Genetics, Osaka University Medical School, Japan
HOOD, LEE, Dept. of Biology, California Institute of Technology, Pasadena
HOWE, MARTHA, Dept. of Bacteriology, University of Wisconsin, Madison
HSU, PEI-LING, Dept. of Biology and Medicine, Brown University, Providence, Rhode Island
HUGHES, D. G., Beatson Institute for Cancer Research, Glasgow, Scotland
HUNTSMAN, MARY, Cetus Corporation, Berkeley, California
IGO, MICHELE, The Biological Laboratories, Harvard University, Cambridge, Massachusetts
IIDA, SHIGERU, Biozentrum, University of Basel, Switzerland
IINO, TETSUO, Dept. of Biology, University of Tokyo, Hongo, Japan
IKEDA, HIDEO, Institute of Medical Science, University of Tokyo, Takanawa, Japan
ILYINA, T. S., Gamaleya Institute, Academy of Sciences USSR, Moscow
ISBERG, RALPH R., Dept. of Microbiology and Molecular Genetics, Harvard Medical School, Boston, Massachusetts
ISING, G., Institute of Genetics, University of Lund, Sweden
JACKSON, JAMES, Institute for Cancer Research, Columbia University, New York, New York
JASKUNAS, S. R., Dept. of Chemistry, Indiana University, Bloomington
JIMENEZ, ANTONIO, Dept. of Biochemistry, University of Wisconsin, Madison
JOHNSON, REID, Dept. of Biochemistry, University of Wisconsin, Madison
JONES, KENNETH W., Dept. of Animal Genetics, University of Edinburgh, Scotland
JORGENSEN, RICHARD, Dept. of Plant Biology, Carnegie Institution of Washington, Stanford, California
KAHMANN, REGINE, Dept. of Biochemistry, Max-Planck-Institut für Biochemie, Munich, Federal Republic of Germany
KAMP, DIETMAR, Dept. of Biochemistry, Max-Planck-Institut für Biochemie, Munich, Federal Republic of Germany
KAPLAN, DONALD K., Dept. of Microbiology, University of California, Los Angeles
KENYON, CYNTHIA, Massachusetts Institute of Technology, Cambridge
KIDWELL, MARGARET, Dept. of Biology and Medicine, Brown University, Providence, Rhode Island
KIM, BYUNG DONG, Dept. of Biochemistry and Molecular Biology, University of Florida, Gainesville
KLAR, AMAR, Cold Spring Harbor Laboratory, New York
KLECKNER, NANCY, The Biological Laboratories, Harvard University, Cambridge, Massachusetts
KOPECKO, DENNIS J., Dept. of Bacterial Immunology, Walter Reed Army Institute of Research, Washington DC
KORNBLUM, JOHN, Public Health Research Institute, New York, New York
KREISWIRTH, BARRY, Public Health Research Institute, New York, New York
KRETSCHMER, PETER, Bethesda Research Laboratories, Inc., Rockville, Maryland
KUROSAWA, YOSHIKAZU, Basel Institute for Immunology, Switzerland
LANDY, ARTHUR, Dept. of Biology and Medicine, Brown University, Providence, Rhode Island
LATT, SAMUEL, Dept. of Genetics, Childrens Hospital and Medical Center, Boston, Massachusetts
LAUX, ROXANNE, Dept. of Microbiology, University of Chicago, Illinois
LEDER, PHILIP, NICHHD, National Institutes of Health, Bethesda, Maryland
LEDERBERG, E. M., Dept. of Medical Microbiology, Stanford University Medical School, California
LEONG, JOHN, Dept. of Biology and Medicine, Brown University, Providence, Rhode Island
LEVIS, ROBERT, Sidney Farber Cancer Institute, Boston, Massachusetts
LEVY, DANIEL, Cold Spring Harbor Laboratory, New York
LEWIN, BENJAMIN, *Cell*, Cambridge, Massachusetts
LEWIS, HERMAN W., National Science Foundation, Washington, DC
LICHTEN, MICHAEL, Dept. of Biology, Massachusetts Institute of Technology, Cambridge
LIM, JOHNG, Dept. of Biology, University of Wisconsin, Eau Claire

LIU, CHIH-PING, Dept. of Genetics, University of Wisconsin, Madison
LOCKETT, TREVOR, Rockefeller University, New York, New York
LONG, ERIC, Dept. of Biochemistry, NCI, National Institutes of Health, Bethesda, Maryland
LOW, BROOKS, Dept. of Radiobiology, Yale Medical School, New Haven, Connecticut
LUSKY, MONIKA, Institute für Biology III, University of Freiburg, Federal Republic of Germany
MACHATTIE, L. A., Dept. of Microbiology, University of Chicago, Illinois
MACHIDA, CHIYOKO, Dept. of Microbiology, State University of New York, Stony Brook
MACHIDA, YASUNORI, Dept. of Microbiology, State University of New York, Stony Brook
MAJORS, JOHN, Dept. of Microbiology, University of California, San Francisco
MAKI, RICHARD, Basel Institute for Immunology, Switzerland
MANS, RUSTY JAY, Dept. of Biochemistry and Molecular Biology, University of Florida, Gainesville
MARCU, KENNETH, Dept. of Biochemistry, State University of New York, Stony Brook
MAXAM, ALLAN M., Sidney Farber Cancer Institute, Boston, Massachusetts
MAZZARA, GAIL, The Biological Laboratories, Harvard University, Cambridge, Massachusetts
MCCLINTOCK, BARBARA, Cold Spring Harbor Laboratory, New York
MCCORMICK, MARY, Dept. of Microbiology, State University of New York, Stony Brook
MCCUSKER, JOHN, Rosenstiel Basic Medical Sciences Research Center, Brandeis University, Waltham, Massachusetts
MCGROGAN, MICHAEL, Dept. of Biological Sciences, Stanford University, California
MEADE, HARRY, Dept. of Microbiology, Merck and Company, Inc., Rahway, New Jersey
MILLER, CHARLES, Dept. of Microbiology, Case Western Reserve University, Cleveland, Ohio
MILLER, HARVEY I., NCI, National Institutes of Health, Bethesda, Maryland
MIRAULT, M. E., Dept. of Molecular Biology, University of Geneva, Switzerland
MIZUUCHI, KIYOSHI, NIAMDD, National Institutes of Health, Bethesda, Maryland
MORITA, CAROL, Cold Spring Harbor Laboratory, New York
MURPHY, ELLEN, Public Health Research Institute, New York, New York
MURRAY, ANDREW, Sidney Farber Cancer Institute, Boston, Massachusetts
NASH, HOWARD, NIMH, National Institutes of Health, Bethesda, Maryland
NASMYTH, KIM, Dept. of Genetics, University of Washington, Seattle
NASSER, DE LILL, Dept. of Genetic Biology, National Science Foundation, Washington, DC
NEVERS, PATRICIA, 3301 Braunschweig-Mascherode, Am Steintore 3, Federal Republic of Germany
NISEN, PERRY, Dept. of Molecular Biology, Albert Einstein College of Medicine, Bronx, New York
NODA, MAKOTO, Dept. of Microbiology, Keio University School of Medicine, Tokyo, Japan
NOVICK, RICHARD, Public Health Institute, New York, New York
NUGENT, MARILYN, Dept. of Bacteriology, Hammersmith Hospital, Royal Postgraduate Medical School, London, England
NYMAN, KATE, Dept. of Microbiology, State University of New York, Stony Brook
OHTSUBO, EIICHI, Dept. of Microbiology, State University of New York, Stony Brook
OHTSUBO, HISAKO, Dept. of Microbiology, State University of New York, Stony Brook
OTA, YOSHIMI, Public Health Research Institute, New York, New York
PATO, MARTIN, National Jewish Hospital Research Center, Denver, Colorado
PERUCHO, MANUEL, Cold Spring Harbor Laboratory, New York
PERRY, ROBERT P., Institute for Cancer Research, Philadelphia, Pennsylvania
PETERSON, PETER A., Dept. of Agronomy, Iowa State University, Ames
PHILIPPSEN, PETER, Biozentrum, University of Basel, Switzerland
PIFFARETTI, J. C., Dept. of Microbiology, University of Geneva, Switzerland
PIRUZIAN, ELEANORA, Dept. of Molecular Genetics, Academy of Sciences USSR, Moscow
POSTLE, KATHLEEN, Dept. of Microbiology, University of California Medical School, Irvine
POTASH, MARY JANE, Basel Institute for Immunology, Switzerland
POTTER, HUNTINGTON, The Biological Laboratories, Harvard University, Cambridge, Massachusetts
PRASAD, ISHWARJ, Dept. of Pathology, Downstate Medical Center, State University of New York, Brooklyn
PRIEFER, URSULA, Dept. of Biology, University of Bielefeld, Federal Republic Germany
PROCTOR, NEAL G., Dept. of Molecular Biology, University of Wisconsin, Madison
PURUCKER, MARY, Dept. of Molecular Biology, Albert Einstein College of Medicine, Bronx, New York
RABBITTS, T. H., Laboratory of Molecular Biology, Medical Research Council, Cambridge, England
RABOY, BIRHA, Dept. of Molecular Biology, Albert Einstein College of Medicine, Bronx, New York
RADDING, CHARLES, Yale University School of Medicine, New Haven, Connecticut
RAK, BODO, Dept. of Biology, University of Freiburg, Federal Republic of Germany
RASMUSON, BERTIL, Dept. of Genetics, University of Umea, Sweden
REED, RANDALL, Yale University, New Haven, Connecticut
REIF, HANS-JORG, Biozentrum, University of Basel, Switzerland

RÉSIBOIS, ANNE, Dept. of Electron Microscopy, University of Brussels, Belgium
REZNIKOFF, WILLIAM, Dept. of Biochemistry, University of Wisconsin, Madison
RICHARDS, HILARY, Dept. of Bacteriology, Royal Postgraduate Medical School, London, England
RILEY, MONICA, Dept. of Biochemistry, State University of New York, Stony Brook
ROBERTS, JAMES, 640 W. 170 St., New York, New York
ROEDER, WILLIAM, Basel Institute for Immunology, Switzerland
ROSNER, J. L., NIAMDD, National Institutes of Health, Bethesda, Maryland
ROSS, WILMA, Dept. of Biology and Medicine, Brown University, Providence, Rhode Island
ROTH, JOHN, Dept. of Biology, University of Utah, Salt Lake City
ROTHSTEIN, RODNEY, Dept. of Microbiology, New Jersey Medical School, Newark
ROYER, HANS D., Sidney Farber Cancer Institute, Boston, Massachusetts
ROYER-POKORA, BRIGITTE, Sidney Farber Cancer Institute, Boston, Massachusetts
RUBIN, GERALD, Dept. of Biological Chemistry, Sidney Farber Cancer Institute, Boston, Massachusetts
RUVKUN, GARY B., The Biological Laboratories, Harvard University, Cambridge, Massachusetts
SAEDLER, HEINZ, Institute of Biology III, University of Freiburg, Federal Republic of Germany
SAGER, RUTH, Sidney Farber Cancer Institute, Boston, Massachusetts
SAIGO, KAORU, Dept. of Life Sciences, Mitsubishi-Kasei Institute, Tokyo, Japan
SAINT GIRONS, ISABELLE, Dept. of Genetics, University of Cologne, Federal Republic of Germany
SAKANO, HITOSHI, Dept. of Immunology, University of Basel, Switzerland
SALAMINI, FANCESCO, Istituto Sperimentale per la Cerealicolture, Bergamo, Italy
SAMBROOK, JOSEPH, Cold Spring Harbor Laboratory, New York
SANG, HELEN, Dept. of Biochemistry and Molecular Biology, Cambridge, Massachusetts
SAPIENZA, CARMEN, Dept. of Biochemistry, Dalhousie University, Halifax, Canada
SARVETNICK, NORA, Cold Spring Harbor Laboratory, New York
SASTRY, G. R. K., Dept. of Genetics, University of Leeds, England
SCHERER, STEWART, Dept. of Biochemistry, Stanford University School of Medicine, California
SCHILDKRAUT, CARL, Albert Einstein College of Medicine, Bronx, New York
SCHMITT, RUDIGER, Dept. of Biochemistry and Genetics, Federal Republic of Germany
SCHOEFFL, FRIEDRICH, Dept. of Botany, University of Georgia, Athens
SCHOPF, THOMAS, Dept. of Geophysical Sciences, University of Chicago, Illinois
SCHWARTZ, HELEN, Rockefeller University, New York, New York
SCOTT, JUNE, Dept of Microbiology, Emory University, Atlanta, Georgia
SEDIVY, JOHN M., Dept. of Microbiology and Molecular Genetics, Harvard Medical School, Boston, Massachusetts
SHAPIRO, J. A., Dept. of Microbiology, University of Chicago, Illinois
SHAPIRO, LUCY, Dept. of Molecular Biology, Albert Einstein College of Medicine, Bronx, New York
SHERMAN, FRED, Dept. of Radiation Biology and Biophysics, University of Rochester, New York
SHERRATT, DAVID J., Institute of Genetics, University of Glasgow, Scotland
SHINNICK, THOMAS, Scripps Clinic and Research Foundation, La Jolla, California
SILVERSTEIN, E., Downstate Medical Center, State University of New York, Brooklyn
SIM, GEK-KEE, Institute for Cancer Research, Columbia University College of Physicians and Surgeons, New York, New York
SIMON, MELVIN, Dept. of Biology, University of California, La Jolla
SINGH, LALJI, Dept. of Animal Genetics, University of Edinburgh, Scotland
SKALKA, A. M., Dept. of Cell Biology, Roche Institute, Nutley, New Jersey
SO, MAGDALENE, Dept. of Biochemistry, University of California, San Francisco
SOAVE, CARLO, Istituto Biosintesi Vegetali, Milano, Italy
SOMMER, HANS, Dept. of Biochemistry, Ruhr-University, Bochum, Federal Republic of Germany
STAHL, FRANKLIN W., Institute of Molecular Biology, University of Oregon, Eugene
STARLINGER, PETER, Institut für Genetics, University of Cologne, Federal Republic of Germany
STERNBERG, NAT, Frederick Cancer Research Center, NCI, Frederick, Maryland
STIBITZ, SCOTT, Dept. of Biochemistry, University of Wisconsin, Madison
STILES, JOHN, Dept. of Radiation Biology and Biophysics, University of Rochester, New York
STROBEL, EDWARD, Dept. of Tumor Biology, Sidney Farber Cancer Institute, Boston, Massachusetts
SUTCLIFFE, J. GREGOR, Research Institute of Scripps Clinic, La Jolla, California
SYMONDS, NEVILLE, Dept. of Biological Sciences, University of Sussex, Brighton, England
SYVANEN, MICHAEL, Dept. of Microbiology and Molecular Genetics, Harvard Medical School, Boston, Massachusetts
SZOSTAK, JACK, Sidney Farber Cancer Institute, Boston, Massachusetts
SZYBALSKI, WACLAW, University of Wisconsin, Madison
TATCHELL, KELLY, Dept. of Genetics, University of Washington, Seattle
TAYLOR, AUSTIN L., Dept. of Microbiology and Immuniology, University of Colorado, Denver
TEMIN, HOWARD M., University of Wisconsin, Madison

THOMAS, CHARLES JR., Scripps Clinic and Research Foundation, La Jolla, California
TLSTY, THEA D., Dept. of Microbiology and Immunology, Washington University School of Medicine, St. Louis, Missouri
TOMASSINI, JOANNE, Merck, Sharp, & Dohme Research Laboratory, West Point, Pennsylvania
TONEGAWA, SUSUMU, Basel Institute for Immunology, Switzerland
TOUSSAINT, ARIAN, Dept. of Genetics, University of Brussels, Belgium
TU, CHEN-PEI, Dept. of Microbiology and Biochemistry, Pennsylvania State University, University Park
UPCROFT, PETER, Queensland Institute for Medical Research, Brisbane, Australia
VAN DE PUTTE, P., Dept. of Molecular Genetics, University of Leiden, The Netherlands
VAN MONTAGU, MARC, Laboratory for Genetics, Gent, Belgium
VANDE WOUDE, GEORGE, NCI, National Institutes of Health, Bethesda, Maryland
VAPNEK, DANIEL, Dept. of Molecular Genetics, University of Georgia, Athens
WALL, RANDOLPH, University of California School of Medicine, Los Angeles
WALLACE, LINDA, Dept. of Bacteriology, University of Bristol, England
WEIFFENBACH, BARBARA, Brandeis University, Waltham, Massachusetts
WEISBERG, ROBERT, National Institutes of Health, Bethesda, Maryland
WIGLER, MICHAEL, Cold Spring Harbor Laboratory, New York
WILKIE, NEIL M., Institute of Virology, Glasgow, Scotland
WILLIAMS, RICHARD, International Laboratory for Research on Animal Diseases, Nairobi, Kenya
YAMAMOTO, KEITH R., Dept. of Biochemistry and Biophysics, University of California, San Francisco
YARMOLINSKY, MICHAEL, Dept. of Cancer Biology, Frederick Cancer Research Center, Frederick, Maryland
YIN, JERRY, Dept. of Biochemistry, University of Wisconsin, Madison
YOUNG, ELIHU, Sidney Farber Cancer Institute, Boston, Massachusetts
YOUNG, MICHAEL, Rockefeller University, New York, New York
ZAMIR, ADA, Boyce Thompson Institute, Cornell University, Ithaca, New York
ZHANG, YONG-DI, Dept. of Biochemistry, University of Wisconsin, Madison
ZINDER, NORTON, Rockefeller University, New York, New York

First row: H. Bentley Glass; S. Brenner; A.M. Campbell; P. Starlinger
Second row: J. Sambrook; R. Harshey; A. Toussaint; M. Casadaban
Third row: P. Borst, W. Szybalski; J.D. Watson, H. Lewis
Fourth row: B. McClintock, I. Herskowitz; A.I. Bukhari, A. Klar, J.B. Hicks

First row: R. Reed; J.G. Sutcliffe; A. Ahmed; M. Yarmolinsky
Second row: G.R.K. Sastry; F. Sherman; K. Marcu; N. Glansdorff
Third row: M. Pato, N. Symonds; D. Botstein, F.W. Stahl
Fourth row: W.F. Doolittle, P.D. Harriman; P. Dunsmuir, R.B. Flavell

Foreword

In the early 1950s, Barbara McClintock's analysis of crosses between genetically marked corn plants led her to postulate that the activity of key genes was under the control of genetic elements that had the capacity to move from one chromosomal site to another. She called them "controlling elements," noting that when one was inserted next to a gene it inhibited that gene's activity. Conversely, when a control element moved away from a gene, the activity of that gene suddenly reappeared. Movement of control elements thus results in abrupt phenotypic changes, and, because of their high mutability, the respective genes they affect were first erroneously thought to be basically different from ordinary genes.

The idea that genetic elements could move with such facility from one chromosomal site to another flew strongly in the face of conventional genetic wisdom. Confirmation in other organisms thus had to occur before the far-reaching consequences of the corn plant message became generally appreciated.

Most important were the independent observations in the late 1960s of Jim Shapiro and Peter Starlinger that certain highly pleiotropic mutations in *E. coli* were the result of the insertion of descrete DNA segments (insertion sequences or IS's) that had the capacity to jump from one chromosomal site to another. Molecular characterization of these elements became possible soon afterward with the arrival of the restriction enzymes and the recombinant DNA cloning procedures. Quickly it became obvious that IS-like elements were to be found not only in bacteria, but perhaps in all organisms. Equally important was the discovery that closely spaced pairs of IS's can move as units ("transposons") carrying along the genes lying between them. Such transposons bore many similarities with the phage Mu, leading Ahmad Bukhari, Jim Shapiro, and Sankar Adhya to convene a meeting on DNA insertion elements at Cold Spring Harbor in May of 1976. With this meeting, IS elements and transposons moved to the center of the genetic world, commencing a frenzy of experimentation that has shown no sign of abatement.

Choosing "Movable Genetic Elements" as our 1980 Symposium topic was a virtually unavoidable decision, one that became even more appropriate after the invitations went out, when rumors began that the structures of integrated retroviruses were remarkably similar to those of transposons. So our most able Symposium organizers, Ahmad Bukhari and Jim Hicks, had great difficulty in keeping the number of presentations within the bounds of sanity. In doing so, they received invaluable advice from Sidney Brenner, Peter Day, Harrison Echols, J. R. S. Fincham, Walter Gehring, Mel Green, Ira Herskowitz, Lee Hood, Amar Klar, Phil Leder, Barbara McClintock, Howard Nash, Heinz Saedler, Jim Shapiro, Peter Starlinger, Jeff Strathern, and Bob Weisberg.

We were obviously most pleased that Barbara McClintock, who typically did not want to give a formal presentation, nevertheless gave an informal summary of her latest ideas. And most appropriately on the opening night, Bentley Glass briefly recapitulated many of Barbara's key contributions to genetics.

The formal program contained 103 presentations, to which the reports of 14 last-minute informal presentations have been added in these volumes. The total attendance at this Symposium was 302, a number which we handled with more ease than before because of great improvements in our food service facilities on the ground level of Blackford Hall. Our Meetings Office staff, Gladys Kist, Winifred Modzeleski, and Barbara Ward, were as usual indispensable in helping make our Symposium an enjoyable as well as a memorable event.

That we could invite so many participants reflects the substantial financial support again provided by the National Institutes of Health, the National Science Foundation, and the Department of Energy. Because of the rapidly escalating air fares, we needed still additional help, and we wish to acknowledge major support from the Cetus Corporation, Bethesda Research Laboratories, and New England BioLabs.

The preparation of the volumes is never a simple matter, and we are greatly indebted to our most effective Publications staff headed by Nancy Ford. In particular, we shall mention the most competent dedication of Douglas Owen, Nadine Dumser, Dorothy Brown, Doris Calhoun, Joan Ebert, Annette Kirk, and Kathleen Horan.

J. D. Watson

Contents

Part 1

Transposable Elements in Bacteria: Structure and Function

Transposable Elements in Bacteria: Factors Affecting Transposition

Transposable Elements in Bacteria: Mechanism of Transposition

Transposable Elements in Bacteria: Mechanism of Mu Transposition

Biochemistry of Recombination

Part 2

Genetic Instability in Plant Systems

Organization of Genes

Rearrangements in Antibody Genes

COLD SPRING HARBOR SYMPOSIA ON QUANTITATIVE BIOLOGY

VOLUME XLV—PART 1

Some General Questions about Movable Elements and Their Implications

A. CAMPBELL
Department of Biological Sciences, Stanford University, Stanford, California 94305

In this introductory chapter I present a highly selective history of research on movable genetic elements, followed by some comments on the molecular mechanisms of insertion and transposition and the possible roles of specific recombination systems in evolution and development. My general purpose is to call attention to some of the questions we should be thinking about and trying to answer.

HISTORY

Transposable Elements in Maize

Barbara McClintock discovered movable elements in maize in the 1940s and proceeded to demonstrate a remarkable array of properties associated with them, such as controlled chromosome breakage, effects on gene expression, localized mutagenicity, etc. (McClintock 1952). In many respects her work was far ahead of its time. It has been admired by geneticists because of its combined content of perception, biological insight, and experimental and analytical virtuosity, but the means to explore its full implications at a finer level are only now becoming available.

Insertion

In the meantime, information obtained from bacterial genetics provided some relevant insights. The first of these was the realization that temperate bacteriophages such as λ and conjugative plasmids such as F can become physically inserted into the chromosome rather than joined to it by some looser type of connection. This idea seems so natural to us now that it is hard to imagine alternatives to it. However, I believe it is true that both in bacterial genetics through the 1950s and in the genetics of movable elements in higher eukaryotes during the same period, most investigators assumed that the mobility of the elements, their ability to add to chromosomes and to be lost from them, indicated that the connection between a movable element and the rest of the chromosome differed in its physical nature from the bonds that held the normal, "immovable" parts of the chromosome together, such as the bonds connecting one gene with another.

Initially, the evidence for insertion came from fine-structure genetics, especially of bacteriophage λ, starting around 1960. Here advantage was taken of the enormous resolving power of bacterial and phage genetics, which permitted the demonstration that the genome of the inserted element is in fact colinear with that of the chromosome. More direct physical demonstration of insertion became possible with improvements in DNA technology in the 1970s. By 1972, Sharp et al. had produced some remarkable electron micrographs of artificial three-way heteroduplexes between the DNAs from λ viral particles, F plasmids, and an F′ plasmid bearing an inserted λ prophage that were sufficiently definitive to convince even the most skeptical and unimaginative biochemist that insertion was a real process and that the participants were tangible objects amenable to experimental manipulation.

Site-specific Recombination

In the late 1960s another feature of λ insertion became evident, which has come to be called site-specific recombination. The underlying facts were, first, physiological evidence that λ insertion is controlled by the λ repressor and, later, the isolation by several investigators (Zissler 1967; Gingery and Echols 1967; Gottesman and Yarmolinsky 1968) of viral mutants that were unable to insert because of their failure to produce a specific gene product, λ integrase. Subsequently, Guarneros and Echols (1970) and Kaiser and Masuda (1970) showed that a second viral product, excisionase, was required for excision from the chromosome, but not for insertion.

The noteworthy implication here was that insertion and excision are processes in which the enzymes of homologous recombination, such as those encoded by the bacterial *rec* genes, play no part. Rather, they represent breakage and joining of DNA at specific sites by enzymes that recognize those sites and act on them alone.

Several more years elapsed before Nash (1975) succeeded in developing a workable in vitro assay for the integrase reaction and before Landy and Ross (1977) determined the nucleotide sequence of the junction points between prophage and host DNA. As a result of the combined efforts of many people, I think it is fair to say that the amount of direct knowledge about site-specific recombination in λ far exceeds that about any other element. Consequently, many of the generalizations that can be made at this time involve extrapolation from the λ work to other systems.

One such generalization is that most movable elements resemble λ in encoding some of the enzymatic machinery needed for the DNA-joining reactions in which they participate and that these element-encoded enzymes recognize specifically the termini of the element itself. This idea is supported by the evidence available on the relative handful of elements where this has been studied in depth. We may anticipate that this generalization is not absolute and depends more on the evolutionary pressures operative on the elements that have been studied to date than on any more basic principle.

Transposition

The horizons of prokaryote geneticists were extended by the discovery (Jordan et al. 1968; Shapiro 1969; Fiandt et al. 1972) of inserted elements that can move about from one chromosomal location to another (like the McClintock factors) and the later expansion of this category to include elements that carry drug-resistance determinants and other genes with direct phenotypic effects. By the time attention became focused on these elements, molecular techniques were sufficiently advanced so that direct physical demonstration of insertion was simple, though not necessarily easy. This was a fortunate circumstance, because insertion sequences and transposons are inherently more difficult to study genetically than something like λ, which has a demonstrable free phase in which its genetics can be investigated and compared with that of the inserted state.

As these studies developed, some prokaryote geneticists, including myself, realized that we should have anticipated that elements of this sort ought to exist. The bacterial inserting elements that had been studied up to that time were temperate phages and plasmids that could replicate extrachromosomally and could also become inserted and replicated as part of the chromosome. Elements with this dual capacity were dubbed "episomes" by Jacob and Wollman (1958). The course of prokaryotic work in the 1960s was strongly influenced by the desire for a unified concept of the interactions between episomes and chromosomes. Impressed by the similarity of episomes to the McClintock factors, at least some of us imagined that they would turn out to be episomes as well.

They still may. However, we already knew enough about the genetic organization of λ more than 10 years ago to appreciate that the ability to replicate autonomously and the ability to insert are independent attributes. Many phages, including mutants of λ, can replicate but cannot insert. Someone might have thought more seriously about the possible existence of natural elements capable of insertion but not of independent replication and interpreted existing data in that light.

The story has an ironic twist. As the analogy between transferable prokaryotic elements and temperate phages like λ was gradually coming to seem natural, comfortable, and familiar, evidence was developing that the mechanism of transposition is, in fact, fundamentally different from the excision-insertion pathway known in λ. In λ infection, viral DNA enters the cell and assumes a closed circular form. Insertion comprises a single reciprocal exchange taking place somewhere within a 15-bp segment of the viral DNA and an identical 15-bp segment that preexists at a unique site on the chromosome. Excision of prophage from the chromosome reverses this process. No DNA is degraded, and there is no new synthesis. The reaction is remarkable in that its representation as a precise reciprocal exchange has continued to hold true through successively finer levels of analysis both in vitro and in vivo.

Transposition of an insertion sequence like IS*1* as observed in vivo is quite different. Like λ, the inserted element is flanked by a direct oligonucleotide repeat, in this case 9 bp long. Translocation to new sites comprises the appearance of the element at the new site without concomitant loss from the old site. Obviously, DNA synthesis must accompany transposition. Not only is the element itself duplicated, but also 9 bp of DNA originally in the recipient, the two copies of which now flank the insert at its new site. Hence, not only does the basic process differ from λ insertion, but in these elements replication and insertion are connected rather than independent.

DNA Rearrangements

The last topic in this brief history is the discovery of special systems in which DNA rearrangements play a demonstrable role in controlling gene expression or cellular differentiation. These include the control of flagellar antigen synthesis in *Salmonella* phase variation, the determination of mating type in yeast, and the differentiation of vertebrate lymphocytes to form specific types of immunoglobulins. The existence of these systems has focused attention on the possible importance of controlled DNA rearrangements throughout the biological world.

MECHANISM

In considering the mechanism of site-specific recombination, if we restrict ourselves to systems in which a fairly well-defined reaction can be studied in vitro, there is only one system available, λ integrase. Rather than review the biochemistry of this one system, I would like, instead, to explore the implications of the foregoing discussion that λ insertion and IS*1* transposition represent basically different processes. One of the tasks of scientists is to search for syntheses that illuminate whatever unity may be discernible beneath diversity. The time for such a synthesis may be premature. However, it is appropriate to focus our attention on what features these two types of reactions may have in common and whether the enzymes carrying them out might be derived by minor modifications from a common progenitor.

First, what do the two have in common? One feature, which to my knowledge is shared by all prokaryotic insertion systems, is that the inserted element ends up flanked by a direct repeat—frequently of oligonucleotide length but sometimes, as in the case of the F factor, much longer. The direct repeats appear to be there for different reasons in various cases.

With elements like λ, which insert preferentially at a unique site or at a small number of sites, the flanking repeat is present in the element itself and appears to be

Table 1. Site-specific Recombination Systems

Conservative reactions	Duplicative reactions
Insertion of λ, P2, P22, etc.	IS and Tn transpositions
Excision of λ, etc.	Mu insertion
Specific inversions:	Replicon fusion
Salmonella phase variation	Retrovirus insertion?
G loop of Mu	
Clean excision of IS and Tn elements	

part of the DNA recognition site for the enzyme. On the other hand, transposable elements like IS*1* insert at many sites, and there is apparently no specific recognition of the recipient DNA oligonucleotide sequence that ends up flanking the inserted element. What is generally found with such elements is an imperfect inverted repeat at the ends of the inserted elements, as though the two ends had evolved so as to look equivalent to some recognition system when viewed from either end.

Table 1 shows a few systems, mostly from prokaryotes, that involve site-specific recombination. These can be classified into two groups depending on whether they resemble λ insertion or IS*1* transposition. I have labeled such systems "conservative" and "duplicative," respectively. A conservative reaction is a simple polynucleotide exchange. Nothing is created, nothing is destroyed; DNA is simply rearranged. In a duplicative reaction, on the other hand, there is a net increase in DNA.

In a very strict sense, the only demonstrated conservative reactions are λ insertion and excision, where the fate of substrate molecules can be followed directly in vitro. Nevertheless, there is a strong presumption that some of the other systems listed are conservative.

Of special interest are the controlled inversions, such as the G segment of Mu and the segment that regulates phase variation in *Salmonella.* Inversion, like λ insertion, is a simple rearrangement in which the net DNA content is unchanged. We can imagine a controlled inversion system evolving directly and immediately from a λ prophage. Synthetic invertible elements can, in fact, be derived from λ by appropriate manipulation (Reyes et al. 1979). The *Salmonella* phase variation system resembles such a λ model at least insofar as the termini of the invertible segment are two identical oligonucleotide sequences in opposite orientation.

The duplicative reactions include transposition by all of the common insertion sequences and transposons of the Enterobacterioceae and also bacteriophage Mu, which utilizes the transposition mechanism not only for inserting into the chromosome and for transposing from one site to another, but also in its own autonomous replication. I have also tentatively placed retroviruses in this category.

To get back to the basic question: Are there any similarities in the two types of processes that might encourage one to look for common aspects of mechanism? If not, is there any way to seek relevant evidence?

To my mind, the most direct evidence that some steps are common to the two processes would be if someone found one element capable of carrying out both types of processes and could show that the same specific gene and/or its enzymatic product is required for both. There are some suggestive cases with known elements, but I am unaware of any for which there is really hard evidence. To put the question in terms of synthetic invertible elements derived from λ, it is clear that such an element is expected to invert at high frequency in the presence of integrase. I would not expect it to transpose in the presence of integrase, but my mind is open to the possibility that it might do so at low frequency.

The general idea is illustrated in Figure 1. The figure represents successive steps in concerted reactions, not independent reactions with free intermediates. The donor is our artificial λ with inverted ends. The scheme has been drawn to incorporate a "sticky-ended" intermediate which has often been postulated. For our pur-

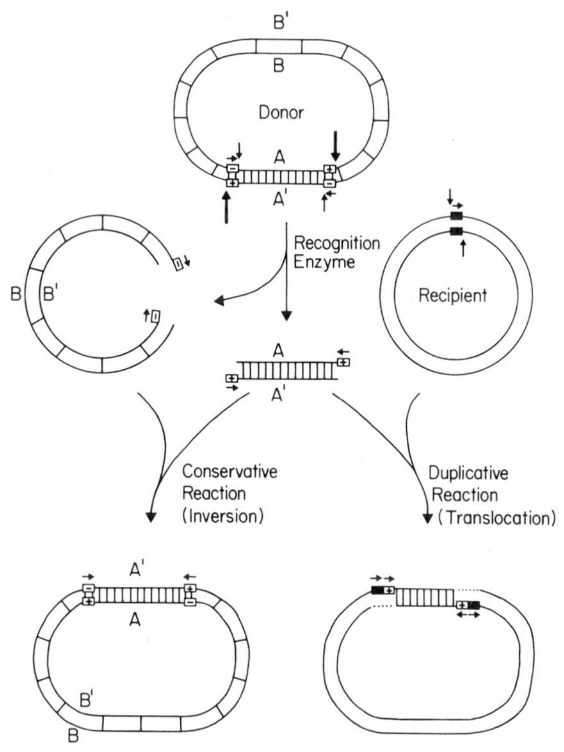

Figure 1. Hypothetical scheme illustrating one possible manner in which a single enzyme might participate in both conservative and duplicative reactions. The donor molecule contains terminal oligonucleotide repeats which, like the 15-bp common core of λ, function both as recognition elements and as sites of action, but which, unlike the λ case, are in inverted orientation. An element-encoded "recognition enzyme" (like λ integrase) makes staggered nicks within the common core regions at both ends (not necessarily simultaneously). The excised molecule then has one of two possible fates: (1) It may, as shown on the left, reinsert into the donor molecule in inverted orientation. This is the standard integrase reaction as applied to a substrate with inverted ends. (2) It may transpose by inserting into a recipient molecule that has been cleaved in the appropriate manner by host enzymes, followed by repair synthesis to generate oligonucleotide repeats. More detailed discussion of the duplicative reaction is given in the text.

poses, such an intermediate provides a convenient, but not necessarily unique, mechanism for promoting homologous recombination within a short DNA segment. We imagine that formation of this intermediate is catalyzed by a sequence-specific, element-encoded enzyme, perhaps acting in conjunction with other enzymes. This first step is then represented as leading to either inversion or transposition depending on what happens next.

In the inversion reaction, the element is reinserted, in opposite orientation, in the donor chromosome. On the transposition side, a recipient chromosome is cleaved by a host enzyme that makes staggered nicks, which must generate ends of the opposite chemical polarity from the original reaction; that is, if the original cuts generated 5′ ends, these must be 3′. The projecting single-stranded ends are then joined, and repair synthesis fills in the gaps.

The purpose of this diagram is to indicate how a single enzyme, or two related variants of the same enzyme, might participate in both types of reactions, depending in part on its association with additional host enzymes. Now I must return to the fact that to simplify things I have cheated quite a lot on the transposition side. Among other things, I have diagrammed a mechanism for transposition but not one that is duplicative. I have, in fact, destroyed the donor chromosome by cleaving it in two.

What can become of it? We know of no mechanism whereby it can rejoin and heal, because its sticky ends are identical rather than complementary. The easiest way to rejoin it is to reinsert the excised DNA. But to both reinsert and transpose, that DNA must be duplicated. How? It could replicate as such, with some special attention to what happens to the ends, to generate one transposable copy and one reinsertable copy.

Somewhat more realistic schemes, some elements of which are incorporated in Figure 1, have been proposed by others (Grindley and Sherratt 1979; Shapiro 1979). Their models invoke a host-enzyme-induced cleavage of recipient DNA to produce sticky ends, which are then ligated to element DNA. Unlike Figure 1, these models do not require production of sticky ends at the ends of the insert. Rather, a cleavage of one of the two strands at each end suffices. (The larger arrows of Fig. 1 represent the cleavage sites in these models.) This is of course simpler than excising the insert, replicating it, and then reinserting a copy. In these schemes, the insert DNA never leaves its original location and hence does not need to be reinserted. There is an additional simplification in that the replication of the insert and the gap-filling that I depicted become part of one replication event rather than two separate ones. Shapiro's scheme postulates, in addition to the enzyme active at the terminus, a second specific enzyme whose site of action is internal to the element and whose mode of action is conservative rather than duplicative.

What do schemes like this predict with respect to the prospects for success in looking for one enzyme that participates in both conservative and duplicative reactions? Even from Figure 1 it is obvious that the recognition enzyme must act in concert with other enzymes and that a given enzyme may therefore be specialized to function in one manner or the other. In that case, Figure 1 may represent not so much the potentialities of a single enzyme as the collective potentialities of a family of closely related enzymes, perhaps with overlapping specificities. Carried a step further, an enzyme that makes two nicks rather than four, as in the Grindley and Shapiro models, would also represent a variant of a family that includes conservative enzymes as well. These considerations encourage a comparative approach and justify skepticism about the likelihood that a single enzyme might function in both manners.

Two further comments on natural transposons and insertion sequences: The termini of the element generally form an imperfect inverted repeat, not a perfect one. This suggests that they are not frequently the sites of conservative reactions, which should depend on precise homology. Second, precise excision does occur (Table 1), although it does not seem to represent a step in transposition. Precise excision, with restoration of gene function, presumably requires recombination between the oligonucleotide repeats of host DNA that flank the element. The reciprocal product is generally not recoverable; so characterization of the reaction as conservative is tentative.

At past Symposia the introductory remarks have sometimes included predictions. It is against my principles to make predictions about science. However, I enjoy the more harmless recreation of making predictions about scientists. I predict that at some stage of the game scientists will dedicate serious effort to understanding the relation between conservative and duplicative reactions. This may happen because the data indicate that such efforts will be auspicious; but even if they are not, the question will be examined in depth because it offers our only hope for a unified picture. Such studies may concentrate either on single totipotent enzymes or on families of recognizably related enzymes.

EVOLUTION

When Barbara McClintock discovered controlling elements in the 1940s, it was obvious that they might play important roles both in evolution and in development. In presenting her results she generally concentrated more attention on the developmental implications than on the evolutionary ones. I believe that this was a wise decision on her part, although I do not know whether her actual reasons for this were the same as mine.

Changes in DNA, ranging from point mutations through gross rearrangements, are the raw material of evolutionary change. We cannot doubt that any mechanism that produces such changes must have occasionally contributed to evolution by creating adaptive gene combinations that have survived. But it is also true that at the time controlling elements were discovered, there was no obvious shortage of known mechanisms (understood at the same level as controlling elements could be understood at the time) that could generate the same end results. Evolutionists did not accord high priority to the search for new mutational mechanisms. The principal outstanding problems in their field did not appear to be at that level.

Although some people may disagree, I believe that, for the most part, the situation remains the same today. We all agree that movable elements can cause deletions, inversions, translocations, etc. Few of us would assert that all deletions or all inversions require their participation. With a few possible exceptions, such as the distribution of dispersed, moderately repetitive DNA within related species, most observable evolutionary change can be adequately (though not necessarily correctly) explained by other mechanisms. Without in any way minimizing the value of some interesting evolutionary speculations that have been made recently, I believe that the case that movable elements provide much of unique explanatory value remains to be made.

One example often cited of the role of specific events in evolution is the bacterial plasmids. Comparisons of the genomic organization of related plasmids show that frequently they differ from one another by insertion of transposons and by replicon fusions catalyzable by specific systems. There can be no reasonable doubt that these events have occurred in nature and have contributed to the diversity of the existing plasmid population.

Accepting that such events have happened and are important, one may still ask whether they really constitute evolution. Let me illustrate the reasoning behind this position with an extreme case. If I survey *Escherichia coli* strains in nature, some are lysogenic for bacteriophage λ and many are not. I could, therefore, speak of the acquisition of a λ as part of the evolution of a strain such as *E. coli* K12. In a sense it is. However, perhaps because of conditioning, I generally regard the transition from phage to prophage and back again as defining the life cycle of the virus, rather than as constituting evolution, which implies (to me) some degree of progressive as well as cyclical change. I tend to consider the relocation of transposons and the insertion of one plasmid into another likewise as constituting part of the life cycle of these elements. Unions are created when convenient and eventually dissolved when their selective value disappears. All this happens within a single bacterial species or a group of species belonging to some Exchanger List, and I suspect it happens at a frequency many times as great as events that lead to a degree of genetic isolation comparable to that separating distinct species of higher organisms.

This is not the time to make a judgment concerning this issue. It may be profitable for bacterial geneticists to consider carefully where they want to draw the line between evolutionary change and variation that is part of the ongoing population biology of the species. They may wish to relegate cyclical processes and pedigrees that are reticulate rather than branching to the latter category. Perhaps, on the other hand, it is impracticable to make the distinction between evolutionary change and nonevolutionary change and we should simply chronicle variation as such. Even in that event, I would prefer to see that decision made consciously and explicitly rather than, as it seems at the moment, implicitly and by default.

DEVELOPMENT

With those remarks, I shall leave evolution and turn to the question that is really the major thrust of the current interest in movable elements, namely, their potentialities for controlling gene action, especially during development.

If evolutionists have generally perceived an abundance of possible mechanisms for generating genetic diversity during evolution, genetically oriented developmentalists contemplating the differentiation of the descendants of a single zygote into more or less permanently distinct types have never enjoyed that luxury. They have, in fact, generally been hard pressed to find any precedents or paradigms in the transmission genetics of organisms, including unicellular ones, that explain the changes in cellular properties during development in a manner that is either satisfactory or satisfying. The classical notion of differentiation as a process that generates changes that are permanent and heritable at the cellular level does not square with the textbook picture of equal distribution of genes between daughter cells whenever a somatic cell divides by mitosis. The demonstration by McClintock that movable elements could translocate to positions adjacent to known genes and permanently modify their expression and that translocation is subject to some degree of temporal control provided a genetic system with many of the features needed to understand how changes could be at once inducible and permanent.

A role of controlling elements in developmental specification implies that the critical relevant changes are nuclear rather than cytoplasmic. On this important question, data from nuclear transplantation experiments have been equivocal. They have sometimes been interpreted as indicating that nuclei from adult cells retain the totipotency of the zygote nucleus, but that conclusion can be questioned on a technical level (McKinnell 1978). I have always found the conclusion conceptually unsatisfying as well, because it seems to beg the question of what is the basis of permanent change, if not genetic.

In any event, there is a growing awareness of the possibility that controlled DNA arrangements are critical in development, and there are a few supporting facts. The most impressive are from antibody-forming cells. Extrapolation from that system to general developmental mechanisms requires some optimism, because part of the motivation for studying immunoglobulin determination is that even among developmental systems, its properties are extraordinary.

There are examples from prokaryotes of the recombinational control of gene expression whose mechanisms are better understood. Paramount among these are the controlled inversions, such as the G region of Mu or the H-antigen control region of *Salmonella*. A gene outside an invertible region transcribed from a promoter inside the region is turned on when the inversion is in one orientation (flip); in the other orientation (flop), the gene is off. Developmental biologists, facing the question of why a given gene is on in one tissue and off in another, though present in both, may well be attracted by the potentialities of such a picture.

I think that most integrative biologists would also agree that, viewed as a model for developmental specification, the system lacks one important feature. It does show us how a specific recombination system can control expression of a gene. That is clearly one aspect of development. It does not include or imply a mechanism by which the recombination itself is controlled so as to fit into the overall developmental program of the organism. Such a program must embody not only the capacity to change, but also the ability to change in a highly directed manner, so that the right switch can be thrown in a prescribed cell at a prescribed time in development. It is in this regard that immunoglobulin synthesis may be atypical, because the system is designed to generate a high level of somewhat random diversity, whereas the typical critical changes in development should be much more tightly controlled.

If we consider the prokaryotic model systems, we can ask whether they are under tight control. The recombinational event in the *Salmonella* switch is controlled by the product of a specific gene within the element. That gene may be subject to some quantitative effects on its rate of action, but there is no evidence that it is tightly regulated. Furthermore, there is no reason to suppose that it should be. The natural role of phase variation seems to be to counter the immunological defenses of the host by creating a more rapid rate of antigenic change than is readily accomplished by mutation alone. That purpose is satisfied by a reversible reaction that occurs constitutively. In development, on the other hand, the critical changes should be regulated and irreversible.

Obviously, there are many mechanisms imaginable whereby site-specific recombination could be controlled. Prokaryotic systems provide some interesting examples of how it is controlled in those systems. I would like to describe some aspects of the regulation of λ insertion and excision in that context.

There is a special reason (other than my own involvement in the work) for talking about λ rather than simpler elements like Tn*3* or Tn*5*. λ represents a genetic element that is sufficiently complex so that it has a regulatory program that includes genes for repression, replication, assembly, and lysis, as well as insertion. The insertion genes must therefore be controlled so as to fit into this program, just as we may expect that analogous developmental events are controlled by the program of the whole organism. It is in this respect that I regard λ regulation as deserving the attention of biologists in general.

As with other aspects of λ biology, the control of site-specific recombination has many facets and has achieved a level of sophistication that may well be unmatched in organisms like higher eukaryotes, which are larger and genetically more sluggish. The whole story is summarized later in this volume (H. Miller et al., this volume). Here I shall discuss only one aspect, namely, the differential control of *int* and *xis* transcription, and only enough of that to indicate some critical features of the story.

For this purpose it may be helpful to look at λ in a particular manner. λ is a virus with a genome of about 50,000 nucleotide pairs. Of this DNA, all of the functions and sites concerned with integration are clustered into one segment about 1500 nucleotides in length. I like to imagine that a virus like λ evolved by the joining together of different modules that previously functioned in isolation or in other contexts. According to that view, these 1500 bp might be derived from and phylogenetically related to simpler insertion elements like IS*1* or IS*2*.

This viewpoint has two relevant consequences. The first relates to the discussion of the relationship between conservative and duplicative reactions. Modular concepts of evolution have become popular in the last few years mainly because the properties of transposons suggested a specific mechanism whereby modules might come together. I find that most of my colleagues are fairly well conditioned to accepting the notion that the insertion region of λ might have a common evolutionary origin with some element like IS*2*. Fewer people seem willing to follow through with what I consider a reasonable corollary, namely, that the common ancestry of the two types should be reflected, at some level, in a commonality of mechanism.

The second feature is the one I wish to address now. Compared with an IS element, λ is a large and complicated entity with a well-ordered life cycle. Hence, the potentialities of the insertion region must at some stage have come under the control of the λ program. How is this accomplished?

Obviously, the relationship between λ and its insertion genes is not the same as that between an organism and a movable element within its genome. The insertion region of λ functions to insert the whole λ genome within the bacterial genome, rather than moving about within the λ genome. Let us ignore that detail for the moment and focus on the regulatory problem common to both cases of subordinating the recombination functions to the program of the whole organism.

The basic problem concerning regulation of λ insertion was already perceptible 10 years ago. Insertion of λ into the *E. coli* chromosome requires only integrase. Excision from the chromosome requires another viral protein, excisionase, as well. The *int* and *xis* genes lie next to each other on the λ DNA and are both included in the major leftward transcript of the virus.

Figure 2 illustrates the lifestyle of λ and its attendant problems. When cells are infected with λ, a fraction of the cell population goes into the productive cycle, eventually dying and liberating virus. In such cells, insertion of viral DNA into the chromosome is a waste, and insertion without excision may cause the infection to

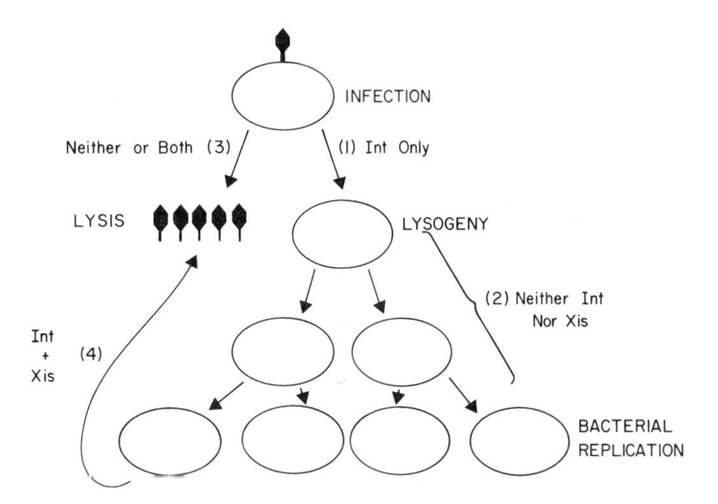

Figure 2. Life cycle of λ, showing the need for three positions of the regulatory switch controlling integrase production: (*1*) *int* on, *xis* off (in those infected cells destined to survive as lysogens); (*2*) both off (in established lysogenic bacteria); (*3*) both on (in induced lysogenic bacteria, and perhaps in infected cells destined for lysis).

abort. Hence, one would like to see *int* and *xis* either both turned on or both turned off. Certainly, integrase alone would be bad. Another fraction of the cell population survives infection because viral functions become repressed. It is in this fraction that insertion, without subsequent excision, is desired. If a surviving cell harbors a repressed, unintegrated phage genome, that genome is eventually diluted out with growth. So here we want integrase alone.

After insertion has taken place and the stable lysogenic condition is established, neither integrase nor excisionase is needed to perpetuate that condition. Integrase alone would not hurt anything; integrase and excisionase together would cause prophage loss. Finally, within the lysogenic population occasional cells become derepressed and go over into the productive cycle. In this case, integrase and excisionase are needed to allow those cells to produce phage.

The mechanism whereby these ends are achieved was worked out between 1970 and 1980 by various people in several laboratories. Figure 3 summarizes the results. The *int* gene can be transcribed from two different promoters—P_L and P_I. P_L is the major leftward promoter. Transcription from P_L is observed after infection and following derepression of a lysogenic cell. In that transcript *int* and *xis* are expressed coordinately. This assures that excision will follow induction and that productively infected cells will not lose viral genomes by burying them in the chromosome. Transcription from P_I, on the other hand, leads to expression only of *int*. The P_I promoter is turned on by the products of genes *c*II and *c*III. These two proteins are also required to turn on the transcription of the *c*I gene, which codes for repressor. Therefore, in those cells that build up a high concentration of *c*II and *c*III, repression becomes established, and it is in those same cells that high levels of integrase are induced. Thus, the two aspects of lysogenization, repression and insertion, are coordinated, not because the two genes are in the same operon, but because two different operons respond to the same control system.

Finally, in the established lysogenic cell, repressor turns off both the transcription of *int* and *xis* from P_L and also the transcription of *c*II and *c*III, which are needed to activate P_I. Thus, both pathways are off, and no energy is wasted on useless synthesis.

Detailed molecular mapping places the P_I promoter, not between the *int* and *xis* genes, but straddling the beginning of the *xis* gene (Abraham et al. 1980). The initiation codon of the *xis* gene is within the Pribnow box

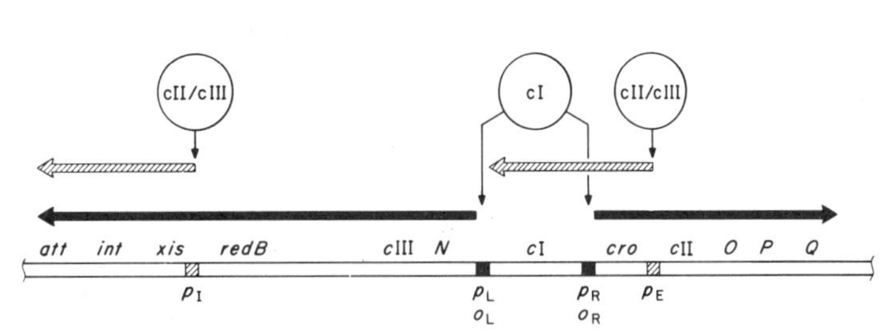

Figure 3. Structure of three-way switch controlling integrase and excisionase production in λ. Position 1 (*int* on, *xis* off) is achieved by turn on of transcription from P_I, which produces a message translated to give integrase but not excisionase. This message is turned on by the *c*II/*c*III proteins, which simultaneously turn on the *c*I gene. The *c*I product (repressor) turns off transcription from P_L and P_R, allowing cell survival and also shutting off excisionase production from P_I. Position 2 (both off) is reached when the *c*I product is elaborated continuously in a lysogenic bacterium. Repressor directly shuts off *int* and *xis* transcription from P_L. It also prevents transcription from P_I, by blocking transcription of the *c*II and *c*III genes. In position 3, the P_L message is made, resulting in coordinate expression of *int* and *xis* without sufficient accumulation of *c*II/*c*III to turn on P_I and P_E.

of this promoter, so that almost the entire *xis* gene is transcribed as a leader for the *int* message. This location of the promoter is intriguing with respect to the evolutionary considerations mentioned earlier. If the insertion region of λ is descended from a free-living IS element, it might seem reasonable that the regulated promoters putting the genes of the element under λ control lie outside the region itself, or at the border between the insertion region and the rest of λ.

With this incomplete and simplified description of λ regulation as an example, I would like now to return to some more general questions concerning developmental specification and differentiation.

If the critical switching event that commits a cell to a particular pathway is a change in DNA involving systems like those known in prokaryotes, at least two requirements must be met. First, the relevant recombinations must, as in the λ example, be regulated so as to function at the proper time. Second, there must be some mechanism that makes the switching event effectively (though not necessarily absolutely) irreversible. A process that allows free switching back and forth, even during a brief window in time, would not do.

Table 2 lists several possible methods for achieving irreversibility. Duplicative reactions, by their nature, are not directly reversible. The basic mechanism creates a new element without destroying an old one. Iteration of that operation will not restore the original condition. Excision events, which apparently occur in immunoglobulin synthesis, are irreversible even if the underlying chemical process is both conservative and symmetrical. This is because the reciprocal products suffer different fates depending on whether or not they contain a centromere (in eukaryotes) or an active replication origin (in prokaryotes).

Differential recognition of the sites involved in the forward and reverse reactions, as exhibited by λ, is another mechanism. The basis for the different catalytic specificities of insertion and excision lies in the differential recognition of the base sequences flanking the phage and bacterial sites. Thus, in the presence of integrase alone the reaction is effectively irreversible.

Finally, the rearrangement event itself may influence the expression of the genes whose products generate the rearrangement. One example of such recombinational control of recombination functions is seen in bacteriophage P2, where the insertion site lies within the *int* operon, which is therefore inactivated by the insertion event (Bertani 1970). That effect, without differential site recognition, would suffice to render insertion irreversible. In bacteriophage λ, also, there is good evidence that *int* expression is affected by downstream DNA that is contiguous to it in the phage, but not the prophage, state, although the nature of the effect in λ is somewhat different from that in P2 (H. Miller et al., this volume).

With respect to developmental programs, the last two mechanisms, especially, can be described with a computer analogy. We may imagine a specific cell receiving an instruction of the form, "Now is the time to move from square one to square two." We may, more realistically, dissect this instruction into two components: External information, both temporal and positional, impinges on the cell and is read as, "If you are now in square one, move to square two." Execution of the order requires internal, propioceptive information that answers the question, "Which square am I in right now?" The last two mechanisms also offer the possible advantage of being effectively irreversible under certain circumstances and potentially reversible under others.

The organizers of this meeting expressed the hope that "the Symposium will focus on the mechanisms by which specific genetic rearrangements occur and the mechanisms by which these rearrangements may control gene expression." We may predict that if and when another Symposium is held covering some of the same ground as this one, the focus will be on a third question as well: It will focus not only on how rearrangements occur and how rearrangements control gene expression, but also on the mechanisms by which the occurrence of specific rearrangements is itself controlled and integrated into the overall biology of the organism. At that point molecular biologists may fairly claim to have come to terms with the major implications of McClintock's classical work, in which these same questions were posed at a different level.

Table 2. Mechanisms for Irreversibility of Site-specific Recombination

Mechanism	Example
Duplicative reaction	transposition of IS*1*
Excision	loss of IS*1*
Site recognition (*int-xis* type)	insertion of λ prophage
Site regulation of enzymes	insertion of P2 prophage

REFERENCES

ABRAHAM, J., D. MASCARENHAS, R. FISCHER, M. BENEDIK, A. CAMPBELL, and H. ECHOLS. 1980. DNA sequence of regulatory region for integration gene of bacteriophage λ. *Proc. Natl. Acad. Sci.* **77:** 2477.

BERTANI, L. 1970. Split operon control of a prophage gene. *Proc. Natl. Acad. Sci.* **65:** 331.

FIANDT, M., W. SZYBALSKI, and M. H. MALAMY. 1972. Polar mutations in *lac, gal* and phage λ consist of a few DNA sequences inserted in either orientation. *Mol. Gen. Genet.* **119:** 223.

GINGERY, R. and H. ECHOLS. 1967. Mutants of bacteriophage λ unable to integrate into the host chromosome. *Proc. Natl. Acad. Sci.* **58:** 1507.

GOTTESMAN, M. and M. YARMOLINSKY. 1968. Integration negative mutants of bacteriophage lambda. *J. Mol. Biol.* **31:** 487.

GRINDLEY, N. D. F. and D. J. SHERRATT. 1979. Sequence analysis at IS*1* insertion sites: Models for transposition. *Cold Spring Harbor Symp. Quant. Biol.* **43:** 1257.

GUARNEROS, G. and H. ECHOLS. 1970. New mutants of bacteriophage λ with a specific defect in excision from the host chromosome. *J. Mol. Biol.* **47:** 565.

JACOB, F. and E. WOLLMAN. 1958. Les épisomes, élements génétique ajoutes. *C. R. Acad. Sci.* **247:** 154.

JORDAN, E., H. SAEDLER, and P. STARLINGER. 1968. O^c and strong polar mutations in the *gal* operon are insertions. *Mol. Gen. Genet.* **102:** 353.

KAISER, A. and T. MASUDA. 1970. Evidence for a prophage excision gene in λ. *J. Mol. Biol.* **47:** 557.

LANDY, A. and W. ROSS. 1977. Viral integration and excision: Structure of the lambda *att* sites. *Science* **197:** 1147.

MCCLINTOCK, B. 1952. Chromosome organization and genic expression. *Cold Spring Harbor Symp. Quant. Biol.* **16:** 13.

MCKINNELL, R. G. 1978. *Cloning—Nuclear transplantation in amphibia.* University of Minnesota Press, Minneapolis.

NASH, H. A. 1975. Integrative recombination of bacteriophage lambda DNA *in vitro. Proc. Natl. Acad. Sci.* **72:** 1072.

REYES, O., M. GOTTESMAN, and S. ADHYA. 1979. Formation of lambda lysogens by IS*2* recombination: *gal* operon-lambda p_R promoter fusion. *Virology* **94:** 400.

SHAPIRO, J. A. 1969. Mutations caused by the insertion of genetic material into the galactose operon of *Escherichia coli. J. Mol. Biol.* **40:** 93.

———. 1979. Molecular model for the transposition and replication of bacteriophage Mu and other transposable elements. *Proc. Natl. Acad. Sci.* **76:** 1933.

SHARP, P., M. HSU, and N. DAVIDSON. 1972. Note on the structure of prophage λ. *J. Mol. Biol.* **71:** 499.

ZISSLER, J. 1967. Integration-negative (*int*) mutants of phage λ. *Virology* **31:** 189.

Trans-acting Genes of Bacteriophages P1 and Mu Mediate Inversion of a Specific DNA Segment Involved in Flagellar Phase Variation of *Salmonella*

T. IINO AND K. KUTSUKAKE
Laboratory of Genetics, Department of Biology, Faculty of Science, University of Tokyo, Tokyo 113, Japan

In genetics, the inversion of a chromosomal region was for a long time thought to be an aberrant event. However, recurrent inversions of specific DNA segments in chromosomes have recently been found in several biological systems, and the phenomena are now being regarded as essential for regulation of the expression of specific genes. Flagellar phase variation in *Salmonella* is an example of such phenomena.

Flagellar phase variation was originally detected as alternative appearances of two distinct types of flagellar antigen in a so-called "diphasic" *Salmonella* clone (Andrews 1922). The phenomenon has attracted the attention of microbial geneticists because it resembles the occurrence of forward and back mutation, but it is distinguished by its extraordinarily high frequency and also by its oscillation between two fixed types. Since the discovery of a genetic exchange system (i.e., phage-mediated transduction) in *Salmonella,* genetic analyses of flagellar phase variation have been proceeding. On the basis of accumulated experimental data, we are now able to describe the controlling system of the phenomenon.

Diphasic *Salmonella* strains possess two nonallelic structural genes, *H1* and *H2,* for flagellin, the component protein of flagellar filaments. These genes are specific in phase 1 and phase 2, respectively (for review, see Iino 1969). Thus, flagellar phase variation occurs by the alternative expression of these two genes (Lederberg and Edwards 1953). The ability to switch from one phase to the other is controlled by the chromosomal region where *H2* is located (Lederberg and Iino 1956). Together with the *rh1* gene, which specifies a repressor of the transcription of *H1*, *H2* constitutes an operon (Pearce and Stocker 1967; Fujita et al. 1973; Suzuki and Iino 1973). When the *H2* operon is active (*H2*-on), phase-2 flagellin is produced and *H1* is repressed by the product of *rh1.* When *H2* changes to the inactive state (*H2*-off), both *H2* and *rh1* stop transcription and *H1* starts the synthesis of phase-1 flagellin. Thus, the frequency of flagellar phase variation depends on the frequency of the change of state of *H2.*

The question was raised as to how the change of state of *H2* is controlled genetically. The possibility of the involvement of a structural anomaly in the *H2* region associated with a recombinational event has been discussed previously (Pearce and Stocker 1967; Iino 1969), and the effect of *recA* and either *recB* or *recC* mutations on flagellar phase variation was looked for (Lederberg and Stocker 1970). However, the final clarification had to await direct analysis of the DNA structure of the *H2* region. This was recently accomplished by cloning the DNA region of the *H2* operon in both the on state and the off state by means of in vitro DNA recombination techniques, and it is now known that the alternate inversion of a specific 900-bp DNA sequence, termed *PD,* adjacent to *H2* causes switch on and switch off of the *H2* operon (Zieg et al. 1977; Silverman et al. 1979).

Before we proceeded with the structural analysis of *PD,* a gene, *vh2,* that controls the frequency of flagellar phase variation was detected in a stable-phase strain of *Salmonella abortusequei* (Iino 1961a). The frequency of flagellar phase variation varies among different *Salmonella* strains. For example, the *S. typhimurium* C77 strain changes from phase 1 to phase 2 and from phase 2 to phase 1 at the frequencies of 1×10^{-5} and 3×10^{-4} per bacterial generation cycle, respectively (Stocker 1949). Conversely, in *S. abortusequi* SL23 the frequencies are as low as $<10^{-7}$ in both directions. Analysis of transduction between such strains indicated that they differ in a gene, *vh2,* that is closely linked to *H2*; ordinary diphasic strains of *S. typhimurium* carry the $vh2^{+}$ allele, whereas *S. abortusequi* SL23 carries $vh2^{-}$ (Iino 1961a).

The finding of the specific DNA inversion at *PD* provoked the question of whether *vh2* and *PD* are identical. The answer was subsequently provided by a *cis-trans* test; i.e., *PD* was shown to act in *cis* position to the *H2* operon (Zieg et al. 1977, 1978; Silverman et al. 1979), whereas *vh2* was shown to act in *trans* as well as in *cis* (Kutsukake and Iino 1980a). Furthermore, with the phase-stable mutants of *S. typhimurium,* a *trans*-acting gene, *hin,* which is presumed to be identical with *vh2,* was assigned in the DNA segment *PD* (Silverman et al., this volume).

From these results, the dual controlling system for flagellar phase variation became apparent: *PD* is the invertible DNA region that switches the activity of the *H2* promoter on and off, and *vh2* produces a *trans*-acting factor that controls the inversion of *PD.* The overall scheme of the phase variation is shown in Figure 1.

The recurrent inversions of specific DNA segments are not unique to flagellar phase variation. Among bacteriophages, the G segment of Mu DNA, containing at least two genes essential for phage growth (Howe et al. 1979), carries out recurrent inversions (Howe and Bade 1975). The specific inversion is correlated with the formation of infectious phage particles (Bukhari and Am-

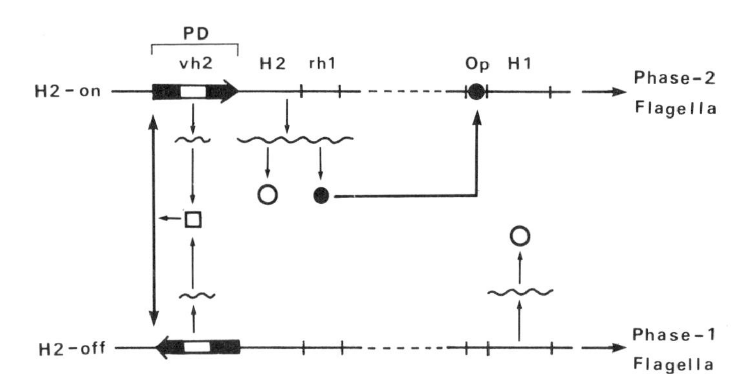

Figure 1. Genetic system involved in flagellar phase variation of *Salmonella.* (○) Flagellin; (●) *H1* repressor; (□) effector of *PD* inversion.

brosio 1978; Kamp et al. 1978). The gene analogous to *vh2* in flagellar phase variation, namely, the *trans*-acting gene for the specific inversion, was detected in the β segment of Mu DNA and termed *gin* (Chow et al. 1977). A similar inversion region also has been reported in bacteriophage P1 DNA (Lee et al. 1974; Chow and Bukhari 1976) and termed the C region (Yun and Vapnek 1977). Furthermore, the G segment of a Mu strain that is defective in a *trans*-acting factor of G inversion was shown to be inverted by P1 DNA, and the existence of a common inversion system between Mu and P1 has been suggested (Kamp et al. 1978, 1979).

We have been deeply interested in the generality of the genetic system for specific DNA inversions and have been investigating the interactions between the inversion system of *Salmonella* and those of the bacteriophages (Kutsukake and Iino 1980a,b). In this paper we report the results obtained on the effects of the *trans*-acting functions of Mu and P1 on flagellar phase variation of *Salmonella.*

EXPERIMENTAL PROCEDURES

Examination of the* trans *effect for flagellar phase variation. The procedure for this examination is to put phage DNA or a DNA segment into a *vh2*⁻ *Salmonella* cell as a plasmid and look for the occurrence of flagellar phase variation in offspring carrying the plasmid. When the phase variation occurred at a frequency significantly higher than in the *vh2*⁻ clone, it was inferred that at least a gene effective in *trans* position for the switch on and switch off of the *H2* operon was present in the plasmid.

To detect the phase variation, it is convenient to use *H2*⁻ bacteria because they are motile in phase 1 but nonmotile in phase 2. The motile and nonmotile clones of *Salmonella* are easily discriminated on semisolid nutrient gelatin-agar media (NGA); the former spread into the media and form large, thin colonies called swarms, whereas the latter grow confined at the area of inoculation and form compact colonies (Iino and Enomoto 1971). Details of the procedure are described in the notes to Table 1.

For these reasons, derivatives of *Salmonella* strain SJW1250 carrying stable *vh2*⁻ and *H2*⁻, namely, KK1251 (*galE*), KK1252 (*rpsL* = *strA*), and KK1253 (*galE*), were chosen as tester strains. They are all nonmotile in phase 2 and motile in phase 1, producing flagella of the gt antigen type specified by *H1*-gt. The fixed phase is phase 1 in KK1253 and phase 2 in the others. In these strains the phase variation occurs at a frequency of less than 10^{-8} per bacterial generation cycle in both directions. The nonmotile mutants, namely, Fla⁻ or Mot⁻, which are liable to be confused with the phase-2 phenotype, appear at a frequency of less than 10^{-7} in these cultures.

Preparation of phage DNA and its hybrid molecule by using vector DNA. The P1 phages used were P1*kc* (Lennox 1955) and P1Cm*clr*100 (Rosner 1972). The latter phage has Tn*9* and is temperature-inducible. Therefore, the chloramphenicol-resistance (Cm^r) marker of the transposon was conveniently used for the detection of bacteria carrying the prophage, and bacteria cured of the prophage were easily obtained by cultivation at 42 °C. Bacteriophage P7, which is closely related to P1 and whose prophage is known to persist in a plasmid state like that of P1 (Hedges et al. 1975), was provided

Table 1. Effects of Prophage Genomes on the Change from Phase 2 to Phase 1

Tester phase-2 bacteria	*vh2*	Plasmids	Production of phase-1 swarms on NGA
KK1251	–	none	–
		P1Cm*clr*100	+
		P1*kc*	+
		P7	+
KK1252	–	none	–
		F′*lac*⁺	–
		F′*lacI*^{s}::Mu*cts*62	+
LT2			
galE	+	none	+
		P1Cm*clr*100	+
LT2 *str*^{r}	+	none	+
		F′*lac*^{s}::Mu*cts*62	+

Cells from a single colony were inoculated onto an NGA plate and incubated at 30 °C for 5 hr, and the change from phase 2 to phase 1 was detected by the appearance of swarms. This procedure has routinely been used to discriminate *vh2*⁺ from *vh2*⁻ strains; swarms appear only when the frequency of phase shift is higher than 10^{-6}. Frequency of the cured cells appearing in the plasmid-carrying culture was less than 10^{-6}. LT2, used as a control, is a diphasic strain of *S. typhimurium* expressing i-type flagella in phase 1 and 1.2-type in phase 2. To immobilize its phase-2 cells, anti-1.2 serum was added to the test plates.

by J. R. Scott (Emory University) and used in some experiments. To examine a segment of P1Cm*clr*100 DNA, it was inserted into plasmid vector pCR1 (Covey et al. 1976) by in vitro DNA recombination. In the recloning experiment, pBR322 (Bolivar et al. 1977) was used as vector.

For the tests with Mu DNA, a hybrid plasmid composed of F′, namely, F′*lacI*^{s}::Mu*cts*62, and a hybrid λ phage containing the β-G segment (Allet and Bukhari 1975), which was provided by A. I. Bukhari (Cold Spring Harbor Laboratory), were used. The hybrid phage was propagated with *Escherichia coli* strain BU8072 (Bukhari and Allet 1975).

Phage DNA was prepared from purified phage particles obtained after thermal induction of a corresponding lysogen or lytic infection. Plasmid DNA was prepared according to the method of Nishimura et al. (1977). Digestion of DNA molecules with restriction endonucleases was performed according to the method of Bächi and Arber (1977). Fractionation of the digested DNA fragments by means of agarose gel electrophoresis and ligation of DNA fragments with vector plasmids were carried out according to the method of Sugiura and Kusuda (1979). The detailed procedures of each step have been described previously (Kutsukake and Iino 1980b).

Construction of prophage- or plasmid-carrying bacteria. F′ factors carrying a *lac* region and their hybrids with Mu were transferred from their carriers into tester bacteria by F-mediated conjugation.

The hybrid plasmids derived from pCR1 were introduced by transformation into a *recA* derivative of an *E. coli* K12 strain, KH802 (F^{-}*galK lacY met* SuII^{+} rK^{-} mK^{+}) (Velten et al. 1976), according to the method of Lederberg and Cohen (1974). F′*lac*$^{+}$ from *E. coli* LC169 (Δ*proB-lac leu thy*/F′*lac*$^{+}$) was then introduced into the transformants, and the hybrid plasmids were transferred from them into the *Salmonella* tester strains by F-mediated conjugation.

The hybrid plasmids derived from pBR322 were introduced directly by transformation into the *Salmonella* tester strains, which carried the *galE* mutation, according to the method of Lederberg and Cohen (1974).

RESULTS

Flagellar Phase Variation of *vh2*$^{-}$ *Salmonella* Carrying Prophage Genomes in Their Cytoplasms

The phase-2-stable (*vh2*$^{-}$) *Salmonella* strains KK1251 and KK1252 produced swarms on NGA when prophages of P1, P7, and Mu were introduced (Table 1). Expression of phase 1 by the swarm-forming cells was confirmed by specific agglutination with anti-gt serum. Successive cultures of the swarm-forming cells resulted in the recurrent production of phase-2, nonmotile cells. When these prophages were introduced into a diphasic (*vh2*$^{+}$) *Salmonella* strain, LT2, subclones predominantly in phase 1 and those predominantly in phase 2 appeared recurrently from a carrier bacterial clone.

Two explanations are possible for these results. One explanation is that the prophage genomes suppress *vh2*$^{-}$ of the carrier bacteria, resulting in the occurrence of frequent flagellar phase variation. The second explanation is that they suppress *rh1*$^{+}$ activity and let the phase-1 phenotype be expressed. To discriminate between these possibilities, flagellar phases of the tester bacteria cured of these prophages were examined. The cured clones originated from KK1251 and KK1252 that carried the plasmids were selected at the nonpermissive temperature for the plasmids and isolated as Cm- and/or P1-sensitive and as lactose-nonfermentative, respectively. The flagellar phases of the independently isolated cured clones were identified by their swarm-forming ability on NGA. If the first explanation is correct, phase-1 and phase-2 cells existing at the time of the curing must be fixed at the existing phase and both phase-1- and phase-2-stable clones must be detected by this procedure. On the other hand, if the second explanation is correct, the cured clones must become phase-2, i.e., nonmotile, due to the release of suppression against *rh1*$^{+}$. As shown in Table 2, clones fixed in phase 1 and in phase 2 were detected, indicating that the first explanation is correct. Segregation of phase-1 and phase-2 clones in the cured KK1251 in which Tn*9* (Cm^{r}) of P1Cm*clr*100 was present as a chromosomal constituent means that the innate phage P1 genome is effective for suppression.

A remarkable result derived from these experiments was that the frequency of change from phase 1 to phase

Table 2. Flagellar Phase of the *vh2*$^{-}$ Strains Cured of Prophage P1 or Mu

Bacteria strains	Selected phenotypes	No. of cured colonies tested	No. of colonies fixed in: motile (phase 1)	No. of colonies fixed in: nonmotile (phase 2)	Frequency of change from phase 1 to phase 2
KK1251/	T^{r}	400	70	330	$>2.8 \times 10^{-2}$
P1Cm*clr*100	T^{r}, Cm^{r}	200	64	136	$>2.3 \times 10^{-2}$
KK1252/	T^{r}	260	140	120	$>1.5 \times 10^{-2}$
F′*lacI*^{s}::Mu*cts*62					

Each motile clone carrying either of the plasmids was grown in Luria (L) broth (Miller 1972) at 30°C for about 30 generations. The cultures were then diluted and spread on indicator plates, and the colonies growing at the nonpermissive temperature (42°C) for the plasmids were each isolated, cultivated as independent clones, and examined for their motilities after the absence of the plasmids was confirmed. The cured clones selected by both temperature resistance (T^{r}) and chloramphenicol resistance (Cm^{r}) were inferred to be KK1251 with Tn*9* (Cm^{r}) of P1Cm*clr*100 inserted in their chromosomes. The frequency of change from phase 1 to phase 2 was estimated according to the method of Stocker (1949), assuming that the ratio of motile and nonmotile clones corresponds to that of the phase-1 and phase-2 cells existing at the time of spreading of the culture.

2 in the prophage-carrying clones was estimated to be as high as $\sim 2 \times 10^{-2}$ per cell-generation cycle. This value is 100-fold higher than those calculated for diphasic *S. typhimurium* strains (Stocker 1949).

Phage P1 Mutants Defective in Both Suppression of *vh2⁻* and Their Own Inversions

To determine whether or not the suppressor activity of the phage P1 genome to $vh2^-$ is associated with the inversion system of its own C region, mutants that failed to suppress $vh2^-$ were isolated. A culture of P1Cm*clr*100 lysogen was mutagenized by *N*-methyl-*N'*-nitro-*N*-nitrosoguanidine and the phage particles were obtained from it. Through infection of the phages and the selection of Cm^r cells, clones of KK1251 that were lysogenic with respect to the mutagen-treated P1Cm*clr*100 were established. Each clone thus obtained was examined for its ability to produce swarms on NGA. Among them, 15 clones were found to fail in swarm formation, i.e., they were mutants that were unable to suppress $vh2^-$ of the host bacteria.

Phage particles were prepared from each of these mutants and a phase-1-stable strain, KK1253, was lysogenized with them. Clones cured of phages were obtained by the procedure described in the notes to Table 2, and the flagellar phases of their subclones were examined. All of the 500 subclones examined from all clones were motile, indicating that these clones were all fixed in phase 1. In a parallel experiment employing wild-type phages, both phase-1- and phase-2-fixed subclones were detected from the individual clones at a ratio of about 1.5:1. Thus, these mutants were inferred to be unable to suppress the change not only from phase 2 to phase 1, but also from phase 1 to phase 2.

Phage DNAs prepared from two representatives of these mutants (P1 no.17 and no.28) were denatured and reannealed by the formamide technique and observed by electron microscopy according to the procedures of Yamagishi et al. (1976). The inversion loop of about 3000 bp specific for the C region was observed in 10 of 51 molecules prepared from the wild-type phages, whereas no inversion loop was detected among more than 50 molecules observed from the mutant phages. This result indicates that the mutants also failed to invert the C region of their own DNAs.

Localization of the *vh2⁻* Suppressors on the Chromosomes of P1 and Mu

To identify the chromosomal localization of the $vh2^-$ suppressors of phages P1 and Mu, methods of restriction mapping were adopted.

For the analysis on P1, the largest fragment (fragment 1) obtained by digestion of P1Cm*clr*100 DNA with *Eco*RI was inserted into the *Eco*RI cleavage site of pCR1. The resulting hybrid plasmid, pKK2, was effective for suppression of $vh2^-$ when it was introduced into KK1252. Plasmids lacking various regions of the P1 DNA of pKK2 were constructed by excision with restriction endonucleases and fusion of the excised termini with ligation. Among these plasmids, those that could not suppress $vh2^-$ in the tester bacteria were commonly missing the *Bgl*II-excised segment of about 5 kb in which the C region resided (pKK22, pKK23, and pKK24 in Fig. 2a). The *Bgl*II-excised segment was then

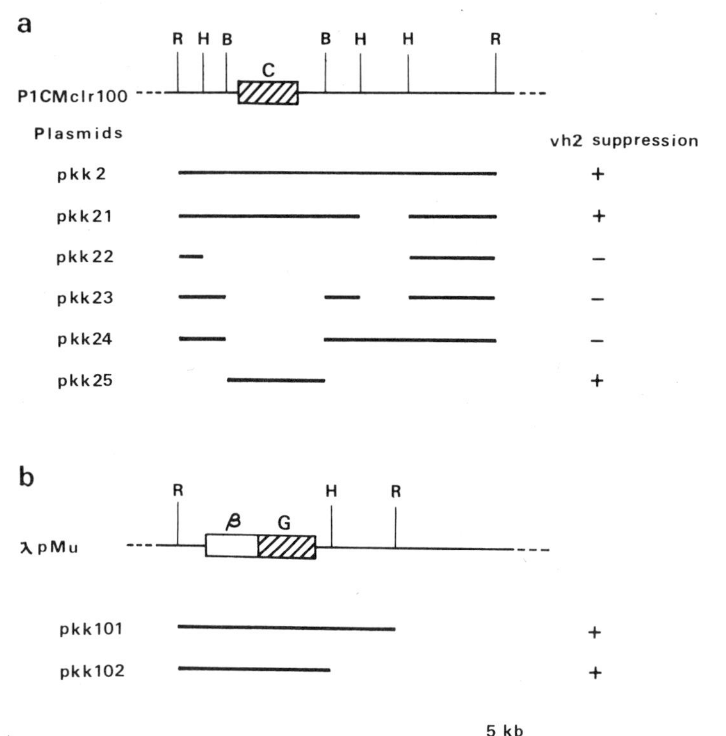

Figure 2. $vh2^-$ suppressor activity of various P1Cm*clr*100 (*a*) and λpMu (*b*) DNA segments cloned with plasmid vectors. (R, H, B) Cutting sites of restriction endonucleases *Eco*RI, *Bam*HI, and *Bgl*II, respectively. (C, β, G) C region, β segment, and G segment of the phage genomes. Lines on each hybrid plasmid denote the DNA regions of the P1 or Mu genome existing in the hybrid DNA.

inserted in the *Bam*HI site of pBR322. This hybrid plasmid (pKK25) was found to suppress *vh2*⁻ in the tester bacteria, although pBR322 itself could not suppress *vh2*⁻. Thus, the location of the *vh2*⁻ suppressor of P1 was confined in the 5-kb segment including the C region.

By using similar procedures we found that the *Eco*RI-excised segment covering the β-G segment of Mu DNA showed suppressor activity (pKK101 in Fig. 2b). A hybrid plasmid of pBR322 in which its own *Eco*RI—*Bam*HI excision segment was substituted by the *Eco*RI—β-G—*Bam*HI segment of pKK101 was also active for the suppression (pKK102 in Fig. 2b). Thus, the *vh2*⁻ suppressor of Mu was assigned to the β-G segment of 5 kb.

DISCUSSION

At least two factors are prerequisite for the inversion of a specific DNA segment. One is a *cis*-acting factor, i.e., a specific DNA sequence in the inverting segment. The other is a *trans*-acting one, i.e., a cytoplasmic factor that recognizes the specific DNA sequence and inverts the DNA segment. In flagellar phase variation of *Salmonella,* the former resides in *PD* and the latter corresponds to the product of *vh2* (Kutsukake and Iino 1980a).

The results reported here clearly show that prophage genomes of P1 and Mu can suppress *vh2*⁻ of *Salmonella* and markedly enhance the frequency of flagellar phase variation in phase-stable (*vh2*⁻) strains. Their effectiveness in their plasmid states indicates that they are active in *trans* position to the chromosome of the host bacteria. Thus, the genomes of those phages were inferred to have genes homologous to *vh2* with regard to the regulation of phase variation. As has been attributed to the function of *vh2* (Kutsukake and Iino 1980a), those genes may produce an effector that inverts *PD.* We termed such activity "din activity" and the genes of the bacteriophages exerting this activity as *din* (Kutsukake and Iino 1980a).

After examining the din activity of cloned DNA segments of P1 and Mu genomes, it was inferred that *din* genes reside in or near the C region of P1 and in the β-G segment of Mu. It is worth noting that the specific inversions have been detected in these segments (Howe and Bade 1975; Bächi and Arber 1977); that is, the *din* genes are located closely associated with the specific inversion segments in these bacteriophages.

In bacteriophage Mu, the *gin* gene, which regulates the frequency of inversion of the G segment, has been shown to be located in the β segment (Chow et al. 1977). The G segment of a *gin* mutant that failed to invert the segment recovered invertibility through the inversion system of P1 (Kamp et al. 1978). In this report it was further demonstrated that *din* mutants that failed to invert *PD* also failed to invert their own C regions. These phenomena strongly suggest that the inversion systems of *Salmonella,* P1, and Mu operate alike and that *gin* and *din* are identical.

The invertible DNA regions of P1 and Mu are homologous not only in size, i.e., about 3000 bp, but also in base sequences (Chow and Bukhari 1976). This size is more than three times larger than that of *PD* in *Salmonella* (Zieg et al. 1977). This difference may not necessarily reflect differences between the inversion systems of bacteriophages and *Salmonella;* rather, it may mean that the specific regions involved in the inversion of a DNA segment are relatively short and consist of only a part of the inversion segment.

With regard to the homology of the specific inversion systems of Mu and P1, a possibility has been discussed that they have evolutionally derived from a common transposable element (Kamp et al. 1979). Similarly, the possibility may exist that the *vh2-PD* system also has derived from the same common origin as the inversion systems of these bacteriophages. It has been assumed that the *H2* locus of *Salmonella* originated by translocation of *H1* (Iino 1961b, 1969). For example, the *H1* gene of an ancestral *Salmonella* may have been picked up by a bacteriophage and inserted together with a transposable element in the phage genome into a new site in the *Salmonella* chromosome. Thereafter, structural differentiation of the inserted segment may have resulted in the establishment of the new gene, *H2,* and its controlling system, which operates through the specific inversion.

ACKNOWLEDGMENT

This work was supported by a Grant-in-Aid (410708) for Scientific Research from the Ministry of Education, Science, and Culture, Japan.

REFERENCES

Allet, B. and A. I. Bukhari. 1975. Analysis of bacteriophage Mu and λ-Mu hybrid DNAs by specific endonucleases. *J. Mol. Biol.* **92:** 529.

Andrews, F. W. 1922. Studies in group-agglutination. I. The *Salmonella* group and its antigenic structure. *J. Pathol. Bacteriol.* **25:** 515.

Bächi, B. and W. Arber. 1977. Physical mapping of *Bgl*II, *Bam*HI, *Eco*RI, *Hin*dIII and *Pst*I restriction fragments of bacteriophage P1 DNA. *Mol. Gen. Genet.* **153:** 311.

Bolivar, F., R. Rodriguez, M. Betlach, H. L. Heyneker, H. W. Boyer, J. Crosa, and S. Falkow. 1977. Construction and characterization of new cloning vehicles. II. A multipurpose cloning system. *Gene* **2:** 95.

Bukhari, A. I. and B. Allet. 1975. Plaque-forming λ-Mu hybrids. *Virology* **63:** 30.

Bukhari, A. I. and L. Ambrosio. 1978. The invertible segment of bacteriophage Mu DNA determines the adsorption properties of Mu particles. *Nature* **271:** 575.

Chow, L. T. and A. I. Bukhari. 1976. The invertible DNA segments of coliphages Mu and P1 are identical. *Virology* **74:** 242.

Chow, L. T., R. Kahmann, and D. Kamp. 1977. Electron microscopic characterization of DNAs of non-defective deletion mutants of bacteriophage Mu. *J. Mol. Biol.* **113:** 591.

Covey, C., D. Richardson, and J. Carbon. 1976. A method for the detection of restriction sites in bacterial plasmid DNA. *Mol. Gen. Genet.* **145:** 155.

Fujita, H., S. Yamaguchi, and T. Iino. 1973. Studies on H-O variants in *Salmonella* in relation to phase variation. *J. Gen. Microbiol.* **76:** 127.

HEDGES, R. W., A. E. JACOB, P. T. BARTH, and N. J. GRINTER. 1975. Compatibility properties of P1 and φAMP prophages. *Mol. Gen. Genet.* **141:** 263.

HOWE, M. M. and F. G. BADE. 1975. Molecular biology of bacteriophage Mu. *Science* **190:** 624.

HOWE, M. M., J. W. SCHUMM, and A. L. TAYLOR. 1979. The *S* and *U* genes of bacteriophage Mu are located in the invertible G segment of Mu DNA. *Virology* **92:** 108.

IINO, T. 1961a. A stabilizer of antigenic phase in *Salmonella abortusequi. Genetics* **46:** 1465.

———. 1961b. Anomalous homology of flagellar phases in *Salmonella. Genetics* **46:** 1471.

———. 1969. Genetics and chemistry of bacterial flagella. *Bacteriol. Rev.* **33:** 454.

IINO, T. and M. ENOMOTO. 1971. Motility. In *Methods in microbiology* (ed. N. R. Norris and D. W. Ribbons), vol. 5A, p. 145. Academic Press, New York.

KAMP, D., R. KAHMANN, D. ZIPSER, T. R. BROKER, and L. T. CHOW. 1978. Inversion of the G DNA segment of phage Mu controls phage infectivity. *Nature* **271:** 577.

KAMP, D., L. T. CHOW, T. R. BROKER, D. KWOH, D. ZIPSER, and R. KAHMANN. 1979. Site-specific recombination in phage Mu. *Cold Spring Harbor Symp. Quant. Biol.* **43:** 1159.

KUTSUKAKE, K. and T. IINO. 1980a. A *trans*-acting factor mediates inversion of a specific DNA segment in flagellar phase variation of *Salmonella. Nature* **284:** 479.

———. 1980b. Inversion of specific DNA segments in flagellar phase variation of *Salmonella* and inversion systems of bacteriophages P1 and Mu. *Proc. Natl. Acad. Sci.* (in press).

LEDERBERG, E. M. and S. N. COHEN. 1974. Transformation of *Salmonella typhimurium* by plasmid deoxyribonucleic acid. *J. Bacteriol.* **119:** 1072.

LEDERBERG, E. M. and B. A. D. STOCKER. 1970. Phase variation in *rec*$^-$ of *Salmonella typhimurium. Bacteriol. Proc.*, p. 35.

LEDERBERG, J. and P. R. EDWARDS. 1953. Serotypic recombination in *Salmonella. J. Immunol.* **71:** 232.

LEDERBERG, J. and T. IINO. 1956. Phase variation in *Salmonella. Genetics* **41:** 744.

LEE, H. J., E. OHTSUBO, R. C. DEONIER, and N. DAVIDSON. 1974. Electron microscopic heteroduplex studies of sequence relations among plasmids of *Escherichia coli.* V. *ilv*$^+$ deletion mutants of F14. *J. Mol. Biol.* **89:** 585.

LENNOX, E. S. 1955. Transduction of linked genetic characters of the host by bacteriophage P1. *Virology* **1:** 190.

MILLER, J. H. 1972. *Experiments in molecular genetics.* Cold Spring Harbor Laboratory, Cold Spring Harbor, New York.

NISHIMURA, Y., Y. TAKEDA, A. NISHIMURA, H. SUZUKI, M. INOUYE, and Y. HIROTA. 1977. Synthetic ColE1 plasmids carrying genes for cell division in *Escherichia coli. Plasmid* **1:** 67.

PEARCE, U. B. and B. A. D. STOCKER. 1967. Phase variation of flagellar antigens in *Salmonella:* Abortive transduction studies. *J. Gen. Microbiol.* **49:** 335.

ROSNER, J. L. 1972. Formation, induction, and curing of bacteriophage P1 lysogens. *Virology* **48:** 679.

SILVERMAN, M., J. ZIEG, M. HILMEN, and M. SIMON. 1979. Phase variation in *Salmonella:* Genetic analysis of a recombinational switch. *Proc. Natl. Acad. Sci.* **76:** 391.

STOCKER, B. A. D. 1949. Measurement of rate of mutation of flagellar antigenic phase in *Salmonella typhimurium. J. Hyg.* **47:** 398.

SUGIURA, M. and J. KUSUDA. 1979. Molecular cloning of tobacco chloroplast ribosomal RNA genes. *Mol. Gen. Genet.* **172:** 137.

SUZUKI, H. and T. IINO. 1973. *In vitro* synthesis of phase-specific flagellin of *Salmonella. J. Mol. Biol.* **81:** 57.

VELTEN, J., K. FUKUDA, and J. ABELSON. 1976. *In vitro* construction of bacteriophage λ and plasmid DNA molecules containing DNA fragments from bacteriophage T4. *Gene* **1:** 93.

YAMAGISHI, H., H. INOKUCHI, and H. OZEKI. 1976. Excision and duplication of su3$^+$-transducing fragments carried by bacteriophage φ80. I. Novel structure of φ80 *sus2* psu3$^+$ DNA molecule. *J. Virol.* **18:** 1016.

YUN, T. and D. VAPNEK. 1977. Electron microscopic analysis of bacteriophages P1, P1CM, and P7. *Virology* **77:** 376.

ZIEG, J., M. HILMEN, and M. SIMON. 1978. Regulation of gene expression by site specific inversion. *Cell* **15:** 237.

ZIEG, J., M. SILVERMAN, M. HILMEN, and M. SIMON. 1977. Recombinational switch for gene expression. *Science* **196:** 170.

Analysis of the Functional Components of the Phase Variation System

M. SILVERMAN, J. ZIEG, G. MANDEL, AND M. SIMON
Department of Biology, University of California, San Diego, La Jolla, California 92093

The oscillation of flagellar phenotypes of *Salmonella*, termed phase variation, is regulated by a genetic rearrangement. The concept that specific recombinational processes may determine the nature of gene expression was elaborated by McClintock on the basis of her work with maize. McClintock (1957) suggested that mobile elements could become associated with specific gene systems and could be modified to regulate gene expression. The advent of new and powerful biochemical and genetic techniques has allowed the identification and physical analysis of transposable elements in prokaryotes (Kleckner 1977) and eukaryotes (Cameron et al. 1979; Potter et al. 1979). Such elements may contribute plasticity to the genome, which is important in regulating gene activity or in stimulating genetic diversity. Interconversion of yeast mating types (Strathern et al. 1980), generation of immunoglobulin diversity in mice (Maki et al. 1980), and control of antigenic variability in trypanosomes (Williams et al. 1979) all involve rearrangement of the structure of the genetic material. In prokaryotes, site-specific inversion of a DNA segment regulates expression of genes of bacteriophages Mu and P1 (Kamp et al. 1978) and of the *Salmonella* flagellar phase system (Zieg et al. 1977). Our objectives have been to dissect the structure of the phase variation system and to test the hypothesis, first formulated by McClintock, that there is a fundamental connection between transposable elements and the recombinational control of gene expression.

In *Salmonella*, two genes code for the major flagellar structural protein flagellin. These two genes, *H1* and *H2*, map in different regions of the *Salmonella* genome (Lederberg and Edwards 1953). The capacity of cells to alternate or switch between expression of *H1* and *H2* is called phase variation, and this variation of antigenicity presumably allows *Salmonella* to evade the host immune response. The frequency with which cells undergo phase transition varies with different *Salmonella* strains from 10^{-3} to 10^{-5} per bacterium per generation (Stocker 1949). The alternative expression of *H1* and *H2* is controlled by the state of a genetic element linked to *H2* (Lederberg and Iino 1956). Another gene, *rh1*, linked to and coordinately expressed with *H2*, codes for a repressor substance that prevents expression of *H1* (Fujita et al. 1973; Silverman et al. 1979a). Thus, when a cell is expressing *H2*, it is also expressing *rh1*. This results in the repression of expression of *H1*, and only *H2*-type flagella are formed. When a cell is in phase 1, neither the *H2* nor *rh1* product is synthesized, and *H1* can be expressed, leading to the formation of the *H1*-type flagella.

To understand the mechanism of phase variation at the molecular level, recombinant molecules containing the regions encoding *H1* and *H2* were constructed and cloned in *Escherichia coli*, where the phase variation effect could be reproduced. Genetic and physical analyses of the recombinant DNA molecules showed that inversion of a 1000-bp region adjacent to *H2* controlled the expression of the *H2* operon; in one orientation the *H2* operon was "on," and in the opposite orientation the *H2* operon was "off" (Zieg et al. 1977). The inversion of this region that contains the *H2* operon promoter was found to be site-specific and independent of the RecA recombination system of *E. coli* (Zieg et al. 1978; Silverman et al. 1979b). To define the mechanism by which the regulatory control element is inverted, we have isolated and characterized mutants defective in the switching process (Silverman and Simon 1980) and have determined the nucleotide sequence of the inversion region (Zieg and Simon 1980). The analysis reported here resulted in the definition of two functions necessary for recombinational gene switching: One function is *trans*-acting and is encoded by the *hin* gene (*H inversion*), which is located within the inversion region, and a second function is *cis*-acting and consists of a 14-bp sequence flanking the inversion region in the inverted repeat configuration. These functions are compared with those necessary for other site-specific recombinational events.

RESULTS

Genetic Definition of Inversion Functions

A hybrid λ phage (λ*fla*157) was constructed by inserting a 3.75-kb *Eco*RI restriction endonuclease fragment that carried the *H2* region derived from *Salmonella* onto a λ cloning vehicle. This hybrid phage was used to study *H2* switching in *E. coli*, which is monophasic and has only one flagellin-specifying gene (*hag*), which is homologous to *H1* of *Salmonella*. A Hag^- *E. coli* strain lysogenized with the hybrid phage alternated between a nonflagellate and a flagellate (*H2* serotype) phenotype. Cells with these two phenotypes could be distinguished conveniently by their susceptibility to flagellotropic phage χ (Silverman et al. 1979b). Cells lysogenized with phage in the *H2*-on configuration were sensitive to this phage, whereas lysogens of phage in the *H2*-off configuration were resistant. The proportion of cells in a population in either state could be measured as a function of the number of generations of growth, and thus the frequency of phase transition could be determined.

λ*H2* phage were mutagenized by insertion of the transposable element Tn*5* (Berg 1977; Kleckner et al. 1977), and the resulting λ*H2*::Tn*5* phage were characterized with regard to *H2* switching and to the location of the Tn*5* insertion (Silverman and Simon 1980). Lysogen clones that were defective in *H2*-on to *H2*-off switching were recognized as those that were sensitive to the flagellotropic phage. Only those lysogen clones harboring λ*H2*::Tn*5* phage with *H2* fixed in the on configuration as a result of a defect in a switching function would appear as phage-sensitive clones. The frequency of *H2*-on to *H2*-off for λ*H2*::Tn*5* lysogens with normal phase transition was approximately 10^{-2} per cell per generation. Thirteen independent λ*H2*::Tn*5* phage with reduced frequencies of *H2* switching were isolated.

Two classes of mutant phenotypes were apparent: One (intermediate class) showed approximately one fifth the frequency of *H2* switching, and a second (null class) showed an approximately 1000-fold reduction in the frequency of phase transition. The effect of the insertion was symmetrical; i.e., mutants defective in *H2*-on to *H2*-off switching showed a similar reduction in transition in the opposite direction, *H2*-off to *H2*-on. The locations of Tn*5* insertions were determined by Southern blot analysis of restriction fragments (Silverman and Simon 1980), and these locations are shown in Figure 1A. Mutants with the null switching phenotype contained Tn*5* insertions in a 500-bp region bounded by Tn*5* insertions λ*fla*242 and λ*fla*245. Mutants with the intermediate switching defects contained Tn*5* insertions in a region defined by insertions λ*fla*250, λ*fla*252, λ*fla*255, and λ*fla*257. Tn*5* stimulates the formation of deletions adjacent to the point of insertion (Berg 1977). The deletions usually have one endpoint within the transposable element and extend in either direction from that point into adjacent sequences. Such deletion mutants of hybrid λ*H2*::Tn*5* phage were obtained by chelating-agent (Na pyrophosphate) selection (Parkinson and Huskey 1971). Deletions originating at Tn*5* insertion points inside the inversion region in λ*fla*250, λ*fla*252, λ*fla*255, and λ*fla*257 would be particularly interesting as these insertions retain most of the phase transition function and the loss of remaining function could be correlated with deletions. Like the λ*H2*::Tn*5* mutants, the deletion mutants were characterized with regard to *H2* switching, and the approximate extent of deletion was determined by Southern blot hybridization analysis (Fig. 1B summarizes these results).

On the basis of our examination of the physical and genetic properties of λ*H2* deletion phage, the following conclusions were drawn. First, deletions past the crossover points for inversion completely abolished inversion; thus, inversion did not occur with hybrids λ*fla*350, λ*fla*356, λ*fla*358, and λ*fla*364. Second, inversion was reduced by almost 1000-fold when the deletion extended into the 500-bp region previously defined as necessary for inversion. Thus, extension of the deletion into the 120-bp or 190-bp fragments, e.g., in λ*fla*378 and λ*fla*381, markedly reduced inversion, and extensive deletion, as in λ*fla*364, completely eliminated inversion. Deletion mutant λ*fla*385 showed inversion. In fact, the *H2* switching phenotype of hybrid λ*fla*385 was similar to that of the wild-type λ*H2* phage (λ*fla*157) rather than to that of the parent λ*H2*::Tn*5* hybrid (λ*fla*255). The reduction in switching frequency observed with the intermediate class of switching mutant (i.e., λ*fla*255) could therefore be attributed to elongation of the inversion region by insertion of the Tn*5* element and not to the interruption of a particular switching function located at the point of Tn*5* insertion. Third, λ*fla*380 and λ*fla*351 showed wild-type switching frequencies, as measured by restriction analysis, but λ*fla*351 had lost the ability to express *H2*. As expected, hybrids λ*fla*350, λ*fla*356, and λ*fla*358 also had the $H2^-$ phenotype. These data suggest that the location of the *H2* promoter is within the first 100 bp of the invertible control region. Figure 1C summarizes the genetic functions necessary for *H2* switching: sites where inversion crossover takes place, IR(L) and IR(R), and a region of about 500 bp, *hin*, located inside the inversion region.

Tn*5* insertions in a region of approximately 500 bp inside the inversion sequence resulted in loss of a function necessary for normal *H2* switching. To test the possibility that this region codes for a *trans*-acting factor necessary for *H2* switching, a genetic complementation test was devised. One such arrangement is shown in Figure 2. Here, a λ*H2*::Tn*5* switching mutant (λ*fla*247) was grown in cells that either were coinfected with another phage (λ*fla*350) or contained a plasmid (pJZ110); the second phage or the plasmid provided the region necessary for inversion in *trans*. Phage grown in such a manner were used to lysogenize a Hag^- *E. coli* strain, and kanamycin-resistant (Km^r) lysogens (those that contained the λ*H2*::Tn*5* hybrid) were examined. DNA from the mutant λ*H2*::Tn*5* phage was fixed in the *H2*-on configuration, and phage in the *H2*-off configuration arose only if the *trans* donor element complemented *H2* switching to give the complementation product shown in Figure 2. Those λ*H2*::Tn*5* phage that had switched to *H2*-off were recognized as Km^r lysogens that were resistant to the flagellotropic phage χ. (For a more detailed account of this complementation test, see Silverman and Simon [1980].) From an examination of the phage recovered from the complementation test as lysogens, it was apparent that a significant proportion of the mutant population switched to the *H2*-off phenotype. When the function was provided in *trans*, the mutant phage showed frequencies of *H2* switching from one eighth to one fourth that of the wild-type control. Thus, the region of DNA provided in *trans* can complement switching defects, and we designate this region the *hin* gene. λ*H2* phage with deletions in the *hin* gene could also be complemented to switch, but only if the crossover points were intact, which suggests that these sites are *cis*-acting. Hybrid λ*H2*::Tn*5* phage (null class) switched at a low, but measurable, frequency in Rec^+ *E. coli* strains but were incapable of any measurable switching in $RecA^-$ hosts. Apparently spontaneous (*hin*-independent) switching can occur through RecA-mediated recombi-

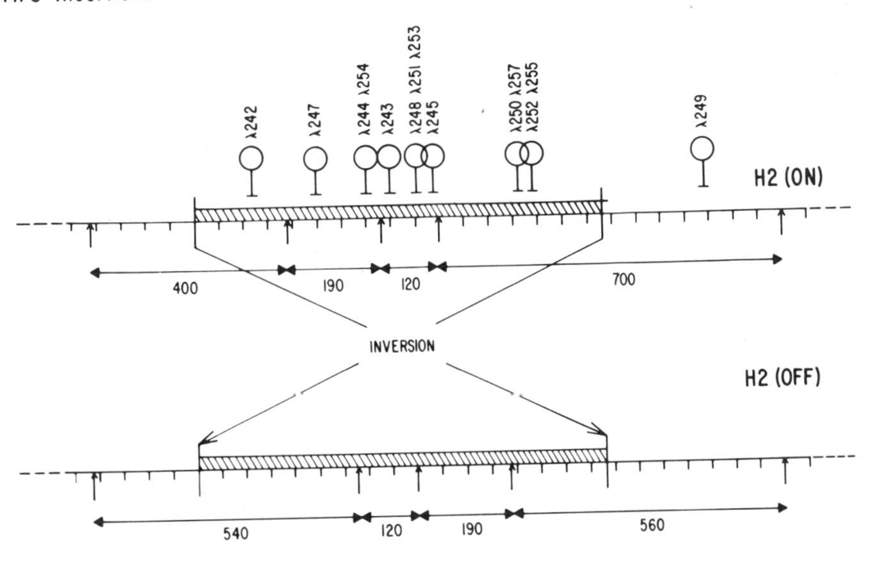

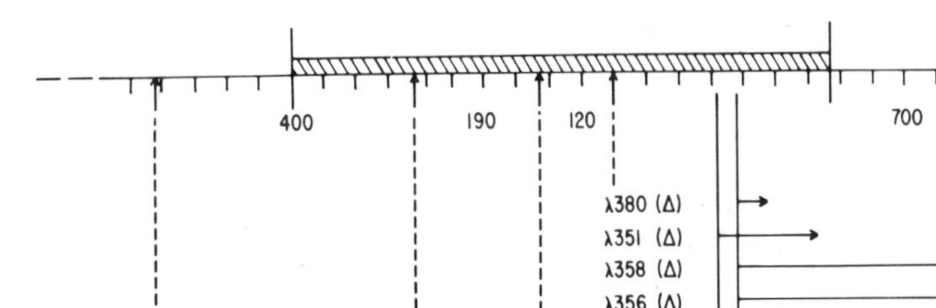

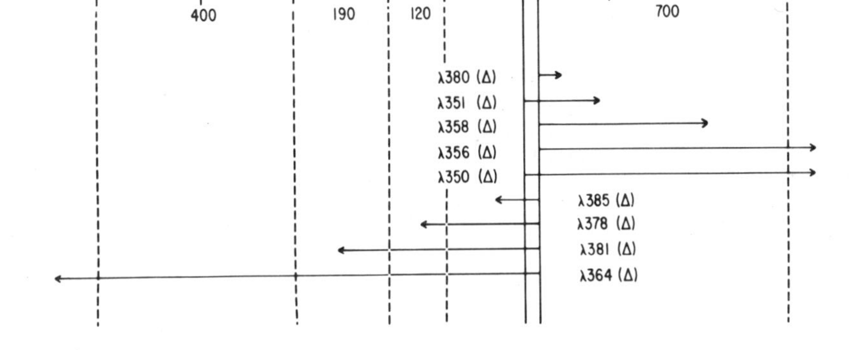

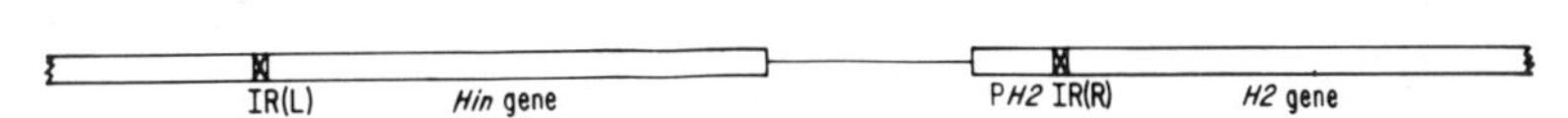

Figure 1. H2 switching mutants. *(A)* Locations of Tn*5* insertions in mutant λ*H2*::Tn*5* phage. The inversion region (▨) and adjacent DNA sequences (*H2* gene on right) are shown in both the *H2*-on and *H2*-off configurations. Restriction of DNA at *Hpa*II sites (vertical arrows) resulted in six DNA fragments. The 700-bp and 400-bp fragments are characteristic of the *H2*-on phase, and the 560-bp and 540-bp fragments are characteristic of the *H2*-off phase. Length of DNA is marked in 50-bp intervals, and positions of Tn*5* insertions are accurate to approximately 25 bp. Tn*5* positions were determined from Southern blot analysis of *Hpa*II restriction fragments of λ*H2*::Tn*5* DNA (see Silverman and Simon 1980). *(B)* Tn*5*-generated deletion mutants derived from four λ*H2*::Tn*5* mutants (intermediate switching class). One terminus of the deletion is within an arm of the Tn*5* element (vertical solid lines), and deletion extends through Tn*5* DNA (kanamycin determinant is deleted) to another terminus in *H2* DNA (horizontal arrows). Deletion endpoints are imprecise but are accurate relative to *Hpa*II sites (vertical dashed lines), to crossover points for inversion (vertical lines at end of inversion sequence), and to other deletion endpoints. Mapping of deletions was done by restriction analysis as in *A*. *(C)* Functions defined by the phenotypes of the mutant phage.

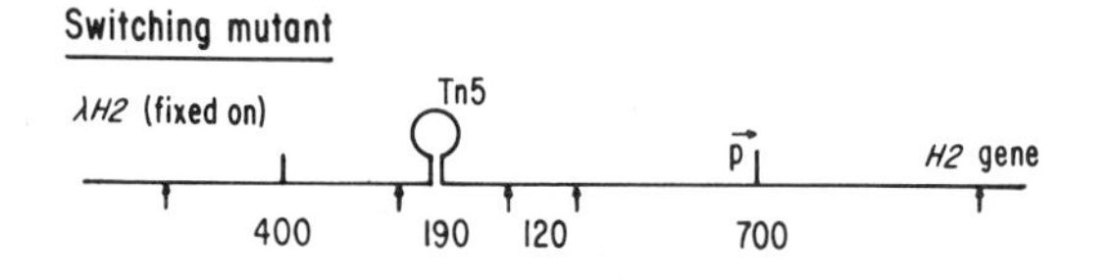

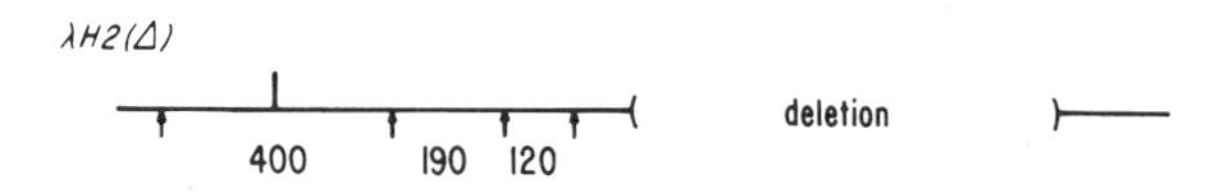

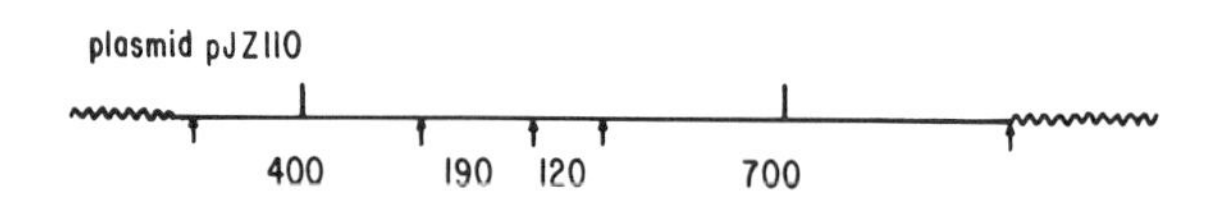

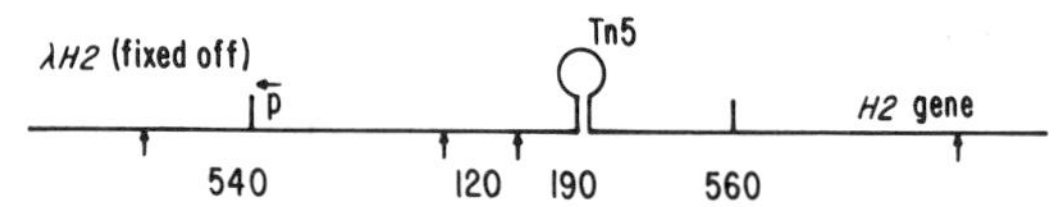

Figure 2. Complementation analysis of switching mutants. Genetic test to determine whether λ*H2*::Tn*5* mutants could be complemented to switch from *H2*-on to *H2*-off by providing a function in *trans*. Null-class λ*H2*::Tn*5* switching mutants (i.e., λ*fla*247) were used to infect cells that were either coinfected with a λ*H2* deletion mutant, such as λ*fla*350 (*trans*-function donor), or already contained a hybrid *H2* plasmid, such as pJZ110 (*trans*-function donor). The resulting phage lysate was used to form λ*H2*::Tn*5* lysogens, and the H2 phenotype of the lysogen was then measured by the $\varkappa$ resistance test. DNA was purified from $\varkappa$-resistant lysogens (complementation product) and subjected to restriction analysis to verify the structure of the λ*H2*::Tn*5* resulting from *trans* complementation.

nation, albeit at very low frequencies (approximately 10^{-6} per cell per generation). The crossover points may contain homologous sequences that can be recognized by the RecA system and lead to inversion. The *hin*-mediated system provides a site-specific, RecA-independent pathway that allows phase transition to occur at much higher frequencies (10^{-2} per cell per generation).

A number of years ago, Iino (1961) found a naturally occurring variant in *Salmonella* that was fixed with respect to phase transition. He identified a genetic locus, *vh2*⁻, that acted as a stabilizer of phase transition. Although the physical nature of *vh2*⁻ is not known, Kutsukaki and Iino (1980) showed recently that this variant could be complemented to switch by the inversion functions specified by Mu and P1. They also showed that a Hin^+ plasmid constructed in our laboratory could complement the *vh2*⁻ variant. Therefore, in complementation tests, the *vh2*⁻ defect was equivalent to a *hin* defect, and, as expected, the Hin^- mutants described in this paper were complemented efficiently in lysogens when Mu or P1 was present. We found that Mu- or P1-complemented switching occurred at approximately one fifth the frequency observed for Hin^+ phage. Using mutants defective in the Mu inversion function *gin* (*G*-loop *in*version), Kamp and Kahmann (this volume) performed the reciprocal complementation test and found that Gin^- mutants were capable of inversion when the *hin* function was provided in *trans*. These results suggest that each of these inversion systems (H2, Mu, and P1) codes for a polypeptide that is required for the inversion event and that these functions, *gin* and *hin*, are homologous, even though the precise structure of the invertible sequence in each system is probably quite different. Tn*3* and Tn*5* did not complement Hin^- mutants and thus do not have a function equivalent to *hin*.

Nucleotide Sequence Determination

To define precisely the functions necessary for *H2* switching, derivatives of plasmid pJZ110, which carries a fragment of the *Salmonella* genome containing the inversion region and flanking sequences, were used to determine the nucleotide sequence (Zieg and Simon 1980). The sequences of the inversion region and the proximal arm segments, determined according to the method of Maxam and Gilbert (1980), are shown in Figure 3. It was possible to identify the precise locations of the ends of the inversion region by comparing DNA sequences of fragments that represent the *H2*-on and *H2*-off orientations. In a plasmid population that could exist in both orientations, two sets of restriction fragments that extended from within the inversion region into the arm sequences could be obtained. Each set represented one orientation of the inversion region. When the sequences of these four end fragments were compared, a 14-bp inverted repeat sequence was found to exist at the boundaries of the inversion region. The recombinational event that resulted in the inversion of the DNA segment must have occurred between these 14-bp inverted repeat sequences, and the homology shared by the inverted repeat sequences apparently was necessary for this process. These repeats also corresponded to the crossover sites defined genetically as *cis*-acting. These inverted repeat sequences (TTATCAAAAACCTT) are indicated in Figures 3 and 4 as IR(L) and IR(R).

An analysis of this 14-bp inverted repeat sequence has revealed that it shares homology with DNA regions in other systems that are involved in recombination and transposition. This sequence resembles the λ*att* core region (7, TTA[G]TA[T]AAAAA[A]GCTG, -9) as well as regions on the P and P′ arms (-64, TTAT[G]CAAAATCTAA, -50 P; 15, TTATAAAAAAGCAT, 28 P′) (Landy and Ross 1977). Two regions within the Tn*10* inverted repeat show some homology (a, TTATCAAAATCATT; b, TT[G]ATAAAAATCATT) (Kleckner 1979), as does a region just outside the inverted repeat of Tn*3*, a noncoding region of the transposon (4889, TTATCAAAAAGGAT, 4902) (Heffron et al. 1979). (Brackets indicate additional bases; underlined bases differ from the *H2* inverted repeat sequence.) These sequence similarities (as well as functional similarities to Mu and P1)

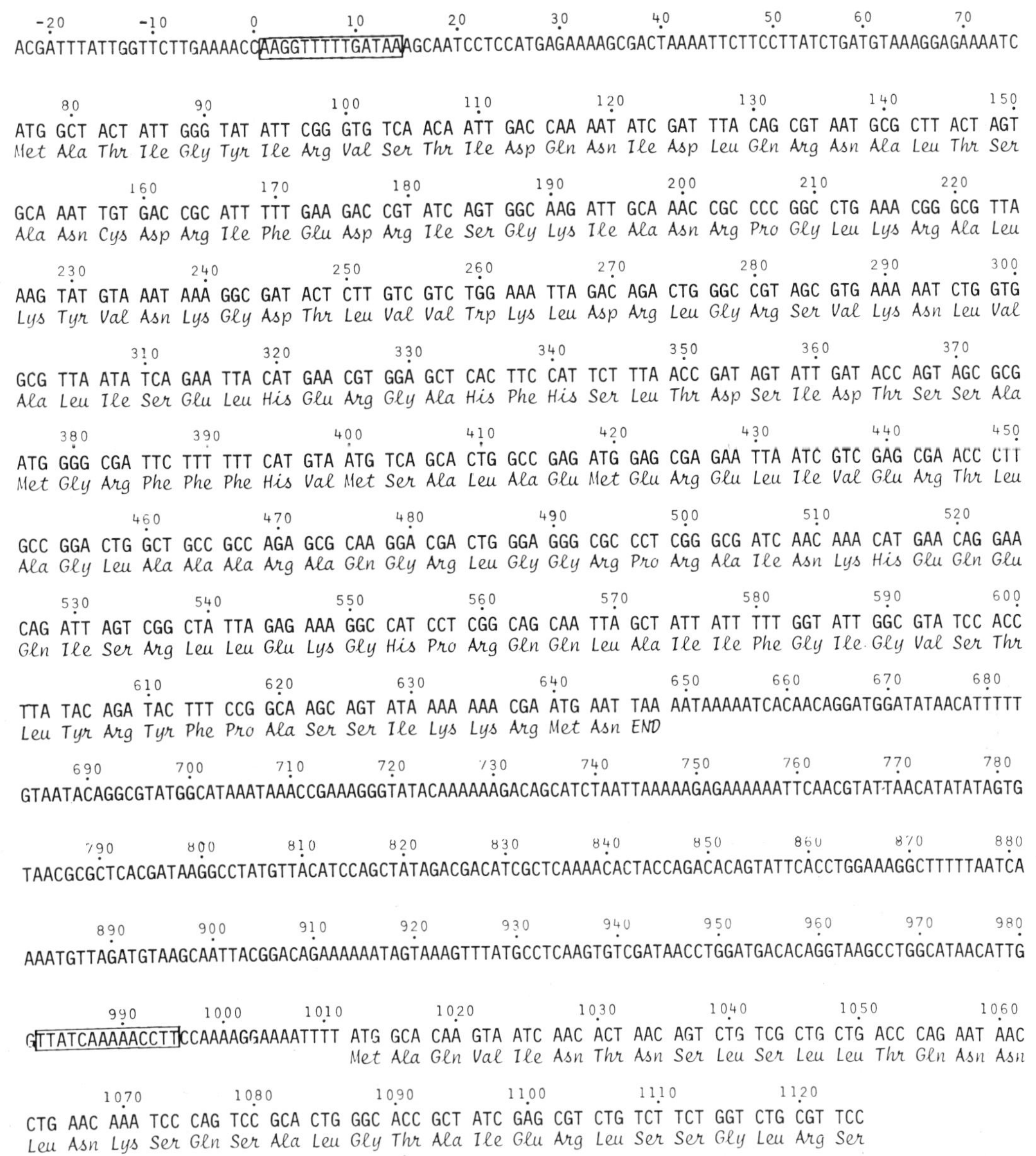

Figure 3. Nucleotide sequence of the inversion region. The 5′→ 3′ strand of the nucleotide sequence of the inversion region and the proximal arm sequences are shown. The inversion region, which extends from +1 to +995, includes flanking inverted repeat sequences (shown within boxes). Amino acid sequences predicted by the nucleotide sequence are shown for the *hin* and *H2* structural genes. The *hin* gene is located between nucleotides 76 and 648. The *H2* gene begins at position 1012 and continues rightward beyond the end of the nucleotide sequence.

suggest an evolutionary relationship among all of these movable genetic elements.

The direction of transcription of *H2* was shown genetically to be left to right, as indicated in Figure 2 (Silverman et al. 1979b). The nucleotide sequence revealed only one possible ATG sequence on the 5′→ 3′ strand (not followed by an in-phase termination codon) that could correspond to the translational start signal for *H2*. Confirmation of this site as the beginning of the *H2* structural gene came from a comparison of the amino acid sequence predicted by the nucleotide sequence with that determined directly from purified *H2* flagellin (*enx* serotype) (Zieg and Simon 1980). The agreement of these two sequences indicates that the *H2* structural gene commences at the ATG sequence at position 1012, 16 bp outside the inversion region to the right of IR(R). This sequence is preceded by the sequence GGAA at position 1002, which resembles part of the sequence proposed for ribosome binding (Shine and Dalgarno 1974). Only the reading frame containing this ATG remained open for the 114 bp that have been sequenced. The 16 bp between IR(R) and the beginning of the *H2* structural gene do not correspond to sequences known to be involved in transcription initiation. In addition, the genetic evidence discussed above indicates that the promoter for *H2* synthesis probably lies within the first 100 bp of the inversion region. Analysis of the nucleotide sequence in this region reveals several possible sites

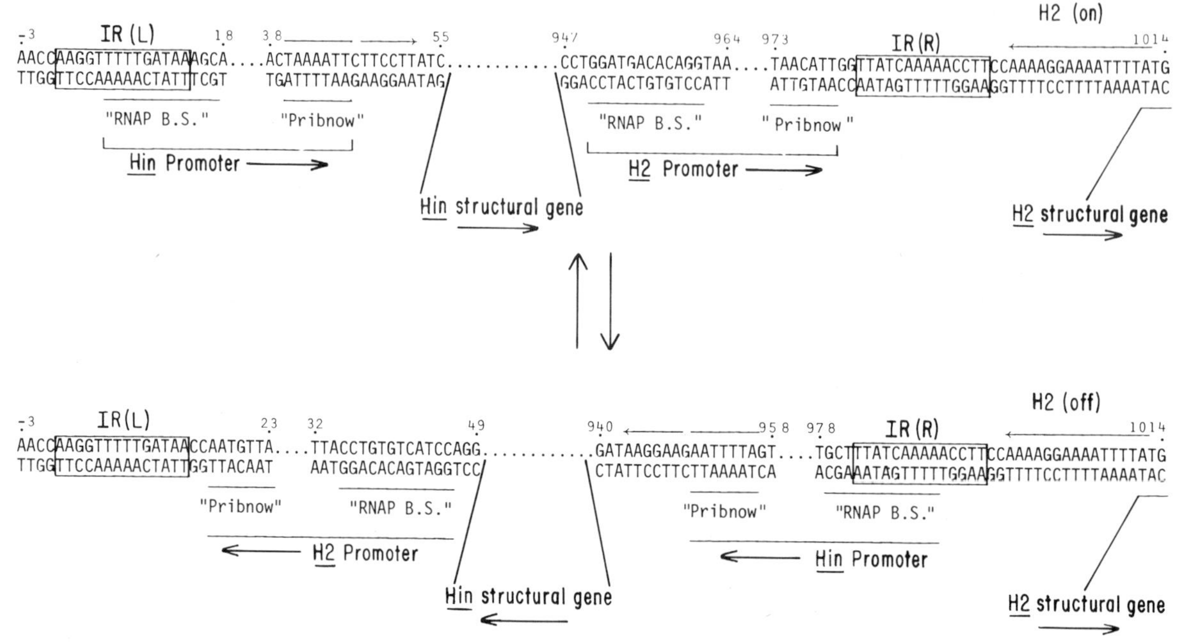

Figure 4. Location of sites for initiation of transcription. The nucleotide sequence of the inversion region is shown in both the *H2*-on and *H2*-off configurations. Flanking inverted repeat sequences are shown within boxes. The locations of the putative *hin* and *H2* promoters are shown below the nucleotide sequence. Sequences that resemble the prototypic RNA polymerase binding site (RNAP B.S.) and the site that precedes transcription initiation (Pribnow) are underlined, as are the ATG start codons for *hin* and *H2*. The sequences between nucleotides 40 and 53 and between nucleotides 1000 and 1012 (numbering as in the *H2*-on phase) constitute an inverted repeat in the *H2*-on configuration and a direct repeat in the *H2*-off configuration.

that resemble prototypic sequences for promoters (Fig. 4). The most homologous sequence is found between positions 950 and 979. The sequence between 973 and 979 resembles the consensus Pribnow sequences (Pribnow 1975). It is preceded at positions 950 to 961 by a region similar to the prototypic sequence for RNA polymerase recognition (Seeburg et al. 1977). However, other sequences in this region also may be recognized as promoter sequences, and the precise location of initiation of transcription of *H2* is currently being determined using enzymatic techniques.

Transcription of *H2* is regulated positively by a variety of *trans*-acting elements (Silverman and Simon 1974). However, the exact sequences that might be involved in the binding of a regulatory protein are not apparent. There is an open translational frame preceding *H2* that would allow for the synthesis of a polypeptide of 56 amino acids encoded by the nucleotides from position 805 to position 975. There is no evidence, however, that such a protein is synthesized from this region of the DNA or that it plays any role in the regulation of *H2* expression. Another sequence that might play a role in *H2* expression is located between the ATG start codon of *H2* and the inverted repeat sequence. This 20-bp region can form a stem-loop structure by intrastrand hydrogen bonding, but there is no evidence that this sequence has a function in gene regulation. Some of the DNA region shown genetically to be nonfunctional both in the inversion process and in the expression of *H2* flagellin lies within the high-A-T-rich region between nucleotides 648 and 981. The DNA segments between nucleotides 648 and 783 and between nucleotides 863 and 928 have an A + T content of approximately 72%. Such sequence compositions are characteristic of regions upstream of promoters (Nakamura and Inouye 1979) and they may be important in enhancing promoter activity and thus in regulating *H2* expression.

The genetic studies discussed above indicate that a region within the inversion is required to encode a product necessary for inversion. The nucleotide sequence provides direct evidence for the existence of the *hin* gene. There is an open translational frame that begins with the ATG at position 76 and terminates with the TAA sequence at position 648 (190 amino acids). The sequence GGAG precedes the ATG at position 66 and may serve as part of the site for ribosome binding. The predicted amino acid sequence of the *hin* protein is shown in Figure 3. Although no DNA homology has been found between this region and other genes coding for proteins involved in recombination and transposition, a striking homology exists between the amino acid sequences of the *hin* protein and the *tnpR* protein of Tn*3*. Computer analysis has revealed that the two proteins have 36% amino acid sequence identity, with two small gaps introduced to align the sequence (Simon et al 1980). Although, as discussed above, the Tn*3* element would not complement *hin* defects, these remarkable similarities most assuredly indicate some evolutionary relationship between the two proteins.

The site of the promoter for the *hin* gene may be defined by sequences at positions 40 to 46, similar to the Pribnow sequence, and at positions 6 to 17, which resembles the consensus sequence for RNA polymerase recognition (Fig. 4). Note that this latter sequence would overlap with the 14-bp inverted repeat sequence but would remain the same in the case of either orientation of the inversion region. Sequences upstream from the RNA polymerase recognition site might affect the activity of the promoter, and these sequences would be different for the *H2*-on and *H2*-off configurations. These sequences would exist outside the inversion region in the arm sequences and may direct different levels of *hin* protein in the *H2*-on and *H2*-off orientations. Such an arrangement would account for the small bias in *H2* switching toward the *H2*-off phase that was observed previously in *Salmonella* (Zieg et al. 1977). Orientation-dependent differences in *hin* expression may explain why particular recombinant plasmids appear to exhibit different frequencies of inversion (J. Zeig and M. Simon, unpubl.). Furthermore, assuming that the *hin* product may interact with the inverted repeat sequences to effect inversion and that the *hin* promoter may be part of one inverted repeat sequence, it is conceivable that the *hin* product participates in the regulation of the *hin* gene. It is apparent that active transcription from the *hin* and *H2* promoters proceeds across IR(L) and IR(R) in either orientation of the inversion region. The effects that these two processes, inversion and transcription, may have on each other are not known. In addition, certain sequence structures are affected by the orientation of the inversion region. As shown in Figure 4, the *H2*-on orientation creates an inverted repeat of the sequences between positions 40 and 53 and between positions 1000 and 1012. In the *H2*-off orientation, these sequences constitute a direct repeat. These sequences not only may play a role in regulating inversion, but also may affect the expression of the adjacent *H2* and *hin* genes.

Identification of Gene Product

Genetic tests and DNA sequence analysis support the definition of a gene (*hin*) necessary for inversion of the *H2* control element and suggest the existence of a small polypeptide encoded by the *hin* gene. We have identified a polypeptide that has the properties of the *hin* product; its presence correlates with genetic function, and its molecular weight agrees with that predicted from the DNA sequence of the *hin* gene. Protein synthesis was programmed in vivo using plasmid-containing, minicell-producing strains (Matsumura et al. 1977) and in vitro using a cell-free transcription-translation system (Zubay et al. 1970) to which plasmid DNA was added. Figure 5 shows the electrophoretic separation of [^{35}S]methionine-labeled polypeptides encoded by a variety of Hin$^+$ and Hin$^-$ plasmids. Plasmid pJZ121, which contains the *hin* gene, directed the synthesis of a 19,000-m.w. polypeptide in minicells (Fig. 5, lane 2). Plasmids pMS8 and pMS19 contain the *H2* insert from λ*H2* Hin$^+$ phage (λ*fla*380 and λ*fla*385) and also directed the synthesis of the 19,000-m.w. polypeptide (Fig. 5, lanes 3 and 4). Plasmids pMS14 and pMS1 contain the *H2* insert from λ*H2* Hin$^-$ mutants (λ*fla*381 and λ*fla*378) and did not program the synthesis of the 19,000-m.w. polypeptide (Fig. 5, lanes 5 and 7). The appearance of this polypeptide coincides with the presence of *hin* activity, but an alternative explanation for the 19,000-m.w. polypeptide is that its synthesis results from transcription by the *H2* promoter in the *H2*-off configuration. The Hin$^+$ plasmids used here invert frequently and have this particular *H2* promoter configuration, whereas the Hin$^-$ plasmids (pMS1 and pMS14) are fixed in the *H2*-on orientation. The Hin$^-$ plasmid pMS14 was switched to the *H2*-off orientation (by *trans* complementation), and the resulting pMS14 derivative did not direct the synthesis of the 19,000-m.w. polypeptide (Fig. 5, lane 6). Therefore, transcription from the *H2* promoter in the *H2*-off orientation does not direct the synthesis of the 19,000-m.w. polypeptide, and its synthesis is best explained as the product of the *hin* gene.

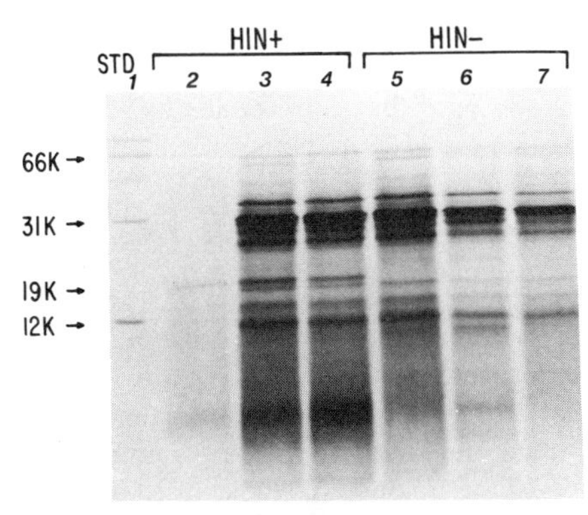

Figure 5. Polypeptide synthesis directed by recombinant plasmids. Synthesis of [^{35}S]methionine-labeled polypeptides was programmed in vivo in minicells containing plasmid pJZ121 or in vitro in a cell-free transcription-translation system with plasmid pMS8, pMS19, pMS14, or pMS1. Polypeptides were electrophoresed on an 18% polyacrylamide slab gel, and labeled bands were visualized by autoradiography. Plasmid pJZ121 contains the *hin* gene (Zieg et al. 1978). *Eco*RI restriction enzyme inserts from a variety of λ*H2* Hin$^+$ and Hin$^-$ deletion mutants were cloned into plasmid pBR322 and used for in vitro programming. Plasmids pMS1 and pMS14 contain the *H2* inserts from λ*H2* Hin$^-$ mutants λ*fla*378 and λ*fla*381 described in Fig. 1B. Plasmids pMS8 and pMS19 contain the *H2* inserts from λ*H2* Hin$^+$ mutants λ*fla*380 and λ*fla*385. (Lanes 5 and 7) Programming with Hin$^-$ plasmids in the *H2*-on configuration; (lane 6) programming with a derivative of plasmid pMS14 in the *H2*-off configuration.

Furthermore, the region of DNA that contains *vh2*$^-$ has been cloned. This clone corresponds to the DNA sequences that are adjacent to *H2,* which includes the inversion region and sequences 3 kb distal to the inversion region. When this DNA was examined in the in vitro protein-synthesizing system, synthesis of the *H2* product, but not the *hin* product, could be detected. This result is also consistent with hybridization studies of *vh2*$^-$ DNA. These studies revealed the absence of the 120-bp sequence corresponding to the middle of the *hin* gene

(E. Szekely and M. Simon, in prep.). All of these results strongly suggest that *vh2*⁻ is a naturally occurring deletion of a sequence that corresponds to part of the *hin* gene and that the *vh2*⁻ mutants are effectively Hin⁻.

DISCUSSION

Alternation of expression of the *H2* operon is controlled by a reversible coupling of the *H2* promoter to this operon. Reversible coupling is achieved by the inversion of a 1000-bp segment of DNA that contains the *H2* promoter and is adjacent to the amino-acid-coding sequences of the *H2* operon. Nucleotide sequence determination allowed precise location of the *H2* promoter and the coding sequences of the *H2* operon with respect to the inverting region.

Inversion is site-specific and independent of the host RecA system and requires at least two functions. One function is *cis*-acting and consists of a pair of 14-bp inverted repeat sequences located at the boundaries of the inversion region. A homologous recombination event between the 14-bp inverted repeat sequences results in the inversion of the DNA segment between them. Deletion of either of these sequences, IR(L) or IR(R), prevents *H2* switching. Since the deletions remove DNA adjacent to the 14-bp sites, neighboring sequences may also influence this recombinational process. A *trans*-acting function encoded by a sequence inside the inversion region is also required for *H2* switching. Definition of this function, the *hin* product, is based on three lines of evidence. (1) Genetic analysis of *H2* switching mutants identified a *trans*-acting function specified by a sequence inside the inversion region (*hin* gene). (2) Examination of the nucleotide sequence revealed the presence of a translational frame that could encode a low-molecular-weight polypeptide (190 amino acids). (3) The location of this sequence coincided with the location of the *hin* gene. Using recombinant plasmids to program the synthesis of polypeptide products, a low-molecular-weight polypeptide (19,000 m.w.) was correlated with the presence of *hin* activity. The size of this polypeptide is consistent with the amino-acid-coding capacity of the open translation frame found in the *hin* region of the nucleotide sequence.

hin-mediated inversion of the *H2* control element is independent of RecA function. However, λ*H2* Hin⁻ mutants do show a very low, but measurable, frequency of *H2* switching when the RecA recombination system is functional. For example, *H2* switching in RecA⁺ or RecA⁻ *E. coli* lysogenized with λ*H2* Hin⁺ hybrids is 10^{-2} per cell per generation, whereas that in RecA⁺ *E. coli* lysogenized with λ*H2* Hin⁻ phage is approximately 10^{-6} per cell per generation. RecA⁻ *E. coli* lysogenized with λ*H2* Hin⁻ phage are incapable of *H2* switching. This suggests that the DNA homology provided by the 14-bp inverted repeat sites can be recognized by the RecA system. However, RecA-mediated *H2* switching is very infrequent and probably is not a significant alternative mode of *H2* switching. It should be emphasized that although the *hin* product is necessary for *H2* switching, it may not be sufficient for this event. We cannot exclude the possibility that functions provided by the *E. coli* host also participate in *H2* switching.

Inversion of a specific region of the genome of Mu regulates the expression of genes that determine the host range of the phage particle (van de Putte et al. 1980). A structurally similar inversion system exists in P1 (Chow and Bukhari 1976). With both Mu and P1, the inversion event requires a phage-encoded function (*gin* function). The Mu and P1 inversion-mediating functions are interchangeable (Kamp et al. 1978). A naturally occurring switching mutant in *Salmonella, vh2*⁻, can be complemented to switch if the Mu or P1 inversion function is provided in *trans* (Kutsukake and Iino 1980). The Hin⁻ mutants described in this paper are also complemented in *trans* by Mu or P1. Thus, *hin* and *gin* encode functionally equivalent proteins. These three inversion systems probably share a common ancestry, although the Mu and P1 inversion regions are structurally different from the *H2* inversion region.

λ*H2*::Tn*5* Hin⁻ mutants used in this study contained the λ genes for integrative recombination (*int, xis*). Any contribution to *H2* switching made by these λ gene products was not detected at the sensitivity of the complementation measurement. Hin⁻ mutants were also not complemented by the introduction of Tn*3* or Tn*5*. However, from a comparison of nucleotide sequences it is apparent that parts of the *H2* inversion region share significant homology with parts of phage λ and several transposable elements. Specifically, the 14-bp inverted repeat sequence necessary for *H2* inversion is similar in sequence to regions of the λ*att* site and to regions of the arms of Tn*10*. The amino acid sequence of the *hin* protein (190 amino acids) was inferred from the nucleotide sequence and was compared with the amino acid sequence of the *tnpR* protein (185 amino acids) of Tn*3* (Heffron et al. 1979). A computer-generated comparison revealed that the *hin* and *tnpR* proteins had 36% amino acid identity, with two small gaps introduced to adjust the sequence alignment. Recent evidence (Arthur and Sherratt 1979) indicates that the *tnpR* protein functions to resolve the cointegrate structure of Tn*3*, which may be an intermediate in the process of transposition. Cointegrate resolution is thought to involve a site-specific recombination event between directly repeated sites in the Tn*3* elements.

There are a number of site-specific recombination events that have the same general features. Figure 6 summarizes the notion that site-specific recombination within a relatively small sequence that is repeated can lead to excision, deletion, or cointegrate resolution (Fig. 6C), as well as to fusion or integration (Fig. 6B) or inversion (Fig. 6A). A polypeptide homologous to the *hin* product could play a role in mediating all of these events. Its function might be to recognize the site at which the recombination event is to occur and to allow other protein factors to catalyze that event. Alternatively, this polypeptide itself might both recognize the specific sites and catalyze the strand exchange necessary for rearrangement. In the case of a complex, highly evolved transposon such as Tn*3*, both a function that catalyzes transposition and another function that could

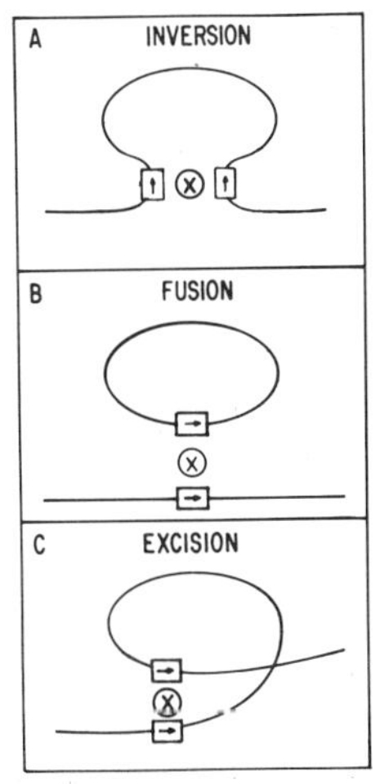

Figure 6. Recombinational systems. Functions essential for inversion of the *H2* control region are, as defined by this study, *cis*-acting elements consisting of 14-bp sequences in inverted repeat configurations and a *trans*-acting function (*hin*) that mediates site-specific DNA-strand exchange at the *cis*-acting sites (*A*). (⟹) *cis*-acting sites; (⊗) *hin*-like function. If the repeat sites are located on separate DNA molecules, *hin*-mediated exchange might result in fusion of the molecules (*B*). If the repeat sites are in direct repeat configuration (*C*), intramolecular exchange might result in excision of the region flanked by the sites. Thus, by extrapolating from *H2* inversion functions, processes that resemble phage integration (*B*) and phage excision or cointegrate resolution (*C*) would result.

participate and mediate site-specific recombination, such as the *tnpR* function, exist on the same transposon. We can develop a scenario in which phase variation may have resulted from an initial event that transposed an ancestral copy of *H1* to another site on the chromosome. The subsequent loss of the ability to transpose fixed the gene, and further mutation brought its expression under control of an adjacent site-specific recombination-mediating system. As further sequence data become available it will be possible to determine the precise relationship between transposition and site-specific recombination and the roles that these two processes play in gene expression in both highly differentiated organisms and the simpler prokaryotes.

ACKNOWLEDGMENTS

This work was supported by grants from the National Science Foundation and the American Cancer Society.

REFERENCES

Arthur A. and D. Sherratt. 1979. Dissection of the transposition process: A transposon-encoded site-specific recombination system. *Mol. Gen. Genet.* **175:** 267.

Berg, D. E. 1977. Insertion and excision of the transposable kanamycin resistance determinant Tn*5*. In *DNA insertion elements, plasmids, and episomes* (ed. A. I. Bukhari et al.), p. 205, Cold Spring Harbor Laboratory, Cold Spring Harbor, New York.

Cameron, J. R., E. Loh, and R. W. Davis. 1979. Evidence for transposition of dispersed repetitive DNA families in yeast. *Cell* **16:** 739.

Chow, L. and A. I. Bukhari. 1976. The invertible DNA segments of coliphages Mu and P1 are identical. *Virology* **74:** 242.

Fujita, H., S. Yamaguchi, and T. Iino. 1973. Studies of H-O variants in *Salmonella* in relation to phase variation. *J. Gen. Microbiol.* **76:** 127

Heffron, F., B. J. McCarthy, H. Ohtsubo, and E. Ohtsubo. 1979. DNA sequence analysis of the transposon Tn*3:* Three genes and three sites involved in transposition of Tn*3*. *Cell* **18:** 1153.

Iino, T. 1961. A stabilizer of antigenic phase in *Salmonella abortusequi. Genetics* **46:** 1465.

Kamp, D., R. Kahmann, D. Zipser, T. R. Broker, and L. T. Chow. 1978. Inversion of the G DNA segment of phage Mu controls phage infectivity. *Nature* **271:** 577.

Kleckner, N. 1977. Translocatable elements in procaryotes. *Cell* **11:** 11.

———. 1979. DNA sequence analysis of Tn*10* insertions: Origin and role of 9 bp flanking repetitions during Tn*10* translocation. *Cell* **16:** 711.

Kleckner, N., J. Roth, and D. Botstein. 1977. Genetic engineering *in vivo* using translocatable drug-resistance elements. *J. Mol. Biol.* **166:** 125.

Kutsukake, K. and T. Iino. 1980. A *trans*-acting factor mediates inversion of a specific DNA segment in flagellar phase variation of *Salmonella. Nature* **284:** 479.

Landy, A. and W. Ross. 1977. Viral integration and excision: Structure of the lambda *att* sites. *Science* **197:** 1147

Lederberg, J. and P. Edwards. 1953. Serotypic recombination in *Salmonella. J. Immunol.* **71:** 323.

Lederberg, J. and T. Iino. 1956. Phase variation in *Salmonella. Genetics* **41:** 743.

Maki, R., A. Traunecker, H. Sakano, W. Roeder, and S. Tonegawa. 1980. Exon shuffling generates an immunoglobulin heavy chain gene. *Proc. Natl. Acad. Sci.* **77:** 2138.

Matsumura, P., M. Silverman, and M. Simon. 1977. Synthesis of *mot* and *che* gene products of *Escherichia coli* programmed by hybrid ColEl plasmids in minicells. *J. Bacteriol.* **132:** 996.

Maxam, A. and W. Gilbert. 1980. Sequencing end-labeled DNA with base-specific chemical cleavages. *Methods Enzymol.* **65:** 499.

McClintock, B. 1957. Controlling elements and the genes. *Cold Spring Harbor Symp. Quant. Biol.* **21:** 197.

Nakamura, K. and M. Inouye. 1979. DNA sequence of the gene for the outer membrane lipoprotein of *E. coli:* An extremely AT-rich promoter. *Cell* **18:** 1109.

Parkinson, J. S. and R. J. Huskey. 1971. Deletion mutations of bacteriophage lambda. *J. Mol. Biol.* **56:** 369.

Potter, S. S., W. J. Borein, P. Dunsmuir, and G. M. Rubin. 1979. Transposition of elements of the *412, copia,* and *297* dispersed repeated families in *Drosophila. Cell* **17:** 415.

Pribnow, D. 1975. Bacteriophage T7 early promoters: Nucleotide sequences of two RNA polymerase binding sites. *J. Mol. Biol.* **99:** 419.

Seeburg, P., C. Nusslein, and H. Schaller. 1977. Interaction of RNA polymerase with promoters from bacteriophage fd. *Eur. J. Biochem.* **74:** 107.

Shine, J. and L. Dalgarno. 1974. The 3′-terminal sequence of *Escherichia coli* 16S ribosomal RNA: Complementarity to nonsense triplets and ribosome binding sites. *Proc. Natl. Acad. Sci.* **71:** 1342.

Silverman, M. and M. Simon. 1974. Characterization of *Escherichia coli* flagellar mutants that are insensitive to catabolite repression. *J. Bacteriol.* **120:** 1196.

———. 1980. Phase variation: Genetic analysis of switching mutants. *Cell* **19:** 845.

Silverman, M., J. Zieg, and M. Simon. 1979a. Flagellar phase variation: Isolation of the *rh1* gene. *J. Bacteriol.* **137:** 517.

SILVERMAN, M., J. ZIEG, M. HILMEN, and M. SIMON. 1979b. Phase variation in *Salmonella:* Genetic analysis of a recombinational switch. *Proc. Natl. Acad. Sci.* **76:** 391.

SIMON, M., J. ZIEG, M. SILVERMAN, G. MANDEL, and R. DOOLITTLE. 1980. Phase variation: Evolution of a controlling element. *Science* **209:** 1370.

STOCKER, B. A. D. 1949. Measurements of rate of mutation of flagellar antigenic phase in *Salmonella typhimurium. J. Hyg.* **47:** 398.

STRATHERN, J. N., E. SPATOLA, C. MCGILL, and J. B. HICKS. 1980. Structure and organization of transposable mating type cassettes in *Saccharomyces* yeasts. *Proc. Natl. Acad. Sci.* **77:** 2839.

VAN DE PUTTE, P., S. CRAMER, and M. GIPHART-GASSLER. 1980. Invertible DNA determines host range specificity of bacteriophage Mu. *Nature* **286:** 218.

WILLIAMS, R. O., J. R. YOUNG, and P. A. O. MAJIWA. 1979. Genomic rearrangements correlated with antigenic variation in *Trypanosoma brucei. Nature* **282:** 847.

ZIEG, J. and M. SIMON. 1980. Analysis of the nucleotide sequence of an invertible controlling element. *Proc. Natl. Acad. Sci.* **77:** 4196.

ZIEG, J., M. HILMEN, and M. SIMON. 1978. Regulation of gene expression by site-specific inversion. *Cell* **15:** 237.

ZIEG, J., M. SILVERMAN, M. HILMEN, and M. SIMON. 1977. Recombinational switch for gene expression. *Science* **196:** 170.

ZUBAY, G., D. A. CHAMBERS, and L. C. CHEONG. 1970. Cell-free studies on the regulation of the *lac* operon. In *The lactose operon* (ed. J.R. Beckwith and D. Zipser), p. 375. Cold Spring Harbor Laboratory, Cold Spring Harbor, New York.

Genesis and Natural History of IS-mediated Transposons

S. Iida, J. Meyer, and W. Arber
Department of Microbiology, University of Basel, Biozentrum, CH-4056 Basel, Switzerland

Transposable elements unable to replicate autonomously have been classified as either transposons (Tn) or insertion sequences (IS), depending on whether or not they carry detectable genes. Transposons are generally flanked by repeated DNA sequences, and some of these repeats are known to be IS elements. In the case of IS*1*-flanked transposons, the terminal repeats can be in either direct (MacHattie and Jackowski 1977; Iida and Arber 1980) or inverted (So et al. 1979) orientation. This, together with considerations based on the knowledge of the nucleotide sequence of IS*1*-flanked chloramphenicol (Cm) transposons (Alton and Vapnek 1979; Marcoli et al. 1980) and on the fact that inverse IS*1*-mediated transposition occurs (Reif 1980; Rosner and Guyer 1980), poses the questions of whether and how any DNA segment can evolve to become a component of an IS*1*-flanked transposon.

Here we will present evidence that, in fact, a determinant for chloramphenicol resistance (Cm^r) located on plasmid pBR325 that does not contain an IS element can be mobilized by the integration of IS*1* in its neighborhood and the subsequent formation of a transposon. The resulting transposons are flanked by either direct or inverted repeats of IS*1* and are found to transpose as stable units, e.g., from λ DNA to P1-15 DNA. It will be shown that these IS*1* elements may also cause decay of the transposons, either by deletion formation, excision, or homologous recombination. Our observations lead to a discussion of the evolutionary implications of IS-mediated transposons.

MATERIALS AND METHODS

Bacteria Strains, Phages, and Plasmids

Escherichia coli K12 strain SI2381 is *hsdR* ($r_K^- m_K^+$) *met supE supF rif*, and strain SI2385 is Δ*lacZ* λ^r. These strains and phage λp*lac5imm*^{21}*c*Its were from our collection. Phage P1-15 hybrid 2 is a recombinant between phage P1 and plasmid p15B and it does not carry an IS*1* (Arber and Wauters-Willems 1970; S. Iida et al., in prep.). The other *E. coli* K12 and phage strains used have been described previously (Iida and Arber 1977; Iida et al. 1978; S. Iida et al., in prep.). Plasmid pBR325 was described by Bolivar (1978) and pJO505 (pJR505 renamed) was described by Reif (1980).

Isolation of pBR325::P1 and pBR325::IS*1*

A phage P1 lysate was prepared by heat induction (Iida and Arber 1977) of WA921(P1*c*1ts225, pBR325). With the resulting lysate, strain WA921 was infected at a multiplicity of 0.1 plaque-forming phage per cell and Cm^r transductants were isolated. The screening of these transductants for ampicillin-resistance (Ap^r) and tetracycline-resistance (Tc^r) phenotypes, for phage production upon heat induction, and for high frequency of Cm^r transduction by the resulting lysates allowed us to classify the transductants. About one tenth carried pBR325::P1 cointegrate plasmids; this was confirmed by physical characterization.

pBR325::IS*1* segregants of pBR325::P1 were isolated as follows. WA921(pBR325::P1) was grown in 50 ml of Luria broth supplemented with 300 μg/ml of chloramphenicol. From this culture, plasmids were prepared (Meyers et al. 1976), and these served to transform WA921 to Cm^r. Since transformation with P1 plasmid DNA is inefficient, most of the Cm^r transformants carried pBR325::IS*1*.

Transposition of Tn*2651* from the Genome of λp*lac5imm*^{21} to That of P1-15

Phage P1-15 hybrid 2 was prepared by heat induction of SI2381(λp*lac5imm*^{21}*c*Its::Tn*2651*, λ*imm*^{21} c^+*b*515*b*519, P1-15 hybrid 2 *c*1ts225). The lysate served to infect SI2385(P1-15 hybrid 2 *c*1ts225::Tn*10*) at a multiplicity of 0.1 plaque-forming phage per cell. Cm^r transductants were selected on EMBLacCm plates. Among Cm^rLac^- colonies producing killer particles upon heat induction, tetracycline-sensitive (Tc^s) cells (substitution of P1-15 prophage as a result of P1 plasmid incompatibility) that produced no λ*imm*^{21} phage but did produce plaque-forming P1-15Cm phage were screened. The same procedure was applied to demonstrate transposition of the other five newly isolated Cm transposons.

Electron Microscopy

The formation of heteroduplex molecules and snapback structures and their mounting for electron microscopy have been described previously (Iida et al. 1978; Meyer and Iida 1979).

RESULTS

Strategy for the Demonstration of In Vivo Genesis of IS*1*-flanked Cm Transposons

To document the in vivo genesis of IS*1*-flanked transposons, we decided to proceed stepwise as outlined in Figure 1. Because of its easy selection, the Cm^r determi-

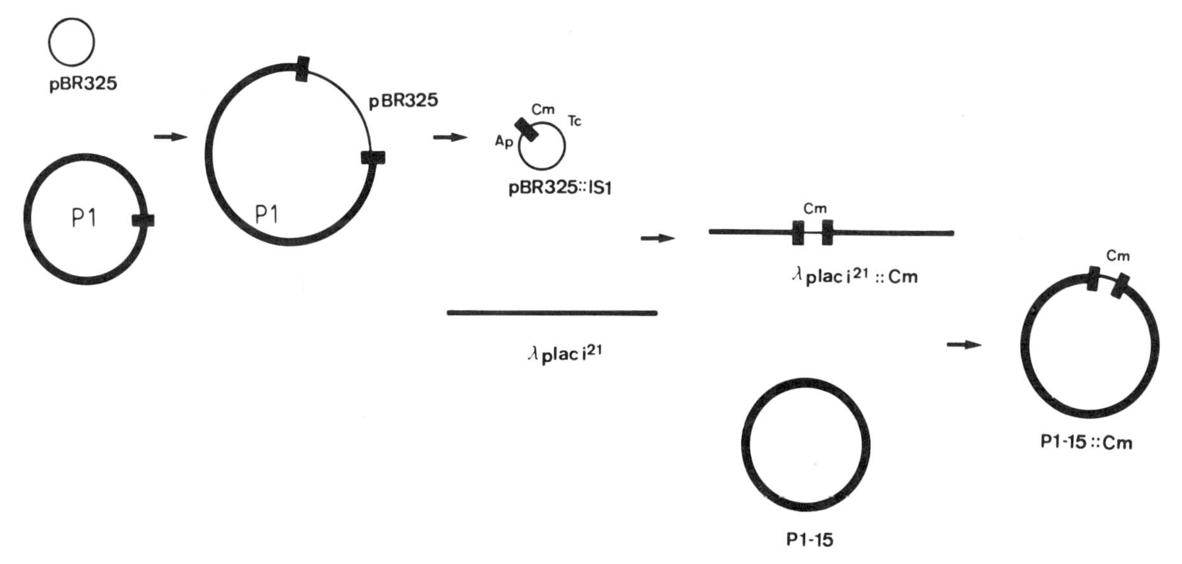

Figure 1. Schematic representation of the strategy used for the in vivo genesis of IS*1*-mediated Cm transposons. See text for explanations.

nant carried on the plasmid pBR325 was chosen as starting material to become a component of an IS*1*-flanked transposon. We planned to integrate the IS*1* element of P1 (Iida et al. 1978) in the vicinity of the Cm marker of pBR325 by a first step of transpositional cointegration of the two plasmids and by subsequent segregation of the P1 part from the cointegrate. We expected that this would yield pBR325::IS*1* molecules. In a next step we planned to propagate bacteriophage λ*plac5imm*^{21} in a cell carrying the pBR325::IS*1* plasmid. By infection of chloramphenicol-sensitive (Cm^s) bacteria with the resulting phage lysate, some Cm^r transductants should be obtained. We expected that some of these might contain a λ prophage carrying the Cm marker derived from pBR325 on an IS*1*-mediated transposon. A study of phenotypic traits of Cm^r transductants and the analysis of the physical structure of the phage genomes isolated from such lysogens could insure the expectation. The confirmation that the Cm marker flanked by two IS*1* elements would indeed be a transposon could come from its transposition to an unrelated replicon, in particular to the DNA of phage P1-15.

Transpositional Cointegration of P1 and pBR325

In principle, IS*1* can mediate the cointegration between two replicons by two different mechanisms. First, the IS*1* element carried on one of the replicons can transpose into the other replicon and thereby yield a cointegrate in which each of the two junctions contains one IS*1* element in the same orientation. This mechanism can act independently of whether or not the second replicon also contains IS*1*. Second, mechanisms of reciprocal recombination between two intact IS*1* elements or between an intact IS*1* element carried in one of the replicons and a segment of IS*1* carried in the other replicon can also bring about cointegration. This mechanism, in contrast to that described first, does not result in any material gain; i.e., the IS*1* element is not replicated. A further difference between the two mechanisms is that junctions obtained by the latter cointegration mechanism are always at the sites of residential IS*1* material, whereas transpositional cointegration can mediate the integration of the IS*1*-carrying replicon into any chosen site of the second replicon (Shapiro and MacHattie 1979; Iida and Arber 1980; Iida 1980; Ohtsubo et al. 1980).

Since it was our aim to demonstrate that a genome segment originally not linked to IS*1* material could become a component of an IS*1*-mediated transposon, it was important to show that pBR325 does not contain any IS*1* sequences. By a thorough physical analysis of the Cm segment of pBR325, the results of which are shown in Figures 2 and 3, and by DNA-DNA hybridization studies with an IS*1* probe, we could not, in fact, detect any evidence for the presence of IS*1* sequences in pBR325. Therefore, this plasmid can only serve as a target in transpositional cointegration and not for IS*1*-mediated cointegration by the reciprocal recombination pathway.

For the isolation of pBR325::P1 cointegrates, we prepared a P1 lysogenic strain carrying the pBR325 plasmid. A culture of this strain was induced for phage P1 production and the resulting lysate served to transduce WA921 into Cm^r. About one tenth of the Cm^r transductants contained a full cointegrate plasmid (Table 1), hence including the determinants for ampicillin (Ap) and tetracycline (Tc) resistance. Upon induction, such transductants gave phage lysates able to transduce Cm^r and other resistances at high frequency. Restriction cleavage analysis of some of the cointegrates isolated confirmed the expected size of transpositional cointegrates as being the sum of the P1 genome, the pBR325 plasmid, and one additional IS*1* element. In addition,

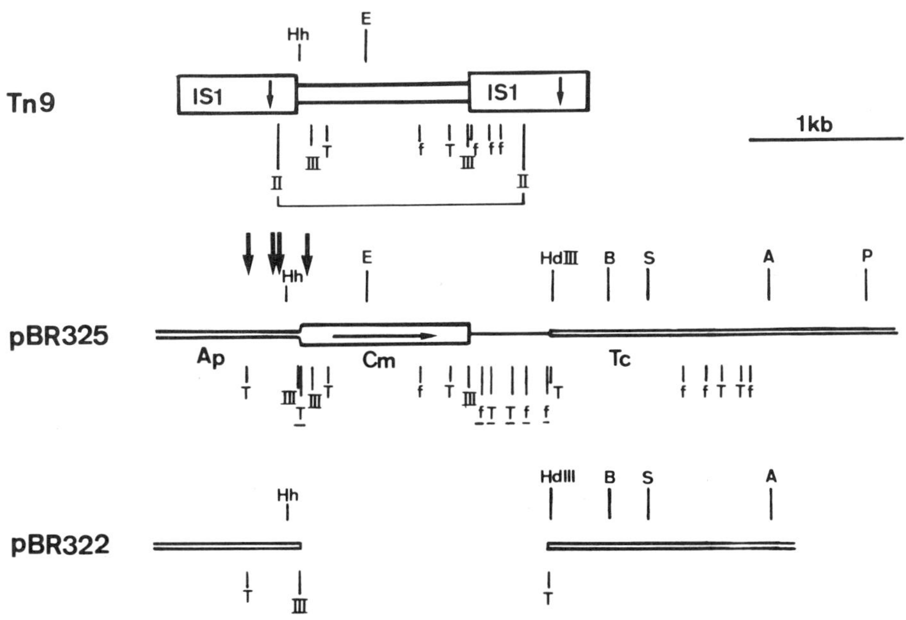

Figure 2. Partial restriction cleavage maps of pBR325, and its ancestors Tn*9* and pBR322, and the location of the sites of IS*1*-mediated cointegration of P1 with pBR325. Plasmid pBR325 had been constructed by cloning the nuclease-S1-treated *Hae*II fragment (shown with brackets) of Tn*9* into the nuclease-S1-treated *Eco*RI site of pBR322 (Bolivar 1978). The drawings of the relevant restriction cleavage sites of Tn*9* (*top*) and of part of pBR322 (*bottom;* drawn interrupted at the *Eco*RI site) are based on published nucleotide sequences (Ohtsubo and Ohtsubo 1978; Alton and Vapnek 1979; Sutcliffe 1979). Restriction cleavage of pBR325 suggested a more complex structure than might have been expected from its origin. The pBR325 map drawn in the middle is the result of our analysis. The *Pst*I sites within IS*1* (*top*) are indicated by small vertical arrows. Other restriction cleavage sites are: (A) *Ava*I; (B) *Bam*HI; (E) *Eco*RI; (f) *Hin*fI; (HdIII) *Hin*dIII; (Hh) *Hha*I; (P) *Pvu*II; (S) *Sal*I; (T) *Taq*I; (II) *Hae*II; (III) *Hae*III. All cleavage sites are shown for the enzymes *Ava*I, *Bam*HI, *Hin*dIII, and *Sal*I, but only the relevant cleavage sites for the other enzymes are drawn. Underlined symbols are for sites on pBR325 that are absent from both Tn*9* and the region of pBR322 adjacent to the *Eco*RI site. At the left junction of pBR322 with the Cm region of Tn*9*, both the *Hae*III and a *Mbo*II (not shown, but to the right of *Hae*III) recognition sites of pBR322 are present, but the *Hha*I site of Tn*9* is absent. Instead, a new *Taq*I site was detected at this junction. At the right junction between the Cm insert and pBR322, the *Hae*III site near the right end of the Cm segment seems to be carried also in pBR325, but the three *Hin*fI sites of IS*1* are absent. The leftmost of these missing *Hin*fI sites is only 7 bp to 11 bp inside the IS*1* of Tn*9*. Hence, these data point to the absence of IS*1* segments in pBR325. Instead, pBR325 carries another insertion (drawn with a thin line) between Cm and Tc. Both the restriction cleavage data (of which only *Taq*I and *Hin*fI sites are drawn) and analysis by electron microscopy (Fig. 3) indicate that this segment is the duplication of a 0.5-kb part of the Tc region of pBR322 in inverted orientation. At the right junction between this insertion and the pBR322 DNA, a *Hae*III site appeared (not shown). It is located between the *Taq*I site of pBR322 and the rightmost *Hin*fI site of the insertion. The four large vertical arrows drawn above the pBR325 map mark the sites of integration of P1 in the four studied pBR325::P1 cointegrates and the sites of IS*1* in their pBR325::IS*1* segregants. These segregant plasmids are, from the leftmost to the rightmost insertion sites, pSHI214, pSHI211, pSHI212, and pSHI213. The insertion site of IS*1* in pSHI214 is within the Ap^r determinant. None of the plasmids studied is affected in its Cm resistance. The extension of the structural gene *cat* for Cm resistance (Alton and Vapnek 1979; Marcoli et al. 1980) is indicated in pBR325 by a horizontal arrow.

the presence of two IS*1* elements in the resulting cointegrates was verified by Southern blotting and hybridization with ^{32}P-labeled IS*1* probes.

The junction made between P1 and pBR325 material occurred at the same site (at the residential IS*1*) of the P1 genome in all four cointegrates, but at four different sites on pBR325 (indicated by arrows in Fig. 2). These latter sites happened to be clustered in the Ap-Cm region. In fact, cointegration at the leftmost site on pBR325 destroyed the Ap^r determinant, whereas all four cointegrates studied still expressed Cm resistance. This variability of integration sites offers independent evidence that cointegration of P1 into pBR325 did not occur via recombination with IS*1* nucleotide sequences hidden in pBR325. Rather, all cointegrates studied must have originated in IS*1*-mediated transposition.

Segregation of pBR325::P1 Yields pBR325::IS*1*

Segregant plasmids of the structure pBR325::IS*1* were isolated from each of the pBR325::P1 plasmids as described in Materials and Methods. The physical structure of the resulting segregants, labeled pSHI211 to pSHI214, was investigated by restriction cleavage analysis and in Southern hybridization experiments. This allowed confirmation and more precise mapping of the sites in pBR325 at which the IS*1* elements are carried (Fig. 2). These experiments also revealed that the IS*1* elements in all four plasmids studied are carried in the same orientation. For pSHI213, the integration site and orientation of IS*1* were confirmed by nucleotide sequence analysis at both junctions (data not shown).

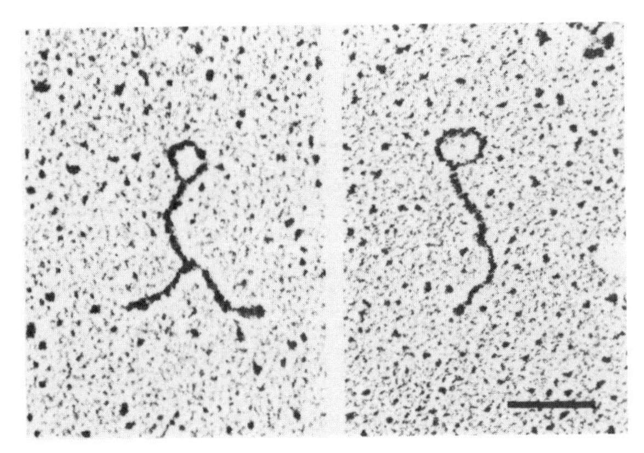

Figure 3. Electron micrographs of the snapback structure observed in single-stranded DNA of pBR325. The left photograph shows the snapback structure observed in single-stranded *Eco*RI:*Pvu*II doubly digested plasmid DNA. The lengths of the short and long single-stranded ends, stem, and loop were measured as 0.52±0.10 kb, 0.70±0.10 kb, 0.51±0.07 kb, and 0.75±0.12 kb, respectively (62 molecules were observed). The photograph on the right shows the snapback structure observed in *Eco*RI:*Ava*I doubly digested plasmid DNA. The lengths of the single-stranded end, stem, and loop were measured as 0.51±0.11 kb, 0.54±0.08 kb, and 0.74±0.12 kb, respectively (64 molecules were observed). The other single-stranded end was too short to be measured properly and is thus less than 50 to 100 bp in size. The location of the relevant *Ava*I, *Eco*RI, and *Pvu*II sites on pBR325 are shown in Fig. 2. This study was made after we had learned from P. Prentki and E. Galley (pers. comm.) that linearized single-stranded pBR325 DNA forms a snapback structure. Our experiment allowed precise mapping of the inserted repeat. The bar indicates the length of 500-bp double-stranded DNA.

Isolation and Characterization of λp*lac5imm*^{21} Phages Carrying New Cm Transposons

In parallel experiments, WA921(pSHI211) and WA-921(pSHI213) were lysogenized with phage λp*lac*5-*imm*^{21}cIts. Cultures of the lysogens were heat-induced and the resulting λ lysates served to infect WA921 at a multiplicity of 1 plaque-forming phage per cell. Cm^r transductants arose at a frequency on the order of 10^{-8} per infected cell (Table 1). On testing of their Tc and Ap phenotypes and their capacity to liberate phage lysates transducing Cm^r, Ap^r, and Tc^r at high frequency, most Cm^r transductants were revealed to contain λ-plasmid cointegrate genomes (Table 1). However, we directed our attention to a minor class of Cm^r transductants that were Ap^s and Tc^s and that, on heat induction, produced plaque-forming λ phages transducing Cm^r at high frequency, as revealed by the plaque center test (Iida and Arber 1977). A total of six such derivatives were found, three in each of the two parallel experiments.

The physical structures of all six new phage genomes were investigated by restriction cleavage and by electron microscopy studies (Fig. 4). These indicated that all six λ strains carried an IS*1*-flanked Cm transposon at a different site (Fig. 5) and that each transposon was different from the other (Fig. 6). Transposons derived from pSHI211 were labeled Tn*2651* to Tn*2653*; those derived from pSHI213, Tn*2654* to Tn*2656*.

Only one of the six transposons, Tn*2651*, carried its flanking IS*1* elements as direct repeats (Figs. 4A–D and 6). It was revealed to be relatively unstable (Table 2), segregating λp*lac5imm*^{21}::IS*1* phages that were not providing Cm resistance. This segregation is likely due to *recA*-mediated reciprocal recombination between the two IS*1* elements. As shown in Table 2, this segregation occurs with a rate similar to that in another IS*1*-mediated Cm transposon, Tn*981*, carried in a similar arrangement.

Transposons Tn*2652* and Tn*2655* carried their flanking IS*1* elements as inverted repeats (Fig. 6). As for Tn*2651*, the DNA structure between the flanking IS*1* elements of Tn*2652* and Tn*2655* pointed to a direct descendence from the parental plasmids pSHI211 and pSHI213, respectively. These two transposons showed a high degree of stability (Table 2), which appears to be a property of transposons carrying their flanking IS*1* elements in an inverted orientation.

Tn*2654* demonstrates that this statement should not be taken as a general rule. Indeed, Tn*2654* is also flanked by inverted IS*1* elements, and its DNA structure is similar to that of Tn*2655* but slightly shorter. Tn*2654* was revealed to be quite unstable (Table 2); λp*lac5imm*^{21}::Tn*2654* produced λp*lac5imm*^{21}Cm^s with high frequency, even after induction of a lysogen. Judging from restriction cleavage and electron microscopy studies of two Cm^s derivatives, these appeared to be caused by precise, or nearly precise, excision; i.e., no detectable deletion or insertion was observed if compared with λp*lac5imm*^{21}. Since λp*lac5imm*^{21}::Tn*2654* produced smaller plaques than λp*lac5imm*^{21}, its Cm^s derivatives carrying no insertion may have a considerable growth advantage. Thus, the apparent high segregation rate is likely to be attributable to a physiological effect of the presence of Tn*2654* at a particular site on the λ genome rather than to be an intrinsic property of Tn*2654* itself.

The two remaining transposons have a rather unexpected structure. Tn*2653* carries the Cm region and the adjacent DNA segment of pBR325 as inverted duplications (Figs. 4 and 6). Together with the flanking IS*1* elements, which are also carried in inverted orientation, this gives Tn*2653* an inverted duplication of about 2.6 kb. Only a 0.8-kb part of the Tc region of pSHI211 is carried as a nonrepeated segment, but it does not provide Tc resistance.

Tn*2656* is also flanked by inverted IS*1* elements. The 2.4-kb DNA segment carried inside of this transposon was identified to have its origin in pSHI213 as expected. But this segment, which does not provide Tc resistance, had been restructured. We interpret the results of the analysis as shown in Figure 6 to mean that a partial duplication similar to the one described for Tn*2653*, with inversion of the duplicated material, was followed by an IS*1*-mediated deletion formation (Reif and Saedler 1975), resulting in the structure of Tn*2656*.

Both Tn*2653* and Tn*2656* are relatively stable (Table 2). Some of the Cm^s segregants were studied and they

Table 1. Estimated Frequencies of IS*1*-mediated DNA Rearrangements

Phage preparation	WA921(P1,pBR325)	WA921(λ*plac5imm*^{21},pSHI211)	WA921(λ*plac5imm*^{21},pSHI213)	WA921(λ*plac5imm*^{21},pJO505)
Multiplicity of infection of WA921 (plaque-forming phage/cell)	0.1	1	1	0.3
Frequency of Cm^r transduction (Cm^r transductants per plaque-forming phage)	2.8×10^{-7}	2.1×10^{-8}	4.7×10^{-8}	5.3×10^{-7}
Cointegrates detected in Cm^r transductants	pBR325::P1	(λ*plac5imm*^{21}::pSHI211)	(λ*plac5imm*^{21}::pSHI213)	λ*plac5imm*^{21}::pJO505
cointegrate frequency	2.1×10^{-8}	1.6×10^{-8}	4.0×10^{-8}	3.1×10^{-9}
Cm^r caused by transposition				
detected structure	—	—	—	λ*plac5imm*^{21}::Tn*981*
frequency				4.0×10^{-7}
Appearance of new Cm transposons				
structure		λ*plac5imm*^{21}::Tn*Cm*	λ*plac5imm*^{21}::Tn*Cm*	
frequency		10^{-9}	2×10^{-9}	

All bacteria were heat-induced. The resulting phage lysates served in the transduction of WA921 for Cm^r. All frequencies are expressed per plaque-forming phage of the transducing lysates. With the exception of the Cm^r transduction frequency, they are estimates rather than measured values since they are based on the interpretation of phenotypic traits (for all structures shown) and of relatively few structural determinations (for structures shown without parentheses). Plasmid pJO505 (not discussed in the text) was studied as a control. It carries three IS*1* elements in the same orientation, two of which form the flanking elements of Tn*981* (Reif 1980). Tn*981* is a Cm transposon originally carried in phage P1Cm204 (Arber et al. 1979), and its nucleotide sequence is known (Marcoli et al. 1980).

A comparison of the given frequencies shows that the formation of λCm^r transducing phage genomes more often occurs by transposition than by cointegration if the Cm^r determinant is already carried on a transposon, as in pJO505. In contrast, in the transduction of Cm^r from pSHI211 and pSHI213, cointegrates appear to be the most abundant Cm^r transducing phage derivatives.

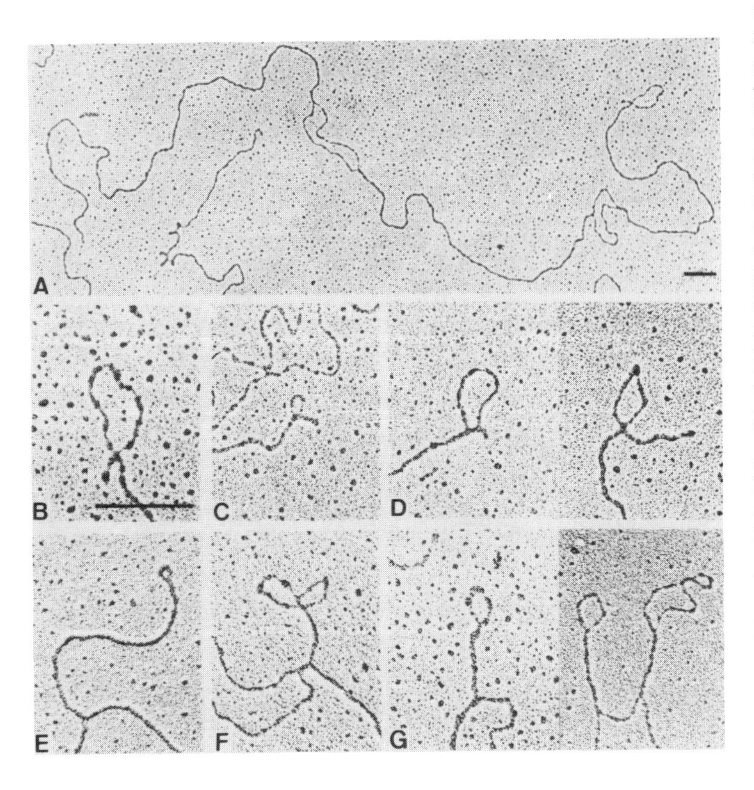

Figure 4. Electron micrographs documenting direct or inverted repeats flanking the Cm transposons. (*A–D*) Evidence for directly repeated terminal IS*1* sequences in Tn*2651*. (*A*) Heteroduplex molecule between λ*plac5imm*^{21}::Tn*2651* and λ DNA. See Fig. 5 for a schematic representation. (*B*) Close up of the insertion loop (Tn*2651*) located at the right end of the molecule shown in *A*. (*C*) The small loop in a heteroduplex molecule between λ*plac5imm*^{21}::IS*1* (a Cm^s derivative of λ*plac5imm*^{21}::Tn*2651*) and λ DNA represents the IS*1* left behind at the original integration site of Tn*2651*. (*D*) The loop in heteroduplex molecules between DNA of λ*plac5imm*^{21}::Tn*2651* and its Cm^s derivative has a fixed size but a variable position within IS*1*. (*E–G*) Representatives of Cm transposons with terminal inverted repeats. (*E*) Snapback representing Tn*2653* in a heteroduplex formed between λ*plac5imm*^{21}::Tn*2653* and λ DNA. (*F*) Snapback representing Tn*2656* in a heteroduplex formed between λ*plac5*-*imm*^{21}::Tn*2656* and λ DNA. The Tn*2656* is carried at a distance of about 0.15 kb from the *lac* nonhomology region, part of which is seen in the left part of the picture. (*G*) Snapback representing Tn*2654* in a heteroduplex formed between λ*plac5imm*^{21}::Tn*2654* and λ DNA (*left*), and in single-stranded P1-15::Tn*2654* DNA (*right*) where Tn*2654* is inserted close to the C-loop. The bars indicate the length of 1-kb double-stranded DNA. *B–G* are printed at the same enlargement.

proved to be products of IS*1*-mediated deletion of the Cm^r determinant (Fig. 7).

Transposition of the Newly Isolated Cm Transposons

To demonstrate that the Cm segments carried on the six Cm^r λ*plac5imm*^{21} phage derivatives described previously are indeed transposons, we grew P1-15 hybrid 2 (S. Iida et al., in prep.) in the presence of these phages as described in Materials and Methods. Plaque-forming P1-15 phages carrying the Cm^r marker were isolated and their DNA subjected to restriction cleavage analysis. As shown in Figure 8, the Cm fragments produced by *Pst*I, which cleaved within flanking IS*1*, had the same size with both λ*plac5imm*^{21} and P1-15 derivatives, i.e., before and after transposition. Southern blotting and hybridization with a ^{32}P-labeled IS*1* probe confirmed that the Cm segment carried on P1-15 contained two copies of IS*1* (Fig. 9). In addition, viewed with the electron microscope, snapback structures of all five Cm transposons, Tn*2652* to Tn*2656*, with inverted IS*1* were the same size on λ*plac5imm*^{21} DNA as on P1-15 DNA (shown in Fig. 4G for Tn*2654*).

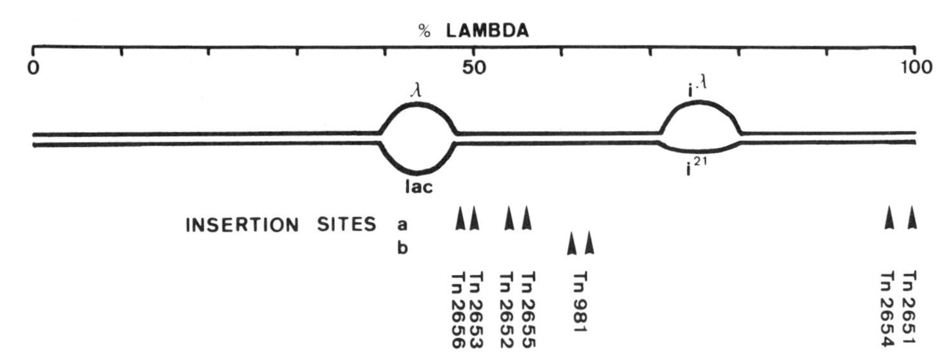

Figure 5. Schematic representation of heteroduplex molecules between λ DNA and DNA of λ*plac5imm*^{21} carrying an IS*1*-flanked Cm transposon. The top line gives the λ map coordinates (Szybalski and Szybalski 1979). (♠) Indicates the insertion sites (*a*) of the various Cm transposons derived from pBR325::IS*1* and (*b*) of Tn*981* as transposed from pJO505, respectively. The localizations of the insertion sites shown are based both on restriction cleavage analysis and electron microscopy studies of heteroduplex molecules. Additional insertion sites for Tn*981* determined by restriction cleavage analysis only and located between the sites shown for Tn*2653* and Tn*2652* are not included.

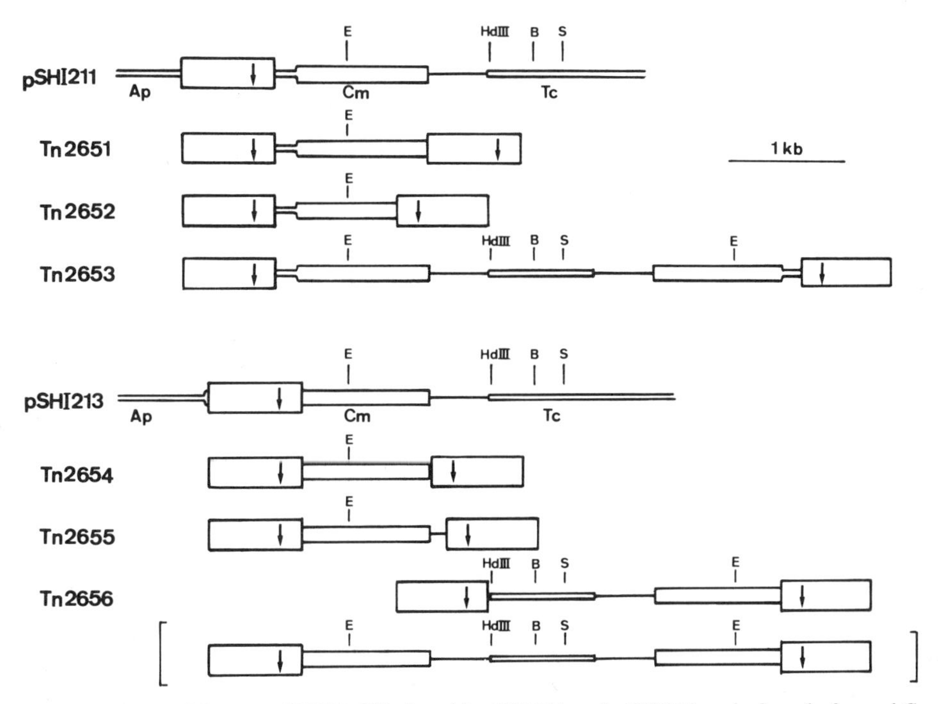

Figure 6. Relevant structure of the two pBR325::IS*1* plasmids pSHI211 and pSHI213 and of newly formed Cm transposons derived from them. The structures of plasmids and Cm transposons are drawn using the same symbols as in Fig. 2. The orientation of the IS*1* elements is seen from the unique *Pst*I cleavage site (arrows) within IS*1*. The structure drawn in brackets at the bottom is a putative precursor having given rise to Tn*2656* by IS*1*-mediated deletion formation.

DISCUSSION

Transposons encompass a class of movable genetic elements the propagation of which depends on their residing on a replicon and that contain one or more detectable genes not obviously related with transposition functions. This latter criterion distinguishes these from IS elements. In the last few years, it has become obvious that naturally occurring transposons fall into distinctive subclasses. One of these classes is represented by Tn*3*, which carries only short inverted repeats at both ends (Heffron et al 1979). Another class is formed by transposons flanked by nucleotide sequences approximately 1 kb in length that, in some instances, are known to be IS elements. Several IS*1*-flanked transposons representing this class have been given much attention recently, e.g., Tn*9*, which contains the Cm^r determinants (MacHattie and Jackowski 1977), and the newly named Tn*2671*, which is identical to the 23-kb-long r-determinant of the R plasmid NR1-Basel (Iida and Arber 1980). Tn*2671* has been shown to incorporate into the genome of bacteriophage P1. This results in a genome that is too big to be packaged in its entirety into phage P1 particles. Therefore, rare plaque-forming derivatives can easily

Table 2. Stability of IS*1*-flanked Cm Transposons on the λp*lac5imm*^{21} Phage Genome

Phage strain	Percentage of plaques containing Cm^r lysogens: induced lysate	plate stock	Major mechanism responsible for instability
λp*lac5imm*^{21}::Tn*981*	99	25	reciprocal recombination between IS*1*
λp*lac5imm*^{21}::Tn*2651*	100	18	reciprocal recombination between IS*1*
λp*lac5imm*^{21}::Tn*2652*	100	100	no Cm^s segregants studied
λp*lac5imm*^{21}::Tn*2653*	100	98	IS*1*-mediated deletion formation
λp*lac5imm*^{21}::Tn*2654*	95	2	excision of the transposon
λp*lac5imm*^{21}::Tn*2655*	100	99	IS*1*-mediated deletion formation
λp*lac5imm*^{21}::Tn*2656*	100	96	IS*1*-mediated deletion formation

Phage λp*lac5imm*^{21} carrying Cm transposons were prepared either by heat induction of lysogenic bacteria or by the confluent lysis plate method. From each lysate, 100 plaques were subjected to the plaque center test as described previously (Iida 1977; Iida and Arber 1977). The percentage of plaques containing Cm^r lysogens is taken as a measure for the stability of the Cm transposon, although we are aware that at least some, if not all, Cm^s segregant phages might have had a growth advantage over their parental phages carrying a Cm transposon. The structures of a few Cm^s segregant phage genomes were analyzed by restriction cleavage and electron microscopy studies of the heteroduplex molecule, and the conclusions are given in the last column. Tn*2651* and Tn*981* carry their IS*1* elements as direct repeats. They segregate Cm^s phages mainly by intramolecular reciprocal recombination between these two IS*1* elements. The other five transposons carry their flanking IS*1* elements in inverted orientation. Their preponderant mechanism of segregation of Cm^s derivatives is IS*1*-mediated deletion formation (Fig. 7). However, efficient excision of the transposon is seen with Tn*2654*, and possible explanations are given in the text.

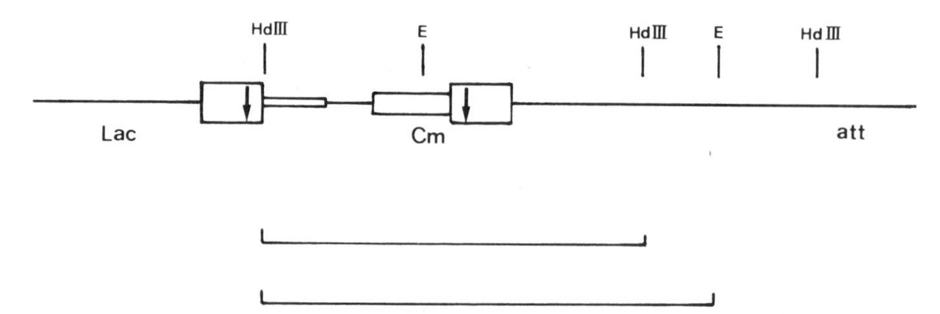

Figure 7. Partial structure of λ*plac5imm*^{21}::Tn*2656* and two of its Cm^s deletion derivatives. The relevant structure of λ*plac5imm*^{21}::Tn*2656* is drawn as in Fig. 6. The 2 brackets (*below*) symbolize IS*1*-mediated deletions on two λ*plac5imm*^{21}Cm^s derivatives. The site and orientation of Tn*2656* on λ*plac5imm*^{21} DNA and the extents of the deletions are based both on restriction cleavage analysis and heteroduplex studies by electron microscopy. The restriction cleavage sites on the λ*plac5imm*^{21} genome are described by Szybalski and Szybalski (1979).

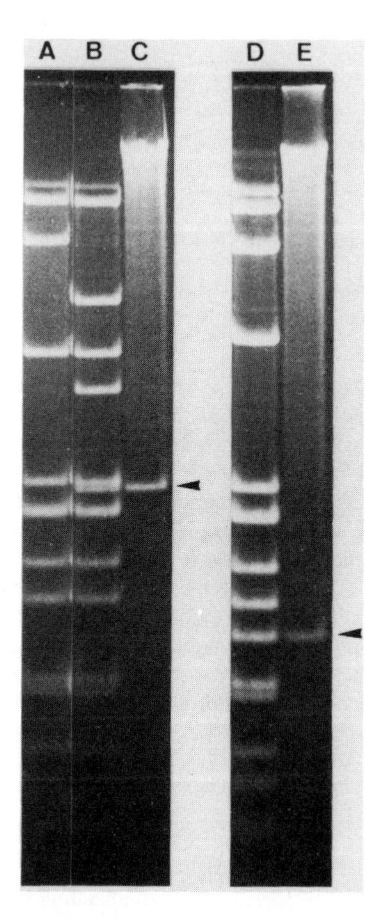

Figure 8. Evidence that Tn*2656* and Tn*2655* are indeed transposons and transpose as discrete units. Preparation of phage DNA, cleavage with restriction endonuclease *Pst*I, and 0.8% agarose gel electrophoresis were carried out as described by Iida and Arber (1980). (*A*) λ*plac5imm*^{21} DNA; (*B*) λ*plac5imm*^{21}::Tn*2656* DNA; (*C*) P1-15::Tn*2656* DNA; (*D* λ*plac5imm*^{21}::Tn*2655* DNA; (*E*) P1-15::Tn*2655* DNA. Phage P1-15 DNA carries no *Pst*I cleavage site (S. Iida et al., in prep.). Therefore, *Pst*I cleavage of P1-15::Tn*Cm* DNA yields one fragment (indicated by arrows) that corresponds in size to the material contained between the two unique *Pst*I sites of the IS*1* elements flanking the Cm transposons. This figure demonstrates that the size of these fragments is maintained after transposition of Tn*2655* and Tn*2656* from λ*plac5imm*^{21} DNA to P1-15 DNA. The same results were obtained with Tn*2651*, Tn*2652*, Tn*2653*, and Tn*2654*.

be selected, and many of these were shown to carry internal deletions to the Tn*2671*. The study of such phages revealed a large variety of IS*1*-mediated Cm transposons (Arber et al. 1979). In the interpretation of structural studies of such P1Cm phage derivatives, several possible pathways for the formation of these phages have been discussed (Iida and Arber 1980). Several of these pathways involve reciprocal recombination and transposition (including IS*1*-mediated deletion formation) in various orders, making use of both IS*1* elements of Tn*2671* and in some instances also of the IS*1* element resident in the P1 genome (Iida et al. 1978). An alternative model involving only a single IS*1* element of Tn*2671* and only transposition mechanisms, but not reciprocal recombination, to form P1Cm derivatives has also been proposed (Iida and Arber 1980). However, the experimental material did not allow determination of whether this latter pathway had ever been used. If it were, the model further postulated that any genome segment located in the vicinity of an IS*1* element could, in principle, become a component of a newly formed transposon, provided that the distal part of this segment presented a target for the transposition of a second IS*1* (Meyer et al. 1980; J. Miller et al.; Saedler et al. both this volume). This hypothesis was the basis of the experimental approach discussed in this paper.

Because of its convenience of selection, we chose a gene (*cat*) responsible for Cm resistance as a candidate to become incorporated into an IS*1*-mediated transposon. This gene is carried on a 6-kb plasmid, pBR325. In a first step, we had to introduce IS*1* into the vicinity of the *cat* gene. Since IS*1* transposition into pBR325 does not offer a selective advantage, we decided to proceed via cointegration of pBR325 with the donor molecule of the IS*1*, the genome of phage P1 that carries one IS*1* element (Iida et al. 1978). Cointegrate pBR325::P1 molecules arise at low frequency (Table 1), but because of the presence of resistance determinants, they are easy to isolate. We had assured ourselves that pBR325 neither contains an intact IS*1* nor any detectable IS*1* sequence. Therefore, most of the cointegrates it might have with P1 DNA were expected to be mediated by the transposition of the IS*1* of P1. Structural analysis of cointegrates confirmed this view. Each of four cointe-

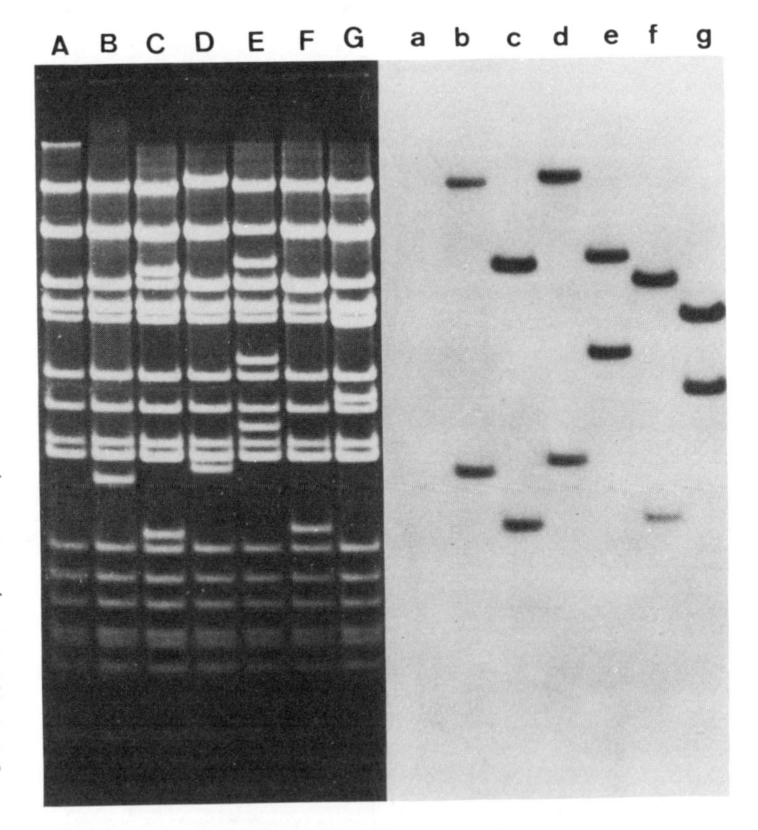

Figure 9. The transposed Cm segments carry two IS*1* elements. Preparation of phage DNA, cleavage with restriction endonuclease, gel electrophoresis, and filter hybridization of *Eco*RI fragments with a ^{32}P-labeled λ::IS*1* probe were performed as described by Iida and Arber (1980). Electrophoresis was carried out in 0.8% agarose. Lanes *A–F* show *Eco*RI fragments, and lanes *a–f* the hybridization of these *Eco*RI fragments with ^{32}P-labeled λ*r*14::IS*1* DNA (Hirsch et al. 1972). (*A, a*) P1-15 DNA; (*B, b*) P1-15::Tn*2654* DNA; (*C, c*) P1-15::Tn*2655* DNA; (*D, d*) P1-15::Tn*2656* DNA; (*E, e*) P1-15::Tn*2653* DNA; (*F, f*) P1-15::Tn*2651* DNA; (*G, g*) P1-15::Tn*2652* DNA. Note that P1-15 hybrid 2 DNA (*a*) contains no IS*1*.

grates studied had the P1 genome integrated at a different site on the pBR325 plasmid, but the sites of the junctions on the P1 genome were in all four cases at its resident IS*1*.

pBR325::P1 cointegrates show properties similar to those of an ordinary P1 prophage. In particular, the vegetative cycle of phage reproduction is heat-inducible, and plaque-forming phages are thereby produced. These are specialized transducing phages for the antibiotic resistances of pBR325.

The cointegrate structures are of considerable stability. However, we knew from previous experiments with P1::Tn*2671* genomes that the two IS*1* elements at the junctions of the P1 genome with the r-determinant insert recombine with each other at a low frequency (Iida and Arber 1977, 1980). By analogy, we expected that in the experiment discussed here a few pBR325::IS*1* plasmids would be segregated out. These segregant molecules could indeed be isolated upon their selective propagation and transformation.

In several pBR325::IS*1* plasmids, an IS*1* element had become a neighbor of the *cat* gene. The question of whether this genome segment could further evolve to become part of an IS*1*-mediated transposon could then be attacked. We have clearly demonstrated that it is indeed possible to find such a transposon carried on a λ-phage derivative that had been grown in the presence of pBR325::IS*1*. From the resulting lysate, λCm^r transducing phages were selected; their structural analysis revealed that in each of six independent cases studied, the *cat* gene was carried in an IS*1*-flanked transposon.

An analysis of the transduction, cointegration, and transposition frequencies shown in Table 1 suggests that with Cm^r plasmids carrying one IS*1* element, transpositional cointegration with λ DNA represents a major source for the formation of specialized Cm^r transducing phages. In contrast, if a plasmid carries an already established Cm transposon, as in pJO505, transposition of the transposon is the most important source of specialized Cm^r transducing λ phages.

A variety of different structures was revealed upon study of the six newly isolated Cm transposons (Fig. 6). Several possible explanations for their formation from pBR325::IS*1* plasmids are given in Figure 10.

Pathways C1 and C2 consist of one IS*1*-mediated transpositional cointegration with the λ target genome and one IS*1*-mediated deletion formation. In principle, these two steps could occur in either order, as was already proposed by Iida and Arber (1980). These pathways involve neither reciprocal recombination nor the transposition of a chromosomal IS*1*.

In pathway D, the second IS*1* could either be a copy of the preexisting IS*1* of pBR325::IS*1* or it could originate from the host chromosome. In either case, transposition with resolution of the intermediate structure (Arthur and Sherratt 1979; Shapiro 1979) would be responsible for the first step shown. The Cm segment flanked by two IS*1* elements would then transpose as a unit into the λ target genome. Depending on the polarity chosen for the transpositional insertion of IS*1* into the target site at the first step, the resulting transposon could be flanked by either directly repeated (D1) or inversely repeated (D2) IS*1* elements.

Pathway E postulates the inversion of the Cm seg-

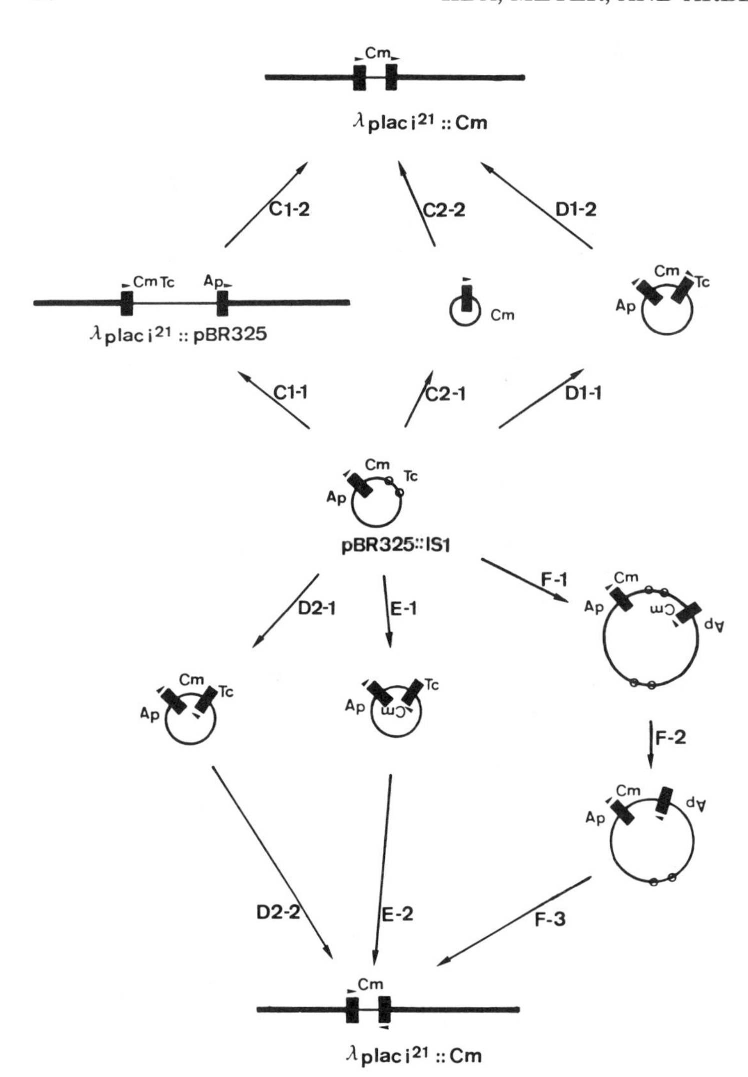

Figure 10. Proposed pathways for the formation of IS*1*-flanked Cm transposons. IS*1* elements are drawn as black boxes, and the orientation is shown by arrowheads. Two small circles on pBR325::IS*1* represent the inverted duplication on pBR325 (Figs. 2 and 3). Drawings of λ*placi*[21]::Cm on the top and bottom represent plaque-forming phages λ*plac5imm*[21] carrying Cm transposons with directly repeated and inversely repeated IS*1* elements, respectively. IS*1*-mediated DNA rearrangements are: IS*1*-mediated transpositional cointegration (C1-1 and C2-2); IS*1*-mediated deletion formation (C1-2, C2-1, and F-2); transposition of Cm segments flanked by two IS*1* elements (D1-2, D2-2, E-2, and F-3); transposition of IS*1* (D1-1 and D2-1); IS*1*-mediated inversion (E-1). Pathway F-1 leads to inverse dimerization of pBR325::IS*1* by recombination between inverted repeats of pBR325. Pathways C, D, and E may apply to any replicon carrying an IS*1* element, whereas pathway F is specific for replicons carrying a partial inverted duplication. Details are discussed in the text. Note that pathway C corresponds to pathway B3 described by Iida and Arber (1980).

ment by intramolecular transposition of IS*1* without its resolution step. This is, in fact, equivalent to the deletion discussed in pathway C2-1, with the difference that the transpositional insertion occurs with inversed polarity (Shapiro 1979). Such inversion mediated by IS*1* has been reported twice (Cornelis and Saedler 1980; Iida et al. 1980).

Any of the pathways C1, C2, or D1 could account for the formation of Tn*2651,* and either pathway D2 or E for the formation of Tn*2652,* Tn*2654,* and Tn*2655.* The two remaining transposons studied, Tn*2653* and Tn*2656,* however, show complex structures, and none of the pathways discussed could easily explain their formation. We would like to propose that their origin is related to the specific structure of pBR325, which carries an inverted duplication (Figs. 2 and 3). By reciprocal recombination, this duplication could give rise to plasmid dimerization in inverted orientation, as shown in pathway F of Figure 10, resulting in the formation of a Cm transposon containing two copies of the *cat* gene in inverted orientation. This is exactly the structure displayed by Tn*2653.* Subsequent IS*1*-mediated deletion formation (F-2) could then give rise to transposons carrying part of their inserts in the same orientation as the parental plasmid and part in inverted orientation. Such a structure is displayed by Tn*2656.* Depending on the extent of the material deleted, pathway F could also explain the formation of Tn*2652,* Tn*2654,* and Tn*2655,* all of which carry their flanking IS*1* elements in inverted orientation.

At present, we have no solid evidence as to which of the pathways shown in Figure 10 is used and with what probability. In a first approximation, it might be expected that the formation of a transposon flanked by directly repeated IS*1* is about as probable as that of a transposon flanked by inversely repeated IS*1*, if the special pathway F is not taken into account. Since Tn*2651* with directly repeated IS*1* has been isolated, we would like to consider that pathways D2 and E may indeed contribute to form the Cm transposons with inverted repeats.

With regard to the question raised concerning the origin of the second IS*1* element in the formation of transposons by pathway D, an observation made by

Arber et al. (1979 and this volume) may be relevant: In the P1 plasmid, intramolecular transposition of the IS*1* resident in the P1 DNA into a linked genome region (usually resulting in deletion formation) appears more frequently than the transposition of an IS*1* element from the *E. coli* chromosome onto the P1-related genome P1-15 that does not yet carry an IS*1*. For this reason, we favor the idea that in the formation of new transposons the second IS*1* might often be a copy of the first.

All newly isolated transposons were shown to transpose as units to a replicon unrelated to the carrier DNA molecule. Most of them also were revealed to be remarkably stable in a test for Cm^s segregants. Reciprocal recombination between the two IS*1* elements was shown to be the most important cause of decay of Tn*2651*, which carries its IS*1* elements as direct repeats. We expect that reciprocal recombination also acts on transposons with inverted IS*1*. But this should not lead to the segregation of the *cat* gene; rather, it should only invert the transposon with regard to the genetic content of the carrier DNA. We expect that such inversion is not accompanied by the loss of Cm resistance, and it was therefore not monitored in our study of transposon stability based on the appearance of Cm^s segregants. It is presumably for this reason that, in general, transposons flanked with inverted repeats appeared to be considerably more stable than those flanked with direct repeats. However, transposons with inversely repeated IS*1* also produced a few Cm^s segregants. These were shown to be mainly caused by IS*1*-mediated deletion formation, deleting in the documented cases the *cat* gene together with some adjacent phage genetic material (Fig. 7). Of course, this mechanism also acts on transposons carrying their flanking IS*1* elements in the same orientation, as was also reported by Muster and Shapiro (this volume).

Tn*2654* as isolated on λ*plac5imm*^{21} was exceptionally unstable, and the two independent Cm^s segregants examined were shown not to contain any IS*1*, indicating that the transposon had been excised. However, we have not investigated further whether this is a particular property of Tn*2654* or rather whether it is related to its particular location on the carrier DNA molecule.

Besides these various processes affecting the stability of IS*1*-mediated transposons, additional mechanisms may also be expected to cause their decay, e.g., the excision of a single IS*1* or the insertion of an unrelated transposable element into a flanking IS*1* or into the *cat* gene, thereby destroying its activity. Some of these mechanisms provide for a direct fusion of unrelated genetic materials, which is of evolutionary importance.

We believe that the conclusions made from studies with IS*1* may have a more general validity. The *E. coli* chromosome contains a number of different IS elements that are able to transpose into the P1 plasmid (Arber et al. 1979 and this volume). It is conceivable that these IS elements might also be able to build up at any time new transposons from chosen genome regions. These new transposons would then have a certain stability and decay again according to rules similar to those outlined here. If one also takes into account that transposons have the opportunity to integrate into phage genomes and into conjugative plasmids, from which they can transpose again to the chromosome of a newly infected host bacterium, it becomes clear that IS elements are potent, although relatively inefficient (because of their low transposition rates), promoters of exchange of practically any gene among different bacterial strains. In addition, IS elements, even if carried in a single copy on a replicon, e.g., on a phage genome or on a conjugative plasmid, can give rise to (1) transpositional cointegration with any other replicon and (2) recombinational cointegration with another replicon also carrying the same IS element (Iida 1980). This provides additional opportunities for spreading of genetic material to other bacterial strains. For all these reasons and in view of the diversity of IS elements found in microorganisms, a particular gene or genome segment could sequentially be taken in charge by a number of different IS elements and thereby be transported through a wide variety of organisms.

SUMMARY

The natural genesis of IS*1*-mediated transposons containing the genetic determinant *cat* for chloramphenicol resistance is documented. First, the small plasmid pBR325 containing the *cat* gene served as a target in IS*1*-mediated transpositional cointegration with the genome of bacteriophage P1, which was the source of the IS*1*. From the resulting pBR325::P1 plasmids, pBR325::IS*1* segregants were isolated. Upon growth of a phage λ derivative in the presence of this plasmid, rare plaque-forming λCm^r specialized transducing phages were formed. In each of six independent λCm^r isolates studied, the *cat* gene was carried between flanking IS*1* elements. In one case, these IS*1* elements were in the same orientation; in the other five cases, they were in opposite orientation. All of these IS*1*-*cat*-IS*1* structures transposed as units to the genome of phage P1-15, pointing to stable maintenance of the transposon. However, appropriate selection allowed us to follow the decay of these transposons. Models to explain the genesis of transposons with directly and inversely repeated IS elements are discussed, as well as the evolutionary implications of these mechanisms.

Note Added in Proof

Absence of any IS*1* sequences on pBR325 was confirmed by DNA sequencing (P. Prentki et al., in prep).

ACKNOWLEDGMENTS

We wish to thank S. Schrickel and M. Stalhammar-Carlemalm for their excellent technical assistance; T. A. Bickle, M. Lusky, and H. J. Reif for providing restriction enzymes and strains; and P. Prentki for communicating results prior to publication. This work was supported by grant no. 3.479.79 from the Swiss National Science Foundation.

APPENDIX I

Spontaneous Mutations in the *Escherichia coli* Prophage P1 and IS-Mediated Processes

W. ARBER, M. HÜMBELIN, P. CASPERS, H. J. REIF, S. IIDA, AND J. MEYER

Department of Microbiology, University of Basel, Biozentrum, CH-4056 Basel, Switzerland

A recent genetic and physical study (Arber et al. 1979) of spontaneous P1 prophage mutants affected in the vegetative phage propagation revealed that at the most a few percent of spontaneous mutations were caused by nucleotide substitution or other small alterations in the nucleotide sequence. About 70% of all mutations studied were relatively long deletions clustering around the single IS*1* element carried in the wild-type P1 genome. The rest, about 27% of the mutants, carried an additional DNA segment most likely representing a transposable IS element from the *Escherichia coli* host chromosome rather than a partial duplication of phage genetic material. On the one hand, these findings point to the importance of IS elements in the production of spontaneous mutations. On the other hand, they open an interesting approach to investigate the diversity of IS elements carried in bacteria serving as hosts for bacteriophage P1.

To study most efficiently the transposition of IS elements from the host chromosome to the prophage, we have now used as a target phage P1-15 hybrid 2 (S. Iida et al., in prep.), which does not carry an IS*1*. With this phage and in contrast to wild-type P1, we expected not to accumulate IS*1*-mediated deletions among spontaneous occurring mutations. It is probable that the P1-15 phage genome used does not initially carry any transposable element.

After the bacteria lysogenic for phage P1-15 *c*1ts had been kept for two months at 30 °C and allowed to grow for about 100 generations, between 100 and 300 single colonies of each of 100 independent clones were tested for production of plaque formers on induction at 42 °C. In this test, one or several colonies not producing active phage were found in 42 of the 100 cultures, and the average frequency of such colonies was about 0.3%. Expression of P1-15-specific restriction (Res$^+$) occurred in 52 out of 61 subclones not producing phage. This points to the presence of a defective plasmid genome. The 9 Res$^-$ subclones were not yet studied further. These could be strains cured from the P1-15 prophage or, as shown in previous experiments (Arber et al. 1979), strains carrying a mutation affecting the Res functions. As a result, a total of 33 clones with independent mutations affecting the production of plaque-forming phage were kept in our collection.

From 17 of these 33 strains, the P1-15 plasmids were isolated; all of the 17 mutants carried an insertion or a partial duplication. The size of each insertion was estimated from the electrophoretic mobility of the restriction fragments carrying the additional material. As seen from Table 3, the insertions were 0.8 to 1.5 kb long, with a few possible exceptions both to lower and slightly higher values. That the insertions are not all identical is also seen from the presence of sites for restriction cleavage in some of the insertions and from hybridization with known IS probes (Table 3). A comparison of restriction sites carried in the insertions with those located on the P1 genome adjacent to the insertions renders the possibility of local partial duplications unlikely for most of the mutated prophages. Rather, the additional genetic material represents transposable elements from the host bacterium.

It is seen in Figure 11 that the site in which an insertion is carried on the mutated P1-15 genome has not been chosen fully at random. Of particular interest is a hot spot or hot region in which 8 of the 17 studied insertions are found. This region is for the time being limited by the approximately 1.5-kb-long overlap between the restriction fragments *Bgl*II-3 and *Bam*HI-5. Most of the remaining insertions are carried at single sites, but a certain clustering to particular regions of the phage genome is obvious. Some regions known to carry also genes essential for vegetative phage development have not yet been found to carry an insertion.

For comparison and as a further support of the above conclusions, the data on 21 previously studied insertions rendering the wild-type P1 genome defective (Arber et al. 1979) are also reported in Figure 11. One of these insertions, 705A, has in the meantime been identified as a $\gamma\delta$ insertion element of 5.7-kb length. The remaining insertions vary in size between an estimated 0.6 kb and 1.6 kb. Seven of the 21 insertions are carried in the above-defined hot region of the target genome. Among these is insertion 520B. This has now been shown to be identical with the 1.25-kb IS element carried in the r-determinant of the R plasmid NR1-Basel (Arber et al. 1979). It is given the name IS*30*, and it has one *Bgl*II site and one *Hin*dIII site.

Both IS-mediated deletion formation and IS-mediated transposition mutagenesis prevail in the formation of spontaneous P1 prophage mutations. In the studies with P1 (Arber et al. 1979), only two spontaneous mutants, or about 3% of those analyzed, did not measurably differ from wild-type mutations in their genome sizes. These mutations could be caused by nucleotide substitution or by very small insertions or deletions. No such mutant was among the 17 P1-15 derivatives studied. Hence, nucleotide substitution must represent at the most a few percent of spontaneously occurring P1 and P1-15 prophage mutations. Because of the observed clustering of the IS-mediated mutations, these conclusions may not be relevant to all of the genes carried on the phage genome or to those that are looked at individually.

The experimental conditions used here, phases of active growth alternating with the stationary phase, may reflect conditions encountered by bacteria in their natural environment. Therefore, our conclusion that

Table 3. Characterization of Insertions Carried in Spontaneous P1-15 Prophage Mutants

Mutant no.	Restriction fragments altered			Estimated size (kb)	Additional restriction cleavage sites	Hybridization with IS probes	
	*Eco*RI	*Bgl*II	*Bam*HI			tested	hybridized with
6	5	1	2	1.2	*Hin*dIII	x	none
8	1	3	5	1.4			
9	19	6	4	1.4	*Hin*dIII	x	none
10	11	1	2	1.4			
28	1	3	5	1.4		x	IS*2*
34B	1	3	5	1.5			
36A	1	3	7	1.4			
41A	1	3	5	1.5			
47	1	3	5	1.2		x	IS*2*
48	1	3	5	1.5			
55	10	3	1	1.0	*Eco*RI, *Bgl*II	x	none
58	1	3	5	1.4			
66	5	1	2	1.2	*Hin*dIII	x	IS*2*
76	1	3	1	1.3	*Eco*RI, *Bgl*II		
80	1	3	5	1.1			
86	6	8	2	1.1	*Hin*dIII	x	IS*2*
96	1	5	5	0.8		x	IS*1*

Mutant plasmid digests were compared in electrophoresis with those of P1-15 DNA. In each instance, one fragment of the P1-15 genome had disappeared and its number is indicated. Instead, either a new fragment of increased size or two new fragments were seen. The size of the insertion was estimated from the electrophoretic mobilities of the respective fragments. These estimates may have an error of ±20%, but the available data clearly show that not all insertions studied have the same size. Where marked, the insertion carried one site each for cleavage by the restriction enzymes indicated, but additional closely linked sites for the same enzymes cannot be excluded. Only mutants numbered 6, 9, 55, 66, and 86 were tested for the presence of *Hin*dIII restriction cleavage sites in the insertion. For the strains indicated (x), hybridization with ^{32}P-labeled probes of IS*1*, IS*2*, and IS*4* was studied by Southern blotting. For this test, plasmids of mutants 9 and 55 were digested with *Bgl*II, all others with *Bam*HI. In addition, hybridization with an IS*5* probe was investigated with mutants 9, 28, 47, and 96.

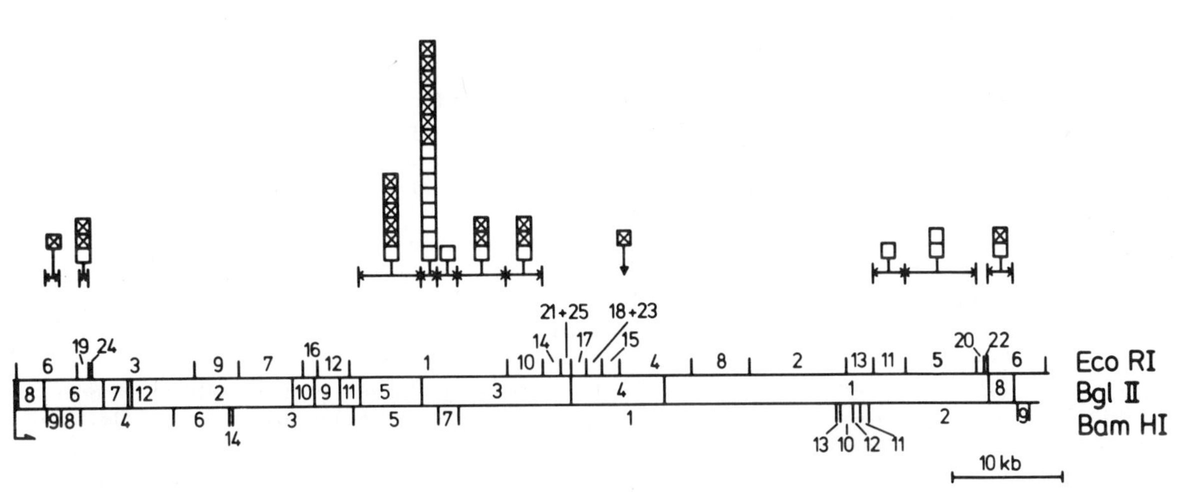

Figure 11. Location of independently isolated insertions on P1-15 and P1 genomes. The *Bgl*II and *Bam*HI restriction maps of the P1-15 hybrid 2 genome are from Iida et al. (in prep.); the *Eco*RI map is from unpublished data of S. Iida. Numbers identify restriction fragments in order of increasing electrophoretic mobility in 0.7% agarose. The fragments are not drawn to scale because emphasis was given to correctly representing the order of the recognition sites for the three enzymes. (□) The sites on the P1-15 genome where the 17 studied insertions are located; (⊠) the sites of a total of 21 analyzed insertions carried in spontaneous prophage mutants of wild-type P1 (Arber et al. 1979; P. Caspers, unpubl). All of these map in the parts of the genome homologous between P1-15 and P1. One insertion was mapped precisely (shown by vertical arrowhead) using electron microscopy to perform heteroduplex analysis. All other insertions were mapped by restriction cleavage analysis (see Table 3 for the 17 insertions in P1-15) so their locations are not precisely known. In particular, it is not known whether sites carrying several insertions represent hot spots or hot regions on the genome. The P1-15 genome region common to fragments *Eco*RI-7, *Bgl*II-2, and *Bam*HI-3 is not totally homologous with P1 DNA; this is the area in which the P1 genome also carries its resident IS*1* element (Iida et al. 1978). The circular map is linearized at the site where DNA packaging into phage particles starts, and the arrow (→) indicates the direction of packaging (Bächi and Arber 1977).

transposition mutagenesis and IS-mediated deletion formation represent the most important sources of spontaneous mutations in P1 plasmids opens the question of whether or not the same mechanisms may prevail in the formation of spontaneous *E. coli* chromosomal mutations. Perhaps they do, but one might expect that processes of IS-mediated restructuring within the bacterial chromosome may often be lethal, as they are in fact to the viral functions of P1 in our experiments, and thus these restructurings might not be monitored by easy means.

SUMMARY

IS transposition is the major cause of spontaneous loss of the ability of P1 prophage and its relative P1-15 to produce plaque-forming phage particles. A number of different IS elements from the *E. coli* host contribute to this effect.

ACKNOWLEDGMENT

We thank the Swiss National Science Foundation for its support (grant no. 3.479.79).

APPENDIX II

Analysis of Transposition of IS*1-kan* and Its Relatives

H. J. Reif and W. Arber

Department of Microbiology, University of Basel, Biozentrum CH-4056 Basel, Switzerland

To investigate sequences of IS*1* that are necessary for transposition, it would be helpful to follow the transposition of individual copies of IS*1*. This is problematic for two reasons: First, IS*1*—like many other IS elements—is phenotypically cryptic; it does not code for selectable genes and this prohibits observation of transposition of IS*1* by genetic means. Second, IS*1* occurs in multiple copies in the *Escherichia coli* chromosome (Saedler and Heiss 1973) and therefore physical analysis of its transposition is difficult. We overcame both problems by splicing a DNA fragment coding for resistance to the antibiotic kanamycin (*kan*) into the *Pst*I site of one copy of IS*1*. We call the resulting IS element IS*1-kan*.

The strategy for splicing the *Pst*I fragment of phage P1-*kan*, containing the gene for kanamycin resistance, into IS*1* and the various plasmids constructed from such cloning experiments are shown in Figure 12. The plasmid pJO505 contains three copies of IS*1* (Tn*982*); two of these flank a gene for chloramphenicol resistance (Tn*981*). Plasmid pJO50 is a derivative of this plasmid and contains only one copy of IS*1*. Into this IS*1*, we cloned the *Pst*I fragment of P1-*kan*, which resulted in pJO51. The *Pst*I fragment of P1-*kan* was also inserted into pBR322. This plasmid, pJO603, was used as a control in studies on IS*1-kan* transposition. Another plasmid constructed was pJO700, which is similar to pJO505 except that one copy of IS*1* is now replaced by IS*1-kan*.

RESULTS

The relevant structures of plasmids pJO603, pJO505, pJO51, and pJO700 and the transposition frequencies of the kanamycin and chloramphenicol determinants in them are shown in Figure 13.

In the control experiment with pJO603, we found that the *kan* fragment has no transposition ability by itself. Indeed, no transposition products were found on screening 10^{10} phages. We also determined transposition of the *tet* gene from pJO603, which is also flanked by the fragments of the direct repeats of the *kan* determinant. With 5.3×10^{-10} transpositions per phage, we found a frequency several orders below those found with IS*1*.

To get a standard value for IS*1*-mediated transposition, we measured the frequencies of transposition of Tn*981* and Tn*982*, the composite transposons on pJO505. We found a high value for transposition of the *cat* gene (4.5×10^{-4}/phage), as had been observed previously in a *recA*$^+$ host (Reif 1980). In a *recA*$^-$ background, essentially the same level of transposition is obtained (Fig. 13).

The third line on Figure 13 shows clearly that IS*1-kan* transposes as frequently as an IS*1*-flanked transposon in a *recA*$^+$ genetic background. We have analyzed several of the phage DNAs of P1-15::IS*1-kan* by restriction cleavage (data not shown). *Pst*I digests of the phage DNAs contained a fragment that migrated during electrophoresis indistinguishable from the *Pst*I fragment of P1-*kan* which was cloned into pJO50. Furthermore, using a *Bgl*II digest, it was easily determined where IS*1-kan* integrated into the P1-15 genome, and several different integration sites were also observed. With each of the isolates analyzed, the new *Bgl*II fragment (IS*1-kan* does not contain a *Bgl*II site) was cut by *Pst*I into three fragments. One of the fragments corresponded to the *kan* insert, and the two others comprised the junction fragments between P1-15 DNA and IS*1-kan*. The addition of the sizes of the two junction fragments yielded values that were always roughly 800 bp longer than the corresponding target fragment of P1-15. This corresponds well to the size of IS*1*, which contains 768 bp (Johnsrud 1979). Thus, it had been concluded that IS*1-kan* had transposed from pJO51.

In contrast to transposition of Tn*981* and Tn*982*, we find a strong *recA* dependence for transposition of IS*1-kan*. In the *recA*$^-$ background, transposition of IS*1-kan* is reduced 100-fold. Instead, most of the transductants turn out to present cointegrates appearing with a frequency of about 10^{-4}/phage, whereas they were a minor class in the *rec*$^+$ host.

Thus, transposition of IS*1-kan* is severely reduced in *recA*$^-$, but cointegrate formation is strongly enhanced. The fourth line in Figure 13 shows transposition properties of an IS*1*-flanked *cat* transposon, in which one of

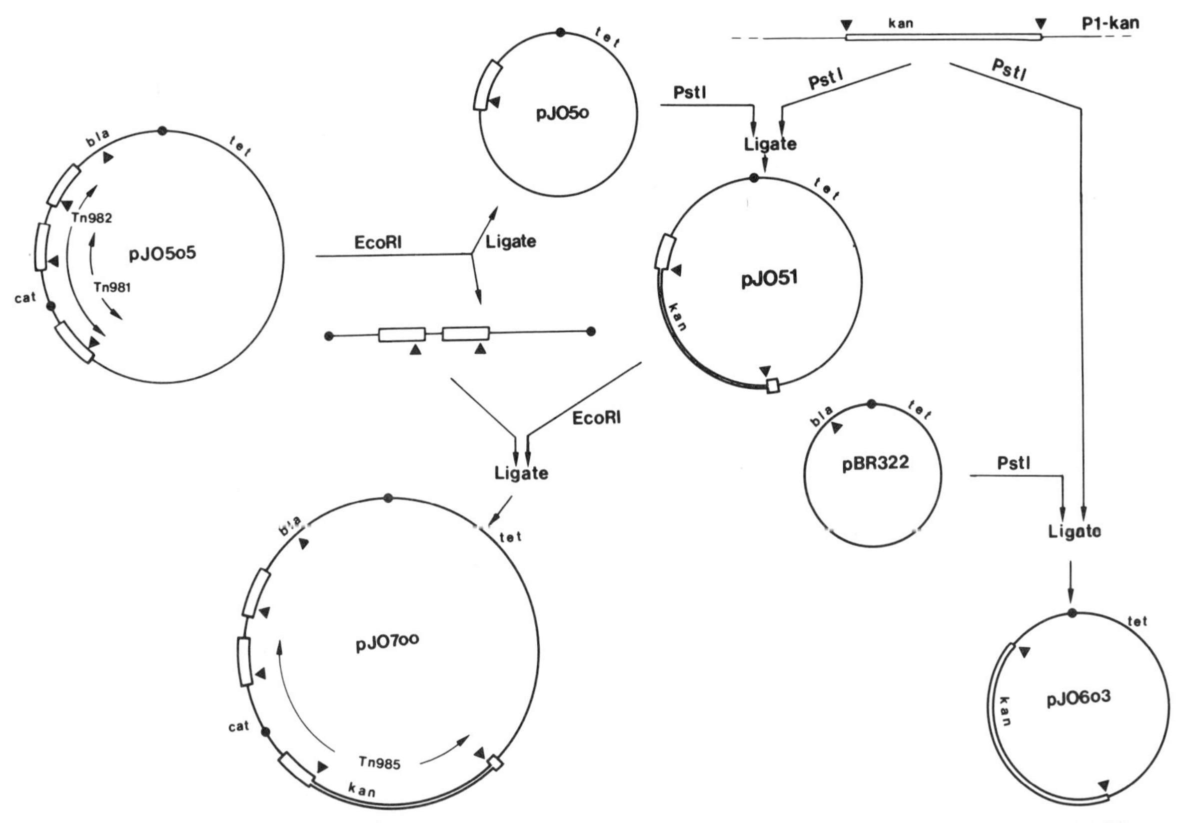

Figure 12. Cloning strategy for construction of the plasmids pJO51, pJO603, and pJO700. *bla*, β-lactamase = ampicillin resistance; *tet* = tetracycline resistance; *kan* = kanamycin resistance; *cat*, chloramphenicolacetyltransferase = chloramphenicol resistance. (●) *Eco*RI site; (▲) *Pst*I site; (▭) IS*1*.

Plasmid	relevant structures	Transposition of: kan		cat	
		rec⁺	recA	rec⁺	recA
pJO6o3	kan tet	$<10^{-10}$	x	–	–
pJO5o5	cat tet	–	–	$4{,}5\times10^{-4}$	$2{,}3\times10^{-4}$
pJO51	kan	$4{,}0\times10^{-4}$	$3{,}5\times10^{-6}$	–	–
pJO7oo	cat kan	$6{,}0\times10^{-6}$	$2{,}0\times10^{-6}$	$8{,}0\times10^{-5}$	$8{,}0\times10^{-6}$

Figure 13. The transposition frequencies of *kan* and *cat* genes from the different plasmids mentioned in the text are given as events per plaque-forming P1-15 phage. In case of plasmid pJO700, transposition of *kan* alone (IS*1-kan*) as well as of *cat* (Tn*985*) could be measured. In the latter case, all chloramphenicol-resistant (Cm^r) isolates were also kanamycin-resistant (Km^r). Thus, *cat* cannot transpose without IS*1-kan*. (x) Not measured; (—) the marker is not on the plasmid; (▲) *Pst*I sites.

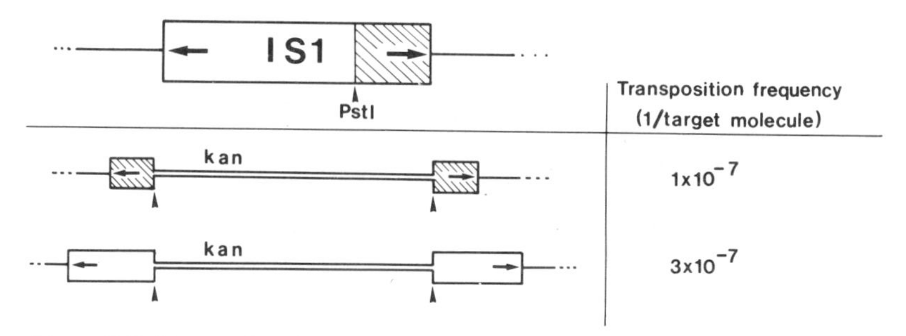

Figure 14. Transposition of Mini-inverted and Maxi-inverted IS*1*-*kan*. The frequencies given are events per P1-15 plaque-forming phage tested. (→) The terminal, imperfect inverted repeats; (♠) *Pst*I sites used for cloning. Experimental details will be described elsewhere.

the flanking repeats has been replaced by IS*1*-*kan*. The *kan* insert reduces *cat* transposition slightly (factor of 5) and a *recA* influence is clearly present.

Plasmid pJO700 also allows analysis of the influence of neighboring IS*1* elements on transposition of IS*1*-*kan*. In contrast to pJO51, where it was the only transposable element on the plasmid, it is part of a transposon on pJO700 and has to transpose out of a transposon. Surprisingly, it was found that IS*1*-*kan* transposes from pJO700 with a frequency reduced 60-fold compared to that of pJO51. The presence of other copies of IS*1* in the vicinity of IS*1*-*kan* apparently partially inhibits its transposition.

The observation that IS*1*-*kan* transposition is reduced about 100-fold in a *recA*$^-$ background and that it mediates cointegrate formation is reminiscent of the behavior of certain mutants of Tn*3* (Gill et al. 1978; Arthur and Sherratt 1979). These Tn*3* mutants are defective in a sequence called *IRS* (*I*nternal *R*esolution *S*tructure).

The results presented here comparing transposition behavior of Tn*981* and IS*1*-*kan* lead us to the conclusion that with respect to cointegrate resolution, IS*1* behaves in a manner similar to Tn*3*. If this conclusion is correct, it would place the *IRS* site of IS*1* in the vicinity of the *Pst*I site, since insertion of a DNA fragment into the *Pst*I site damages the presumptive *IRS* site and reduces resolution 100-fold.

A more puzzling observation is the decrease of transposition of IS*1*-*kan* from pJO700 in a *recA*$^+$ host. Apparently, the presence of other copies of IS*1* on the same plasmid in the vicinity of IS*1*-*kan* affects its transposition negatively. This could be the result of the various cointegrate structures that can be formed by a plasmid with more than one transposable element.

The assumption that an *IRS* site for IS*1*-specific transposition is close to the *Pst*I site of IS*1* led us into a preliminary search for sequence homologies. Several other sequences for site-specific recombination, such as the ends of insertion sequences and transposable elements, the *IRS* region of Tn*3*, and the attachment sites for phage λ integration, are fairly well defined. We noticed that close to the left side of the *Pst*I site of IS*1* is a sequence (Johnsrud 1979), AGTGCAACT, that not only has strong homologies to the inverted repeats at the ends of IS*1*, but also is homologous to the border between *P′* and *O* of the λ attachment site (Landy and Ross 1979) AGTCCAACT.

The 5-bp sequence, CAACT, common to all four sequences mentioned also shows up in the region where the *IRS* sequence of Tn*3* is located (Gill et al. 1978; Arthur and Sherratt 1979; Heffron et al. 1979). The probability for the sequence homologies to occur at all the sites by chance is rather low. One could speculate about some functional significance of the 5-bp sequence mentioned above in an environment for nonhomologous recombination.

The result that initiation of transposition of IS*1*-*kan* (cointegrate formation) is possible in a *recA*$^-$ host and that in a *recA*$^+$ host complete transposition of IS*1*-*kan* is very efficient allows us to test transposition properties of structural relatives of IS*1*-*kan*. In particular, mutant IS*1*-*kan* can serve to analyze which of the sequences of IS*1* is necessary in *cis* for transposition. It has been known for some time that intact inverted ends are necessary for transposition of Tn*3* and that deletions of the inverted repeats at the ends cannot be complemented in *trans* (Heffron et al. 1977). Furthermore, Saedler et al. (this volume) found that in case of IS*1* mutations within the terminal, imperfect inverted repeats abolish IS*1*-dependent deletion formation (Reif and Saedler 1975), a process believed to be an abortive intermediate of the transposition process (Arthur and Sherratt 1979; Shapiro 1979). We tested for other possible *cis*-like sequences acting on IS*1* and that would be important for transposition by analyzing Mini-inverted IS*1*-*kan* and Maxi-inverted IS*1*-*kan*. These sequences have perfect repeats at their ends but lack considerable parts of the IS*1* sequences. Both derivatives still show transposition, though only to a reduced level (Fig. 14). The imperfections of the repeats at the ends of IS*1* do not seem to have an important functional role in transposition. So et al. (1979) and Iida et al. (this volume) described transposons from *E. coli* that are flanked by inverted repeats of IS*1*. They seem to transpose as efficiently as a transposon flanked with direct repeats. Thus, the reduction of transposition observed with Mini-inverted IS*1*-*kan* and Maxi-inverted IS*1*-*kan* must have other causes and involve different *cis*-acting sites.

Note Added in Proof

Transposition of IS*1*-*kan* and Mini-inverted IS*1*-*kan* have now been proved by hybridization experiments.

ACKNOWLEDGMENTS

We are thankful to Drs. T. Bickle, S. Iida, J. Shephard, and H. Eibel for helpful discussions; Drs. G. Hobom, S. Iida, R. Matthes for strains and unpublished information; and Dr. T. Bickle for generously supplying us with restriction enzymes. We are glad to acknowledge R. Hiestand for technical assistance and I. Reif for typing the manuscript. This work was supported by grant no. 3.479.79 from the Swiss National Science Foundation.

REFERENCES

ALTON, N. K. and D. VAPNEK. 1979. Nucleotide sequence analysis of the chloramphenicol resistance transposon Tn*9*. *Nature* **282:** 864.

ARBER, W. and D. WAUTERS-WILLEMS. 1970. Host specificity of DNA produced by *Escherichia coli*. XII. The two restriction and modification systems of strain 15T⁻. *Mol. Gen. Genet.* **108:** 203.

ARBER, W., S. IIDA, H. JÜTTE, P. CASPERS, J. MEYER, and C. HÄNNI. 1979. Rearrangements of genetic material in *Escherichia coli* as observed on the bacteriophage P1 plasmid. *Cold Spring Harbor Symp. Quant. Biol.* **43:** 1197.

ARTHUR, A. and D. SHERRATT. 1979. Dissection of the transposition process: A transposon-encoded site-specific recombination system. *Mol. Gen. Genet.* **175:** 267.

BÄCHI, B. and W. ARBER. 1977. Physical mapping of *BglII*, *Bam*HI, *Eco*RI, *Hind*III, and *Pst*I restriction fragments of bacteriophage P1 DNA. *Mol. Gen. Genet.* **153:** 311.

BOLIVAR, F. 1978. Construction and characterization of new cloning vehicles. III. Derivatives of plasmid pBR322 carrying unique *Eco*RI sites for selection of *Eco*RI generated recombinant DNA molecules. *Gene* **4:** 121.

CORNELIS, G. and H. SAEDLER. 1980. Deletions and an inversion induced by a resident IS*1* of the lactose transposon Tn*951*. *Mol. Gen. Genet.* **178:** 367.

GILL, R., F. HEFFRON, G. DOUGAN, and S. FALKOW. 1978. Analysis of sequences transposed by complementation of two classes of transposition deficient mutants of Tn*3*. *J. Bacteriol.* **136:** 742.

HEFFRON, F., P. BEDINGER, J. CHAMPOUX, and S. FALKOW. 1977. Deletions affecting transposition of an antibiotic resistance gene. In *DNA insertion elements, plasmids, and episomes* (eds. A. J. Bukhari et al.), p. 161. Cold Spring Harbor Laboratory, Cold Spring Harbor, New York.

HEFFRON, F., B. J. MCCARTHY, H. OHTSUBO, and E. OHTSUBO. 1979. DNA sequence analysis of the transposon Tn*3*: Three genes and three sites involved in transposition of Tn*3*. *Cell* **18:** 1153.

HIRSCH, H. J., P. STARLINGER, and P. BRACHET. 1972. Two kinds of insertions in bacterial genes. *Mol. Gen. Genet.* **119:** 191.

IIDA, S. 1977. Directed integration of an F′ plasmid by integrative suppression. Isolation of plaque forming λ transducing phage for the *dna*C gene. *Mol. Gen. Genet.* **155:** 153.

———. 1980. A cointegrate of the bacteriophage P1 genome and the conjugative R plasmid R100. *Plasmid* **3:** 278.

IIDA, S. and W. ARBER. 1977. Plaque forming specialized transducing phage P1: Isolation of P1CmSmSu, a precursor of P1Cm. *Mol. Gen. Genet.* **153:** 259.

———. 1980. On the role of IS*1* in the formation of hybrids between the bacteriophage P1 and the R plasmid NR1. *Mol. Gen. Genet.* **177:** 261.

IIDA, S., J. MEYER, and W. ARBER. 1978. The insertion element IS*1* is a natural constituent of coliphage P1 DNA. *Plasmid* **1:** 357.

———. 1980. IS*1* mediated inversion observed in phage P1CmTc1 DNA. *Experientia* **36:** 748.

JOHNSRUD, L. 1979. DNA sequence of the transposable element IS*1*. *Mol. Gen. Genet.* **169:** 213.

LANDY, A. and W. ROSS. 1977. Viral integration and excision: Structure of the lambda *att* sites. *Science* **197:** 1147.

MACHATTIE, L. A. and J. B. JACKOWSKI. 1977. Physical structure and deletion effects of the chloramphenicol resistance element Tn*9* in phage lambda. In *DNA insertion elements, plasmids, and episomes* (ed. A. I. Bukhari et al.), p. 219. Cold Spring Harbor Laboratory, Cold Spring Harbor, New York.

MARCOLI, R., S. IIDA, and T. A. BICKLE. 1980. The DNA sequence of an IS*1*-flanked transposon coding for resistance to chloramphenicol and fusidic acid. *FEBS Lett.* **110:** 11.

MEYER, J. and S. IIDA. 1979. Amplification of chloramphenicol resistance transposons carried by phage P1Cm in *Escherichia coli*. *Mol. Gen. Genet.* **176:** 209.

MEYER, J., S. IIDA, and W. ARBER. 1980. Does the insertion element IS*1* transpose preferentially into A+T-rich DNA segments? *Mol. Gen. Genet.* **178:** 471.

MEYERS, J. A., P. SANCHEZ, L. P. ELWELL, and S. FALKOW. 1976. Simple agarose gel electrophoretic method for the identification and characterization of plasmid deoxyribonucleic acid. *J. Bacteriol.* **127:** 1529.

OHTSUBO, H. and E. OHTSUBO. 1978. Nucleotide sequence of an insertion element, IS*1*. *Proc. Natl. Acad. Sci.* **75:** 615.

OHTSUBO, E., M. ZENILMAN, and H. OHTSUBO. 1980. Plasmids containing insertion elements are potential transposons. *Proc. Natl. Acad. Sci.* **77:** 750.

REIF, H. J. 1980. Genetic evidence for absence of transposition functions from the internal part of Tn*981*, a relative of Tn*9*. *Mol. Gen. Genet.* **177:** 667.

REIF, H. J. and H. SAEDLER. 1975. IS*1* is involved in deletion formation in the *gal* region of *E. coli* K12. *Mol. Gen. Genet.* **137:** 17.

ROSNER, J. L. and M. S. GUYER. 1980. Transposition of IS*1*-λBIO-IS*1* from a bacteriophage λ derivative carrying the IS*1*-*cat*-IS*1* transposition (Tn*9*). *Mol. Gen. Genet.* **178:** 111.

SAEDLER, H. and B. HEISS. 1973. Multiple copies of insertion DNA sequences IS*1* and IS*2* in the chromosome of *E. coli* K12. *Mol. Gen. Genet.* **122:** 267.

SHAPIRO, J. A. 1979. Molecular model for the transposition and replication of bacteriophage Mu and other transposable elements. *Proc. Natl. Acad. Sci.* **76:** 1933.

SHAPIRO, J. A. and L. A. MACHATTIE. 1979. Integration and excision of prophage λ mediated by the IS*1* element. *Cold Spring Harbor Symp. Quant. Biol.* **43:** 1135.

SO, M., F. HEFFRON, and B. J. MCCARTHY. 1979. The *E. coli* gene encoding heat stable toxin is a bacterial transposon flanked by inverted repeats of IS*1*. *Nature* **277:** 453.

SUTCLIFFE, J. G. 1979. Complete nucleotide sequence of the *Escherichia coli* plasmid pBR322. *Cold Spring Harbor Symp. Quant. Biol.* **43:** 77.

SZYBALSKI, E. H. and W. SZYBALSKI. 1979. A comprehensive molecular map of bacteriophage lambda. *Gene* **7:** 217.

Transposons Encoding Trimethoprim or Gentamicin Resistance in Medically Important Bacteria

N. Datta, M. Nugent, and H. Richards
Department of Bacteriology, Royal Postgraduate Medical School, London W12 OHS, England

An understanding of the epidemiology of antibiotic resistance depends on the ability to recognize particular bacterial strains, their plasmids, and the resistance genes that they carry. We have collected and studied bacteria resistant to trimethoprim (Tp^r) and/or gentamicin (Gm^r) with the aim of elucidating the spread of resistance. Our study dates back to the introduction of these drugs to medicine in the late 1960s. Trimethoprim resistance is important in bacteria that cause infections in the community as well as in hospitals, whereas the problems presented by gentamicin resistance are encountered chiefly in hospitals.

The first trimethoprim R plasmids to be identified all belonged to incompatibility group W (IncW) and were found in several bacterial species that infected or colonized patients in a group of London hospitals (Datta and Hedges 1972; Fleming 1972). They conferred resistance to trimethoprim and sulfonamides (Tp^{r}Su^r). R388 is an example of these plasmids (Jacob et al. 1977). The origin of their Tp^r gene is not known, and no evidence has been found for its transposability.

Later, another kind of trimethoprim R plasmid, of IncIα, was identified in *Escherichia coli* strains isolated from calves that had been given high doses of trimethoprim on an experimental farm (Hedges et al. 1972; Fleming 1973; Datta and Barth 1976). One example, R483, carried a 9-megadalton (MD) sequence that encoded resistance to trimethoprim and streptomycin/spectinomycin (Tp^r,Sm^r/Sp^r) and that was readily transposable between plasmids or into the chromosome of *E. coli* K12 (Barth et al. 1976). This sequence is designated transposon 7 (Tn*7*) (Campbell et al. 1977). Subsequently, trimethoprim R plasmids belonging to 12 different incompatibility groups have appeared (Jobanputra and Datta 1974; Acar et al. 1977; Towner 1979; Datta et al. 1980), partly as the result of dissemination of Tn*7* (Barth and Datta 1977; Richards and Nugent 1979). The frequency of Tn*7* transposition is very high, $>10^{-2}$ per cell (Barth et al. 1976).

Trimethoprim R plasmids determine the production of dihydrofolate reductase (DHFR). The host's own DHFR is inhibited by trimethoprim, and the plasmid provides a substitute that is resistant to such inhibition (Amyes and Smith 1974; Sköld and Widh 1974).

The earliest gentamicin R plasmids identified belonged to three different groups: IncC, IncM, and IncFII (Witchitz and Chabbert 1972; Chabbert et al. 1979; Datta et al. 1979a). Those of IncC and IncFII determined the same enzyme for gentamicin adenylylation, ANT(2″), which also confers tobramycin resistance, whereas the earliest-reported gentamicin (not tobramycin) R plasmid of IncM, pIP135, determined an acetylating enzyme, AAC(3)1 (Witchitz and Gerbaud 1972). The linkage of gentamicin resistance (Gm^r) genes to their respective types of plasmids has been very stable, and currently most of the gentamicin R plasmids that we identify in Enterobacteriaceae are IncC or IncFII, which determine ANT(2″). The IncM plasmid pTH1 (described in Results because it includes Tn*7*) resembles pIP135, which dates from nearly 10 years earlier, in its incompatibility group and its gentamicin-modifying enzyme, AAC(3)1. Although some of its other resistance genes are different, pTH1 is very similar to pIP135 in its pattern of cleavage by *Eco*RI (Chabbert et al. 1979; J. W. Witchitz, pers. comm.).

Gm^r genes, although they made their appearance in plasmids of medically important bacteria soon after gentamicin was first used medically, have not been disseminated among different classes of plasmids to the same extent as Tp^r genes. There is now evidence, however, of changing linkages and transpositions.

MATERIALS AND METHODS

Collection of strains. Non-hospital bacteria were isolates from human urinary tract infections, *Salmonella* species from man and animals, and isolates of *E. coli* from animal sources. Hospital bacteria were obtained during outbreaks of infection (Casewell et al. 1977; Rennie and Duncan 1977) and during a continuous survey in our own hospital (Datta et al. 1980). Bacterial species and strains were identified by conventional biochemical and serological methods. The *E. coli* K12 strains and plasmids used are described in Table 1.

Plasmid transfer and classification of plasmids by incompatibility. These were done as described previously (Datta et al. 1980).

Transposition. Transposition was indicated if a resistance gene (or genes) was retained after the plasmid that had carried it was eliminated during incompatibility tests. Resistance-transfer tests were then used to show the new linkage of the retained gene or genes. Serial transposition from plasmid to plasmid, plasmid to chromosome, or chromosome to plasmid was looked for in *recA* K12 host strains. Having thus identified a transposon, we then determined its molecular mass (on the basis of the increase in molecular weight upon its acquisition by another plasmid) and its susceptibility to cutting by restriction enzymes (from altered restriction

Table 1. Strains of *E. coli* K12 and Plasmids

E. coli

Strain	Characteristics	Reference or source
J53-1	*pro met nal*^{r}	Bachmann (1972)
J62-2	*pro his trp rif*^{r}	Bachmann (1972)
M259	*met* $r^- m^+$	N. Murray
PB1455	*thr leu* B1 *recA*	P. Bergquist
PB1474	*ilv recA*	P. Bergquist

Plasmids

No.	Resistance pattern	Inc group	Reference
R388	Tp Su	W	Jacob et al. (1977)
R389a	Tp Su Cm Sm	W	Jacob et al. (1977)
R483	Tp Sm	Iα	Jacob et al. (1977)
pIP135	Sm Su Tc Gm	M	Jacob et al. (1977)
pTH1	Ap Sm Su Tc Cm Km Tp Gm Hg	M	Datta et al. (1979b)
R721	Tp Sm	Iδ	Barth and Datta (1977)
pBW1	Tp Sm	Iδ	Barth and Datta (1977)
R751	Tp	P	Jacob et al. (1977)
R27	Tp Su	W	Pattishall et al. (1977)
pHH500	Tp Su	—	Datta et al. (1979c)
pHH502	Tp Tc Cm	P	Datta et al. (1979c)
pBR322	Ap Tc	—	Bolivar et al. (1977)
pHH1313b	Ap Tc Cm Gm Tb	FII	Nugent et al. (1979)
pHH1307	Tp Su Gm Ap Hg	W	Datta et al. (1980)

fragments yielded by a plasmid upon acquisition of the new gene[s]).

DNA extractions. Plasmid DNA was extracted according to the method of Barth and Grinter (1974).

Restriction enzyme analysis. The restriction enzymes *Bam*HI, *Eco*RI, *Hin*dIII, *Pst*I, and *Sal*I were obtained from BRL, Inc., and the digestion buffers and conditions used were as recommended by BRL. Gel electrophoresis was performed as described by Datta et al. (1979b).

Plasmid molecular weights. These were estimated either by restriction enzyme analysis of purified DNA or by the single-colony lysis method described by Eckhardt (1978).

Cloning of the internal fragments of Tn7. Approximately 10 μg each of ColE1::Tn7 and pBR322 DNA were together digested with *Hin*dIII, according to the method of Barth (1979), except that they were ligated for only 5 hours. Ampicillin-resistant (Ap^r) transformants were tested for sensitivity to tetracycline (Tc), and any Ap^{r}Tc^s clones were analyzed further.

Preparation of radioactive probes for identification of transposon sequences. ^{32}P-labeled probes were prepared according to the method of Rigby et al. (1977). The DNAs used in the probes were small plasmids into which the transposon had inserted itself in vivo, small plasmids into which restriction fragments of transposons had been inserted in vitro, and vector plasmids without added sequences.

Hybridization of plasmid DNA with transposon probes. This was carried out according to the method of Heinaru et al. (1978).

RESULTS

Trimethoprim Resistance

The spread of Tn7 in naturally occurring plasmids and bacteria. We have recognized the Tn7 sequence by (1) its ability to transpose linked genes for high-level trimethoprim resistance and low-level streptomycin/spectinomycin resistance; (2) its molecular mass (9 MD); (3) its having one cut site for restriction enzyme *Eco*RI, two for *Bam*HI, and three for *Hin*dIII at characteristic positions and giving fragments of predictable sizes (Fig. 1); and (4) its hybridization with radioactive probes derived from the prototype Tn7.

Until 1977, Tn7 had been recognized in plasmids from *E. coli* strains isolated from non-hospital sources where selection for trimethoprim resistance was known, i.e., from calves treated with trimethoprim and from a human subject on long-term trimethoprim-sulfonamide therapy. It has since appeared in more and more environments. In a London hospital in 1977, there was an outbreak of infection with multiply resistant *Klebsiella* sp. serotype K16. The epidemic bacterial strain carried a plasmid of IncM that included Tn7. The same IncM plasmid was found in epidemiologically related isolates of *E. coli* and *Citrobacter koseri,* and in all three bacterial hosts Tn7 was present in at least two loci: the IncM plasmid and the host chromosome (Datta et al 1979b). We could not determine where the transposon came from, but something is known of the origin of its vector IncM plasmid pTH1. It determines, among other things, gentamicin resistance and resembles gentamicin R plasmids found in Paris before 1972 (see above). The evidence suggests that by 1977 the IncM plasmid was widely dispersed and had recently and/or locally acquired Tn7. With Tn7 included, the IncM plasmid was

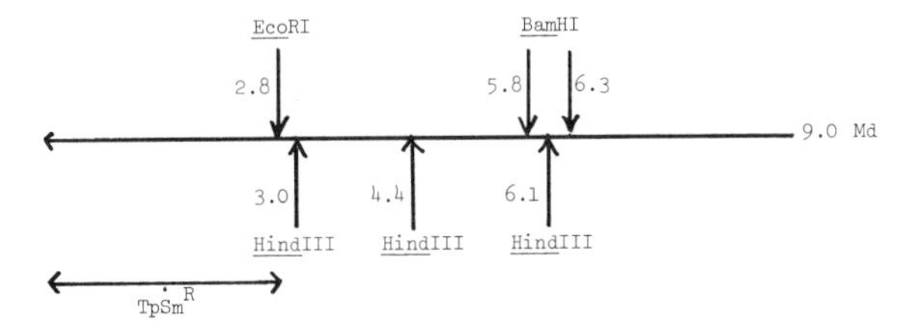

Figure 1. Restriction enzyme map of Tn7.

found in bacteria that caused another outbreak of hospital infection about 25 miles north of London. In this case, there was evidence of the transposon in other naturally occurring plasmids in the same bacterial host that contained the IncM plasmid (V. M. Hughes, pers. comm.). Tn7 was also identified in plasmids of at least three groups (IncC, IncIδ, and IncX), during a survey of hospital bacteria (Datta et al. 1980), and it has appeared in plasmids of *Salmonella typhimurium,* which infects man and animals (Richards et al. 1978).

Hybridization tests using radioactive probes. A ^{32}P-labeled probe of ColE1::Tn7 was made in vitro and tested for hybridization with unlabeled plasmid DNA that had been transferred to nitrocellulose filters by the Southern blot technique. We wanted to know whether the method would identify (1) Tn7 and (2) Tp^r genes encoding DHFR but not part of Tn7. We tested plasmids that were known to include Tn7, plasmids that were suspected to contain Tn7, and trimethoprim R plasmids without Tn7. The latter group consisted of R388, R389a, R751, R27, pHH500, and pHH502, which we have shown, by genetic techniques and restriction enzyme analysis, not to contain Tn7. Figure 2 and Table 2 show the results of hybridization tests.

We demonstrated hybridization of Tn7 only with plasmids that we knew or suspected to contain Tn7 itself. Plasmid-determined DHFRs are of two immunologically distinct types, DHFR I and DHFR II (Pattishall et al. 1977; Fling and Elwell 1980). Tn7 encodes type I. Type II (or DHFR immunologically related to type II) is determined by R27, R388, and R751 (Fling and Elwell 1980); R389a is related to R388 in incompatibility group and origin and therefore is also likely to specify type II. The DHFRs of plasmids pHH500 and pHH502 have not been studied. Our results, therefore, support the findings of Fling and Elwell (1980) in showing no homology between the genes for DHFR I and DHFR II. Further studies will show whether the genetic sequence for the type-I enzyme is found somewhere other than as part of Tn7.

We have cloned the 1.4-MD internal *Hin*dIII fragment of Tn7 into pBR322. This fragment does not contain the gene for DHFR (Fig. 1); no gene product is known for it. We have made ^{32}P-labeled probes of this plasmid and have shown that it hybridizes specifically to plasmids containing or suspected of containing Tn7. It does not hybridize with RP4 and therefore is not homologous to any equivalent sequence in Tn*1*. We are using this probe as a quick method of identifying Tn7 in our collection of plasmids.

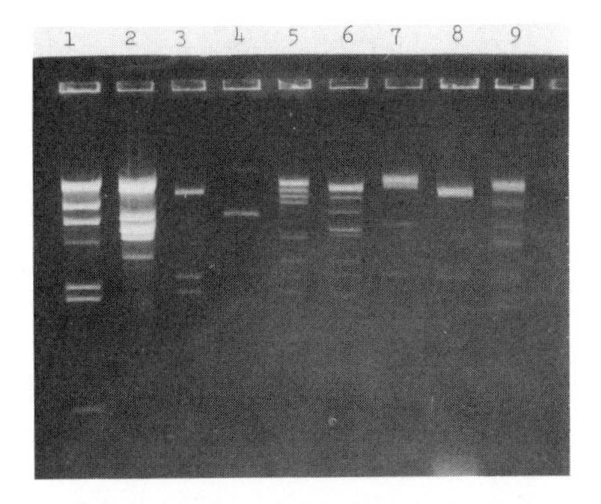

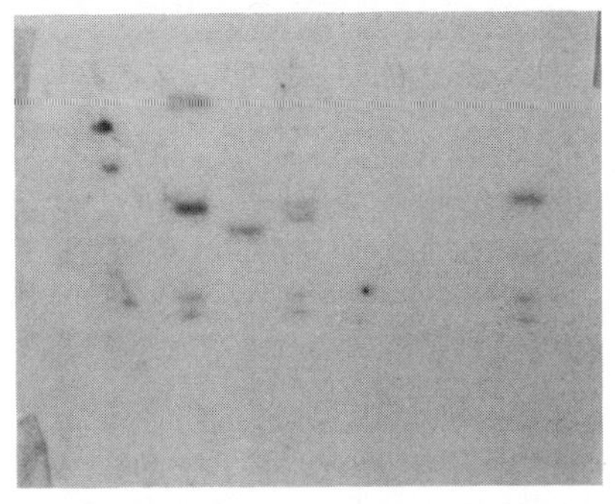

Figure 2. Restriction enzyme analysis of trimethoprim R plasmids. (*Top*) Gels contain: (*1*) λ *Hin*dIII; (*2*) λ *Eco*RI; (*3*) ColE1::Tn7 *Hin*dIII; (*4*) ColE1 *Hin*dIII; (*5*) R483 *Hin*dIII; (*6*) pHH502 *Hin*dIII; (*7*) R389a *Hin*dIII; (*8*) pHH500 *Hin*dIII; (*9*) pTH1 *Hin*dIII. (*Bottom*) Hybridization of the ColE1::Tn7 ^{32}P-labeled probe to the gel above.

Gentamicin Resistance

Transposon Tn732. The gentamicin-tobramycin resistance (Gm^r/Tm^r) genes determine various enzymes that modify and inactivate the drugs, and there is an association between particular Gm^r/Tm^r genes and plasmids of particular incompatibility groups (Witchitz and Gerbaud 1972; Datta et al. 1979a). This implies that the acquisition of such a gene by any plasmid is a rare event.

A transposon, Tn*732*, determining the gentamicin/tobramycin adenylylating enzyme ANT (2″) was identified in an IncFII plasmid, pHH1313b, from bacteria

Table 2. Hybridization of ColE1::Tn7 ^{32}P-labeled Probe

Plasmid DNA	Inc group	Hybridization	Transposon present
ColE1::Tn7		+	Tn7
RP4	P	–	Tn*1*
RP4::Tn7	P	+	Tn*1* + Tn7
R483	Iα	+	Tn7
R721	Iδ	+	Tn7
pBW1	Iδ	+	Tn7
pTH1	M	+	Tn7
pHH502	P	–	—
pHH500		–	—
R27	W	–	—
R388	W	–	—
R389a	W	–	—
R751	P	–	Tn*402*

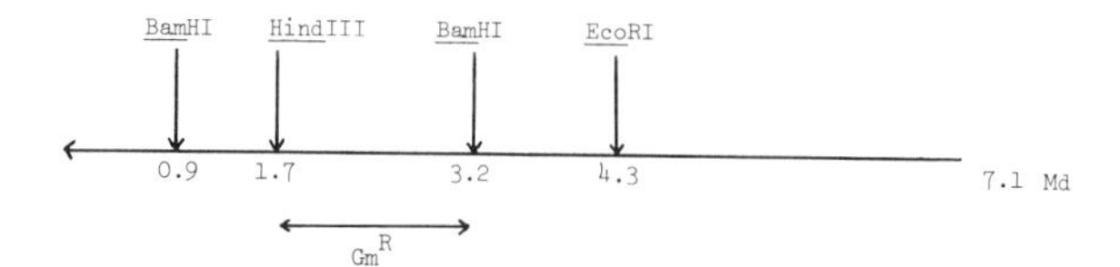

Figure 3. Restriction enzyme map of Tn*732*.

Table 3. IncW Plasmids in Bacteria from Patients in Hammersmith Hospital

Year	Resistance pattern
1974	Tp Su
1977–1980	Tp Su
	Tp Su Km
	Tp Su Km Gm Ap Hg
	Tp Su Gm Ap Hg
	Gm Su Ap Hg

isolated in Canada (Nugent et al. 1979). Tn*732* has a molecular mass of 7.1 MD and contains one cut site each for *Eco*RI and *Hin*dIII and two sites for *Bam*HI (Fig. 3). From the cloning of Tn*732* fragments onto pBR322, the gene for the gentamicin adenylylating enzyme has been located within a 1.5-MD portion of the transposon (Fig. 3).

We have examined the insertion of Tn*732* into plasmids of different incompatibility groups and so far have only seen transposition to plasmids of groups FII and P1. The frequency of transposition to R1 was 1 in 10^6, and to RP4 it was 1 in 10^5. These figures are low compared with those of other transposons, although within the reported range (Kleckner 1977). Restriction enzyme maps of 12 insertions of Tn*732* into the IncP plasmid RP4 show only two insertion sites, and these are only 0.1 MD apart. Restriction maps of eight insertions of Tn*732* into R1 show seven of these to be at the same site. The low frequency of transposition and the specificity of insertion of Tn*732* may explain why it is not as widespread as Tn*7*.

In our initial characterization of Tn*732*, we showed that it inserts into the IncFII plasmid R1 in such a way that an existing *Eco*RI site of R1 is duplicated (Nugent et al. 1979). Transposition of DNA elements leads to the duplication of an existing short sequence at the site of insertion in the recipient molecule (Grindley 1978). Therefore, our results with Tn*732* suggest that in one case the duplicated sequence happened to contain an *Eco*RI site.

Transposons encoding the gentamicin acetylating enzymes. The first trimethoprim R plasmids, such as R388, conferred resistance to trimethoprim and sulfonamides only. The spread of such plasmids among bacteria in London hospitals was observed by Jobanputra and Datta (1974) and Grey et al. (1979), and they were still found in our recent survey at Hammersmith Hospital, London (Datta et al. 1980). In the latter survey, however, there was more variation in the resistance patterns of IncW plasmids (Table 3), and many of them, e.g., pHH1307, determined the pattern TpSuGmApHg. Their gentamicin resistance was associated with an acetylating enzyme. We postulated, therefore, that R388-like plasmids had acquired new genes by transposition. Ampicillin and mercuric chloride resistance were known to be, at least sometimes, transposable (Campbell et al. 1977; Stanisich et al. 1977), but at that time we had not heard of transposable AAC(3) genes (Rubens et al. 1979a,b). Because gentamicin is such an important antibiotic in hospital medicine, we were particularly interested in the possibility of a gentamicin R transposon, but in our first experiments with pHH1307 no such transposon was detected. Recently, however, we have found transposition of 5.8 MD of DNA (including the gentamicin acetylating gene) from pHH1307 to R389a in *recA* and also in Rec^+ hosts. We have called the transposed sequence Tn*733*. Upon digestion with *Eco*RI, it has a 2.4-MD internal fragment that we can use for the identification of this transposon by hybridization with other plasmids.

Rubens et al. (1979a,b) have identified three transposons (Tn*1696*, Tn*1699*, and Tn*1700*) that confer resistance to multiple drugs, including gentamicin and/or tobramycin, and determine gentamicin acetylation. The relationship of Tn*733* to these transposons has not been investigated.

Incidence of trimethoprim or gentamicin resistance in bacteria of medical importance. Trimethoprim resistance in bacteria that cause human infections outside the hospital remains relatively infrequent, as compared with resistance to antibacterial drugs that have been used in medicine for longer periods. Table 4 shows the figures for ampicillin resistance and trimethoprim resistance in *E. coli* that caused urinary tract infections in three groups of patients in England and Wales. The incidence in hospitals of infections caused by bacteria resistant to trimethoprim and/or gentamicin varies very much from one hospital to another and in the same hospitals at different times. The incidence depends, at least in part, on the local usage of these drugs. Table 5 shows figures for trimethoprim and gentamicin resistance for Hammersmith Hospital (Datta et al. 1980).

DISCUSSION

In the past decade we have seen the very successful worldwide spread of β-lactamase transposons (e.g., Tn*1*) in plasmids of many classes and bacteria of many gen-

Table 4. *E. coli* Isolated from Urinary Tract Infections in the Community

	Percentage resistant to	
Year	ampicillin	trimethoprim
1960	0	0
1970	5	1
1980	25	1

Data from Harkness et al. (1975), Grüneburg (1976 and pers. comm.), and our own unpublished results summarize findings in several groups of patients in England and Wales.

Table 5. Enterobacteria Causing Infections in Hammersmith Hospital

	Percentage resistant to	
Year	trimethoprim	gentamicin
1970	5.6	0
1975	11	2.4
1979	15	5.9

era, including *Haemophilus influenzae* and *Neisseria gonorrhoeae* (Table 4) (Laufs et al. 1979). This appears to be an evolutionary response to the selective pressure of the intensive use of broad-spectrum, semisynthetic penicillins, especially ampicillin. Trimethoprim (usually sold in combination with a sulfonamide) resembles ampicillin in having a broad spectrum of antibacterial activity and negligible toxicity; it can also be taken in convenient oral form as tablets or suspension. A spread of Tp^r genes in the community and in hospitals, similar to that of β-lactamase genes, thus appears possible or even probable. To date, however, the frequency of trimethoprim resistance in bacteria isolated from human sources outside hospitals is low (Table 4). In hospitals, the frequency is higher (Table 5). Towner et al. (1980) report that in Nottingham, England, the frequency of trimethoprim resistance in bacteria from clinical specimens did not rise between 1978 and 1979, but among Tp^r strains the proportion carrying transmissible trimethoprim R plasmids had increased significantly. The range of incompatibility groups known to include trimethoprim R plasmids is shown in Table 6.

The range of groups that include gentamicin R plasmids is smaller (Table 6). The gentamicin adenylylating enzyme, ANT(2″), is often found in plasmids of IncC and IncFII in hospital bacteria. Perhaps the early acquisition of the gene by plasmids of these two groups may have resulted from transposition of the Tn*732* sequence from one to the other, or to both from an unknown donor replicon. The report by Chabbert et al. (1979) of IncM plasmids (from bacteria isolated in New York) that determine ANT(2″) may represent another example of this transposition, since AAC(3)1 rather than ANT(2″) is the Gm^r determinant typical of IncM.

The frequency with which a transposable genetic element is found in naturally occurring plasmids is presumably influenced by the readiness with which it transposes between replicons and also by the degree of selective pressure exerted in its favor in the environment. In the case of transposons encoding β-lactamase genes, both factors have been operative: Transposition is very efficient, and selection is intense. The transposons we have discussed here vary greatly in efficiency of transposition. The trimethoprim-resistance transposon Tn*7* transposes at high frequency; for example, when IncM plasmid pTH1, which includes Tn*7*, was transferred to *E. coli* K12 and subsequently eliminated, the transposon was retained in the K12 chromosome in 37% of the clones tested (Datta et al. 1979b). Another trimethoprim-resistance transposon, Tn*402*, recognized in IncP plasmid R751 by Shapiro and Sporn (1977), transposes at frequencies too low to be detected in our studies. But Tn*402* determines plasmid DHFR II, an enzyme that is carried by naturally occurring plasmids of at least four different incompatibility groups (Fling and Elwell 1980), which is indirect evidence of the spread of Tn*402* in nature.

Table 6. Range of Plasmid Incompatibility Groups Known to Include Examples of Resistance to Trimethoprim or Gentamicin

Trimethoprim	Gentamicin
B, C, FII, FIV, Iα, Iδ, K, M, N, P, W, X	C, FII, M, N, P, W

The frequency of transposition and the intensity of selection may not be unrelated. Tomich et al. (1980) have made the interesting observation that an erythromycin R transposon in a plasmid of *Streptococcus faecalis* is induced to transpose at increased frequency on exposure to low concentrations of erythromycin. Analogous mechanisms for control of the transposition property in plasmids of gram-negative bacilli remain to be discovered.

The evolution of multiple resistance in medically important bacteria has evidently resulted from the acquisition by plasmids of many kinds of DNA inserts encoding resistance genes. Our studies are aimed toward finding out the extent to which particular transposable elements are disseminated among different plasmids and different bacteria. Thus, we know that Tn*1* and its close relatives, such as Tn*3*, have spread on a worldwide scale. Other β-lactamase-determining transposons (Hedges et al. 1977; Nugent and Hedges 1979) have been followed less closely. At least two different trimethoprim R transposons and two or more gentamicin R transposons have been discussed here. We are preparing additional radioactively labeled DNA probes for the identification of their (1) resistance genes and (2) transposon sequences, other than the resistance genes, with which we plan to study the occurrence and spread of these transposons in bacteria isolated from many sources.

ACKNOWLEDGMENT

This work was supported by the Medical Research Council of England.

REFERENCES

ACAR, J. F., F. W. GOLDSTEIN, G. R. BERBAUD, and Y. A. CHABBERT. 1977. Plasmides de résistance au triméthoprime: Transférabilité et groupes d'incompatibilité. *Ann. Microbiol.* **128A:** 41.

AMYES, S. G. B. and J. T. SMITH. 1974. R-factor trimethoprim resistance mechanism: An insusceptible target site. *Biochem. Biophys. Res. Comm.* **58:** 412.

BACHMANN, B. J. 1972. Pedigrees of some mutant strains of *Escherichia coli* K12. *Bacteriol. Rev.* **36:** 525.

BARTH, P. T. 1979. Plasmid RP4, with *Escherichia coli* DNA inserted *in vitro,* mediates chromosomal transfer. *Plasmid* **2:** 130.

BARTH, P. T. and N. DATTA. 1977. Two naturally occurring

transposons indistinguishable from Tn*7*. *J. Gen. Microbiol.* **102:** 129.

Barth, P. T. and N. J. Grinter. 1974. Comparison of the deoxyribonucleic acid molecular weights and homologies of plasmids conferring linked resistance to streptomycin and sulfonamides. *J. Bacteriol.* **120:** 618.

Barth, P. T., N. Datta, R. W. Hedges, and N. J. Grinter. 1976. Transposition of a deoxyribonucleic acid sequence encoding trimethoprim and streptomycin resistances from R483 to other replicons. *J. Bacteriol.* **125:** 800.

Bolivar, F., R. L. Rodriguez, P. J. Greene, M. C. Betlach, H. L. Heyneker, H. W. Boyer, J. H. Crosa, and S. Fialkow. 1977. The circular restriction map of pBR322. In *DNA insertion elements, plasmids, and episomes* (ed. A. I. Bukhari et al.), p. 686. Cold Spring Harbor Laboratory, Cold Spring Harbor, New York.

Campbell, A., D. Berg, E. Lederberg, P. Starlinger, D. Botstein, R. P. Novick, and W. Szybalski. 1977. Nomenclature of transposable elements in prokaryotes. In *DNA insertion elements, plasmids, and episomes* (ed. A. I. Bukhari et al.), p. 15. Cold Spring Harbor Laboratory, Cold Spring Harbor, New York.

Casewell, M. W., M. T. Dalton, M. Webster, and I. Phillips. 1977. Gentamicin-resistant *Klebsiella aerogenes* in a urological ward. *Lancet* **II:** 444.

Chabbert, Y.-A., A. Roussel, J. L. Witchitz, M. J. Sanson-Le Pors, and P. M. Courvalin. 1979. Restriction patterns of plasmids belonging to incompatibility groups I1, C, M and N: Application to plasmid taxonomy and epidemiology. In *Plasmids of medical, environmental and commercial importance* (ed. K. T. Timmis and A. Pühler). Elsevier/North Holland, Amsterdam.

Datta, N. and P. T. Barth. 1976. Compatibility properties of R483, a member of the I plasmid complex. *J. Bacteriol.* **125:** 796.

Datta, N. and R. W. Hedges. 1972. Trimethoprim resistance conferred by W plasmids in Enterobacteriaceae. *J. Gen. Microbiol.* **72:** 349.

Datta, N., V. Hughes, and M. Nugent. 1979a. Gentamicin resistance plasmids. In *Plasmids of medical, environmental and commercial importance* (ed. K. T. Timmis and A. Pühler). Elsevier/North Holland, Amsterdam.

Datta, N., V. M. Hughes, M. E. Nugent, and H. Richards. 1979b. Plasmids and transposons and their stability and mutability in bacteria isolated during an outbreak of hospital infection. *Plasmid* **2:** 182.

Datta, N., M. Nugent, S. G. B. Amyes, and P. McNeilly. 1979c. Multiple mechanisms of trimethoprim resistance in strains of *Escherichia coli* from a patient treated with long-term co-trimoxazole. *J. Antimicrob. Chemother.* **5:** 399.

Datta, N., S. Dacey, V. Hughes, S. Knight, H. Richards, G. Williams, M. Casewell, and K. P. Shannon. 1980. Distribution of genes for trimethoprim and gentamicin resistance in bacteria and their plasmids in a general hospital. *J. Gen. Microbiol.* **118:** 495.

Eckhardt, T. 1978. A rapid method for identification of plasmid DNA in bacteria. *Plasmid* **1:** 584.

Fleming, M. P. 1973. Trimethoprim-resistance and its transferability in *E. coli* isolated from calves treated with trimethoprim-sulphadiazine: A two year study. *J. Hyg.* **71:** 669.

Fleming, M. P., N. Datta, and R. N. Grüneberg. 1972. Trimethoprim resistance determined by R factors. *Brit. Med. J.* **1:** 726.

Fling, M. E. and L. P. Elwell. 1980. Protein expression in *Escherichia coli* minicells containing recombinant plasmids specifying trimethoprim-resistant dihydrofolate reductase. *J. Bacteriol.* **141:** 779.

Grey, D., J. M. T. Hamilton-Miller, and W. Brumfitt. 1979. Incidence and mechanisms of resistance to trimethoprim in clinically isolated gram-negative bacilli. *Chemotherapy* **25:** 147.

Grindley, N. D. F. 1978. IS*1* insertion generates duplication of a nine base pair sequence at its target site. *Cell* **13:** 419.

Grüneberg, R. N. 1976. Susceptibility of urinary pathogens to various antimicrobial substances: A four year study. *J. Clin. Pathol.* **29:** 292.

Harkness, J. L., F. M. Anderson, and N. Datta. 1975. R factors in urinary tract infection. *Kidney Int.* **8:** S130.

Hedges, R. W., N. Datta, and M. P. Fleming. 1972. R factors conferring resistance to trimethoprim but not sulfonamides. *J. Gen. Microbiol.* **73:** 573.

Hedges, R. W., M. Matthew, D. I. Smith, J. M. Cresswell, and A. E. Jacob. 1977. Properties of a transposon conferring resistance to penicillins and streptomycin. *Gene* **1:** 241.

Heinaru, A., C. J. Duggleby, and P. Broda. 1978. Molecular relationships of degradative plasmids determined by in situ hybridisation of their endonuclease-generated fragments. *Mol. Gen. Genet.* **160:** 347.

Jacob, A. E., J. A. Shapiro, L. Yamamoto, D. I. Smith, S. N. Cohen, and D. Berg. 1977. Plasmids studied in *Escherichia coli* and other enteric bacteria. In *DNA insertion elements, plasmids, and episomes* (ed. A. I. Bukhari et al.), p. 607. Cold Spring Harbor Laboratory, Cold Spring Harbor, New York.

Jobanputra, R. S. and N. Datta. 1974. Trimethoprim resistance factors in Enterobacteria from clinical specimens. *J. Med. Microbiol.* **7:** 169.

Kleckner, N. 1977. Translocatable elements in procaryotes: Review. *Cell* **11:** 11.

Laufs, R., P.-M. Kaulfers, G. Jahn, and U. Teschner. 1979. Molecular characterisation of a small *Haemophilus influenzae* plasmid specifying β-lactamase and its relationship to R factors from *Neisseria gonorrhoeae. J. Gen. Microbiol.* **111:** 223.

Nugent, M. E. and R. W. Hedges. 1979. The nature of the genetic determinant for the SHV-1 β-lactamase. *Mol. Gen. Genet.* **175:** 239.

Nugent, M. E., D. H. Bone, and N. Datta. 1979. A transposon, Tn*732,* encoding gentamicin/tobramycin resistance. *Nature* **282:** 422.

Pattishall, K. H., J. Acar, J. J. Burchall, F. W. Goldstein, and R. J. Harvey. 1977. Two distinct types of trimethoprim-resistant dihydrofolate reductase specified by R plasmids of different compatibility groups. *J. Biol. Chem.* **252:** 2319.

Rennie, R. P. and I. B. R. Duncan. 1977. Emergence of gentamicin-resistant *Klebsiella* in a general hospital. *Antimicrob. Agents Chemother.* **11:** 179.

Richards, H. and M. Nugent. 1979. The incidence and spread of transposon 7. In *Plasmids of medical, environmental and commercial importance* (ed. K. N. Timmis and A. Pühler). Elsevier/North Holland, Amsterdam.

Richards, H., N. Datta, C. Wray, and W. J. Sojka. 1978. Trimethoprim resistance plasmids and transposons in *Salmonella. Lancet* **II:** 1194.

Rigby, P. W. J., M. Dieckmann, C. P. Rhodes, and P. Berg. 1977. Labelling deoxyribonucleic acid to high specific activity *in vitro* by nick translation with DNA polymerase I. *J. Mol. Biol.* **113:** 237.

Rubens, C. E., W. F. McNeill, and W. E. Farrar. 1979a. Transposable plasmid deoxyribonucleic acid sequence in *Pseudomonas aeruginosa* which mediates resistance to gentamicin and four other antimicrobial agents. *J. Bacteriol.* **139:** 877.

———. 1979b. Evolution of multiple-antibiotic-resistance plasmids by transposable plasmid deoxyribonucleic acid sequences. *J. Bacteriol.* **140:** 713.

Shapiro, J. A. and P. Sporn. 1977. Tn*402:* A new transposable element determining trimethoprim resistance that inserts in bacteriophage λ. *J. Bacteriol.* **129:** 1632.

SKÖLD, O. and A. WIDH. 1974. A new dihydrofolate reductase with low trimethoprim sensitivity induced by an R factor mediating high resistance to trimethoprim. *J. Biol. Chem.* **249:** 4324.

STANISICH, V., P. M. BENNETT, and M. H. RICHMOND. 1977. Characterization of a translocation unit encoding resistance to mercuric ions that occurs on a nonconjugative plasmid in *Pseudomonas aeruginosa. J. Bacteriol.* **129:** 1227.

TOMICH, P. K., F. Y. AN, and D. B. CLEWELL. 1980. Properties of erythromycin-inducible transposon Tn*917* in *Staphylococcus faecalis. J. Bacteriol.* **141:** 1366.

TOWNER, K. J. 1979. Classification of transferable plasmids conferring resistance to trimethoprim isolated in Great Britain. *FEMS Microbiol. Lett.* **5:** 319.

TOWNER, K. J., N. J. PEARSON, P. A. PINN, and F. O'GRADY. 1980. Increasing importance of plasmid-mediated trimethoprim resistance in Enterobacteriaceae: Two six-month clinical surveys. *Br. Med. J.* **280:** 517.

WITCHITZ, J. L. and Y. A. CHABBERT. 1972. Résistance transférable à la gentamicine. II. Transmission et liaisons du caractère de résistance. *Ann. Inst. Pasteur* **122:** 367.

WITCHITZ, J. L. and G. R. GERBAUD. 1972. Classification de plasmides conférant la résistance à la gentamicine. *Ann. Inst. Pasteur* **123:** 333.

A Study of the Dissemination of Tn*1681*: A Bacterial Transposon Encoding a Heat-stable Toxin among Enterotoxigenic *Escherichia coli* Isolates

M. So,*[†] R. Atchison,[†] S. Falkow,[‡] S. Moseley,[‡] and B. J. McCarthy[†]
[†] *Department of Biochemistry and Biophysics, University of California, San Francisco, California 94143;* [‡] *Department of Microbiology, University of Washington, Seattle, Washington 98195*

The gene encoding the STI toxin is located within a transposon (Tn*1681*) flanked by inverted repeats of IS*1*. Using a 300-bp *Hin*dIII fragment subclone of the STI gene that was free of IS*1* sequences, we carried out molecular studies to determine the occurrence of the STI gene in enterotoxigenic *Escherichia coli* isolated from a variety of animal sources: human, bovine, porcine and ovine. Such studies revealed that the same gene occurs in STI$^+$ enterotoxigenic *E. coli*, producing the STI toxin isolated from all of the aforementioned sources. However, not all STI producers cross-reacted with the probe, indicating the presence of another distinctly different STI gene. The enterotoxigenic strains isolated from animals in which both ST and LT were produced did not cross-react with the probe, whereas those strains isolated from humans did. Although some of the cross-reactive STI genes have the same sequence organization as Tn*1681*, others appear to have entirely different flanking sequences, suggesting a different evolutionary history and possibly a different mechanism of dissemination.

E. coli is a normal inhabitant of the gastrointestinal tracts of humans and some warm-blooded animals, but, under certain circumstances, it can turn against its host and become a pathogen. One group of intestinal diseases caused by *E. coli* includes syndromes known as traveler's diarrhea and infantile diarrhea. Infantile diarrhea, as the name implies, affects the newborn. It is a devastating and sometimes fatal disease, and it affects a variety of agricultural animals as well as humans (Smith and Linggood 1971, 1972; Gorbach and Khurana 1972; Sack 1975; Ryder et al. 1976). The livestock industry suffers a sizable loss each year due to this disease. Several other microorganisms can cause infantile diarrhea in humans; however, in nations with a low socioeconomic standing, the majority of cases of infantile diarrhea is largely *E. coli*-mediated (D. G. Evans, pers. comm.).

The symptoms of diarrhea are elicited by the action of toxins produced by these *E. coli* (Smith and Gyles 1970; Gyles 1971). One toxin (LT) is a heat-labile protein antigenically and functionally related to cholera toxin: It has similar binding (B) and active (A) subunits (Gyles 1971; Smith and Sack 1973; Dallas and Falkow 1979), and its primary mode of action is the stimulation of adenylate cyclase in the target-cell membrane (Evans et al. 1972). Its activity is assayed by induction of morphological changes of Chinese hamster ovary (CHO) (Guerrant et al. 1974) or Y-1 (Donta et al. 1973) cells exposed to LT. The other toxin (ST) is heat-stable, low in molecular weight, and only slightly immunogenic. It stimulates guanylate cyclase (Field et al. 1978; Hughes et al. 1978) and does not bind to the target-cell membrane. Recent studies in Brazil and Egypt indicate that the majority of *E. coli*-mediated diarrhea is largely due to ST producers (D. G. Evans, pers. comm.).

The genes encoding ST and LT lie on plasmids. They can occur separately or on the same plasmid. There are at least two genetically distinct ST toxins: STI is assayable in suckling mice (Dean et al. 1972), and STII is assayable in the ileal loops of weaned rabbits and piglets (Smith and Halls 1967). Earlier studies on ST + LT producers isolated from animals indicate the ST from these strains is strongly positive in weaned piglets and does not react well in suckling mice (Gyles 1979). The STI gene is located on a transposon (Tn*1681*) flanked by inverted repeats of IS*1*, and its nucleotide sequence has been determined (So et al. 1979). The corresponding amino acid sequence is cysteine-rich and its amino-terminal portion bears a striking resemblance to the signal sequence of the phage fd minor coat protein. That the STI toxin is first made as a pretoxin containing a signal sequence is supported by the observation that two suckling mouse active proteins of sizes varying approximately by the size of the hydrophobic amino-terminal portion of the amino acid sequence can be precipitated from an in vitro transcription-translation system by anti-STI antibodies (Lathe et al. 1980).

It appears, then, that there are several clinical syndromes affecting a variety of animals that are due to these enterotoxigenic *E. coli.* Molecular analyses of the ST gene have so far been done with a series of subclones from an enterotoxigenic *E. coli* responsible for an epidemic in calves. Several questions with regard to ST-induced infantile diarrhea remain to be answered: (1) How many different ST genes do enterotoxigenic *E. coli* harbor? (2) What is the genetic nature of these ST toxins, and how are they disseminated from *E. coli* to *E. coli?* (3) What is the relationship, if any, between the ST and LT genes? (4) Are traveler's diarrhea and infantile diarrhea related syndromes? In this paper we report the results of several epidemiological experiments based primarily on Southern hybridizations designed to answer some of these questions.

* Present address: Cold Spring Harbor Laboratory, Cold Spring Harbor, New York 11724.

MATERIALS AND METHODS

Clinical isolates of *E. coli* from human sources were obtained from D. G. Evans, R. B. Sack, M. Gurwith, and S. Gorbach. Isolates from animal sources were from C. Gyles and H. W. Smith. All isolates tested were positive for ST and/or LT, using either the suckling mouse assay or the ileal loop assay for ST and the CHO or Y-1 cell assay for LT. These strains and other pertinent information are listed in Table 1.

Plasmid DNA was isolated from the wild-type *E. coli* using a modification of a procedure of M. Yasu and H. Ohtsubo (pers. comm.). From a log-phase culture of cells grown in L broth, 1.5 ml were spun and the supernate discarded. The pellet was resuspended in 0.5 ml of TE buffer (20 mM Tris-HCl [pH 8], 5 mM EDTA). An equal volume of a phenol-ethanol mixture was added (75% ethanol, 2% phenol, 20 mM Tris-HCl [pH 8], 10 mM EDTA) and it was mixed briefly using a Vortex mixer. The cells were pelleted in a Brinkman microfuge for 2 minutes and the supernatant completely removed by suction using a drawn-out pasteur pipette. The pellet was resuspended in 0.1 ml of TE buffer and 10 μl of RNase (10 mg/ml, heat-treated). After the addition of 5 μl of 10% SDS to the sample, the tube was gently inverted two or three times to mix. The sample was incubated 5–10 minutes at 37 °C, at which time slight clearing was observed. The debris was spun down in a Brinkman microfuge for 30 minutes and removed with a pipette. The supernatant was extracted with phenol-chloroform and the DNA precipitated two or three times with 0.1 M sodium acetate, pH 5 (final concentra-

Table 1. List of Toxin-producing *E. coli* Clinical Isolates Used in This Study and Their Homology to the STIa Gene

Strain	Toxin produced	Source	Homology with STIa probe
(H)M443C1	ST	R. B. Sack[1]	+
(H)M109C2	ST		–
(H)M326C3	ST		+
(H)A237C4	ST		–
(H)A407C4	ST		–
(H)M626C5	ST		+
(H)M408C1	ST/LT		–
(H)M411C1	ST/LT		+
(H)M111C5	ST/LT		+
(H)447C5	ST/LT		–
(H)CP74	ST	D. Evans[2]	–
(H)H10407P	ST	C. Gyles[3]	+
(H)H10407	ST		+
(H)431	ST		+
(H)284-1	ST/LT	M. Gurwith[4]	–
(H)342	ST/LT		+
(H)A39-5	ST/LT		–
(H)I273	ST		+
(H)C674	ST		–
(H)H470	ST (unstable)		–
(H)TX1	ST	S. Gorbach[5]	+
(H)052005-74	ST		+
(H)214-4	ST		+
(H)193-4	ST/LT		–
(H)K326C5	ST/LT		–
(H)B2C	ST/LT		–
(H)TD514C2	ST	E. Gangarosa[6]	+
(H)TD471C2	ST		+
(P)P3	ST	C. Gyles and H. W. Smith[7]	+
(P)P122d	ST		+
(P)P334	ST		+
(P)P155	ST		+
(P)P16M	ST		+
(P)P16	ST		–
(P)P95	ST		–
(P)307	ST/LT		–
(O)S13	ST		+
(B)B42	ST		+
(B)B44	ST		+
(B)B117	ST		+

The letters in parentheses refer to the type of animal from which the *E. coli* was isolated: H, human; B, bovine; O, ovine; P, porcine. LT, heat-labile; ST, stable.

[1] Johns Hopkins Medical School, Baltimore, Md.; [2] Univ. of Texas, Houston, Tex.; [3] Univ. of Guelph Veterinary School, Guelph, Ont., Canada; [4] Veterans' Administration, Wadsworth Hospital Center, Los Angeles, Calif.; [5] Tufts Univ. Medical School, Medford, Mass.; [6] Center for Disease Control, Atlanta, Ga.; [7] Houghton Poultry Research Station, Houghton, Huntingdon, England.

tion). The vacuum-dried pellet was dissolved in 50 μl of Tris buffer (5 mM Tris-HCl [pH 8], 0.25 mM EDTA); part of the DNA was loaded directly onto an agarose gel and part was first restricted with restriction enzyme.

*Pst*I was obtained from Boehringer Mannheim and T4 polymerase from B. Alberts and J. Barry (Dept. of Biochemistry, Univ. of California, San Francisco, Calif.). Plasmid DNA was transferred from agarose gels to nitrocellulose filters by the method of Southern (1975). The 330-bp Tc1 probe was labeled with [α-^{32}P]dATP using T4 polymerase, according to the method of O'Farrell (pers. comm.). All hybridizations were done in 30% formamide, 0.1× SSC at 37 °C. The technique for visualizing inverted-repeat structures in plasmid DNA was taken from Ohtsubo (1976), with the modifications that the final reaction volume was 100 μl and the denaturation was done by boiling the samples for 5 minutes. Agarose gel electrophoresis was carried out in a horizontal apparatus of our own design and construction. When the same DNA samples were to be hybridized to two different probes, the samples were electrophoresed in parallel in the same gel and the agarose strips cut out after straining in ethidium bromide for transfer to nitrocellulose filters. All agarose gels were 1% in Tris-borate-EDTA buffer (100 mM Tris-base, 100 mM boric acid, 20 mM EDTA).

RESULTS AND DISCUSSION

We have previously described the isolation and characterization of the STI gene from B41, a plasmid of 65 MD in an *E. coli* strain causing diarrhea in calves (So et al. 1979; So and McCarthy 1980). A series of STI$^+$ subclones from B41 was generated, and the nucleotide sequence of the STI gene was determined. Among these was a 330-bp fragment of DNA (Tc1) cloned with *Hin*dIII linkers containing only the STI-coding sequence and no IS*1* sequences. This insert was initially used as a probe to determine the extent of nucleotide sequence homology between STI and STII. We have now used Tc1 as the probe for all subsequently described Southern blotting experiments. A series of enterotoxigenic *E. coli* strains were used in our studies. These were isolated from humans and animals from widely divergent geographical locations and at various times. All clinical isolates from human sources were tested and shown to be ST-positive in the suckling mouse assay and/or LT-positive in CHO cells. The animal isolates were tested and shown to be ST-positive in either the mouse assay or the ileal loop assay in weaned piglets. Some were also LT-positive.

DNA was prepared from clinical isolates using a procedure that yielded both plasmid and a small amount of chromosome. The presence of chromosomal DNA in the preparations was desired because we wished to detect any chromosomally located STI sequences if they existed. In one series of experiments, native plasmid DNA from these wild-type isolates was analyzed by agarose gel electrophoresis and transferred to nitrocellulose filters. Figure 1 shows the banding pattern of the DNA from a series of wild-type enterotoxigenic *E. coli* purified by this quick screening procedure. Most strains harbored two or more plasmids, the majority of which were of a high molecular weight. These are typical observations for wild-type *E. coli* strains, whether they are pathogens or not, and the functions encoded by most of these plasmids are unknown.

In a separate series of experiments, the DNAs were first restricted with either *Hin*dIII or *Pst*I before electrophoresis. The STI gene in Tc1 is flanked by inverted repeats of IS*1*. In the entire transposon there are only two *Pst*I sites, one in each of the IS sequences. The nucleotide sequence arrangement of Tn*1681* predicts that if the same transposon is present in the DNA of all these isolates, a 1.7-kb *Pst*I fragment will hybridize to the Tc1 probe. Figure 2 shows a representative Southern blot in which Tc1 was blotted to several STI-producing *E. coli* strains of animal origin. These strains were isolated from piglets (P334), lamb (S13), and calves (B42, B44, B117). The DNAs were first restricted with *Pst*I before electrophoresis. Most of the strains contained a 1.7-kb fragment that hybridized to the Tc1 probe. The one exception (S13) contained a 1.2-kb cross-hybridizing fragment. *Pst*I-restricted DNA from STI-producing strains of human origin have also been examined. Figure 3 shows that not all human STI$^+$ strains hybridized to the probe, indicating the presence of yet another genetically distinct STI toxin. In some of the strains that cross-react with the Tc1 probe, the *Pst*I fragments that hybridize are not 1.7 kb, but instead range in size from 550 bp to 5 kb. In one strain, M443C1, three cross-hybridizing *Pst*I fragments appear. The presence of three such fragments is not the result of partial digestion by *Pst*I, as all three bands are of the same intensity and when pBR322 DNA is included in the sample it is digested to completion.

Since the cross-hybridizing *Pst*I fragments in these wild-type isolates are so different in size from the one expected for Tn*1681*, it is expected that their flanking sequences are different and that these STI genes are not at all similar in sequence arrangement to Tn*1681*. (In Tn*1681*, the IS sequences flank a 550-bp segment in which 330 bp are coding sequences for the toxin.) A 408-bp *Pst*I-*Alu*I fragment located entirely within IS*1* was used as a probe in hybridizations to these isolates. Lanes marked a in Figures 4 and 5 show whole-plasmid DNA hybridized to the Tc1 probe and lanes marked b show the same DNA hybridized to IS*1*. DNAs from *E. coli* isolated from animals are shown in Figure 4 and DNAs from *E. coli* isolated from human sources are shown in Figure 5. IS*1* and Tc1 hybridized to the same plasmid in strains that showed the cross-reacting 1.7-kb *Pst*I fragment. However, the IS*1* probe did not hybridize to the same plasmid as Tc1 when the cross-reactive fragment differed in size from 1.7 kb. The exception is strain S13, in which Tc1 hybridized to a 1.2-kb *Pst*I fragment. That IS*1* hybridized to the same plasmid as Tc1 can most probably be explained by the orientation of the IS*1* sequences flanking ST. Directly repeating IS*1* sequences on either side of the ST gene will yield a 1.2-kb *Pst*I fragment that should hybridize to both Tc1 and IS*1*. In strain 203, which contains three different size cross-hybridizing *Pst*I fragments, Tc1 hybridized to only one plasmid band, indicating that these three STI genes

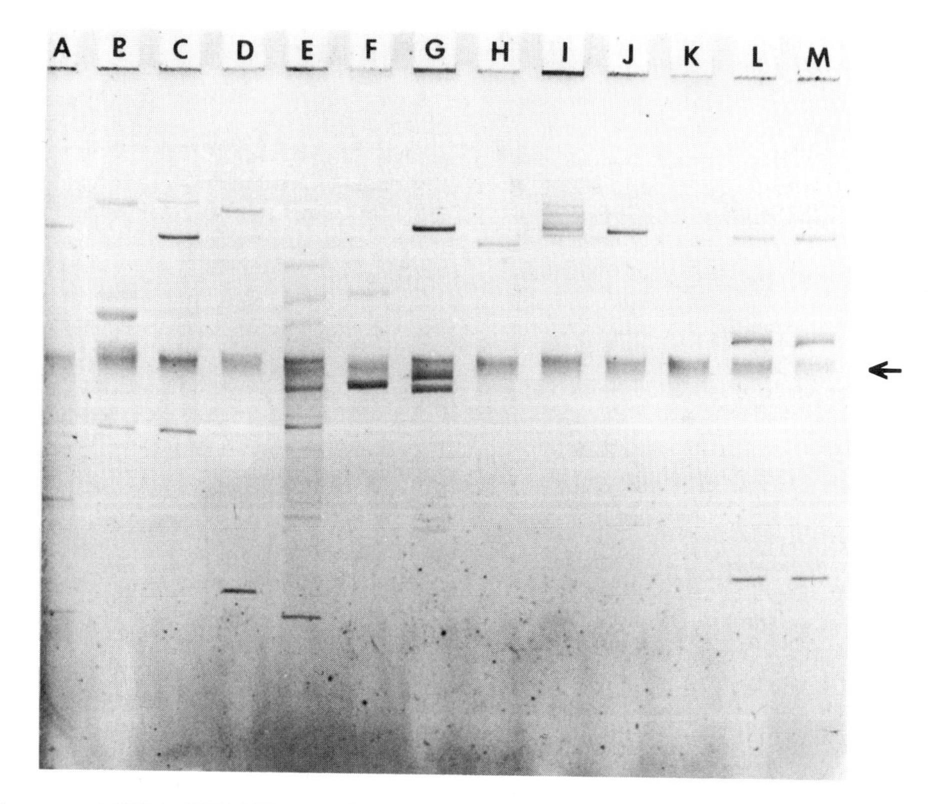

Figure 1. Agarose gel (1%) in TBE (100 mM Tris-base, 100 mM boric acid, 2 mM GCTA) of plasmid DNA isolated from clinical isolates by a quick screening procedure using 1.5 ml of cells (see Materials and Methods). (*A*) P334; (*B*) S13; (*C*) B42; (*D*) B44; (*E*) B117; (*F*) I273; (*G*) H10407P; (*H*) M443C1; (*L*) M411C1; (*M*)M111 C5. The smeary fluorescent region in the middle of the gel is chromosomal DNA (arrow).

are located on one plasmid. In all these strains, IS*1* hybridizes to the chromosomal DNA because the chromosome contains several copies of IS*1*. The Tc1 probe also hybridizes to the chromosomal fraction, since, under our lysis conditions, linearized plasmid DNA banded at that location on the agarose gels. We have never observed Tc1 to hybridize solely to the chromosomal DNA and not to a plasmid band.

Table 1 shows the results of all hybridizations of the Tc1 probe to whole-plasmid DNA from all wild-type enterotoxigenic isolates. Tc1 did not hybridize to strains from animal sources known to produce LT as well as ST, whereas it did cross-react with two of four ST/LT-producing strains of human origin. In these latter strains, it appears that the same plasmid encodes both STI and LT. The basis for this difference in plasmid specificity for strains is not clear. Because the genes encoding the plasmid components required for full enteropathogenicity by *E. coli* are plasmid-borne, most clinical isolates tend to be of certain serotypes. It is possible that some ST/LT plasmids have a compatibility with or specificity for certain host genes existing in some *E. coli* serotypes. For instance, some plasmid incompati-

P334
S13
B42
B44
B117
1.7kb—
1.2kb—

Figure 2. Southern blot of *Pst*I-restricted plasmid DNA from the quick-screen isolation, with ^{32}P-labeled 330-bp Tc1 fragment as probe. The *E. coli* strains used were ST producers isolated from diseased animals.

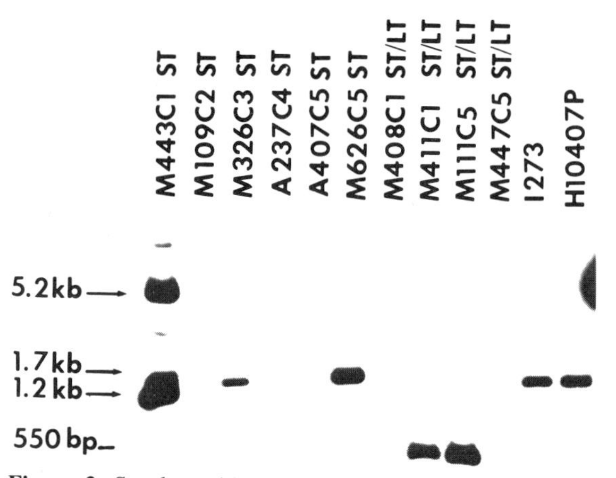

Figure 3. Southern blot of *Pst*I-restricted plasmid DNA from the quick-screen procedure, with ^{32}P-labeled 330-bp Tc1 fragment as probe. The *E. coli* strains used were ST producers isolated from humans suffering from diarrhea.

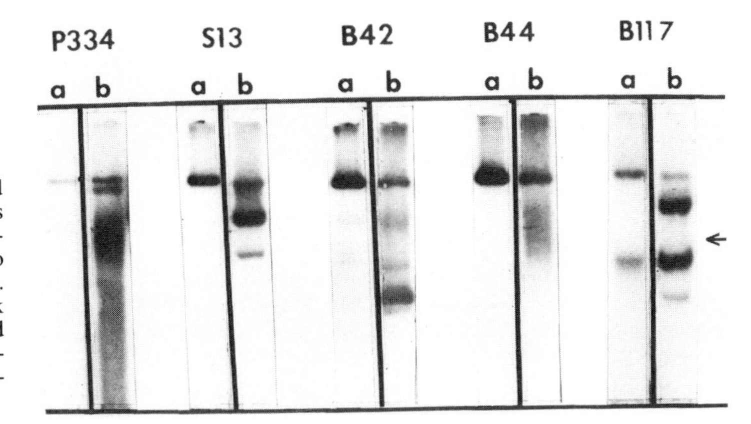

Figure 4. Southern blots of whole-plasmid DNA with ^{32}P-labeled Tc1 (*a*) and IS*1* probes (*b*). The IS*1* probe consists of a 408-bp *Pst*I-*Alu*I internal fragment (Ohtsubo and Ohtsubo 1978) isolated from pMS3 (So et al. 1979). Cold plasmid DNA was isolated by the quick screening method described in Materials and Methods. (←) Approximate location of chromosome. The strains shown are of animal origin.

bility groups are more efficient in conjugal transfer than others. In the natural environment, an *E. coli* that is a particularly good recipient might often be found harboring a conjugally proficient plasmid. Another explanation for these results is that enterotoxigenic *E. coli* from animal sources tend to be isolated and characterized by laboratories that are closely associated with veterinary medicine. Such laboratories have the facilities and the expertise to handle larger animals and would therefore not tend to prefer one type of ST assay over the other. Indeed, the ST/LT strains used in this study were first shown to be ST producers by their reactivity in the ileal loop assay using weaned rabbits or piglets. Still, it is intriguing that these ST/LT strains all produce STIb exclusively.

From these results and from the results of a broader epidemiological study we reported earlier, it is clear that there are at least two other heat-stable toxins that can be assayed in suckling mice. The genes encoding one of these ST toxins, which here we designate STIa, have been extensively studied and are part of Tn*1681*. The genetic nature of STIb is unknown. We have been studying a plasmid encoding STIb and have found that plasmid DNA from this strain does not produce snapback structures that can be visualized by gel electrophoresis when the plasmid DNA is subjected to denaturation and renaturation followed by treatment with exonuclease S1. When B41 is analyzed in this manner, a discreet 768-bp fragment of DNA (the snapback structure containing IS*1*) can be visualized in the agarose gel (data not shown). This result suggests that the nucleotide sequences surrounding the STIb gene are not arranged in inverted repeats like the majority of other bacterial transposons. We are now analyzing various subclones of this gene to determine the presence of direct repeats.

Many clinical isolates do contain an STIa gene homologous to Tc1. However, it is clear that the same toxin gene can have entirely different flanking sequences. The latter are not likely to be located within Tn*1681* since their surrounding sequences are not IS*1* and their cross-hybridizing *Pst*I fragments differ in size. Yet there seems to be some pattern to the flanking sequences of these STIa genes since they are located on *Pst*I fragments of certain sizes. It is possible that these sets of STIa-containing *Pst*I fragments are the products of recombination within the IS*1* sequences since IS*1* is commonly found in the *E. coli* chromosome and some plasmids. Whether these other STIa genes are located on their own transposons or are a part of a larger set of

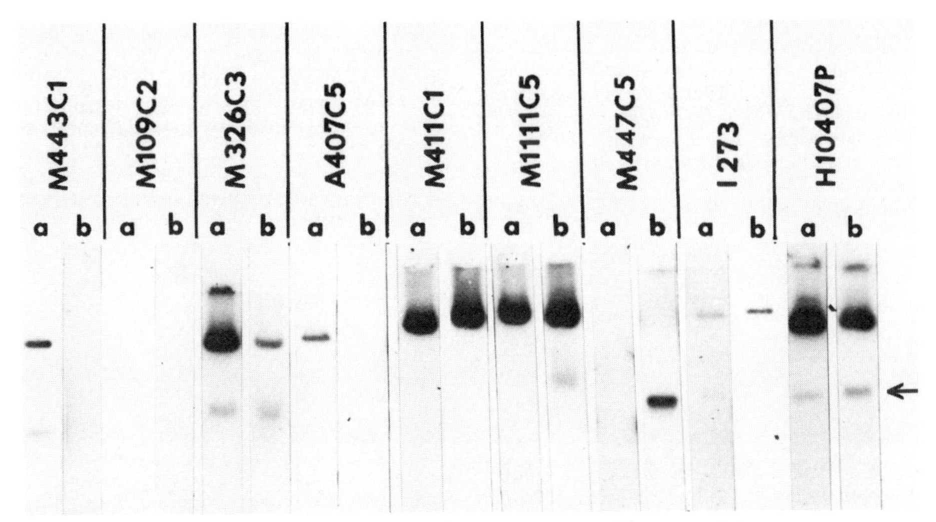

Figure 5. The same method as in Fig. 4, except that the strains shown are of human origin.

other transposons is not known. At present, we are studying their ability to transpose.

Pathogenicity in *E. coli* is a multifaceted phenomenon, and the interactions and interrelationships of the genes encoding pathogenicity and virulence factors are complicated and constantly evolving. This set of molecular epidemiological experiments has shown that the *E. coli* heat-stable toxins are a genetically diverse group of proteins. So far, we have been able to identify three different ST toxins on the basis of Southern hybridizations and animal models: STIa and STIb can be assayed in suckling animals, and STII can be assayed in weaned animals. The difference in susceptibility of animals to these toxins suggests the possibility of a change in receptor sites that is manifest during the development of the animal and indicates that infantile diarrhea and traveler's diarrhea are syndromes that may be differentiated, at least in part, on the basis of the toxins responsible for the symptoms. Moreover, the same toxin may be responsible for the disease in animals as well as in humans.

In all cases examined, the STIa gene is located on a plasmid. It is part of the Tn*1681* transposon in some, but not all, of the isolates. In the latter cases, the restriction fragments containing the STIa gene appear to fall into certain size classes. Whether these classes of sequences arose from the Tn*1681* transposon or from independent recombinational events is not known. The dissemination of these other STIa genes may depend on mechanisms other than transposition. Further molecular and epidemiological studies are needed to answer this question.

REFERENCES

DALLAS, E. W., D. M. GILL, and S. FALKOW. 1979. Cistrons encoding *Escherichia coli* heat labile toxin. *J. Bacteriol.* **139:** 850.

DEAN, A. G., Y. C. CHING, R. G. WILLIAMS, and L. B. HARDEN. 1972. Test for *Escherichia coli* enterotoxin using infant mice: Application in a study of diarrhea in children in Honolulu. *J. Infect. Dis.* **125:** 407.

DONTA, S.T., H. W. MOON, and S. C. WHIP. 1974. Detection of heat labile *Escherichia coli* enterotoxin with the use of adrenal cells in tissue culture. *Science* **183:** 334.

FIELD, M., L. H. GRAF, JR., W. J. LAIRD, and P. L. SMITH. 1978. Heat stable enterotoxin of *Escherichia coli: In vitro* effects on gualinate cyclase activity, cyclic GMP concentration, and ion transportation in small intestine. *Proc. Natl. Acad. Sci.* **75:** 2800.

GORBACH, S. L. and C. M. KHURANA. 1972. Toxigenic *Escherichia coli:* A cause of infantile diarrhea in Chicago. *N. Engl. J. Med.* **287:** 791.

GUERRANT, R. L., L. L. BRUNTON, T. C. SCHNAITMAN, L. I. REBHUN, and A. G. GILMAN. 1974. Cyclic adenosine monophosphate and alteration of Chinese hamster ovary cell morphology: A rapid sensitive in vitro assay for the enterotoxins of *Vibrio cholera* and *Escherichia coli. Infect. Immun.* **10:** 320.

GYLES, C. L. 1971. Heat labile and heat stable forms of the enterotoxin from *E. coli* strains enteropathogenic for pigs. *Ann. N.Y. Acad. Sci.* **176:** 314.

———. 1974. Immunological study of the heat labile enterotoxins of *Escherichia coli* and *Vibrio cholera. Infect. Immun.* **129:** 277.

———. 1979. Limitations of the infant mouse test for *Escherichia coli* heat stable enterotoxin (ST). *Can. J. Comp. Med.* **43:** 371.

HUGHES, J. M., F. MURAD, B. CHANG, and R. L. GUERRANT. 1978. Role of cyclic GMP in the action of heat-stable enterotoxin of *Escherichia coli. Nature* **271:** 755.

LATHE, R., P. HIRTH, M. DEWILDE, N. HARFORD, and J. P. LECOQ. 1980. Cell free synthesis of enterotoxin of *Escherichia coli* from a cloned gene. *Nature* **284:** 473.

OHTSUBO, H. and E. OHTSUBO. 1976. Isolation of inverted repeat sequences, including IS1, IS2, and IS3, in *Escherichia coli* plasmids. *Proc. Natl. Acad. Sci.* **73:** 2316.

RYDER, R. W., I. K. WACHSMUTH, A. BUXTON, D. G. EVANS, H. L. DUPONT, E. MASON, and F. F. BARRETT. 1976. Heat stable enterotoxigenic *E. coli* in a newborn nursery: Relation to infantile diarrhea. *N. Engl. J. Med.* **295:** 849.

SACK, R. B. 1975. Human diarrheal disease caused by enterotoxigenic *Escherichia coli. Annu. Rev. Microbiol.* **29:** 333.

SMITH, H. W. and C. L. GYLES. 1970. The relationship between two apparently different enterotoxins produced by enteropathogenic strains of *Escherichia coli* of porcine origin. *J. Med. Microbiol.* **3:** 419.

SMITH, H. W. and S. HALLS. 1967. Studies on *Escherichia coli* enterotoxin. *J. Pathol. Bacteriol.* **93:** 531.

SMITH, H. W. and M. LINGGOOD. 1971. Observations on the pathogenic properties of the K88, Hly and Ent plasmids of *E. coli* with particular reference to porcine diarrhea. *J. Med. Microbiol.* **4:** 467.

———. 1972. Further observations on *E. coli* enterotoxins with particular regard to those produced by atypical piglet strains and by calf and lamb strains: The transmissible nature of these enterotoxins and of a K antigen possessed by calf and lamb strains. *J. Med. Microbiol.* **5:** 243.

SO, M. and B. J. MCCARTHY. 1980. Nucleotide sequence of a bacterial transposon, Tn*1681,* encoding an *E. coli* heat stable toxin (ST) and its identification in enterotoxigenic *E. coli* strains. *Proc. Natl. Acad. Sci.* **77:** 4011.

SO, M., F. HEFFRON, and B. J. MCCARTHY. 1979. The *E. coli* gene encoding heat stable toxin (ST) is a bacterial transposon flanked by inverted repeats of IS1. *Nature* **277:** 453.

SOUTHERN, E. M. 1975. Detection of specific sequences among DNA fragments separated by agarose gel electrophoresis. *J. Mol. Biol.* **98:** 503.

Basis of Transposition and Gene Amplification by Tn*1721* and Related Tetracycline-resistance Transposons

R. Schmitt, J. Altenbuchner, K. Wiebauer, W. Arnold,* A. Pühler,* and F. Schöffl*
*Lehrstuhl für Genetik, Universität Regensburg, D-8400 Regensburg, Federal Republic of Germany; *Institut für Mikrobiologie, Universität Erlangen, D-8520 Erlangen, Federal Republic of Germany*

A novel property of transposable elements has been detected in the closely related tetracycline-resistance (Tc^r) transposons Tn*1721* and Tn*1771*, i.e., the ability to amplify a 5.3-kb portion comprising the drug-resistance genes (*tet* region). A model accounting for both *recA*-dependent gene amplification and *recA*-independent transposition to other replicons is shown in Figure 1 (Schmitt et al. 1979; Schöffl and Pühler 1979). The salient features include three identical repeats, two of which bracket the repetitious *tet* region in direct orientation; the third, inversely oriented, is located at one terminus of the minor transposon. Each repeat contains an *Eco*RI restriction site. These repeat sequences divide Tn*1721* into two domains, one responsible for translocation and the other responsible for drug resistance. This paper includes (1) results in support of this model and a discussion of (2) the possible origin and (3) physiological significance of transposons like Tn*1721*.

RESULTS AND DISCUSSION

The 5.3-kb *Eco*RI fragment designated as the *tet* region has been cloned and shown to encode genes for inducible tetracycline resistance (Mattes et al. 1979). The *tet* genes comprise a region of less than 2 kb located between restriction coordinates 5.4 and 7.5 kb (Fig. 1), as analyzed by insertion mutagenesis (P. M. Bennett, pers. comm.). In *rec*$^+$, but not in *recA* cells, this *tet* region is capable of forming multiple tandem repeats (Mattes et al. 1979). Likewise, spontaneous losses of the *tet* region are dependent on the host *recA* function. This suggests the presence of homologous sequences that flank the resistance genes in direct orientation and provide the basis for homologous recombination. The 5.4-kb minor transposon[1] has structural features of an insertion element; i.e., it is bounded by inverted-repeat sequences and does not encode known genes unrelated to transposition. The minor transposon has also been found as an insertion in plasmid pRSD2 encoding genes for raffinose utilization (Burkardt et al. 1978). It is capable of forming deletions that originate at either end of the element. Independent translocation of the minor transposon has recently been examined after in vitro insertion of a kanamycin-resistance (Km^r) marker into its single *Pst*I site (R. Mattes and R. Schmitt, in prep.).

Direct evidence for the proposed model has been obtained by sequence analyses of the DNA region surrounding the three *Eco*RI sites located in the postulated repeats. By pursuing different strategies in sequencing the homologous Tn*1721* (isolated in India; Schmitt et al. 1979) and Tn*1771* elements (isolated in Germany; Schöffl and Pühler 1979), identical results were obtained for corresponding regions. The sequences shown in Figure 2 reveal three 38-bp repeats that are identical except for a mismatch of three bases at the inverted ends. Translocation of these elements generates 5-bp direct repeats in the recipient replicon. This places Tn*1721* into a class with Tn*3* (Ohtsubo et al. 1979), Tn*501* (Brown et al. 1980), Tn*551* (Khan and Novick 1980), $\gamma\delta$ (Reed et al. 1979), Mu (Allet 1978), and the 200-bp insertion in pSC101 (Ravetch et al. 1979), all of which generate a 5-bp duplication of recipient sequences.

Our model (Fig. 1), initially proposed on the basis of indirect data, has been validated by these sequence analyses. Repeats II and III, bounding the repetitious *tet* region in direct orientation, are 38 bp long, as defined on the basis of their homology with inverted-repeat I. There is, however, more extensive sequence homology between DNA regions adjacent to direct repeats II and III (Fig. 2). It comprises the 87 nucleotides sequenced thus far and may extend much farther. This notion is corroborated by a homoduplex analysis. As diagramed in Figure 3, in vitro inversion of the minor transposon at the *Eco*RI sites has been accomplished (pJOE112). In this configuration, homologous sequences flanking the *tet* genes can be detected by intramolecular duplex formation. As seen in Figure 3 (inset), the double-stranded portion amounts to 1.8 kb (1.16 megadaltons [MD]). This homology adjacent to the direct repeats (pJOE105) provides an excellent basis for the intramolecular recombination of *tet* (Yagi and Clewell 1976; Meyer and Iida 1979). Furthermore, no amplification of *tet* genes has been observed in this inverted configuration (pJOE112). The residual 38-bp homology flanking *tet* after inversion of the minor transposon is apparently not sufficient for homologous recombination. On the other hand, transposition of Tn*1721* with the inverted minor transposon is normal. This finding has an interesting implication. If during the transposition process the inverted ends have to be brought together, the 38-bp homology is sufficient for the *tnp* function(s) involved,

[1] The 5.4-kb transposable DNA region has been termed the minor transposon to distinguish it from the entire 10.7-kb transposon, which includes both the 5.4-kb region and the amplifiable 5.3-kb *tet* region. In addition, the term minor transposon distinguishes this region from the smaller IS elements (Starlinger and Saedler 1976).

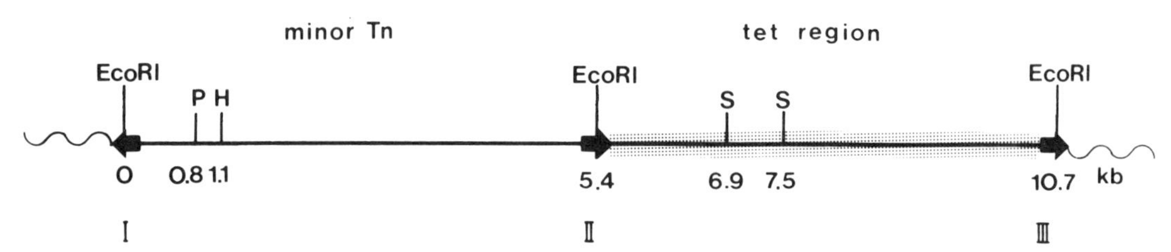

Figure 1. Schematic model of Tn*1721* and Tn*1771* with three identical repeats (arrows), each containing an *Eco*RI restriction site. Arrowheads indicate the orientation of the repeats; II and III bound the repetitious *tet* region (encoding tetracycline resistance) in direct orientation; repeat I is inversely oriented and flanks the minor transposon (encoding transposition functions) at its distal end. Restriction coordinates are marked in kb (H = *Hin*dIII; P = *Pst*I; S = *Sma*I).

whereas it cannot be used efficiently by the *recA*-dependent host recombination system.

Similarities in the restriction maps of regions surrounding the *tet* genes of Tn*1721* and of R plasmid RP4 prompted us to test for possible interrelationships. The Tc^r determinant of RP4 is inducible but shows neither gene amplification nor transposition. A heteroduplex analysis of the corresponding regions, using appropriate restriction fragments of Tn*1721* (inserted into pRSF2124; So et al. 1975) and RP4, is shown in Figure 4. The regions of duplex DNA correspond to the *tet* genes and to Tn*A*, respectively, which are present in both of these plasmids. The homology in the *tet* region amounts to 1.97 MD (3 kb), thus indicating the close relationship and possible identity of the resistance genes of Tn*1721* and RP4.

The evolutionary route leading to formation of Tn*1721* and related transposons is hypothesized below. One could imagine that the nucleus of the evolving Tc^r transposon is an insertion element like the minor transposon. Following two insertions of this element into RP4 (or a related plasmid), as diagramed in Figure 5, we arrive at two minor transposons bounding the *tet* region in direct orientation. A partial deletion of the right-hand

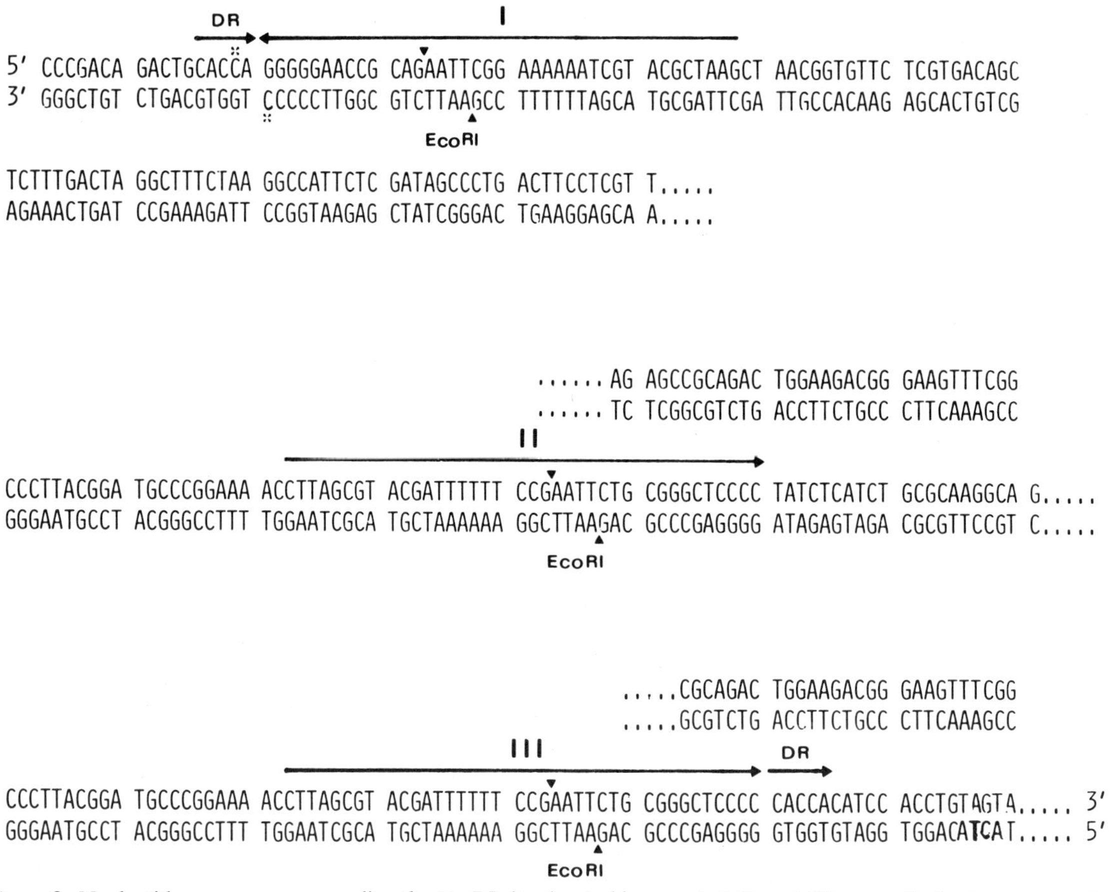

Figure 2. Nucleotide sequences surrounding the *Eco*RI sites located in repeats I, II, and III, respectively. Long arrows indicate the extent and orientation of the 38-bp repeats (with the *Eco*RI sites marked). Short arrows mark the 5-bp direct repeats (DR) flanking the transposon at the site of insertion.

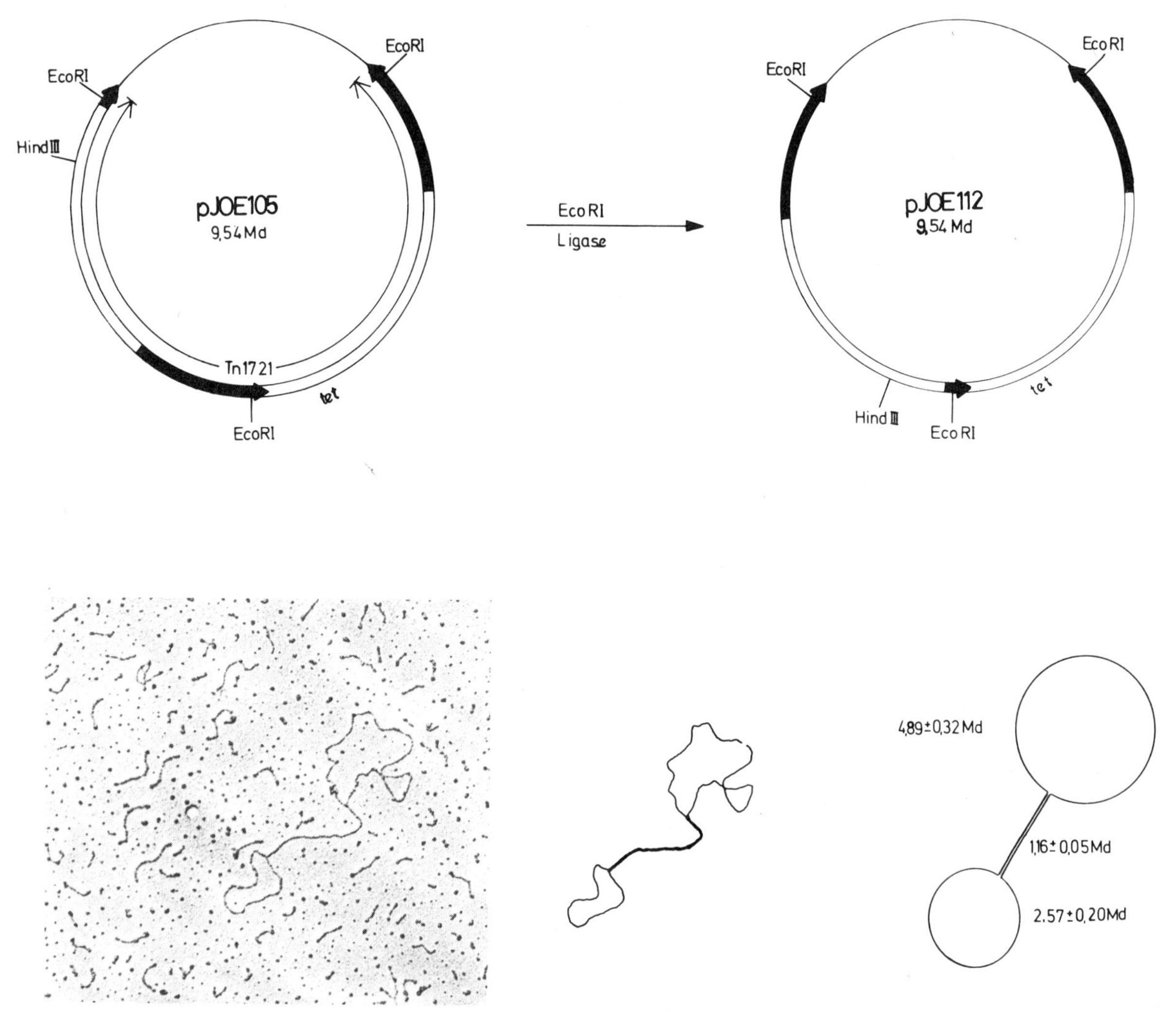

Figure 3. Diagram showing Tn*1721* inserted into the high-copy vector pJOE105 and in vitro inversion of the minor transposon at the *Eco*RI sites. The homoduplex (inset and tracing) shows hybridization of two 1.16-MD (1.8-kb) regions in the single-stranded circular molecule. According to contour-length measurements, this homology corresponds to the portions adjacent to the direct repeats of Tn*1721*.

insertion, as illustrated in Figure 5, would generate Tn*1721*. Both the insertion of transposable elements in direct orientation bracketing a DNA region and site-specific deletion extending from one terminus of an inverted-repeat sequence are events commonly observed with many other transposable DNA segments. Thus, the association of a resistance determinant with a minor transposon can lead to a new element capable of gene amplification and transposition, mechanisms that must have been invented by nature long ago. Although we are just at the beginning of an epidemiological study, present data indicate a widespread distribution of similar Tc^r transposons; i.e., they have now been found in India, Germany, and Belgium.

Finally, the physiological role of a Tc^r determinant that, under appropriate conditions, can be amplified and/or transposed deserves further emphasis. We have isolated plasmids containing from one to eight copies of the *tet* region and have separately transformed them into *recA E. coli* K12 cells. These strains, each carrying one plasmid type, were used to study the resistance levels of noninduced cells. As shown in Figure 6, the number of cells resistant to discrete concentrations of tetracycline increases with the number of *tet* genes present on the plasmid, thus demonstrating a gene-dosage effect of *tet* on resistance levels. It is important to note that only about 10^{-8} cells of a population bear amplified plasmids. Therefore, without taxing the bacterium, amplification ensures survival of the species at high doses of the drug, and transposition of the *tet* genes ensures their maintenance in case of plasmid loss.

ACKNOWLEDGMENTS

We are grateful to Dennis Kopecko for his critical reading of the manuscript. This investigation was supported by grants from the Deutsche Forschungsgemeinschaft.

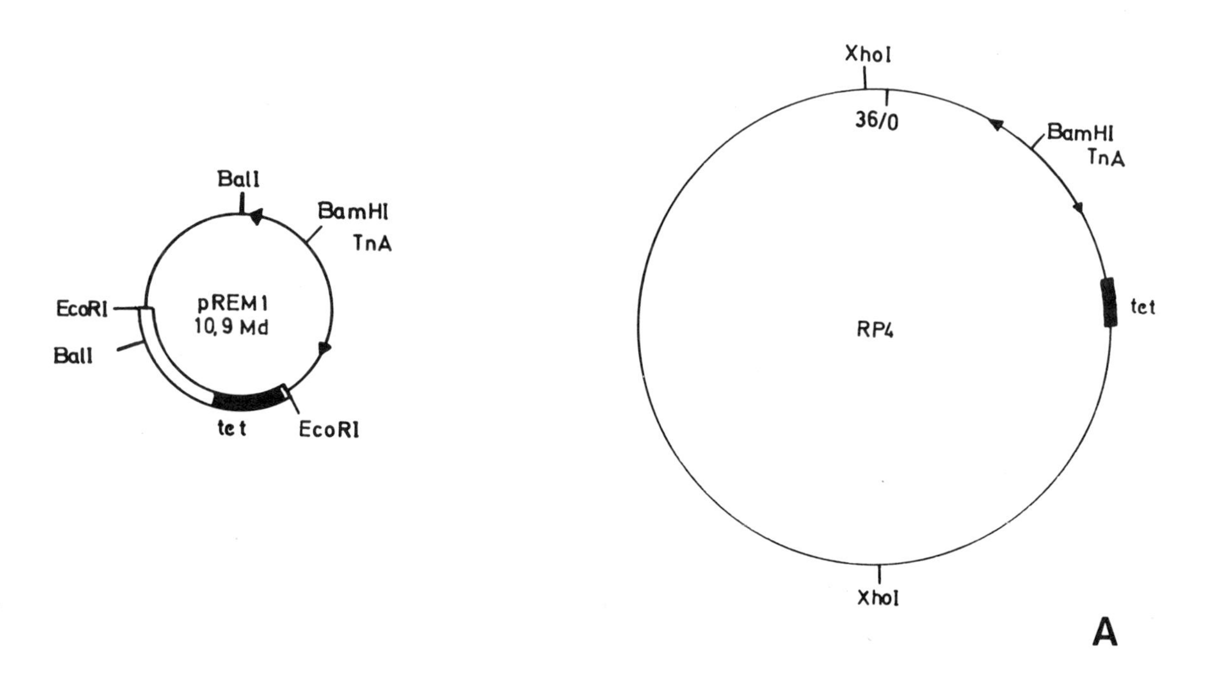

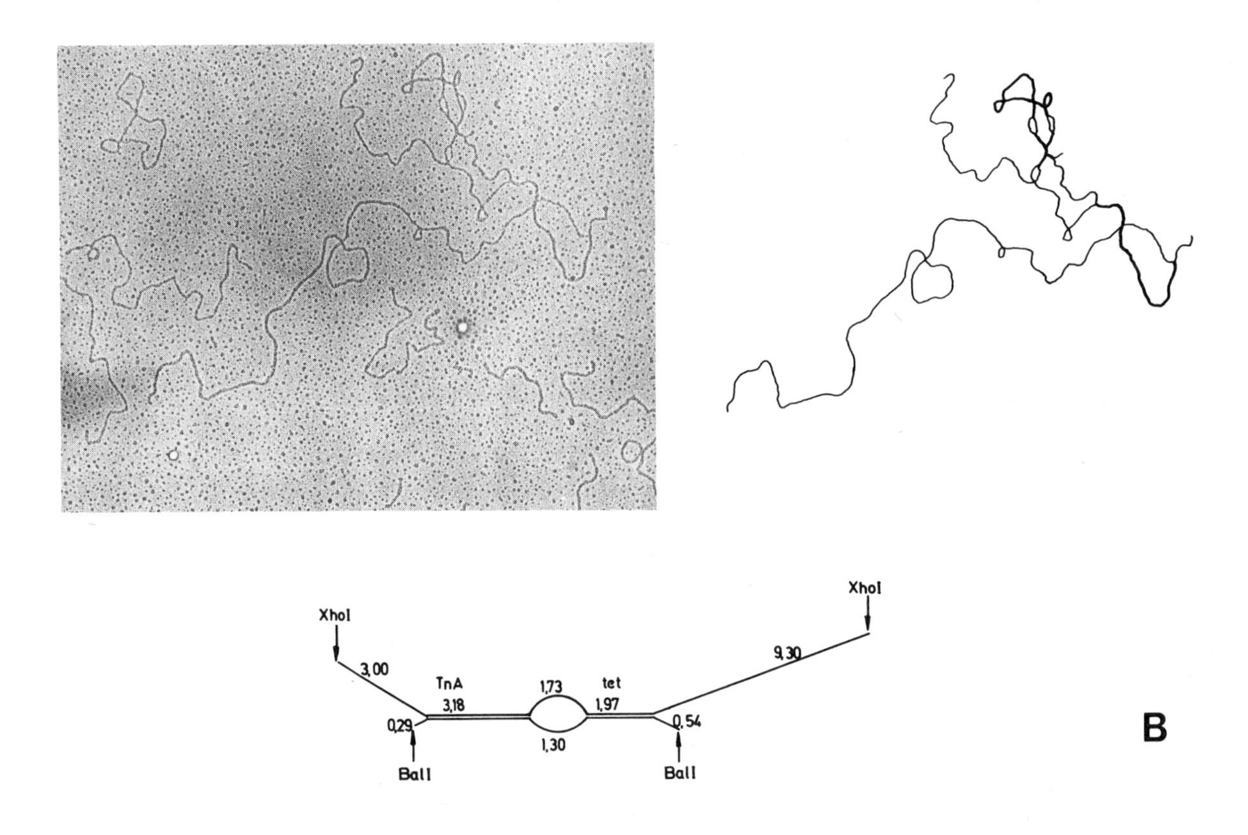

Figure 4. (*A*) Diagram of pREM1 (Mattes et al. 1979) and RP4. The locations of Tn*A*, Tc^r determinants (*tet*), and relevant restriction sites are drawn according to scale. (*B*) Heteroduplex formed between the 19.2-MD *Xho*I fragment of RP4 and the 7.3-MD *Bal*I fragment of pREM1, seen in *A*. Whereas Tn*A* forms an underwound loop, typical for elements flanked by inverted repeats, there is a linear duplex of 1.97 MD (3 kb) between regions encoding *tet*, thus indicating homology.

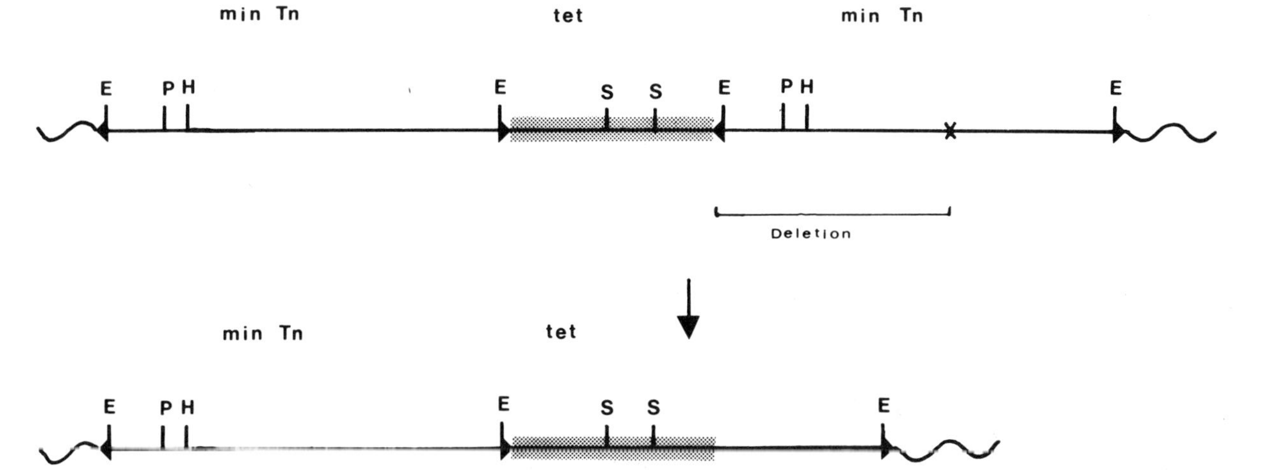

Figure 5. Diagram showing the possible evolution of Tn*1721*. (*Top*) Two directly oriented minor transposons (min Tn) flanking the *tet* region of RP4. (*Bottom*) Partial deletion of the right-handed element, as indicated, generates a Tn*1721*-like element.

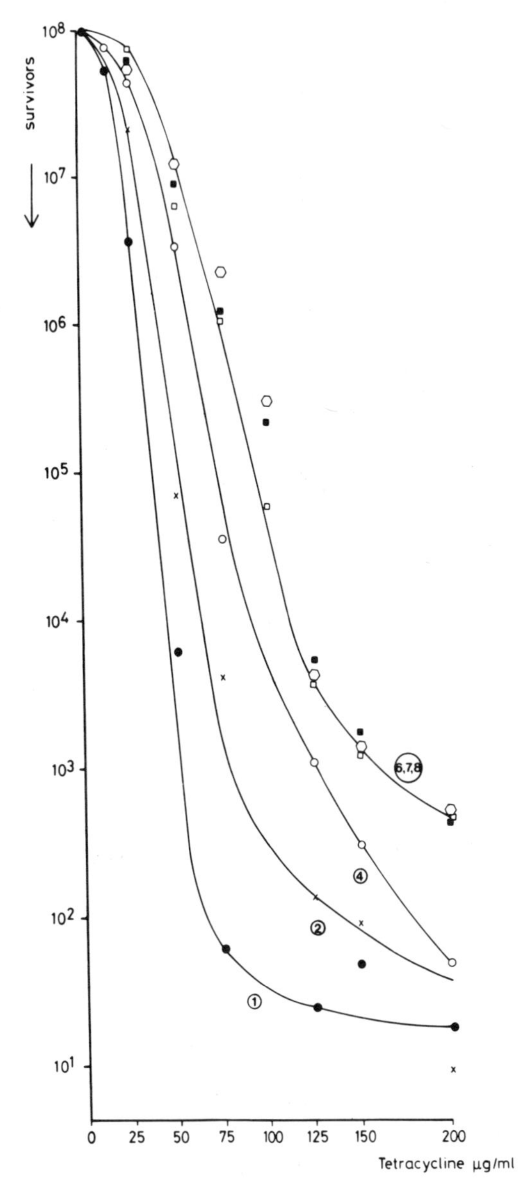

Figure 6. Resistance curves of *E. coli recA* strains harboring pRSD1 (R388::Tn*1721*) with either one, two, four, six, seven, or eight copies of the *tet* region, as marked by circled numbers. Cells growing exponentially without tetracycline were plated on solid media containing increasing concentrations of tetracycline as indicated on the abscissa.

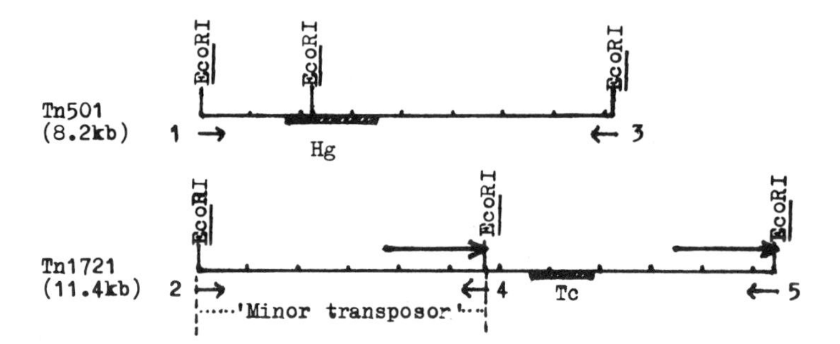


```
                EcoRI
1)  GGGGGAACCGCAGAATTCGGAAAAAATCGTACGCTAAGCTAACGGTGTTCTCGTGACAGCTCTTTGA--
    CCCCCTTGGCGTCTTAAGCCTTTTTTAGCATGCGATTCGATTGCCACAAGAGCACTGTCGAGAAACT--

2)  GGGGGAACCGCAGAATTCGGAAAAAATCGTACGCTAAGCTAACGGTGTTCTCGTGACAGC
    CCCCCTTGGCGTCTTAAGCCTTTTTTAGCATGCGATTCGATTGCCACAAGAGCACTGTCG

3)  GGGGGGCTCGCAGAATTCGGAAAAAATCGTACGCTAAGGTTTTCCGGGCATCCGTAGGGGCCGAAAG--
    CCCCCCGAGCGTCTTAAGCCTTTTTTAGCATGCGATTCCAAAAGGCCCGTAGGCATCCCCGGCTTTC--
        *  *
4)  GGGGAGCCCGCAGAATTCGGAAAAAATCGTACGCTAAGGTTTTCCGGGCATCCGT
    CCCCTCGGGCGTCTTAAGCCTTTTTTAGCATGCGATTCCAAAAGGCCCGTAGGCA
        *  *
5)  GGGGAGCCCGCAGAATTCGGAAAAAATCGTACGCTAAGGTTTTCCGGGCA
    CCCCTCGGGCGTCTTAAGCCTTTTTTAGCATGCGATTCCAAAAGGCCCGT
```

Figure 7. Maps and terminal sequences of Tn*501* and Tn*1721*. In the maps of the transposons, the hatched regions show the locations of the genes that code for the resistances, the heavy arrows indicate the direct repeats in Tn*1721*, and the small, numbered arrows indicate the origins and directions of the sequences shown. Asterisks indicate base pairs in Tn*1721* that are different from those in the corresponding Tn*501* sequence. The sequence data for Tn*501* are from Brown et al. (1980 and unpubl.) and those for Tn*1721* are from Schmitt et al. (this volume).

APPENDIX

Transposons Tn*501* and Tn*1721* Are Closely Related

C.-L. Choi,* J. Grinsted,*
J. Altenbuchner,†
R. Schmitt,† and M. H. Richmond

** Department of Bacteriology, University of Bristol, Bristol BS8 lTD, England; † Lehrstuhl für Genetik, Universität Regensburg, D-8400 Regensburg, Federal Republic of Germany.*

The terminal sequences of Tn*501,* which codes for mercuric ion resistance (Hg^r), and Tn*1721,* which codes for tetracycline resistance (Tc^r), are almost identical, the similarity also extending to sequences adjoining the inverted-repeat sequences (Fig. 7). It should be noted that Tn*1721* contains a second copy of one of the terminal sequences about halfway along its length (Fig. 7); this is the terminus of the minor-transposon portion of the transposon, which can transpose independently of the rest of the element.

There are *Eco*RI sites within the inverted repeats of both elements (see Fig. 7), so they can be almost exactly excised from host replicons by the action of this enzyme. Thus, the sequences comprising the two transposons were isolated using *Eco*RI and the fragments compared by hybridization. The only homology between the left ends of Tn*501* and Tn*1721* resides at the extremity of each and must correspond to the identity seen in the sequences (Fig. 7; the extent of this identity is in fact 82 bp; F. Schöffl et al., in prep.). The minor-transposon portion of Tn*1721* and the larger of the *Eco*RI fragments of Tn*501* are essentially homologous. However, the two are not completely homologous: First, the sizes of the two do not correspond (Fig. 7), and, second, the detailed restriction enzyme maps are quite different. When radioactive Tn*1721* was hybridized with Tn*501* that had been fragmented with *Bgl*I, some of the fragments from the large *Eco*RI fragment of Tn*501* did not hybridize with the minor transposon. The regions of the two transposons that show considerable homology are those sequences that contain genes necessary for transposition. This suggests that a common ancestor of Tn*501* and Tn*1721* may have been a transposable element that was similar to the minor transposon. The fact that there is some heterogeneity in the regions of homology suggests that the two transposons diverged some time ago.

In Figure 8, the inverted-repeat sequences of Tn*501* and Tn*1721* are compared with three other transposons. All of these transposons cause duplication of 5 bp of host DNA at the site of insertion. That the inverted-repeat sequences of all of these elements are related is prima facie evidence for a common origin for this group

```
Tn3     GGGGTCTGACGCTCAGTGGAACGAAAACTCACGTTAAG
γδ      GGGGTTTGAGGGCCAATGGAACGAAAACGTACGTT
Tn551   GGGGTCCGAGCGCACGAGAAATTTGTATCGATAAG
Tn501   GGGGGGCTCGCAGAATTCGGAAAAAATCGTACGCTAAG
Tn1721  GGGGAGCCCGCAGAATTCGGAAAAAATCGTACGCTAAG
```

Figure 8. Inverted-repeat sequences of transposons that generate a 5-bp direct repeat of host DNA. Underlined letters show the homology with Tn*3* when the sequences are simply aligned as shown. Data from Reed et al. (1979) (γδ), Takeya et al. (1979) (Tn*3*), Brown et al. (1980) (Tn*501*), Khan and Novick (1980) (Tn*551*), and Schmitt et al. (this volume) (Tn*1721*).

of transposons. The minor transposon discussed here might have been the progenitor of these transposons.

ACKNOWLEDGMENTS

The work described was financed by grants from the Medical Research Council to J. G. and M. H. R.

REFERENCES

ALLET, B. 1978. Nucleotide sequence at the ends of bacteriophage Mu DNA. *Nature* **257:** 553.

BROWN, N. L., C.-L. CHOI, J. GRINSTED, M. H. RICHMOND, and P. R. WHITEHEAD. 1980. Nucleotide sequences at the ends of the mercury-resistance transposon Tn*501*. *Nucleic Acids Res.* **8:** 1933.

BURKARDT, H. J., R. MATTES, K. SCHMID, and R. SCHMITT. 1978. Properties of two conjugative plasmids mediating tetracycline resistance, raffinose catabolism and hydrogen sulphide production in *Escherichia coli. Mol. Gen. Genet.* **166:** 75.

KHAN, S. A. and R. P. NOVICK. 1980. Junction DNA sequences of an erythromycin transposon (Tn*551*) from *S. aureus:* Homology with Tn*3*. *Plasmid* **4:** 148.

MATTES, R., H. J. BURKARDT, and R. SCHMITT. 1979. Repetition of tetracycline resistance determinant genes on R. plasmid pRSD1 in *Escherichia coli. Mol. Gen. Genet.* **168:** 173.

MEYER, J. and S. IIDA. 1979. Amplification of chloramphenicol resistance transposons carried by phage P1Cm in *Escherichia coli. Mol. Gen. Genet.* **176:** 209.

OHTSUBO, H., H. OHMIRO, and E. OHTSUBO. 1979. Nucleotide-sequence analysis of the Tn*3* (Ap): Implications for insertion and deletion. *Cold Spring Harbor Symp. Quant. Biol.* **43:** 1269.

RAVETCH, J. V., M. OHSUMI, P. MODEL, G. F. VOVIS, D. FISCHHOFF, and N. D. ZINDER. 1979. Organization of a hybrid between phage f1 and plasmid pSC101. *Proc. Natl. Acad. Sci.* **76:** 2195.

REED, R. R., R. A. YOUNG, J. A. STEITZ, N. D. F. GRINDLEY, and M. S. GUYER. 1979. Transposition of the *Escherichia coli* insertion element $\lambda\delta$ generates a five-base-pair repeat. *Proc. Natl. Acad. Sci.* **76:** 4882.

SCHMITT, R., E. BERNHARD, and R. MATTES. 1979. Characterization of Tn*1721*, a new transposon containing tetracycline resistance genes capable of amplification. *Mol. Gen. Genet.* **172:** 53.

SCHÖFFL, F. and A. PÜHLER. 1979. Intramolecular amplification of the tetracycline resistance determinant of transposon Tn*1771* in *Escherichia coli. Genet. Res.* **33:** 253.

SO, M., R. GILL, and S. FALKOW. 1975. The generation of a ColE1-Ap^r cloning vehicle which allows detection of inserted DNA. *Mol. Gen. Genet.* **142:** 239.

STARLINGER, P. and H. SAEDLER. 1976. IS-elements in microorganisms. *Curr. Top. Microbiol. Immunol.* **75:** 111.

TAKEYA, T., H. NOMIYAMA, J. MIYOSHI, K. SHIMADA, and Y. TAKAGI. 1979. DNA sequences of the integration sites and inverted repeated structure of transposon Tn*3*. *Nucleic Acids Res.* **6:** 1831.

YAGI, Y. and D. CLEWELL. 1976. Plasmid-determined tetracycline resistance in *Streptococcus faecalis:* Tandemly repeated resistance determinants in amplified forms of pAMα1 DNA. *J. Mol. Biol.* **102:** 583.

Hitchhiking Transposons and Other Mobile Genetic Elements and Site-specific Recombination Systems in *Staphylococcus aureus*

R. P. NOVICK, S. A. KHAN, E. MURPHY, S. IORDANESCU,* I. EDELMAN, J. KROLEWSKI,† AND M. RUSH†

*Department of Plasmid Biology, The Public Health Research Institute of the City of New York, New York, New York 10016; *Institute Cantacuzino, Bucharest 35, Rumania; †Department of Biochemistry, New York University Medical Center, New York, New York 10016*

There appear to be two major classes of site-specific recombination systems in bacteria, all presumably involving one or more specific recombination enzymes. One of these consists of reversible, reciprocal exchanges between the recombining segments and is typified by the insertion and excision of temperate phages such as coliphage λ. In this case, the recombining sites have a common core (Landy and Ross 1977) and nonidentical flanking sequences, and the recombination is orientation-specific as well as site-specific. Other examples include site-specific insertion of penicillinase plasmids into the φ11-prophage–chromosome junction in *Staphylococcus aureus* (Schwesinger and Novick 1975) and the site-specific recombination between two P1 prophages described recently by Hoess and Sternberg (1980).

Reversible inversions or "flip-flops" are probably similar in mechanism to reversible insertions. These have been identified in eukaryotes as well as in prokaryotes, and in several cases they have been shown to be involved in genetic control. In *S. aureus,* one of the penicillinase plasmids has recently been found to contain an invertible region that is bounded by an inverted repeat of 650 bp (Murphy and Novick 1979). The inversion-recombination apparently occurs at a specific site within the inverted-repeat segment. A related plasmid, pI258, contains only one copy of this segment, which serves as a locus for high-frequency, site-specific recombination between the two plasmids (Murphy and Novick 1980) and establishes the connection between intra- and intermolecular site-specific, reciprocal recombination. Recombinations of this type are reversible and may be regarded as disruptive in the sense that the preexisting configurations of the recombining elements are not preserved.

The other class of site-specific recombination involves the insertion of heterologous segments, referred to as transposable elements, into nonhomologous loci. These events may be regarded as duplicative (Campbell, this volume): A copy of the inserting segment is retained by the donor genome (Bennett et al. 1977; Novick 1978), and therefore replication is required. Transpositions are also irreversible: The enzymes that are responsible for the insertion do not seem to catalyze excision. Typical elements of this type include the various transposons and insertion sequence (IS) elements as well as coliphage Mu (for review, see Kopecko 1980).

In this paper we present recent results with site-specific recombination systems in *S. aureus,* including a transposon that is very similar to Tn*3,* namely, Tn*551* (Em^r), and two new systems that seem somewhat different from those in gram-negative bacteria, namely, site-specific, apparently irreversible cointegrate formation and a high-frequency hitchhiking transposon, Tn*554.*

Tn*551*

Tn*551* is a 5.2-kb transposon specifying resistance to erythromycin (Em) and other macrolide antibiotics (Pattee et al. 1977; Novick et al. 1979b). It was first encountered as a resistance determinant on plasmid pI258 (see Table 1 for a list of plasmids) and was identified as having been responsible for the integration of Em resistance into the host chromosome following transduction with UV-irradiated transducing phage (Novick 1967). Many chromosomal insertions, some mutagenic, have been mapped (Pattee et al. 1977), and secondary plasmid insertions have been isolated. These were obtained most readily from a chromosomal donor of the transposon, and a series of these have been mapped in plasmid pI6187 by restriction analysis. The results of this mapping, shown in Figure 1, illustrate a clear regional specificity of the insertions, which possibly is similar to that seen with Tn*3.*

Examination of the sequences surrounding the junctions of the transposons with the flanking DNA has revealed three other significant features of this transposon. These sequences (Khan and Novick 1980) are presented in Figure 2 and Table 2, and they show that Tn*551* contains an inverted-repeat sequence of 40 bp at its termini and generates a 5-bp direct repeat upon insertion. Moreover, an examination of the inverted-repeat sequences shows that 14 of the first 18 nucleotides are identical with the corresponding region of Tn*3,* a similarity that is not likely to have occurred by chance. Possibly, the transposon is evolutionarily very ancient, and its basic molecular format evolved before it acquired or developed genetic determinants of resistance to antibiotics and before the divergence of gram-positive and gram-negative bacteria. Finally, one of the plasmids that has been used for sequence determination is a derivative of pI258 from which there has been a precise deletion of Tn*551.* This deletion and six others that are indistin-

Table 1. Plasmids

Strain	Plasmid	Description
RN1978	pI258*repA18*	28-kb penicillinase plasmid containing Tn*551*, Tsr mutant (Novick 1974)
RN647	pI6187	naturally occurring plasmid identical to pI258 but lacking Tn*551* (Novick et al. 1979a)
RN1306	pI524*repA2*	32-kb penicillinase plasmid with reversible inversion (Murphy and Novick 1979), Tsr mutant
RN1687	pII147*repA38*	32-kb penicillinase plasmid, Tsr mutant
SA190	pT181	4.4-kb Tc^r plasmid (Iordanescu et al. 1978)
SA233	pE194	3.6-kb Em^r plasmid
SA231	pC194	2.9-kb Cm^r plasmid
UB4011	pUB110	4.5-kb Km^r plasmid
SA268	pS194	4.5-kb Sm^r plasmid
RN2951	pRN4166	pRN1107 Ω171 (Tn*554*) (pRN1107 is a derivative of pI524*repA2*) (Phillips and Novick 1979)
RN3146	pRN4169	pI524*repA2* Ω172 (*Eco*RI-A::φ11Ω177[Tn*554*])
RN3240	pRN4171	pI258*repA18* Ω174 (*Eco*RI-A::Tn*554*)
RN3665	pRN4174	pII147*repA38* Ω175 (*Eco*RI-B::Tn*554*)
RN3666	pRN4175	pII147*repA38* Ω176 (*Eco*RI-D::Tn*554*)

guishable from it by restriction analysis (Novick et al. 1979a) were generated by UV-irradiation of transducing phage propagated on a strain containing the parental plasmid, followed by transduction with selection for plasmid-determined cadmium (Cd) resistance (Novick 1967). It is concluded from this observation that precise excision of the transposon can be induced by UV-irradiation, and in this behavior it seems dissimilar to Tn*3*.

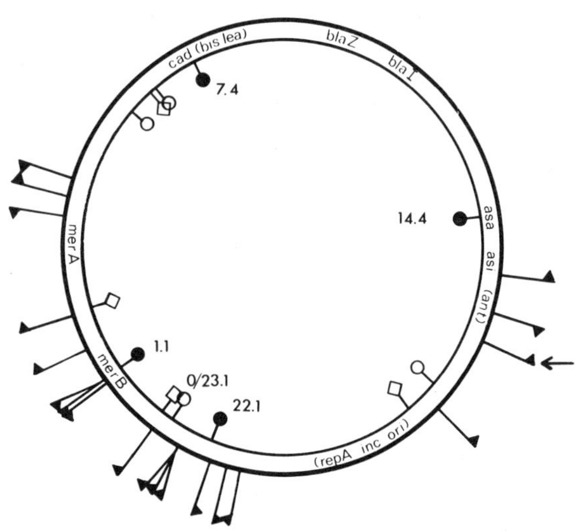

Figure 1. Regional specificity of Tn*551* insertions. Insertions of the transposon were obtained by transduction from a donor strain containing pI6187. Selection was for the plasmid Cd^r and transposon Em^r markers. Insertions were mapped by restriction endonuclease digestion as described previously (Novick et al. 1979b). Insertions are indicated by flags; direction of flag indicates orientation of the transposon (Novick et al. 1979b). Plasmid markers indicated include resistances to cadmium ions (*cad*), bismuth ions (*bis*), lead ions (*lea*), arsenate ions (*asa*), arsenite ions (*asi*), antimony ions (*ant*), mercuric ions (*merA*), organic mercurials (*merB*), penicillin (*bla*), replication (*repA*), incompatibility (*inc*), and replication origin (*ori*). As reported previously, pI6187 is apparently identical with pI258, except that the latter contains the naturally occurring Tn*551* insertion (Novick et al. 1979a), which is indicated by the arrow. Restriction enzyme recognition sites are indicated on the inside of the map with kilobase coordinates as noted (Novick et al. 1979b). (●) *Eco*RI; (○) *Bgl*II; (◊) *Hpa*I; (□) *Bam*HI.

Site-specific Cointegrates

Several small heterologous (1.8–3 megadaltons [MD]) *S. aureus* R plasmids that have been found to undergo cotransduction are included in Table 1. Cotransduction can involve two or three of these plasmids. It occurs at a frequency of several percent of the singles (Kasuga and Mitsuhashi 1968; Iordanescu et al. 1978) and is often accompanied by recombinational exchange, in which the roles of base-sequence homology and general recombination functions are unclear (Iordanescu 1977). Occasionally, cointegrates are formed during cotransduction (Iordanescu 1975); Table 3 lists some of these and the combinations of plasmids that have given rise to them. The cointegrates apparently are stable to secondary transfer either by transduction or by transformation (Iordanescu 1977), and we describe here our studies of the conditions under which they are formed and the sites that are involved in their formation.

All of the cointegrates thus far obtained have been isolated following transduction; for the four pairs of plasmids tested, they occurred at a frequency of 1×10^{-4} to 4×10^{-4} per singly selected transductant. Because of their rarity, they have been detected only where one of the components was a conditional lethal (temperature sensitive for replication [Tsr]) mutant. An attempt to determine whether the transduction step was actually involved in their formation or was simply serving to select them from among a very large population of singles involved the construction of a strain with a Tsr derivative of pT181 (pSA0301) (Iordanescu 1976) and a stable plasmid (pC194). Selection for tetracycline (Tc) resistance at the restrictive temperature gave rise to thermoresistant colonies at a frequency of about 10^{-6}. Twenty of these were analyzed by agarose gel electrophoresis, and all contained both original plasmids in their independent configurations. Since the copy number of pSA0301 is about 60 (S. Carleton and R. Novick, unpubl.), if cointegrates are formed by this route, they must occur at a frequency of less than 10^{-8}. Once formed, however, the cointegrates are transduced at about the same frequency as their component plasmids, and so we conclude that the phage plays a direct role in their formation. Although the cointegrates are stable to

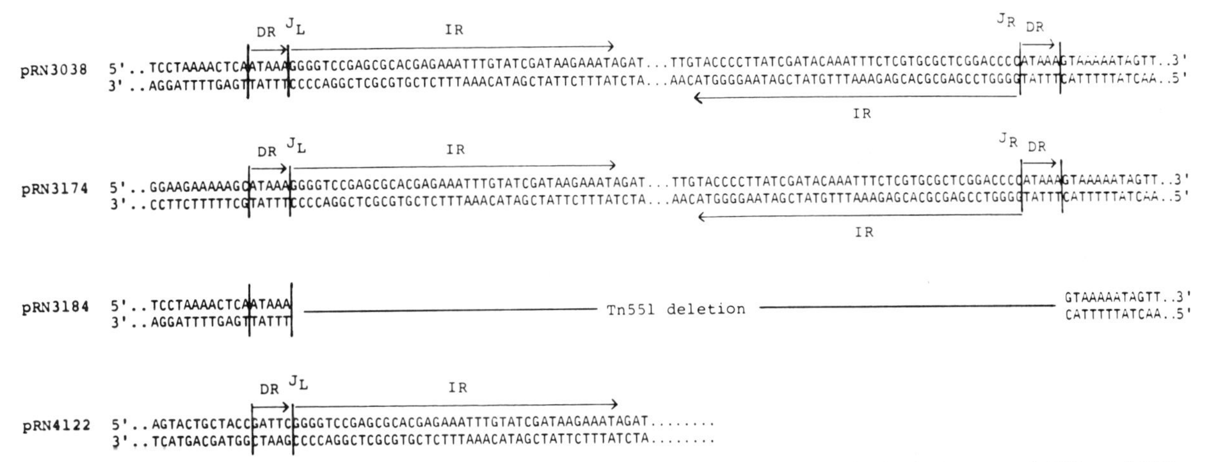

Figure 2. Tn*551* junction sequences. Sequences were determined according to the method of Maxam and Gilbert (1977); details are reported elsewhere (Khan and Novick 1980). Plasmids included are pRN3038, a derivative of the naturally occurring plasmid containing Tn*551* (pI258); pRN3174, a spontaneous deletion that terminates at the outer end of the 5-bp direct repeat; pRN3184, a UV-induced precise excision of Tn*551* from pI258; and one junction of pRN4122, one of the insertions of Tn*551* into pI6187 shown in Fig. 1. DR indicates direct repeat, IR indicates inverted repeat, and J_L and J_R indicate the left and right junctions of the transposon and the parent plasmid pI258, left being the counterclockwise end of the transposon on the map shown in Fig. 1.

secondary transduction, it is not clear whether the frequency of dissociation is actually lower than the frequency of formation, since 10^5 to 10^6 secondary transductants would have to be scored by replica-plating to provide a valid estimate.

Figure 3 shows restriction patterns obtained with several series of cointegrates involving different pairs of plasmids. In each series the various cointegrates were isolated following transduction with a phage lysate prepared on a different clone of the donor culture. These may therefore be regarded as independent isolates. It was noted that in most cases all of the examples of cointegrates between a given pair of plasmids are indistinguishable by restriction analysis. In one case, that of pT181::pE194, nine cointegrates were examined, and there were two different patterns: two of one type and seven of the other. On the basis of these results, it is suggested that each plasmid has one or more specific sites with which it recombines with other plasmids to form cointegrates and that the recombination process is orientation- as well as site-specific. We next investigated whether a given plasmid uses the same site to recombine with more than one other plasmid. For this analysis, we chose three cointegrates, representing the three possible pairwise combinations of three of the plasmids—pE194, pT181, and pC194 (only one of the two cointegrate types involving pT181 and pE194 was studied). Each of the cointegrates (pSA4500, pSA5600, and pSA5710) was digested with a restriction enzyme for which it has a single recognition site, and heteroduplexes were formed in the three possible pairwise combinations and analyzed by electron microscopy. The results, listed in Table 4 and shown in Figures 4 and 5, support the conclusion that in each case a given plasmid uses the same site for the formation of cointegrates with at least two other plasmids. In addition, since the sums of the double-stranded portions of the heteroduplexes were in each case equal to the length of the common plasmid, it can be concluded that if the recombination sites are homologous, then they cannot be longer than 50–100 bp (the limit of resolution in these electron micrographs). We have designated these recombination sites RS and have given them identification notations corresponding to the plasmid of origin, as listed in Table 3. Two additional features of the cointegrates can be discerned from an examination of the diagrammatic representation in Figure 5. First, pT181 and pE194 share a homologous segment of about 0.5 kb that is not involved in the formation of cointegrates, including those making use of the second pair of RS, as typified by pSA4510. This homologous region, however, has been found to be the site of spontaneous excision of a segment of a cointegrate

Table 2. Inverted-repeat Sequences of Tn Elements Generating 5-bp Direct Repeats

Element		5′-End sequence	Reference
Tn*551* (J_L)	5′	GGGGTCCGA^{G}CGCAC^{G}AGAAATTTGTATCGATAAGAAATA	Khan and Novick (1980)
Tn*551* (J_R)	5′	GGGGTCCGA^{G}CGCAC^{G}AGAAATTTGTATCGATAAGGGGTA	Khan and Novick (1980)
Tn*3*	5′	GGGGTCTGA CGCTC AGTGGAACGAAAACTCACGTTAAG	Ohtsubo et al. (1979)
$\gamma\delta$	5′	GGGGTTTGA GGGCC AATGGAACGAAAACGTACGTT	Reed et al. (1979)
pSC101 (J_L)	5′	GGGGTTTGA GGTCC AACCGTACGAAAACGTACGGTAAG	Ravetch et al. (1979)
pSC101 (J_R)	5′	GGGGTCTGA GGGCC AATGGAACGAAAACGTACGTTAGT	Ravetch et al. (1979)
Tn*501*		GGGGGAACC GCAGA ATTCGGAAAAAATCGTACGCTAAG	J. Grinsted (pers. comm.)

Table 3. Cointegrates

Plasmid	Derivation	kb	Phenotype	Rs usage[a]
pSA4500	pSA0301::pE194[b]	8.0	Tc^rEm^r Inc3 Inc11	RS-T_B × RS-E_B
pSA4510	pSA0301::pE194	8.0	Tc^rEm^r Inc3 Inc11	$T_A \times E_A$
pSA4520	pSA0301::pE194	8.0	Tc^rEm^r Inc3 Inc11	$T_B \times E_B$
pSA4540	pSA301::pE194	8.0	Tc^rEm^r Inc3 Inc11	$T_A \times E_A$
pSA4550	pSA301::pE194	8.0	Tc^rEm^r Inc3 Inc11	$T_A \times E_A$
pSA4570	pSA301::pE194	8.0	Tc^rEm^r Inc3 Inc11	$T_A \times E_A$
pSA4590	pSA301::pE194	8.0	Tc^rEm^r Inc3 Inc11	$T_A \times E_A$
pSA5500	pT181::pE194	8.0	Tc^rEm^r Inc3 Inc11	$T_A \times E_A$
pSA5510	pT181::pE194	8.0	Tc^rEm^r Inc3 Inc11	$T_A \times E_A$
pSA5520	pT181::pE194	8.0	Tc^rEm^r Inc3 Inc11	n.t.
pSA5600	pSA0301::pC194	7.3	Tc^rCm^r Inc3 Inc8	$T_B \times C$
pSA5610	pSA0301::pC194	7.3	Tc^rCm^r Inc3 Inc8	$T_B \times C$
pSA5620	pSA0301::pC194	7.3	Tc^rCm^r Inc3 Inc8	$T_B \times C$
pSA5660	pSA0301::pC194	7.3	Tc^rCm^r Inc3 Inc8	$T_B \times C$
pSA5700	pE194::pC194	6.5	Em^rCm^r Inc11 Inc8	$E_B \times C$
pSA5710	pE194::pC194	6.5	Tc^rCm^r Inc3 Inc8	$E_B \times C$

[a] RS indicates recombination site. Sites are designated according to plasmid on which they occur. If more than one site has been demonstrated, uppercase letters are used to differentiate them. Thus, RS-T_A represents the first site identified on PT181. n.t. indicates not tested.

[b] pSA0301 is a Tsr mutant of pT181 (Iordanescu 1976).

between these same two plasmids formed by restriction-ligation in vitro. In this case, deletions were observed in only one of the two possible orientations of this cointegrate clone (G. Adler and R. Novick, unpubl.). Second, the RS observed so far always participate in recombination in the same orientation with respect to one another, suggesting that they are asymmetric. In the case of pT181::pE194, the two pairs of RS seem to interact in only two different ways: RS-T_A × RS-E_A and $T_B \times E_B$. Neither of the cross combinations has been observed among the nine examples examined so far. These interactions suggest a modular organization for the small *S. aureus* plasmids and may serve as an alternative to transposon insertion in the evolution of plasmid complexity.

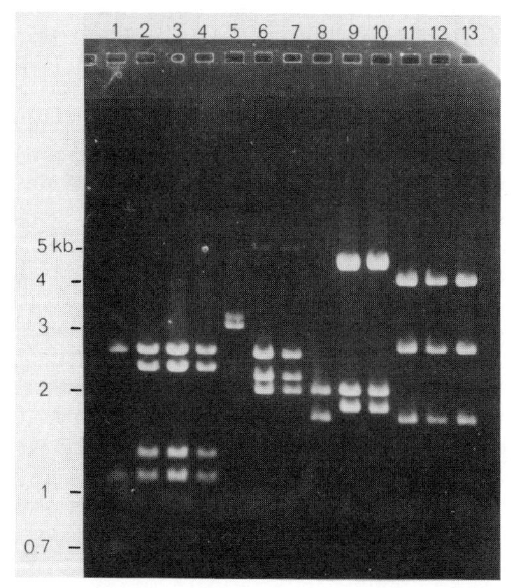

Figure 3. Restriction patterns of cointegrate plasmids. Restriction digestions were performed by dye-cesium centrifugation as described previously (Novick et al. 1979a). Digests were analyzed by horizontal agarose gel electrophoresis and were stained with ethidium bromide and photographed with UV transillumination. (Tracks *1-5*) *Hpa*I-*Hpa*II double digests; (tracks *6-13*) *Hpa*II digests. (*1*) pT181; (*2-4*) pT181::pC194 cointegrates; (*5*) pC194 (upper band represents undigested DNA); (*6, 7*) pC194::pE194 cointegrates; (*8*) pE194; (*9, 10*) pT181::pE194 cointegrates, using RS-T_B and RS-E_B; (*11-13*) pT181::pE194 cointegrates, using RS-T_A and RS-E_A.

Tn*554*

A common phenotype in *S. aureus* is due to linked chromosomal determinants of spectinomycin (Sp) resistance and inducible Em resistance (Lacey 1975). The latter, like that of Tn*551*, is of the well-studied macrolide-lincosamide-streptogramin (MLS) phenotype (Weisblum and Demohn 1969). Early studies from this laboratory (Novick and Bouanchaud 1971; Wyman et al. 1974) revealed that a highly mobile genetic element is involved; more recent studies (Phillips and Novick

Table 4. Electron Microscopy of Heteroduplexes

Combination tested	Common plasmid	D1[a]	D2	D1 and D2	N[b]
pSA4500 × pSA5710	pE194 (3.6)[d]	3.3 ± .2[c]	0.38 ± .06	3.7 ± .2 (3.6)[d]	12
pSA4500 × pSA5600	pT181 (4.4)	2.3 ± .2	2.1 ± .1	4.4 ± .3 (4.4)	27
pSA5600 × pSA5710	pC194 (2.9)	1.3 ± .2	1.7 ± .2	3.0 ± .4 (2.9)	15

[a] D1 and D2 represent the two double-stranded arms flanking the substitution loop (see Figs. 4 and 5).

[b] N indicates the number of molecules measured.

[c] Values are given in kilobases ± S.D.

[d] Numbers in parentheses represent the known molecular length of the common plasmid in each case.

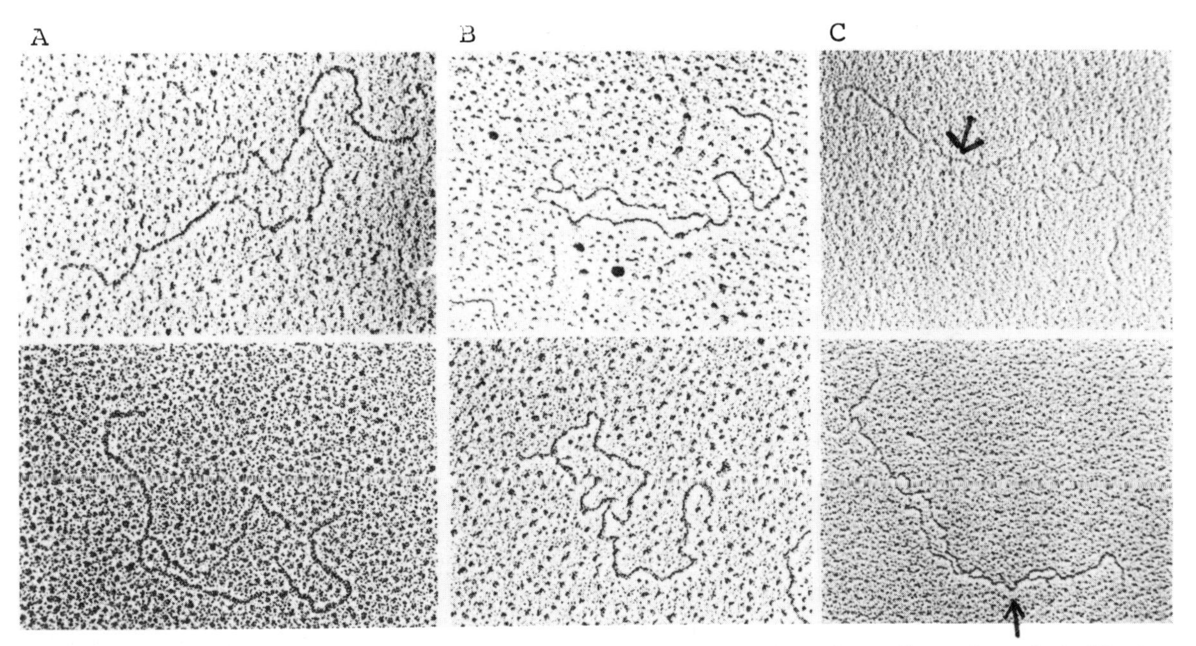

Figure 4. Heteroduplex analysis of cointegrate plasmids. Heteroduplexes were prepared according to the method of Davis et al. (1971) and were photographed with a Philips EM 300 electron microscope. (*A*) pSA4500 × pSA5600, digested with *Kpn*I; (*B*) pSA4500 × pSA5710, digested with *Pst*I; (*C*) pSA5600 × pSA5710, digested with *Bgl*II (see Fig. 5). Note region of homology between pT181 and pE194 (arrow).

1979) have shown that this element is a transposon with several distinctive properties. The prototype has been designated Tn*554*, and its behavior has been analyzed chiefly by means of transduction with ϕ11, a well-characterized general transducing phage from *S. aureus.* Tn*554* is located at a specific map position on the staphylococcal chromosome, and we have found that when it is introduced into a new recipient, it returns to this site with a high degree of preference.

Figure 6 shows the results of in situ hybridization

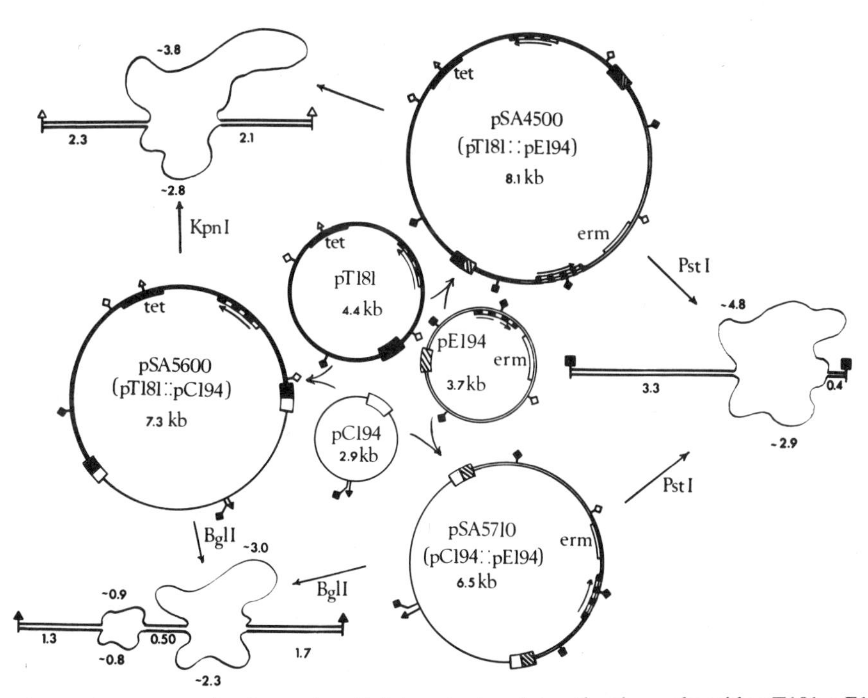

Figure 5. Formation and structure of cointegrates. Cointegrates involving the three plasmids pT181, pE194, and pC194 (smaller circles) are diagramed. Recombination sites are represented by closed boxes, and recombination is depicted as bisecting these, by analogy with λ integration. Relevant restriction sites and recombination sites are indicated. (△) *Kpn*I; (◇) *Hpa*I; (◆) *Hpa*II; (■) *Pst*I; (▲) *Bgl*II; (▬) RS-T_B; (▨) RS-E_B; (□) RS-C_A. The heteroduplexes shown in Fig. 4 are also diagramed.

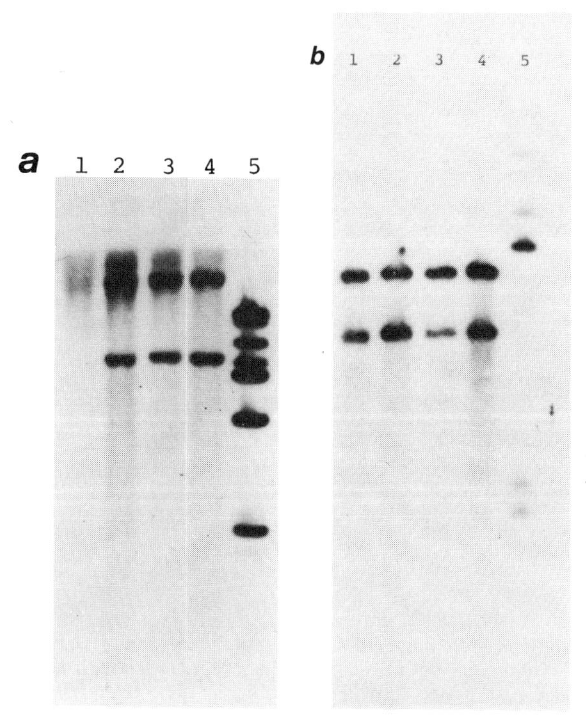

Figure 6. Chromosomal insertion of Tn*554*, as demonstrated by in situ hybridization. Chromosomal DNA was prepared from a series of strains containing Tn*554* insertions and was analyzed by the in situ hybridization method of Southern (1975), as modified by Basilico et al. (1979). Plasmid pRN4166 DNA was labeled with ^{32}P by nick translation in vitro (Rigby et al. 1977) and used as a probe for Tn*554* sequences. (*a*) (Tracks *1-4*) *Xba*I digests; (track *5*) *Bgl*II digest. (*1*) Tn*554*$^-$ recipient strain; (*2-4*) independent insertions obtained by transducing Tn*554* from a chromosomal donor to a *rec*$^-$ recipient; (*5*) pRN4166 plasmid DNA. (*b*) All *Eco*RI digests. (*1-4*) Chromosomal insertions of Tn*554* obtained from the various 2° plasmid insertions, pRN4166, pRN4171, pRN4174, and pRN4175, respectively, after transfer to a Tn*554* strain; (*5*) plasmid pRN4166 DNA. The control samples (tracks *5* in *a* and *b*) contained an amount of plasmid DNA corresponding to what would be expected for one chromosome equivalent.

analysis (Southern 1975) of chromosomal insertions either from a chromosomal donor into a *rec*$^-$ recipient (Fig. 6a) or from a plasmid donor into a *rec*$^+$ recipient (Fig. 6b). All seven insertions are at the same site, at least to the limits of resolution of the blotting analysis. A key finding is that if this site (referred to as the primary site) is vacant, transposition into it, following transduction from a chromosomal donor occurs at frequencies approaching 100%; but if the site is occupied, the apparent transposition frequency is reduced by at least 1000-fold (Phillips and Novick 1979). Table 5 shows the results of transduction into *rec*$^+$ and *rec*$^-$ hosts with and without a (suitably marked) copy of the transposon at the primary site. The more informative results are obtained with the *rec*$^-$ recipient, as conventional recombination tends to mask the effect in the *rec*$^+$ recipient. The frequency of transduction of Tn*554* into the *rec*$^-$ recipient is, if anything, even higher than that of a plasmid; transduction of the conventional transposon Tn*551* from a chromosomal site in the donor to the *rec*$^-$ recipient has not been observed and is therefore at least 10^4-fold less frequent. It has also been found that the few transductants of Tn*554* into the *rec*$^-$ recipient with an occupied primary site do not represent either secondary-site insertions or tandems at the primary site; rather, they appear to be recombinants between the incoming and resident copies (Phillips 1978).

Attempts to obtain plasmid insertions by selecting for cotransduction of a plasmid and the transposon to a Tn*554*$^-$ recipient have been uniformly unsuccessful; although cotransductants are obtained at a low frequency, they have always turned out to have Tn*554* reinserted at its primary site and the plasmid unchanged (Phillips 1978). Plasmid insertions, however, have been obtained by transfer to a recipient in which the primary site is occupied by the transposon. These insertions are rare, occurring at frequencies on the order of 10^{-5} of the Tn*554* transduction frequency, and we have thus far succeeded in isolating only five insertions (Table 1). Three of these (pRN4171, pRN4174, and pRN4175) were isolated by selecting for cotransduction from a donor containing the target plasmid plus a chromosomal Tn*554* to a recipient containing a suitably marked copy of the transposon. As revealed by restriction analysis of plasmid DNA isolated from these strains, these three are all simple 6.2-kb insertions into the respective target plasmids. Examination of one of these by heteroduplex analysis revealed an insertion loop corresponding to the transposon and lacking any observable inverted-repeat structure (not shown). Of the remaining two of the five insertions isolated, which were obtained by transduction to a strain containing both the target plasmid and a resident copy of the transposon, one (pRN4166) is a simple insertion, and the other (pRN4169) is a complex plasmid consisting of the transposon, the target plasmid, and the genome of the transducing phage ϕ11 (see Fig. 7). The transposon is actually inserted into the phage genome, and, as a consequence, the phage is defective, but it still expresses immunity. The transposon-containing defective phage is inserted into the plasmid, apparently using its normal attachment site, and the phage-plasmid complex dissociates upon UV induction of the phage (R. P. Novick, unpubl.). The site on the plasmid that is involved results in the inactivation of the plasmid gene for arsenate resistance (Asa^r); insertions of this same plasmid into the prophage-chromosome junctions, also inactivating the Asa^r determinant and also dissociated by phage induction, have been encountered previously (Schwesinger and Novick 1975).

Although three different plasmids were involved in the secondary insertions, the three are closely related (Novick et al. 1979a), and, in fact, the insertions occurred at clearly different sites in regions in which the three plasmids share homology. This observation, in conjunction with the phage insertion, suggests that there is no particular site specificity for secondary insertions.

Properties of Secondary Tn*554* Insertions

The five secondary insertions vary strikingly with respect to several key properties, which illuminates the uniqueness of Tn*554* as a transposon. These properties

Table 5. Transductional Analysis of Tn*554* with ϕ11

Donor strain containing	Recipient strain *recA*	Recipient strain Tn*554* (*erm*$^-$*spc*$^+$)	Em^r transductants tr/10^7 pfu	Em^r transductants Cd^r (%)	Cd^r transductants tr/10^7 pfu	Cd^r transductants Em^r (%)	Ratio Em^r → Tn*554*$^-$ / Em^r → Tn*554*$^+$	Ratio Em^r / Cd^r-Em^r
Primary chromosomal Tn*554*								
8325-4 (*chr*::Tn*554* [*erm*$^+$*spc*$^-$])	+	−	10,420					
	+	+	570				25	
	−	−	704					
	−	+	1				~700	
Secondary plasmid Tn*554*								
8325-4 (pRN4166)	+	−	272	16	66	82		6
	+	+	47	57	27	87	6	<2
8325-4 (pRN4169)	+	−	2,000,000					
	+	+	13,000				154	
8325-4 (pRN4171)	+	−	900	22	300	60	~1	4.5
	+	+	1,000	26	350	~100		4
8325-4 (pRN4174)	+	−	11,000	90	9,200	~100	~1	~1
	+	+	9,400	94	8,500	~100		~1
8325-4 (pRN4175)	+	−	13,000	92	11,000	~100	~1	~1
	+	+	13,000	96	11,000	~100		~1
Standard plasmid								
8325-4 (pI258)	+	−	6,000	100	6,100	100	1	1
	+	+	6,100	100	6,250	100		1
	−	−	500	100	525	100	1	1
	−	+	550	100	600	100		1

include the ability to serve as a transposon donor, the frequency of apparently precise excision during growth of ϕ11, and the extent to which transduction of the excised transposon is inhibited in a Tn*554*$^+$ recipient.

Transducing phage (ϕ11) lysates were prepared on strains containing each of the five plasmid-Tn*554* complexes and used to transduce a pair of recipients, one (RN2897) containing a suitably marked copy of Tn*554* at the primary site and the other (RN27) without such a copy. Transductants were selected for the Tn*554* Em^r marker, for the plasmid Cd^r marker, or for both. As all of the target plasmids are Tsr (Novick 1974), the frequency of transposition following transduction could be evaluated readily by selecting for thermal elimination of the plasmid from doubly resistant transductants and then testing for loss or retention of the Tn*554* Em^r marker. A similar experiment was performed by means of protoplast transformation (Chang and Cohen 1979) with pRN4166 DNA purified by dye-cesium centrifugation of a cleared lysostaphin lysate (Novick et al. 1979a). These experiments confirmed the previous observation that a copy of Tn*554* inhibits transposition (Phillips and Novick 1979); in fact, when the recipient contained a copy of the transposon, no transpositions were observed in either several hundred doubly resistant transductants or 100 doubly resistant pRN4166 transformants, as scored by the coelimination test (see Table 6). A few transductants with thermostable Em resistance were obtained with pRN4171; these have been shown to be due to conventional recombination between the plasmid and chromosomal Tn*554* genomes. Transfer to a Tn*554*$^-$ recipient, however, unexpectedly revealed that the secondary plasmid insertions varied widely in their ability to act as transposon donors. Thus, one of the secondaries (pRN4166) gave a transposition frequency of 100%; transposition to the chromosome occurred in all of 35 doubly resistant transductants (Phillips and Novick 1979) and in all of 50 transformants, as scored by the coelimination test. Two of the others (pRN4174 and pRN4175) gave frequencies of less than 10%; the Cd^r and Em^r markers, with rare exceptions, were always coeliminated from doubly resistant transductants. In the case of pRN4171, the fourth plasmid insertion, the frequency of transposition was intermediate between these two extremes, i.e., about 20% by the coelimination test

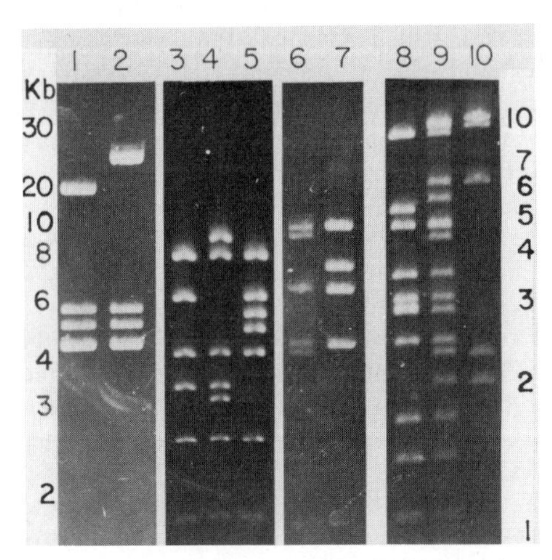

Figure 7. Restriction patterns of Tn*554* 2° insertions. Plasmid DNA was digested with commercially obtained restriction endonucleases and analyzed by agarose gel electrophoresis as described previously (Novick et al. 1979a). Restriction endonuclease digests of Tn*554* insertions. (*1, 2*) *Bcl*I digests of pRN1107 (target) and pRN4166, respectively; (*3, 4, 5*) *Eco*RI digests of pII147*repA38* (target), pRN4174, and pRN4175, respectively; (*6, 7*) *Hpa*I digests of pRN4171 and pI258*repA18* (target), respectively; (*8, 9, 10*) *EcoRI* digests of ϕ11 DNA, pRN4169, and pI524*repA2* (target), respectively. The kilobase scale at left refers to tracks *1-7*, and the scale at right refers to tracks *8-10*.

Table 6. Excision and Transposition of Tn*554* from Plasmids with 2° Insertions

Plasmid	Excision frequency[a] (%)	Transposition frequency[b] recipient (%)	
		Tn*554*$^-$	Tn*554*$^+$
Tn*554*			
pRN4166 (transduction)	84	100	<1
pRN4166 (transformation)	<1	100	<1
pRN4169 (transduction)	10	$>90^c$	$<10^c$
pRN4171 (transduction)	78	~20	2
pRN4174 (transduction)	15	<10	<1
pRN4175 (transduction)	<10	<10	<1
Tn*551*			
pI258*repA18* (transduction)	$<<1$	$\sim 10^{-4}$	—

[a] Number of singly Em^r transcipients as percentage of doubly resistant transcipients; recipient strain Tn*554*$^-$.

[b] Percentage of doubly resistant transcipients that show thermal curing of the Cd^r marker, but not of the Em^r marker.

[c] Percentage of Em^r transductants that are not defective lysogens.

(see Table 6). The frequency of transposition with pRN4169 has been more difficult to measure because growth of the transducing phage automatically results in dissociation of the phage-plasmid complex. The estimate in Table 6 is based on the relative frequency of nonlysogens versus defective lysogens among the Em^r transductants.

Examination of the transduction data in Table 5 reveals a second unexpected result, namely, that the growth of phage ϕ11 induces excision of the transposon. This can be deduced from the fact that the frequency of singly Em^r transductants is always higher than that of the singly Cd^r or doubly resistant transductants; this effect is also variable and is directly correlated with the Tn*554* transposition activity of the plasmid. Thus, for pRN4166, the highest-frequency transposon donor, the cotransduction frequency of Cd and Em resistances was 82% of the singly Cd^r transductants and about 16% of the singly Em^r ones, whereas for pRN4174 and pRN4175, the two low-frequency transposon donors, there was only a slight excess of singly Em^r transductants. The secondary plasmid insertion with the intermediate frequency (pRN4171) also gave an intermediate cotransduction frequency (see Table 5). With a stable Cd^r-Em^r plasmid such as pI258, however, the frequencies of singly and doubly resistant transductants and transformants are always identical, so that the results seen with Tn*554* cannot be attributed to marker effects.

Another unexpected, and thus far unexplained, finding in this experiment is that there also seemed to be a substantial variation in the degree of reduction, due to a resident Tn*554*, of the frequency of singly Em^r transductants. Thus, with one of the insertions (pRN4169) there was a reduction of 150-fold, whereas with another one (pRN4171) there was no detectable reduction. Since establishment of the Em^r determinant in these singly resistant transductants seems to occur predominantly by transposition, this latter result is inconsistent with the usual inhibition seen in other similar crosses.

Some of these results have been verified by an examination of plasmid DNA isolated from singly and doubly resistant transductants and transformants involving a high-frequency secondary Tn*554* donor (pRN4166) and a Tn*554*$^-$ recipient. Three Cd^rEm^s transductants had plasmid restriction patterns that were indistinguishable from that of the original plasmid without the transposon, suggesting precise excision of the latter. Similarly, a Cd^rEm^s transductant of pRN4171 had also lost the transposon. Plasmid DNAs isolated from three pRN4166 Cd^rEm^r transductants gave somewhat different results. On the basis of the coelimination test, there had been a transposition to the primary chromosomal site in all three cases. In one the transposon had excised concomitantly, and apparently precisely, from the plasmid, whereas in the other two the transposon-plasmid complex was still intact. Finally, 11 doubly resistant transformants were also analyzed. In all of these there had been a transposition of Tn*554* to the chromosome, but, on the basis of its restriction pattern, the plasmid-transposon complex was intact. These results suggest that Tn*554* ordinarily transposes by a duplicative mechanism similar to that of other transposons. However, this may not always be the case; as noted above, several attempts to obtain plasmid insertions by cotransduction to a Tn*554*$^-$ recipient always yielded clones in which the transposon was at its primary site and the plasmid did not contain any insertion (Phillips 1978). This result, in conjunction with the above-mentioned example of excision following cotransduction, raises the interesting possibility that Tn*554* can also transpose by an excisive mechanism that could be related to the phenomenon of phage-induced excision.

CONCLUSION

Tn*554* is considered to be the prototype of a new class of mobile genetic elements, and experiments to illuminate its precise modus operandi are in progress. In gram-negative organisms, plasmids such as ColE1 have evolved a means of genetic transfer that makes use of conjugative machinery provided by larger Tra^+ plasmids. This is appropriate in view of the prevalence of the latter. In *S. aureus,* however, it appears that phages serve as the usual, if not the only, vector of natural genetic transfer, and it is therefore reasonable to imagine that certain genetic elements have evolved a mechanism of specific mobilization that makes use of phages. Our results suggest that certain Tn*554* insertions have junctions that are recognized for excision by the phage, and, when this occurs, a unique transduction mechanism comes into play that greatly favors the "free" transposon over other gene segments. Presumably, this would involve attachment to some other DNA, as the transposon is only 6.2 kb in length, whereas the ϕ11 genome is about 45 kb. One attractive candidate is the phage itself, and it is possible that one of our secondary insertions (pRN4169) represents a trapped transposon-carrying phage intermediate.

In this context, it is noted that site-specific plasmid cointegrates have been described in *Escherichia coli* in connection with conjugative mobilization (Broome-Smith 1980), and, as described above, similar cointegrates in *S. aureus* seem to occur in connection with transductional mobilization and could reflect an additional example of phage-specific effects on other genetic elements.

Finally, when an element such as Tn*554* inserts into an active secondary site on a replicon that is then transferred to a Tn*554*⁻ recipient, the transposon will come immediately to occupy its normal primary site in the recipient organism. We refer to this sequence of events as "hitchhiking" and predict that it will be found to be a common pathway for special genetic elements in gram-positive bacteria.

ACKNOWLEDGMENTS

This work was funded by grants from the National Science Foundation (PCM77-25476) and the National Institutes of Health (GM-27253).

REFERENCES

Basilico, C., S. Gattoni, D. Zouzias, and G. Della Valle. 1979. Loss of integrated viral DNA sequences in polyoma-transformed cells is associated with an active viral *A* function. *Cell* **17:** 645.

Bennett, P. M., J. Grinsted, and M. H. Richmond. 1977. Transposition of Tn*A* does not generate deletions. *Mol. Gen. Genet.* **154:** 205.

Broome-Smith, J. 1980. *RecA* independent, site-specific recombination between ColE1 or ColK and a miniplasmid they complement for mobilization and relaxation: Implications for the mechanisms of DNA transfer during mobilization. *Plasmid* **4:** 51.

Chang, S. and S. N. Cohen. 1979. High frequency transformation of *Bacillus subtilis* protoplasts by plasmid DNA. *Mol. Gen. Genet.* **168:** 111.

Davis, R. W., M. Simon, and N. Davidson. 1971. Electron microscope heteroduplex methods for mapping regions of base sequence homology in nucleic acids. *Methods Enzymol.* **21:** 413.

Hoess, R. and N. Sternberg. 1980. P1 site-specific recombination. *J. Supramol. Struct.* (in press).

Iordanescu, S. 1975. Recombinant plasmid obtained from two different, compatible staphylococcal plasmids. *J. Bacteriol.* **124:** 597.

———. 1976. Temperature-sensitive mutant of a tetracycline resistance staphylococcal plasmid. *Arch. Roum. Pathol. Exp. Microbiol.* **35:** 257.

———. 1977. Relationships between cotransducible plasmids in *Staphylococcus aureus. J. Bacteriol.* **129:** 71.

Iordanescu, S., M. Surdeanu, P. Della Latta, and R. Novick. 1978. Incompatibility and molecular relationships between small staphylococcal plasmids carrying the same resistance marker. *Plasmid* **1:** 468.

Kasuga, T. and S. Mitsuhashi. 1968. Drug resistance of staphylococci. VIII. Genetic properties of resistance to tetracycline in *Staphylococcus aureus* E169. *Jpn. J. Microbiol.* **12:** 269.

Khan, S. and R. P. Novick. 1980. Terminal nucleotide sequences of Tn*551,* a transposon specifying erythromycin resistance in *Staphylococcus aureus:* Homology with Tn*3*. *Plasmid* **4:** 148.

Kopecko, D. J. 1980. Specialized genetic recombination systems in bacteria: Their involvement in gene expression and evolution. *Prog. Mol. Subcell. Biol.* **7:** (in press).

Lacey, R. W. 1975. Antibiotic resistance plasmids of *Staphylococcus aureus* and their clinical importance. *Bacteriol. Rev.* **39:** 1.

Landy, A. and W. Ross. 1977. Viral integration and excision: Structure of the lambda *att* sites. *Science* **197:** 1147.

Maxam, A. and W. Gilbert. 1977. A new method for sequencing DNA. *Proc. Natl. Acad. Sci.* **74:** 560.

Murphy, E. and R. P. Novick. 1979. Physical mapping of *Staphylococcus aureus* penicillinase plasmid pI524: Characterization of an invertible region. *Mol. Gen. Genet.* **175:** 19.

———. 1980. Site-specific recombination in plasmids of *Staphylococcus aureus. J. Bacteriol.* **141:** 316.

Novick, R. P. 1967. Penicillinase plasmids of *Staphylococcus aureus. Fed. Proc.* **26:** 29.

———. 1974. Studies on plasmid replication. III. Isolation and characterization of replication-defective mutants. *Mol. Gen. Genet.* **135:** 131.

———. 1978. The mechanism of translocation in bacteria. *Brookhaven Symp. Biol.* **29:** 272.

Novick, R. P. and D. Bouanchaud. 1971. Extrachromosomal nature of drug resistance in *Staphylococcus aureus. Ann. N.Y. Acad. Sci.* **182:** 279.

Novick, R. P., E. Murphy, T. J. Gryczan, E. Baron, and I. Edelman. 1979a. Penicillinase plasmids of *Staphylococcus aureus:* Restriction-deletion maps. *Plasmid* **2:** 109.

Novick, R. P., I. Edelman, M. Schwesinger, A. Gruss, E. Swanson, and P. A. Pattee. 1979b. Genetic translocation in *Staphylococcus aureus. Proc. Natl. Acad. Sci.* **76:** 400.

Ohtsubo, H., H. Ohmori, and E. Ohtsubo. 1979. Nucleotide-sequence analysis of Tn*3* (Ap): Implications for insertion and deletion. *Cold Spring Harbor Symp. Quant. Biol.* **43:** 1269.

Pattee, P. A., N. E. Thompson, D. Haubrich, and R. P. Novick. 1977. Chromosomal map locations of integrated plasmids and related elements in *Staphylococcus aureus. Plasmid* **1:** 38.

Phillips, S. 1978. "Transposon 554: A novel genetic element encoding erythromycin-spectinomycin resistance in *Staphylococcus aureus.*" Ph.D. thesis, New York University, New York.

Phillips, S. and R. Novick. 1979. Tn*554:* A repressible site-specific transposon in *Staphylococcus aureus. Nature* **278:** 476.

Ravetch, R. R., R. A. Young, J. A. Steitz, N. D. F. Grindley, and M. S. Guyer. 1979. Transposition of the *Escherichia coli* insertion element γ generates a five-base pair-repeat. *Proc. Natl. Acad. Sci.* **76:** 4882.

Reed, J. V., M. Ohsumi, P. Model, G. F. Vovis, D. Fischhoff, and N. D. Zinder. 1979. Organization of a hybrid between phage f1 and plasmid pSC101. *Proc. Natl. Acad. Sci.* **76:** 2195.

Rigby, P. W. J., M. Diceckman, C. Rhodes, and P. Berg. 1977. Labeling deoxyribonucleic acid to high specific activity *in vitro* by nick translation with DNA polymerase I. *J. Mol. Biol.* **113:** 237.

Schwesinger, M. D. and R. P. Novick. 1975. Prophage-dependent plasmid integration in *Staphylococcus aureus. J. Bacteriol.* **123:** 724.

Southern, E. M. 1975. Detection of specific sequences among DNA fragments separated by gel electrophoresis. *J. Mol. Biol.* **98:** 503.

Weisblum, B. and V. Demohn. 1969. Erythromycin-inducible resistance in *Staphylococcus aureus:* Survey of antibiotic classes involved. *J. Bacteriol.* **98:** 447.

Wyman, L., R. Goering, and R. P. Novick. 1974. Genetic control of chromosomal and plasmid recombination in *Staphylococcus aureus. Genetics* **76:** 681.

Evidence for Conjugal Transfer of a *Streptococcus faecalis* Transposon (Tn*916*) from a Chromosomal Site in the Absence of Plasmid DNA

A. E. FRANKE AND D. B. CLEWELL

Departments of Oral Biology and Microbiology and the Dental Research Institute, The University of Michigan Schools of Dentistry and Medicine, Ann Arbor, Michigan 48109

Streptococcus faecalis strain DS16, a multiply drug-resistant clinical isolate obtained from St. Joseph's Mercy Hospital (Ann Arbor), was previously shown to harbor two plasmids (Clewell et al. 1978; Tomich et al. 1979b). One of the plasmids, pAD1 (35 megadaltons [MD]), is conjugative and determines hemolysin and bacteriocin activities; the other plasmid, pAD2 (15 MD), is nonconjugative and determines resistance to streptomycin (Sm), kanamycin (Km), and erythromycin (Em) (inducible). The last resistance is located on a transposon (Tn*917*) that can be induced to transpose by exposing the cells to a low (0.5 μg/ml) concentration of erythromycin (Tomich et al. 1979a, 1980). A tetracycline-resistance (Tc^r) determinant was presumed to be chromosome-borne, since derivatives cured of pAD1 and pAD2 and devoid of plasmid DNA (on the basis of physical analysis) maintained a resistance to Tc (minimum inhibitory concentration was 30 μg/ml). We report here that the Tc resistance is located on a transposon (Tn*916*) and, interestingly, is capable of conjugal transfer in the absence of plasmid DNA.

METHODS

Most of the materials and methods used in the experiments reported in this paper have been described previously (Clewell et al. 1974; Dunny and Clewell 1975). The strains used are described in Table 1.

Broth matings were carried out as described previously (Dunny and Clewell 1975). Filter mating experiments were performed as follows. Overnight cultures were mixed in a ratio of one donor to ten recipients, diluted tenfold in N2GT broth (Franke et al. 1978) (total volume 5 ml), and collected on a Millipore membrane (type GS) filter. The filters were placed on N2GT–horse-blood–agar plates and incubated at 37 °C for 20 hours. The cells were then resuspended in 1 ml of medium, and the suspension was spread on plates containing the appropriate selective drugs. Controls consisting of donor or recipient cells alone were treated similarly to determine the frequency of spontaneously appearing drug-resistant mutants in the population. Transfer-frequency measurements were obtained by determining the number of transconjugants per recipient at the end of the mating. (In view of the low frequencies encountered and a concern regarding the frequency of spontaneous mutants of recipients, it was more meaningful to express frequencies as a function of recipient concentration rather than donor concentration.)

RESULTS

Identification of Tc Resistance on a Transposon

Filter matings between DS16 and the plasmid-free recipient JH2-2 gave rise to Tc^r transconjugants at a frequency of about 10^{-6} per recipient. Analyses of the transconjugants revealed three different classes. About 90% or more of the transconjugants were hemolytic, contained pAD1, and harbored the Tc^r determinant on the chromosome. Another class (1–3%) was nonhemolytic and plasmid-free; Tc resistance was chromosome-borne. A third class represented cases where Tc resistance was now linked to pAD1. Many of the latter strains exhibited a hyperexpression of hemolysin, giving rise to zones of hemolysis three to four times the usual diameter, whereas others were "typically" hemolytic or nonhemolytic. Examination of plasmid DNAs from such strains by sucrose density gradient centrifugation and agarose gel electrophoresis consistently showed that pAD1 had increased in size, reflecting an insertion of approximately 10 MD.

Strain DJ2 is a Tc^r transconjugant of JH2-2 that is plasmid-free. Introduction of pAD1 into this strain followed by its use as a donor in matings with the isogenic JH2SSp again gave rise to Tc^r transconjugants of all three classes.

Figure 1 shows the *Eco*RI restriction products of pAD1, as well as two cases where Tc resistance was linked with the plasmid. pAD1 gives rise to eight fragments, designated A through H (Fig. 1a). The plasmid pAM210 came from a strain having hyperexpression for hemolysin; it appears to have the Tc-resistance element inserted into fragment D, as the latter has been replaced by a fragment about 10 MD larger (Fig. 1b). pAM211 shows an insertion in fragment F (Fig. 1c), which is a nonhemolytic derivative.

These data indicate that the chromosome-borne Tc^r determinant is located on a 10-MD transposon capable of inserting at different sites on pAD1. This element will be referred to subsequently as Tn*916*. Additional experiments (data not shown) have indicated that Tn*916* will also transpose to the conjugative plasmids pAMγ1 (Dunny and Clewell 1975) and pOB1 (Oliver et al.

Table 1. *S. faecalis* Strains Used in This Study

Strain	Comments/reference
DS16	contains pAD1 and pAD2 (Tomich et al. 1979b)
DS16C3	plasmid-free; cured of pAD1 and pAD2
JH2-2	plasmid-free; resistant to Rif and Fa (Jacob and Hobbs 1974)
JH2SSp	plasmid-free; resistant to Sm and Sp
DJ2	plasmid-free Tc^r transconjugant from DS16 × JH2-2
UV202	Rec$^-$ derivative of JH2-2 (Yagi and Clewell 1980)
UV202-T23	plasmid-free Tc^r transconjugant from DS16C3 × UV202
OG1SSp	plasmid-free; resistant to Sm and Sp (Dunny et al. 1979)
OG1SSpT1	plasmid-free Tc^r transconjugant from DJ2 × OG1SSp

1977). Transposition from the chromosome to pAD1 occurred at an unaltered frequency (data not shown) in strain UV202 (Yagi and Clewell 1980), a Rec$^-$ derivative of JH2-2.

Transferability of Tn*916* in the Absence of Plasmid DNA

Strain DS16C3 is a derivative of DS16 that has been cured of both pAD1 and pAD2. When used as a donor in filter matings with JH2-2 or the Rec$^-$ strain UV202, Tc^r transconjugants could be selected at low frequency (Table 2). Similarly, using the plasmid-free DJ2 as a donor, Tc^r transconjugants were derived. In the latter case, transfer to the isogenic JH2SSp, as well as to the nonisogenic plasmid-free OG1SSp, was observed. A transconjugant of the DS16C3 × UV202 mating, designated UV202-T23, was capable of acting as a donor when crossed with JH2SSp. Platings of donors or recipients alone on the selective plates yielded no colonies, and "heavy" platings of recipients did not reveal Tc^r mutants ($<10^{-10}$ per recipient). In addition, examination of unselected markers (where possible) confirmed the identity of the transconjugants.

Transfer of the "complete" Tn*916* occurred in the above matings, as transconjugants were capable of inserting the transposon into pAD1 plasmids introduced subsequently. This was observed even in the case of UV202-T23.

It was of interest to determine whether the transfer of Tn*916* from a plasmid-free host was unique to the transposon or a common property of any chromosomal determinant in *S. faecalis*. The only available selectable chromosomal markers, however, are resistances to Sm, spectinomycin (Sp), fusidic acid (Fa), and rifampicin (Rif). The frequency of spontaneous mutation to resistance to these drugs is 10^{-8} to 10^{-7}, a level that would interfere with detecting transconjugants in this frequency range. To circumvent this problem, we took advantage of our earlier observation that the chromosomal Sm^r, Sp^r, and Fa^r markers are closely linked (roughly 90% cotransferable [Franke et al. 1978]). Thus, we examined the cotransfer of the Sm^r and Sp^r markers with the notion that two simultaneous recipient mutations would not occur at a detectable frequency. Plasmid-free strain OG1SSp-T1, with Tn*916* located on the chromosome, and a derivative containing the conjugative plasmid pAMγ1 were used as donors in filter matings with the recipient JH2-2. (The donor and recipient are not isogenic; the donor is actually a *liquifaciens* subspecies of *S. faecalis* and can be distinguished from the recipient by its production of "gelatinase.") In the absence of plasmid DNA, no transfer of SmSp resistance was observed under conditions where Tc^r transconjugants were obtained (Table 2). When the conjugative plasmid was present, Sm^{r}Sp^r transconjugants appeared at low frequency as expected (Franke et al. 1978); the frequency of Tc^r transconjugants increased by 2 orders of magnitude. Tc-resistance transfer was not accompanied by SmSp resistance, and vice versa.

Transfer from a plasmid-free donor, therefore, appears to be specific for Tn*916*. Failure to transfer SmSp resistance tends to rule out transformation or generalized transduction as the basis for transfer. (It is noteworthy that transformation has never been reported for *S. faecalis*, and deliberate efforts to transform JH2-2

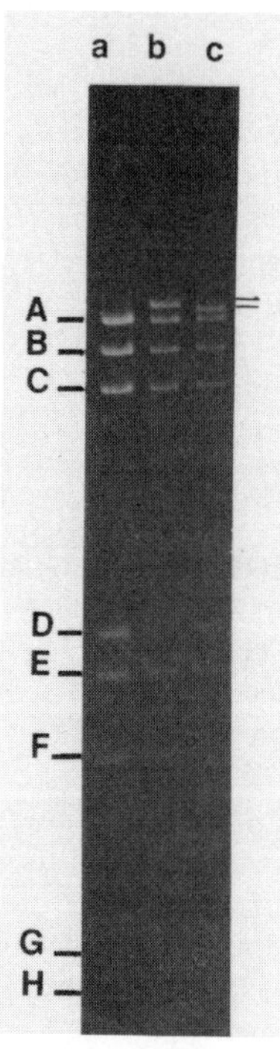

Figure 1. Comparison of the DNA fragments generated by *Eco*RI digestion of pAD1 and plasmid DNA enlarged by Tn*916* insertion. *Eco*RI digestion and electrophoresis of the resulting DNA fragments were carried out essentially as described previously (Yagi and Clewell 1976). The sizes of the *Eco*RI fragments were determined on 0.7% agarose gels by comparison with the *Eco*RI fragments of bacteriophage λ. (*a*) pAD1 plasmid DNA from a "typically" hemolytic isolate. Restriction enzyme cleavage generates eight fragments (A–H) with the following molecular weights: 11.8, 9.8, 7.6, 2.5, 2.2, 1.6, 0.7, and 0.5 MD. (*b*) Plasmid pAM210 DNA isolated from a strain exhibiting hyperexpression of the plasmid-encoded hemolysin. (*c*) Plasmid pAM211 DNA from a nonhemolytic strain.

Table 2. Transfer of Tc Resistance in the Absence of a Conjugative Plasmid

Donor	Recipient	Frequencies of transconjugants acquiring markers[a]	
		Tc^r	Sm^rSp^r
DS16C3	JH2-2	3×10^{-8}	
DS16C3	UV202	6×10^{-9}	
DJ2	JH2SSp	2×10^{-8}	
DJ2	OG1SSp	4×10^{-8}	
UV202-T23	JH2SSp	3×10^{-8}	
OG1SSp-T1	JH2-2	4×10^{-8}	$<3 \times 10^{-10}$
OG1SSp-T1 (pAMγ1)	JH2-2	2×10^{-6}	2×10^{-8}

The indicated donor strains were mixed with plasmid-free recipient cells and spread on appropriate selective plates, as described in text. Depending on the recipient strain, Tc^r transconjugants were selected on plates containing 10 μg of tetracycline per ml and either 25 μg each of rifampicin and fusidic acid per ml or 1 mg of streptomycin per ml. Transfer of Sm and Sp resistance was selected on plates containing 50 μg of rifampicin per ml and 500 μg each of streptomycin and spectinomycin per ml.

[a] The frequency of transconjugants is expressed relative to the number of recipients at the end of the mating. The transfer frequencies given are an average of at least two mating experiments.

have been unsuccessful [P. K. Tomich et al., unpubl.].) Furthermore, the frequency of Tn*916* transfer was unaltered in matings between DS16C3 and JH2-2 carried out in the presence of pancreatic DNase (50 μg/ml present in the broth during mixing and also in the plate during incubation). In addition, when recipients were mixed with cell-free donor filtrate (in place of donor cells) for 1 hour prior to collection and incubation on filters, Tc^r derivatives could not be obtained. Exposure of donors to mitomycin C (0.5 μg/ml) for 2 hours or to UV-irradiation (15 or 90 sec using a General Electric G8T5 lamp at a distance of 90 cm) prior to mating had no effect on the frequency of Tn*916* transfer. Exposure of donor cells to chloroform prior to mating eliminated the transfer of Tc resistance. Thus, the transfer of Tn*916* from a plasmid-free donor appears to require direct contact between donors and recipients.

DISCUSSION

There is good support for the view that the Tc^r determinant of strain DS16 is located on a transposon, which we have designated Tn*916*. Having a size of approximately 10 MD, Tn*916* is capable of Rec-independent transposition from a chromosomal location to several conjugative plasmids and to different sites on pAD1. Insertion into the latter also may affect hemolysin expression.

A most interesting aspect of Tn*916* is its ability to transfer in the apparent absence of plasmid DNA, a process that occurs equally well in three different host systems and under conditions where either the donor or the recipient is Rec$^-$. Direct contact between donors and recipients appears to be necessary; extensive efforts to demonstrate the involvement of transformation or transduction were unsuccessful. Thus, transfer appears to occur by a conjugationlike process.

Possible interpretations of these data are as follows. It is conceivable that the *S. faecalis* chromosome has a fertility potential capable of mobilizing determinants that reside on it. It is not clear what the nature of this potential might be; possibly, it could reflect an integrated, perhaps defective, conjugative plasmid with residual fertility functions. The fact that other chromosomal resistance markers were not observed to mobilize under conditions where Tn*916* transferred would tend to argue against this possibility. It is conceivable, however, that these other markers could, in fact, enter the recipient with equal efficiency, but their recombination with the recipient chromosome was less efficient than in the case of Tn*916,* which could have supplied its own recombination (integrase or transposase) enzyme(s). In addition, it would have to be assumed that the three nonisogenic strains from which Tn*916* was observed to transfer with equal efficiency all contained this fertility potential. This seems unlikely, unless the segment of DNA encoding the fertility potential always accompanied the transposon.

Another interpretation of these data considers the possibility that Tn*916* encodes its own transfer functions. At a size of 10 MD, there is room for significantly more genetic information than that required for Tc resistance. Conceivably, the transposon is a plasmidlike (and/or phagelike) element that lacks replicative autonomy but maintains specific information for transfer. Transfer could be envisioned as a specific duplication and/or excision of the element followed by an expression of transfer functions. Once the DNA has passed into the recipient, a "zygotic induction" of an integrase or a transposase may facilitate insertion into the chromosome.

A seemingly remote alternative is that Tn*916* might actually be located on a nonchromosomal element, which, for reasons unknown, has escaped physical detection. In this regard, however, it is noteworthy that extensive efforts to irreversibly "cure" the cells of the Tc^r determinant have been unsuccessful.

There is a precedent for the notion of an excision-insertion process involving a resistance transposon. Phillips and Novick (1979) have observed that upon entry into *Staphylococcus aureus* cells via a transducing fragment, Tn*554* (Em^r) is excised and specifically inserted into the chromosome. The transposon is believed to undergo a "zygotic induction" upon entry into the recipient.

Recently, Shoemaker et al. (1980) showed that certain strains of *Streptococcus pneumoniae* containing chromosome-borne Tc^r and chloramphenicol-resistance (Cm^r) determinants transfer these markers by a DNase-resistant process (on filters) at low frequency in the absence of conjugative plasmids; however, efforts to transfer the determinants to *S. faecalis* (JH2-2) were unsuccessful. It is not known whether these determinants are located on transposons.

The identification of a chromosomal drug-resistance transposon capable of conjugal transfer has, to our knowledge, not been reported previously. In view of the widespread appearance of drug-resistant clinical isolates of streptococci, it will be of interest to see the extent to which resistance elements similar to Tn*916* are involved.

ACKNOWLEDGMENTS

We thank Paul Tomich, Yoshihiko Yagi, and M. Cindy Gawron for helpful discussions. This work was supported by U.S. Public Health Service research grants from the National Institute of Dental Research (DE-02731) and the National Institute of Allergy and Infectious Diseases (AI-10318). A. E. F. is a predoctoral genetics trainee supported by a National Research Service Award from the National Institutes of Health. D. B. C. is the recipient of a National Institutes of Health Research Career Development Award (AI-00061).

REFERENCES

Clewell, D. B., Y. Yagi, S. P. Damle, and F. An. 1978. Plasmid-related gene amplification in *Streptococcus faecalis*. In *Microbiology—1978* (ed. D. Schlessinger), p. 29. American Society for Microbiology, Washington, D.C.

Clewell, D. B., Y. Yagi, G. M. Dunny, and S. K. Schultz. 1974. Characterization of three plasmid DNA molecules in a strain of *Streptococcus faecalis:* Identification of a plasmid determining erythromycin resistance. *J. Bacteriol.* **117:** 283.

Dunny, G. M. and D. B. Clewell. 1975. Transmissible toxin (hemolysin) plasmid in *Streptococcus faecalis* and its mobilization of a noninfectious drug resistance plasmid. *J. Bacteriol.* **124:** 784.

Dunny, G. M., R. A. Craig, R. L. Carron, and D. B. Clewell. 1979. Plasmid transfer in *Streptococcus faecalis:* Production of multiple sex pheromones by recipients. *Plasmid* **2:** 454.

Franke, A. E., G. M. Dunny, B. Brown, F. An, D. Oliver, S. P. Damle, and D. B. Clewell. 1978. Gene transfer in *Streptococcus faecalis:* Evidence for mobilization of chromosomal determinants by transmissible plasmids. In *Microbiology—1978* (ed. D. Schlessinger), p. 45. American Society for Microbiology, Washington, D.C.

Jacob, A. E. and S. J. Hobbs. 1974. Conjugal transfer of plasmid-borne multiple antibiotic resistance in *Streptococcus faecalis* var. *zymogenes. J. Bacteriol.* **117:** 360.

Oliver, D., B. Brown, and D. B. Clewell. 1977. Characterization of a plasmid determining hemolysin and bacteriocin production in *Streptococcus faecalis* strain 5952. *J. Bacteriol.* **130:** 948.

Phillips, S. and R. Novick. 1979. Tn*554*—A site-specific repressor-controlled transposon in *S. aureus. Nature* **278:** 476.

Shoemaker, N., M. Smith, and W. Guild. 1980. DNase-resistant transfer of chromosomal *cat* and *tet* insertions by filter mating in *Pneumococcus. Plasmid* **3:** 80.

Tomich, P. K., F. Y. An, and D. B. Clewell. 1979a. A transposon (Tn*917*) in *Streptococcus faecalis* that exhibits enhanced transposition during induction of drug resistance. *Cold Spring Harbor Symp. Quant. Biol.* **43:** 1217.

———. 1980. Properties of erythromycin-inducible transposon Tn*917* in *Streptococcus faecalis. J. Bacteriol.* **141:** 1366.

Tomich, P. K., F. Y. An, S. P. Damle, and D. B. Clewell. 1979b. Plasmid-related transmissibility and multiple drug resistance in *Streptococcus faecalis* subsp. *zymogenes* strain DS16. *Antimicrob. Agents Chemother.* **15:** 828.

Yagi, Y. and D. B. Clewell. 1976. Plasmid-determined tetracycline resistance in *Streptococcus faecalis:* Tandemly repeated resistance determinants in amplified forms of pAMα1 DNA. *J. Mol. Biol.* **102:** 583.

———. 1980. Recombination-deficient mutant of *Streptococcus faecalis. J. Bacteriol.* **143:** 966.

Inverted-repeat Nucleotide Sequences in *Escherichia coli* and *Caulobacter crescentus*

P. NISEN AND L. SHAPIRO
Department of Molecular Biology, Division of Biological Sciences, Albert Einstein College of Medicine, Bronx, New York 10461

The nucleotide sequences of several prokaryotic insertion sequence (IS) elements have been determined (Ohtsubo and Ohtsubo 1978; Ghosal et al. 1979). Hybridization experiments and computer analysis of reported nucleotide sequences have shown that mature *Escherichia coli* ribosomal RNA (rRNA) contains sequences homologous to portions of IS elements IS*1* and IS*2* (Nisen and Shapiro 1979). Evidence has accumulated that suggests that IS elements can modulate gene expression at the DNA level, e.g., by promoting chromosomal rearrangements, providing promoters, or inducing polarity (Starlinger and Saedler 1976). The sequence homologies between IS elements and *E. coli* rRNA raise the possibility that they play an additional role.

We describe here a computer analysis of published nucleotide sequences that reveals additional homologies involving IS*1* and IS*2*, and the localization of these homologies in both the IS element and precursor and mature rRNAs is shown. Sequences in the inverted-repeat termini of IS*1* and IS*2* exist in both the 16S and 23S rRNA precursor stem structures and in both cases encompass the RNase-III processing sites reported by Bram et al. (1980). The nucleotide sequence spanning the RNase-III site in precursor 16S rRNA exists not only in the inverted-repeat terminus of IS*2*, but also in the terminus of *E. coli* 6S RNA.

Inverted-repeat termini are a common feature of prokaryotic transposable elements. They have been shown to be required for transposition (Chow et al. 1979; Gill et al. 1979). The known transposability of inverted-repeat-terminated IS elements and transposons (Tn) and the potential importance and ubiquity of these sequences led us to analyze the inverted-repeat DNA (IR-DNA) population of *Caulobacter crescentus* (Nisen et al. 1979b). This prokaryote undergoes a cell cycle composed of morphologically and biochemically distinct cell types, namely, swarmer cells and stalked cells, which are easily separable (Shapiro 1976). IR DNA was observed to hybridize to different regions of stalked- and swarmer-cell-derived chromosomal DNA, suggesting that IR-DNA sequences in *Caulobacter* may rearrange.

METHODS

The computer analysis of nucleotide sequences was carried out using the Stanford Molgen Project at the National Institutes of Health Sumex-Aim Facility. The computer program was originally developed by Korn et al. (1977) and modified by P. Friedland, D. Brutlag, and L. Kedes at Stanford University. The parameters and statistical probabilities employed were the same as those defined and developed by these workers. In all cases, matches with a probability of <0.1 were considered significant, taking into account the sizes and locations of the sequences compared.

Procedures for the isolation of RNA and DNA, use of restriction endonucleases, agarose gel electrophoresis, blotting to filter paper, and hybridizations have been described elsewhere (Nisen et al. 1979a,b).

RESULTS AND DISCUSSION

Distribution of IS Elements and Homology between IS*1* and IS*2*

Certain IS elements have been identified on various bacterial, phage, and plasmid DNAs by either hybridization or electron microscopic heteroduplex analysis (Saedler and Heiss 1973; Deonier and Hadley 1976; Starlinger and Saedler 1976; Chow 1977; Deonier et al. 1979). To determine the extent of the distribution of IS elements, we first characterized several plasmid and phage DNAs containing IS*1*, IS*2*, and IS*5*, and then hybridized these labeled probes to DNAs from 15 different bacterial and eukaryotic genomes. During characterization of these probes, limited regions of homology between IS*1* and IS*2* were observed by hybridization to Southern blots. Computer analysis of the nucleotide sequences of IS*1* and IS*2* revealed statistically significant regions of homology that can account for the hybridization results. One of the sequence matches detected by computer analysis is shown in Figure 1. This particular homology contains a tetranucleotide sequence tandemly repeated three times. This resembles the repeated sequence at the hotspot for spontaneous mutations in the *lacI* gene (Farabaugh et al. 1978).

Homology between the IS elements and various genomes was determined by hybridizing labeled DNA containing IS*1*, IS*2*, and IS*5* sequences to Southern blots of chromosomal DNA cleaved with restriction endonucleases. IS*1* and IS*5* appear to be limited to the enteric bacteria, whereas IS*2* sequences can also be detected in *Pseudomonas putida*, *Pseudomonas aeruginosa*, and *Serratia marcescens*. As described below, sequence homologies between these IS elements and rRNA could contribute to part of the observed hybridization of IS probes. Sequences homologous to IS*1*, IS*2*, or IS*5* were

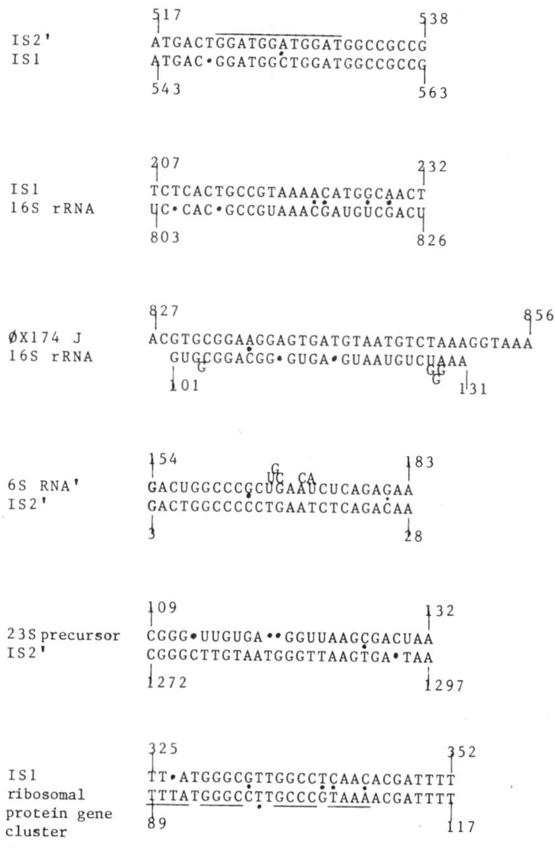

Figure 1. A sample of the types of nucleotide sequence homologies detected. A complete compilation of the nucleotide sequence homologies will be reported elsewhere (P. Nisen and L. Shapiro, in prep.). The identity of each sequence is listed on the left. The numbers correspond to the coordinates from the published sequence. Primes (′) correspond to the complementary strand of the reported sequence in inverted order; dots indicate mismatches. The sequences shown are all considered statistically significant by the criteria and calculations developed by D. Brutlag, P. Friedland, and L. Kedes.

not detected in *Saccharomyces cerevisiae, Dictyostelium discoideum,* or calf thymus DNA.

E. coli rRNA Contains Sequences Homologous to IS*1* and IS*2*

IS*1* and IS*2* are present in multiple copies in the *E. coli* chromosome (Saedler and Heiss 1973; Deonier and Hadley 1976; Chow 1977). They can modulate gene expression at the DNA level, and they generate various chromosomal rearrangements (Starlinger and Saedler 1976). To test the possibility that IS elements could exert control through partial RNA transcripts, we hybridized λ*NNc*I857*r*14 (carrying IS*1*) and the plasmid pBR322 (carrying two thirds of IS*2*) to Northern blots (Alwine et al. 1977) containing whole-cell *E. coli* RNA, which was separated by agarose gel electrophoresis. The probe containing IS*1* hybridized predominantly to 23S rRNA, and, to a lesser extent, to 16S rRNA. IS*2*, however, hybridized equally well to both 23S and 16S rRNA (Nisen et al. 1979a). pBR322 and λ DNA without IS elements did not hybridize to the filters. Hybridization to messenger RNA (mRNA) would not likely have been detected by this experiment because so little of it (in relation to rRNA) is present on the filter. Since the nucleotide sequences for IS*1* (Ohtsubo and Ohtsubo 1978), IS*2*, 23S rRNA (Brosius et al. 1980), and 16S rRNA (Brosius et al. 1978) are known, we used a computer program to detect homologous regions between these sequences to confirm our hybridization results. As indicated in Methods, the computer calculates the probability that a particular sequence match would not occur randomly given the sizes of the sequences compared.

The locations of nucleotide sequences in mature 23S and 16S rRNA that are homologous to sequences in IS*1* and IS*2* are summarized in schematic form in Figure 2. Examples of a few of these sequence matches are listed in Figure 1. These extensive regions of homology between the IS elements and mature 16S and 23S rRNA can account for the hybridization observed between these sequences. Since IS probes can hybridize to rRNA, it is likely that they also hybridize with the DNA encoding rRNA (rDNA). Therefore, as described above, hybridization of IS probes to chromosomal DNA may, in part, represent hybridization to rDNA. To test this possibility, we hybridized probes containing IS element and rDNA sequences to Southern blots of *E. coli* K12 (strain AE2000) chromosomal DNA. The probe that contained the rDNA sequences hybridized to three bands in both *Eco*RI- and *Bam*HI-digested chromosomal DNA. One of these bands is located at the same region of the blot to which IS-containing probes also hybridized. This suggests that some of the hybridization by IS probes (IS*1*, IS*2*, and IS*5*) to chromosomal DNA may involve rDNA sequences.

To ascertain that homologies between IS elements and rRNA equivalent in statistical significance to those detected by the computer do not exist between any random pairs of DNA or RNA sequences, the nucleotide sequences of pBR322 (Sutcliffe 1978), SV40 (Fiers et al. 1978), two histone genes (Sures et al. 1978), φX174, and several ribosomal protein gene clusters (Post et al. 1979) were compared with 16S and 23S rRNA, IS*1*, and IS*2*. Statistically significant homologies were detected only with a palindrome in the leader sequence for a ribosomal protein gene cluster and with the ribosome-binding sites in φX174 DNA (Fig. 1). A region of homology between the φX174 gene-*J* ribosome-binding site and the 5′ end of the 16S rRNA overlaps the region of homology between IS*2* and 16S rRNA. Thus, there is a nucleotide sequence shared between a ribosome-binding site in φX174 DNA, IS*2*, and the 5′ end of 16S rRNA. This observation led us to compare sequences of other known ribosome-binding sites with those of rRNA and IS elements. By means of computer analysis, we compared 30 prokaryotic mRNA ribosome-binding sites (Steitz 1979). Twenty of these ribosome-binding sites were found to have statistically significant homology with 16S rRNA. Of these 20, 15 are clustered at the 5′ end of the molecule. Some examples of these sequence matches are shown in Figure 1.

It should be noted that in contrast to the model of

Shine and Dalgarno (1974), in which the 3′ end of 16S rRNA is complementary to a segment in the 5′ end of a message, these matches represent homology between the 5′ end of an mRNA (i.e., the ribosome-binding site) and rRNA. Therefore, the region of rRNA that is homologous to an mRNA ribosome-binding site is complementary, in an antiparallel fashion, to the DNA strand that has just been transcribed. This raises the possibility that interaction between the rRNA and the newly transcribed DNA serves to couple transcription and translation. Alternatively, the observed homologous sequences in rRNA, the ribosome-binding sites, and IS elements may function as common recognition sites for a protein or they can be viewed as nucleotide sequences that have become stably integrated over the course of evolution. IS elements, or sequences like them, could at one time have interrupted DNA sequences coding for rRNA; the sequences remaining in the rRNA could represent remnants left behind by imprecise excision or duplication.

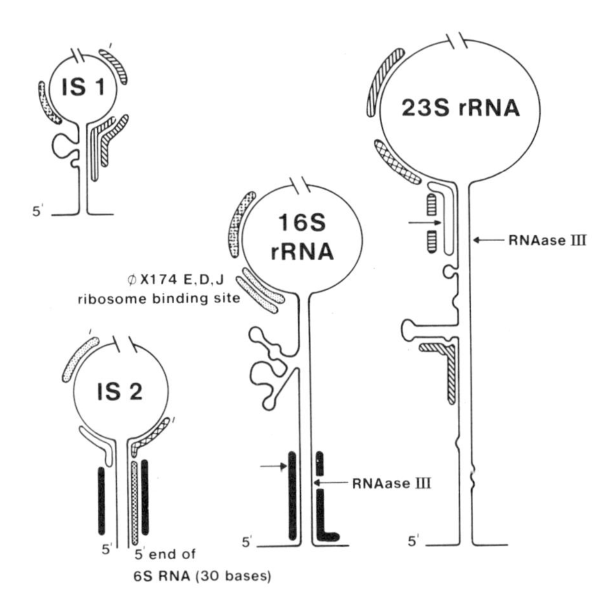

Figure 2. Schematic of the locations of homologous regions between IS*1*, IS*2*, and precursor and mature rRNAs. Segments of IS elements and rRNA homologous with portions of ϕX174 DNA (Sanger et al. 1978) and 6S RNA (Bailey and Apirion 1979) are also depicted. Each nucleotide sequence corresponds to a single strand of the molecule. The structures have been drawn so as to show the inverted-repeat termini of IS*1* (Ohtsubo and Ohtsubo 1978) and IS*2* (Ghosal et al. 1979) as well as the putative stem structures of the 16S and 23S rRNA precursors proposed by Bram et al. (1980). The sites of RNase-III cleavage are indicated. Bars of identical pattern indicate homologous sequences. A prime (′) next to a homology bar indicates that it is the inverted complement of the sequence represented.

Sequence Similarity between Inverted-repeat Termini of IS*1* and IS*2* and the Precursor Stems of 16S and 23S rRNA

Homologies with the inverted-repeat termini of IS*1* and IS*2* were found in the precursor stems of 16S and 23S rRNA and in the terminus of stable 6S RNA. The locations of these homologies are depicted in schematic form in Figure 2. Some of these sequence matches are shown in Figure 1. Sequences homologous to the inverted-repeat termini of IS*1* and IS*2* are present in the RNase-III processing sites in the stems of the 16S and 23S rRNA precursors. Numerous other RNA sequences were analyzed for similar homologies, e.g., 4.5S RNA, 5S RNA, and several tRNAs, and none were found. The similarity between the sequences in the inverted-repeat termini of IS*1* and IS*2* and those in the RNase-III processing stems of 23S and 16S rRNA suggests that these sequences may perform a common function. For example, they could provide a means for the removal of transcripts of IS elements that would otherwise adversely affect essential cellular functions.

Cell-cycle-associated Rearrangement of *Caulobacter* IR DNA

The potential importance and ubiquity of inverted-repeat sequences led us to analyze them in *Caulobacter*. At least 3% of the *Caulobacter crescentus* genome is comprised of IR-DNA sequences (Wood et al. 1976; Nisen et al. 1979b). They exist in two size classes: one 100–600 bp long and the other 1500–3000 bp long. Experiments with exonuclease VII and endonuclease S1 revealed that at least 4.5% of the *C. crescentus* chromosome is contained between the IR-DNA sequences (Nisen et al. 1979b).

Each of the different cell types of the *Caulobacter* cycle can be isolated from synchronous cell cultures (Shapiro 1976). Chromosomal DNA was isolated from different cell types, digested with restriction endonuclease *Bam*HI, and analyzed by agarose gel electrophoresis. The pattern of restriction fragments appeared to be identical for each of the different DNA preparations. Similarly, stalked- and swarmer-cell-derived DNAs yielded indistinguishable patterns when cleaved with restriction endonucleases *Eco*RI and *Hin*dIII, and, as expected, these patterns differed from those obtained with *Bam*HI (Nisen et al. 1979b).

To identify the locations of IR DNA in the stalked and swarmer cell chromosomes, whole-cell populations and individual clones of IR DNA were hybridized to restriction fragments of the genomic DNA (Nisen et al. 1979b). Hybridization of unfractionated IR DNA to Southern blots of stalked and swarmer cell DNA treated with either *Bam*HI or *Eco*RI showed that, in general, the IR DNA was homologous to a limited number of similar sites in the two genomes. However, some differences were consistently noted. To localize these differences, ^{32}P-labeled λWES•IR-DNA clones were hybridized individually to portions of the same Southern blot of *Bam*HI digests of stalked- and swarmer-cell-derived DNA. It was found that for two clones tested, designated IR1 and IR2, both hybridized to the same four regions in stalked- and swarmer-cell-derived DNA; however, IR1 also hybridized to two additional bands in the swarmer-cell-derived DNA. This experiment was repeated with different stalked and swarmer cell DNA preparations that had been treated with *Bam*HI; IR1

DNA consistently hybridized to the same two additional regions in the swarmer cell DNA. When ^{32}P-labeled λWES•IR1 DNA was hybridized to a Southern blot of *Eco*RI-generated fragments of stalked and swarmer cell DNA, IR1 DNA also hybridized to additional regions of the swarmer cell DNA (Nisen et al. 1979b). These preliminary results suggest that a sequence rearrangement of a part of the IR1 DNA may have occurred.

This observation may have regulatory significance since the cell cycle of *C. crescentus* involves temporally regulated differentiation of specific cell types. Swarmer cells and stalked cells, for example, differ with respect to their abilities to synthesize and assemble flagella and pili, their patterns of cytoplasmic and membrane protein synthesis (Shapiro and Maizel 1973; Cheung and Newton 1977; Iba et al. 1978; Lagenaur and Agabian 1978; Agabian et al. 1979; Evinger and Agabian 1979), their patterns of membrane phospholipid synthesis, and their abilities to initiate DNA replication.

ACKNOWLEDGMENTS

We wish to thank the Stanford Molgen Project at the National Institutes of Health Sumex-Aim Program for allowing us to use their facility for our computer analysis of nucleotide sequences. In particular, we are grateful to D. Brutlag, P. Friedland, and L. Kedes, who have developed this program. We especially thank John Sninsky for his assistance with the computer analysis and helpful discussions.

This investigation was supported by grant GB42545X from the National Science Foundation and by U.S. Public Health Service grant GM 11301 from the National Institutes of Health. P. N. is a Medical Science Training Program Fellow (U.S. Public Health Service grant 5T5-GM1674 from the National Institutes of Health), and L. S. is the recipient of a Hirschl Trust Award.

REFERENCES

Agabian, N., M. Evinger, and G. Parker. 1979. Generation of asymmetry during development: Segregation of type-specific proteins in *Caulobacter*. *J. Cell Biol.* **81:** 123.

Alwine, J., D. Kemp, and G. Stark. 1977. Method for detection of specific RNA's in agarose gels by transfer to diazobenzyloxymethyl paper and hybridization with DNA probes. *Proc. Natl. Acad. Sci.* **74:** 5350.

Bailey, S. and D. Apirion. 1979. Repetitive DNA in *E. coli:* Multiple sequences complementary to small stable RNA's. *Mol. Gen. Genet.* **172:** 339.

Bram, R. J., R. A. Young, and J. A. Steitz. 1980. The ribonuclease III site flanking 23S sequences in the 30S ribosomal precursor RNA of *E. coli. Cell* **19:** 393.

Brosius, J., T. Dull, and H. Noller. 1980. Complete nucleotide sequence of a 23S rRNA gene from *E. coli. Proc. Natl. Acad. Sci.* **77:** 201.

Brosius, J., M. C. Palmer, P. J. Kennery, and H. Noller. 1978. Complete nucleotide sequence of a 16S rRNA gene from *E. coli. Proc. Natl. Acad. Sci.* **75:** 4801.

Cheung, R. R. and A. Newton. 1977. Patterns of protein synthesis during development in *Caulobacter crescentus. Dev. Biol.* **56:** 184.

Chou, J., P. Lemaux, P. M. Casadaban, and S. N. Cohen. 1979. Transposition protein of Tn*3:* Identification and characterization of an essential repressor-controlled gene product. *Nature* **282:** 801.

Chow, L. 1977. The organization of putative insertion sequences of the *E. coli* chromosome. *J. Mol. Biol.* **113:** 611.

Deonier, R. and R. Hadley. 1976. Distribution of inverted IS length sequences in the *E. coli* K-12 chromosome. *Nature* **264:** 191.

Deonier, R., R. Hadley, and M. Hu. 1979. Enumeration and identification of IS*3* elements in *E. coli* strains. *J. Bacteriol.* **137:** 1421.

Evinger, M. and N. Agabian. 1979. *Caulobacter crescentus* nucleoid: Analysis of sedimentation behavior and protein composition during the cell cycle. *Proc. Natl. Acad. Sci.* **76:** 175.

Farabaugh, P., M. Schmeissner, M. Hofer, and J. Miller. 1978. Genetic studies of the *lac* repressor VII: On the molecular nature of spontaneous hotspots in the *lac* I genes of *E. coli. J. Mol. Biol.* **126:** 847.

Fiers, W., G. Contreras, R. Haegeman, R. Rogiers, H. Van de Voorde, H. Van Heuverswyn, J. Van Herreweghe, and M. Ysebaert. 1978. Complete nucleotide sequence of SV40 DNA. *Nature* **273:** 113.

Ghosal, D., H. Sommer, and H. Saedler. 1979. Nucleotide sequence of the transposable element IS*2. Nucleic Acids Res.* **6:** 111.

Gill, R., F. Heffron, and S. Falkow. 1979. Identification of the protein encoded by the transposable element Tn*3* which is required for its transposition. *Nature* **282:** 797.

Iba, H., A. Fukuda, and Y. Okada. 1978. Rate of major protein synthesis during the cell cycle of *Caulobacter crescentus. J. Bacteriol.* **135:** 642.

Korn, L. J., C. L. Queen, and M. N. Wegman. 1977. Computer analysis of nucleic acid regulatory sequences. *Proc. Natl. Acad. Sci.* **74:** 4401.

Lagenaur, C. and N. Agabian. 1978. *Caulobacter* flagella organelle: Synthesis, compartmentation and assembly. *J. Bacteriol.* **135:** 1062.

Nisen, P. and L. Shapiro. 1979. *E. coli* ribosomal RNA contains sequences homologous to insertion sequences IS*1* and IS*2. Nature* **282:** 872.

Nisen, P., M. Purucker, and L. Shapiro. 1979a. DNA sequence homologies among bacterial insertion sequence elements and genomes of various organisms. *J. Bacteriol.* **140:** 588.

Nisen, P., R. Medford, J. Mansour, M. Purucker, A. Skalka, and L. Shapiro. 1979b. Cell cycle-associated rearrangement of inverted repeat DNA sequences. *Proc. Natl. Acad. Sci.* **76:** 6240.

Ohtsubo, H. and E. Ohtsubo. 1978. Nucleotide sequence of an insertion element IS*1. Proc. Natl. Acad. Sci.* **75:** 615.

Post, L., G. Strycharz, M. Nomura, H. Lewis, and P. Dennis. 1979. Nucleotide sequence of the ribosome protein gene cluster adjacent to the gene for RNA polymerase β in *E. coli. Proc. Natl. Acad. Sci.* **76:** 1697.

Saedler, H. and B. Heiss. 1973. Multiple copies of insertion sequence DNA sequences IS*1* and IS*2* on the *E. coli* chromosome. *Mol. Gen. Genet.* **122:** 266.

Sanger, F., A. Couljon, T. Friedman, G. Air, B. Barrell, N. Brown, J. Fiddes, C. Hutchinson III, P. Slocombe, and M. Smith. 1979. The nucleotide sequence of bacteriophage ϕX174. *J. Mol. Biol.* **125:** 225.

Shapiro, L. 1976. Differentiation in the *Caulobacter* cell cycle. *Annu. Rev. Microbiol.* **30:** 377.

Shapiro, L. and J. V. Maizel. 1973. Synthesis and structure of *Caulobacter crescentus* flagella. *J. Bacteriol.* **113:** 478.

Starlinger, P. and H. Saedler. 1976. IS elements in microorganisms. *Curr. Top. Microbiol. Immunol.* **75:** 111.

Steitz, J. A. 1979. Genetic signals and nucleotide sequences in mRNA. In *Biological regulation and develop-*

ment (ed. R. Goldberger), p. 478. Plenum Press, New York.

Sures, I., J. Lowry, and L. Kedes. 1978. The DNA sequence of the sea urchin (*S. purpuratus*) H2A and H2B histone coding and spacer regions. *Cell* **15:** 1033.

Sutcliffe, J. 1978. pBR322 restriction map derived from the DNA sequence: Accurate DNA size markers up to 4361 nucleotide pairs long. *Nucleic Acids Res* **5:** 2721.

Wood, N., A. Rake, and L. Shapiro. 1975. Structure of *Caulobacter* DNA. *J. Bacteriol.* **126:** 1305.

ISR1: An Insertion Element Isolated from the Soil Bacterium *Rhizobium lupini*

U. B. PRIEFER,* H. J. BURKARDT,† W. KLIPP,* AND A. PÜHLER*
* *Lehrstuhl für Genetik, Fakultät für Biologie, Universität Bielefeld, D-4800 Bielefeld, Federal Republic of Germany;*
† *Institut für Mikrobiologie und Biochemie, Lehrstuhl für Mikrobiologie, Universität Erlangen-Nürnberg, D-8520 Erlangen, Federal Republic of Germany*

This report deals with an insertion element (IS*R1*) that was found in the soil bacterium *Rhizobium lupini* and that was originally isolated by Heumann (1968).

We made the observation that the IncP plasmid RP4 (Datta et al. 1971; Datta and Hedges 1972), which confers resistance against ampicillin (Ap), kanamycin (Km), and tetracycline (Tc), is quite unstable in *R. lupini* in contrast to other host bacteria (Datta et al. 1971; O'Gara and Dunican 1973; Beringer 1974). Loss of the whole plasmid occurs at a strikingly high frequency, in some cases 50–60%. Clones that harbor mutated plasmids arise at a rate of up to 10%, or even more, and can easily be identified by simple plating techniques. We could isolate Ap^s, Km^s, and Tra^- mutants in various combinations, but for some reason we never observed loss of the Tc resistance.

Identification of IS*R1*

It seemed rather unlikely to us that the extraordinarily high instability of plasmid RP4 in *R. lupini* was due to spontaneous mutation events. However, molecular analysis of the RP4 mutations in *R. lupini* was impossible due to the instability of the mutant molecules. We therefore introduced the RP4 mutants into *E. coli,* where they proved to be nearly absolutely stable. Thus, we had the possibility to isolate homogeneous plasmid DNA from *E. coli* and to study the molecular background of the RP4 mutations generated in *R. lupini.*

We analyzed different RP4 mutants by contour-length measurements, heteroduplex experiments, and restriction analysis. It turned out that all RP4 mutations that had arisen in *R. lupini* correlated to the insertion of a defined DNA sequence. This DNA element, denoted IS*R1* (*i*nsertion *s*equence of R*hizobium lupini 1*), is shown in Figure 1. It carries single recognition sites for restriction enzymes *Hin*dIII, *Pst*I, and *Bam*HI and is 1.15 kb long, which is very common for IS elements.

General Properties of IS*R1* Insertions

The molecular study of IS*R1* insertions in 16 different RP4 mutants revealed the following general properties: (1) IS*R1* can integrate in both orientations; (2) IS*R1* can generate mutations; (3) IS*R1* preferentially inserts into three different regions on plasmid RP4.

Mutations caused by IS*R1* can be generated in two different ways: either by integration into the corresponding gene region or by generating deletions of variable length. These deletions can extend in one or in both directions, starting from the ends of the IS*R1* insertion.

According to the mutant phenotypes (Ap^s, Km^s, Tra^-), we could identify three different regions on plasmid RP4 into which IS*R1* can insert (Fig. 2). All RP4 mutants investigated so far carry an IS*R1* sequence next to the Ap-resistance gene encoded by transposon Tn*1* (Hedges and Jacob 1974; Campbell et al. 1977). This insertion within Tn*1* is very common for RP4 plasmids isolated from *R. lupini* and is present even in those molecules that show no alteration in phenotype. The insertion is very specific with respect to the orientation and integration site and does not affect the expression of β-lactamase (*bla*). Loss of Ap resistance is observed only in connection with deletions.

Most of the mutant RP4 molecules harbor a second IS*R1* insertion in the neighborhood of the single *Hin*dIII site on plasmid RP4. These insertions are less specific than those within Tn*1*. IS*R1* is found to be inserted in both orientations and into different positions, either within or in the neighborhood of the Km-resistance coding region. Again there occur associated deletions of variable lengths.

In three mutants, IS*R1* insertions were identified within the transfer region Tra3, inactivating the transferability of the RP4 plasmid. In all three cases, the insertions proved to be identical with respect to their orientations and integration sites.

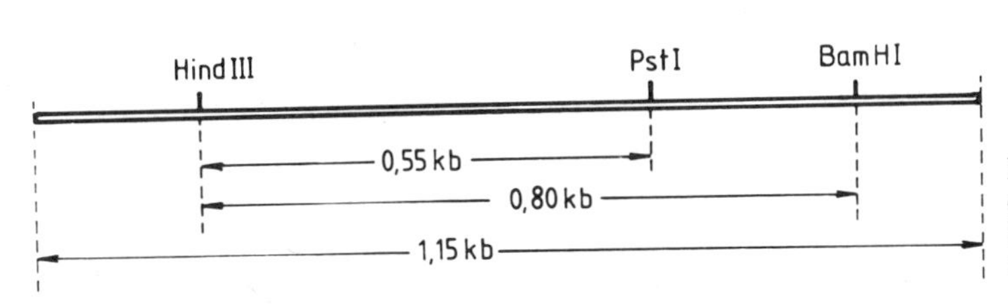

Figure 1. Map of the insertion element IS*R1* detected in *R. lupini.* IS*R1* has a length of 1.15 kb and carries recognition sites for the restriction enzymes *Hin*dIII, *Pst*I, and *Bam*HI.

The Structure of Two Typical RP4::IS*R1* Mutants

Figure 3 shows an RP4 mutant that is Ap^s, Km^s, and Tra$^-$. This mutant (RP4-5.4) carries two IS*R1* insertions: one in the area of Tn*1* and the second in the opposite region. Both insertions are accompanied by deletions. The deletion within Tn*1* has a length of 6.1 kb and eliminates a part of the Ap-resistance gene. The other one starts within the Km^r determinant and extends through the transfer region Tra2 into the transfer region Tra3. Altogether, 24.6 kb of RP4 DNA are removed, and the molecule is reduced to 57% of its original length.

This structure, obtained by restriction analysis, was confirmed by a homoduplex experiment. The fact that the two IS*R1* sequences are oriented to opposite directions results in the typical homoduplex molecule shown in part B of Figure 3. The two IS*R1* elements form a short, double-stranded snapback structure of 1.15 kb in length, and the single-stranded nonhomologous regions loop out.

An RP4 mutant carrying three IS*R1* insertions is shown in Figure 4. This plasmid (RP4-5.82) again is Ap^s, Km^s, and Tra$^-$. Ap resistance is lost by a deletion of 4.65 kb, caused by the IS*R1* insertion within Tn*1*. Km sensitivity and transfer deficiency are due to IS*R1* insertions into the corresponding gene regions.

This triple arrangement should result in two characteristic homoduplex molecules: The insertion within Tra3 can hybridize either with the inverted IS*R1* sequence within Tn*1* or with the IS*R1* element located within the Km^r determinant. Both structures could be found in the electron microscope (data not shown).

The Proteins of Tn*1*::IS*R1*

As mentioned above, all RP4 mutants tested so far carry an IS*R1* element within Tn*1* that does not affect the expression of β-lactamase. To test whether IS*R1* affects Tn*1* in some other way, we analyzed the polypeptides produced by Tn*1*::IS*R1* in *E. coli* minicells. Simultaneously, this study should demonstrate whether or not IS*R1* itself codes for some protein or carries transcription or translation signals.

The investigations (data not shown) revealed that the pattern of Tn*1*::IS*R1* differs from that of wild-type Tn*1* (Dougan et al. 1979) in two proteins, one with a molecular weight of 16,500 and the other with a molecular weight of greater than 100,000.

The interpretation of these two peptides is given in Figure 5. They can be explained by analogy to the model of Tn*3* (Chou et al. 1979a,b; Gill et al. 1979; Heffron and McCarthy 1979) since Tn*1* and Tn*3* possess about 85% base-sequence homology (Rubens et al. 1976). As we know from restriction analysis, IS*R1* is integrated about 150 bp to the right of the *Bam*HI site. We therefore assume that the insertion maps within the far right end of gene *tnpR*. This gene codes for a 19,000-m.w. protein that represses both its own transcription and the transcription of *tnpA*. We suggest that this insertion causes a premature termination of the *tnpR*-encoded repressor protein. This means that IS*R1* carries a termination signal for either transcription or translation. Instead of the 19,000-m.w. repressor, a truncated, inactive *tnpR* product with a molecular weight of 16,500 is synthesized (denoted in Fig. 5 as R*) that no longer reduces transcription of *tnpR* and *tnpA*. Thus, both gene

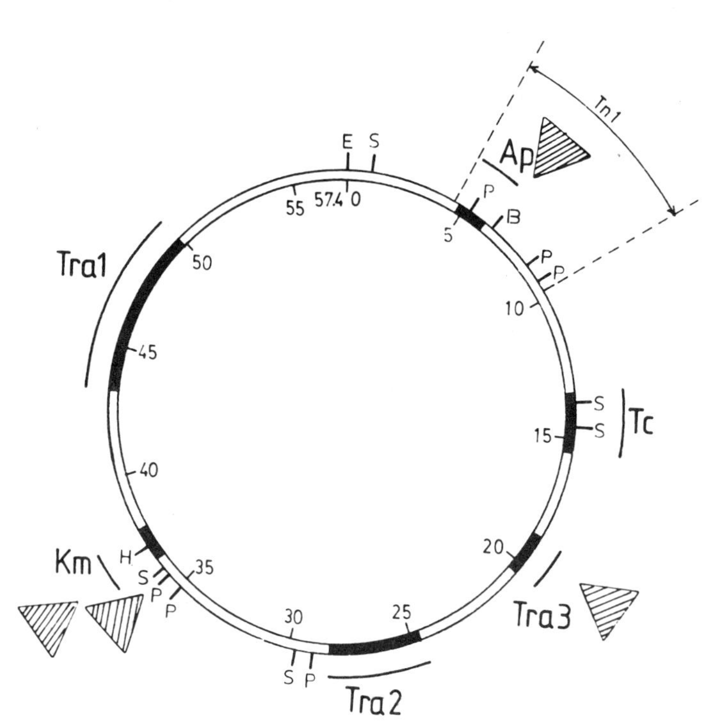

Figure 2. Map of plasmid RP4 and distribution of IS*R1* insertions. The map of RP4 is calibrated from 0 to 57.4 kb (Burkardt et al. 1979). Genetic data are taken from Figurski et al. (1976), Stanisich and Bennett (1976), Barth and Grinter (1977a,b), Barth et al. (1978), and Barth (1979). Ap, Km, and Tc refer to genes encoding resistance to ampicillin, kanamycin, and tetracycline, respectively; genes involved in conjugal transfer are indicated by Tra1, Tra2, and Tra3. The location of Tn*1* is based on the data of Heffron et al. (1975a,b) and R. Simon (pers. comm.). The positions of restriction sites for the enzymes *Eco*RI (E), *Bam*HI (B), *Hin*dIII (H), *Pst*I (P), and *Sma*I (S) are based on the data of DePicker et al. (1977), Grinsted et al. (1977), Meyer et al. (1977), and Spitzbarth (1978). The crosshatched triangles show the sites on plasmid RP4 where we have mapped IS*R1* insertions.

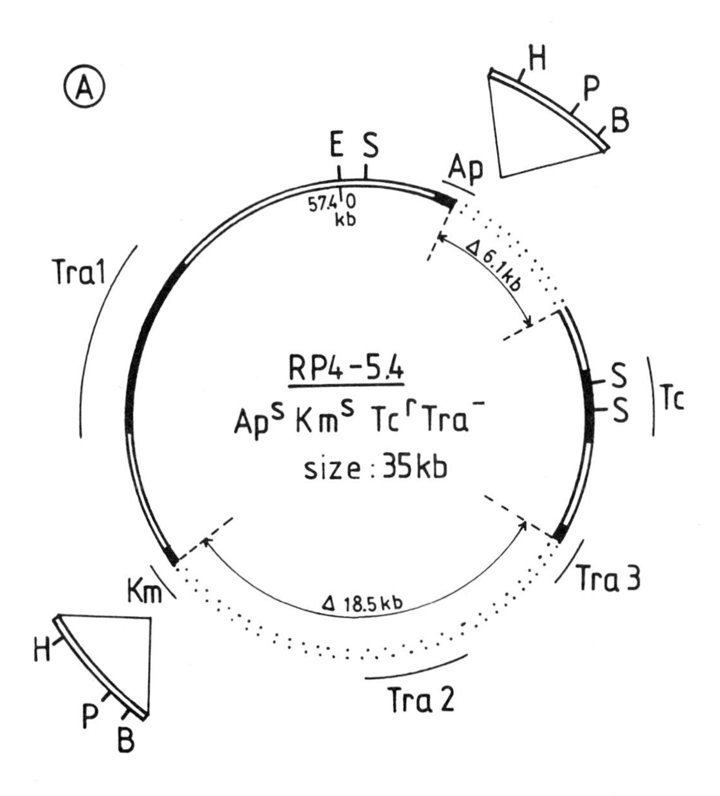

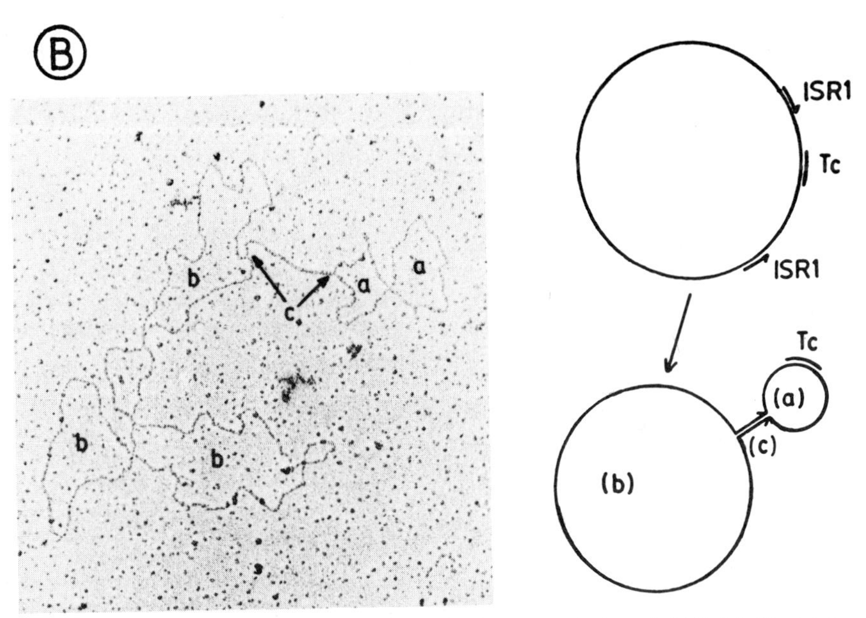

Figure 3. Structure of the mutant plasmid RP4–5.4. (*A*) The physical map of plasmid RP4-5.4. (Restriction sites and genetic markers are those of Fig. 2). Two IS*R1* sequences are integrated into the RP4 molecule, generating deletions of various extents (dotted lines). The IS*R1* elements are inserted in opposite directions, which is documented by the order of their restriction sites. (*B*) A typical homoduplex molecule of RP4-5.4 and its interpretation. The insertion elements form a stretch of double-stranded DNA (c) with two single-stranded loops (a and b).

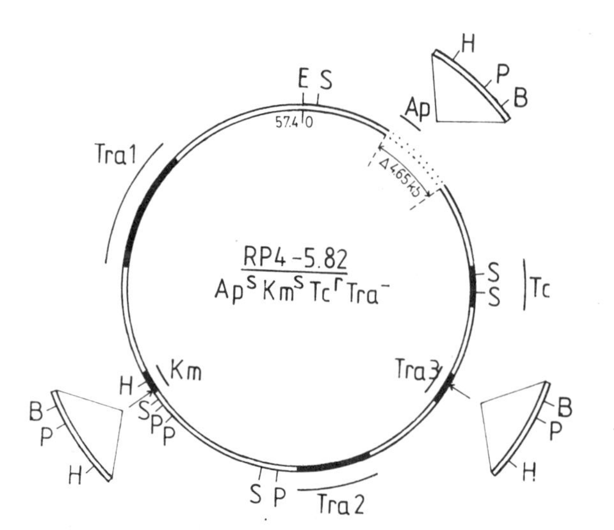

Figure 4. Structure of the mutant plasmid RP4-5.82. This mutant, which is Ap^s, Km^s, and Tra$^-$, carries three IS*R1* insertions, as determined by restriction enzyme analysis. The insertion within Tn*1* is correlated with a deletion of 4.65 kb (dotted line).

products, the truncated repressor (m.w. 16,500) and the transposase (m.w. 120,000), are synthesized in increased amounts.

This interpretation is in accord with the observation that the frequency of Tn*1*::IS*R1* transposition is 15–20-fold higher than that of wild-type Tn*1* (data not shown).

SUMMARY

The insertion element IS*R1* was isolated from the soil bacterium *R. lupini.* In this strain, IS*R1* shows a very strong affinity to plasmid RP4. It causes RP4 mutations at the strikingly high frequency of 10^{-2} to 10^{-1}, either by the integration itself or by generating deletions. In *E. coli,* IS*R1* seems to be inactive. No evidence could be obtained for a promoter site on IS*R1* or for an IS*R1*-encoded protein. Our results indicate, however, an IS*R1*-specific termination signal for either transcription or translation.

ACKNOWLEDGMENTS

This investigation was financed by grants from the Deutsche Forschungsgemeinschaft. U. P. was supported by stipends from the Studienstiftung des Deutschen Volkes. We would like to thank Mrs. Agnes Schulte for photographic aid.

REFERENCES

BARTH, P. T. 1979. RP4 and R300B as wide host-range plasmid cloning vehicles. In *Plasmids of medical, environmental, and commercial importance* (ed. K. N. Timmis and A. Pühler), p. 399. Elsevier/North-Holland Biomedical Press, Amsterdam.

BARTH, P. T. and N. GRINTER. 1977a. Map of plasmid RP4 derived by insertion of transposon C. *J. Mol. Biol.* **113:** 455.

———. 1977b. A Tn*7* insertion map of RP4. In *DNA insertion elements, plasmids, and episomes* (ed. A. I. Bukhari et al.), p. 675. Cold Spring Harbor Laboratory, Cold Spring Harbor, New York.

BARTH, P. T., N. J. GRINTER, and D. E. BRADLEY. 1978. Conjugal transfer system of plasmid RP4: Analysis by transposon *7* insertion. *J. Bacteriol.* **133:** 43.

BERINGER, J. E. 1974. R factor transfer in *Rhizobium leguminosarum. J. Gen. Microbiol.* **84:** 188.

BURKARDT, H. J., U. PRIEFER, A. PÜHLER, G. RIESS, and P. SPITZBARTH. 1979. Naturally occurring insertion mutants of broad host range plasmids RP4 and R68. In *Plasmids of medical, environmental, and commercial importance* (ed. K. N. Timmis and A. Pühler), p. 387. Elsevier/North-Holland Biomedical Press, Amsterdam.

CAMPBELL, A., D. E. BERG, D. BOTSTEIN, E. M. LEDERBERG, R. P. NOVICK, P. STARLINGER, and W. SZYBALSKI. 1977. Nomenclature of transposable elements in prokaryotes. In *DNA insertion elements, plasmids, and episomes* (ed. A. I. Bukhari et al.), p. 15. Cold Spring Harbor Laboratory, Cold Spring Harbor, New York.

CHOU, J., M. J. CASADABAN, P. G. LEMAUX, and S. N. COHEN. 1979a. Identification and characterization of a self-regulated repressor of translocation of the Tn*3* element. *Proc. Natl. Acad. Sci.* **76:** 4020.

CHOU, J., P. G. LEMAUX, M. J. CASADABAN, and S. N. COHEN. 1979b. Transposition protein of Tn*3:* Identification and characterisation of an essential repressor-controlled gene product. *Nature* **282:** 801.

DATTA, N. and R. W. HEDGES. 1972. Host ranges of R factors. *J. Gen. Microbiol.* **70:** 453.

DATTA, N., R. W. HEDGES, E. J. SHAW, R. B. SYKES, and M. H. RICHMOND. 1971. Properties of an R factor from *Pseudomonas aeruginosa. J. Bacteriol.* **108:** 1244.

DEPICKER, A., M. VAN MONTAGU, and J. SCHELL. 1977. Physical map of RP4. In *DNA insertion elements, plasmids, and episomes* (ed. A. I. Bukhari et al.), p. 678. Cold Spring Harbor Laboratory, Cold Spring Harbor, New York.

DOUGAN, G., M. SAUL, A. TWIGG, R. GILL, and D. SHERRATT. 1979. Polypeptides expressed in *Escherichia coli*

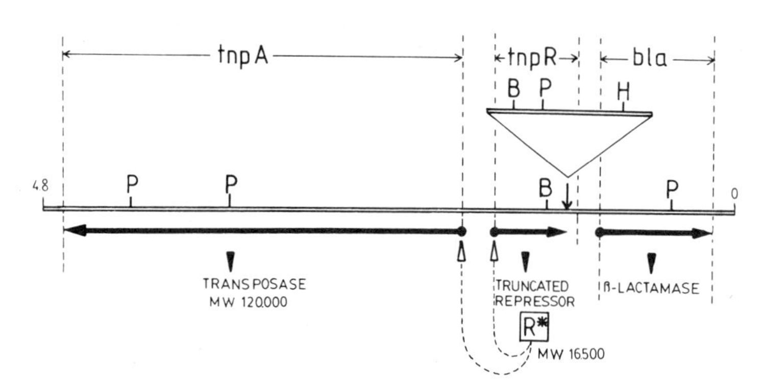

Figure 5. Model for the genetic structure of Tn*1* carrying an IS*R1* insertion (Tn*1*::IS*R1*). The length of Tn*1* is based on the data of Heffron et al. (1975a,b). The locations of the coding regions for the *β*-lactamase (*bla*), the repressor (*tnpR*), and the transposase (*tnpA*) are taken from the model of Tn*3* (Chou et al. 1979a,b; Gill et al. 1979; Heffron and McCarthy 1979). (↓) The insertion site of IS*R1*. Restriction sites: (P) *Pst*I; (B) *Bam*HI; (H) *Hin*dIII.

K-12 minicells by transposition elements Tn*1* and Tn*3*. *J. Bacteriol.* **138:** 48.

FIGURSKI, D., R. MEYER, D. S. MILLER, and D. R. HELINSKI. 1976. Generation in vitro of deletions in the broad host range plasmid RK2 using phage Mu insertions and a restriction endonuclease. *Gene* **1:** 107.

GILL, R. E., F. HEFFRON, and S. FALKOW. 1979. Identification of the protein encoded by the transposable element Tn*3* which is required for its transposition. *Nature* **282:** 797.

GRINSTED, J., P. M. BENNETT, and M. H. RICHMOND. 1977. A restriction enzyme map of R-plasmid RP1. *Plasmid* **1:** 34.

HEDGES, R. W. and A. E. JACOB. 1974. Transposition of ampicillin resistance from RP4 to other replicons. *Mol. Gen. Genet.* **132:** 31.

HEFFRON, F. and B. J. MCCARTHY. 1979. DNA sequence analysis of the transposon Tn*3*: Three genes and three sites involved in transposition of Tn*3*. *Cell* **18:** 1153.

HEFFRON, E., C. RUBENS, and S. FALKOW. 1975a. Translocation of a plasmid DNA sequence which mediates ampicillin resistance: Molecular nature and specificity of insertion. *Proc. Natl. Acad. Sci.* **72:** 3623.

HEFFRON, F., R. SUBLETT, R. HEDGES, A. JACOB, and S. FALKOW. 1975b. Origin of the TEM beta-lactamase gene found on plasmids. *J. Bacteriol.* **122:** 250.

HEUMANN, W. 1968. Conjugation in starforming *Rhizobium lupini. Mol. Gen. Genet.* **102:** 132.

MEYER, R., D. FIGURSKI, and D. R. HELINSKI. 1977. Restriction enzyme map of RK2. In *DNA insertion elements, plasmids, and episomes* (ed. A. I. Bukhari et al.), p. 680. Cold Spring Harbor Laboratory, Cold Spring Harbor, New York.

O'GARA, F. and L. K. DUNICAN. 1973. Transformation and physical properties of R-factor RP4 transferred from *Escherichia coli* to *Rhizobium trifolii. J. Bacteriol.* **116:** 1177.

RUBENS, C., F. HEFFRON, and S. FALKOW. 1976. Transposition of a plasmid deoxyribonucleic acid sequence that mediates ampicillin resistance: Independence from host *rec* functions and orientation of insertion. *J. Bacteriol.* **128:** 425.

SPITZBARTH, P. 1978. "Molekulare Charakterisierung von Insertionsmutanten des RP4-Plasmids." Ph.D. thesis, Universität Erlangen-Nürnberg.

STANISICH, V. A. and P. M. BENNETT. 1976. Isolation and characterization of deletion mutants involving the transfer genes of P group plasmids in *P. aeruginosa. Mol. Gen. Genet.* **149:** 211.

IS*1*-mediated DNA Rearrangements

H. SAEDLER, G. CORNELIS,* J. CULLUM, B. SCHUMACHER, AND H. SOMMER
Max-Planck-Institut für Züchtungsforschung, Egelspfad, 5 Köln-Vogelsang, Federal Republic of Germany;
**Unité de Microbiologie, Université de Louvain UCL30.58, B 1200 Brussels, Belgium*

When IS*1* is present in a particular DNA sequence, numerous DNA rearrangements occur in its vicinity at frequencies several orders of magnitude higher than when IS*1* is absent (Reif and Saedler 1975, 1977; Cornelis and Saedler 1980). This paper reports on investigations of IS*1*-induced deletions, transpositions of a given IS*1* to nearby sites in both orientations, and IS*1*-mediated inversions. In addition, an analysis of the regions within IS*1* and in the target sites involved in these rearrangements is presented that leads to a model for the mechanism of IS*1* integration.

DNA Rearrangements Generated by IS*1*

Many years ago, Reif and Saedler (1975) reported the formation of IS*1*-mediated deletions. In this process, IS*1* is retained and thus can promote further rounds of increased deletion formation (Reif and Saedler 1977). DNA sequence analysis showed that deletions terminate precisely at the end of the integrated IS*1* element (Ohtsubo and Ohtsubo 1978).

Recently, we showed that IS*1* can also promote inversion of adjacent genetic material (Cornelis and Saedler 1980). Such a structure is, however, more complex and seems to involve transposition of a given IS*1* element to a nearby site in inverse orientation, with concomitant inversion of the genetic material between the two inversely oriented copies of IS*1*. Figure 1 outlines a model for the formation of the observed structures. The diagram is based on suggestions made by Faelen et al. (1975), Bennett et al. (1977), and Grindley and Sherratt (1979) and is outlined more explicitly by Shapiro (1979) and Arthur and Sherratt (1979). The model of Figure 1 gives rise to four different products. However, because IS*1* occurs in several copies in the *Escherichia coli* chromosome (Saedler and Heiss 1973) and because it is not possible to differentiate between these copies, it is difficult to interpret unambiguously the relative frequencies of the different IS*1*-mediated rearrangements.

Recently, however, we described a system that avoids this difficulty by using an IS*1* foreign to the *E. coli* genome and that is distinguishable from the *E. coli* IS*1* elements by its lack of a *Pst*I cleavage site (Cornelis and Saedler 1980). This system exploits transposon Tn*951*, first isolated in *Yersinia enterocolitica* (Cornelis et al. 1978). As shown in Figure 2, the 16.6-kb Tn*951* transposon harbors an IS*1* element at position 3.1–3.9 kb, immediately in front of the *lacI, Z,* and *Y* genes. Tn*951* was transposed to plasmid pDG1 (Ghosal and Saedler 1977), and one particular isolate was chosen to transform a Δ(*lac*) strain (Cornelis et al. 1980). In this strain, Y^- mutations in the *lac* operon of Tn*951* can be isolated with a frequency of about 10^{-4}. The majority of these were shown to be deletions by their inability to revert to Lac$^+$. In addition, five random Lac$^-$ isolates from Tn*951* were checked by DNA restriction analysis and found to contain deletions extending from the resident IS*1* of Tn*951* into the *lac* operon and beyond (Cornelis and Saedler 1980). However, with appropriate screening (using Xgal in the selective plates), numerous Y^- derivatives were found that were still *lacZ*$^+$. Plasmids isolated from these derivatives exhibit three further classes of IS*1*-induced rearrangements: direct transpositions, inverse transpositions, and inversion coupled with inverse transposition. Ten *lacZ*$^+$ Y^- isolates were picked at random. They all reverted to Lac$^+$, although at different frequencies, and in only one isolate could reversion be induced by ICR191A. This particular isolate therefore probably contains a frameshift mutation, this being supported by the fact that no change in molecular weight is seen in the DNA fragments generated by triple digests with *Hin*dIII, *Eco*RI, and *Bam*HI or in *Pvu*II digests. Each of the other nine isolates showed an altered restriction pattern, and Southern blotting experiments revealed that each carried a second IS*1* element. The second IS*1* element in the Tn*951*-*lacY*$^-$ derivatives lacked a *Pst*I restriction site, like the parental IS*1* on Tn*951*. Hence, they seem to derive from the parental IS*1*-*951* allele. Transposition of IS*1* in inverse orientation relative to the resident IS*1*-*951* is most easily demonstrated by electron microscopic analysis of snapback structures. The length of the double-stranded stem is consistent with the length of IS*1*, and the loop length corresponds to the distance between IS*1*-*951* and the position of the inverted IS*1* copy in *lacY*.

Transposition of IS*1* in direct orientation does not give snapback structures, but the integration site of IS*1* can easily be mapped by restriction analysis and Southern blots. Structures involving simple transposition of IS*1* in inverse orientation could be distinguished from those in which the segment between the two IS*1* elements was inverted with the help of the single assymmetric *Eco*RI restriction site at 8.5 kb; in the latter instance, the molecular weights of two DNA fragments are affected (see Fig. 2), whereas only one fragment is affected in the former case. Of the nine *lacZ*$^+$$Y^-$ mutations studied, five were due to the integration of IS*1* in direct orientation, two were due to integration in inverse orientation, and the remaining two showed IS*1*-mediated inversions (Fig. 2). This indicates that at least those three classes of IS*1*-mediated rearrangements occur with similar frequencies; IS*1*-induced deletions are usually much more frequent. However, the area in which the deletion end points can occur to give a Y^-

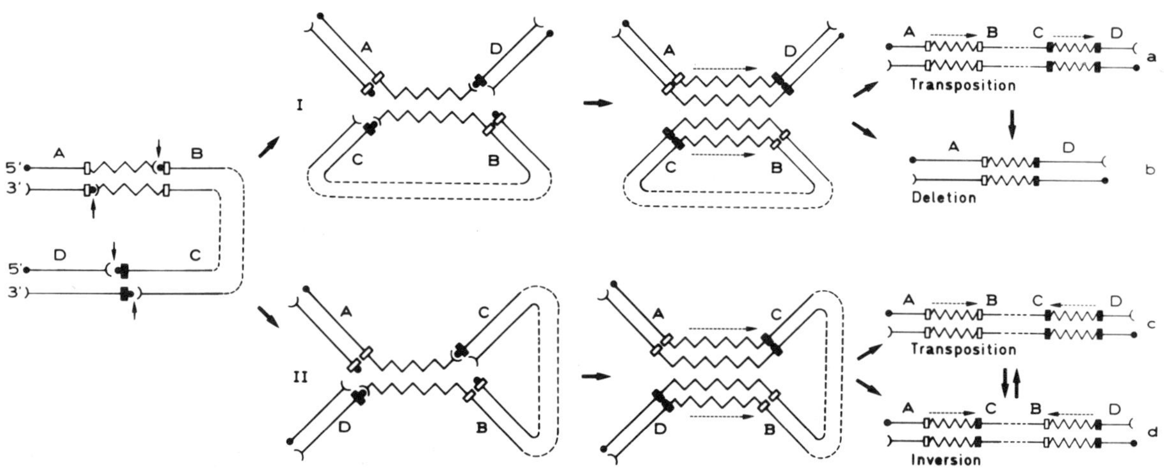

Figure 1. Intramolecular model for transposition, deletion formation, and inversion (based on model of Shapiro [1979]). The first step involves single-strand endonucleolytic cuts at both ends of IS*1* and a 9-bp staggered nick at the new target site. The resulting free IS*1* ends are then joined to the DNA ends produced at the target site; this is possible in either of two orientations (upper and lower diagrams). Replication across IS*1* produces the structures shown in the third step. A site-specific recombination across the two IS*1* copies then produces transposition in direct (*a*) or inverse (*c*) orientation. In the absence of this recombination step, the upper diagram shows loss of the BC segment to produce a deletion (*b*); in the lower diagram an inversion results (*d*).

phenotype is not limited to the *lacY* gene, and therefore a higher frequency of deletions, as compared with other events, is to be expected.

Among six *lac*⁻ mutations isolated previously, five turned out to be due to deletions, two of which terminated around the only *Eco*RI site within Tn*951*. The remaining *lac*⁻ mutation was shown to be due to an IS*1*-mediated inversion close to the *Eco*RI site (Cornelis and Saedler 1980). Therefore, deletions with end points in the same area as the other IS*1*-mediated events are probably no more abundant than these other events.

To summarize, the four products described in Figure 1 are formed with similar frequencies in a given system.

To study various parameters in the reactions, the for-

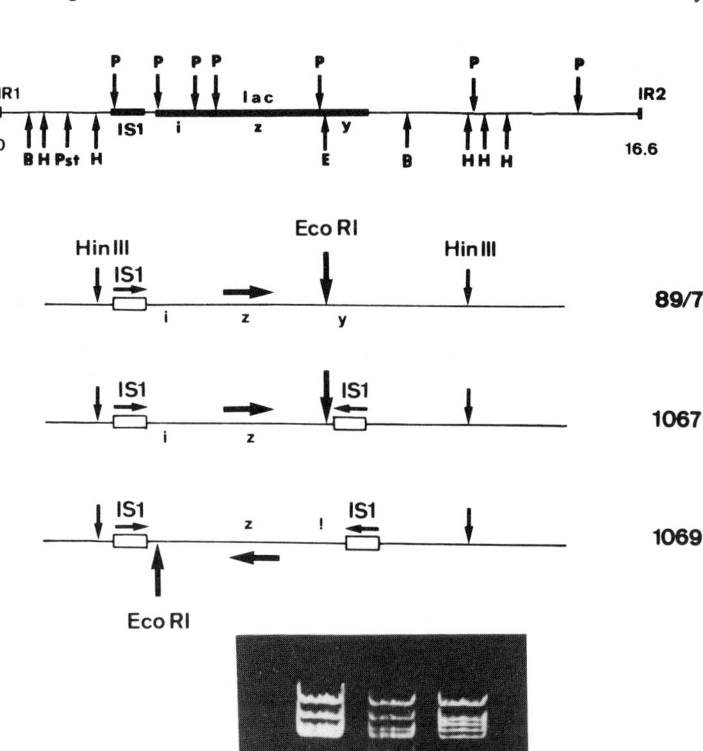

Figure 2. IS*1* transpositions and IS*1*-mediated inversions in Tn*951*. The top line shows a restriction map of Tn*951* for the enzymes *Pvu*II (P), *Bam*HI (B), *Hin*dIII (H), *Pst*I (Pst), and *Eco*RI (E). The *Hin*dIII and *Eco*RI cuts in the vicinity of the *lac* operon are shown for the parent strain 89/7 (pDG1-7::Tn*951*) and two derivatives carrying an IS*1* either in inverse orientation (1067) or in inverse orientation with an inversion (1069). The *Hin*dIII, *Eco*RI double-digest fragments of these three strains are shown in the photograph of the agarose gel at the bottom of the figure.

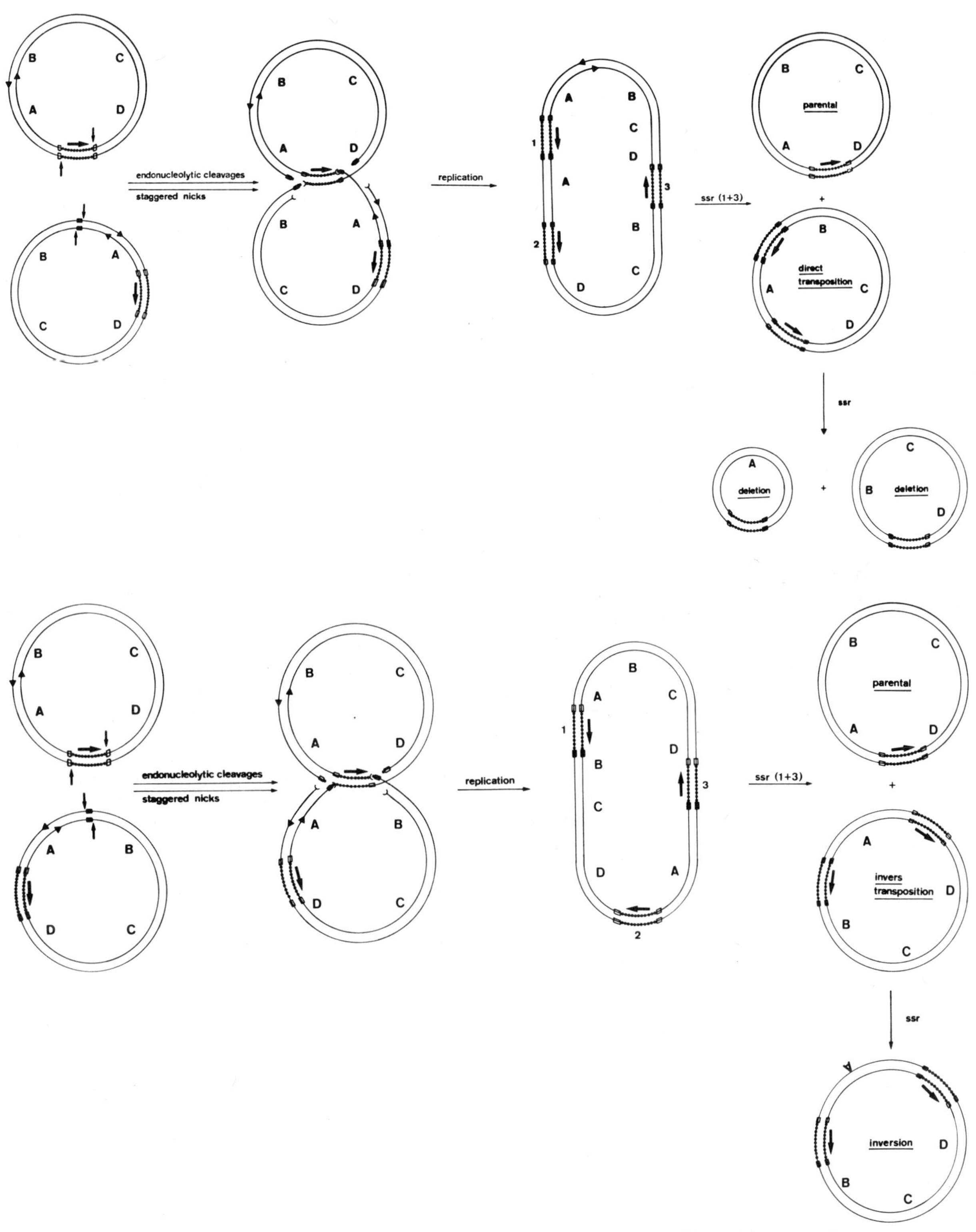

Figure 3. Intermolecular model for transposition, deletion formation, and inversion. This model assumes the same steps as those in Fig. 1. However, the IS*1* and the target site are assumed to be carried on two copies of the same plasmid rather than on the same DNA molecule. (*Top*) The steps leading to transposition in direct orientation; (*bottom*) those for inverse orientation.

mation of deletions was chosen as a representative reaction because it is easy to monitor. The system used for these studies employs IS*1* elements integrated into the control region of the *gal* operon in either possible orientation. As with Tn*951*, the substrate for IS*1*-induced deletion formation is located on plasmid pDG1.

Sites within IS*1* Required for Deletion Formation

Since pDG1 is a multicopy plasmid, both intermolecular reactions between identical molecules (Fig. 3) and intramolecular reactions (Fig. 1) can occur. Although the same end products are formed, the pathways of intra- and intermolecular reactions differ. For example, it can be seen from Figure 1 that no recombinational step seems to be involved in deletion formation by the intramolecular pathway, whereas for intermolecular reactions the fusion products have to be resolved by recombination. In both pathways deletion formation requires endonucleolytic cleavages at both ends of the integrated IS*1* element. Therefore, one could expect that disruption of the distal end of IS*1*, which from the structure of the deletions is not obviously involved, would prevent IS*1*-mediated deletions.

Figure 4 gives the frequencies of deletions extending into the *galE* gene in an IS*1* derivative in which IS*2* had integrated in orientation II about 20 bp inward from the *galE* distal end of IS*1*. This results in a considerable reduction in deletion formation.

Also included in Figure 4 are the deletion frequencies observed in a strain carrying an internal deletion in IS*1* that removes the *Pst*I site; both ends are intact, as revealed by DNA sequence analysis, but, nonetheless, IS*1*-mediated deletion formation is severely reduced.

Deletion formation, therefore, seems to require not only the end of IS*1* at which the new fusion sequence is formed, but also the distal end of the element. This is in agreement with the models outlined in Figures 1 and 3. The effect of the internal IS*1* deletion could be due to the loss of a site needed for recombination in the model (Fig. 3). If this interpretation is true, then the majority of deletions in this system, where IS*1* is carried on the plasmid pDG1, are probably via the intermolecular pathway, which is recombination-dependent.

DNA Guiding Sequences in IS*1* Integration

IS*1*-induced deletions terminate precisely at the end of the integrated element (Ohtsubo and Ohtsubo 1978). This recombinational event is independent of *recA* (Reif and Saedler 1975) and does not occur between short homologous sequences; nonetheless, homologous sequences seem to be involved in this process (see Fig. 5). A (*chlD-pgl*) deletion (128/9) was isolated from pDG1 *galOP*-128::IS*1*. Both the junction DNA sequence and the sequence of the target site in the parental strain were determined.

As shown in Figure 5, the deletion starts precisely at the end nucleotide of IS*1* and does not involve recombination between short homologous sequences. However, there do exist noteworthy sequence relations between the target site and both ends of IS*1*, as indicated by the arrows in Figure 5. A direct repeat of a sequence in the right end of IS*1* occurs to the left of the cutting site in the target area. In addition, two other sequences in the left end of IS*1* are seen as direct repeats to the right of the cutting site in the target area. Thus, these target sequences might be used as guiding sequences in the deletion event that results in the 128/9 sequence, which still retains one pair of the crosswise repeats at the target site and the distal end of IS*1*. These findings led us to suggest that, in the transposition process, small sequence homologies between the target area and the ends of IS*1* guide this element to its new position of integration (see Fig. 6). Since the lengths of the sequence homologies are only 4–6 bp, the specificity of this process is low, which is in agreement with the observed low specificity of IS*1* integrations. The model for IS*1* integration given in Figure 6 also accounts for the observation that IS*1* apparently prefers AT-rich regions for integration (Devos et al. 1979; Meyer et al. 1980), since the IS*1* termini are themselves AT-rich (60–70%). Although the model predicts the orientation of integrations at a given target site, the sequence homologies are small, and therefore other pairs might exist at the target under question that can guide IS*1* into the opposite orientation of integration also. The 11 published DNA sequences for IS*1* and Tn*9* integration sites and the 2 Tn*9*-induced deletion sequences (Calos and Miller 1980) are compatible with the guiding-sequence model. However, further testing is needed, for example, by sequencing numerous independent isolates from genuine hotspots.

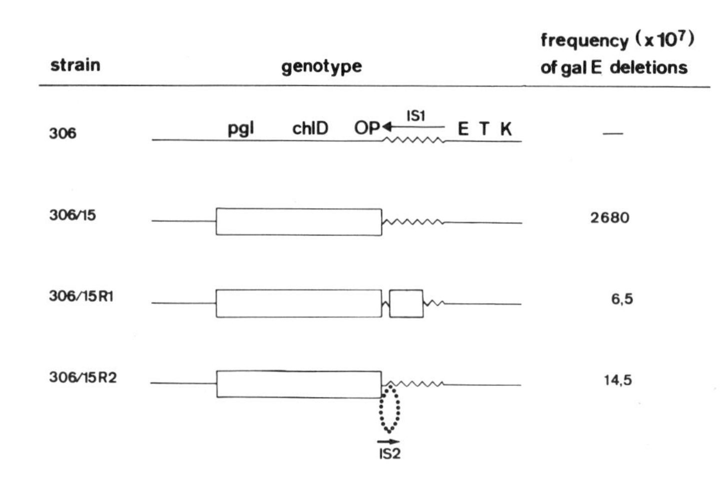

Figure 4. IS*1*-mediated deletion formation; effect of mutations in IS*1*. All of the derivatives shown are carried on the plasmid pDG1. The starting IS*1* insertion was *galOP*-306::IS*1* (Saedler et al. 1972). The (*chlD-pgl*) deletion 306/15 was selected; this makes secondary deletions into *galE* at a high frequency. Two derivatives of this strain, 306/15R1 and 306/15R2, had IS*1* mutations caused by an internal deletion in IS*1* and an IS*2* insertion into IS*1*, respectively. These formed deletions at a much lower frequency than 306/15 as shown.

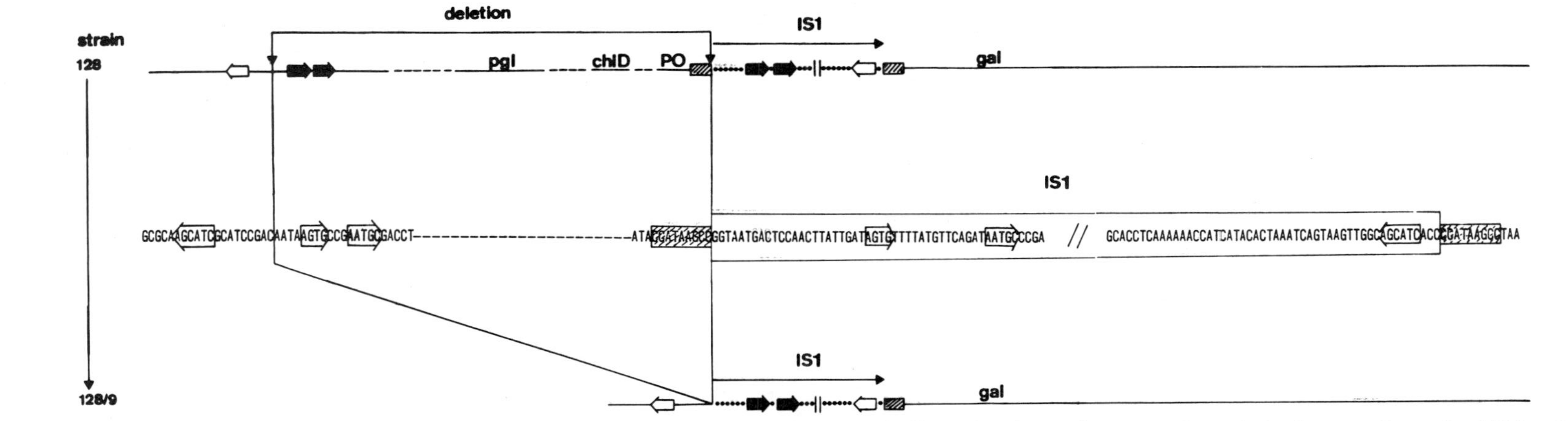

Figure 5. Short sequence homologies observed in the formation of the deletion 128/9. The arrows indicate the observed sequence homologies between the ends of IS*1* and the end point of the deletion.

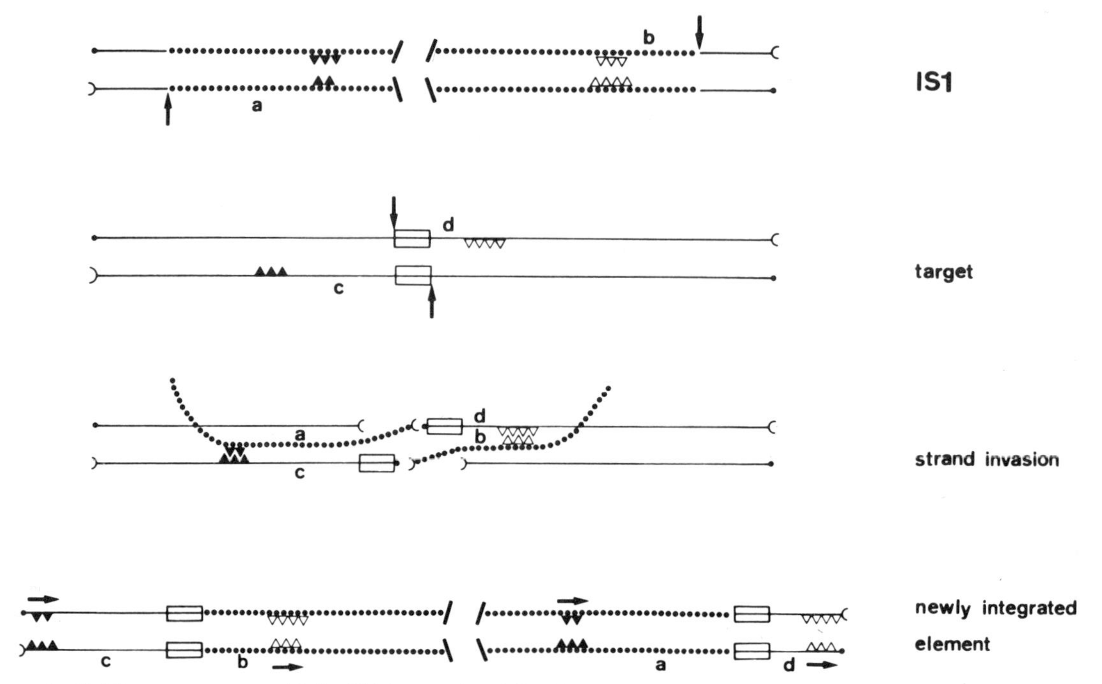

Figure 6. Guiding-sequence model for IS*1* integration. The filled and open triangles indicate the homology between the ends of IS*1* and sequences to the left and right of the new target. Possible base-pairing at these sites during strand invasion would lead to selection of the target site as shown. (*a* and *b, c* and *d*) The distances of the small homologies from the ends of IS*1* and from the 9-bp region at the target site, respectively. These distances can vary. However, the first pair are within the inverted-repeat termini of IS*1,* whereas the others can overlap the 9-bp region.

ACKNOWLEDGMENTS

We thank Dr. Pat Nevers for many useful discussions. This work was supported by the Deutsche Forschungsgemeinschaft, a NATO research grant (no. 1564), and Fonds de Developpement Scientifique of the University of Louvain. J. C. was supported by a Royal Society European Programme Fellowship.

REFERENCES

Arthur, A. and D. Sherratt. 1979. Dissection of the transposition process: A transposon-encoded site-specific recombination system. *Mol. Gen. Genet.* **175:** 267.

Bennett, P. M., J. Grinsted, and M. H. Richmond. 1977. Transposition of TnA does not generate deletions. *Mol. Gen. Genet.* **154:** 205.

Calos, M. P. and J. H. Miller. 1980. Molecular consequences of deletion formation mediated by the transposon Tn*9*. *Nature* **285:** 38.

Cornelis, G. and H. Saedler. 1980. Deletions and an inversion induced by a resident IS*1* of the lactose transposon Tn*951*. *Mol. Gen. Genet.* **178:** 367.

Cornelis, G., D. Ghosal, and H. Saedler. 1978. Tn*951:* A new transposon carrying a lactose operon. *Mol. Gen. Genet.* **160:** 215.

Cornelis, G., U. Meier, G. Hoeksma, D. Ghosal, J. Cullum, and H. Saedler. 1980. The lactose transposon Tn*951:* Characterization of transposition. *Ann. Microbiol.* **131A:** 233.

Devos, R., R. Contreras, J. van Emmelo, and W. Fiers. 1979. Identification of the translocatable element IS*1* in a molecular chimera constructed with plasmid pBR322 DNA into which a bacteriophage MS2 DNA copy was inserted by the poly(dA)·poly(dT) linker method. *J. Mol. Biol.* **128:** 621.

Faelen, M., A Toussaint, and J. de la Fonteyne. 1975. Model for the enhancement of λ-*gal* integration into partially induced Mu-1 lysogens. *J. Bacteriol.* **121:** 873.

Ghosal, D. and H. Saedler. 1977. Isolation of the mini-insertions IS*6* and IS*7* of *E. coli. Mol. Gen. Genet.* **158:** 123.

Grindley, N. D. F. and D. Sherratt. 1979. Sequence analysis at IS*1* insertion sites: Model for transposition. *Cold Spring Harbor Symp. Quant. Biol.* **43:** 1257.

Meyer, J., S. Iida, and W. Arber. 1980. Does the insertion element IS*1* transpose preferentially into A+T-rich DNA segments? *Mol. Gen. Genet.* **178:** 471.

Ohtsubo, H. and E. Ohtsubo. 1978. Nucleotide sequence of an insertion element, IS*1*. *Proc. Natl. Acad. Sci.* **75:** 615.

Reif, H.-J. and H. Saedler. 1975. IS*1* is involved in deletion formation in the *gal* region of *E. coli* K12. *Mol. Gen. Genet.* **137:** 17.

———. 1977. Chromosomal rearrangements in the *gal* region of *E. coli* K12. In *DNA insertion elements, plasmids, and episomes* (ed. A. I. Bukhari et al.), p. 81. Cold Spring Harbor Laboratory, Cold Spring Harbor, New York.

Saedler, H. and B. Heiss. 1973. Multiple copies of the insertion-DNA sequences IS*1* and IS*2* in the chromosome of *E. coli* K12. *Mol. Gen. Genet.* **122:** 267.

Saedler, H., J. Besemer, B. Kemper, B. Rogenwirth, and P. Starlinger. 1972. Insertion mutations in the control region of the *gal* operon of *E. coli.* I. Biological characterisation of the mutations. *Mol. Gen. Genet.* **115:** 258.

Shapiro, J. A. 1979. Molecular model for the transposition and replication of bacteriophage Mu and other transposable elements. *Proc. Natl. Acad. Sci.* **76:** 1933.

Genetic Organization of Tn*5*

S. J. Rothstein,* R. A. Jorgensen,† J. C.-P. Yin, Z. Yong-di,‡
R. C. Johnson, and W. S. Reznikoff
Department of Biochemistry, College of Agricultural and Life Sciences, University of Wisconsin, Madison, Wisconsin 53706

A number of bacterial transposable elements have been found that encode resistance to various antibiotics. They are of interest both for their importance in the proliferation of bacterial populations resistant to therapeutically useful antibiotics and as model systems for studying the process of transposition itself. To facilitate the understanding of the transposition process, we have chosen to analyze the gene organization of one transposon, the transposable drug-resistance element Tn*5*.

The transposon Tn*5* is a genetic complex approximately 5700 bp in length. It is composed of two inverted repeats, each 1500 bp long, which bound a unique 2700-bp central region (Berg et al. 1975; Jorgensen et al. 1979; E. A. Auerswald and H. Schaller, pers. comm.). Tn*5* DNA is known to encode the following functions: (1) the enzyme neomycin phosphotransferase II (NPT II),[1] which results in resistance to several aminoglycoside antibiotics such as kanamycin (Km) and neomycin (Nm) (Berg et al. 1978), and (2) transposition functions (Meyer et al. 1979; Rothstein et al. 1980). In describing the organization of Tn*5*, we extend results presented previously (Rothstein et al. 1980) and present a sequence analysis of matched 130-bp regions of both inverted repeats known to be important for determining the level of NPT-II production and the level of transposition. The complete sequence of Tn*5* has been determined by E. A. Auerswald and H. Schaller (pers. comm.).

The existence of a detailed restriction map of Tn*5* (Fig. 1) has allowed us to use recombinant DNA technology to generate a variety of specific mutations. Many of these are described in Figure 2. Phenotypic analysis of antibiotic resistance and transposition properties of these mutations can be useful in determining where different functions are located. They can also be analyzed with regard to protein-coding functions by using the minicell system. The restriction map is also useful in that it allows us to localize in vitro the RNA polymerase binding sites and transcription initiation points. Finally, it allows the sequence analysis of selected regions. All of these approaches have been utilized to achieve a definition of the genetic organization of Tn*5*.

Present address: * Plant Breeding Institute, Maris Lane, Trumpington, Cambridge CB2 2LQ, England; † Carnegie Institution of Washington, 290 Panama Street, Stanford, California 94305. ‡ Chengtu Institute of Biology, Chengtu Branch, Chinese Academy of Sciences, Chengtu, Sichuan, The People's Republic of China.

[1] NPT II is also named aminoglycoside 3′-phosphotransferase II.

RESULTS

Genetic Determinants for Nm Resistance

The structural gene for NPT II. The structural gene for NPT II lies within a 1275-bp region bounded by the *Hin*dIII site in the left[2] inverted repeat and the *Sma*I site in the unique central region. This has been determined by studying the Nm^r phenotype of cells carrying various mutant Tn*5* DNAs and by examining the ability of mutant DNAs to code for NPT-II protein in minicells. The left-hand side of the NPT-II gene is defined by the deletion plasmid pRZ111 and the insertion plasmid pRZ172. Cells carrying these DNAs manifest wild-type levels of Nm resistance (Table 1) and code for equivalent amounts of the NPT-II protein (Fig. 3). On the basis of the localization of the NPT-II promoter (discussed below) and the low, but significant, levels of Nm resistance encoded by the *Bgl*II insertion plasmid pRZ135, we believe that the NPT-II structural gene actually starts to the right of the *Bgl*II site, but this has not been shown rigorously. The right-hand boundary of the NPT-II gene must be to the left of the *Sma*I site (shown in Fig. 1), since the deletion plasmid pRZ152 codes for a normal level of Nm resistance (Table 1) and for an NPT-II protein of the correct size (see Fig. 3).

The promoter for the NPT-II gene. Mutant plasmids that contain insertions or other genetic alterations starting at the *Bgl*II site in the left inverted repeat (e.g., pRZ135) code for significantly lower levels of Nm resistance (Table 1). This observation is most easily explained by assuming that the promoter for the NPT-II gene lies between the *Hin*dIII and *Bgl*II sites of the left inverted repeat. The RNA polymerase binding sites have been mapped in Tn*5* through the use of a filter-binding assay that measures the retention of restriction fragments on nitrocellulose by RNA polymerase (Rothstein et al. 1980) (see Fig. 1). These studies indicate that the *Hin*dIII-*Bgl*II fragment of the left inverted repeat binds to RNA polymerase in a specific heparin-resistant fashion, as would be expected for a promoter-containing fragment.

It is of interest to determine whether the comparable sequence from the other inverted repeat might not also

[2] To facilitate our description of Tn*5* gene organization, we will use the convention of naming the inverted repeats as "left" or "right" depending on their orientations in Figures 1, 2, 4, and 5.

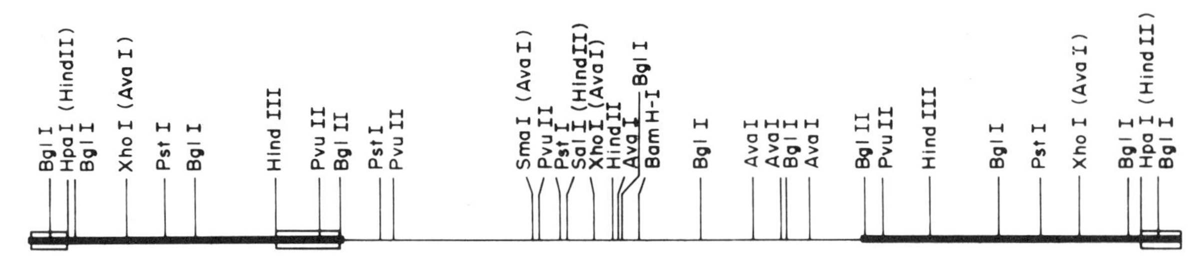

Figure 1. Restriction map of Tn*5*, described previously by Jorgensen et al. (1979). (▭) DNA fragments that were found to form heparin-resistant complexes with RNA polymerase (Rothstein et al. 1980).

be able to perform the same function if suitably placed. This has been tested by examining the properties of the *Bgl*II inversion pRZ141. This mutation codes for substantially lower levels of Nm resistance and protein NPT II (see Table 1 and Fig. 3), suggesting that the right inverted repeat cannot substitute for the left inverted repeat in its promoter function. In agreement with this, RNA polymerase binding studies fail to show any binding activity for the right *Bgl*II-*Hin*dIII fragment.

The difference in promoter activity between the matched *Hin*dIII-*Bgl*II regions has also been studied by sequence analysis. We have determined the sequence of 130 bp proceeding outward from the *Bgl*II sites in both inverted repeats. The results are shown in Figure 4. This sequence agrees in its entirety with that determined by E. A. Auerswald and H. Schaller (pers. comm.). Starting at 105 bp from the *Bgl*II site and continuing up to 56 bp from the *Bgl*II site in the left inverted repeat, one can identify a sequence that contains many of the conserved sites found for promoters, as described by Rosenberg and Court (1979) (see Fig. 4b). We propose that this is the NPT-II promoter. The right inverted repeat contains the same sequence with a single-base-pair difference. This difference (a G/C instead of a T/A) is also indicated in Figure 4, a and b. As shown, one of the most highly conserved sites in the model promoter sequence has been altered by this difference (see Fig. 4b); thus, it

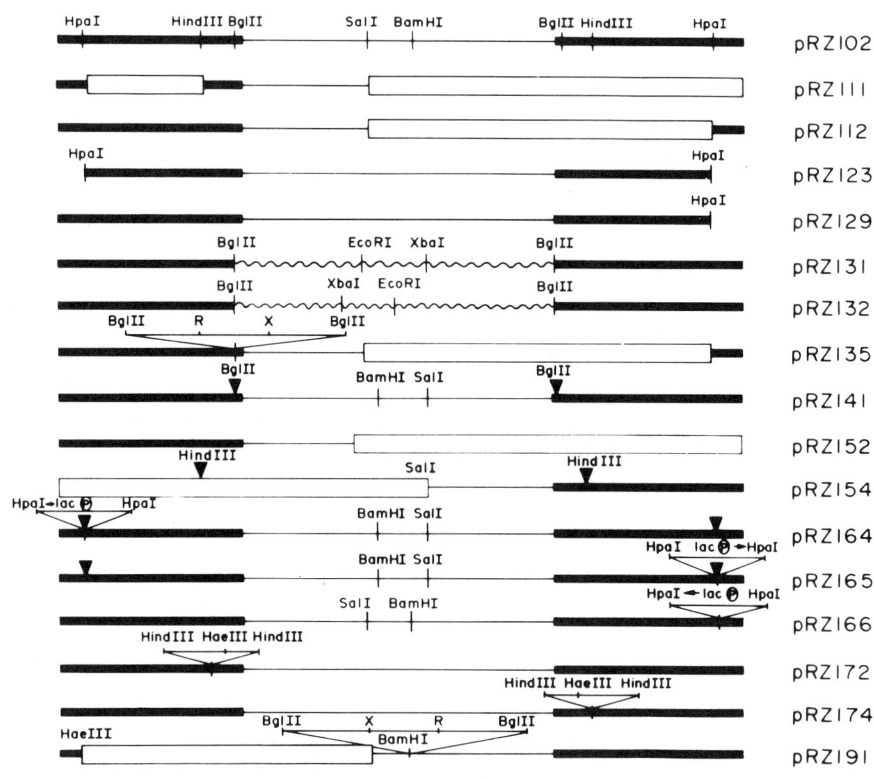

Figure 2. Structures of Tn*5* mutations. These mutations were constructed using recombinant DNA techniques. The techniques used and the physical analyses of the resulting structures have been described previously (Jorgensen et al. 1979; Rothstein et al. 1980; S. J. Rothstein et al., in prep.). There are four general classes of mutations: deletions (▭), substitutions (∿∿∿), insertions (▽), and inversions (▼▼). For the inversions, the closed triangles denote the inversion end points, and the *Bam*HI and *Sal*I sites are in reverse order compared with the wild-type Tn*5*. The directions of transcription of the lactose promoter in insertion mutations pRZ164, pRZ165, and pRZ166 are indicated by arrows. The insertion in pRZ131, pRZ132, pRZ135, and pRZ191 is a *Bgl*II fragment from Tn*10*, which encodes tetracycline (Tc) resistance.

Table 1. Nm Resistance of Tn*5* Mutants in C600-SF8

Plasmid present	EOP_{50} (μg/ml neomycin)
pRZ102	90
pRZ112	75
pRZ135	2
pRZ141	10
pRZ152	85
pRZ172	130

Nm^r phenotypes encoded by Tn*5* mutant plasmids. EOP_{50} values are the concentrations of neomycin required to kill 50% of cells carrying the indicated plasmids, which were determined as described by Rothstein et al. (1980).

is not surprising that the *Hin*dIII-*Bgl*II fragment of the left inverted repeat can function as a better promoter than the *Bgl*II-*Hin*dIII fragment of the right inverted repeat.

Genetic Determinants of Transposition Functions and Proteins Encoded by the Inverted Repeats

The right inverted repeat codes for functions required for transposition and for two proteins. All mutations that alter the integrity of the right inverted repeat eliminate detectable transposition of Tn*5*. These mutations include pRZ112 and pRZ174, which are described in Figure 2. They were tested by examining the frequency of transposition of the Nm-resistance gene from the relevant plasmids onto λ*bbnin.* The data are presented in Figure 5. Meyer et al. (1979) have found that the transposition defect in a mutation similar to pRZ112 can be complemented in *trans;* therefore, these mutations prevent the synthesis of a diffusible product.

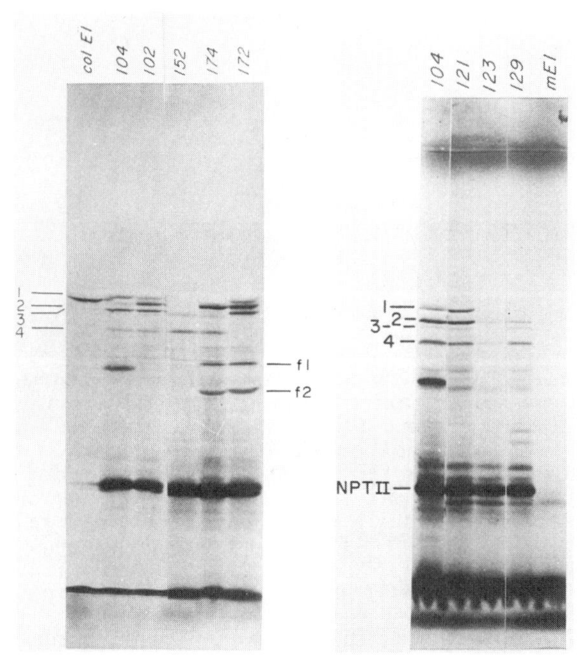

Figure 3. Minicell analysis of Tn*5* proteins. Minicells containing the indicated plasmids were prepared as described previously (Roozen et al. 1971; Rothstein et al. 1980). [^{35}S]Methionine-labeled polypeptides were analyzed by electrophoresis on a 12.5% SDS-polyacrylamide gel (Laemmli 1970; Rothstein et al. 1980). (1, 2, 3, 4) Proteins encoded by the inverted repeats. Polypeptide 3 is often very faint and at times difficult to see. The colicin protein in ColE1 is shortened in extracts from strains containing Col E1::Tn*5*, since the transposon is inserted in the colicin gene. In pRZ102, the colicin protein electrophoreses between polypeptides 1 and 2, whereas in pRZ104 it is considerably smaller. Fusion proteins synthesized in pRZ172 and pRZ174 are denoted f1 and f2.

When examined in the minicell system (see Fig. 3), plasmids pRZ112 and pRZ174 fail to code for two Tn*5*-specific peptides, proteins 1 and 2, with apparent molecular weights of 58,000 and 54,000. Thus, these proteins are obvious candidates for the required transposition function lost in thcsc mutations. Sincc thcsc protcins arc modulated in concert in most mutations affecting the right inverted repeat (including the *Hin*dIII-inversion–*Sal*I-deletion pRZ154) and since their combined molecular weight greatly exceeds the coding capacity of the inverted repeat, we conclude that these proteins are encoded by overlapping sequences. It is possible that one protein is a processed product of the other, although no evidence exists to suggest this.

The minicell analysis of various mutant plasmids suggests that the proteins encoded by the right inverted repeat (Tn*5* proteins 1 and 2) are translated from right to left, with their N termini being near the *Hpa*I site near the outside edge of the inverted repeat and the difference in their molecular weights being due to differences in their N termini. Deletions such as those carried on plasmids pRZ123 and pRZ129, which cut from the outside edge of the transposon up to the *Hpa*I site, appear to abolish synthesis of protein 1 but permit low, but detectable, levels of protein-2 synthesis (Fig. 3). Although the failure to synthesize protein 1 is compatible with either orientation for the relevant gene, the altered level of protein-2 synthesis is most easily explained by assuming that the controlling elements for its gene have been removed by these deletions. Analysis of the *lacP-O* insertion mutation carried on pRZ166 confirms this hypothesis. In this case, a 789-bp *Hin*dII fragment has been inserted into the *Hpa*I site with the promoter facing inward. Protein 2 and a fusion protein are made, and their syntheses are under the control of *lac* regulatory signals (i.e., sensitive to catabolite repression and *lac* repressor control) (data not shown).

The genetic and protein-coding data described above predict the existence of a promoter between the *Hpa*I site and the end of the right inverted repeat. In fact, RNA-polymerase–DNA filter-binding studies have shown that a potential promoter does exist in this region (Rothstein et al. 1980) (see Fig. 1) and that DNA fragments from this region do stimulate transcription initiation in a purified in vitro system.

The C termini of proteins 1 and 2 must lie beyond the *Bgl*II cut site, since plasmids carrying substitutions of the internal *Bgl*II fragment (pRZ131 and pRZ132) synthesize longer fusion peptides instead of proteins 1 and 2 (data not shown). Mutation pRZ132 is one exception to the generalization that synthesis of proteins 1

(a)

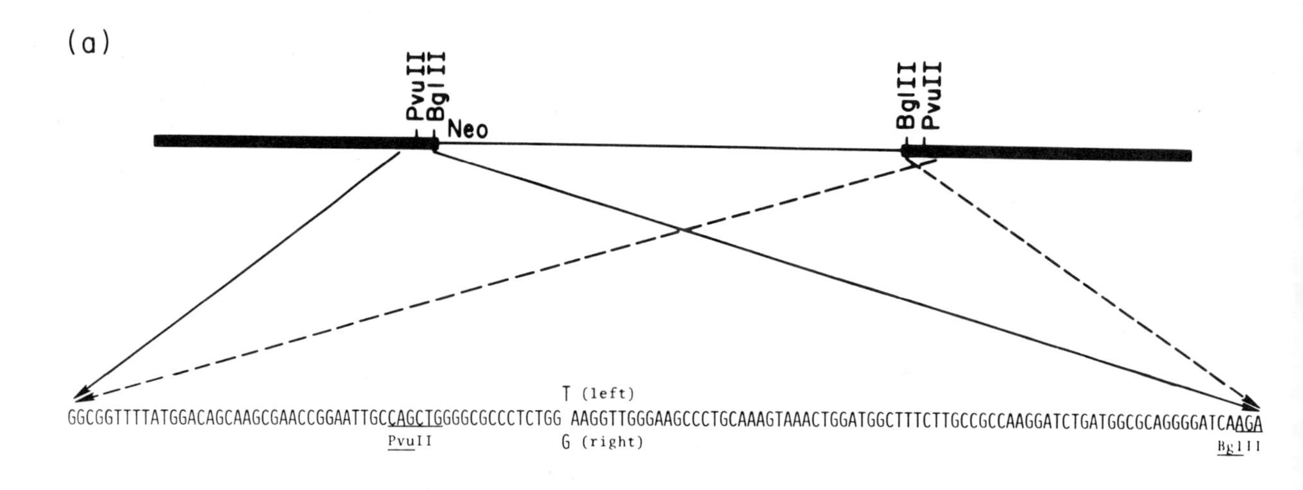

(b)

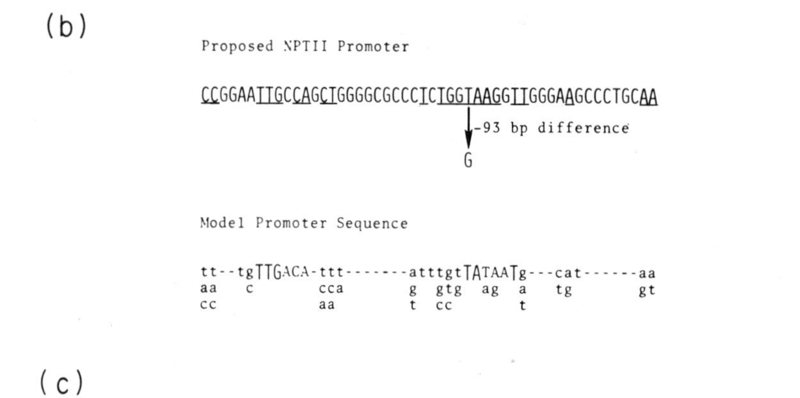

(c)

IR Proteins and the -93 bp Difference

```
          UAA (left)
... CUC UGG    GGU UGG GAA GCC CUG CAA AGU AAA CUG GAU GGC UUU CUU GCC GCC AAG GAU CUG AUG GCG CAG GGG AUC AAG AUC U(GA) ...
          GAA (right)

          STOP(left)
... Leu Trp    Gly Trp Glu Ala Leu Gln Ser Lys Leu Asp Gly Phe Leu Ala Ala Lys Asp Leu Met Ala Gln Gly Ile Lys Ile STOP
          Glu (right)
```

Figure 4. (*a*) Sequence of the *Pvu*II-*Bgl*II regions from the left and right inverted repeats. The sequences of 130 bp proceeding outward from the *Bgl*II sites in both inverted repeats were determined using techniques developed by Maxam and Gilbert (1977) and Sanger et al. (1977) and will be described elsewhere by Rothstein et al. (in prep.). For simplicity, only one strand is shown. From extended sequence data determined by E. A. Auerswald and H. Schaller (pers. comm.), the centers of the *Bgl*II sites are known to be 17 bp from the internal ends of the inverted repeats. There is a single-base-pair nonhomology 76 bp from the *Bgl*II sites or (as indicated in the figure) 93 bp from the internal ends of the inverted repeats. (*b*) Proposed NPT-II promoter. Within the 130-bp left inverted repeat shown in *a*, one can identify a sequence that bears a strong resemblance to the model promoter sequence described by Rosenberg and Court (1979). This model sequence is indicated, and its similarities to the proposed NPT-II promoter are underlined. The −93-base pair T/A-G/C nonhomology destroys the similarity to a highly conserved site in the model sequence. (*c*) Peptide-coding sequence. A proposed reading frame for the proteins encoded by the inverted repeats is deduced from the sequence data of E. A. Auerswald and H. Schaller (pers. comm.), as are the 2 bp beyond the *Bgl*II site (in parentheses). The −93 T/A base pair on the left generates a UAA nonsense codon, whereas on the right the comparable position has a GAA (Glu) codon. The termination codon for the proteins encoded by the right inverted repeat is deduced from the E. A. Auerswald and H. Schaller (pers. comm.) sequence data.

and 2 is required for transposition, since it does show significant transposition (J. Shapiro; D. Berg; S. J. Rothstein et al.; all unpubl.). Presumably, the fusion peptides maintain some enzymatic activity. The exact end point for these peptides can be deduced from the sequence data of E. A. Auerswald and H. Schaller (pers. comm.) and is presented schematically in Figure 4c. The stop codon for the proteins encoded by the right inverted repeat (proteins 1 and 2) includes the last base pair of the *Bgl*II recognition sequence and is within the inverted repeat. This conclusion corrects our previous suggestion (Rothstein et al. 1980) that the genes for the right inverted repeat might extend into the unique central region.

The left inverted repeat differs from the right inverted repeat in protein-coding capacity and transposition functions. The transposition and protein-coding functions

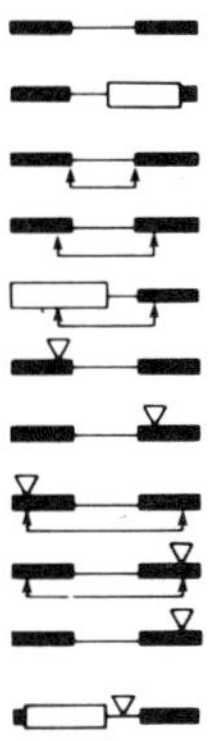

Plasmid present	Structure	Tn5 proteins synthesized					Fusion proteins synthesized	% Transposition of wt
		1	2	3	4	NPTII		
102		+	+	+	+	+	—	100%
112		-	-	+	+	+	—	<0.5%
141		+	+	+	+	<	—	n.d.
143		+	+	+	+	+	—	100%
154		-	-	+	+	+	—	n.d.
172		+	+	-	-	+	38K + 34K	27%
174		-	-	+	+	+	38K + 34K	<0.5%
164		-	<	+	+	+	—	<0.5%
165		+	+	-	-	+	—	21%
166		-	<	+	+	+	~79K	<0.5%
191		+	+	-	-	-	—	50%*

Figure 5. Proteins encoded by and transposition frequencies of Tn*5* mutants.

of the left inverted repeat differ from those of the right inverted repeat in the following ways:

1. Mutations altering the integrity of the left inverted repeat (but leaving the terminus of the inverted repeat intact) perturb the frequency of transposition but do not prevent its occurrence. (The data for plasmid pRZ172 are shown in Fig. 5.) It is not known why these mutations alter transposition frequencies (preliminary structural studies suggest that it is not due to a failure to resolve cointegrate structures and that the transposed structures are identical with the starting structures), but it is clear that transposition can occur in the absence of an intact left inverted repeat. This is most obvious with regard to pRZ191, which deletes all but the terminal 117 bp of the left inverted repeat and encodes resistance to tetracycline rather than to neomycin. It still transposes at high levels (see Fig. 5).
2. Although many features of the protein-coding functions of the left inverted repeat are similar to those of the right inverted repeat (e.g., the existence of two proteins encoded by overlapping sequences, the location of the promoter, and the location of the N termini), there is an obvious difference: They have lower apparent molecular weights (53,000 and 49,000 as opposed to 58,000 and 54,000; see results for pRZ172 in Fig. 3). As described below, the difference in transposition function is most probably explained by the differences in protein-coding functions, and these differences are related to different C termini for the two groups of proteins. Furthermore, this difference can be explained by the same single-base-pair nonhomology between the two inverted repeats that generates a promoter found on the left but not on the right.

One approach to understanding the difference between the two inverted repeats is to analyze the properties of double mutations that include an inversion originating at defined sites in the two inverted repeats in combination with an insertion or a deletion mutation in one of the new "hybrid" inverted repeats. Examples of plasmids carrying such mutations include pRZ154, pRZ164, and pRZ165. Analyses of their properties are shown in Figures 3 and 5. The conclusion is that inversions at the *Hpa*I or *Hin*dIII sites invert the protein-coding and transposition functional differences between the two inverted repeats (the hybrid left repeat codes for proteins 1 and 2 and is required for transposition, whereas the hybrid right repeat codes for proteins 3 and 4 and is not required for transposition). Thus, the sequences from the ends of the inverted repeats to the *Hin*dIII sites are interchangeable.

The difference in protein-coding capacity between the two inverted repeats can be localized to the matched *Hin*dIII-*Bgl*II fragments and is due to the fact that the C termini of proteins 3 and 4 occur in this region, as distinct from where the termini of proteins 1 and 2 are located. In the case of the *Bgl*II inversion carried on plasmid pRZ141, the minicell analysis indicates that the inversion has not altered the protein-coding capacity. In plasmids carrying substitutions of the internal *Bgl*II fragment (pRZ131 and pRZ132), proteins 3 and 4 are still made, indicating that they terminate prior to the *Bgl*II site (unpubl.).

Analysis of the sequence data presented in Figure 4 suggests an explanation for the observed differences. The same base-pair difference that generates a promoter in the left inverted repeat also generates a UAA nonsense codon not found in the right inverted repeat (see Fig. 4c). From their extended sequence data, E. A. Auerswald and H. Schaller (pers. comm.) suggested that this nonsense codon would be located in the correct reading frame for terminating proteins 3 and 4. A prediction of this hypothesis is that mutations affecting the right inverted repeat should be transposition-proficient in an ochre suppressor (*supB*) genetic background. Preliminary results indicate that this prediction is correct.

CONCLUSION

A model for the genetic organization of Tn*5* is shown in Figure 6. There are five polypeptides that have been found to be encoded by Tn*5*, two being translated from

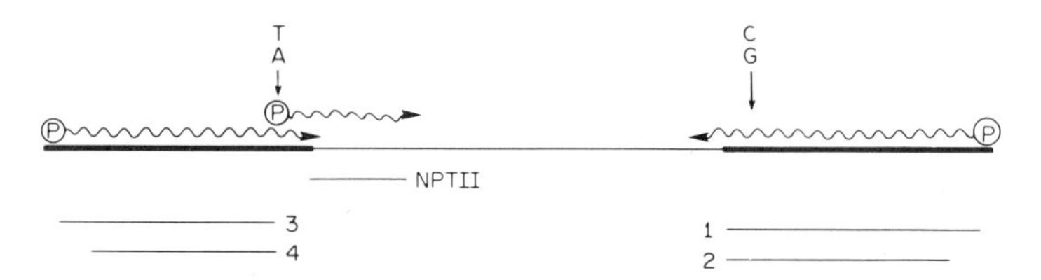

Figure 6. The genetic organization of Tn*5*. The locations of the inverted repeats and NPT-II promoters, the proteins encoded by Tn*5*, and the single-base-pair nonhomology are shown. One or both of the proteins encoded by the right inverted repeat (1 and 2) are known to be required for the transposition process.

each inverted repeat. The fifth is NPT II. The difference in the coding capacity between the two repeats is caused by the single-base-pair difference shown in Figures 4 and 6. The T/A base pair in the left repeat results in the presence of a nonsense codon, accounting for the foreshortening of its coded polypeptides. The active polypeptide(s) for transposition are encoded by the right inverted repeat.

The other major difference between the two inverted repeats involves their abilities to promote resistance to high levels of neomycin. When an inversion places the resistance gene adjacent to the right inverted repeat, then the resistance of the cell to neomycin is lowered considerably. This difference in the repeats can be explained by the same base-pair difference that affects their polypeptide-coding capacity. The inside edge of the left repeat has a sequence that matches the model promoter sequence of Rosenberg and Court (1979) quite well, as shown in Figure 4b. The single-base-pair difference in the two repeats occurs at one of the most highly conserved bases in the generalized promoter sequence.

It is possible to construct a tranposon from Tn*5* that is of the Tn*3* "type" with short inverted repeats (Heffron et al. 1975). An example is pRZ191, which has 117-bp inverted repeats (most of the left repeat is deleted) and still transposes at a high frequency. The only regions of the left repeat that appear to be absolutely necessary for the functions of the transposon are its two ends. The outside edge is presumably needed to provide a substrate for the transposase enzyme; the inside end carries the promoter for the Nm-resistance gene.

The genetic structure of Tn*5* as it is known today may tell us something about its genetic evolution. It has been suggested that the inverted repeats associated with a transposon are derived originally from a pair of homologous insertion sequences (MacHattie and Jackowski 1977; Kleckner 1977). If this were the case in Tn*5*, then it would appear that Tn*5* must have undergone at least two changes. One change would be a point mutation in the left repeat, creating a promoter sequence. The second change would be a deletion starting at the inside edge of the left repeat, which would remove the original promoter for the NPT-II gene and thereby place this gene under the control of the "new" inverted-repeat promoter.

There are several areas of inquiry with respect to Tn*5* that we are actively pursuing. One involves analyzing the Tn*5* promoters for their exact in vitro start sites and for the possibility that one or more of them are regulated in vivo. Another involves finding out exactly where each translation start site is located. We are also searching for point mutations in Tn*5* that will alter the level of transposition and that will be useful in further defining its genetic structures. Finally, we wish to determine how protein 1 and/or protein 2 acts in the transposition process.

ACKNOWLEDGMENTS

We are grateful to Heinz Schaller for communicating his results prior to publication. This work was supported by grant GM-19670 from the National Institutes of Health and grant PCM79-10686 from the National Science Foundation. J. Y. was supported by National Institutes of Health training grant GM-07133. Z. Y. is a visiting scholar from The People's Republic of China. R. C. J. was supported by National Institutes of Health training grant GM-07215. W. S. R. was supported in part by a career development award (GM-30970) from the National Institutes of Health and was the recipient of a Harry and Evelyn Steenbock Career Development Award.

REFERENCES

Berg, D. E., R. Jorgensen, and J. Davies. 1978. Transposable kanamycin-neomycin resistance determinants. In *Microbiology—1978* (ed. D. Schlessinger), p. 13. American Society for Microbiology, Washington, D.C.

Berg, D. E., J. Davies, B. Allet, and J. Rochaix. 1975. Transposition of R-factor genes to bacteriophage λ. *Proc. Natl. Acad. Sci.* **72:** 3628.

Heffron, F., R. Sublett, R. Hedges, A. Jacob, and S. Falkow. 1975. Origin of the TEM beta-lactamase gene found on plasmids. *J. Bacteriol.* **122:** 250.

Jorgensen, R. A., S. J. Rothstein, and W. S. Reznikoff. 1979. A restriction enzyme cleavage map of Tn*5* and location of a region encoding neomycin resistance. *Mol. Gen. Genet.* **177:** 65.

Kleckner, N. 1977. Translocatable elements in procaryotes. *Cell* **11:** 11.

Laemmli, U. 1970. Cleavage of structural proteins during the assembly of the head of bacteriophage T4. *Nature* **227:** 680.

MacHattie, L. and J. B. Jackowski. 1977. Physical structure and deletion effects of the chloramphenicol resistance element Tn*9* in phage lambda. In *DNA insertion elements, plasmids, and episomes* (ed. A. I. Bukhari et al.), p. 219. Cold Spring Harbor Laboratory, Cold Spring Harbor, New York.

Maxam, A. and W. Gilbert. 1977. A new method for sequencing DNA. *Proc. Natl. Acad. Sci.* **74:** 560.

Meyer, R., G. Boch, and J. Shapiro. 1979. Transposition of DNA inserted into deletions of the Tn*5* kanamycin resistance element. *Mol. Gen. Genet.* **171:** 7.

Roozen, D., R. Fenwick, and R. Curtiss III. 1971. Synthesis of ribonucleic acid and protein in plasmid-con-

taining minicells of *Escherichia coli* K-12. *J. Bacteriol.* **107:** 21.
ROSENBERG, M. and D. COURT. 1979. Regulatory sequences involved in promotion and termination of RNA transcription. *Annu. Rev. Genet.* **13:** 319.
ROTHSTEIN, S. J., R. A. JORGENSEN, K. POSTLE, and W. S. REZNIKOFF. 1980. The inverted repeats of Tn*5* are functionally different. *Cell* **19:** 795.
SANGER, F., S. NICKLEN, and A. R. COULSON. 1977. DNA sequencing with chain-terminating inhibitors. *Proc. Natl. Acad. Sci.* **74:** 5463.

Structural Analysis of Tn*5*

E.-A. AUERSWALD,* G. LUDWIG, AND H. SCHALLER
Mikrobiologie, Universität Heidelberg, Im Neuenheimer Feld 230, 6900 Heidelberg, Federal Republic of Germany

Transposon Tn*5* is a movable DNA element of about 5.3 kbp that carries resistance to the aminoglycoside antibiotics kanamycin (Km) and neomycin (Nm) (Berg et al. 1975) and that also seems to encode proteins that participate in the transposition reaction (Rothstein et al. 1980). Tn*5* has two 1.5-kb terminal inverted-repeat (IR) sequences that flank a central unique region of about 2.3 kb. From an evolutionary point of view, this structure can be envisaged to have evolved from two identical insertion sequence (IS) elements that were originally able to move independently but that lost this ability after fusing with the resistance gene to give rise to a larger transposable unit. To obtain more insight into its structural organization and to understand the rules governing the recombinational events that lead to integration and loss of the transposon, we have analyzed the structure of Tn*5*, in particular the inverted repeats, and the adjacent host DNA sequences by DNA sequencing.

To facilitate this analysis, we transferred Tn*5* into bacteriophage fd, which has a small, well-defined genome (6.408 kb; Beck et al. 1978) and which, as a filamentous phage, tolerates the insertion of relatively large DNA elements (Herrman et al. 1978). The large fraction of transposon DNA in the chimeric phage is best visualized by an electron micrograph of fd::Tn*5* DNA (Fig. 1). fd viral DNA is single-stranded. Therefore, the transposon DNA forms a stem-loop structure, which covers close to half of the fd::Tn*5* genome. Three fd::Tn*5* isolates that were able to multiply without fd helper phage were used in this study.

Sites of Tn*5* Integration in Phage fd

Mapping and sequence analysis of the fd DNA–Tn*5* junctions (Auerswald 1979, as quoted in Schaller 1979) showed that Tn*5* had integrated into fd DNA at two closely spaced sites in the intergenic region (IG) of the phage genome: twice at position 5550/51 (fd::Tn*5*-33; fd::Tn*5*-42) and once at the nearby position 5545/46 (fd::Tn*5*-30). Since there are also other positions in the intergenic region of fd that tolerate insertion of foreign DNA fragments (Schaller 1979; Neugebauer 1980; Herrman et al. 1980), this suggested that this part of fd DNA may be a hotspot for Tn*5* integration in the IG of phage fd. Sequence comparison of this region with the ends of Tn*5* (Auerswald 1979) and with integration sites of Tn*5* in plasmid pBR322 (E.-A. Auerswald and J. Collins, unpubl.) did not reveal any significant sequence homology that could be responsible for an enhanced probability of Tn*5* integration.

Comparison of the nucleotide sequences at the two different integration sites (Fig. 2) allowed an unequivocal definition of the ends of Tn*5*. It also showed that 9 bp from fd DNA, next to the integration sites, had been duplicated, a result that at the time of its discovery contributed to the general acceptance of the phenomenon of host DNA duplication at the site of the integration of a movable DNA element (Plasmid Workshop, Berlin, April 1978).

Loss of Tn*5* from fd::Tn*5*

Tn*5*, like other transposons, is "lost" from genomes with high frequency (Berg et al. 1975; Kleckner 1977). Neither the relation of this reaction to transposition nor its mechanism is understood. The fd system provides the possibility for rapid enrichment of such revertants, since phages with shortened genomes multiply at increased rates (Herrman et al. 1978, 1980). Km-sensitive revertants are therefore present in high proportion in cultures of fd::Tn*5* after multiple passages. Sixty such revertants were isolated from separate cultures of an *E. coli recA* strain. When analyzed by electrophoresis on 1% agarose gels for the size of their viral DNA, all contained DNA molecules very similar to that of fd wild type. This result indicates that essentially all Tn*5* sequences were deleted with the loss of the Km-resistance gene from fd::Tn*5* and suggests that a precise excision of the transposon might have occurred. Further fine analysis was therefore carried out with 20 of these pseudo-wild-type phages by electrophoresis of *Hae*III-digested viral (single-stranded) DNA on 6% polyacrylamide gels. As shown in Figure 3, at this higher level of resolution distinct differences were observed between the various isolates with respect to the sizes of the DNA fragments that carried the fd DNA–Tn*5* junction (fragments *Hae* E+H and *Hae* E+H+D). Out of 20 revertants analyzed by this method, 7 showed a slight increase in size (up to 100 nucleotides), indicating that Tn*5* sequences were still present. Five showed an overall loss of DNA of up to 80 nucleotides, indicating that fd sequences had been deleted, whereas 8 revertants again could not be distinguished clearly from fd wild type. However, when 3 of these latter strains (R30-8, R30-32, and R33-3) were analyzed further by DNA sequencing, no evidence for a precise excision event was found. As shown in Figure 4, each DNA molecule still contained a short but variable piece from the very end of the left inverted repeat of

* Present address: Institut für Biochemie, Bayer AG Forschungszentrum, 5600 Wuppertal 1, Federal Republic of Germany.

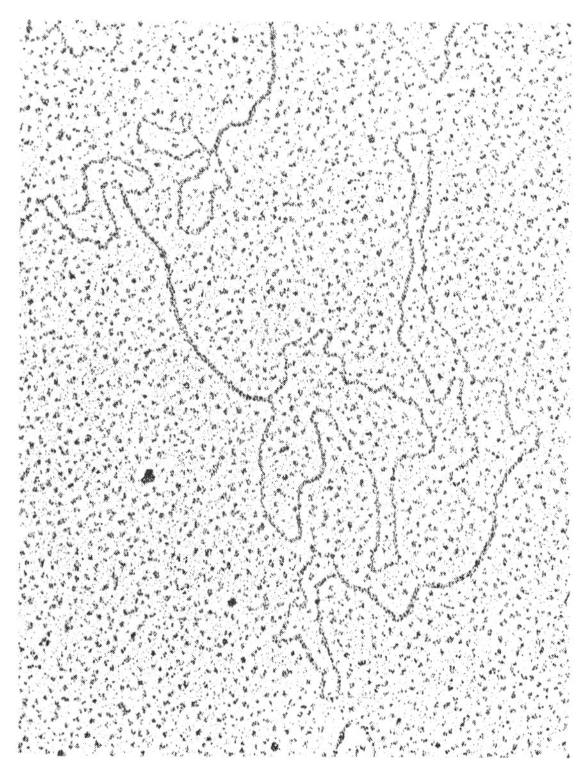

Figure 1. Electron micrograph of DNA from phage fd:Tn*5*-30. The large single-stranded loop is fd DNA; the double-stranded stem with the small loop is formed from Tn*5* DNA by annealing of the two inverted-repeat sequences. Courtesy of H. Zentgraf (German Cancer Research Center, Heidelberg).

Tn*5*, an increase that was partly compensated by a loss of DNA sequences from the phage DNA. A similar structure was also observed in phage R30-42, which contained a shortened genome as already indicated by restriction analysis (Fig. 3). A quantitative evaluation of the overall analysis of the reversion products thus allows us to conclude that precise excision is at best a rare event in the pathway(s) that leads to a loss of Tn*5* from fd DNA.

A more detailed analysis of the results above gives some insight into the rules governing the excision of Tn*5* from fd DNA. In all four revertants (R30-8, R30-32, R33-3, and R30-42), loss of Tn*5* is accompanied by the deletion of the right fd DNA–Tn*5* junction. These deletions include the 9-bp duplication of the host DNA and extend as far as 71 bp into the fd genome. As indicated in Figure 4, new junctions between fd DNA and the remaining Tn*5* sequences are formed only at regions with substantial sequence homology. The same octanucleotide AGCGYCCT common to positions 5560 in fd and 19′ in Tn*5* is utilized for recombination in both R30-8 and R33-3, although their parent phages differ in the site of Tn*5* integration (Fig. 4).

Other Tn*5* sequences from around position 40′ are linked at homologous sequences in two different sites in fd DNA: one around position 5590, giving rise to R30-32; the other at position 5620, yielding R30-42. The latter phage shows the highest extent of sequence homology (7 nucleotides) at the site where crossover between Tn*5* and fd occurs. Others, like R30-32, show as few as three consecutive homologous bases at the exact site of crossover but are framed by further homologous nucleotide sequences. Thus, loss of Tn*5* from fd DNA seems to be caused by a *recA*-independent recombination system that can utilize a low level of sequence homology and that could also lead to a precise excision event by recombination between the terminal 9-bp duplications of host DNA. Such a recombination system may be activated by a sequence-specific nicking by a transposase protein that acts close to the ends of the transposon. Analogous recombination events between two homologous heptanucleotides in fd DNA that lead to amplification and loss of the origin of DNA replication also seem to be initiated by the introduction of single-strand breaks at these sites (Schaller 1979).

Nucleotide Sequences from Tn*5*

Sequence analysis of Tn*5* was carried out preferentially with the inverted repeats, since these seem to carry the essential information for transposition, i.e., the terminal nucleotide sequence(s) recognized by the transposase system and possibly also the transposase gene(s). The double-stranded snapback DNA from fd::Tn*5* single strands provided a ready source for the DNA sequencing, which was done according to the method of Maxam and Gilbert (1977). It was used either after cleavage by restriction endonucleases (e.g., *Hae*III, *Hin*fI, *Hin*dII, *Hin*dIII) in the intact, otherwise single-stranded fd::Tn*5* molecule or after removal of the single-stranded parts of the fd::Tn*5* DNA by extensive digestion with nuclease S1 using the conditions described by Shenk et al. (1975). This approach yields nucleotide

```
fd::Tn5-30

5531              5551      10           10'           5551      5571
GGTTACGCGCAGCGTGA CCGCTACAC CTGACTCTTA.... Tn5 ....TAAGAGTCAG CCGCTACAC TTGCCAGCGCCCTAGCGC

fd::Tn5-33 und fd::Tn5-42

5531                 5551      10          10'           5551      5571
GGTTACGCGCAGCGTGACCGCT ACACTTGCC CTGACTCTTA... Tn5 ...TAAGAGTCAG ACACTTGCC AGCGCCCTAGCGCC
```

Figure 2. Duplication of 9 bp at two integration sites of Tn*5* in fd DNA (Auerswald 1979). The positions in the fd sequence are taken from Beck et al. (1978).

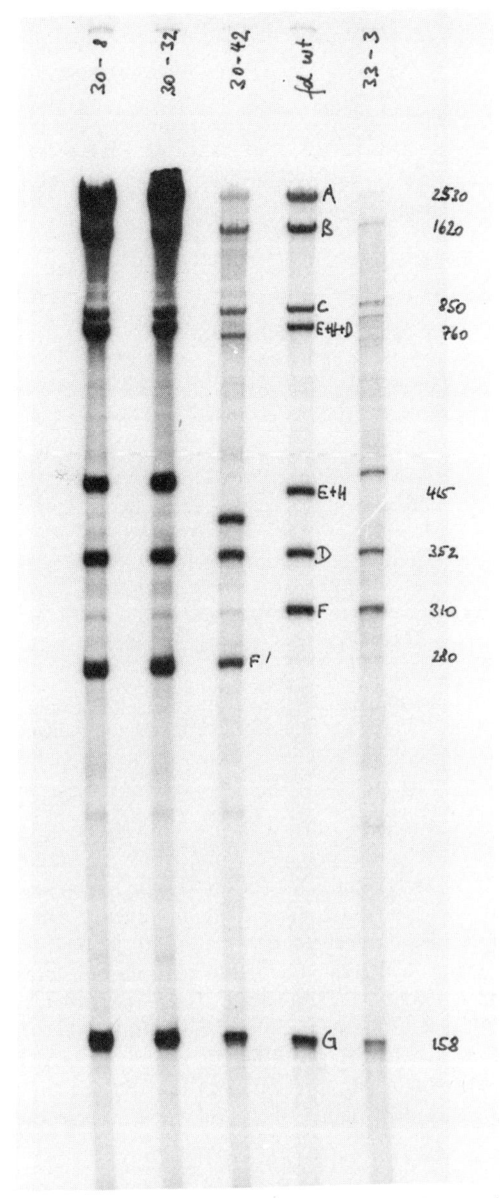

Figure 3. Size changes in the DNAs from Km-sensitive revertants of fd::Tn*5* phages relative to fd wild type. Single-stranded viral DNAs were digested with restriction endonuclease *Hae*III and separated by polyacrylamide gel electrophoresis on a 6% sequencing gel. Fragments E+H and E+H+D contained the Tn*5* integration sites. The change of fragment F to F′ in derivatives from fd::Tn*5*–30 is caused by the deletion of 30 nucleotides in viral gene III and is unrelated to the loss of Tn*5* from the intergenic region (Auerswald 1979).

sequences from one strand of each IR sequence only, and it is therefore based on the assumption that both IRs contain identical nucleotide sequences in an inverted array. This seemed indeed to be the case since no fragmentation or internal nicking of the IR snapback DNA was observed during nuclease-S1 digestion, which was so extensive as to "nibble" about 50 bp from either end of the IR and to degrade base-complementary hairpin structures in fd DNA that are fairly resistant to nuclease-S1 digestion (Gray et al. 1978). Nevertheless, a single-base change was detected between the two IR sequences, indicating that in contrast to what has been suggested by Shenk et al. (1975), nuclease S1 is unable to act on single-base mismatches in a DNA double strand. To obtain sequences from either DNA strand of a single IR, we have recently started to analyze deletion mutants from fd::Tn*5* that contain only one of the two IRs. The sequence runs obtained and most of the restriction sites used for sequencing are presented in Figure 5, and the resulting composite IR sequences are given in Figure 6. These data also include neighboring sequences from either side of the unique central region.

According to this analysis (Figs. 5 and 6), the length of the Tn*5* IR sequence is 1534 bp. The size of the unique central region is now estimated to be about 2.7 kb, which increases the total size of Tn*5* to about 5.7 kb. For the 85% of the IR sequences that can be compared at present, both IR sequences are identical except for a single-base change at position 1442, where a T/A base pair in IR_L is replaced by a C/G base pair in IR_R. This change, which was first noted by Rothstein et al. (1980), leads to a difference in the coding capacities of the two IRs (see below) and also seems to activate a cryptic promoter at this site in IR_L that enhances the level of expression of the nearby gene for Km resistance (Rothstein et al. 1980). Thus, this single-base change can account for both of the functional differences between IR_L and IR_R observed experimentally by Rothstein et al. (1980).

Translational Reading Frames in the Inverted Repeats of Tn*5*

When the nucleotide sequences of the two IRs are analyzed for their coding capacities, open translational reading frames of substantial length are found only in the direction toward the unique region. Of these, one very long frame, which covers almost all of each inverted repeat, is of particular interest because its length correlates well with the sizes of the proteins encoded by the Tn*5* IR sequences in a minicell system (Rothstein et al. 1980). As indicated in Figure 6, this translational frame starts with an ATG codon at position 92 (that the start is at position 80 is less likely as this ATG is preceded by a poor Shine-Dalgarno [1974] sequence) and is terminated in the left IR by a TAA stop codon at position 1442. In the right IR this stop codon is changed to a GAA glutamine codon, so that a translational stop is only encountered after a further 26 codons, at position 1520, just before the end of the IR sequence. This finding predicts that the two IRs encode very similar large polypeptides that differ only by the additional 26 amino acids at the CO_2H terminus of the IR_R protein(s). The sizes of the proteins predicted from the nucleotide sequence (51K for IR_L and 54K for IR_R) are in good agreement with the apparent molecular weights of the major polypeptides encoded in IR_L and IR_R, which were determined by Rothstein et al. (1980) to be 53K (49K) and 58K (54K), respectively. With the nucleotide sequences of the IRs at hand, these correlations can now be tested specifically. Having this knowledge of the sequences and exact cleavage maps for many restriction

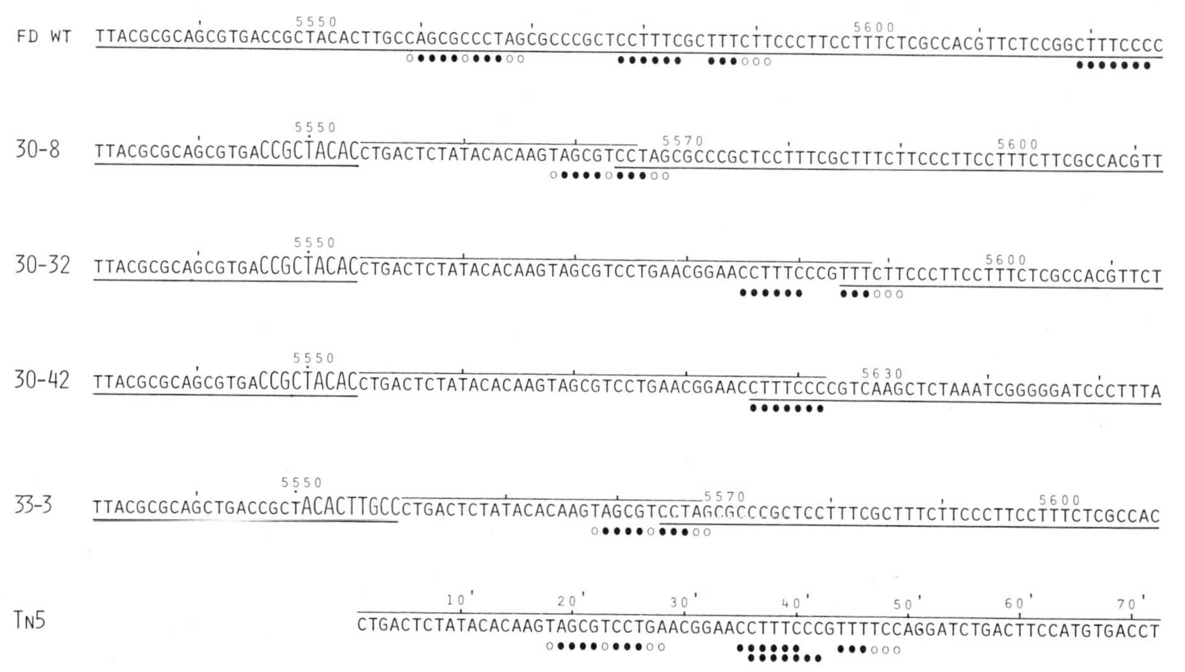

Figure 4. Nucleotide sequences formed after loss of Tn*5* from fd DNA. Sequences from Tn*5* are overlined, those from fd are underlined. Bases present in both fd DNA and Tn*5* DNA at the newly formed junction are marked by dots, and common pyrimidine or purine nucleotides by open circles. The 9 nucleotides duplicated in the parental fd::Tn*5* phages are indicated by large capital letters.

enzymes will allow manipulation of the putative transposase gene(s) in a predictable manner and the identification and alteration of the signals that control its (their) expression. So far, no promoter signal has been located in the nucleotide sequence that precedes the presumed beginning of the structural gene.

The nucleotide sequence determined also covers the beginning of the structural gene for the aminoglycoside-3′-*O*-phosphotransferase, which had been located previously in the left half of the central region of Tn*5* (Jorgensen et al. 1979; Rao and Rogers 1979). As indicated in Figure 6, this gene starts with the ATG codon at position 1551, which is part of a good ribosome-binding sequence in messenger RNA (mRNA). The amino acids derived from the succeeding coding sequence are in good agreement with the determination of the NH_2-terminal amino acid sequence of the phosphotransferase protein (J. Davies, pers. comm.). Thus, the structural gene for this enzyme starts only 17 nucleotides after the end of IR_L, which suggests that its expression is controlled at least in part by signals in the IR_L sequence. One such control signal seems to be the promoter

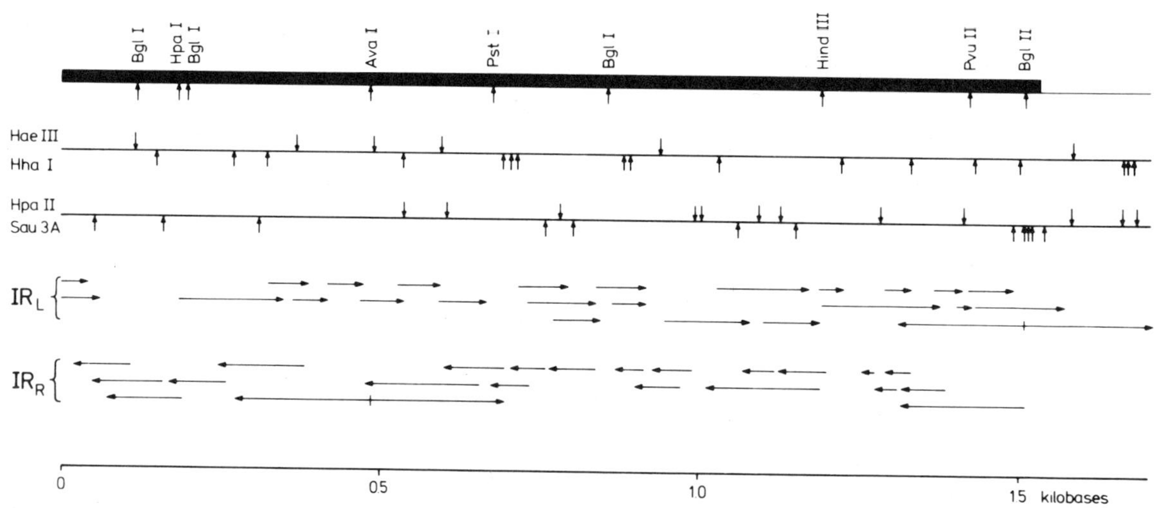

Figure 5. Strategy for nucleotide sequence determination of the inverted-repeat DNA from Tn*5*. Cleavage maps of the restriction endoncleases used are presented. Horizontal arrows indicate the directions and extents of the sequencing runs.

```
            10          20          30          40          50          60          70          80          90         100

   0 CTGACTCTTA TACACAAGTA GCGTCCTGAA CGGAACCTTT CCCGTTTTCC AGGATCTGAC TTCCATGTGA CCTCCTAACA TGGTAACGTT CATGATAACT
 100 TCTGCTCTTC ATCGTGCGGC CGACTGGGCT AAATCTGTGT TCTCTTCGGC GGCGCTGGGT GATCCTCGCC GTACTGCCCG CTTGGTTAAC GTCGCCGCCC
 200 AATTGGCAAA ATATTCTGGT AAATCAATAA CCATCTCATC AGAGGGTAGT GAAGCCATGC AGGAAGGCGC TTACCGATTT TACCGCAATC CCAACGTTTC
 300 TGCCGAGGCG ATCAGAAAGG CTGGCGCCAT GCAAACAGTC AAGTTGGCTC AGGAGTTTCC CGAACTGCTG GCCATTGAGG ACACCACCTC TTTGAGTTAT
 400 CGCCACCAGG TCGCCGAAGA GCTTGGCAAG CTGGGCTCTA TTCAGGATAA ATCCCGCGGA TGGTGGGTTC ACTCCGTTCT CTTGCTCGAG GCCACCACAT
 500 TCCGCACCGT AGGATTACTG CATCAGGAGT GGTGGATGCG CCCGGATGAC CCTGCCGATG CGGATGAAAA GGAGAGTGGC AAATGGCTGG CAGCGGCCGC
 600 AACTAGCCGG TTACGCATGG GCAGCATGAT GAGCAACGTG ATTGCGGTCT GTGACCGCGA AGCCGATATT CATGCTTATC TGCAGGACAG GCTGGCGCAT
 700 AACGAGCGCT TCGTGGTGCG CTCCAAGCAC CCACGCAAGG ACGTAGAGTC TGGGTTGTAT CTGATCGACC ATCTGAAGAA CCAACCGGAG TTGGGTGGCT
 800 ATCAGATCAG CATTCCGCAA AAGGGCGTGG TGGATAAACG CGGTAAACGT AAAAATCGAC CAGCCCGCAA GGCGAGCTTG AGCCTGCGCA GTGGGCGCAT
 900 CACGCTAAAA CAGGGGAATA TCACGCTCAA CGCGGTGCTG GCCGAGGAGA TTAACCCGCC CAAGGGTGAG ACCCCGTTGA AATGGTTGTT GCTGACCGGC
1000 GAACCGGTCG AGTCGCTAGC CCAAGCCTTG CGCGTCATCG ACATTTATAC CCATCGCTGG CGGATCGAGG AGTTCCATAA GGCATGGAAA ACCGGAGCAG
1100 GAGCCGAGAG GCAACGCATG GAGGAGCCGG ATAATCTGGA GCGGATGGTC TCGATCCTCT CGTTTGTTGC GGTCAGGCTG TTACAGCTCA GAGAAAGCTT
1200 CACGCTGCCG CAAGCACTCA GGGCGCAAGG GCTGCTAAAG GAAGCGGAAC ACGTAGAAAG CCAGTCCGCA GAAACGGTGC TGACCCCGGA TGAATGTCAG
1300 CTACTGGGCT ATCTGGACAA GGGAAAACGC AAGCGCAAAG AGAAAGCAGG TAGCTTGCAG TGGGCTTACA TGGCGATAGC TAGACTGGGC GGTTTTATGG
1400 ACAGCAAGCG AACCGGAATT GCCAGCTGGG GCGCCCTCTG GTAAGGTTGG GAAGCCCTGC AAAGTAAACT GGATGGCTTT CTTGCCGCCA AGGATCTGAT
1500 GGCGCAGGGG ATCAAGATCT GATCAAGAGA CAGGATGAGG ATCGTTTCGC ATGATTGAAC AAGATGGATT GCACGCAGGT TCTCCGGCCG CTTGGGTGGA
1600 GAGGCTATTC GGCTATGACT GGGCACAACA GACAATCGGC TGCTCTGATG CCGCCGTGTT CCGGCTGTCA GCGCAGGGGC GCCGGTTCTT TTTGTCAAGA
            10          20          30          40          50          60          70          80          90         100

1400' ACAGCAAGCG AACCGGAATT GCCAGCTGGG GCGCCCTCTG GGAAGGTTGG GAAGCCCTGC AAAGTAAACT GGATGGCTTT CTTGCCGCCA AGGATCTGAT
1500' GGCGCAGGGG ATCAAGATCT GATCAAGAGA CAGGGGCCGG CCCACGCTGT CGTCCAATCT CCCAAGACAC GCCGCCACCG CGCACCGTCG CGGCGAGCTG
```

Figure 6. Nucleotide sequences of the inverted repeats of Tn5 and of the adjacent unique central region sequences. The complete sequence is presented for IR_L, but only the terminal sequence is presented for IR_R. Potential start and stop codons for large open reading frames are boxed; the ends of the IR sequence are indicated by brackets. The heavy arrows overline short inverted repeats at the ends of each IR, which may be involved in the independent transposition of the IR elements. Recognition sites for restriction endonucleases are overlined and are listed in Table 1.

Table 1. Recognition Sites for Restriction Endonucleases in Nucleotide Sequences of Tn*5* Inverted Repeats and in Adjacent Sequences of the Unique Central Region

Restriction endonuclease	Sequence	Position
*Alu*I	AGCT	420, 429, 875, 1185, 1196, 1299, 1352, 1378, 1424
*Ava*I	CYCGRG	485
*Bal*I	TGGCCA	369
*Bgl*I	GCCNNNNNGGC	119, 197, 863
*Bgl*II	AGATCT	1515
*Eco*RII	CCAGG	49, 406
*Hae*II	RGCGCY	151, 266, 323, 705, 1430, 1678
*Hae*III	GGCC	118, 370, 490, 595, 940, 1586
*Hga*I	GCGTC	21, 1032
*Hha*I	GCGC	152, 267, 324, 538, 695, 706, 718, 886, 895, 1030, 1223, 1333, 1431, 1502, 1671, 1679
*Hin*dII	GTYRAC	185
*Hin*dIII	AAGCTT	1195
*Hin*fI	GANTC	4, 746, 1010
*Hpa*I	GTTAAC	185
*Hpa*II	CCGG	542, 607, 785, 996, 1004, 1092, 1127, 1286, 1413, 1584, 1661, 1683
*Hph*I	GGTGA	158, 965
*Mbo*II	GAAGA, TCTTC	106, 143, 416, 775
*Mst*I	TGCGCA	885
*Pst*I	CTGCAG	680
*Pvu*II	CAGCTG	1423
*Sau*3A	GATC	53, 161, 310, 763, 805, 1063, 1153, 1493, 1510, 1516, 1521, 1540
*Taq*I	TCGA	486, 765, 856, 1008, 1038, 1065, 1151
*Tha*I	CGCG	455, 656, 839, 931, 1031
*Xho*I	CTCGAG	485

See Fig. 6 for complete sequences.

created by the base change at position 1442. A further signal of unknown functional significance seems to be located at the *Bgl*II site at position 1515, as gene expression can be affected by the cloning of foreign DNA into this site (Rao and Rogers 1979; B. Reiss and R. Herrmann, unpubl.).

SUMMARY

Nucleotide sequences have been determined for the 1.5-kb inverted repeats of Tn*5* and for their junctions with the central unique region and with host DNA. The primary findings stemming from this analysis are:

1. Integration of Tn*5* is accompanied by the duplication of 9 bp of host DNA.
2. Loss of Tn*5* occurs by crossover between short, homologous nucleotide sequences near the junction between Tn*5* and host DNA.
3. The IR sequences contain long, open translational reading frames that may code for transposase proteins.
4. The two IR sequences differ by a single-base change. This alteration accounts for the two functional differences observed between IR_L and IR_R: It shortens the reading frame for the transposase gene in IR_L, and it improves the efficiency of a promoter for the nearby Km-resistance gene.
5. The NH_2 terminus of the structural gene for the Km-resistance gene maps at the very left border of the unique region, i.e., very close to the end of IR_L.

These results support the view that the inverted repeats of Tn*5* stem from two identical copies of an originally independently moving DNA element. In the transposon, one of these, IR_L, seems to have evolved toward a close physical and functional linkage with the antibiotic-resistance gene.

ACKNOWLEDGMENTS

We thank W. S. Reznikoff for the exchange of unpublished information, E. Beck for contributing to the sequence determination, and the Deutsche Forschungsgemeinschaft for financial support (Scha 134/9).

REFERENCES

Auerswald, E.-A. 1979. "Struktur des Transposons Tn*5* und Analyse seiner Integration und Exzision im Bakteriophagen fd." Ph.D. thesis, University of Heidelberg, Federal Republic of Germany.

Beck, E., R. Sommer, E.-A. Auerswald, C. Kurz, B. Zink, G. Osterburg, H. Schaller, K. Sugimoto, H. Sugisaki, T. Okamoto, and M. Takanami. 1978. Nucleotide sequence of bacteriophage fd DNA. *Nucleic Acids Res.* **5:** 4495.

Berg, D. E., J. Davies, B. Allet, and J. Rochaix. 1975. Transposition of R-factor genes to bacteriophage λ. *Proc. Natl. Acad. Sci.* **72:** 3628.

Gray, C. P., R. Sommer, C. Polke, E. Beck, and H. Schaller. 1978. Structure of the origin of DNA replication of bacteriophage fd. *Proc. Natl. Acad. Sci.* **75:** 50.

Herrmann, R., K. Neugebauer, H. Zentgraf, and H. Schaller. 1978. Transposition of a DNA sequence determining kanamycin resistance into the single-stranded genome of bacteriophage fd. *Mol. Gen. Genet.* **159:** 171.

Herrmann, R., K. Neugebauer, E. Pirkl, H. Zentgraf, and H. Schaller. 1980. Conversion of bacteriophage fd into an efficient single-stranded DNA vector system. *Mol. Gen. Genet.* **177:** 231.

JORGENSEN, R. A., S. J. ROTHSTEIN, and W. S. REZNIKOFF. 1979. A restriction enzyme cleavage map of Tn5 and location of a region encoding neomycin resistance. *Mol. Gen. Genet.* **177:** 65.

KLECKNER, N. 1977. Translocatable elements in procaryotes. *Cell* **11:** 11.

MAXAM, A. M. and W. GILBERT. 1977. A new method for sequencing DNA. *Proc. Natl. Acad. Sci.* **74:** 560.

NEUGEBAUER, K. 1980. "Untersuchungen zur Konstruktion von Enzelstrangvektoren in *Escherichia coli* und eines Exportvektors für *Bacillus subtilis*." Ph.D. thesis, University of Heidelberg, Federal Republic of Germany.

RAO, R. N. and S. G. ROGERS. 1979. Plasmid pKC7: A vector containing ten restriction endonuclease sites suitable for cloning DNA segments. *Gene* **7:** 79.

ROTHSTEIN, S. J., R. A. JORGENSEN, K. POSTLE, and W. S. REZNIKOFF. 1980. The inverted repeats of Tn5 are functionally different. *Cell* **19:** 795.

SCHALLER, H. 1979. The intergenic region and the origins for filamentous phage DNA replication. *Cold Spring Harbor Symp. Quant. Biol.* **43:** 401.

SHENK, T. E., C. RHODES, P. W. J. RIGBY, and P. BERG. 1975. Biochemical method for mapping mutational alterations in DNA with S1 nuclease: The location of deletions and temperature sensitive mutations in simian virus 40. *Proc. Natl. Acad. Sci.* **72:** 989.

SHINE, J. and J. DALGARNO. 1974. The 3′-terminal sequence of *Escherichia coli* 16S ribosomal RNA: Complementary to nonsense triplets and ribosome binding sites. *Proc. Natl. Acad. Sci.* **71:** 1342.

Insertion, Excision, and Inversion of Tn*5*

D. E. BERG, C. EGNER, B. J. HIRSCHEL,* J. HOWARD, L. JOHNSRUD, R. A. JORGENSEN,†‡ AND T. D. TLSTY

Department of Microbiology and Immunology and Department of Genetics, Washington University School of Medicine, St. Louis, Missouri 63110; †Department of Biochemistry, University of Wisconsin, Madison, Wisconsin 53706

The defining characteristic of bacterial transposons is their ability to move to new loci in the absence of extensive DNA sequence homology or the *recA* protein (for review, see Calos and Miller 1980; Starlinger 1980). Transposons increase the fitness of bacterial populations by facilitating gene flow among different species and by changing the arrangement and the control of expression of genes within a single species. Transposons that contain genes for antibiotic resistance appear to be responsible for the recurrence of the same resistance genes in otherwise unrelated R-factor plasmids (see Falkow 1975). In the laboratory, such resistance transposons can be used to mutate host genes, to probe and alter the control of gene expression, and to move genes to new locations.

In this paper we focus on the properties of Tn*5*, a 5700-bp transposon that contains a gene encoding kanamycin resistance (kan^r) bracketed by terminal 1500-bp inverted repeats (Berg et al. 1975; Berg 1977). Tn*5* is a natural component of an R-factor plasmid obtained from *Klebsiella* and has no homology with sequences normally present in the chromosome of *Escherichia coli* K12 (Berg and Drummond 1978). Tn*5* inserts into many different sites in a genome. It causes mutations that are polar on the expression of distal genes in the operon and can revert by excision of Tn*5* in both $recA^+$ and $recA^-$ cells (Berg 1977; Shaw and Berg 1979; Berg et al. 1980). The excision of Tn*5* is not correlated with its movement to new sites. Conversely, transposition leaves one copy of Tn*5* at the original site (Berg 1977). Thus, transposition and excision result from two distinct processes.

Restriction endonuclease (Jorgensen et al. 1979a) and DNA sequence (E.-A. Auerswald and H. Schaller, pers. comm.) analyses have shown that the left and right inverted repeats of Tn*5* are nearly identical. However, analyses of insertion and substitution derivatives of Tn*5* have shown that the right inverted repeat encodes a *trans*-acting function necessary for transposition of Tn*5* to new sites (a transposase) and that the left inverted repeat does not encode a transposase but contains the promoter responsible for the expression of the kan^r gene in the central portion of the transposon (Meyer et al. 1979; Rothstein et al. 1980).

The experiments presented in this paper provide insights into (1) the functional organization of Tn*5*, by showing that its inverted repeats are themselves transposons and that a Tn*5* derivative with direct repeats can transpose; (2) the mechanism of excision, by assessing the roles of the Tn*5*-encoded transposase and the inverted repeats of Tn*5* in excision; and (3) the flexibility in the control of gene expression, by characterizing a derivative of Tn*5* that functions as a mobile recombinational switch.

RESULTS AND DISCUSSION

Insertions and Substitutions within Tn*5* That Affect Transposition Proficiency

To define the roles of the left and right inverted repeats of Tn*5* in transposition, we inserted a DNA fragment containing a gene that encodes tetracycline resistance (tet^r) into the *Bgl*II restriction sites near the interior ends of the left and right inverted repeats of a Tn*5* element. These and the other Tn*5* derivatives shown in Figure 1 were moved by homologous recombination to sites in the *lac* operon of an F′ *lac* episome (see Berg et al. 1980; Berg 1980). The F′ episomes were transferred to a strain lysogenic for λ*b*515*b*519*c*I857*nin*5*S*am7, and the frequencies of transposition to λ were determined after thermal induction of the λ prophage and selection of tet^r or kan^r transducing phage.

We obtained the following approximate frequencies: 2×10^{-6} to 5×10^{-6} for Tn*5*-wild type, Tn*5*-132, Tn*5*-134, and Tn*5*-145; 1×10^{-8} to 2×10^{-8} for Tn*5*-131 and Tn*5*-133; and less than 10^{-9} for Tn*5*-112 and Tn*5*-410. The results with Tn*5*-112, in which most of the right inverted repeat is deleted, and with Tn*5*-131 and Tn*5*-133, in which the tet^r fragment is fused in one orientation with the internal end of the right inverted repeat, support the conclusion (Rothstein et al. 1980) that only the right inverted repeat of Tn*5* encodes a functional transposase protein. Since the transposase gene appears to be transcribed and translated from sites near the exterior end of the right inverted repeat (Rothstein et al. 1980), it is likely that these fusions result in transposases with altered carboxyl termini and that the transposase encoded by Tn*5*-132 and Tn*5*-134 retains more activity than that encoded by Tn*5*-131 and Tn*5*-133.

Inverted Repeats of Tn*5* Can Transpose by Themselves

Tn*5* probably arose after insertion of a pair of identical insertion sequence (IS) elements—small transposons

Present addresses: *Departement de maladies infectieuses, Hôpital Cantonal, CH-1211, Gèneve, Switzerland; ‡Carnegie Institution, Stanford, California 94305.

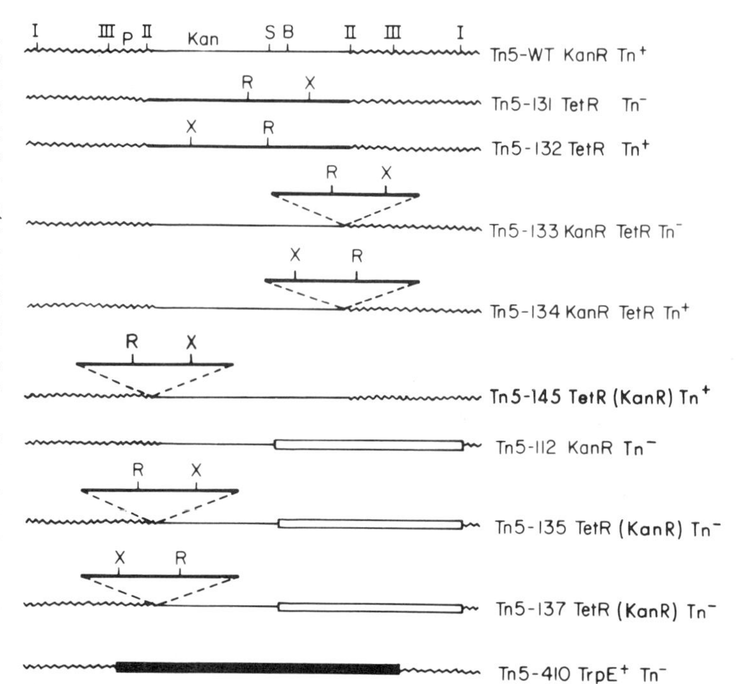

Figure 1. Map of Tn*5* and its derivatives. (〰) Inverted repetitions; (▭) deletions; (P) promoter for expression of kanamycin resistance. Restriction endonuclease cleavage sites: (I) *Hpa*I; (II) *Bgl*II; (III) *Hin*dIII; (R) *Eco*RI; (X) *Xba*I; (B) *Bam*HI. Tn^+ and Tn^- indicate transposition-proficient and transposition-deficient, respectively, as defined in the text. Tn*5*-112 was generated by partial digestion of ColE1::Tn*5* with *Hin*dII and contains an internal deletion extending from the central *Sal*I (*Hin*dII) site to the *Hpa*I (*Hin*dII) site in the right inverted repeat. Tn*5*-131 and Tn*5*-132 contain, as a substitute for the central *Bgl*II fragment of Tn*5*-wild type (WT), a 2700-bp fragment, which encodes tet^r, generated by *Bgl*II digestion of Tn*10* (Jorgensen et al. 1979b). In Tn*5*-133 and Tn*5*-134 the same *Bgl*II tet^r fragment is inserted into the *Bgl*II site in the right inverted repeat. Tn*5*-135 and Tn*5*-137 contain the tet^r fragment in the left inverted repeat of Tn*5*-112. The tet^r fragments of Tn*5*-131 and Tn*5*-133 are in one orientation, and these elements are transposition-deficient. The tet^r fragments of Tn*5*-132 and Tn*5*-134 are in the other orientation, and these elements are transposition-proficient (see text). Tn*5*-145 was generated in vivo by homologous recombination between Tn*5*-wild type in the *lac* operon and Tn*5*-135 in a λ*b*221*c*I857 phage, essentially as outlined in Fig. 6. Tn*5*-135, Tn*5*-137, and Tn*5*-145 confer resistance to kanamycin when grown in the presence of low levels of tetracycline. Tn*5*-410 was generated by Meyer et al. (1979) and contains a 5300-bp fragment containing the *trp* operon *E* and *D* genes in place of the central *Hin*dIII fragment of Tn*5*-wild type; it is transposition-deficient.

carrying only the genes and sites necessary for their own transposition—on either side of a previously immobile kan^r gene. Support for this view includes the findings that one of the terminal repeats of Tn*5* encodes a transposase (see above), that IS*1* elements bracket the cam^r gene of Tn*9* and the *ent* gene of Tn*1681* (MacHattie and Jackowksi 1977; So et al. 1979), and that one of the terminal repeats of Tn*10* functions as an IS element (Sharp et al. 1973; Ross et al. 1979).

We used two tests to determine whether the repeats of Tn*5* could transpose by themselves. The first test was based on the finding of Guyer (1978) that fertility factor F mediates the transfer of the amp^rtet^r plasmid pBR322 to recipient *E. coli* strains at a frequency of 10^{-6} and that the transferred pBR322 plasmids invariably contain insertions of one of the F factor's IS elements. We found that FΔ::Tn*5*, a derivative of F lacking all known IS elements (M. Guyer, pers. comm.) and to which we had transposed Tn*5*, mediates the transfer of the amp^rtet^r traits of pBR322 at a frequency of 10^{-9} to 10^{-10} when either $recA^+$ or $recA^-$ donors and recipients are used.

Sixty-five of the transferred plasmids were analyzed genetically and by restriction endonuclease digestion and gel electrophoresis. Forty-four encoded kanamycin resistance; each of 12 representatives of this class analyzed was 5.7 kb larger than pBR322 and contained an insertion of one copy of Tn*5*. Fifteen of 17 kanamycin-sensitive transformants analyzed contained plasmids the same size as pBR322 and were not analyzed further. Two contained plasmids 1.5 kb larger than pBR322. Restriction endonuclease digestions such as depicted in Figure 2 showed that this 1.5-kb insert corresponded to one of the inverted repeats of Tn*5*, which we will designate IS*50*. In subsequent experiments, we identified four additional insertions of IS*50* into pBR322, one of which inactivated pBR322's ampicillin-resistance gene.

We have determined the DNA sequence at the pBR322–IS*50* junction for an insertion between nucleotides 4320 and 4321 of pBR322. The results show that IS*50*, like Tn*5*, generates 9-bp direct duplications of the target sequence (D. E. Berg et al., in prep.).

To show that both inverted repeats could transpose, we characterized λtet^r phage generated by the transposition from Tn*5*-134 (in which tet^r marks the right inverted repeat). Three of 23 phage from Tn*5*-134 and 4 of 45 phage from Tn*5*-145 did not contain a kan^r determinant. Restriction endonuclease analyses confirmed that the λtet^rkan^s phage resulted from insertion of a single inverted repeat into λ and that the λtet^rkan^r phage resulted from transposition of the entire Tn*5*-134 or Tn*5*-145 element (data not shown). We conclude, therefore, that both inverted repeats, which we designate IS*50*-R and IS*50*-L, can transpose from Tn*5* to new locations. R. Laux (pers. comm.) has observed independently the transposition of the right inverted repeat from Tn*5*-134. Because the results described above showed that the left repeat lacks a functional transposase gene, we are testing the prediction that only IS*50*-R will be capable of continued transposition in a cell lacking other Tn*5* sequences.

A Transposon with Direct Repeats of IS*50*

To analyze the importance of the orientation of the terminal repeat in transposition and excision, we constructed two Tn*5* derivatives (designated Tn*5*-DR1 and Tn*5*-DR2) containing direct repeats of IS*50* in the two possible orientations. Plasmid pBR322, which con-

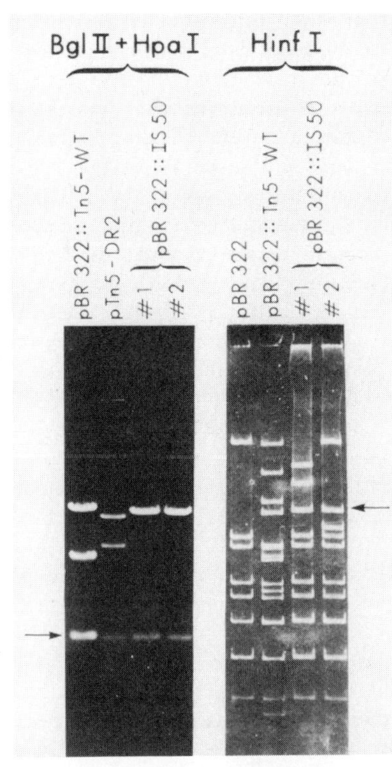

Figure 2. Restriction endonuclease analysis of pBR322 containing insertions of IS*50* and Tn*5*. Insertions of Tn*5* and IS*50* into pBR322 were generated as described in the text. Plasmid DNA was extracted according to the method of Birnboim and Doly (1979), digested with restriction endonucleases as recommended by the suppliers (New England Biolabs and Bethesda Research Labs), and electrophoresed in Tris-acetate-buffered agarose gels as described previously by Berg and Drummond (1978). (*Right*) *Hin*fI digestion. The internal fragment extending from position 3 to position 732 in the inverted repeats of Tn*5* (Schaller 1979 and pers. comm.) is indicated by the arrow. (*Left*) *Bgl*II + *Hpa*I digestion. The internal 1300-bp fragment generated by digestion of IS*50* and the inverted repeats of Tn*5* is indicated by the arrow. pBR322 DNA is not cleaved by these enzymes.

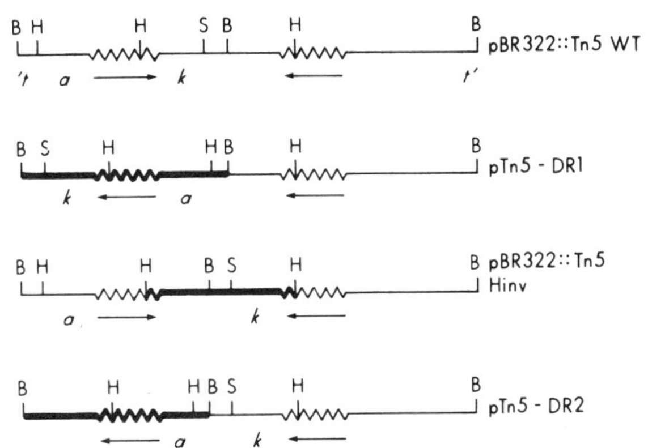

Figure 3. Strategy for construction of plasmids containing direct repetitions of IS*50:* pTn*5*-DR1 and pTn*5*-DR2. Restriction endonuclease sites: (B) *Bam*HI; (H) *Hin*dIII; (S) *Sma*I. *a, t,* and *k* indicate resistance to ampicillin, tetracycline, and kanamycin, respectively. (—) Region inverted at each step; (∿∿) repeats of IS*50*. The circular map of pBR322::Tn*5* and its derivatives is arbitrarily linearized at the *Bam*HI site in pBR322. pTn*5*-DR1 was generated from the pBR322::Tn*5* plasmid indicated by *Bam*HI digestion, ligation, and selection of $amp^r kan^r tet^s$ transformants. pTn*5*-DR2 was generated from the same pBR322::Tn*5* plasmid by inversion of the central *Hin*dIII fragment of Tn*5*, following partial *Hin*dIII digestion to generate the plasmid labeled Hinv. This plasmid was digested with *Bam*HI and ligated, and a plasmid with the structure indicated as pTn*5*-DR2 was found among the transformants. These plasmids were characterized by digestion with *Hin*dIII, *Bam*HI, and *Sma*I.

tains an insertion of Tn*5* near its amp^r gene, was digested with restriction endonucleases and ligated, and recombinant DNA molecules were selected as indicated in Figure 3. The structures of plasmids pTn*5*-DR1 and pTn*5*-DR2 were confirmed by digestion with restriction endonucleases *Bam*HI, *Sma*I, and *Hin*dIII (data not shown).

We found that pTn*5*-DR1 is stable in $recA^-$ cells but extremely unstable in $recA^+$ cells. Cells carrying intact pTn*5*-DR1 grow very slowly, whereas rare derivatives that have lost ampicillin resistance grow normally. Consequently, less than 2% of the plasmid molecules extracted from cultures grown for 100 generations without ampicillin contain an amp^r gene. amp^s derivatives of pTn*5*-DR2 were also found but were not favored strongly (<5% after 100 generations). The genetic basis for the selective disadvantage conferred by Tn*5*-DR1 is under investigation.

The sizes of the amp^s derivatives of pTn*5*-DR1 and pTn*5*-DR2 plasmids, estimated by agarose gel electrophoresis, were 6000 bp and 5600 bp, respectively (data not shown), as expected from homologous recombination between the directly repeated IS*50* sequences (see Figs. 4 and 3). Digestion with restriction enzymes *Taq*I and *Hin*fI confirmed that all Tn*5*-DR sequences, except for one copy of the IS*50* element, were missing from the amp^s plasmids (data not shown). Because these segregants are only formed in $recA^+$ cells, they probably arise by homologous recombination between the direct repeats of the pTn*5*-DR plasmids (Fig. 4).

We emphasize that the consequences of intramolecular recombination between repeated sequences depend on their orientation. For direct repeats, recombination leads to loss of the intervening segment, as shown in Figure 4. For inverted repeats, recombination leads to reversal of the orientation of the intervening segment (see A Mobile Recombinational Switch and Fig. 8 below). Both events can have profound effects on the phenotype of an organism.

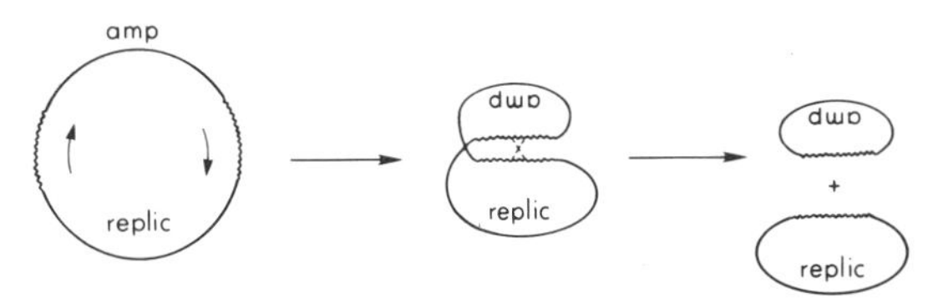

Figure 4. Recombination between the direct repeats in a pTn*5*-DR plasmid. (replic) Replication region of the plasmid. (amp) Segment within Tn*5*-DR1 encoding ampicillin resistance and segment within Tn*5*-DR2 encoding ampicillin and kanamycin resistances. (∿∿) Direct repetitions (IS*50*).

Figure 5. Physical map of representative Tn*5*-DR2 insertions in *lacZ*. *lacZ*::Tn*5*-DR2 alleles were generated by transposition from the bacterial chromosome to an F′ *lac* episome, as described in the text, and were recombined in vivo into the ColE1-*lac* plasmid pMC4 (Calos et al. 1978) as described by Berg et al. (1980). The resulting *lacZ*::Tn*5*-DR2 plasmids were analyzed by digestion with *Eco*RI, *Bam*HI, and *Taq*I. 0, 1, 2 and 3 refer to distances, measured in kilobases, from the *Eco*RI site, which is approximately 50 bp from the carboxyl terminus of *lacZ*. Symbols below the line indicate insertions of Tn*5*-DR2 in which the kan^r gene is transcribed from left to right; symbols above the line indicate insertions of Tn*5*-DR2 in the opposite orientation.

We studied the transposition of Tn*5*-DR2 because it is more stable than Tn*5*-DR1 and because the resistances to both kanamycin and ampicillin, which it encodes, facilitate the selection of rare transposition events. The frequency of transposition of Tn*5*-DR2 from plasmid pTn*5*-DR2 to phage λ was approximately 10^{-9} per phage; this is comparable to the frequency of transposition of Tn*5*-wild type from plasmid pBR322::Tn*5*. An F′ *lac* strain was lysogenized with a λTn*5*-DR2 phage, and transpositions of Tn*5*-DR2 from the chromosomal site to the F′ *lac* episome were selected by crossing the F′ lac^+ lysogen with a $recA^-$ lac^- recipient. Approximately 2% of $amp^r kan^r$ exconjugants were lac^- and resulted from insertion of Tn*5*-DR2 into the *lac* operon. Of 117 independent *lac*::Tn*5*-DR2 mutants examined, 90 were $lacY^-$ and 27 were $lacZ^-$. This preferential insertion of Tn*5*-DR2 into *lacY* is similar to that observed earlier with Tn*5*-wild type (Berg et al. 1980).

To determine physically the orientations and positions of insertions, representative *lacZ*::Tn*5*-DR2 alleles were recombined onto a ColE1-*lac* plasmid as described by Berg et al. (1980), and each of the resulting *lac*::Tn*5*-DR2 plasmids was analyzed by digestion with *Eco*RI, *Bam*HI, and *Taq*I. The results (summarized in Fig. 5) showed that insertion had occurred in either orientation and that each insertion was at a different site. Thus, Tn*5*-DR2, like Tn*5*-wild type, can insert into many different sites in a gene.

Transposase-independent Excision of Tn*5*

The vast majority of mutations induced by insertion of Tn*5* revert by precise excision. Revertants are Kan^s, and thus excision involves loss of the element from the cell, not movement to new sites (Berg 1977). In our analyses of *lac*::Tn*5* insertion mutations, we found that lac^+ revertant frequencies depended on the site of insertion and on linkage to the F factor but did not depend strongly on the *recA* (homologous recombination) function. Revertant frequencies ranged from 3×10^{-8} to 2×10^{-4} in a screen of 150 mutants in an F′ *lac* strain. Although revertant frequencies of representative Tn*5* insertion mutations were not changed when the mutations were introduced into the chromosomes of Hfr strains, they were reduced approximately 100-fold when the mutations were introduced into the chromosomes of F^+ or F^- strains (C. Egner and D. E. Berg, in prep.). Tests to determine the basis of *cis* enhancement of Tn*5* excision by F are in progress.

To test whether precise excision was essentially an abortive transposition event reflecting an unusual action of the Tn*5*-encoded transposase at the Tn*5*–bacterial gene junction or whether it was a form of illegitimate recombination similar to the formation of spontaneous deletions (Berg 1977), we examined the frequencies of excision of the Tn*5* elements shown in Figure 1, some of which are deficient in transposition. We determined the effects of loss of transposition functions on excision, independently of effects of the site of insertion, by generating sets of strains containing different Tn*5* elements at the same site, using the procedure outlined in Figure 6. This procedure involves homologous recombination between the inverted repeats of a Tn*5* element inserted in the F′ *lac* episome and a second element brought into the cell as part of a λ phage. Restriction endonuclease analyses of *lac*::Tn*5* alleles have verified the precise replacement of one Tn*5* element by another in either orientation, as predicted by Figure 6 (Berg 1980; Berg et al. 1980).

Table 1 lists representative frequencies of excision. Although these frequencies are influenced by the size or sequence organization of the Tn*5* element, comparison of the Tn*5* elements that differ in their abilities to transpose shows that excision does not require transposition proficiency. We therefore conclude that excision is a transposase-independent process.

We also examined the excision of Tn*5*-DR2 and IS*50* from five representative sites in *lacZ* in the same F′ *lac* episome. The average frequency of excision of Tn*5*-DR2 was 3×10^{-9} and that of a single copy of IS*50* generated by recombination within Tn*5*-DR2 (Fig. 4) was 8×10^{-9}, in sharp contrast to the average frequency of 10^{-6} obtained for the excision of Tn*5*-wild type (Table 1 and unpubl.). Since Tn*5*-DR2 transposes with a frequency and specificity similar to Tn*5*-wild type, these results indicate that transposase is not necessary for excision and that the inverted repeats of Tn*5*-wild type play a role in the excision process.

Imprecise excision is the removal of much or all of the transposon without restoration of the ancestral gene sequence. Since Tn*5* insertions are polar, imprecise excision can be detected in *lacZ*::Tn*5* mutants by selecting for a $LacY^+$ phenotype. Two kinds of $LacY^+$ derivatives are obtained. In one type, the $lacZ^-$ mutant allele does not revert because part of *lacZ* is deleted; the other type can revert to $lacZ^+$ because part of Tn*5* remains inserted in *lacZ* and no *lacZ* sequences are deleted (Berg 1977). The results of C. Egner and D. E. Berg (in prep.) indicate that imprecise excision is also independent of the transposase function and that it does depend on the terminal inverted repeats of Tn*5*.

A Mobile Recombinational Switch

Recombinational switches that control the expression of particular genes (Silverman and Simon 1980; van de Putte et al. 1980) appear to play important roles in bacterial populations by maintaining heterogeneity in the

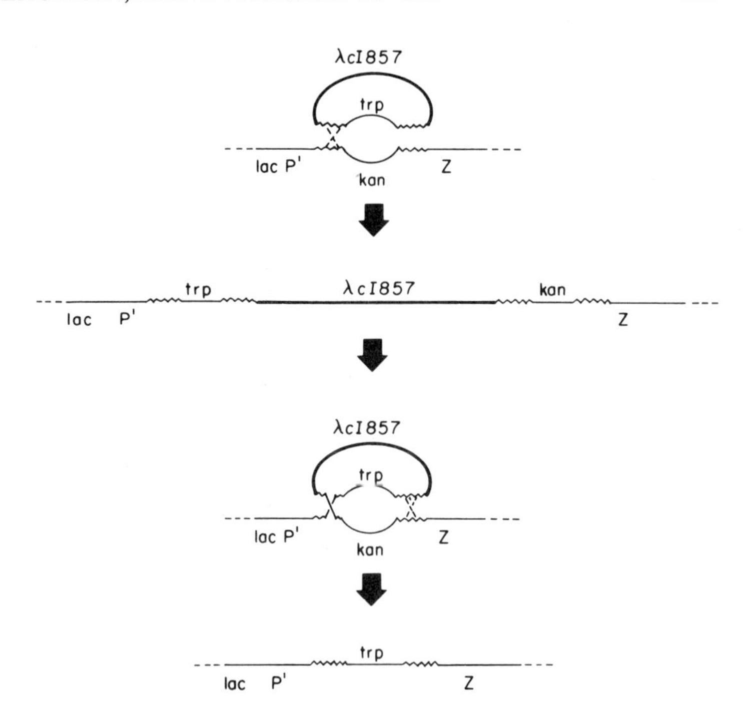

Figure 6. Replacement of Tn*5* (*kan*^{r}). Shown here is the replacement of Tn*5*-wild type (*kan*^{r}) in the *lac* promoter by Tn*5*-410 (*trp*$^+$). After infection of Trp$^-$ DB1506 cells containing a *lac*::Tn*5* insertion mutation with λ*b*221*c*I857 Tn*5*-410, lysogens were selected as Trp$^+$ λ-immune colonies at 32 °C. Since the *b*221 deletion makes this λ phage unable to integrate into the bacterial attachment site, stable lysogens generally resulted from a crossover between homologous portions of inverted repetitions of Tn*5*-wild type in *lac* and Tn*5*-410 in λ. Nonlysogenic segregants were selected by their ability to grow at 41 °C, a temperature that induces λ*c*I857 prophage development. The heat-resistant Trp$^+$ colonies were found at frequencies of 10^{-2} to 10^{-3}, and they arose by a second crossover on the other side of the nonhomology between Tn*5*-wild type and Tn*5*-410. To ensure a high yield of pure Trp$^+$Kan^s clones, the lysogenic cultures were routinely grown at 41 °C for five to ten generations in tryptophan-free media containing glucose to inhibit adsorption of λ particles. Analogous procedures result in the replacement of Tn*5*-wild type with Tn*5*-131 (*tet*^{r}) and then replacement of Tn*5*-410 with Tn*5*-wild type, Tn*5*-112 (*kan*^{r}), Tn*5*-135 (*tet*^{r}*kan*^{r}), etc.

phenotypes of cells growing in a uniform environment. Experiments reported here show that Tn*5*-112 can function as such a mobile recombinational switch. Tn*5*-112 was derived from Tn*5* by deletion of an internal 3-kb segment, which removes all but the terminal 186 bp of the right inverted repeat and brings the *kan*^{r} gene near the right end of the element as shown in Figure 1. The *kan*^{r} gene is transcribed from left to right (Jorgensen et al. 1979a).

We found that both Tn*5* and IS*50,* when inserted in *lacZ,* are polar on *lacY.* Therefore, terminators in Tn*5*-wild type that block progression of the *kan*^{r} transcript into nearby bacterial genes are in the repeats and are likely to be deleted in Tn*5*-112. The *kan*^{r} promoter within Tn*5*-112 therefore should stimulate expression of distal genes when the element is in the correct orientation. This was tested by placing Tn*5*-112 in the *lac*-operon-promoter region (*lacP*) and analyzing its effects on β-galactosidase synthesis.

Strains carrying Tn*5*-112 in both orientations in *lacP* were generated in two steps. First, Tn*5*-wild type in *lacP* was replaced by Tn*5*-410 (Trp) as outlined in Figure 6. Then, in an analogous cross, the Tn*5*-410 element was replaced by Tn*5*-112 (*kan*^{r}). One isolate from each of the 33 independent replacements of Tn*5*-410 by Tn*5*-112 was tested for *lac* expression on MacConkey agar. Sixteen isolates formed pale-pink colonies, which were indistinguishable from isogenic *lacP*::Tn*5*-wild type strains, and 17 isolates formed dark-pink colonies, which is indicative of much stronger lactose fermentation (Table 2).

β-Galactosidase assays confirmed the existence of two classes of Tn*5*-112 replacements at this site (Table 2). The strains designated OFF, like the ancestral Tn*5*-wild type strain, make no detectable β-galactosidase. In contrast, the ON strains synthesize β-galactosidase constitutively at a rate close to the basal level of expression of a wild-type *lac* operon.

Restriction endonuclease analyses were used to test whether the ON and OFF phenotypes depend on the orientation of Tn*5*-112, as predicted if it is the *kan*^{r} transcript that continues into distal bacterial genes. Four independent ON and four independent OFF *lacP*::Tn*5*-112 recombinant alleles were crossed into the ColE1-*lac* plasmid, and the plasmids that resulted were analyzed by digestion with restriction endonuclease

Table 1. Excision Frequencies of Tn*5* Derivatives

Insertion site	Lac$^+$ revertant frequencies ($\times 10^{-6}$) Tn*5*-wild type Tn$^+$	Tn*5*-131 Tn$^-$	Tn*5*-132 Tn$^+$	Tn*5*-133 Tn$^-$	Tn*5*-134 Tn$^+$	Tn*5*-410 Tn$^-$	Tn*5*-112 Tn$^-$
*Z*155	0.4 ± .02	0.1 ± .01	0.1 ± .01	0.05 ± .01	0.06 ± .01	0.07 ± .01	0.01 ± .02
*Z*202	10 ± 2	5 ± 1	5 ± 1	2 ± 0.2	3 ± 1	3 ± 0.6	3 ± 1
*Y*148	30 ± 10	20 ± 4	20 ± 5	10 ± 4	10 ± 2	10 ± 4	5 ± 1

Cultures of derivatives of DB1506 (F′ *proB*$^+$ *lac*/Δ[*proB-lac*] Δ*trpE5*) containing the indicated *lac*::Tn*5* insertion mutations were streaked on broth agar without antibiotics and incubated at 37 °C. Individual colonies were suspended in 11 ml of broth and shaken at 37 °C for 10 hr. The stationary-phase cultures were centrifuged, washed once with 10 mM $MgSO_4$, and suspended in 10 mM $MgSO_4$; aliquots were spread on minimal Difco Noble agar lactose plates and incubated at 37 °C for 36 hr. Each revertant frequency (±S.D.) was based on five separate cultures. Tn$^+$ indicates transposition-proficient; Tn$^-$ indicates transposition-deficient.

Table 2. Expression of *lacZ* in Cultures of *lacP*204::Tn*5* Insertion Mutants

	Lactose fermentation[a]	β-Galactosidase level[b]	Phenotype designation	Orientation[c]
Bacterial strain				
lac$^+$	+++	1	Lac$^+$	
*lacZ*124::Tn*5*	0	<0.05	LacZ$^-$	
*lacP*204::Tn*5*-wild type				
orientation I	+	<0.05	OFF	I
orientation II	+	<0.05	OFF	II
*lacP*204::Tn*5*-112				
1	++	0.75	ON	II
2	+	<0.05	OFF	I

The *lac*$^+$ bacterial strain was DB1506 (F′ *proB*$^+$ *lac*$^+$/Δ[*proB-lac*] Δ*trpE5*). Other strains were derived from it by introduction of the indicated Tn*5* element.

[a] Lactose fermentation was scored on MacConkey agar.

[b] β-Galactosidase was assayed as described by Miller (1972). The values are expressed relative to the basal (noninduced) level for *lac*$^+$ strain DB1506 set at 1.

[c] The orientation was determined from restriction endonuclease digestion as shown in Fig. 7.

*Sma*I (Fig. 7a). One pattern of fragments was in each case associated with the ON phenotype, and a second pattern was associated with the OFF phenotype. This outcome indicates that the ON phenotype results from Tn*5*-112 in the orientation in which transcription of *kan*^{r} proceeds toward distal *lac* operon genes.

To extend the concept of Tn*5*-112 as a mobile recombinational switch, we asked whether Tn*5*-112 in the OFF orientation could turn on distal genes by inversion (Fig. 8). Derivatives of a *lacZ*155::Tn*5*-112 OFF mutant that expressed *lacY* were selected on minimal melibiose agar at 41 °C and recovered at a frequency of approximately 10^{-7}. Although a small fraction of the LacY$^+$ isolates were Kan^s (10 of 83 tested) and thus resulted from excision, the majority (88%) were Kan^r, as expected for inversion of Tn*5*-112. To test whether they did result from inversion, nine independent ON alleles were recombined into the ColE1-*lac* plasmid and analyzed by digestion with *Sma*I (Fig. 7b). The fragments of each ON isolate were identical, and their sizes indicated that Tn*5*-112 can invert at its site of insertion. Since the 186-bp terminal repetition of Tn*5*-112 is close to the size limit for homologous pairing in generalized recombination, it seemed possible that the low frequency of Tn*5*-112 inversion might be *recA*-independent and due to the action of an "invertase" analogous to those that change *Salmonella* flagellar antigens and the host range of phage Mu (Silverman and Simon 1980; van de Putte et al. 1980). Consequently, we also selected LacY$^+$Kan^r derivatives of a *recA*$^-$*lacZ*155::Tn*5*-112 OFF mutant strain. Approximately equal frequencies of ON derivatives were seen in the *recA*$^+$ and *recA*$^-$ strains, which might have indicated that Tn*5*-112 inversion is not *recA*-independent. Further analysis indicated that the ON derivatives of the *recA*$^-$ strains did not result from the type of simple crossover depicted in Figure 8. Genetic and molecular tests showed that these F′ epi-

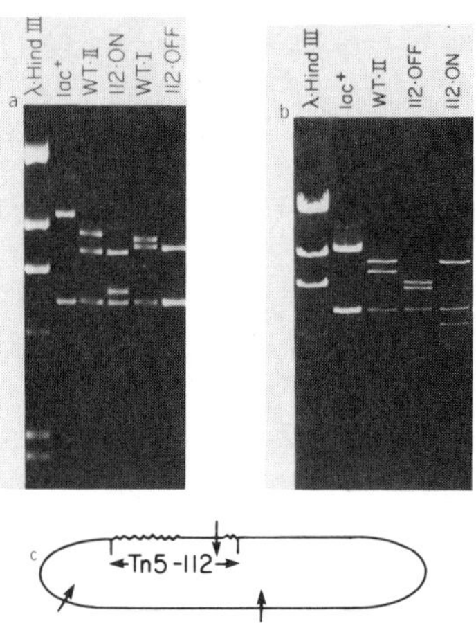

Figure 7. Agarose gel electrophoresis of *Sma*I digests of pMC4::Tn*5*-112 and pMC4::Tn*5*-wild type DNAs. These plasmids were generated as described in the legend to Fig. 5. Lac$^+$ designates the parental pMC4 plasmid that does not contain Tn*5*. The sizes of reference fragments in the lanes marked λ·*Hind*III are (*top to bottom*) 23, 9.8, 6.6, 4.5, 2.5, and 2.2 kb (Murray and Murray 1975). (*a*) *lacP*204::Tn*5*-wild type (WT), *lacP*204::Tn*5*-112, and *lacZ*155::Tn*5*-112 OFF alleles were generated in F′ *lac*::Tn*5* episomes by the replacement procedure described in the legend to Fig. 6. (*b*) The *lacZ*155::Tn*5*-112 ON allele shown is representative of nine characterized alleles that resulted from selection for the ON phenotype in *recA*$^+$ cells containing an F′ *lacZ*155::Tn*5*-112 OFF episome. (*c*) Pattern of *Sma*I cleavages of a pMC4::Tn*5*-112 plasmid. The sizes of fragments from pMC4 *lac*$^+$ were approximately 11 kb and 5.5 kb (common to all plasmids analyzed). The fusion fragments resulting from insertion into the 11-kb pMC4 fragment are 8.1 kb and 5.9 kb for *lacP*204::Tn*5*-112 ON and 5.5 kb and 8.8 kb for *lacP*204::Tn*5*-112 OFF; 8.1 kb and 9.6 kb for *lacP*204::Tn*5*-wild type II and 8.4 kb and 8.8 kb for *lacP*204::Tn*5*-wild type I; 9.2 kb and 8.2 kb for *lacZ*155::Tn*5*-wild type, 7.1 kb and 6.6 kb for *lacZ*155::Tn*5*-112 OFF, and 9.2 kb and 4.8 kb for *lacZ*155::Tn*5*-112 ON.

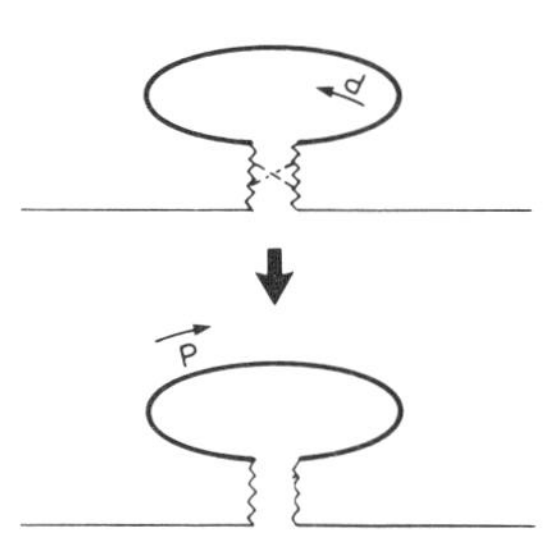

Figure 8. Crossing-over between inverted repeats. The orientation of the intervening segment (designated P) relative to the rest of the molecule is reversed by recombination. (∿∿) Inverted repeats.

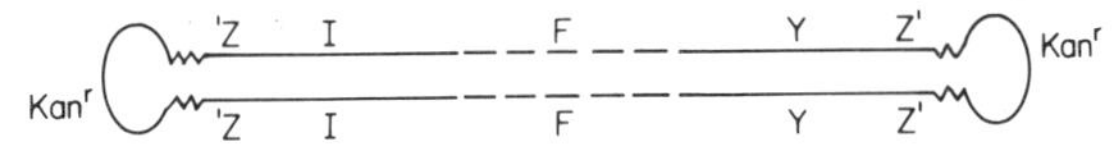

Figure 9. The structure of inverted dimer F′ episomes generated as *lacZ*::Tn*5*-112 ON derivatives of a *recA*⁻ F′ *lacZ*::Tn*5*-112 OFF strain. I and Y refer to the *lacI* and *lacY* genes; ′Z and Z′ refer to the amino terminal and carboxyterminal portions of *lacZ*, respectively.

somes have undergone a major DNA sequence rearrangement. We are currently testing the hypothesis that they are complete inverted dimers containing two copies of Tn*5*-112, one flanked by the *lac*-operator-proximal part of *lacZ* and the other flanked by the distal part of *lacZ* (Fig. 9). Since the simple inversion of Tn*5*-112 occurs only in *recA*⁺ cells, we conclude that it is due to homologous crossing-over, not site-specific exchange.

Kutsukake and Iino (1980) demonstrated that a product encoded by prophage Mu (probably its invertase) can stimulate variation in the flagellar antigens of *Salmonella*, a process known to occur by inversion of a specific DNA segment containing terminal inverted repeats. We found that prophage Mu did not enhance Tn*5*-112 inversion. This result supports the view that the Mu invertase recognizes specific DNA sequences rather than a general feature of an inverted repetition.

We have shown previously that inverted dimers similar to those described above predominate among newly isolated λdv plasmids, generated as deletion derivatives of the phage λ*nin*5. The type of replication-error model we invoked to explain the formation of λdv inverted dimers (Berg 1974; Chow et al. 1974) can also account for the formation of F′ inverted dimers.

The Promoter in Tn*5*-112 That Controls Distal Gene Expression

To test whether the *kan*ʳ promoter in Tn*5*-112 turns on distal gene expression, we replaced Tn*5*-112 present in the ON orientation in the *lac* promoter with Tn*5*-135 and Tn*5*-137, essentially as outlined in the legend to Figure 6. Tn*5*-135 and Tn*5*-137 are derivatives of Tn*5*-112 that contain insertions of the polar *tet*ʳ fragment between the *kan*ʳ promoter and the *kan*ʳ structural gene (Fig. 1) (Jorgensen et al. 1979a). Expression of the *tet*ʳ gene is induced by tetracycline (Beck 1979), and bidirectional transcription of this *tet*ʳ segment occurs in vitro (K. Bertrand, pers. comm.). We found that in each of ten independent replacements (five by Tn*5*-135 and five by Tn*5*-147), the cells were Kanˢ and Lac⁻ in media lacking tetracycline but partially Kanʳ and Lac⁺ in media containing 2 μg/ml of tetracycline. From the Kanˢ Lac⁻ phenotype conferred by Tn*5*-135 and Tn*5*-137 in the absence of tetracycline, we conclude that the *kan*ʳ promoter is responsible for the distal gene expression instigated by Tn*5*-112. The tetracycline-induced kanamycin resistance and *lac* expression probably represent bidirectional transcription initiated at the *tet* promoter in Tn*5*-135 and Tn*5*-137.

Model for the Excision of Tn*5*

We propose that excision of Tn*5* involves a failure to replicate Tn*5* and that Tn*5* excision is similar to the formation of spontaneous deletions. Spontaneous deletion formation involves interactions of closely spaced direct repeats (Farabaugh et al. 1978; Galas 1978). A 9-bp duplication of target sequences is present at the sites of Tn*5* insertion (Schaller 1979). We have found that Tn*5* excision does not require the *recA* product (Berg 1977), that transposase is not necessary for excision, and that terminal repeats in inverted orientations facilitate Tn*5* excision (this paper). Our proposal is therefore an attempt to show how interactions between inverted repeats of Tn*5* can bring the 9-bp flanking repeats into juxtaposition so that the rare but normal process of spontaneous deletion formation can operate.

As shown in Figure 10, we envision that double-stranded DNA is occasionally denatured in vivo, perhaps by the action of the products of *recBC* or *rep*, which can cause extensive DNA denaturation in vitro, provided excess single-strand-binding protein is present (Rosamond et al. 1979; Yarronton and Gefter 1979). Intramolecular reannealing between complementary sequences of the inverted repeats of Tn*5* would create hairpin structures that would bring the flanking direct repeats close together. Provided there is preferential copying of single-strand templates and retardation of chain growth at the single-strand–double-strand junc-

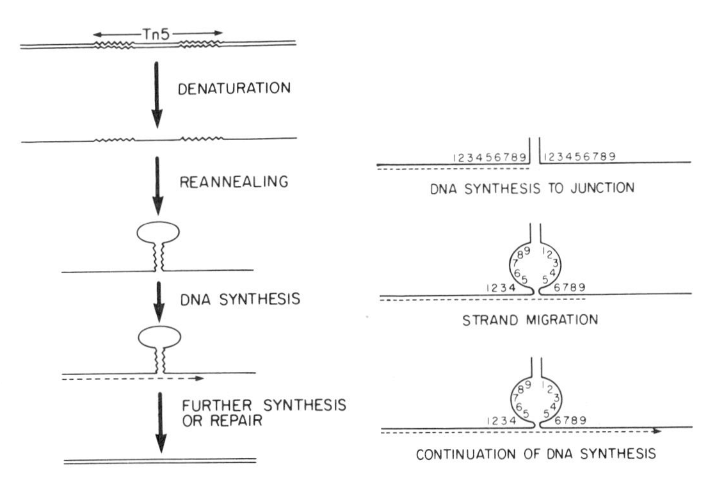

Figure 10. Model for transposase-independent excision of Tn*5*. (*a*) DNA denaturation (perhaps catalyzed by the *recBC* or *rep* proteins) is followed by intramolecular reannealing between complementary sequences in the inverted repeats. (*b*) Detailed view of DNA synthesis after reannealing as in *a*. (1–9) Juxtaposed 9-base repeats of the target sequence. (----) A new polynucleotide chain whose elongation will be slowed at the junction between single- and double-strand regions, which may facilitate strand migration, a local denaturation of the end of the nascent chain, and renaturation with the second copy of the 9-base sequence. The commitment to the excision event is fixed by the continuation of chain growth after strand migration.

tions in vivo (as seen in vitro with DNA polymerase I [Richardson et al. 1964; Masamune and Richardson 1971]), this configuration would promote slippage between the juxtaposed 9-base repeats by facilitating dissociation of the end of the nascent strand, hydrogen bonding with the second copy of the 9-base repeat, and then the continuation of DNA synthesis.

Since imprecise excision is stimulated by the inverted repeats and is independent of the transposase function, it can also be interpreted in the context of the model depicted in Figure 10. Deletion of part of the gene into which Tn*5* inserted may result when repeated sequences other than those that immediately bracket Tn*5* are brought into proximity by DNA denaturation and intramolecular renaturation. Other imprecise excision events in which much of Tn*5*, but none of the adjacent sequences, is deleted may occur by template switching–a copying of part of one strand in the foldback structure and then continuation of synthesis on the complementary strand–essentially as proposed by us for the formation of λdv plasmids from phage λ (Berg 1974; Chow et al. 1974) and as has been seen by Backman et al. (1979) during the replication of small plasmids.

We propose that any DNA segment containing long terminal inverted repeats can be excised by copy errors following denaturation and intramolecular reannealing as outlined here. This would include Tn*10* and Tn*1681*, which have terminal 1400-bp and 768-bp inverted repeats, respectively (Kleckner et al. 1975; So et al. 1979), and might also include sequences such as IS*1* and Tn*3* if their short (<40 bp) terminal inverted repeats are sufficiently stable. Analogous excision mechanisms may operate in eukaryotes, as recent analyses have shown that specific DNA segments are excised from the variable regions of antibody genes during the differentiation of the immune system and that these segments contain terminal inverted repeats (Early et al. 1980). The programmed excision of these antibody gene segments might also be independent of DNA cleavage and may involve the directed copying of a single-stranded-DNA template following DNA denaturation and intrastrand reannealing at specific times during development, essentially as proposed here for the more random events involving Tn*5*.

SUMMARY AND CONCLUSIONS

The results described here provide insights into the transposition, excision, and control of gene expression by the kanamycin-resistance transposon Tn*5*. We described *tet*^{r} insertion and substitution derivatives of Tn*5* in which the ability to transpose depends on the orientation of the *tet*^{r} fragment, probably because of differences in the activity of the transposase protein encoded by the Tn*5*-*tet* derivatives. We found that both 1500-bp inverted repeats of Tn*5* (IS*50*-L and IS*50*-R) are themselves transposons and that a Tn*5* derivative with direct (instead of inverted) repeats transposes with the same frequency and specificity as Tn*5*-wild type. These findings support the views that Tn*5* was formed by the association of a previously immobile *kan*^{r} determinant and a pair of IS elements that furnished the genes and sites for transposition and that perhaps any DNA segment is potentially transposable once sandwiched between a pair of transposons.

Excision of Tn*5* from sites of insertion is independent of its ability to transpose to new sites but it does depend on the terminal inverted repeats of Tn*5*. We propose that "excision" results from the slippage of a nascent DNA strand between the two copies of the 9-bp direct repeats of the target sequence that flank Tn*5* and from the failure to copy the Tn*5* sequence during DNA synthesis. This slippage is likely to be facilitated, or even instigated, by denaturation and intrastrand reannealing of the region containing Tn*5*.

In Tn*5*-112, a terminator that stops transcription initiated at the *kan*^{r} promoter of Tn*5* is deleted. Consequently, the effect of Tn*5*-112 on the expression of nearby genes depends on its orientation. When Tn*5*-112 is in the OFF orientation in an operon, *recA*$^+$-dependent homologous recombination inverts Tn*5*-112 and permits transcription from its *kan*^{r} promoter to stimulate distal gene expression. In *recA*$^-$ cells, another type of event results in a major DNA sequence rearrangement, possibly the formation of an inverted dimer of the entire F′ episome. Thus, Tn*5*-112 provides a model for a mobile recombinational switch of the type likely to be important in modulating the expression of particular genes and it may also provide insights into the role of the *recA* protein in the maintenance of the integrity of prokaryotic genomes.

ACKNOWLEDGMENTS

We are grateful to Dr. C. M. Berg for stimulating discussion and critical review of the manuscript. This work was supported by U.S. Public Health Service grant AI 14267 to D. E. Berg and grant GM 19670 to W. R. Reznikoff.

REFERENCES

Backman, K., M. Betlach, H. W. Boyer, and S. Yanofsky. 1979. Genetic and physical studies on the replication of ColE1-type plasmids. *Cold Spring Harbor Symp. Quant. Biol.* **43:** 69.

Beck, C. F. 1979. A genetic approach to analysis of transposons. *Proc. Natl. Acad. Sci.* **76:** 2376.

Berg, D. E. 1974. Genetic evidence for two types of gene arrangements in new λ*dv* plasmid mutants. *J. Mol. Biol.* **86:** 59.

———. 1977. Insertion and excision of the transposable kanamycin resistance determinant Tn*5*. In *DNA insertion elements, plasmids, and episomes* (ed. A. I. Bukhari et al.), p. 205. Cold Spring Harbor Laboratory, Cold Spring Harbor, New York.

———. 1980. Control of gene expression by a mobile recombinational switch. *Proc. Natl. Acad. Sci.* **77:** 4880.

Berg, D. E. and M. Drummond. 1978. Absence of DNA sequences homologous to transposable element Tn*5* (Kan) in the chromosome of *Escherichia coli* K12. *J. Bacteriol.* **136:** 419.

Berg, D. E., A. Weiss, and L. Crossland. 1980. The polarity of Tn*5* insertion mutations in *Escherichia coli*. *J. Bacteriol.* **142:** 439.

BERG, D. E., J. DAVIES, B. ALLET, and J. D. ROCHAIX. 1975. Transposition of R factor genes to bacteriophage λ. *Proc. Natl. Acad. Sci.* **72:** 3628.

BIRNBOIM, H. C. and J. DOLY. 1979. A rapid alkaline extraction procedure for screening recombinant plasmid DNA. *Nucleic Acids Res.* **7:** 1513.

CALOS, M. and J. H. MILLER. 1980. Transposable elements. *Cell* **20:** 579.

CALOS, M., L. JOHNSRUD, and J. H. MILLER. 1978. DNA sequence at the integration sites of the insertion element IS*1*. *Cell* **13:** 411.

CHOW, L. T., N. DAVIDSON, and D. E. BERG. 1974. Electron microscope study of the structures of λ*dv* DNAs. *J. Mol. Biol.* **86:** 69.

EARLY, P., H. HUANG, M. DAVIS, K. CALAME, and L. HOOD. 1980. An immunoglobulin heavy chain region gene is generated from three segments of DNA: V_H, D, and J_H. *Cell* **19:** 981.

FALKOW, S. 1975. *Infectious multiple drug resistance.* Pion, London.

FARABAUGH, P. I., U. SCHMEISSNER, M. HOFER, and J. H. MILLER. 1978. Genetic studies on *lac* repressor. VII. On the molecular nature of spontaneous hotspots in the *lacI* gene of *Escherichia coli. J. Mol. Biol.* **126:** 847.

GALAS, D. 1978. An analysis of sequence repeats in the *lacI* gene of *Escherichia coli. J. Mol. Biol.* **126:** 858.

GUYER, M. 1978. The γ-δ sequence of F is an insertion sequence. *J. Mol. Biol.* **126:** 347.

JORGENSEN, R. A., S. J. ROTHSTEIN, and W. S. REZNIKOFF. 1979a. A restriction endonuclease cleavage map of Tn*5* and location of a region encoding neomycin resistance. *Mol. Gen. Genet.* **177:** 65.

JORGENSEN, R. A., D. E. BERG, B. ALLET, and W. S. REZNIKOFF. 1979b. Restriction endonuclease cleavage map of Tn*10*, a transposon which encodes tetracycline resistance. *J. Bacteriol.* **137:** 681.

KLECKNER, N., R. CHAN, B.-K. TYE, and D. BOTSTEIN. 1975. Mutagenesis by insertion of a drug resistance element carrying an inverted repetition. *J. Mol. Biol.* **97:** 561.

KUTSUKAKE, K. and T. IINO. 1980. A *trans*-acting factor mediates inversion of a specific DNA segment in flagellar phase variation in *Salmonella. Nature* **284:** 479.

MACHATTIE, L. and J. B. JACKOWSKI. 1977. Physical structure and deletion effects of the chloramphenicol resistance element Tn*9* in phage lambda. In *DNA insertion elements, plasmids, and episomes* (ed. A. I. Bukhari et al.), p. 219. Cold Spring Harbor Laboratory, Cold Spring Harbor, New York.

MASAMUNE, Y. and C. C. RICHARDSON. 1971. Strand displacement during deoxyribonucleic acid synthesis at single strand breaks. *J. Biol. Chem.* **246:** 2692.

MEYER, R., G. BOCH, and J. SHAPIRO. 1979. Transposition of DNA inserted into deletions of the Tn*5* kanamycin resistance element. *Mol. Gen. Genet.* **171:** 7.

MILLER, J. H. 1972. *Experiments in molecular genetics.* Cold Spring Harbor Laboratory, Cold Spring Harbor, New York.

MURRAY, K. and N. MURRAY. 1975. Phage lambda receptor chromosomes for DNA fragments made with restriction endonuclease III of *Haemophilus influenzae* and restriction endonuclease I of *Escherichia coli. J. Mol. Biol.* **98:** 551.

RICHARDSON, C., R. B. INMAN, and A. KORNBERG. 1964. Enzymic synthesis of deoxyribonucleic acid. XVIII. The repair of partially single stranded DNA templates by DNA polymerase. *J. Mol. Biol.* **9:** 46.

ROSAMOND, J., K. M. TELANDER, and S. LINN. 1979. Modulation of the action of the *recBC* enzyme of *Escherichia coli* K12 by Ca^{2+}. *J. Biol. Chem.* **254:** 8646.

ROSS, D. G., P. GRISAFI, N. KLECKNER, and D. BOTSTEIN. 1979. The ends of Tn*10* are not IS*3*. *J. Bacteriol.* **139:** 1097.

ROTHSTEIN, S. J., R. A. JORGENSEN, K. POSTLE, and W. S. REZNIKOFF. 1980. The inverted repeats of Tn*5* are functionally different. *Cell* **19:** 795.

SCHALLER, H. 1979. The intergenic region and the origins for filamentous phage DNA replication. *Cold Spring Harbor Symp. Quant. Biol.* **43:** 401.

SHARP, P. A., S. N. COHEN, and N. DAVIDSON. 1973. Electron microscope heteroduplex studies of sequence relations among plasmids of *Escherichia coli.* II. Structure of drug resistance (R) factors and F factors. *J. Mol. Biol.* **75:** 235.

SHAW, K. J. and C. M. BERG. 1979. *Escherichia coli* auxotrophs induced by insertion of the transposable element Tn*5*. *Genetics* **92:** 741.

SILVERMAN, M. and M. SIMON. 1980. Phase variation: Genetic analysis of switching mutants. *Cell* **19:** 845.

SO, M., F. HEFFRON, and B. MCCARTHY. 1979. The *E. coli* gene encoding heat stable toxin is a bacterial transposon flanked by inverted repeats of IS*1*. *Nature* **277:** 453.

STARLINGER, P. 1980. IS elements and transposons. *Plasmid* **3:** 241.

VAN DE PUTTE, P., S. CRAMER, and M. GIPHART-GASSLER. 1980. Invertible DNA determines host specificity of bacteriophage Mu. *Nature* **286:** 218.

YARRONTON, G. T. and M. L. GEFTER. 1979. Enzyme catalyzed DNA unwinding: Studies on *Escherichia coli rep* proteins. *Proc. Natl. Acad. Sci.* **76:** 1658.

Analysis of the Structure and Function of the Kanamycin-resistance Transposon Tn*903*

N. D. F. GRINDLEY AND C. M. JOYCE
Department of Molecular Biophysics and Biochemistry, Yale University School of Medicine, New Haven, Connecticut 06510

The kanamycin-resistance transposon Tn*903* (Nomura et al. 1978) was first identified as a constituent of the F-like R plasmid R6 (Sharp et al. 1973) and is a component of the cloning vector pML21 (Hershfield et al. 1976). It is one of the class of prokaryotic transposable elements that generate duplication of a 9-bp host sequence upon integration (Oka et al. 1978; for reviews, see Grindley 1979; Starlinger 1980; Calos and Miller 1980). Tn*903* contains a 1000-bp unique region flanked by a pair of inverted-repeat sequences of 1050 bp. This structure is characteristic of many of the transposons that generate 9-bp repeats, e.g., Tn*5*, Tn*10*, and Tn*1681* (Berg et al. 1975; Kleckner et al. 1975; So et al. 1979). In the case of at least two of these, the inverted-repeat sequence either is known to transpose (Tn*10*) (Sharp et al. 1973) or has been identified as a transposable element characterized previously (Tn*1681*) (So et al. 1979).

We have undertaken a genetic and DNA sequence analysis of Tn*903*, and our results are summarized in this paper. A 1050-bp inverted repeat of Tn*903* is shown to act as an insertion sequence (IS) element, which we have designated IS*903*. We have constructed a derivative of Tn*903* (Tn*903*Δ1) that lacks a 520-bp segment in one of its 1050-bp inverted repeats but that can still transpose. Insertion of a 10-bp fragment at several sites within the nondeleted arm resulted in an inability to transpose. Correlation of the DNA sequence of the inverted repeat with the sites of the insertion mutations suggests that translation of a 921-bp coding frame, contained entirely within the 1050-bp repeat, is required for transposition. These studies are described in detail elsewhere (Grindley and Joyce 1980). In addition, we have started an analysis of replicon fusions mediated by the partially deleted derivative Tn*903*Δ1. Preliminary results suggest that the Tn*903* "transposase" is probably a *cis*-acting protein. We discuss the observed structures of the cointegrates in relation to current molecular models for transposition.

RESULTS AND DISCUSSION

Deletion Derivatives of Tn*903*

Restriction mapping of Tn*903* indicated that each 1050-bp inverted repeat contains two *Pvu*II sites separated by about 520 bp (see Figs. 1 and 2). To use these sites to create deletions in Tn*903*, we first cloned the 5000-bp *Eco*RI-*Sal*I fragment from pML21 into the *Eco*RI-*Sal*I sites of pNG16, a derivative of pBR322 that lacks the *Pvu*II site (see Fig. 1). The resulting plasmid, pNG18, was digested with *Pvu*II, and the DNA mixture was religated. Kanamycin-resistant (Km^r) transformants were selected and their plasmid DNAs analyzed. Derivatives were obtained that had lost the 520-bp *Pvu*II fragment from right (pNG35), left (pNG24), or both (pNG23) inverted repeats of Tn*903* but had retained the central region encoding kanamycin resistance in its original orientation. Kanamycin resistance could still transpose from pNG35 or pNG24 to an F plasmid, but no transposition from pNG23 could be detected. Transposition from pNG35 was as efficient as that from the parental plasmid (pNG18), whereas transposition from pNG24 was reduced about tenfold (the reason for this reduction is not yet clear). These results indicate that the 1050-bp flanking repeat supplies a function (or a site within the central 520-bp region) necessary for transposition of Tn*903* and that only one of the repeats needs to be intact.

Insertion Mutations within Tn*903*Δ1

The transposon present on pNG35, which we have designated Tn*903*Δ1, has been subjected to a further genetic analysis using in vitro mutagenesis (Heffron et al. 1978). We have generated mutations in the intact 1050-bp flanking sequence by inserting into several specific restriction sites a 10-bp DNA fragment that contains a *Bam*HI recognition site (for details, see Grindley and Joyce 1980). The sites of insertion were the *Pvu*II sites at positions 180 and 700, the *Sma*I sites at positions 450 and 650, and the *Ava*II site at position 950 (see Fig. 2). At all five sites we have obtained insertion mutations that cause a loss of kanamycin-resistance transposition. Surprisingly, one mutant with an insert at *Sma*I-650 (from a total of five at this site) could still transpose, but with a frequency reduced by about tenfold.

We have checked the structure of several mutants by DNA sequencing. Significantly, the insertion at *Sma*I-650 that does not prevent transposition is 9 bp long, whereas another insertion at the same site that does abolish transposition is the expected 10 bp long; only the latter would cause a shift in a coding frame. This result suggests very strongly that the DNA region at position 650 is translated into protein. All the other insertions analyzed would result in frameshifts.

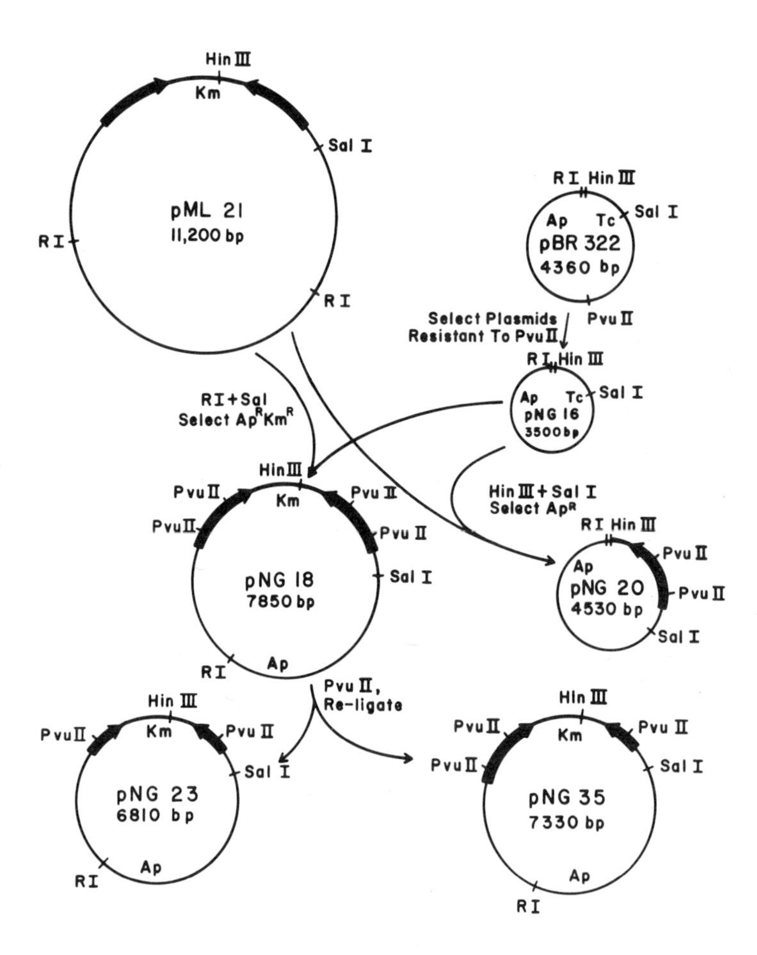

Figure 1. Relationships between the plasmids used in this study. Tn*903* is represented by two heavy arrows and the intervening segment marked Km. Km, Ap, and Tc indicate resistances to kanamycin, ampicillin, and tetracycline, respectively.

The 1050-bp Repeat of Tn*903* Is an IS Element

A nonconjugative, nonmobilizable plasmid can be transferred by a conjugative plasmid if one of them carries a transposable element that can mediate fusion with the other (Guyer 1978). Cointegrates formed in this manner contain two copies of the element, each in the same orientation and one at each junction between the replicons (see Fig. 3, pathway i) (Faelen et al. 1975; Gill et al. 1978; Shapiro and MacHattie 1979). We have constructed a plasmid (pNG20; see Fig. 1) that contains the right 1050-bp repeat of Tn*903,* carried on a 1650-bp *Hin*III-*Sal*I fragment from pML21, cloned into the *Hin*III-*Sal*I sites of pNG16 (a nonconjugative, nonmobilizable plasmid). The ampicillin resistance (Ap^r) encoded by pNG16 was not transferred at a detectable rate from a $RecA^-$ strain that carries pNG16 and FΔ(0–15) (a derivative of F that lacks the active transposable element

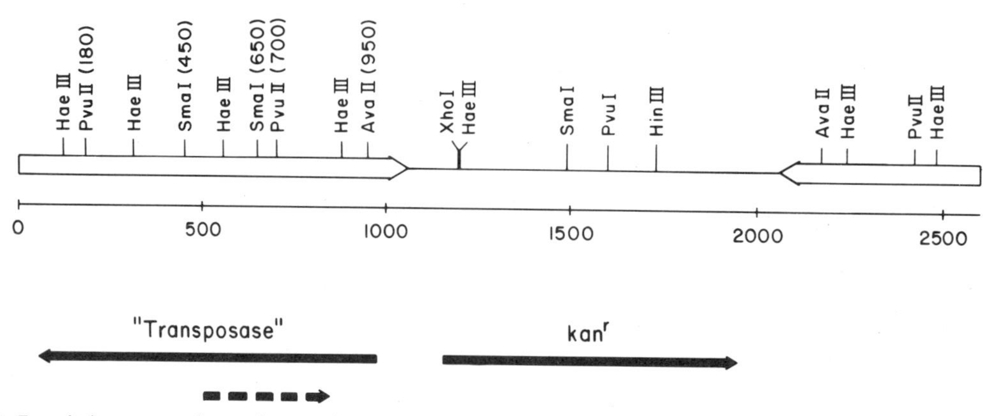

Figure 2. Restriction map and genetic organization of Tn*903*Δ1. The numbers in parentheses are distances in base pairs from the left end of the transposon. The restriction sites so marked are those at which mutations have been constructed by insertion of a *Bam*HI linker fragment (see text). Both polypeptides potentially encoded by the inverted repeat are shown, although we have no evidence supporting the existence of the smaller one.

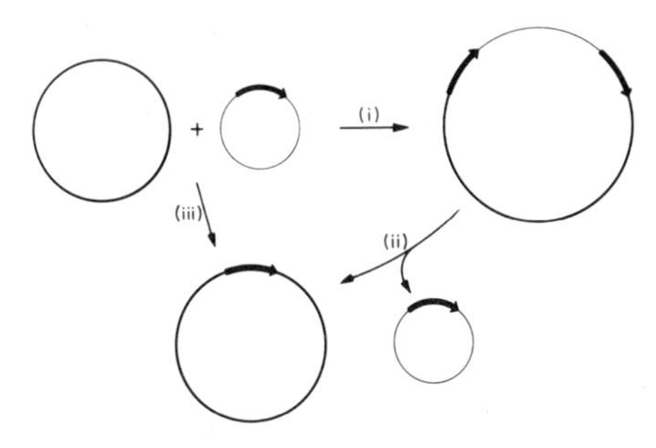

Figure 3. Replicon fusion and transposition mediated by an insertion element. The insertion element is represented by the heavy arrow. Replicon fusion (*i*) results in formation of a cointegrate that contains two copies of the element, both in the same orientation and one at each junction between the plasmids. A cointegrate can dissociate (*ii*) by recombination between the two copies of the element. The result of dissociation is equivalent to a simple transposition of the element (*iii*). It is not known whether transposition can occur as a single-step process as shown in pathway *iii* or whether a combination of pathways *i* and *ii* is required.

$\gamma\delta$) (Sharp et al. 1972). In contrast, ampicillin resistance was transferred at a low rate (about 10^{-6} to 10^{-7} per F^+ transconjugant) from a $RecA^-$ strain carrying pNG20 and FΔ(0–15). This result suggested that the cloned fragment in pNG20 contained an insertion element.

Analysis of the F^+Ap^r transconjugants showed that they contained a single, large plasmid. Digestions with *Eco*RI and with *Sma*I showed that pNG20 had inserted into many sites in F and that the fusion was accompanied by duplication of the 1050-bp repeat sequence of Tn*903* (Grindley and Joyce 1980) (see Fig. 4). This repeated sequence is therefore an insertion sequence,

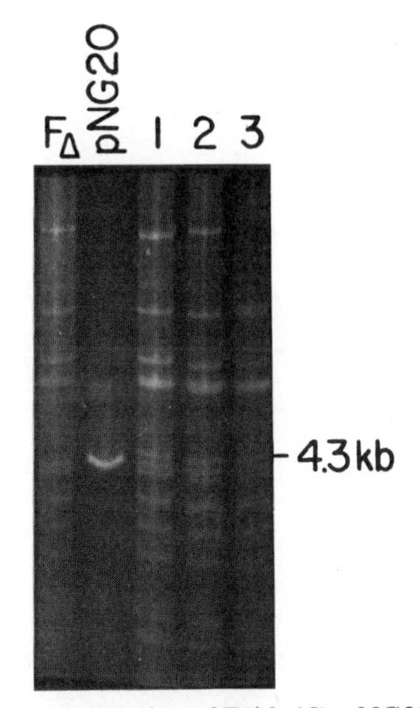

Figure 4. *Sma*I digestion of F(Δ0–15)::pNG20 cointegrates. Digestion of each of three independent FΔ(0–15)::pNG20 cointegrates (*1–3*) produces the 4300-bp fragment that is also produced by digestion of pNG20. As *Sma*I cuts pNG20 only within the 1050-bp inverted repeat of Tn*903*, this result shows that the pNG20 sequences are flanked by direct repeats of the 1050-bp element IS*903*.

which we have designated IS*903*. It should be noted that Ohtsubo et al. (1980) have detected a 1000-bp insertion sequence, designated IS*102*, as a natural component of pSC101. By the criteria of partial DNA sequencing and restriction mapping, IS*102* is highly homologous, but not identical, to IS*903*.

DNA Sequencing of Tn*903*

We have determined the nucleotide sequence of the intact 1050-bp repeat of Tn*903*Δ1 and also of all the regions of Tn*903* carried on pNG20. The strategy for sequencing Tn*903*Δ1 and for defining the internal ends of the inverted repeats is described in detail elsewhere (Grindley and Joyce 1980). The DNA sequence of the left (intact) arm of Tn*903*Δ1 is identical with that of pNG20 (the right arm of Tn*903*). However, we do not know whether the 520-bp *Pvu*II fragment in the intact arm of Tn*903*Δ1 was derived from the left or right repeat of Tn*903*. It is possible, therefore, that the natural left repeat of Tn*903* differs in sequence from the right repeat within the region between the two *Pvu*II sites. The deduced sequence of Tn*903* (Fig. 5) shows the following features:

1. Structure of IS*903*. The inverted repeats of Tn*903* are 1057 bp long. The 18 bp at the inside end of each is an exact inverted repeat of the 18 bp at the outside end. Such terminal inverted repeats are characteristic of other IS elements and presumably constitute the recognition sites for transposase action.
2. Potential coding regions in IS*903*. The IS*903* sequence has two long, uninterrupted coding regions, one from each DNA strand. One runs from the potential initiation codon GTG at position 980 to the TGA stop codon at position 59, a stretch of 307 codons. The other partially overlaps the first but runs in the reverse direction, from the ATG at position 507 to the TGA at position 849, a distance of 114 codons. These are the only two coding regions that include the *Sma*I-650 site, through which it appears that translation must proceed for transposition to occur (see above). They are also the only translatable frames with potential products longer than 70 amino acids. We believe that the longer of the two frames encodes the required protein. This frame includes all five sites of transposition-defective insertion mutations, whereas the shorter frame includes only two of these insertion sites. The existence of an insertion mutation at *Ava*II-950 suggests that the gene product, called here the transposase, is the full 307 amino acids long and does not start at a subsequent initiation codon (see Grindley and Joyce 1980). Translation of this coding frame is shown in Figure 5. The amino acid composition is strongly basic: 53 arginine and lysine residues, compared with 28 aspartate and glutamate. This contrasts with the transposase of Tn*3*, which is essentially neutral (Heffron et al. 1979).
3. Transcription signals in IS*903*. Although we believe we have deduced the location of the translational start of the transposase gene, examination of the

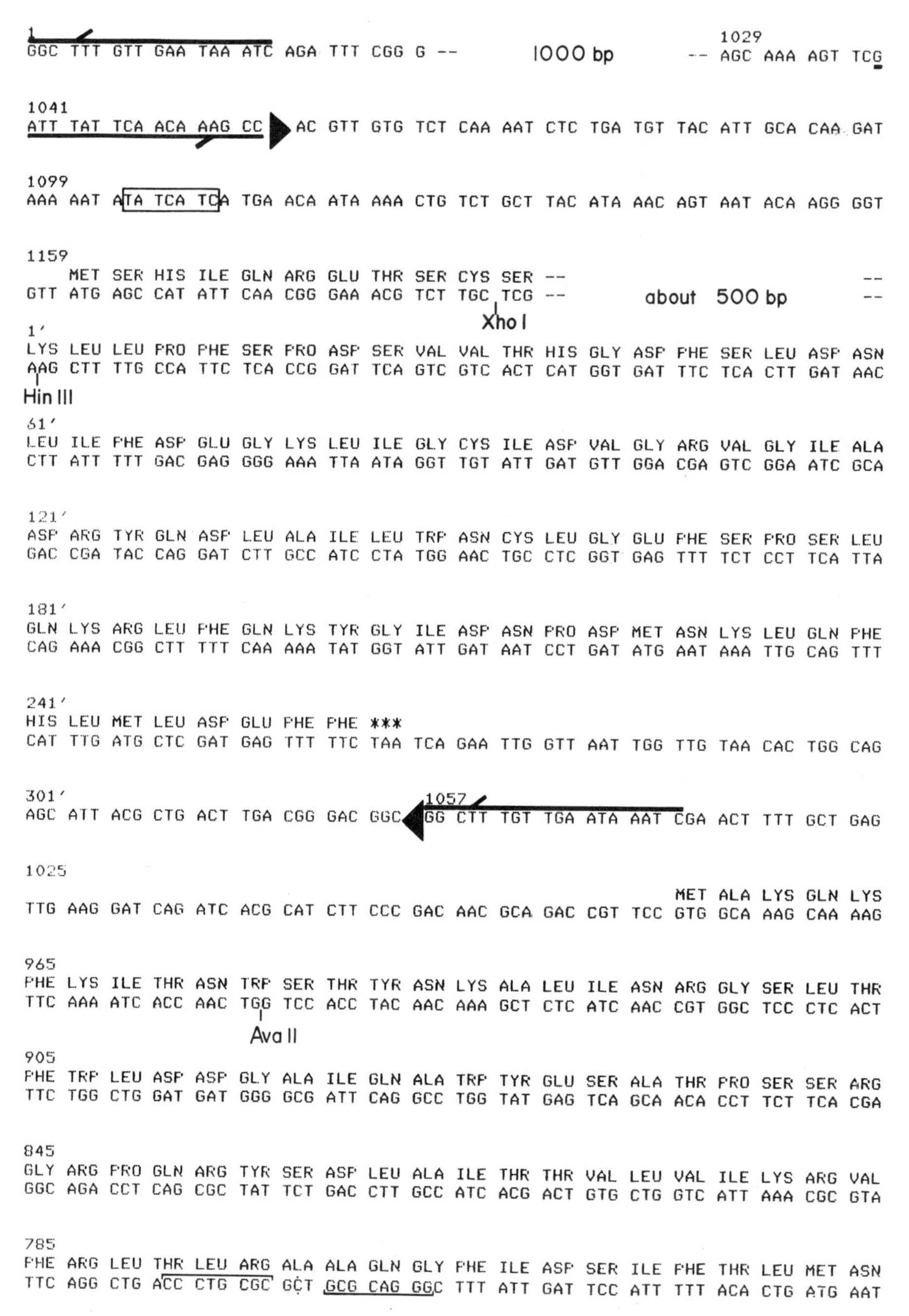

Figure 5. DNA sequences of Tn*903*. The transposon is flanked by inverted repeats of the 1057-bp IS*903*; these are numbered 1–1057 and 1057–1. The inside ends of each are indicated by arrowheads, and the 18-bp terminal repeats are indicated by heavy lines. The unique region containing the *kan* gene runs from position 1058 to the *Xho*I site at position 1195, across an unsequenced stretch of 500 bp to the *Hind*III site at 1′, and ends at position 326′. Regions presumed to encode the kanamycin phosphotransferase and the transposase are shown translated.

DNA sequence upstream shows no Shine-Dalgarno sequence (Shine and Dalgarno 1974) and no obvious promoter sequences. This may indicate that the gene is expressed only at a low rate. Consistent with this interpretation, we, using the maxicell system of Sancar et al. (1979), and J. Davies (pers. comm.), using minicells (Roozen et al. 1971), have not been able to detect a polypeptide encoded by Tn*903* other than the kanamycin phosphotransferase. (None of the alternative initiation sites in the 921-bp reading frame or in the 342-bp coding frame on the opposite strand have Shine-Dalgarno sequences.) Possible sites for termination of transcription (Rosenberg and Court 1979) occur in both orientations within IS*903*. The strongest candidate occurs within the presumed transposase-encoding sequence and consists of an 8-base inverted repeat (which contains seven G + C residues), followed by a stretch rich in T residues

```
725
VAL PRO LEU ARG CYS PRO ASP TYR SER CYS VAL SER ARG ARG ALA LYS SER VAL ASN ILE
GTT CCG TTG CGC TGC CCG GAT TAC AGC TGT GTC AGC AGG CGG GCA AAG TCG GTT AAT ATC
                                  |
                                PvuII

665
SER PHE LYS THR PHE THR ARG GLY GLU ILE ALA HIS LEU VAL ILE ASP SER THR GLY LEU
AGT TTC AAA ACG TTC ACC CGG GGT GAA ATC GCG CAT CTG GTG ATT GAT TCC ACC GGG CTG
                         |
                        SmaI

605
LYS VAL PHE GLY GLU GLY GLU TRP LYS VAL LYS LYS HIS GLY GLN GLU ARG ARG ARG ILE
AAG GTC TTT GGT GAA GGC GAA TGG AAA GTC AAA AAG CAT GGC CAG GAA CGC CGC CGT ATC

545
TRP ARG LYS LEU HIS LEU ALA VAL ASP SER LYS THR HIS GLU ILE ILE CYS ALA ASP LEU
TGG CGT AAG CTG CAT CTC GCC GTT GAC AGT AAA ACA CAT GAA ATC ATC TGC GCT GAC CTG

485
SER LEU ASN ASN VAL THR ASP SER GLU ALA PHE PRO GLY LEU ILE ARG GLN THR HIS ARG
TCG CTG AAC AAT GTG ACC GAC TCA GAA GCC TTC CCG GGT CTT ATC CGG CAG ACT CAC AGA
                                             |
                                            SmaI

425
LYS ILE ARG ALA ALA SER ALA ASP GLY ALA TYR ASP THR ARG LEU CYS HIS ASP GLU LEU
AAA ATC AGG GCA GCA TCG GCA GAC GGC GCT TAC GAC ACC CGG CTC TGT CAC GAT GAA CTG

365
ARG ARG LYS LYS ILE SER ALA LEU ILE PRO PRO ARG LYS GLY ALA GLY TYR TRP PRO GLY
CGG CGT AAG AAA ATC AGC GCG CTT ATC CCT CCC CGA AAA GGT GCG GGT TAC TGG CCC GGT

305
GLU TYR ALA ASP ARG ASN ARG ALA VAL ALA ASN GLN ARG MET THR GLY SER ASN ALA ARG
GAA TAT GCA GAC CGT AAC CGT GCA GTG GCT AAT CAG CGA ATG ACC GGG AGT AAT GCG CGG

245
TRP LYS TRP THR THR ASP TYR ASN ARG ARG SER ILE ALA GLU THR ALA MET TYR ARG VAL
TGG AAA TGG ACA ACA GAT TAC AAC CGT CGC TCG ATA GCG GAA ACG GCG ATG TAC CGG GTA

185
LYS GLN LEU PHE GLY GLY SER LEU THR LEU ARG ASP TYR ASP GLY GLN VAL ALA GLU ALA
AAA CAG CTG TTC GGG GGT TCA CTG ACG CTG CGT GAC TAC GAT GGT CAG GTT GCG GAG GCT
       |
     PvuII

125
MET ALA LEU VAL ARG ALA LEU ASN LYS MET THR LYS ALA GLY MET PRO GLU SER VAL ARG
ATG GCC CTG GTA CGA GCG CTG AAC AAA ATG ACG AAA GCA GGT ATG CCT GAA AGC GTG CGT

65
ILE ALA ***
ATT GCC TGA AAA CAC AAC CCG CTA CGG GGG AGA CTT ACC CGA AAT CTG ATT TAT TCA ACA

5
AAG CC
```

Figure 5. *(Continued)*

(positions 775 to 750; see Fig. 5). Termination at this site would result in attenuation of the transposase transcript.

4. Kanamycin-resistance gene (*kan*). Our own results (N. Grindley, unpubl.) and those of others (Casadaban and Cohen 1980) suggest that *kan* is expressed from left to right (Fig. 5) and includes the *Xho*I site. There is a potential initiation codon (ATG) at position 1162 that is preceded by the Shine-Dalgarno sequence AAGGGG at position 1152. Upstream lies a region with the characteristics of a transcriptional promoter (Rosenberg and Court 1979): TATCATC at the Pribnow box (positions 1106 to 1112) and TGTTACATTG at the −35 region. We suggest that transcription and translation of the kanamycin phosphotransferase start at these sites. The gene product has been identified as a polypeptide with a molecular weight of 27,500 (Andrés et al. 1979) and would therefore require about 250 codons. About 500 bp separate the *Xho*I site from the *Hind*III site (see Figs. 2 and 5). Only one of the three possible reading frames starting at the *Hind*III site is long enough to encode the C-terminal portion of kanamycin phosphotransferase; this is shown translated in Figure 5.

Preliminary Study of Replicon Fusions Mediated by Tn*903*Δ1

In Figure 3, pathway i illustrates replicon fusion mediated by an insertion element. Most current molecular models for transposition can explain the fusion of two replicons by a transposable element (Arthur and Sherratt 1979; Grindley and Sherratt 1979; Shapiro 1979; Harshey and Bukhari, this volume). However, the model proposed by Shapiro (1979) and Arthur and Sherratt (1979) requires that fusion of two replicons be an intermediate in transposition, i.e., transposition occurs exclusively via pathways i and ii together and not by the direct pathway (pathway iii) allowed in the other models. A corollary of the Shapiro–Arthur-Sherratt model is that resolution of cointegrates (pathway ii) in a $RecA^-$ strain must be transposon-mediated. (Clearly, resolution could also be accomplished by the *recA* recombination system, as it involves a homologous reciprocal recombination.)

We have undertaken a study of replicon fusions mediated by Tn*903*Δ1. This is a complex transposon that is composed of two distinguishable IS elements, IS*903* and the deleted IS*903*Δ1. These two, in combination with the *kan* gene, form a third element, Tn*903*Δ1. As shown in Figure 6, one might expect any one of these three elements to mediate the fusion and, consequently, be duplicated in the resulting cointegrate. The three structures, A, B, and C, are the intermediates postulated for transposition of IS*903*, IS*903*Δ1, and Tn*903*Δ1, respectively, according to the Shapiro–Arthur-Sherratt model.

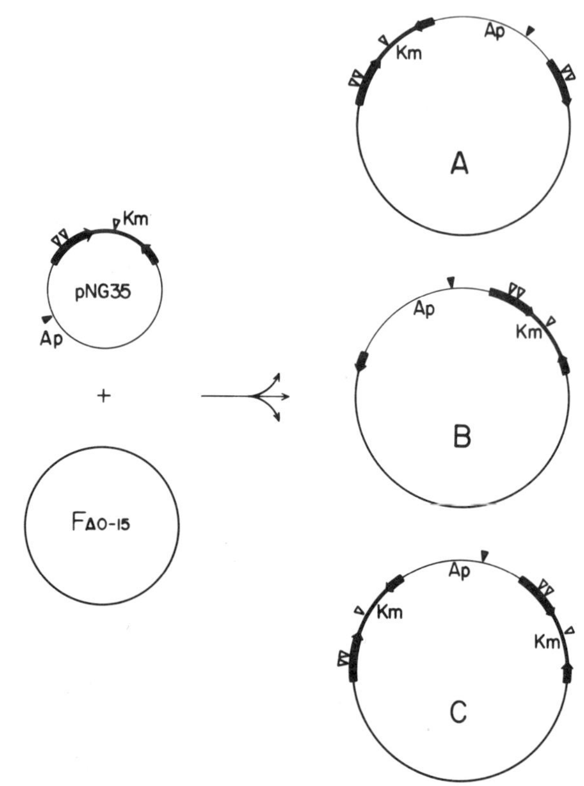

Figure 6. Replicon fusion mediated by Tn*903*Δ1: alternative structures of cointegrates. Replicon fusion results in a direct repeat of the element that mediated the fusion (see Fig. 3). A complex transposon such as Tn*903*Δ1 actually contains three transposable elements: the intact IS*903*, the deleted IS*903*Δ1, and the complete Tn*903*Δ1. Each of these three presumably could mediate the fusion resulting in the three structures shown in *A*, *B*, and *C*, respectively. *Tn903*Δ1 is represented by the heavy arrows and the intervening segment marked Km (kanamycin resistance). Ap (ampicillin resistance) is a marker on pNG35 outside the transposon. (▽) *Sma*I sites on the pNG35 sequences; (▼) single *Eco*RI site. FΔ(0–15) contains several *Sma*I and *Eco*RI sites, but these are not shown.

We have obtained 20 independent cointegrates between FΔ(0–15) and pNG35 by selecting for simultaneous transfer of Ap^r and Km^r from a $RecA^-$ strain carrying FΔ(0–15) and pNG35 to a $RecA^-$ recipient. The cointegrates are stable in the recipient strain, an observation that contrasts with the extreme instability of Tn*3*-mediated or γδ-mediated cointegrates (Guyer 1978; Sherratt et al., this volume). Analysis of all 20 cointegrates shows them to have structure A, i.e., fusion was mediated by the intact copy of IS*903*. Digestion with *Eco*RI (which cuts pNG35 just once; see Fig. 1) showed that each cointegrate had lost a single F fragment and had gained two new fragments (data not shown). These are equal in total size to the missing F fragment plus pNG35 plus an additional 1100 bp. (In place of the additional 1100 bp, structure B would give an additional 550 bp, and structure C an extra 2600 bp.) Furthermore, digestion with *Sma*I (which cuts once in the *kan* gene and makes two cuts in the intact IS*903*, but none in IS*903*Δ1) produced a novel fragment of 6300 bp, identical with that produced from pNG35 itself (Fig. 7). This fragment would only be produced from structures A and C (see Fig. 6). Finally, by passage through a $RecA^+$ host, we have obtained the dissociated F^+Ap^s plasmids from ten separate cointegrates. In each case, kanamycin resistance was also lost (dissociation of structure C would lead to retention of Km^r on the F plasmid), and *Eco*RI digestion showed that the missing F fragment was replaced by a single fragment 1100 bp larger. Therefore, all ten dissociated plasmids are F::IS*903* and were derived from structure A.

Why are all 20 cointegrates formed by replicon fusion mediated by IS*903* and not by IS*903*Δ1 or Tn*903*Δ1? We had anticipated that IS*903*Δ1 would be as effective (or nearly so) as IS*903* because we believed that the transposase produced by the intact element would act at the terminal sites provided by either intact or deleted arms. IS*903*Δ1 consists of 180 bp from the outside end and 360 bp from the inside end of IS*903* and therefore probably contains the terminal sites necessary for transposition. That Tn*903*Δ1 transposes with the same efficiency as Tn*903* demonstrates that the 180-bp outside end is sufficient. Transposition of Ap^r from pNG35 in the absence of Km^r demonstrates that the inside end of the deleted IS*903*Δ1 is also an active site (N. Grindley, unpubl.).

The observation that IS*903* is strongly preferred over IS*903*Δ1 as the mediator of replicon fusion can be explained in two ways. First, the asymmetry could be

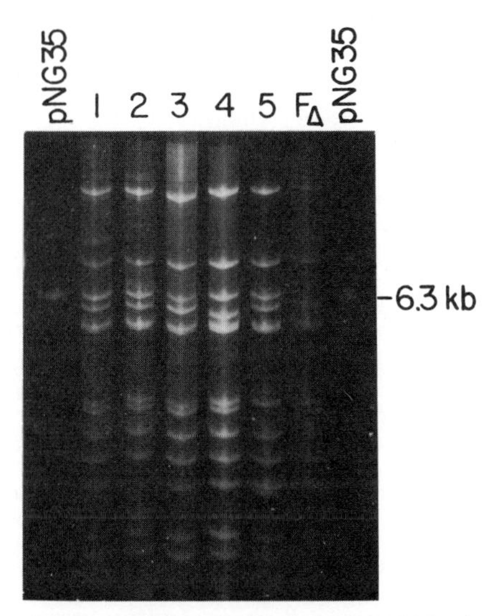

Figure 7. *Sma*I digestion of FΔ(0–15)::pNG35 cointegrates. Digestion of five independent FΔ(0–15)::pNG35 cointegrates produces the 6800-bp fragment that is also produced by digestion of pNG35 but is absent from the digest of FΔ(0–15). As *Sma*I cuts pNG35 only within the intact IS*903* (twice) and the *kan* gene (once) but not in the deleted IS*903*Δ1, this result shows that a directly repeated copy of IS*903* or Tn*903*Δ1 (but not of IS*903*Δ1) lies at each junction between the fused replicons.

caused by external influences. For example, replication or transcription of pNG35 might run into Tn*903*Δ1 from one direction only and thereby activate (or inhibit) the initiation of transposition from that end.

An alternative explanation, and one that we prefer, is that the transposase of Tn*903* is *cis*-acting and acts first at the terminal site immediately adjacent to the region encoding the C terminus of the protein, i.e., the outside end of the intact IS*903* of Tn*903*Δ1 (see Fig. 2). Consistent with the idea of a *cis*-acting transposase for Tn*903*, we have been unable to complement the transposition deficiency of two Tn*903* mutants. Using the plasmid pSC101, which contains IS*102*, an element highly homologous to IS*903* and with identical terminal repeats (Ohtsubo et al. 1980), we find no increase in Km^r transposition from pNG23 (which contains the doubly deleted derivative of pNG18; see Fig. 1) or from pNG35 with the insertion mutation at the *Ava*II-950 site. Failure to observe complementation of transposition mutants has also been noted by Kleckner et al. (this volume) with Tn*10* and by Ohtsubo et al. (this volume) with IS*1*. These two elements and Tn*903* belong to the class that generate 9-bp repeats upon integration. These results contrast with the complementation that is readily observed with Tn*3* and γδ mutants (Heffron et al. 1977; R. Reed, pers. comm.) and with the element IS*101* from pSC101 (Ravetch et al. 1979; Miller and Cohen 1980), all of which generate 5-bp repeats upon integration. The *cis* action of a transposase could be readily explained by assuming that the protein is usually present in very low amounts (<1 copy/cell) and has a very high affinity for its recognition site, perhaps binding to this site even before completion of protein synthesis. A precedent for a *cis*-acting, site-specific endonuclease is provided by the cistron-*A* protein of φX174 (Tessman 1966; Franke and Ray 1972).

To determine whether the asymmetry is caused by a *cis*-acting transposase or by external influences, we need to study cointegrates formed by a plasmid, such as pNG24, in which the left arm of Tn*903* contains the deletion. These experiments are in progress.

Are Cointegrates Intermediates in Transposition?

Although a *cis*-acting transposase could explain our failure to detect IS*903*Δ1-mediated fusions (Fig. 6B), it does not explain the lack of Tn*903*Δ1-mediated fusions (Fig. 6C). Experimentally, we find that Tn*903*Δ1 transposes at about the same frequency as IS*903* mediates replicon fusion. Therefore, to postulate cointegrates with two copies of the entire Tn*903*Δ1 (Fig. 6C) as intermediates in transposition (as required by the Shapiro–Arthur-Sherratt model), one would have to argue that Tn*903*Δ1 cointegrates are very short-lived. This could be either because they are unstable even in the $RecA^-$ hosts used (unlikely because of the observed stability of IS*903* cointegrates; Fig. 6A) or because resolution of cointegrates to give transposition (Fig. 3, pathway ii) generally is coupled to the initial replicon fusion. A simpler explanation for the lack of cointegrates containing duplicated Tn*903*Δ1 is that these are not intermediates in Tn*903*Δ1 transposition and that for elements such as Tn*903*, replicon fusion represents a recombinational pathway that is an alternative to simple transposition (Grindley and Sherratt 1979).

On the other hand, in the case of Tn*3* there is strong evidence that cointegrates are intermediates in transposition and that their resolution is transposon-mediated. Cointegrates containing a direct repeat of the complete Tn*3* transposon are the only products of transposition if Tn*3* contains a mutation in the *tnpR* gene (Arthur and Sherratt 1979) or lacks the internal resolution site (Gill et al. 1978). These are the only conditions under which Tn*3* cointegrates are observed, although transient fusions mediated by a nonmutant transposon may be inferred from plasmid-mobilization experiments (Sherratt et al., this volume). Apparently, cointegrates involving two copies of the nonmutant Tn*3* are resolved rapidly.

Recent results of Sherratt and coworkers (Arthur and Sherratt 1979; Sherratt et al., this volume) indicate that resolution of Tn*3* cointegrates requires the product of *tnpR* and not the transposase (the product of *tnpA*). At first, this was unexpected, since the product of *tnpR* had been characterized as the repressor of *tnpA* (and *tnpR*) expression (Chou et al. 1979a,b; Gill et al. 1979). However, when one considers the difference between transposition and resolution of cointegrates, it is not so surprising. Transposition is a replicative recombination involving insertion of DNA at a nonhomologous target, whereas cointegrate resolution is a conservative recombination between two homologous sites, analogous to integration and excision in λ (Nash 1977) or phase variation by DNA inversion in *Salmonella* (Silverman et al., this volume). It appears that the internal resolution site in Tn*3* (the site required for *recA*-independent resolution of cointegrates) lies very close to (and may overlap) the site(s) at which the product of *tnpR* effects control of *tnpA* and *tnpR* expression (Heffron et al. 1979). The product of *tnpR* would normally occupy this site and would therefore be on hand to resolve cointegrates formed during a transposition.

The apparent saturation of the coding capacity of IS*903* with the postulated 307-amino-acid transposase suggests that IS*903*, unlike Tn*3*, has no equivalent of the *tnpR* gene and therefore has no separate pathway for resolving cointegrates. This suggestion is supported by the stability of IS*903* cointegrates in $RecA^-$ hosts. The difference in genetic organization between Tn*903* and Tn*3*, the contrasting stabilities of their cointegrates, and the difference in their abilities to resolve cointegrates by a transposon-encoded pathway may reflect a basic difference between the mechanism of transposition of Tn*3*-like elements and that of elements like Tn*903* (e.g., IS*1* and Tn*10*). We feel that the apparently *cis*-acting transposase of Tn*903* (and probably of IS*1* and Tn*10*) favors a model for transposition of these elements in which the transposase acts at a single end of the element and finds the second end (and recognition site) by threading the transposing DNA of the element through the protein-DNA transposition complex. Simple transposition or replicon fusion would then be an alternative consequence of the recombinational process, as described previously (Grindley and Sherratt 1979). This would contrast with the Shapiro–Arthur-Sherratt model, which involves the simultaneous joining of both ends of an element to the new integration site.

ACKNOWLEDGMENTS

We thank Allison Levy for excellent technical assistance. This work was supported by grants GM-25227 and GM-28470 from the U.S. Public Health Service.

REFERENCES

Andrés, I., P. M. Slocombe, F. Cabello, J. K. Timmis, R. Lurz, H. J. Burkardt, and K. N. Timmis. 1979. Plasmid replication functions. II. Cloning analysis of the *repA* replication region of antibiotic resistance plasmid R6-5. *Mol. Gen. Genet.* **168:** 1.

Arthur, A. and D. J. Sherratt. 1979. Dissection of the transposition process: A transposon-encoded site-specific recombination system. *Mol. Gen. Genet.* **175:** 267.

Berg, D., J. Davies, B. Allet, and J.-D. Rochaix. 1975. Transposition of R factor genes to bacteriophage lambda. *Proc. Natl. Acad. Sci.* **72:** 3628.

Calos, M. P. and J. H. Miller. 1980. Transposable elements. *Cell* **20:** 579.

Casadaban, M. J. and S. N. Cohen. 1980. Analysis of gene control signals by DNA fusion and cloning in *E. coli. J. Mol. Biol.* **138:** 179.

Chou, J., M. J. Casadaban, P. G. Lemaux, and S. N. Cohen. 1979a. Identification and characterization of a self-regulated repressor of translocation of the Tn*3* element. *Proc. Natl. Acad. Sci.* **76:** 4020.

Chou, J., P. G. Lemaux, M. J. Casadaban, and S. N. Cohen. 1979b. Transposition protein of Tn*3*: Identification and characterization of an essential repressor-controlled gene product. *Nature* **282:** 801.

Faelen, M., A. Toussaint, and J. de Lafonteyne. 1975. Model for the enhancement of λ-*gal* integration into partially induced Mu-1 lysogens. *J. Bacteriol.* **121:** 873.

Franke, B. and D. S. Ray. 1972. *cis*-limited action of the gene-*A* product of bacteriophage φX174 and the essential bacterial site. *Proc. Natl. Acad. Sci.* **69:** 475.

Gill, R., F. Heffron, and S. Falkow. 1979. Identification of the protein encoded by the transposable element Tn*3* which is required for its transposition. *Nature* **282:** 797.

Gill, R., F. Heffron, G. Dougan, and S. Falkow. 1978. Analysis of the sequences transposed by complementation of two classes of transposition-deficient mutants of Tn*3*. *J. Bacteriol.* **136:** 742.

Grindley, N. D. F. 1979. Integration of transposable DNA elements: Analysis by DNA sequencing. *Proceedings of the 1979 ICN-UCLA Symposium: Extrachromosomal DNA* (ed. D. J. Cummings et al.), p. 155. Academic Press, New York.

Grindley, N. D. F. and C. M. Joyce. 1980. A genetic and DNA sequence analysis of the kanamycin resistance transposon, Tn*903*. *Proc. Natl. Acad. Sci.* (in press).

Grindley, N. D. F. and D. J. Sherratt. 1979. DNA sequence analysis at IS*1* insertion sites: Models for transposition. *Cold Spring Harbor Symp. Quant. Biol.* **43:** 1257.

Guyer, M. 1978. The γδ sequence of F is an insertion sequence. *J. Mol. Biol.* **126:** 347.

Heffron, F., M. So, and B. J. McCarthy. 1978. In vitro mutagenesis using synthetic restriction sites. *Proc. Natl. Acad. Sci.* **75:** 6012.

Heffron, F., P. Bedinger, J. J. Champoux, and S. Falkow. 1977. Deletions affecting the transposition of an antibiotic resistance gene. *Proc. Natl. Acad. Sci.* **74:** 702.

Heffron, F., B. J. McCarthy, H. Ohtsubo, and E. Ohtsubo. 1979. DNA sequence analysis of the transposon Tn*3*: Three genes and three sites involved in transposition of Tn*3*. *Cell* **18:** 1153.

Hershfield, V., H. W. Boyer, L. Chow, and D. R. Helinski. 1976. Characterization of a mini-ColE1 plasmid. *J. Bacteriol.* **126:** 447.

Kleckner, N., R. K. Chan, B.-K. Tye, and D. Botstein. 1975. Mutagenesis by insertion of a drug resistance element carrying an inverted repetition. *J. Mol. Biol.* **97:** 561.

Miller, C. A. and S. N. Cohen. 1980. F plasmid provides a function that promotes *recA*-independent site specific fusions of pSC101 replicon. *Nature* **285:** 577.

Nash, H. A. 1977. Integration and excision of bacteriophage λ. *Curr. Top. Microbiol. Immunol.* **78:** 171.

Nomura, N., H. Yamagishi, and A. Oka. 1978. Isolation and characterization of transducing coliphage fd carrying a kanamycin resistance gene. *Gene* **3:** 39.

Ohtsubo, H., M. Zenilman, and E. Ohtsubo. 1980. Insertion element IS*102* resides in the plasmid pSC101. *J. Bacteriol.* **144:** 131.

Oka, A., N. Nomura, K. Sugimoto, H. Sugisaki, and M. Takanami. 1978. Nucleotide sequence at the insertion sites of a kanamycin transposon. *Nature* **276:** 845.

Ravetch, J. V., M. Ohsumi, P. Model, G. F. Vovis, D. Fischhoff, and N. D. Zinder. 1979. Organization of a hybrid between phage fl and plasmid pSC101. *Proc. Natl. Acad. Sci.* **76:** 2195.

Roozen, K. J., R. G. Fenwick, and R. Curtiss III. 1971. Synthesis of ribonucleic acid and protein in plasmid-containing minicells of *E. coli. J. Bacteriol.* **107:** 21.

Rosenberg, M. and D. Court. 1979. Regulatory sequences involved in the promotion and termination of RNA transcription. *Annu. Rev. Genet.* **13:** 319.

Sancar, A., A. M. Hack, and W. D. Rupp. 1979. A simple method for identification of plasmid-coded proteins. *J. Bacteriol.* **137:** 692.

Shapiro, J. A. 1979. Molecular model for the transposition and replication of bacteriophage Mu and other transposable elements. *Proc. Natl. Acad. Sci.* **76:** 1933.

Shapiro, J. A. and L. A. MacHattie. 1979. Prophage insertion and excision mediated by IS*1*. *Cold Spring Harbor Symp. Quant. Biol.* **43:** 1135.

SHARP, P. A., S. N. COHEN, and N. DAVIDSON. 1973. Electron microscopy heteroduplex studies of sequence relations among plasmids of *E. coli.* II. Structure of drug resistance (R) factors and F factors. *J. Mol. Biol.* **75:** 235.

SHARP, P. A., M.-T. HSU, E. OHTSUBO, and N. DAVIDSON. 1972. Electron microscope heteroduplex studies of sequence relations among plasmids of *E. coli.* I. Structure of F-prime factors. *J. Mol. Biol.* **71:** 471.

SHINE, J. and L. DALGARNO. 1974. The 3′-terminal sequence of *E. coli* 16S ribosomal RNA: Complementarity to nonsense triplets and ribosome binding sites. *Proc. Natl. Acad. Sci.* **71:** 1342.

SO, M., F. HEFFRON, and B. J. MCCARTHY. 1979. The *E. coli* gene encoding heat stable toxin is a bacterial transposon flanked by inverted repeats of IS*1*. *Nature* **277:** 453.

STARLINGER, P. 1980. IS elements and transposons. *Plasmid* **3:** 241.

TESSMAN, E. S. 1966. Mutants of bacteriophage S13 blocked in infectious DNA synthesis. *J. Mol. Biol.* **17:** 218.

Identification of a Sex-factor-affinity Site in *E. coli* as $\gamma\delta$

M. S. GUYER,*† R. R. REED,‡ J. A. STEITZ,‡ AND K. B. LOW§

* *Laboratory of Molecular Biology, National Institute of Arthritis, Metabolism and Digestive Diseases, Bethesda, Maryland 20205;* ‡ *Department of Molecular Biophysics and Biochemistry and* § *Radiobiology Laboratories, Yale University, New Haven, Connecticut 06510*

Many conjugative bacterial plasmids are capable of promoting the conjugal transfer of normally physically unlinked replicons, including the main bacterial chromosome (for review, see Holloway 1979). Such behavior has been termed mobilization (Novick et al. 1976) or conduction (Clark and Adelberg 1962; Clark and Warren 1979). One model proposed to describe at least some such interactions involves recombination between regions of extensive nucleotide sequence homology on both the conjugative and nonconjugative replicons (Clark and Warren 1979). Such recombination-mediated processes occur in the cases of some experimentally altered plasmids (Wilkins 1969; Kleckner et al. 1977; Barth 1979).

Insertion sequence (IS) elements (for review, see Starlinger 1980 and this volume) are resident in many bacterial replicons, including the main chromosome as well as plasmids (Saedler and Heiss 1973; Deonier et al. 1979). Much attention has been focused on the ability of these genetic elements to mediate specialized recombination processes in which nucleotide sequence homology plays little, if any, role; transposition and other types of genetic rearrangements are examples. In addition to these processes, the presence of an identical IS element on each of two different replicons creates an increased opportunity for recombination between those replicons in a recombination-proficient host. Thus, Davidson et al. (1975) have proposed that recombination between any of several IS elements located both on F and on the main chromosome of *Escherichia coli* K12 results in the integration of F and the formation of the Hfr-type chromosome. The formation of stable Hfr chromosomes has been shown to be a largely *recA*-dependent process (Cullum and Broda 1979).

The translocatable element $\gamma\delta$ (Guyer 1978), also known as Tn*1000,* is one of three IS elements on F that was identified as an integration sequence in the studies of Davidson et al. (1975). The Hfr chromosomes formed in which $\gamma\delta$ is located at the F–chromosome junction appear to be quite unstable (e.g., Ra-2) (Low 1967; Palchaudhuri et al. 1976). We have found recently that, as predicted by the hypothesis of Davidson et al., $\gamma\delta$ can be detected on the chromosome of a number of strains of *E. coli* K12, although not in all strains (R. Reed, unpubl.). Here we report the results of experiments designed to demonstrate the involvement of recombination between $\gamma\delta$ elements in certain cases of F-promoted chromosomal transfer.

METHODS

Plasmid constructions. Plasmid pOX38 was isolated as a conjugative replicon composed of the largest *Hin*dIII digestion fragment of F. F was completely digested with *Hin*dIII. The DNA was treated with T4 DNA ligase (New England BioLabs) at a low DNA concentration, and the ligation mixture was then used to transform an F^- host. The transformed cell population was incubated at 37 °C until it reached saturation, and the culture was then diluted and reincubated. The subculturing regimen was repeated ten times to allow for the spread of the conjugative plasmid throughout the population. Samples of the culture were streaked out, and individual colonies were then tested for sensitivity to sex-specific phages.

Plasmid pOX67 is a derivative of FΔ(0–15) in which $\gamma\delta$ has been inserted, by transposition, at about F coordinate 20 (within *Eco*RI fragment 14 of F). The plasmid was isolated from a conjugal mating in which the donor strain (MG1184) carried both FΔ(0–15) and pOX14 (see Table 1). $Tc^rAp^r[Sm^r$, bacteriophage $T6^r]$ transconjugants from such a cross were found to contain two plasmids, pOX14 and FΔ(0–15)::$\gamma\delta$. The plasmid from one such transconjugant was designated pOX67.

Plasmid pOX119 is a derivative of pOX38 in which $\gamma\delta$ has been inserted into F at approximately coordinate 42–44 in a procedure analogous to that described above.

Quantitative matings. Equal numbers of exponentially growing donor and recipient cells were mixed, incubated for 1 hour at 37 °C, mixed in a Vortex mixer, and plated on selective medium. Samples were plated on Luria agar containing 50 μg/ml of streptomycin (Sm) to recover the recipient cells. To score for the transfer of the donor plasmid, 80 Sm^r colonies were each placed into Luria broth containing Sm and 5 mM $CaCl_2$. The cultures were incubated overnight; samples were spotted onto EMB-0 agar containing 5 mM $CaCl_2$, which had been spread with more than 10^{10} plaque-forming units (pfu) of phage f2. $f2^s$ strains gave dark spots in this test, whereas $f2^r$ strains gave light spots. The fraction of recipients that were $f2^s$ was then used to calculate the total number of $f2^s$ transconjugants. The number of recombinants for any given marker was then normalized to the number of $f2^s$ transconjugants.

† Present address: The Genex Corporation, Bethesda, Maryland 20205.

Table 1. Bacterial Strains and Plasmids

Bacterial strain	Relevant genotype	Reference
MG1043[a]	F^- *thyA*$^-$; contains three chromosomal copies of $\gamma\delta$	Guyer (1978)
MG1655[b]	F^-; contains no detectable chromosomal copies of $\gamma\delta$	Bachmann (1972)
MG1656	same as MG1655, but F^+; -W1485 (LED)	
AB1157		Bachmann (1972)
MG1343	same as MG1043, but F^+	
MG1347	same as MG1343, but *recA56*	

Plasmid	Reference
F	Sharp et al. (1972
FΔ(0–15)	Hu et al. (1975)
pOX38	this paper
pOX67 (FΔ(0–15)::$\gamma\delta$)	this paper
pOX119 (pOX38::$\gamma\delta$)	this paper
pOX14 (pBR322::$\gamma\delta$)	Guyer (1978)

[a] MG1043 is an F^- derivative of a strain designated W1485 obtained from N. Davidson, Department of Chemistry, California Institute of Technology.
[b] MG1655 is an F^- derivative of W1485 (CGSC#5024) that contains no $\gamma\delta$, in contrast to other strains denoted W1485 or W1485E.

RESULTS

Requirement for Both Plasmid and Chromosomal Copies of $\gamma\delta$

MG1043 is a strain that carries three chromosomal copies of $\gamma\delta$ (R. Reed, unpubl.). An F^+ derivative, MG1343, was found to transfer chromosomal markers, such as *argE*$^+$ or *proA*$^+$, at a frequency of about 10^{-4} per F^+ transconjugant (Table 2) in matings with different recipients. This is a rather high level of transfer from an F^+ donor; typical values reported in the literature are 10 to 100 times lower (Reeves 1960; Curtiss and Renshaw 1969a,b; Cullum and Broda 1979). We have found that the efficient transfer in the case of MG1343 is dependent on the presence of $\gamma\delta$ on both F and the main chromosome of this strain.

The requirement for the presence of a chromosomal copy (or copies) of $\gamma\delta$ is inferred from a comparison of the transfer frequencies of chromosomal markers obtained in matings with MG1343 and with MG1656. The latter is an F^+ derivative of MG1655 (Table 1), a strain that carries no chromosomal copy of $\gamma\delta$. MG1656 transfers *leu*$^+$ and other markers only about 1% as well as MG1343 (data not shown).

The effect of the deletion of $\gamma\delta$ from F on the ability of the plasmid to promote transfer of chromosomal markers indicates a requirement for the presence of the IS element on the plasmid. FΔ(0–15) is a spontaneous deletion mutant of F (Sharp et al. 1972) that lacks the region 0–14.5F. Included in the deletion are $\gamma\delta$ (2.8–8.5F) and $\alpha_2\beta_2$ (IS*3*b, 13.2–14.5F), as well as other regions (0–2.8F and 8.5–13.2F) of unidentified function. MG1652, which is MG1043 (FΔ[0–15]), transfers chromosomal markers considerably less efficiently than does MG1343, the F^+ isogenic strain (Fig. 1; Table 2). Another deletion mutant of F, pOX38 (Table 1), also has reduced ability to promote chromosomal transfer from the MG1043 genetic background (Fig. 1; Table 2). Plasmid pOX38 carries a deletion extending from the leftmost *Hin*dIII site within $\alpha_1\beta_1$ (at about 93.5F) to 41.3F. Thus, except for the region of IS*3*a from α_1 (93.2F) to the first *Hin*dIII site of IS*3*, none of the known IS elements of F are present on this conjugative plasmid. Plasmid pOX38 may be slightly more defective (about two- to threefold) than FΔ(0–15) in its ability to effect the transfer of a $\gamma\delta$-containing chromosome (Table 2).

Evidence that the deletion of $\gamma\delta$ is primarily responsible for the failure of both FΔ(0–15) and pOX38 to transfer chromosomal markers at high efficiency comes from the observation that reintroduction of $\gamma\delta$ into each of the deleted plasmids restores much of the ability to mobilize the chromosome. Thus, FΔ(0–15)::$\gamma\delta$ (pOX67; Table 1) transfers *thr*$^+$ *leu*$^+$ from MG1043 almost as well as F (Table 2). Similarly, pOX38::$\gamma\delta$ (pOX119) promotes chromosomal transfer much better than pOX38 and apparently as well as F (Fig. 1; Table 2).

Chromosomal transfer in this system is largely *recA*-dependent. A *recA56* derivative of MG1343 (MG1367) is about a 1000-fold poorer donor for chromosomal markers than is the Rec$^+$ donor (data not shown). Thus, the requirement for copies of the $\gamma\delta$ element on both replicons appears to result from a requirement for sequence homology. On the basis of these results, we conclude that most of the transfer of chromosomal DNA from the $\gamma\delta$-containing background of strain MG1043 comes from cells that contain a cointegrate, Hfr-like

Table 2. Chromosomal Transfer in Various Strains

Donor strain	Plasmid	$TL^+[Sm^rThy^+]$	$Arg^+[Sm^rThy^+]$
MG1343	F	1	1
MG1652	FΔ(0–15)	0.05	0.07
MG1650	pOX67	0.2	0.4
MG1345	pOX38	0.01	0.03
MG1689	pOX119	1.3	3.0

The genetic background of each strain is MG1043 (Table 1), which contains three chromosomal copies of $\gamma\delta$. With MG1343, the average number of recombinants/f2^s transconjugant obtained were $TL^+[Sm^rThy^+] = 6.0 \times 10^{-4}$ and $Arg^+[Sm^rThy^+] = 1.1 \times 10^{-3}$ ($TL^+ = Thr^+Leu^+$). The data are values relative to those shown for MG1343.

Figure 1. Photograph of replica-mating selecting for Arg$^+$[Sm^{r}Thy$^+$] recombinants, using donors described in Table 2.

chromosome generated by *recA*-dependent recombination between the γδ sequence on F and a chromosomal copy of γδ (Clark and Warren 1979). Thus, in this instance, chromosomal transfer by F appears to be an example of plasmid (F)-mediated conduction (Clark and Adelberg 1962; Clark and Warren 1979) of a nontransmissible replicon (the main chromosome).

It should be noted, however, that our conclusion must be somewhat tempered because of two problems with the experimental design. First, the genetic backgrounds of MG1043 and MG1655 may not be isogenic (see Table 1). Therefore, some genetic difference between the two strains other than γδ could affect F-mediated chromosomal transfer. Second, the deletions in both FΔ(0–15) and pOX38 remove F sequences in addition to γδ. The fact that reinsertion of γδ largely restores the ability of the plasmid to conduct the chromosome demonstrates specifically that at least part of the defect in the deleted plasmid is due to the deletion of γδ. However, pOX67 does not display a completely wild-type ability to transfer chromosomal markers in liquid matings; thus, there may be some contribution to transfer efficiency from non-γδ-encoded functions in the deleted regions of F. We are currently examining both of these possibilities.

Identification of γδ in *sfa* Strains

Richter (1957), Adelberg and Burns (1960), and others have described derivatives of Hfr strains of *E. coli* that no longer carry an integrated F factor but in which some "memory" of the integrated plasmid is retained. Reintroduction of wild-type F into such strains yields a donor in which the extrachromosomal plasmid can effect transfer of chromosomal markers from the point of origin and with the orientation of the integrated plasmid of the original Hfr strain. Such strains are considered to have a "sex-factor-affinity" (*sfa*) site (Adelberg and Burns 1960), which appears at the original site of F integration and with which an autonomous F element interacts preferentially to promote chromosomal transfer.

We have examined the chromosomal DNAs of a number of strains of *E. coli* K12 for the presence of sequences that hybridize to γδ (R. Reed, unpubl.). Of the 20 strains examined, there were only four in which homology to the γδ-containing probe (pOX14; see Table 1) could be detected. MG1043, as noted above, contains three chromosomal copies of γδ. At least one of these is complete and is capable both of transposition (M. S. Guyer, unpubl.) and of complementing in *trans* a deletion mutant of γδ affecting transposition (R. Reed, unpubl.).

Two of the γδ-containing strains were already known to be *sfa* strains: RaF$^-$ (Low 1967), which contains two chromosomal copies of γδ, and P4X-1 (Adelberg and Burns 1960), which contains a single copy. The fourth strain, W1485E (Bachmann 1972), an F$^+$ that also contains a single chromosomal copy of γδ, was not known to be an *sfa* strain, but it transfers chromosomal markers at high efficiency in an oriented manner (Figs. 2 and 3). The point of origin (counterclockwise of the *ilv* operon) corresponds to that of the Hfr strain KL25 (Low 1972), which is derived from W1485E. In other words, W1485E (F$^+$) behaves as an *sfa* strain, although it had not been identified previously as such. Thus, a complete

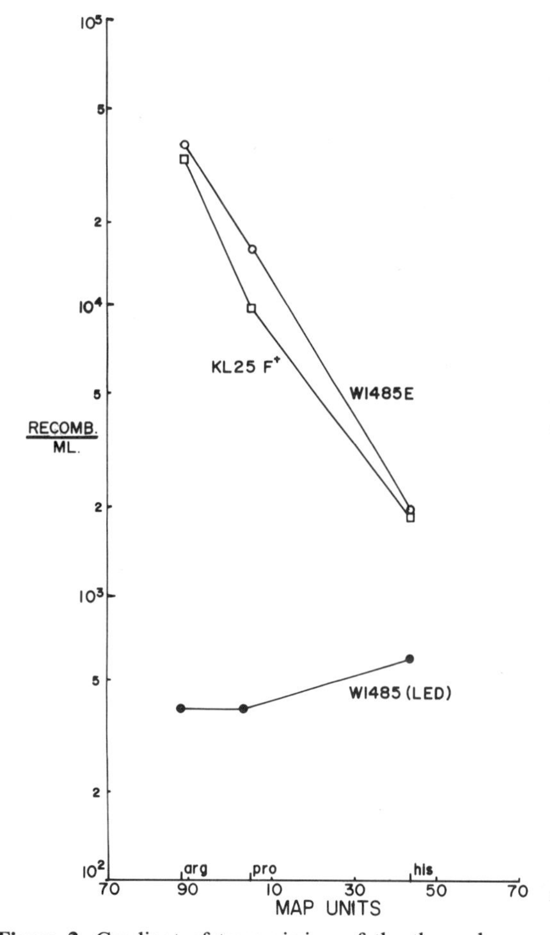

Figure 2. Gradient of transmission of the three chromosomal markers *arg, pro,* and *his* in mating with AB1157, demonstrating oriented transfer at elevated frequency in the cases of W1485E and KL25 F$^+$, as described in the text. W1485(LED) is the parent of MG1655 and carries no γδ on the chromosome.

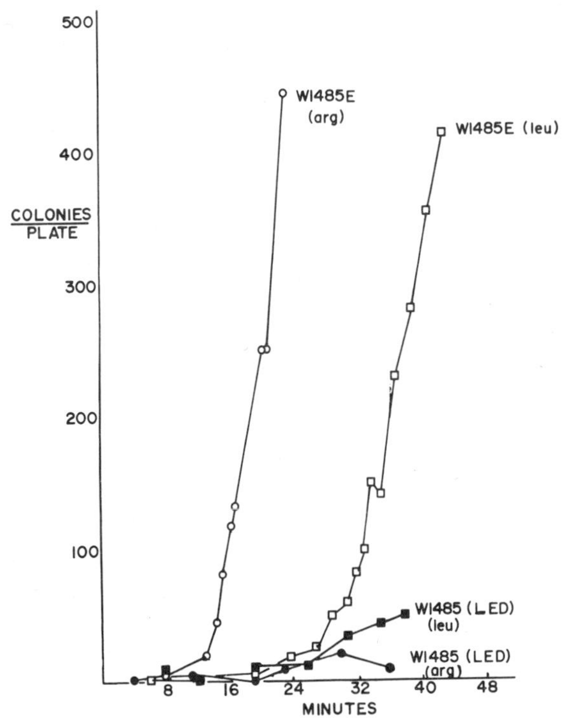

Figure 3. Time of entry of two chromosomal markers in the $\gamma\delta^+$ and the $\gamma\delta^-$ strains (see text).

correlation (4 of 4) between the presence of $\gamma\delta$ and the presence of an *sfa* site in the chromosome of *E. coli* K12 exists. Conversely, the absence of an *sfa* site in strain MG1655 can be correlated with the absence of $\gamma\delta$ in the chromosome of that strain.

DISCUSSION

Our results indicate that certain discrete DNA sequences, such as IS elements, can provide regions of nucleotide sequence homology that are used by the generalized recombination system to recombine different prokaryotic replicons. Since IS elements, by virtue of their ability to engage in homology-independent recombination, are able to translocate from one replicon to another, they can truly be considered "portable regions of homology." This property of IS elements has been exploited as a powerful new manipulative tool for the geneticist (Kleckner et al. 1977; Bachmann and Low 1980). Similarly, they must be considered, at least potentially, as important agents of horizontal transmission of genetic information among prokaryotes (Reanney 1976; Holloway 1979).

F can promote the conjugal transmission of pBR322 by a *recA*-independent process that results in the acquisition of $\gamma\delta$ by the R plasmid (Guyer 1978). In an analogous manner, conjugal transmission of pOX14 (pBR322::$\gamma\delta$) by FΔ(0–15) leads to the appearance of a copy of $\gamma\delta$ in the conjugative plasmid (see derivation of pOX67 in Table 1). In both cases, a region of homology between the two plasmids is produced during the transposition event. The homology thus provided by $\gamma\delta$ can then be used as a site of interaction between the two plasmids; subsequently, in a Rec$^+$ host, F can mobilize pOX14 considerably more efficiently (ca. 5×10^{-4} per F$^+$ transconjugant) than it can mobilize pBR322 (ca. 1×10^{-6} per F$^+$ transconjugant) (M. S. Guyer, unpubl.). Regions of homology exist on both replicons after transposition regardless of whether the IS element was originally present on the conjugative or nonconjugative plasmid. Indeed, it does not even matter if the IS element is not present on either replicon. We have found that pBR322 can be transferred with FΔ(0–15) from the MG1043 genetic background at a very low, but detectable, level (10^{-11} to 10^{-9} per F$^+$ transconjugant). Analysis of the transferred plasmids in these rare transconjugants has shown that both contain $\gamma\delta$ (R. Reed, unpubl.). Since neither of the plasmids in the donor strain initially contained $\gamma\delta$, the transposable element must have come from the chromosome. It is therefore possible for two replicons to acquire a region of homology that was originally present on a third replicon.

Given these considerations, a situation can easily be imagined in which the conjugative plasmid is not one with a limited host range, such as F, but one with a much broader host range, such as RP1. If the plasmids were to acquire an IS element from the chromosome of any particular host, the transfer of genetic information of that host to any other species with which the plasmid was capable of promoting conjugation could occur. An example of this phenomenon may be the case of plasmid R68 and its variant R68.45, which was isolated on the basis of increased ability to promote chromosomal transfer in *Pseudomonas aeruginosa* (Haas and Holloway 1976) and was shown to have suffered an insertion of an 1800-bp sequence (Reiss et al. 1978).

Our results raise again the question of the basis for the F-promoted transfer of chromosomal markers from strains that do not contain chromosomal copies of $\gamma\delta$. We have found that most *E. coli* K12 F$^-$ sublines do not possess $\gamma\delta$. Furthermore, the frequency of chromosomal transfer with one $\gamma\delta$-free F$^+$ strain, MG1655, is about 10^{-6} per F$^+$ transconjugant. This is similar to the frequencies classically reported for F-mediated chromosomal mobilization. Therefore, we assume that most of the original studies of the ability to promote chromosomal transfer utilized $\gamma\delta$-free strains and therefore cannot be interpreted as being due to the recombination of F with $\gamma\delta$ *sfa* sites. Older studies have shown that F-promoted chromosomal transfer is largely *recA*-dependent (Clowes and Moody 1966; Cullum and Broda 1979) and can be accounted for only partially by the presence of stable Hfr chromosomes in the F$^+$ population. The existence of "unstable Hfr" chromosomes or "incomplete integration" of F has been suggested (Curtiss and Stallions 1969; Evenchick et al. 1969; Holloway 1979). In light of current information, several possibilities can be suggested. All *E. coli* K12 strains that have been examined do contain several copies each of IS*2* and IS*3* (Deonier et al. 1979). These elements could act as weak *sfa* sites by recombining with the $\alpha\beta$ and $\epsilon\zeta$ sequences of F. (Our results [Table 2] indicate that the ability of $\gamma\delta$ to promote chromosomal transfer is considerably greater than that of $\alpha\beta$ [IS*3*] or $\epsilon\zeta$ [IS*2*]. It is not clear whether this is due to the fact that $\gamma\delta$ is significantly larger than either of the other sequences or to some other property of $\gamma\delta$.) Furthermore, it must be considered that there may be other regions of homology between the *E. coli* chromosome and F, in addition to the three IS elements that have been described.

A different possibility is suggested by various models for the transposition of IS elements (Arthur and Sherratt 1979; Grindley and Sherratt 1979; Shapiro 1979). These models postulate the existence of a cointegrate or fused-replicon structure as an intermediate in, or as a by-product of, transposition of a translocatable element. The cointegrate is proposed to consist of a hybrid molecule containing the (transpositional) donor and recipient replicons linked at each end by a copy of the IS element. In the case of an IS element being transposed from F to the chromosome, such a cointegrate would be formally equivalent to an Hfr chromosome, i.e., a copy of F integrated into the *E. coli* chromosome with a direct duplication of the IS element at each junction (Davidson et al. 1975). Speculating further, we can suggest that $\gamma\delta$ would transpose much more often from F to pBR322, and at least 100 times more frequently than IS*2* or IS*3* (Guyer 1978). Also, the cointegratelike structures of the F′ plasmids F14 and F105 (in which direct duplications of $\gamma\delta$ are present at the junctions of F and chromosomal sequences) are quite unstable and excise F by *recA*-independent recombination between the duplicated $\gamma\delta$ sequences (Ohtsubo et al. 1974; Palchaudhuri et al. 1976). Therefore, we would expect that Hfr-like cointegrates formed as the product of transposition and containing duplicated $\gamma\delta$ sequences would also be unstable. This could account for the unstable Hfr chromosomes postulated to be involved in the F-promoted transfer of chromosomal markers (Curtiss and Renshaw 1969a; Curtiss and Stallions 1969). On the other hand, the generation of cointegrates during transposition of $\gamma\delta$ would be expected to be *recA*-independent, whereas chromosomal mobilization by F is *recA*-dependent (Clowes and Moody 1966; Cullum and Broda 1979). Questions raised by these points are currently being examined further.

REFERENCES

Adelberg, E. A. and S. N. Burns. 1960. Genetic variation in the sex factor of *Escherichia coli. J. Bacteriol.* **79:** 321.

Arthur, A. and D. Sherratt. 1979. Dissection of the transposition process: A transposon-encoded site-specific recombination system. *Mol. Gen. Genet.* **175:** 267.

Bachmann, B. J. 1972. Pedigrees of some mutant strains of *Escherichia coli* K12. *Bacteriol. Rev.* **36:** 525.

Bachmann, B. J. and K. B. Low. 1980. Linkage map of *Escherichia coli* K-12. *Microbiol. Rev.* **44:** 1.

Barth, P. T. 1979. Plasmid RP4, with *E. coli* DNA inserted *in vitro,* mediates chromosomal transfer. *Plasmid* **2:** 130.

Clark, A. J. and E. A. Adelberg. 1962. Bacterial conjugation. *Annu. Rev. Microbiol.* **16:** 289.

Clark, A. J. and G. J. Warren. 1979. Conjugal transmission of plasmids. *Annu. Rev. Genet.* **13:** 99.

Clowes, R. C. and E. E. M. Moody. 1966. Chromosomal transfer from "recombination-deficient" strains of *Escherichia coli* K12. *Genetics* **53:** 717.

Cullum, J. and P. Broda. 1979. Chromosome transfer and Hfr formation by F in *rec*$^+$ and *recA* strains of *Escherichia coli* K12. *Plasmid* **2:** 358.

Curtiss, R., III and J. Renshaw. 1969a. F$^+$ strains of *Escherichia coli* K12 defective in Hfr formation. *Genetics* **63:** 7.

———. 1969b. Kinetics of F transfer and recombination production in F$^+$ × F$^-$ matings in *Escherichia coli* K12. *Genetics* **63:** 39.

Curtiss, R., III and D. R. Stallions. 1969. Probability of F integration and frequency of stable Hfr donors in F$^+$ populations of *Escherichia coli* K12. *Genetics* **63:** 27.

Davidson, N., R. C. Deonier, S. Hu, and E. Ohtsubo. 1975. Electron microscope heteroduplex studies of sequence relations among plasmids of *Escherichia coli.* X. Deoxyribonucleic acid sequence organization of F and F-primes, and the sequences involved in Hfr formation. In *Microbiology—1974* (ed. D. Schlessinger), p. 56. American Society for Microbiology, Washington, D.C.

Deonier, R. C., R. G. Hadley, and M. Hu. 1979. Enumeration and identification of IS*3* elements in *Escherichia coli* strains. *J. Bacteriol.* **137:** 1421.

Evenchick, A., K. A. Stacey, and W. Hayes. 1969. Ultraviolet induction of chromosomal transfer by autonomous sex factors in *E. coli. J. Gen. Microbiol.* **56:** 1.

Grindley, N. D. F. and D. J. Sherratt. 1979. Sequence analysis of IS*1* insertion sites: Models for transposition. *Cold Spring Harbor Symp. Quant. Biol.* **43:** 1257.

Guyer, M. 1978. The $\gamma\delta$ sequence of F is an insertion sequence. *J. Mol. Biol.* **126:** 347.

Haas, D. and B. W. Holloway. 1976. R factor variants with enhanced sex factor activity in *Pseudomonas aeruginosa. Mol. Gen. Genet.* **144:** 243.

Holloway, B. W. 1979. Plasmids that mobilize bacterial chromosome. *Plasmid* **2:** 1.

Kleckner, N., J. Roth, and D. Botstein. 1977. Genetic engineering *in vivo* using translocatable drug-resistance elements: New methods in bacterial genetics. *J. Mol. Biol.* **116:** 125.

Low, K. B. 1967. Inversion of transfer modes and sex factor-chromosome interactions in conjugation in *Escherichia coli. J. Bacteriol.* **93:** 98.

———. 1972. *Escherichia coli* K-12 F-prime factors, old and new. *Bacteriol. Rev.* **36:** 587.

Novick, R. P., R. C. Clowes, S. N. Cohen, R. Curtiss III, N. Datta, and S. Falkow. 1976. Uniform nomenclature for bacterial plasmids: A proposal. *Bacteriol. Rev.* **40:** 168.

Ohtsubo, E., R. C. Deonier, H. J. Lee, and N. Davidson. 1974. Electron microscope heteroduplex studies of sequence relations among plasmids of *Escherichia coli.* IV. The sequences in F14. *J. Mol. Biol.* **89:** 565.

Palchaudhuri, S., W. K. Maas, and E. Ohtsubo. 1976. Fusion of two F-prime factors in *Escherichia coli* studied by electron microscope heteroduplex analysis. *Mol. Gen. Genet.* **146:** 215.

Reanney, D. 1976. Extrachromosomal elements as possible agents of adaptation and development. *Bacteriol. Rev.* **40:** 552.

Reeves, P. 1960. Role of Hfr mutants in F$^+$ × F$^-$ crosses in *Escherichia coli* K-12. *Nature* **185:** 265.

Reiss, G., H. Burkhardt, and A. Pickler. 1978. Molecular characterization of R68.45, a plasmid with chromosomal donor ability. *Hoppe-Seylers Z. Physiol. Chem.* **359:** 1139.

Richter, A. 1957. Complementary determinants on an Hfr phenotype in *E. coli* K-12. *Genetics* **42:** 391.

Saedler, H. and B. Heiss. 1973. Multiple copies of the insertion DNA sequences IS*1* and IS*2* in the chromosome of *E. coli* K-12. *Mol. Gen. Genet.* **122:** 267.

Shapiro, J. A. 1979. Molecular model for the transposition and replication of bacteriophage Mu and other transposable elements. *Proc. Natl. Acad. Sci.* **76:** 1933.

Sharp, P. A., M.-T. Hsu, E. Ohtsubo, and N. Davidson. 1972. Electron microscope studies of sequence relations among plasmids of *E. coli.* I. Structure of F-prime factors. *J. Mol. Biol.* **71:** 471.

Starlinger, P. 1980. IS elements and transposons. *Plasmid* **3:** 241.

Wilkens, B. M. 1969. Chromosome transfer from F-*lac*$^+$ strains of *E. coli* K-12 mutant at *recA, recB,* or *recC. J. Bacteriol.* **98:** 599.

Internal Rearrangements of IS*2* in *Escherichia coli*

A. Ahmed,* K. Bidwell,† and R. Musso†

Department of Genetics, University of Alberta, Edmonton, Canada T6G 2E9; †Cancer Biology Program, National Cancer Institute, Frederick Cancer Research Center, Frederick, Maryland 21701

Mutations caused by the insertion of movable genetic elements are often accompanied by regional destabilization of the chromosome. This instability is a direct consequence of processes (such as inversions, deletions, or excisions) promoted by these elements that constitute the bases of the seemingly diverse phenomena discussed at this Symposium. Although most of these phenomena stem from alterations of the adjoining regions of the chromosome, these elements also promote internal rearrangements that contribute to their novel, and often bewildering, properties. In this article we describe some internal sequence alterations of IS*2* that were detected as revertants of the *gal3* mutation of *Escherichia coli.* We explain the unusual behavior of these revertants in terms of special features of their DNA sequences.

The *gal3* System

The *gal3* mutation (for review, see Ahmed 1977) was caused by the insertion of a 1327-bp IS*2* element into the promoter region of the *gal* operon of *E. coli* (Fig. 1). In this orientation (designated I), IS*2* prevents transcription of the three structural genes *E, T,* and *K* of the *gal* operon which code for the epimerase, transferase, and kinase enzymes, respectively. As a result, *gal3* acts as an extreme-polar mutation, but the polarity can be suppressed by mutations altering the transcription-termination factor rho. In the opposite orientation (designated II), IS*2* is nonpolar and appears to promote transcription.

gal3 reverts spontaneously to produce three different kinds of gal^+ revertants. The first class consists of stable revertants that are inducible for the *gal* operon and normally transduced by phage λ. The second class consists of revertants that are stable but exhibit constitutive enzyme synthesis at a level (~150%) exceeding that of a fully induced wild type. These revertants are not normally transduced by λ unless additional genetic events (deletions) are introduced. The third class consists of unstable revertants that show constitutive enzyme synthesis at a relatively low level (65%) and may, or may not, be transducible by λ. These revertants typically segregate gal^- and gal^{wc} (weak-constitutive) colonies at a high rate (~10^{-2} segregants/cell/division). The gal^{wc} segregants show constitutive enzyme synthesis at low levels (7–20%) and segregate gal^- colonies at a rate of 10^{-3}. The gal^- segregants arising from either the unstable revertants or the gal^{wc} segregants are identical to the original *gal3* parent. None of these events require the *recA* function. A summary of the reversion and segregation behaviors of *gal3* and its revertants is presented in Fig. 2.

DNA heteroduplex analysis of λ*gal* phages bearing various *gal3* revertants indicated that the inducible revertants remove IS*2* from the promoter by precise excision, whereas the constitutive revertants (gal^c) retain IS*2* in altered forms at the original site of insertion (Ahmed 1977; Ahmed and Johansen 1978). The unstable revertant gal^c331, and gal^{wc} segregants derived from it, were found to retain the complete IS*2* element in the original orientation. On the other hand, the stable revertant gal^c200 (harboring a *chlD-pgl* deletion Δ*31*) was found to carry a partially deleted IS*2* element that retained only 200 bp from IS*2* in the *gal* promoter. In the following sections we describe results of our sequence studies on the altered forms of IS*2* present in these revertants. This work was simplified considerably by the determination of the complete nucleotide sequence of IS*2* by Ghosal et al. (1979b).

RESULTS

Sequence of the Unstable Reversion gal^c331

gal^c331 is a typical representative of the unstable constitutive class of revertants. Overnight cultures of this revertant, grown from single colonies, typically contain ~15% gal^- and gal^{wc} segregants, which appear as deep red colonies on tetrazolium-galactose plates (Fig. 3). This reversion was successfully transferred to a λ*gal* genome, and a permanent λgal^c331 transducing line was established (Ahmed and Johansen 1978).

A comparison of the electrophoretic mobilities of fragments generated by *Hin*dIII digestion of λ*gal3* and λgal^c331 DNA revealed that a 879-bp fragment, originating from the IS*2*–*galE* junction in *gal3,* had increased in size to 987 bp in gal^c331. Since the other *Hin*dIII fragments had remained unchanged, it was concluded that the *gal3*→ gal^c331 reversion was caused by the addition of a 108-bp sequence to the 879-bp fragment of *gal3.* This conclusion was strengthened by the observation that the *Hin*dIII digests of gal^c331 invariably produced not only the 987-bp fragment, but also 933-bp and 879-bp fragments, all in submolar yields. As shown later, this heterogeneity arises from the instability of gal^c331, which always contains gal^{wc} and *gal3* segregants.

The 879-bp *Hin*dIII fragment from λ*gal3* and the 987-bp *Hin*dIII fragment from λgal^c331 were cloned separately in the *Hin*dIII site of pBR322. Recombinant

gal
attλ pgl chlD OP IS2 E T K

Figure 1. Map of the *gal* region of the *E. coli* K12 chromosome. *O* and *P* designate the operator and promoter, and *E, T,* and *K* designate the epimerase, transferase, and kinase genes of the *gal* operon. A 1327-bp IS*2* element (bold line) was inserted in orientation I in the *gal* promoter to produce the *gal3* mutation.

plasmids were identified by the ampicillin-resistant and tetracycline-sensitive phenotypes of the transformants. The presence of the correct IS*2-galE* fragment in each recombinant plasmid was verified by restriction analysis with *Hin*dIII. The plasmid containing the *gal3* fragment produced a sharp band corresponding to the 879-bp fragment (in addition to the *Hin*dIII-cleaved 4362-bp pBR322 band). In contrast, the plasmid carrying the *gal*^{c}*331 Hin*dIII fragment produced one sharp 987-bp band plus two faint bands 933 bp and 879 bp in length (in addition to the linearized pBR322 band). The faint bands correspond, in length, to the bands from the two segregation products of *gal*^{c}*331*. These observations confirmed that the correct fragment displaying instability had been cloned and that the presence of the entire IS*2* element on the plasmid was not necessary for the expression of instability. The faint bands arising due to instability were seen even when the plasmids carrying the *gal*^{c}*331* fragment were amplified in a *recA*$^-$ host.

A restriction map of the 879-bp *Hin*dIII fragment carrying the IS*2–galE* junction from *gal3* is shown in Figure 4. By comparing the restriction patterns of *Hin*dIII fragments from *gal3* and *gal*^{c}*331*, the alteration responsible for the *gal*^{c}*331* reversion (i.e., addition of 108 bp) was localized between the two *Hae*III cuts. The DNA sequence of this region was determined by the method of Maxam and Gilbert (1977). The *Hha*I fragment from the *gal3*- or *gal*^{c}*331*-containing plasmid was isolated, cut with *Hpa*II or *Hin*fI, ^{32}P-labeled, and sequenced (Fig. 5). The sequences for *gal3* and *gal*^{c}*331*, together with the wild-type *gal*$^+$ sequence (Musso et al. 1977), are presented in Figure 6.

In *gal*$^+$, transcription starts at position +1 and proceeds to the right into the structural genes of the *gal* operon. The sequence to the left of the startpoint contains, in order, the Pribnow boxes of the two *gal* promoters *P1* and *P2*, the "−35" or RNA polymerase recognition region, and the *gal* operator (Adhya and Miller 1979; Di Lauro et al. 1979). The *gal3* sequence shows that IS*2* is inserted at the transcription startpoint. Because of a 5-bp duplication at the other end of IS*2*, the site of insertion in *gal3* may be regarded as from +1 to +5. The IS*2* residues have been assigned negative numbers, so that the right (*galE*) end is numbered −1 and the left (*chlD-pgl*) end is −1327. This is done to avoid possible confusion arising from some minor dif-

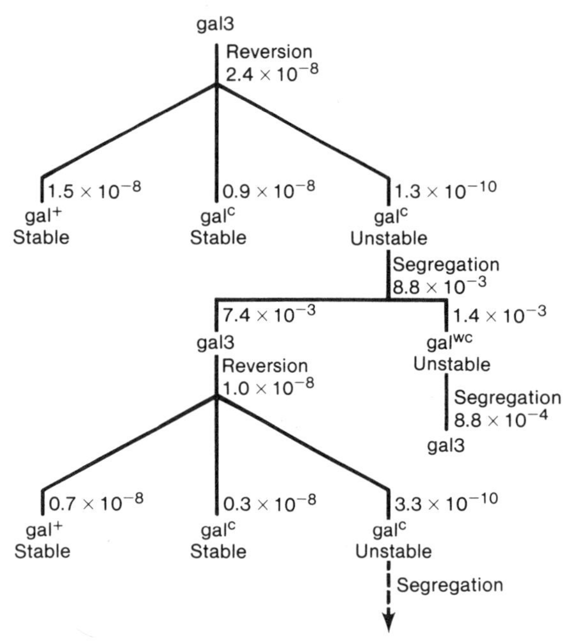

Figure 2. Reversion behavior of the *gal3* mutation. *gal3* reverts spontaneously to produce inducible (*gal*$^+$) or constitutive (*gal*^{c}) revertants. The latter can be stable (e.g., *gal*^{c}*200*) or unstable (e.g., *gal*^{c}*331*). The unstable revertants segregate *gal*^{wc} and *gal*$^-$ colonies at a high frequency. The *gal*^{wc} segregants also are unstable and produce *gal*$^-$ colonies at high frequency. The *gal*$^-$ segregants are identical to the original *gal3* mutation. Rates are expressed as revertants (or segregants) per cell per division. (Reprinted, with permission, from Ahmed and Johansen 1978.)

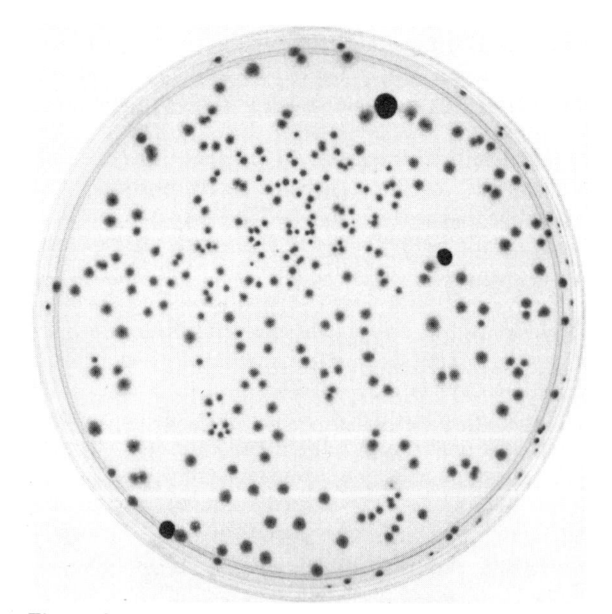

Figure 3. Instability of *gal*^{c}*331*. An overnight culture of *gal*^{c}*331 recA*$^-$, grown from a single colony, was spread on a tetrazolium-galactose plate and incubated. The *gal*$^-$ and *gal*^{wc} segregants appear on this indicator medium as deep red sectors or colonies against a background of pink *gal*^{c}*331* colonies.

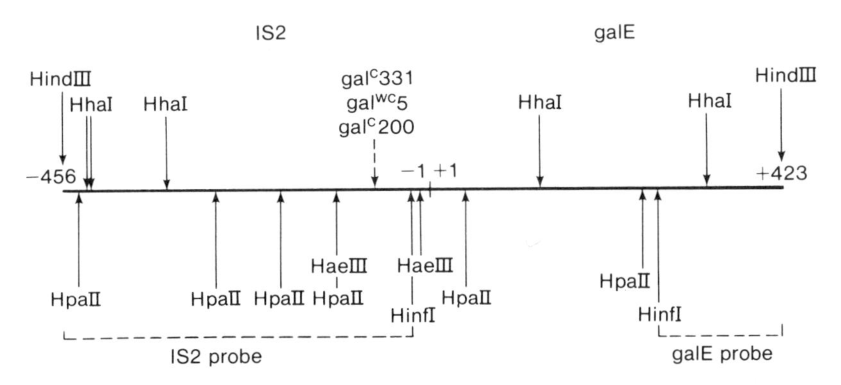

Figure 4. Restriction map of the cloned 879-bp *Hin*dIII fragment of *gal3* originating from the IS*2–galE* junction. Positive and negative numbers refer to *galE* and IS*2* residues, respectively. Restriction sites of various enzymes used for sequencing are indicated. The site of the 108-bp or 54-bp addition in *gal*^{c}*331* or *gal*^{wc}*5* is indicated by a vertical dashed arrow. This is also the site for IS*2*(II) insertion in *gal*^{c}*200*. Sequence analyses were performed using fragments 5′ ^{32}P-labeled at the *Hin*fI and *Hpa*II sites bracketing this site. Dashed lines at the bottom show the IS*2* and *galE* probes used for Southern analysis.

ferences from the sequence determined by Ghosal et al. (1979b).

A comparison of the *gal3* and *gal*^{c}*331* sequences shows that the 108-bp addition responsible for the *gal*^{c}*331* reversion occurred at position −66/−67 of the *gal3* sequence. Since this change results in repeating directly the sequence from positions −33 to −66 at positions −141 to −174, the 108-bp addition can be considered to have occurred anywhere between positions −32/−33 and −66/−67 of *gal3*. The added sequence is similar, but not identical, to the IS*2-6* sequence reported by Ghosal and Saedler (1978), who pointed out that the latter can be derived by copying parts of both strands of IS*2* between positions −33 and −78. As a consequence, large inverted repeats exist in the DNA sequences of IS*2-6* and *gal*^{c}*331*. The *gal*^{c}*331* sequence also contains three sets of partially overlapping direct repeats: a and b (34 bp); c and d (52 bp); and e, f, and g (23 bp). The significance of these direct repeats in generating instability is discussed later.

Sequences of Segregants

λ*gal* phages bearing *gal*$^{-}$ and *gal*^{wc} segregants can be readily obtained from an *E. coli* strain deleted for the *gal* operon and lysogenic for λ*gal*^{c}*331* and λ phages (Ahmed and Johansen 1978). Restriction analyses of the DNA of several independent λ*gal*^{wc} and λ*gal*$^{-}$ segregants indicated that the 987-bp *Hin*dIII fragment present in *gal*^{c}*331* is replaced by a 933-bp fragment in *gal*^{wc} or an 879-bp fragment in *gal*$^{-}$ segregants. This result shows that the *gal*^{c}*331* → *gal*^{wc} or *gal*$^{-}$ segregation is accompanied by the loss of 54 bp or 108 bp, respectively, from the *gal*^{c}*331 Hin*dIII fragment. The 933-bp fragment from a *gal*^{wc} segregant (λ*gal*^{wc}*5*) was cloned in the *Hin*dIII site of pBR322. Restriction analyses of the recombinant plasmids showed that *gal*^{wc}*5* plasmid produces a 933-bp band and a faint 879-bp band (in addition to the pBR322 band) by *Hin*dIII digestion. This instability agrees well with the segregation of *gal*$^{-}$ colonies by the *gal*^{wc}*5* parent. On the other hand, the plasmid carrying the *Hin*dIII fragment of the *gal*$^{-}$ segregant produced only one 879-bp band, which was indistinguishable from that produced by *gal3*.

Restriction analysis of the purified *Hin*dIII fragment from *gal*^{wc}*5* localized the alteration to the same *Hae*III fragment that contains the *gal*^{c}*331* alteration (Fig. 4). The sequence of *gal*^{wc}*5* in this region is presented in Figure 6. It is clear that the *gal*^{wc}*5* segregant arose by the loss of a 54-bp segment from *gal*^{c}*331* at position −95/−96. Because of the presence of direct repeats, however, the loss can be considered to have occurred anywhere between positions −66/−67 and −95/−96 of *gal*^{c}*331*. Removal of this internal 54-bp segment results in bringing the direct repeats a and b closer. Thus, the *gal*^{wc}*5* sequence also possesses two 34-bp direct repeats, labeled h and i in Figure 6.

The sequence of the *gal*$^{-}$ segregant (Fig. 6) was obtained by analysis of the 879-bp *Hin*dIII fragment isolated from *gal*^{c}*331* DNA. It is identical to *gal3* and is free of large direct or inverted repeats. This sequence clearly originated by the loss of an 108-bp segment from *gal*^{c}*331* at position −66/−67 or, in the strict sense, from anywhere between positions −32/−33 and −66/−67.

Sequence of *gal*^{c}*200* Δ*31*

gal^{c}*200* is a stable revertant of *gal3* that exhibits constitutive synthesis of *gal* enzymes at a high level. Like other revertants of its class, *gal*^{c}*200* fails to establish permanent λ*gal* transducing lines. However, it was possible to construct a stable λ*gal* line after a *chlD-pgl* deletion (Δ*31*) had occurred between *gal* and prophage λ. A physical map of λ*gal*^{c}*200* Δ*31* DNA, based on heteroduplex analysis, has been published (Ahmed and Johansen 1975; Ahmed and Scraba 1978). These studies showed that this DNA retains only an ~200-bp fragment of IS*2* (responsible for the *gal*^{c} phenotype), adjacent to a 6300-bp *chlD-pgl* deletion (Δ*31*).

Restriction analysis of *gal*^{c}*200* Δ*31* indicated that the partial deletion of IS*2* has removed the *Hin*dIII cutting site located on IS*2*. A 6600-bp *Hin*dIII-*Eco*RI fragment of this DNA, extending from the *Hin*dIII site in *galE* to the *Eco*RI site between Δ*31* and *att*B • P′, was purified

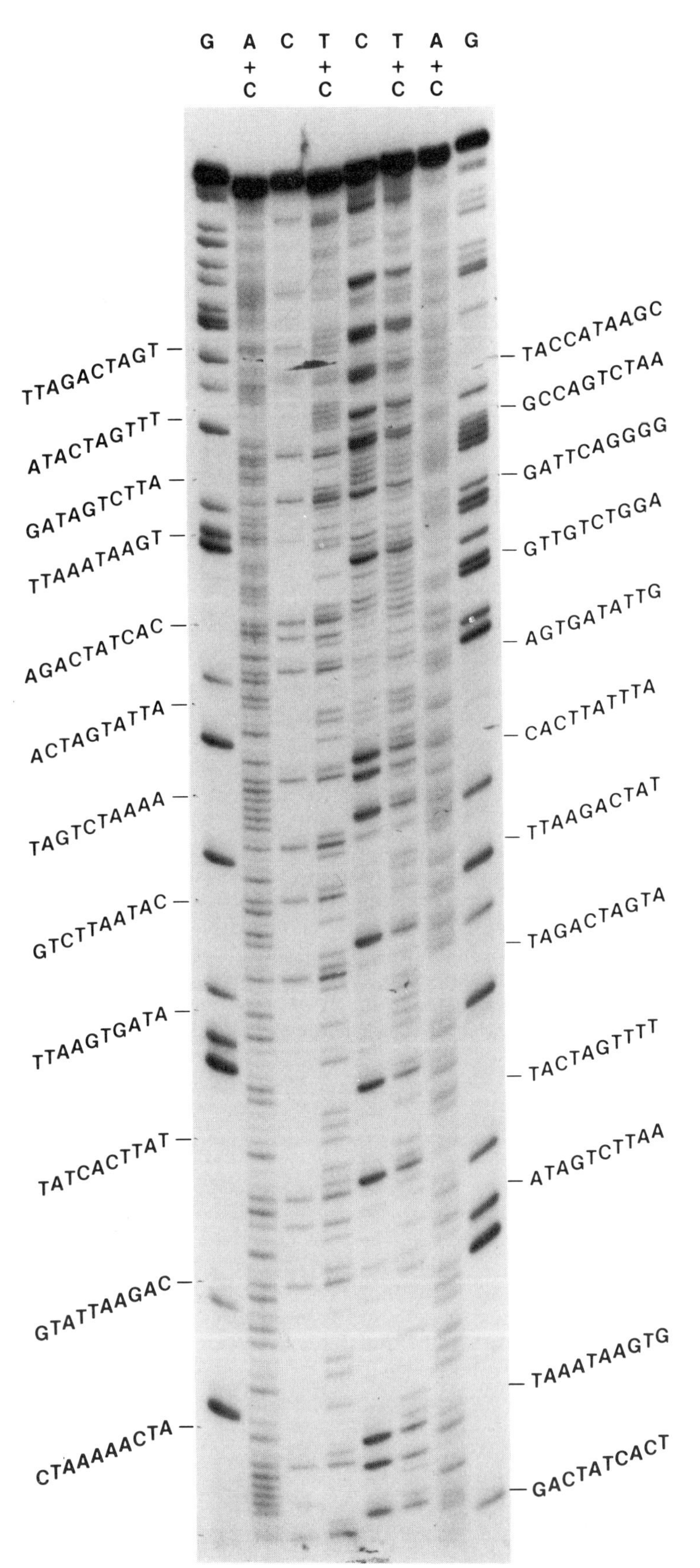

Figure 5. Sequence analysis of *gal*^{c}*331*. The 264-bp *Hpa*II fragment spanning the addition in *gal*^{c}*331* was 5′-end-labeled and the DNA stands separated by gel electrophoresis. Each strand was subjected to base-specific chemical degradation according to the method of Maxam and Gilbert (1977). The products were fractionated by electrophoresis on an 8% acrylamide-7M urea gel, and the first 110 residues were run off the gel. The right four lanes correspond to the DNA strand whose sequence is shown in Fig. 6, whereas the left four lanes are derived from the complementary strand.

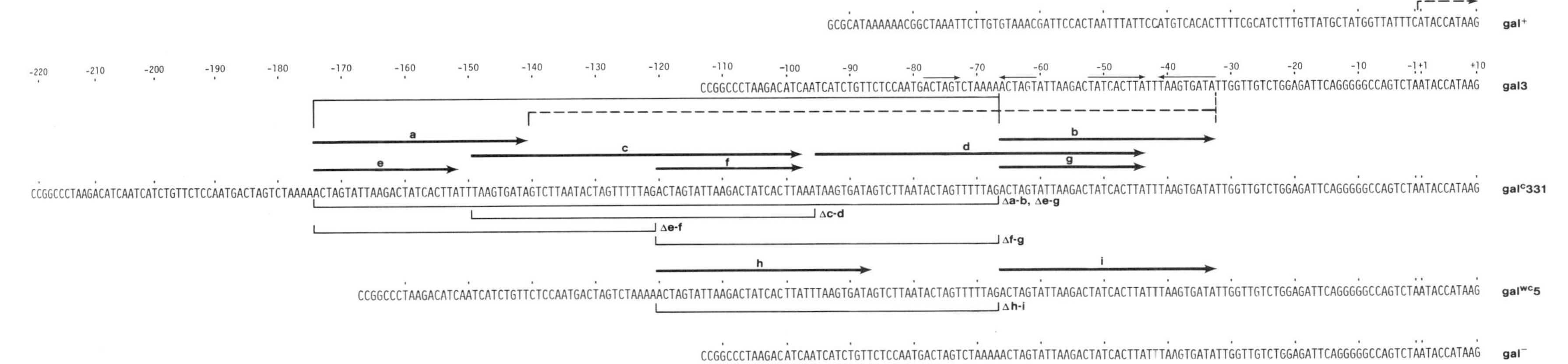

Figure 6. DNA sequences of gal^+, *gal3*, gal^c331, $gal^{wc}5$, and a gal^- segregant. Only one strand, in which the sequence is homologous to the *gal* mRNA sequence, is shown. The gal^+ sequence shows the transcription start point at +1, to the left of which are the two *gal* promoters (*P1* and *P2*) and the *gal* operator (Musso et al. 1977; Di Lauro et al. 1979). In *gal3*, IS*2*(I) is inserted just before +1. Negative numbers refer to IS*2* residues. Two pairs of inverted repeats are shown by arrows above the *gal3* sequence. The 108-bp segment, added to *gal3* between positions −32/−33 and −66/−67 to give rise to gal^c331, is indicated by solid and dashed lines. Three sets of direct repeats (a,b; c, d; and e,f,g), responsible for gal^c331 instability, are shown by bold arrows. Deletions Δa-b and Δe-g remove 108 bp from gal^c331 to generate the *gal3* sequence, whereas Δc-d, Δe-f, and Δf-g remove 54 bp to generate the gal^{wc} sequences (Fig. 10). Deletions are shown as occurring from arrowtail to arrowtail only, but they can occur anywhere between two homologous sites in any set of direct repeats. The $gal^{wc}5$ sequence contains the direct repeats h and i; deletions between these two generate a gal^- segregant. The observed sequence of a gal^- segregant of gal^c331 is presented in the bottom line. Also present in these sequences are inverted repeats of 60 bp and 35 bp centered at −123/−124 and −150/−151 in gal^c331 and of 23 bp centered at −69/−70 in $gal^{wc}5$.

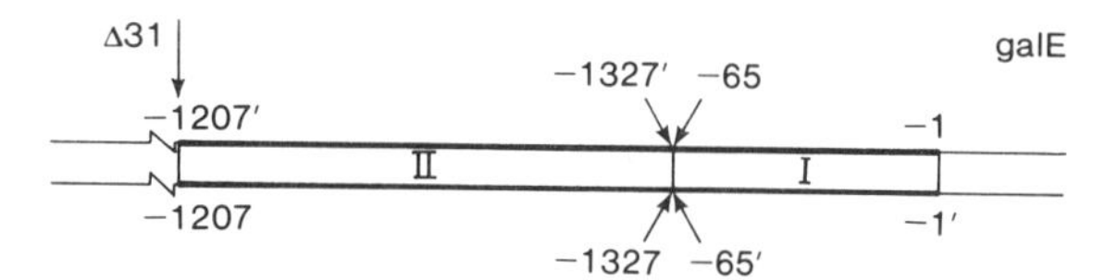

Figure 7. Structure of IS*2* in *gal*[c]*200 Δ31*. The structure is based on sequence determination of a 674-bp *Xho*I–*Hpa*II segment spanning the IS*2* fragment retained in *gal*[c]*200 Δ31*. The 186-bp fragment contains 65 bp (−1 to −65) from the right end of IS*2*(I), and 121 bp (−1327′ to −1207′) from the left end of IS*2*(II). Primed numbers indicate complementary-strand sequences. *Δ31* fuses position −1207′ of IS*2*(II) with a sequence derived from the far end of the *pgl* region.

and cloned between the *Eco*RI and *Hin*dIII sites of pBR322. In this configuration, the *galE* end is ligated to the *tet* gene through the *Hin*dIII site. Colonies harboring these recombinant plasmids exhibited a moderate level of tetracycline resistance but could be easily distinguished from the higher level of resistance conferred by the parent plasmid. The tetracycline resistance of the recombinant plasmid suggests that a promoter exists within the IS*2* fragment present in *gal*[c]*200 Δ31*.

The sequence of 674 bp from an *Xho*I site at the left of *Δ31* to the first *Hpa*II site in *galE* has been determined. The essential features of this sequence are summarized in Figure 7. *gal*[c]*200 Δ31* contains 186 bp of IS*2* sequence, located between the new joint created by *Δ31* and *galE*. In this sequence, the 65 bp at the right are derived from the right end of IS*2* (−1 to −65), the 121 bp at the left are derived from the complementary strand of the left end of IS*2* (−1327′ to −1207′), and the remaining IS*2* sequences are missing. Thus, the 65-bp segment at the right is in orientation I and the 121-bp segment at the left is in orientation II. This rearranged sequence must harbor the promoter responsible for constitutive expression of the *gal* operon. The sequence to the left of −1207′ is derived from the far end of the *pgl* region, which has become fused to the IS*2* fragment by *Δ31*. Apparently, this deletion extended up to residue −1207′ in IS*2* and, therefore, does not show the characteristic endpoint specificity of IS*2*-promoted deletions.

Structure of the Stable Reversion *gal*[c]*200*

Despite repeated attempts, the *gal*[c]*200* reversion could not be transferred directly to a λ*gal* genome unless a deletion, such as *Δ31*, was first introduced (Ahmed and Johansen 1975). Similar difficulties were encountered in transferring *gal*[c]*200* to a plasmid by in vitro cloning or by homogenote formation. As discussed later, this unusual property likely reflects a unique structural feature of *gal*[c]*200* and also poses a problem in determining its sequence. The sequence of *gal*[c]*200 Δ31*, however, provided a clue and allowed us to consider two alternative schemes for the origin of *gal*[c]*200* (Fig 8). According to the first scheme, the *gal3* → *gal*[c]*200* reversion occurred by the inversion of a segment (−66 to −1327) within IS*2* (Fig. 8b). Therefore, *gal*[c]*200* would contain residues −1 to −65 in orientation I and residues −1327′ to −66′ in orientation II. The deletion *Δ31* would extend up to residue −1207′ to generate the observed *gal*[c]*200 Δ31* sequence (Fig. 8d). According to the second scheme, *gal*[c]*200* arose by the insertion of a second IS*2* in orientation II into the IS*2* already present in *gal3* in orientation I. Insertion of an IS*2*(II) at position

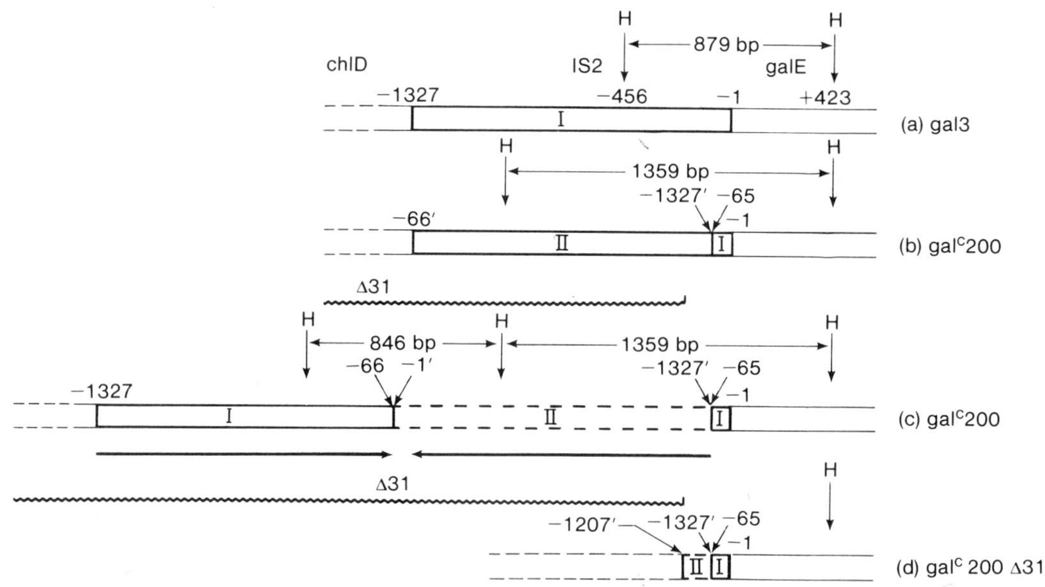

Figure 8. Proposed schemes for the origin of *gal*[c]*200*. The IS*2* element in *gal3* (*a*) underwent inversion of the −66 to −1327 segment to produce the *gal*[c]*200* structure shown in *b*. Alternatively, an IS*2* in orientation II (----) was inserted at position −65/−66 of the IS*2* present in *gal3* (——) in orientation I (*c*). The resulting structure creates a 1261-bp inverted repeat (bold arrows under *c*). Both structures can generate the *gal*[c]*200 Δ31* structure (*d*) by deleting (~~~) up to −1207′ and eliminating the inverted repeat. Various *Hin*dIII (H) fragments predicted by the two schemes, and identified by Southern hybridization (Fig. 9), are indicated. The fragment sizes shown in *c* do not take into account the 5-bp duplication normally created during the insertion of IS*2*.

−65/−66 of IS*2*(I) would generate the *gal*^{c}*200* structure shown in Figure 8c. The *gal*^{c}*200* Δ*31* structure (Fig. 8d) can easily be derived from this structure by assuming that Δ*31* extends up to residue −1207′ within IS*2*(II). Both schemes predict structures that would be stable and constitutive, because IS*2* is believed to harbor a functional promoter in orientation II (Saedler et al. 1974).

The two structures proposed for *gal*^{c}*200* can be distinguished experimentally on the basis of the predicted sizes of *Hin*dIII fragments. The inversion structure in Figure 8b predicts that the 879-bp fragment, a characteristic of *gal3,* would be replaced by a new 1359-bp fragment in *gal*^{c}*200.* On the other hand, the insertion structure (Fig. 8c) predicts that the 879-bp *gal3* fragment would be replaced by two new fragments, 1359 bp and 846 bp in length. These predictions were tested by Southern blot hybridization (Southern 1975) of electrophoretically separated *Hin*dIII fragments of total *E. coli gal3* and *gal*^{c}*200* DNA with labeled probes representing IS*2* or *galE* sequences (Fig. 9). The probe for IS*2* sequences was a 438-bp *Hin*dIII–*Hin*fI fragment from IS*2,* and the *galE* probe was an ~150-bp *Hin*fI–*Hin*dIII fragment from *galE* (Fig. 4). Normally, the IS*2* probe shows homology to five different fragments (size range ~950 bp to 20,000 bp) generated by *Hin*dIII cleavage of *E. coli gal*$^+$ DNA (not included in Fig. 9). A sixth band, approximately 900 bp in size, showing homology to the IS*2* probe, appears in *E. coli gal3* DNA. This new band was identified as the 879-bp *Hin*dIII fragment originating from the IS*2*–*galE* junction of *gal3* by comparison with *Hin*dIII-cleaved λ*gal490* DNA (which carries an IS*2* insertion 1 bp to the right of the site of *gal3* insertion) (Fig. 9c, lanes 1 and 2). The hybridization pattern of *gal*^{c}*200* (lane 3) shows that the 879-bp fragment of *gal3* is replaced by two new bands, corresponding closely in size to the 1359-bp and 846-bp fragments expected from the structure shown in Figure 8c. This finding is supported by results with the *galE* probe (Fig. 9d). As expected, this probe hybridizes with one 879-bp *Hin*dIII-generated fragment present in λ*gal490* (lane 1) or *E. coli gal3* (lane 2) DNA. In *gal*^{c}*200* (lane 3), the *galE* probe hybridizes with a 1359-bp fragment that corresponds to only one of the two fragments detected by the IS*2* probe. The structure in Figure 8c shows that this is precisely what is expected: The 1359-bp fragment should contain the *galE* sequence, whereas the 846-bp fragment should not. These results exclude the inversion scheme (Fig. 8b) and strongly support the secondary IS*2* insertion proposed in Figure 8c.

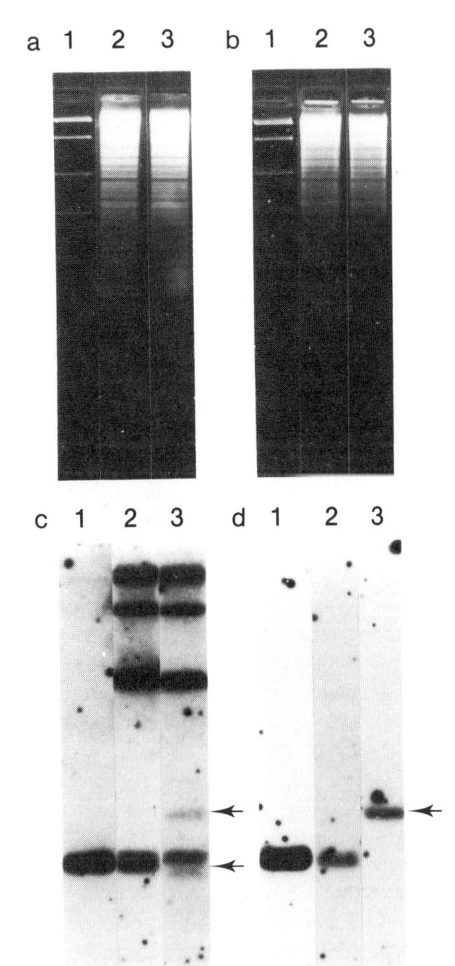

Figure 9. Southern analysis. *Hin*dIII digests of DNA extracted from λp*gal490* (lane *1*), *E. coli gal3* (lane *2*), and *E. coli gal*^{c}*200* (lane *3*) were fractionated by electrophoresis through 1% agarose slab gels, stained with ethidium bromide, and illuminated with a short-wave-length UV lamp for photography (*a* and *b*). The DNA was transferred to nitrocellulose filters and hybridized against ^{32}P-labeled probes representing the IS*2* sequences (*c*) or *galE* sequences (*d*), derived by *Hin*fI cleavage of the 3′-end-labeled 879-bp *Hin*dIII fragment (Fig. 4). Arrows point to the new 1359-bp and 846-bp fragments appearing in *gal*^{c}*200.* The new bands are faint in *c* probably because the 1359-bp fragment hybridizes with the 438-bp IS*2* probe by only 47 bp (positions −19 to −65), whereas the 846-bp fragment undergoes self-annealing, causing poor hybridization with the IS*2* probe. In contrast, the 1359-bp fragment shows strong hybridization in *d* because it contains the sequence homologous to the entire length of the ~150-bp *galE* probe (Figs. 4 and 8).

DISCUSSION

The Site of IS*2* Insertion

The sequence studies described above show that an IS*2* element was inserted at the *gal* promoter to produce the *gal3* mutation. The site of insertion in *gal3* is, therefore, identical with that found in *gal308* (Ghosal and Saedler 1978) and 1 bp to the left of the site of *gal490* insertion (Musso and Rosenberg 1977). Since these are the only IS*2* insertions known in the *gal* region, it appears that a sequence near the transcription startpoint constitutes a preferred target for IS*2* insertion.

The presence of IS*2* in the *gal* promoter provides an opportunity to monitor spontaneous sequence alterations in the insertion element as *gal*$^+$ revertants. In the simplest case, precise excision of IS*2* produces inducible revertants by restoring the wild-type *gal*$^+$ sequence. Internal rearrangements in IS*2,* however, produce constitutive revertants that provide valuable information about the role of direct and inverted repeats in generating slippage errors during repair or replication. These events provide explanations for some peculiar phenomena that have been known to be associated with IS*2*-induced mutations.

gal^{c}*331:* An Error Caused by Inverted Repeats

The unstable revertant *gal*^{c}*331* arose by the addition of a 108-bp segment at a point between positions −32/−33 and −66/−67 of the *gal3* sequence (Fig. 6). The added sequence is not new; it can be derived by copying along both strands of DNA of the region covered by the two pairs of inverted repeats shown above the *gal3* sequence. Ghosal and Saedler (1978) proposed an ingenious explanation based on replication slippage caused by inverted repeats to account for this mode of DNA addition. The newly replicated strand may fold back at the inverted repeat, thereby forming a new template, so that synthesis proceeds in the reverse direction along the newly synthesized strand for a short distance, before resuming replication along the correct template strand. A series of such errors, occurring at the two inverted repeats, can generate the 108-bp addition found in *gal*^{c}*331.* Such errors of replication must be rare events, occurring at a rate of $\sim 10^{-10}$ to 10^{-9}/cell/division.

Segregation: Errors Caused by Direct Repeats

Addition of the 108-bp sequence causes constitutive expression of the *gal* operon and introduces genetic instability in *gal*^{c}*331.* Although the promoter sequence has not yet been identified, the basis of instability is clearly established. The added segment generates a cluster of partially overlapping direct repeats labeled a and b, c and d, or e, f, and g in Figure 6. Removal of the intervening sequence between homologous sites in a and b (shown as Δa−b in Figure 6) or e and g (Δe−g) would result in a precise loss of 108 bp, generating a sequence identical to the original *gal3* sequence (Fig. 10). Sequence analysis of one such segregation product has confirmed this expectation. Therefore, these deletions explain the production by *gal*^{c}*331* of *gal*$^{-}$ segregants that are identical with *gal3.* On the other hand, removal of a segment between homologous sites in the direct repeats c and d (Δc−d) or f and g (Δf−g) would result in a loss of 54 bp, producing a sequence identical with *gal*^{wc}*5.* The Δe−f event would also remove 54 bp from *gal*^{c}*331,* but the resulting sequence differs from the *gal*^{wc}*5* sequence at positions −96 and −97 (underlined in Fig. 10). Although we have not sequenced additional *gal*^{wc} segregants, it is likely that the different enzyme levels (7% and 20%) found in different *gal*^{wc} segregants (Ahmed and Johansen 1978) originate from this minor sequence difference. The level of constitutive enzyme synthesis exhibited by *gal*^{wc}*5* is ~20% of the wild-type level.

Like *gal*^{c}*331,* the *gal*^{wc} segregants are also unstable and produce *gal*$^{-}$ colonies. The *gal*^{wc}*5* sequence in Figure 6 shows that a Δh−i event, occurring between the direct repeats h and i, would remove 54 bp to produce a sequence identical with *gal3* (Fig. 10).

Removal of the intervening sequence between direct repeats can occur by homologous recombination or deletion. The *recA* independence of *gal*^{c}*331* segregation suggests that these are true deletions arising by strand slippage along direct repeats, according to the scheme first proposed by Streisinger (Streisinger et al. 1967; Miller 1978). After the first component of a direct repeat (e.g., b in Fig. 6) has been replicated, the newly replicated strand may pair with the second component (a) and continue onward, generating a deletion Δa−b. As indicated by *gal*^{c}*331* segregation, these deletions must occur at an exceptionally high rate (10^{-3} to 10^{-2}/cell/division). In fact, this high rate is rather surprising for a replication slippage mechanism and we cannot exclude a novel *recA*-independent recombination mechanism. The Δa−b and Δe−g events (which produce *gal*$^{-}$ segregants) seem to occur more frequently than Δc−d, Δe−f, or Δf−g (which produce *gal*^{wc} segregants), suggesting that the frequency of these slippage errors may depend on the distance between the direct repeats. The presence of large inverted repeats in *gal*^{c}*331* and *gal*^{wc}*5* sequences might facilitate slippage between the direct repeats and thus contribute to the high rate of deletions. Since the frequency of *gal*^{c}*331* segregation is increased in the presence of a temperature-sensitive (*ts*) DNA ligase mutation (A. Ahmed, unpubl.), it appears that prolonged accumulation of nicks may favor errors of strand slippage.

gal^{c}*200:* Insertion of an IS*2* into IS*2*

Results of Southern analysis show that the constitutive revertant *gal*^{c}*200* arose by the insertion of an IS*2*(II) into IS*2*(I) present in *gal3* (Fig. 8c). This observation supports the proposal (Saedler et al. 1974) that IS*2* may carry a promoter in orientation II. Enhancement of gene expression by the insertion of IS*2*(II) has also been reported for *int-c* mutants of λ (Zissler et al. 1977) and for *trp5* and *his3* genes of yeast cloned in *E. coli* (Walz et al. 1978; Brennan and Struhl 1980). Since the second IS*2* is inserted in an orientation opposite to the first, no large direct repeats of the kind observed in *gal*^{c}*331* are created. Consequently, *gal*^{c}*200* is stable, and *gal*$^{-}$ segregants have never been observed.

The sequence at the insertion site (−65/−66) in *gal*^{c}*200* bears no extensive similarity to the sequence at the preferred insertion site at +1, although both are adenine- and thymine (AT)-rich regions of DNA. It seems more likely that selection of IS*2* rearrangements as *gal*^{c} revertants strongly favors changes at position −65 or −66, as found in *gal*^{c}*200, gal*^{c}*331,* and IS*2-7* (Ghosal et al. 1979a). Moreover, J. Besemer (pers. comm.) has found that a constitutive reversion of *gal308* occurred by the fusion of an IS*2*(II) element to position −65 of IS*2*(I), producing a junction (−1327′/−65) identical with that found in *gal*^{c}*200* Δ*31* (Fig. 7). Therefore, internal rearrangements of IS*2* occurring to the left of this region may not be detectable as *gal*^{c} revertants due to

CCGGCCCTAAGACATCAATCATCTGTTCTCCAATGACTAGTCTAAAAACTAGTATTAAGACTATCACTTATTTAAGTGATATTG **Deletion Product** (-30)

Δa-b, Δe-g, Δh-i

CCGGCCCTAAGACATCAATCATCTGTTCTCCAATGACTAGTCTAAAAACTAGTATTAAGACTATCACTTATTTAAGTGATATTG **gal3** (-30)

CCGGCCCTAAGACATCAATCATCTGTTCTCCAATGACTAGTCTAAAAACTAGTATTAAGACTATCACTTATTTAAGTGATAGTCTTAATACTAGTTTTTAGACTAGTATTAAGACTATCACTTATTTAAGTGATATTG **Deletion Products** (-30)

Δc-d Δf-g

CCGGCCCTAAGACATCAATCATCTGTTCTCCAATGACTAGTCTAAAAACTAGTATTAAGACTATCACTTA<u>AA</u>TAAGTGATAGTCTTAATACTAGTTTTTAGACTAGTATTAAGACTATCACTTATTTAAGTGATATTG (-30)

Δe-f

CCGGCCCTAAGACATCAATCATCTGTTCTCCAATGACTAGTCTAAAAACTAGTATTAAGACTATCACTTATTTAAGTGATAGTCTTAATACTAGTTTTTAGACTAGTATTAAGACTATCACTTATTTAAGTGATATTG **$gal^{wc}5$** (-30)

Figure 10. Comparison of observed and expected sequences of *gal*^{c}*331* segregants. The actual *gal3* sequence is identical with the sequence of *gal*$^-$ segregants expected from Δa–b or Δe–g events in *gal*^{c}*331* and Δh–i event in *gal*^{wc}*5* (Fig. 6). The actual *gal*^{wc}*5* sequence is identical with the Δc–d or Δf–g product from *gal*^{c}*331* but differs from the Δe–f product by 2 bp (underlined). The sequence of only one strand of DNA is shown.

either the lack of promoter activity or the presence of a transcription-termination site.

Inhibition of λ*gal* Formation

A remarkable property of *gal*^{c}*200* is the total suppression of the production of normal λ*gal* transducing particles. A lysate of a wild-type *gal*$^+$ (λ) strain, prepared by induction, contains a high titer of λ particles (10^9 to 10^{10}/ml) and a few λ*gal* transducing particles formed by defective excision. The λ*gal*:λ ratio of such low-frequency-transducing (LFT) lysates is typically around 5×10^{-7}. These particles produce normal *gal*$^+$ transductants, as judged by the segregation of *gal*$^-$ colonies and production of high-frequency-transducing (HFT) lysates. In sharp contrast, induction of *gal*^{c}*200*(λ) yields lysates that contain normal titers of λ, but very few transducing particles (λ*gal*:λ $= < 9 \times 10^{-9}$). Even these few particles are abnormal as judged by the properties of their transductants. These transductants are constitutive (indicating that *gal*^{c}*200* is transduced) but do not produce *gal*$^-$ segregants or HFT lysates. These transductants, therefore, do not arise by λ*gal* addition but by substitution via *gal-gal* exchange. This phenomenon is not restricted to *gal*^{c}*200,* because a large number of stable constitutive revertants tested were found to behave in the same manner. This drastic inhibition of λ*gal* production can also be demonstrated by density gradient centrifugation of HFT lysates from *gal*$^-$ (λ*gal*^{c}) (λ) strains, in which case only λ, but not λ*gal*^{c}, particles are found (Ahmed 1977).

How does *gal*^{c}*200* selectively block the production of normal λ*gal* particles? The answer to this question may be provided by the structure of *gal*^{c}*200* drawn in Figure 8c. Insertion of an IS*2* into an IS*2,* in the opposite orientation, creates a giant 1261-bp inverted repeat (shown as bold arrows in Fig. 8c) in which the two arms are separated by a short 66-bp segment. The presence of such a large inverted repeat might act as a strong barrier to the progress of λ*gal* DNA replication. Because of the extensive sequence homology, the two arms of the inverted repeat are expected to fold back in a newly replicated strand, so that synthesis would proceed backwards using the newly synthesized strand as a template. This would result in the creation of complex branched structures of bizarre genetic constitution, which would not be suitable for packaging. Therefore, these lysates would be devoid of normal λ*gal* particles. The rare particles capable of *gal* transduction found in these lysates would be the ones whose formation is independent of replication. Such particles are λ*doc*R, which are generated by cutting at the far side of the *gal* operon followed by a second cutting at the prophage *cos* site by *ter.* As a result, these aberrant particles carry only one cohesive end of λ and, often, the *gal* operon (Little and Gottesman 1971; Sternberg and Weisberg 1975). The abnormal properties of the *gal*$^+$ transductants arising from the λ*doc*R particles (Gingery and Echols 1969) are strikingly similar to those exhibited by the rare transductants obtained from *gal*^{c}*200*(λ) lysates. We infer that the only transducing particles present in these lysates are λ*doc*R. The replicational block may also explain the failure to transfer *gal*^{c}*200* to a plasmid. However, the possibility that *gal*^{c}*200* acts by blocking a step in the packaging process only cannot be eliminated.

Δ*31*: Loss of Deletion Endpoint Specificity

IS*2* promotes the formation of specific deletions (Ahmed and Johansen 1975). Heteroduplex analysis shows that these deletions extend from an endpoint fixed at the left terminus of IS*2* (Fig. 1) to various points in the *chlD-pgl* region, thus retaining the entire IS*2* element (Ahmed and Scraba 1978). Results of DNA sequence analysis have confirmed these observations and show that the fixed endpoint corresponds precisely to position −1327 at the left terminus of IS*2* (Peterson et al. 1979; R. Musso et al., unpubl.).

The specificity noted above is exhibited by *chlD-pgl* deletions isolated from *gal3* or *gal*^{c}*331.* However, the behavior of *gal*^{c}*200* is quite different. First, the frequency of *chlD-pgl* deletions is slightly increased. Second, the majority (19/23) of these deletions are *gal*$^-$, indicating that they destroy the *gal*^{c} promoter and likely enter the *gal* operon by deleting both IS*2* elements present in *gal*^{c}*200.* Third, some of those deletions that retain *gal*^{c}*200* allow its packaging to form normal transducing particles. Δ*31* is among the few deletions recovered that retain partial *gal*^{c}*200* phenotype and can be packaged as λ*gal*^{c}*200* Δ*31* (Ahmed and Johansen 1975). Analysis of Δ*31* (Fig. 8 c and d) shows that rather than extending up to the termini of either one of the two IS*2* elements, this deletion extended into the interior of IS*2*(II) and removed all sequences to the left of position −1207′. Consequently, this deletion retains only a 65-bp fragment from IS*2*(I) and a 121-bp fragment from IS*2*(II).

The occurrence of deletions showing altered specificity suggests some kind of interference in this process by the *gal*^{c}*200* structure. A common feature of these deletions is that they eliminate, either partially or completely, the two IS*2* elements harbored by *gal*^{c}*200.* Therefore, the lack of specificity may be attributed to the presence in *gal*^{c}*200* of a replication barrier in the form of a massive inverted repeat. Removal of this barrier by deletions (Fig. 8 c and d) would provide selective advantage by allowing DNA replication. Consequently, deletions exhibiting an apparent lack of endpoint specificity would be observed frequently in *gal*^{c}*200.*

To conclude, our studies show that the presence of direct or inverted repeats in specialized DNA sequences (such as insertion elements) and, perhaps, normal genomic sequences, can have a profound influence on the genetic behavior of the organism.

ACKNOWLEDGMENTS

Financial support for the research described in this paper was provided by the Natural Sciences and Engineering Research Council of Canada and the National Cancer Institute (under contract NO1-CO-75380 with

Litton Bionetics, Inc.). We thank N. Sternberg, J. Burckhardt, and M. Whiteway for critically reviewing this manuscript.

REFERENCES

ADHYA, S. and W. MILLER. 1979. Modulation of the two promoters of the galactose operon of *Escherichia coli. Nature* **279:** 492.

AHMED, A. 1977. The *gal3* mutation of *E. coli.* In *DNA insertion elements, plasmids, and episomes* (ed. A. I. Bukhari et al.), p. 37. Cold Spring Harbor Laboratory, Cold Spring Harbor, New York.

AHMED, A. and E. JOHANSEN. 1975. Reversion of the *gal3* mutation of *Escherichia coli:* Partial deletion of the insertion sequence. *Mol. Gen. Genet.* **142:** 263.

———. 1978. The basis of instability of a revertant of the *gal3* insertion of *Escherichia coli. J. Mol. Biol.* **121:** 269.

AHMED, A. and D. SCRABA. 1978. Nature of deletions formed in response to IS*2* in a revertant of the *gal3* insertion of *E. coli. Mol. Gen. Genet.* **163:** 189.

BRENNAN, M. and K. STRUHL. 1980. Mechanisms of increasing expression of a yeast gene in *Escherichia coli. J. Mol. Biol.* **136:** 333.

DI LAURO, R., T. TANIGUCHI, R. MUSSO, and B. DE CROMBRUGGHE. 1979. Unusual location and function of the operator in the *Escherichia coli* galactose operon. *Nature* **279:** 494.

GHOSAL, D. and H. SAEDLER. 1978. DNA sequence of the mini-insertion IS*2-6* and its relation to the sequence of IS*2. Nature* **275:** 611.

GHOSAL, D., J. GROSS, and H. SAEDLER. 1979a. DNA sequence of IS*2-7* and generation of mini-insertions by replication of IS*2* sequences. *Cold Spring Harbor Symp. Quant. Biol.* **43:** 1193.

GHOSAL, D., H. SOMMER, and H. SAEDLER. 1979b. Nucleotide sequence of the transposable DNA-element IS*2. Nucleic Acids Res.* **6:** 1111.

GINGERY, R. and H. ECHOLS. 1969. Integration, excision, and transducing particle genesis by bacteriophage λ. *Cold Spring Harbor Symp. Quant. Biol.* **33:** 721.

LITTLE, J. and M. GOTTESMAN, 1971. Defective lambda particles whose DNA carries only a single cohesive end. In *The bacteriophage lambda* (ed. A. D. Hershey), p. 371. Cold Spring Harbor Laboratory, Cold Spring Harbor, New York.

MAXAM, A. and W. GILBERT. 1977. A new method for sequencing DNA. *Proc. Natl. Acad. Sci.* **74:** 560.

MILLER, J. 1978. The *lacI* gene: Its role in *lac* operon control and its use as a genetic system. In *The operon* (ed. J. Miller and W. Reznikoff), p. 31. Cold Spring Harbor Laboratory, Cold Spring Harbor, New York.

MUSSO, R. and M. ROSENBERG. 1977. Nucleotide sequences at two sites for IS*2* DNA insertion. In *DNA insertion elements, plasmids, and episomes* (ed. A. I. Bukhari et al.), p. 597. Cold Spring Harbor Laboratory, Cold Spring Harbor, New York.

MUSSO, R., R. DI LAURO, M. ROSENBERG, and B. DE CROMBRUGGHE. 1977. Nucleotide sequence of the operator-promoter region of the galactose operon of *Escherichia coli. Proc. Natl. Acad. Sci.* **74:** 106.

PETERSON, P., D. GHOSAL, H. SOMMER, and H. SAEDLER. 1979. Development of a system useful for studying the formation of unstable alleles of IS*2. Mol. Gen. Genet.* **173:** 15.

SAEDLER, H., H. REIF, S. HU, and N. DAVIDSON. 1974. IS*2*, a genetic element for turn-off and turn-on of gene activity in *E. coli. Mol. Gen. Genet.* **132:** 265.

SOUTHERN, E. 1975. Detection of specific sequences among DNA fragments separated by gel electrophoresis. *J. Mol. Biol.* **98:** 503.

STERNBERG, N. and R. WEISBERG. 1975. Packaging of prophage and host DNA by coliphage λ. *Nature* **256:** 97.

STREISINGER, G., Y. OKADA, J. EMRICH, J. NEWTON, A. TSUGITA, E. TERZAGHI, and M. INOUYE. 1967. Frameshift mutations and the genetic code. *Cold Spring Harbor Symp. Quant. Biol.* **31:** 77.

WALZ, A., B. RATZKIN, and J. CARBON. 1978. Control of expression of a cloned yeast (*Saccharomyces cerevisiae*) gene (*trp5*) by a bacterial insertion element (IS*2*). *Proc. Natl. Acad. Sci.* **75:** 6172.

ZISSLER, J., E. MOSHARRAFA, W. PILACINSKI, M. FIANDT, and W. SZYBALSKI. 1977. Position effects of insertion sequences IS*2* near the genes for prophage λ insertion and excision. In *DNA insertion elements, plasmids, and episomes* (ed. A. I. Bukhari et al.), p. 381. Cold Spring Harbor Laboratory, Cold Spring Harbor, New York.

Activation of Gene Expression by IS*2* and IS*3*

N. Glansdorff, D. Charlier, and M. Zafarullah
Département de Microbiologie, Vrije Universiteit Brussel, and Research Institute of the CERIA, B-1070 Brussels, Belgium

Insertion of an IS element between the promoter of a bacterial gene and the gene itself can turn the gene either off or on. Until now, the "on" effect has been observed exclusively in the case of IS*2* elements inserted in orientation II with respect to the gene considered; the activating IS*2* is either fused to the gene by a deletion, as in high constitutive mutants of the galactose operon (Saedler et al. 1974), or inserted in front of it, as in integrase constitutive mutants of bacteriophage λ (Pilacinski et al. 1977). In all instances reported, the activating effect has been strongly selected for.

The purpose of this paper is: (1) to discuss evidence indicating that IS*2* in orientation II does not invariably promote active DNA transcription into adjacent genes, especially when no strong selection is applied (Boyen et al. 1978; Charlier et al. 1978); (2) to show that the on-off effect originally associated with IS*2* can be observed with IS*3* as well; (3) to discuss what bearing these observations have on the idea that some IS elements constitute mobile promoters and to present the alternative hypothesis of formation of new promoters at the site of insertion.

METHODS

All bacterial and phage strains used in this work have been described previously (Charlier et al. 1978). Genetic techniques, enzyme assays, and culture conditions were as described by Elseviers et al. (1972). Heteroduplex mapping and the strategy followed to identify IS*2* elements were as described previously (Charlier et al. 1978). The identification of IS*3* elements is discussed in the text. The genetic nomenclature follows that used by Bachmann and Low (1980).

RESULTS

Isolation of Insertion Mutants

We have looked for IS elements inserted in the control region of the divergent *argECBH* operon (Fig. 1). Using a divergent gene cluster rather than a classical operon makes it possible to investigate the influence on gene expression exerted by an IS element on the genes on either side of the element.

The control region (Fig. 1) is delimited to the right by a set of deletions that strongly reduce the transcription of *argE* but not of *argH*, this effect being *cis*-dominant (Elseviers et al. 1972; G. Bény and N. Glansdorff, unpubl.). Hence, the conclusion that the promoters for *argE* and *argCBH* face their cognate structural genes over the operator region, a situation that is now thought to prevail for the *bio* operon as well (Otsuka and Abelson 1978). The relevant DNA sequence has been worked out recently and appears to be in keeping with the genetic evidence (Piette et al. 1980).

To select for mutant derivatives displaying higher *argE* activity, we have used a strain (P4XSUP102; Elseviers et al. 1972) in which a deletion affecting part of *argB*, the whole of *argC*, and the promoter for *argE* results in very weak expression of *argE*. The deletion (Fig. 1) has been recombined into transducing phage λ13d*ppc argECBH* (Mazaïtis et al. 1976) to permit the screening of mutants harboring DNA rearrangements by isopycnic centrifugation of lysates in CsCl and by electron microscopy.

For a relatively weak selection, we have plated out cells on the natural substrate of the *argE* enzyme *N*-α-acetylornithinase, i.e., L-acetylornithine (50 μg/ml), as described previously (Elseviers et al. 1972). For a stronger selection, the lower-affinity substrate L-acetylarginine (200 μg/ml; Baumberg 1970) was used. The overall frequency of mutants was around 10^{-6} in the former case and 10^{-7} in the latter; about half of them appeared to harbor insertions, and the rest consisted mainly of point mutations and tandem or inverted duplications (Charlier et al. 1979). Enzyme assays (Table 1) were performed on derivatives obtained by transducing the mutant λd*ppc arg* at low multiplicity into a *recA* derivative of strain MN42 λ^- (Charlier et al. 1978) that harbors a *ppc argECBH* deletion; transductants were selected, and subsequent cultures grown, on minimum essential medium supplemented with glucose (selecting for Ppc^+) and arginine (100 μg/ml).

Characterization of Insertion Mutants

Selection on acetylornithine frequently provides mutants with low *argE* activities (see Table 1, AO series). Mutants AO7 and AO10 have been mentioned in previous publications (Boyen et al. 1978; Charlier et al. 1978). In the five AO strains, acetylornithinase remains well below the normal repressed level of that enzyme; this limited production is nevertheless growth-supporting, providing 20 times as much enzyme as the parental deletion strain. Hybridization assays for *argE* mRNA (data not shown) are in keeping with the enzyme assays.

The five strains harbor an IS*2* element in either orientation II (AO7, AO249, AO255) or orientation I (AO10, AO245) with respect to *argE*. Their respective locations are given in Figure 1. The sequence of the insertion site has been determined for AO7 (Piette et al. 1980). At

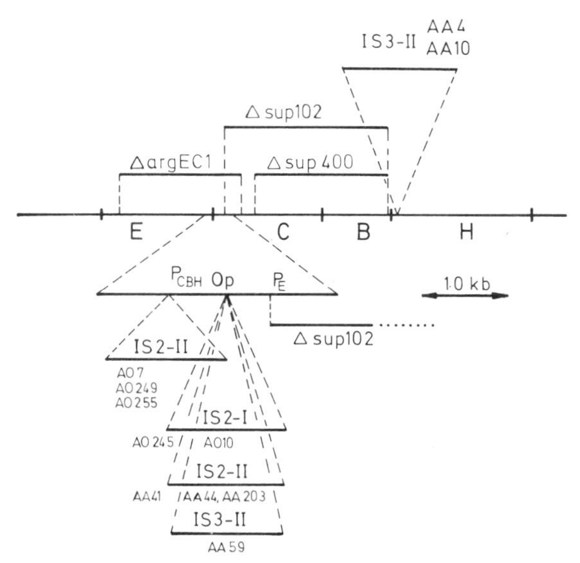

Figure 1. Physical map of the *argECBH* bipolar operon drawn to scale according to published evidence (Charlier et al. 1978; Linn et al. 1979). Orientation of insert is with respect to *argE*. (P) Promoter; (Op) operator region; (Δ) deletion.

Table 1. Specific Activities of Enzymes E and H

Strain	Enzyme E[a]	Enzyme H[a]	Insert[b]
P4X (wt)	6.0	0.16	
P4X*argR*	86	9.0	
SUP102	≤0.2	1.0	
SUP102*argR*	≤0.2	5.1	
AO7	1.9	0.55	IS*2*-II
AO7*argR*	2.8	1.62	
AO249	1.6	1.0	IS*2*-II
AO249*argR*	2.3	1.31	
AO255	2.9	0.60	IS*2*-II
AO10	1.8	0.44	IS*2*-I
AO10*argR*	2.0	0.52	
AO245	1.35	0.81	IS*2*-I
AO245*argR*	2.26	0.95	
AA41	6.0	0.10	IS*2*-II
AA41*argR*	17.8	0.42	
AA44	7.8	0.10	IS*2*-II
AA44*argR*	20.4	0.34	
AA203	8.3	<0.02	IS*2*-II
AA4	32.0	<0.02	IS*3*-II
AA4*argR*	30.0	<0.02	
AA10	31.0	<0.02	IS*3*-II
AA10*argR*	25.0	<0.02	
AA59	17.5	0.07	IS*3*-II
AA59*argR*	15.4	0.07	

Strains grown in minimum essential medium supplemented with 100 μg of L-arginine/ml.
[a] Sp. act. = μmole(s)/hr/mg of protein.
[b] IS orientation indicated is with respect to *argE*.

least two of the five base pairs adjacent to the element on the left-hand side (ATTGT) are known to be duplicated on the other side because their apposition to the sequence TGAC creates a *Hinc*II site, which has been detected. Up to nucleotide 92, at least, the sequence is identical with that of the IS*2* present in *gal* mutant 308 (Ghosal et al. 1979) and in the IS*2*(II)*gal* fusion analyzed by J. Besemer et al. (in prep.). As predicted from an earlier analysis (Boyen et al. 1978), the element borders on the putative promoter for *argCBH* (Fig. 1), the functioning of which it appears to impede to a certain extent (Boyen et al. 1978; Table 1). The most remarkable fact is that none of the five IS*2* elements promotes high expression of the gene with respect to which it happens to be in orientation II: *argE* for AO7, AO249, and A0255 (as determined by enzyme and mRNA assays) and *argH* for the other two (compare in Table 1 enzyme levels of genetically derepressed [*argR*] derivatives of the mutants and of the parental P4XSUP102 strain). In the latter case, the lower level of *argH* activity cannot be attributed to the fortuitous insertion of the IS element in a ribosome-binding site that would be present in the control region; indeed, in *argEC* deletion mutants, *argH* is known to be translated efficiently from mRNA transcribed from a weak secondary promoter located in the distal part of *argB* (Elseviers et al. 1972; Cunin et al. 1975). A relatively weak selection thus leads to recovery of IS*2* elements in orientation I or II with respect to the gene reactivated.

Selection on acetylarginine yields mutants with higher rates of *argE* expression (see the AA series in Table 1). This substrate offers the advantage of bypassing the subsequent steps of arginine biosynthesis (*argF, G,* and *H*), and it is thus possible to recover mutants that have lost *argH*, whereas on acetylornithine an activity of at least 0.1 unit is necessary to give a visible colony after 24 hours of incubation at 32°C.

Mutants AA41, AA44, and AA203 were found to harbor IS*2* elements in orientation II with respect to *argE*. Heteroduplex mapping and, for AA44, preliminary sequence data indicate that these IS*2* elements are located to the right of the promoter for *argCBH*, very close to the SUP102 deletion (see Fig. 1). Being in orientation I with respect to *argH*, these IS*2* elements exert the expected polar effect on the production of argininosuccinase. Acetylornithinase activities, though higher than in the AO series, remain low to moderate compared with fully derepressed levels of that enzyme.

The inserts present in mutants AA4, AA10, and AA59 were found to display homology, not with IS*2*, but rather with the IS*3* elements inserted into *lacZ* on the transducing phage λ*plac*MS505 (Malamy et al. 1972) (see Figs. 2 and 3). With respect to *argE*, the IS*3* of all three mutants is in the orientation opposite to that found in the *lacZ* mutant, where it is known to be strongly polar on *lacY* (Malamy et al. 1972). If we consider the orientation in *lacZ* as orientation I with respect to *lacY*, then these IS*3* elements are in orientation II relative to *argE*.

The very low levels of *argH* in mutants AA4 and AA10 make it possible to select for $ArgH^+$ derivatives by plating cells on minimum essential medium supplemented with ornithine (100 μg/ml), thus without applying any selective pressure on *argE*. This again illustrates the advantage of using a bipolar operon to investigate the effect of IS elements on gene expression. Ornithine-utilizing derivatives were obtained at a frequency of about 10^{-6}: More than 95% appeared to have lost the insertion and to have regained the low *argE* activity and

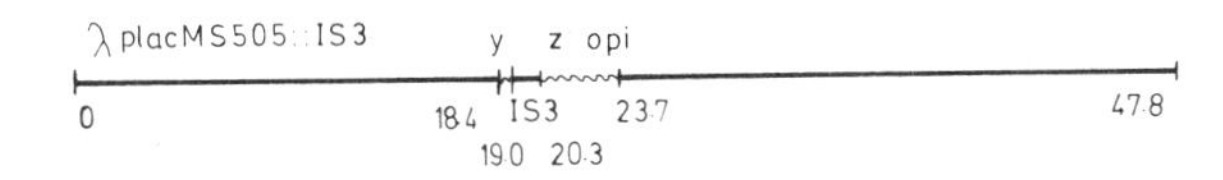

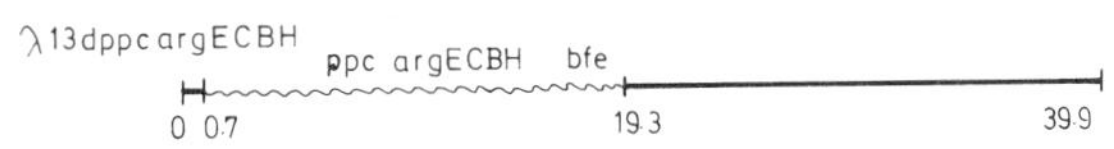

Figure 2. Genetic maps of λ*plac*MS505 (Malamy et al. 1972) and λ13d*ppc argECBH* (Mazaïtis et al. 1976). (—) Phage DNA; (〰) bacterial DNA. Coordinates are given in kb.

the *argH* activity proper to the parental P4XSUP102 strain. The insertion itself thus appears at least partly responsible for the reactivation of *argE*. On the other hand, mutant AA10 has been found to segregate, spontaneously (at a rate of about 0.5×10^{-4}/cell/generation), strains growing slowly on acetylarginine and displaying about 30% of the enzyme-E activity observed in AA10 itself; these derivatives appear to have kept all or most of the insertion element. It seems conceivable that the AA10 reactivation may not be accounted for fully by a sequence present on the IS itself.

DISCUSSION

At first sight, a straightforward interpretation of the properties of the mutants of the AA series (Table 1) would have appeared to be that the reactivation of *argE* is due to a promoter present on the IS element and geared in orientation II, as assumed a few years ago by Saedler et al. (1974) in the case of IS*2*. IS*3* would therefore appear as another insertosome able to turn on an adjacent gene, provided it is inserted in the required orientation. However, this interpretation is not tenable anymore, at least for IS*2*, without making the additional assumption that there exist different molecular species of that element. Indeed:

1. It is striking that the IS*2* elements present in our mutants of the AO series (Boyen et al. 1978) display only a very weak genetic activity in orientation II; even in mutants of the AA series the activity remains low to moderate.
2. There is no obvious candidate for a promoter in orientation II in the now-available sequence of that element (Ghosal et al. 1979).
3. W. Szybalski (pers. comm.) has found a spontaneous isolate of an IS*2* element inserted in bacteriophage λ that does not appreciably transcribe into adjacent genes in orientation II; this is in keeping with our observations on the low *argH* activity in strains of the AO series (Table 1) where the selection was applied on *argE*.
4. J. Besemer et al. (in prep.) have repeated the observation that originally led Saedler et al. (1974) to advocate the mobile-promoter hypothesis: the fusion between an IS*2* in orientation II and the *gal* operon previously inactivated by an IS*2* in orientation I inserted within the *gal* control region. A relic of the IS*2*(I) was found at the junction, confirming Rak's elaboration (1976) of the original observation. J. Besemer et al. (pers. comm.) have acquired strong evidence that in those strains the capacity to transcribe *gal* is determined by the nature of the new junction or in some cases by unstable rearrangements within the leftover segment in orientation I but not by the IS*2*(II) alone; most significantly, low-activity segregants obtained from such an unstable strain can become reactivated by an unidentified insertion at about 800 nucleotides from the extremity of the IS*2*(II), indicating that loss of activity in the segregant is not due to further rearrangement of the IS*2*(I) relic into a stop signal that could interrupt a hypothetical transcription coming from the IS*2*(II) element.

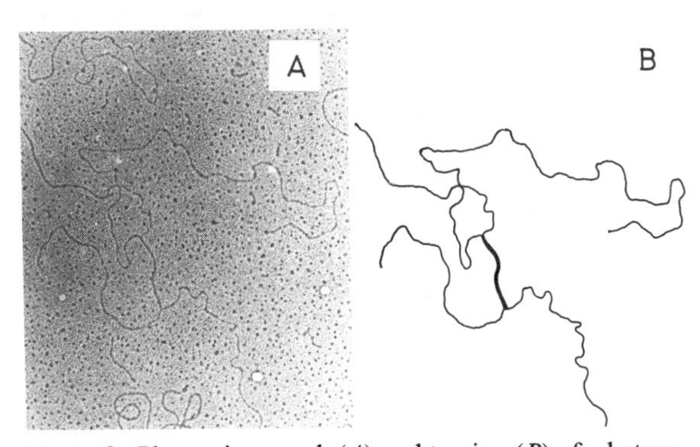

Figure 3. Photomicrograph (*A*) and tracing (*B*) of a heteroduplex molecule between DNA from λ*plac*MS505 and DNA from λ13d*ppc argECBH* carrying the SUP102 deletion and the IS*3* element of strain AA4.

It might be envisioned (Ghosal et al. 1979) that some copies of IS*2* do not carry promoters, whereas others do. However, the interpretation of the very instance that gave rise to the mobile-promoter hypothesis is now called into doubt by the work of J. Besemer et al. (in prep.). Although the association of certain insertosomes (presently IS*2* and IS*3*) with the restoration of gene activity is a fact of great biological significance, on the basis of our studies with *arg* mutants (Boyen et al. 1978; Charlier et al. 1978) and the work of J. Besemer et al. (in prep) on *gal* mutants, it must be concluded that a convincing case for a strong mobile promoter has yet to be produced. At any rate, because gene-activating IS*2* or IS*3* elements are found preferentially in one orientation, it is tempting to suggest that these elements carry a sequence that is inactive by itself but can become part of a promoter after the proper junction or the proper rearrangement is made. The behavior of the ornithine-utilizing derivatives of AA4 and AA10 indeed suggests that part of an IS sequence is involved in the promoter. The most obvious possibility would be that IS*2* carries an RNA polymerase recognition site that could become part of a functional promoter after insertion at the appropriate distance from a putative binding site (Pribnow box) present on the chromosome. There is a sequence around nucleotide 20 on the IS*2* elements that is present in *gal* mutant 308 (Ghosal et al. 1979), in the strain described by J. Besemer et al. (in prep.), and in mutants AO7 and AA44 (Piette et al. 1980; this paper) and that resembles previously reported recognition sites. Instances of IS*2*-promoted gene activations that can be accounted for by the present hypothesis will be reviewed elsewhere by D. Charlier and J. Besemer (in prep.). In this respect, it may be significant that in mutant AO7, which displays low levels of acetylornithinase, there is no obvious RNA polymerase binding site close to the end of the cognate IS element (Piette et al. 1980).

The formation of promoters by the apposition of two otherwise inactive sequences is an interesting possibility from the point of view of cell differentiation; indeed, rather than moving around a potentially harmful, "unbridled" promoter, new ones could be created at specific junctions. Moreover, one may wonder (and test) whether the occurrence of rearrangements at the very time of insertion could play a role in the formation of new promoters by providing part or all of their sequences. Indeed, it is known (Ghosal and Saedler 1978) that new promoters can be created near the I-end of IS*2* by what appears to be mistakes in DNA replication; we suggest that the transposition of an insertosome or transposon could, by virtue of requiring DNA replication (see Arthur and Sherratt 1979; Shapiro 1979), constitute an event favorable to the selection of promoter-like rearrangements due to replication errors.

ACKNOWLEDGMENTS

This work was supported by the Belgian Fonds voor Kollektief en Fundamenteel Onderzoek. M. Z. acknowledges a fellowship of the Cultural Agreements between Belgium and Pakistan.

REFERENCES

ARTHUR, A. and D. SHERRATT. 1979. Dissection of the transposition process: A transposon-encoded site-specific recombination system. *Mol. Gen. Genet.* **175:** 267.

BACHMANN, B. J. and K. B. LOW. 1980. Linkage map of *Escherichia coli* K-12. *Microbiol. Rev.* **44:** 1.

BAUMBERG, S. 1970. Acetylhistidine as substrate for acetylornithinase: A new system for the selection of arginine regulation mutants in *Escherichia coli. Mol. Gen. Genet.* **106:** 162.

BOYEN, A., D. CHARLIER, M. CRABEEL, R. CUNIN, S. PALCHAUDHURI, and N. GLANSDORFF. 1978. Studies on the control region of the bipolar *argECBH* operon of *Escherichia coli.* I. Effect of regulatory mutations and IS*2* insertions. *Mol. Gen. Genet.* **161:** 185.

CHARLIER, D., M. CRABEEL, R. CUNIN, and N. GLANSDORFF. 1979. Tandem and inverted repeats of arginine genes in *Escherichia coli.* Structural and evolutionary considerations. *Mol. Gen. Genet.* **174:** 75.

CHARLIER, D., M. CRABEEL, S. PALCHAUDHURI, R. CUNIN, and N. GLANSDORFF. 1978. Heteroduplex analysis of insertion mutations in the bipolar *argECBH* operon of *Escherichia coli. Mol. Gen. Genet.* **161:** 175.

CUNIN, R., A. BOYEN, P. POUWELS, N. GLANSDORFF, and M. CRABEEL. 1975. Parameters of gene expression in the bipolar *argECBH* operon of *Escherichia coli* K-12. The question of translational control. *Mol. Gen. Genet.* **140:** 51.

ELSEVIERS, D., R. CUNIN, N. GLANSDORFF, S. BAUMBERG, and E. ASHCROFT. 1972. Control regions within the *argECBH* gene cluster of *Escherichia coli* K-12. *Mol. Gen. Genet.* **117:** 349.

GHOSAL, D. and H. SAEDLER. 1978. DNA sequence of the mini-insertion IS*2-6* and its relation to the sequence of IS*2*. *Nature* **275:** 611.

GHOSAL, D., H. SOMMER, and H. SAEDLER. 1979. Nucleotide sequence of the transposable DNA-element IS*2*. *Nucleic Acids Res.* **6:** 1111.

LINN, T., M. GOMAN, and J. SCAIFE. 1979. Lambda transducing bacteriophage carrying deletions of the *argECBH-rpoBC* region of the *Escherichia coli* chromosome. *J. Bacteriol.* **140:** 479.

MALAMY, M. H., M. FIANDT, and W. SZYBALSKI. 1972. Electron microscopy of polar insertions in the *lac* operon of *Escherichia coli. Mol. Gen. Genet.* **119:** 207.

MAZAÏTIS, A. J., S. PALCHAUDHURI, N. GLANSDORFF, and W. K. MAAS. 1976. Isolation and characterization of λ*dargECBH* transducing phages and heteroduplex analysis of the *argECBH* cluster. *Mol. Gen. Genet.* **143:** 185.

OTSUKA, A. and J. ABELSON. 1978. The regulatory region of the biotin operon in *Escherichia coli. Nature* **276:** 689.

PIETTE, J., R., CUNIN, M. CRABEEL, A. BOYEN, N. GLANSDORFF, C. SQUIRES, and C. L. SQUIRES. 1980. Nucleotide sequence of the control region of the *argECBH* bipolar operon in *Escherichia coli. Arch. Int. Physiol. Biochim.* **88:** (in press).

PILACINSKI, W., E. MOSHARRAFA, R. EDMUNDSON, J. ZISSLER, M. FIANDT, and W. SZYBALSKI. 1977. Insertion sequence IS*2* associated with *int*-constitutive mutants of bacteriophage lambda. *Gene* **2:** 61.

RAK, B. 1976. *gal* mRNA initiated within IS*2*. I. Hybridization studies. *Mol. Gen. Genet.* **149:** 135.

SAEDLER, H., H.-J. REIF, S. HU, and N. DAVIDSON. 1974. IS*2*, a genetic element for turn-off and turn-on of gene activity in *E. coli. Mol. Gen. Genet.* **132:** 265.

SHAPIRO, J. A. 1979. Molecular model for the transposition and replication of bacteriophage Mu and other transposable elements. *Proc. Natl. Acad. Sci.* **76:** 1933.

IS*1*-promoted Events Associated with Drug-resistance Plasmids

M. CHANDLER, M. CLERGET, AND L. CARO
Department of Molecular Biology, University of Geneva, 1211 Geneva 4, Switzerland

The plasmids R1*drd*19, R6, and R100-1 are members of a class of bacterial plasmids that confer multiple antibiotic resistance on the host cell. They have the striking feature that most of their resistance genes are clustered in a unique region called the r determinant (r-det), which is flanked by two directly repeated copies of the insertion element IS*1* (Hu et al. 1975; Ptashne and Cohen 1975). The r-det generally occurs as a cointegrate with a second element, RTF, which contributes both *r*eplication and *t*ransfer *f*unctions to the plasmid (Fig. 1A). The clustering of the resistance genes between the IS*1* elements suggests that the r-det may function in the translocation of these genes as a unit. Indeed, it has been shown that the entire r-det can transpose from R100-1 to bacteriophage P1, albeit at very low frequency (Arber et al. 1979).

Other lines of evidence also indicate that r-det structures behave as an integral unit. For example, growth of *Proteus mirabilis* carrying R100-1 in the presence of chloramphenicol (Cm), an antibiotic to which R100-1 specifies resistance, results in amplification of the entire r-det unit as monomeric and multimeric circles and as tandem multimers attached to the RTF (Hashimoto and Rownd 1975); in *Salmonella typhimurium,* the r-det-associated drug resistances are rapidly lost as a unit during growth in nonselective media (Watanabe and Ogata 1970). This type of behavior is not observed when the plasmid is carried by *Escherichia coli* and therefore must be determined by host-specified factors.

We present here results demonstrating that functions present on the R plasmid are also important in determining r-det behavior. We have isolated a mutant derivative of R100-1 that produces supernumerary covalently closed, circular (CCC) copies of the r-det in *E. coli,* and we show that this phenotype can be repressed in *trans* by the presence of the closely related plasmids R1*drd*16 and R1*drd*19 (Meynell and Data 1967). We also present evidence that the formation of circles by this mechanism can provide an alternative pathway to direct transposition for the translocation of r-det elements from one replicon to another. In this study we have used a small r-det derivative that carries resistance to kanamycin (Km^r), which we have isolated from the plasmid R1*drd*19.

Finally, we have begun to study in detail the transposition properties of elements like the r-det that are flanked by IS*1* sequences. We have compared the frequencies of transposition of a series of IS*1*-flanked transposons varying in length between 2.6 kb and 10.4 kb. We find that the frequency of transposition decreases dramatically as the length of the DNA between the IS*1* elements is increased.

The Appearance of r-det in *E. Coli*

Normally, the plasmid R100-1 is maintained in *E. coli* as a single CCC species (Fig. 1a). The appearance of CCC r-det copies is rare. R100-1 can integrate into the *E. coli* chromosome to form an Hfr strain and in so doing is able to suppress the defect in the initiation of chromosome replication caused by the temperature-sensitive *dnaA* mutation (Nishimura et al. 1973). It does so by providing the host chromosome with a functional replication system and an origin of replication (Bird et al. 1976). Integration of R100-1 can result in the appearance of CCC r-det units (Chandler et al. 1977c) by recombination between the directly repeated copies of IS*1* (Chandler et al. 1977a) (Fig. 1b). In the Hfr strain that we have studied in greatest detail, LC2633, integration occurred by a process that we have called inverse transposition (Fig. 1a,b), in which the ends of the IS*10* inverted repeats proximal to the tetracycline-resistance (Tc^r) genes of Tn*10* are employed for the integration. This results in transposition of the R plasmid into the chromosome, with concomitant loss of the region of Tn*10* that specifies Tc^r (Chandler et al. 1979b).

The copy number of r-det relative to RTF units has been measured by DNA-DNA hybridization using total Hfr DNA and has been found to be approximately 1.4 (Chandler et al. 1979a; Silver et al. 1980). In other words, an excess of r-det sequences is produced. The level of overproduction as measured by the hybridization assay is similar in both *recA*$^+$ and *recA*$^-$ derivative Hfr strains, but the appearance of r-det in its circular form (Fig. 1b) is dependent on the products of both the *recA*$^+$ and *recC*$^+$ genes (Chandler et al. 1979a; Silver et al. 1980). These observations, taken together with our inability to establish purified r-det in *E. coli* using a variety of methods (Chandler et al. 1977c; Lane et al. 1979), has led us to propose a working model for circular r-det production (Silver et al. 1980). We have proposed that r-det is produced continuously from the integrated plasmid by what might be called a conservative excision, which involves a plasmid-determined overreplication of the r-det region followed by a host-specified circularization and excision of the supernumerary r-det copy. The molecular mechanism of r-det production, however, is at present unclear. To clarify this phenomenon, we have begun to examine the contribution of plasmid-specified functions.

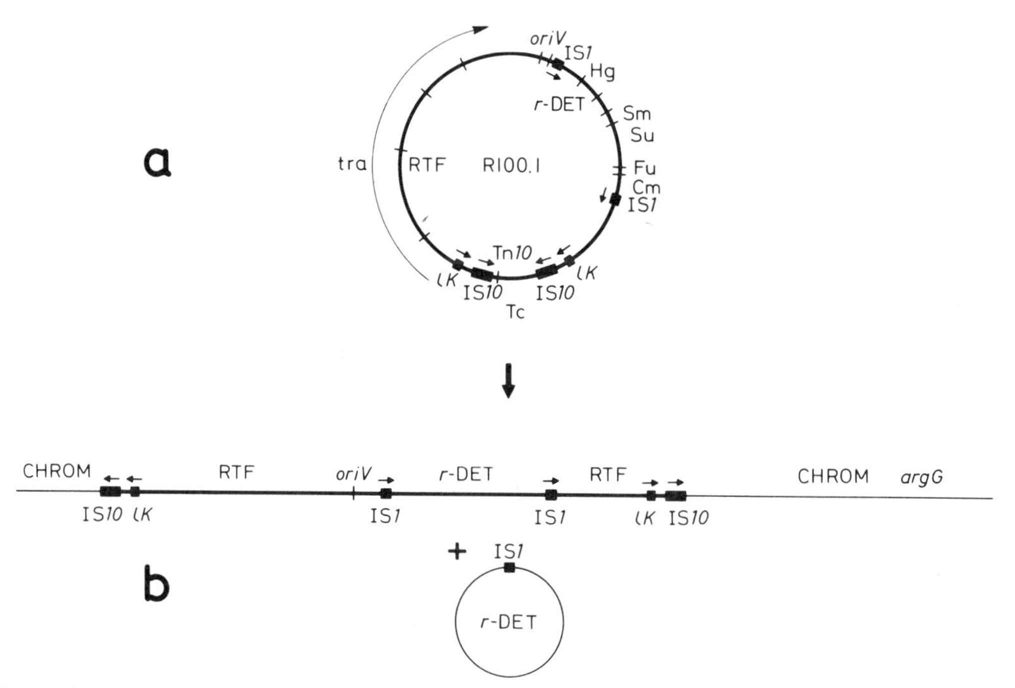

Figure 1. (*a*) Physical map of R100-1 showing *Eco*RI-generated DNA fragments (Tanaka et al. 1976), location of the transfer operon (*tra*), origin of vegetative replication (*oriV*), Tn*10*, and r-det. The positions of the inverted repeats ικ (Ohtsubo and Ohtsubo 1977) are also shown. The approximate positions of resistances to mercury (Hg), streptomycin (Sm), sulfonamide (Su), fusidic acid (Fu), chloramphenicol (Cm), and tetracycline (Tc) are indicated. (*b*) Structures of the integrated R100-1 in Hfr strain LC2633 (Chandler et al. 1979b) and of the supernumerary r-det. Insertion has occurred using the internal ends of the IS*10* elements with loss of the Tc^r region of Tn*10*.

Isolation of an r-det-producing Derivative of R100-1

We have previously reported the isolation of an R′ derivative plasmid from the r-det-producing Hfr strain LC2633 (Chandler et al. 1979b). This R′ (pLC115) carries a segment of bacterial DNA that includes the *argG* locus and has a molecular length of approximately 240 kb, as determined by electron microscopy.

Electron microscopy of denatured and rapidly renatured preparations of the plasmid shows a single-stranded loop of the size expected for R100-1 (78.0 ± 5.5 kb) separated from a second larger loop (139.9 ± 13.1 kb) of bacterial DNA by a double-stranded stem of 1.34 ± 0.17 kb (Fig. 2a). The stem results from intramolecular hybridization of the inverted IS*10* repeats carried by the plasmid (Chandler et al. 1979b).

To determine whether the ability to produce r-det is retained by pLC115, we have transferred the plasmid from its original $recA^-$ host into a $recA^+$ recipient. Plasmid DNA isolated from the single exconjugant analyzed was found to contain DNA circles of the size expected for the r-det (data not presented). Subsequent passage of the plasmid resulted in the loss of the bacterial *argG* marker. When one such derivative plasmid, pLC117, was transferred to a $recA^-$ recipient strain, the plasmid DNA was found to have a unique molecular length of 84 kb, as determined by electron microscopy. Comparison of the *Eco*RI digestion products of pLC117 (Fig. 3, lane 2) with those of R100-1 (Fig. 3, lane 1) shows that they are similar and indicates that most, if not all, of the bacterial sequences have been deleted.

We have determined the extent of the deletion more accurately by examining in the electron microscope heteroduplex molecules formed between pLC117 and the RTF unit of R100-1. An electron micrograph of a typical heteroduplex is shown in Figure 4. A large single-stranded loop can be observed. It has a molecular length of 22 kb, which is consistent with it being the r-det region of pLC117. The characteristic loop-stem structure of Tn*10* associated with the RTF molecule can also be seen. In addition, there is a deletion loop extending in both directions from Tn*10* that terminates in a small double-stranded stem of approximately 0.18 kb. The deletion that gave rise to pLC117 therefore seems to have occurred at the level of two small sequences that are inverted with respect to each other and that flank Tn*10* in R100-1. Their size and position suggest that they are identical to the ικ elements described by Ohtsubo and Ohtsubo (1977) and shown by them to promote deletions in R100-1. The stability and smaller size of pLC117 make it simpler to manipulate than the large R′. We have therefore used pLC117 in subsequent studies. Although circular r-det cannot be detected in pLC117 plasmid DNA preparations from a $recA^-$ background, either by agarose gel electrophoresis of purified plasmid DNA (Fig. 3, lane 5) or by electron microscopy of cleared lysates of these strains, transfer of pLC117 into a $recA^+$ recipient leads to the appearance of CCC

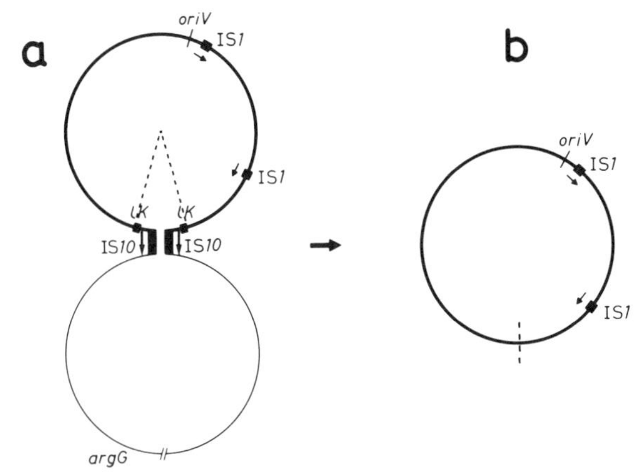

Figure 2. (*a*) Structure of an R′ pLC115 (derived from LC2633) molecule in a DNA preparation that had been denatured and rapidly renatured (Davis et al. 1971). The heavy region shows R100-1 sequences; the light region shows bacterial DNA carrying *argG*. Measurements were made using ϕX174 viral and RFII DNAs as internal standards (see text) (Barrell et al. 1976). (----) Extent of the deletion that gives rise to pLC117. (*b*) Structure of pLC117, a plasmid derived from pLC115 by spontaneous deletion at the ιK inverted repeats of pLC115 (see Fig. 4). Its contour length is given in the text.

r-det molecules. This can be seen clearly as an additional plasmid band following agarose gel electrophoresis of undigested plasmid DNA (Fig. 3, lane 4). *Eco*RI digestion of these DNA preparations (Fig. 3, lanes 2 and 3) reveals that the preparation of pLC117 isolated from the *recA*$^+$ genetic background carries a

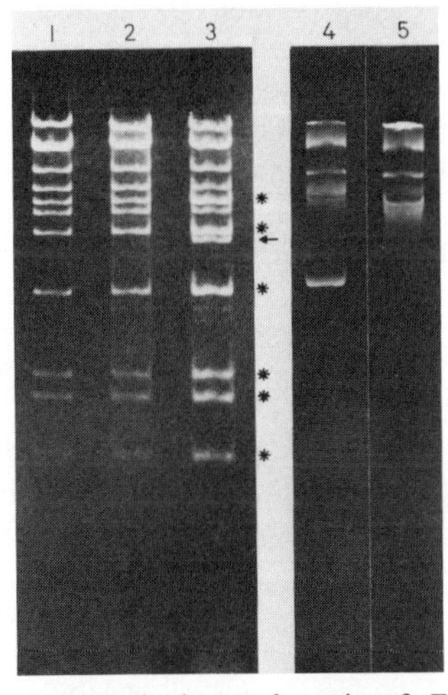

Figure 3. Agarose gel electrophoresis of *Eco*RI-digested plasmid DNA. (*1*) R100-1; (*2*) pLC117 DNA isolated from a *recA*$^-$ host; (*3*) pLC117 DNA isolated from a *recA*$^+$ host. Asterisks indicate fragments present in higher molar quantities. They are all present in the r-det region of the plasmid (Chandler et al. 1977c). Arrow indicates the fragment derived from r-det excision by IS1/IS1 recombination. (*4*) Agarose gel electrophoresis (0.6%) of an undigested preparation of pLC117 isolated from a *recA*$^+$ host (note the presence of an extra major band); (*5*) agarose gel electrophoresis (0.6%) of an undigested preparation of pLC117 isolated from a *recA*$^-$ host.

subset of fragments in higher molar quantity than when these fragments are carried by the parent plasmid. These fragments are those characteristic of the r-det and include a fragment not found in the parent plasmid but which is generated by recombination between the flanking copies of IS*1* during excision of the r-det (Chandler et al. 1977a,c).

The property of r-det production persists through many passages of the plasmid alternately into *recA*$^-$ and *recA*$^+$ strains having a variety of genetic backgrounds. Integration of R100-1 into the chromosome has therefore resulted in a permanent change in plasmid behavior. In addition to r-det production, this mutation renders pLC117 unstable. In several experiments we have observed the appearance of RTF units at low frequency (Fig. 5, lane 4). These arise by deletion of the r-det.

To determine whether the mutated function of pLC117 is *trans*-acting, we have investigated the effect of the related plasmids R1*drd*16 and R1*drd*19 on r-det production. R1*drd*19 carries resistances to Km and ampicillin (Ap), in addition to most of the resistances associated with the r-det of R100-1 (see Figs. 1 and 7). Insertion of the Ap^r genes of Tn*3* has occurred within the Hg^r genes of the r-det, with the result that R1*drd*19 does not express Hg^r (see Fig. 7). The plasmid R1*drd*16 is a derivative of R1*drd*19 that has lost all resistances except for Km^r (Meynell and Data 1967).

The effect of introducing R1*drd*16 into the Hfr strain LC2633 is shown in Figure 5. This figure shows the results of agarose gel electrophoresis of cleared lysates of LC2633 (Fig. 5, lane 1), of LC2633/R1*drd*16 (Fig. 5, lane 2), and of a derivative of LC2633/R1*drd*16 that has lost the R1*drd*16 plasmid (Fig. 5, lane 3). As shown, the characteristic r-det plasmid band present in LC2633 disappears with the introduction of R1*drd*16 (which appears on the gel as a specific band of high molecular weight). Loss of R1*drd*16 from the Hfr strain results in the reappearance of the r-det. In the experiment shown, we used cultures that had been grown at 37° C. Similar results have been obtained with cultures grown both at 30°C and at 42°C. Figure 5 also shows the results of agarose gel electrophoresis of purified plasmid DNA isolated from *recA*$^+$ strains carrying pLC117 (Fig. 5, lane 4) and both pLC117 and R1*drd*19 (Fig. 5, lane 5). The introduction of the latter results in the disappearance of the r-det band. The results indicate that production of the r-det is due to a plasmid mutation that can be repressed in *trans*.

The existence of mechanisms that allow overproduction, circularization, and excision of the r-det suggests an alternative pathway to that of transposition for its translocation from replicon to replicon (Fig. 6a). Supernumerary r-det circles resulting from this process should be able to form a cointegrate with a suitable recipient replicon either by transposition using its single copy of IS*1* (Fig. 6c) or by recombination between this IS*1* element and a resident IS*1* element in the recipient (Fig. 6b). These pathways have recently been proposed from an analysis of cointegrates formed between the prophage P1 and R100-1 (Iida and Arber 1980).

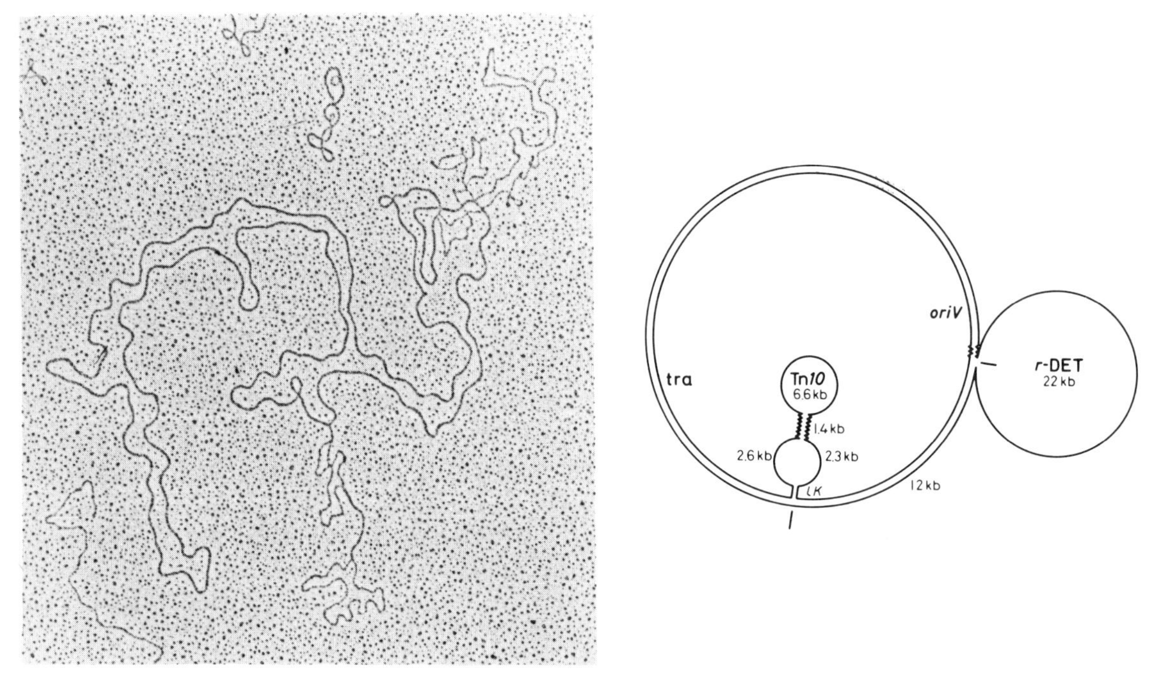

Figure 4. Structure of pLC117. The electron micrograph shows a typical heteroduplex molecule formed between pLC117 and the RTF unit of R100-1. It shows a large, single-stranded loop representing the r-det region of pLC117 and the typical loop-stem structure of the RTF-associated Tn*10* flanked by two unequal single-strand regions. The deletion loop terminates in a short, double-stranded stem representing the ικ inverted repeats that flank Tn*10* in the plasmid (Ohtsubo and Ohtsubo 1977). Measurements were made using φX174 viral and RFII DNAs as internal standards.

Isolation of a Replicating r-det Structure from R1*drd*19

Three major problems are encountered in studying the relationship between r-det transposition and its translocation by circle formation followed by cointegrate formation: (1) The frequency of r-det transposition is very low (Arber et al. 1979). (2) r-det is large (22 kb) and therefore difficulties arise in selecting a suitable recipient replicon. (3) The properties of the circular r-det cannot be analyzed in the absence of the parent R plasmid, since it cannot be established on its own. Ideally, we would like a smaller r-det derivative that exhibits an increased frequency of transposition and that can be established as a circle independent of the parent R plasmid.

Recently, we have isolated just such a derivative from the r-det region of R1*drd*19. The results of analysis of heteroduplex molecules formed among R1*drd*19, R6, and R100-1 and their derivatives (Sharp et al. 1973) indicate that these plasmids are closely related. The r-det regions of R100-1 and R1*drd*19 differ from each other in only three respects (Fig. 7a,b) (Kopecko et al. 1976). R1*drd*19 carries two insertion loops: one with a molecular length of 4.6 ± 0.62 kb, consistent with its being the Ap^r transposon Tn*3*, and a second loop of 9.56 ± 1.04 kb that carries the Km^r gene(s) (Kopecko et al. 1976). The r-det of R100-1 carries a single insertion (1.45 ± 0.27 kb) of unknown function. The r-det region of a heteroduplex between R1*drd*19 and R100-1 is shown in Figure 7a. Figure 7b shows the results of measurements made on nine such molecules, and in Figure 7c we have aligned the *Eco*RI, *Hin*dIII, and *Sal*I restriction maps of the r-det regions of both plasmids (Tanaka et al. 1976; Ohtsubo and Ohtsubo 1977; Blohm and Goebel 1978; Arber et al. 1979; Chandler et al. 1979a; M. Clerget et al., in prep.). When *Eco*RI digests of the two plasmids are separated by agarose gel electrophoresis and the separated fragments are transferred to a nitrocellulose filter (Southern 1975) and challenged with an IS*1*-specific probe, R100-1 shows two homologous IS*1* bands corresponding to fragments at either extremity of the r-det (Chandler et al. 1977a). In the case of R1*drd*19, three fragments exhibit IS*1* homology (M. Clerget et al., in prep.). Further mapping of these homologies has shown that in addition to the two flanking IS*1* elements, R1*drd*19 carries a third copy of IS*1* within the r-det. This is located at one extremity of the Km^r loop (Fig. 7c) (M. Clerget et al., in prep.). The 9.6-kb Km^r region is able to transpose from R1*drd*19 to a suitable λ phage (λ*b*515*b*519*nin*5) using standard techniques (Berg et al. 1975). Transposition is mediated by the flanking copies of IS*1* to yield an insertion of 10.4 kb (the 9.6-kb insertion with an additional IS*1*; Fig. 7c). It can also transpose from a λ lysogen to a resident RTF plasmid at a frequency of about 10^{-7} (see Fig. 9). We have given this transposon the number Tn*2350*.

Like R100-1, R1*drd*19 can integrate into the *E. coli* chromosome, but it gives rise to three r-det circular species (J. Frey, pers. comm.), representing all three possible IS*1*/IS*1* recombination products (Fig. 7c), including a 9.6-kb circular form of Tn*2350*.

In contrast to the purified circular r-det of R100-1, which we have consistently been unable to establish in-

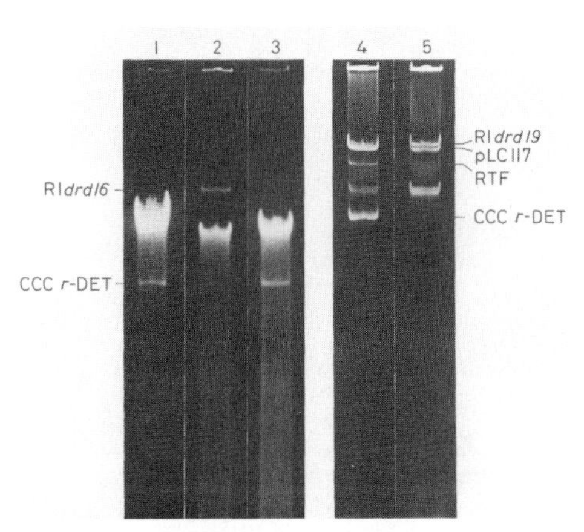

Figure 5. Effect of R1*drd*16 and R1*drd*19 on r-det production in the Hfr strain LC2633 and in a strain carrying pLC117. R1*drd*16 (Km^r) was introduced by conjugation into LC2633 ($Cm^rFu^rSm^rHg^r$) from *E. coli* C600 by selection for Km^rCm^r. To remove R1*drd*16 subsequently, LC2633/R1*drd*16 was grown in medium without Km, and a Km^s segregant was isolated. R1*drd*19 ($Km^rAp^rCm^rFu^r$-Sm^rSu^r) was introduced by conjugation from C600 into LC690 (= C600 Nal^rP1^r) carrying pLC117. Selection was made for Nal^rHg^r (pLC117) and both Ap^r and Km^r. Selection was maintained at all times. (Lanes *1-3*) Agarose gel electrophoresis (0.6%) of cleared lysates (Clewell and Helinski 1969). (*1*) LC2633 showing an r-det band in addition to the chromosomal DNA; (*2*) LC2633/R1*drd*16 showing loss of the circular r-det and appearance of a band of high molecular weight corresponding to R1*drd*16; (*3*) an R1*drd*16 segregant of LC2633/R1*drd*16 showing loss of the R1*drd*16 band and reappearance of the r-det. (Lanes *4, 5*) Agarose gel electrophoresis (0.6%) of plasmid DNA isolated by CsCl–ethidium-bromide density centrifugation (Clewell and Helinski 1969). (*4*) pLC117 DNA showing the r-det; (*5*) pLC117/R1*drd*19 DNA showing the presence of two large plasmid species and the disappearance of r-det.

dependently in *E. coli,* the circular form of Tn*2350* (pTn*2350*) can be introduced by transformation and maintained in the absence of the parent plasmid. It appears as a distinct plasmid band on agarose gels of cleared lysates of Km^r transformants, has a molecular length of 9.6 kb, and has a restriction pattern consistent with its production by recombination between the two flanking IS*1* elements. Figure 8 shows the *Eco*RI digestion products of the parent plasmid R1*drd*19, pTn*2350,* λ*bbin*::Tn*2350,* and λ*bbnin,* respectively, for comparison. Each Tn*2350*-carrying replicon has two fragments in common (indicated by asterisks). The junction fragments of Tn*2350* in both R1*drd*19 and λ*bbnin*::Tn*2350* are also indicated. In addition to the two common fragments, pTn*2350* carries a larger fragment with a molecular length of 6.2 kb, consistent with its formation by recombination between the flanking IS*1* elements (Fig. 7c).

Although we have not yet studied the replication of this transposon in detail, it is clear that it is unstable. It is lost from approximately 60% of cells in an overnight culture grown under nonselective conditions and which had undergone about 18 generations of growth.

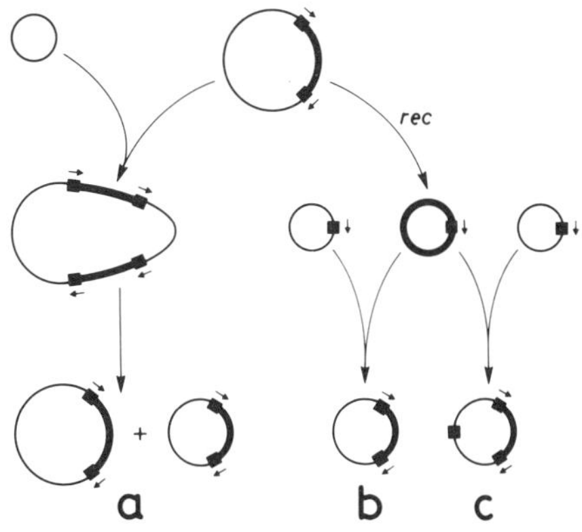

Figure 6. (*a*) Cointegrate model (Shapiro 1979). A cointegrate is formed that includes two copies of the entire r-det. This is then resolved to give R100-1 and a copy of the transposon inserted into the recipient plasmid. An alternative would be to imagine a cointegrate formed with a single IS*1* element instead of the entire r-det. Resolution would then result in either a single copy of IS*1* in the recipient replicon, with the reformation of R100-1, or a transfer of the r-det to the recipient molecule, leaving the RTF unit with a single copy of IS*1*. (*b, c*) An alternative pathway involving the production of supernumerary copies of the r-det and their excision by means of host recombination functions. The circular forms carrying one copy of IS*1* can translocate to a recipient replicon either by homologous recombination with a copy of IS*1* resident on the replicon (*b*) or by cointegrate formation using its own IS*1* (*c*).

In addition to its interest as a second, previously unknown replication system for R1*drd*19, we can use pTn*2350* to determine whether such circular r-det structures can be trapped efficiently on a suitable recipient replicon. Initial results indicate that it can be trapped efficiently as a stable cointegrate with the RTF unit of R100-1. We have crossed *E. coli* strain C600 carrying both pTn*2350* and RTF with a Nal^r recipient LC799 (= C600 $Nal^rP1^r\lambda^r$) with selection for Tc^rNal^r for transfer of RTF and for $Km^rTc^rNal^r$ for transfer of RTF::Tn*2350.* The frequency of cotransfer of Km^rTc^r compared with that of Tc^r alone was determined to be approximately 10^{-5}. The linkage between RTF and Tn*2350* is stable: Clones having received Tc^rKm^r subsequently transfer both resistances at high frequency to a secondary recipient. This demonstrates a rather efficient translocation of pTn*2350* onto RTF, but the experiments do not distinguish between transposition of pTn*2350* and homologous recombination between its resident IS*1* and the IS*1* element located on RTF.

Frequency of IS*1*-mediated Transposition as a Function of Transposon Size

The observation that the frequency of transposition of the r-det of R100-1, a 22-kb segment of DNA, is low (Arber et al. 1979), whereas that of Tn*2350* (10.4 kb) is several orders of magnitude higher, and that of Tn*9,* an IS*1*-flanked transposon of 2.6 kb, is higher still, prompted us to investigate in a systematic way whether transposon size has an influence on transposition frequency. In these experiments, we have used a series of

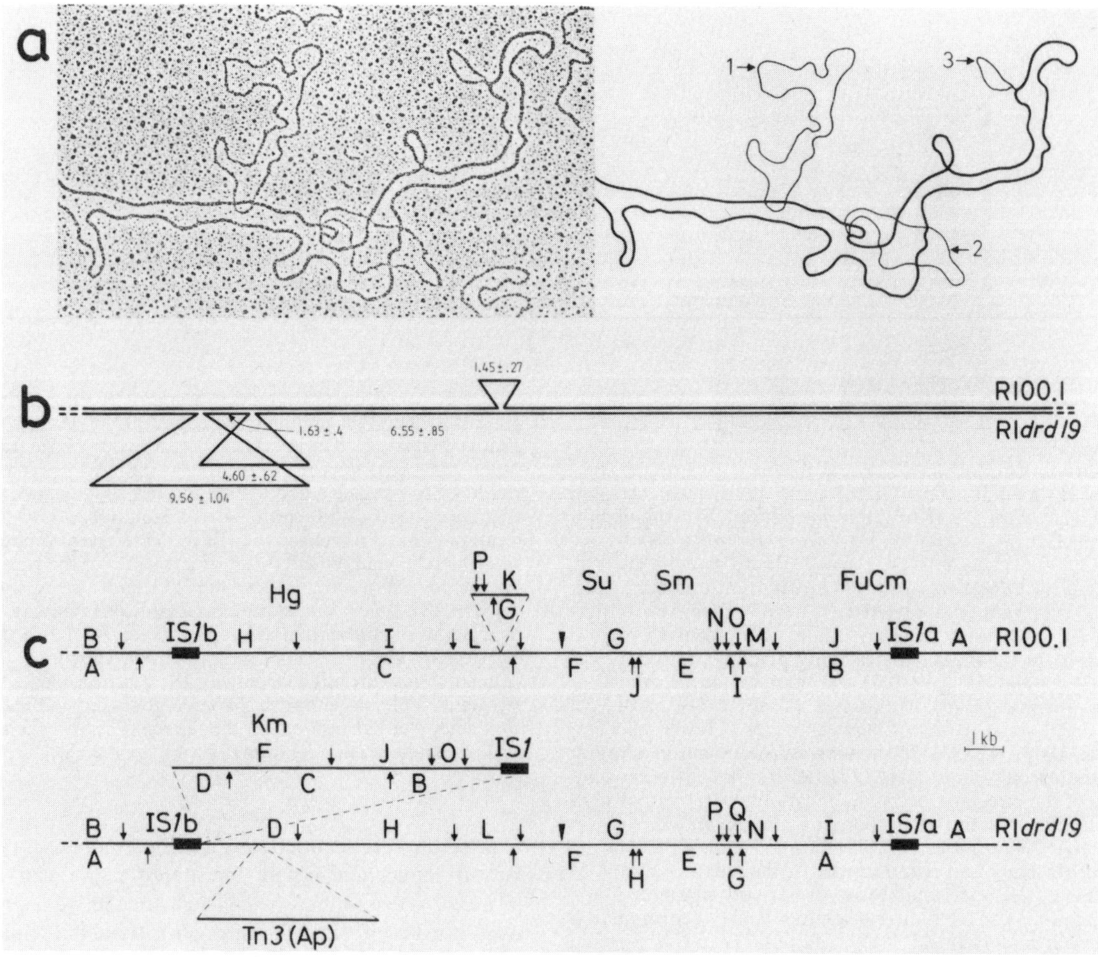

Figure 7. (*a*) A typical heteroduplex formed between the r-det regions of the plasmids R100-1 and R1*drd*19. It shows three single-stranded insertion loops: (*1*) the Km^r transposon Tn*2350,* (*2*) Tn*3* associated with R1*drd*19, and (*3*) one of unknown function associated with R100-1. (*b*) Measurements of these features made on nine molecules. (*c*) A comparison of the restriction maps of these regions: (↓) *Eco*RI; (↑) *Hin*dIII; (▼) *Sal*I. For R100-1, the results of several authors have been combined (Tanaka et al. 1976; Ohtsubo and Ohtsubo 1977; Arber et al. 1979; Chandler et al. 1979a). The positions of the IS*1* elements were obtained from Chandler et al. (1977a); those of the resistance genes were obtained from Lane and Chandler (1977). The map of the r-det region of R1*drd*19 will be published elsewhere (M. Clerget et al., in prep.). It has been aligned with the map of R100-1 and with the heteroduplex map shown in *b* and shows good agreement. The third copy of IS*1* was located at one end of the Km insertion loop (Tn*2350*) by analysis of λ::Tn*2350* derivative phage (M. Clerget et al., in prep.).

IS*1*-flanked transposons derived from the r-det of R100-1 by deletion following transposition to the prophage P1 (Arber et al. 1979). These transposons, which all specify Cm^r, were moved from P1 to λ*bbnin* by induction of a double lysogen carrying λ*bbnin* and the respective P1 derivative and subsequent infection of a P1^{r}λ^s stain (LC690). Phage λ stocks prepared from a Cm^r clone of LC690 were used to infect LC543, a *recA*$^-$ derivative of W1485 carrying the RTF plasmid. We also made LC543 lysogens of λ::Tn*9* (λ*cam*1 was obtained from L. Rosner, Laboratory of Molecular Biology, National Institute of Arthritis and Metabolic Diseases, National Institutes of Health) and λ::Tn*2350.* Transposition frequencies of this series of IS*1*-flanked transposons (ranging in size from 2.6 kb to 10.4 kb) were measured by the conjugation assay described above, using LC799 (P1^{r}λ^{r}Nal^r) as recipient. For Tn*9* and the r-det-derived transposons, selection was made for Cm^r, and for Tn*2350,* selection was for Km^r. The results of these experiments, presented in Figure 9, show that there is a dramatic reduction in the frequency of transposition of these elements as their sizes increase. Moreover, the decline in transposition frequency seems to be exponential, decreasing by half for each additional 4 kb of DNA. Since we are limited in this type of experiment by the size of the DNA that can be inserted into the λ*bbnin* phage, we have not attempted to analyze transposons larger than Tn*2350.* Extrapolation of this curve to transposons having the size of r-det, however, gives a frequency of transposition of 10^{-9} to 10^{-10}, consistent with the observations of others (Arber et al. 1979).

DISCUSSION

The results we have presented show that the plasmid R100-1 carries a function involved in the production of CCC copies of its r-det. The results of a physical analysis of an r-det-producing derivative of R100-1 (pLC117),

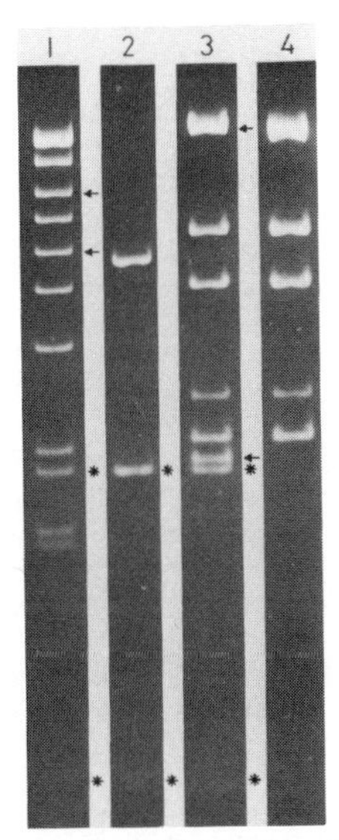

Figure 8. *Eco*RI-digested R1*drd*19 (*1*), pTn*2350* (*2*), λ*bbnin*::Tn*2350* (*3*), and λ*bbnin* (*4*). Two fragments (*) are present in *1, 2,* and *3.* They represent fragments internal to Tn*2350.* In the pTn*2350* DNA (*2*), a single, large band can also be observed. This is the result of recombination between its two flanking IS*1* elements (Fig. 7c) in circularization. As expected, it carries homology to IS*1* (data not shown). The junction fragments both in the parent plasmid (*1*) and in the λ derivative (*3*) are indicated by arrows.

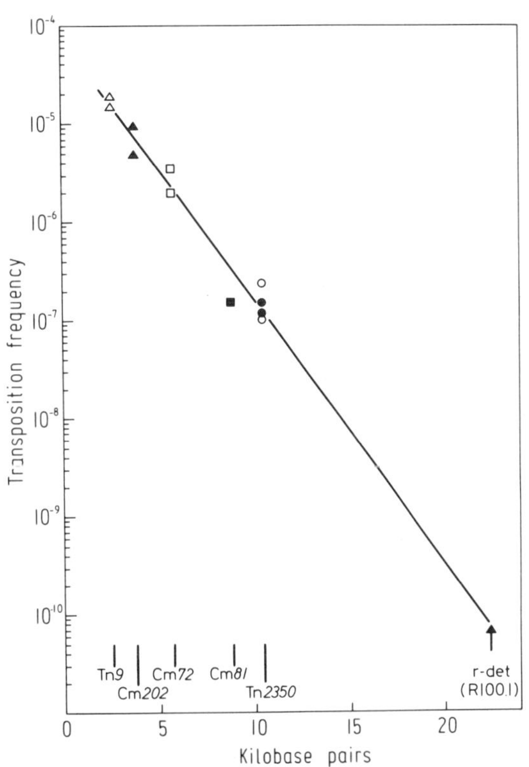

Figure 9. Transposition frequency of IS*1*-flanked transposons. In addition to Tn*9* (▲) (2.6 kb) and Tn*2350* (○, ●) (10.4 kb) (M. Clerget et al., in prep.), three transposons of intermediate length were employed: Cm*72* (□), Cm*81* (■), and Cm*202* (▲). They have contour lengths of 5.7 kb, 8.8 kb, and 3.7 kb, respectively (Arber et al. 1979). They were described by Arber et al. (1979) as deletion products of the r-det of R100-1 and were isolated by them on the phage P1. Cm*72* and Cm*202* confer the Cm^rFu^r phenotype, and, in addition, Cm*81.1* confers Sm^r. The transposons were introduced into a λ*bbnin* phage as described in the text, and these derivative phages were introduced into a *recA*⁻ recipient (LC543) carrying the RTF unit of R100-1. Transposition was measured as the ratio of cells having received RTF (Tc^r) and the transposon (Cm^r for Tn*9*, Cm*72*, Cm*81*, and Cm*202*: Km^r for Tn*2350*) to cells having received the RTF in a mating with LC790 (= C600 $P1^r\lambda^rNal^r$). All matings were performed in an identical manner at 33 °C overnight, and exconjugants were selected at 42 °C. In the case of Tn*2350*, both *recA*⁺ (○) and *recA*⁻ (●) donors were used.

using restriction enzymes and electron microscopy of heteroduplex molecules formed between the plasmid and an RTF unit of R100-1, indicate that the property of r-det production results from a mutation rather than from a change in structure. The location of the mutation on the plasmid has not yet been determined. The function leading to r-det appearance seems to be subject to repression in the parent plasmid, since the presence of a closely related plasmid in the original Hfr cell or in the same cell as the mutant pLC117 results in the disappearance of r-det circles. Since the mutation arises following integration of R100-1 into the chromosome and the selection procedure requires that the integrated plasmid be able to function in replicating the host chromosome, it is possible that a change in replication properties of the plasmid is involved. In this regard, it should be noted that replication of the parent R100-1 plasmid is largely unidirectional when in the autonomous state (Silver et al. 1977), whereas replication from the plasmid origin in the Hfr cell is largely bidirectional (Chandler et al. 1977b). Experiments to determine the origin and mode of replication of pLC117 are in progress.

Recently, it has been observed that the plasmid F, which has a large region of homology with R100-1 (Sharp et al. 1973), carries a function called *fer.* Mutants have been isolated (*ferB*) that stimulate both precise excision of Tn*5* and Tn*10* in *cis* and in *trans* and *recA*-dependent recombination between two directly repeated copies of the insertion element IS*3* in the F′ F128 (Hopkins et al. 1980). It is believed that *ferB* is a repressor that acts on the structural gene *ferA*. The effect of *ferB* on *recA*-promoted recombination between the IS*3* elements may be similar to the *recA*-promoted recombination between the IS*1* elements, which results in r-det production from R100-1. In this light, it is interesting to note that the *fer* functions have been located at the end of the transfer operon of F (Hopkins et al. 1980), a region that shares sequence homology with R100-1 (Sharp et al. 1973).

The plasmid R1*drd*19 also possesses a mechanism that results in circular r-det production. Closer examination of this plasmid has revealed that it carries three copies of the IS*1* element within the r-det region (M. Clerget et al., in prep.). The third IS*1* copy is located at one extremity of a Km^r transposon, which we have designated Tn*2350*. This transposon can circularize in the same way as the r-det of R100-1 but, in contrast, can be maintained in the absence of the parent R plasmid. R1*drd*19 therefore carries a second, previously uncharacterized, replication system.

It has been proposed that circularization of r-det structures may provide an additional pathway to transposition for the translocation of antibiotic-resistance genes from one replicon to another (Iida and Arber 1980). The existence of plasmid-specified mechanisms for r-det circularization tends to support this view. Our observation that the small circular form of Tn*2350* (pTn*2350*) can be trapped efficiently by the RTF unit of R100-1 provides more direct evidence for this type of pathway. Our experiments do not distinguish between translocation by transposition of pTn*2350* using its single IS*1* copy and translocation by recombination between its resident IS*1* and that of RTF, but the relatively high frequency we observe (10^{-5}) suggests that homologous recombination is involved. Ohtsubo et al. (1980) have recently presented results demonstrating that a single copy of IS*1* on a plasmid can indeed be employed in transposition (cointegrate formation) onto a second plasmid.

What factors govern the frequency of transposition of IS*1*-flanked transposable elements such as the r-det and Tn*2350?* We have already indicated that the presence of directly repeated flanking insertion elements provides additional pathways, via circle formation, for their translocation. In the absence of circle formation, e.g., in *recA*$^-$ cells, translocation of these elements must occur by direct transposition. Presumably, all IS*1*-flanked elements employ the same mechanism for transposition. It is surprising, therefore, that the frequency of transposition of a series of these elements whose overall length varies from 2.6 kb to 10.4 kb decreases by two orders of magnitude. Two possible explanations for this behavior are apparent. It is possible that transposition requires that both ends of the element be brought into apposition and that the probability of this decreases as the distance between the IS*1* elements increases. The probability of ligating the ends of a linear DNA molecule depends on its length in a qualitatively similar manner (Dugaiczyk et al. 1975). An alternative explanation is that the initial steps involved in transposition are processive; i.e., transposition requires that a complex initiate at one end of the transposon and progress to the opposite extremity. Such a process could involve replication of the element. It could generate the observed exponential decrease in transposition were the probability that the complex aborted constant per unit length.

ACKNOWLEDGMENTS

We thank G. Churchward, H. Krisch, and P. Prentki for critically reading the manuscript, A. Bruschi for technical assistance, and E. Boy de la Tour and E. Gallay for electron microscopy. This work was supported by grant 3.245.77 from the Swiss National Science Foundation.

REFERENCES

Arber, W., S. Iida, H. Jütte, P. Caspers, J. Meyer, and C. Hänni. 1979. Rearrangements of genetic material in *Escherichia coli* as observed on the bacteriophage P1 plasmid. *Cold Spring Harbor Symp. Quant. Biol.* **43:** 1197.

Barrell, B. G., G. M. Air, and C. A. Hutchison. 1976. Overlapping genes in bacteriophage φX174. *Nature* **264:** 34.

Berg, D. E., J. Davies, B. Allet, and J.-D. Rochaix. 1975. Transposition of R factor genes to bacteriophage λ. *Proc. Natl. Acad. Sci.* **72:** 3628.

Bird, R. E., M. Chandler, and L. Caro. 1979. Suppression of an *E. coli dnaA* mutation by the integrated R factor R100.1: Change of chromosome replication origin in synchronized cultures. *J. Bacteriol.* **126:** 1215.

Blohm, D. and W. Goebel. 1978. Restriction map of the antibiotic resistance plasmid R1*drd*-19 and its derivatives pKN102 (R1*drd*-19B2) and R1*drd*-16 for the enzymes *Bam*HI, *Hin*dIII, *Eco*RI and *Sal*I. *Mol. Gen. Genet.* **167:** 119.

Chandler, M., L. Silver, and L. Caro. 1977a. Suppression of an *Escherichia coli dnaA* mutation by the integrated R factor R100.1: Origin of chromosome replication during exponential growth. *J. Bacteriol.* **131:** 421.

Chandler, M., L. Silver, J. Frey, and L. Caro. 1977b. Suppression of an *Escherichia coli dnaA* mutation by the integrated R factor R100.1: Generation of small plasmids after integration. *J. Bacteriol.* **130:** 303.

Chandler, M., L. Silver, D. Lane, and L. Caro. 1979a. Properties of an autonomous r-determinant from R100.1. *Cold Spring Harbor Symp. Quant. Biol.* **43:** 1223.

Chandler, M., B. Allet, E. Gallay, E. Boy de la Tour, and L. Caro. 1977c. Involvement of IS*1* in the dissociation of the r-determinant and RTF components of the plasmid R100.1 *Mol. Gen. Genet.* **153:** 289.

Chandler, M., E. Roulet, L. Silver, E. Boy de la Tour, and L. Caro. 1979b. Tn*10* mediated integration of the plasmid R100.1 in the bacterial chromosome: Inverse transposition. *Mol. Gen. Genet.* **173:** 23.

Clewell, D. B. and D. R. Helinski. 1969. Supercoiled circular DNA-protein complex in *E. coli:* Purification and induced conversion to an open circular form. *Proc. Natl. Acad. Sci.* **62:** 1159.

Davis, R. W., M. Simon, and N. Davidson. 1971. Electron microscope heteroduplex methods for mapping regions of base sequence homology in nucleic acids. *Methods Enzymol.* **21:** 413.

Dugaiczyk, A., H. W. Boyer, and H. M. Goodman. 1975. Ligation of *Eco*RI endonuclease-generated DNA fragments into linear and circular structures. *J. Mol. Biol.* **96:** 171.

Hashimoto, H. and R. Rownd. 1975. Transition of the R factor NR1 in *Proteus mirabilis:* Level of drug resistance of nontransitioned and transitioned cells. *J. Bacteriol.* **123:** 56.

Hopkins, J. D., M. B. Clements, T. Y. Lang, R. R. Isberg, and M. Syvanen. 1980. Recombination genes on the *Escherichia coli* sex factor specific for transposable elements. *Proc. Natl. Acad. Sci.* **77:** 2814.

Hu, S., E. Ohtsubo, N. Davidson, and H. Saedler. 1975. Electron microscope heteroduplex studies of sequence relationships among bacterial plasmids: Identification and mapping of the insertion sequences IS*1* and IS*2* in F and R plasmids. *J. Bacteriol.* **122:** 764.

Iida, S. and W. Arber. 1980. On the role of IS*1* in the formation of hybrids between the bacteriophage P1 and the R plasmid NR1. *Mol. Gen. Genet.* **177:** 261.

Kopecko, D. J., J. Brevet, and S. N. Cohen. 1976. Involvement of multiple translocating DNA segments and recombinational hotspots in the structural evolution of bacterial plasmids. *J. Mol. Biol.* **108:** 333.

LANE, D. and M. CHANDLER. 1977. Mapping of the drug resistance genes carried by the r-determinant of the R100.1 plasmid. *Mol. Gen. Genet.* **157:** 17.

LANE, D., M. CHANDLER, L. SILVER, A. BRUSCHI, and L. CARO. 1979. The construction and replication properties of hybrid plasmids composed of the r-determinant of R100.1 and the plasmids pCR1 or pSC201. *Mol. Gen. Genet.* **168:** 337.

MEYNELL, E. and N. DATTA. 1967. Mutant drug resistance factors with high transmissibility. *Nature* **214:** 885.

NISHIMURA, A., Y. NISHIMURA, and L. CARO. 1973. Isolation of Hfr strains from R^+ and $ColV2^+$ strains of *Escherichia coli* and derivation of an R'*lac* factor by transduction. *J. Bacteriol.* **116:** 1107.

OHTSUBO, H. and E. OHTSUBO. 1977. Repeated DNA sequences in plasmids, phages and bacterial chromosomes. In *DNA insertion elements, plasmids, and episomes* (ed. A. I. Bukhari et al.), p. 49. Cold Spring Harbor Laboratory, Cold Spring Harbor, New York.

OHTSUBO, E., M. ZENILMAN, and H. OHTSUBO. 1980. Plasmids containing insertion elements are potential transposons. *Proc. Natl. Acad. Sci.* **77:** 750.

PTASHNE, K. and S. N. COHEN. 1975. Occurrence of insertion sequence (IS) regions on plasmid deoxyribonucleic acid as direct and inverted nucleotide duplications. *J. Bacteriol.* **122:** 776.

SHAPIRO, J. A. 1979. Molecular model for the transposition and replication of bacteriophage Mu and other transposable elements. *Proc. Natl. Acad. Sci.* **76:** 1933.

SHARP, P. A., S. N. COHEN, and N. DAVIDSON. 1973. Electron microscope heteroduplex studies of sequence relations among plasmids of *E. coli.* II. Structure of drug resistance (R) factors and F factors. *J. Mol. Biol.* **75:** 235.

SILVER, L., M. CHANDLER, E. BOY DE LA TOUR, and L. CARO. 1977. Origin and direction of replication of the drug resistance plasmid R100.1 and of a resistance transfer factor derivative in synchronized cultures. *J. Bacteriol.* **131:** 929.

SILVER, L., M. CHANDLER, D. LANE, and L. CARO. 1980. Production of extrachromosomal r-determinant circles from integrated R100.1: Involvement of the *E. coli* recombination system. *Mol. Gen. Genet.* (in press).

SOUTHERN, E. M. 1975. Detection of specific sequences among DNA fragments separated by gel electrophoresis. *J. Mol. Biol.* **98:** 503.

TANAKA, N., J. H. CRAMER, and R. H. ROWND. 1976. *Eco*RI restriction endonuclease map of the composite R plasmid NR1. *J. Bacteriol.* **127:** 619.

WATANABE, T. and Y. OGATA. 1970. Genetic stability of various resistance factors in *Escherichia coli* and *Salmonella typhimurium. J. Bacteriol.* **102:** 363.

Intramolecular Transposition of a β-Lactamase Sequence and Related Genetic Rearrangements

R. C. CLOWES, P. L. HOLMANS, AND S. J. CHIANG
Programs in Biology, The University of Texas at Dallas, Richardson, Texas 75080

Transposable genetic elements mediating ampicillin resistance (Ap^r) were the first identified in prokaryotes (Hedges and Jacob 1974) and have been among those investigated most extensively. The most recent studies of these transposons, especially of Tn*3* (Chou et al. 1979; Cohen et al. 1979; Heffron et al. 1979), have had considerable influence on the development of molecular models (see Shapiro 1979) that can plausibly accommodate the salient observations derived in the main from studies of transpositions from one replicon to another. From these models, the events that result following transposition from one site to another on the same replicon can also be predicted. The observations reported below were made on such a system of intramolecular transposition and confirm a number of these predictions while providing some new evidence leading to suggested modifications to the model proposed.

The system under study is the plasmid R6K, on which is located the Ap^r transposon Tn*2660*, shown in early reports from heteroduplex experiments to be homologous to Tn*3* (Heffron et al. 1975). R6K is a 38.5-kb conjugative plasmid of repressed fertility found at about 13 copies per chromosomal equivalent; it specifies streptomycin resistance (Sm^r) in addition to Ap^r (Kontomichalou et al. 1970).

Tn*2660* can be localized on R6K by use of the restriction enzymes *Eco*RI, *Bam*HI, and *Sma*I. As shown in Figure 1A, the single *Sma*I site is situated 1.5 kb from the unique *Bam*HI site, inferred from Tn*3* studies to be located asymmetrically on Tn*2660* about 1379 bp from the nearest terminus (Heffron et al. 1979), thus orienting Tn*2660* as shown in Figure 1A (Holmans et al. 1978).

Characteristics of Deletions

Following the observation that a mutant plasmid that had lost the ability to determine Sm^r had a greatly increased conjugal transfer proficiency, a number of similar mutants were isolated from a recipient strain following conjugal transfer. When restriction enzyme cleavage products of the mutant DNAs were examined following agarose gel electrophoresis, they were all found to have lost the unique *Sma*I site, one of the *Eco*RI sites, and between about 9 kb and 12 kb of DNA. When heteroduplexes of each mutant DNA with R6K DNA were examined by electron microscopy following *Bam*HI cleavage, the distances of the sites of the single-stranded deletion loops from the nearest end were found to be uniformly 1.36 kb. Since this distance corresponded closely to that of the *Bam*HI site from the nearest terminus of Tn*3*, it was concluded that one end of the deletion loops was probably sited at the corresponding Tn*2660* terminus and that the deletions extended past the *Sma*I site to a series of clustered distal termini as shown in Figure 1A (Holmans and Clowes 1979).

The deletion mutants were originally isolated at a frequency of about 7×10^{-4} following transformation of *recA*$^+$ recipients using R6K DNA. They were later derived from crosses of donors and recipients that both carried the *recA*$^+$ allele, and although similar mutants could be isolated when either donor or recipient was *recA56*, none were found when both parents were *recA56*. The conclusion was drawn that one of the events leading to deletion required *recA*$^+$ host activity. However, our more recent preliminary studies indicate that when a donor strain with a *recA16* allele is crossed with the same *recA56* recipient, Sm^s deletion mutants of R6K can be isolated at a low frequency.

A Duplicate of Tn*2660* Is Always Inversely Oriented with Respect to the Resident Tn*2660*

Insertions of a duplicate copy of Tn*2660* in R6K were achieved by selection of host colonies that were hyperresistant to ampicillin (Ap^{rr}). These colonies were selected from *recA*$^-$ hosts, either following growth in liquid culture or by direct plating in ampicillin concentrations greater than 2 mg/ml; the strains carrying R6K were inhibited to less than 0.1% survival of very small colonies at 2 mg/ml. Plasmid DNA was isolated from these Ap^{rr} strains. Although in a number of cases the DNA appeared to be of the same size as R6K and was produced in an increased number of copies to the chromosomal equivalent (multicopy mutants), in many cases there was no increase in the copy number, but the plasmid DNA had increased by about 5 kb in size. The increase in plasmid size was concluded to be due to the insertion of a second copy of Tn*2660* some distance from the first when it was found that *Bam*HI cleavage of mutant DNAs gave rise to two fragments of sizes adding to 43.5 kb. This supposition was confirmed by electron microscopic examination of the mutant plasmid DNAs, following denaturation and rapid renaturation, in which were seen double-strand "snapback" regions of approximately 5 kb separating two single-stranded loops. This evidence also led to the conclusion that in all (22) instances, the duplicate Tn*2660* is inversely oriented with respect to the resident transposon. From measurements of the sizes of restriction enzyme fragments and of snapback structures, the loca-

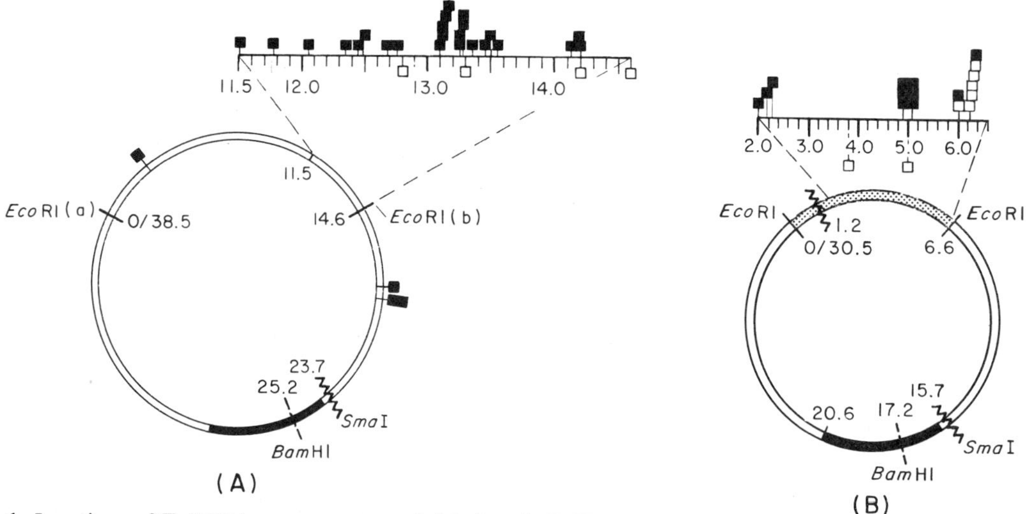

Figure 1. Locations of Tn*2660* inverse repeats and deletions in R6K (*A*) and pSJC102 (*B*). Restriction enzyme sites are shown as intercepts of the double circle representing the plasmid molecule. Cleavage sites: (——) *Eco*RI; (----) *Bam*HI; (〰) *Sma*I. The numbers inside the circle represent site coordinates (in kb) (the *Eco*RI site most distal to the *Bam*HI site being designated coordinate 0) measured in the direction toward the second *Eco*RI site. The black area represents the location of transposon Tn*2660*. The horizontal scale represents an expanded segment of the molecule; each box above the line or outside the circle represents the position of an inverse repeat of Tn*2660*. (■) Inversions of the DNA between the inverse repeats; (□) parental orientation. Each box below this line represents one terminus of a deletion with the other endpoint at the transposon terminus at coordinate 23.8 (R6K) or at coordinate 20.6 (pSJC102). pSJC102 is a hybrid comprised of the larger 23.9-kb *Eco*RI segment of R6K inserted into the ColE1 replicon (shown stippled).

tions of these inverse repeats were concluded to be clustered, as shown in Figure 1A, in the same localized region of R6K as the distal termini of the previously mapped deletions (Chiang and Clowes 1980).

Duplicate Transposons Arise Predominantly by Intramolecular Transposition

A discrimination between inter- and intramolecular transposition as the mechanism for the origin of the inverse-repeat transposon sequences was made by studies of a host strain harboring two plasmids, one incorporating a wild-type Tn*2660* and the other incorporating a mutant Tn*2660* in which Ap^r was expressed only at a low temperature. The mutated Tn*2660* (Tn*2662*) was selected on an R6K plasmid following in vitro mutagenesis of R6K DNA with hydroxylamine. The wild-type Tn*2660* was carried on a hybrid plasmid (pSJC102) on which the smaller (14.6-kb) *Eco*RI segment of R6K, carrying the replication loci (Crosa et al. 1976, 1978; Kolter and Helinski 1978a,b; Kolter et al. 1978), had been replaced by the ColE1 replicon (Fig. 1B). The two plasmids are thus completely compatible, and the biplasmid host strain shows additive Ap^r of the strains carrying either plasmid independently. Similar selection for Ap^{rr} clones from this biplasmid host led to the finding that duplications of the Tn*2660* sequence occurred in either plasmid at about the same frequency (of 26 Ap^{rr} clones, 14 resulted from Tn*2660* duplications in the mutant R6K plasmid [pSJC301] and 12 from duplications in the ColE1 hybrid plasmid [pSJC102]). The two plasmid DNAs were separated for further study by transformation into independent recipient cells, and the Ap^r phenotypes of these hosts were measured at 30 °C and 42 °C. In all 18 cases studied where the Tn*2660* duplication had occurred in the hybrid pSJC102 plasmid harboring the wild-type Tn*2660* sequence, the resistance at 30 °C or at 42 °C was twice that of the host carrying pSJC102. From this it was concluded that the duplicated transposons also carried wild-type sequences. In all six cases where a second transposon sequence was found in the mutant R6K plasmid, the host cells showed twice the level of pSJC301 resistance at 30 °C, but at 42 °C no measurable resistance was found. From this it was concluded that the second transposon copy also determined temperature-sensitive Ap^r. Thus, unless the two plasmids are located in different regions of the host cell, these results are consistent with the conclusion that the probability of transposition from one plasmid to another is considerably less than transposition from one plasmid to a second location in the same molecule (Chiang and Clowes 1980).

Intramolecular Transposition Is Frequently Accompanied by Inversions of the DNA between the Transposons

As found with Tn*2660* insertions in the wild-type R6K, the formation of snapback structures by mutant plasmid DNAs indicated that in all 6 pSJC301::Tn*2662* molecules and in all 18 pSJC102::Tn*2660* molecules examined, the second transposon was oriented inversely with respect to the resident transposon. Restriction enzyme analysis of the pSJC301::Tn*2662* plasmid DNAs was consistent in all cases with an inversion of the DNA sequences between the two transposons. Thus, in one plasmid with a duplicated Tn*2662* sequence at coordinate 13.11, the *Eco*RI fragments measured about 16.2

kb and 27 kb, instead of the expected 19.5 kb and 23.9 kb. This was confirmed by heteroduplex measurements between the mutant plasmid and R6K, which, instead of showing a 38.5-kb double-strand region with a 5-kb single-stranded loop, produced molecules with double-strand regions of approximately either 28 kb or 14.5 kb, each with two single-stranded loops greatly in excess of 5 kb. Similar restriction enzyme analysis of the pSJC102::Tn*2660* mutants showed that of 18 molecules examined, 12 had inversions of the DNA between the two Tn*2660* sequences, whereas 6 did not. (For example, cleavage of one inverted molecule with an inverse Tn*2660* repeat at coordinate 2.15 gave *Eco*RI fragments of about 19 kb and 16.2 kb, instead of the expected 11.4-kb and 23.9-kb fragments, and *Sma*I cleavage gave two fragments of 5.85 kb and about 29.4 kb, instead of the expected 14.75-kb and 19.35-kb fragments [see Fig. 1B].)

The digestion products of the 22 R6K::Tn*2660* mutants isolated earlier were also consistent with an inversion of the DNA between the two Tn*2660* sequences. Thus, although duplicate copies of the Tn*2660* sequence are presumed to be transposed from the resident Tn*2660* sequence with about equal frequency in either replicon and to produce exclusively inverse repeats, all 28 cases of transposition into R6K are accompanied by inversion of the intervening DNA, whereas inversion accompanies only 12 out of a total of 18 transpositions into the hybrid plasmid pSJC102.

DISCUSSION

These experiments confirm previous conclusions that transposition is not accompanied by excision of the original transposon (Berg 1977) and that the sites of insertion of Ap^r transposons such as Tn*3* are clustered in specific regions (Tu and Cohen 1980). Other results of this study—that the proximal endpoints of deletions extend from one or the other terminus of Tn*2660* and the distal endpoints are clustered in the same regions as those into which the majority of transpositions of a second copy of Tn*2660* occur, and that all the duplicated Tn*2660* sequences are oriented as inverse repeats with respect to the resident Tn*2660*—conform to those predictable from the Shapiro (1979) model as applied to intramolecular transposition; i.e., they arise from the same primary molecular events, in which ligation of the 3′-OH end of a transposon sequence with the 5′-PO_4 end of a target sequence may take place either between the same strands of the molecule, leading to deletions of the DNA of those segments that do not have replication loci, or between opposite strands, resulting in duplication of Tn*2660* as an inverse repeat with respect to the parental structure (see Fig. 2). In intermolecular transposition, this model proposes that the early events result in replicon fusion to produce an intermediate structure that is a cointegrate of the two plasmids carrying direct repeats of the transposon at the two junctions; this intermediate is then resolved into the two plasmid products by a site-specific *recA*-independent recombination event. The structure corresponding to this cointegrate, which would arise from intramolecular transposition, is a molecule with inverse repeats of the transposon between which the DNA is inverted (see Fig. 2D), the site-specific recombination event then leading to a restoration of the parental DNA orientations. This event seems not to be a necessary consequence following intramolecular transposition of Tn*2660* and, more importantly, appears to be replicon-dependent. Moreover, the inversions are stable, and the presence of molecules with the alternative orientations could not be detected in DNAs from plasmids with or without inversions of the parental sequence, even after extensive subculture in *recA*⁻ hosts. When transferred to *recA*⁺ hosts, the inversions were

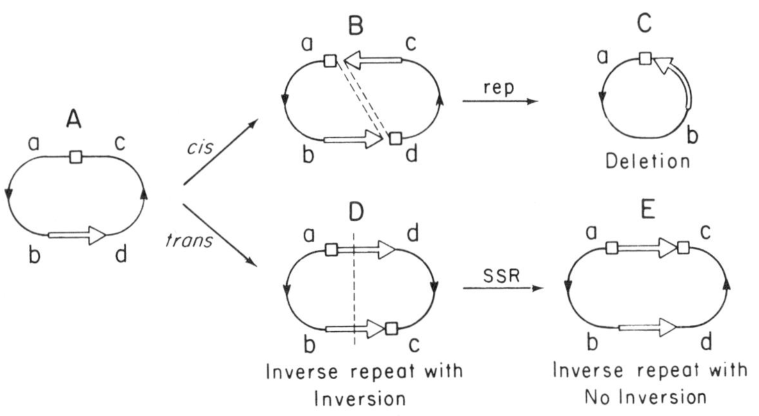

Figure 2. Diagrammatic representation of events in intramolecular transposition following the proposal of Shapiro (1979). (*A*) Molecule with sequences *a, b, c,* and *d,* carrying a transposon (arrow) and a target site (□). Following single-strand nicking at 3′-OH ends of transposon sequence and 5′-PO_4 ends of target sequence, ligation occurs between these ends, either on the same strand (*cis*) or on opposite strands (*trans*). *cis* ligation leads to separation of the molecule into two segments (*B*), each with a transposon sequence. If replication loci are situated only on segment *ab*, a deletion of the region *cd* will result (*C*). *trans* ligation (*D*) leads to insertion of a duplicate transposon as an inverse repeat, together with an inversion of the sequence *cd* with respect to *ab*. Site-specific, *recA*-independent recombination (SSR) between the two transposons (----) would produce structure *E* with parental DNA orientations restored.

also stable, and only after approximately 100 generations of growth could the reverse orientation be detected in about 1% of the molecules.

Since the positions of the duplicated Tn*2660* are approximately the same distance from the resident Tn*2660* in both pSJC102 and R6K, which are of similar size, it seems unlikely that steric effects can account for differences in the incidence of inversions in the two replicons. However, they differ considerably in their mechanisms of replication. ColE1 (and pSJC102; see Chiang and Clowes 1980) replicates unidirectionally from a unique origin (Inselburg 1974; Lovett et al. 1974; Tomizawa et al. 1974), is dependent on host DNA polymerase I enzyme (Kingsbury and Helinski 1973), and is not inhibited by chloramphenicol (Clewell 1972). R6K replication is bidirectional and asymmetric from two origins (Lovett et al. 1975; Crosa et al. 1976), is independent of host *polA* function, and is inhibited by chloramphenicol (Arai and Clowes 1975). Therefore, it seems possible that the differences observed in the incidence of DNA inversions associated with transposition are due to replication differences, from which it may be inferred that the limited replication proposed by Shapiro (1979) to be initiated from either or both of the unpaired ends of transposon and target sequences (or from an internal site) may not occur but that normal vegetative replication of the plasmid may be necessary to achieve the duplication of the transposon and target sequences.

The possibility that Tn*2660* is somehow defective in the site-specific recombination step, either by site or functional defect (see Arthur and Sherratt 1979), seems unlikely from the absence of inversion in a number of hybrid pSJC102 molecules. This conclusion has been confirmed by determining that intermolecular transposition of R6K is "normal." Thus, when DNA was isolated from a host carrying two plasmids (ColE1::Tn*2660* and a hybrid plasmid, pSJC101, in which the 14.6-kb *Eco*RI replication fragment of R6K is coupled to the *Eco*RI kanamycin-resistance [Km^r] fragment from pML2 [Hershfield et al. 1974]) and transformed into a $polA^-$ recipient strain, 1% of the Km^r transformants were also Ap^r (and inferred to be due to transposition of Tn*2660* to pSJC101), whereas no Ap^r ColE1-immune colonies (of phenotypes characteristic of a cointegrate plasmid) were found in 10^5 Km^r colonies tested. (Tests of plasmid DNAs from several Km^rAp^r colonies confirmed this conclusion, and transformation controls of both ColE1::Tn*2660* and pSJC101 into a $polA^+$ recipient occurred with normal frequencies.)

The frequency of intramolecular transposition of Tn*2660* into R6K is difficult to evaluate from the selection method chosen, but from the results presented, it is clear that intermolecular transposition of a duplicate copy occurs at least one order of magnitude less frequently and is considerably lower in frequency than that expected from the preceding experiments. This is consistent with previous reports of transpositional immunity, in which intermolecular transposition was measured and intramolecular events would not have been identified (Robinson et al. 1977). Tn*2660* appears to produce immunity to intermolecular transposition to the same extent as Tn*3*, as judged by preliminary experiments. Intermolecular transposition was studied by testing plasmid DNA from a biplasmid host carrying pCR1::Tn*2662* (a Km^r ColE1 hybrid plasmid, pCR1 [Covey et al. 1976], into which had been transposed the temperature-sensitive Ap^r transposon Tn*2662*) and R6K, using DNA from the biplasmid host carrying pCR1 and R6K as control. At 42 °C, the frequency of Ap^r clones among the Km^r transformants in the control (which were also Sm^s and thus characteristic only of transposition of Tn*2660* from R6K into pCR1) was 8×10^{-3}, whereas in the test system using pCR1::Tn*2662*, Ap^rSm^s transformants were found at 42 °C at a frequency of 10^{-6} among the Km^r transformants. Thus, the presence of Tn*2662* in pCR1 reduced the (intermolecular) transposition of Tn*2660* to 10^{-4} the normal level. It therefore appears that transposition immunity has less effect on transposition from another site within the same plasmid than on transposition from a different plasmid.

ACKNOWLEDGMENTS

We thank Elzora Jordan and Teresa Hayes for excellent technical assistance and Barbara Bruton-Davis for some of the electron microscopy studies, and we are grateful to Walter Dempsey, Donald Helinski, and Barbara Bachmann for providing bacterial strains. The studies were supported by a U.S. Public Health Service research grant from the National Institute of Allergy and Infectious Diseases (AI-10468).

REFERENCES

ARAI, T. and R. C. CLOWES. 1975. Replication of stringent and relaxed plasmids. In *Microbiology—1974* (ed. D. Schlessinger), p. 141. American Society for Microbiology, Washington, D.C.

ARTHUR, A. and D. SHERRATT. 1979. Dissection of the transposition process: A transposon-encoded site-specific recombination system. *Mol. Gen. Genet.* **175:** 267.

BERG, D. 1977. Insertion and excision of the transposable kanamycin resistance determinant Tn*5*. In *DNA insertion elements, plasmids, and episomes* (ed. A. I. Bukhari et al.), p. 205. Cold Spring Harbor Laboratory, Cold Spring Harbor, New York.

CHIANG, S. J. and R. C. CLOWES. 1980. Intramolecular transposition and inversion in plasmid R6K. *J. Bacteriol.* **142:** 668.

CHOU, J., M. J. CASADABAN, P. G. LEMAUX, and S. N. COHEN. 1979. Identification and characterization of a self-regulated repressor of translocation of the Tn*3* element. *Proc. Natl. Acad. Sci.* **76:** 4020.

CLEWELL, D. 1972. Nature of ColE1 plasmid replication in *Escherichia coli* in the presence of chloramphenicol. *J. Bacteriol.* **110:** 667.

COHEN, S. N., M. J. CASADABAN, J. CHOU, and C.-P. TU. 1979. Studies of the specificity and control of transposition of the Tn*3* element. *Cold Spring Harbor Symp. Quant. Biol.* **43:** 1247.

COVEY, C., D. RICHARDSON, and J. CARBON. 1976. A method for the deletion of restriction sites in bacterial plasmid deoxyribonucleic acid. *Mol. Gen. Genet.* **145:** 155.

CROSA, H., K. LUTTROPP, and S. FALKOW. 1976. Mode of replication of the conjugative R-plasmid RSF1040 in *Escherichia coli. J. Bacteriol.* **126:** 454.

———. 1978. Molecular cloning of replication and incompatibility regions from the R-plasmid R6K. *J. Mol. Biol.* **124:** 443.

Hedges, R. W. and A. Jacob. 1974. Transposition of ampicillin resistance from RP4 to other replicons. *Mol. Gen. Genet.* **132:** 31.

Heffron, F., B. J. McCarthy, H. Ohtsubo, and E. Ohtsubo. 1979. DNA sequence analysis of the transposon Tn*3*: Three genes and three sites involved in transposition of Tn*3*. *Cell* **18:** 1153.

Heffron F., R. Sublett, R. Hedges, A. Jacob, and S. Falkow. 1975. Origin of the TEM β-lactamase gene found on plasmids. *J. Bacteriol.* **122:** 250.

Hershfield, V., H. W. Boyer, C. Yanofsky, M. A. Lovett, and D. R. Helinski. 1974. Plasmid ColE1 as a molecular vehicle for cloning and amplification of DNA. *Proc. Natl. Acad. Sci.* **71:** 3455.

Holmans, P. L. and R. C. Clowes. 1979. Transposition of a duplicate antibiotic resistance gene and generation of deletions in plasmid R6K. *J. Bacteriol.* **137:** 977.

Holmans, P., G. C. Anderson, R. C. Clowes. 1978. TnA-directed deletions and translocations within the R6K plasmid. In *Microbiology—1978* (ed. D. Schlessinger), p. 38. American Society for Microbiology, Washington, D.C.

Inselburg, J. 1974. Replication of colicin El plasmid DNA in minicells from a unique replication initiation site. *Proc. Natl. Acad. Sci.* **71:** 2256.

Kingsbury, D. T. and D. R. Helinski. 1973. Temperature-sensitive mutants for the replication of plasmids in *Escherichia coli:* Requirement for deoxyribonucleic acid polymerase I in the replication of the plasmid ColE1. *J. Bacteriol.* **114:** 1116.

Kolter, R. and D. R. Helinski. 1978a. Construction of plasmid R6K derivatives *in vitro:* Characterization of the R6K replication region. *Plasmid* **1:** 571.

———. 1978b. Activity of the replication terminus of plasmid R6K in hybrid replicons in *Escherichia coli. J. Mol. Biol.* **124:** 425.

Kolter, R., M. Inuzuka and D. R. Helinski. 1978. Trans-complementation-dependent replication of a low molecular weight origin fragment from plasmid R6K. *Cell* **15:** 1199.

Kontomichalou, P., M. Mitani, and R. C. Clowes. 1970. Circular R-factor molecules controlling penicillinase synthesis, replicating in *Escherichia coli* under either relaxed or stringent control. *J. Bacteriol.* **104:** 34.

Lovett, M. A., L. Katz, and D. R. Helinski. 1974. Unidirectional replication of plasmid ColE1 DNA. *Nature* **251:** 337.

Lovett, M. A., R. B. Sparks, and D. R. Helinski. 1975. Bidirectional replication of plasmid R6K DNA in *Escherichia coli;* correspondence between origin of replication and position of single-strand break in relaxed complex. *Proc. Natl. Acad. Sci.* **72:** 2905.

Robinson, M. K., P. M. Bennett, and M. H. Richmond. 1977. Inhibition of TnA translocation by TnA. *J. Bacteriol.* **129:** 407.

Shapiro, J. A. 1979. Molecular model for the transposition and replication of bacteriophage Mu and other transposable elements. *Proc. Natl. Acad. Sci.* **76:** 1933.

Tomizawa, J.-I., Y. Sakakibara, and T. Kakefuda. 1974. Replication of colicin El plasmid DNA in cell extracts: Origin and direction of replication. *Proc. Natl. Acad. Sci.* **71:** 2260.

Tu, C. P-D. and S. N. Cohen. 1980. Translocation specificity of the Tn*3* element: Characterization of sites of multiple insertions. *Cell* **19:** 151.

Detection of Replicational Inceptor Signals in IS5

M. LUSKY,* M. KRÖGER, AND G. HOBOM
Institut für Biologie III, der Universität Freiburg, D-7800 Freiburg i.Br., Federal Republic of Germany

In hybrid plasmid dissection and reconstruction studies, we have shown that two different signal structures are involved in initiation of DNA replication in various lambdoid phages (Lusky and Hobom 1979a,b). The origin of replication (*ori*) serves as the initiation site for leftward primer RNA synthesis, whereas the replicational inceptor (*ice*), located in the λ*c*II gene, controls termination of primer RNA synthesis, as well as initiation of daughter-strand DNA synthesis (see Fig. 1). Hybrid plasmids containing only *ori* are unable to initiate lambdoid replication, whereas plasmids containing *ice* in connection with a leftward promoter such as λp_o do show replicational activity (minimal replication system). Initiation in the *ori-ice* maximal replication system is dependent on transcriptional activation of *ori*, which is stabilized by interaction with phage-specific *O* proteins. Initiation in the p_o-*ice* minimal replication system is *cis*-repressed in *ori*-p_o-*ice* hybrid plasmids. This replicational repression effect is only dependent on the left section of *ori*, on transcriptional *ori* activation, and on the NH_2-terminal domain of *O* protein. This was observed by employing CO_2-terminal variants or amber mutants of *O* that also retain fully repressive control over the cryptic minimal replication system (Hobom and Lusky 1980).

Further studies with lambdoid *ice*⁻ mutants obtained by site-directed mutagenesis in vitro support our interpretation that the *ice* signal structures consist of a single-stranded hairpin loop and an attached oligo(G) tail sequence, resembling transcription terminator signals.

In our search for substitute nonlambdoid inceptor signals, we have detected functionally equivalent signals at or near both ends of the IS*5* insertion element in phage λKH100. These insertion-specific inceptor signals are located in opposite orientation. Only if connected with an external promoter fragment for outside → in transcription will either of the IS*5* inceptor signals be able to sustain high-copy plasmid replication in the absence of any other initiation system.

MATERIALS AND METHODS

Phage λKH100 (Blattner et al. 1974) was provided by W. Szybalski (Univ. of Wisconsin, Madison, Wis.). The plasmid pEX-*lac*110-9T was obtained from H. Weiher, B. Zink, and H. Schaller (Univ. of Heidelberg, Heidelberg, FRG). *Escherichia coli* strain K12 SC294 *polAts* was used for transformation. All experimental procedures used throughout this study have been described previously (Lusky and Hobom 1979a).

RESULTS

Cloning of IS5

In λKH100 DNA, the IS*5* element, which is 1350 bp long, is integrated at position λ:38149, as determined by DNA sequence analysis (M. Kröger, unpubl.), in between the two *Hin*dIII restriction sites located at λ:37770 and λ:38291. The IS*5*-containing *Hin*dIII fragment was isolated from λKH100 DNA and cloned into the *Hin*dIII restriction site of pHL113, which is *Hin*dIII-*Bam*HI-deleted pBR313 DNA containing a λ*c*II*am*41-*am*60 *Hin*dIII-*Bgl*II fragment (λ:38291–38935). The λ fragment in pHL113 carries the λp_R and λp_{rm} promoters. As shown in Figure 2, the resulting plasmid, pHL118 (complete IS*5* in its λKH100 location together with the flanking λ DNA segments carrying the λp_R and λp_{rm} promoters), shows autonomous replication at 42 °C in a *polAts* strain, as does its *Eco*RI deletion derivative pHL132. In contrast, the plasmid pHL133, which has the IS*5*-containing *Hin*dIII fragment inserted into the *Hin*dIII site of pBR313, does not show replicational activity at 42 °C (data not shown).

For more detailed analysis, this *Hin*dIII fragment was recleaved by *Bgl*II at an IS*5* internal *Bgl*II site. Both *Hin*dIII-*Bgl*II subfragments containing either the right or left end of IS*5* have been inserted into a *lac*UV5 promoter-carrying vector plasmid. This vector plasmid is a pBR322 derivative containing the *lac* promoter fragment inserted between its *Eco*RI and *Hin*dIII sites; due to the presence of two *Bam*HI sites in a proximal and a distal position to a single *Hin*dIII site, both IS*5* subfragments were inserted immediately distal to the inducible promoter and in both orientations (see Fig. 2). The results obtained prove that the IS*5* DNA contains nonlambdoid inceptor signals at or near both ends of the insertion element. These insertion-specific *ice* signals are able to sustain multicopy hybrid plasmid replication. Similar to the lambdoid minimal replication systems (Lusky and Hobom 1979a), both IS*5* *ice* signals have orientation-dependent activity and require RNA-polymerase-catalyzed primer-RNA synthesis for initiation of replication. Only if provided with an outside → in transcription initiated at an external promoter, such as p_{lac} or λp_{rm}, will either of them show replicational activity. The results also show that the IS*5* *ice* signals must be located at or near the ends of IS*5*, between the left end and the *Bgl*II site (180 bp from the left end) and between the right end and the *Eco*RI site (104 bp from the right end).

* Present address: Department of Molecular Biology, University of California, Berkeley, California 94720.

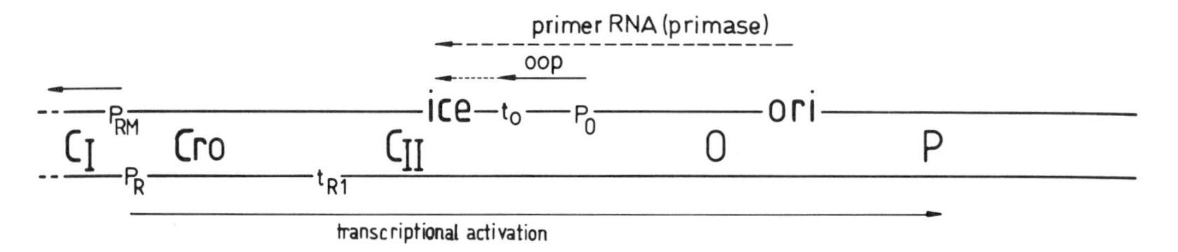

phage	mode of initiation		primer RNA synthesis	initiation proteins NH_2-O	CO_2-O	P	transcriptional activation
λ,21:	MINI-system	P_0 → ice	RNA polymerase	–	–	+	–
434:	MINI-system	P_0 → ice	RNA polymerase	–	–	–	–
λ,21:	MAXI-system	ori → ice	(primase)	+	+	+	+
434:	MAXI-system	ori → ice	(primase)	+	+	+	+
λ,21,434:	replicative repression	ori	—	+	–	–	+

Figure 1. Schematic representation of the replication regions of lambdoid phages. Genes are indicated in the central line; mRNAs transcribed rightward are indicated below this line, together with their transcription signals; mRNAs and signals oriented leftward are shown above this line. The table below summarizes the phage-coded conditions that are necessary (+) or not necessary (−) for initiation of replication in the maximal (MAXI) and in the minimal (MINI) initiation systems.

Any further conclusion as to their location or structure (in comparison to lambdoid *ice* signals) has to await DNA sequence determination.

DNA Sequence at the Ends of IS*5*

Figure 3 shows the DNA sequence at the ends of IS*5* and its integration site in λKH100 DNA (M. Kröger, unpubl.), which was determined as described by Maxam and Gilbert (1977). Upon integration of IS*5*, 4 bp of the λ DNA target site are duplicated. By sequencing one integration site, a 3-bp duplication cannot be excluded; but the comparison of the IS*5* integration sites in λKH100 and in phage Mu (Kahmann and Kamp 1979) shows that the 4-bp duplication is correct. In contrast to all other known IS elements, this is the first case where, upon integration, an even number of base pairs at the target site is duplicated. Whether the number of duplicated base pairs has any function in the mechanism of transposition is still unknown. Furthermore, two nearly perfect inverted repeats of 16 bp are found at the ends of IS*5*, as is the case with other IS elements.

DISCUSSION

Significance of IS-specific Inceptor Signals

The detection of IS*5*-specific inceptor (*ice*) signals, which are able to initiate DNA replication (at least on the hybrid plasmid level), leads to the question of whether other IS elements or transposons might also contain insertion-specific *ice* signals. The detection of replicational activity in a Tn*903*-containing λ DNA fragment after cloning (λpk35; Moore et al. 1980) also supports this hypothesis.

In the DNA sequence of IS*4*, structures resembling transcriptional stop signals have been found near both of its termini (Klaer et al., this volume). It remains to be seen whether these structures, when activated by an outside → in transcription, could initiate plasmid DNA replication.

So far, nothing is known about the significance of inceptor signals in IS elements or transposons. Their possible functions are discussed below from two slightly different aspects.

1. DNA replication is a prerequisite of transposition, because transposition of an element from one chromosomal site to another does not result in loss of the element at the original locus (Bennett et al. 1977; Ljungquist and Bukhari 1977; Shapiro 1979). Therefore, one could assume that insertion-specific inceptor signals might be useful for DNA amplification of an IS element before the transposition event. Due to the activation of such *ice* signals by transcription initiated at an outside (of the IS element)-located promoter, the region containing the IS element or transposon preferentially might be replicated. After such a segmental amplification step, a fusion of the target site to the newly synthesized DNA could take place according to the transposition model reported by Coelho et al. (this volume).

   In the case of IS*5*, the integration properties of plasmids carrying the left part of phage Mu with an IS*5* insertion in the Mu part seems to be dependent on the insertion element (van de Putte et al., this volume). A comparison of the DNA sequences of the ends of Mu and IS*5* shows that they have some homology (R. Kahmann, pers. comm.). It remains to be seen, by the creation of mutants, for example,

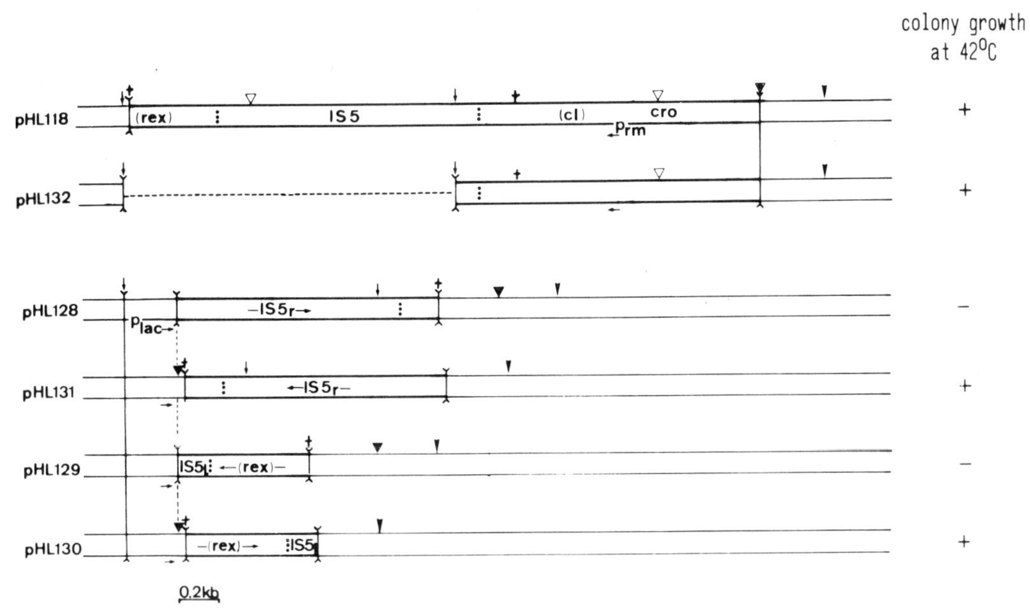

Figure 2. Cloning analysis of IS*5*-containing hybrid plasmids. The first series describes the cloning of the IS*5*-containing *Hin*dIII fragment into pHL113 and the *Eco*RI deletion derivative of that plasmid. The second series shows hybrid plasmids with cloned right and left fragments of the IS*5* element together with short flanking λ DNA segments in either orientation, as indicated by the fragment internal arrows that refer to the normal orientation in λKH100 DNA. p_{lac} refers to the *lac*UV5 promoter with an additional promoter up mutation (H. Weiher, B. Zink, and H. Schaller, pers. comm.). Three fragment internal dots indicate the IS*5*/λ DNA boundaries. Relevant genes and signal structures are drawn in full for the first member of a series; related arrows below the lines mark the positions and orientations of these signal structures in other members of that series. Key to restriction symbols: small arrow, *Eco*RI; larger arrowhead, *Sal*I; cross, *Hin*dIII; open triangle, *Bgl*II; closed triangle, *Bam*HI; dot inside triangle, *Bgl*II/*Bam*HI fusion (resistant to either digestion). Cloning boundaries are indicated by fusion signs (Ⅰ).

whether the ends of IS*5* could substitute for the ends of Mu during transposition and whether a replication step due to the IS*5*-specific *ice* signals activated by outside → in transcription is involved in the creation of cointegrates during the transposition event.

2. No IS element has yet been shown to replicate or to exist in a separate, physically autonomous state. By biochemically splicing a known small replicon into the central part of a transposable element, Cohen et al. (1979) have isolated a self-replicating transposition intermediate. Due to the ability of IS-specific inceptor signals to initiate DNA replication, IS elements also could exist as autonomously replicating units, but only when the IS element is flanked by sequences that carry a promoter as initiation site for primer-RNA synthesis. Such an IS element could then serve as a transposable *ice* element in a way that, upon transposition, new replicons could be created.

ACKNOWLEDGMENTS

We thank G. Schlingmann for her expert technical assistance and R. Kahmann, P. van de Putte, and I. Herskowitz for valuable discussions. This work has been supported by the Deutsche Forschungsgemeinschaft.

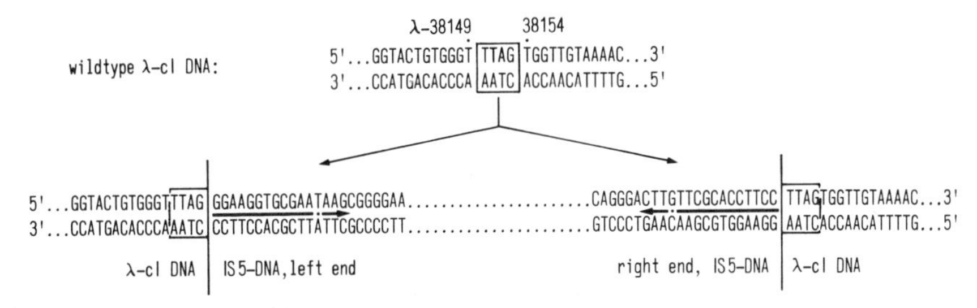

Figure 3. The ends of insertion sequence IS*5* in λKH100. The first line shows the wild-type λ*c*I DNA around the integration site of IS*5*. The boxed areas indicate base pairs that are duplicated upon integration of IS*5*. The second line shows the sequence at the ends of IS*5* together with flanking λ DNA. Arrows between the upper and the lower strand indicate the inverted repeats, which contain one mismatched base pair (•). The nucleotide positions indicated refer to an "on"-specific *Eco*RI recognition sequence = 40,000.

REFERENCES

BENNETT, P. M., J. GRINSTED, and M. H. RICHMOND. 1977. Transposition of TnA does not generate deletions. *Mol. Gen. Genet.* **154:** 205.

BLATTNER, F. R., M. FIANDT, K. K. HASS, P. A. TWOSE, and W. SZYBALSKI. 1974. Deletions and insertions in the immunity region of coliphage lambda: Revised measurement of the promoter-startpoint distance. *Virology* **62:** 458.

COHEN, S. N., M. J. CASADABAN, J. CHOU, and C.-P. D. TU. 1979. Studies of the specificity and control of transposition of the Tn*3* element. *Cold Spring Harbor Symp. Quant. Biol.* **43:** 1247.

HOBOM, G. and M. LUSKY. 1980. Origin and inceptor of DNA replication in bacteriophage lambda. *ICN-UCLA Symp. Mol. Cell. Biol.* **19:** 231.

KAHMANN, R. and D. KAMP. 1979. Nucleotide sequences of the attachment sites of bacteriophage Mu DNA. *Nature* **280:** 247.

LJUNGQUIST, E. and A. I. BUKHARI. 1977. State of prophage Mu DNA upon induction. *Proc. Natl. Acad. Sci.* **74:** 3143.

LUSKY, M. and G. HOBOM. 1979a. Inceptor and origin of DNA replication in lambdoid coliphages. I. The λ DNA minimal replication system. *Gene* **6:** 137.

———. 1979b. Inceptor and origin of DNA replication in lambdoid coliphages. II. The λ DNA maximal replication system. *Gene* **6:** 173.

MAXAM, A. M. and W. GILBERT. 1977. A new method for sequencing DNA. *Proc. Natl. Acad. Sci.* **74:** 560.

MOORE, D. D., K. DENNIS-THOMPSON, and F. R. BLATTNER. 1980. Organization of sequence at lambdoid bacteriophage origins of replication. *ICN-UCLA Symp. Mol. Cell. Biol.* **19:** 221.

SHAPIRO, J. A. 1979. Molecular model for the transposition and replication of bacteriophage Mu and other transposable elements. *Proc. Natl. Acad. Sci.* **76:** 1933.

Genes Are Things You Have Whether You Want Them or Not

C. SAPIENZA AND W. F. DOOLITTLE
Department of Biochemistry, Dalhousie University, Halifax, Nova Scotia, Canada B3H 4H7

The statement we have chosen as the title for this paper is that of an unidentified elementary-school child quoted in an article ("Digestion is best accomplished on an empty stomach; original views on health from kids") appearing in the April 1979 issue of *Self* magazine. It neatly summarizes a notion recently elaborated in detail by Orgel and Crick (1980) and by us (Doolittle and Sapienza 1980), which was alluded to earlier by Dawkins (1976), Bodmer (see Walker 1978), and Crick (1979) and which we suspect has lain dormant in the minds of many molecular biologists and evolutionary theorists for several years without being articulated explicitly.

That notion, which follows logically from a consideration of the essential natures of natural selection, nucleic acids, and organisms, is this: Natural selection operates on any entities capable of self-replication and heritable variation to produce descendants more fit for survival within their environments. Nucleic acids are such entities, and organisms are environments in which nucleic acids can evolve. DNA sequences whose expressions significantly affect organismal phenotype evolve in response to phenotypic selection. They may best be understood in terms of their contribution either to short-term fitness of individual organisms or to long-term fitness (evolutionary adaptability) of populations of organisms. However, most organisms harbor additional DNA sequences whose behaviors and very existences are exceedingly difficult to rationalize in these terms. We and Orgel and Crick proposed that many of these may be the product of what we call nonphenotypic selection. Nonphenotypic selection operates within genomes, independently of organismal phenotype, to produce DNAs whose only function is to ensure their own preservation within subsequent generations of the organisms that bear them. We believe it to be true, on logical grounds, that if there are mechanisms by which DNA sequences can ensure their own perpetuation, then such self-perpetuating DNAs (selfish DNAs) will of necessity arise and be more-or-less well maintained by nonphenotypic selection.

A logical corollary of this position is that the existence of DNAs that have evolved mechanisms for self-perpetuation need not be rationalized in any other terms. Thus, much molecular biological speculation on the function of such DNAs is unnecessary and is founded on premises that are false.

Our proposals can only be correct in more than a formal sense if indeed there are mechanisms by which DNA sequences can ensure their own preservation within genomes without operating through phenotype. Transposition is one such mechanism (Doolittle and Sapienza 1980; Orgel and Crick 1980).

Transposable Elements in Prokaryotes and Eukaryotes

There is direct or indirect genetic and/or physical evidence for the existence of transposable elements in the genomes of gram-negative and gram-positive eubacteria (Bukhari et al. 1977; Stuttard and Rozee 1980), blue-green algae (Lau et al. 1980), and at least one archaebacterium (Weidinger et al. 1979). There has also long been genetic evidence for the existence of transposable elements in eukaryotes (McClintock 1957; Peterson 1977). Physical evidence for transposition has come more recently from analyses of individually cloned members of dispersed middle-repetitive-DNA families of yeast and *Drosophila* (Cameron et al. 1979; Potter et al. 1979; Strobel et al. 1979; Wensink et al. 1979); these cloned middle repetitive DNAs bear certain structural similarities to bacterial transposable elements. Most striking are the data of Young (1979), which indicate that many, and perhaps all, of the middle-repetitive-DNA families of *Drosophila* are extremely mobile. We predict that most middle repetitive DNAs of most eukaryotes will ultimately be proven to be transposable elements, or the degenerate and immobilized descendants of such elements. We do not mean to imply by this that mechanisms of transposition of eukaryotic DNAs will in all cases be the same or necessarily similar to mechanisms involved in the transposition of prokaryotic transposons and insertion sequences (Arthur and Sherratt, 1979; Grindley and Sherratt 1979; Shapiro 1979). For instance, cycles of transcription, reverse transcription, and chromosomal reinsertion could affect transposition (and genomic survival) of noninfectious retrovirus-like elements.

Acceptance of the ubiquity of transposable elements leads one to ask what these elements are for, or, in less teleological terms, what are the selection pressures responsible for the origin and maintenance of these unusual DNA sequences. In general, two kinds of functions, both arising through phenotypic selection, have been proposed. The first is the specific regulation of gene expression through precise transpositional juxtaposition of regulatory sequences and/or protein-coding sequences. Antigenic variation in *Salmonella* (Silverman et al. 1979), adsorption and infectivity of phage

Mu (Bukhari and Ambrosio 1978; Kamp et al. 1978), mating-type variation in yeast (Nasmyth and Tatchell 1980), antigenic variation in trypanosomes (Hoeijmakers et al. 1980), and the immunoglobulin system (Maki et al. 1980) all provide good examples of such "recombinational switching." We do not deny that such specific transpositional regulatory mechanisms have on occasion arisen and been maintained by phenotypic selection operating through organismal fitness. However, it seems unlikely that the majority of prokaryotic transposable elements or eukaryotic mobile middle repetitive DNAs act as specific regulatory recombinational switches. In recognition of this, most authors have suggested a second (and predominant) function for transposable elements: the promotion of evolutionary adaptability (Cohen 1976; Kleckner 1977; Nevers and Saedler 1977; Cameron et al. 1979; Strobel et al. 1979; Kopecko 1980).

The insertions, deletions, inversions, and genomic rearrangements that often result from transposition will indeed inevitably increase the genetic variability of a population and may optimize its response (in terms of the generation of types that are more fit) to new environmental challenges. However, we question the assumption (often only implicit) that transposable elements arise and are maintained by phenotypic selection for genetic variability. The selective advantage offered by increased genetic variability is a very long-term one and should only be invoked with caution (Dawkins 1976; Maynard Smith 1978). Furthermore, it is unnecessary to invoke such an advantage, as transposable elements may, by their very nature, be ensured of self-perpetuation.

Transposable Elements as Selfish DNAs

A single copy of a DNA sequence of no phenotypic benefit to the organism that carries it risks deletion. But a DNA that can spawn additional identical copies of itself elsewhere in a genome is ensured of survival; its eradication requires multiple, independent, and simultaneous deletion events.

Prokaryotic transposable elements exhibit four traits that are more easily interpreted as selfish behavior patterns arising through nonphenotypic selection than as "altruistic" behavior patterns (beneficial to their hosts) arising through phenotypic selection (Bukhari et al. 1977; Stuttard and Rozee 1980).

1. Transposition requires replication of the element and results in retention or regeneration of the maternal copy. This is difficult to prove in all cases, but maternal copy survival is an essential feature of all recent transposition models (Arthur and Sherratt 1979; Grindley and Sherratt 1979; Shapiro 1979). It is also essential for self-perpetuation: Simple translocation from one site to another would increase genetic variability but provides no insurance against deletion.
2. Daughter elements resulting from transposition are independently transposable and do not require maternal assistance. Maternally dependent daughters would promote genetic variability as well as independent daughters, but the entire family could be lost through deletion of the mother element, unless this was maintained by phenotypic selection.
3. Transposition-specific functions are often encoded by the transposable elements themselves. Dependence on host functions is dangerous only for a genetic element whose presence is not phenotypically beneficial to the host. Some smaller prokaryotic transposable elements may indeed require for their transposition proteins they do not encode (Reif 1980). We view these elements as degenerate and comparable to defective viruses; their transposition may depend on functions of autonomous elements from which they descend, and their long-term survival is not assured.
4. Transposable elements of one family generally do not promote the transposition of members of unrelated families. If transposable elements are selfish, then unrelated elements compete for space within the genome; we would not expect them to cooperate. The fact that the generation of 3-, 5-, 9-, or 11-bp flanking repeats is element-specific implies that transposition mechanisms are themselves element-specific, and there is other, more direct, evidence for specificity in transposition (Kopecko 1980; Shapiro 1980). On the other hand, if transposable elements were beneficial to their hosts and maintained by phenotypic selection, cooperation between them might be anticipated.

It is not immediately clear whether any of the above traits, which seem to be generally characteristic of prokaryotic transposable elements and which are consistent with the notion that these are selfish molecular parasites, has yet been shown to be characteristic of eukaryotic transposable elements. It can therefore be taken as a testable prediction of the selfish-DNA hypothesis that eukaryotic transposable elements will behave in ways that are formally, if not mechanistically, identical.

Middle Repetitive DNAs, Transposition, and the Regulation of Gene Expression

If it proves to be true that most eukaryotic middle repetitive DNAs are transposable elements (or the degenerate descendants of once-transposable elements) and if the transposition mechanisms employed by them are formally similar (in the ways described above) to transposition mechanisms in prokaryotes, then the origin and existence of middle repetitive DNAs in eukaryotes require no other explanations; they are the inevitable outcome of nonphenotypic selection.

This view is in strong contrast to those of Britten, Davidson, and coworkers (Britten and Davidson 1969; Davidson et al. 1973; Graham et al. 1974; Crain et al. 1976; Davidson and Britten 1979), who assign regulatory roles to the middle repetitive DNAs that are interspersed with unique-sequence DNA and evolutionary significance (the generation of new, adaptive regulatory interactions) to the movements of such elements. It is certainly true that transposable DNAs involved in recombinational switching can best be viewed in this way and certainly plausible that some originally selfish middle repetitive DNAs could be recruited to function by phenotypic selection. However, the supposition that middle repetitive DNAs are selfish elements vitiates one of the major appealing facets of the models of Britten and Davidson; namely, that these models can account for the existence and relative sequence homogeneity of these otherwise mysterious components of the eukaryotic genome.

We know of no data that are consistent with the Britten-Davidson models (Davidson and Britten 1979), and that are inconsistent with our alternative view, and we know of at least one major body of observation that seems more easily rationalized in terms of nonphenotypic selection mechanisms. Accumulating evidence indicates that the patterns of interspersion of middle repetitive DNA and unique-sequence DNA vary widely in a way that makes little phylogenetic sense. Until recently, most organisms examined exhibited the "*Xenopus* pattern" of interspersion of short (<500 bp) middle repetitive units and short (1000–3000 bp) stretches of unique-sequence DNA. *Drosophila,* in which much longer (>5000 bp) middle repetitive elements are separated by much longer (>13,000 bp) unique-sequence stretches, was an exception, but this organism also has an exceptionally small genome size (Manning et al. 1975). It is now clear that short- and long-period interspersion patterns are scattered among the eukaryotes in a way that correlates only poorly with genome size (Smith et al. 1980) and not at all with phylogenetic position (Fig. 1). Furthermore, some eukaryotes show no middle-repetitive-DNA component at all (Timberlake 1978). The conclusion of Krumlauf and Marzluf (1980) that, for *Neurospora,* "very short repeated sequences, undetected by renaturation experiments, may serve as recognition sites for eukaryotic gene regulation," although a testimonial to the heuristic impact of Britten-Davidson models, seems unduly strained.

If, however, middle repetitive DNAs are to be viewed as self-perpetuating elements that, in general, play no role (regulatory or otherwise) in organismal phenotype, then new ways of looking at the observed variation in interspersion patterns can be entertained. One could speculate that the spacing (but not the existence) of middle repetitive elements is determined both by nonphenotypically selected properties of the elements themselves and by phenotypic selection pressures related to chromosome-pairing and recombination mechanisms. There may, for instance, be constraints on element proximity imposed by the transposition mechanisms themselves and by the "acceptability" of transposable-element transcripts to RNA:RNA splicing enzymes (which would determine whether insertions into introns were lethal). Adjacent repetitive elements of the same family can generate inversions or deletions of unique-sequence DNAs between them. The frequency of such events, and the likelihood of their being lethal, would depend on (1) the fraction of unique-sequence DNA that is essential for organismal survival, (2) the number of different middle-repetitive-DNA families

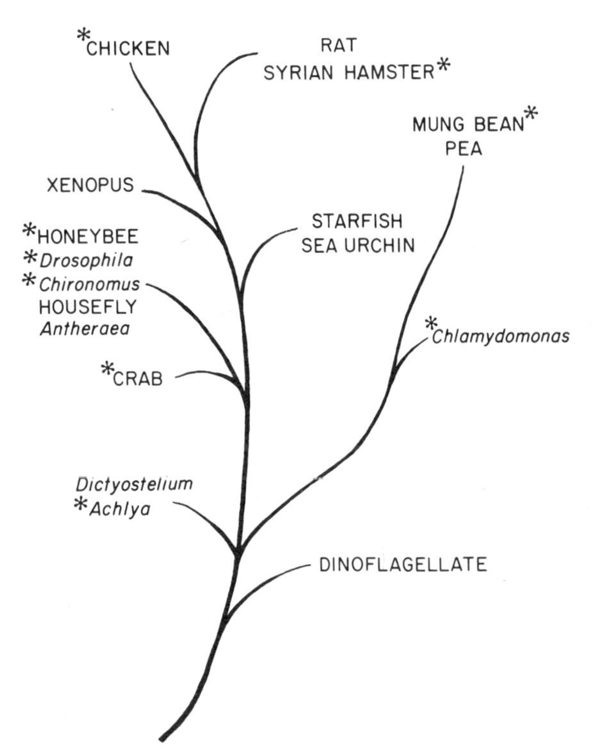

Figure 1. Phylogenetic tree of some organisms in which genome organization has been studied. Asterisk denotes that the middle repetitive DNA and unique-sequence DNA is arranged as in the *Drosophila,* or long-period interspersion, pattern; others have genome organization characteristic of the *Xenopus,* or short-period interspersion, pattern. Data from Davidson et al. (1973), Graham et al. (1974), Firtel and Kindle (1975), Manning et al. (1975), Crain et al. (1976), Efstratiadis et al. (1976), Howell and Walker (1976), Wells et al. (1976), Galau et al. (1977), Hudspeth et al. (1977), Moyzis et al. (1977), Arthur and Strauss (1978), Smith and Boal (1978), Christie and Skinner (1979), Hinnebusch et al. (1980), Smith et al. (1980), and Thompson and Murray (1980).

and, hence, the likelihood of finding close copies of identical elements, and (3) the frequency of inter- and intrachromosomal crossover events.

Evolutionary Behavior of Selfish Transposable Elements

If transposable elements are selfish molecular parasites, then their evolution within genomes should be comprehensible in the same terms as the evolution of organisms within their environments. Identical daughter elements generated by transposition of a single maternal element can experience a variety of evolutionary fates. The four most obvious ones are shown in Figure 2.

1. Those daughters that faithfully retain the maternal DNA sequence will survive by the mechanisms that preserve that sequence. Drift will be limited to nucleotide sequence changes that do not significantly alter element-encoded products involved in transposition or specific DNA regions recognized by these products and/or required for transposition.

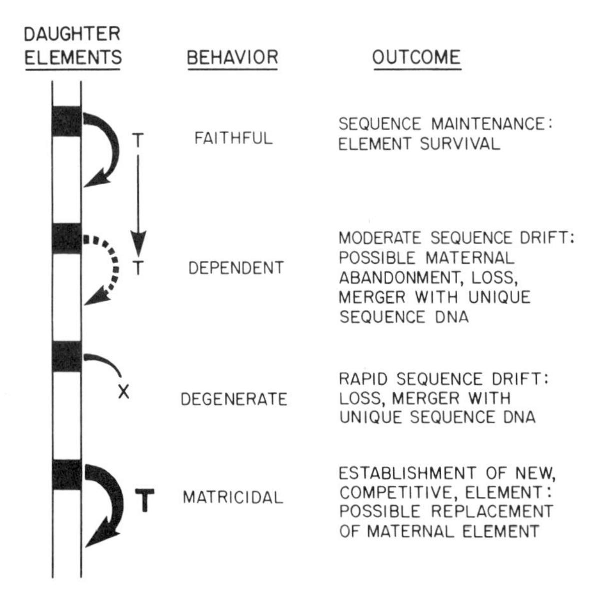

Figure 2. Possible behavior of daughter transposable elements and evolutionary fate associated with that behavior.

2. Some daughters may lose the ability to promote their own transposition but retain the ability to be transposed by maternal gene products acting in *trans.* (Mutant transposable elements of this type are well known [Meyer et al. 1979].) These will drift more rapidly in overall sequence and risk "maternal abandonment," i.e., mutations of the maternally encoded transposase that result in failure to transpose such dependent elements. Abandonment is probably selectively advantageous to the mother element, since the propagation of non-self-transposable daughters does nothing to ensure the preservation of the mother element itself.
3. Some daughters may lose the ability to be transposed altogether. There are no constraints on sequence drift in such degenerate daughters. They may be deleted or become indistinguishable from unique-sequence DNA. Their rates of elimination should be related to phenotypic pressures for reduced genome size (*r* selection in the terminology of Cavalier-Smith [1978]). When these pressures are high, one would expect degenerate (or dependent) daughters to be eliminated rapidly, and thus that the degree of sequence homogeneity within members of a single middle-repetitive-DNA family would be high. When phenotypic selection pressures for the elimination of excess DNA are low, middle-repetitive-DNA families should contain a higher proportion of drifting dependent or degenerate daughters and show less sequence homogeneity. This may explain why middle repetitive DNAs of *Drosophila,* which has the lowest genome size of any well-studied higher eukaryote, are more homogeneous in sequence than are middle repetitive DNAs of most other species (Wensink 1978).
4. Some daughters may become more readily transposable than the maternal element that generated them. This could occur either by mutation in DNA sequences directly involved in the transposition process, by mutations resulting in more efficient transposase functions, or by complementary mutations in both functions. Such matricidal elements would be in competition with, and might ultimately replace, the maternal element and its more faithful daughters.

Eukaryotic middle-repetitive-DNA families are known to change in sequence, sequence homogeneity, and copy number during the evolutionary divergence of the organisms that bear them (Britten and Kohne 1968; Galau et al. 1977; Klein et al. 1978; Moore et al. 1978). We suggest that most of the data on middle-repetitive-DNA-family size, sequence homogeneity, and phylogenetic distribution can be explained by processes of the sort suggested here, which are most easily understood if the assumption is made that these middle repetitive DNAs arc (or were) selfish transposable elements.

Limitations to Selfishness

Obviously, it is not to the advantage of any selfish transposable element to spawn unlimited numbers of daughter copies. All that is required is that enough copies be present to balance the inevitable loss of some by deletion and to ensure representation in each and every descendant of a sexually reproducing population. Both the prokaryotic transposable elements (Heffron et al. 1979) and the demonstrably transposable middle repetitive DNAs of *Drosophila* (Potter et al. 1979; Strobel et al. 1979; Young 1979) seem to show some sort of copy-number control, although not all elements (e.g., the *Alu*I family of mammalian repetitive DNAs [Rubin et al. 1980]) appear to exert reproductive restraint. In any event, we can envision no self-imposed limitation on the total number of different selfish transposable elements competing for space within the genome; limits must be imposed by organismal physiology.

Transposable middle repetitive DNAs and their dependent or degenerate descendants can be considered part of the excess DNA whose presence and variability constitute the C-value paradox. In general, we would expect the amount of excess DNA carried by an organism to be determined by the relative intensities of the two types of natural selection, phenotypic and nonphenotypic (Fig. 3). Nonphenotypic selection should, as a rule, increase the amount of excess DNA. The replication (and in the case of active, self-transposing elements, transcription, and translation) of excess DNA must represent an energetic burden that, although difficult to estimate, is undeniably real (Orgel and Crick 1980); phenotypic selection should, in general, favor the elimination of such DNA. The intensity of nonphenotypic selection should be relatively independent of organismal physiology; nonphenotypic evolution is an inherent property of the genetic material. The intensity of phenotypic selection, however, is very much a function of cellular physiology. Cavalier-Smith (1978) has recently reviewed the extensive data that correlate C values with cell and nuclear volume, cell-cycle length, and minimum generation time. In general, species (*r* strategists) under intense

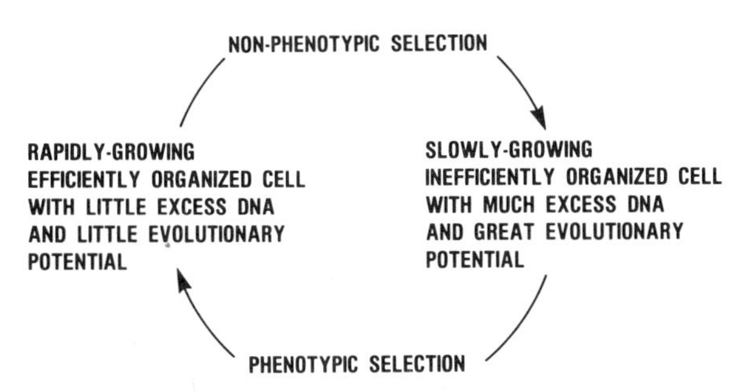

Figure 3. Two opposing types of natural selection operating on organisms. Excess DNA is increased by nonphenotypic selection and decreased by phenotypic selection.

r selection (which favors short generation times, early reproduction, and small size) show low C values. Organisms under *K* selection (*K* strategists) show longer generation times, delayed reproduction, slow development, and high C values. Cavalier-Smith (1978) suggests that it is the bulk of excess DNA itself that controls developmental rates and that *r* and *K* selections exert their effects by modulating C values. It remains arguable whether high C values are the cause or the consequence of the adoption of *K* strategies, but, in any event, the views of Cavalier-Smith are not incompatible with ours. Nonphenotypic selection provides a mechanism by which at least one component of the excess DNA of eukaryotic cells can be expanded rapidly, either when phenotypic selection pressures against the accumulation of excess DNA are relaxed or when *K* selection requires such accumulation.

REFERENCES

Arthur, A. and D. J. Sherratt. 1979. Dissection of the transposition process: A transposon-encoded site-specific recombination system. *Mol. Gen. Genet.* **175:** 257.

Arthur, R. R. and N. A. Strauss. 1978. DNA-sequence organization in the genome of the domestic chicken (*Gallus domesticus*). *Can. J. Biochem.* **56:** 257.

Britten, R. J. and E. H. Davidson. 1969. Gene regulation for higher cells: A theory. *Science* **165:** 349.

Britten, R. J. and D. E. Kohne. 1968. Repeated sequences in DNA. *Science* **161:** 529.

Bukhari, A. I. and L. Ambrosio. 1978. The invertible segment of bacteriophage Mu DNA determines the adsorption properties of Mu particles. *Nature* **272:** 575.

Bukhari, A. I., J. A. Shapiro, and S. L. Adhya, eds. 1977. *DNA insertion elements, plasmids, and episomes.* Cold Spring Harbor Laboratory, Cold Spring Harbor, New York.

Cameron, J. R., E. Y. Loh, and R. W. Davis. 1979. Evidence for transposition of dispersed repetitive DNA families in yeast. *Cell* **16:** 739.

Cavalier-Smith, T. 1978. Nuclear volume control by nucleoskeletal DNA, selection for cell volume and cell growth rate, and the solution of the DNA C-value paradox. *J. Cell Sci.* **34:** 247.

Christie, N. T. and D. M. Skinner. 1979. Interspersion of highly repetitive DNA with single copy DNA in the genome of the red crab, *Geryon quinquedens. Nucleic Acids Res.* **6:** 781.

Cohen, S. N. 1976. Transposable genetic elements and plasmid evolution. *Nature* **263:** 731.

Crain, W. R., E. H. Davidson, and R. J. Britten. 1976. Contrasting patterns of DNA sequence arrangement in *Apis mellifera* (honeybee) and *Musca domestica* (housefly). *Chromosoma* **59:** 1.

Crick, F. H. C. 1979. Split genes and RNA splicing. *Science* **204:** 264.

Davidson, E. H. and R. J. Britten. 1979. Regulation of gene expression: Possible role of repetitive sequences. *Science* **204:** 1052.

Davidson, E. H., B. R. Hough, C. S. Amenson, and R. J. Britten. 1973. General interspersion of repetitive with non-repetitive sequence elements in the DNA of *Xenopus. J. Mol. Biol.* **77:** 1.

Dawkins, R. 1976. *The selfish gene.* Oxford University Press, England.

Doolittle, W. F. and C. Sapienza. 1980. Selfish genes, the phenotype paradigm and genome evolution. *Nature* **284:** 601.

Efstratiadis, A., W. R. Crain, R. J. Britten, E. H. Davidson, and F. C. Kafatos. 1976. DNA sequence organization in the lepidopteran *Antheraea pernyi. Proc. Natl. Acad. Sci.* **73:** 2289.

Firtel, R. A. and K. Kindle. 1975. Structural organization of the genome of the cellular slime mold *Dictyostelium discoideum:* Interspersion of repetitive and single-copy DNA sequences. *Cell* **5:** 401.

Galau, G. A., M. E. Chamberlin, B. R. Hough, R. J. Britten, and E. H. Davidson. 1977. Evolution of repetitive and non-repetitive DNA. In *Molecular evolution* (ed. F. J. Ayala), p. 200. Sinauer, Sunderland, Massachusetts.

Graham, D. E., B. R. Neufeld, E. H. Davidson, and R. J. Britten. 1974. Interspersion of repetitive and non-repetitive DNA sequences in the sea urchin genome. *Cell* **1:** 127.

Grindley, N. D. F. and D. J. Sherratt. 1979. Sequence analysis at IS*1* insertion sites: Models for transposition. *Cold Spring Harbor Symp. Quant. Biol.* **43:** 1257.

Heffron, F., B. J. McCarthy, H. Ohtsubo, and E. Ohtsubo. 1979. DNA sequence analysis of the transposon Tn*3:* Three genes and three sites involved in the transposition of Tn*3*. *Cell* **18:** 1153.

Hinnebusch, A. G., L. C. Klotz, E. Inmergut, and A. R. Loeblich III. 1980. Deoxyribonucleic acid sequence organization in the genome of the dinoflagellate *Crypthecodinium cohnii. Biochemistry* **19:** 1744.

Hoeijmakers, J. H. J., A. C. C. Frasch, A. Bernards, P. Borst, and G. A. M. Cross. 1980. Novel expression—linked copies of the genes for variant surface antigens in trypanosomes. *Nature* **284:** 78.

Howell, S. H. and L. L. Walker. 1976. Informational complexity of the nuclear and chloroplast genomes of *Chlamydomonas reinhardi. Biochim. Biophys. Acta* **418:** 249.

Hudspeth, M. E. S., W. E. Timberlake, and R. B. Goldberg. 1977. DNA sequence organization in the water mold *Achlya. Proc. Natl. Acad. Sci.* **74:** 4332.

Kamp, D., R. Kahmann, D. Zipser, T. R. Broker, and L. T. Chow. 1978. Inversion of the G DNA segments of phage Mu controls phage infectivity. *Nature* **271:** 577.

Kleckner, N. 1977. Translocatable elements in prokaryotes. *Cell* **11:** 11.

Klein, W. H., T. L. Thomas, C. Lai, R. H. Scheller, R. J. Britten, and E. H. Davidson. 1978. Characteristics of individual repetitive sequence families in the sea urchin genome studied with cloned repeats. *Cell* **14:** 889.

Kopecko, D. 1980. Involvement of specialized recombination in the evolution and expression of bacterial genes. In *Plasmids and transposons: Environmental effects and maintenance mechanisms* (ed. C. Stuttard and K. R. Rozee), p. 165. Academic Press, New York.

Krumlauf, R. and G. A. Marzluf. 1980. Genome organization and characterization of the repetitive and inverted repeat DNA sequences in *Neurospora crassa. J. Biol. Chem.* **255:** 1138.

Lau, R. H., C. Sapienza, and W. F. Doolittle. 1980.

Cyanobacterial plasmids: Their widespread distribution and evidence for regions of homology between plasmids in the same and different strains. *Mol. Gen. Genet.* **178:** 203.

MAKI, R., A. TRAUNECKER, H. SAKANO, W. ROEDER, and S. TONEGAWA. 1980. Exon shuffling generates an immunoglobulin heavy chain gene. *Proc. Natl. Acad. Sci.* **77:** 2138.

MANNING, J. E., C. W. SCHMID, and N. DAVIDSON. 1975. Interspersion of repetitive and non-repetitive DNA sequences in the *Drosophila melanogaster* genome. *Cell* **4:** 141.

MAYNARD SMITH, J. 1978. *The evolution of sex.* Cambridge University Press, England.

MCCLINTOCK, B. 1957. Controlling elements and the gene. *Cold Spring Harbor Symp. Quant. Biol.* **21:** 197.

MEYER, R., G. BOCH, and J. SHAPIRO. 1979. Transposition of DNA inserted into deletions of the Tn*5* kanamycin resistance element. *Mol. Gen. Genet.* **171:** 7.

MOORE, G. P., R. H. SCHELLER, E. H. DAVIDSON, and R. J. BRITTEN. 1978. Evolutionary change in the repetition frequency of sea urchin DNA sequences. *Cell* **15:** 649.

MOYZIS, R., J. BONNET, and P. O. P. TS'O. 1977. DNA sequence organization of the Syrian hamster. *J. Cell Biol.* **75:** 103a.

NASMYTH, K. A. and K. TATCHELL. 1980. The structure of transposable yeast mating type loci. *Cell* **19:** 753.

NEVERS, B. and H. SAEDLER. 1977. Transposable genetic elements as agents of gene instability and chromosomal rearrangements. *Nature* **268:** 109.

ORGEL, L. E. and F. H. C. CRICK. 1980. Selfish DNA: The ultimate parasite. *Nature* **284:** 604.

PETERSON, P. A. 1977. The position hypothesis for controlling elements in maize. In *DNA insertion elements, plasmids, and episomes.* (ed. A. I. Bukhari et al.), p. 429. Cold Spring Harbor Laboratory, Cold Spring Harbor, New York.

POTTER, S. S., W. J. BOREIN, JR., P. DUNSMUIR, and G. M. RUBIN. 1979. Transposition of elements of the 412, *copia* and 297 dispersed repeated gene families in *Drosophila. Cell* **17:** 415.

REIF, H. F. 1980. Genetic evidence for absence of transposition functions from the internal part of Tn*981,* a relative of Tn*9. Mol. Gen. Genet.* **177:** 667.

RUBIN, C. M., C. M. HOUCK, P. L. DEININGER, T. FRIEDMANN, and C. W. SCHMID. 1980. Partial nucleotide sequence of the 300-nucleotide interspersed repeated human DNA sequences. *Nature* **284:** 372.

SHAPIRO, J. A. 1979. Molecular model for the transposition and replication of bacteriophage Mu and other transposable elements. *Proc. Natl. Acad. Sci.* **76:** 1933.

———. 1980. A model for the genetic activity of transposable elements involving DNA replication. In *Plasmids and transposons: Environmental effects and maintenance mechanisms* (ed. C. Stuttard and K. R. Rozee), p. 229. Academic Press, New York.

SILVERMAN, M., J. ZIEG, M. HILMEN, and M. SIMON. 1979. Phase variation in *Salmonella:* Genetic analysis of a recombinational switch. *Proc. Natl. Acad. Sci.* **76:** 391.

SMITH, M. J. and R. BOAL. 1978. DNA sequence organization in the common Pacific starfish *Pisaster ochraceous. Can. J. Biochem.* **56:** 1048.

SMITH, M. J., A. LUI, K. K. GIBSON, and J. K. ETZKORN. 1980. DNA sequence organization in the starfish *Dermasterias imbricata. Can. J. Biochem.* **58:** 352.

STROBEL, E., P. DUNSMUIR, and G. M. RUBIN. 1979. Polymorphisms in the chromosomal locations of elements of the 412, *copia* and 297 dispersed repeated gene families in *Drosophila. Cell* **17:** 429.

STUTTARD, C. and K. R. ROZEE, eds. 1980. *Plasmids and transposons: Environmental effects and maintenance mechanisms.* Academic Press, New York.

THOMPSON, W. F. and M. G. MURRAY. 1980. Contrasting patterns of DNA sequence organization in plants. In *Genome organization and expression in plants* (ed. C. J. Leaver), p. 1. Plenum Press, New York.

TIMBERLAKE, W. E. 1978. Low repetitive DNA content in *Aspergillus nidulans. Science* **202:** 973.

WALKER, P. M. B. 1978. Genes and non-coding DNA sequences. *Ciba Found. Symp.* **66:** 25.

WEIDINGER, G., G. KLOTZ, and W. GOEBEL. 1979. A large plasmid from *Halabacterium halobium* carrying genetic information for gas vacuole formation. *Plasmid* **2:** 377.

WELLS, R., H. D. ROYER, and C. P. HOLLENBERG. 1976. Non-*Xenopus*-like DNA sequence organization in the *Chironomus tentans* genome. *Mol. Gen. Genet.* **147:** 45.

WENSINK, P. C. 1978. Sequence homology within families of *Drosophila melanogaster* middle repetitive DNA. *Cold Spring Harbor Symp. Quant. Biol.* **42:** 1033.

WENSINK, P. C., S. TABATA, and C. PACHL. 1979. The clustered and scrambled arrangement of moderately repetitive elements in *Drosophila* DNA. *Cell* **18:** 1231.

YOUNG, M. W. 1979. Middle repetitive DNA: A fluid component of the *Drosophila* genome. *Proc. Natl. Acad. Sci.* **76:** 6274.

Transposition Immunity

L. J. WALLACE, J. M. WARD, P. M. BENNETT, M. K. ROBINSON, AND M. H. RICHMOND
Department of Bacteriology, University of Bristol, Bristol, BS8 1TD, England

A plasmid that carries a Tn*A* transposon cannot acquire another copy of Tn*A* by transposition, a phenomenon known as transposition immunity (Robinson et al. 1977). To date, no other transposon has been shown to confer to its host replicon immunity to further transposition of a like transposon. To our knowledge, only two transposons other than Tn*A* have been examined for transposition immunity: Tn*10* (Bennett et al. 1977) and Tn*7* (Brevet and Hassan, pers. comm.). Neither Tn*10* nor Tn*7* shows transposition immunity with respect to the like transposon; but Tn*10* does confer to its host replicon some immunity to transposition of Tn*A* (Bennett et al. 1977).

Plasmids That Carry Two Copies of Tn*A* Are Stable

Transposition immunity is not the result of an inherent instability created by the same replicon carrying two Tn*A* elements. Stable plasmids that carry two copies of Tn*A* have been constructed by in vitro recombination techniques (Robinson et al. 1978). Furthermore, plasmids that have acquired two copies of Tn*A* by transposition have been isolated (Robinson et al. 1978; Tu and Cohen 1980). However, all of these naturally occurring plasmids that carry two copies of Tn*A* acquired both Tn*A* copies in the same experiment, i.e., the two transposition events could not be separated in time. In contrast, all attempts to transpose, sequentially, two copies of wild-type Tn*A* onto the same replicon, by first establishing one Tn*A* and then seeking to introduce a second copy of Tn*A* by transposition, have failed (Bennett et al. 1977). These observations lead us to conclude that a replicon may acquire a second copy of Tn*A* by transposition provided the second event occurs before the chemical or physical change that results in immunity is established. Once immunity is established, however, no further transposition of wild-type Tn*A* to that replicon can occur.

Transposition Immunity Is *cis*-acting

Tn*A* can transpose to a nonimmune plasmid at normal frequencies in the presence of an immune plasmid, indicating that transposition immunity is *cis*-acting (Robinson et al. 1977). Indeed, it is essential that transposition immunity does act only in *cis,* because *trans*-acting transposition immunity would block all Tn*A* transposition. Therefore, a site that is normally carried on Tn*A* must be involved in the establishment of immunity.

We have approached the investigation of transposition immunity in two ways. The first objective was to identify those regions of Tn*A* that are required to confer immunity on the host replicon. The second objective was to identify Tn*A*-encoded proteins involved in the establishment, or recognition, of immunity.

Tn*A* transposons from four different origins have been used in this study: Tn*1* from RP4, Tn*801* from RP1, Tn*802* from RP1-1 (Grinsted et al. 1978), and Tn*3* from R1 (Cohen 1976). Tn*1*, Tn*801*, and Tn*802* appear to be identical and may indeed have come from the same source (Grinsted et al. 1978). They differ slightly from Tn*3* (C.-L. Choi, pers. comm.), but as far as expression of transposition immunity is concerned, they are indistinguishable.

EXPERIMENTAL PROCEDURES

The bacterial strains used in this study were *recA* strains (*recA56*) and so were deficient in classical (homologous) recombination, with the exception of *Escherichia coli* C600, which was used for the initial screening of the plasmids used in the studies reported here. *E. coli* strain UB1780, which carries a copy of Tn*A* (Tn*802*) on the chromosome (Bennett and Richmond 1976), was used to test the ability of plasmids to acquire Tn*A* by transposition. Plasmids to be tested were introduced into strain UB1780 by conjugation. Plasmid-containing progeny from these crosses, once purified, were left for at least 5 days on Dorset egg slopes. This procedure allows transposition events to accumulate within the population (P. M. Bennett, unpubl.). The frequency of transposition of Tn*A* to a particular plasmid was estimated by outcrossing plasmids from the UB1780 host to an appropriate *recA* recipient strain and determining the proportion of plasmid::Tn*A* recombinants in the outcrossed plasmid population. In practice, plasmid::Tn*A* recombinants were selected by incorporating carbenicillin (Cb) (500 μg/ml) into the selective medium, and parent plasmids were selected by incorporating into the selective medium a drug to which the plasmid confers resistance (trimethoprim [Tp], 25 μg/ml, for R388 and derivatives; kanamycin [Km], 30 μg/ml, for pUB1601). Transposition frequencies are recorded as the ratio of plasmid::Tn*A* recombinants detected in the outcross to the total number of plasmids detected.

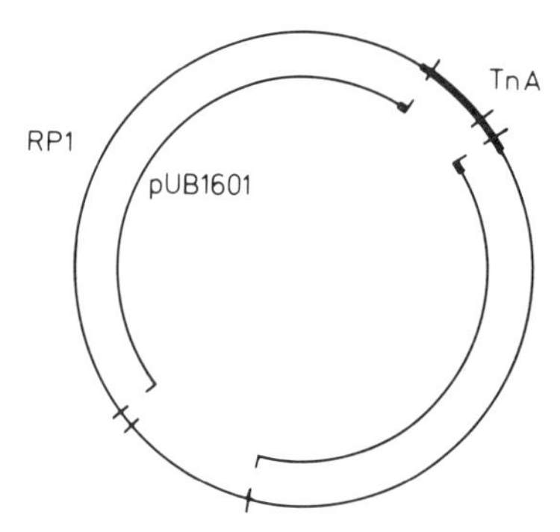

Figure 1. RP1 ($Ap^rKm^rTc^r$) and a deletion derivative pUB1601 ($Km^rTc^rAp^s$). The thick line represents Tn*A* (Tn*801*) sequences, and the short vertical bars indicate *Pst*I sites. The two large *Pst*I fragments present in pUB1601 are oriented as in RP1. pUB1601 was constructed by Dr. J. Grinsted (University of Bristol).

RESULTS

Location of the Sequences of TnA Necessary for the Establishment of Transposition Immunity

The first piece of evidence that the whole of Tn*A* is not required to confer immunity on the host replicon came from experiments with pUB1601, a derivative of the IncP resistance plasmid RP1 (Grinsted et al. 1972) (Fig. 1). Plasmid pUB1601 consists of two large *Pst*I fragments of RP1 ligated together so that they retain their normal orientations with respect to each other. This plasmid contains only 990 bp from the end of Tn*A* distal to the *bla* gene, i.e., the gene that encodes β-lactamase, and 450 bp from the end of Tn*A* proximal to *bla*. The ends of the deletion are defined by the extreme *Pst*I sites within Tn*A*. pUB1601 no longer confers resistance to ampicillin (Ap; or carbenicillin) but is still capable of conjugal transfer. Despite the fact that most of Tn*A* has been deleted to form pUB1601, Tn*A* will not transpose onto this plasmid. We concluded, therefore, that one or both of the ends of Tn*A* are required to confer to the host replicon immunity to further Tn*A* transpositions.

To determine whether either end of Tn*A* by itself is sufficient to establish immunity, we have used Tn*A* derivatives of the IncW resistance plasmid R388, a conjugal plasmid that confers resistance to trimethoprim and sulfonamides (Su) (Datta and Hedges 1972). The particular R388::Tn*A* plasmids used are more easily manipulated than RP1 or its derivatives because the genetic and physical maps have been aligned (Ward and Grinsted 1978) and it is known that R388 contains only two sites for the restriction endonuclease *Pst*I and that these sites are located close together in the region that contains the resistance genes, which is situated away from regions essential for plasmid maintenance, replication, and transfer. Furthermore, R388 acquires Tn*A* by transposition at relatively high frequencies ($\sim 2 \times 10^{-2}$) (Robinson et al. 1977).

The first derivative of R388 used was pUB501 (Fig. 2), a plasmid isolated and characterized by Robinson et al. (1977). It carries a mutation that prevents expression of active β-lactamase and, as a consequence, no longer confers resistance to ampicillin (or carbenicillin). Plasmid pUB501 shows transposition immunity, i.e., a second copy of Tn*A* will not transpose onto pUB501. Two deletion derivatives of pUB501, specifically, pUB5574 and pUB5575, were constructed using the *Pst*I restriction endonuclease (Fig. 2). Derivative pUB5774, which contains only the end of Tn*A* proximal to the *bla* gene, is not immune to Tn*A* transposition and acquires Tn*A* at frequencies similar to those observed for R388 itself ($\sim 2 \times 10^{-2}$). In contrast, the derivative of pUB501 analogous to pUB1601 (Fig. 1), specifically, pUB5575 (Fig. 2), which contains both ends of Tn*A*, is immune and will not accept Tn*A* by transposition. We concluded, therefore, that the end of Tn*A* distal to *bla* is necessary to confer immunity, whereas the end of Tn*A* proximal to *bla* is not sufficient but could be necessary to confer immunity.

The second R388 derivative used was pUB1621 (Fig. 3). This plasmid was one of a number of plasmids isolated for this study that have Tn*A* inserted into the sulfonamide-resistance gene(s) of R388 (Ward and

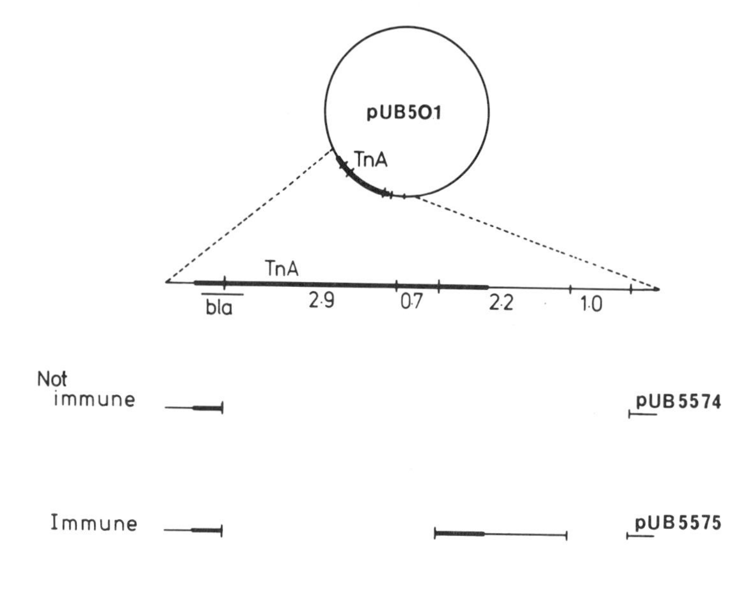

Figure 2. pUB501 (R388::Tn*A* [$Tp^rSu^rAp^s$]) and deletion derivatives. (——) R388 sequences; (▬▬) Tn*A* sequences. *Pst*I sites are indicated by short vertical bars and sizes are given in kb. Sequences not deleted in derivatives pUB5574 and pUB5575 are shown. Construction of the deletion derivatives was as follows: pUB501 DNA was digested with *Pst*I restriction endonuclease. When the reaction was complete, the *Pst*I was heat-inactivated. The DNA fragments were ligated with T4 DNA ligase, and the ligated DNA was used to transform *E. coli* C600 to trimethoprim resistance. Plasmids constructed and isolated in this way were tested for their ability to acquire Tn*A* by transposition (see Experimental Procedures).

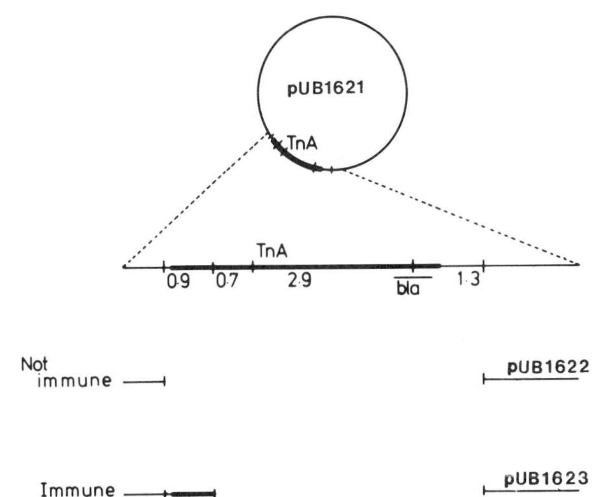

Figure 3. pUB1621 (R388::Tn*A* [$Tp^rSu^sAp^r$]) and deletion derivatives. (——) R388 sequences; (▬▬) Tn*A* sequences. Tn*A* is inserted into the gene(s) that confers resistance to sulfonamide. *Pst*I sites are indicated by short vertical bars, and sizes are given in kb. Sequences not deleted in the derivatives pUB1622 and pUB1623 are shown. These plasmids were constructed as described in the legend to Fig. 2.

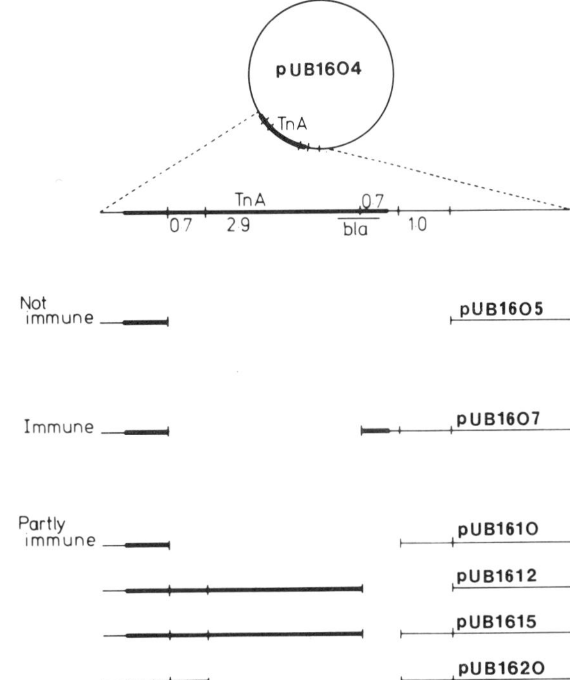

Figure. 4. pUB1604 (R388::Tn*A* [$Tp^rSu^sAp^r$]) and deletion derivatives. (——) R388 sequences; (▬▬) Tn*A* sequences. Tn*A* is inserted into the gene(s) conferring resistance to sulfonamide. *Pst*I sites are indicated by short vertical bars, and sizes are given in kb. Sequences not deleted in derivatives pUB1605, pUB1607, pUB1610, pUB1612, pUB1615, and pUB1620 are shown. Deletion derivatives were constructed as follows: pUB1604 DNA was partially digested with *Pst*I endonuclease. The resulting fragments were used to transform *E. coli* C600 to trimethoprim resistance. There was no ligation step. Plasmids constructed and isolated in this way were tested for their ability to acquire Tn*A* by transposition (see Experimental Procedures).

Grinsted 1978). This ensures that all the *Pst*I sites of the R388::Tn*A* plasmids are relatively close together, which facilitates deletion generation using *Pst*I restriction. Two derivatives, pUB1622 and pUB1623, were constructed from pUB1621 (Fig. 3). pUB1622 contains no Tn*A* sequences and acquires Tn*A* by transposition at normal frequencies, i.e., it is not immune. On the other hand, the derivative that contains only the end of Tn*A* distal to *bla*, pUB1623 (Fig. 3), does not accept Tn*A* by transposition, i.e., it is immune. These data show that the end of Tn*A* distal to the *bla* gene, as defined by the *Pst*I restriction site at that end, can confer immunity on its host replicon in the absence of any other Tn*A* sequence.

We have found, however, that the end of Tn*A* distal to *bla*, although necessary to confer immunity, is not always sufficient to ensure it. Derivatives of plasmid pUB1604 demonstrate this fact (Fig. 4). Plasmid pUB1604 is another R388::Tn*A* recombinant plasmid with Tn*A* located in the sulfonamide-resistance gene(s) of R388. The site of the insertion is about 300 bp away from the site of insertion of the Tn*A* element carried by pUB1621 (cf. Figs. 3 and 4). However, pUB1605, the derivative of pUB1604 that contains only the end of Tn*A* distal to *bla*, does not display transposition immunity, as might have been expected from previous results, but accepts Tn*A* at normal frequencies, i.e., similar to those shown by R388 itself. In contrast, pUB1607, which carries sequences from both ends of Tn*A* (and so is analogous to plasmids pUB1601 [Fig. 1] and pUB5575 [Fig. 2]), does not accept Tn*A* by transposition and thus is immune.

A number of derivatives of pUB1604 that acquire Tn*A* by transposition, but at reduced frequencies ($\sim 1 \times 10^{-3}$), were isolated. Four plasmids representative of this class, pUB1610, pUB1612, pUB1615, and pUB1620, are shown in Figure 4. Like pUB1605 (Fig. 4), each plasmid contains the end of Tn*A* distal to *bla*, and, in addition, each plasmid carries other sequences derived from R388, Tn*A*, or both, but no one additional sequence is present in all cases. This analysis has been carried one stage further. A 3200-bp fragment that carries a gene specifying resistance to chloramphenicol (Cm) and that is derived from the IncW resistance plasmid S-a was cloned into the single *Pst*I site of the nonimmune plasmid pUB1605 (Fig. 5). The derivative so generated, pUB1629 (Fig. 5), is completely immune to Tn*A* transposition, despite the fact that plasmid S-a, from which the chloramphenicol-resistance fragment was derived, shows no immunity itself, but instead acquires Tn*A* at frequencies similar to those of R388. We conclude, therefore, that sequences of the host replicon can affect the expression of transposition immunity even though the appropriate nucleotide sequence (a 990-bp sequence at the end of Tn*A* distal to *bla*, as defined by the *Pst*I site at that end) is carried on the host replicon. Other sequences, not necessarily derived from Tn*A*, inserted adjacent to the immunity sequence can relieve

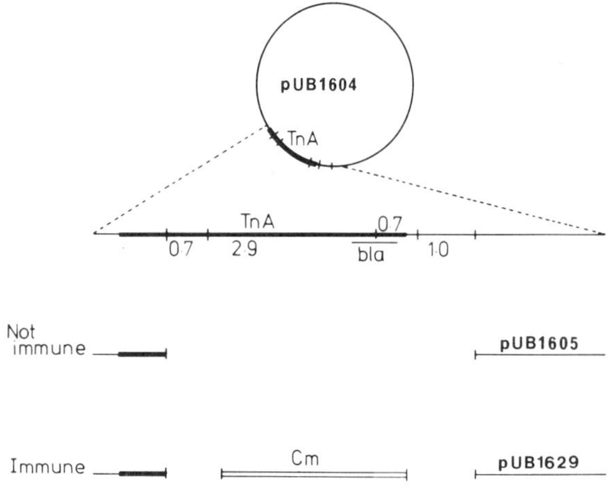

Figure 5. pUB1604 (R388::Tn*A* [$Tp^rSu^sAp^r$]) and derivatives pUB1605 and pUB1629. (——) R388 sequences; (▬) Tn*A* sequences; (▭) sequences derived from the plasmid S-a. *Pst*I sites are indicated by short vertical bars, and sizes are given in kb. pUB1604 and pUB1605 are described in the legend to Fig. 4. pUB1629 was constructed by ligating together the products of a *Pst*I digest of both pUB1605 and the IncW plasmid S-a (Hedges and Datta 1971), which confers resistance to streptomycin, sulfonamide, chloramphenicol, and kanamycin; the gene conferring resistance to chloramphenicol is located on a 3.2-kb *Pst*I fragment of S-a (J. M. Ward, unpubl.). The ligation mixture was used to transform *E. coli* C600 to trimethoprim and chloramphenicol resistance. Plasmid pUB1629, constructed and isolated in this way, was tested for its ability to acquire Tn*A* by transposition (see Experimental Procedures).

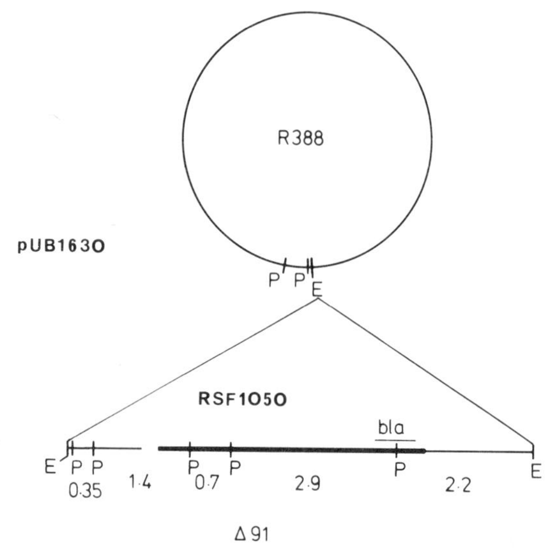

Figure 6. pUB1630 ($Tp^rSu^rAp^s$). pUB1630 was constructed from plasmids R388 and RSF1050 Δ91 (see text) by using the single *Eco*RI restriction site present in each molecule. *Eco*RI-generated linear fragments of each plasmid were mixed, ligated with T4 DNA ligase, and then used to transform *E. coli* C600 to amphenicol and trimethoprim resistance. Transformants contained a single plasmid species comprised of both plasmids. Plasmid DNA was isolated from one of these, treated with hydroxylamine as described by Humphreys et al. (1976), and then used to transform *E. coli* C600 to trimethoprim resistance. These Tp^r transformants were screened to detect Ap^s clones. One of these carries plasmid pUB1630. The circle represents R388 sequences; the linear sequence inserted at the *Eco*RI site (E) of R388 is RSF1050 Δ91. (▬) Tn*A* (Tn*3*) sequences present on the molecule. The break between the thin and thick lines on the DNA insert represents deletion 91. P indicates *Pst*I restriction sites (sizes are given in kb).

this block to the expression of transposition immunity either wholly or partly. However, the most important observation is that the other end of Tn*A*, i.e., the end proximal to *bla*, relieves the block completely. Thus, the end of Tn*A* distal to *bla* (as defined by the *Pst*I site at that end) is necessary for expression of transposition immunity but is not always sufficient to ensure that expression, and therefore the end of Tn*A* proximal to *bla*, may, on occasion, be necessary to guarantee transposition immunity.

We have examined one other deletion of Tn*A* (Tn*3*), which was originally characterized by Heffron et al. (1977). Deletion 91 is a deletion of RSF1050 (PMB8:: Tn*3*) that extends 250 bp into the end of Tn*3* distal to *bla* (Heffron et al. 1979), the end of Tn*A* that we have found to carry immunity sequences. We have cloned plasmid RSF1050 Δ91 into R388 by taking advantage of the single *Eco*RI site of each plasmid and have constructed plasmid pUB1630 (Fig. 6).

Plasmid pUB1630 (see legend to Fig. 6) was tested and shown to be immune to Tn*A* transposition. From this result, we conclude that the first 250 bp at the end of the immunity end of Tn*A*, i.e., the end distal to *bla*, are not required; consequently, the sequence that constitutes the inverted repeat at the ends of Tn*A* is not involved in the expression of immunity. These data locate (to ~750 bp of Tn*A*) a site necessary, but not always sufficient, for the establishment of immunity.

Mutants of Tn*A* That No Longer Recognize Immune Molecules

Plasmids carrying deletions of Tn*A* but that are still immune to further Tn*A* transposition events have proved useful not only in locating the sequences of Tn*A* involved in immunity, but also in isolating mutants of Tn*A* that fail to recognize an immune molecule. Three Tn*A*-encoded proteins have been reported: a β-lactamase, which is the product of the *bla* gene; a protein with a molecular weight of 120,000, which is required for transposition and is the product of the *tnpA* gene; and a protein with a molecular weight of 19,000, the product of the *tnpR* gene, which acts to repress the synthesis of itself and the product of the *tnpA* gene (Chou et al. 1979; Gill et al. 1979; Heffron et al. 1979). The product of the *tnpR* gene may also play an active role in transposition (Arthur and Sherratt 1979). The extent and direction of transcription of these genes are known, and these genes, plus their control regions, account for the great majority of Tn*A* sequences (Chou et al. 1979; Gill et al. 1979). Plasmids such as pUB1601 (Fig. 1), pUB5575 (Fig. 2), pUB1623 (Fig. 3), and pUB1607 (Fig. 4), which have large internal deletions of Tn*A*, cannot produce any of the three gene products. It is possible, therefore, to search for mutants of Tn*A* that do not recognize a molecule such as pUB1601 as an immune molecule. The rationale for this is that the remnants of Tn*A* on pUB1601 (and like molecules) furnish the immunity site but no longer code for the protein(s) that act at this site. Therefore, to establish immunity, these proteins must be provided by a second copy of Tn*A* that can provide the missing functions, normally the Tn*A* element one is trying to transpose onto the plasmid. In such a situation, transposition immunity would be established, and transposition of Tn*A*, e.g., to pUB1601, could not occur. However, if the second copy of Tn*A* carries a mutation within the gene(s) that codes for the *trans*-acting immunity function(s), then transposition immunity could not be established, and Tn*A* transposition to pUB1601 would occur, provided, of course, that such an immunity mutant were still capable of self-mediated transposition.

pUB1601::Tn*A* derivatives can be isolated at very low frequencies ($\leq 10^{-7}$). The two Tn*A* elements we have found in this way have been transposed to the bacterial chromosome and can be shown to be mutants of Tn*A* that do not recognize that a molecule such as pUB1601 is immune to Tn*A* transposition. These mutants transpose onto pUB1601, as well as onto nonimmune plasmids, at normal frequencies ($\sim 2 \times 10^{-2}$). These mutant Tn*A*s will not transpose onto immune molecules that produce wild-type *tnpA*- and *tnpR*-gene products but have a mutation within *bla*. Therefore, wild-type proteins encoded by Tn*A* can complement the Tn*A* immunity mutants in *trans*. We conclude from these results that the mutations are likely to be located within either of the genes known to code for transposition functions, i.e., *tnpA* or *tnpR*, although the possibility that Tn*A* carries overlapping genes has not yet been ruled out.

We have begun experiments to determine which of the Tn*A*-coded proteins will complement the mutant Tn*A*s. To this end, we have used a derivative of pUB1621, specifically, pUB1624 (Fig. 7), that carries only one complete Tn*A* gene, the *tnpR* gene, and approximately half of the *tnpA* gene. The promoters for both of these genes remain intact. pUB1624 was introduced into strains that carried one of the mutant Tn*A*s on the chromosome and the plasmid pUB1601. In one case, the mutant Tn*A* continues to transpose onto pUB1601 at normal frequencies (about 2×10^{-2}). In the other case, however, the mutant Tn*A* would no longer transpose onto pUB1601, although it would do so in the absence of pUB1624. We believe, therefore, that one of the two mutant Tn*A*s is complemented by the product of gene *tnpR*.

The second molecule used in these complementation studies is pUB1625. This plasmid is a cointegrate of the two plasmids R388 and RSF103 (also called ΔAp; Heffron et al. 1977). RSF103 is an RSF1010::Tn*1* recombinant plasmid with a deletion within Tn*1* that terminates at one end in *tnpR* and at the other end in *bla*. The deleted Tn*1* of RSF103 is transposition-proficient, but transposition leads to the production of cointegrates of

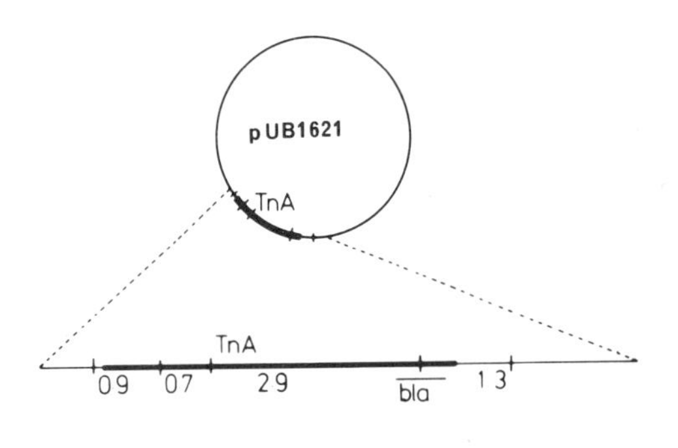

pUB1624

Figure 7. pUB1621 (R388::Tn*A* [$Tp^rSu^sAp^r$]) and derivative pUB1624. (——) R388 sequences; (▬▬) Tn*A* sequences. *Pst*I sites are indicated by short vertical bars, and sizes are given in kb. pUB1624 was constructed as described in the legend to Fig. 3.

RSF103 and the recipient replicon with two copies of the deleted transposon (Arthur and Sherratt 1979). RSF103 has been used extensively in complementation studies to provide the product of gene *tnpA* in the absence of that of gene *tnpR* (Heffron et al. 1977; Gill et al. 1978; Arthur and Sherratt 1979). We took advantage of this property of RSF103, namely, formation of cointegrates, to construct pUB1625, which is the product of transposition of RSF103 onto R388.

pUB1625 was introduced into strains that carried one of the mutant Tn*A*s on the chromosome and the plasmid pUB1601. In the case of the mutant Tn*A*, which was apparently complemented by the product of the *tnpR* gene of pUB1624, no complementation was observed with plasmid pUB1525. This mutant Tn*A* transposed onto pUB1601 at normal frequencies in the presence of pUB1625. This result is consistent with our conclusion that the mutation carried by this Tn*A* is located within the *tnpR* gene. We believe that the remaining mutant Tn*A* is likely to be complemented by the product of gene *tnpA*, unless, of course, Tn*A* carries a series of overlapping genes. Our tentative conclusion, however, is that both the product of gene *tnpA* and that of gene *tnpR* are involved in transposition immunity.

DISCUSSION

We have located a nucleotide sequence at one end of Tn*A*, specifically, the end distal to *bla* (as defined by the *Pst*I site at that end), that is necessary, and in some cases sufficient, for the expression of transposition immunity. However, this is not always so. Although several different sequences partly or wholly restore expression of immunity when they are located next to the inactive immunity sequence, the most significant finding is that a 450-bp sequence from the other end of Tn*A* (the end proximal to *bla*) will completely restore transposition immunity when inserted next to an inactive immunity sequence. It would appear, therefore, that Tn*A* has a built-in safeguard to guarantee expression of transposition immunity.

In addition, we have identified two *trans*-acting components of immunity, both encoded by Tn*A*. We believe that the likeliest explanation for the involvement of the *tnpA*-gene product is that this protein is involved in transposition in the initial interaction between donor and recipient replicons and therefore may be the component of the transposition mechanism that recognizes the immunity block. A mutation in this protein might well result in an altered protein that still mediates transposition but no longer recognizes the block. It is more difficult to suggest an explanation for the involvement of the *tnpR*-gene product. It is possible that this protein is involved in mediating whatever change occurs that establishes transposition immunity either directly or by subverting a host-cell protein to produce the change. We believe it is likely that a host function is involved, and we have begun experiments to examine this possibility. It seems likely that as the components of immunity and their interactions are identified, then the nature of the change involved in the establishment of transposition immunity will become apparent. Finally, it would seem that transposition immunity is a complex effect that is mediated by several components, both *cis*- and *trans*-acting, and, furthermore, that the mechanism is intimately associated with the transposition process itself.

ACKNOWLEDGMENT

This work was supported in part by a Programme Grant to M. H. R. from the Medical Research Council.

REFERENCES

ARTHUR, A. and D. SHERRATT. 1979. Dissection of the transposition process: A transposon-encoded site-specific recombination system. *Mol. Gen. Genet.* **175:** 267.

BENNETT, P. M. and M. H. RICHMOND. 1976. Translocation of a discrete piece of deoxyribonucleic acid carrying an *amp* gene between replicons in *Escherichia coli. J. Bacteriol.* **126:** 1.

BENNETT, P. M., M. K. ROBINSON, and M. H. RICHMOND. 1977. R-factors: The properties and possible control. In *Topics in infectious diseases* (ed. J. Drews and G. Högenauer), vol. 2, p. 81. Springer-Verlag, New York.

CHOU, J., P. G. LEMAUX, M. J. CASADABAN, and S. N. COHEN. 1979. Transposition protein of Tn*3:* Identification and characterization of an essential repressor-controlled gene product. *Nature* **282:** 801.

COHEN, S. N. 1976. Transposable genetic elements and plasmid evolution. *Nature* **263:** 731.

DATTA, N. and R. W. HEDGES. 1972. Trimethoprim resistance conferred by W plasmids in Enterobacteriaceae. *J. Gen. Microbiol.* **72:** 349.

GILL, R. E., F. HEFFRON, and S. FALKOW. 1979. Identification of the protein encoded by the transposable element Tn*3* which is required for its transposition. *Nature* **282:** 797.

GILL, R., F. HEFFRON, G. DOUGAN, and S. FALKOW. 1978. Analysis of sequences transposed by complementation of two classes of transposition-deficient mutants of Tn*3. J. Bacteriol.* **136:** 742.

GRINSTED, J., P. M. BENNETT, S. HIGGINSON, and M. H. RICHMOND. 1978. Regional preference of insertion of Tn*501* and Tn*802* into RP1 and its derivatives. *Mol. Gen. Genet.* **166:** 313.

GRINSTED, J., J. R. SAUNDERS, C. INGRAM, R. B. SYKES, and M. H. RICHMOND. 1972. Properties of an R-factor which originated in *Pseudomonas aeruginosa* 1822. *J. Bacteriol.* **110:** 529.

HEDGES, R. W. and N. DATTA. 1971. fi$^-$ R factors giving chloramphenicol resistance. *Nature* **234:** 220.

HEFFRON, F., P. BEDINGER, J. J. CHAMPOUX, and S. FALKOW. 1977. Deletions affecting the transposition of an antibiotic resistance gene. *Proc. Natl. Acad. Sci.* **74:** 702.

HEFFRON, F., B. J. MCCARTHY, H. OHTSUBO, and E. OHTSUBO. 1979. DNA sequence analysis of the transposon Tn*3:* Three genes and three sites involved in transposition of Tn*3. Cell* **18:** 1153.

HUMPHREYS, G. O., G. A. WILLSHAW, H. R. SMITH, and E. S. ANDERSON. 1976. Mutagenesis of plasmid DNA with hydroxylamine: Isolation of mutants of multi-copy plasmids. *Mol. Gen. Genet.* **145:** 101.

ROBINSON, M. K., P. M. BENNETT, and M. H. RICHMOND. 1977. Inhibition of Tn*A* translocation by Tn*A. J. Bacteriol.* **129:** 407.

ROBINSON, M. K., P. M. BENNETT, J. GRINSTED, and M. H. RICHMOND. 1978. The stable carriage of two Tn*A* units on a single replicon. *Mol. Gen. Genet.* **160:** 339.

TU, C-P. D. and S. N. COHEN. 1980. Translocation specificity of the Tn*3* element: Characterization of sites of multiple insertions. *Cell* **19:** 151.

WARD, J. M. and J. GRINSTED. 1978. Mapping of functions in the R-plasmid R388 by examination of deletion mutants generated in vitro. *Gene* **3:** 87.

Regulation of Tn5 Transposition

D. BIEK AND J. R. ROTH
Department of Biology, University of Utah, Salt Lake City, Utah 84112

The transposable drug-resistance element Tn*5* consists of a unique sequence of 2.4 kb (Berg et al. 1975) flanked by inverse repeats of 1.45 kb (Berg et al. 1975). The Tn*5* *neo* gene encodes an enzyme that inactivates neomycin (Nm) and kanamycin (Km), providing the host cell with resistance to these antibiotics (Berg et al. 1978; Jorgensen et al. 1979). Evidence has been presented (Meyer et al. 1979; Postle and Reznikoff 1979; Rothstein et al. 1980) that the flanking inverted-repeat sequences encode proteins important to the transposition ability of the Tn*5* element. In this paper evidence is presented that the transposition activity of Tn*5* is subject to regulation.

Previously, a different transposable element, Tn*3*, was shown to possess a mechanism for regulating its transposition frequency (Gill et al. 1978; Heffron et al. 1978). This regulation occurs by repression of transposase synthesis (Chou et al. 1979a,b; Gill et al. 1979; Heffron et al. 1979). However, the biological role of this mechanism is still uncertain. A genetically intact Tn*3* element has not been shown to vary its transposition frequency in response to any biologically relevant signal. In the case of Tn*5*, the mechanism of control is unknown, but the regulation can be demonstrated under natural conditions. This regulatory ability may be of adaptive significance to the Tn*5* element.

RESULTS

Initial Observations

The suggestion that Tn*5* might regulate its frequency of transposition was first proposed on the basis of observations of the behavior of Tn*5* insertion auxotrophs in P22-mediated generalized transductional crosses in *Salmonella typhimurium*. When a Tn*5* insertion auxotroph is used as a donor, a high percentage (approximately 10%) of the kanamycin-resistant (Km^r) transductants arise by transposition of Tn*5* from the transduced fragment to new sites in the recipient chromosome (Biek and Roth 1980). This surprisingly high frequency of transposition is not seen when Tn*5* is established in a strain. When Tn*5* insertion auxotroph mutants are used as recipients and are transduced to prototrophy, virtually all transductants lose kanamycin resistance; apparently, less than about 0.1% of the cells in the Tn*5*-carrying recipient population have acquired secondary Tn*5* elements by transposition even though these cells have divided many times since last being cloned. This variation in transposition frequency suggested that Tn*5* might regulate this process. According to this hypothesis, Tn*5* might actively transpose immediately after introduction into a naive host; this transposition activity might be inhibited or repressed once Tn*5* is established in the new host. Unfortunately, these initial observations were based on comparisons of results from rather dissimilar crosses performed under different selective conditions. Differences in the state (linear or supercoiled) of the Tn*5*-carrying DNA in the two situations being compared are likely to be important.

A Demonstration of Transposition Regulation

As a better test of the possibility of regulation of transposition, Tn*5* was introduced into *Salmonella* recipients that already harbored a Tn*5* element. The transposition frequency was compared with that seen when Tn*5* is introduced into a naive recipient (i.e., one that does not contain a Tn*5* element). In both cases, the donor Tn*5* element being observed is present in a linear transduced fragment, and selection is made for kanamycin resistance under conditions that require inheritance of the donor Tn*5* element by transposition. (The recipient Tn*5* carries a *neo* mutation that inactivates its kanamycin-resistance determinant. Recipient cells carry a *recA* mutation that prevents inheritance of the donor Tn*5* element by standard recombination.) Typical data from such crosses are presented in Table 1. The presence of a Tn*5* element in the recipient results in a 12- to 70-fold reduction in the transposition frequency of the donor Tn*5* element. Certain deletion mutants of Tn*5* are unable to reduce transposition.

Regulatory Effects Are Exerted in *trans*

Apparently, the recipient Tn*5* element encodes a product that inhibits or represses transposition activity of the incoming Tn*5* element. The inhibitory effect is exerted in *trans* on a Tn*5* element introduced into a recipient cell on a transduced fragment. This conclusion is supported by the fact that a Tn*5* element present on an F′ plasmid inhibits Tn*5* transposition both to the chromosome and to the F′ plasmid (compare TT6147 and TT6148 in Table 1). (Hence, this cannot be a *cis*-acting mode of inhibition as is the case for Tn*A* [Robinson et al. 1977].)

A Genetic Map of Tn5

To determine what region of the Tn*5* element is required for production of the inhibitory effect, a variety of mutations affecting the Tn*5* element were isolated and used to construct a genetic map (Fig. 1). The mutant Tn*5* elements were placed in (*recA*$^-$) recipient cells

Table 1. Frequency of Tn*5* Transposition in Tn*5* (*neo*^{s}) Recipients

Recipient	*neo* allele	Relevant genotype	Donor Tn*5* insertion *hisG*::Tn*5*	*hisD*::Tn*5*
TT4630	—	without Tn*5*	1.0	1.0
TT4090	5	Tn*5* neo_{am}	0.01 ± 0.01	0.01 ± 0.01
TT4093	8	Tn*5* neo_{pt}	0.06 ± 0.04	0.05 ± 0.03
TT4150	15	Tn*5* ▽	0.93 ± 0.36	0.94 ± 0.41
TT4151	16	Tn*5* ▽	0.37 ± 0.16	0.34 ± 0.10
TT6147	—	F′ without Tn*5*	1.22 ± 0.46	1.29 ± 0.22
TT6148	5	F′ Tn*5* neo_{am}	0.10 ± 0.01	0.13 ± 0.02

Phage P22 grown on two Tn*5* insertion auxotrophs (*hisG9653*::Tn*5* in TT2677 and *hisD9644*::Tn*5* in TT2715) was used to transduce *recA1* recipients containing various mutant Tn*5* elements (*neo*^{s}) to kanamycin resistance. Frequencies of transposition of the donor Tn*5* elements are expressed relative to their transposition in a recipient that does not contain Tn*5* (TT4630). Transductions were performed as described by Biek and Roth (1980). To test whether Tn*5* transposition inhibition is a *trans* or a *cis* effect, strains containing an F′ *his* episome with or without Tn*5* present on it were employed (TT6147 and TT6148). All recipients have isogenic chromosomes, and the transposition frequencies listed above have not been corrected for slight differences in ability to plate phage P22 (which varies by no more than 10% between strains). The results represent mean and S.D. for four or more determinations for each cross (except for TT6147 and TT6148, for which only two determinations were made).

and tested for their abilities to inhibit transposition of a nonmutant donor Tn*5* element from transduced fragments. By correlating the map position of a Tn*5* mutation with its effect on the production of the inhibitory factor, it should be possible to identify the region of Tn*5* required for production of that factor.

The mutant Tn*5* elements include *neo* mutants selected for loss of the drug-resistance determinant. These mutants were isolated after localized mutagenesis (as described by Hong and Ames [1971]) and appear to carry point mutations, including nonsense and temperature-sensitive mutations, within the *neo* gene. Deletion mutants of Tn*5* were selected as derivatives of a Tn*5* element inserted in the *hisG* gene that have lost the polarity effect on the promoter-distal *hisD* gene. Most of these deletions result from fusion of the *hisD* transcript to a transcript originating within the Tn*5* element. A few deletions, including those extending to the left of Tn*5* in Figure 1, have probably removed sites within Tn*5* that normally prevent the *his*-specific transcript from crossing the inserted Tn*5* element (*hisG*::Tn*5*).

Localization of the Regulatory Determinant

When these mutant Tn*5* elements were checked for their abilities to produce active inhibitory factor, it was found that all Tn*5* elements with *neo* point mutations still produce inhibitor. The group tested included a Tn*10* insertion within the resistance gene (*neo*81::Tn*10*) as well as several nonsense mutations (*neo*5, *neo*6). These results suggest that the *neo* gene is not itself involved in transposition inhibition. Furthermore, since the *neo* Tn*10* insertion (*neo*81::Tn*10*) did not eliminate the inhibitory activity by virtue of this insertion's polar

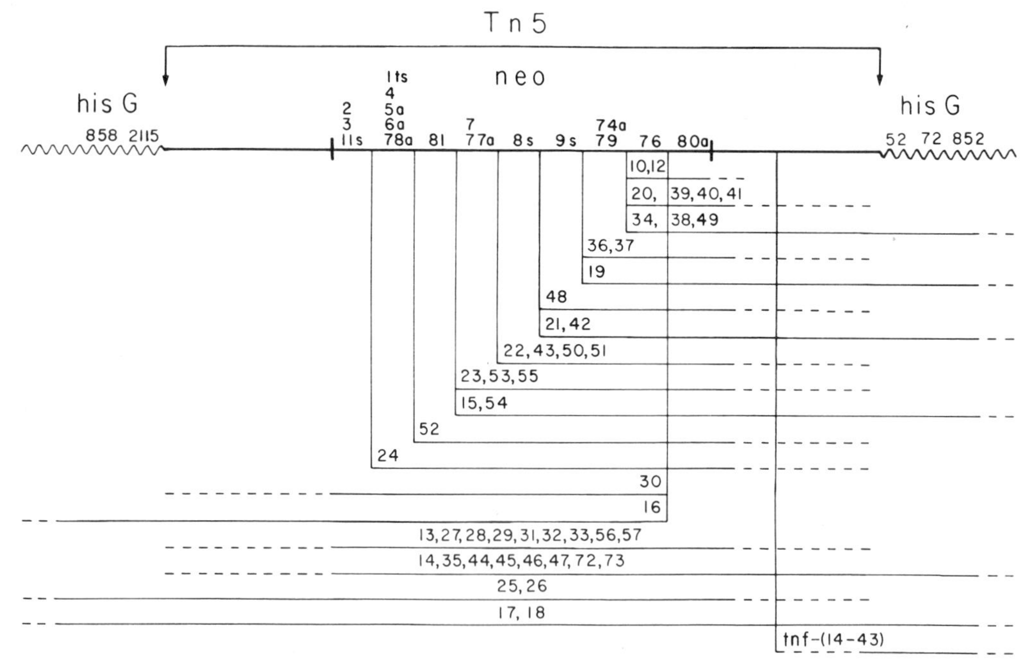

Figure 1. Deletion map of the Tn*5* gene (*neo*) conferring kanamycin-neomycin resistance. (———) Tn*5* sequences: (〰) adjacent chromosomal *hisG* sequences. Neither the position of the *neo* gene with respect to the entire element nor its size is drawn to scale. Refer to Biek and Roth (1980) for a detailed description of the isolation and characterization of *neo* mutants as well as of the construction of the genetic map. Where known, the types of mutations are indicated (*a* = amber, *ts* = temperature sensitive, and *s* = stable). Tn*5* mutants *neo*1–*neo*12 and *neo*81 (*neo*::Tn*10*) are able to produce the transposition inhibitory factor. (Strains containing Tn*5* *neo*74–*neo*80 have not yet been checked for this property.) The following *neo* deletions have lost the inhibitory activity, 13, 15–18, 20, 22–33, 36, 37, 39–41, 43, 48–50, 52, 53, 55–57, 72, 73, as have *tnf*14–*tnf*43.

nature, the *neo* transcript must not include sequences distal to the *neo* gene that provide message for the inhibitory factor. (The *neo* transcript starts immediately to the left of the *neo* gene and proceeds rightward across it as presented in Fig. 1 [modified from Rothstein et al. 1980].)

The behavior of deletion mutants was unexpected. All *neo* deletions tested (except the small deletions *neo*10 and *neo*12) have lost the ability to produce inhibitor. The deletions tested include those extending in either direction from the *neo* gene of Tn*5* into adjacent chromosomal (*hisG*) sequences. These deletions lack one or the other of Tn*5*'s inverted repeats, and they overlap within the *neo* gene (e.g., deletions *neo*15 and *neo*16). Since it seems unlikely that the region of overlap is critical to inhibition, we are led to the conclusion that sequences on both sides of the *neo* gene are required for inhibitor production. This conclusion is reinforced by the existence of other deletion pairs that extend to the left or right outside of Tn*5* but do not overlap (e.g., deletions *neo*16 and *tnf*14). Tn*5* elements containing either of these deletions lack inhibitory activity even though they affect noncontiguous parts of Tn*5*. Results such as these are most easily accounted for if sequences flanking both sides of the *neo* gene are essential for the production of the transposition inhibitor.

DISCUSSION

It is possible that the inhibitor will prove to be the transposase itself. Work by Rothstein et al. (1980) has demonstrated the involvement of both inverted-repeat sequences in the transposition of Tn*5*. Disruption of either sequence reduces (to different extents) the transposition ability of the element. Each of the two inverse repeats has been shown to encode (probably from overlapping genes) two different polypeptides. The left repeat (orientation of Tn*5* as shown in Fig. 1) encodes proteins with molecular weights of 53,000 and 49,000 (Postle and Reznikoff 1979; Rothstein et al. 1980). The right repeat encodes proteins with molecular weights of 58,000 and 54,000 (Rothstein et al. 1980). We propose that a complex of proteins from the left and right repeats may act as the inhibitor (or repressor) of Tn*5* transposition. This complex could prove to be identical with the transposase. If the inhibitor is the native transposase, the inhibitory effect would likely be produced by a mechanism affecting transposase synthesis. If the inhibitory complex is not the transposase, the inhibitory effect on transposition could be produced either by repression of transposition functions or by direct interaction with proteins or DNA sites critical to transposition.

ACKNOWLEDGMENTS

This work was supported by grant GM 27068 from the National Institutes of Health (to J. R.) and by grant 1/T32/GM 07464 from the U.S. Public Health Service (to D. B.).

REFERENCES

Berg, D. E., R. Jorgensen, and J. Davies. 1978. Transposable kanamycin-neomycin resistance determinants. In *Microbiology—1978* (ed. D. Schlessinger), p. 13. American Society for Microbiology, Washington, D.C.

Berg, D. E., J. Davies, B. Allet, and J. Rochaix. 1975. Transposition of R factor genes to bacteriophage λ. *Proc. Natl. Acad. Sci.* **72:** 3628.

Biek, D. and J. R. Roth. 1980. Regulation of Tn*5* transposition in *Salmonella typhimurium. Proc. Natl. Acad. Sci.* (in press).

Chou, J., M. J. Casadaban, P. G. Lemaux, and S. N. Cohen. 1979a. Identification and characterization of a self-regulated repressor of translocation of the Tn*3* element. *Proc. Natl. Acad. Sci.* **76:** 4020.

Chou, J., P. G. Lemaux, M. J. Casadaban, and S. N. Cohen. 1979b. Transposition protein of Tn*3*: Identification and characterization of an essential repressor-controlled gene product. *Nature* **282:** 801.

Gill, R. E., F. Heffron, and S. Falkow. 1979. Identification of the protein encoded by the transposable element Tn*3* which is required for its transposition. *Nature* **282:** 797.

Gill, R., F. Heffron, G. Dougan, and S. Falkow. 1978. Analysis of sequences transposed by complementation of two classes of transposition-deficient mutants of Tn*3*. *J. Bacteriol.* **136:** 742.

Heffron, F., M. So, and B. J. McCarthy. 1978. *In vitro* mutagenesis of a circular DNA molecule by using synthetic restriction sites. *Proc. Natl. Acad. Sci.* **75:** 6012.

Heffron, F., B. J. McCarthy, H. Ohtsubo, and E. Ohtsubo. 1979. DNA sequence analysis of the transposon Tn*3*: Three genes and three sites involved in transposition of Tn*3*. *Cell* **18:** 1153.

Hong, J. S. and B. N. Ames. 1971. Localized mutagenesis of any specific small region of the bacterial chromosome. *Proc. Natl. Acad. Sci.* **68:** 3158.

Jorgensen, R. A., S. J. Rothstein, and W. S. Reznikoff. 1979. A restriction enzyme cleavage map of Tn*5* and location of a region encoding neomycin resistance. *Mol. Gen. Genet.* **177:** 65.

Meyer, R., G. Boch, and J. Shapiro. 1979. Transposition of DNA inserted into deletions of the Tn*5* kanamycin resistance element. *Mol. Gen. Genet.* **171:** 7.

Postle, K. and W. S. Reznikoff. 1979. Identification of the *Escherichia coli tonB* gene product in minicells containing *tonB* hybrid plasmids. *J. Mol. Biol.* **131:** 619.

Robinson, M. K., P. M. Bennett, and M. H. Richmond. 1977. Inhibition of Tn*A* translocation by Tn*A*. *J. Bacteriol.* **129:** 407.

Rothstein, S. J., R. A. Jorgensen, K. Postle, and W. S. Reznikoff. 1980. The inverted repeats of Tn*5* are functionally different. *Cell* **19:** 795.

Mutants of *Escherichia coli* Affected in the Processes of Transposition and Genomic Rearrangements

G. B. Smirnov, T. S. Ilyina, Y. M. Romanova, A. P. Markov, and E. V. Nechaeva
Department of General Medical Microbiology, Institute of Epidemiology and Microbiology, N. F. Gamalei AMS USSR, Moscow 123098, USSR

Recombination associated with the transposable genetic elements results in (1) transposition of movable genetic elements themselves, (2) formation of different genomic rearrangements promoted by the elements, and (3) precise excision of the transposable elements from the sites at which they happen to insert. Little is known about the genetic control of these processes. Nevers et al. (1977) have isolated a chromosomal mutation that reduces the frequency of IS*1*-induced deletions. It has also been demonstrated that some Tn elements contain particular genes necessary for transposition and that presumably encode recombination enzymes (Heffron et al. 1977; Chou et al. 1979; Gill et al. 1979). The requirements for these genes for the processes that generate genomic rearrangements promoted by Tn elements have not been determined. Other Tn elements, e.g., Tn*9*, Tn*981*, and Tn*601*, possess only genes coding for drug resistance and flanking repeated sequences and lack genes necessary for transposition, at least in the internal coding part of the Tn element (Berg et al. 1978; Marcoli et al. 1980; Reif 1980; Rosner and Guyer 1980).

One can suggest that migration of the Tn elements of the latter type (and probably even of the former) is determined by the host chromosomal genes. This assumption is supported by the recent findings showing that several genes required for integration of phage λ into the *att*BOB′ site exist on the chromosome of *Escherichia coli* K12 (Miller and Friedman 1977; Williams et al. 1977; Miller et al. 1979).

On the basis of these arguments, we have performed a search of bacterial mutants having (1) altered ability to carry out transposition of Tn elements and (2) reduced frequency of formation of genomic rearrangements promoted by the Tn elements. The mutants have been isolated using two selective procedures. In this paper we present the genetic analysis and some properties of the mutants. The results obtained show that the transposition of a given Tn element and the genomic rearrangements promoted by the same Tn element are under the genetic control of different genes. Furthermore, the transpositions of different Tn elements share common steps when affected by the same chromosomal mutations. The same is true for the genomic rearrangements promoted by the different Tn elements.

The results summarized in this paper are presented in detail elsewhere (Ilyina et al. 1980a,b; Smirnov and Markov 1980).

RESULTS

Isolation of Mutants Deficient in Tn-element Migration or Genomic Rearrangements

Different schemes of selection were used for isolation of the mutants defective in *Tn*-element *m*igration (designation *tnm*) and in *g*enomic *r*earrangements (designated *ger*). The bacterial and phage strains used in the study are listed in Table 1.

Selection A. This selection was used for isolation of the mutants deficient in Tn*9* transposition. We have shown previously that Tn*9*, being integrated in one particular chromosomal site called *att*Tn9A, is very unstable and capable of transposition into the genome of phage λ or P1*vir* with a very high frequency. These phages, grown lytically on the host carrying Tn*9* in the *att*Tn*9*A site, were able to "transduce" Tn*9* into the recipient genome with frequencies of about 1×10^{-4} to 4×10^{-4} per infecting particle (Ilyina et al. 1978, 1980c). To select mutants deficient in transposition of Tn*9*, phage P1 grown on the *E. coli* strain KS7201 carrying Tn*9* in the site *att*Tn*9*A was spread onto the surface of nutrient agar containing chloramphenicol (Cm). Drops of suspensions of individual bacterial clones that survived mutagen treatment were spotted into the marked areas of the same plate (for details, see Ilyina et al. 1980a). Nonmutant clones formed a considerable number of Cm-resistant (Cm^r) colonies in the spot, whereas the zones of mutant clones were empty. Further quantitative assays of transposition ability were performed that took into account the exact number of bacteria and infecting phage. The differences in transposition frequencies between mutants and wild-type bacteria were more than 4 orders of magnitude.

Five tnm^- mutations that mapped at the same definite position on the *E. coli* chromosome were isolated by this selection procedure.

Selection B. This procedure selects for temperature-resistant (T^r) survivals of lysogens that carry prophage λ*att*80*c*I857*S*7. It was shown that the formation of T^r survivors in the population lysogenic for λ*c*I857 was associated with deletions (Neubauer 1970; Marchelli et al. 1976). It seems that other genomic rearrangements could also be involved. Deletions and inversions are readily induced by Tn elements (see Kleckner et al. 1979; Ross et al. 1979; Shapiro and MacHattie 1979). We have compared the frequency of T^r survivors in the

Table 1. Strains of *E. coli* K12 and Bacteriophages

Strain	Relevant character	Source[a]
	(*a*) E. coli	
AB259	HfrH; *thi*	W. Hayes
K10	HfrC; *ilv relA*	L. Cavalli
Ra-2	Hfr; *sfa*4 *sfa*5	B. K. Low[b]
C600	F$^-$; *thr leu thi supE tonA*	lab. collection
BT1000	F$^-$; *polA met thy strA*	W. Goebel[c]
AB1157	F$^-$; *thr leu proA2 his thi argE lac gal ara xyl mtl tsx strA*	P. Howard-Flanders[d]
GC146	F$^-$; *pro his metA malB lac gal strA*	M. Castellazzi[e]
KS874	as GC146, but *recA*	lab. collection
KS707	*proA*$^+$ transductant of AB1157	lab. collection
KS706	as KS707, but *recA*	lab. collection
KS837	F$^-$; *ara leu tonA lacZ proC tsx trp lys mtl metE thy strA*	lab. collection
KS8372	*tnm-2 dna*$_{ts}$ *gerA* mutant of KS837	
KS8373	as KS8372, but *recA*	
KS8374	T^r revertant of KS8372	
KS7071	*tnm*1 mutant of KS707	
KS7072	as KS7071, but *recA*	
KS7073	*tnm*3 mutant of KS707	
KS7074	*tnm*4 mutant of KS707	
KS7075	*tnm*5 mutant of KS707	
KS7076	as KS7073, but *recA*	
KS7077	as KS7075, but *recA*	
KS898	AB1157(λ*att*80*c*I857*S*7[Tn*9*])	
KS1016	*gerB* mutant of KS898	
KS1013	*gerC* mutant of KS898	
KS1101	as KS898, but possesses Tn*10* in prophage	
KS1102	as KS1016, but possesses Tn*10* in prophage	
KS1055	*gerC* mutant of KS898	
KS1117	as KS898, but possesses Tn*601* in prophage	
KS1119	as KS1016, but possesses Tn*601* in prophage	
KS1160	as KS1013, but possesses Tn*601* in prophage	
KS1161	as KS1013, but possesses Tn*10* in prophage	
KS1162	as KS1055, but possesses Tn*601* in prophage	
KS1163	as KS1055, but possesses Tn*10* in prophage	
KS1401	Hly152 (Tn*3*)/*tnm*1 *pro his strA* (λ*att*80*c*I857*S*7 [Tn*9*])	
KS914	*metA*$^+$ *gerA* transductant of GC146	
KS1331	GC146(λ*b*221*c*I857 [Tn*601*])	
KS1332	KS914(λ*b*221*c*I857 [Tn*601*])	
KS1333	GC146(λ*att*80*c*I857*S*7 [Tn*10*])	
KS1334	KS914(λ*att*80*c*I857*S*7 [Tn*10*])	
KS1335	GC146(λ*att*80*c*I857*S*7)	
KS1336	KS914(λ*att*80*c*I857*S*7)	
KS918	GC146(λ*att*80*cI857S7* [Tn*9*])	
KS916	KS914(λ*att*80*c*I857*S*7 [Tn*9*])	
KS919	GC146(λ*att*80*c*I857*S*7 [Tn*601*])	
KS920	KS914(λ*att*80*c*I857*S*7 [Tn*601*])	
KS1337	pEN25 Tc Km Ap Mu*cts metA malB tnm*5 Mu*cts*/KS874	
KS1338	pEN25/GC146, but *malB*$^+$ *tnm*1	
KS1339	pEN25/KS874, but *malB*$^+$ *tnm*2	
KS1340	pEN25/KS7076	
KS1341	pEN25/KS7077	
KS1342	KS1339, cured of plasmid	
	(*b*) *Bacteriophages*	
P1*vir*	*virA*	lab. collection
P1*vir* (Tn*9*)	*virA cat*	
λ*att*80 (Tn*9*)	*att*80*c*I857*S*7*cat*	
λ*b*221 (Tn*3*)	*b*221*c*I857*bla*	
λ*b*221 (Tn*601*)	*b*221*c*I857*aphA*	
λ*b*221 (Tn*10*)	*b*221*c*I857*N*$^-$*rex*173::Tn*10*	N. Kleckner[f]
λ*b*221 (Tn*5*)	*b*221*c*I857*aph*	D. Berg[g]
λ*imm*^{21} (Tn*10*)	*imm*^{21}*tet*	S. Syneoky[h]

[a] Unless otherwise indicated, the source is this paper.
[b] Yale University, New Haven, Conn.
[c] Institut für Genetik und Microbiologie, Universitat Wurzburg, F.R.G.
[d] Yale University, School of Medicine, New Haven, Conn.
[e] Institut de Biologie Moleculaire, Paris, France.
[f] Harvard University, Cambridge, Mass.
[g] Washington University, St. Louis, Mo.
[h] Institute of Genetics of Industrial Microorganisms, Moscow, U.S.S.R.

populations of two strains: KS902, lysogenic for phage λ*att*80*c*I857*S*7, and KS898, lysogenic for phage λ*att*80*c*I857*S*7 (Tn*9*). The strains were isogenic for all markers except for Tn*9* in the prophage in strain KS898. Strain KS898 produces T^r survivors with a frequency of about 2×10^{-4}, whereas T^r derivatives of strain KS902 were formed with 50 times lower frequency. Thus, the predominant fraction of events leading to the production of T^r survivors in the population of strain KS898 were Tn*9*-dependent.

To isolate mutants deficient in the formation of genomic rearrangements, bacterial suspensions of the individual clones of strain KS898 that survived after mutagenic treatment were dropped into marked areas on the

Table 2. General Properties of *tnm*⁻ and *ger*⁻ Mutants

Strain	Phenotype		Frequency of lysogenization (%) by phage		RecA phenotype[a]	Sensitivity to phage Mu-1[b]	Temperature sensitivity (42 °C)
	Tnm	Ger	P1*kc* (Tn9)	λ*att*80 (Tn9)			
KS707	+	+	3.0	28	+	s	r
KS837	+	+	4.0	24	+	s	r
KS7071	−	+	5.1	18.4	+	r	r
KS8372	−	−	4.9	26.3	+	r	s
KS7073	−	+	0.9	18.8	+	s	r
KS7074	−	+	2.3	27.6	+	r	r
KS7075	−	+	1.0	20.5	+	s	r
KS898	+	+	n.t.	n.t.	+	s	s
KS1016	+	−	n.t.	n.t.	+	s	s
KS1013	+	−	n.t.	n.t.	+	s	s
KS1055	+	−	n.t.	n.t.	+	n.t.	s

[a] The RecA phenotype was determined by measuring the frequency of transconjugants in the crosses with Hfr K10 and the frequency of phage P1 transductants. + indicates proficiency.
[b] s indicates sensitive, r indicates resistant, and n.t. indicates not tested.

surfaces of two nutrient agar plates. One plate was incubated at 42 °C; the other was incubated at 32 °C. Those clones that formed fewer colonies in the drop areas at 42 °C than formed on an untreated control plate were tested further. Eight *ger*⁻ mutants were isolated by this method.

General Properties of Mutant Strains

The general properties of the mutant strains isolated are presented in Table 2. We have found that, in general, *recA*-dependent recombination was not changed in the mutants. The *tnm*⁻ and *ger*⁻ mutants tested were unaffected in their ability to be lysogenized by phage λ (Table 2) and P1*clr cml*100 and in their ability to be transformed with DNAs of the plasmids Rsc11 and ColE1::Tn*3* (data not presented). Efficiency of plating (EOP) of phages P1 and λimm^{21} was not affected by either the *tnm*⁻ or *ger*⁻ mutation. (EOP of λimm^{21} on the KS1016 host was 0.001; however, mapping data indicated that reduction of EOP had nothing to do with the *ger*⁻ defect of this strain.) These data show that the *tnm*⁻ and *ger*⁻ mutations did not alter site-specific recombination of phage λ and argue against the possibility of abnormal DNA breakdown, which destroys incoming donor DNA in the mutant cells. Some of the *tnm*⁻ mutants were unable to propagate phage Mu-1. The nature of their resistance to Mu is unknown.

Strain KS8372 possesses both the Tnm⁻ and Ger⁻ phenotypes and carries a mutation that confers temperature sensitivity (T^s). Strain KS1016, in addition to having the properties indicated in Table 2, was unable to produce mature plaque-forming phage after temperature or UV induction.

Genetic Mapping

Mutants isolated using selection-A procedure. All five mutations found in the mutants isolated using the selection-A procedure described earlier were mapped in the same chromosomal region. Therefore, we will describe mapping experiments with one representative: strain KS8372 *tnm*2, which was T^s in addition to having the Tnm⁻ and Ger⁻ phenotypes. *tnm*2 and *ts* were found to be different mutant alleles, since revertants to the T^r phenotype always retained the Tnm⁻ property. However, the *ts* and *tnm*2 mutations were very tightly linked: More than 99% of the T^r clones obtained after P1*vir* transduction of the *ts*⁻ *tnm*2 strain KS8372 were also Tnm⁺. Approximately 70% of Leu⁺ [Str^r] transconjugants obtained in the cross Hfr Ra-2 × KS8374 were T^r and Tnm⁺. The same linkage of *tnm*2 and *ts* to *metE* was observed in the cross Hfr K10 × KS8372. These data placed *ts* and *tnm*2 mutations in the interval between *metE* and *leu* (Fig. 1).

Accordingly, *ts* and *tnm*2 mutations were found to be cotransducible with *metA* in 3.5% of the cases and with *malB* in 46.5%. The linkage of *metA* with *malB* was higher than that of *metA* with *tnm*2 (*ts*); at the same time, *tnm*2 (*ts*) was more closely linked to *malB* than to *metA*. In all cases, with only one exception, the *tnm*2 and *ts* mutations were cotransducible. Of 160 *malB*⁺ transductants obtained in the cross P1 [*malB*⁺ *ts*⁻ *tnm*2] × GC146 *malB*⁻ *ts*⁺ *tnm*⁺, one had the phenotype Mal⁺ T^s Tnm⁺, which was in agreement with the gene order *malB-ts-tnm*. Taken together, the results of transductional mapping argue in favor of the gene order *metA-malB-ts-tnm*2.

The *ger*⁻ mutation of strain KS8372 (designated *ger*2) was closely linked to the *tnm*2 mutation. Phage P1

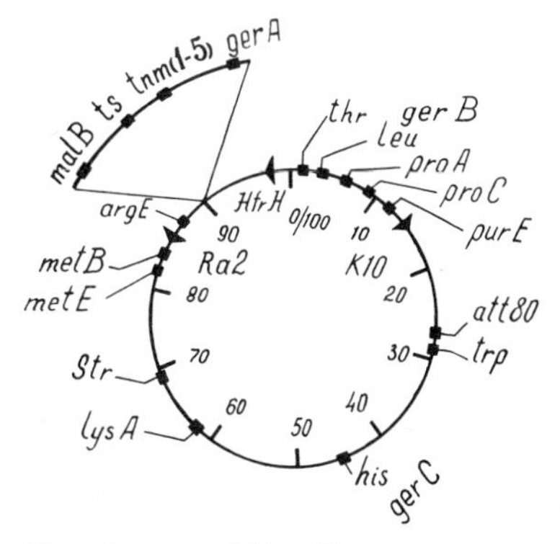

Figure 1. Genetic map of *E. coli*.

grown on the *metA*$^+$ *tnm*2 *ger*2 strain KS8374 was used to transduce *metA*$^-$ strain GC146 to Met$^+$. Among the 51 MetA$^+$ transductants, 4 were Tnm$^-$ and 1 was Tnm$^-$ and Ger$^-$.

To determine whether the *tnm*$^-$ mutations were recessive or dominant in the tnm$^+$/tnm$^-$ merodiploids, the F′ plasmids F112 and F118, which carry the *tnm*$^+$ allele, were introduced into the *tnm*1 and *tnm*2 mutants. Both merodiploids were Tnm$^+$, indicating the recessiveness of the *tnm*1 and *tnm*2 alleles.

To determine whether the *tnm*$^-$ mutations isolated belonged to one or more complementation groups, complementation analysis of four *tnm*$^-$ mutations was carried out. The plasmid RP4-Mu*cts* was introduced into the *metA*$^+$ *tnm*5 transductant of strain GC146. After partial thermoinduction of the Mu*cts* prophage (Faelen and Toussaint 1976), this strain was crossed with the *metA*$^-$ recipient. Selected Met$^+$ transconjugants possessed complex plasmids consisting of the RP4 genome and chromosomal markers *argE*$^+$, *metA*$^+$, *malB*, and *tnm*5, presumably surrounded by the duplicated Mu genomes. One such plasmid was crossed into the *metA*$^-$ *tnm*1 and *metA*$^-$ *tnm*2 strains using *metA*$^+$ as a selective marker. The frequencies of Tn*9* transposition from the phage P1 genome into the chromosome, and probably into the plasmid of merodiploids, were measured (Table 3). One can see that the *tnm*2 and *tnm*5 mutations belong to different complementation groups, whereas the *tnm*3 mutations affect the same complementation group as the *tnm*5 mutation.

Mutants isolated using selection B. Here, we present mapping and some properties of three of eight *ger*$^-$ mutants isolated by the procedure described earlier as selection B.

To determine the relationship of the Ger$^-$ phenotype of strain KS1016 to the prophage λ*att*80 and/or to the Tn*9* inserted into this prophage, we have transduced prophage λ*att*80*c*I857*S*7 (Tn*9*) by phage P1 to strain AB1157, selecting for Cm^r progeny. All 32 Cm^r λ^r T^s transductants obtained were indistinguishable from the original lysogenic strain KS898, i.e., Ger$^+$. Thus, the *ger*$^-$ mutation of strain KS1016 was located outside the prophage. To map this mutation, strain KS1016 was crossed with the donor strain Hfr K10 (see Fig. 1). Prototrophic recombinants of different classes were tested for their Ger phenotype. The following linkage of donor *ger*$^+$ marker to the selected markers was found: *proA*, 1/47 (2%); *thr*$^+$ *leu*$^+$, 43/46 (93%); *argE*$^+$, 16/46 (35%); and *his*$^+$, 2/46 (4%). These data placed the *ger*$^-$ mutation in the vicinity of the *thr* and *leu* genes. When strain KS1016 was crossed with the Hfr H donor, 75% of the Thr$^+$ Leu$^+$ transconjugants were Ger$^+$, indicating the presence of the *ger*$^-$ mutation in the region corresponding to the leading end of the Hfr H chromosome.

Thus, the *ger*$^-$ mutations of strains KS8374 and KS1016 were mapped at different locations and therefore affect the functions of different genes. We have named the gene affected in strain KS8374 *gerA* and the gene affected in strain KS1016 *gerB*. Transductional mapping showed 7% linkage betweeen *gerB1* and *leu*, which is in agreement with the conjugation mapping data.

In the conjugal cross of Hfr K10 with another *ger*$^-$ strain, KS1013, 28% of His$^+$ progeny were Ger$^+$. In a similar cross with the recipient KS1055, 50% linkage between *ger* and *his* markers was found. No linkage of *ger* to other recipient markers was found in either case. Thus, the *ger*$^-$ alleles of strains KS1013 and KS1055 seem to locate in the *his* region of the bacterial chromosome (Fig. 1). The symbol *gerC* is used for the gene mapped in this region.

Table 3. Complementation of *tnm* Alleles

Strain	*tnm* genotype	Tn*9* transposition frequency[a]
KS1337	*tnm*5/*tnm*$^+$	3.0×10^{-6}
KS1341	*tnm*5/*tnm*5	$<10^{-8}$
KS1339	*tnm*5/*tnm*2	7.5×10^{-6}
KS1340	*tnm*5/*tnm*3	$<10^{-8}$
KS1342[b]	*tnm*2	$<10^{-8}$

[a] Quantitative assay of Tn*9* transposition from phage P1 was carried out as described in the notes to Table 4.

[b] To prove the presence of the *tnm*2 mutation in the chromosome of KS1339, the strain was cured of the pEN25 plasmid, and the resulting derivative was designated KS1342 and was found to be Tnm$^-$.

Quantitative Studies on Transposition

Two methods were used to study the effects of *tnm*$^-$ mutations on transposition frequency. According to the first method, transposition of Tn elements from the infecting phage genome to the bacterial chromosome was measured. The second approach was to score the transposition of Tn elements from the nonconjugative, nonmobilizable plasmid to the conjugative plasmid. The results of these experiments are presented in Tables 4 and 5.

Transposition from the infecting phage genome to the bacterial chromosome. Transpositions of Tn*3*, Tn*9*, and Tn*10* were *recA*-independent and gave rise to T^r transposants, whereas the transposition of Tn*601* was partially *recA*-dependent (Table 4). Moreover, kanamycin-resistant (Km^r) clones obtained after infection of the recipients with λ*b*221*c*I857*S*7 (Tn*601*) were always T^s and λ^r, indicating that the integration of the λ genome carrying Tn*601* into the bacterial chromosome had occurred. The integration process probably used Tn*601* as an analog of an attachment site for integration of the λ genome, which lacks its own *att* site. This result is in agreement with earlier observations on the particular features of Tn*601* integration (Berg et al. 1978).

The *tnm*1 and *tnm*2 mutations blocked almost completely the transposition of Tn*9* and caused a sharp decrease in the frequency of Tn*3*, Tn*10*, and Tn*601* transposition. Ampicillin-resistant (Ap^r) clones appeared with low frequency after infection of the *tnm*1 or *tnm*2 mutants with λ*b*221*c*I857*S*7 (Tn*3*), were highly unstable, and were ultimately lost during subsequent purification on the selective medium. Thus, even those rare

Table 4. Effect of *tnm*⁻ Mutations on Transposition of Different Tn Elements from the Infecting Phages to the Bacterial Chromosome

Bacterial strain	Relevant genotype	Transposition frequency[a] Tn*9*	Tn*3*	Tn*10*	Tn*601*
KS707	*tnm*$^+$	1.5×10^{-4}	7.9×10^{-3}	2.0×10^{-5}	1.0×10^{-2}
KS706	*tnm*$^+$ *recA*$^-$	1.1×10^{-4}	n.t.	1.6×10^{-5}	3.4×10^{-3}
KS7071	*tnm*1	$<1.0 \times 10^{-9}$	5.0×10^{-8}	1.0×10^{-8}	4.0×10^{-5}
KS7072	*tnm*1 *recA*$^-$	$<1.0 \times 10^{-9}$	n.t.	1.0×10^{-8}	1.0×10^{-8}
KS8372	*tnm*2 *ts*	$<1.0 \times 10^{-9}$	5.0×10^{-8}	1.4×10^{-7}	8.0×10^{-5}
KS8373	*tnm*2 *ts recA*$^-$	$<1.0 \times 10^{-9}$	n.t.	1.0×10^{-7}	1.6×10^{-6}
KS8374	*tnm*2 *ts*$^+$	$<1.0 \times 10^{-9}$	5.0×10^{-8}	1.0×10^{-7}	4.0×10^{-5}
KS7073	*tnm*3	$<1.0 \times 10^{-9}$	5.9×10^{-3}	2.2×10^{-5}	1.6×10^{-2}
KS7076	*tnm*3 *recA*$^-$	$<1.0 \times 10^{-9}$	3.1×10^{-3}	1.1×10^{-5}	6.3×10^{-3}
KS7074	*tnm*4	$<1.0 \times 10^{-9}$	4.2×10^{-3}	7.0×10^{-7}	1.9×10^{-4}
KS7075	*tnm*5	1.0×10^{-6}	9.3×10^{-3}	5.4×10^{-6}	2.5×10^{-3}
KS7075	*tnm*5 *recA*$^-$	1.0×10^{-6}	1.5×10^{-3}	n.t.	n.t.

The strains were grown to a density of 5×10^8 to 8×10^8/ml, washed, and resuspended in solution containing 0.85% NaCl, 2.5×10^{-3} M $CaCl_2$, and 1×10^{-2} M $MgSo_4$. Bacteria were infected with the following phages: P1 grown on strain KS7201 for transposition of Tn*9*, λ*b*221 (Tn*3*), λ*b*221 (Tn*601*), and λ*b*221 (Tn*10*). Adsorption times were 30 min for P1 and λ (Tn*3*) and 60 min for the remaining phages. Suspensions were plated in appropriate dilutions on the selective media containing antibiotics required for selection. The plates were incubated at 32 °C.

[a] The frequency of transposition was calculated as the ratio of the number of drug-resistant clones/ml to the titer of plaque-forming infecting phages.

Ap^r clones formed in *tnm*1 or *tnm*2 mutants could not be attributed to the true transposants. Unlike the *tnm*1 and *tnm*2 mutations, the *tnm*3 mutation only affected transposition of Tn*9* (Table 4).

Transposition from the nonconjugative plasmids to the conjugative plasmid genome. In contrast to the data presented in Table 4, mutation *tnm*1 had no effect on transposition of Tn*3* from nonconjugative plasmids to the R1*drd*16 plasmid (Table 5). The donor strains used in these experiments retained the *tnm*1 allele, since they still exhibited the Tnm⁻ phenotype in the test with infecting phage P1*vir* (Tn*9*). Thus, the *tnm*1 mutation abolished transposition of Tn*9*, Tn*10*, Tn*3*, and Tn*601* from the infecting phage to the bacterial chromosome, but it had no effect on the transposition of Tn*3* from nonconjugative plasmids to the conjugative plasmid.

Effect of *ger*⁻ Mutations on Genomic Rearrangements

The ability of bacteria lysogenic for λ*att*80cI857*S*7, which carries a particular Tn element, to form T^r derivatives was taken as a measure of genomic rearrangements. The effects of different *ger*⁻ mutations on the frequency of T^r derivatives are demonstrated in Table 6. Neither of the *tnm*⁻ mutations influenced the formation of T^r derivatives. Mutations *gerA2*, *gerB1*, and *gerC*, located in different regions of the bacterial chromosome, affect the formation of T^r survivors to different extents. It is of interest that the frequency of T^r derivatives in the lysogenic populations carrying prophages with Tn elements other than Tn*9* was also affected by the *ger*⁻ mutations tested.

DISCUSSION

In this paper we have presented the results of a search for bacterial mutants defective in Tn-element transposition or Tn-element-promoted genomic rearrangements, which was undertaken in an effort to define the genetic control and interrelationship of these processes. Mutations affecting transposition of Tn*9*, Tn*10*, Tn*601*, and Tn*3* were mapped in a definite region of the bacterial chromosome close to the *malB* locus. Mutations affecting genomic rearrangements promoted by Tn*9*, Tn*10*, and Tn*601* were mapped at different chromosomal locations: *gerA* was linked to *malB* and *gerB* was linked to *leu* and *gerC* in the *his*⁻ region.

The existence of chromosomal genes necessary for the transposition of Tn elements and for genomic rearrangements promoted by Tn elements is analogous to the existence of host genes required for site-specific recombination of λ (see Miller et al. 1979). Thus, one

Table 5. Transposition of Tn*3* from Plasmid ColE1::Tn*3* or Rsc11::Tn*3* to Plasmid R1*drd*16 in *tnm*$^+$ and *tnm*1 Strains

Cross no.	Donor	Recipient	Tn*3* transposition frequency
1	KS870 *tnm*$^+$ ColE1::Tn*3*, R1*drd*16	BT1000 *polA*$^-$	10^{-2}
2	KS871 *tnm*1 ColE1::Tn*3*, R1*drd*16	BT1000 *polA*$^-$	7.3×10^{-3}
3	KS872 *tnm*$^+$ Rsc11::Tn*3*, R1*drd*16	KS707	5.5×10^{-2}
4	KS873 *tnm*1 Rsc11::Tn*3*, R1*drd*16	KS707	4.2×10^{-2}

To measure the frequency of Tn*3* transposition from ColE1 to R1*drd*16 (crosses 1 and 2), donor strains that carry both plasmids were crossed with a *polA* recipient unable to maintain ColE1. The donor strains KS872 and KS873 (crosses 3 and 4) possess nonconjugative *mob*⁻ plasmid Rsc11::Tn*3*. Ap^r progeny in crosses could appear, therefore, only as a result of transposition of Tn*3* to R1*drd*16. The ratio of Ap^r progeny to Km^r progeny was taken as a measure of transposition frequency. The protocol for these experiments was suggested by Dr. W. Goebel (Institut für Genetik und Microbiologie, Universitat Wurzburg, F.R.G.), in whose laboratory the donor strains were constructed.

Table 6. Effect of *ger*⁻ Mutations on the Frequency of T^r Clones Formed in the Lysogenic Populations That Carry Prophage *att*80*c*I857 with Different Tn Elements

Bacterial strain	*ger* allele	Frequency of T^r survivors[a]		
		Tn*9*	Tn*10*	Tn*601*[b]
KS918	+	10^{-4}	—	—
KS916	*A2*	$<5.0 \times 10^{-9}$	—	—
KS1333	+	—	3.0×10^{-4}	—
KS1334	*A2*	—	$<2.0 \times 10^{-9}$	—
KS919	+	—	—	1.2×10^{-5}
KS920	*A2*	—	—	$<5.0 \times 10^{-10}$
KS898	+	2.0×10^{-4}	—	—
KS1016	*B1*	5.2×10^{-6}	—	—
KS1013	*C3*	6.0×10^{-8}	—	—
KS1055	*C4*	$<7.0 \times 10^{-7}$	—	—
KS1101	+	—	5.0×10^{-6}	—
KS1102	*B1*	—	$<2.0 \times 10^{-7}$	—
KS1161	*C3*	—	$<5.8 \times 10^{-7}$	—
KS1163	*C4*	—	$<2.7 \times 10^{-7}$	—
KS1117	+	—	—	3.0×10^{-3}
KS1119	*B1*	—	—	1.6×10^{-6}
KS1160	*C3*	—	—	$<4.2 \times 10^{-9}$
KS1162	*C4*	—	—	8.7×10^{-7}

Bacteria were grown overnight at 32 °C in Lennox (1955) broth containing the appropriate antibiotic, diluted 1:10 in fresh broth without drugs, and incubated 2 hr at 32 °C with shaking. Suspensions were plated with and without dilution on the brain-heart-infusion agar plates in duplicate. Plates seeded with diluted suspensions were incubated at 32 °C for determination of viable count. Colonies were counted after 48 hr of incubation at 43 °C and 32 °C.

[a] Frequency of T^r derivatives was determined as the ratio of viable counts at 43 °C to that at 32 °C.

[b] Transposable elements inserted into the prophage λ*att*80*c*I857 of a particular strain are indicated.

could not consider Tn elements to be self-integrating, independent travelers along the target replicons. We were faced with the situation that neither of the *tnm*⁻ mutations affected genomic rearrangements promoted by the Tn element whose transposition was blocked in the mutants, and that neither of the *ger*⁻ mutations led to the drastic reduction in the transposition frequency. The data show that the genes controlling transposition are different from the genes controlling genomic rearrangements. However, the existence of common genes is not excluded. Transposition and genomic rearrangements could be the results of two different recombination processes that lack common steps or possess them. If the latter is true, we have isolated mutations affecting specific steps only.

First, the mutation affecting genomic rearrangements promoted by transposable element IS*1* was isolated by Nevers et al. (1977). The *del*⁻ mutation was mapped close to *lysA*, ouside the regions where the mutations described in this paper were mapped (Nevers and Saedler 1978). The *del*⁻ mutation and the *ger*⁻ mutations have no effect on Tn*9* transposition from the phage P1 genome to the bacterial chromosome. However, in contrast to the *ger*⁻ mutations, the *del*⁻ mutation did not affect the frequency of T^r derivatives in the population lysogenic for λ*c*I857*S7* (Tn*9*) (P. Nevers, pers. comm.; T. S. Ilyina, unpubl.). Thus, the *del*⁻ mutation alters even another recombination step associated with IS*1*.

The idea about the similarity of transposition and genomic rearrangements promoted by Tn elements has now become widely distributed. Weinstock and Botstein (1979), taking into account the regional specificity characteristic of both insertion of Tn*1* and deletions promoted by Tn*1*, conclude that the processes "share some common mechanistic feature." Similarly, Noel and Ames (1978) had found the site within the *Salmonella* gene *hisJ* that was highly preferred for both Tn*10* insertion and recombination producing deletions. Kleckner and Ross (1979) have also discussed the similarity between insertion of Tn*10* and Tn*10*-promoted rearrangements; however, they have pointed out that the first process utilizes outer ends, whereas the second utilizes both ends of insertion sequences that flank the Tn element. Finally, Ohtsubo et al. (1979) proposed a model that presumed the transposition of the Tn element as a step in Tn-induced deletion formation. As already mentioned, the mutations isolated in the course of this study affect *either* transposition *or* genomic rearrangements. As soon as transposition is believed to be a complex multistep process, then the *tnm*⁻ mutations studied may affect the steps specific for transposition but not for genomic rearrangements. These steps have nothing to do with restriction or other phenomena that lead to DNA breakdown of donor DNA, since the *tnm*⁻ mutants, which exhibited unchanged efficiency of plating phages λ and P1, could be easily lysogenized with phage λ or transformed with plasmid DNA. The data available do not allow us to draw the final picture, but they do indicate at least the existence of different steps in the processes of transposition and genomic rearrangements.

Mutations that affect transposition (isolated in our experiments) fall into two different classes. The representatives of the first class (mutations *tnm*1 and *tnm*2) cause almost complete block of Tn*9* transposition from the phage P1 genome to the bacterial chromosome and reduce significantly the frequency of transposition of Tn*3*, Tn*10*, Tn*601* (Table 4), and Tn*5* (data not presented) tested in the same system. The *tnm*3 mutation belongs to another class, since it specifically abolished Tn*9* transposition. Therefore, the transposition of Tn*9*, on the one hand, and Tn*3*, Tn*5*, Tn*10*, and Tn*601*, on the other, share common features affected by *tnm*1 and *tnm*2 mutations, as well as some specific features. One of the latter, specific for Tn*9* transposition, is affected by the *tnm*3 mutation. It is interesting that the transposable elements Tn*3*, Tn*5*, Tn*9*, and Tn*10*, all restricted in their migration by the same mutations, *tnm*1 and *tnm*2, differ in their mode of insertion. Namely, Tn*3* was found to be bracketed by direct repeats of five target base pairs, whereas the other Tn elements mentioned were found to be bracketed by nine target base pairs (Johnsrud et al. 1978; Kleckner 1979; Ohtsubo et al. 1979).

The features of transposition of Tn*3* common to those of Tn*9*, affected by the *tnm*1 mutation, appeared to be nonessential when transposition from the nonconjugative plasmid to the conjugative plasmid, rather than from the infecting phage genome to the chromosome, took place. The nature of this phenomenon is obscure. Recently, the models for transposition were created, showing the complexity of the transposition process and predicting the formation of joint replicons as an intermediate step of transposition (Gill et al. 1978; Shapiro 1979). Integration of phage λ, directed by Tn*9* instead of its own Int system, also required some additional intermediate steps, e.g., λ (Tn*9*) dimer formation (MacHattie and Shapiro 1978; see also Kleckner and Ross 1979). One can speculate that the degree of requirements in some additional steps of transposition is different, depending on the type and state of donor and/or recipient structures.

The *ger*$^-$ mutations caused reduction or complete block of genomic rearrangements promoted by Tn*9*, Tn*10*, and Tn*601* and led to the formation of T^r survivors in populations lysogenic for λ*att*80*c*I857. It seems, therefore, that the common steps of rearrangements promoted by different Tn elements exist irrespective of some differences in the structure of their ends. The next step of this work will consist of studying the effects of *ger*$^-$ mutations on genomic rearrangements independent of Tn and IS elements.

Genetic-mapping data show that transposition and genomic rearrangements are under the control of different genes. However, close linkage between the group of *tnm*$^-$ mutations and one of the *ger*$^-$ mutations studied argue in favor of some interrelationship of these processes. The *tnm*$^-$ mutations of this cluster were found to be tightly linked to the novel mutation affecting DNA replication (Ilyina et al. 1980b). The very close linkage between the *tnm*$^-$ and the dna_{ts} mutations raises the possibility that they might be mutations in the same structural gene. Thus, very probably, the processes of transposition and genomic rearrangements are related to the processes involved in DNA replication.

ACKNOWLEDGMENTS

We thank Dr. Goebel for bacterial strains and for suggestions of some experiments, Dr. Saedler for helpful discussions, and Dr. Nevers for generously providing us with unpublished data.

REFERENCES

Berg, D. E., R. Jorgensen, and J. Davies. 1978. Transposable kanamycin-neomycin resistance determinants. In *Microbiology—1978* (ed. D. Schlessinger), p. 13. American Society for Microbiology, Washington, D.C.

Chou, J., P. G. Lemaux, M. J. Casadaban, and S. N. Cohen. 1979. Transposition protein of Tn*3*: Identification and characterization of an essential repressor-controlled gene product. *Nature* **282:** 801.

Faelen, M. and A. Toussaint. 1976. Bacteriophage Mu-1: A tool to transpose and to localize bacterial genes. *J. Mol. Biol.* **104:** 525.

Gill, R., F. Heffron, and S. Falkow. 1979. Identification of the protein encoded by the transposable element Tn*3* which is required for its transposition. *Nature* **282:** 797.

Gill, R., F. Heffron, G. Dougan, and S. Falkow. 1978. Analysis of sequences transposed by complementation of two classes of transposition-deficient mutants of Tn*3*. *J. Bacteriol.* **136:** 742.

Heffron, F., P. Bedinger, J., Champoux, and S. Falkow. 1977. Deletions affecting transposition of an antiobiotic resistance gene. In *DNA insertion elements, plasmids, and episomes* (ed. A. I. Bukhari et al.), p. 151. Cold Spring Harbor Laboratory, Cold Spring Harbor, New York.

Ilyina, T. S., E. V. Nechaeva, and G. B. Smirnov. 1980a. The method of isolation and initial characterization of bacterial mutants with altered ability to transpose Tn*9*. *Genetika* **16:** 957.

Ilyina, T. S., E. V. Nechaeva, N. G. Belikov, and G. B. Smirnov. 1978. The specificity of insertion of the transposon controlling chloramphenicol-resistance (Tn*9*) into *Escherichia coli* K12 chromosome. *Genetika* **14:** 2071.

Ilyina, T. S., Y. M. Romanova, E. V. Nechaeva, and G. B. Smirnov. 1980b. The genetic control of the transposition process in *Escherichia coli*. *Genetika* (in press).

Ilyina, T. S., E. V. Nechaeva, L. N. Pasynkova, N. I. Smirnova, and G. B. Smirnov. 1980c. Tn*9* integration sites and their influence on the transposon properties. *Genetika* **16:** 46.

Johnsrud, L., M. Calos, and J. H. Miller. 1978. The transposon Tn*9* generates a 9 bp repeated sequence during integration. *Cell* **15:** 1209.

Kleckner, N. 1979. DNA sequence analysis of Tn*10* insertions: Origin and role of 9 bp flanking repetitions during Tn*10* translocation. *Cell* **16:** 711.

Kleckner, N. and D. G. Ross. 1979. Translocation and other recombination events involving the tetracycline-resistance element Tn*10*. *Cold Spring Harbor Symp. Quant. Biol.* **43:** 1233.

Kleckner, N., K. Reichardt, and D. Botstein. 1979. Inversions and deletions of the *Salmonella* chromosome generated by the translocatable tetracycline resistance element Tn*10*. *J. Mol. Biol.* **127:** 89.

MacHattie, L. A. and J. A. Shapiro. 1978. Chromosomal integration of phage λ by means of a DNA insertion element. *Proc. Natl. Acad. Sci.* **75:** 1490.

Marchelli, C., P. Ghelardini, and S. Nasi. 1976. Deletions induced by heat treatment of *E. coli* K12 lysogenic for λ prophages. *Genetics* **82:** 161.

Marcoli, R., S. Iida, and T. Bickle. 1980. The DNA sequence of an IS*1*-flanked transposon coding for resistance to chloramphenicol and fusidic acid. *FEBS Lett.* **110:** 11.

Miller, H. J. and D. J. Friedman. 1977. Isolation of *Escherichia coli* mutants unable to support lambda integrative recombination. In *DNA insertion elements, plasmids, and episomes* (ed. A. I. Bukhari et al.), p. 349. Cold

Spring Harbor Laboratory, Cold Spring Harbor, New York.

MILLER, H. J., A. KIKUCHI, H. A. NASH, R. A. WEISBERG, and D. I. FRIEDMAN. 1979. Site-specific recombination of bacteriophage λ: The role of host gene product. *Cold Spring Harbor Symp. Quant. Biol.* **43:** 1121.

NEUBAUER, L. 1970. Prophage deletions selected by heat induction of bacteria lysogenic for λ*int6 c*I857. *Virology* **42:** 225.

NEVERS, P. and H. SAEDLER. 1978. Mapping and characterisation of an *E. coli* mutant defective in IS-*1*-mediated deletion formation. *Mol. Gen. Genet.* **160:** 209.

NEVERS, P., H. J. REIF, and H. SAEDLER. 1977. A mutant of *E. coli* defective in IS1-mediated deletion formation. In *DNA insertion elements, plasmids, and episomes* (ed. A. I. Bukhari et al.), p. 125. Cold Spring Harbor Laboratory, Cold Spring Harbor, New York.

NOEL, K. D. and D. F.-L. AMES. 1978. Evidence for a common mechanism for the insertion of the Tn*10* transposon and for the generation of Tn*10* stimulated deletions. *Mol. Gen. Genet.* **166:** 217.

OHTSUBO, H., H. OHMORI, and E. OHTSUBO. 1979. Nucleotide sequence analysis of Tn*3*(Ap): Implication for insertion and deletion. *Cold Spring Harbor Symp. Quant. Biol.* **43:** 1269.

REIF, H. J. 1980. Genetic evidence for absence of transposition functions from the internal part of Tn*981*, a relative of Tn*9*. *Mol. Gen. Genet.* **177:** 667.

ROSNER, J. L. and M. S. GUYER. 1980. Transposition of IS*1*-*bio*-IS*1* from a bacteriophage λ derivative carrying the IS*1*-*cat*-IS*1* transposon (Tn*9*). *Mol. Gen. Genet.* (in press).

ROSS, D. G., J. SWAN, and N. KLECKNER. 1979. Physical structures of Tn*10*-promoted deletions and inversions: Role of 1400 bp inverted repetitions. *Cell* **16:** 721.

SHAPIRO, J. A. 1979. Molecular model for the transposition and replication of bacteriophage Mu and other transposable elements. *Proc. Natl. Acad. Sci.* **76:** 1933.

SHAPIRO, J. A. and L. A. MACHATTIE. 1979. Integration and excision of prophage mediated by the IS*1* element. *Cold Spring Harbor Symp. Quant. Biol.* **43:** 1135.

SMIRNOV, G. B. and A. P. MARKOV. 1980. Isolation and genetic analysis of *Escherichia coli* mutants deficient in formation of genomic rearrangements. *Genetika* (in press).

WEINSTOCK, G. M. and D. BOTSTEIN. 1979. Regional specificity of illegitimate recombination associated with the translocatable ampicillin-resistance element Tn*1*. *Cold Spring Harbor Symp. Quant. Biol.* **43:** 1209.

WILLIAMS, J. G. K., D. L. WULFF, and H. A. NASH. 1977. A mutant of *Escherichia coli* deficient in a host function required for phage lambda integration and excision. In *DNA insertion elements, plasmids, and episomes* (ed. A. I. Bukhari et al.), p. 357. Cold Spring Harbor Laboratory, Cold Spring Harbor, New York.

Isolation of a *polA* Mutation That Affects Transposition of Insertion Sequences and Transposons

M. B. Clements and M. Syvanen
Department of Microbiology and Molecular Genetics, Harvard Medical School, Boston, Massachusetts 02115

The bacterial transposable elements include the insertion sequences and the transposons. In addition to undergoing transposition, these elements mediate a number of other genetic rearrangements, including deletions, inversions, and inverted duplications (Botstein and Kleckner 1977; Kleckner et al. 1979; Ross et al. 1979; Syvanen 1980). They excise precisely or imprecisely in a reaction that appears to be independent of the transposition reaction (Rubens et al. 1976; Berg 1977).

Genetic studies of transposable elements in *E. coli* have focused primarily on the elements themselves. The transposons Tn*3* (Chou et al. 1979; Gill et. al. 1979) and Tn*5* (Rothstein et al. 1980) have been shown to carry structural genes for proteins required for their own transposition. Little has been discovered, however, about host functions involved in transposition. A number of host functions that interact with other transposon-mediated genetic rearrangements have been identified. A mutation designated *del* (Nevers and Saedler 1978) reduces the frequency of IS*1*-mediated deletion formation in the *gal* operon and also inhibits precise excision of IS*1*. Mutations in *himA*, isolated by their inhibition of bacteriophage λ integration, also cause deficiencies in precise excision of transposons and growth of bacteriophage Mu (Miller et al. 1979). The *ferB* mutations, which map in the *E. coli* sex factor F, stimulate precise excision of a number of different transposable elements (Hopkins et al. 1980). None of these mutations affect the transposition process itself.

One reason that host mutants affecting transposition have not been isolated is that screening for such mutants has been difficult. We have developed a new transposition screen and have isolated a number of mutant strains carrying mutations in the *E. coli* chromosome that inhibit transposition. One of these mutations is in *polA*, the structural gene for DNA polymerase I.

MATERIALS AND METHODS

Bacterial strains. All are derivatives of *E. coli* K12. Mutants were isolated in SY203, which is F^- (*lac pro*)ΔXIII $argE_{am}$ *araC nalA* rif^r *thi.* Male derivatives of SY203 contained F′ *lac proAB* (F′128) carrying an insertion of either Tn*5* or Tn*10* in *lacZ*. The recipient in all F′ crosses from SY203 was SY410, which is F^- *strA araC* (*lac pro*)ΔXIII *recA1*. Mutant strains are numbered consecutively beginning at SY601 (e.g., SY634 carries the *tnd34* mutation).

Genetic methods. Media used, generalized transduction with phage P1, and ethylmethane sulfonate mutagenesis were as described by Miller (1972). Mutagenized cells were screened on MacConkey indicator medium containing 1% salicin.

Transposition assays. Mutants and wild-type control were assayed for transposition of Tn*5* and Tn*10* in two ways: (1) The strains were grown to mid log phase and infected at a multiplicity of 0.2 with phage λ*b*221*O*am29*P*am80*rex*::Tn*5* or with phage λ*b*221cIII::Tn*10* *O*am29. Cells were plated on media containing either kanamycin (Km) or tetracycline (Tc) to select for transposants. (2) The strains carrying F′ *lac*::Tn*5 pro* and a chromosomal insertion of Tn*10* were grown to mid log phase and mated with SY410 (F^-*strA recA*). Streptomycin (Sm) was used to select against the donor. The rate of transposition is the ratio of the number of F′ *lac*::Tn*5 pro* Tn*10* exconjugants ($Tc^rSm^rKm^r$) versus the number of parental F′ exconjugants (Sm^rKm^r). Similar experiments were performed using F′ *lac*::Tn*10* and a chromosomal insertion of Tn*5* to measure the frequency of transposition of Tn*5* onto F.

Enzyme assays. Assays of the DNA polymerase I activity of the mutant SY634 were performed according to the method of Lehman and Chien (1973). All assay mixtures contained 10 mM N-ethylmaleimide to insure inhibition of polymerases II and III.

RESULTS

The Salicin Papillation Assay for Transposition Mutants

Wild-type *E. coli* K12 is unable to ferment β-glucosides (Bgl^-), including the sugar salicin (Prasad and Schaefler 1974). However, *E. coli* K12 mutates spontaneously to Bgl^+ at frequencies varying from 10^{-6} to 10^{-7}. Recently, J. Felton et al. (in prep.) have shown that the most common mutational event giving Bgl^+ mutants is a transposition of either IS*1* or IS*5* into *bglR*, the controlling region of the *bgl* operon. Point mutations to Bgl^+ occur at a much lower frequency. Therefore, when Bgl^- bacteria are plated on MacConkey indicator medium containing salicin, the red Bgl^+ papillae that appear on the surface of each white colony after several days of growth represent individual transposition events

Table 1. Relative Transposition Frequencies of Strains Carrying *tnd* Mutations

	(a) Transposition from λ to the chromosome		(b) Transposition from the chromosome to F	
Mutation	Tn*5*	Tn*10*	Tn*5*	Tn*10*
wild type	1.00	1.00	1.00	1.00
tnd2	0.25	0.06	—	1.05
tnd5	0.03	0.05	0.80	—
tnd7	0.05	0.11	—	1.25
tnd8	0.12	0.47	—	1.05
tnd36	0.10	0.02	1.91	—
tnd34	0.07	0.05	0.20	0.27

The transposition frequencies in column a are defined as the number of drug-resistant colonies that arise per cell infected by nonreplicating λ phage deleted for the attachment site and carrying the indicated transposon. The frequencies in column b are defined in Materials and Methods. Transposition frequencies are given relative to a parental value of 1.00.

occurring during the growth of the colony. We have used this papillation screening technique to isolate mutants with altered frequencies of IS*1* and IS*5* transposition.

A Bgl$^-$ strain was mutagenized with ethylmethane sulfonate and cells were plated on MacConkey indicator medium containing salicin. Approximately 150,000 colonies were screened for mutants giving altered numbers of Bgl$^+$ papillae. Mutants obtained in this way were of three types: (a) those giving no papillae; (b) those giving more papillae than the wild-type strain; and (c) those giving fewer papillae than wild type but still able to mutate to Bgl$^+$. Mutants of type a, which were seen at a frequency of 10^{-2}, were assumed to be deficient in some structural gene of the *bgl* system rather than in transposition and were not tested further. Mutants of type b were isolated at a frequency of 10^{-3}. Because Bgl$^+$ papillae may arise by spontaneous point mutation of the *bglR* locus (J. Felton et al., in prep.) as well as by transposition, the high rate of papillation associated with this mutant class might be the result of either a moderate increase in transposition frequency or a large increase in the frequency of spontaneous point mutation. Therefore, 60 type-b mutants were tested for their rates of mutation to streptomycin resistance, which occurs only by point mutation. All were found to undergo this event at rates 50- to 1000-fold higher than the wild type, indicating that they are mutator mutants. Therefore, this mutant class has not been analyzed further.

Mutants of type c, those with reduced but detectable papillation rates, were isolated at a frequency of 10^{-3}. Since we were interested in identifying genes whose products affect transposition of a broad range of transposable elements, these mutants were tested for their ability to transpose Tn*5* and Tn*10*. This was done by transposing either Tn*5* or Tn*10* from λ into the chromosome of the type c mutant. Of more than 100 mutants tested, 12 were found to be deficient in transposition by this assay, the magnitude of the effect varying from 4- to 50-fold (Table 1, col. a). We designate the mutations carried by these transposition-deficient strains *tnd.* All strains with *tnd* mutations reduce the transposition frequency from λ to the chromosome for both Tn*5* and Tn*10* while not affecting the plating efficiency of the phage. However, another assay, measuring transposition of either Tn*5* or Tn*10* from the chromosome to the sex factor F by mating out of the sex factor (see Materials and Methods) gives a reduced rate for only one mutant, SY634, which carries *tnd34* (Table 1, col. b).

tnd34 Is an Allele of *polA*

SY634 (*tnd34*) is unable to grow at 43 °C, is sensitive to ultraviolet light and to methylmethane sulfonate, and is unable to support the growth of phage λ carrying the *red* mutation. None of the other *tnd* mutant strains exhibit this phenotype. Since all of these properties are associated with certain alles of *polA,* a number of genetic tests were performed to determine whether *tnd34* maps near the *polA* locus. SY634 was transduced to kanamycin resistance with phage P1 grown on a strain carrying a Tn*5* insertion known to be linked to *polA*. Approximately 50% of the transductants were found to be temperature-resistant. All temperature-resistant transductants tested were also resistant to ultraviolet light and gave wild-type transposition frequencies, as is shown in Table 2. Similar results were obtained with a Tn*10* insertion known to be linked to *polA*. This genetically

Table 2. *tnd34* Is an Allele of *polA*

Number of Tc^r transductants	Temperature sensitivity	UV sensitivity	Transposition frequency from λ to the chromosome
8	ts$^+$	resistant	wild type
8	ts	sensitive	reduced

The strain SY604 was transduced with phage P1 selecting for transduction of a Tn*5* transposon known to be cotransducible with *polA*. These transductants were then tested for several of the *tnd34* properties, including the transposition property. All these properties are linked to each other and to the transposon.

Table 3. Effects of Known *polA* Mutations on Transposition

Allele	Polymerase I defect	Relative transposition frequencies: from λ to the chromosome	from the chromosome to F
wild type	—	1.00	1.00
polA1	polymerase	0.30	0.30
polAex1	exonuclease	0.61	0.21
tnd34	—	0.10	0.14

The mutations *polA1* and *polAex1* were transduced into the SY203 background by linkage to a Tn*10* insertion cotransducible with *polA*. Wild-type and *tnd34* strains also carried this Tn*10* insertion.

maps *tnd34* in the *polA* region of the *E. coli* chromosome.

DNA polymerase I of *E. coli* has both a polymerase and a 5′-3′ exonuclease activity (Lehmen and Uyemura 1976). Cleavage of the enzyme yields two fragments, one carrying the polymerase activity and the other exonuclease. Genetic work also shows the independence of the two activities. The "Cairns mutant," *polA1* (De Lucia and Cairns 1969), is deficient in the polymerase but not in the exonuclease. Another mutant, *polAex1,* is deficient in the exonuclease but not in the polymerase (Konrad and Lehman 1974). We tested derivatives of SY203 carrying the *polA1* and *polAex1* mutations for transposition of Tn*5* and Tn*10* from lambda to the chromosome and from the chromosome to F. The results of these tests, shown in Table 3, indicate that both mutations reduce transposition frequencies somewhat but that *tnd34* has a stronger effect.

To demonstrate further that *tnd34* affects *polA,* we have assayed the enzyme activities associated with DNA polymerase I. In crude extracts, SY634 shows only 0.6% as much polymerase activity as does its ts^+ transductant when the substrate is nicked salmon-sperm DNA (Table 4). With other DNA substrates there is no detectable activity. The residual activity in SY634 is present when assays are carried out at 42°C (data not shown).

DISCUSSION

Our finding that a mutant defective in measured transposition frequency is due to a mutant DNA polymerase I may be of significance with regard to the suspected role of DNA synthesis in the transposition process. Current thought on the mechanism of transposition is that the entire transposable element is replicated, the resulting replica is transposed, and the transposable element in the donor is conserved. We would like to suggest that DNA polymerase I plays a role in this replication process.

Alternative roles for polymerase I must be considered. One possibility is that the immediate product of transposition is a nicked or gapped DNA structure, and that active polymerase is needed for repair to avoid a lethal break or rapid excision of the element. Another possibility is that the mutant polymerase is an inhibitor.

tnd34 causes a more pronounced reduction in transposition frequency than does *polA1* or *polAex1,* indicating that *tnd34* is a different type of mutation in the *polA* gene. This conclusion is supported by the fact that extracts of a strain carrying *tnd34* are deficient not only in the polymerase activity (Table 4), but also in the 5′-3′ exonuclease activity of DNA polymerase I (M. Syvanen, unpubl.).

Other *tnd* strains isolated by the salicin-papillation screening technique appear to affect transposition from bacteriophage λ to the chromosome but not from the chromosome to the sex factor F. This may be because the assay of transposition onto F, which involves mating out of the sex factor, allows expression of the transposon-encoded resistance to occur in a wild-type background, whereas the assay of transposition onto the chromosome does not. Mutant strains may be interfering with expression of the phenotype normally conferred by the transposable element. Alternatively, the transposition events measured by the two assays, which involve different replicons, may also involve different processes. Some gene products involved in one type of event may not be involved in the other. If so, analysis of other *tnd* mutations will identify host factors in transposition that are replicon-specific.

Table 4. A Strain Carrying *tnd34* Is Missing the Polymerase Activity of DNA Polymerase I

Strain	Percentage polymerase activity with template: nicked salmon-sperm DNA	poly[dA: $(dT)_{12}$]	in vitro ColEl replication
SY634(*tnd34*)	0.6	<1	<0.3
Wild type	100	100	100

100% is 1.3 pmoles of ^{32}P incorporated into acid-insoluble material per mg of protein per hr. Nicked salmon-sperm DNA was prepared by partial DNase digestion (Aphosian and Kornberg 1962). ^{32}P-labeled deoxyadenosine triphosphate (sp. act. 500 Ci/mmole) was purchased from New England Nuclear Corp.

ACKNOWLEDGMENTS

We are grateful to J. Hopkins, R. Isberg, C. Kaiser, and J. Sedivy for their advice and assistance, and to A. Wright for his help in developing the screening technique. Our work was supported by National Institutes of Health grant GM-28142-01.

REFERENCES

Aposhian, H. and A. Kornberg. 1962. Enzymatic synthesis of deoxyribonucleic acid. IX. The polymerase formed after T2 bacteriophage infection of *Escherichia coli*. *J. Biol. Chem.* **237:** 519.

Berg, D. E. 1977. Insertion and excision of the transposable kanamycin determinant Tn*5*. In *DNA insertion elements, plasmids, and episomes* (ed. A. I. Bukhari et al.), p. 205. Cold Spring Harbor Laboratory, Cold Spring Harbor, New York.

Botstein, D. and N. Kleckner. 1977. Translocation and illegitimate recombination by the tetracycline resistance element Tn*10*. In *DNA insertion elements, plasmids, and episomes* (ed. A. I. Bukhari et al.), p. 185. Cold Spring Harbor Laboratory, Cold Spring Harbor, New York.

Chou, J., P. G. Lemaux, M. J. Casadaban, and S. N. Cohen. 1979. Transposition protein of Tn*3*: Identification and characterization of an essential repressor-controlled gene product. *Nature* **282:** 801.

De Lucia, P. and J. Cairns. 1969. Isolation of an *E. coli* strain with a mutation affecting DNA polymerase. *Nature* **224:** 1164.

Gill, R. E., F. Heffron, and S. Falkow. 1979. Identification of the protein encoded by the transposable element Tn*3* which is required for its transposition. *Nature* **282:** 797.

Hopkins, J. D., M. B. Clements, T-Y. Liang, R. R. Isberg, and M. Syvanen. 1980. Recombination genes on the *Escherichia coli* sex factor specific for transposable elements. *Proc. Natl. Acad. Sci.* **77:** 2814.

Kleckner, N., K. Reichardt, and D. Botstein. 1979. Inversions and deletions of the *Salmonella* chromosome generated by the translocatable tetracycline-resistance element Tn*10*. *J. Mol. Biol.* **127:** 89.

Konrad, E. B. and I. R. Lehman. 1974. A conditional lethal mutant of *Escherichia coli* K12 defective in the 5′3′ exonuclease associated with DNA polymerase I. *Proc. Natl. Acad. Sci.* **71:** 2048.

Lehman, I. R. and J. R. Chien. 1973. Persistence of deoxyribonucleic acid polymerase I and its 5′3′ exonuclease activity in *polA* mutants of *Escherichia coli* K12. *J. Biol. Chem.* **248:** 7717.

Lehman, I. R. and D. G. Uyemura. 1976. DNA polymerase I: Essential replication enzyme. *Science* **193:** 963.

Miller, H. I., A. Kikuchi, H. A. Nash, R. A. Weissberg, and D. I. Friedman. 1979. Site-specific recombination of bacteriophage λ: The role of host gene products. *Cold Spring Harbor Symp. Quant. Biol.* **43:** 1121.

Miller, J. 1972. *Experiments in molecular genetics.* Cold Spring Harbor Laboratory, Cold Spring Harbor, New York.

Nevers, P. and H. Saedler. 1978. Mapping and characterization of an *E. coli* mutant defective in IS*1*-mediated deletion formation. *Mol. Gen. Genet.* **160:** 209.

Prasad, I. and S. Schaefler. 1974. Regulation of the β-glucoside system in *Escherichia coli* K12. *J. Bacteriol.* **120:** 638.

Rothstein, S. J., R. A. Jorgensen, K. Postle, and W. S. Reznikoff. 1980. The inverted repeats of Tn*5* are functionally different. *Cell* **19:** 795.

Ross, D. G., J. Swan, and N. Kleckner. 1979. Physical structures of Tn*10*-promoted deletions and inversions: Role of 1400 bp inverted repetitions. *Cell* **16:** 721.

Rubens, C., F. Heffron, and S. Falkow. 1976. Transposition of a plasmid deoxyribonucleic acid sequence that mediates ampicillin resistance: Independence from host *rec* functions and orientation of insertion. *J. Bacteriol.* **128:** 425.

Syvanen, M. 1980. Tn*903* induces inverted duplications in the chromosome of bacteriophage lambda. *J. Mol. Biol.* **139:** 1.

Substitution of Silent Bacterial Genes by a Bacteriophage λ Variant Carrying IS*1*

E. Olson, P. Tomich, C. Parsons, K. Leason, D. Jackson, and D. Friedman
Department of Microbiology and Immunology, University of Michigan, Ann Arbor, Michigan 48109

Temperate bacteriophage of the lambdoid family can exist in either of two states, as free phage particles or as prophage in which the DNA is integrated in the bacterial chromosome (Weisberg et al. 1977). In the prophage state, all lytic functions are repressed by the interaction of a unique, phage-specified repressor with operator sites on the phage chromosome (Ptashne et al. 1976). In addition to functional regulation by a repressor, there is structural regulation based on genome organization. Because phage genes are arranged in clusters, groups of genes can be transcribed in sequence and regulated by a cascade mechanism, in which the expression of one set of genes requires the prior expression of another (Herskowitz 1973). In this way, a repressor can bind in one small region and influence the expression of all phage functions.

The relatedness of lambdoid phages is evidenced by the high degree of homology between their DNAs (Davidson and Szybalski 1971) as well as by a striking similarity in organization (Hershey 1971). The quintessential feature distinguishing lambdoid phages from each other is that each encodes a unique repressor-operator system. The details of this regulation have been worked out for phage λ (reviewed in Ptashne et al. 1976) and the following is a brief discussion of this work (see also Fig. 1).

The *c*I gene encodes for the repressor and can be transcribed from two promoters, p_{RE} and p_{RM}. The former functions to establish, and the latter functions to maintain, the synthesis of repressor. Binding of the repressor to the two operators, o_L and o_R, is both necessary and sufficient to maintain repression, since this prevents transcription from the two early promoters, p_L and p_R. These promoters control expression of functions involved in lytic growth as well as those regulating repressor synthesis. The latter class include the *c*II and *c*III gene products that stimulate transcription from p_{RE} and the *cro* gene product that acts to turn off repressor synthesis. The small region of the phage genome that encodes these operators, promoters, and structural genes is called the immunity region. This complicated circuitry ultimately results either in transcription from p_{RM} or from p_L and p_R. Thus, a bacterium that carries the immunity region and adjacent genes of a lambdoid phage would be expected to express either the repressor or adjacent lytic functions.

The homology observed between the DNA of the various lambdoid phages is reflected in vivo by efficient recombination between these phages. A number of hybrids have been isolated that carry varying extents of genetic information from different lambdoid phages (Kaiser and Jacob 1957; Franklin et al. 1965; Liedke-Kulke and Kaiser 1967; Szpirer et al. 1969). Recombination also occurs between these phages and the *Salmonella* phage P22 (Gemski et al. 1972; Botstein and Herskowitz 1974) under appropriately engineered conditions. Hybrid phages are identified by two names: the name of the phage whose immunity is expressed, preceded by the name of the other contributing phage; e.g., λ*imm*P22 is a λ-P22 hybrid that expresses the immunity of P22.

Studies with λ*imm*P22 have revealed a striking exception to the exclusive immunity rule discussed above. *Salmonella* phage P22 and coliphage 21 have functionally equivalent immunities, and heteroduplex analysis reveals significant homology between the immunity regions of these two phages (Botstein and Herskowitz 1974).

We have isolated a variant of phage λ, λ*alt*SF, that shows high-level exchange of sets of genetic information (Friedman et al. 1980). Although the source of the alternate sets of genes has not been determined for each substitution, one form of λ*alt*SF, λ*alt*SF22, acquired a set of genes, including the immunity region and replication genes of *Salmonella* phage P22, from the *E. coli* chromosome. Paradoxically, these P22 genes are silent on the *E. coli* chromosome even though they apparently have all the information necessary to permit expression. In this paper we report the details of λ*alt*SF isolation and the characterization of an insertion sequence that plays an important role in the substitution process.

MATERIALS AND METHODS

Media. Tryptone media used have been described previously (Friedman et al. 1973).

Bacterial strains. All bacterial strains used are derivatives of *E. coli* K12. K213 is a derivative of C600 that carries the *groP* mutation (Georgopoulos and Herskowitz 1971). K99 is strain C600 that carries a SuII suppressor (Appleyard 1954). K1544 carries a number of auxotrophic and fermentative markers (Robert-Baudouy and Portalier 1974). K96 is a derivative of K99 that carries a defective λ prophage. K369, which carries a defective λimm^{21} prophage, was obtained from D. Berg (Washington Univ. School of Medicine, St. Louis, Mo.). K37 is a nonlysogenic strain used routinely in studies of λ and was derived from the standard laboratory strain W3102 (Gottesman and Yarmolinsky 1968). K450 carries the *nusB*-5 mutation (Friedman et al.

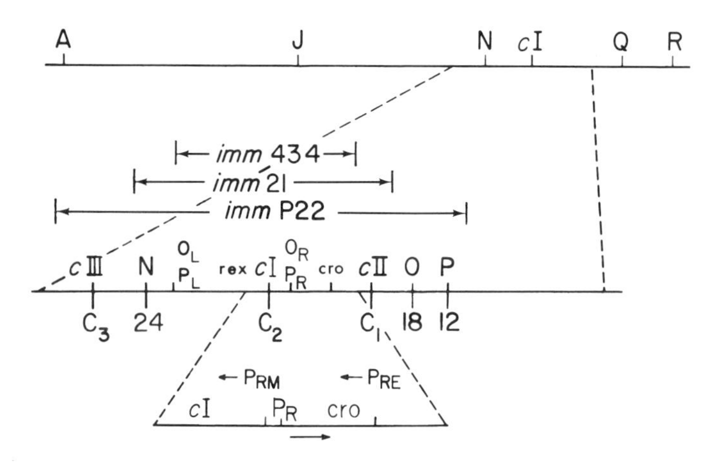

Figure 1. Genetic map of λ and hybrid phages. The top line shows representative markers on the total λ genetic map. The arrows between the vertical bars show the extent of the substitutions corresponding to the immunity regions of the various hybrid phages. The first expansion shows the immunity region of λ and P22. The markers listed above the line represent genes of λ and those below the line analogous genes of P22. The second expansion shows the promoters controlling establishment, p_{RE}, and maintenance, p_{RM}, of *c*I synthesis as well as the promoter controlling rightward transcription, p_R. Arrows indicate directions of transcription.

1976). K1091 carries the *sip* mutation, which selectively inhibits λ*imm*P22 growth. The following lysogens were constructed from K37: K124 carries a prophage; K160 carries a λimm^{21} prophage. K100 is a derivative of K37 carrying a *recA* mutation (Clark 1973). The pSM1 plasmid (Ohtsubo and Ohtsubo 1978) was obtained from H. Ohtsubo (SUNY, Stony Brook, N.Y.). K985 is *attB*-deleted and was obtained from M. Gottesman (National Cancer Institute, NIH, Bethesda, Md.).

Phages. λ*c*I*b*2 and λimm^{21}*c*I*xis* were obtained from M. Gottesman. λ*c*I857*r*14 was derived from λ*N*am7 53*c*I857*r*14 (Brachet et al. 1970), which was obtained from W. Szybalski (University of Wisconsin, Madison, Wisc.). λ*imm*P22hy7 was obtained from D. Botstein (Massachusetts Institute of Technology, Cambridge, Mass.). λ171, whose genotype is *b*221 *c*I171::Tn*10c*I857, was obtained from N. Kleckner (Harvard University, Cambridge, Mass.) (Kleckner et al. 1978). λ*imm*P22-*cam*1 and λ*imm*P22*int*6$c_2$30*12*am$H_4$$cam^r$ were constructed in this laboratory.

Heteroduplex studies. The techniques outlined by Davis et al. (1971) were used. Molecules were measured using a graphics digitizer.

Restriction enzyme analysis of DNA was performed using the methods outlined by Deleys and Jackson (1976). DNA fragment transfer and hybridization was performed employing the method of Southern (1975; see Fig. 5.) CsCl gradients of phage lysates were fractionated, collected, and analyzed as outlined by Brachet et al. (1970).

RESULTS

λ*alt*

alt phages are variants of coliphage λ that exhibit the unusual property of substituting or alternating blocks of genes (Friedman et al. 1980). λ*alt* was isolated from a lysate of λ*c*I857*r*14 (Tomich and Friedman, unpubl.). The *r*14 phage carries an IS*1* element in the *c*II gene (Brachet et al. 1970). J. Zissler and F. Goetz (pers. comm.) have independently isolated an *alt* phage starting with a derivative of λ that carries an IS*2* in the *xis* gene.

Substitution of different sets of genes has been observed with λ*alt* phage (*alt* stands for alternates). However, change of immunity is the alternation reaction easiest to observe and thus has been the focus of our studies. As an example, we look at λ*alt*λ. Normally, λ does not form plaques on a λ lysogen. The rare mutants that do form plaques on the lysogen still carry the λ immunity region but have mutations that interfere with repressor binding (Ptashne et al. 1976). In contrast, lysates of λ*alt*λ contain a highly variable fraction (10^{-6}–10^{-2}) that form plaques on a λ lysogen. The majority of these phage carry the immunity of phage 21. Phage of the latter class are called λalt^{21}, because, at a low and variable frequency, lysates of these phage contain phages with a different immunity. *alt* phages also show variation in the range of hosts they will infect, appearing at times to have the host range of λ and at other times to have that of phage φ80 (Hershey and Dove 1971). Lysogens formed with either of the *alt* phages are stable and fail to exhibit the immunity characteristics of the alternate phage. However, on induction, these lysogens release, at a low frequency, phage of the alternate immunity. Similar growth of either of the non-*alt* phages, λ or λimm^{21}, fails to produce any alternation, showing that alternation is a characteristic unique to *alt* phages.

The *alt* phages tend to be extremely unstable, losing the ability to alternate after a few growth cycles. Moreover, the extreme variation in the alternation frequency made meaningful studies of the phenomenon very difficult. We therefore selected for an Alt phage that showed high levels of alternation.

Isolation of λ*alt*SF

A variant of λ*alt*λ that substitutes immunity regions at high frequency was selected for its ability to form relatively normal appearing plaques on a bacterial lawn composed of two lysogens. The lysogens carry either a defective λ or λimm^{21} prophage. This phage variant is called λ*alt*SF (SF stands for *s*ubstitutes *f*requently).

The strategy underlying this selection is based on the following considerations. (1) Lysogens do not support growth of phage with an immunity homologous to that of the prophage. (2) Since *alt* phages change immunity at very low frequency, they will not form plaques on lysogens carrying a prophage with the same immunity. (3) When a mixed lawn containing two bacteria, one that supports the growth of a phage and the other a lysogen immune to infection by that phage, is used for plating the phage, plaque formation is poor. This presumably occurs because the lysogen is unaffected by the growth of a phage carrying an immunity homologous to that of the prophage and thus overgrows the region of the plaque. (4) A variant of λ*alt* that changes immunity at high frequency would be expected to produce phages of both immunity types following single infection. (5) If the phage initially infects the lysogen that exhibits the heterologous immunity, phage growth will occur resulting in the production of some phage with the alternate immunity. Thus, both bacterial types in the lawn will be productively infected and a normal-appearing plaque should be formed.

The efficacy of this strategy is proved by the following observation. When a lysate of *alt* is plated on a mixed lawn formed from a lysogen carrying a defective λ prophage (K96) and a lysogen carrying a defective λimm^{21} prophage (K369), plaque formation is extremely poor. However, phages forming relatively normal plaques on this mixed lawn are observed at a frequency of approximately 10^{-8}. One phage isolated in this manner was selected for further characterization and was plaque-purified twice. Lysates of this phage exhibited the characteristic expected for a rapidly alternating derivative, that of substituting immunity regions at high frequency.

Biological Properties of λ*alt*SF

The two forms of λ*alt*SF display phages λ or 21 immunity, the same characteristics as the original Alt phage. A number of experiments outlined below demonstrate conclusively that the λ*alt*SF phage exhibiting phage 21 immunity characteristics carries the immunity region and associated genes of P22. Although P22 and 21 share the same immunity characteristics, they have different early regulation and replication genes (Botstein and Herskowitz 1974). Moreover, phage 21 is *E. coli*-specific, whereas P22 is *Salmonella*-specific.

Both forms of λ*alt*SF show substantially higher substitution of immunity regions than that observed with the original *alt* phage. However, lysates of the two forms show different levels of alternate phage production; 10–30% for the λ form and 1% or less for the P22 form. Substitution is not limited to these two immunity regions. Other derivatives that exhibit neither phage λ nor phage 22 immunity have been isolated but, as of yet, they have not been characterized. In addition to substitution of immunity specificities, λ*alt*SF phage show associated changes in early regulation and replication genes.

These substitution reactions can also be demonstrated when the λ*alt*SF phage are grown in K100, a strain isogenic with K37 except that it is *recA*⁻ (Clark 1973). Thus, substitution occurs independently of the *recA*-mediated recombination system.

Associated with the immunity regions of lambdoid phages are genes encoding positive regulation functions. In the case of λ, the function is the product of the *N* gene (Herskowitz 1973). Analogous genes are found in both λimm^{21}, N_{21} (Friedman et al. 1973), and λ*imm*P22, gene *24* (Hilliker and Botstein 1976). These genes all map to the left of their respective repressor genes. Their products act to stimulate gene expression by permitting transcription complexes to transcend transcription-termination sites (Roberts 1969).

Mutants of *E. coli* have been isolated that interfere with phage growth by failing to support the activity of these *N*-like functions. One example is strain K450, which carries the *nusB*-5 mutation (Friedman et al. 1976). As shown in Table 1, λ, λimm^{21}, and λ*imm*P22 can all be distinguished from each other by their plating patterns on K450. At 32 °C, λimm^{21} plates poorly, λ plates moderately well, and λ*imm*P22 plates well. At 40 °C, λ and λimm^{21} completely fail to plate, whereas λ*imm*P22 plates on K450. As expected for a phage with the immunity-*N* region of λ, λ*alt*SFλ plates poorly at 32 °C and fails to plate at 40 °C on K450. The alternate form that exhibits the immunity character of phages 21 and P22 plates at both temperatures, suggesting that this form of *alt*SF has gene *24* of P22.

P22 differs from λ and 21 in its replication genes. Hybrids that have the replication genes of P22 can be distinguished from phages that carry the analogous region from λ by their ability to plate on a bacterial mutant with a mutation, *groP*, in the *dnaB* gene (Hilliker and Botstein 1976). As shown in Table 1, λ*alt*SFλ fails to plate on the *groP* strain K213, whereas the λ*alt*SF22 phage plates well on this host. This suggests that al-

Table 1. Growth of λ*alt*SF Phage on Various *E. coli* Strains

Bacteria	Relevant genotype	λ	λimm^{21}	λ*imm*P22	λ*alt*SFλ	λ*alt*SF22
K37	wild type (nonlysogen)	+	+	+	+	+
K124	λ lysogen	−	+	+	−	+
K160	21 lysogen	+	−	−	+	−
K450	*nusB*-5	±(−)	−(−)	+(+)	−(−)	+(+)
K1091	*sip*	+	+	−	+	−
K213	*groP*	−	−	+	−	+

Each of the listed strains was cultivated in tryptone broth (TB) at 32 °C, and overnight cultures were used to form bacterial lawns on TB plates. The seeded plates were spotted with dilutions of each of the listed phages and then incubated overnight at 32 °C. Phage lysates that contained low percentages of phage exhibiting the alternate immunity were used. A second plate seeded with K450 was spotted in a similar manner but incubated at 40 °C. (+) Normal growth; (−) substantial reduction in growth; (±) moderate reduction in growth; symbols enclosed in parentheses are the results of incubation at 40 °C. In the cases where the *alt*SF phage are listed as not plating, the small number of phage that formed plaques were shown to carry the alternate immunity.

though the λ form has λ replication genes, the alternate form of λ*alt*SF has P22 replication genes.

Further evidence that λ*alt*SF22 carries P22 genes derives from experiments with another *E. coli* mutant, K1091. This strain has the *sip* mutation, which causes selective inhibition of growth of phage with the immunity and replication regions from P22 (M. Strauch and D. Friedman, unpubl.). We find that λ*alt*SFλ plates on K1091 at 32°C, whereas λ*alt*SF22 does not. Collectively, these findings indicate that λ*alt*SF22 has P22 genes. This assumption was proven by experiments analyzing λ*alt*SF DNA.

Structure of λ*alt*SF

λ*alt*SF genomes were analyzed by electron microscopy of heteroduplexed DNA and gel electrophoresis of restriction-enzyme-treated DNA. A schematic representation of molecules analyzed by DNA heteroduplexing techniques is shown in Figure 2. In the following discussion we will locate points on the chromosome as percentages according to the scale shown on the map in Figure 2.

Three phages with deletions, additions, or substitutions of DNA as markers were used to analyze the λ*alt*SF DNA. λc*Ib*2 has a 14% deletion of DNA between the 44–58% points (Davidson and Szybalski 1971). The λ*imm*[21] used in these experiments has a substitution on the far right side that is seen as two loops (Fig. 2). A hybrid λ phage carrying the *Q, S,* and *R* genes from φ80 (Davidson and Szybalski 1971) has an identical substitution. λ*imm*P22*cam* carries the Tn*9* transposon (which encodes resistance to chloramphenicol [Cm^r]) in the *b*2 region (Rosner and Gottesman 1977). The Tn*9* insertion can be used for orienting heteroduplexed molecules.

Comparison of the heteroduplex analysis of DNA from λ*alt*SFλ with that of λc*Ib*2 reveals that in addition to the *b*2 deletion observable as a large loop near the 50% point, there is a small loop of DNA near the 75% point. This loop, referred to as the SF loop, could represent either a deletion or addition of DNA. A comparison of λ and λ*alt*SFλ DNA by *Eco*RI digestion and

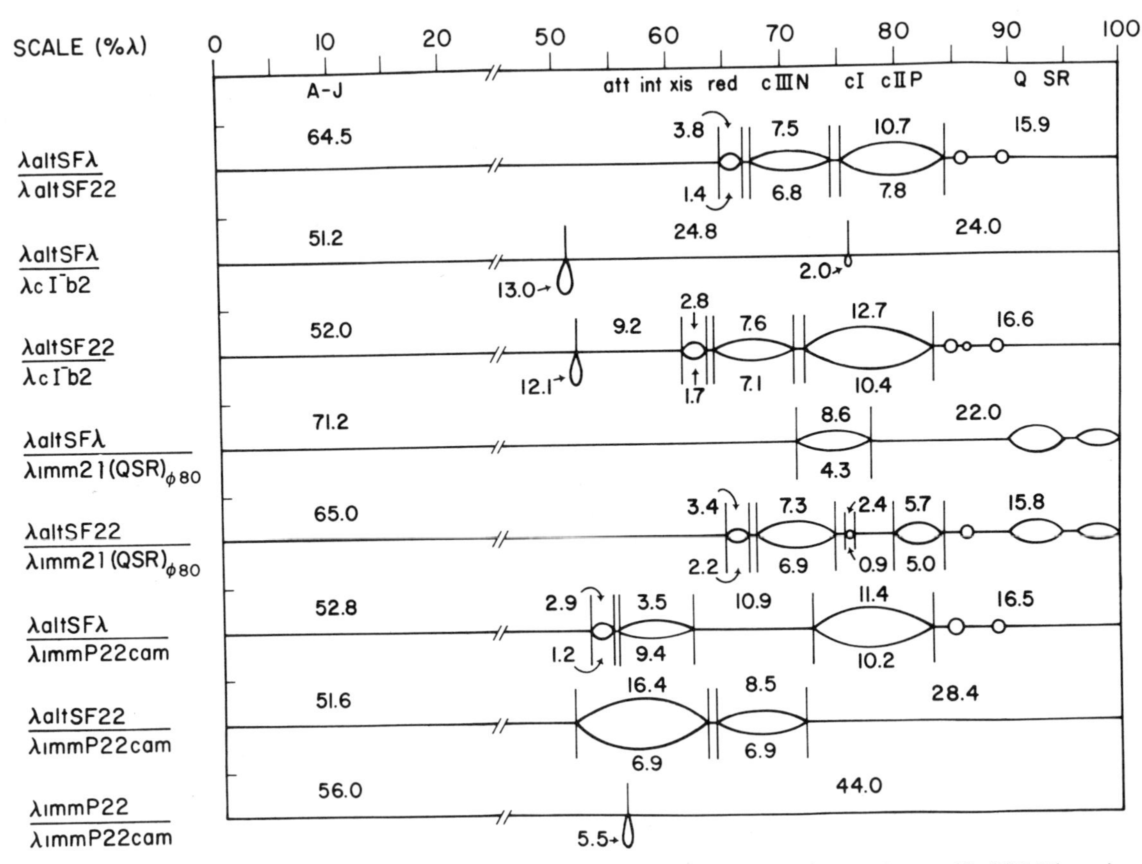

Figure 2. Schematic representation of DNA heteroduplexes defining the structure of the two forms of λ*alt*SF. The pairs of DNA used are listed in the column on the left. The molecules are aligned with respect to the heteroduplex linear map derived for λ (Davidson and Szybalski 1971), which is shown at the top. For reference, several λ genes are placed on the map. The heteroduplexed molecules were prepared from purified phages according to the methods of Davis et al. (1971). The distances between the various sites on the molecules are given as percentages of the measured heteroduplex molecule. Because of the variations within the substitutions, as well as the presence of deletions and insertions, the representations do not always align perfectly. For these studies, a set of derivatives of the *alt* phages that showed lower levels of alternation were used in order to have a more homogeneous population (e.g., see Fig. 4A, lane c). It should be noted that comparisons of the heteroduplexes made with the higher-substituting parental phages showed no obvious differences in structure. Therefore, although we have no evidence as to the cause of the reduction of substitution at this time, these phages can be used for structural studies.

subsequent gel electrophoresis (see Fig. 4) shows that fragment 2 from the λ*alt*SFλ DNA is larger than that from λ DNA, and thus the λ*alt*SFλ DNA has an insertion. Fragment 2 is known to contain the *c*I and surrounding genes (Helling et al. 1974), the region where the SF insertion is located by the heteroduplex analysis. We estimate the SF insertion to be ~1000 bp.

To locate the SF loop more precisely, DNA from λ*alt*SFλ was heteroduplexed with DNA isolated from a λ derivative that carries a Tn*10* element in the region of the *c*I gene, λ171 (Kleckner et al. 1978). As shown in Figure 3, the SF loop is located immediately to the left of the Tn*10* element. The fact that λ*alt*SFλ forms a clear plaque is consistent with the location of the *alt*SF loop within or near the *c*I gene.

The structure of λ*alt*SF22 is more complicated. As shown in Figure 2, this phage has the *24* gene, immunity region, and replication genes of P22. This was directly demonstrated by heteroduplex analysis with λ*imm*P22 DNA carrying a Tn*9* (*cam*) insertion. As seen in Figure 2, there is homology between λ*alt*SF22 DNA and λ*imm*P22*cam*1 DNA from the right end to the 72% point, a region of the genome that includes the replication, immunity, and *24-N* genes. Confirming our conclusion as to the nature of λ*alt*SF22 genes is the heteroduplex comparison of λ*alt*SF22 DNA and λimm^{21} DNA. In this case, we observe a region of homology at the 72% point corresponding to the c_2-*c*I genes. This region of homology is flanked by two regions of nonhomology that correspond to the *N-24* genes to the left and the replication genes to the right. This nonhomology is to be expected since the λimm^{21} phage has the *N* gene of 21 and the replication genes of λ. One striking finding is that λ*alt*SF22 DNA carries a substituted region of DNA in the 64–74% region. This DNA is not homologous with either the P22 or λ DNA normally found in this region.

A derivative of λ called λ*rev*, which, like λ*alt*SF22, carries a substitution of bacterial genes in the 65–75% region, has been isolated (Zissler et al. 1971). Relevant to the alternation process, the reverse DNA carries an alternate set of recombination genes in this region (Gottesman et al. 1974). However, heteroduplex analysis shows that λ*alt*SF22 and λ*rev* are not homologous in the regions of the substitutions (data not shown).

The *alt*SF Insertion Is IS*1*

The *alt*SF insertion was estimated to be approximately 1000 bp in length, which places it in the size range of IS*1* (Ohtsubo and Ohtsubo 1978). Moreover, λ*r*14, the apparent progenitor of λ*alt*SF, carries an IS*1* element in the *c*II gene (Brachet et al. 1970). It should be noted that this is not the location of the *alt*SF insertion, which is in the *c*I gene. However, the former considerations led us to determine whether the *alt*SF insertion is IS*1*. Southern transfer experiments represented in Figure 4 show that the λ*alt*SFλ insertion is IS*1* and

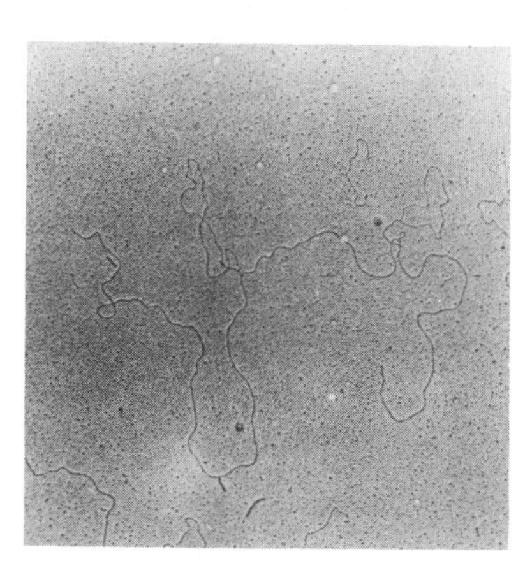

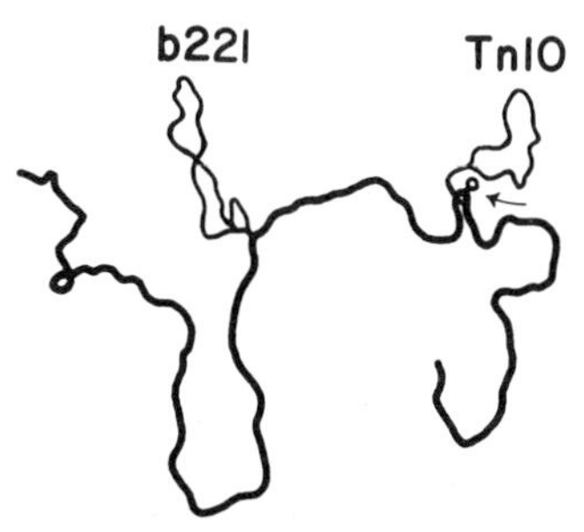

Figure 3. Heteroduplex molecule formed between λ*alt*SFλ and λ171. λ171 carries a Tn*10* element in the left side of the *c*I gene as well as the *b*221 deletion (Kleckner et al. 1978). The location of these two markers is indicated on the diagram. The arrow points to the location of the SF insertion.

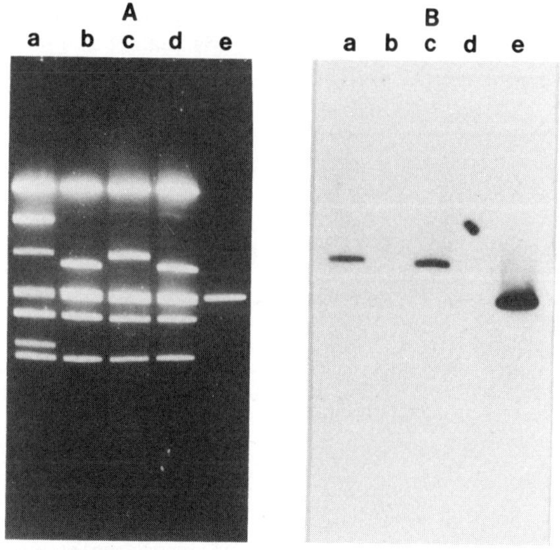

Figure 4. Restriction enzyme and Southern transfer analysis of *alt*SF DNA. (*A*) The following phage and plasmid DNAs were digested with *Eco*RI and analyzed by electrophoresis on a 0.7% agarose gel. λ*alt*SFλ⁻ denotes derivatives of λ*alt*SFλ that fail to show substitution. (*a*) λ*alt*SFλ and λ*alt*SF22 DNA (0.5 μg); (*b*) $\lambda alt\mathrm{SF}\lambda_1^-$ (0.5 μg); (*c*) $\lambda alt\mathrm{SF}\lambda_2^-$ (0.5 μg); (*d*) λ (0.5 μg); (*e*) pSM1 (0.1 μg). (*B*) The DNA fragments shown in *A* were denatured and transferred to nitrocellulose paper (see Materials and Methods). IS*1*-containing plasmid pSM1 was ^{32}P-labeled (0.3 μg, 3 × 10^6 cpm/μg) and used as a hybridization probe to DNA bound to the filter. Hybridization solution contained 5× SSC and 1% Sarkosyl. After 18 hr at 65 °C, the filter was washed at 65 °C in 2× SSC, 0.5 M glycine, and 0.5% SDS. After autoradiography, hybridization was detected in lanes *a*, *c*, and *e*.

demonstrate that this insertion is, at least in part, responsible for the unusual recombination events exhibited by the λ*alt*SFλ phage. We focus on the λ*Eco*RI fragment 2, which is known to contain the *c*I gene (Helling et al. 1974). This is fragment 3 in Figure 4A, lane a. In this case, the digested DNA shows extra fragments that result from the subpopulation of λ*alt*SF22 found in lysates of λ*alt*SFλ. This fragment is obviously larger in the case of λ*alt*SFλ DNA (Fig. 4A, lanes a and c) and hybridizes with the IS*1*-containing probe pSM1 (Fig. 4B). The functional significance of the *alt*SF insertion was inferred from the analysis of DNA from a revertant of λ*alt*SFλ that concomitantly lost the ability to substitute genes and regained the ability to form turbid plaques. Since the *alt*SF insertion maps in the *c*I gene, this suggests that the *alt*SF insertion was precisely excised from the phage genome. The experiments shown in the b lanes of Figure 4 A and B confirm this conclusion. Fragment 2 from the revertant DNA is the same size as that from λ wild-type DNA, and there is no hybridization with the IS*1* probe. We note that mutants of λ*alt*SFλ that retain the IS*1* insertion but fail to alternate can also be isolated (c lanes of Fig. 4 A and B). The nature of these mutations is not as yet known.

Location of Cryptic P22 Sequences

Genetic and biochemical studies have previously been used to demonstrate that P22 material, including the immunity region and replication genes, is located on the *E. coli* K12 chromosome (Friedman et al. 1980). We now present evidence based on zygotic induction that locates the cryptic genes between minutes 89 and 100 on the K12 genetic map.

A specially constructed derivative of P22, λ*imm*-P22*int*6$c_2$30*12*amH_4*cam*^{r}, was used as an in vivo probe for the cryptic genes. We will first discuss the placement of the phage in the cryptic genes. This P22 hybrid, called λ*imm*P22*int*$^-$*12*$^-$*cam*^{r}, was used to lysogenize K985, a derivative of K37 that is deleted for the normal λ attachment site (*att*B). The fact that the phage is *int*$^-$ (it cannot make the phage-determined product necessary for integrative recombination [Gottesman and Yarmolinsky 1968]) and the bacterium is deleted for *att*B insures that the phage will not integrate by the usual site-specific recombination mechanism (Shimada et al. 1973). Therefore, lysogens should be found in which the P22 hybrid has integrated by homologous recombination with cryptic prophage genes (see Fig. 5). Since this should be a rare event, we included the *cam*^{r} marker to give another selecting mechanism, Cm^r, in addition to P22 immunity. Thus, lysogens were identified as bacteria resistant both to infection by λ*imm*P22 and to Cm.

We next determined whether the prophage was located in the cryptic P22 sequences by inducing the lysogens and scoring the type of phage released. The strategy underlying this test is based on the following arguments and is schematically represented in Figure 5. If the infecting phage integrated into the resident P22 sequences that contained the P22 replication genes *12* and *18*, we would expect a reasonable percentage of the lysate to carry the *12*$^+$ allele. This follows from the assumption that the phages have integrated by homologous recombination and will be excised by homologous

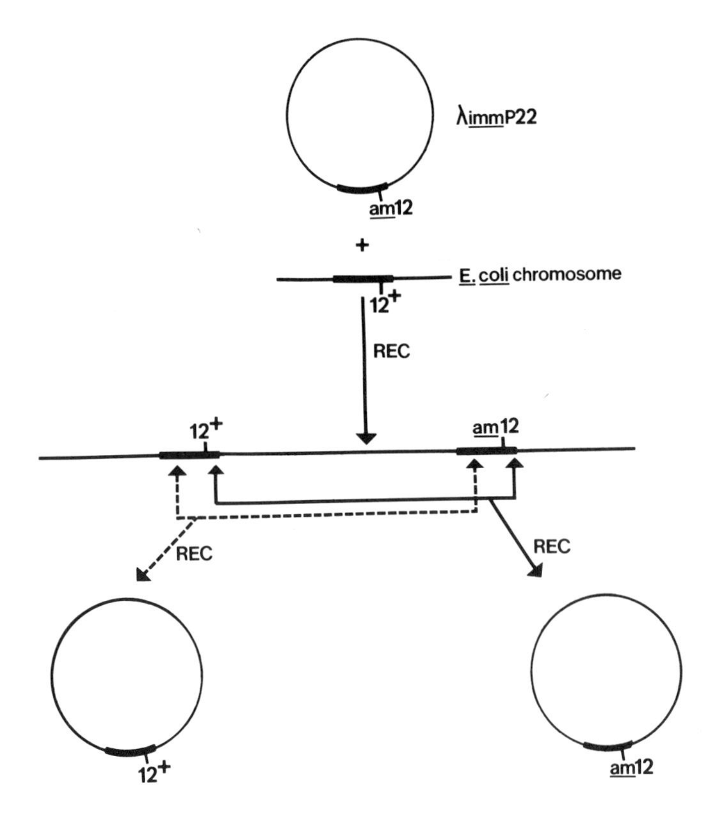

Figure 5. Integration and excision of λ*imm*P22 by homologous recombination. The P22 hybrid phage is *int*$^-$ and the bacterium is *att*B-deleted; therefore, the phage must integrate by homologous recombination. The lysogens carry a duplication of the P22 sequences. Excision of the prophage can also occur by homologous recombination. This recombination can occur anywhere in the region of homology. Therefore, phages carrying either the am*12* or *12*$^+$ allele should be formed depending on the region where recombination occurs.

recombination. Therefore, if the recombination can occur with equal probability in any part of the regions of homology, some of the phages will have been produced in a manner to yield genomes with the amber$^+$ allele. Sixteen lysogens resulting from infection with λ*imm*P22*int*$^-$*12*$^-$*cam*^{r} were tested for production of amber$^+$ and amber$^-$ phages following UV induction. The amber$^+$ phages plate on both *su*II (K99) and *su*^{o} hosts (K37), whereas the amber$^-$ phages plate only on the *su*II host. One isolate, K1471, was arbitrarily chosen for further study. Analysis of phage yield revealed that approximately equal numbers of amber$^+$ and amber$^-$ λ*imm*P22 phages were produced following induction.

K1471 was made into an Hfr donor of the P4X type, which donates from minute 7 in a counterclockwise direction (Bachmann and Low 1980). This Hfr strain was mated with the recipient K1544, and various auxotrophic markers were used for selection. The results of the mapping are illustrated in Figure 6. We find that *proB* (min 6) and *thr* (min 100) markers can be transferred from the Hfr-P22 lysogen efficiently, but distal markers, including *argB* (min 89), are transferred inefficiently. This break in linkage suggests that zygotic induction has occurred and the transfer of the λ*imm*P22 prophage follows transfer of *thr* but occurs prior to transfer of *argB*. This places the cryptic P22 genes in the region covering minutes 89 to 100 of the *E. coli* K12 chromosome (Bachmann and Low 1980).

We have no evidence as to the origin of the λ genes available for substitution. However, the presence of λ sequences on the *E. coli* chromosome of supposed nonlysogens has previously been reported (Berg and Drummond 1978).

DISCUSSION

The high-frequency substitution of blocks of genes as exhibited by λ*alt*SF phage reflects an obviously efficient recombination process. Our studies have focused on the λ rather than the P22 form because the physical structure of λ*alt*SFλ DNA appeared to be more amenable to analysis. The only difference distinguishable by heteroduplex analysis between λ and λ*alt*SFλ DNAs is the small insertion in the *c*I gene. The isolation of variants of λ*alt*SFλ that have lost the insertion and concomitantly lose the ability to substitute genes implicates the insertion in the substitution process. The demonstration that the *alt*SF insertion is an IS*1* element is consistent with such a recombinational role.

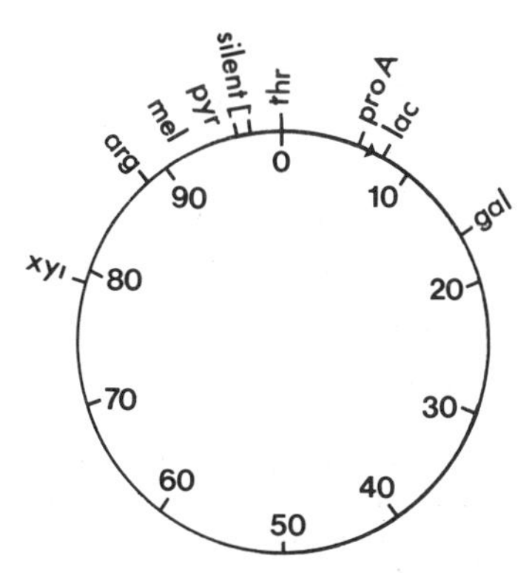

Figure 6. Map of *E. coli* K12 showing region where P22 genes have been located. The arrow indicates site of origin of transfer of P4X Hfr. Appropriate markers are indicated.

IS*1* elements have been shown to participate in a number of recombination-related processes, including transposition, excision, inversion-deletion formation, duplication, and plasmid cointegration (reviewed by Starlinger and Saedler 1976; Kleckner 1977; Kopecko 1980). Studies by MacHattie and Shapiro (1978) suggested that IS*1* encodes genes whose products are involved in the IS-mediated recombination events. In addition to promoting their own movement, IS*1* elements appear to promote the recombinational activity of a number of transposons, e.g., Tn*9*, Tn*1681*, and Tn(R-det) (see Kopecko 1980). In these cases, the IS*1* elements are carried at each end of the transposable segment. Transposition of Tn*9* is a relatively infrequent event; Rosner and Gottesman (1977) reported a frequency of 2×10^{-5} for transposition from the λ genome to the bacterial chromosome.

The substitution of phage genes generated by the *alt*SF-IS*1* element resembles transposition in two ways. First, the substitution occurs only when the IS*1* is present. Second, the substitution occurs in the absence of the host homologous recombination functions. Substitution differs from translocation in three major ways. First, substitution is a high-frequency event, with 30% of the lysate containing phage with substituted genes. Second, substitution occurs during lytic growth of the *alt*SF phage, whereas transposition of Tn*9* to the phage genome does not occur during lytic growth (Gottesman and Rosner 1975). Third, the IS*1* element of λ*alt*SFλ is located internally at the ends of the substituted genes, whereas the IS*1* elements of Tn*9* are located at the ends of the transposon (MacHattie and Jackowski 1977; Rosner and Gottesman 1977).

Two alternate explanations for the extraordinary activity of the *alt*SF-IS*1* element are compatible with the studies presented in this paper: (1) The element could be a variant endowed with high activity. Such an IS*1* could have arisen de novo during the selection for *alt*SF or represent a preexisting class of IS*1*s found on the *E. coli* chromosome. Since multiple copies of IS*1* have been identified on the *E. coli* chromosome (Saedler and Heiss 1973), it is formally possible that there is such IS*1* heterogeneity. Consistent with this assumption is the recent finding of an IS*1* element that is missing a restriction site usually found in this element (Cornelis and Saedler 1980). (2) The position of the IS*1* element in the *c*I gene of λ might permit high-level expression of IS*1*-encoded proteins. For example, the *c*I gene can be transcribed from the λ p_{RE} promoter (Fig. 1), which, in the presence of *c*II and *c*III functions, shows high levels of transcription of *c*I (Schmeissner et al. 1980). Upon infection, the *c*II-*c*III-stimulated transcription from p_{RE} might result in high-level expression of *alt*SF-IS*1*-encoded functions. Studies by MacHattie and Shapiro (1978) were interpreted as showing that expression of IS*1* functions is enhanced if the IS*1* is located near a λ promoter.

These two models are experimentally distinguishable, and current studies are directed toward elucidating the mechanism of *alt*SF-directed recombination.

The ability of λ*alt*SF to substitute genes en bloc in an efficient manner could play an obvious role in rapid evolution of the virus. It has been suggested that evolution by exchange of functional units offers a means for local diversification, whereas overall functional homology is maintained (Fisher 1930; Hershey 1971; Botstein and Herskowitz 1974). Linkage of genes in functional groups is commonly observed and has been referred to as orthotopic linkage (Stahl and Murray 1966). The isolation of λ*alt*SF suggests that there may be a special system for promoting exchanges of linkage groups.

Rearrangements of blocks of genes have been shown to be important ontologically in eukaryotic cells. For instance, studies with yeast have led Herskowitz and coworkers (Hicks et al. 1977) to propose that variation in mating type is determined by exchange of blocks of genetic information. Although the genetic information for both mating types, α and **a**, is carried by each yeast cell, only certain sets of genes are expressed, depending on which block of genes is at the mating-type locus. "Silent" copies of each mating type are carried at other sites located at some distance from the mating-type locus.

Temperate phages would appear to be ideally suited to participate in a process of "reshuffling" functional units since they have potential to exist as prophage. A bacterial host carrying prophage or even parts of prophage could serve as a reservoir for the exchangeable functional units. Indeed, the widespread occurrence of defective prophages in the bacterial world (Garro and Marmur 1970) shows that a range of potential donors is available.

ACKNOWLEDGMENTS

The authors thank David Botstein, Ira Herskowitz, Eric Flamm, Max Gottesman, Gary Gussin, J. L. Rosner, Harvey Miller, Mark Ptashne, and L. S. Baron for their helpful discussion and encouragement. James Zissler and Fred Goetz are thanked for conveying unpublished information. Emma Williams is thanked for her patient care in typing the manuscript. Lisa Mashni is thanked for her care in preparing the figures. These studies were supported by grants from the National Institutes of Health to D.F.

REFERENCES

Appleyard, R. K. 1954. Segregation of new lysogenic types during growth of the doubly lysogenic strain derived from *Escherichia coli* K-12. *Genetics* **26:** 440.

Bachmann, B. J. and K. B. Low. 1980. Linkage map of *Escherichia coli* K12, edition 6. *Microbiol. Rev.* **44:** 1.

Berg, D. E. and M. Drummond. 1978. Absence of DNA sequences homologous to transposable element Tn*5* (Kan) in the chromosome of *Escherichia coli* K-12. *J. Bacteriol.* **136:** 419.

Botstein, D. and I. Herskowitz. 1974. Properties of hybrids between *Salmonella* phage P22 and coliphage. *Nature* **251:** 584.

Brachet, P., H. Eisen, and A. Rambach. 1970. Mutations of coliphage λ affecting the expression of replicative functions O and P. *Mol. Gen. Genet.* **108:** 266.

Clark, A. 1973. Recombination deficient mutants of *E. coli* and other bacteria. *Annu. Rev. Genet.* **7:** 67.

Cornelis, G. and H. Saedler. 1980. Deletions and an inversion induced by a resident IS*1* of the lactose transposon Tn*951*. *Mol. Gen. Genet.* **178:** 367.

Davidson, N. and W. Szybalski. 1971. Physical and chemical characteristics of lambda DNA. In *The bacteriophage lambda* (ed. A. D. Hershey), p. 45. Cold Spring Harbor Laboratory, Cold Spring Harbor, New York.

Davis, R. W., M. Simon, and N. Davidson. 1971. Electron microscope heteroduplex methods for mapping regions of base sequence homology in nucleic acids. *Methods Enzymol.* **21:** 413.

Deleys, R. J. and D. A. Jackson. 1976. Electrophoretic analysis of covalently closed SV40 DNA: Boltzmann distributions of DNA species. *Nucleic Acids Res.* **3:** 641.

Fisher, R. A. 1930. *The genetical theory of natural selection.* Clarendon Press, Oxford.

Franklin, N. C., W. F. Dove, and C. Yanofsky. 1965. The linear insertion of a prophage into the chromosome of *E. coli* shown by deletion mapping. *Biochem. Biophys. Res. Comm.* **18:** 910.

Friedman, D. I., M. Baumann, and L. S. Baron. 1976. Cooperative effects of bacterial mutations affecting N gene expression. I. Isolation and characterization of a *nus*B mutant. *Virology* **73:** 119.

Friedman, D. I., G. S. Wilgus, and R. J. Mural. 1973. Gene N regulator function of phage λ*imm*21: Evidence that a site of N action differs from a site of N recognition. *J. Mol. Biol.* **81:** 505.

Friedman, D. I., P. Tomich, C. Parsons, E. Olson, E. L. Flamm, and R. Deans. 1980. λ*alt*SF: A phage variant that acquired the ability to substitute sets of genes at high frequency. *Proc. Natl. Acad. Sci.* **78:** 410.

Garro, A. J. and J. Marmur. 1970. Defective bacteriophages. *J. Cell. Physiol.* **76:** 253.

Gemski, P., Jr., L. S. Baron, and N. Yamamoto. 1972. Formation of hybrids between coliphage and *Salmonella* phage P22 with a *Salmonella typhimurium* hybrid sensitive to these phages. *Proc. Natl. Acad. Sci.* **69:** 3110.

Georgopoulos, C. P. and I. Herskowitz. 1971. *Escherichia coli* mutants blocked in lambda DNA synthesis. In *The bacteriophage lambda* (ed. A. D. Hershey), p. 553. Cold Spring Harbor Laboratory, Cold Spring Harbor, New York.

Gottesman, M. M. and J. L. Rosner. 1975. Acquisition of a determinant for chloramphenicol resistance by coliphage lambda. *Proc. Natl. Acad. Sci.* **72:** 5041.

Gottesman, M. and M. Yarmolinsky. 1968. Integration-negative mutants of bacteriophage lambda. *J. Mol. Biol.* **31:** 487.

Gottesman, M. M., M. E. Gottesman, S. Gottesman, and M. Gellert. 1974. Characterization of bacteriophage reverse as an *Escherichia coli* phage carrying a unique set of host-derived recombination functions. *J. Mol. Biol.* **88:** 471.

Helling, R. B., H. M. Goodman, and H. W. Boyer. 1974. Analysis of endonuclease R·*Eco*RI fragments of DNA from lambdoid bacteriophages and other viruses by agarose-gel electrophoresis. *J. Virol.* **14:** 1235.

Hershey, A. D. 1971. Comparative molecular structure among related phage DNA's. *Carnegie Inst. Wash. Year Book* **70:** 3.

Hershey, A. D. and W. Dove. 1971. Introduction to lambda. In *The bacteriophage lambda* (ed. A. D. Hershey), p. 3. Cold Spring Harbor Laboratory, Cold Spring Harbor, New York.

Herskowitz, I. 1973. Control of gene expression in bacteriophage lambda. *Annu. Rev. Genet.* **7:** 289.

Hicks, J. B., J. N. Strathern, and I. Herskowitz. 1977.

The cassette model of mating-type interconversion. In *DNA insertion elements, plasmids, and episomes* (ed. A. I. Bukhari et al.), p. 457. Cold Spring Harbor Laboratory, Cold Spring Harbor, New York.

HILLIKER, S. and D. BOTSTEIN. 1976. Specificity of genetic elements controlling regulation of early functions in temperate bacteriophages. *J. Mol. Biol.* **106:** 537.

KAISER, A. D. and F. JACOB. 1957. Recombination between related temperate bacteriophages and the genetic control of immunity and prophage localization. *Virology* **4:** 509.

KLECKNER, N. 1977. Translocatable elements in prokaryotes. *Cell* **11:** 11.

KLECKNER, N., D. F. BARKER, D. G. ROSS, D. BOTSTEIN, J. A. SWAN, and M. ZABEAU. 1978. Properties of the translocatable tetracycline-resistance element Tn*10* in *Escherichia coli* and bacteriophage lambda. *Genetics* **90:** 427.

KOPECKO, D. J. 1980. Specialized genetic recombination systems in bacteria: Their involvement in gene expression and evolution. *Prog. Mol. Subcell. Biol.* **7:** 135.

LIEDKE-KULKE, M. and A. D. KAISER. 1967. Genetic control of prophage insertion specificity in bacteriophage λ and 21. *Virology* **32:** 465.

MACHATTIE, L. A. and J. B. JACKOWSKI. 1977. Physical structure and deletion effects of the chloramphenicol resistance element Tn*9* in phage lambda. In *DNA insertion elements, plasmids, and episomes* (ed. A. I. Bukhari et al.), p. 219. Cold Spring Harbor Laboratory, Cold Spring Harbor, New York.

MACHATTIE, L. A. and J. A. SHAPIRO. 1978. Chromosomal integration of phage by means of DNA insertion element. *Proc. Natl. Acad. Sci.* **75:** 1490.

OHTSUBO, H. and E. OHTSUBO. 1978. Nucleotide sequence of an insertion element, IS*1*. *Proc. Natl. Acad. Sci.* **75:** 615.

PTASHNE, M., K. BACKMAN, M. Z. HUMAYUN, A. JEFFREY, R. MAURER, B. MEYER, and R. T. SAUER. 1976. Autoregulation and function of a repressor in bacteriophage lambda. *Science* **194:** 156.

ROBERT-BAUDOUY, J. M. and R. C. PORTALIER. 1974. Studies of mutations in glucuronate catabolism in *E. coli* K12. *Mol. Gen. Genet.* **131:** 31.

ROBERTS, J. W. 1969. Termination factor for RNA synthesis. *Nature* **224:** 1168.

ROSNER, J. L. and M. M. GOTTESMAN. 1977. Transposition and deletion of Tn*9*: A transposable element carrying the gene for chloramphenicol resistance. In *DNA insertion elements, plasmids, and episomes* (ed. A. I. Bukhari et al.), p. 213. Cold Spring Harbor Laboratory, Cold Spring Harbor, New York.

SAEDLER, H. and B. HEISS. 1973. Multiple copies of the insertion-DNA sequences IS*1* and IS*2* in the chromosome of *E. coli* K-12. *Mol. Gen. Genet.* **122:** 267.

SHIMADA, K., R. A. WEISBERG, and M. E. GOTTESMAN. 1973. Prophage lambda at unusual chromosomal locations. II. Mutations induced by bacteriophage lambda in *Escherichia coli. J. Mol. Biol.* **80:** 297.

SCHMEISSNER, U., D. COURT, S. SHIMATAKE, and M. ROSENBERG. 1980. The promoter for the establishment of repressor synthesis in bacteriophage. *Proc. Natl. Acad. Sci.* **77:** 3191.

SOUTHERN, E. M. 1975. Detection of specific sequences among DNA fragments separated by gel electrophoresis. *J. Mol. Biol.* **98:** 503.

STAHL, F. W. and N. E. MURRAY. 1966. The evolution of gene clusters and genetic circularity in microorganisms. *Genetics* **53:** 569.

STARLINGER, P. and H. SAEDLER. 1976. IS-elements in microorganisms. *Curr. Top. Microbiol. Immunol.* **75:** 11.

SZPIRER, J., R. THOMAS, and C. M. RADDING. 1969. Hybrids of bacteriophage and ϕ80. A study of nonvegetative functions. *Virology* **37:** 585.

WEISBERG, R. A., S. GOTTESMAN, and M. E. GOTTESMAN. 1977. Bacteriophage λ: The lysogenic pathway. In *Comprehensive virology* (ed. H. Frankel-Conrat and R. Wagner), vol. 3, p. 197. Plenum Press, New York.

ZISSLER, J., E. SIGNER, and F. SCHAEFER. 1971. The role of recombination in growth of bacteriophage lambda. I. The gamma gene. In *The bacteriophage lambda* (ed. A. D. Hershey), p. 455. Cold Spring Harbor Laboratory, Cold Spring Harbor, New York.

Studies on Transposition Mechanisms and Specificity of IS*4*

R. KLAER, S. KÜHN,* H.-J. FRITZ, E. TILLMANN, I. SAINT-GIRONS,† P. HABERMANN, D. PFEIFER, AND P. STARLINGER

Institut für Genetik, Universität Köln, Federal Republic of Germany

Insertion sequence (IS) elements are transposable DNA elements that are normal constituents of the *Escherichia coli* chromosome. They differ in size from the larger transposons and they lack known genes encoding proteins with enzymatic function. Known effects are exerted in *cis* position only, and usually their presence cannot be selected for (for reviews, see Cohen 1976; Starlinger and Saedler 1976; Bukhari et al. 1977; Kleckner 1977; Starlinger 1980).

In the case of the larger transposons Tn*5* (Rothstein et al. 1980) and Tn*10* (N. Kleckner, pers. comm.), the large inverted repeats flanking the transposon encode functions necessary for transposition. For Tn*10*, the independent transposition of the flanking IS*10* has been demonstrated (Ross et al. 1979). Thus, structures resembling known IS elements can carry the gene(s) for their own transposition.

Is this also true for *E. coli* IS elements? This question cannot be answered easily for IS*1*, IS*2*, and IS*3*, because all of these elements are present in several copies in the *E. coli* chromosome (Saedler and Heiss 1973; Deonier et al. 1979) and cannot be altered simultaneously. Sequence analysis of IS*1* and IS*2* did not show large open reading frames, and therefore we turned to IS*4*, which is present in one copy only in the *E. coli* chromosome and thus may be more amenable to functional alterations.

In this paper we report our studies on the transposition of IS*4*, on the specificity of integration, and on the DNA sequence, and we try to relate our sequence to functional studies. In addition, we describe a new kind of chromosomal aberration observed with high frequency in the vicinity of IS*4*.

IS*4* Is Presently Known in Three Different Sites

Several strains of *E. coli* K12 carry only a single copy of IS*4*. Only a single band is seen when the DNAs of these strains are digested with either *Eco*RI, *Hin*dIII, *Hin*dII, *Kpn*I, or *Pvu*II and hybridized by the Southern blotting technique against purified IS*4* DNA or interior restriction fragments encompassing most of this element. These *Hin*dIII fragments from five strains have been cloned in pBR322, and restriction maps have been obtained. These maps showed unambiguously that only one copy of IS*4* is carried on the fragment.

IS*4*-containing fragments from different strains were found to be of different sizes. However, this is not due to IS*4* being located in different regions of the chromosome in these strains. The cloned fragments are derived from the same region of the chromosome. We showed this by obtaining Southern blots against restriction fragments from the cloned DNA of one strain that are located outside IS*4*. This seemed to indicate that IS*4* resided in the same site and that the DNA surrounding it had undergone some evolution. This assumption was proved to be correct by sequence analysis showing that IS*4* resides in exactly the same position in these strains (see below). This (yet unmapped) site is called the common site of IS*4*.

Strain F165 carries two copies of IS*4*. One copy is in the common site and the other copy is in a DNA fragment of different size that shares no homology with the DNA segments surrounding IS*4* in the common site that were used in the experiments mentioned above. We call this site the F165 site of IS*4* (Klaer and Starlinger 1980). The third site of IS*4* is found in *galT*, where insertions of this element have been observed repeatedly (Pfeifer et al. 1977).

Although these observations do not yet allow conclusions to be drawn about the distribution of IS*4* integration sites, they seem to indicate that the integration requirements may be more stringent than those for IS*1*. IS*4* in its "common site" is more conservatively retained than the DNA surrounding it. It will be interesting to see whether this is due to a selective value of IS*4* at this position.

IS*4* Generates 11-bp or 12-bp Duplications at the Site of Integration

At all of the three known sites of IS*4*, a short sequence is found duplicated at the termini. For the common site and for the F165 site, this sequence is 12 base pairs (bp) long. In the case of the common site, three different isolates showed the same integration site and the same duplication (Habermann et al. 1979; Klaer and Starlinger 1980).

At the integration site of IS*4* in *galT*, an 11-bp duplication is formed. This has been found in three independent isolates with IS*4*, being present in both of its possible orientations. In this case, the sequence of the *gal* operon in the integration region is known (Bidwell and Landy 1979). Prior to IS*4* integration, the 11-bp sequence is present only once. The same was found for one revertant to Gal$^+$ that had lost both IS*4* and one of the 11-bp duplications.

Present addresses: * Max-Planck-Institut für Zellphysiologie 6802 Ladenburg, Federal Republic of Germany; † Institut Pasteur, F-75724 Paris, Cedex 15, France.

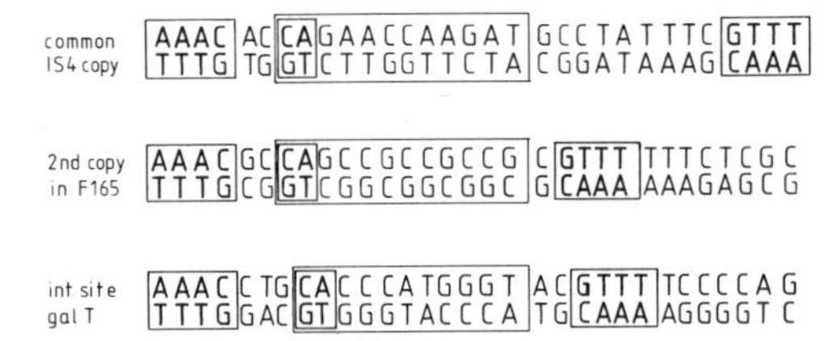

Figure 1. Integration sites of IS*4*. The common site of IS*4* on the chromosome is located in a *Hin*dIII fragment, the size of which differs in different *E. coli* K12 strains (Klaer and Starlinger 1980). The *Hin*dIII fragment from strain JB239 was cloned in pBR322 from a total DNA digest, and the appropriate colonies were identified by colony hybridization (Grunstein and Hogness 1975) against IS*4* DNA purified according to the method of Schmidt et al. (1976) or against internal restriction fragments of IS*4*. The F165 site is located in a 10.5-kb *Eco*RI fragment that shares no sequence homology with the *Hin*dIII fragment at the common site apart from IS*4*. IS*4* from *galT* was cloned from the DNA of the appropriate phage λd*galT*::IS*4*. Sequence determinations were done according to the method of Maxam and Gilbert (1977). The 5′ termini used for the determinations were generated by the cleavage within IS*4* of either an *Ava*I site at position 79 from one end or an *Hin*dII site located at position 26 from the other end. The sequence around IS*4* in the common site was confirmed by sequencing from the opposite strand, whereas the sequence of the integration site in *galT* is in agreement with the sequence determined by Bidwell and Landy (1979). The sequences are written assuming that the sequence in the absence of IS*4* would contain only one of the repeated sequences flanking the element. This was shown to be true in the case of *galT* only.

The three duplicated sequences flanking IS*4* in the three known sites are different (Fig. 1). Specifically, the sequences surrounding IS*4* in the F165 site and in *galT* are very different from the surrounding sequence in the common site. This shows that the duplications are generated during integration and cannot be created by a recombination event between one sequence carried along with the element and another sequence in the host chromosome. In this respect, IS*4* resembles IS*1* and IS*2* (Calos et al. 1978; Grindley 1978; Ghosal et al. 1979). It is distinguished from these, however, not only by the length of the duplication, but also by the fact that the length of the duplicated sequence is variable; with all other elements investigated so far, a duplication of fixed length was found for the same element at different sites. If the duplications are generated by the introduction of staggered nicks (Calos et al. 1978; Grindley 1978), the enzyme responsible must have a more flexible way to interact with the target DNA.

Specificity of IS*4* Integration

The 11-bp sequence duplicated at the integration site of IS*4* in *galT* shares a CA dinucleotide at one terminus with the duplications of 12 bp observed at the other two integration sites of the IS element. A dinucleotide clearly is not sufficient to specify a single site in *galT*, where the integration of IS*4* could be detected by our selection procedure. We therefore looked for sequence similarities outside the duplicated sequences. In all three sites we observed inverted AAA trinucleotides on either side of the duplication at a nearly fixed distance to the region duplicated upon integration. At a slightly more variable position, an AAAC tetranucleotide is found at both sides of IS*4* in all positions. We do not know whether these sequence similarities have any meaning for the specificity of integration. If they do, they parallel the slightly variable kind of interaction typical of RNA polymerase, with its variable distance between the Pribnow box and the −35-bp site (Rosenberg and Court 1979), rather than the very precise recognition sites found with those restriction enzymes recognizing symmetrical sites on the target DNA. Possibly, it may be more than a superficial analogy that restriction enzymes recognizing an asymmetrical site, like *Hph*I (Kleid et al. 1976), also show more variability in choosing the distance at which they cleave the DNA.

IS*4* May Be Replicated upon Transposition

Experiments with phage Mu (Ljungquist and Bukhari 1977) and with Tn*801* (Bennett et al. 1977) have indicated that transposons may be replicated upon transposition. Although the experiments done with elements present in many copies in the cell must be of a quantitative nature, IS*4* lends itself to a more direct experiment. Upon transposition of IS*4* from the common site to *galT*, a new band hybridizing to this element is expected; its size can be predicted from the known maps of the *gal* operon and from the length of IS*4*. This new band can be found in mutants such as those in Figure 2. In addition, the old band is still present, indicating a duplication of IS*4* during or after transposition (Klaer et al. 1980). However, it cannot be ruled out by this type of experiment that the chromosome arm donating the copy of IS*4* transferred to *galT* was subsequently lost and that the copy still residing in the common site did not participate in the transposition event (Berg 1977). The unambiguous demonstration of replication upon transposition will have to await experiments with appropriate replication mutants. However, this is difficult when an IS element is to be tested for which no direct selection procedure is known.

IS*4*::Cm Does Not Transpose

To facilitate transposition studies, we have introduced a *Pst*I fragment cleaved from a Tn*9*-like transposon (Iida and Arber 1977), and thus carrying permuted fragments of IS*1* and the gene encoding chloramphenicol (Cm) transacetylase, into the *Pst*I site of IS*4*. This site is located in the central part of IS*4*, and we hoped that it would not destroy *cis*-acting sequences, as we thought that these would be more likely to be found at the termini. The loss of a *trans*-acting function was thought not to be deleterious, since the intact IS*4* present in the chromosome was assumed to be able to complement any *trans*-acting function. An analogous experiment, in which we had introduced a kanamycin-resistance gene from Tn*5* into IS*2*, had yielded a structure perfectly capable of transposition. We used a *recA* strain lysogenic for both λ and λd*gal*. λd*gal* carries a known integration site for IS*4*, and as the DNA of this isolate is shorter

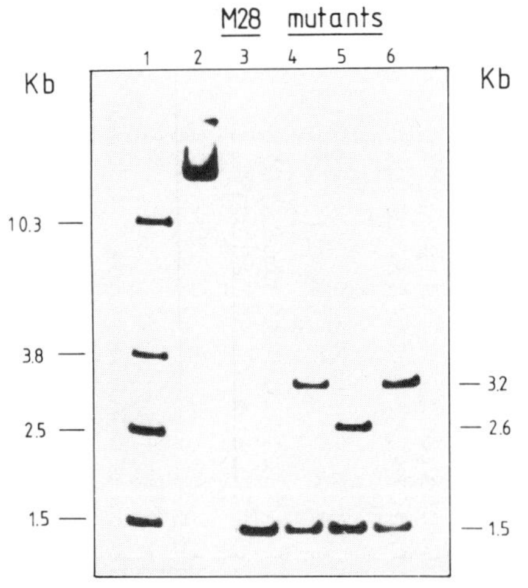

Figure 2. IS*4* is present in one additional copy after transposition to *galT*. Hybridization of ^{32}P-labeled IS*4* DNA against *Hin*dII fragments of the DNAs of the parental strain M28 (= *E. coli* K12 *galU*) (lane 3) and three *gal*^{r} mutants (lanes 4–6). These *gal*^{r} mutants had been tentatively characterized as being caused by the insertion of IS*4* by transducing them with an LFT λ lysate into a *galU*$^+$ background, mapping them by crosses with appropriate phage λd*gal* into deletion group 9 of *galT* (Pfeifer et al. 1977), and showing that they revert to Gal$^+$ at a frequency of 10^{-9}/cell plated. The molecular-weight marker (lane 1) consisted of appropriate digests of pKS51 (Klaer and Starlinger 1980), each of which yielded a fragment of different size capable of hybridizing to IS*4*. Lane 2 contains undigested bacterial DNA as a control to show that the transfer of large DNA to the nitrocellulose filter was sufficient (the method of Wahl et al. [1979] was used). The lengths of the fragments hybridizing to IS*4* in the three mutants are compatible with expectations from the restriction map of λ under the assumption that the IS*4* copies are integrated in both possible orientations. When larger fragments are used (cleavage with *Hin*dIII, which does not cleave within IS*4*), three bands of identical size and compatible with expectations from the restriction map are found (data not shown). The 1.5-kb band found in the parental strain is unaltered in the mutant, as is the 6.7-kb band after *Hin*dIII digestion (not shown).

than λ DNA, the transposition of IS*4*::Cm was not expected to impair phage maturation. The cells carried IS*4*::Cm on a pBR322 derivative. After induction of the phage, we found no (less than 10^{-11}) Cm transducers among the *gal* transducers. The inability of IS*4*::Cm to transpose was not due to the overproduction of an IS*4* repressor from this structure on the multicopy plasmid, since the transposition of the Cm-less IS*4* from the common site to *galT* was not measurably reduced, and no transposition of IS*4*::Cm was observed. Whether *cis*-acting proteins encoded by IS*4* are responsible for transposition, or whether sequences located centrally are required in *cis* for the transposition reaction, or whether other factors are responsible is not known.

The Sequence of IS4 Contains Terminal Inverted Repeats, a Large Open Reading Frame, and Possibly Also Control Sequences

To obtain some clues as to the functions carried out by IS*4*, we determined the DNA sequence, starting from a cloned fragment that carries IS*4* in the common site (Fig. 3). IS*4* is 1426 bp long. It carries at its termini 18-bp inverted repeats that have two mismatches in positions 8 and 9. In this respect, IS*4* is similar to all other insertion elements and transposons. Although the function of these inverted repeats is not known, it is conceivable that they make the two ends of the IS*4* molecule similar for an enzyme scanning the DNA sequence in the direction either into or out of IS*4*. This symmetry indicates that the transposable elements behave as rods with two nearly identical ends rather than as circles, as all plasmids and many bacteriophages do.

Like other IS elements, IS*4* is strongly polar in either orientation. In the case of IS*4*, structures are easily recognized from the sequences that resemble known, rho-independent transcription stop signals (Rosenberg and Court 1979). Moving into IS*4* in the 5′→3′ direction from either end, a potential stem-loop structure followed by a stretch of T residues is encountered soon after the termini of the element (Fig. 4). Although functional studies still have to be carried out to show that these structures are solely responsible for the strong polarity, the similarity to known signals is striking. This is different from the situation for IS*1* and IS*2*, which are also very polar, but do not show sequences that can be identified easily as polarity signals. It is conceivable that different polarity signals are used in different IS elements. If this were found to be true, it might be asked whether a polarity signal of one kind or another has a selective value for the survival of IS elements.

Most of the interior of IS*4* is occupied by a long open reading frame starting with an ATG codon at position 85 and ending with TAA at position 1413. A sequence of this length could encode 442 amino acids, and the polypeptide would have a molecular weight of 54,000, as deduced from the amino acid sequence encoded by the open frame.

Counting from the putative ATG start codon, the sequence GGA occupies positions −8 to −10, where such a sequence would be expected in a ribosome-binding site. In the case of the f1 coat-protein gene, the binding site consists of the same three nucleotides, which are even displaced 2 nucleotides to the right when compared with the usual position of this trinucleotide relative to the initiator codon (Steitz 1979). We conclude that the requirements for a ribosome-binding site may be met.

The positions for a possible promoter for the large IS*4* gene are rather restricted, because transcription starting upstream relative to the putative transcription stop signal would most probably come to a halt here. As the potential ribosome-binding site is located within the stem of this structure, the starting nucleotide of the leader sequence must also be sought within this structure, possibly at a purine far enough inside the stem-loop structure to preclude the formation of the RNA secondary structure. We suggest that A59 or G61 fulfills these requirements. A Pribnow box must then be found at the appropriate distance and can be tentatively identified as AAGAATC starting at position 48. The fit with the canonical Pribnow box is not very good, but it becomes better when the sequence immediately preceding

```
        10        20        30        40        50        60        70        80        90       100
         .         .         .         .         .         .         .         .         .         .
TAATGCCGATCAGTTAAGGATCAGTTGACCGATCCAGTGGCTGTGTAAGAATCCGGAAACGCTCACTTGTTTCCGGATTTTTTTATGCACATTGGACAGG
ATTACGGCTAGTCAATTCCTAGTCAACTGGCTAGGTCACCGACACATTCTTAGGCCTTTGCGAGTGAACAAAGGCCTAAAAAAATACGTGTAACCTGTCC

       110       120       130       140       150       160       170       180       190       200
         .         .         .         .         .         .         .         .         .         .
CTCTTGATCTGGTATCCCGTTACGATTCTCTGCGTAACCCACTGACTTCTCTGGGGGATTACCTCGACCCCGAACTCATCTCTCGTTGCCTTGCCGAATC
GAGAACTAGACCATAGGGCAATGCTAAGAGACGCATTGGGTGACTGAAGAGACCCCCTAATGGAGCTGGGGCTTGAGTAGAGAGCAACGGAACGGCTTAG

       210       220       230       240       250       260       270       280       290       300
         .         .         .         .         .         .         .         .         .         .
AGGTACTGTAACGCTACGCAAGCGCCGTCTTCCCCTCGAAATGATGGTCTGGTGTATTGTTGGCATGGCGCTTGAGCGTAAAGAACCTCTTCACCAGATT
TCCATGACATTGCGATGCGTTCGCGGCAGAAGGGGAGCTTTACTACCAGACCACATAACAACCGTACCGCGAACTCGCATTTCTTGGAGAAGTGGTCTAA

       310       320       330       340       350       360       370       380       390       400
         .         .         .         .         .         .         .         .         .         .
GTGAATCGCCTGGACATCATGCTGCCGGGCAATCGCCCCTTCGTTGCCCCCAGTGCCGTTATTCAGGCCCGCCAGCGCCTGGGAAGTGAGGCTGTCCGCC
CACTTAGCGGACCTGTAGTACGACGGCCCGTTAGCGGGGAAGCAACGGGGGTCACGGCAATAAGTCCGGGCGGTCGCGGACCCTTCACTCCGACAGGCGG

       410       420       430       440       450       460       470       480       490       500
         .         .         .         .         .         .         .         .         .         .
GCGTGTTCACGAAAACAGCGCAGCTCTGGCATAACGCCACGCCGCATCCGCACTGGTGCGGCCTGACCCTGCTGGCCATCGATGGTGTGTTCTGGCGCAC
CGCACAAGTGCTTTTGTCGCGTCGAGACCGTATTGCGGTGCGGCGTAGGCGTGACCACGCCGGACTGGGACGACCGGTAGCTACCACACAAGACCGCGTG

       510       520       530       540       550       560       570       580       590       600
         .         .         .         .         .         .         .         .         .         .
ACCGGATACACCAGAGAACGATGCAGCCTTCCCCCGCCAGACACATGCCGGGAACCCGGCGCTCTACCCGCAGGTCAAAATGGTCTGCCAGATGGAACTG
TGGCCTATGTGGTCTCTTGCTACGTCGGAAGGGGGCGGTCTGTGTACGGCCCTTGGGCCGCGAGATGGGCGTCCAGTTTTACCAGACGGTCTACCTTGAC

       610       620       630       640       650       660       670       680       690       700
         .         .         .         .         .         .         .         .         .         .
ACCAGCCATCTGCTGACGGCTGCAGCCTTCGGCACGATGAAGAACAGCGAAAATGAGCTTGCTGAGCAACTTATAGAACAAACCGGCGATAACACTCTGA
TGGTCGGTAGACGACTGCCGACGTCGGAAGCCGTGCTACTTCTTGTCGCTTTTACTCGAACGACTCGTTGAATATCTTGTTTGGCCGCTATTGTGAGACT

       710       720       730       740       750       760       770       780       790       800
         .         .         .         .         .         .         .         .         .         .
CGTTAATGGATAAAGGTTATTACTCACTGGGACTGTTAAATGCCTGGAGCCTGGCGGGAGAACACCGCCACTGGATGATACCTCTCAGAAAGGGAGCGCA
GCAATTACCTATTTCCAATAATGAGTGACCCTGACAATTTACGGACCTCGGACCGCCCTCTTGTGGCGGTGACCTACTATGGAGAGTCTTTCCCTCGCGT
```

```
       810       820       830       840       850       860       870       880       890       900
         .         .         .         .         .         .         .         .         .         .
ATATGAAGAGATCAGAAAACTGGGTAAAGGCGATCATCTGGTGAAGCTGAAAACCAGCCCGCAGGCACGAAAAAAGTGGCCGGGACTGGGAAATGAAGTG
TATACTTCTCTAGTCTTTTGACCCATTTCCGCTAGTAGACCACTTCGACTTTTGGTCGGGCGTCCGTGCTTTTTTCACCGGCCCTGACCCTTTACTTCAC

       910       920       930       940       950       960       970       980       990      1000
         .         .         .         .         .         .         .         .         .         .
ACTGCCCGCCTGCTGACCGTGACGCGCAAAGGAAAAGTCTGCCATCTGCTGACGTCGATGACGGACGCCATGCGCTTCCCCGGAGGAGAAATGGGGGATC
TGACGGGCGGACGACTGGCACTGCGCGTTTCCTTTTCAGACGGTAGACGACTGCAGCTACTGCCTGCGGTACGCGAAGGGGCCTCCTCTTTACCCCCTAG

      1010      1020      1030      1040      1050      1060      1070      1080      1090      1100
         .         .         .         .         .         .         .         .         .         .
TGTACAGTCATCGCTGGGAAATCGAACTGGGATACAGGGAGATAAAACAGACGATGCAACGGAGCAGGCTGACGCTGAGAAGTAAAAAGCCGGAGCTTGT
ACATGTCAGTAGCGACCCTTTAGCTTGACCCTATGTCCCTCTATTTTGTCTGCTACGTTGCCTCGTCCGACTGCGACTCTTCATTTTTCGGCCTCGAACA

      1110      1120      1130      1140      1150      1160      1170      1180      1190      1200
         .         .         .         .         .         .         .         .         .         .
GGAGCAAGAGCTGTGGGGTGTCTTACTGGCTTATAATCTGGTGAGATATCAGATGATTAAAATGGCGGAACATCTGAAAGGTTACTGGCCGAATCAACTG
CCTCGTTCTCGACACCCCACAGAATGACCGAATATTAGACCACTCTATAGTCTACTAATTTTACCGCCTTGTAGACTTTCCAATGACCGGCTTAGTTGAC

      1210      1220      1230      1240      1250      1260      1270      1280      1290      1300
         .         .         .         .         .         .         .         .         .         .
AGTTTCTCAGAATCATGCGGAATGGTGATGAGAATGCTGATGACATTGCAGGGCGCTTCACCGGGACGTATACCGGAGCTGATGCGCGATCTTGCAAGTA
TCAAAGAGTCTTAGTACGCCTTACCACTACTCTTACGACTACTGTAACGTCCCGCGAAGTGGCCCTGCATATGGCCTCGACTACGCGCTAGAACGTTCAT

      1310      1320      1330      1340      1350      1360      1370      1380      1390      1400
         .         .         .         .         .         .         .         .         .         .
TGGGACAACTTGTGAAATTACCGACAAGAAGGGAAAGGGCCTTCCCGAGAGTGGTAAAGGAGAGGCCCTGGAAATACCCCACAGCCCCGAAAAAGAGCCA
ACCCTGTTGAACACTTTAATGGCTGTTCTTCCCTTTCCCGGAAGGGCTCTCACCATTTCCTCTCCGGGACCTTTATGGGGTGTCGGGGCTTTTTCTCGGT

      1410      1420
         .         .
GTCAGTTGCTTAACTGACTGGCATTA
CAGTCAACGAATTGACTGACCGTAAT
```

Figure 3. The sequence of IS*4*. The sequence was determined according to the method of Gilbert and Maxam (1977). Restriction fragments were produced with different enzymes. Both strands were sequenced completely, and complementarity was established.

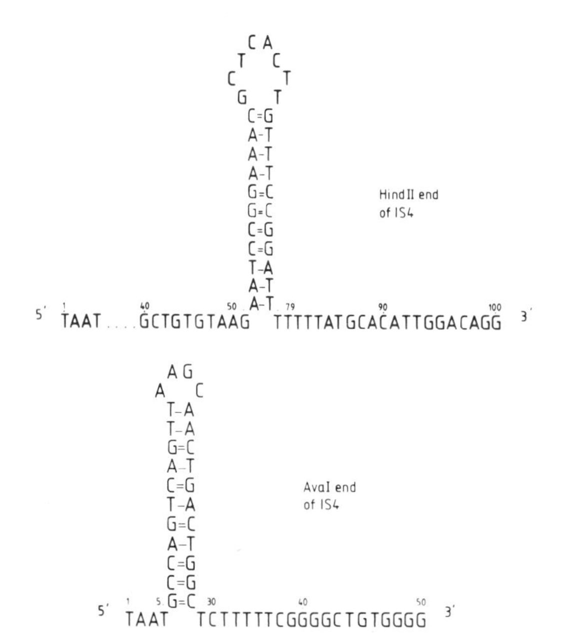

Figure 4. Two possible transcription stop signals near the termini of IS*4*.

the heptanucleotide is taken into consideration (Rosenberg and Court 1979) (Fig. 5).

The sequence GTTGAC at positions 24 to 28 qualifies very well for a −35 site. Considering these features of the sequence, it seems possible that the large open frame can be transcribed and translated. However, this does not seem to occur in vivo. Any considerable transcription of IS*4* extending all the way through the element until at least the TAA that terminates the open reading frame would have no signal left within IS*4* to terminate here. It would have to be assumed that transcription should extend into adjacent sequences. Within *galT*, this should lead to readthrough transcription into the transferase and kinase genes. It is conceivable that the lack of translation of the distal part of *galT* would allow the stop of transcription within *galT*, but this

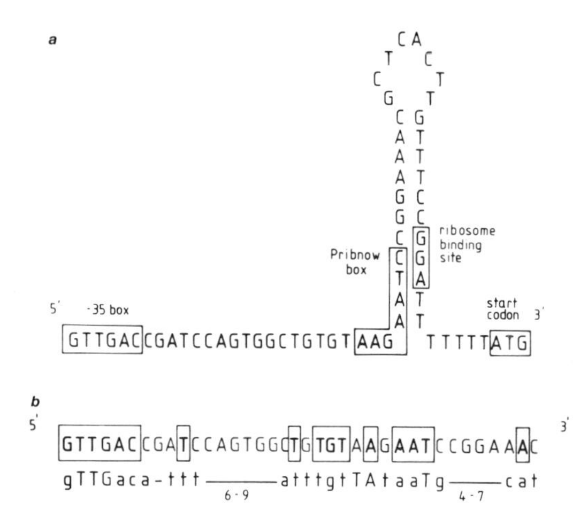

Figure 5. (*a*) Hypothetical promoter and ribosome-binding site preceding the large open reading frame of IS*4*; (*b*) comparison of the hypothetical promoter with the consensus sequence of other promoters (Rosenberg and Court 1979).

would not be more effective than has been found with amber mutations in *galT* (Jordan and Saedler 1967). As these mutations allow a residual synthesis of galactokinase of more than 10% of the induced wild-type level, we would expect the same rate of synthesis for the IS*4* oriented in orientation II. However, in both of the orientations the residual kinase is not measurable and thus is at least a factor of 100 lower than expected (Besemer and Herpers 1977). We must conclude either that the putative promoter is not active, though very similar structures can initiate transcription effectively, or that the IS*4* gene is regulated.

When we look for genes involved in repression, either we can consider autoregulation or we must look for another gene. We cannot rule out a priori the possibility that this gene is located outside IS*4*. If it is in IS*4*, it cannot be in another frame in the same orientation, as no other decent open frames are found here. The longest open reading frame next to the large one is detected in orientation I. It would encode 131 amino acids, starting with ATG in position 609 (reading backward). Here, no decent ribosome-binding site is present, but in the case of the λcI maintenance promoter, this is also lacking (Ptashne 1978). The sequences CAAGCTC at positions 661 to 655 and TTTGT at positions 682 to 678 might qualify marginally for a Pribnow box and a −35 site, respectively. Again, the argument regarding the polarity of IS*4*, this time in orientation I, does not call for pronounced transcription in vivo, and as no transcription terminators are found reading outward in this direction, again, either the gene is not transcribed efficiently or it must be under (auto) regulation. If it were to encode a repressor, it might be expressed only weakly.

To test the assumptions made about the sequences found in IS*4*, two approaches can be used. One approach is to look for the products of transcription and translation predicted by the sequence in vivo (minicells) and in vitro (pure transcription system and cell-free system for protein synthesis). This is presently being done. The other approach would be to construct mutants in the putative genes, e.g., by insertion or removal of fragments, and to study the functional consequences. This latter approach has only been started and, in the case of studies with *trans*-acting functions, will require the construction of an IS*4*-less bacterium. This is presently being attempted.

Bidirectional Deletions around IS*4*

Chromosomal aberrations near IS elements have often been described. The prototype is the adjacent deletion at IS*1* that terminates exactly at the element and has the other terminus a variable distance from it (Reif and Saedler 1975). Similar, as well as more complicated, aberrations have been described near Tn*3* (Nisen et al. 1978; Weinstock et al. 1979), Tn*10* (Kleckner et al. 1979; Ross et al. 1979), and phage Mu (Toussaint et al. 1977). In all these cases, the resulting structure originated from the fusion of one terminus of the element with non-IS DNA.

Here, we report preliminary observations on a type of

deletion that not only removes IS*4*, but also removes bacterial DNA sequences adjacent to both termini of IS*4*. Thus, in these deletions, two non-IS termini are linked. These deletions were first observed in heteroduplex studies with DNA extracted from phage λd*galT*::IS*4*, carrying IS*4* in both orientations. Although these observations indicated that the deletions occurred on only those DNAs that carry IS*4*, they were not sufficient to prove the point. We therefore constructed plasmids that were derived from pBR322 and carried DNA segments of various lengths, including the *gal* operon with IS*4* inserted in *galT*. Isogenic variants were prepared by selection of Gal^+ revertants shown to differ from the original plasmids only by the absence of IS*4*.

The results of experiments with two of these plasmids are presented in Table 1. The deletions of the type described above are found at a frequency of about 10^{-3} of the cells tested. The revertants that have lost IS*4* do not yield deletions at a comparable frequency. From these observations, we conclude that the formation of the deletions is dependent on the presence of IS*4* between the endpoints of the deletion, because the Gal^+ revertants still carry IS*4* in the bacterial chromosome. However, initial experiments with two other plasmids and with IS*4* in the chromosome have yielded either no deletions at all or deletions at a reduced frequency even in the IS*4*-free plasmids. We are continuing our experiments on this point in order to clarify the dependence of the deletions observed on the presence of IS*4*. We do not have a convincing hypothesis for the creation of deletion junctions removed from the IS*4* sequence and still dependent on it. Chow et al. (1974) described the formation of λdv particles and explained the frequent observation of head-to-head dimers with the assumption that bidirectional replication starting within λ DNA in some cases comes to a halt within this DNA and that the θ structure thus formed can be excised from the nonreplicated DNA and be replicated as such. If the two ends of the nonreplicated DNA could be joined, a bidirectional deletion would result. We have no evidence, however, for a replication initiated within IS*4*. More experiments are needed to show in which sites the deletions occur and to obtain clues on the mechanism.

Table 1. Bidirectional Deletions Produced by pBR322 Derivatives Carrying Bacterial DNA Sequences, Including the *gal* Operon with IS*4*

Plasmid	No. of colonies screened	No. of white colonies found	Frequency
pKS106	12,340	36	2.9×10^{-3}
pKS106-R	9,841	0	
pKS104	7,232	9	1.2×10^{-3}
pKS104-R	7,854	0	
pKS105	7,831	18	2.3×10^{-3}
pKS105-R	12,080	9	9.0×10^{-4}

Two of the plasmids (pKS104 and pKS105) carry in the *Pst*I site of pBR322 a *Pst*I fragment of λd*galT*75::IS*4*. The latter phage carries in addition to IS*4* in *galT* a copy of IS*1*, 11.7 kb upstream of the *gal* operon (R. Klaer, unpubl.). The *Pst*I site in IS*1* is one terminus of the fragments inserted in all three plasmids. pKS104 and pKS105 differ in fragment orientation relative to the plasmids. pKS104-R and pKS105-R are derived from their parent plasmid by selection for Gal^+. The precise excision of IS*4* was verified by restriction analysis (data not shown). pKS106 is derived from pKS104-R by replacement of a DNA fragment, extending from the *Hin*dIII site in *galE* through the *gal* operon to the *Eco*RI site of pBR322, with a fragment derived from pKS107, extending from the *Hin*dIII site in *galE* through *galT* with IS*4* to an *Eco*RI site distal to the *gal* operon. pKS104 and pKS106 are thus similar, but the latter carries approximately 5.4 kb of bacterial DNA that is not present in pKS104, and it also lacks ~700 bp of pBR322. The total lengths of the plasmids are 20.4 kb and 25.1 kb, respectively. The plasmids were propagated in a Δ*gal recA* strain. DNAs of the plasmids were prepared and used to transform *galE-N*10 to tetracycline resistance (Tc^r). Most of the colonies became Gal^+, due to the complementation of the *galE* mutation by the functional epimerase gene in the plasmid. Bidirectional deletions were assumed to extend into the epimerase gene and thus to give rise to white Gal^- Tc^r colonies. White colonies were picked, and, after purification, they were hybridized against a *Hin*dIII-*Pvu*II fragment (^{32}P-labeled by nick translation), which extends from the center of *galE* to the end of *galT* and contains IS*4*. All white colonies failed to hybridize to this fragment, indicating a lack in the plasmids not only of IS*4*, but also of sequences adjacent to the IS element. The residual hybridization due to the sequences in the chromosome was readily distinguished from the stronger signal due to sequences carried by the multicopy plasmid. No deletions were found in plasmids pKS106-R and pKS104-R, which are derived from their parental plasmids by reversion to Gal^+ by excision of IS*4*. The failure to detect deletions can be ascribed to the lack of IS*4*, to the decrease in length, or to an unknown factor. As pKS104 is shorter than pKS106-R, we do not believe that length per se is important. Deletions are observed in pKS105-R, though at a lower frequency than in the parental plasmid. All but one of these carry an additional copy of IS*1*, a feature not observed in those deletions tested for IS*1* that were obtained from pKS105. We believe that the deletions obtained from pKS105-R are of a different kind and need further study.

DISCUSSION

At present, IS*4* is known in three sites, but the transposition into only one of them has been observed repeatedly. This site is located in *galT*. As the *gal* operon has also been seen as the target for the transposition of IS*1* and IS*2*, their transposition specificities can be compared. Clearly, IS*4* is different in this respect from the other two. IS*2* has been seen in the leader sequence and *galE* only, and IS*1* is found there preferentially, although it has been seen in various other positions in *galE* and *galT* (Fiandt et al. 1972; Hirsch et al. 1972).

As IS*1* junctions with surrounding DNA have been sequenced both in *galT* (Grindley 1978) and in the leader sequence (Kühn et al. 1979), and in many other sites (reviewed by D. J. Galas et al., in prep.), the integration sites of IS*1* and IS*4* can be compared at the sequence level. The following differences are observed:

1. In the leader sequence, four isolates of IS*1* occupy three narrowly spaced positions. Three isolates of IS*4* in *galT* (in both orientations) are found in the same position.
2. Of the 9-bp duplications flanking IS*1*, a GC pair is found at both ends in 23 of 28 cases (D. J. Galas et al., in prep.); of the exceptions, two come from the *gal* operon. IS*4* in its three sites has an 11- or 12-bp duplication that carries a CA(G) at one end and no similarity at the other end.

3. Near the integration sites of IS*1*, a limited homology with the termini of IS*1* is often observed (D. J. Galas et al., in prep.). This is similar to the situation observed with Tn*3* (C.-P. D. Tu and S. N. Cohen, in prep.). Very little homology is seen between sequences near the three integration sites of IS*4* and their inverted termini.
4. IS*1* prefers AT-rich sequences (Meyer et al. 1980; D. J. Galas et al., in prep.). Again, this is similar to the situation with Tn*3* (C.-P. D. Tu and S. N. Cohen, in prep.). For IS*4*, no AT preference is seen with regularity. Although the 110 bp around the chromosomal site of IS*4* have an AT content of 64%, the two sites at which transposition has been recently observed have AT contents of 33% (F165 site) and 46% (*galT*), respectively.

Is it possible to formulate hypotheses about recognition on the basis of this limited information? In the case of IS*4*, the data seem compatible with recognition of a site on double-stranded DNA. The length of the region would be about 25 bp to 26 bp; the sites of cleavage would be on one side of the DNA molecule. The inverted AAA sequences would also be approximately above each other, but at an angle to the cleavage sites. Still, the enzyme would mainly make contact with IS*4* from one side.

In the case of IS*1*, the situation seems to be more complicated. Although the introduction of staggered nicks exactly 9 bp apart might occur also on the double strand with its exact geometry, the combination of AT-rich regions and of homology between the IS element and the integration site suggests that partial melting and pairing of single strands play a role. One way to explain these data would be to assume that, independent of IS insertion, staggered nicks are introduced unspecifically by an unknown enzyme (topoisomerase?) that would hold the cleaved DNA together until ligation of at least one strand had taken place. Only at AT-rich sites would partial melting and pairing to a nicked IS*1* occasionally lead to the ligation of the host DNA termini to this element rather than to their previous junctions. The situation must be more complicated than this, however, because random staggered nicks, followed by partial denaturation and pairing to an IS element, should occasionally place the IS*1* integration site next to a 5-bp nick, and the creation of a 5-bp duplication next to IS*1* should be the consequence. However, even in the leader sequence of the *gal* operon, where IS*1* and IS*2* are found in nearly identical positions, IS*1* is always flanked by a 9-bp duplication and IS*2* by a 5-bp duplication. It must therefore be postulated that specificity is already involved at the time the staggered nicks are formed. It is difficult to reconcile these data without further information. Perhaps a partially melted region where IS*1* invades the DNA induces a 9-bp staggered nick in neighboring double-stranded DNA.

With regard to the mechanism of transposition, our experiments are compatible with the replication models of Grindley and Sherratt (1979) and Shapiro (1979), without proving them. Another problem is the location of the transposition genes and their regulation. From the data on Tn*5* (Meyer et al. 1979; Rothstein et al. 1980) and on Tn*10* (N. Kleckner, pers. comm.), the possibility emerges that the functions necessary for transposition are encoded by the IS elements themselves. The data on IS*4* are compatible with this hypothesis. The failure of IS*4*, carrying a DNA segment encoding chloramphenicol resistance in its *Pst*I site, to transpose from the chromosomal site carried in a plasmid to λd*gal* suggests a function of IS*4* that can be insertionally inactivated. As the cells carry an additional copy of IS*4* in the chromosome, this might even be a *cis*-acting function. It is not impossible that this function is similar to the recombination site found in Tn*3* (Gill et al. 1978; Arthur and Sherratt 1979), since the experiments have been done in a *recA* background. It is thus conceivable that the 12-kb plasmid carrying the marked IS*4* has been fused into λ DNA (rendering it too large to be packaged) without elimination of the rest of the plasmid. This possibility is being tested.

It will be interesting to study the control of transposition genes, as these have to encode functions that are expressed very rarely. It will have to be determined whether the products of these genes are synthesized in a burst in rare cells and whether several consecutive or simultaneous transposition events occur in these cells. In maize, the simultaneous action of a *trans*-acting factor encoded by the controlling element *Ac* on several copies of the transposable element *Ds* in a small minority of endosperm cells has been reported (McClintock 1949).

The sequences of IS*1* and IS*2* do not show as much easily recognizable genetic information as do those of IS*50* and IS*10* (the IS elements of Tn*5* and Tn*10*, respectively) and IS*4*. These latter elements are present in two copies and one copy only in most bacterial cells; of the two copies of the elements flanking the transposons, only one seems to be fully active (Rothstein et al. 1980; N. Kleckner, pers. comm.). IS*1* and IS*2*, on the other hand, are present in several copies each. Is it conceivable that many of the IS*1* and IS*2* copies are devoid of (some) transposition information and depend on a master IS element? This again would be similar to the observation on controlling elements in maize, which often consist of both transposition-active and transposition-passive elements (McClintock 1965).

A further point deserving attention is the uniform size of IS elements. IS*2* through IS*5*, IS*8*, and the elements flanking Tn*5* and Tn*10* are all in the range of 1.2 kb to 1.5 kb (Bukhari et al. 1977; Depicker et al. 1980). Only two of them are larger (Tn*3*, which by the structure of its inverted repeats is more similar to IS elements than to the larger composite transposons [Ohtsubo et al. 1979] and which contains transposition information of approximately 4.5 kb [Chou et al. 1979; Gill et al. 1979], and γδ [Reed et al. 1979]), and two are smaller (IS*1* and the short IS-like sequence found in pSC101 [Ravetch et al. 1979]).

Of the normal-size IS elements, two (IS*50* and IS*4*) seem to be so densely packed with genetic information that even overlapping functions are seen (e.g., between the genes encoding peptides 3 and 4 and the *neo* promoter in IS*50* or between the inverted repeats and one of the putative transcription stop sites on IS*4*). It is not likely that this coincidence is fortuitous. If it has a reason, this must probably be looked for in functions common to most IS elements (but different for Tn*3*). As recombination is involved, a length dependence might indicate the need for a certain amount of superhelicity at a certain stage of the process.

The study of IS elements and transposons has shown that transpositions of genetic material and other chromosomal aberrations can be produced at a rate equal to or higher than that of point mutations. In light of these findings, it is astonishing that the genetic map of *E. coli* is relatively stable, as can be seen from the comparison of the maps of *E. coli* and *Salmonella typhimurium*. Although these two bacteria differ considerably in base sequence, their maps are still rather similar. It is hard to avoid the conclusion that, in nature, a powerful selection must be at work that conserves the general DNA arrangements against the dispersive forces of transposition. It will be interesting to learn what the selective values of the prevailing genetic map are.

ACKNOWLEDGMENT

This work has been supported by Deutsche Forschungsgemeinschaft through SFB 74.

REFERENCES

ARTHUR, A. and D. SHERRATT. 1979. Dissection of the transposition process: A transposon-encoded site-specific recombination system. *Mol. Gen. Genet.* **175:** 267.

BENNETT, P., J. GRINSTED, and M. RICHMOND. 1977. Transposition of TnA does not generate deletions. *Mol. Gen. Genet.* **154:** 205.

BERG, D. E. 1977. Insertion and excision of the transposable kanamycin resistance determinant Tn*5*. In *DNA insertion elements, plasmids, and episomes,* (ed. A. I. Bukhari et al.), p. 205. Cold Spring Harbor Laboratory, Cold Spring Harbor, New York.

BESEMER, J. and M. HERPERS. 1977. Suppression of polarity of insertion mutations within the *gal* operon of *E. coli. Mol. Gen. Genet.* **151:** 295.

BIDWELL, K. and A. LANDY. 1979. Structural features of λ site-specific recombination of a secondary *att* site in *galT. Cell* **16:** 397.

BUKHARI, A. I., J. A. SHAPIRO, and S. L. ADHYA, eds. 1977. *DNA insertion elements, plasmids, and episomes.* Cold Spring Harbor Laboratory, Cold Spring Harbor, New York.

CALOS, M. P., L. JOHNSRUD, and J. H. MILLER. 1978. DNA sequences at the integration sites of the insertion element IS*1*. *Cell* **13:** 411.

CHOU, J., P. LEMAUX, M. CASADABAN, and S. N. COHEN. 1979. Identification and characterization of a bi-directional genetic unit regulating translocation of the Tn*3* element. *Nature* **282:** 801.

CHOW, L., D. E. BERG, and N. DAVIDSON. 1974. Electron microscope study of the structures of λdv DNAs. *J. Mol. Biol.* **86:** 69.

COHEN, S. N. 1976. Transposable genetic elements and plasmid evolution. *Nature* **263:** 731.

DEONIER, R., R. HADLEY, and M. HU. 1979. Enumeration and identification of IS*3* elements in *Escherichia coli* strains. *J. Bacteriol.* **137:**1421.

DEPICKER, A., M. DEBLOCK, D. INZE, M. VAN MONTAGU, and J. SCHELL. 1980. IS-like element IS*8* in RP4 plasmid and its involvement in cointegration. *Gene* **10:** 329.

FIANDT, M., W. SZYBALSKI, and M. H. MALAMY. 1972. Polar mutations in *lac, gal* and phage lambda consist of a few DNA sequences inserted with either orientation. *Mol. Gen. Genet.* **119:** 223.

GHOSAL, D., H. SOMMER, and H. SAEDLER. 1979. Nucleotide sequence of the transposable element IS*2*. *Nucleic Acids Res.* **6:** 1111.

GILL, R., F. HEFFRON, and S. FALKOW. 1979. Identification of the protein encoded by the transposable element Tn*3* which is required for its transposition. *Nature* **282:** 797.

GILL, R., F. HEFFRON, G. DOUGAN, and S. FALKOW. 1978. Analysis of sequences transposed by complementation of two classes of transposition-deficient mutants of Tn*3*. *J. Bacteriol.* **136:** 742.

GRINDLEY, N. D. F. 1978. IS*1* insertion generates duplication of a nine base sequence at its target site. *Cell* **13:** 419.

GRINDLEY, N. D. F. and D. SHERRATT. 1979. Sequence analysis at IS*1* insertion sites: Models for transposition. *Cold Spring Harbor Symp. Quant. Biol.* **43:** 1257.

GRUNSTEIN, M. and D. S. HOGNESS. 1975. Colony hybridization: A method for the isolation of cloned DNAs that contain a specific gene. *Proc. Natl. Acad. Sci.* **72:** 3961.

HABERMANN, P., R. KLAER, S. KÜHN, and P. STARLINGER. 1979. IS*4* is formed between eleven or twelve base pair duplications. *Mol. Gen. Genet.* **175:** 369.

HIRSCH, H. J., P. STARLINGER, and P. BRACHET. 1972. Two kinds of insertions in bacterial genes. *Mol. Gen. Genet.* **119:** 191.

IIDA, S. and W. ARBER. 1977. Plaque-forming specialized transducing phage P1: Isolation of P1 Cm Sm Su, a precursor of P1 Cm. *Mol. Gen. Genet.* **153:** 259.

JORDAN, E. and H. SAEDLER. 1967. Polarity of amber mutations and suppressed amber mutations in the galactose operon of *E. coli. Mol. Gen. Genet.* **100:** 282.

KLAER, R., and P. STARLINGER. 1980. IS*4* chromosomal site in *E. coli. Mol. Gen. Genet.* **178:** 285.

KLAER, R., D. PFEIFER, and P. STARLINGER. 1980. IS*4* is still found in its chromosomal site after transposition to *galT. Mol. Gen. Genet.* **178:** 281.

KLECKNER, N. 1977. Translocatable elements in procaryotes. *Cell* **11:** 11.

KLECKNER, N., K. REICHARDT, and D. BOTSTEIN. 1979. Inversions and deletions of the *Salmonella* chromosome generated by the translocatable tetracycline-resistance element Tn*10*. *J. Mol. Biol.* **127:** 89.

KLEID, D., Z. HUMAYUN, A. JEFFREY, and M. PTASHNE. 1976. Novel properties of a restriction endonuclease isolated from *Haemophilus parahaemolyticus. Proc. Natl. Acad. Sci.* **73:** 293.

KÜHN, S., H.-J. FRITZ, and P. STARLINGER. 1979. Close vicinity of IS*1* integration sites in the leader sequence of the *gal* operon of *E. coli. Mol. Gen. Genet.* **167:** 235.

LJUNGQUIST, E. and A. I. BUKHARI. 1977. State of prophage Mu DNA upon induction. *Proc. Natl. Acad. Sci.* **74:** 3143.

MAXAM, A. and W. GILBERT. 1977. A new method for sequencing DNA. *Proc. Natl. Acad. Sci.* **74:** 560.

MCCLINTOCK, B. 1949. Mutable loci in maize. *Carnegie Inst. Wash. Year Book* **48:** 142.

———. 1965. The control of gene action in maize. *Brookhaven Symp. Biol.* **18:** 162.

MEYER, R., G. BOCH, and J. A. SHAPIRO. 1979. Transposition of DNA inserted into deletions of the Tn*5* kanamycin resistance element. *Mol. Gen. Genet.* **171:** 7.

MEYER, J., S. IIDA, and W. ARBER. 1980. Does the insertion element IS*1* transpose preferentially into A+T-rich DNA segments? *Mol. Gen. Genet.* (in press).

NISEN, P. D., D. J. KOPECKO, J. CHOU, and S. N. COHEN. 1978. Site-specific DNA deletions occurring adjacent to the termini of transposable ampicillin resistance element (Tn*3*). *J. Mol. Biol.* **117:** 975.

OHTSUBO, H., H. OHMORI, and E. OHTSUBO. 1979. Nucleotide sequence analysis of Tn*3* (Ap): Implications for insertion and deletion. *Cold Spring Harbor Symp. Quant. Biol.* **43:** 1269.

PFEIFER, D., D. KUBAI-MARONI, and P. HABERMANN. 1977. Specific sites for integration of IS elements within the transferase gene of the *gal* operon of *E. coli* K12. In *DNA insertion elements, plasmids, and episomes* (ed. A. I. Bukhari et al.), p. 31. Cold Spring Harbor Laboratory, Cold Spring Harbor, New York.

PTASHNE, M. 1978. Gene control by the lambda phage repressor. In *Integration and excision of DNA molecules* (ed. P. H. Hofschneider and P. Starlinger), p. 1. Springer-Verlag Berlin.

RAVETCH, J., M. OHSUMI, P. MODEL, G. VOVIS, D. FISCHOFF, and N. ZINDER. 1979. Organization of a hybrid between phage f1 and plasmid pSC101. *Proc. Natl. Acad. Sci.* **76:** 2195.

REED, R. R., R. A. YOUNG, J. A. STEITZ, N. GRINDLEY, and M. S. GUYER. 1979. Transposition of the *E. coli* insertion element $\gamma\delta$ generates a five base pair repeat. *Proc. Natl. Acad. Sci.* **76:** 4882.

REIF, H. J. and H. SAEDLER. 1975. IS*1* is involved in deletion formation in the *gal* region of *E. coli* K12. *Mol. Gen. Genet.* **137:** 17.

ROSENBERG, M. and D. COURT. 1979. Regulatory sequences involved in the promotion and termination of RNA transcription. *Annu. Rev. Genet.* **13:** 319.

ROSENBERG, M., D. COURT, H. SHIMATAKE, C. BRADY, and D. L. WULFF. 1978. The relationship between function and DNA sequence in an intercistronic regulatory region in phage lambda. *Nature* **272:** 414.

ROSS, D. G., J. SWAN, and N. KLECKNER. 1979. Nearly precise excision: A new type of DNA alteration associated with the translocatable element Tn*10*. *Cell* **16:** 721.

ROTHSTEIN, S., R. JORGENSEN, K. POSTLE, and D. REZNIKOFF. 1980. The inverted repeats of Tn*5* are functionally different. *Cell* **19:** 795.

SAEDLER, H. and B. HEISS. 1973. Multiple copies of the insertion-DNA sequences IS*1* and IS*2* in the chromosome of *E. coli* K12. *Mol. Gen. Genet.* **122:** 267.

SCHMIDT, F., J. BESEMER, and P. STARLINGER. 1976. The isolation of IS*1* and IS*2* DNA. *Mol. Gen. Genet.* **145:** 145.

SHAPIRO, J. A. 1979. A molecular model for the transposition and replication of bacteriophage Mu and other transposable elements. *Proc. Natl. Acad. Sci.* **76:** 1933.

STARLINGER, P. 1980. IS elements and transposons. *Plasmid* **3:** 241.

STARLINGER, P. and H. SAEDLER. 1976. IS-elements in microorganisms. *Curr. Top. Microbiol. Immunol.* **75:** 111.

STEITZ, J. A. 1979. Translational control at the molecular level. In *Biological regulation and development* (ed. R. Goldberger), vol. 1, p. 349. Plenum Press, New York.

TOUSSAINT, A., M. FAELEN, and A. I. BUKHARI. 1977. Mu-mediated illegitimate recombination as an integral part of the Mu life cycle. In *DNA insertion elements, plasmids and episomes* (ed. A. I. Bukhari et al.), p. 275. Cold Spring Harbor Laboratory, Cold Spring Harbor, New York.

WAHL, G. M., M. STERN, and G. R. STARK. 1979. Efficient transfer of large DNA fragments from agarose gels to diazobenzyloxymethyl-paper and rapid hybridization by using dextran sulfate. *Proc. Natl. Acad. Sci.* **76:** 3683.

WEINSTOCK, G., M. SUSSKIND, and D. BOTSTEIN. 1979. Regional specificity of illegitimate recombination by the translocatable ampicillin-resistance element Tn*1* in the genome of phage P22. *Genetics* **92:** 685.

Genetic Organization of Tn*10* and Analysis of Tn*10*-associated Excision Events

N. Kleckner, T. J. Foster,* M. A. Davis, S. Hanley-Way, S. M. Halling, V. Lundblad, and K. Takeshita

The Biological Laboratories, Harvard University, Cambridge, Massachusetts 02138

The transposable tetracycline-resistance (Tc^r) element Tn*10* moves as a discrete unit from one place to another in DNA. Tn*10* is 9300 bp in length. The ends of the element are 1400-bp inverted repeats, and the intervening 6500 bp of nonrepeated material include the asymmetrically located 2500-bp Tc^r determinant, shown in Figure 1 (Kleckner et al. 1975; Jorgensen and Reznikoff 1979).

Tn*10* promotes a variety of other DNA rearrangements besides transposition; most notably, deletions and inversions (Kleckner et al. 1979a; Ross et al. 1979a; Ross and Kleckner 1980). Each of these Tn*10*-promoted recombination events involves specific transposon-encoded sites and functions, and each results in the joining of one or more transposon (Tn*10* or IS*10*) ends to a new target DNA site (Kleckner and Ross 1979).

Insertions of Tn*10* are flanked by short (9-bp) direct repeats of a target DNA sequence. The repeat appears to arise by duplication of target sequences during insertion, and the presence of such a repeat is not required for subsequent transposition of Tn*10* to a new site. These observations have suggested that the occurrence of the short repeat simply reflects the way in which the target DNA molecule is interrupted and joined to Tn*10* material during transposition: The target duplex must be broken at positions on the two strands that are 9 nucleotide pairs apart (Kleckner 1979).

When Tn*10* was first identified, it was suggested that the 1400-bp inverted repeats of the element might actually be individual insertion sequences analogous to IS*1* or IS*2*. In Tn*10*, two copies of the insertion sequence would be cooperating to effect transposition of the intervening 6500-bp segment (Chan 1974; Kleckner et al. 1975; Botstein and Kleckner 1977). Physical analysis has previously shown that both of the inverted-repeat segments of the element are *structurally* intact IS-like sequences; the inside ends of the two segments are both active sites in Tn*10*-promoted inversions and deletions (Ross et al. 1979a).

In this paper we report the isolation and genetic analysis of deletion mutants and structural variants of Tn*10*. This analysis has established the following additional features of the inverted-repeat segments. First, the two repeat segments together encode all of the genetic determinants, both sites and functions, that are important for Tn*10* transposition. Second, the two repeat segments (designated IS*10*-right and IS*10*-left in Fig. 1) are not functionally equivalent. IS*10*-right is fully functional and is capable by itself of promoting Tn*10* transposition. IS*10*-left is vestigial and promotes transposition only at a very low frequency when IS*10*-right is deleted. Third, each of the individual IS*10* sequences is functionally symmetrical; the inside ends of the IS*10* segments are functionally equivalent to the outside ends.

Complementation analysis (summarized below) has also established four additional features of Tn*10* function. First, IS*10*-right encodes at least one *trans*-acting function that works at the ends of the element. Second, all of the sites specifically required for normal Tn*10* transposition have been localized to the outermost 70 bp at each end of the element; there is no evidence that specific sites internal to the element play an essential role. Third, although transposition function(s) can be seen to work in *trans*, at least one of them appears to act preferentially in *cis*. Fourth, some aspect of Tn*10* function renders wild-type Tn*10* immune or insensitive to the action of incoming functions being provided in *trans* from another Tn*10* element.

During our analysis of Tn*10*-promoted DNA rearrangements, we have also identified three DNA alterations that are associated with Tn*10* and involve excision of all or part of the element. All three excision events involve specific sequences at or near the ends of the transposon; however, unlike transposition, deletions, and inversions, these events do not result in new joints between the end(s) of the transposon and some new target DNA site. As discussed below, all three events are structurally analogous: Excision occurs between two short, direct-repeat sequences and results in deletion of all intervening material plus one copy of the direct repeat. In all three cases, the direct repeats in question occur at either end of an inverted repeat.

The results summarized below suggest that all three types of excision events occur by pathways that are fundamentally separate from the pathway(s) for Tn*10*-promoted transposition. Furthermore, Tn*10* elements that have shorter inverted repeats give excision at reduced frequencies. These and other observations suggest that the inverted repeats may play an important structural role in excision.

Present address: * Department of Microbiology, Trinity College, University of Dublin, Dublin 2, Ireland.

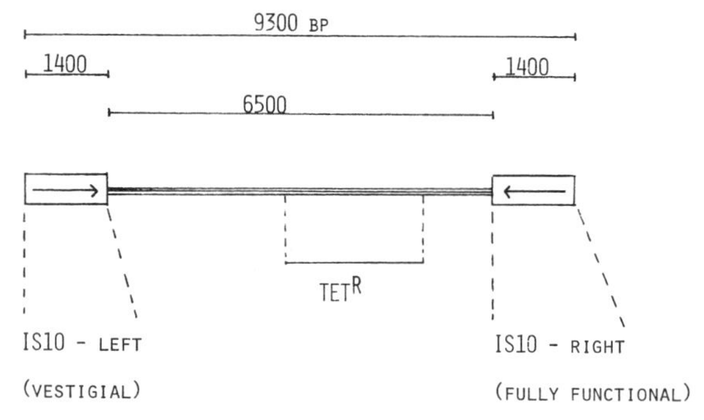

Figure 1. Structure of Tn*10*.

RESULTS

Genetic Organization of Tn*10*

***A transposon containing only the IS*10 *segments of Tn*10.** To assess the role of the IS*10* segments in Tn*10* transposition, we have isolated by genetic methods a new transposon that contains none of the internal 6500 bp of Tn*10* but consists of a new drug-resistance determinant flanked by inverted repeats of IS*10*-left and IS*10*-right. The new transposon was isolated from a multicopy *amp*^{r} *his*::Tn*10* plasmid (see Excision of Tn*10* for details). As shown in Figure 2, previous observations predicted that such a plasmid should give rise not only to normal Tn*10* transpositions, but also to a second symmetrically related type of transposition involving the two "inside" ends of the IS*10* segments and resulting in movement of an IS*10*(R)-*amp*^{r}-*ori*-IS*10*(L) segment. If the two IS*10* elements contain all of the genetic information important for transposition, and if the "inside" ends can function equally as well as the "outside" ends, then this newly transposed segment should itself be a functioning transposon capable of subsequent cycles of insertion into new positions in the absence of a wild-type Tn*10* element.

Transpositions of the IS*10*-*amp*^{r}-*ori*-IS*10* segment into bacteriophage λ were isolated by directly selecting phages that had acquired the plasmid origin. Integration of a multicopy plasmid origin into a replication-defective (*P*am) λ phage will suppress the replication defect, allowing the phage to make a distinctive, small clear plaque on a normally nonpermissive host (K. Takeshita and N. Kleckner, unpubl.). For one such IS*10*-*amp*^{r}-*ori*-IS*10* phage, a deletion variant was subsequently isolated in which the plasmid origin was deleted but the *amp*^{r} determinant and the flanking IS*10* segments were retained intact. The structures of this phage and its parent have been confirmed by restriction mapping and heteroduplex analysis (Foster et al. 1981a).

The ability of the IS*10*-*amp*^{r}-IS*10* segment to transpose further in a host that does not contain any other Tn*10* or IS*10* segment was measured by a conjugational mating-out assay analogous to those described in detail in Figure 3. Transposition of ampicillin resistance from a λ::IS*10*-*amp*^{r}-IS*10* prophage onto an F plasmid lacking other IS elements or $\gamma\delta$ occurs at the same frequency as transposition of tetracycline resistance from a λ::Tn*10* prophage, as measured in parallel experiments. Approximately 3×10^{-7} drug-resistant exconjugants are obtained per total F^+ exconjugants.

We conclude from these results that together the two IS*10* segments contain all of the information necessary to specify Tn*10* transposition. Further evidence leading to the same conclusion is presented below. Furthermore, we conclude that the two ends of each individual IS*10* segment must be functionally symmetrical. Ends that are normally the inside ends in wild-type Tn*10* can function efficiently as outside ends in the IS*10*-*amp*^{r}-IS*10* transposon.

Deletion mutants of Tn*10*: Nonidentity of IS*10*-left and IS*10*-right. To probe in more detail the role of the IS*10* segments, we have isolated and characterized deletion mutants spanning the entire length of Tn*10*, excluding the tetracycline-resistance gene. The starting point for the deletion analysis was a multicopy *amp*^{r} plasmid carrying the *hisOGD* region of the *Salmonella* chromosome into which was crossed by genetic recombination a genetically characterized *Salmonella hisG*::Tn*10* insertion. Lesions were introduced into the resulting *amp*^{r} *his*::Tn*10* plasmid by recombinant DNA techniques. Manipulations were carried out in such a way (see Foster et al. 1981a) that lesions could be introduced separately into regions to the left side and to the right side of the *tet*^{r} determinant of Tn*10*, and desired pairs of left-side lesions and right-side lesions could then be combined to give doubly altered elements.

The resulting series of altered Tn*10* elements were then moved, again by genetic recombination, into both a λ-*hisOGD* transducing phage, where they were tested for Tn*10* transposition, and back into the *Salmonella* chromosome, where they were tested for several other previously characterized Tn*10*-related chromosomal rearrangements. By using this approach, we have been able to move Tn*10* elements from one type of genome to another and from one organism to another regardless of the transposition proficiency of the mutant and without

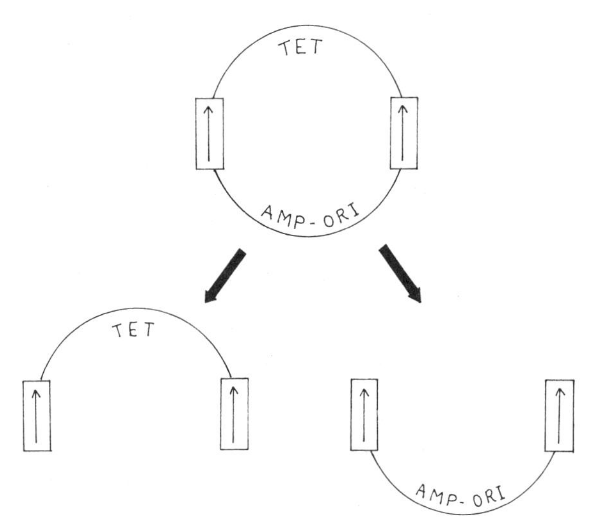

Figure 2. Multicopy *amp*^{r} plasmid containing Tn*10* can give rise to two types of transposition products.

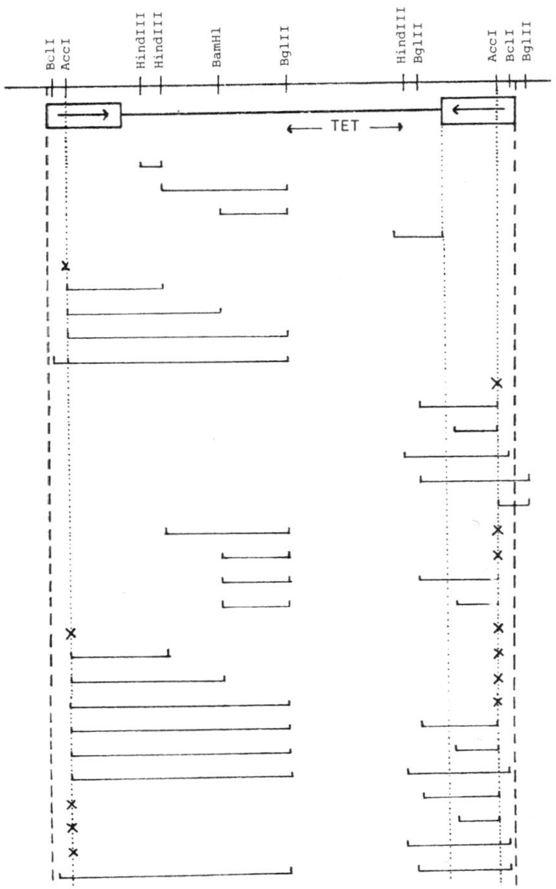

BclI AccI HindIII HindIII BamHI BglII HindIII BglII AccI BclI BglII

TET

TRANSPOSITION FREQUENCY

	"LAMBDA HOP" ASSAY TET^R COLONIES PER INPUT LAMBDA::Tn10 x 10^{-6}	"MATING-OUT" ASSAY TET^R PRO$^+$ EXCONJUGANTS PER PRO$^+$ EXCONJUGANT x 10^{-6}
WILD TYPE	2.2	1.4
DEL 1	4.1	
DEL 6	2.9	1.0
DEL 7	2.5	
DEL 9	0.4**	1.0
INS 2	3.5	0.5
DEL 2	2.5	2.0
DEL 3	3.5	
DEL 4	1.4	2.2
DEL 12*	6.0	1.7
INS 1	0.05	0.005
DEL 10	0.1	0.005
DEL 11	0.1	0.01
DEL 13	0.07	0.07
DEL 8	<0.01	<0.008
DEL 14	<0.01	<0.008
DEL 6 INS 1	0.09	
DEL 7 INS 1	0.08	
DEL 7 DEL 10	0.03	
DEL 7 DEL 11	0.12	
INS 2 INS 1	<0.01	<0.009
DEL 2 INS 1	<0.01	
DEL 3 INS 1	<0.01	
DEL 4 INS 1	<0.01	
DEL 4 DEL 10	<0.01	
DEL 4 DEL 11	<0.01	
DEL 4 DEL 13	<0.01	
INS 2 DEL 10	<0.01	
INS 2 DEL 11	<0.01	
INS 2 DEL 13	<0.01	
DEL 16 DEL 17	<0.01	

Figure 3. Structures and transposition frequencies of altered Tn*10* elements. Tn*10* elements containing deletions (⌴) and/or small insertions (x) were constructed on a multicopy *his*::Tn*10* plasmid and crossed by genetic recombination into a λ-*his* transducing phage (see Foster et al. 1981a). The altered elements were tested for transposition proficiency by two assays. Lambda-hop assay: Tc^s cells are infected with phage λ-*his*::Tn*10* under conditions where the phage cannot replicate, repress, integrate, or kill the host cell; Tc^r bacteria arising from such an infection contain insertions of Tn*10* in the host chromosome (Kleckner et al. 1978). Mating-out assay: Lysogens of each phage λ-*his*::Tn*10* made in NK5830 (= *recA*$^-$ Δ(*lac pro*)XIII/F′ *lac pro*). Each lysogen was mated with NK6641 (= *recA*$^-$ λ^{r}Δ(*lac pro*)XIII Sm^r), and Pro$^+$ Sm^r and Pro$^+$ Tet^r Sm^r exconjugants were selected individually. The latter class results from mating of donor cells in which Tn*10* has transposed into the F′ *lac pro* element prior to or during conjugation.

any change in the chromosomal sequences immediately adjacent to the element.

The structures of the Tn*10* variants isolated and the phenotypes of these variants in two different transposition tests are shown in Figure 3 and can be summarized as follows:

1. If the right side of Tn*10* is intact, lesions affecting the left side do not lower transposition frequencies. The longest left-side deletion, *del*12, removes all Tn*10* material from the left edge of the *tet*ʳ segment outward to a point only 70 bp from the left end of the element.
2. In contrast, every lesion that alters IS*10*-right reduces transposition 25- to 100-fold, depending on the type of assay used. However, as expected, if all transposition-related information lies within the IS*10* elements, the right-side deletion *del*9, which extends inward from the inside end of IS*10*-right, leaves transposition proficiency unimpaired.
3. The combination of any left-side lesion affecting IS*10*-left with any right-side lesion affecting IS*10*-right results in total abolition of Tn*10* transposition. However, again as expected, left-side deletions that do not enter IS*10*-left have no effect when combined with right-side lesions.
4. Deletions that extend across the outside end of the element (*del*8 and *del*14) also abolish transposition completely.

These data again suggest that the IS*10* inverted repeats encode all of the genetic information required for Tn*10* transposition. Furthermore, the data show that IS*10*-right is fully functional and can promote Tn*10* transposition at normal levels, even when IS*10*-left is almost completely deleted; on the other hand, IS*10*-left is only partially active and can promote transposition at only a very low level when IS*10*-right is inactivated or deleted. Both of these conclusions are further supported by data in the following section and elsewhere (Foster et al. 1981a).

The deletion mutants in Figure 3 have also been tested for their ability to give rise to Tn*10*-promoted Tc^s inversions. Inversions are identified genetically as a particular type of polarity-relief revertant in the *Salmonella hisG*::Tn*10* system described previously. Physical experiments have suggested that formation of these Tc^s inversions involves an interaction between some DNA target site and the two inside ends of the IS*10* segments, shown in Figure 4. In accordance with this suggestion, mutants lacking one (or both) inside ends never give rise to inversions, even in cases such as *del*4 where the mutant is perfectly normal in its transposition frequency. For mutants that retain both IS*10* inside ends intact, inversion proficiency parallels transposition proficiency, consistent with the idea that transposition and inversion events are promoted by the same Tn*10* function(s). These data are presented elsewhere (Foster et al. 1981a).

Complementation of Tn*10* mutants. The Tn*10* mutant *del*16 *del*17 is deleted for all of IS*10*-right and all of IS*10*-left except for the outermost 70 bp at each end of

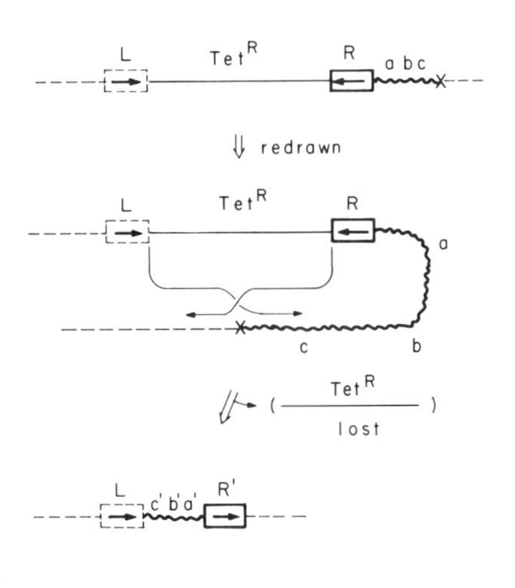

Figure 4. Proposed explanation for Tn*10*-promoted inversions. The two inside ends of the IS*10* sequences interact with an adjacent target DNA site (with or without concomitant DNA replication).

the Tn*10* element (see Fig. 3). When this mutant is present in a multicopy plasmid, it will transpose efficiently into an F plasmid if functions are provided in *trans* from a chromosomally inserted IS*10* element, as shown by the data in Table 1. These results demonstrate directly that IS*10* (in fact, IS*10*-right, see below) encodes at least one diffusible function that is essential for Tn*10* transposition and that acts on the ends of the Tn*10* element. The results also show that Tn*10* *del*16 *del*17 contains all of the sites required in *cis* for normal Tn*10* transposition and hence that all of the relevant sites lie in the outermost 70 bp at each end of the element. There do not seem to be sites internal to the element that are essential for normal transposition.

Complementation analysis of Tn*10* *del*16 *del*17 and of other Tn*10* mutants has also been carried out under an

Table 1. Complementation of Tn*10* *del*16 *del*17 by IS*10*-*amp/cam*-IS*10*

	Transposition frequency ($\times 10^{-5}$) (Tet^r Pro$^+$ exconjugants/ total Pro$^+$ exconjugants)	
	Tn*10* element on multicopy plasmid	
IS*10* element in chromosome	Tn*10*[a]	Tn*10* *del*16 *del*17
None	4.8	0.0001
IS*10*-*amp/cam*-IS*10* (1)[b]	4.1	1.2
IS*10*-*amp/cam*-IS*10* (2)[b]	2.3	1.2

Transposition of Tn*10* from the multicopy plasmid into F was measured by the mating-out assay described in Fig. 3.

[a] Tn*10* element used as a control in this experiment is the equivalent of *del*12 (Fig. 3); it contains an extensive deletion of IS*10*-left analogous in structure to *del*16.

[b] IS*10*-*amp/cam*-IS*10* element is a derivative of the IS*10*-*amp*ʳ-IS*10* element described in the text; the *cam*ʳ genes from Tn9 have been inserted into the *amp*ʳ gene of the element.

Table 2. Differential Action of "High-hopper" IS*10* Functions on Wild-type Tn*10* and Tn*10* *del*16 *del*17 *kan*^{r}

Tn*10* Element(s) in Chromosome	Transposition frequency (× 10^{-7}) (drug^r Pro$^+$ exconjugants/ total Pro$^+$ exconjugants)			
	no plasmid[a]		plasmid containing IS*10*-right from Tn*10* "high-hopper" mutant	
	tet^r	*kan*^{r}	*tet*^{r}	*kan*^{r}
Wild-type Tn*10* (*tet*^{r})	3.2	<0.005	16.2	n.t.[c]
Tn*10* *del*16 *del*17 *kan*^{r} [b]	<0.005	<0.005	n.t.	2000
Wild-type Tn*10* (*tet*^{r}) and Tn*10* *del*16 *del*17 *kan*^{r}	n.t.	n.t.	16.2	2400

Transposition was measured by the mating-out assay described in Fig. 3.

[a] The same results are obtained if a plasmid not encoding IS*10*-right is used as control.

[b] Tn*10* *del*16 *del*17 *kan*^{r} is a derivative of Tn*10* *del*16 *del*17 (Fig. 3) in which the kanamycin-resistance gene of Tn*5* has been substituted for the *tet*^{r} determinant of Tn*10*. The *kan*^{r} derivative has been characterized extensively and behaves identically to the original *tet*^{r} deletion mutant in every test.

[c] n.t. indicates not tested.

experimental situation where a single copy of the mutant element is integrated into the bacterial chromosome (as a λ-*his*::Tn*10* prophage), and the function-donor is either a single-copy IS*10-amp/cam*-IS*10* insertion in the chromosome or a multicopy plasmid carrying IS*10*-right. Such experiments reveal an additional feature of Tn*10* function. We have identified a situation in which functions from a multicopy IS*10* plasmid can be shown to work efficiently in *trans* on a chromosomal Tn*10* *del*16 *del*17 insertion but not on a second, wild-type Tn*10* element present elsewhere in the same chromosome of the same cell. This situation (Table 2) involves as the function donor a multicopy plasmid carrying IS*10*-right from a Tn*10* "high-hopper" transposition-up mutant (M. A. Davis, unpubl.). The chromosome of the cell contains Tn*10* *del*16 *del*17 at one location and wild-type Tn*10* at another location. In such a strain, the complemented Tn*10* *del*16 *del*17 element transposes at an exceptionally high frequency, several hundred fold higher than when the functions are provided by wild-type IS*10*-right, and comparable to the transposition frequency of the original intact Tn*10* high-hopper mutant. Apparently, the high-hopper IS*10*-right produces more and/or better transposition functions that can work in *trans* on Tn*10* *del*16 *del*17. In sharp contrast to the behavior of *del*16 *del*17, the wild-type Tn*10* element present in the same chromosome transposes at a frequency that is not very different from its normal level in the absence of any IS*10* plasmid and that is actually 100-fold lower than the frequency of the complemented Tn*10* *del*16 *del*17 element in the same cell.

This result suggests that wild-type Tn*10* is not acted on by incoming plasmid-encoded functions, even in a situation where in the same cell those functions are working efficiently on Tn*10* *del*16 *del*17. It would appear that the functions are present, available, and active but simply fail to act effectively on the normal Tn*10* element.

In experiments similar to those presented in Table 2, a plasmid carrying wild-type IS*10*-right has been used as function donor for several different hydroxylamine-induced, transposition-defective point mutants of Tn*10* (M. A. Davis and N. Kleckner, unpubl.). None of the point mutants are complemented by the wild-type plasmid, although in parallel experiments Tn*10* *del*16 *del*17 is complemented fully. This finding is consistent with the notion that incoming wild-type functions fail to act on the mutant elements. Full *cis/trans* tests to rule out *trans* dominance of the mutant phenotype are in progress.

The failure of Tn*10* to be acted upon by incoming functions must reflect some interesting feature(s) of the way in which transposition functions are made and act and/or the way transposition is regulated. Particular scenarios are considered in the Discussion.

Excision of Tn*10*

Identification of three Tn10-associated excision events. Recombination events actively promoted by Tn*10* (transposition, inversions, and deletions) involve specific transposon-encoded sites and functions and always result in the joining of one or more transposon (Tn*10* or IS*10*) ends to a new target DNA site. We have identified three additional DNA alterations associated with Tn*10* that also involve specific sequences at or near the ends of the transposon but that do not result in new transposon-target DNA joints. These three events, diagramed in Figure 5 and discussed below, are precise excision of Tn*10*, nearly precise excision of Tn*10*, and precise excision of the nearly precise excision remnant.

Precise excision of Tn*10* is defined genetically as reversions of a Tn*10* insertion mutation, usually as reversion of an insertion auxotroph to prototrophy (Kleckner et al. 1975). Transposition of Tn*10* to a new site results in duplication of a 9-bp target DNA sequence and integration of the transposon segment between the 9-bp repeats. The DNA sequences of two independent His$^+$ revertants of a *Salmonella hisG*::Tn*10* insertion have been determined. This analysis shows directly that precise excision involves a deletion be-

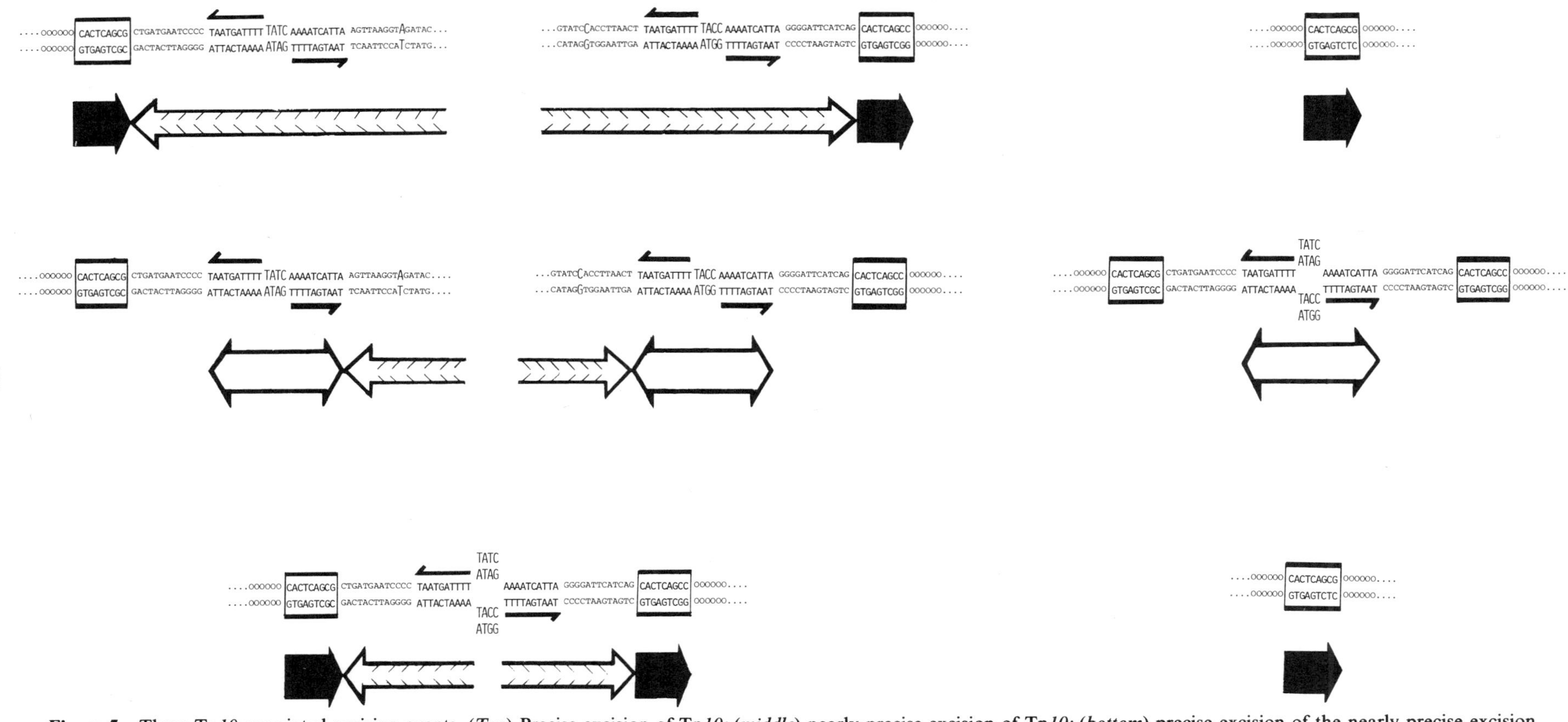

Figure 5. Three Tn*10*-associated excision events. (*Top*) Precise excision of Tn*10;* (*middle*) nearly precise excision of Tn*10;* (*bottom*) precise excision of the nearly precise excision remnant. Sequences shown are for excision products of *Salmonella hisG*9424::Tn*10.*

tween the short direct repeats of target DNA sequence that flank an inserted Tn*10* element. This deletion restores the original wild-type target DNA sequence (Foster et al. 1981b).

Nearly precise excision is a similar event involving short repeat sequences that occur internal to the element, once near each end. Deletion of material between these repeats, analogous to excision of material between the 9-bp repeats in precise excision, results in excision of all but 50 bp of Tn*10*. Nearly precise excisions of Tn*10* were identified during physical mapping and DNA sequence analysis of Tn*10*-related rearrangements of phage λ (Ross et al. 1979a,b). We have been able to show that these nearly precise excisions correspond to a genetically identified class of Tn*10* polarity-relief revertants, which have the distinguishing feature that they give rise to precise-excision full revertants at very high frequencies. DNA sequence analysis of two independently isolated polarity-relief, high-frequency-reverting revertants of a *hisG*::Tn*10* insertion showed that these revertants are, in fact, nearly precise excisions of the type described previously (Foster et al. 1981b). This correlation provides us with a genetic assay for nearly precise excision and has made possible further functional and mechanistic analyses, as well as revealing the third type of excision event, precise excision of the nearly precise excision remnant. Nearly precise excision derivatives still contain sequences of both the left and right ends of Tn*10*, flanked by their original 9-bp target DNA repeats. We presume that precise excision of this material, detected genetically as a reversion event, results in restoration of the original wild-type target DNA sequence, but this has not been directly confirmed by sequence analysis.

As indicated in Figure 5, the three Tn*10*-associated excision events are structurally analogous. In each case, excision occurs between two short, direct-repeat sequences and results in deletion of all intervening material plus one copy of the direct repeat. In all three cases, the direct repeats in question occur at either end of an inverted repeat.

The genetic properties of the three excision events are compared in Table 3. All three events occur in a Rec^- host at frequencies comparable to or higher than the frequencies in a Rec^+ host. It is likely that processes other than homologous recombination are involved (see below).

Precise-excision frequencies differ dramatically for Tn*10* insertions at different sites. As shown, Tn*10* insertions at two sites in *hisG* differ by at least 20-fold in their reversion frequencies. A parallel site-dependence is observed for precise excision of nearly precise excision derivatives at these two sites. In contrast, the frequency of nearly precise excision itself is essentially the same for Tn*10* insertions at both sites. These results may reflect the fact that both precise-excision events involve the 9-bp flanking repeats of the target DNA sequence, which differ for insertions at different sites (including the two *hisG* sites) (S. M. Halling, unpubl.). Nearly precise excision, on the other hand, involves the same short repeat sequences within Tn*10*, regardless of the position of the element with respect to target DNA sequences.

These three excision events have been of interest partly because of their potential relationship to Tn*10* transposition and partly because of their unusual structural consequences. It was of interest to determine whether these events depend on transposon-encoded functions or sites and whether the host functions that promote these events are the same as or different from the host functions involved in Tn*10*-promoted deletions, inversions, or transpositions. Precise excision occurs at a much lower frequency than transposition; however, its consequences for the DNA target are the reverse of transposon integration. Nearly precise excision involves sites within and very near the ends of the element and occurs at frequencies comparable to the frequencies of

Table 3. Genetic Characterization of Tn*10*-associated Excision Events in *Salmonella*

Strain	*hisG*::Tn*10* insertion	Site of insertion[a]	Orientation of insertion[b]	RecA phenotype[c]	Frequency of Tn*10* precise excisions[d]	Relative frequency of NPE[e]	Frequency of precise excisions from NPE[d]
				Experiment 1			
NK616	*hisG*9424	*G*1	A	$RecA^+$	2.7×10^{-8}	1	3.6×10^{-5}
NK690	*hisG*9424	*G*1	A	$RecA^-$	4.2×10^{-8}	3.8	3.6×10^{-5}
				Experiment 2			
NK120	*hisG*9424	*G*1	A	$RecA^+$	3.1×10^{-8}	1 (2×10^{-6})	2.8×10^{-5}
NK159	*hisG*9427	*G*1	B	$RecA^+$	2.9×10^{-8}	0.5	2.3×10^{-5}
NK1198	*hisG*9433	*G*2	A	$RecA^+$	1.2×10^{-9}	0.5	1.4×10^{-7}
NK1199	*hisG*8671	*G*2	B	$RecA^+$	1.0×10^{-9}	1.1	1.2×10^{-7}

[a] *G*1 is a major hotspot for Tn*10* insertion (Kleckner et al. 1979b). All insertions at *G*1 occur at exactly the same 9-bp target site; all insertions at *G*2 occur at a second 9-bp target site (S. M. Halling, unpubl.).

[b] Orientation means the orientation of the nonrepeated central portion of Tn*10* relative to the surrounding chromosomal material. In orientation A, the tet^r genes are close to the *his* promoter.

[c] NK616 and NK690 are isogenic except for the *recA* allele.

[d] Frequency of His^+ revertants per viable cell plated; plates counted after 3 days at 37 °C. Most revertants arise on the plate so that revertant frequencies can be compared without interference from preexisting clonal jackpots.

[e] NPE indicates nearly precise excision. Strains were plated in parallel for polarity-relief revertants, and individual revertants were then tested to identify NPEs (see text).

Tn*10*-promoted deletions and inversions. It was therefore relevant to understand whether any of these events are mechanistically related to Tn*10* transposition and, if not, to understand what types of cellular-DNA-handling processes might be involved.

Effects of Tn*10 *mutations on excision. The series of biochemically generated deletion and insertion mutations of Tn*10* that have been tested for their consequences on Tn*10*-promoted transposition (above) have also been tested for their effects on Tn*10*-associated excision events. Data from representative experiments are shown in Table 4. The following three generalizations can be made on the basis of such experiments.

First, there is no correlation between the transposition proficiency of a mutant and its capacity to undergo precise excision or nearly precise excision. Within a group of mutants having the same structure, excision frequencies are essentially invariant, whereas transposition frequencies vary over several orders of magnitude.

Second, there is a consistent correlation between the length of the element's terminal inverted repeat and the frequencies of both precise excision and nearly precise excision. All mutants with inverted repeats reduced from 1400 bp to 400 bp have excision frequencies that are five- to tenfold lower than for wild type or for mutants retaining normal-length inverted repeats. This correlation holds regardless of whether the mutant with reduced repeats carries a deletion in IS*10*-right, IS*10*-left, or both. Similarly, mutants with only a 70-bp inverted repeat are even further reduced in excision over mutants with 400-bp repeats.

Third, a mutant (*del*16 *del*17) that retains only the outermost 70 bp of IS*10*-right and of IS*10*-left still retains the capacity to give both precise excisions and nearly precise excisions at measurable frequencies. This property of *del*16 *del*17 demonstrates that pathway(s) exist for precise excision and for nearly precise excision that are totally independent of Tn*10*-encoded functions. The possibility that sites at the very ends of the element are involved is not excluded. None of the results obtained thus far give any indication that Tn*10* transposition functions are involved in any way in Tn*10* excision, and the simplest idea is that one or, more likely, several pathways for excision involving only host-encoded functions are available.

The dependence of excision on the lengths of the inverted repeats provides support for the idea that the inverted repeats actually play a structural role in stimulating or facilitating both precise excision and nearly precise excision. Particular models for how excision might occur are considered below.

Additional support for the notion that the inverted repeats play a structural role in excision is provided by analysis of many independent Tn*10* insertions into a single site, the *G*1 "hotspot" in the *Salmonella hisG* gene. As indicated in Figure 5, DNA sequence analysis has revealed that the left and right inverted repeats of wild-type Tn*10* differ at several nucleotide positions in the outermost 200 bp at the ends of the element (Kleckner 1979; S. M. Halling, unpubl.). These positions of difference include a cluster of three nucleotide pairs about 26 bp in from each end, a fourth nucleotide pair about 50 bp in from each end, and nucleotide pairs at three other positions farther in. DNA sequence analysis of *hisG*::Tn*10* insertions at the *hisG*1 insertion hotspot (Kleckner et al. 1979b) has revealed that several of the insertions no longer show the characteristic differences between IS*10*-left and IS*10*-right. Instead, both inverted-repeat segments have the sequence characteristic of IS*10*-right.

The relevance of these observations for the present discussion derives from the additional finding, shown in Figure 6, that the two types of Tn*10* insertions identified by DNA sequence analysis also correspond to two different classes with regard to their precise-excision frequencies. Insertions whose inverted repeats are exactly homologous to one another have a reversion frequency approximately threefold higher than do insertions of the wild-type element. Both types of insertions lie at exactly the same 9-bp target site in *hisG;* hence, the differences in excision frequencies must reflect differences in the Tn*10* elements themselves. The orientation of the ele-

Table 4. Effects of Tn*10* Mutations on Excision

Exp.	Tn*10* mutant	Length of inverted repeat (bp)	Transposition[a]	Relative frequency: precise excision	Relative frequency: nearly precise excision
1	wild type	1400	1	1	
	*ins*2	1400	1.5	0.85	
	*ins*1	1400	0.02	0.9	
	*ins*1 *ins*2	1400	0.005	1.3	
	*del*4	400	0.6	0.12	
	*del*4 *ins*1	400	0.005	0.11	
2	*ins*2	1400	1.5	1	1
	*ins*1	1400	0.02	1.1	1
	*del*10	400	0.04	0.2	0.3
	*del*11	400	0.04	0.2	0.09
	*del*13	70	0.04	0.08	0.003

Precise excision and nearly precise excision were measured following reintroduction of the mutant Tn*10* elements into the histidine region of the *Salmonella* chromosome (for details, see Foster et al. 1981b).

[a] See Fig. 3.

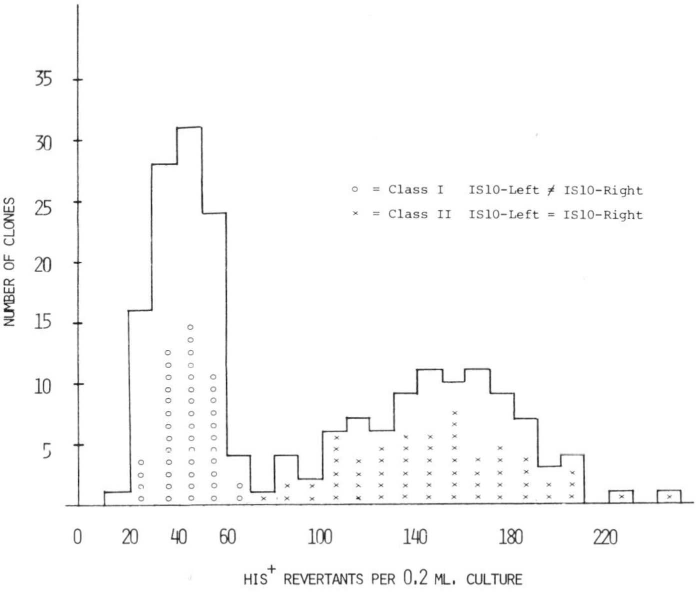

Figure 6. Precise-excision frequencies for two types of Tn*10* insertions at a single site in *hisG*. Eleven independent Tn*10* insertions at a single site in *hisG* were analyzed for reversion to His$^+$. For each strain, 15 independent clones were plated for revertants (Kleckner et al. 1979b). The distribution of the numbers of colonies on each of the 165 plates is shown. Seven of these insertions have been shown by DNA sequence analysis to occur at exactly the same 9-bp target site and to be of two types. In one type, IS*10*-right and IS*10*-left differ in sequence; in the other type, they are identical (see text). The distributions of revertant colonies from clones of these seven insertions are indicated by x and o.

ment with respect to the histidine operon is not relevant; both groups of insertions contain representatives inserted in both possible orientations. This result is consistent with the idea that the inverted repeats of Tn*10* play an important structural role in precise excision. If the inverted repeats do actually interact during excision, it is reasonable that the more similar the sequences involved, the higher the frequency of excision. Other interpretations, however, are in no way excluded.

Host mutants affecting Tn*10 *excision. To probe the role of host-encoded functions in Tn*10*-associated excision events, we have isolated mutants of *Escherichia coli* K12 that exhibit increased frequencies of precise excision for a particular *lacZ*::Tn*10* insertion. Mutants were identified according to the procedure of Hopkins et al. (1980) by plating for individual colonies on MacConkey lactose indicator plates and screening for colonies containing increased numbers of Lac$^+$ papillae. A total of 39 independent mutants (ethylmethanesulfonate-induced or nitrosoguanidine-induced) have been identified and tested for all Tn*10*-associated excision events, for Tn*10* transposition, and for a number of other phenotypes that might result from mutations in cellular-DNA-handling functions. The phenotypes of these mutants are described in Table 5.

The mutants identified on the basis of their papillation phenotype also exhibit increased frequencies of Lac$^+$ revertants in liquid-culture reversion tests. Increases range from 10-fold to 10,000-fold. On the basis of tests not involving Tn*10* (i.e., mutator activity and plating of mutant λ phages), the host mutants exhibit a wide range of different phenotypes: Many are mutators, many are altered for plating of one or another type of λ

Table 5. Thirty-nine Host Mutants Increased for Precise Excision of Tn*10*

				Other Tn*10* excision events		Non-Tn*10* phenotypes	
	No. of mutants	Precise excision of Tn*10*	Tn*10* transposition	nearly precise excision (NPE) of Tn*10*	precise excision of NPE	mutator or antimutator	altered plating of λ mutants
Class I							
transposition normal							
A	15	up 5–20×	normal	up 2–30×	normal	6/15	6/15
B	14	up 20–200×	normal	up 10–70×	normal	11/14	10/14
C	4	up 200–10,000×	normal	up 25–400×	normal	3/4	4/4
Total	33						
Class II							
transposition altered							
A	4	up 10–30×	up 10–20×	up 5–200×	normal	4/4	3/4
B	2	up 10–20×	down 10–20×	up 5–80×	normal	2/2	2/2
Total	6						

Precise excision: reversion of *lacZ*2900::Tn*10* to Lac$^+$. Nearly precise excision: high-reverting Hol$^+$ polarity-relief revertants of *hisG*5107::Tn*10* (see text and Foster et al. 1981b). Precise excision of nearly precise excisions: reversion of NPEs to His$^+$. Mutator: Nal^r and Sm^r mutants selected. λ mutants tested: λ$^+$; λ*gam*210; λ*red*270*gam*210; λ*red*3*gam*210; λO^{ts}21; λcI857*cro*27. Reversion and mutator phenotypes determined after growth of several independent clones to saturation in broth medium at 37 °C. λ phages tested for efficiency of plating and for plaque morphology. Tn*10* transposition measured by the lambda-hop assay described in Fig. 3.

derivative, some are altered in both types of tests, and some are wild type for both. Thus, it would appear that mutations that affect Tn*10* excision do also affect other fundamental DNA-handling processes. The significance of these additional phenotypes is being investigated further.

The range of phenotypes observed in these tests contrasts sharply with the general uniformity of excision and transposition phenotypes (see below). This finding, plus preliminary mapping data on some of the mutants (V. Lundblad, unpubl.), suggests that the 39 mutants are distributed among at least several and perhaps many different genetic loci.

Only three mutants exhibit significant increases in the frequency of Tn*10* transposition, three other mutants exhibit slight decreases, and the remaining 33 mutants are wild type for transposition. If the mutations are indeed distributed among a variety of different genetic loci, these findings provide additional support for the view that the pathway(s) for Tn*10* excision events are generally separate from the pathway for Tn*10* transposition.

All of the mutants analyzed exhibit increased frequencies of nearly precise excision as well as of precise excision, and none of the mutants are altered more than a few fold for second-step excision of nearly precise excisions. This finding strongly suggests that precise excision and nearly precise excision occur by closely related pathways and that these pathways are generally distinct from the pathway(s) for the third type of excision event.

DISCUSSION

Genetic Organization of Tn*10*

The combination of genetic and structural information presented here and elsewhere makes it clear that Tn*10* is indeed a composite element whose inverted repeats are structurally intact IS-like sequences. The inverted repeats encode all of the information required for normal Tn*10* transposition, with IS*10*-right being fully functional and IS*10*-left providing functions at only 1% the level of IS*10*-right. IS*10*-left and IS*10*-right also differ at the nucleotide sequence level; i.e., of the outermost 200 bp at each end of Tn*10*, the two sequences differ at seven positions (Kleckner 1979; S. M. Halling, unpubl.).

Since the functions of a single IS*10* element are sufficient for normal Tn*10* transposition, there should be no selective pressure for retention of both IS*10* segments intact. It is therefore reasonable that one element should retain or improve in function while the second element accumulates first one and then more and more defects. We presume that IS*10*-left, which barely functions at all, has already accumulated one or more deleterious mutations and has thus become vestigial. Further dissection of the IS*10* segments has provided a reasonably simple picture of this element, as presented below.

First, each IS*10* element is functionally symmetrical; the two ends of an IS*10* element need not be identical, but they are functionally equivalent. This finding places constraints on transposition mechanisms. For example, if transposition involves DNA replication originating at the end(s) of the element, such replication must be able to proceed from either or both ends.

Second, all of the sites needed for normal Tn*10* transposition lie in the outermost 70 bp at each end of the element. For Tn*3*, evidence has been presented that a site internal to the element affects the nature and stability of transposition and cointegration products (Gill et al. 1978; Heffron et al. 1979 and this volume; Sherratt et al., this volume). There is thus far no evidence for such an internal site in the case of Tn*10*.

It is conceivable that IS*10*/Tn*10* and Tn*3* work by substantially different mechanisms, but it seems more likely that they are two variations on fundamentally the same theme. For example, all of the essential transposition functions of Tn*3* lie between one end of the element and the internal site (Heffron et al. 1979). Perhaps the internal site actually used to be an end, and Tn*3* is now carrying out in the middle of the element processes that in simpler elements occur at the ends, or perhaps Tn*3* has modified the processes that normally occur at ends to a form more suitable to an internal site. Another, not mutually exclusive, possibility is that the internal site of Tn*3* is not the actual physical site of a recombination event but is regulating at a distance events that actually occur at the end(s) of the element. In this case, IS*10* and Tn*3* might differ only in regulatory features rather than in mechanistic details of recombination/replication processes.

Third, IS*10* encodes at least one function that is essential for its own transposition and that acts at the ends of the element.

Fourth, a chromosomal insertion of wild-type Tn*10* is not acted on by incoming IS*10* functions provided in *trans* from a multicopy plasmid, even in a situation where in the same cell those functions are working efficiently on a chromosomal insertion of Tn*10* *del*16 *del*17. One particular explanation for these results would be that a *cis*-acting regulatory function (not encoded by Tn*10* *del*16 *del*17) is always present in a Tn*10*-containing cell, bound to the ends of the element in such a way as to directly (sterically) exclude incoming functions from interacting with the ends.

Complementation data and other observations have also led us to consider the possibility that Tn*10* transposition might actually require some activation process (analogous to transcriptional activation of phage λ replication?). Such a possibility is attractive because it could provide additional potential for complex regulatory responses and because it could separate the regulatory processes from the actual replication/recombination events themselves. If transposition did require activation, then a Tn*10*-encoded regulatory function present at all times could prevent action of incoming functions indirectly by blocking the activation step.

The notion of activation raises yet another possible explanation for the complementation data. If activation of the element occurred only infrequently and were very tightly coupled to production and *cis* action of transposition function(s), functions diffusing to the element in *trans* might have a much lower chance of seeing the activated element than do functions encoded in *cis*. The activation process could be, for example, a single round of transcription/translation of transposition functions, which then would act locally on the activated element. By such an hypothesis, the susceptibility of Tn*10 del*16 *del*17 to incoming functions could mean that it is fortuitously activated by the adjacent *tet*^{r} region (in an unregulated way) and/or that the two deletion mutations have eliminated the requirement for activation and have turned the ends of the element into simple sites to be acted upon by incoming functions.

Both Tn*3* and Tn*5* have been shown to control negatively their own transposition (Chou et al. 1979; Gill et al. 1979; Biek and Roth 1980 and this volume). It is too early to tell whether Tn*10* differs fundamentally from these elements or whether the different elements have thus far been investigated from different points of view. Particularly in the case of Tn*5*, which is structurally similar to Tn*10* (Berg et al. 1975; Rothstein et al., this volume), the negative regulation of Tn*5* and the blocking and/or activation of Tn*10* may converge into a common process once both are understood.

Finally, when a transposition-defective point mutant of Tn*10* is integrated in the bacterial chromosome, it cannot be complemented for transposition by a wild-type IS*10*-right. Whether the complementing element is present in a single copy or in multiple copies, the mutant element apparently is insensitive to any wild-type functions that do reach it. For Tn*903*, IS*1*, and Tn*5*, it has also been found that local lesions (usually filled-in restriction sites or insertion mutations) are not complementable (Grindley and Joyce 1980 and this volume; Saedler et al., this volume; J. Roth, pers. comm.). In these cases, it is not possible to be certain that the failure of complementation is due to some functional aspect of the element rather than to the fact that essential sites have been inactivated. However, it seems most likely that noncomplementation of point mutations is, in fact, a general property of many transposons.

This unexpected property has one interesting implication: Even in a cell that harbors at different sites many different copies of the same element, a single point mutation in any one element will be sufficient to kill it completely. The transposition frequencies of most endogenous IS elements under normal laboratory conditions are comparable to or lower than the frequency of spontaneous point mutations. Thus, in many cases, a given IS element may well die of a point mutation before it has a chance to (presumably) replicate and transpose further. This may provide some modest constraint on the takeover of the bacterial genome by active transposable elements. It might also mean that many of the IS insertions present in prokaryotic genomes are functionally defective "corpses" rather than active transposable elements.

Overview

It has been suggested that transposable elements are molecular parasites, which replicate and transpose to new sites and thus render themselves immune from elimination even in the absence of any direct selective pressure for their survival (Doolittle and Sapienza 1980; Orgel and Crick 1980). This is a particularly attractive description of the IS elements. They carry out their replication/transposition events at the low frequency befitting an unselected process and compactly carry only those sites and functions that are relevant for their own perpetuation. Eventually, of course, these simple elements may be coupled to other determinants (phages, drug resistance) that may, in turn, make possible direct phenotypic selection for, among other things, more and better transposition.

By this view, we would consider Tn*10* still to be a very simple element. The transposition frequency is as low as that of any insertion element known, and the individual IS*10* sequence still retains a fundamental symmetry and simplicity of organization. IS*10* is essentially a pair of functionally equivalent ends plus a limited number of functions, at least one of which must recognize and act at the ends of the element. Complementation studies reinforce the picture of a very self-contained functional unit, with functions acting poorly in *trans* and with transposition quite possibly under tight, perhaps *cis*-acting, control. It is not yet clear which host functions are involved in Tn*10* transposition, but we would speculate that IS*10* may parasitize host functions to a considerable extent, more perhaps than would elements with larger coding capacities, such as Tn*3*, $\gamma\delta$, or bacteriophage Mu.

Tn*10*-associated Excision Events

Four lines of evidence suggest that Tn*10*-associated excision events occur by pathways that are fundamentally different from Tn*10* transposition: (1) A deletion mutant of Tn*10* that lacks all but 70 bp of the two IS*10* segments still gives rise to both precise excisions and nearly precise excisions. There must be one or more pathways for these events that do not depend on Tn*10*-encoded transposition functions. (2) Tn*10* elements that are structurally intact but defective for transposition give precise excisions and nearly precise excisions at the same frequency as transposition-proficient elements having the same structure. (3) Precise excision and nearly precise excision occur at decreased frequencies in Tn*10* deletion mutants that have shorter inverted repeats. Transposition frequencies do not vary depending on the length of the inverted repeats. (4) Host mutants isolated on the basis of their stimulation of Tn*10* precise excision also turn out to stimulate nearly precise excision. The vast majority of these mutants have little or no effect on Tn*10* transposition.

It is still possible, however, that the pathways for excision and for transposition do overlap in certain respects. It would, in fact, be quite remarkable if the two types of DNA rearrangements did not involve a single, common cellular-DNA-handling protein. The properties of host mutants that affect both pathways should be informative in this regard.

As stressed previously, all three Tn*10*-associated excision events involve analogous structures. In each case, excision occurs between two short, direct-repeat sequences and results in deletion of all intervening material plus one copy of the direct repeat. In all three cases, the direct repeats in question occur at either end of an inverted repeat.

The available information suggests that precise excision and nearly precise excision are mechanistically very closely related. Every host mutant that is increased for precise excision is also increased for nearly precise excision; furthermore, deletions that reduce the length of the inverted repeats of Tn*10* reduce the frequencies of both events to roughly the same extent.

This close correlation in behavior of precise excision and nearly precise excision could have a trivial explanation. It is conceivable that all precise-excision revertants actually arise by the two-step pathway involving first nearly precise excision and then precise excision of the nearly precise excision remnant; in this case, there would be no single-step event that gives precise excision directly from intact Tn*10*. Three observations argue against this possibility, although they do not totally exclude it. First, the frequencies of the two individual excision events in the two-step pathway would appear to be substantially too low for their product to account for the observed frequency of precise excision (see Table 1). However, it might still be that the two steps in the pathway occur in some coupled fashion, rather than independently, to give precise excision. Second, none of the host mutants that are increased for precise excision are increased in the second step of the two-step pathway. Some such mutants should have been found if all precise excisions occur in two steps, unless the second step is far from rate-limiting in the two-step pathway. Third, several of the host mutants isolated because they increase precise excision of a *lacZ*::Tn*10* insertion turn out not to increase precise excision of a *hisG*::Tn*10* insertion. The reason for this specificity is not understood. However, these same mutants do increase nearly precise excision of Tn*10* from the same *hisG*::Tn*10* site where precise excision is unaffected. If all precise-excision events involve nearly precise excision as an intermediate, all mutants that increase nearly precise excision (and leave the second step in the pathway unaltered) should increase precise excision as well.

It seems most likely that precise excision and nearly precise excision respond similarly to mutational perturbations because they are structurally analogous events that occur by the same pathway.

The fact that a reduction in the length of the inverted repeats of Tn*10* reduces the frequencies of precise excision and nearly precise excision suggests that the inverted repeats of the element may play some structural role in the excision events. The simplest idea would be that some interaction between the inverted repeats serves to bring into proximity the flanking direct-repeat sequences, which are actually involved in the excision itself and which are normally separated by up to thousands of base pairs in thc casc of wild-type Tn*10*. In principle, the inverted repeats at the two ends of Tn*10* could come together and interact with each other either as intact double-stranded DNA duplexes or in intrastrand pairings as single-strand snapbacks involving one or both strands. These possibilities are shown in Figure 7.

There are no data available for Tn*10* or for any other element that distinguishes between these two classes of possibilities. The *recA* independence of excision argues against the involvement of this particular function; however, other activities in the cell (synaptases? [Potter and Dressler 1980]) might be capable of carrying out the appropriate reaction.

We are more inclined, however, to favor the possibility that excision involves or is stimulated by intrastrand snapback pairing. Any snapbacks that do form in a cell could reasonably be expected to persist as stable structures. As such, they will certainly pose a severe block to the progress of DNA replication forks (Gefter and Sherman 1977), and excision events could arise as the consequence of abortive or aberrant replication past the block. Alternatively, the structures by themselves could serve as targets for nucleases or, if ends are present, for

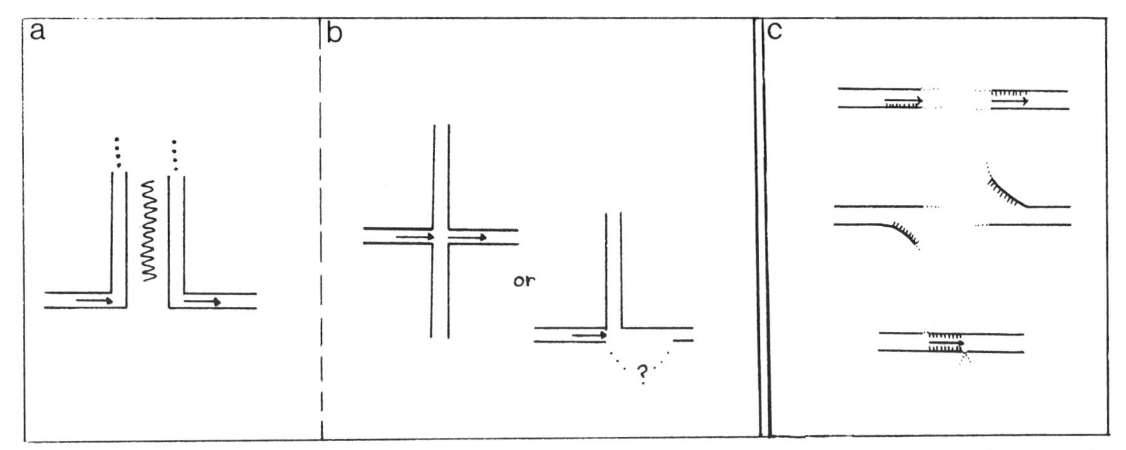

Figure 7. Possible components of precise and nearly precise excision. (*a,b*) Inverted repeats of Tn*10* may interact as intact double-stranded duplexes or as intrastrand snapbacks on one or both strands to create or to stabilize a structure that can give rise to excision products. (*c*) Involvement of direct repeats in excision suggests that "repairing" or "slippage" of nucleotides between the two repeat sequences could be the determining process in formation of a novel joint during excision.

polymerases that could somehow generate the appropriate excision product.

Scenarios for excision must also take into account the role of the flanking direct repeats, which are the physical site of the recombination event. Regardless of how the event is initiated or finally resolved, the involvement of direct repeats suggests that "slippage" or "repairing" of the 9-bp sequence on one strand from its normal partner to the corresponding sequence on the other side of Tn*10* is responsible for formation of the novel joint created in the deletion event (see Fig. 7c). One attractive feature of excision models involving snapback structures is that such structures readily explain how the two 9-bp repeat sequences are brought close together under circumstances where nicking, excising, slippage, and polymerizing could easily occur. It is less obvious exactly how such processes would be facilitated or promoted by pairing between the inverted repeats as intact duplexes.

At least some spontaneous deletion mutations in *E. coli* involve excision of DNA segments between short direct repeats (Farabaugh et al. 1978). It is therefore possible that the pathways for precise excision and nearly precise excision are related to the pathway(s) for spontaneous deletion formation.

None of the host mutants exhibiting increased precise-excision frequencies were significantly altered in the second-step precise excision of the nearly precise excision. Thus, it would appear that this third excision event occurs by a different pathway. This is not unreasonable in view of the fact that the total length of the nearly precise excision remnant is 50 bp plus the flanking 9-bp repeats, and the inverted repeats in this remnant are only 23 bp in length. It is quite plausible that the potential for internal secondary structure should play a role in precise excision of this remnant, but it is also reasonable that quite different factors will be involved than those affecting inverted repeats 1400 bp in length and separated by up to several thousand base pairs.

ACKNOWLEDGMENTS

This work was supported by grants to N. K. from the National Science Foundation and the National Institutes of Health. T. J. F. was supported by a National Institutes of Health Fogarty Foundation Fellowship, and S. M. H. was supported by a National Institutes of Health postdoctoral traineeship. We are grateful to Jeff Way and Denise Roberts for comments, criticisms, and helpful discussions.

REFERENCES

Berg, D., J. Davies, B. Allet, and J.-D. Rochaix. 1975. Transposition of R-factor genes to bacteriophage lambda. *Proc. Natl. Acad. Sci.* **72:** 3628.

Biek, D. and J. R. Roth. 1980. Regulation of Tn*5* transposition in *Salmonella typhimurium. Proc. Natl. Acad. Sci.* **77:** 6047.

Botstein, D. and N. Kleckner. 1977. Translocation and illegitimate recombination by the tetracycline-resistance element Tn*10*. In *DNA insertion elements, plasmids, and episomes* (ed. A. I. Bukhari et al.), p. 185. Cold Spring Harbor Laboratory, Cold Spring Harbor, New York.

Chan, R. K. 1974. "Specialized transduction by bacteriophage P22." Ph.D thesis, Massachusetts Institute of Technology, Cambridge.

Chou, J., M. Casadaban, P. Lemaux, and S. N. Cohen. 1979. Identification and characterization of a self-regulated repressor of translocation of the Tn*3* element. *Proc. Natl. Acad. Sci.* **76:** 4020.

Doolittle, W. F. and C. Sapienza. 1980. Selfish genes, the phenotype paradigm and genome evolution. *Nature* **284:** 601.

Farabaugh, P. J., U. Schmeissner, M. Hofer, and J. R. Miller. 1978. Genetic studies of the *lac* repressor. VII. On the molecular nature of spontaneous hotspots in the *lacI* gene of *Escherichia coli. J. Mol. Biol.* **126:** 847.

Foster, T. J., M. A. Davis, D. E. Roberts, K. Takeshita, and N. Kleckner. 1981a. Genetic organization of transposon Tn*10*. *Cell* **23:** (in press).

Foster, T. J., V. Lundblad, S. M. Halling, S. Hanley-Way, and N. Kleckner. 1981b. Tn*10*-associated excision events: Relationship to transposition and role of direct and inverted repeats. *Cell* **23:** (in press).

Gefter, M. L. and L. A. Sherman. 1977. The role of DNA structure in DNA replication. In *The organization and expression of the eukaryotic genome.* (ed. E. M. Bradbury and K. Javaherian), p. 233. Academic Press, New York.

Gill, R., F. Heffron, and S. Falkow. 1979. Identification of the protein encoded by the transposable element Tn*3* which is required for its transposition. *Nature* **282:** 797.

Gill, R., F. Heffron, G. Dougan, and S. Falkow. 1978. Analysis of sequences transposed by complementation of two classes of transposition-deficient mutants of Tn*3*. *J. Bacteriol.* **136:** 742.

Grindley, N. D. F. and C. Joyce. 1980. A genetic and DNA sequence analysis of the kanamycin-resistance transposon Tn*903*. *Cell* (in press).

Heffron, F., B. J. McCarthy, H. Ohtsubo, and E. Ohtsubo. 1979. DNA sequence analysis of the transposon Tn*3*: Three genes and three sites involved in transposition of Tn*3*. *Cell* **18:** 1153.

Hopkins, J. D., M. B. Clements, T-Y. Liang, R. R. Isberg, and M. Syvanen. 1980. Recombination genes on the *E. coli* sex factor specific for transposable elements. *Proc. Natl. Acad. Sci.* **77:** 2814.

Jorgensen, R. A. and W. S. Reznikoff. 1979. Organization of structural and regulatory genes that mediate tetracycline-resistance in transposon Tn*10*. *J. Bacteriol.* **138:** 705.

Kleckner, N. 1979. DNA sequence analysis of Tn*10* insertions: Origin and role of 9-bp flanking repetitions during Tn*10* translocation. *Cell* **16:** 711.

Kleckner, N. and D. G. Ross. 1979. Translocation and other recombination events involving the tetracycline-resistance transposon Tn*10*. *Cold Spring Harbor Symp. Quant. Biol.* **43:** 1233.

Kleckner, N., K. Reichardt, and D. Botstein. 1979a. Inversions and deletions of the *Salmonella* chromosome generated by the translocatable tetracycline-resistance element Tn*10*. *J. Mol. Biol.* **127:** 89.

Kleckner, N., D. Barker, D. Ross, and D. Botstein. 1978. Properties of the translocatable tetracycline-resistance element Tn*10* in *E. coli* and bacteriophage λ. *Genetics* **90:** 427.

Kleckner, N., R. K. Chan, B. K. Tye, and D. Botstein. 1975. Mutagenesis by insertion of a drug resistance element carrying an inverted repetition. *J. Mol. Biol.* **97:** 561.

Kleckner, N., D. Steele, K. Reichardt, and D. Botstein. 1979b. Specificity of insertion by the translocatable tetracycline-resistance element Tn*10*. *Genetics* **92:** 1023.

ORGEL, L. E. and F. H. C. CRICK. 1980. Selfish DNA: The ultimate parasite. *Nature* **284:** 604.

POTTER, H. and D. DRESSLER. 1980. DNA synaptase: An enzyme that fuses DNA molecules at a region of homology. *Proc. Natl. Acad. Sci.* **77:** 2390.

ROSS, D. G., J. SWAN, and N. KLECKNER. 1979a. Physical structures of Tn*10*-promoted deletions and inversions: Role of 1400 base pair inverted repetitions. *Cell* **16:** 721.

———. 1979b. Nearly-precise excision: A new type of recombination event associated with the translocatable tetracycline-resistance element Tn*10*. *Cell* **16:** 733.

Recombination Involving Transposable Elements: On Replicon Fusion

C. J. Muster and J. A. Shapiro
Department of Microbiology, The University of Chicago, Chicago, Illinois 60637

In bacterial cells, transposable elements participate in several related nonhomologous recombination events, including transposition, replicon fusion, adjacent deletion, adjacent inversion, and transposition of random DNA segments. As has been shown previously (Cohen and Shapiro 1980), all of these recombination events can be summarized under two headings—full transposition and genome rearrangements—if we examine only the regions directly involved in the recombination between DNA segments (Fig. 1). The principal features of these two groups of recombination events are the following: (1) ligation of transposable-element termini to nonhomologous target sequences in the cellular genome; (2) duplication of a short oligonucleotide sequence from the target molecule; and (3) duplication of the transposable element (for a recent review of the data, see Calos and Miller 1980). In a few cases, it is known that transposable-element mutants lacking a gene product necessary for transposition are also defective for genome rearrangements (Heffron et al. 1977; Faelen et al. 1978; N. Kleckner and T. J. Foster, pers. comm.). Thus, the biochemical steps leading to the formation of both classes of recombination products appear to involve at least one reaction catalyzed by the same protein.

A major focus of interest in studies of nonhomologous recombination is the elucidation of the molecular events leading to the formation of different recombinant products. On the basis of the two classes of recombination events illustrated in Figure 1, studies of Tn*3* mutants (Heffron et al. 1977; Gill et al. 1978), and analysis of bacteriophage Mu replication (Ljungquist and Bukhari 1977; Waggoner and Pato 1978), we proposed a sequential molecular model for nonhomologous recombination and replication of transposable elements (Fig. 2) (Shapiro and MacHattie 1979; Shapiro 1979, 1980). Two novel features of this model merit emphasis: (1) specific semiconservative replication of the transposable element after the formation of replication-fork structures at its extremities by single-strand cleavage and ligation reactions; and (2) reciprocal exchange between daughter transposable-element duplexes to convert genome arrangement products to full transposition products. In this paper we summarize some recent experiments on replicon fusions mediated by a mutant Tn*3* element. They support our model by demonstrating that a monomeric replicon carrying a single copy of a transposable element will recombine with another replicon to yield a fused product carrying two copies of the transposable element (Fig. 3). These experiments indicate that transposable-element duplication occurs during the fusion reaction itself (which must therefore involve specific replication) and not prior to recombination as a result of passive replication by the carrier molecule.

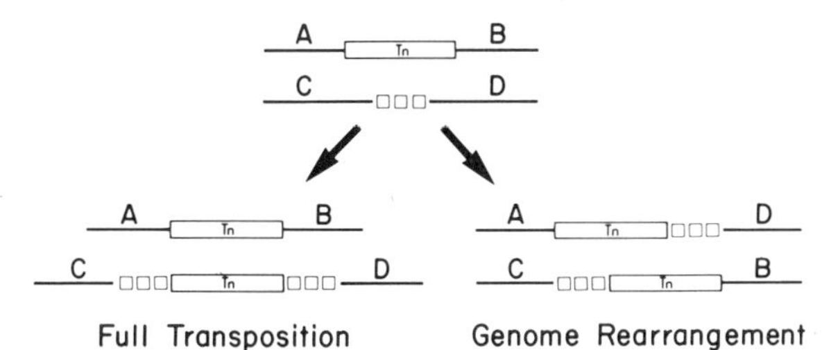

Figure 1. Nonhomologous recombination events involving transposable elements. Genome rearrangements include replicon fusions, adjacent deletions, excision of circular structures carrying deleted DNA plus a copy of the transposable element, and adjacent inversions (Shapiro 1979, 1980; Cohen and Shapiro 1980).

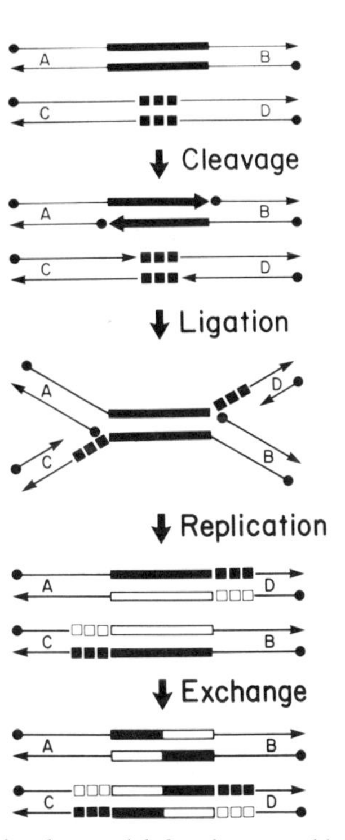

Figure 2. Molecular model for the recombination events summarized in Fig. 1 (details are explained in Shapiro 1979, 1980).

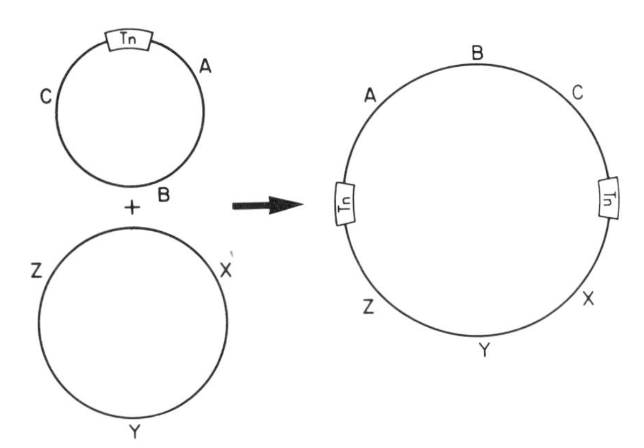

Figure 3. Replicon fusion.

Origin and Structure of λ*Cb*1

Heffron and his colleagues (1977) have isolated deletion mutants of Tn*3* (and of the closely related Tn*1* element) that have multiple transposition defects. These include a deletion, Tn*3*Δ596, resulting in lack of transposase synthesis (directed by the *tnpA* gene [Chou et al. 1979]), lack of repressor synthesis (directed by *tnpR*), and lack of a site needed for some step leading to full transposition but not needed for replicon fusion (*tnpS* = IRS of Heffron et al. 1979). Thus, mutant elements like Tn*3*Δ596 (Fig. 4) will mediate only replicon fusion events when complemented by an active $tnpA^+$ gene. In the absence of Rec functions, the fusion products are stable, but homologous recombination will resolve them to final transposition products (Gill et al. 1978). A mutant element like Tn*3*Δ596 offers two important experimental advantages in studies of nonhomologous recombinations: (1) only a single recombination product is formed in Rec^- cells, and (2) the need for $tnpA^+$ complementation makes it possible to determine whether recombination events are catalyzed by transposase (i.e., are specific).

To test Tn*3*Δ596-mediated replicon fusion as a mechanism for inserting a λ prophage into the bacterial genome, we isolated λcI857*red3b2*::Tn*3*Δ596 phage. The phage containing the transposon were isolated from lysates of λcI857*red3b2* grown in a cell carrying RSF1596 (pMB8::Tn*3*Δ596) and RSF103 (RSF1010::Tn*1*ΔAp, $tnpA^+tnpR^-$). We are confident that Tn*3*Δ596 is inserted into the λ genome via a replicon-fusion intermediate because the phage particles in this lysate that initially transduced the transposon bla^+ marker conferring carbenicillin resistance (Cb^r) also transduced the colicin immunity marker (Iel^+) from pMB8. After incubation in a Rec^+ transductant, phage could be isolated that transduced Cb^r but not Iel^+. The structure of one Iel^- phage, labeled λ*Cb*1, is depicted in Figure 5.

Table 1. Transduction by λ*Cb*l

Rec^- strain	Transductants/10^6 infecting phage[a]
MV12 (R388)	6.2[b]
MV12 (R388, RSF103)	328.0
MV12 (λ) (R388)	0.05
MV12 (λ) (R388, RSF103)	78.8

λ*Cb*l infections at moi <0.01.

[a] Transduction levels are the means of four experiments. The values of TnpA-mediated transduction of cultures carrying RSF103 can vary because of Tn*1*ΔAp insertions into R388 before infection (unpubl.). The inserts exert transposition immunity (Robinson et al. 1977) and therefore lower transduction frequencies as they accumulate in a culture. We have measured spontaneous induction in λ-resistant derivatives of the lysogenic cultures. As expected for *recA* strains (Hertman and Luria 1967), the level of spontaneous induction is very low ($\leq 10^{-7}$ pfu/cell in a culture at infected density). Since this level is no more than 1% of the number of infected $TnpA^+$ cells that yield transductants, the spontaneous derepression of the prophage could not have contributed significantly to the number of transduced clones obtained in our experiments.

[b] Most of these transductants are unstable, which indicates that they arose by abortive lysogenization (Kellenberger et al. 1961).

Genome Insertion of λ*Cb*1

Because of the *b2* deletion, λ*Cb*1 will not lysogenize *Escherichia coli* strains by the normal prophage insertion mechanism. Low levels of lysogenization can be measured very sensitively by monitoring the formation of Cb^r transductants immune to superinfection by λcI. Table 1 gives the results of λ*Cbl* infections of various related *recA* strains at low multiplicities of infection

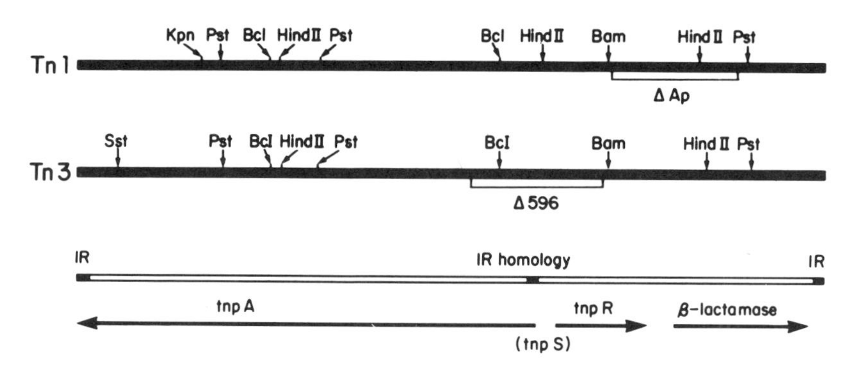

Figure 4. Physical and transcriptional maps of Tn*1* and Tn*3*. The Δ596 and ΔAp deletions are indicated. The inverted repeat (IR) homology may represent the site necessary for cointegrate resolution (Heffron et al. 1979), which is here called *tnpS*. Tn*1*ΔAp is $tnpA^+S^+R^-$ bla^-. Tn*3*Δ596 is $tnpA^-S^-R^-$ bla^+. Note the *Kpn*I-*Sst*I/*Hin*dII polymorphisms in these two closely related elements.

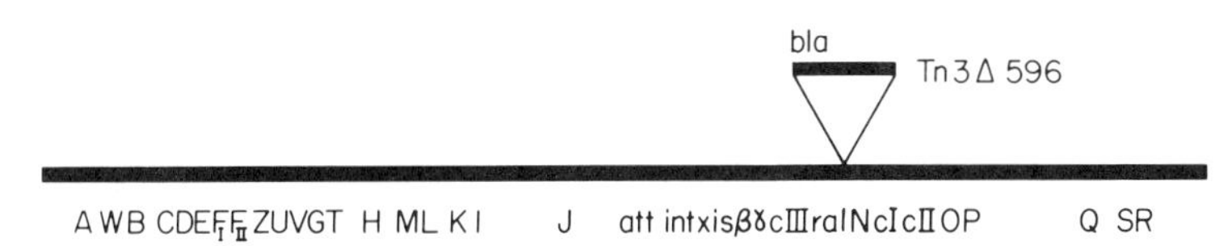

Figure 5. Structure of nondefective λ*Cb*1. The λcI857*red*3*b*2::Tn*3*Δ596 was constructed as described in the text. The site of Tn*3*Δ596 insertion and its orientation were confirmed by genetic and restriction enzyme mapping.

(moi). The results demonstrate that stable transduction is greatly facilitated by TnpA activity (i.e., by the presence in the recipient of the *tnpA*$^{+}$*tnpR*$^{-}$ RSF103 plasmid), indicating that a large majority of insertions occur by Tn*3*-specific recombination. In the absence of TnpA activity, prophage immunity greatly reduces transduction (by preventing abortive lysogenization), but immunity does not greatly reduce TnpA-dependent transduction. Since prophage immunity blocks the replication of superinfecting phage (Thomas and Bertani 1964), the latter observation shows that TnpA activity will recombine a monomeric, nonreplicating λ*Cb*1 molecule with some segment of the recipient genome to yield a stably inserted λ*Cb*1 prophage. About 80–90% of the insertions of λ*Cb*1 occur in the R388 plasmid, yielding either Tra^{-} plasmid mutants or Tra^{+} plasmids carrying R388 and λ*Cb*1 markers together. This preference for the plasmid over the bacterial chromosome parallels earlier results of other investigators (Kretschmer and Cohen 1977; P. M. Bennett et al.; R. Gill; both pers. comm.).

To confirm the nature of the replicon fusion event involved in lysogenization by λ*Cb*1, we extracted R388-λ*Cb*1 plasmids after transduction of both sensitive and immune cells. In both cases, the plasmid molecules contained viral sequences sandwiched between direct repeats of Tn*3*Δ596. Figure 6 shows two examples where the plasmid contained a complete λcI857*red*3*b*2 genome. In other cases (after transfer to sensitive cells), some of the viral sequences were missing in the R388-λ*Cb*1 plasmids. But here, too, the remaining viral sequences were flanked by Tn*3*Δ596 repeats, and the deletions appeared to extend from one end of a Tn*3*Δ596. Thus, it seems that the transfer selected for adjacent deletion derivatives of a complete fusion product, as these deleted molecules would not undergo zygotic induction.

Since Tn*3* does not carry its own origin of theta replication (Cohen et al. 1979), the duplication of Tn*3*Δ596 must have arisen during the recombination event.

DISCUSSION

Experiments with Tn*9* and Tn*10* previously showed that de novo DNA synthesis almost certainly creates the oligonucleotide duplications that flank the transposed element at its new site, as only a single copy of this sequence (at the target site) was involved in the recombination event (Johnsrud et al. 1978; Kleckner 1979). However, the origin of two copies of the transposable element in the genome after transposition (Bennett et al. 1977; Faelen and Toussaint 1978; Fennewald and Shapiro 1979) or in the product of replicon fusion (Toussaint and Faelen 1973; Gill et al. 1978; Shapiro and MacHattie 1979) was not so clear. Two possibilities were suggested to account for this apparent duplication: (1) the donor suicide model, in which the donor replicon that participates in the transposition reaction is lost from the genome and is replaced by a sister replicon; and (2) the dimer donor model, in which replicon fusions are generated by nonreplicative transposition of a transposon-carrier replicon-transposon structure from a dimeric donor molecule. We believe that neither of these alternative explanations is tenable. The suicide model cannot account for replicon fusions and other genome rearrangements where donor DNA (i.e., segments initially adjacent to the transposable element) are covalently linked to target DNA in the final products of recombination. Our results (and analogous unpublished experiments performed by M. Faelen with λ::mini-Mu phages transducing immune cells) exclude the dimer model. Thus, it appears that replication of the transpos-

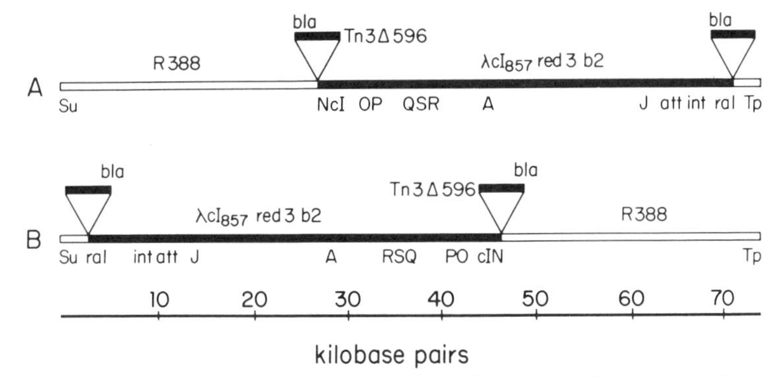

Figure 6. Structure of R388::λ*Cb*1 cointegrates. (*A*) Structure of pXJS488/752, isolated from MV12 (R388, RSF103) transduced to carbenicillin resistance by λ*Cb*1. (*B*) Structure of pXJS700/701, isolated from MV12 (λ) (R388, RSF103) transduced to carbenicillin resistance by λ*Cb*1. The orientation of the R388 Tn*3*Δ596 and λ sequences are indicated by the genetic markers shown, and the cointegrates have been opened into linear structures at the sole *Eco*RI site in R388.

able element during replicon fusion is the only reasonable explanation for the duplication found in the final product. It seems very likely that the same conclusion is also valid for complete transposition.

ACKNOWLEDGMENTS

We thank Gary Boch for excellent technical assistance, R. Gill and M. Fennewald for bacterial strains, and M. Benedik for phage. This research was supported by a grant from the U.S. Public Health Service (NIGMS-24960). C. J. M. was the recipient of a postdoctoral traineeship from the National Institutes of Health (GM-07543), and J. A. S. was the recipient of a Research Career Development Award (1-K04-AI-00118).

REFERENCES

BENNETT, P. M., J. GRINSTED, and M. H. RICHMOND. 1977. Transposition of Tn*A* does not generate deletions. *Mol. Gen. Genet.* **154:** 205.

CALOS, M. P. and J. H. MILLER. 1980. Transposable elements. *Cell* **20:** 579.

CHOU, J., P. G. LEMAUX, M. J. CASADABAN, and S. N. COHEN. 1979. Transposition protein of Tn*3*: Identification and characterization of an essential repressor-controlled gene product. *Nature* **282:** 801.

COHEN, S. N. and J. A. SHAPIRO. 1980. Transposable genetic elements. *Sci. Am.* **242:** 40.

COHEN, S. N., M. J. CASADABAN, J. CHOU, and C.-P. D. TU. 1979. Studies of the specificity and control of transposition of the Tn*3* element. *Cold Spring Harbor Symp. Quant. Biol.* **43:** 1247.

FAELEN, M. and A. TOUSSAINT. 1978. Stimulation of deletions in the *Escherichia coli* chromosome by partially induced Mu *cts*62 prophages. *J. Bacteriol.* **136:** 477.

FAELEN, M., O. HUISMAN, and A. TOUSSAINT. 1978. Involvement of phage Mu-1 early functions in Mu-mediated chromosomal rearrangements. *Nature* **271:** 580.

FENNEWALD, M. A. and J. A. SHAPIRO. 1979. Transposition of Tn*7* in *P. aeruginosa* and isolation of *alk*::Tn*7* mutations. *J. Bacteriol.* **139:** 264.

GILL, R. E., F. HEFFRON, and S. FALKOW. 1979. Identificaton of the protein encoded by the transposable element Tn*3* which is required for its transposition. *Nature* **282:** 797.

GILL, R., F. HEFFRON, G. DOUGAN, and S. FALKOW. 1978. Analysis of the sequences transposed by complementation of two classes of transposition-deficient mutants of Tn*3*. *J. Bacteriol.* **136:** 742.

HEFFRON, F., P. BEDINGER, J. CHAMPOUX, and S. FALKOW. 1977. Deletions affecting the transposition of an antibiotic resistance gene. *Proc. Natl. Acad. Sci.* **74:** 702.

HEFFRON, F., B. J. MCCARTHY, H. OHTSUBO, and E. OHTSUBO. 1979. DNA sequence analysis of the transposon Tn*3*: Three genes and three sites involved in transposition of Tn*3*. *Cell* **18:** 1153.

HERTMAN, I. and S. E. LURIA. 1967. Transduction studies on the role of a *rec*$^+$ gene in the ultraviolet induction of prophage lambda. *J. Mol. Biol.* **23:** 117.

JOHNSRUD, L., M. CALOS, and J. MILLER. 1978. The transposon Tn*9* generates a 9 bp repeated sequence during integration. *Cell* **15:** 1209.

KELLENBERGER, G., M. ZICHICHI, and J. WEIGLE. 1961. A mutation affecting the DNA content of bacteriophage λ and its lysogenic properties. *J. Mol. Biol.* **3:** 399.

KLECKNER, N. 1979. DNA sequence analysis of Tn*10* insertions: Origin and role of 9 bp flanking repetitions during Tn*10* translocation. *Cell* **16:** 711.

KRETSCHMER, P. J. and S. N. COHEN. 1977. Selected translocation of plasmid genes: Frequency and regional specificity of translocation of the Tn*3* element. *J. Bacteriol.* **130:** 888.

LJUNGQUIST, E. and A. I. BUKHARI. 1977. State of prophage Mu DNA upon induction. *Proc. Natl. Acad. Sci.* **74:** 3143.

ROBINSON, M. K., P. M. BENNETT, and M. H. RICHMOND. 1977. Inhibition of TnA translocation by TnA. *J. Bacteriol.* **129:** 407.

SHAPIRO, J. A. 1979. Molecular model for the transposition and replication of bacteriophage Mu and other transposable elements. *Proc. Natl. Acad. Sci.* **76:** 1933.

———. 1980. A model for the genetic activity of transposable elements involving DNA replication. In *Plasmids and transposons* (ed. C. Stuttard and K. R. Rozee), p. 229. Academic Press, New York.

SHAPIRO, J. A. and L. A. MACHATTIE. 1979. Insertion and excision of λ prophage mediated by the IS*1* element. *Cold Spring Harbor Symp. Quant. Biol.* **43:** 1135.

THOMAS, R. and L. R. BERTANI. 1964. On the control of the replication of temperate bacteriophages superinfecting immune hosts. *Virology* **24:** 241.

TOUSSAINT, A., and M. FAELEN. 1973. Connecting two unrelated DNA sequences with a Mu dimer. *Nat. New Biol.* **242:** 1.

WAGGONER, B. and M. L. PATO. 1978. Early events in the replication of Mu prophage DNA. *J. Virol.* **27:** 587.

Genetic and Sequencing Studies of the Specificity of Transposition into the *lac* Region of *E. coli*

J. H. MILLER, M. P. CALOS, AND D. J. GALAS
Département de Biologie Moléculaire, Université de Genève, Genève, Suisse

Although the cardinal property of transposable elements is their ability to translocate from one position on a donor chromosome or plasmid to one of many different positions on the same or a different DNA molecule, a wide range of insertion specificities has been noted for different elements. At one extreme are elements that insert with little recognizable specificity. A well-studied example in this category is phage Mu (Bukhari and Zipser 1972; Daniell et al. 1972). At the other end of the spectrum is the element IS*4*, which has been observed at only one position in the *gal* region (Habermann et al. 1979). All 20 occurrences of IS*4* in *galT* are inseparable by genetic methods, and the three that have been sequenced are at the identical nucleotide position. The majority of elements display specificities in between these two extremes. For example, Tn*10* also shows a marked preference for certain hotspots, but it integrates at other points as well. Thus, in the ten-gene *his* operon of *Salmonella typhimurium*, about 40% of the Tn*10* insertions are, at the nucleotide level, at the same point, whereas the others are distributed among at least 21 other sites, as determined by genetic analysis (Botstein and Kleckner 1977; Kleckner et al. 1979; S. Halling and N. Kleckner, unpubl.). Genetic analysis of Tn*10* insertions in *lacZ* shows a similar pattern (Foster 1977).

Certain elements integrate preferentially into regions of DNA, and yet they have many integration sites within these stretches. This phenomenon has been observed for the integration of Tn*3* (Kretschmer and Cohen 1977) and a similar transposon, Tn*802* (Grinsted et al. 1978), into a number of different plasmids; for Tn*1* into phage P22 (Weinstock et al. 1979); for IS*1* into the *gal* control region (Kühn et al. 1979), and for Tn*9* into certain parts of the *lac* region (Johnsrud et al. 1978; Miller et al. 1980) (see below).

We have studied the specificity of integration of several transposable elements into the same target region, *lacI,Z,Y*, of *Escherichia coli*. The DNA sequence of much of this 5600-bp region is known, enabling correlation of the frequency of insertion with aspects of the DNA sequence. We report here the genetic analysis of insertions of Tn*5*, Tn*9*, and Tn*10* into the *lac* region (Table 1 reviews some of their properties) and sequence studies of many of the Tn*9* insertions. Several specificity determinants emerge that can be applied to explain the integration specificity of Tn*9* as well as other elements.

MATERIALS AND METHODS

Media. All media have been described previously (Miller 1972).

Detection of transposition. Overnight cultures of the relevant donor strain (see Fig. 1 and text) were grown at 30 °C, 34 °C, or 37 °C and subcultured in Luria-Bertani (LB) broth at 34 °C or 37 °C until a density of 2×10^8 cells/ml was reached. A mating mixture was prepared by mixing 0.5 ml of the donor cells with 0.5 ml of a freshly saturated culture of the recipient cells (usually Q90C unless otherwise noted). After 30 minutes at 34 °C or 37 °C without aeration, the cells were shaken at 37 °C for 30 minutes at 30 rpm. At the end of this time, the cells were titered for the presence of diploids carrying the *lacproB* episome in the recipient background by plating on medium containing streptomycin and no proline, and samples of the mating mixture were plated on selective medium to allow the detection of transpositions. The selective medium contained 5-bromo-4-chloro-3-indolyl-β-D-galactosidase (Xgal), streptomycin (Sm), and either chloramphenicol (Cm), kanamycin (Km), or tetracycline (Tc). These antibiotics were present at 20 μg/ml. After 48 hours at 37 °C, Z^- colonies were identified as white colonies and I^- colonies as deep-blue colonies. Because the starting strain has a higher basal level of β-galactosidase than normal, all other colonies were light blue (due to the increased basal level). The plates were then replicated onto lactose MacConkey plates with the relevant antibiotic. Lac^- colonies are white on this medium. Lac^- colonies that were not white (Z^+) on the Xgal plates were picked and shown to be Y^- by subsequent analysis.

Determination of transposition frequencies. Transpositions were detected as described above, but a large number (50–100) of 0.5-ml cultures were seeded with approximately 100 cells of the donor and allowed to grow without aeration at 34 °C until a predetermined time, at which point 0.2 ml of a freshly saturated culture of Q90C was added. After the mating, the entire contents of the tube was spread onto two selective plates. The numbers of I^-, Z^-, and Y^- colonies were scored from each culture and were analyzed to verify that transpositions were involved. The fraction of cultures (P_0) having no mutants of a particular type was used to calculate the frequency of transposition from the formula $P_0 = e^{-m}$, where m equals the average number of mutational events per sample. (All mutants were verified by reversion and mapping experiments.) Five cultures were titered at the moment of mating, and the average titer was used to compute the mutation or transposition rate per generation. Under the conditions of the experiment, the efficiency of the mating ap-

Table 1. Some Properties of Tn*5*, Tn*9*, and Tn*10*

Transposon	Resistance marker	Length (bp)	Length of inverted repeat (bp)	Length of repeat generated upon insertion (bp)
Tn*5*	Km	~5400	~1500	9
Tn*9*	Cm	2638	18/23[a]	9
Tn*10*	Tc	~9300	~1400	9

References: Tn*5*, Berg et al. (1975) and Schaller (1979); Tn*9*, Gottesman and Rosner (1975), MacHattie and Jackowski (1977), Johnsrud et al. (1978), Ohtsubo and Ohtsubo (1978), and Alton and Vapnek (1979); and Tn*10*, Foster et al. (1975), Kleckner et al. (1975), and Kleckner (1979).

[a] Flanked by direct repeat of IS*1* (768 bp).

proaches 100%. The change in cell number during the mating is within one doubling, and the small correction necessary would not significantly affect the relative numbers given in Table 3. Moreover, by scoring I⁻, Z⁻, and Y⁻ colonies in the same experiment, the ratios of the frequencies of transposition for a given element should represent their true values.

Mapping. Mapping experiments were done essentially as described previously (Schmeissner et al. 1977a; Johnsrud et al. 1978). Phenocopy crosses, as described by Foster (1977), were employed for insertion × insertion crosses. The self-cross served as a control. In a number of cases, the high reversion of particular Tn*5* or Tn*10* insertions limited the resolution of the mapping.

Strains for mapping. The strains 5217, 5203, 5165, 5131, and 5079 (Δ*lac his strA*) carry an F′ *lac* episome harboring the respective deletions *274, 250, 209, 169,* and *103*. These strains were kindly provided by H. Khatoon (Cold Spring Harbor Laboratory).

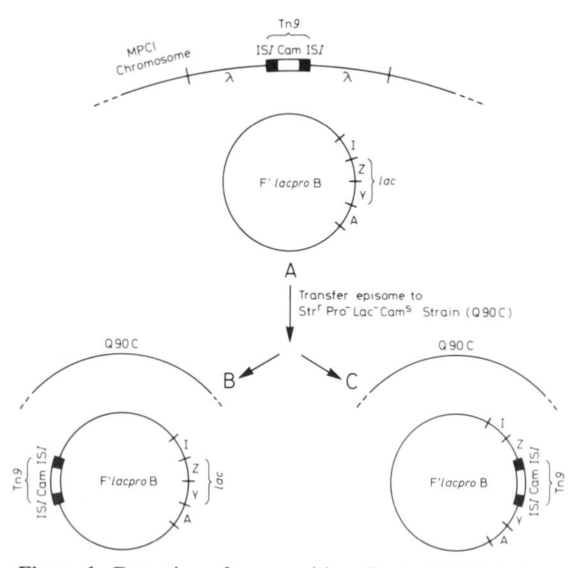

Figure 1. Detection of transposition. Strain MPC1 harbors a λ::Tn*9* integrated in the chromosome and an F′ *lacproB* episome. Transposition of Tn*9* from the chromosome to the episome occurs during growth and is detected after transfer of F′ *lacproB* to a recipient strain, Q90C. By selecting for Pro⁺Cm^{r}Sm^r colonies on different *lac* indicator plates, clones containing Tn*9* in different parts of the *lac* region of the episome can be identified (see Materials and Methods).

Construction of plasmids containing the* lac *operon; derivatives of pBR322 carrying Tn9 in the* lac *operon. We cloned DNA fragments containing the *lac* region by using DNA from strains carrying Tn*9* insertions in *lacZ* on an F′ *lacproB* episome. This procedure allowed direct selection for Cm^r clones, because Tn*9* determines resistance to chloramphenicol. We took advantage of the fact that neither the *lac* region (Teather et al. 1978) nor Tn*9* (MacHattie and Jackowski 1977) contains a *Hin*dIII cleavage site. Episomal DNA, prepared from three different mutants carrying different *lacZ*::Tn*9* insertions, was digested with *Hin*dIII and ligated to *Hin*dIII-cleaved pBR322 DNA. Ap^{r}Cm^r clones were selected after transformation into strain Q90C. (Δ*lacproB strA malB*). Several colonies were recovered for each transformation and purified. Plasmid DNA from each of the candidates was analyzed by restriction analysis. In all three cases, the resulting pBR322-derived plasmids carried both the *lac* region and the *lacZ*::Tn*9* insertion. The identity of each insertion was verified by backcrosses and shown to be at the same position as the original insertion in the episome.

Revertant plasmids were obtained that had lost the Tn*9* insert in vivo (Galas et al. 1980), and these were used to prepare, by genetic recombination (Galas et al. 1980), plasmids containing each insertion to be analyzed.

RESULTS

Tn*9* Transposition

Tn*9* carries the determinants for chloramphenicol resistance flanked by single copies of the insertion sequence IS*1* in the same orientation (Gottesman and Rosner 1975; MacHattie and Jackowski 1977). The entire element is 2638 bp long (Alton and Vapnek 1979) and, like IS*1* (Calos et al. 1978; Grindley 1978), generates a 9-bp repeated sequence during integration (Johnsrud et al. 1978).

Figure 1 shows the design of the experiments used in this work to detect transposition. For the transposon Tn*9*, we used the strain MPC1 (see Table 2), which carries Tn*9* on a heat-inducible λ prophage in a Sm^s background. The strain also harbors a *lacproB* deletion and carries an F′ *lacproB* episome. During growth of a culture of MPC1, transpositions of Tn*9* onto the episome occur and can be detected by transferring the F′ factor to a Sm^r recipient strain and monitoring the Cm^r colonies (see Fig. 1 and Materials and Methods). Transposition into the *I* gene can be scored on Xgal indicator plates, since the I⁻ colonies are blue on this medium (see Miller 1972), whereas Z⁻ colonies remain white after 48 hours. Y⁻ mutants are light blue after 48 hours on Xgal medium, and they score Lac⁻ on lactose MacConkey indicator plates.

Frequency of Transposition

In the MPC1-strain background (consult Table 2), approximately 1 in 10^5 of the episomes transferred to

Table 2. Bacterial Strains

Strain	Sex	Genotype	Source
Q90C	F⁻	*ara* Δ(*lacproB*) *strA malB*	this paper
Q90C-1	F⁻	*ara* Δ(*lacproB*) *strA malB recA*	this paper
MPC1	F′ *lacproB*(I^qL8)	*ara* Δ(*lacproB*) (λ::Tn*9*,cI857)	this paper
MPC3	F′ *lacproB*(I^qL8)	*ara* Δ(*lacproB*) (λ::Tn*9*,cI857) *rif recA*	this paper
MPC4	F′ *lacproB*(I^qL8)	*ara val*ʳ Δ(*lacproB*)::Tn*5*	derived from DB1000
MPC5	F′ *lacproB*(I^qL8)	*ara val*ʳ Δ(*lacproB*)::Tn*10 gal his trp lys strA*	derived from DU3083
DU3083	F⁻	*lac gal his trp lys strA*::Tn*10*	M. Kehoe and T. J. Foster
DB100	F⁻	*trpR tna* Δ*trpE5*::Tn*5*	D. Berg
CSH63	HfrH	*ara val*ʳ Δ(*lacproB*)	Miller (1972)

The bacterial strains used for detection of transposition are described. (For other strains used in mapping or constructions, see Fig. 2 and Materials and Methods.) Strains MPC4 and MPC5 were constructed from DB100 and DU3083, respectively, by crossing in the *lac-proB* deletion from strain CSH63. Except where otherwise indicated, the *thi* marker is uncertain. Strains MPC1 and MPC3 carry the λ*cam*1. Stocks of λ*cam*1 were provided by L. Rosner (Laboratory of Molecular Biology, National Institute of Arthritis, Metabolism, and Digestive Diseases, National Institutes of Health).

the recipient cells (in the experiment diagramed in Fig. 1) carried the transposon Tn*9*. We determined the frequency of transposition into the *lac* region by examining a large number of individual mating cultures and by employing the method of Luria and Delbrück (1943). From each culture, the total number of I⁻, Z⁻, and Y⁻ colonies resulting from the transposition of Tn*9* was measured, and the fraction of cultures having no transposition into *I, Y,* or *Z* was used to compute the respective frequencies of transposition per cell per generation (for further details see Materials and Methods). As Table 3 shows, under the conditions of the experiment, Tn*9* integrates into *lacY* at a frequency of 0.8×10^{-8}/cell/generation, into *lacZ* at 1×10^{-9}/cell/generation, and into *lacI* at 4×10^{-10}/cell/generation. The *I* and *Y* genes are 1080 bp and 1251 bp long, respectively (Farabaugh 1978; Büchel et al. 1980), whereas the *Z* gene consists of 3063 bp (Fowler and Zabin 1978). (Each of these genes is preceded by an untranslated region into which insertions would generate a detectable phenotype. The lengths of these regions are 60–70 bp.) Thus, these tests indicate that Tn*9* integrates into the *Y* gene about 15- to 20-fold more frequently than it integrates into the *I* and *Z* genes, when equal lengths of DNA are considered.

To show that the preferential integration into the *Y* gene is not an artifact of the method, we also collected 50 *lac*⁻ insertions of Tn*9* from independent cultures, without determining whether they were in *Z* or *Y* until afterwards. Forty eight of these insertions were in the *Y* gene and only two were in the *Z* gene, further substantiating the preferential integration into *Y*.

Distribution of Tn*9* Insertions in the *lac* Region

We collected I⁻ and Z⁻ mutants from more than 200 individual cultures of MPC1 and MPC3 (a *recA rif* derivative of MPC1), and we analyzed no more than one I⁻ or Z⁻ colony per culture. Y⁻ colonies were isolated from MPC1 cultures. We mapped the insertions against deletions and by insertion × insertion crosses. The resulting distributions are shown separately in Figure 2. (Because of the different frequencies of transposition into each gene, to consider all of the data together requires a normalization step; see below.)

It is apparent that there are numerous insertion sites throughout the *lac* region. However, there are regions that are favored for insertion. In addition to much of the *Y* gene (Table 3), the very end of the *Z* gene and a small segment within the *I* gene are also favored relative to

Table 3. Transposition Frequencies

Strain	Transposable element	*lac* target gene	Total no. of cultures	No. of cultures without transposition in target gene	Average no. of cells in cultures	Transposition frequency into target gene ($\times 10^9$/cell/generation)[a]
MPC1	Tn*9*	*I*	95	88	2.0×10^8	0.38
	Tn*9*	*Z*	95	76	2.0×10^8	1.1
	Tn*9*	*Y*	76	16	2.0×10^8	7.8
MPC4	Tn*5*	*I*	95	49	3.0×10^7	22
	Tn*5*	*Z*	90	19	3.0×10^7	52
	Tn*5*	*Y*	—	—	—	77*
MPC5	Tn*10*	*I*	97	97	6.5×10^7	<.16
	Tn*10*	*Z*	97	88	6.5×10^7	1.5
	Tn*10*	*Y*	—	—	—	1.0*

[a] Transposition frequencies were determined per cell per generation (see Lea and Coulson 1949; Moore and Sherman 1975). An asterisk indicates the cases of Tn*5* and Tn*10* in which the frequencies of integration in the *Y* gene were calculated by multiplying the frequencies of integration into the *Z* gene by the ratio of *Y*:*Z* integrations found among integrations into *lac* in independent experiments (see text). It is not ruled out that the presence of the F factor may affect transposition frequencies. However, all of the frequencies were measured in the presence of F and under similar experimental conditions.

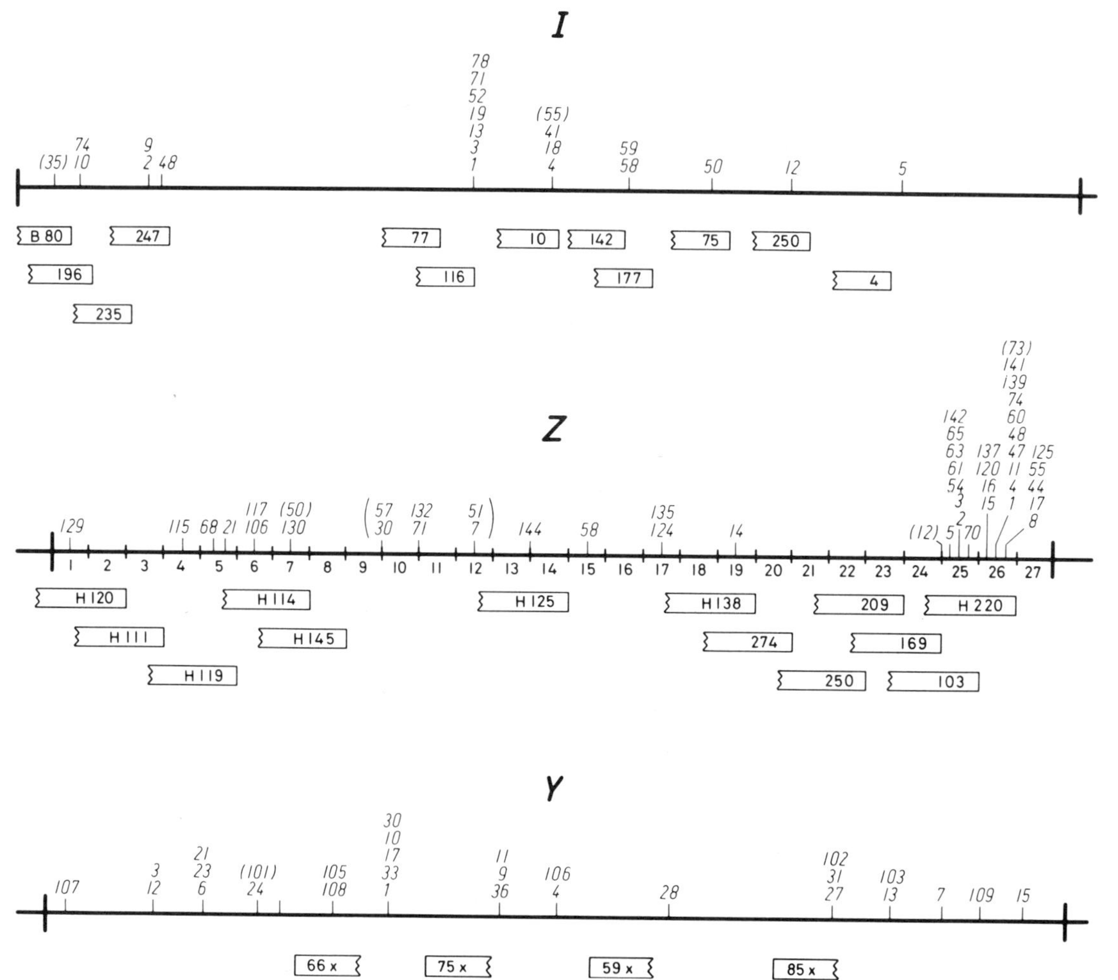

Figure 2. Tn*9* integration points in *lac*. The distributions of Tn*9* insertions in *lacI, Z,* and *Y* are shown. Different sample sizes were considered in each case. In *lacI,* the mutations were mapped against the deletions described previously by Schmeissner et al. (1977a,b). Although, by convention, insertion mutations are referred to as, e.g., *lacI2*::Tn*9,* for simplicity they are listed here only by allele numbers, such as *2.* The positions of insertions *1, 3, 13, 19, 52, 71,* and *78* have been established by sequencing or analysis of restriction digests (Johnsrud et al. 1978; M. P. Calos, unpubl.), which separates the genetically indistinguishable cluster in the middle of the gene into several distinct integration sites. Although the position of each *I*-gene deletion group is known with a high degree of precision, sequencing results show that a reduction of recombination frequencies occurs as the deletion approaches the end of the insertion, so that the insertion lies 50–75 bp beyond the point estimated from deletion mapping in the *I* gene. The position of each deletion is therefore increased by 50–75 bp on the *I*-gene map shown here. In the *Z* gene, the deletions *H120,* etc., were described by Gho and Miller (1974) and are harbored in strains CSH13 through 20 (Miller 1972). The deletions *274, 250, 209, 169,* and *103* are from strains described in Materials and Methods. The *Z* gene is divided into 27 sections, each corresponding to a segment defined by Zipser and coworkers (1970). The number above the line corresponds to one of 48 *lacZ*::Tn*9* insertions. All mutations with numbers below 100 were derived in strain MPC1, whereas those with numbers greater than 100 were derived in the *recA* strain MPC2 (as were the five *lacI*::Tn*9* insertions *1, 2, 3, 4,* and *35*). The cluster at the end of the *Z* gene has been separated into more than the six to eight groups shown here by sequencing studies reported in the accompanying figures and in Galas et al. (1980), which establish 15 sites for the mutations mapping past insertion *5.* The insertions in the *Y* gene have been mapped against some of the deletions reported by Hobson et al. (1977). As in the *Z* gene, those insertions depicted in different clusters are at different sites based on genetic crosses described by Miller et al. (1980). Some of these clusters may be subdivided further by sequence analysis.

other regions within the same gene. Yet within each of these regions there are many different insertion sites, as established by genetic crosses (Miller et al. 1980), restriction mapping, and sequencing (see the following sections) (Johnsrud et al. 1978; Galas et al. 1980). Moreover, after considering Figure 2 and Table 3, we conclude that the preferred region of integration at the end of the *Z* gene can be viewed as an extension of a much larger region that includes most or all of the *Y* gene. Integrations into the *Y* gene are seven times more frequent than integrations into the *Z* gene (Table 2) in MPC1, where two thirds of the *Z*-gene insertions are at the end of *Z.* Thus, Tn*9* inserts into the *Y* gene about tenfold more frequently than into the end of the *Z* gene. Since the preferred region at the end of the *Z* gene is very close to 10% of the size of the *Y*-gene region (Büchel et al. 1980; Galas et al. 1980), the frequency of integration per unit length is the same.

Comparison of Different Regions of *lac*

The ends of the *Z* gene and the *Y* gene constitute a region into which Tn*9* integrates approximately 25- to 50-fold more frequently per unit length of DNA than into most of the rest of *lacZ* and *lacI*, whereas a short stretch in the middle of the *I* gene is favored by about a factor of 10. These estimates are derived from Table 3 and Figure 2. Since only one third of the Z^- insertions in MPC1 are in the remainder of *Z* (~2950 bp), the frequency of insertion into the *Y* gene (~1300 bp, if one includes the short region between *Z* and *Y*) (Büchel et al. 1980) is $7 \times 3 \times 2950/1300$, or approximately 48 times more frequent per unit length than into the rest of the *Z* gene. A similar treatment for the major portion of the *I* gene yields a factor of approximately 25–30 for the ratio of insertions into the *Y* gene versus insertions into most of the *I* gene. Figure 3 summarizes these data by showing the relative Tn*9* integration index for different parts of the *lac* region. This diagram identifies two "hot" regions for Tn*9* integration (enclosed in boxes) within which there are many different integration points.

Does the DNA sequence of regions favored for integration reflect any pattern? Inspection of the sequence reveals a good correlation between AT richness and a high frequency of integration of Tn*9*, as shown in Figure 3. The region in the *I* gene that is the most favored for integration is the most AT-rich segment (when intervals of 50 nucleotides are considered), and the *Y* gene is significantly richer than the *I* gene.

Tn*5*

Figure 4 shows the distribution of Tn*5* insertions, which were detected by the same method shown in Figure 1 but using strain MPC4 and selective medium containing kanamycin (see Table 2 and Materials and Methods). Tn*5* integrates into *lac* more frequently than any of the other elements studied (Table 3). To determine the relative frequency of insertion into the *Y* and *Z* genes, we collected 50 independent lac^- insertions using indicator media not distinguishing between Z and Y^- phenotypes. Twenty-one were Z^- insertions and 31 were Y^- insertions. Multiplying this 3:2 ratio by the inverse ratio of the respective gene lengths gives a preference for the *Y* gene of four- to fivefold per unit distance. (Since the distribution in the *Z* gene is slightly skewed toward the end of the gene, the preference for the region consisting of the *Y* gene and the end of the *Z* gene is somewhat higher.) Yet again, as with Tn*9*, there are many integration points within these regions.

Tn*10*

Foster (1977) has described a set of *lac*::Tn*10* insertions in which Lac⁻ mutants were selected on an F′*lac* episome. Of the 35 Tn*10* insertions in the *lac* region, 21 are in the *Z* gene and 14 are in the *Y* gene. Of the 21 *Z*-gene insertions, 18 map in one small deletion interval at the end of the *Z* gene. We further characterized the Tn*10* insertions of Foster by insertion × insertion crosses and by mapping the insertions in the *Y* gene, as well as detecting additional insertions, using strain MPC5. These results are shown in Figure 5. Within the region at the end of the *Z* gene, the Tn*10* insertions form a tight cluster and cannot be separated (with one possible exception) by the genetic techniques that resolve the Tn*9* insertions in this region into numerous subgroups.

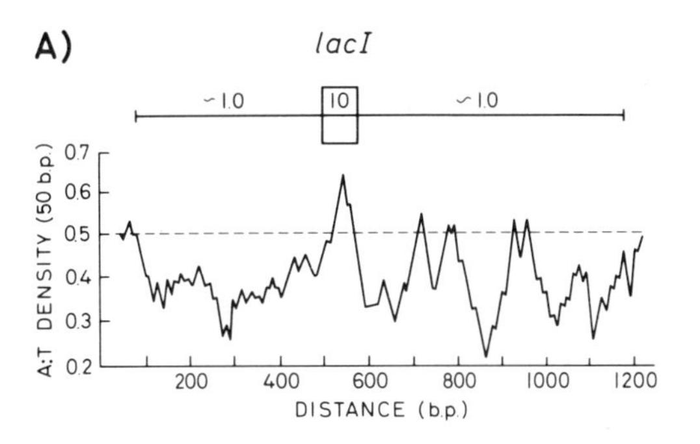

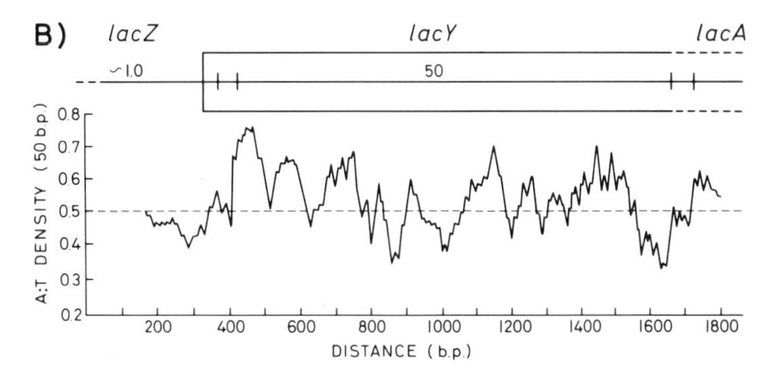

Figure 3. Correlation of Tn*9* insertion frequency with AT richness. The approximate frequencies of integration of Tn*9* (per unit distance) in *lacI*, part of *lacZ*, and *lacY* are shown at the tops of *A* and *B*. The AT density of the corresponding regions, based on the sequences of 50-bp intervals, is shown at the bottom of *A* and *B*. (The remainder of the *Z* gene has a normalized frequency of 1.0 on this scale, but the sequence is incomplete, so this region is not shown here.)

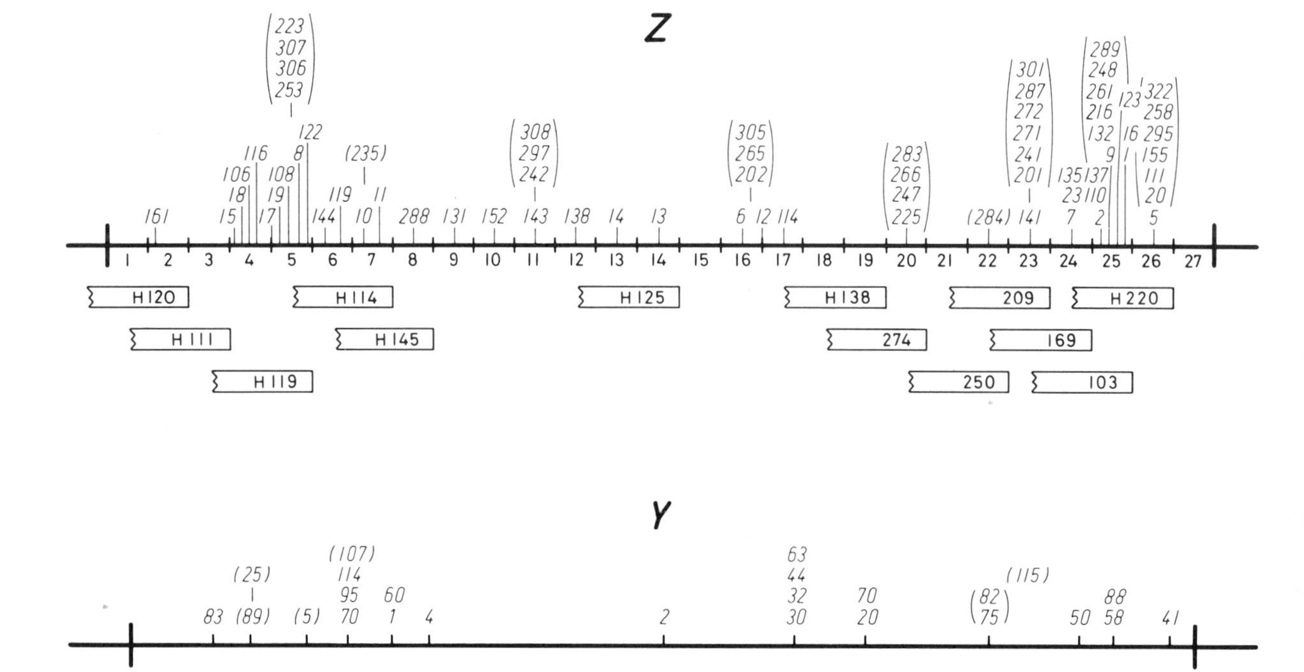

Figure 4. Tn*5* insertions in *lacZ* and *Y*. (*Top*) Distribution of 69 Tn*5* insertions in *lacZ;* (*bottom*) position of 25 Tn*5* insertions in *lacY*. The different sample sizes do not reflect the fact that insertions are more frequent into *Y* (Table 3), and the distributions should therefore be considered separately. Parentheses indicate clusters that have not been mapped further. (The clusters in *Z* segments 25 and 26 have been subdivided further, although this is not indicated here.) For other details, see legend to Fig. 2.

Sequence Analysis of Insertion Sites

From a collection of 49 Tn*9* insertions in *lacZ*, we chose to sequence the sites of 27 insertions that mapped in the last two of 27 deletion groups (Miller et al. 1980). Of the insertions resulting in a Z^+Y^- phenotype, we also sequenced the site mapping closest to the *Z*-gene insertions. This last insertion was found to be just within the *Z* gene itself.

We constructed derivatives of the plasmid pBR322 containing the entire *lac* operon (see Materials and Methods) (Galas et al. 1980). The respective Tn*9* insertion was crossed genetically into the *lac* region of these plasmids. The *Eco*RI restriction site, near the end of *lacZ*, provided a convenient origin for sequencing in both directions, with 5′-end or 3′-end labeling. The labeling and sequencing procedures used in each case are described by Galas et al. (1980).

The approximate position of each insertion site was determined by analysis of restriction digests. Because the flanking IS*1* elements of Tn*9* are in the same orientation and have no *Eco*RI site but do have a single *Pst*I site (Grindley 1977; Ohtsubo and Ohtsubo 1977), the *Z*-gene–Tn*9* juncture point will be carried on a fragment bounded by an *Eco*RI and a *Pst*I site. The size of this fragment is determined by both the distance of the insertion from the *Eco*RI site and the orientation on the Tn*9*. Since all insertions analyzed were within 120 bp of the *Eco*RI site, the size differences are such that both pieces of information can be derived from the size of the fragment.

The positions of all 28 insertions sequenced are shown in Figure 6. It is evident that they are distributed at numerous sites within the last 160 bp of the *Z* gene. In Figure 6, each arrow indicates the point of insertion, with the convention that the 9 bp to the left of the arrow are present on both sides of the inserted element as a consequence of the insertion process (Calos et al. 1978; Grindley 1978). (In this manner, it is easier to compare the positions of different insertions, although it should be kept in mind that this is a convention and does not imply a mechanism of insertion.) Diagrams illustrating this convention are shown at the top and bottom of the figure. An arrow on top of the sequence indicates the orientation of Tn*9* as shown in the bottom diagram, and an arrow on the bottom indicates the opposite orientation as shown in the top diagram.

Sixteen different sites are defined in this collection, and six of these are represented by more than one occurrence. There is no evident orientation bias. However, half of the insertions are clustered about 30 bp from the end of the *Z* gene.

Büchel et al. (1980) have sequenced the end of the *Z* gene, the intergenic region, and the *Y* gene. We have

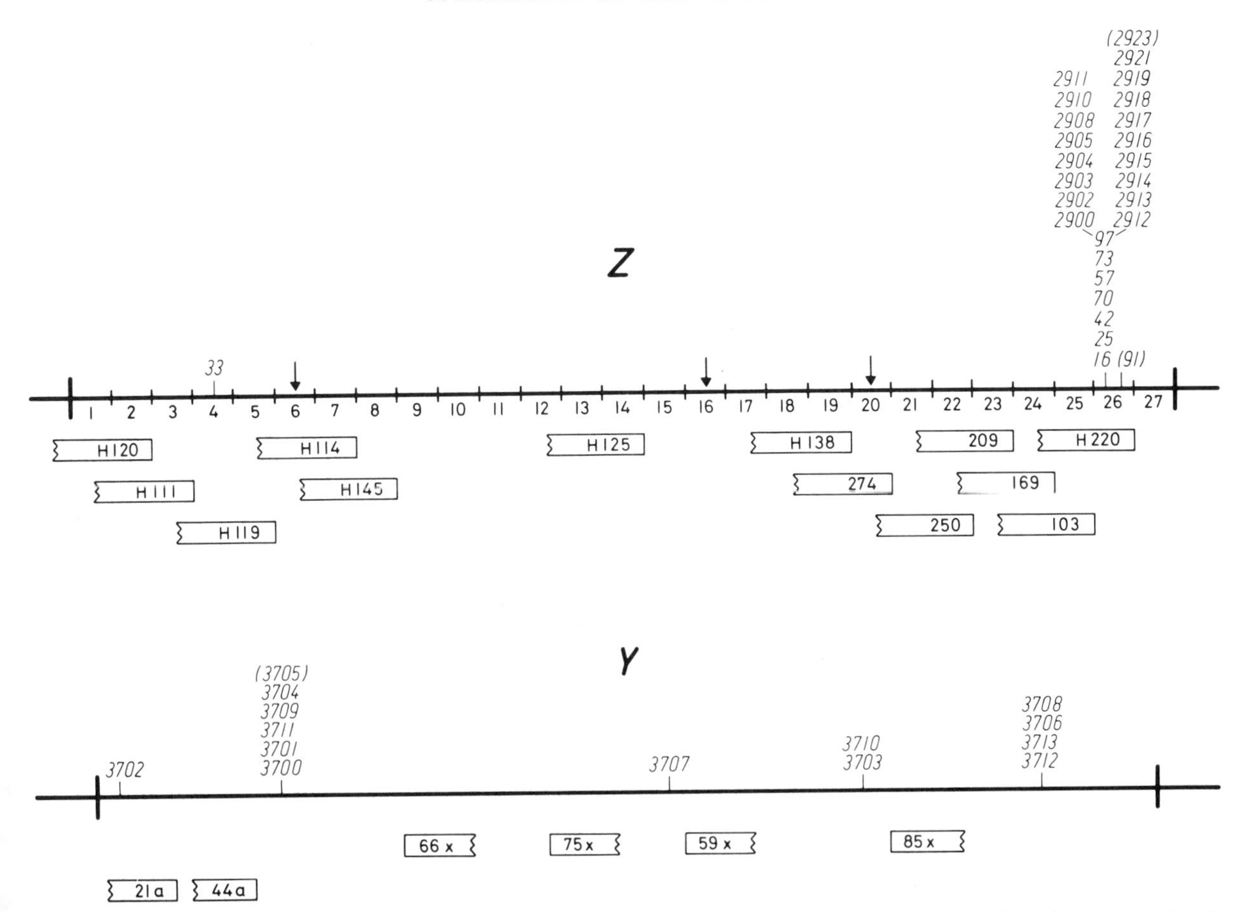

Figure 5. Tn*10* insertions in *lacZ* and *Y*. The top portion of the figure shows the map positions of 21 insertions detected and mapped by Foster (1977) (the 2900 series of mutations and the three arrows) and Tn*10* insertions detected in this work. All but one of the insertions at the end of the *Z* gene fail to be separated by genetic mapping experiments (see Materials and Methods). For further details, see legend to Fig. 2.

verified, by sequencing, that the region of our plasmid from the last *Hin*dII site in the *Z* gene to the beginning of the *Y* gene is the same.

Sequence Specificity of the Insertion Sites

The data described above enable us to examine the DNA sequences surrounding 28 independent insertions for specific patterns that correlate with the selection of an insertion site. Upon examination of the 9 bp repeated by each insertion event and the 6 bp on either side (a 21-bp interval; see Table 4), one striking feature becomes evident. In most, but not all, of the 9-bp repeats themselves, the end base pairs are G:C pairs without apparent regard to orientation. Of the 28 insertion events, 23 have this feature, including all sites where multiple occurrences were observed (see Fig. 7).

To appreciate the statistical significance of the apparent preference for G:C base pairs at the end points, consider that if a set of 28 9-bp sites were chosen completely at random (from DNA with a normal base composition), the probability that any 23 (or more) of these sites would have G:C base pairs at both ends is less than 1 in 10^9! (See Galas et al. 1980 for details.) Likewise, the probability that 11 (or more) of 16 randomly chosen sites would have G:C base pairs at each end is about 3 in 10^4.

Homologies with IS*1* Ends

On examination of the sequence at the end of *lacZ*, we find that there is an exact homology to the final 7 bp of IS*1*. This homology lies in the middle of the region most dense with insertions. The proximity of this sequence to many sites of multiple insertion could be responsible, at least in part, for the observed clustering. Figure 8 displays the insertion pattern, with this homology and several partial homologies indicated on the sequence.

If a homology with the ends of the elements can increase the insertion frequency in its surrounding region, then we might expect to find such homologies near a number of previously reported insertion sites. Table 5 shows the set of insertions from all previous studies of IS*1* and Tn*9*, indicating homologies with the ends of IS*1*. One region known to have an enhanced insertion

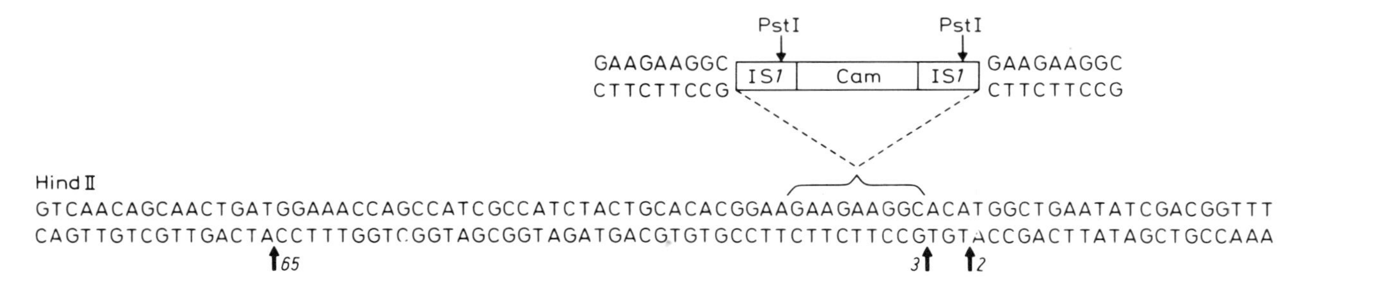

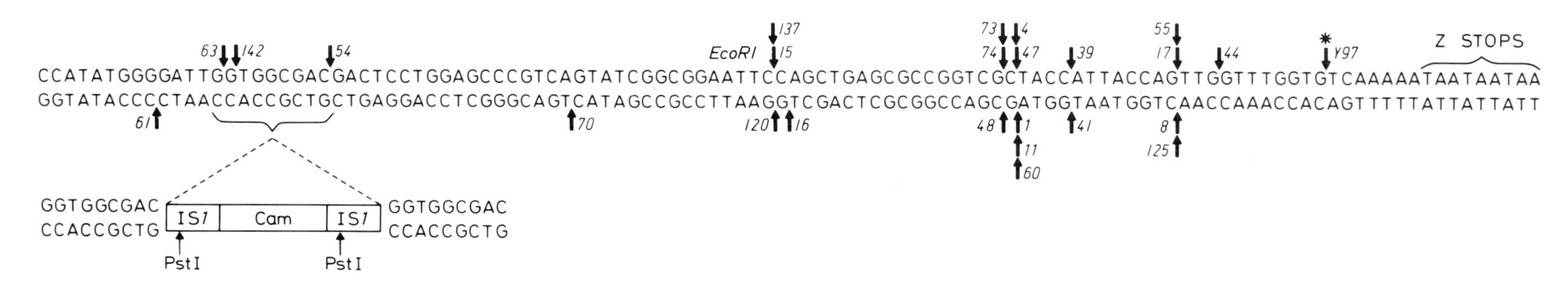

Figure 6. The positions and orientations of 28 insertions of Tn9 into the *Z* gene between the last *Hin*dII site and the end of the gene. The arrows indicating the insertion sites are drawn according to the convention illustrated by the diagrams at the top and bottom of the figure, showing an integrated element. Each arrow is drawn so that the 9-bp repeat is to the left. The bracket in the examples indicates this sequence. An arrow above the sequence indicates that the Tn*9* is oriented like the one in the diagram at the bottom (*Z54*). An arrow below the sequence indicates the opposite orientation, as illustrated in the upper diagram (*Z3*). The rightmost arrow (marked with an asterisk) indicates the position of the Tn*9* for *lac Y97*::Tn*9*, which exhibits a $Z^{+}Y^{-}$ phenotype (see text).

Table 4. Sequences at Tn9 Insertion Sites

lacZ::Tn9 insertion	Insertion site	Orientation	No. of insertions
65	TCAACA GCAACTGAT GGAAAC	←	1
3	CACACG GAAGAAGGC ACATGG	←	1
2	GCGGAA GAAGGCACA TGGCTG	←	1
61	CGGTTT CCATATGGG GATTGG	←	1
63	TCCATA TGGGGATTG GTGGCG	→	1
142	CCATAT GGGGATTGG TGGCGA	→	1
54	GGGATT GGTGGCGAC GACTCC	→	1
70	TCCTGG AGCCCGTCA GTATCG	←	1
120, 137, 15	GTATCG GCGGAATTC CAGCTG	←→→	3
16	TATCGG CGGAATTCC AGCTGA	←	1
73, 74, 48	GCTGAG CGCCGGTCG CTACCA	→→←	3
4, 47	CTGAGC GCCGGTCGC TACCAT	→→	5
1, 11, 60		←←←	
139, 141	GCGCCG GTCGCTACC ATTACC	→←	2
17, 55	CGCTAC CATTACCAG TTGGTT	→→	4
8, 125		←←	
44	TACCAT TACCAGTTG GTTTGG	→	1
lacY97::Tn9	CCAGTT GGTTTGGTG TCAAAA	→	1

The sequence in the target DNA of the 9-bp repeat and 6-bp flanking sequence is presented for each of the Tn9 insertions sequenced in this paper. The sequence is given for the upper strand, oriented as in Fig. 6. The mutation numbers in the first column are as in Fig. 6. The orientation of the element is indicated in the third column, with the convention that a right-pointing arrow represents the orientation shown at the bottom of Fig. 6 (with the *Pst*I site nearest the left end).

frequency for Tn9 lies in the *I* gene near the insertion *lacI1*::Tn9 (Johnsrud et al. 1978; Miller et al. 1980). Here we find that a 6-bp out of 7-bp match with an end of IS*1* falls within the 9-bp repeated sequence itself. Note, also, that of the three IS*1* insertions in the *galT* gene sequenced by Grindley (1978), one is next to a 7-bp exact homology and one overlaps a 6-bp out of 7-bp homology. The probability that such an ensemble of insertions would fall as close to such homologies by chance is calculated to be about 1 in 10^4 (Galas et al. 1980).

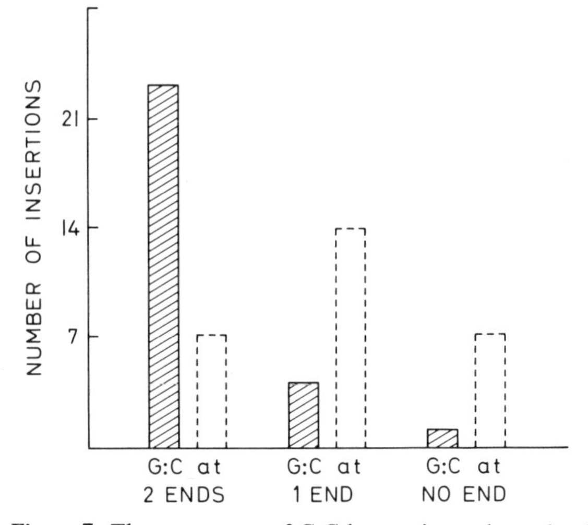

Figure 7. The occurrence of G:C base pairs at the ends of the 9-bp repeats. Hatched bars represent the observed number of the 28 Tn9 insertions sequenced in this work (see Fig. 6 and Table 4) that have G:C base pairs at both ends (first column), one end only (second column), and at neither end (last column). Dashed bars represent the expected number in each of these categories for a completely random sample of 28 9-bp sequences.

DISCUSSION

Since the size and nature of the target DNA can influence the observed integration pattern of transposable elements, we have examined the specificity of insertion of several elements, using the same DNA target—the 5600-bp *lacI,Z,Y* region carried on an F′ *lacproB* episome.

Tn*9*, which contains an IS*1* element at each end, shows a marked preference for a region extending from approximately the last 100 bp of the *Z* gene through much or all of the *Y* gene, a stretch including more than 1000 bp. Within this region, insertions occur 25- to 50-fold more frequently (per unit length of DNA) than in the rest of the *Z* and *I* genes, except for a short stretch in the middle of the *I* gene. The preference for these two regions is not due to the presence of one or even several hotspots, since Tn*9* integrates at many positions throughout the *Y* gene, at the end of the *Z* gene, and in the middle of the *I* gene. Examination of the DNA sequence reveals a good correlation between a high frequency of integration of Tn9 and AT richness (see Fig. 3). The importance of AT-rich regions for integration of IS*1* and Tn*9* is supported by data from other authors (Devos et al. 1979; Federoff 1979; Meyer et al. 1980). Moreover, we note that the *gal* control region, which serves as a preferential region for IS*1* integration (see Starlinger 1977; Kühn et al. 1979), is also extraordinarily AT-rich (Musso et al. 1977).

This type of strong preference for regions of DNA within which there are multiple insertion points has also been shown for Tn*1* in phage P22 (Weinstock et al. 1979). Smaller effects of a similar nature have been described for Tn*3* integration into several different plasmids (Kretschmer and Cohen 1977; Tu and Cohen 1980).

Tn*5* insertions are more uniformly distributed in *lac*

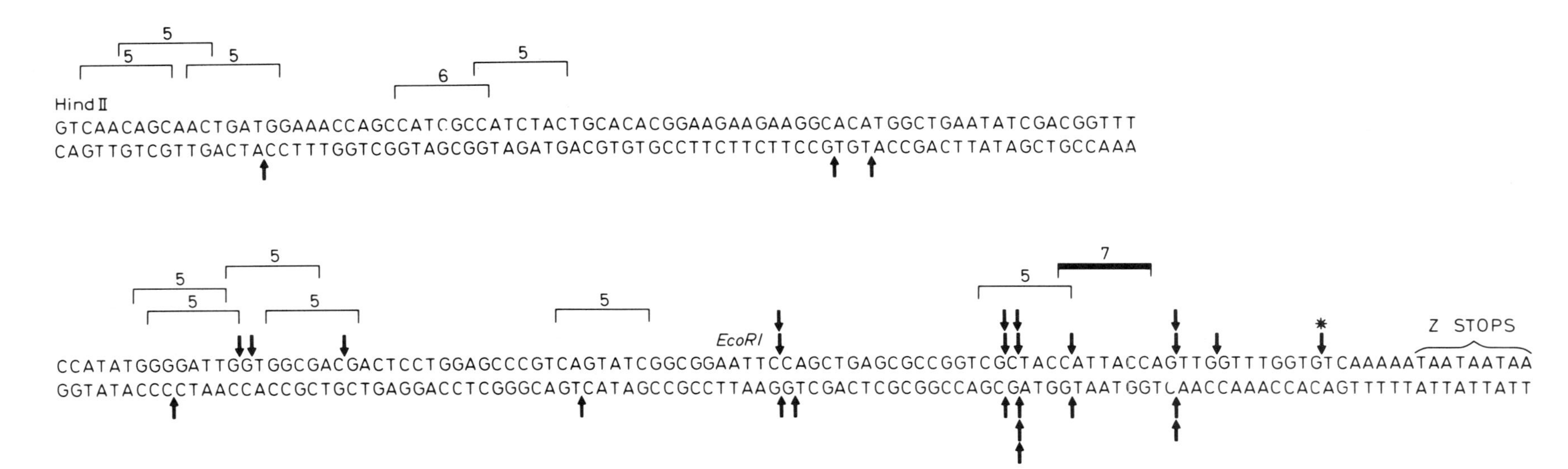

Figure 8. Homologies with the ends of IS*1* at the end of the *Z* gene. The homologies with the terminal 7 bp of IS*1* (see Table 5) with a 5-bp match or greater are indicated here by brackets, with the extent of the homology indicated above the bracket. The position of each of the insertions is marked by an arrow, according to the convention used in Fig. 6.

Table 5. Homolgies with IS*1* Ends Near the Sites of Insertion

IS*1* ends	5′GGTGATG.....CATTACC3′ 3′CCACTAC.....GTAATGG5′
gal control	CATACCATAAGCCTAATGGAG GTATGGTATTCGGATTACCTC
S104	TCCTCTGCGCAGGTAATGT AGGAGACGCGTCCATTACA
S188	TAAGGAACGACCATGACGC ATTCCTTGCTGGTACTGCG
N102	CCACATCGCCGCTACAA GGTGTAGCGGCGATGTT
MS2	GCAGGCATCGTGGTGTCA CGTCCGTAGCACCACAGT
lacI1::Tn*9*	CACCCATCAACAGTATT GTGGGTAGTTGTCATAA
lacI2::Tn*9*	AAGTGGAAGCGGCGATGGC TTCACCTTCGCCGCTACCG
lacZ1::Tn*9*	TGATTACGGATTCACTGGCCG ACTAATGCCTAAGTGACCGGC
S114 S58	AGTAACGTTATACGATGTCGCAGAGTATGC TCATTGCAATATGCTACAGCGTCTCATACG

The first line shows the 7 bp at either end of IS*1*. They are inverted repeats with one purine degeneracy at position 4. The remaining entries show the sequence of the region surrounding each of the insertions that have been reported previously: *gal* control region, Kühn et al. (1979); S104, S108, and N102, Grindley (1978); MS2, Devos et al. (1979); *lacI1, lacI2,* and *lacZ1*, Johnsrud et al. (1978); and S58 and S114, Calos et al. (1978). The boxes indicate homology with the ends of IS*1*. The underlined 9 bp are those repeated on insertion of the element.

than are Tn*9* integration points (Fig. 4). Although the end of the *Z* gene and the entire *Y* gene are still favored by Tn*5*, the five- to sevenfold preference (per unit distance) is less pronounced than the effects seen with Tn*9*. (Berg et al. [1980] have reported similar results with Tn*5* in *lac*.) Determination of transposition frequencies indicates that Tn*5* is much more efficient than Tn*10* or Tn*9* in transposing into the *lac* region (Table 3).

Previous studies of Tn*10* indicate a much smaller number of integration sites than for Tn*9* or Tn*5*. Thus, 135 insertions in the ten-gene *his* operon of *S. typhimurium* are distributed among 22 sites (Botstein and Kleckner 1977; Kleckner et al. 1979). Moreover, one prominent hotspot in *hisG* accounts for 50% of all of the insertions found. All of the members of the hotspot are inseparable by genetic methods, and ten of ten have been shown to be at the same nucleotide position (S. Halling and N. Kleckner, unpubl.). The Tn*10* insertions in *lacZ* and *Y* follow this pattern. Foster (1977) showed that 18 of 21 insertions in the *Z* gene are in the same small deletion group (1977), and this cluster is inseparable by genetic methods (this paper). Also, the 15 *Y*-gene insertions are grouped into several tight clusters (Fig. 4). The hotspot at the end of the *Z* gene and the insertions in the *Y* gene constitute 33 of the 36 *lac*::Tn*10* insertions and thus suggest that Tn*10* is clustering in the same large region as Tn*9*, also by a preferential factor of 30 to 50 per unit distance. However, since only a few sites are involved, this conclusion is still tentative.

Sequence Specificity of IS*1* and Tn*9* Insertions

The work reported here shows that, at the sequence level, the insertions are not randomly distributed within favored regions. From 28 insertions of Tn*9* at the end of the *Z* gene, analyzed by DNA sequencing, two types of patterns emerge.

Cutting-site preference. Upon integration, IS*1* and Tn*9* generate a repeated sequence of 9 bp of target DNA on each side of the inserted element (Calos et al. 1978; Grindley 1978; Johnsrud et al. 1978). By inspecting the 9-bp repeat at each insertion site, we can define a strong preference for G:C base pairs at each end, without apparent regard to orientation (see Table 4). In fact, 23 of 28 insertion events result in a G:C base pair at each end of the 9-bp repeat. (This would occur fortuitously with a probability of less than 10^{-9}.) This 4.6:1 preference overall occurs despite a 1:3 ratio in the DNA target region of 9-bp stretches with a G:C base pair at each end to those without. The average preference per available site or 9-bp stretch is therefore approximately 14:1 for G:C base pairs at each end. A larger set of data, obtained by considering the results from this study together with other published IS*1* and Tn*9* insertion-site sequences, strengthens these conclusions (Galas et al. 1980).

This type of specificity is reminiscent of that seen for the topoisomerase DNA gyrase (Gellert et al. 1976; for review, see Wang and Liu 1979). In its DNA-cleavage activity, gyrase has the ability to act at many sites but exhibits marked preferences for certain sites (Gellert et al. 1979; Peebles et al. 1979). A study of the DNA-sequence basis of this preference reveals that gyrase makes a 4-bp staggered cut such that one strand is cut between a TG dinucleotide (Morrison and Cozzarelli 1979). The G component appears to be common to most or all gyrase cleavage sites, whereas the T component is common to at least all of the preferred sites. No other invariant elements were detected, although additional determinants must be present to account for the infrequency of cleavage. We propose a diverse recognition sequence with common elements, in analogy to interaction of RNA polymerase and DNA (for review, see Rosenberg and Court 1979). Whatever the enzyme responsible for the Tn*9*/IS*1* specificity at the nucleotide level, its preference for G:C base pairs at either end of the site is the only strongly preferred element.

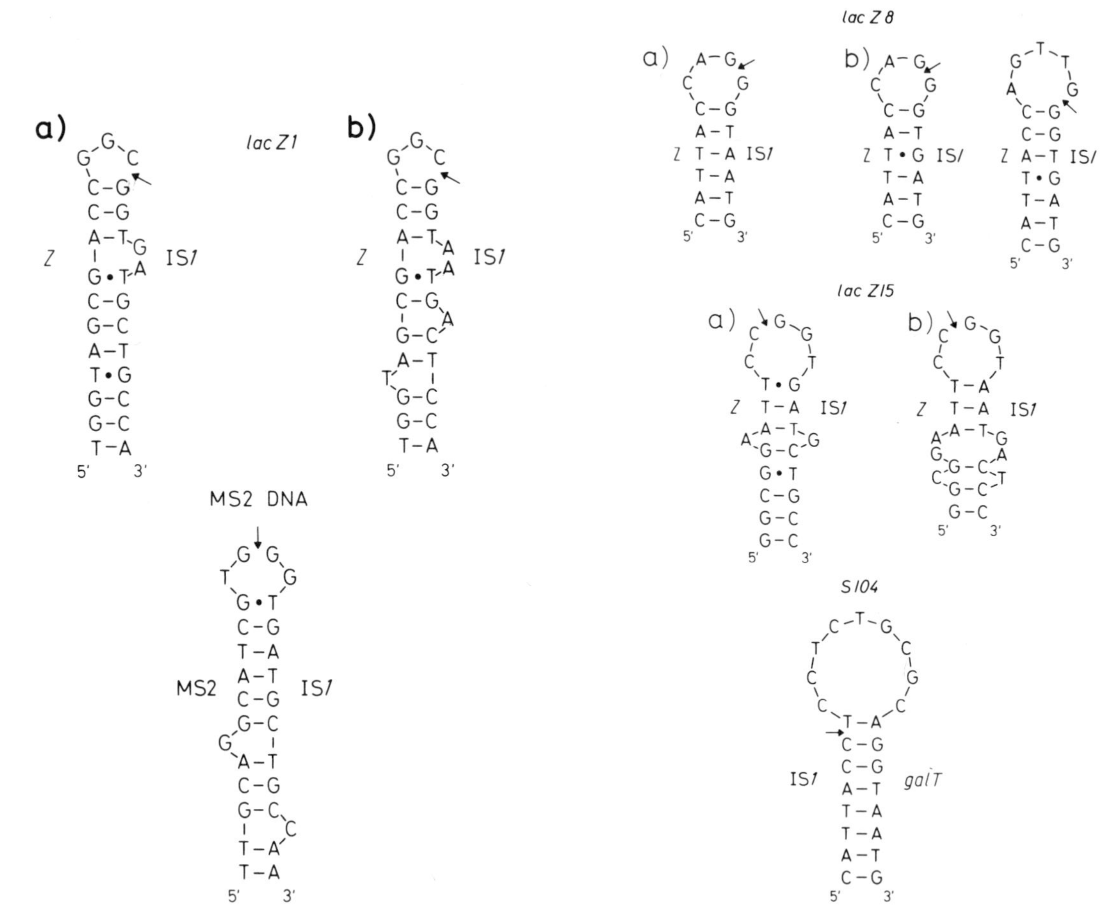

Figure 9. Potential duplex structures between target and transposon DNA. We have indicated here secondary structures that could be involved in an intermediate step in the transposition process, perhaps in the stabilization of the preligation complex (see text). The arrow in each case indicates the boundary between the target and IS*1* DNAs, which are labeled in each drawing. The sets of structures labeled with a *lacZ* mutation number indicate a cluster of insertions (as shown in Fig. 6). In each set, *a* and *b* indicate the structures formed with the two ends of IS*1*, respectively. For the *lacZ8* group, we have also indicated a particularly stable structure for *Z44* (*right*). S104 and MS2 are described in Table 5.

Homologies. Several sites of multiple insertions cluster within a 16-bp region near the end of the *Z* gene. Within this stretch is a 7-bp sequence identical to the last 7 bp of IS*1* (see Fig. 8). An inspection of other published IS*1* and Tn*9* insertion sites (see Table 5 and Galas et al. 1980) reveals a statistically significant presence of sequences homologous to the ends of IS*1* at or very near the sites of insertion. This finding raises the possibility that such homologies are involved in site selection during insertion or at least render some sites more favorable than others for insertion.

One can envision that an early step in transposition site selection results in a staggered cut at the target site. If the same enzyme that recognizes the ends of IS*1* during transposition also generates the staggered cuts at the target site, then we might expect the enzyme to recognize and bind to sites on the target DNA similar to the ends of IS*1*. These stretches would serve as primary binding sites, with nearby stretches that have less homology operating as secondary binding sites.

Alternatively, the observed homologies might reflect the fact that secondary structures of the type depicted in Figure 9 can form between strands of target DNA and the ends of IS*1*. These secondary structures could stabilize a preligation complex during the process of transposition. Reasonable structures can be drawn for many of the favored sites, as indicated in Figure 9.

Tu and Cohen (1980) have sequenced 65 insertions of Tn*3* into a small plasmid. They find many integration sites in the plasmid but detect a fourfold preference for an AT-rich segment, within which multiple occurrences at the same nucleotide position are found in a number of cases. One hotspot of this type has 26 occurrences. The respective 5-bp repeats are mostly A:T base pairs. These authors also postulate that a combination of AT richness and the presence of a sequence showing a strong homology to the ends of Tn*3* is responsible for the large number of insertions in this particular segment.

In analogy to what we find for Tn*9*, it is possible that a "cutting-site preference" for an A:T base pair at each end of the 5-bp repeat is also reflected in the data of Tu and Cohen. Of the 31 different integration sites (including all seven at which multiple insertions have been found), 22 have an A:T base pair at each end. This could be, however, simply a consequence of an overall preference for A:T base pairs within the 5-bp repeat.

CONCLUSIONS

Several factors probably act in concert to produce the final observed distribution of IS*1* and Tn*9* insertion sites. On the larger scale, AT richness (favoring denaturability) results in more frequent integration within regions of DNA.

AT richness is also characteristic of RNA-polymerase-binding sites (Jones et al. 1977; Nakamura and Inouye 1979) and apparently reflects a need for denaturing DNA during transcription initiation. Perhaps an analogous step in transposition is facilitated by denaturability. Among AT-rich regions, the presence of sequences homologous to the ends of IS*1* may result in more frequent integration. The concerted action of these two factors might explain the large-scale distribution of Tn*9* insertions within *lacI*. As Figure 10 shows, there are seven points in the *I*-gene sequence with a homology to the ends of IS*1* of 6 bp out of 7 bp and one perfect 7 bp out of 7 bp. In only one case does a match of 6 bp out of 7 bp coincide with an AT-rich stretch, and this region is the most prominent for Tn*9* integration into *lacI* (Miller et al. 1980), with an insertion frequency tenfold above the rest of the gene (per unit distance). (The one match

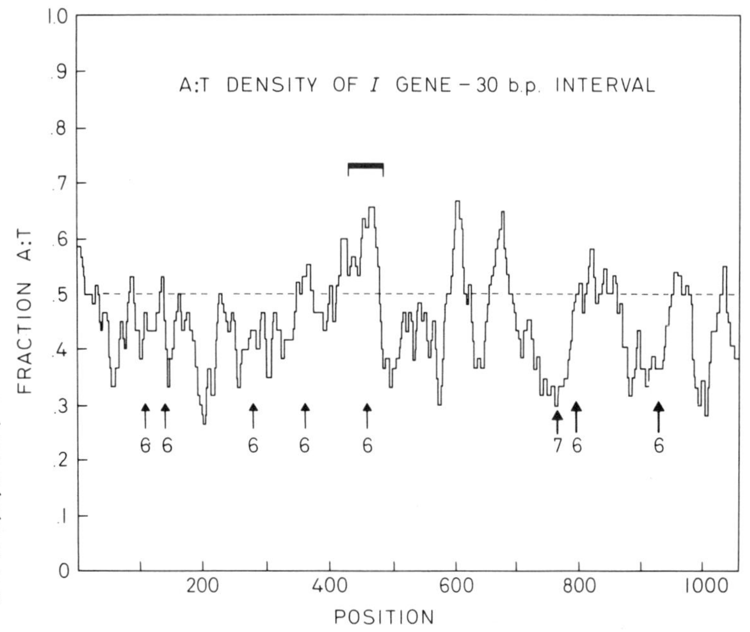

Figure 10. Correlation of AT richness and IS*1* homologies in the *lacI* gene. The fraction of A:T base pairs in a 30-bp interval is plotted here against the position of the center of the interval. The arrows indicate the locations of the 6-bp or 7-bp homology with the ends of IS*1*. The bracket indicates the region in which Tn*9* insertion is about tenfold enhanced over the average for the remainder of the gene (Miller et al. 1980).

of 7 bp out of 7 bp occurs in the least AT-rich stretch in the gene.)

At the nucleotide level, it is clear that a strong cutting-site preference is superimposed on these other determinants, so that G:C base pairs are greatly preferred at each end of the 9-bp sequence that is repeated during integration. The enzyme responsible for this preference is probably element-specific.

ACKNOWLEDGMENTS

We thank M. Hofer and M. Campiche for expert technical assistance. This work was supported by a grant from the Swiss National Fund (3.493.0.79).

REFERENCES

ALTON, N. K. and D. VAPNEK. 1979. Nucleotide sequence analysis of the chloramphenicol resistance transposon Tn*9*. *Nature* **282:** 864.

BERG, D. E., A. WEISS, and L. CROSSLAND. 1980. The polarity of Tn*5* insertion mutations in *E. coli. J. Bacteriol.* (in press).

BERG, D. E., J. DAVIES, B. ALLET, and J.-D. ROCHAIX. 1975. Transposition of R factor genes to bacteriophage lambda. *Proc. Natl. Acad. Sci.* **72:** 3628.

BOTSTEIN, D. and N. KLECKNER. 1977. Translocation and illegitimate recombination by the tetracycline resistance element Tn*10*. In *DNA insertion elements, plasmids, and episomes* (ed. A. I. Bukhari et al.), p. 185. Cold Spring Harbor Laboratory, Cold Spring Harbor, New York.

BÜCHEL, D. E., B. GRONENBORN, and B. MÜLLER-HILL. 1980. Sequence of the lactose permease gene. *Nature* **283:** 541.

BUKHARI, A. I. and D. ZIPSER. 1972. Random insertion of Mu-1 DNA within a single gene. *Nat. New. Biol.* **236:** 240.

CALOS, M. P., L. JOHNSRUD, and J. H. MILLER. 1978. DNA sequence at the integration sites of the insertion element IS*1*. *Cell* **13:** 411.

DANIELL, E., R. ROBERTS, and J. ABELSON. 1972. Mutations in the lactose operon caused by bacteriophage Mu. *J. Mol. Biol.* **69:** 1.

DEVOS, R., R. CONTRERAS, J. VAN EMMELA, and W. FIERS. 1979. Identification of the translocatable elements IS*1* in a molecular chimera constructed with plasmid pBR322 DNA into which a bacteriophage MS2 DNA copy was inserted by the poly(dA)·poly(dT) linker method. *J. Mol. Biol.* **128:** 621.

FARABAUGH, P. J. 1978. Sequence of the *lacI* gene. *Nature* **274:** 765.

FEDEROFF, N. 1979. Deletion mutants of *Xenopus laevis* 5S ribosomal DNA. *Cell* **16:** 551.

FOSTER, T. J. 1977. Insertion of the tetracycline resistance translocation unit Tn*10* in the *lac* operon of *E. coli* K12. *Mol. Gen. Genet.* **154:** 305.

FOSTER, T. J., T. HOWE, and K. RICHMOND. 1975. Translocation of the tetracycline resistance determinant from R100-1 to the *E. coli* K12 chromosome. *J. Bacteriol.* **124:** 1153.

FOWLER, A. V. and I. ZABIN. 1978. Amino acid sequence of beta-galactosidase. XI. Peptide ordering procedures and the complete sequence. *J. Biol. Chem.* **253:** 5521.

GALAS, D., M. P. CALOS, and J. H. MILLER. 1980. Sequence analysis of Tn*9* insertions in the *lacZ* gene. *J. Mol. Biol.* (in press).

GELLERT, M., K. MIZUUCHI, M. H. O'DEA, and H. A. NASH. 1976. DNA gyrase: An enzyme that introduces superhelical turns into DNA. *Proc. Natl. Acad. Sci.* **73:** 3872.

GELLERT, M., K. MIZUUCHI, M. H. O'DAE, H. OHMORI, and J. TOMIZAWA. 1979. DNA gyrase and DNA supercoiling. *Cold Spring Harbor Symp. Quant. Biol.* **43:** 35.

GHO, D. and J. H. MILLER. 1974. Deletions fusing the *i* and *lac* regions of the chromosome in *E. coli. Mol. Gen. Genet.* **131:** 137.

GOTTESMAN, M. M. and J. L. ROSNER. 1975. Acquisition of a determinant for chloramphenicol resistance by coliphage lambda. *Proc. Natl. Acad. Sci.* **72:** 5041.

GRINDLEY, N. D. F. 1977. Physical mapping of IS*1* by restriction endonucleases. In *DNA insertion elements, plasmids, and episomes* (ed. A. I. Bukhari et al.), p. 115. Cold Spring Harbor Laboratory, Cold Spring Harbor, New York.

———. 1978. IS*1* insertion generates duplication of a nine base pair sequence at its target site. *Cell* **13:** 419.

GRINSTED, J., P. M. BENNETT, S. HIGGINSON, and M. H. RICHMOND. 1978. Regional preference of insertion of Tn*501* and Tn*802* into RP1 and its derivatives. *Mol. Gen. Genet.* **166:** 313.

HABERMANN, P., R. KLAER, S. KÜHN, and P. STARLINGER. 1979. IS*4* is found between eleven or twelve base pair duplications. *Mol. Gen. Genet.* **175:** 369.

HOBSON, A. C., D. GHO, and B. MÜLLER-HILL. 1977. Isolation, genetic analysis, and characterization of *E. coli* mutants with defects in the *lacY* gene. *J. Bacteriol.* **131:** 830.

JOHNSRUD, L., M. P. CALOS, and J. H. MILLER. 1978. The transposon Tn*9* generates a 9 base pair repeated sequence during integration. *Cell* **15:** 1209.

JONES, B. B., H. CHAN, S. ROTHSTEIN, R. D. WELLS, and W. S. REZNIKOFF. 1977. RNA polymerase binding sites in λp*lac*5 DNA. *Proc. Natl. Acad. Sci.* **74:** 4914.

KLECKNER, N. 1979. DNA sequence analysis of Tn*10* insertions: Origin and role of 9 base pair flanking repetitions during Tn*10* translocation. *Cell* **16:** 711.

KLECKNER, N., R. K. CHAN, B.-K. TYE, and D. BOTSTEIN. 1975. Mutagenesis by insertion of a drug-resistance element carrying an inverted repetition. *J. Mol. Biol.* **97:** 561.

KLECKNER, N., D. STEELE, K. REICHARDT, and D. BOTSTEIN. 1979. Specificity of insertion by the translocatable tetracycline-resistance element Tn*10*. *Genetics* **92:** 1023.

KRETSCHMER, P. J. and S. N. COHEN. 1977. Selected translocation of plasmid genes: Frequency and regional specificity of translocation of the Tn*3* element. *J. Bacteriol.* **130:** 888.

KÜHN, S., H.-J. FRITZ, and P. STARLINGER. 1979. Close vicinity of IS*1* integration sites in the leader sequence of the *gal* operon of *E. coli. Mol. Gen. Genet.* **167:** 235.

LEA, D. E. and C. A. COULSON. 1949. The distribution of the numbers of mutants in bacterial populations. *J. Genet.* **49:** 264.

LURIA, S. E. and M. DELBRÜCK. 1943. Mutations of bacteria from virus sensitivity to virus resistance. *Genetics* **28:** 491.

MACHATTIE, L. A. and J. B. JACKOWSKI. 1977. Physical structure and deletion effects of the chloramphenicol resistance element Tn*9* in phage lambda. In *DNA insertion elements, plasmids, and episomes* (ed. A. I. Bukhari et al.), p. 219. Cold Spring Harbor Laboratory, Cold Spring Harbor, New York.

MEYER, J., S. IIDA, and W. ARBER. 1980. Does the insertion element IS*1* transpose preferentially into AT-rich DNA segments? *Mol. Gen. Genet.* **178:** 471.

MILLER, J. H. 1972. *Experiments in molecular genetics.* Cold Spring Harbor Laboratory, Cold Spring Harbor, New York.

MILLER, J. H., M. P. CALOS, D. GALAS, M. HOFER, D. E. BÜCHEL, and B. MÜLLER-HILL. 1980. Genetic analysis of transpositions in the *lac* region of *E. coli. J. Mol. Biol.* (in press).

MOORE, C. W. and F. SHERMAN. 1975. Role of DNA sequences in genetic recombination in the iso-1-cytochrome *c* gene of yeast. *Genetics* **79:** 397.

MORRISON, A. and N. R. COZZARELLI. 1979. Site-specific cleavage of DNA by *E. coli* DNA gyrase. *Cell* **17:** 175.

MUSSO, R. E., R. DI LAURO, M. ROSENBERG, and B. DE CROMBRUGGHE. 1977. Nucleotide sequence of the promoter region of the galactose operon of *E. coli. Proc. Natl. Acad. Sci.* **74:** 106.

NAKAMURA, K. and M. INOUYE. 1979. DNA sequence of the gene for the outer membrane lipoprotein of *E. coli:* An extremely AT-rich promoter. *Cell* **18:** 1109.

OHTSUBO, H. and E. OHTSUBO. 1977. Sequence of the ends of IS*1* element. In *DNA insertion elements, plasmids, and episomes* (ed. A. I. Bukhari et al.), p. 591. Cold Spring Harbor Laboratory, Cold Spring Harbor, New York.

———. 1978. Nucleotide sequence of an insertion element, IS*1*. *Proc. Natl. Acad. Sci.* **75:** 615.

PEEBLES, C. L., N. P. HIGGINS, K. N. KREUZER, A. MORRISON, P. O. BROWN, A. SUGINO, and N. R. COZZARELLI. 1979. Structure and activities of *E. coli* DNA gyrase. *Cold Spring Harbor Symp. Quant. Biol.* **43:** 41.

ROSENBERG, M. and D. COURT. 1979. Regulatory sequences involved in the promotion and termination of RNA transcription. *Annu. Rev. Genet.* **13:** 319.

SCHALLER, H. 1979. The intergenic region and the origins for filamentous phage DNA replication. *Cold Spring Harbor Symp. Quant. Biol.* **43:** 401.

SCHMEISSNER, U., D. GANEM, and J. H. MILLER. 1977a. Genetic studies of the *lac* repressor. II. Fine structure deletion map of the *lacI* gene and its correlation with the physical map. *J. Mol. Biol.* **109:** 303.

———. 1977b. Revised gene-protein map for the *lacI*-gene–*lac* repressor system. *J. Mol. Biol.* **117:** 572.

STARLINGER, P. 1977. Mutations caused by the integration of IS*1* and IS*2* into the *gal* operon. In *DNA insertion elements, plasmids, and episomes* (ed. A. I. Bukhari et al.), p. 25. Cold Spring Harbor Laboratory, Cold Spring Harbor, New York.

TEATHER, R. M., B. MÜLLER-HILL, U. ABRUTSCH, G. AICHELE, and P. OVERATH. 1978. Amplification of the lactose carrier protein in *E. coli* using a plasmid vector. *Mol. Gen. Genet.* **159:** 239.

TU, C.-P. D. and S. N. COHEN. 1980. Translocation specificity of the Tn*3* element: Characterization of sites of multiple insertions. *Cell* **19:** 151.

WANG, J. C. and L. F. LIU. 1979. DNA topoisomerases: Enzymes that catalyze the concerted breaking and rejoining of DNA backbone bonds. In *Molecular genetics* (ed. J. H. Taylor), part III, p. 65. Academic Press, New York.

WEINSTOCK, G., M. SUSSKIND, and D. BOTSTEIN. 1979. Regional specificity of illegitimate recombination by the translocatable ampicillin resistance element Tn*1* in the genome of phage P22. *Genetics* **92:** 685.

ZIPSER, D., S. ZABELL, J. ROTHMAN, T. GRODZICKER, and M. WENK. 1970. Fine structure of the gradient of polarity in the *Z* gene of the *lac* operon of *E. coli. J. Mol. Biol.* **49:** 251.

Tn*3* Encodes a Site-specific Recombination System: Identification of Essential Sequences, Genes, and the Actual Site of Recombination

F. Heffron,* R. Kostriken,* C. Morita,* and R. Parker
Department of Biochemistry and Biophysics, University of California, San Francisco, California 94143

In the 13 years since their discovery, prokaryotic insertion sequences (IS) have changed from a scientific curiosity (Jordan et al. 1967; Malamy 1967; Saedler and Starlinger 1967a,b) to a potent tool for bacterial geneticists (Kleckner et al. 1975; Botstein and Kleckner 1977; Chumley et al. 1979). Their significance in prokaryotic evolution is now well established: Insertion sequences and their cousins, transposons (Tn), promote rapid evolution by providing plasticity to the prokaryotic genome via deletions and rearrangements. They account for the emergence of new pathogenic bacteria by the acquisition of traits, such as toxins and antibiotic resistance, that contribute to their pathogenicity (So et al. 1979). The sequence organization of movable elements in eukaryotes is surprisingly analogous to that of bacterial transposons. *copia* in *Drosophila* (Potter et al. 1979), Ty1 in yeast (Cameron et al. 1979), and retroviruses in mammalian cells (Hughes et al. 1978; Sabran et al. 1979) all contain large directly repeated sequences at either end. Insertion of these elements generates short direct repeats of sequences found once in the original DNA, just as insertion elements in bacteria generate such repeats (Grindley 1978; Johnsrud et al. 1978; Fink et al. Majors et al. Rubin et al.; all this volume). Aside from the inherent fascination that transposons hold for us, understanding movable elements in bacteria may prove critical to understanding similar movable elements in higher organisms.

As a first step toward dissecting prokaryotic transposition, we are determining the complete organization of the transposon Tn*3*. Tn*3* is 4957 bp in length and encodes the most common β-lactamase found among gram-negative bacteria (Heffron et al. 1975b). The structural and functional anatomy of Tn*3* has been determined by constructing and analyzing a large number of mutations within this transposon. The mutations were constructed in vitro either by (1) treatment of the plasmid with a combination of nucleases to introduce deletions (Heffron et al. 1977) or (2) random insertion of synthetic *Eco*RI restriction sites (Heffron et al. 1978). The overall structure of Tn*3*, together with many of the mutations and their phenotypes, is shown in Figure 1.

Our dissection of Tn*3* is based on: (1) complementation of the various transposition-negative mutations shown in Figure 1 with an intact transposon or with many of the mutations shown; (2) determination of the DNA sequence of the complete transposon and many of the mutations within it; (3) identification of Tn*3*-specific polypeptides in minicells and the alteration of these polypeptides by the mutations; and (4) our attempt to determine the regulation of these genes by analysis of transcriptional fusions.

The results of this study reveal that Tn*3* encodes three sequences required in *cis* for its transposition in addition to three gene products (Heffron et al. 1977, 1979; Gill et al. 1978, 1979; Chou et al. 1979a,b). Aside from the β-lactamase, Tn*3* encodes a high-molecular-weight transposase, the product of the *tnpA* gene (see Fig. 1), that is absolutely required for its transposition. Mutations in a second gene (*tnpR*) located adjacent to the transposase show an increase in transposition frequency and a corresponding increase in the amount of transposase produced. From these and additional results, we have determined that the product of this gene negatively regulates its own synthesis and that of the transposase. The ends of Tn*3* form a perfect 38-bp inverted repeat (Ohtsubo et al. 1979). We have shown that both of these ends are required in *cis* for transposition (Heffron et al. 1977; Gill et al. 1979). This paper reports experiments identifying a site-specific recombination function encoded by Tn*3* which includes a third site necessary for Tn*3* transposition.

A functional subset of the internal deletions (see Fig. 1) invariably transposes to form fusions between the donor and recipient plasmid when complemented rather than transforms to form precise insertions (Gill et al. 1978; Arthur and Sherratt 1979). The fusions are cointegrates containing the donor plasmid linked to the recipient via two direct repeats of Tn*3* (see Fig. 2). Our explanation for this result has been that cointegrates could be intermediates in transposition and that the function deleted is required for recombination between the direct repeats of Tn*3* in the cointegrate. When we first carried out these experiments, we observed the same result regardless of the complementing Tn*3* derivative (Gill et al. 1978). Therefore, we hypothesized that these deletions removed a sequence essential for this recombination; we termed this region the internal resolution site, or IRS (Heffron et al. 1979). This paper demonstrates that Tn*3* does encode a site-specific recombination function and reports that this recombination requires the product of the *tnpR* gene. In addition, we have shown that the recombination is reciprocal and site-specific (at least to within 19 nucleotides), taking place through the sequence CGAAA-

* Present address: Cold Spring Harbor Laboratory, Cold Spring Harbor, New York 11724.

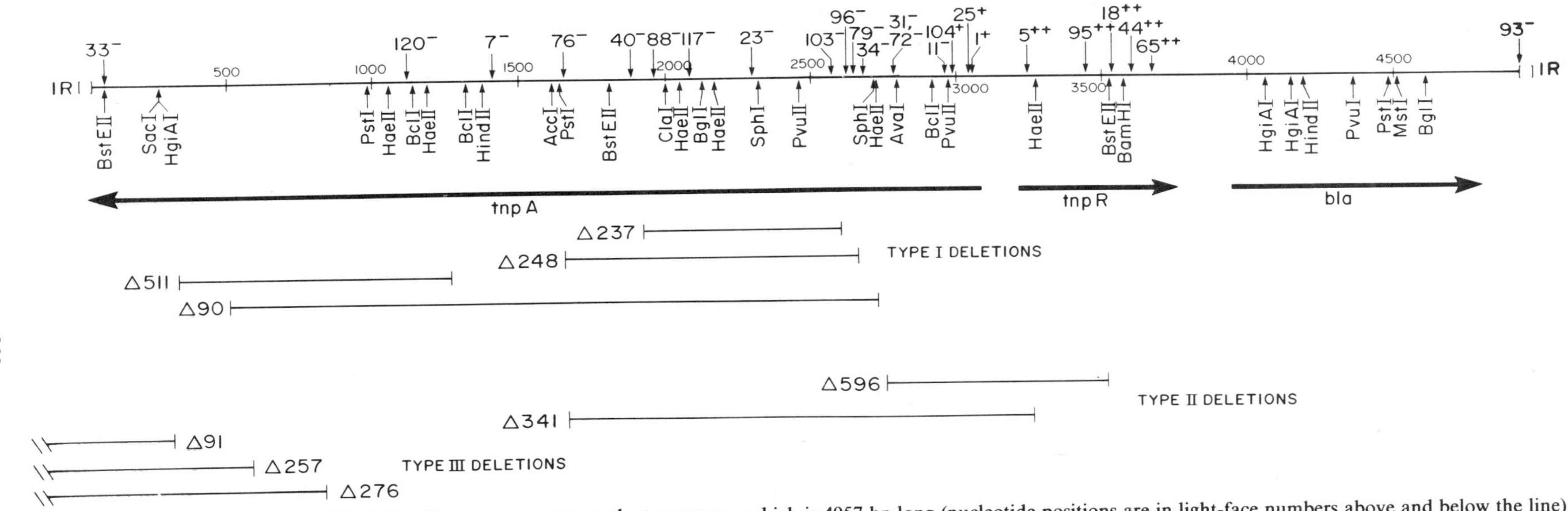

Figure 1. Sequence organization of Tn*3*. The line at top represents the transposon, which is 4957 bp long (nucleotide positions are in light-face numbers above and below the line). Numbers above the line represent the *Eco*RI insertion mutations (Heffron et al. 1978) that have been sequenced (Heffron et al. 1979); the − or + adjacent to the number corresponds to the transposition phenotype of the mutation. Some of the more useful restriction sites are shown below the line; (→) position of proteins. *bla*, the gene for the β-lactamase encoded by Tn*3*, is part of pBR322 and was first sequenced by Sutcliffe (1978). *tnpA* and *tnpR* stand for transposase and transposon repressor/recombinase, respectively. A number of deletions isolated in the initial study (Heffron et al. 1977) are shown at the bottom of the figure. The deletions were divided into three classes on the basis of complementation tests. Type-I deletions give normal transposition when complemented, whereas type-II deletions (which all extend farther to the right than type-I) form cointegrates when complemented. Type-III deletions all remove the left-hand inverted repeat and are *cis*-dominant, transposition-negative.

TYPE I DELETIONS: SIMPLE INSERTION

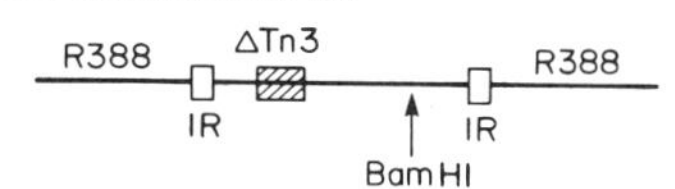

TYPE II DELETIONS: COINTEGRATE FORMATION

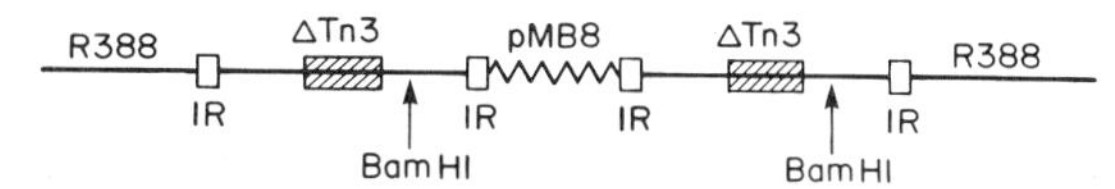

Figure 2. Structure of cointegrates. This figure compares the results observed for complementation of different transposition-negative deletions (Gill et al. 1978). Transposition has occurred from pMB8::Tn*3* to the plasmid R388. Type-I deletions (as described in Fig. 1) show normal insertion. Type-II deletions form cointegrates containing two complete copies of the original transposon as a direct repeat.

TATTATAAATTATC located between the amino terminus of the transposase and the amino terminus of the repressor.

MATERIALS AND METHODS

Bacterial strains and plasmids. All bacterial strains and plasmids are given in Table 1.

Isolation of plasmid DNA. Large-scale preparation of plasmid DNA was carried out using a gentle lysis procedure followed by dye-isopycnic centrifugation (Clewell and Helinski 1969). The ethidium bromide or propidium diiodide was removed by repeated isopropanol extractions, and, as a final step, the DNA was purified away from small-molecular-weight RNA and DNA fragments by chromatography over an A-50 column (Heffron et al. 1978). Small-scale rapid preparation of plasmid DNA was carried out according to a "quick screen" procedure (So et al., this volume).

Digestion with restriction enzymes and agarose gel electrophoresis. All enzymes employed in this work were purchased from New England Biolabs, and digestions were carried out according to its recommendation. Agarose and acrylamide gel electrophoreses were carried out according to the methods of Greene et al. (1974) and Sharp et al. (1973), respectively.

Constructions. Most constructions were carried out using purified restriction fragments isolated from acrylamide gels. After staining, the band was excised, placed in a dialysis sack with Tris-borate-EDTA buffer (100 mM Tris base, 100 mM boric acid, 2 mM EDTA; TBE), and electrophoresed in TBE for several hours. The supernatant was removed, concentrated with butanol (Stafford and Bieber 1975), and spun dialyzed (Neal and Florini 1973). DNA concentrations were determined by electrophoresis of a small sample and visual comparison with standard DNAs. The technique of spin dialysis was used whenever it was necessary to change buffers or remove enzymes in the construction. Ligation efficiency was improved by phosphatasing the vector with calf intestinal phosphatase (Boehringer-Mannheim) (Ullrich et al. 1977). Blunt-end ligations were carried out using a 25-fold excess of linkers at a total end concentration of 1 μM as described previously (Heffron et al. 1978). Sticky-end ligations were carried out at much lower DNA concentrations, usually about 50 μg/ml.

Table 1. List of Plasmids Employed in This Work

Bacterial strains or plasmids	Phenotype
SF800[a]	*nal*^{r} *polA*
MV12[a]	*recA*$^-$
RSF103[a]	*tnpR*$^-$*tnpA*$^+$ derivative of a Tn*1* insertion into RSF1010; used for complementation
pBR322[b]	Ap^r, Tc^r, derivative of ColE1
pACYC184[c]	Cm^r Tc^r derivative of the miniplasmid from *E. coli* 15
pCM10	contains a single copy of the Tn*3* IRS cloned in pBR322
pCM11	
pCM102	contains 2 copies of the IRS of Tn*3* as a direct repeat cloned in pACYC184
pCM202	pBR322 derivative that includes the complete *tnpR* gene
pRK100	contains a large internal fragment of γδ in pCM100 (orientation 1 = inverted repeat)
pRK101	contains a large internal fragment of γδ in pCM100 (orientation 2 = direct repeat)
pCM72-104	derivatives of pBR322, constructed as shown in Fig. 3, that contain an intact transposon missing the region between the *Eco*RI insertion mutations indicated
pCM72-1	same as pCM72-104
pCM72-5	same as pCM72-104
pCM72-65	same as pCM72-104
pCM25-1	same as pCM72-104
pCM104-65	same as pCM72-104
pCM25-65	same as pCM72-104
pCM1-65	same as pCM72-104
pCM5-65	same as pCM72-104

[a] Heffron et al. 1977.
[b] Bolivar et al. 1977.
[c] Chang and Cohen 1978.
All others are from this work.

Detection of transposition and cointegrates. Transposition frequencies were determined, following mating to W3110, as the proportion of F carrying a transposed copy of Tn*3* (Heffron et al. 1977; So et al. 1978). Transposition frequency was calculated as the frequency of ampicillin-resistant transconjugants per F$^+$ transconjugant. The F derivative RSF2001 (γδ$^+$), which is kanamycin-resistant, was used (Heffron et al. 1977). The presence of cointegrates in the transconjugants was checked by toothpicking colonies onto plates containing either colicin or tetracycline. The experiments to examine transposition without γδ were carried out using a newly constructed derivative of FΔ (0–15) that carries Tn*903* and is thus kanamycin-resistant, like RSF2001, but is γδ. This plasmid is called pRP101.

DNA sequencing. DNA sequencing was carried out using the chemical modification procedure developed by Maxam and Gilbert (1980).

RESULTS

Deletion Mapping the Internal Resolution Site

In previous work we have postulated that recombination between direct repeats of Tn*3* takes place through an internal site removed by a subset of the Tn*3* deletions. Our current deletion data delineated the IRS to within the *Hae*II fragment, as shown in Figure 1. The new deletion mapping was carried out using the *Eco*RI insertion mutants described previously (Heffron et al. 1978).

As shown in Figure 3, it was possible to construct deletions between any two of these *Eco*RI sites. This construction was carried out using pBR322 (Bolivar et al. 1977), a commonly used cloning vehicle that contains a contiguous region of Tn*3* between nucleotides 3748 and 4957 (Heffron et al. 1979; Sutcliffe 1979). By constructing the transposon in pBR322, we would not only generate the intact transposon, but we would also be able to make use of the gene for tetracycline resistance already present in pBR322. It was thus possible to check for cointegrates (with the appropriate controls) merely by checking for the presence of tetracycline resistance following transposition onto F. The results of the deletion mapping are summarized in Figure 4.

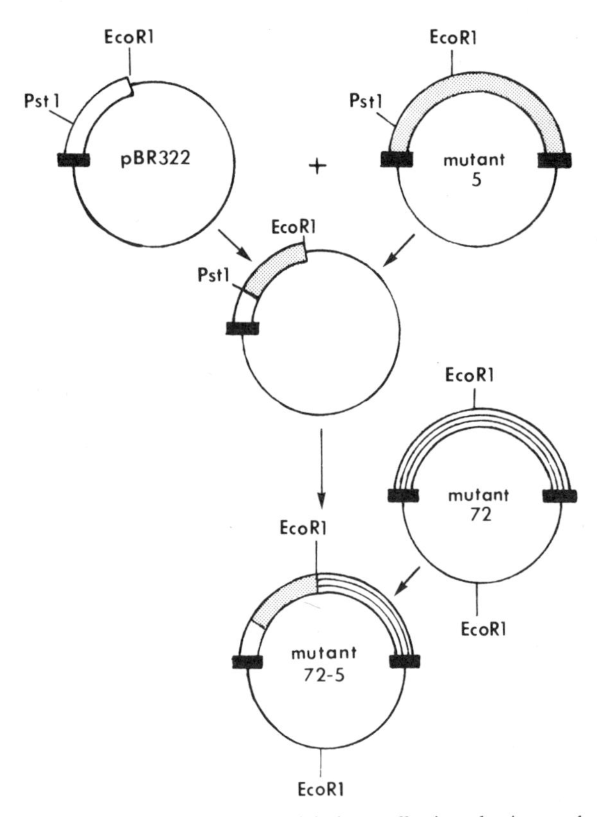

Figure 3. Construction of deletions affecting the internal resolution site. Deletions were constructed in pBR322, which contains the right-hand end of Tn*3* between nucleotides 3748 and 4957, by first exchanging the *Pst*I/*Eco*RI fragment from pBR322 with an analogous fragment from one of the mutants. The complete transposon could then be reconstructed by cloning in the *Eco*RI fragment from another mutant and checking for the appropriate orientation in the recombinant.

As shown in Figure 4, deletions that remove the entire repressor do not give cointegrates in this assay system, which is $\gamma\delta$. However, deletions that remove the intracistronic region between the transposase and the repressor result exclusively in cointegrates. Deletion mapping from the opposite direction gave a surprising result. Deletions that remove an 80-bp sequence (between mutations 104 and 1 in Fig. 4) reproducibly give a high percentage of cointegrates. A deletion that extends past mutation 1 results in cointegrates almost exclusively. Because mutations 1 and 25 both lie in the center of a region essential for resolution, we expected that they would be IRS$^-$ as well. In, fact, we have found that these two mutations have no phenotype other than a slight derepression of transposition frequency. These results suggest that at least two separate sequences are required for site-specific recombination between copies of Tn*3*: a sequence to the left of mutation 25 and another to the right of mutation 25.

Cloning the IRS

The identification of the sequences required for recombination allowed us to subclone a fragment that contains all the essential sequences, as shown in Figure 5. Since we were interested in testing the various mutations and deletions within Tn*3* for their ability to catalyze resolution of cointegrates, we constructed pCM102, an artificial cointegrate containing two copies of the IRS. pCM102 contains two *Eco*RI sites, one within the plasmid and the other flanked by two copies of the IRS. *recA*-independent recombination between these two sequences could be detected merely by the loss of one *Eco*RI site, as shown in Figure 6.

To test whether any transposon functions are required for the recombination, we complemented pCM102 with many of the mutations in Tn*3* shown in Figure 1. Figure 7 shows an agarose gel of *Eco*RI-cleaved plasmid DNA from strains containing pCM102 by itself and pCM102 complemented with various Tn*3* and $\gamma\delta$ derivatives. Lane A shows *Hin*dIII-cleaved λ DNAs used as molecular-weight markers. Lane B shows an *Eco*RI digest of the "tester" plasmid pCM102, which contains two copies of the IRS flanking an *Eco*RI restriction site (the second *Eco*RI site is present on the vector). Recombination between these two copies of the IRS will delete one *Eco*RI restriction site. As shown in lane B, no recombination has taken place for the uncomplemented plasmid. Lane C shows that recombination does take place when a complete copy of Tn*3* is present in the cell. The two bands present in this lane correspond to the recombined form of pCM102, which has lost one *Eco*RI site, and the complementing plasmid, which contains a complete copy of Tn*3* (the uppermost band). This result shows that resolution of cointegrates must depend on a function or functions encoded by the transposon. Lanes D through I show *Eco*RI-restricted plasmid DNA from strains containing pCM102 and various mutants of Tn*3* (*Eco*RI linker insertion mutants). The upper two bands in each of these lanes originate from cleavage of the complementing plasmid

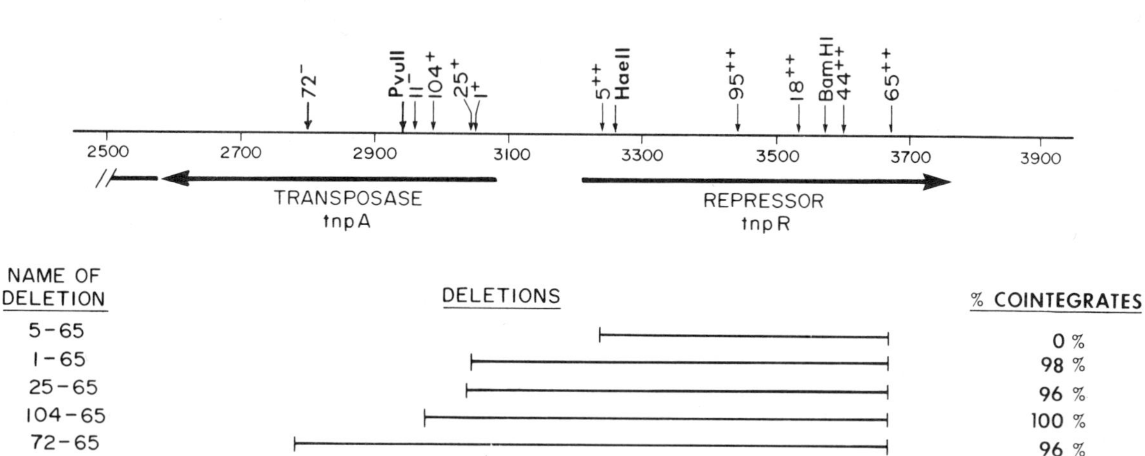

NAME OF DELETION	% COINTEGRATES
5-65	0 %
1-65	98 %
25-65	96 %
104-65	100 %
72-65	96 %
72-104	0 %
72-1	56 %
72-5	100 %
25-1	0 %

Figure 4. Location of deletions affecting the IRS and their phenotypes in Tn*3* (2500–4000 bp). The locations of the translational starts of the transposase and the repressor are shown. *Eco*RI mutations and their phenotypes are shown above the line; all these mutations have been sequenced. We have detected transposition onto RSF2001 (a kanamycin-resistant wild-type F factor containing an insertion of Tn*901;* Heffron et al. 1977) by mating into W3110 and checking for the antibiotic resistance carried outside the transposon. Transposition frequencies and percentages of cointegrates were determined with and without a complementing plasmid (RSF103). Deletion 5–65, which removes the entire repressor, does not result in cointegrates, but the next deletion, 1–65, forms cointegrates exclusively. The actual recombinational junction has been identified in this region at nucleotides 3095–3113. Deletion mapping from the opposite direction yielded a surprise: A deletion that extends up to mutation 104 was not IRS⁻; however, 72–1, only 80 bp larger, results in at least 50% cointegrates. Since mutations 25 and 1 by themselves have wild-type phenotype, these deletion results suggest that two separate regions of the transposon are required for resolution.

(one *Eco*RI site is at the site of the mutation and the second is present in the vector). Lane D shows that a mutation in the transposase will still complement for resolution. The lower band corresponds to the recombined form of pCM102. Lanes E through H show that all *Eco*RI linker insertion mutations that fall within the *tnpR* gene do not complement for resolution. Since there is only a single open reading frame that includes all the mutations that fail to complement for resolution—the *tnpR* gene itself—the repressor must be required for resolution. This small protein regulates its own synthesis as well as that of the transposase and, as these results demonstrate, acts either directly or indirectly as a recombinase.

One additional experiment was carried out to demonstrate that *tnpR* is the only Tn*3*-encoded gene required for resolution. As shown in lane J, pBR322 by itself does not complement for resolution. However, when a fragment from Tn*3* that contains the *tnpR* gene is cloned into pBR322, this recombinant plasmid is now capable of complementing for resolution of pCM102 (lane K). Moreover, we conclude that the 3000 nucleotides of Tn*3* outside the *tnpR* gene are not necessary for recombination.

Interaction between $\gamma\delta$ and Tn*3*

The results of the deletion study shown in Figure 1 suggested that resolution was independent of *tnpR*. Thus, the contradictory results from experiments with pCM102 were surprising. The pCM102 experiments demonstrated that resolution also took place in an F^+ strain (data not shown). The strong structural similarity between Tn*3* and the insertion sequence $\gamma\delta$ (R. Reed and S. Shibuga, in prep.) and the fact that $\gamma\delta$ is present on F suggested that a function encoded by $\gamma\delta$ could be responsible for the recombination. To test this, we complemented pCM102 with pOX14, a derivative of pBR322 that contains $\gamma\delta$ (Guyer 1978). As shown earlier, pBR322 alone will not complement pCM102. The results of the complementation are shown in Figure 7 (lane L). We observe that resolution takes place, and thus, $\gamma\delta$ can supply a function that will complement pCM102 for resolution. We have also observed that $\gamma\delta$ can act in *trans* to repress transcription from the *tnpA* and *tnpR* promoters, as can Tn*3*'s own repressor (R. Kostriken and C. Morita, unpubl.). These two observations led us to postulate that $\gamma\delta$ can supply a repressor/recombinase function analogous to that of Tn*3*. This would also explain why cointegrates were not observed in earlier experiments with mutations in *tnpR*, since these experiments were carried out in a $\gamma\delta$-containing F^+ strain.

The results of the $\gamma\delta$ experiments suggested that the repressor might not only be essential for site-specific recombination, but also could play a direct role in transposition, since it was possible that in all previous experiments we were supplying an equivalent function from $\gamma\delta$ resident within the cell or on F. We have therefore repeated the transposition results for *tnpR*⁻ mutations in

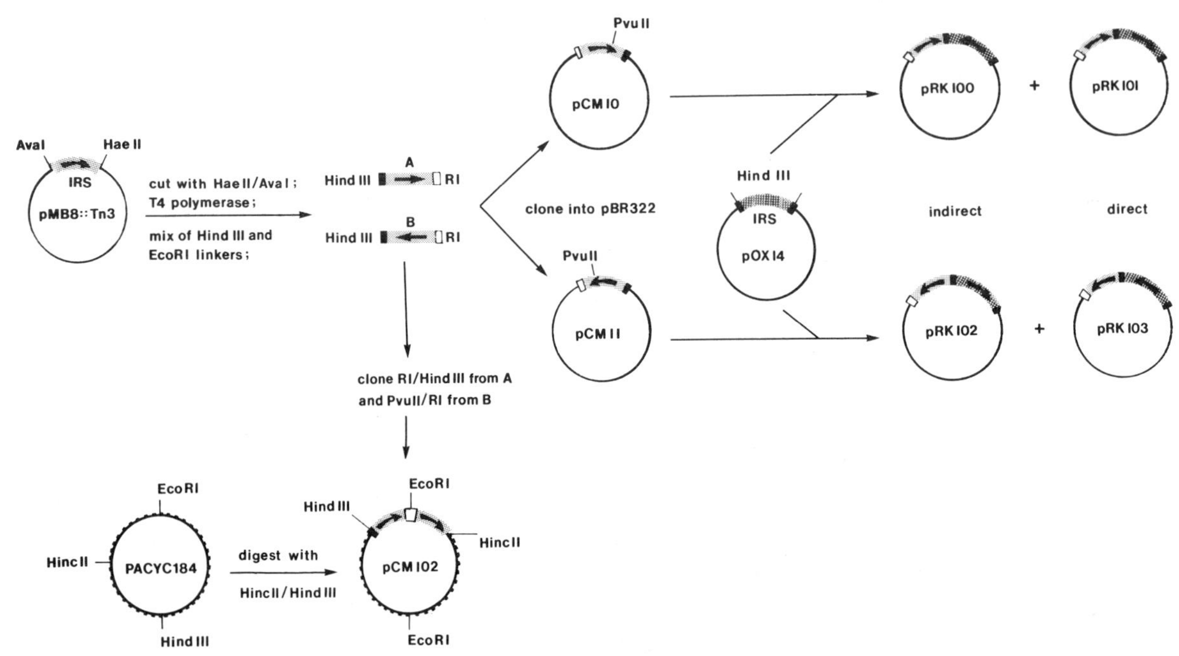

Figure 5. Cloning the IRS and construction of plasmids containing two copies of the IRS. The IRS was cloned out using an *Hae*II/*Ava*I fragment from Tn*3* slightly larger than our deletion results would suggest is necessary. The ends of the purified fragment were made flush with T4 polymerase and ligated to a mixture of *Hin*dIII and *Eco*RI linkers (see Materials and Methods). The ligated mixture was treated with both *Eco*RI and *Hin*dIII and cloned into *Hin*dIII/*Eco*RI-digested pBR322 to generate pCM10 and pCM11. pCM102 was constructed from these two cloned fragments, using pACYC184 (Chang and Cohen 1978), by digestion with *Hin*cII and *Hin*dIII and ligation to purified *Pvu*II/*Eco*RI-digested IRS from pCM101 and purified *Hin*dIII/*Eco*RI-digested IRS from pCM10. The construction of pRK100 and pRK101 is also shown; these contain copies of the IRSs from both Tn*3* and $\gamma\delta$ as a direct or inverted repeat, respectively.

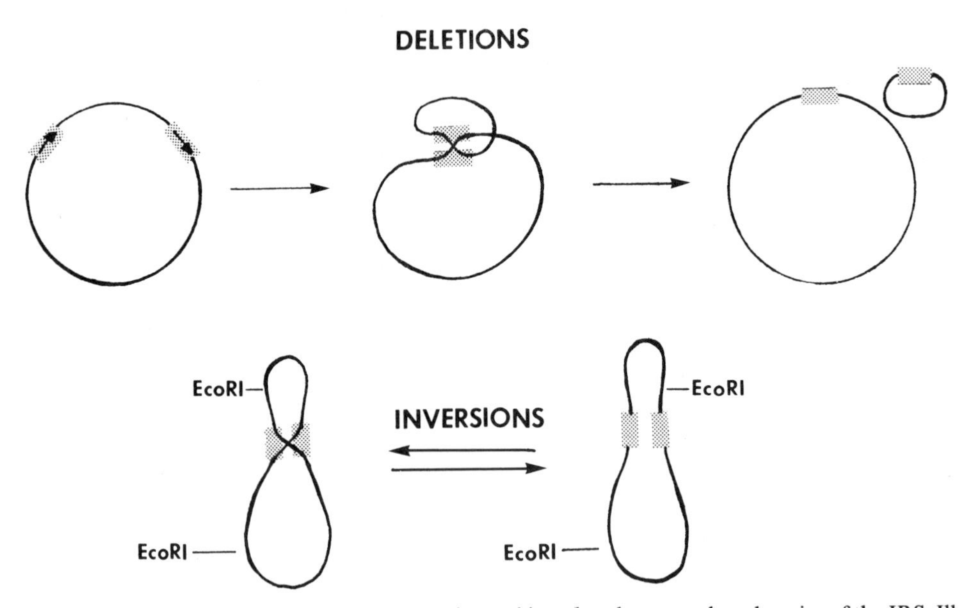

Figure 6. Schematic diagram showing deletions and inversions taking place between cloned copies of the IRS. Illustrated are the types of recombination we would expect from the plasmids described in Fig. 5. (*Top*) In pCM102 or pRK100, which contain direct repeats of the IRS, recombination between these two copies will result in the loss of one copy and the intervening DNA. The deletion can be easily detected in pCM102 since intervening DNA contains an additional *Eco*RI site. (*Bottom*) In pRK101, which contains an inverted repeat of the IRS and two asymmetrically located restriction sites, recombination will invert the DNA between the two IRSs and produce a new restriction pattern.

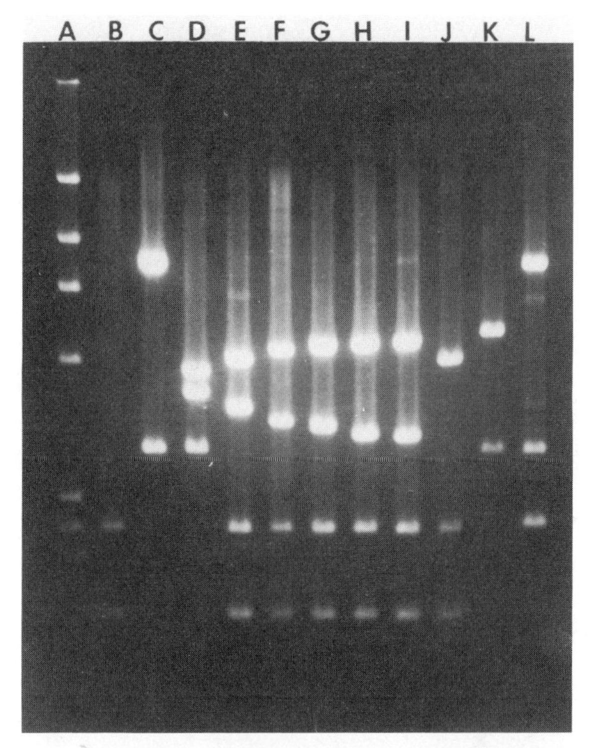

Figure 7. Agarose gel electrophoresis of the resolution plasmid pCM102 complemented with various Tn*3* derivatives. (*A*) *Hin*dIII-digested phage λ DNA; (*B*) pCM102 by itself. In the *recA*⁻ strain MV12 we observe no recombination between the two copies of the IRS even after propagating this plasmid for more than 100 generations. (*C*) The result obtained when pCM102 is complemented with plasmid RSF1050, which contains a complete copy of Tn*3*. Clearly, RSF1050 complements the recombination. (*D–I*) The results for various *Eco*RI linker mutations in Tn*3*. (*D*) An early frameshift mutation in the transposase (mutation 11); (*E–I*) *Eco*RI insertion mutations within *tnpR* (mutations 5, 95, 18, 44, 65). (*J*) Complementation with pBR322; (*K*) complementation by a construct of pBR322, pCM202, in which the complete *tnpR* gene has been reconstituted. pBR322 contains a contiguous fragment from the right end of Tn*3* that includes the last five amino acids of the repressor. pCM202 was constructed by replacing the *Pst*I/*Eco*RI fragment of Tn*3* with an analogous fragment from mutation number 1, thus adding 695 bp and reconstituting the complete *tnpR* gene. (*L*) Complementation by pOX14, a pBR322 derivative that contains γδ (Guyer 1978).

the absence of a γδ and observed that transposition still takes place and at an even higher frequency than previously reported for $tnpR^-$ mutations. The normal frequency for Tn*3* transposition onto F is 10^{-3} ampicillin-resistant transconjugants per F^+ transconjugant. The frequency we observe in the absence of both Tn*3* and γδ repressor is at least 100 times this (these results are a "steady state" value, whereas the actual per generation frequency is considerably lower). Furthermore, transposition in the absence of both *tnpR* and the analogous function from γδ results in essentially 100% cointegrates, as we would predict.

***Site-specific recombination between γδ and Tn3 takes place in* cis.** There are a large number of nucleotide differences between Tn*3* and γδ (R. Reed and S. Shibuga, in prep.); however, because γδ could supply a repressor that would work in *trans* on Tn*3*, it was possible that site-specific recombination could take place in *cis* between a copy of Tn*3* and a copy of γδ. This should allow us to identify the site of recombination by sequencing across the recombinational junction. This experiment was carried out by cloning an internal fragment from γδ into pCM10 and pCM11. γδ contains a large internal *Hin*dIII fragment that encodes a portion of its transposase, a repressor in an analogous location, and other unidentified functions (R. Reed and S. Shibuga, in prep). The inserted *Hin*dIII fragment could be in either of two orientations to give inversions or deletions between the two IRSs. We observe that both types of recombination take place.

Recombination between copies of the IRS is reciprocal and equally efficient in both directions. Recombination between inverted repeats of the IRS should invert the intervening DNA (see Fig. 6) if that recombination is reciprocal and equally efficient in both directions. Inversion resulting from recombination between the two copies of the IRS can be detected by cleaving with a restriction enzyme that cuts asymmetrically. Figure 8 shows the results when inversion takes place between the cloned Tn*3* IRS in pCM10 and the IRS in the large internal *Hin*dIII fragment from γδ. The plasmid has been digested with *Hin*dIII, and thus inversion would result in four bands instead of the expected two. As is obvious in Figure 8, recombination has resulted in an equal proportion of the two possible recombinants, which suggests that the recombination is indeed reciprocal.

Identification of the precise recombinational locus. Figure 8 also shows a recombinant plasmid in which deletion has taken place between the Tn*3* and γδ IRSs. This chimeric plasmid should contain a hybrid IRS, and because of the large number of nucleotide differences between these two transposons, the actual site of recombination within Tn*3* should be determinable to within a few base pairs by sequencing across the recombinational junction. We isolated the DNA from one of the recombinants (in which recombination took place between the IRSs of Tn*3* and γδ). A fragment containing the hybrid IRS together with an analogous fragment from γδ were both end-labeled and sequenced. An autoradiograph of the DNA sequencing gel is shown in Figure 9. Comparison of the sequence of Tn*3* with those of γδ and the chimera suggests that recombination takes place within a 19-bp sequence present in both Tn*3* and γδ. This result, together with the sequence results from other independently isolated recombinants, suggests that the recombination must be site-specific at least to within 19 bp. This sequence is shown in Table 2 and is compared with the ends of the invertible loop from *Salmonella* and the bacteriophage λ *att* site. The sequence is amazingly A/T-rich, containing 16/19 As and Ts.

Figure 8. Recombination in *cis* between Tn*3* and $\gamma\delta$. This gel shows the result when we construct a plasmid with inverted and direct repeats of the $\gamma\delta$ and Tn*3* IRSs. As shown in Fig. 5, we have cloned a *Hin*dIII fragment from $\gamma\delta$ containing a second copy of the IRS into pCM10, which contains the IRS from Tn*3*. This figure shows an agarose gel of recombinants containing an inverted repeat and a direct repeat of the IRS digested with *Hin*dIII. Relative to the IRS, the two *Hin*dIII sites are asymmetrically located, and inversion through the IRS should invert the intervening DNA and change the restriction pattern. Provided that the recombination is reciprocal and equally efficient in both directions, we would expect an equilibrium to be reached containing equal quantities of the two plasmids. This result is shown in *B*. (*A*) A deletion between the Tn*3* and $\gamma\delta$ IRSs resulting in a recombinant that contains a hybrid IRS.

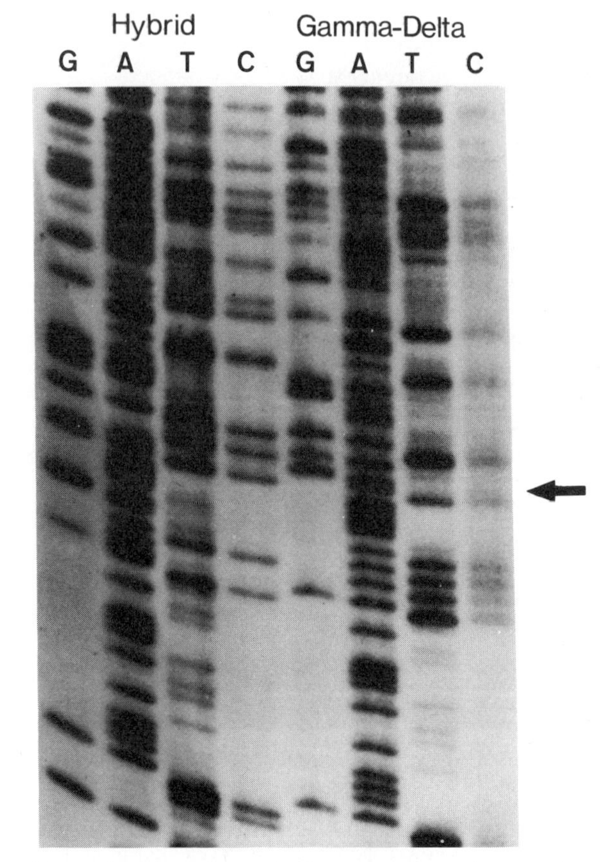

Figure 9. Autoradiograph of a sequencing gel comparing the sequence of a Tn*3*-$\gamma\delta$ recombinant to $\gamma\delta$. (←) The position of the first mismatch. There are no mismatches between Tn*3* and $\gamma\delta$ in the preceding 19 nucleotides in which recombination must have occurred.

DISCUSSION

The overall structure of Tn*3*, as shown in Figure 1, derives from sequence studies (Heffron et al. 1979), direct identification of proteins in minicells (Gill et al. 1979), and from mutational analysis (Heffron et al. 1977, 1978; Gill et al. 1979). The product of the *tnpR* gene acts to regulate negatively its own synthesis and that of the transposase, presumably by binding at one or more operators in this region of the transposon (Gill et al. 1979; Chou et al. 1979a,b). The work presented in this paper clearly demonstrates that Tn*3* encodes a site-specific recombination function as first hypothesized by Gill et al. (1978). This recombination also requires the *tnpR* gene product and takes place within the region in which *tnpR* must be acting to regulate negatively expression of itself and the transposase. Arthur and Sherratt (1979) have come to this same conclusion and have proposed that *tnpR* or some other gene encoded in this region is required for resolution, based on the phenotype of RSF103 (Heffron et al. 1977). Our results have ruled out the possibility that any other gene is encoded here, because the region has only very short open reading frames, except for *tnpR*, and only *tnpR* encompasses all the mutations unable to complement pCM102. Our initial assignment of the acronym *tnpR* stood for transposon repressor, but in view of these experiments, it might be better named the transposon recombinase since this could be considered its most important function.

When we initially identified the *tnpR*⁻ mutations, we tested for formation of cointegrates following transposition onto F. In these experiments, mutations in *tnpR* gave a high proportion of cointegrates, as do mutations of the IRS itself; however, unlike IRS⁻ mutations, these cointegrates are very unstable (F. Heffron, unpubl.). Because the donor cells always contained a copy of $\gamma\delta$, cointegrates that are generated should be resolved via its recombinase. The unstable cointegrates may simply indicate that $\gamma\delta$ does not complement resolution as efficiently as does Tn*3*'s own recombinase.

At least two separate sequences are required for this recombination, although a 23-bp sequence between them (deleted in mutant 25) appears to be nonessential.

Table 2. Comparison of Recombination Sites

Sequence	Site
C G A A A T A T T A T A A A T T A T C	IRS
G C T T T A T A A T A T T T A A T A G	
A A G G T T T T T G A T A A	phase
T T C C A A A A A C T A T T	variation
G C T T T T T T A T A C T A A	*att* core
C G A A A A A A A T A T G A T T	

This table compares the sequence of the IRS in Tn*3* to those of the ends of the invertible loop identified in *Salmonella* phase variation and the phage λ attachment site.

In this respect, this recombination resembles λ integration (Hsu et al. 1980). Like λ integration, the recombination takes place through an A/T-rich region (see Table 2) but requires sequences immediately outside the actual recombinational site. There is no obvious homology between these sequences, however, nor is there any obvious homology with the λ*int* gene (Hoess et al. 1980). Finally, resolution takes place independent of *himA* (R. Parker, unpubl.), a function absolutely required for λ to integrate (Miller and Friedman 1980).

There is another well-documented site-specific recombination system to which the Tn*3* functions might be compared. Phase variation in *Salmonella* results from the expression of different flagellar proteins. Expression of these two proteins is mediated by an invertible loop that can promote expression from an internal promoter in either of two orientations (Silverman et al., this volume). This inversion requires the *hin* gene product encoded within the loop itself. The loop is flanked at either end by a 14-bp inverted repeat through which the recombination actually takes place (see Table 2). As shown in Table 2, the only obvious homology between the IRS and this sequence is that it is also A/T-rich. The *hin* protein itself is strikingly similar to the *tnpR* protein! As first observed by M. Silverman et al. (this volume), this protein shares 33% overall homology with *tnpR*. The phase-variation system and the site-specific recombination system encoded by Tn*3* must surely have been derived from the same original gene(s). It is surprising that the recognition sequences for the recombination no longer bear any obvious relationship, although perhaps when other binding sites for *tnpR* have been identified, we will be able to identify new homologies.

This site-specific recombinational system was first hypothesized on the basis of the phenotype of some of the internal deletions in Tn*3* (Gill et al. 1978). It is very clear now that Tn*3* contains a site-specific recombination system that will very efficiently resolve cointegrates independent of *recA*. The basic question as to whether cointegrates are intermediates in transposition remains unanswered. A few of the possibilities are: (1) Cointegrates might be formed at high frequency as a result of some other aspect of transposon biology that has not yet been identified and are resolved through the IRS; (2) the resolution functions themselves might play an alternate role in transposition so that by their removal transposition results exclusively in cointegrates; and (3) cointegrates might be the predominant or exclusive end products of transposition that are efficiently resolved by this internal recombination system. Although no firm conclusions can be drawn at this time, the last possibility appears the most reasonable. It should be noted, however, that there is some evidence that Tn*10* does not encode a site-specific recombination system and yet does not form cointegrates during transposition (N. Kleckner, pers. comm.). It therefore seems most likely that more than one mechanism will be employed by different transposons. In support of this hypothesis, bacteriophage Mu appears to require different host functions for lysogeny than for lytic growth (Miller and Friedman 1980; M. Howe, pers. comm.), processes that have been thought to be both identical and analogous to transposition.

ACKNOWLEDGMENTS

We thank Bruce Alberts, in whose laboratory this work was carried out, and Ira Herskowitz, without whose interest and enthusiasm this work would not have been done. This work was supported by National Science Foundation grant PC77923858 and, in part, by National Institutes of Health grant GM-24020.

REFERENCES

Arthur, A. and D. Sherratt. 1979. Dissection of the transposition process: A transposon-encoded site-specific recombination process. *Mol. Gen. Genet.* **175:** 267.

Bolivar, F., R. L. Rodriquez, M. Betlach, and H. W. Boyer. 1977. Construction and characterization of new cloning vehicles. I. Ampicillin resistant derivatives of the plasmid pMB9. *Gene* **2:** 75.

Botstein, D. and N. Kleckner. 1977. Translocation and illegitimate recombination by the tetracycline resistance element Tn*10*. In *DNA insertion elements, plasmids, and episomes* (ed. A. I. Bukhari et al.), p. 185. Cold Spring Harbor Laboratory, Cold Spring Harbor, New York.

Cameron, J. R., E. Y. Loh, and R. W. Davis, 1979. Evidence for transposition of dispersed repetitive DNA families in yeast. *Cell* **16:** 739.

Chang, A. C. Y. and S. N. Cohen. 1978. Construction and characterization of amplifiable multicopy DNA cloning vehicles derived from the P15A cryptic miniplasmid. *J. Bacteriol.* **134:** 1141.

Chou, J., M. Casadaban, P. Lemaux, and S. N. Cohen. 1979a. Identification and characterization of a self-regulated repressor of translocation of the Tn*3* element. *Proc. Natl. Acad. Sci.* **76:** 4020.

Chou, J., P. Lemaux, M. Casadaban, and S. N. Cohen. 1979b. Transposition protein of Tn*3*: Identification and characterization of an essential repressor-controlled gene product. *Nature* **282:** 801.

Chumley, G. G., R. Menzel, and J. R. Roth. 1979. Hfr formation directed by Tn*10*. *Genetics* **91:** 639.

Clewell, D. B. and D. R. Helinski. 1969. Supercoiled circular DNA-protein complex in *Escherichia coli:* Induced conversions to an open circular DNA form. *Proc. Natl. Acad. Sci.* **62:** 1159.

Gill, R., F. Heffron, and S. Falkow. 1979. Identification of the protein encoded by the transposable element Tn*3* which is required for its transposition. *Nature* **282:** 797.

Gill, R., F. Heffron, G. Dougan, and S. Falkow. 1978. Analysis of sequences transposed by complementation of two classes of transposition deficient mutants of Tn*3*. *J. Bacteriol.* **136:** 742.

Greene, P. J., M. C. Betlach, H. M. Goodman, and H. Boyer. 1974. The *Eco*RI restriction endonuclease. *Methods Mol. Biol.* **9:** 87.

Grindley, N. D. F. 1978. IS1 insertion generates duplications of a nine base pair sequence at its target site. *Cell* **13:** 419.

Guyer, M. S. 1978. The gamma-delta sequence of F is an insertion sequence. *J. Mol. Biol.* **126:** 347.

Heffron, F., M. So, and B. McCarthy. 1978. In vitro mutagenesis using synthetic restriction sites. *Proc. Natl. Acad. Sci.* **75:** 6012.

Heffron, F., C. Rubins, and S. Falkow. 1975a. The translocation of a plasmid DNA sequence which mediates ampicillin resistance: Molecular nature and specificity of insertion. *Proc. Natl. Acad. Sci.* **72:** 3623.

Heffron, F., P. Bedinger, J. Champoux, and S. Falkow. 1977. Deletions affecting the transposition of an antibiotic resistance gene. *Proc. Natl. Acad. Sci.* **74:** 702.

HEFFRON, F., B. MCCARTHY, H. OHTSUBO, and E. OHTSUBO. 1979. DNA sequence analysis of the transposon Tn*3*: Three genes and three sites involved in transposition of Tn*3*. *Cell* **18:** 1153.

HEFFRON, F., R. SUBLETT, R. HEDGES, A. JACOB, and S. FALKOW. 1975b. Origin of the TEM beta-lactamase gene found on plasmids. *J. Bacteriol.* **122:** 250.

HOESS, R. H., C. FOELLER, K. BIDWELL, and A. LANDY. 1980. Site-specific recombination functions of bacteriophage lambda: DNA sequence of the regulatory regions and overlapping structural genes for Int and Xis. *Proc. Natl. Acad. Sci.* **77:** 2482.

HSU, P-L., W. ROSS, and A. LANDY. 1980. The lambda phage *att* site: Functional limits and interaction with *int* protein. *Nature* **285:** 85.

HUGHES, S., P. SHANK, D. SPECTOR, H. J. KUNG, J. M. BISHOP, and H. E. VARMUS. 1978. Provirus of avian sarcoma virus are terminally redundant, coextensive with unintegrated linear DNA and integrated at many sites. *Cell* **15:** 1397.

JOHNSRUD, L., M. P. CALOS, and J. H. MILLER. 1978. The transposon Tn*9* generates a 9 bp repeated sequence during integration. *Cell* **15:** 1209.

JORDAN E., H. SAEDLER, and P. STARLINGER. 1967. Strong-polar mutations in the transferase gene of the galactose operon of *E. coli. Mol. Gen. Genet.* **100:** 296.

KLECKNER, N., R. K. CHAN, B.-K. TYE, and D. BOTSTEIN. 1975. Mutagenesis by insertion of a drug-resistance element carrying an inverted repeat. *J. Mol. Biol.* **97:** 561.

MALAMY, M. 1967. Frameshift mutations in the lactose operon of *E. coli. Cold Spring Harbor Symp. Quant. Biol.* **31:** 189.

MAXAM, A. and W. GILBERT. 1980. Sequencing end-labeled DNA with base-specific chemical cleavages. *Methods Enzymol.* **65:** 499.

MILLER, H. I. and D. I. FRIEDMAN. 1980. An *E. coli* gene product required for lambda site-specific recombination. *Cell* **20:** 711.

NEAL, M. W. and J. R. FLORINI. 1973. A rapid method for desalting small volumes of solution. *Anal. Biochem.* **55:** 328.

OHTSUBO, H., H. OHMORI, and E. OHTSUBO. 1979. Nucleotide-sequence analysis of Tn*3* (Ap): Implications for insertion and deletion. *Cold Spring Harbor Symp. Quant. Biol.* **43:** 1269.

POTTER, S., W. BROREIN, P. DUNSMUIR, and G. RUBIN. 1979. Transposition of elements of *412, copia* and *297* dispersed repeated gene families in *Drosophila. Cell* **17:** 415.

SABRAN, J. L., J. W. HSU, C. YEATER, A. KAJI, W. I. MASON, and J. M. TAYLOR. 1979. Analysis of integrated avian RNA tumor virus DNA in transformed chicken, duck and quail fibroblasts. *J. Virol.* **29:** 170.

SAEDLER, H. and P. STARLINGER. 1967a. O^c mutations in the galactose operon of *E. coli.* I. Genetic characterization. *Mol. Gen. Genet.* **100:** 178.

———. 1967b. O^c mutations in the galactose operon. II. Physical identification. *Mol. Gen. Genet.* **100:** 190.

SHARP, P., B. SUGDEN, and J. SAMBROOK. 1973. Detection of two restriction endonuclease activities in *Haemophilus parainfluenza* using analytical agarose-ethidium bromide electrophoresis. *Biochemistry* **12:** 3055.

SO, M., F. HEFFRON, and S. FALKOW. 1978. A genetic method for labeling cryptic plasmids. *J. Bacteriol.* **133:** 1520.

SO, M., F. HEFFRON, and B. J. MCCARTHY. 1979. The ST toxin of *E. coli* is a transposon flanked by inverted repeats of IS1. *Nature* **277:** 453.

STAFFORD, D. and D. BIEBER. 1975. Concentration of DNA solutions by extraction with 2-butanol. *Biochim. Biophys. Acta* **378:** 18.

SUTCLIFFE, G. 1978. Nucleotide sequence of the ampicillin resistance gene of *E. coli* plasmid pBR322. *Proc. Natl. Acad. Sci.* **75:** 3737.

———. 1979. Complete nucleotide sequence of the *Escherichia coli* plasmid pBR322. *Cold Spring Harbor Symp. Quant. Biol.* **43:** 77.

ULLRICH, A., J. SHINE, J. CHIRGWIN, R. PICTET, E. TISCHER, W. J. RUTTER, and H. W. GOODMAN. 1977. Rat insulin genes: Construction of plasmids containing the coding sequence. *Science* **196:** 1313.

Tn3: Transposition and Control

M. J. Casadaban,* J. Chou,* P. Lemaux, C.-P. D. Tu,[†] and S. N. Cohen
Departments of Genetics and Medicine, Stanford University School of Medicine, Stanford, California 94305

Gene fusions that bring *lac* gene expression under the control of transcriptional and/or translational regulatory signals within the Tn*3* element (Casadaban and Cohen 1980; Casadaban et al. 1980) have been used to study the mechanism and control of transposition in *Escherichia coli* (Chou et al. 1979a,b; Cohen et al. 1979, 1980). A bidirectional genetic unit consisting of the *tnpA* gene for the transposase protein and the *tnpR* gene for the repressor has been identified and characterized, and the role of these proteins in transposition has been defined (Fig. 1). The *tnpR* gene encodes a repressor for transcription of both genes, thereby accomplishing autoregulatory feedback control of its own expression and modulation of the level of transposition by repressing the synthesis of transposase. In our experiments, *lac* gene fusions have been used to identify, locate, and measure the expression and regulation of the bidirectional promoters and other regulatory elements of the *tnpA* and *tnpR* genes and to isolate mutants of the repressor gene that no longer repress or that are no longer repressed. We have also used *lac* gene fusions to isolate mutants of the *tnpA* gene that result in increased synthesis of the transposase protein and consequently also yield an increased rate of transposition (M. J. Casadaban et al., in prep.).

The *tnpR* (Repressor) Gene

The identification, genetic mapping, DNA sequencing, and mutational analysis of the *tnpR* gene and characterization of its regulatory mechanisms have been described in detail (Chou et al. 1979a; Cohen et al. 1979). The *lac* genes, without any promoter of their own, were inserted at the *Bam*HI site of Tn*3* (as pMC959, Fig. 2) and thereby fused to the promoter for the *tnpR* gene. As a result of this fusion, *lac* expression is repressed when a wild-type copy of *tnpR* is present in the cell. This *lac* gene fusion plasmid was also used to isolate mutants that did not repress the fused *lac* genes.

Among these mutants were amber-suppressible nonsense and missense mutations of *tnpR*, as well as *cis*-constitutive, presumably operator, mutations of the *tnpR* control region that had been fused to *lac*. The DNA sequences of some of these mutants were determined and the location of a protein-chain-terminating amber mutation was found to correlate with the length of a truncated *tnpR* peptide on polyacrylamide gels (Fig. 3). Overproduction of the nonfunctional mutated *tnpR* repressor protein indicated that its gene is autoregulated (Chou et al. 1979a).

Present addresses: * Department of Biochemistry, University of Chicago, Chicago, Illinois 60637. [†] Department of Microbiology, Cell Biology, Biochemistry and Biophysics, The Pennsylvania State University, University Park, Pennsylvania 16802.

A Possible Fourth Gene on Tn*3*

When promoterless *lac* gene fusions were inserted into the *Bam*HI site of Tn*3* in the reverse orientation to the *tnpR* gene (e.g., pMC945, Fig. 2), they were expressed, thus implying that Tn*3* contains a fourth transcriptional unit (Chou et al. 1979b), which we tentatively call *tnpF*. No protein product has been identified for this region of Tn*3* and the DNA sequence does not fit any protein of reasonable size (Chou et al. 1979a; Heffron et al. 1979). However, an apparently *trans*-acting genetic function necessary for DNA transposition and for resolution into separate DNA molecules of an intermediate in the transposition process (Heffron et al. 1977; Gill et al. 1978) has been attributed to this region (Arthur and Sherratt 1979); this function may be assigned to either the *tnpR* gene or to the possible *tnpF* gene. Experiments to distinguish between these possibilities are currently underway.

The *tnpA* Gene

lac fusions were also used to identify and study the *tnpA* gene (Chou et al. 1979b). The protein product of *tnpA* was detected in SDS-polyacrylamide gels only when the *tnpR* gene was mutated, implying that its expression was repressed by *tnpR* (Fig. 3). *lac* gene fusions formed by inserting promoterless *lacZ* gene fragments (Casadaban et al. 1980) into the *Bcl*I sites of Tn*3* (Fig. 2) were used to investigate repressor regulation of *tnpA* and to localize the *tnpA*-controlling elements. All of the controlling elements for *tnpA* and *tnpR*, including promotion, translational initiation, and repressibility, are contained within a small (359 nucleotide) *Sau*3A restriction-endonuclease-generated DNA fragment (Fig. 2).

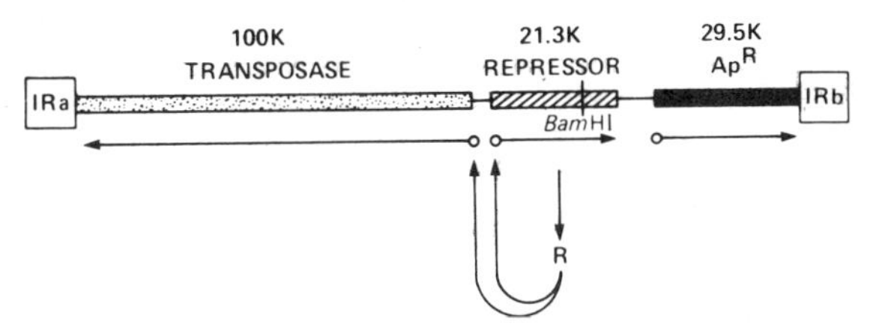

Figure 1. Schematic diagram showing locations of Tn*3* genes and their regulatory signals. A 21K repressor peptide product of *tnpR* acts to reduce transcription from the promoters of the *tnpA* transposase and *tnpR* repressor genes, which are expressed in divergent directions (Chou et al. 1979b).

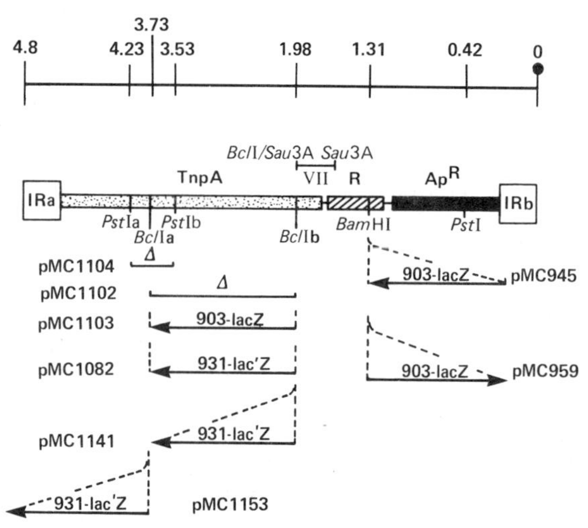

Figure 2. Schematic map showing structure of *lac* insertions into Tn*3* (Chou et al. 1979b). (TnpA) The transposase gene *tnpA*; (R) the repressor gene *tnpR;* (Ap^R) the β-lactamase gene *bla;* (IRa and IRb) the 38-bp-long inverted-repeat termini of Tn*3* (not drawn to scale). The kilobase coordinates of restriction endonuclease cleavage sites are indicated, relative to the right terminus of Tn*3* at IRa; additional *Sau*3A sites within Tn*3* are not shown. The position of *Sau*3A fragment VII, which contains promoters and translational start signals for genes that are encoded in divergent directions is shown; the left terminus of this fragment is also the *Bcl*I$_b$ cleavage site. The segments deleted in the pMC1104 plasmid (Δ*Pst*I) and the pMC1102 plasmid (Δ*Bcl*I) are indicated. The underlying arrows indicate the direction of insertion of 7-kb *Bam*HI-generated fragments containing *lacZ* and *lac'Z* genes that lack promoters (Casadaban et al. 1980). 903-*lacZ* is missing only the promoter for *lacZ* transcription, whereas 931-*lac'Z* lacks the promoter and a translational start site. (For details of constructions, see Chou et al. 1979b.)

This fragment, in one orientation for *tnpA* and the other for *tnpR*, was shown to provide all functions necessary for expression of *lac* genes that had been fused to either end of the fragment; moreover, both the *tnpA* and *tnpR* genes were repressed, as measured by inhibition of expression of the fused *lac* genes by a wild-type repressor protein provided by Tn*3* in *trans* (Chou et al. 1979b).

Overproduction of the Transposase

The *tnpA* transposase protein is synthesized at such a low rate that it was not seen on polyacrylamide gels of

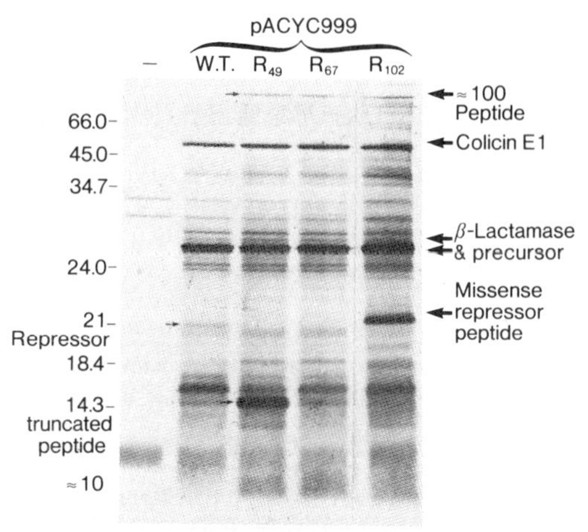

Figure 3. One-dimensional autoradiograph showing ^{35}S-labeled polypeptides synthesized by minicells containing a plasmid carrying Tn*3* or its mutant derivatives. Experimental details have been published elsewhere (Chou et al. 1979a). The locations of molecular-weight standards are shown to the left of the gel. Arrows point to the 21K repressor polypeptide product of *tnpR* observed in extracts of wild-type Tn*3* (W.T.) and to related peptides made by Tn*3* derivatives carrying amber (R$_{49}$) and missense (R$_{102}$) mutations in the repressor gene. The 100K peptide is the transposase product of *tnpA*. The positions of colicin E1 and of β-lactamase and its precursors are indicated. Numbers at left are $\times 10^{-3}$. The track marked — is the minicell strain alone.

^{35}S-labeled plasmid proteins, even when the gene was carried by a high-copy-number plasmid. Only when the *tnpR* gene function was removed, resulting in derepression of *tnpA*, could the transposase be detected as a faint band (Fig. 3) (Chou et al. 1979b; Gill et al. 1979). To increase the synthesis of this protein for biochemical studies of its function, *lac* fusions were used to isolate mutations of Tn*3* that overproduce the transposase.

The pMC1141 fusion (Fig. 2) was used to isolate mutants that increased *lac* expression enough for the cell to be able to utilize lactose for growth. (Details of these experiments will be provided elsewhere.) The promoter region of the *tnpA* gene was examined for five of these mutants by DNA sequence analysis and in each instance was found to have a single base change (Fig. 4). All of the mutations examined are located upstream from the ATG triplet that we believe is the initiation codon for translation of transposase (Chou et al. 1979b);

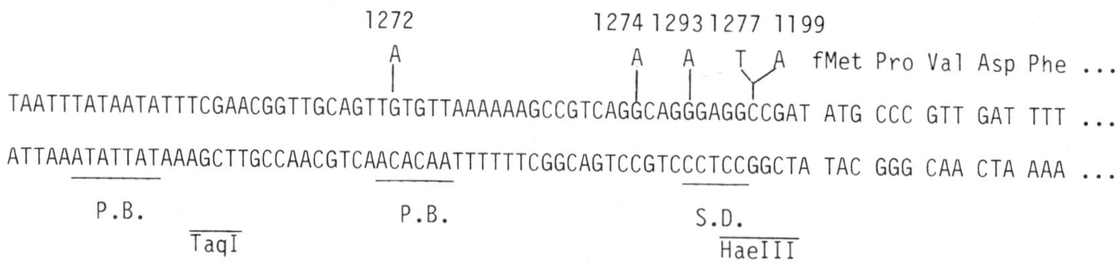

Figure 4. *tnpA* overproduction mutations. The sequence of five different mutational changes that increase the expression of the *tnpA* gene are shown. The amino acids given represent the predicted amino acid sequence of the aminoterminal end of the transposase. P.B. and S.D. are possible Pribnow (1975) box sequences for the promoter and Shine and Delgarno (1975) sequences for initiation of translation. Note that the 1199 and 1277 mutations both inactivate a *Hae*III restriction site and that mutation 1272 changes a GTG sequence into ATG, which is possibly an alternative translation start site for the transposase.

Table 1. Transposition Frequency

Donor plasmid	Transposition frequency	Ratio
pSC101::Tn*3*	8.0×10^{-4}	1.0
pSC101::Tn*3 tnpR*49	3.7×10^{-3}	4.6
pSC101::Tn*3 tnpR*49 *tnpA*^{q}1199	2.1×10^{-2}	27.0
As above, but different insertion site	3.0×10^{-2}	37.0
pSC101::Tn*3 tnpA*Δ*Pst*I	2.6×10^{-6}	0.0033

The transposition frequency was the frequency at which an F′ episome received the Ap^{r} marker from the pSC101 donors following mating into a new cell. *tnpR*49 is an amber mutation in the repressor gene, and *tnpA*^{q}1199 is an overproducing mutation for transposase (Fig. 4). *tnpA*Δ*Pst*I is an interval deletion in the transposase gene (Fig. 2). The procedures used for determining transposition frequency have been described elsewhere (Chou et al. 1979a).

from the positions of the mutations, we suspect that they may improve the efficiency of translational initiation for *tnpA*. We know from our studies of fusions of *tnpA* with the *lac* genes that translational initiation of the *tnpA* protein is much less efficient than translational initiation for the wild-type *lacZ* gene (Chou et al. 1979b); this was determined by comparing the amount of β-galactosidase expression seen with *tnpA-lac* fusions that have *lacZ* gene translational initiation signals with the level of β-galactosidase production by fusions that do not have *lacZ* translational initiation signals and that rely on the translational initiation signals of the *tnpA* gene for *lac* expression.

The most upstream of the five mutations that showed increased expression of *tnpA*, as measured by increased β-galactosidase production by the *tnpA-lacZ* gene fusions (i.e., the 1272 mutation), changed a GTG sequence upstream from the transcriptional control region into an ATG sequence. Heffron et al. (1979) have proposed that this GTG codon may be the initiation site for the transposase protein. If this interpretation is correct, changing the GTG into the more efficient translational start codon ATG would increase production of the *tnpA* protein—as observed. However, the other four mutations that affect expression are located *downstream* from the GTG proposed by Heffron et al. (1979) to be the initiation site for what would be the structural gene for the transposase, and this placement suggests that initiation may not occur at the GTG triplet. An alternative explanation for the observed increased expression of *tnpA* by the 1272 mutation is that this mutation is located in the Pribnow box of the transcriptional control region and that it creates a better RNA polymerase interaction site, resulting in more efficient transcription of the gene (Pribnow 1975).

An additional argument supporting the view that initiation of the transposase peptide occurs at the ATG and not the GTG triplet results from comparison of the DNA sequence of Tn*3* with that of the γδ transposon (Guyer 1978; Reed et al. 1979), which has extensive overall sequence homology with Tn*3* (Reed 1981). γδ has an equivalent ATG sequence but no GTG sequence, and the location where the GTG would line up is in a different phase due to deletions or insertions that apparently have occurred during the evolutions of the γδ and Tn*3* transposons.

To show that these overproducing mutations do indeed overproduce functional transposase, the first mutation isolated (1199) was joined to a functional *tnpA* gene that had been separated by homologous recombination from the *lac* insertion segment used to assay it (details to be published elsewhere). The resulting Tn*3* element showed an increased frequency of Tn*3* transposition (Table 1), which was accompanied by the appearance of a new protein band in polyacrylamide gels on electrophoresis of total protein from bacterial cells containing the mutant element (Fig. 5). This band was the same size as the originally identified transposase (*tnpA*) protein, which ordinarily is not synthesized in quantities large enough to be detected among all the chromosomally encoded proteins of *E. coli* (Chou et al. 1979b).

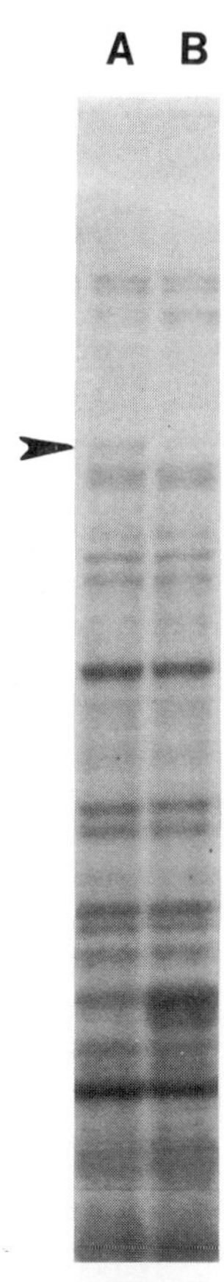

Figure 5. Overproduced transposase protein. The presence of the 1199 overproducing *tnpA* mutation (*A*) results in the synthesis of enough transposase to be seen as a new band (arrowhead) among total *E. coli* protein on a polyacrylamide gel. The band is not seen in an extract of Tn*3*-containing cells carrying a wild-type *tnpA* gene (*B*). (Full details will be published elsewhere.)

Currently, this overproducing mutant is being used as a source of transposase in experiments directed at purifying the *tnpA* gene product for biochemical studies (M. Fennewald et al., unpubl.).

Translocation Specificity of the Tn*3* Element

To investigate the specificity of translocation of the Tn*3* element, we have constructed a small plasmid (pTU4) having a defined structure and containing a DNA segment that previously had been sequenced in its entirety, and we have used this plasmid to study the DNA sequence in the vicinity of multiple independent transposition events. Our analysis of almost 250 independent isolates of insertions of Tn*3* into the pTU4 plasmid has indicated that multiple insertions occur at specific nucleotides located within certain regions of the plasmid (Tu and Cohen 1980a). Moreover, our findings indicate that the clustering of such insertions at certain DNA locations is determined by the nucleotide sequence of the recipient genome.

A description of the construction of the pTU4 plasmid and a map of the plasmid have been presented elsewhere (Tu and Cohen 1980a). Our analysis of Tn*3* insertions into this plasmid indicated that 25% of 247 independent insertion events occurred within the 222-bp *Hha*I fragment 6 of the pTU4 plasmid. DNA sequence analysis of these insertions indicated the presence of several distinct hotspots (i.e., nucleotide positions that had received multiple insertions of Tn*3*): A total of 36 independent insertions (15%) of these examined occurred at three separate locations within an 11-nucleotide region. Moreover, 31 of these 36 insertions were in the same orientation. Determination of the DNA sequence in the vicinity of the hotspot region for Tn*3* insertion (Fig. 6) indicates that major homology with a sequence within Tn*3* exists. Fifteen of 17 nucleotides (mismatch at three locations) in a DNA segment near the hotspot region are identical to 15 of 18 nucleotides within the sequence determined previously for the Tn*3* inverted-repeat termini (Cohen et al. 1979; Ohtsubo et al. 1979; Takeya et al. 1979; Tu and Cohen 1980b).

Analysis of sites of multiple insertions also showed that the previously proposed regional specificity (Kretschmer and Cohen 1977) of Tn*3* insertion is associated with a strong preference for AT-rich segments.

35 40 45 50 55 60
5' A-T-C-T-T-A-T-T-A-A-T-C-A-G-A-T-A-A-A-A-T-A-T-T-T-C-T-A-G-A-
3' T-A-G-A-A-T-A-A-T-T-A-G-T-C-T-A-T-T-T-T-A-T-A-A-A-G-A-T-C-T-

65 70 75 80 85 90
T-T-T-C-A-G-T-G-C-A-A-T-T-T-A-T-C-T-C-T-T-C-A-A-A-T-G-T-A-G-
A-A-A-G-T-C-A-C-G-T-T-A-A-A-T-A-G-A-G-A-A-G-T-T-T-A-C-A-T-C-

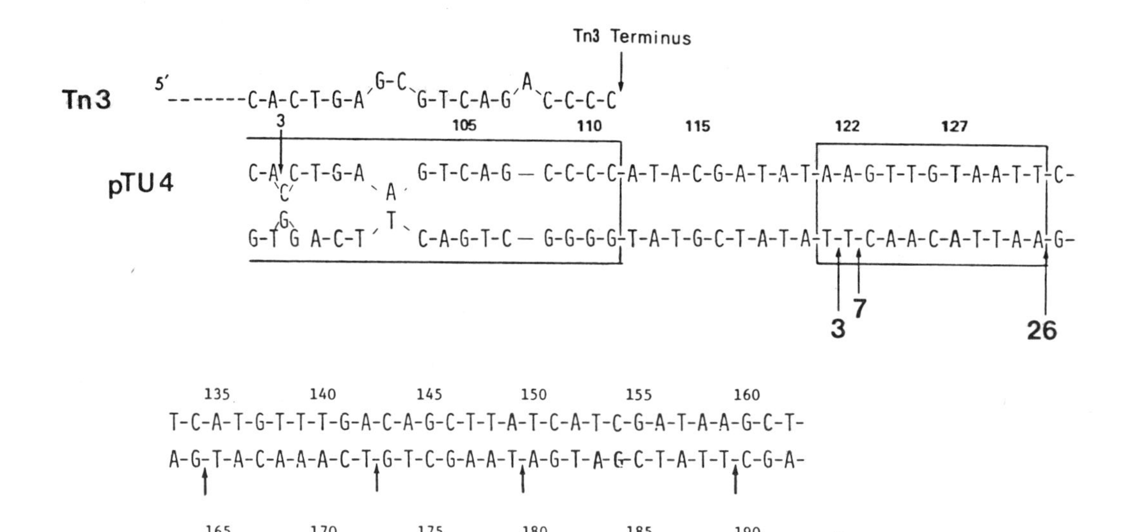

165 170 175 180 185 190
T-T-A-A-T-G-C-G-G-T-A-G-T-T-T-A-T-C-A-C-A-G-T-T-A-A-A-T-T-G- 3'
A-A-T-T-A-C-G-C-C-A-T-C-A-A-A-T-A-G-T-G-T-C-A-A-T-T-T-A-A-C- 5'

Figure 6. DNA sequence homology between terminal nucleotides of Tn*3* and DNA segments in the vicinity of insertional hotspots. Thirty-six multiple insertions of Tn*3* into the recipient genome occur within an 11-bp region near a segment of the recipient genome showing homology with the terminal nucleotides of Tn*3*. Sites of single insertions are indicated by arrows; in instances of multiple insertion, the number of insertions observed at that site is indicated. For details, see Tu and Cohen (1980a).

Furthermore, we have identified within Tn*3* a direct repeat of a 10-bp sequence that is located within the Tn*3* termini: It is of some interest that the preferred orientation of insertion of Tn*3* brings this 10-bp sequence into a position near the Tn*3*-like sequence previously identified on the recipient genome (Tu et al. 1980b).

These findings indicate that the site and orientation of insertion of Tn*3* are at least partly determined by the primary nucleotide sequence of the recipient genome, and they suggest that insertional hotspots may result from the combined effects of AT-richness plus homology of the recipient genome with terminal sequences of Tn*3*. Analogous findings have been published recently by J. H. Miller (this volume) for the Tn*9* transposon. Thus, despite the *recA* independence of transposition of Tn*3* (Kopecko and Cohen 1975), DNA sequence homology may have an important role in such "illegitimate" recombinational events.

REFERENCES

ARTHUR, A. and D. SHERRATT. 1979. Dissection of the transposition process: A transposon-encoded site-specific recombination system. *Mol. Gen. Genet.* **175:** 267.

CASADABAN, M. and S. N. COHEN. 1980. Analysis of gene control signals by DNA fusion and cloning in *Escherichia coli. J. Mol. Biol.* **138:** 179.

CASADABAN, M., J. CHOU, and S. N. COHEN. 1980. *In vitro* gene fusions that join an enzymatically active β-galactosidase segment to amino terminal fragments of exogenous proteins: *E. coli* plasmid vectors for the detection and cloning of translational initiation signals. *J. Bacteriol.* **143:** 971.

CHOU, J., M. CASADABAN, P. LEMAUX, and S. N. COHEN. 1979a. Identification and characterization of a self-regulated repressor of translocation of the Tn*3* element. *Proc. Natl. Acad. Sci.* **76:** 4020.

CHOU, J., P. LEMAUX, M. CASADABAN, and S. N. COHEN. 1979b. Transposition protein of Tn*3*: Identification and characterization of an essential repressor-controlled gene product. *Nature* **282:** 801.

COHEN, S. N., M. CASADABAN, J. CHOU, and D. TU. 1979. Studies of the specificity and control of transposition of the Tn*3* element. *Cold Spring Harbor Symp. Quant. Biol.* **43:** 1247.

COHEN, S. N., M. CASADABAN, J. CHOU, P. LEMAUX, C. MILLER, and D. TU. 1980. Regulation of Tn*3* transposition and specificity of its insertion sites. *Miami Winter Symp.* (in press).

GILL, R. E., F. HEFFRON, and S. FALKOW. 1979. Identification of the protein encoded by the transposable element of Tn*3* which is required for its transposition. *Nature* **282:** 797.

GILL, R., F. HEFFRON, G. DOUGAN, and S. FALKOW. 1978. Analysis of sequences transposed by complementation of two classes of transposition-deficient mutants of Tn*3*. *J. Bacteriol.* **136:** 742.

GUYER, M. S. 1978. The γδ sequence of F is an insertion sequence. *J. Mol. Biol.* **126:** 347.

HEFFRON, F., P. BEDINGER, J. CHAMPOUX, and S. FALKOW. 1977. Deletions affecting the transposition of an antibiotic resistance gene. *Proc. Natl. Acad. Sci.* **74:** 702.

HEFFRON, F., B. J. MCCARTHY, H. OHTSUBO, and E. OHTSUBO. 1979. DNA sequence analysis of the transposon Tn*3*: Three genes and three sites involved in transposition of Tn*3*. *Cell* **18:** 1153.

KOPECKO, D. J. and S. N. COHEN. 1975. Site-specific RecA-independent recombination between bacterial plasmids: Involvement of palindromes at the recombinational loci. *Proc. Natl. Acad.Sci.* **72:** 1373.

KRETSCHMER, P. J. and S. N. COHEN. 1977. Selected translocation of plasmid genes: Frequency and regional specificity of translocation of the Tn*3* element. *J. Bacteriol.* **130:** 888.

OHTSUBO, H., H. OHMORI, and E. OHTSUBO. 1979. Nucleotide-sequence analysis of Tn*3* (Ap): Implications for insertion and deletion. *Cold Spring Harbor Symp. Quant. Biol.* **43:** 1269.

PRIBNOW, D. 1975. Nucleotide sequence of an RNA polymerase binding site at an early T7 promoter. *Proc. Natl. Acad. Sci.* **72:** 784.

REED, R. 1981. Resolution of cointegrate between transposons γδ and Tn*3* defines the recombination site. *Proc. Natl. Acad. Sci.* (in press).

REED, R., R. YOUNG, J. STEITZ, N. GRINDLEY, and M. GUYER. 1979. Transposition of the *E. coli* insertion element γδ generates a five base pair repeat. *Proc. Natl. Acad. Sci.* **76:** 4882.

SHINE, J. and L. DELGARNO. 1975. Determinant of cistron specificity in bacterial ribosomes. *Nature* **254:** 34.

TAKEYA, T., H. NOMIYAMA, J. MIYOSHI, K. SHIMADA, and Y. K. TAKAGI. 1979. DNA sequences of the integration sites and inverted repeated structure of transposon Tn*3*. *Nucleic Acids Res.* **6:** 1831.

TU, C.-P. D. and S. N. COHEN. 1980a. Translocation specificity of the Tn*3* element: Characterization of sites of multiple insertions. *Cell* **19:** 151.

———. 1980b. Effect of DNA sequences adjacent to the termini of Tn*3* on sequential translocation. *Mol. Gen. Genet.* **177:** 597.

Transposon-specified, Site-specific Recombination Systems

D. Sherratt, A. Arthur,* and M. Burke
*Department of Genetics, University of Glasgow, Glasgow G11 5JS, Scotland; *School of Biological Sciences, University of Sussex, Falmer, Brighton BN1 9QG, England*

Transposable genetic elements are unique, nonpermuted DNA sequences that can insert into genomes at sites that lack substantial DNA homology with the element. At least in *Escherichia coli,* this insertion occurs in the absence of a functional homologous recombination system. Because they are always found joined to nonelement sequences, transposable elements appear to have no autonomous existence; however, since transposition appears to be accompanied by specific element replication, they can be considered as replicons. Transposable elements have now been characterized in a number of organisms, and there is no reason to think that they are any less common than viruses.

Transposons are transposable elements that carry a phenotypic marker. Most of the transposons that have been characterized have been found on plasmids, where they encode many of the phenotypic markers. The 4957-bp ampicillin-resistance (Ap^r) transposon Tn*3* has been subjected to detailed structural and functional analyses (Heffron et al. 1977, 1979; Gill et al. 1978; Arthur and Sherratt 1979; Chou et al. 1979). Transposon Tn*1,* derived from RP4, is very closely related to Tn*3,* and they have interchangeable functions. Figure 1 shows a map of transposon Tn*3* and the mutants described in this paper.

The pioneering experiments of Heffron et al. (1977), in which the transposition properties of a set of in-vitro-generated deletions of Tn*3* were studied, indicated not only that the ends of Tn*3* were required for transposition, but also that internal regions specifying diffusible gene products were required for normal transposition. Three Tn*3* genes have since been identified: *bla,* which specifies β-lactamase; *tnpA,* whose product is necessary for transposition; and *tnpR,* which specifies a product that regulates the frequency of transposition and is necessary for the production of normal transposition products.

We have shown previously that the Tn*103* transposon, which is derived from Tn*1,* lacks the product of gene *tnpR* and is transposition-proficient (Arthur and Sherratt 1979; Dougan et al. 1979); it transposes between replicons at high frequency to generate a cointegrate intermediate of the same basic structure as those obtained by Gill et al. (1978) for Tn*3* deletions.

On the basis of these results, we suggested a transposition model (outlined in Fig. 2), similar to that of Shapiro (1979), that has cointegrates as obligatory intermediates in interreplicon transposition. As support for this model, we were able to show that a Tn*103*-containing cointegrate molecule could be complemented efficiently by $tnpR^+$ (but not by $tnpR^-$) Tn*3* derivatives to generate the normal transposition end products. This *tnpR*-dependent resolution of cointegrates requires for its action the presence of an internal resolution site (IRS) close to the N-terminal end of the *tnpR* gene (Arthur and Sherratt 1979; Heffron et al. 1979). Transposon replication is directly implicated in transposition if cointegrates are a normal intermediate. The models also predict that intrareplicon transposition can generate deletions and inversions directly and can lead to duplications and translocations of regions adjacent to transposable elements. The end points for such events are the element ends and the sites preferred for element insertion. In this paper we characterize the Tn*3* site-specific recombination system further and present some other data obtained with other transposons. For convenience, in this paper we refer to Tn*3* and Tn*3* site-specific recombination even when we are describing the properties of Tn*103*-containing molecules. All of our data are consistent with the idea that Tn*1* and Tn*3* are interchangeable as far as transposition and recombination functions are concerned.

Requirements for Cointegrate Resolution

We have shown previously that a Tn*103*-containing cointegrate between pAC184 and RSF1010 (pAA131; Arthur and Sherratt 1979) is stably maintained in $recA^-$ strains, slowly recombines to give pAC184::Tn*103* and RSF1010::Tn*103* in $recA^+$ strains, and is efficiently resolved to the above products in a $recA^-$ strain containing Tn*3*. We also inferred that the product of gene *tnpR,* but not that of gene *tnpA* or one of the inverted repeats, is required for this resolution step. Here we directly confirm those conclusions and show that Tn*3* *tnpR*-dependent recombination occurs equally readily at 30 °C, 37 °C, or 42 °C, and that the product of Tn*1* gene *tnpR* is interchangeable with that of Tn*3*. Although we see both resolution products in a DNA population, at present we do not know whether a single recombination event generates both recombination products.

The results shown in Figure 3 are typical of what is observed when pAA131 cointegrate DNA is propagated in a $recA^-$ strain with and without various Tn*3* derivatives. The $recA^-$ profiles shown were obtained from pAA131 cells that had been transformed with the indicated Tn*3* derivatives. Transformant colonies were regrown once on selective medium, and lysates were prepared from cells derived from a single transformant clone. The $recA^+$ profile shown was obtained after repeated subculture. $recA^+$ profiles obtained soon after transformation show substantial amounts of cointegrate

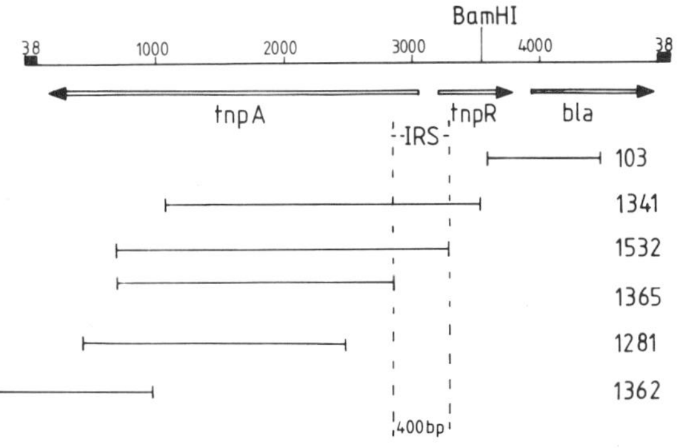

Figure 1. Map of the 4957-bp transposon Tn*3*, indicating the deletion mutants used in this study. All deletions except 103 were isolated in plasmid pMB8: Deletion 103 was isolated in RSF1010 and is derived from Tn*1*. The IRS contains the sequence necessary in *cis* for cointegrate resolution.

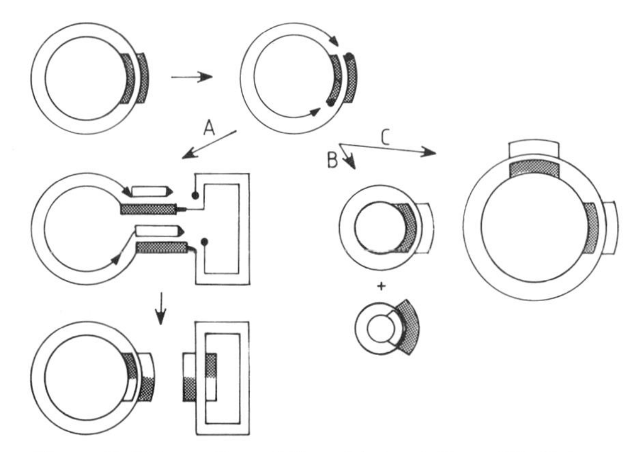

Figure 2. Interreplicon (*A*) and intrareplicon (*B, C*) transpositions of a transposable element (▩) from a circular replicon. Transposition is initiated by recognition of, and nicking adjacent to, a specific sequence at the end of the element. For Tn*3*, this nicking protein could well be the product of gene *tnpA* (transposase), and we propose that it has nicking-closing activity similar to that of the ϕX174 gene-*A* protein (Eisenberg et al. 1977) and to that proposed to be involved in plasmid conjugal transfer (Warren et al. 1978), i.e., during nicking the phosphodiester-bond energy is conserved by covalent joining of the protein to the 5′ phosphate of the phosphodiester chain. The two 5′-end–protein complexes then act in concert to cause a staggered nick at the target site (in the same [*B,C*] or another [*A*] DNA molecule) and to ligate the 5′ ends of the element to the 3′-OH ends of the target DNA. The phosphodiester-bond energy released by the second nicks is conserved by creating further protein-DNA bonds and is used for the final ligation. The nick:nick-ligate:ligate cycle would most easily be achieved by two monomers of the "transpose" acting at each transposon end (see, e.g., Eisenberg et al. 1977). Preferred target sites appear to be AT-rich and close to the insertion point have DNA homology with the element ends (Tu and Cohen 1980). Concomitant element replication, initiated from the two 3′-OH ends at each of the nicks and followed by ligation, yields the structures shown; i.e., a cointegrate for interreplicon transposition (*A;* although replication and ligation are not yet complete in the diagram) and either a deletion (*B*) or an inversion (*C*) for intrareplicon transposition, depending on whether the insertion events occur in the same or opposite strands. Subsequent site-specific (or homologous) recombination will yield the normal transposition end product for interreplicon transposition and will result in repeated flipping of the inverted segment in *C*. The two circular products in *B* (only one of which will normally be able to replicate) will be relatively poor substrates for recombination: If it occurs, the product, containing two directly repeated copies of the element, would be expected to be immediately subject to reexcision. Integration of the smaller (nonreplicating) circle (*B*) into other regions of element homology would result in translocation of other markers on the circle to new positions on the chromosome, whereas integration of the nonreplicating circle via nonelement homology (e.g., into a sister chromosome) would generate a duplication.

DNA (see Arthur and Sherratt 1979). Although the profiles shown were obtained after growth at 37 °C, essentially identical profiles were seen with lysates derived from cells grown at 30 °C and 42 °C, although in the latter case there is a substantial background of degraded DNA in the *recA*$^-$ strains, which is presumably derived from "*rec*less" breakdown.

Although a complementing product of gene *tnpR* is sufficient to produce efficient pAA131 cointegrate resolution, we do not know whether it alone is responsible for the resolution. It could act in association with a chromosomal gene product or with the product of gene *tnpA* (provided by pAA131). The latter possibility seems unlikely from our previous experiments (Arthur and Sherratt 1979) and from the observation that resolution but not transposition occurs at 42 °C.

Tn*3*-mediated Conjugal Transmission

We have used Tn*3*-mediated conjugal transmission in an attempt to determine whether Tn*3* site-specific recombination can act in *trans* to recombine separate molecules containing Tn*3* and to characterize the requirements for this site-specific recombination further.

Plasmid pACYC184 (here abbreviated pAC184) (Chang and Cohen 1978) is not mobilized by conjugation that is promoted by a number of different conjugative plasmids (Twigg and Sherratt 1980): It lacks mobility proteins and the site at which they act. This plasmid, then, can only be transmitted to recipients during conjugation by recombination with the conjugative plasmid and transfer as a cointegrate. This process has been termed conjugal conduction (Clark and Warren 1979). We have therefore used the IncW conjugative plasmid R388 to conjugally transmit pAC184 when various Tn*3* derivatives are present in each plasmid. Figure 4 shows the results of such experiments. pAC184 itself (assayed by transfer of chloramphenicol [Cm] resistance) is not detectably transferred ($<10^{-9}$) by R388 (not shown). Similarly, the derivative of pAC184 that contains the transposition-defective Tn*3* deletion 1341 is not detectably transferred. In contrast, both the Tn*3* and Tn*103* pAC184 derivatives are transferred: pAC184::Tn*103* efficiently, because cointegrates are produced at high frequency and cannot be resolved; and pAC184::Tn*3* at a much reduced frequency, because of the short half-life of cointegrates in a *tnpR*$^+$ cell (compare with the frequency of transfer, i.e., transposition, of Ap resistance). Analysis of the Cm^r transconjugants resulting from the

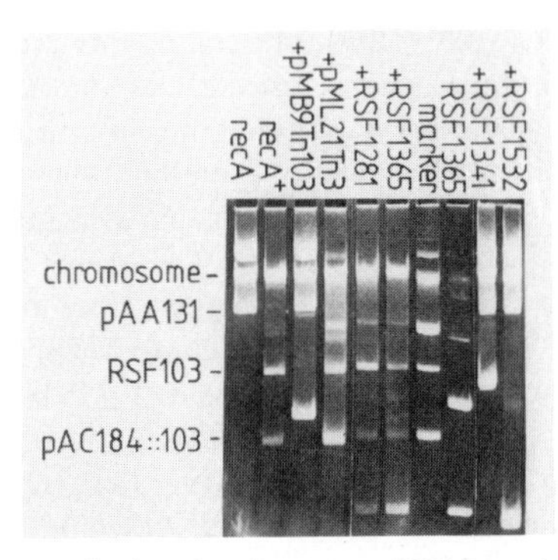

Figure 3. Analysis of cointegrate DNA and its recombination products in single transformed clones. Plasmid DNAs from single clones of a cointegrate-containing (pAA131) *recA*$^-$ strain, a *recA*$^+$ strain, and a series of *recA*$^-$ strains containing Tn*103* or Tn*3* deletion derivatives were analyzed on 0.8% agarose gels. The marker is a mixture of pAA131 resolution products (RSF103 and pAC184::Tn*103*), and the adjacent track contains RSF1365 DNA. RSF103 is RSF1010::Tn*103* and the other RSF plasmids are the pMB8::Tn*3* deletion derivatives referred to in the legend to Fig. 1. RSF1362$^+$ and RP4$^+$ (containing Tn*1*) *recA*$^-$ strains give profiles similar to those for Tn*3*$^+$, RSF1281$^+$, and RSF1365$^+$ *recA*$^-$ cells (not shown). RSF1341 DNA is dimeric.

pAC184::Tn*103*×R388 cross showed them to be cointegrates that could be resolved subsequently in a *tnpR*$^+$ *recA*$^-$ cell or a *recA*$^+$ cell. Transfer of pAC184::Tn*1341* or pAC184::Tn*103* by R388 from a donor cell containing wild-type Tn*3* (pML21::Tn*3*) gave results consistent with our previous observation that Tn*1341* cointegrates cannot be resolved by complementation, whereas Tn*103* cointegrates can be resolved (Arthur and Sherratt 1979).

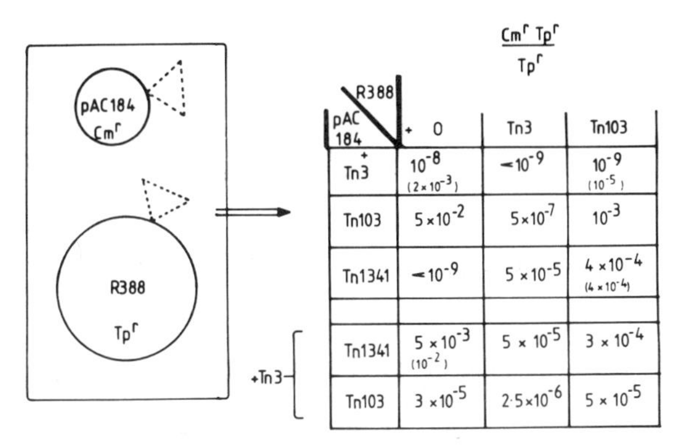

pAC184 \ R388	+ 0	Tn3	Tn103
Tn3$^+$	10^{-8} (2×10^{-3})	$\sim10^{-9}$	10^{-9} (10^{-5})
Tn103	5×10^{-2}	5×10^{-7}	10^{-3}
Tn1341	$\sim10^{-9}$	5×10^{-5}	4×10^{-4} (4×10^{-4})
+Tn3: Tn1341	5×10^{-3} (10^{-2})	5×10^{-5}	3×10^{-4}
+Tn3: Tn103	3×10^{-5}	2.5×10^{-6}	5×10^{-5}

Figure 4. Conjugal transfer of pAC184 and its derivatives by transposon-mediated cointegrate formation. The IncW conjugative plasmid R388 (Tp^r) and its Tn*3* and Tn*103* derivatives were used to transfer pAC184::Tn*3*, pAC184::Tn*103*, and pAC184::Tn*1341*. In the latter two cases, experiments were done both in the presence and absence of the complementing Tn*3*$^+$ plasmid pML21::Tn*3*. pAC184 transfer can occur only when it is a cointegrate with R388, formed either by site-specific recombination between the Tn*3*s of each plasmid or by transposition between the plasmids. The numbers in large type are the fraction of Tp^r transconjugants that are Cm^r (i.e., Cm^{r}Tp^r/Tp^r). The numbers in smaller type represent the fraction of Tp^r transconjugants that are Ap^r (i.e., Ap^{r}Tp^r/Tp^r for the crosses indicated; they are a measure of the frequency of transposition into R388).

Surprisingly, we detected no pAC184::Tn*3* mobilization by R388::Tn*3*. This indicates either that Tn*3* recombination cannot be used to join two Tn*3*-containing molecules, or, if it can, that the equilibrium of the reaction is so far in the direction of cointegrate breakdown that cointegrates arising through recombination are difficult to detect. We favor the latter explanation and suggest that the protein(s) involved in the recombination, once bound at the IRS, acts preferably on other *cis* IRS sequences.

Evidence that Tn*3* recombination can join separate Tn*3*-containing molecules came from analysis of the pAC184::Tn*103* × R388::Tn*3* and pAC184::Tn*3* × R388::Tn*103* crosses. Partial characterization of plasmid DNAs isolated from the Cm^{r}Tp^r (trimethoprim-resistance) transconjugants showed that they generally contain plasmids of the same size and structure as those in the donor; i.e., cointegrate formation had resulted from recombination rather than from transposition, which would have increased the size of the cointegrate by one transposon unit. Occasionally, we have obtained recombination products that are inconsistent with two reciprocal recombination events in which the two crossovers occur at the same position. For example, in the pAC184::Tn*3*×R388::Tn*103* cross we have observed Tp^{r}Cm^{r}Ap^s transconjugants containing pAC184::Tn*103* and R388::Tn*103*, i.e., recombination appears to have been accompanied by genetic conversion. In another cross (pAC184::Tn*103*×R64::Tn*3*) we obtained some transconjugants that contained pAC184::Tn*3*×R64::Tn*103*. The Tn*3* in pAC184 was inserted at precisely the same position and in the same orientation as the original Tn*103*, indicating that the two crossovers had occurred on different sides of the 103 deletion. As yet, we cannot be absolutely sure that all of these plasmids result from Tn*3*-mediated recombination events. If they do, then Tn*3* recombination is not always reciprocal and is not always confined to the IRS.

Analysis of plasmids from the pAC184::Tn*1341* × R388::Tn*3* cross indicated that transfer of chloramphenicol resistance had resulted from transposition of pAC184::Tn*1341* into R388::Tn*3*, which yielded a cointegrate that could not be resolved by Tn*3* recombination in the recipient. Similarly, transposition cointegrates, but not recombination cointegrates, appear to be responsible for transfer of chloramphenicol resistance in the pAC184::Tn*1341*×R388::Tn*103* cross. These data also illustrate the transposition "immunity" observed by Robinson et al. (1977). For example, transposition of Tn*3* from pAC184::Tn*3* into R388 is 200 times higher than transposition into R388::Tn*103*. Similarly, complemented cointegrate formation resulting from Tn*1341* transposition from pAC184::Tn*1341* into R388 is 100 times higher than transposition into R388::Tn*3* and 17 times higher than transposition into R388::Tn*103*. Transposition of Tn*103* from pAC184::Tn*103* into R388 is 50 times higher than transposition into R388::Tn*103*. Since the apparent immunity is observed where there is no functional Tn*3* recombination system and cointegrates are the transpositional end product, then it cannot require the site-specific recombination for its effect.

The reasons for this apparent immunity remain elusive, although it could serve to minimize the frequency of intrareplicon transposition events in vivo; i.e., Tn*3* interreplicon transposition events might be much more frequent than intrareplicon transposition events, which would generate deletions and inversions if the model is correct.

There is no very obvious mechanism for a true *cis*-acting immunity; yet other explanations, e.g., site-specific recombination, or the idea that a cell in which one Tn*3* transposes is in fact induced for all Tn*3* transpositions do not easily clarify the observations. In the latter case, the induction of a Tn*3*-containing recipient replicon might prevent isolation of additional Tn*3* insertion products because (1) it is destroyed (which is unlikely since cointegrate formation indicates that transposing replicons are not destroyed); or (2) its own participation in transposition in some way prevents the isolation of identifiable insertion products (possibly because of cointegration and/or site-specific recombination).

Cointegrate Formation and Site-specific Recombination with Transposons Tn*501* and Tn*7*

We have attempted to determine whether transposons Tn*501* (Hg^r [mercury resistance] 8.2 kb) and Tn*7* (Tp^{r}Sp^r [spectinomycin resistance]/Sm^r [streptomycin resistance]; 12.8 kb) transpose via cointegrate intermediates, which can then be resolved by transposon-specified, site-specific recombination. For the assay, a population of *recA*$^-$ donor cells that contained both the conjugative plasmid R388 and a nonmobilizable plasmid containing either of the transposons was mated with a *recA*$^-$ recipient. Initial selection was for recipients containing either R388 (Tp^r) or R388::Tn*501/7* (Tp^{r}Hg^r or Tp^{r}Sm^r/Sp^r). The latter exconjugants were then scored for the nontransposon markers of the nonmobilizable plasmid (i.e., for presumptive cointegrates). The results of these experiments are shown in Table 1 and Figure 5.

It can be seen clearly that transposition of Tn*7* from the Mob$^-$ ColE1::Tn*7* derivative pDS1702 into R388 is rarely accompanied by stable cointegrate formation: None of 107 Tp^{r}Sp^r/Sm^r transconjugants inherited the ColE1 immunity gene. This suggests that either Tn*7* does not transpose via a cointegrate intermediate or it has an efficient site-specific recombination system like that of Tn*3*. Preliminary experiments also indicate that complemented transposition of a Tn*7* derivative deleted for a 4.7-kb internal *Hin*dIII fragment does not result in cointegrate formation (D. Hodge and D. Sherratt, unpubl.). In contrast, in experiment 1 (Table 1) with Tn*501*, all (137 of 137) Tp^{r}Hg^r transconjugants were immune to colicin (Iea$^+$), indicating that transposition of Tn*501* from the Mob$^-$ ColE1::Tn*501* derivative pDS1509 into R388 generates a stable cointegrate. Surprisingly, when the plasmid DNAs from Tp^{r}Hg^r (Iea$^+$) transconjugants were analyzed on agarose gels (Fig. 5, gel 1), DNA bands corresponding in size to pDS1509 and R388::Tn*501* were seen rather than the expected R388::Tn*501*::pDS1509::Tn*501* cointegrate. Further analysis of these DNAs confirmed this observation. Since no Tp^{r}Hg^r (Iea$^-$) transconjugants were observed, we conclude that transposition in the *recA*$^-$ donor cell generates a stable cointegrate. After conjugal transfer of this to a *recA*$^-$ recipient, resolution to pDS1509 and R388::Tn*501* occurs efficiently. We have analyzed the transfer ability of a number of the R388::Tn*501* plasmids isolated from the recipients: Most are Tra$^+$, indicating that had any such plasmids formed in the donor cell after cointegrate resolution, they could have transferred to the recipient in the absence of pDS1509. Moreover, R388::Tn*501* does not promote the efficient conjugal transfer of pDS1509. It therefore appears that the ColE1::Tn*501*::R388::Tn*501* cointegrate resolution occurs as a consequence of its transfer to a recipient.

Our initial idea was that cointegrate resolution resulted either from derepression of a Tn*501* recombination system on transfer (a sort of zygotic induction) or because of transfer-dependent, site-specific recombination. To test these ideas and to repeat our initial observation, we constructed a pAC184::Tn*501*×R388 *recA*$^-$ strain. This time, transfer was to both a *recA*$^-$ strain (Table 1, exp. 2) and a *recA*$^-$ strain carrying Tn*501* on an RP1 derivative deleted for Tn*801* (Table 1, exp. 3). Tp^{r}Hg^r transconjugants in the *recA*$^-$ strain were se-

Table 1. Tn*7* and Tn*501* Interreplicon Transposition

Exp.	Plasmids in donor	Transconjugants selected
	ColE1::Tn*7* + R388	$\frac{Tp^rSm^r/Sp^r}{Tp^r} = 2.5 \times 10^{-5}$; 0/107 Iea$^+$
1	ColE1::Tn*501* + R388	$\frac{Tp^rHg^r}{Tp^r} = 4.0 \times 10^{-5}$; 137/137 Iea$^+$
2	pAC184::Tn*501* + R388	$\frac{Tp^rHg^r}{Tp^r} = 10^{-3}$; 54/56 Cm^r
3	pAC184::Tn*501* + R388	$\frac{Tp^rCm^r}{Tp^r} = 2.1 \times 10^{-5}$

The donor and recipient strains were both *recA*$^-$. Streptomycin or rifampicin were used to counterselect the donors. In experiment 3, the recipient also carried Tn*501* on an RPl derivative deleted for the Tn*801* transposon (pUB932 from P. Bennett, University of Glasgow). Concentrated donor and recipient cultures were mated on the surfaces of nutrient agar plates.

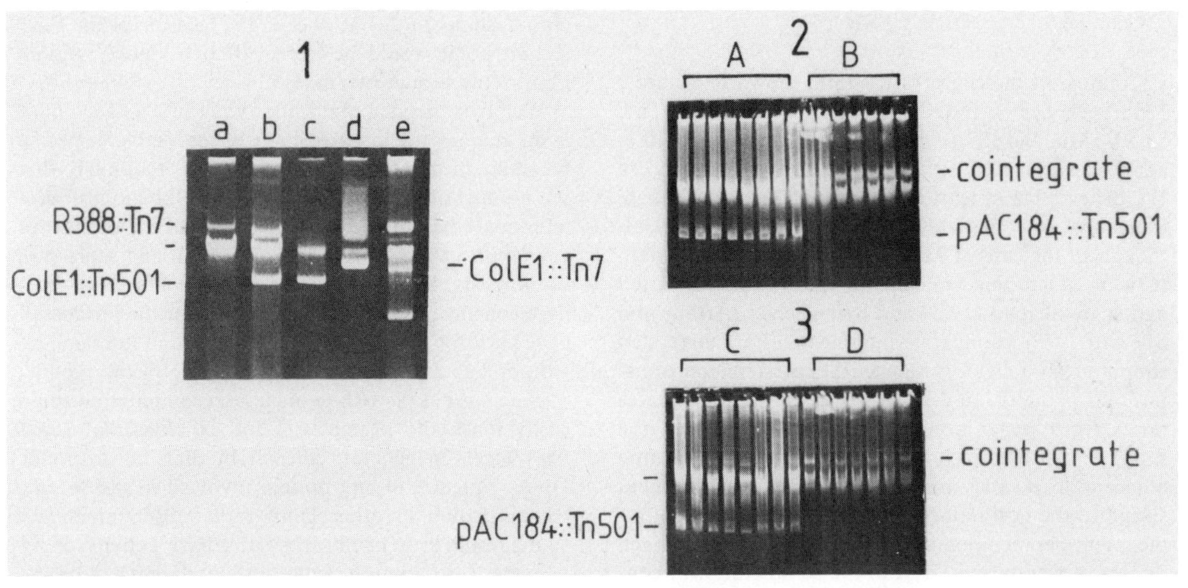

Figure 5. Analysis of plasmids resulting from Tn*7* and Tn*501* transpositions (see Table 1). (Gel *1*) Partially purified plasmid DNAs from either Tp^rSm^r/Sp^r (Iea^-) transconjugants or from Tp^rHg^r (Iea^+) transconjugants resulting from ColE1::Tn*7*×R388 and ColE1::Tn*501*×R388 crosses. (*a*) DNA from Tp^rSm^r/Sp^r (Iea^-) transconjugant; (*b*) DNA from Tp^rHg^r (Iea^+) transconjugant showing R388::Tn*501*×ColE1::Tn*501*; (*c*) ColE1::Tn*501*; (*d*) ColE1::Tn*7*; (*e*) marker: R388::Tn*103*::pAC184::Tn*103*×pAC184::Tn*103*. (Gel *2A*) Plasmids in total lysates of single clones of $Tp^rHg^r(Cm^r)$ transconjugates in experiment 2 in Table 1; (gel *2B*) plasmids in total lysates of Tp^rCm^r transconjugants in experiment 3 in Table 1. (Gel *3*) Plasmids in total lysates of Tp^rCm^r cells subcultured from experiment 3 (Table 1 and gel *2B*); (gel *3C*) DNA from Tp^rCm^r cells regrown on TpCmHg; (gel *3D*) DNA from the same cells regrown on TpCm. All gels contained 0.8% agarose. The unlabeled lines indicate chromosomal DNA.

lected; 54 of 56 were Cm^r, again indicating that most transposition events in the donor had generated cointegrates. Analysis of DNAs from 16 $Tp^rHg^rCm^r$ clones showed that all had resolved their cointegrate DNA (Fig. 5, gel 2A). In contrast, Tp^rCm^r transconjugants in the *recA*$^-$ Tn*501*$^+$ strain (exp. 3) showed no evidence of cointegrate breakdown (Fig. 5, gel 2B). Cells from the colonies shown in Figure 5 (gel 2A,B) were used as donors in further crosses to a *recA*$^-$ recipient. Whereas those shown in gel 2A retransferred Tp resistance, but not Cm resistance, efficiently (2 of 163 Tp^r transconjugants were Cm^r), those in gel 2B generally cotransferred Cm resistance with Tp resistance (17 of 19 Tp^r transconjugants tested were Cm^r). Transformation with DNAs isolated from the clones in gel 2 gave similar results: DNAs from clones in gel 2A generally gave either Tp^rCm^s or Cm^rTp^s transformants, whereas DNAs isolated from the clones in gel 2B gave transformants that were predominantly Tp^rCm^r.

Although this evidence is consistent with the idea of zygotic induction of a Tn*501* recombination function, we also realized that the transconjugants in experiment 3 had been directly selected on TpCm plates (we could not select for Hg^r transfer because the recipient was already Hg^r). Since synthesis of the Hg reductase of Tn*501* (which confers Hg resistance) is known to be induced by Hg salts, it occurred to us that Tn*501* transposition functions, in particular the recombination enzyme, could also be induced by Hg salts. R388×ColE1::Tn*501* and R388×pAC184::Tn*501* donor cells had not been grown on Hg plates prior to the mating: Although some transposition has occurred, there has been little or no cointegrate resolution. After transfer of the cointegrate into a recipient, transconjugants are selected on Hg plates. Perhaps Tn*501* recombination is induced and the cointegrates are resolved. Thus, the nonresolution of cointegrates in the *recA*$^-$ Tn*501*$^+$ recipient could simply be a consequence of our failure to use a Hg selection. To test this, we took from experiment 3 the transconjugants that contained cointegrates (Table 1 and Fig. 5 gel 2B) and regrew them on TpCm and TpCmHg plates. Analysis of the plasmids from the cells grown on the different plates is shown in gel 3. Reculture on TpCm medium results in no observable cointegrate breakdown, whereas growth on TpCmHg results in the efficient appearance of pAC184::Tn*501*. Therefore, we conclude that Hg induces a Tn*501* recombination enzyme that resolves Tn*501* cointegrates efficiently. To confirm this, we repeated experiment 2 and directly selected for transconjugants that contained cointegrates (Tp^rCm^r, in the absence of Hg). Analysis of the plasmid DNA from these transconjugants showed it to contain primarily cointegrates, although on repeated subculture in the absence of Hg some resolution was observed (data not shown).

As yet, we do not know whether Tn*501* transposition is induced by Hg; clearly, some transposition can occur in its absence. What is certain is that efficient resolution of the cointegrates can occur only in cells grown on Hg. Transposition and erythromycin resistance of transposon Tn*917* are both inducible in *Streptococcus faecalis* by erythromycin (Tomich et al. 1980).

DISCUSSION

Cointegrate molecules of the form shown in Figure 2 have been demonstrated with at least Tn*1*, Tn*3*, Tn*5*, Tn*501*, Mu, and IS*1* (Arthur and Sherratt 1979; Gill et al. 1979; Meyer et al. 1979; Shapiro 1979; So et al. 1979). We believe that at least in the case of Tn*501*, these are obligatory intermediates in genetic transposition, although in the case of Mu this seems much less certain (Coelho et al., this volume). In the models suggested earlier by Shapiro (1979) and by ourselves (Arthur and Sherratt 1979), cointegrate intermediates are generated simply as an obligatory step in the interreplicon transposition pathway. In essence, the route to cointegrates arises from recognition (on opposite strands) of the same sequence at each end of the transposon, nicking adjacent to it, and transfer of a preexisting element strand to the new replicon (i.e., generating covalent attachment between donor and recipient replicons). Each duplex of transposon DNA in the cointegrate will contain one parental strand and one newly replicated strand. Similarly, after resolution, the final cointegrate molecules will contain both parental and newly replicated transposon DNA. Furthermore, in these models, the base-pair repeat observed at transposon insertion sites is generated by resolution of the cointegrate; transposon DNA in cointegrates should not be bounded by the small, direct repeats.

A feature of our model is that replication is initiated at the 3′-OH end of the nonelement strand created by the presumptive transposase, and the replicated section consists of the element plus the x-bp repeat. If the model is correct, then transposable elements can be considered as replicons whose specificity of replication resides in the "transposase." Such replication can be as simple and efficient as that of any virus or plasmid.

If transposition between replicons often or always proceeds via cointegrate intermediates, do all transposable elements specify a recombination system to resolve these intermediates? IS*1* apparently does not (So et al. 1979), and it seems fairly unlikely that other IS elements do. In the absence of such site-specific recombination, *recA*-dependent homologous recombination can act, although slowly as compared with the Tn*3* system.

We presume that transposons, which reside predominantly on plasmids and have evolved to transpose between plasmids efficiently, have acquired (or will acquire) site-specific recombination if their transposition proceeds through cointegrate intermediates. IS elements, which are commonly found on chromosomes, may have less opportunity for interreplicon transposition. Their intrareplicon transposition has little use for site-specific recombination. Production of most Hfrs from F^+ strains is *recA*-dependent; however, about 1% are *recA*-independent (Cullum and Broda 1979). The *recA*-dependent Hfrs presumably result from homologous recombination between transposable-element sites on F (e.g., IS*2*, IS*3*, and $\gamma\delta$) and the chromosome, whereas the *recA*-independent Hfrs could well result from transposition of chromosomal transposable elements onto F or transposition of F elements onto the chromosome. In the latter case, the location of the F insertion point would be expected to be more random than in the former two cases.

We favor the idea that the product of Tn*3* gene *tnpR* is the major protein participant in the Tn*3* site-specific recombination. Our earlier data make it unlikely that the product of gene *tnpA* is involved, although host proteins could be. The possible dual roles of the product of gene *tnpR* in site-specific recombination and repression of synthesis of the products of genes *tnpA*/*tnpR* are easily reconciled, since the sites of action of the product of gene *tnpR* for both of these activities are in the same region of Tn*3* and could overlap. Binding of the product of gene *tnpR* at the IRS to mediate recombination could easily mask the promoter(s) for expression of genes *tnpA*/*tnpR*. In general, efficient binding, to a specific DNA sequence, of any protein involved in site-specific recombination or other DNA metabolism/interaction could result in it having the subsidiary activity of repressing transcription from an overlapping/adjacent promoter. Such a suggestion has also been made by Simon (this volume) for *hin*-promoted inversion during switched expression of flagella type in *Salmonella typhimurium.*

The observation that Tn*501* transpositional cointegrates are efficiently resolved to final transposition products only after growth of Tn*501*-containing cells in the presence of Hg salts is intriguing, particularly in view of the derepression of Hg reductase synthesis by Hg salts. We are currently comparing Tn*501* transposition frequencies in the presence and absence of Hg. If transposition is induced by Hg, then the adaptive response of Tn*501*-containing cells would appear to be a derepression of reductase synthesis, an increase in gene copies resulting from transposition, and a spread of the transposon to different replicons. Whatever the detailed nature of Tn*501* gene organization and expression turns out to be, the presence of an inducible cointegrate "resolvase" activity highlights the importance of cointegrates and their resolution in transposon interreplicon transposition.

ACKNOWLEDGMENTS

We thank our colleagues in Brighton and Glasgow for their help and for many useful and stimulating discussions. A. A. is supported by an MRC studentship for training in research methods.

REFERENCES

ARTHUR, A. and D. J. SHERRATT. 1979. Dissection of the transposition process: A transposon-encoded site-specific recombination system. *Mol. Gen. Genet.* **175:** 267.

CHANG, A. C. Y. and S. N. COHEN. 1978. Construction and characterization of amplifiable multicopy DNA cloning vehicles derived from the P15A cryptic miniplasmid. *J. Bacteriol.* **134:** 1141.

CHOU, J., M. J. CASADABAN, P. LEMAUX, and S. N. COHEN. 1979. Identification and characterization of a self-regulated repressor of translocation of the Tn*3* element. *Proc. Natl. Acad. Sci.* **76:** 4020.

Clark, A. J. and G. J. Warren. 1979. Conjugal transmission of plasmids. *Annu. Rev. Genet.* **13:** 99.

Cullum, J. and P. Broda. 1979. Chromosome transfer and Hfr formation by F in *rec*$^+$ and *recA* strains of *Escherichia coli* K12. *Plasmid* **2:** 358.

Dougan, G., M. Saul, A. Twigg, R. Gill, and D. J. Sherratt. 1979. Polypeptides expressed in *Escherichia coli* K-12 minicells by transposition elements Tn*1* and Tn*3*. *J. Bacteriol.* **138:** 48.

Eisenberg, S., J. Griffith, and A. Kornberg. 1977. ϕX174 cistron A protein is a multifunctional enzyme in DNA replication. *Proc. Natl. Acad. Sci.* **74:** 3198.

Gill, R., F. Heffron, and S. Falkow. 1979. Identification of the transposase and transposon-specific repressor encoded by Tn*3*. *Nature* **282:** 20.

Gill, R., F. Heffron, G. Dougan, and S. Falkow. 1978. Analysis of sequences transposed by complementation of two classes of transposition-deficient mutants of Tn*3*. *J. Bacteriol.* **136:** 742.

Heffron, F., B. J. McCarthy, H. Ohtsubo, and E. Ohtsubo. 1979. DNA sequence analysis of the transposon Tn*3*: Three genes and three sites involved in transposition of Tn*3*. *Cell* **18:** 1153.

Heffron, F., P. Bedinger, J. J. Champoux, and S. Falkow. 1977. Deletions affecting the transposition of an antibiotic resistance gene. *Proc. Natl. Acad. Sci.* **74:** 702.

Meyer, R., G. Boch, and J. Shapiro. 1979. Transposition of DNA inserted into deletions of the Tn*5* kanamycin resistance element. *Mol. Gen. Genet.* **171:** 7.

Robinson, M. K., P. M. Bennett, and M. H. Richmond. 1977. Inhibition of Tn*A* translocation by Tn*A*. *J. Bacteriol.* **129:** 407.

Shapiro, J. A. 1979. Molecular model for the transposition and replication of bacteriophage Mu and other transposable sequences. *Proc. Natl. Acad. Sci.* **76:** 1933.

So, M., F. Heffron, and B. McCarthy. 1979. The *E. coli* gene encoding heat-stable toxin is a bacterial transposon flanked by inverted repeats of IS*1*. *Nature* **277:** 453.

Tomich, P. J., F. Y. An, and D. B. Clewell. 1980. Properties of erythromycin-inducible transposon Tn*917* in *Streptococcus faecalis*. *J. Bacteriol.* **141:** 1366.

Tu, C.-P. D. and S. N. Cohen. 1980. Translocation specificity of the Tn*3* element: Characterization of sites of multiple insertions. *Cell* **19:** 151.

Twigg, A. J. and D. J. Sherratt. 1980. Trans-complementable copy-number mutants of plasmid ColE1. *Nature* **283:** 216.

Warren, G. J., A. J. Twigg, and D. J. Sherratt. 1978. ColE1 plasmid mobility and relaxation complex. *Nature* **274:** 259.

Mechanism of Insertion and Cointegration Mediated by IS*1* and Tn*3*

E. OHTSUBO, M. ZENILMAN, H. OHTSUBO, M. MCCORMICK, C. MACHIDA, AND Y. MACHIDA
Department of Microbiology, School of Medicine, State University of New York, Stony Brook, New York 11794

Insertion sequence (IS) and transposable (Tn) elements are discrete segments of DNA that move from one site to another on a different or the same genome and cause structural rearrangements of the genomes (Bukhari et al. 1977). IS*1* is a typical insertion element. It was found originally as an inserted sequence about 700–800 bp in length in various bacterial operons, resulting in strong polar effects on these operons (Fiandt et al, 1972; Hirsch et al. 1972; Jaskunas and Nomura 1977). IS*1* also appears as repeated DNA sequences in a group of resistance plasmids, such as R100 and R1, in phage P1, and in various bacterial chromosomes (Hu et al. 1975b; Arber et al. 1979; Nyman et al. 1981). It is involved in rearrangements of plasmid genomes such that it promotes deletion mutations (Reif and Saedler 1975; Mickel et al. 1977), transposition of antibiotic-resistance and enterotoxin genes (MacHattie et al. 1977; So et al. 1979), and fusion of two different plasmid genomes (Meyer et al. 1979; E. Ohtsubo et al. 1979, 1980).

The nucleotide sequence of IS*1* has been determined to be 768 bp long (Ohtsubo and Ohtsubo 1978; Johnsrud 1979). A sequence of about 35 bp at each end of IS*1* is in an inverted orientation. Upon insertion of IS*1* into any site, it has been shown to generate the duplication of a 9-bp sequence of the target site (Calos et al. 1978; Grindley 1978). The DNA sequence shows that it could code for some small proteins; however, the existence of the genes for these proteins has not yet been substantiated.

Tn*3* is a typical transposable element responsible for resistance to ampicillin (Kopecko et al. 1975). The molecular length of this transposon has been determined to be 4957 bp (Heffron et al. 1979). Genetic studies and the DNA sequence of Tn*3* predict three gene products: a transposase, which is a protein required for transposition; a repressor, which is a protein that regulates its own transcription as well as that of the transposase; and β-lactamase, which confers resistance to ampicillin (Heffron et al. 1978, 1979). The flanking sequence of Tn*3*, a 38-bp inverted repeat, is also required for transposition (Cohen et al. 1978; Heffron et al. 1979; H. Ohtsubo et al. 1979). Tn*3*, like IS*1*, promotes deletion mutations (Nisen et al. 1977; H. Ohtsubo et al. 1979). Furthermore, it has been shown that a specific deletion mutant of Tn*3* mediates the fusion of two different plasmid genomes (Gill et al. 1978).

The subject of this paper is the molecular mechanism of IS*1* insertion and the genetic fusion of two plasmid genomes, mediated by IS*1*. First, we describe a simple genetic system that has recently been developed in our laboratory for the isolation of cointegrated plasmids (E. Ohtsubo et al. 1980), and we show that a plasmid containing one copy of IS*1* is a transposon and that it integrates into a second plasmid containing no IS*1* sequence in a characteristic way: IS*1* is duplicated at the recombinational junctions. The nucleotide sequence results show that a 9-bp sequence at the target site appears as a direct duplication at the junctures between each IS*1* and the recipient plasmid. This suggests that both the IS*1*-insertion event and cointegration mediated by IS*1* occur by a common molecular mechanism, since IS*1* insertion also generates a 9-bp direct duplication of the target site. Thus, analysis of cointegration is a useful approach in the study of the insertion of IS*1* itself.

Second, we demonstrate that mutations occurring inside IS*1* markedly reduce or abolish the frequency of cointegration, suggesting that IS*1* may code for a protein(s) that is required for cointegration, and therefore for the insertion of IS*1*. We also demonstrate that cointegration mediated by IS*1* occurs in several bacterial strains carrying mutations such as *recA, recBC sbcB,* and *polA*.

The genetic system to be described is also useful for the study of cointegration between two different plasmids, both of which carry the IS*1* sequence. We demonstrate that cointegration occurs either by homologous recombination at the IS*1* sequences carried by the two plasmids, even in a *recA*$^-$ strain, or by transposition of one plasmid into another by the same mechanism as that for cointegration between two different plasmids in which only one contained a copy of IS*1*. On the basis of our results, models of IS*1* insertion and cointegration mediated by IS*1* are also discussed.

Finally, we describe how our system is useful in screening new insertion sequences carried by a plasmid and in the study of plasmid cointegration mediated by the known transposon Tn*3*. We demonstrate that only Tn*3* carrying mutations in the repressor gene can form stable cointegrated plasmids with two copies of the Tn*3* mutants at the recombinational junctions. Wild-type Tn*3* and other mutants do not form such cointegrates but instead generate various smaller-size recombinants with one intact Tn*3*. The interpretation of these results is that cointegrated plasmids carrying two Tn*3* elements are formed at first but are subjected to spontaneous deletions eliminating one of the Tn*3* sequences and adding to the stability of the recombinant plasmid.

Isolation and Characterization of Cointegrated Plasmids from Two Different Plasmids, One Carrying IS*1*

The genetic system for the isolation of cointegrates between two different plasmids is straightforward. The key feature in this system is the use of a temperature-sensitive replication mutant plasmid, pHS1, which carries the tetracycline-resistance (Tc^r) gene. The Tc^r marker on pHS1 is rescued at 42 °C by recombination with a second plasmid, which can replicate at the restrictive temperature.

Figure 1 summarizes the genetic and physical properties of the parental plasmids actually used in the isolation of the recombinants. pHS1 is a mutant of pSC101, isolated by Hashimoto-Gotoh and Sekiguchi (1976). Its replication ceases immediately after the temperature shifts from 30 °C to 42 °C within a bacterial cell. The second plasmid, pSM1, is derived from the resistance plasmid R100 and can replicate at 42 °C. It carries no resistance genes but retains one copy of IS*1* (Mickel et al. 1977; Ohtsubo and Ohtsubo 1978). These two plasmids belong to different incompatibility groups and therefore can coexist within the same cell.

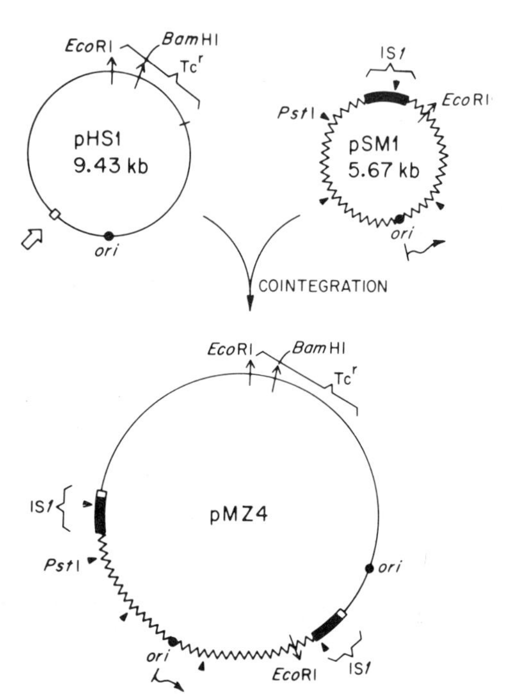

Figure 1. Physical structures of the parental plasmids (pHS1 and pSM1) and one of the cointegrate plasmids (pMZ4). Cleavage sites for the restriction endonucleases *Eco*RI, *Bam*HI, and *Pst*I are shown by arrows and arrowheads. One of four *Pst*I sites on pSM1 is inside IS*1* (■). Note that pHS1 is not cut by *Pst*I. (Another enzyme, *Bst*EII, cleaves in the middle of IS*1* but not in the plasmid sequences. The site of *Bst*EII is not shown.) pSM1 containing a copy of IS*1* recombines with pHS1 (open arrow) to give rise to a cointegrate plasmid pMZ4 (see text). The resulting pMZ4 contains parental plasmid sequences that are flanked by direct repeats of IS*1*. The 9-nucleotide sequence at the target site (□) is also duplicated in a direct orientation in pMZ4 as indicated (see Fig. 4). Approximate locations of origins of replication of the plasmids and the direction of replication of pSM1 (Cabello et al. 1976; Ohtsubo et al. 1977) are shown.

When cells harboring pSM1 and pHS1 were grown overnight at 30 °C, spread on plates containing tetracycline at 10 μg/ml, and incubated at 42 °C for 24 hours, we observed Tc^r cells. Examination of the plasmid DNAs present in these Tc^r cells by isolation of the plasmid DNA molecules indicated that they contained plasmid DNAs larger than parental plasmids. Cells carrying the large plasmids could be isolated from both *recA*$^+$ and *recA*$^-$ strains. Analysis of the frequency of appearance of these cells is described in a later section.

We have analyzed eight kinds of recombinant DNAs that were isolated from independent colonies; six were from *recA*$^+$ and two were from *recA*$^-$. We named these pMZ plasmids (pMZ1–9). Their physical structures, determined by electron microscopy, restriction endonuclease analysis, and nucleotide sequencing, are shown in Figures 1 and 2 (see E. Ohtsubo et al. 1980). The general characteristics of their structure can be summarized as follows: (1) The sequences of both parental plasmids pSM1 and pHS1 are present in the pMZ plasmids. Direct repeats of IS*1* are at the junctions between the parental plasmid sequences. (2) The site of recombination of pSM1 is always at IS*1*, whereas for pHS1 it is different in each pMZ. (3) The orientations of the pHS1 and pSM1 components in each of the pMZ plasmids are identical. (4) Nucleotide sequence analysis of recombinant plasmids and their parental plasmid DNAs revealed that a 9-bp sequence at a target site in the pHS1 plasmid was duplicated at the junction of each IS*1*.

Among the results obtained by various techniques, it is important to describe here the restriction endonuclease analysis used to show the direct duplication of IS*1* in all the cointegrates. As shown in Figure 3, when

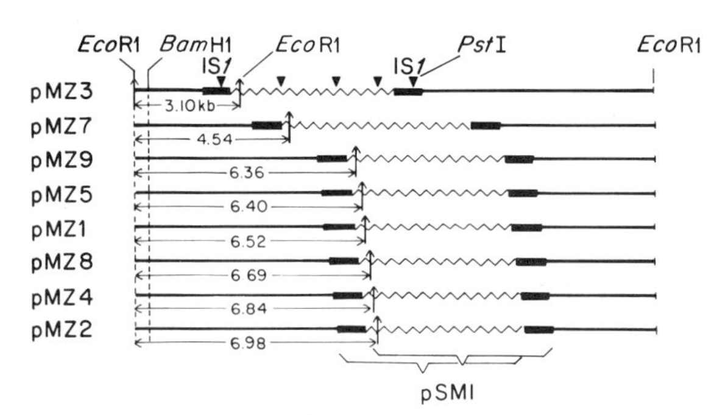

Figure 2. Linear representations of the circular structures of pMZ plasmids (E. Ohtsubo et al. 1980). DNA sequences of parental plasmids pHS1 (——) and pSM1 (∿∿) in each pMZ plasmid are indicated. Restriction endonuclease cutting sites for *Eco*RI and *Bam*HI (arrows) and for *Pst*I (▲) are assigned to indicate the sequence orientation of parental plasmid sequences. Note that IS*1* is located at each junction in the same orientation. Note also that the orientations of pHS1 and pSM1 remain the same in each pMZ plasmid. Numbers are assigned in kilobases for the lengths of the *Eco*RI fragment containing the *Bam*HI site.

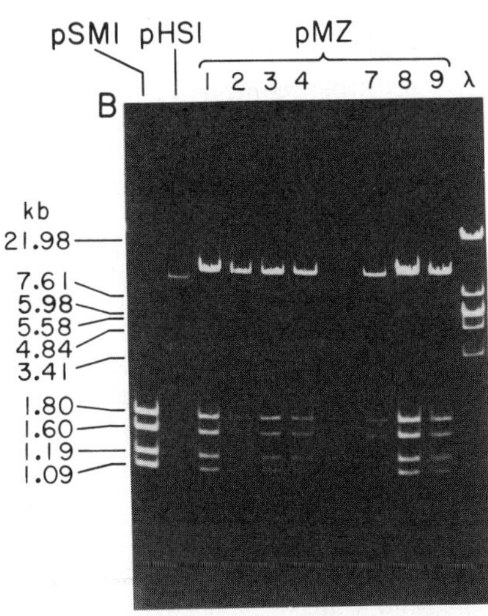

Figure 3. Agarose gel (0.7%) showing *Pst*I digests of pSM1 and of pMZ1–9 isolated from parental plasmids pSM1 and pHS1. Linear pHS1, obtained from pHS1 digested with *Bam*HI, is also shown. The kilobase markers are pSM1 bands, whose lengths were determined from sequence analysis (Ohtsubo and Ohtsubo 1978; J. Rosen et al. 1979 and unpubl.), and *Eco*RI-digested λ DNA (Blattner et al. 1977).

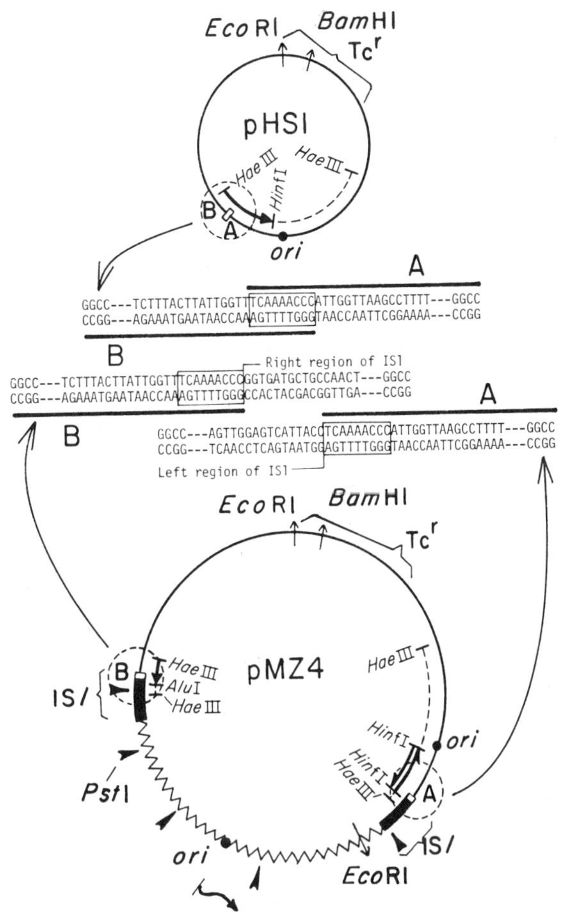

Figure 4. Summary of the sequence analysis of pMZ4 (E. Ohtsubo et al. 1980). The circles denote the region sequenced, and the letters (*A*, *B*) indicate where each sequence appears. The rectangular boxes drawn in the plasmids and around certain sequences denote the 9-base direct repetition generated during cointegration. Arrows parallel to the circle indicate direction of sequencing.

pMZ plasmid DNAs were digested with *Pst*I, which cleaves pSM1 at four sites (including one in IS*1*) but not pHS1 (see Fig. 1), five DNA fragments were generated whose gel patterns were identical for each pMZ (see Fig. 3). Four of these fragments were identical with those of *Pst*I-pSM1 digests; the other one was larger than the linear pHS1 DNA by about 700–800 bp. When pMZ plasmid DNAs were digested with *Eco*RI, which cleaves both pHS1 and pSM1 once, two fragments unique in size for each pMZ plasmid were generated (data not shown, but see Fig. 2). The fact that cleavage with *Pst*I shows identical gel patterns and constancy of pSM1 bands, whereas clear differences exist after cleavage with *Eco*RI, can be explained by a structure of the pMZ plasmids in which IS*1* is duplicated at the junctions of the two parental plasmid sequences in the same orientation; the pSM1 sequence is permuted with the redundant sequence of the IS*1*, as shown in Figures 1 and 2. *Bst*EII is an enzyme that cleaves pSM1 once inside IS*1* and does not cleave pHS1. When this enzyme was used to cleave the pMZ plasmids, two fragments with identical lengths in each pMZ were generated (data not shown). This also supports the duplication of IS*1* in pMZ plasmids. Note that *Pst*I as well as *Bst*EII can be used to identify and characterize the cointegrate plasmid structure mediated by IS*1*.

Figure 4 summarizes the results obtained by DNA sequencing analysis and shows in particular the junction and target sequences for pMZ4. The results indicate that in the recombinant pMZ4, we see the A region of pHS1 leading into one end of IS*1* and the B region leading into the other end. Therefore, IS*1* is directly repeated at the region of recombination. The results also show a 9-base sequence present in pHS1 directly repeated at the junctions with the IS*1* sequences. In the analysis of pMZ2, we obtained identical results showing direct repeats of IS*1* at the junctions. However, the 9 bases duplicated were

5′ G-T-G-T-T-T-T-T-C 3′
3′ C-A-C-A-A-A-A-A-G 5′

which appeared 207 bp away from the integration site of pMZ4. The region containing the integration sites of pMZ2, pMZ4, pMZ8, and probably pMZ1 is rich in A and T bases (66.9%).

It is known that upon insertion IS*1* generates a direct repeat of 9 bases at a target site (Calos et al. 1978; Grindley 1978). This fact and the sequence results obtained above indicate that insertion of IS*1* and cointegration mediated by IS*1* share a common molecular mechanism.

In addition, we observed that pSM1 integration inactivated a gene(s) essential for pHS1 replication (which is normal at 30°C), except for the cases of pMZ3 and pMZ7 (see Fig. 2), which had a functional pHS1 replicon (E. Ohtsubo et al. unpubl.). These data are reminiscent of the facts that an IS*1* insertion inactivates genes and causes strong polar effects in operons.

Thus, the results obtained from analysis of the physical structures of cointegrate plasmids indicate that pSM1, carrying a copy of IS*1*, acts like a transposon and is inserted into pHS1 in its entirety. As a result, the temperature-sensitive replication mutation in pHS1 was suppressed by integration of pSM1.

Effect of Bacterial Host Mutations on IS*1*-mediated Cointegration between pHS1 and pSM1

The frequency of cointegration in vivo between pHS1 and pSM1 can be measured by a fluctuation test, described by Luria and Delbrück (1943). In a *recA*$^+$ strain, the frequency was 4.9×10^{-7} per division cycle, whereas in *recA*$^-$, it was found to be 1.4×10^{-8} per division cycle (see Table 1). This indicates that the cointegration event mediated by IS*1* is independent of *recA* function, although the frequency is lowered in the *recA*$^-$ strain.

We examined the cointegration of these two plasmids in several mutant bacteria (listed in Table 1) to determine which host-gene products affect the recombination event. As shown in Table 1, we found that cointegration can occur in all mutants tested. Note that although cointegration occurs in a *polA*$^-$ mutant, this does not necessarily mean that cointegration mediated by IS*1* does not require polymerase I, since it is known that a reduced amount of polymerase I is still present in the mutant cells and that these cells are still capable of chromosomal replication (Lehman and Chien 1973).

Analysis of Cointegration between Two Different Plasmids, Both Carrying the IS*1* Sequence

We examined the cointegration between two plasmids, both carrying IS*1*. The parental plasmids used were pSM1 and a pHS1 derivative containing a copy of IS*1*, designated pMZ71 (or pHS1::IS*1*). pMZ71 is still temperature-sensitive in replication and carries the Tc-resistance gene. Using the assay system described above, we expect to isolate cointegrates that may be of three different types, as represented in Figure 5. Type I is a recombinant that could be formed by recombination at the IS*1* homologies carried by each parental plasmid. Therefore, the type-I recombinants will contain two IS*1* elements in a direct orientation. Type II can be formed by transposition of pSM1 at its IS*1* site into any site on pMZ71, as observed in the analysis in the previous section of cointegration between pSM1 and pHS1. Finally, type III can be formed by transposition of pMZ71 at its IS*1* site into any site on pSM1. The latter two types of cointegrates will contain three copies of IS*1*, since, upon transposition of a plasmid, an IS*1* originally present in the donor plasmid will be duplicated in the recipient plasmid in a direct orientation at the site of transposition (see Fig. 5). These three types of recombinants will be identified by the use of the restriction endonucleases *Pst*I and *Bst*EII, which cut the IS*1* sequence once.

Cells carrying recombinants between pSM1 and pMZ71 were isolated from *recA*$^+$ and *recA*$^-$ by selecting for Tc resistance at 42 °C. The frequency of recombination was measured as 7.70×10^{-4} per division cycle in the *recA*$^+$ strain, whereas it was measured as 3.31×10^{-8} per division cycle in the *recA*$^-$ strain. Note that these frequencies are higher than those obtained in the recombination between pSM1 and pHS1, in which only pSM1 carried a copy of IS*1*.

Fourteen independent recombinants were isolated from *recA*$^+$. The molecular lengths of the recombinant DNAs, determined by agarose gel electrophoresis, indicated that all are of the same size as the eight pMZ plas-

Table 1. Frequency of IS*1*-mediated Cointegration of the Plasmids pHS1 and pSM1 in *E. coli* Mutants

Mutation	Strain	Concentration of Tc in selective plates (μg/ml)	Frequency/ division cycle
recA	JE5519	20	1.4×10^{-8}
recA$^+$	JE5507	25	4.9×10^{-7}
recBC sbcB	JC7263	15	2.4×10^{-7}
recBC$^+$ *sbcB*$^+$	AB1157	25	4.7×10^{-8}
*polA*1	JG112	20	1.4×10^{-7}
polA$^+$	JG113	30	5.9×10^{-8}

The frequency of cointegration was determined by a fluctuation test as follows. Purified cells containing pHS1 and pSM1 were grown in 5 ml of L broth overnight at 30 °C; 0.1 ml of a 1:10^6 dilution was then inoculated into 10–20 tubes containing 5 ml of broth and incubated overnight at 30 °C. Then, 0.1 ml of the contents from each tube was plated on a prewarmed 10–30 μg/ml Tc plate and incubated at 42 °C for 24 hr. The Tc^r colonies were scored and examined for the presence of large plasmids by the crude lysis method (Y. Machida et al., in prep.), and the frequency of cointegration was calculated in units of resistant bacteria containing cointegrates per division cycle, according to the equation described by Luria and Delbrück (1943). Note that in some strains residual growth of cells was observed even at 42 °C when the Tc plates containing lower doses of Tc were used; thus, we used plates containing higher concentrations of Tc to clear out such cells that do not carry IS*1*-mediated cointegrate plasmids. Under any conditions for the selection of Tc^r cells, the efficiency of plating (eop) of cells containing the standard cointegrate (we used pMZ2) differed in each strain. Thus, to calculate the actual frequency of cointegration, the value obtained by the fluctuation test has been normalized by the value of eop. The cointegration frequency obtained in the *recA*$^+$ and *recA* strain is higher than that reported previously (E. Ohtsubo et al. 1980), since we noticed that the strain has a very low eop on the Tc-containing plate.

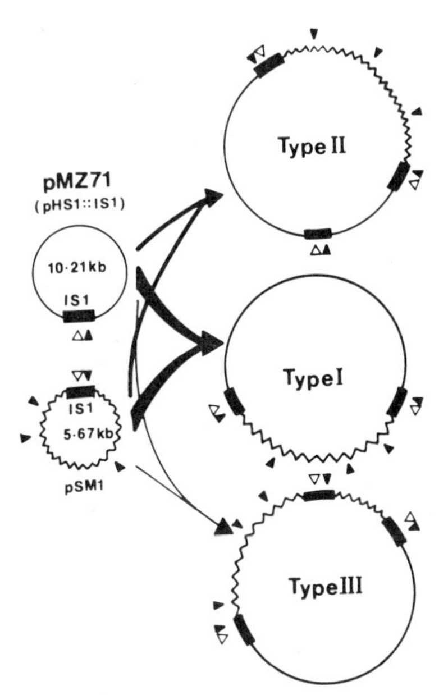

Figure 5. Three types of possible cointegrates that are derived from the parental plasmids pMZ71 (or pHS1::IS*1*) and pSM1, each carrying a copy of IS*1*. Note that the type-I cointegrate contains two copies of IS*1*, whereas type-II and type-III cointegrates contain three copies of IS*1*, two of which will be generated during transposition of either pSM1 or pMZ71 mediated by IS*1*. These three types of cointegrates can be identified readily by the use of the restriction endonucleases *Pst*I, which cleaves IS*1* once and the rest of pSM1 three times (▲), and *Bst*EII, which cleaves IS*1* once in both pSM1 and pMZ71 (△). Type-I cointegrate plasmid digested with *Pst*I will generate five fragments that correspond to *Pst*I-digested fragments of parental pSM1 and pMZ71. *Bst*EII digestion will also generate two fragments that are the same as *Bst*EII fragments of the parental plasmids. Type-II cointegrate plasmids digested with *Pst*I will generate six fragments, four of which are identical to the *Pst*I fragments of the parent pSM1. The *Pst*I fragment of pMZ71 will not be observed; instead, two new bands characteristic for each independent cointegrate will appear, since the sites of transposition of pSM1 into pMZ71 are different. *Bst*EII digests of type-II cointegrates will generate three fragments: One is identical to the *Bst*EII fragment of pSM1, and two specific fragments are smaller than the *Bst*EII fragment of pMZ71. *Pst*I digests of the type-III cointegrates will generate six fragments: One is identical to the *Pst*I fragment of pMZ71, and, of the other five fragments, three correspond to the *Pst*I fragments of pSM1 and two are characteristic for each independent cointegrate, since sites of transposition of pMZ71 into pSM1 are different. The actual cleavage results are shown in Fig. 6.

mids described in the previous section that contain two IS*1* elements. Forty-four independent recombinants were isolated from *recA*$^-$. The molecular lengths of these plasmid DNAs indicated that three of the recombinants were of the same size as those isolated from *recA*$^+$ and thus of the same size as the eight pMZ plasmids. The rest of the recombinants are the same in size but are larger than those obtained from *recA*$^+$. Restriction analysis of these recombinants with *Pst*I and *Bst*EII showed that all the recombinants isolated from *recA*$^+$ and the three of those isolated from *recA*$^-$ identical in size to those from *recA*$^+$ were type-I recombinants (see Fig. 5), indicating that these recombinants resulted from homologous recombination between the IS*1* elements carried by the parental plasmids. Among the 40 recombinants that were isolated from *recA*$^-$ but that showed larger size, 4 were found to be type-II recombinants, whereas 36 were type-III recombinants, indicating that these recombinants were formed by transposition of either one of the parental plasmids into a recipient plasmid, generating a duplication of the IS*1* sequence. This results in plasmids that are larger than the recombinants isolated from *recA*$^+$. Figure 6 shows the results of gel electrophoresis of DNAs of typical recombinants that were digested with *Bst*EII and *Pst*I.

These results suggest that in *recA*$^+$ the increase in the frequency of recombination between pSM1 and pMZ71, when compared with that between pHS1 and pSM1, is due to a relatively high frequency of homologous recombination between the IS*1* elements in pMZ71 and pSM1. In *recA*$^-$, the higher frequency of cointegration between pMZ71 and pSM1, compared with that between pHS1 and pSM1, is explained by the higher frequency of formation of the type-III recombinants, compared with that of type II, since the transposition of pMZ71 into pSM1 probably contributes significantly to the frequency. It is interesting to note that cointegration frequencies depend on which plasmid vehicle carries IS*1* and on which recipient plasmid is used, as discussed in a later section.

We believe that the experiments described in this section provide a model to interpret the Hfr formation by integration of the F plasmid into a host chromosome. Heteroduplex analysis of F′ plasmid molecules has shown that IS*2*, IS*3*, or the $\gamma\delta$ sequence is always at the junctions between the F genome and the host chromosome (Ohtsubo et al. 1974; Davidson et al. 1975; Hu et al. 1975a; Deonier and Davidson 1976). It has been theorized that F is integrated into a host chromosome to form Hfr cells by using the homology of an insertion sequence originally present in both genomes (Ohtsubo et al. 1974; Hu et al. 1975a; Ohtsubo and Hsu 1978). The results obtained from the study of cointegration between two plasmids carrying IS*1* support the theory above, since IS*1*-IS*1* recombination occurred between two plasmid genomes. However, as demonstrated above and in the previous section, our observation of recombination occurring at a site where no insertion sequence originally existed suggests that some Hfr molecules (e.g., HfrOR7 and HfrOR54; see Hu et al. [1975a]) could be formed at a chromosomal site by transposition of F, resulting in the duplication of an insertion sequence within F. In particular, this type of cointegration could occur in a *recA*$^-$ strain in which IS*1*-IS*1* recombination was found to occur with a much lower frequency than that of cointegration accompanied with the duplication of IS*1*.

Analysis of Cointegration between pHS1 and the Second Plasmid ColE1: Discovery of an Insertion Sequence, IS*102*, Residing in pHS1

We studied the recombination between pHS1 and ColE1 (6.65 kb), both of which are supposed to have no

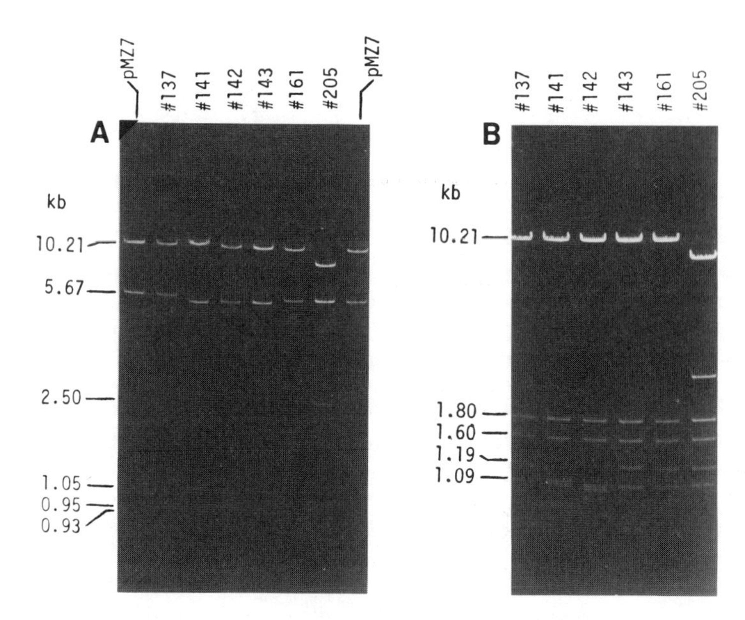

Figure 6. Agarose gels (0.7%) showing *Bst*EII (*A*) and *Pst*I (*B*) digests of various cointegrate plasmids isolated from the parental plasmids pSM1 and pMZ71 (pHS1::IS*1*), except for pMZ7, which has been derived by cointegration between pHS1 and pSM1 and carries two IS*1* elements in a direct orientation. No. 137 was the type-I cointegrate formed in *recA*$^+$; nos. 141, 142, and 143 were the type-III cointegrates formed in *recA*$^-$. No. 161 is the type-I cointegrate formed in *recA*$^-$, and no. 205 is the type-II cointegrate formed in *recA*$^-$. Note that the gel bands of the type-I cointegrates show the same pattern as that of pMZ7 carrying a direct repeat of two copies of IS*1*. The type-III cointegrates lack one of the pSM1 bands in gel *A;* the 5.67-kb fragment is pSM1 DNA. In gel *B,* the 1.80-kb, 1.60-kb, 1.19-kb, and 1.09-kb fragments are from pSM1 DNA and generate two unique extra fragments in each cointegrate, when compared with the pattern of pMZ7. On the other hand, the type-II cointegrate lacks the pMZ71 fragment (10.21 kb) and generates two unique extra fragments, when compared with the gel pattern of pMZ7 (see Fig. 5).

insertion sequences and no sequence homology with each other. We found, however, that recombination also occurred between them at a very low frequency after selection for Tc resistance at 42 °C. Extensive analysis of the physical structures of three of these recombinant plasmids, using restriction endonucleases and the electron microscope heteroduplex method, revealed that the plasmid pHS1 was integrated into different sites of ColE1 (H. Ohtsubo et al. 1980). The recombinant plasmids contained a duplication of a unique 1-kb sequence of pHS1 in a direct orientation at the junctions between the two parental plasmid sequences. In Figure 7, the structure of one recombinant, pHO1, is shown. The location of the 1-kb sequence has been mapped on pHS1 by electron microscopy. This was confirmed by comparing the nucleotide sequences of the recombinants and their parental plasmids. Nucleotide sequence analysis further revealed that a 9-bp sequence at the site of recombination in ColE1 was duplicated at the junction of each of the 1-kb sequences (see Fig. 7). The formation of recombinants is independent of *recA* function. On the basis of our previous finding that a plasmid containing a DNA IS element can recombine with a second plasmid to generate a duplication of the IS element, we conclude that the 1-kb sequence is an insertion sequence, which we designated IS*102.* For convenience, we have also denoted the IS*102* sequence as $\eta\theta$ to assign the orientation of the sequence. Eighteen nucleotides at one end (η end) were found to be repeated in an inverted orientation at the other end (θ end) of IS*102* (H. Ohtsubo et al. 1980).

Nomura et al. (1978) and Oka et al. (1978) reported that a transposable element, Tn*903,* contained inverted repeats of about 1 kb flanking the gene responsible for resistance to kanamycin. It is interesting that the nucleotide sequence at one end (η end) of IS*102* was found to be the same as those at both ends of Tn*903.* From this observation, it is reasonable to assume that IS*102* is the same as the inverted-repeat sequence in Tn*903.* In fact, this sequence is similar to the stem sequence of Tn*903,* which has been sequenced by A. Oka et al. (pers. comm.) and by Grindley and Joyce (this volume), although a few of the nucleotide sequences were different.

Although we have used pHS1 in our previous studies of recombination, we have never observed the involvement of the IS*102* present in pHS1; only the IS*1* present in a plasmid was involved in the formation of recombinant plasmids. This is probably due to the frequency of transposition, which is much higher when using IS*1* than when using IS*102.* To support this idea, we have examined the cointegration between pHS1::IS*1* (i.e., pMZ71; see previous section) and ColE1. The recombination frequency was about 1000 times higher than that for pHS1 and ColE1 (see Table 2). The resulting cointegrates examined were exclusively IS*1*-mediated cointegrates, determined by cleavage analysis with restriction endonucleases, such as *Pst*I and *Bst*EII, as described previously. Note here that pHS1 can be cleaved with the restriction endonuclease *Sst*II or *Sma*I inside the IS*102* sequence. Therefore, *Sst*II and *Sma*I are useful in identifying and characterizing the cointegrated plasmids mediated by IS*102,* using the same principle as that for the identification of IS*1*-mediated cointegrates.

Ohsumi et al. (1978) have reported a case of in vivo recombination between the filamentous phage f1 and the plasmid pSC101. Restriction endonuclease and sequence analyses of two of these derivatives suggest that the recombinant contained the two genomes flanked by a 200-nucleotide sequence in a direct orientation. This sequence, thought to be present in pSC101, generates a 5-base direct repeat at the f1 target site (Fischhoff et al. 1980). Using pHS1, a temperature-sensitive mutant of pSC101, we have not found this sequence to be specifi-

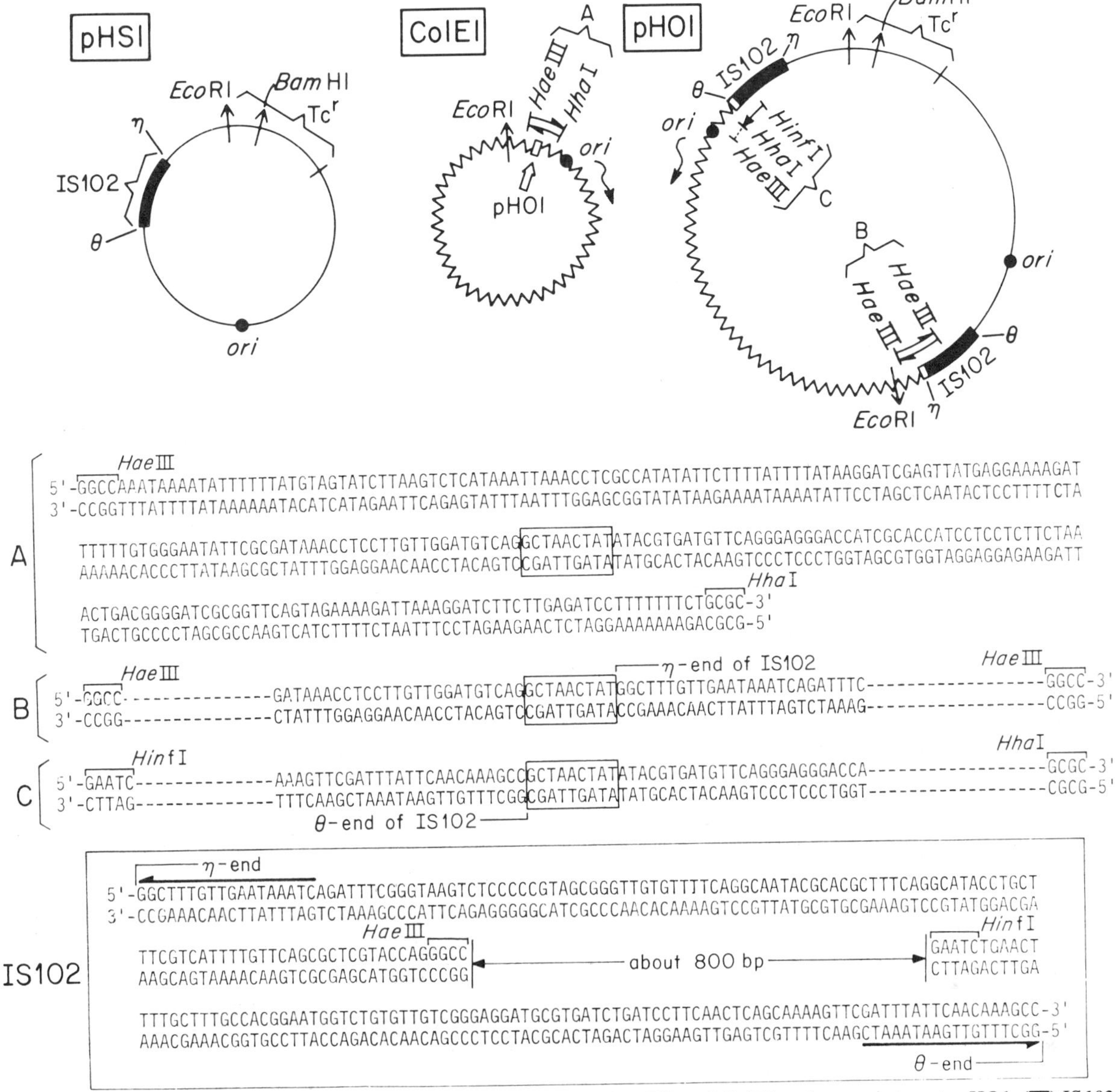

Figure 7. (*Top*) Physical structures of the circular parental plasmids pHS1 and ColE1 and one cointegrate pHO1. (■) IS*102* (or ηθ) sequence, located 1.33 kb away from the *Eco*RI site (H. Ohtsubo et al. 1980). pHO1 was formed by transposition of pHS1 at IS*102* into ColE1 (open arrow). Cleavage sites on these plasmids for the restriction endonucleases *Eco*RI and *Bam*HI are indicated by arrows through the circles. Approximate locations of the origins of replication of the parental and cointegrate plasmids and the direction of replication of ColE1 are shown (Inselburg 1974; Lovett et al. 1974; Tomizawa et al. 1974; Cabello et al. 1976). Locations of the DNA fragments prepared for sequencing by restriction endonucleases such as *Hae*III, *Hin*fI, and *Hha*I are indicated by *A* on ColE1 and by *B* and *C* on pHO1. The DNA strands actually sequenced by the Maxam-Gilbert method (1977) are indicated by solid arrows parallel to the circles. (*Bottom*) The complete nucleotide sequence of region *A* and the nucleotide sequences of the critical regions *B* and *C* are summarized. The nucleotide sequence of IS*102* was determined as shown. The target site for cointegration was a 9-base sequence, which appeared as direct repeats in the pHO1 plasmid.

cally involved in the formation of pHO plasmids; IS*102*, present originally in pHS1, is the only one involved.

Mutations Occurring Inside IS*1* Reduce or Abolish the Formation of Cointegrated Plasmids

In the preceding section, we described a genetic system used to study plasmid cointegration mediated by IS*1* and showed that cointegration and IS*1* insertion events share common molecular mechanisms. Thus, analysis of cointegration is a useful approach in the study of the insertion of IS*1* itself. This section deals with the functional analysis of the IS*1* sequence by mutagenesis of specific sites within IS*1*. The isolation of the IS*1* mutants and their resulting effects on the cointegration ability using the plasmid-cointegration system are discussed.

The genetic cointegration system used here is slightly

Table 2. Effect of Mutations Occurring in the IS*1* Sequence of pHS1::IS*1* on Cointegration with ColE1

Plasmid[a]	Cointegration frequency/ division cycle[b] (%)	No. of examined cointegrates mediated by IS*1*(or mutant IS*1*)[c]	IS*102*[c]
pHS1	1.1×10^{-9}	0	22
pHS1::IS*1*	7.3×10^{-7}	30	0
pHS1::IS*1-1*	6.9×10^{-10}	0	11
pHS1::IS*1-2*	6.7×10^{-10}	5	5
pHS1::IS*1-3*	7.0×10^{-9}	7	16
pHS1::IS*1-7*	$<4.8 \times 10^{-10}$	5	8
pHS1::IS*1-14*	$<4.1 \times 10^{-10}$	1	2
pHS1::IS*1-13*	$<4.5 \times 10^{-10}$	0	6

[a] Mutant plasmids having deletions in the IS*1* sequence; the locations of deletion mutations are shown in Fig. 8.

[b] The frequencies were obtained as described in Table 1. We used JE5507 *recA*$^+$ as a host and 25 μg/ml Tc plates for selection in all the present experiments.

[c] IS*1*- or mutant-IS*1*-mediated cointegrates were characterized by cleavage analysis with the restriction endonuclease *Pst*I or *Bst*EII; IS*102*-mediated cointegrates were readily characterized by cleavage analysis with the restriction endonuclease *Sst*II or *Sma*I, which cleave IS*102* present in the pHS1 sequence.

different from that described previously. In this system, we used two plasmids, pHS1::IS*1* (i.e., pMZ71) and ColE1. pHS1::IS*1* is the derivative of pHS1 that carries a copy of IS*1* and is still temperature-sensitive for replication. This plasmid can form cointegrated plasmids with ColE1 by selection on Tc at 42 °C at a frequency as shown in Table 2. The cointegration was mediated exclusively by the IS*1* in pHS1::IS*1*.

The plasmid pHS1::IS*1* contains a single *Pst*I site and a single *Bst*EII site, both within IS*1*. We obtained deletion mutants at these restriction endonuclease sites by nuclease-S1 treatment followed by ligation. Figure 8 represents the mutants isolated and the locations of the deletion mutations. Note that with nuclease-S1 and ligase treatment, mutants with large deletions, some of which extend into either end of the IS*1* sequence, were also obtained.

Using these mutants, we have analyzed the frequency of the formation of cointegrates between each pHS1::IS*1* deletion mutant and ColE1 and compared it with the frequency of cointegration between pHS1::IS*1* and ColE1. As shown in Table 2, for each mutant the frequency was 100 to 1000 times lower than that for pHS1::IS*1*. The Tc^r cells isolated contained large plasmids, which were possible cointegrates.

We examined these large plasmids to see whether they were IS*1*-mediated, mutant IS*1*-mediated, or IS*102*-mediated cointegrates. As shown in Table 2, when the large plasmids from cells that carried ColE1 and the mutants deleting either end of the IS*1* sequence (pHS1::IS*1-1* and pHS1::IS*1-13*) were examined using several restriction endonucleases, all the large plasmids were found to be cointegrates mediated by the IS*102* sequence. No IS*1*-mediated cointegrate was seen, implying that the ends of IS*1* are vital for IS*1*-mediated cointegration. When large plasmids from cells that carried ColE1 and any one of the other pHS1::IS*1* mutants having a deletion within IS*1* were examined, we observed that about half of the large plasmids were formed by cointegration mediated by IS*102* and half were formed by cointegration mediated by the mutant IS*1* (see Table 2). It should be emphasized here that the *Escherichia coli* K12 strain JE5507 used for the experiments contained ten copies of the IS*1* sequence (Nyman et al. 1981). Therefore, the result of cointegrate formation mediated by mutant IS*1* elements in cells of JE5507 is consistent with the hypothesis that IS*1* codes for a protein(s) required for its own translocation, as predicted by its nucleotide sequence (Ohtsubo and Ohtsubo 1978; Johnsrud 1979); namely, the protein products from the IS*1* elements in the host chromosome complemented the above deletion mutants of IS*1* at a very low efficiency. It is possible that the protein(s) encoded by the IS*1* element is a *cis*-acting protein, similar to the gene-*A* protein of the phage φX174, which shows poor complementation efficiency in vivo (Sinsheimer 1968) perhaps due to favored binding of the mutant protein to the replication origin of φX174 DNA as discussed by Kornberg (1980).

Cointegration Mediated by Tn*3*: Only Repressor-negative Mutants of Tn*3* Can Form Stable Cointegrate Plasmids with Two Tn*3* Sequences

In view of our knowledge concerning IS*1*-mediated and IS*102*-mediated cointegration, it is of great interest to determine whether the transposable element Tn*3* also mediates cointegration and to determine which Tn*3* mutations affect this recombination.

As shown schematically in Figure 9, Tn*3* (4957 bp) contains regions encoding transposase (*tnpA*), repressor (*tnpR*), and β-lactamase (*bla*). It also contains a 38-bp inverted repeat at both ends (IR-R and IR-L) (Heffron et al. 1979; H. Ohtsubo et al. 1979). Mutants of Tn*3* with deletions and/or insertions in these regions are available (see Fig. 9). Tn*3* and all of its mutants are carried in plasmid pMB8 (Heffron et al. 1977, 1979).

Using the system described previously, we have examined recombination between pHS1 and pMB8 carrying wild-type and mutant Tn*3* elements. Table 3 sum-

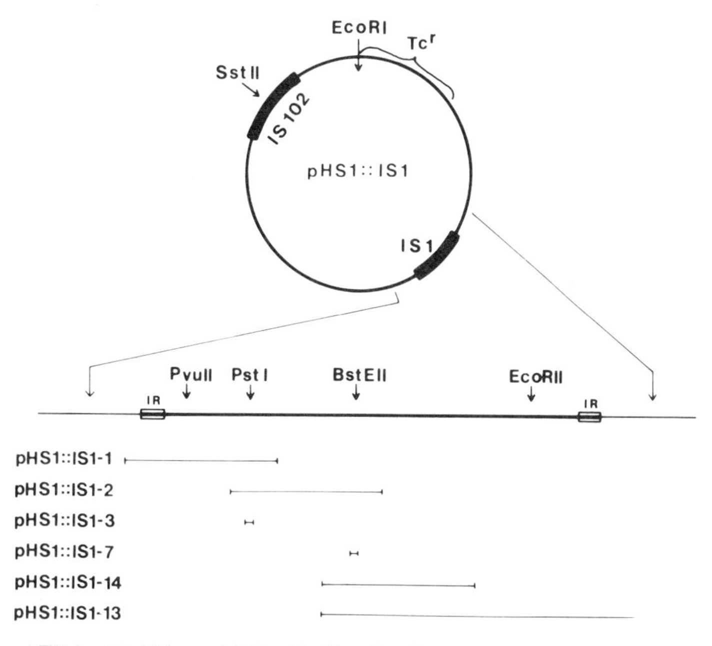

Figure 8. Physical structure of pHS1::IS*1* (i.e., pMZ71) and derivatives with deletions inside IS*1*. Approximate sizes and locations of deletions in independent mutants are indicated at the bottom of the figure, where the expanded IS*1* region is shown. The deletion mutants were isolated as follows. pMZ71 DNA (2 μg) was completely digested with *Pst*I (BRL) or *Bst*EII (BioLabs). DNA was extracted with phenol and precipitated with ethanol after six ether treatments. The DNA precipitated was suspended in an appropriate volume of distilled water and digested with nuclease S1 (Miles). The reaction mixture (30 μl) for nuclease-S1 digestion contained 30 mM sodium acetate buffer (pH 4.6), 4 mM $ZnCl_2$, 0.3 M NaCl, and 400–1200 units of nuclease S1. The reaction mixture was incubated at 25 °C for 60 min, and 1/10 volume of 1 M Tris-HCl (pH 8.0) and 1/20 volume of 0.25 M EDTA were added. After phenol and ether treatments, the sample was dialyzed against 10 mM Tris-HCl (pH 8.0) and 0.1 mM EDTA. The DNA was precipitated with ethanol and suspended in 10 mM Tris-HCl (pH 8.0) and 0.1 mM EDTA. Ligation of blunt-end molecules so prepared was carried out in 20 μl of 60 mM Tris-HCl (pH 8.0), 10 mM $MgCl_2$ 10 mM dithiothreitol, and 1 mM ATP for 12–16 hr at 14 °C with 10 units of T4 DNA ligase (BRL). After incubation, the reaction mixture was diluted with an appropriate volume of distilled water, and then the DNA solution was used for transformation of *E. coli* C600. After selection for Tc resistance, individual transformants were picked. The plasmid DNAs were isolated and examined for absence of the restriction sites.

marizes the frequencies of recombination between the plasmids and the physical characteristics of resulting recombinants isolated from Tc^r cells grown at 42 °C. The results can be summarized as follows:

1. Repressor-negative mutants (*tnpR*$^-$) of Tn*3* mediated cointegration with pHS1 at a very high frequency. Almost all cells harboring pHS1 and pMB8::Tn*3* *tnpR*$^-$ contained recombinants in the absence of any selection. The plasmids isolated from independent Tc^r cells were the same in size. Figure 10 shows the results of gel electrophoresis of plasmid DNA

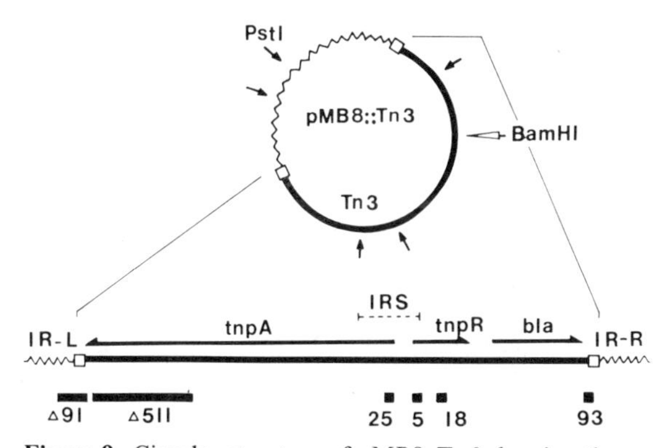

Figure 9. Circular structure of pMB8::Tn*3* showing the sequences of pMB8 and Tn*3* and the approximate locations of cleavage sites for the restriction endonucleases *Pst*I and *Bam*HI. The Tn*3* sequence is expanded at the bottom and is shown by the functional genes such as *tnpA, tnpR, bla,* IRS, IR-R, and IR-L (see text). Locations of mutations occurring in the Tn*3* sequence of pMB8::Tn*3* are indicated below the normal Tn*3* sequence.

cleaved with restriction endonuclease *Pst*I, indicating that the plasmids contain a duplication of the Tn*3* sequence in a direct orientation at the junctions of pHS1 and pMB8.

2. The Tc^r colonies that were independently isolated from cells harboring pHS1 and pMB8::Tn*3* (wild type) contained plasmids heterogeneous in size, larger than parental plasmids, but smaller than the cointegrated plasmids mediated by the Tn*3* *tnpR*$^-$ mutants. All of the plasmids contained all or part of the sequence of their parent, but, unlike the cointegrates mediated by Tn*3* *tnpR*$^-$, they contained only one intact Tn*3* (see Fig. 10). Note also that the frequency of recombinant formation here is lower when compared with that of cointegrates mediated by Tn*3* *tnpR*$^-$.
3. The mutant Tn*3* (no. 25) behaves like wild-type Tn*3*, although the location of the mutation has been mapped in the N-terminus coding region of the transposase gene (see Fig. 9). The frequency of the generation of heterogeneous sizes of recombinants was slightly lower than that for wild-type Tn*3*.
4. When we used pMB8 carrying Tn*3* mutants Δ511 and Δ91, having deletions in the middle or C-terminus coding region of the transposase gene, the resulting recombinants examined were all cointegrated plasmids whose formations were mediated by the IS*102* sequence. This confirms that the transposase is essential not only for Tn*3* transposition, but also for cointegration mediated by Tn*3*. The mutant Δ91 also deletes one end (IR-L) of Tn*3*. This indicates IR-L is vital for plasmid cointegration.
5. The mutant Tn*3* (no. 93) deletes within the inverted-repeat sequence IR-R but retains 14 bp of the origi-

Table 3. Frequency of Recombination between pHS1 and pMB8 Carrying Tn*3* or Its Mutants

Mutant[a]	Mutation deleted in[a]	Frequency/ division cycle[b]	Type of recombinants[c]
Tn*3*	wild type	4.5×10^{-6}	various sizes; carry one Tn*3*
91	*tnpA;* IR-L	2.2×10^{-9}	none
511	*tnpA*	1.2×10^{-9}	none
No. 93	IR-R	7.3×10^{-9}	various sizes; carry one mutant Tn*3*
No. 25	*tnpA*	6.3×10^{-7}	various sizes; carry one mutant Tn*3*
No. 5	*tnpR*	10^{-1}	stable cointegrates; carry two mutant Tn*3*
No. 18	*tnpR*	10^{-1}	stable cointegrates; carry two mutant Tn*3*

[a] Mutants of Tn*3* and the locations of the mutations are summarized in Fig. 9.

[b] Frequencies were determined as described in Table 1. JE5519 *recA*⁻ was used as a host strain. Selective plates contained 10 μg/ml of Tc and no ampicillin.

[c] Δ91 and Δ511 gave rise to cointegrates mediated by only the IS*102* sequence present in pHS1, although in this table it is stated as none. In the no. 93 mutant, about one third of the recombinants were cointegrates mediated by the IS*102* sequence. Recombinants that were not mediated by IS*102* were analyzed with the restriction endonucleases *Pst*I, *Eco*RI, and *Bam*HI.

nal 38-bp sequence at the end (Heffron et al. 1979). The frequency of large-plasmid formation was markedly decreased (see Table 3). Analysis of the large plasmids showed that about one third of them were cointegrates mediated by the IS*102* sequence but that the rest were recombinants of heterogeneous size carrying one copy of Tn*3*. This result suggests that even when the inside of the inverted-repeat sequence of IR-R is deleted, the mutant Tn*3* still retains the ability to generate recombinants, but with a very low efficiency compared with that of wild-type Tn*3*.

An interesting question is why only repressor-negative Tn*3* mediates the formation of stable cointegrates. We wish to interpret the results as follows: Wild-type Tn*3* and mutants such as Tn*3* (no. 25) *tnpA*⁻ and Tn*3* (no. 93) ΔIR-R can probably mediate the formation of cointegrated plasmids containing two direct repeats of Tn*3*, as seen in the case of Tn*3* *tnpR*⁻ mutants. However, the *tnpR*⁺ cointegrated plasmids may be unstable, with two copies of Tn*3* on the same genome, probably due to some effect of the repressor molecules. A similar interpretation of repressor action is described by Sherratt et al. (this volume). Thus, when we select the Tc-resistance gene on pHS1 and the temperature-resistance replication gene of pMB8, one copy or at least a portion of one copy of Tn*3* was eliminated from the selected recombinants containing two sites of action of repressor molecules by a spontaneous deletion to achieve stability. This provides an explanation for the observed heterogeneous sizes of recombinants, smaller than cointegrated plasmids with two Tn*3* *tnpR*⁻.

Heffron et al. (1979 and this volume) proposed that Tn*3* contains a site, called the internal resolution site (IRS), that would resolve cointegrated plasmids with two copies of Tn*3* by a specific recombination within Tn*3*. The result would be two plasmids, each carrying a single copy of Tn*3*. If the repressor is responsible for resolution, the IRS could be the site of action for repressor molecules. The repressor is thought to act on the promoter and operator regions for both the transposase gene and the repressor gene itself (Chou et al. 1979; Gill et al. 1979). Therefore, it is likely that the IRS, if it exists, could be the promoter and operator region that regulates the transcription of the two genes.

Mechanism of Cointegration

To explain how cointegrated plasmids carrying duplications of IS*1* are formed between pHS1 and pSM1, we have proposed the following two models (E. Ohtsubo et al. 1980). One model is based on the fact that the insertion of IS*1* generates a 9-base repetition at the site of insertion (Calos et al. 1978; Grindley 1978). Thus, if insertion of IS*1* first occurred into pHS1, recombination could then take place at the region of IS*1* homology, yielding the cointegrated plasmids carrying two copies of IS*1* and the duplication of the 9-bp sequence at the target site. We think that this is one possible pathway in the formation of cointegrated plasmids, since we have observed homologous recombination between IS*1* elements in the study of cointegration between two plasmids, both carrying IS*1*. The frequency of recombination between two IS*1* elements was high in *recA*⁺, whereas it was very low in *recA*⁻. The frequency in *recA*⁻ is even lower than the frequencies of cointegration, suggesting that the proposed pathway is likely, but almost negligible, particularly in *recA*⁻, unless the frequency of an IS*1* insertion into the plasmid is extremely high.

The formation of the recombinants between pSM1 and pHS1 may also be explained by the following simple model. This model is based on the fact that segments of DNA flanked by a direct repeat of IS*1* are transposable, as demonstrated by Tn*9* (MacHattie and Jackowski 1977) and the r-determinant region of R100 (Arber et al. 1979). The plasmid containing IS*1* (e.g., pSM1) may be generating structures similar to the transposon by polymerization of the genome, which can then act as a transposon, since the polymers contain direct repeats of IS*1* flanking the replication region of pSM1. This could also generate the 9-base repetition upon insertion into pHS1. It is noteworthy that Tn*9*, which contains a direct repeat of IS*1* at its two ends, has

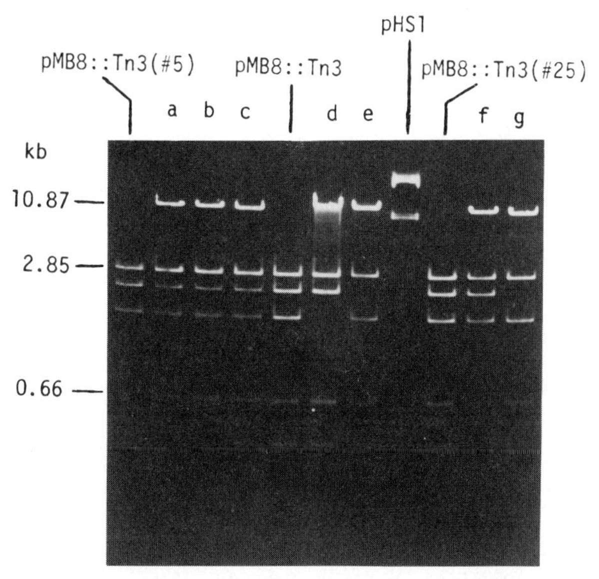

Figure 10. Agarose gels (0.7%) showing *Pst*I digests of recombinants of pHS1 with pMB8 carrying wild-type Tn*3* or mutant Tn*3*. (pHS1 DNA cannot be cleaved with *Pst*I; therefore, in the gel it appears as open and closed circles, but also as linear molecules in a small amount in between open and closed circles.) (*a–c*) Cleavage results of the independent recombinants formed between pHS1 and pMB8::Tn*3* (no. 5) defective in *tnpR* (see Fig. 9). Recombinants were isolated between pHS1 and pMB8::Tn*3* (wild type) (*d, e*) and between pHS1 and pMB8::Tn*3* (no. 25) (*f, g*) having a defect in the N terminus of the *tnpA* gene (see Fig. 9). In *a* through *c,* all the recombinants show the identical gel-band patterns (six bands); five fragments are identical with *Pst*I-digested pMB8::Tn*3* (no. 5), although they showed different patterns when digested with the other enzymes, such as *Eco*RI and *Bam*HI, which cleave both parental plasmids (data not shown). Note that *Pst*I cleaves the Tn*3* sequence at three sites and generates two fragments, 2.85 kb and 0.66 kb, from inside Tn*3* (Heffron et al. 1979) and that in the gel the bands of these fragments appear in a double amount, as judged by the relative brightness of the bands. These results suggest that these recombinants were cointegrates with two copies of Tn*3* (no. 5) in a direct orientation between the pHS1 and pMB8 sequences. In *d* through *g,* on the other hand, independent recombinants were heterogeneous in size and smaller than those formed between pHS1 and pMB8::Tn*3* (no. 5). They show the different patterns of gel bands (five fragments) when compared among themselves. The two Tn*3*-specific fragments generated from these recombinants appear as singlets, suggesting that they contain one copy of the Tn*3* sequence.

been shown to generate 9-base duplications (Johnsrud et al. 1978). This model suggests that any plasmid containing a translocatable DNA element can be a transposon and thereby transpose its entire genome to another different genome. A criterion crucial for this model would then be whether a polymer structure is formed in *recA*$^+$ and *recA*$^-$ backgrounds. Dimer structures of pSM1 and ColE1 have commonly been observed to form in *recA*$^+$ strains. In this case, poor cointegration frequency in *recA*$^-$ may suggest that the frequency of polymerization of pSM1 in *recA*$^-$ may be very low.

Neither model, however, explains the actual molecular mechanism of the IS*1* insertion and IS*1*-mediated cointegration, which generate repetition of 9 bp at the target site. Shapiro (1979) recently proposed a model for the mechanism of insertion in which the cointegrated genome is an intermediate that eventually segregates into one complete donor-genome sequence carrying the insertion sequence and a second genome (recipient) that acquires the same insertion sequence. His model is based on the assumption that the duplication of a target-site sequence and the insertion sequence (or transposable element) occurs by DNA synthesis. Our results, obtained by systematic analysis, including DNA sequencing, demonstrate that such cointegrate structures exist. This seems to support Shapiro's hypothesis. However, the hypothesis that these cointegrates are intermediates in the transposition of DNA elements is still in question, since the recombinant pMZ plasmids carrying two copies of IS*1* in a direct orientation are very stable in *recA*$^+$ as well as in *recA*$^-$ strains and do not generate segregants, even when cultured successively for several days.

We propose that the initial steps of both insertion of IS*1* and cointegration mediated by IS*1* share common molecular mechanisms. These mechanisms involve the assembly of the two inverted-repeat ends of IS*1* at a target site: Nicking and joining between donor and recipient strands give rise to a staggered cut at the target-site sequence, with each end of the insertion element covalently linked to each end of the staggered cut. This results in the formation of a forked molecule containing the donor and recipient linked together as shown in Figure 11A. If displacement DNA synthesis occurs in the $5' \rightarrow 3'$ direction from the $3'$ OH of the primer to completion, a cointegrated molecule will be found, as suggested by Shapiro (1979). If DNA synthesis occurs, but terminates prematurely, and then the strands of the donor molecule (initially carrying the translocatable element) are cleaved, the resulting molecule will be the recipient carrying the translocatable element. A critical point of this model is that the insertion event is not reciprocal, whereas cointegration is reciprocal, as shown in Figure 11. To date, no proof of reciprocity in the insertion event is available.

DISCUSSION

We have described a simple genetic system to study cointegration mediated by IS*1*. Analysis of cointegrated plasmids by various techniques, including nucleotide sequencing, revealed that cointegration mediated by IS*1* and insertion of IS*1* share common molecular mechanisms. The system is also useful for the study of cointegration mediated by Tn*3* and can be applied to the study of the other known insertion and transposable elements. Since the selection of cointegrated plasmids in our system relies on the markers present in the parental plasmids and not those in IS*1* or Tn*3,* the system is useful in the study of IS*1*, which bears no resistance genes. Furthermore, this system allows us to screen for new insertion sequences and, in fact, has resulted in the discov-

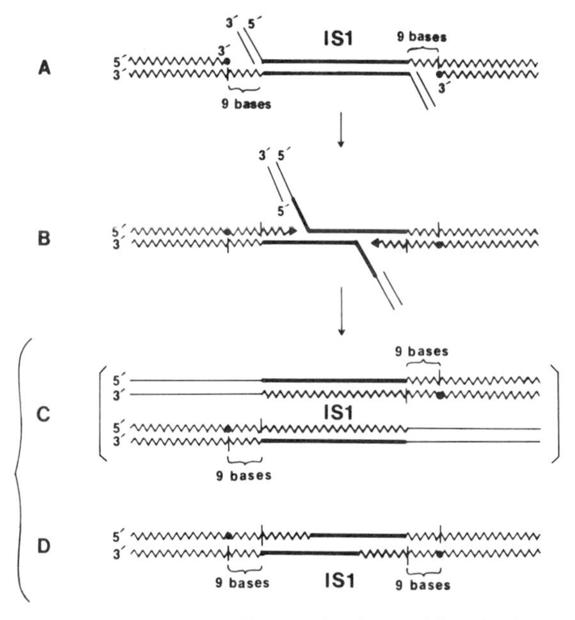

Figure 11. (*A*) Intermediate molecule resulting from recombination of a donor plasmid (——) containing the IS*1* sequence (————) with a recipient plasmid (~~~~) (see text). The donor and recipient plasmids are actually circular, but only portions of them are shown here. Note that a 9-bp sequence in the recipient strands has been subjected to a staggered cut. (*B*) Structure formed when displacement DNA synthesis proceeds from the 3′ ends of the staggered cut of the recipient strands in the 5′→3′ direction across IS*1*. (*C*) Cointegrate structure formed when the DNA synthesis proceeds completely across IS*1* and the newly synthesized strands are joined with the 5′ ends of the donor strands. (*D*) Recipient molecule, which received an IS*1* insertion. This molecule could be formed from the molecule in *B* by removing donor strands during displacement DNA synthesis. Note that in both *C* and *D* a 9-base duplication has been generated.

ery of an insertion sequence, IS*102*, present in pHS1 (i.e., pSC101). A further advantage is that the frequency of cointegration mediated by an element can be assayed by a fluctuation test, enabling us to analyze the effects on cointegration of mutations in the host chromosome or in the element itself. Using this system, further analysis on the genetic and molecular level of cointegration mediated by IS*1* and Tn*3* will extend the knowledge of the molecular events involved in cointegration, as well as in insertion of these translocatable elements. Currently, we are attempting to use the cointegration system between pHS1 and pMB8 carrying repressor-negative Tn*3* to establish an in vitro cointegration system.

ACKNOWLEDGMENTS

We thank F. Heffron for the generous gift of Tn*3* mutants, A. Oka for sending his results on Tn*903* before publication, and D. Davison for help with manuscript preparation. This work was supported by U.S. Public Health Service research grants GM-22007 to E. O. and GM-26779 to H. O. and partly by support from National Institutes of Health training grant CA-09176.

REFERENCES

Arber, W., S. Iida, H. Jütte, P. Caspers, J. Meyers, and C. Hanni. 1979. Rearrangements of genetic material in *Escherichia coli,* as observed on the bacteriophage P1 plasmid. *Cold Spring Harbor Symp. Quant. Biol.* **43:** 1197.

Blattner, F. R., B. G. Williams, A. E. Blechl, K. Denison-Thompson, H. E. Fabor, L.-A. Furlong, D. J. Grunwald, D. O. Kiefer, D. D. Moore, J. W. Schumm, E. L. Sheldon, and O. Smithies. 1977. Charon phages: Safer derivatives of bacteriophage lambda for cloning. *Science* **196:** 161.

Bukhari, A. I., J. A. Shapiro, and S. L. Adhya, eds. 1977. *DNA insertion elements, plasmids, and episomes.* Cold Spring Harbor Laboratory, Cold Spring Harbor, New York.

Cabello, F., K. Timmis, and S. N. Cohen. 1976. Replication control in a composite plasmid constructed by *in vitro* linkage of two distinct replicons. *Nature* **254:** 285.

Calos, M. P., L. Johnsrud, and J. H. Miller. 1978. DNA sequence at the integration sites of the insertion element IS*1*. *Cell* **13:** 411.

Chou, J., P. G. Lemaux, M. J. Casadaban, and S. N. Cohen. 1979. Transposition protein of Tn*3:* Identification and characterization of an essential repressor-controlled gene product. *Nature* **282:** 801.

Cohen, S. N., M. J. Casadaban, J. Chou, and C.-P. D. Tu. 1978. Studies on the specificity and control of transposition of the Tn*3* element. *Cold Spring Harbor Symp. Quant. Biol.* **43:** 1247.

Davidson, N., R. C. Deonier, S. Hu, and E. Ohtsubo. 1975. Electron microscope heteroduplex studies of sequence relations among plasmids of *Escherichia coli.* X. Deoxyribonucleic acid sequence organization of F and of F-primes, and the sequences involved in Hfr formation. In *Microbiology—1974* (ed. D. Schlessinger), p. 56. American Society for Microbiology, Washington, D.C.

Deonier, R. C. and N. Davidson. 1976. The sequence organization of the integrated F plasmid into two Hfr strains of *Escherichia coli. J. Mol. Biol.* **107:** 207.

Fiandt, M., W. Szybalski, and M. H. Malamy. 1972. Polar mutations in *lac, gal,* and phage λ consist of a few IS DNA sequences inserted with either orientation. *Mol. Gen. Genet.* **119:** 223.

Fischhoff, D. A., G. F. Vovis, and N. D. Zinder. 1980. Organization of chimeras between filamentous bacteriophage f1 and plasmid pSC101. *J. Mol. Biol.* **144:** 246.

Gill, R. E., F. Heffron, and S. Falkow. 1979. Identification of the protein encoded by the transposable element Tn*3* which is required for its transposition. *Nature* **282:** 797.

Gill, R., F. Heffron, G. Dougan, and S. Falkow. 1978. Analysis of the sequences transposed by complementation of two classes of transposition-deficient mutants of Tn*3*. *J. Bacteriol.* **136:** 742.

Grindley, N. D. F. 1978. IS*1* insertion generates duplication of a nine base pair sequence at its target site. *Cell* **13:** 419.

Hashimoto-Gotch, T. and M. Sekiguchi. 1976. Isolation of temperature-sensitive mutants of R plasmid by *in vitro* mutagenesis with hydroxylamine. *J. Bacteriol.* **127:** 1561.

Heffron, F., M. So, and B. McCarthy. 1978. *In vitro* mutagenesis of a circular DNA molecule by using synthetic restriction sites. *Proc. Natl. Acad. Sci.* **75:** 6012.

Heffron, F., P. Bedinger, J. Champoux, and S. Falkow. 1977. Deletions affecting transposition of an antibiotic resistance gene. In *DNA insertion elements, plasmids, and episomes* (ed. A. I. Bukhari et al.), p. 161. Cold Spring Harbor Laboratory, Cold Spring Harbor, New York.

HEFFRON, F., B. McCARTHY, H. OHTSUBO, and E. OHTSUBO. 1979. DNA sequence analysis of the transposon Tn*3*: Three genes and three sites involved in transposition of Tn*3*. *Cell* **18:** 1153.

HIRSCH, H.-J., P. STARLINGER, and P. BRACHET. 1972. Two kinds of insertions in bacterial genes. *Mol. Gen. Genet.* **119:** 191.

HU, S., E. OHTSUBO, and N. DAVIDSON. 1975a. Electron microscope heteroduplex studies of sequence relations among plasmids of *Escherichia coli:* Structure of F13 and related F-primes. *J. Bacteriol.* **122:** 749.

HU, S., E. OHTSUBO, N. DAVIDSON, and H. SAEDLER. 1975b. Electron microscope heteroduplex studies of sequence relations among bacterial plasmids. XII. Identification and mapping of the insertion sequences IS*1* and IS*2* in F and R plasmids. *J. Bacteriol.* **122:** 764.

INSELBERG, J. 1974. Replication of colicin E1 plasmid DNA in minicells from a unique replication initiation site. *Proc. Natl. Acad. Sci.* **71:** 2256.

JASKUNAS, S. R. and M. NOMURA. 1977. Mapping ribosome protein gene in *E. coli* by means of insertion mutations. In *DNA insertion elements, plasmids, and episomes* (ed. A. I. Bukhari et al.), p. 487. Cold Spring Harbor Laboratory, Cold Spring Harbor, New York.

JOHNSRUD, L. 1979. DNA sequence of the transposable element IS*1*. *Mol. Gen. Genet.* **169:** 213.

JOHNSRUD, L., M. P. CALOS, and J. H. MILLER. 1978. The transposon Tn*9* generates a 9-bp repeated sequence during integration. *Cell* **15:** 1209.

KOPECKO, D. J. and S. N. COHEN. 1975. Site-specific *recA* independent recombination between bacterial plasmids: Involvement of palindromes at the recombinational loci. *Proc. Natl. Acad. Sci.* **72:** 1373.

KORNBERG, A. 1980. *DNA replication,* p. 511. W. H. Freeman, San Francisco.

LEHMAN, I. R. and J. R. CHIEN. 1973. Persistence of deoxyribonucleic acid polymerase I and its 5′→3′ exonuclease activity in *polA* mutants of *Escherichia coli* K12. *J. Biol. Chem.* **248:** 7717.

LOVETT, M. A., D. G. GUINEY, and D. R. HELINSKI. 1974. Relaxation complexes of plasmids ColE1 and ColE2: Unique site of the nick in the open circular DNA of the relaxed complexes. *Proc. Natl. Acad. Sci.* **71:** 3854.

LURIA, S. E. and M. DELBRÜCK. 1943. Mutations of bacteria from virus sensitivity to virus resistance. *Genetics* **28:** 491.

MACHATTIE, L. A. and J. B. JACKOWSKI. 1977. Physical structure and deletion effects of the chloramphenicol resistance element Tn*9* in phage lambda. In *DNA insertion elements, plasmids, and episomes* (ed. A. I. Bukhari et al.), p. 219. Cold Spring Harbor Laboratory, Cold Spring Harbor, New York.

MAXAM, A. and W. GILBERT. 1977. A new method for sequencing DNA. *Proc. Natl. Acad. Sci.* **74:** 560.

MEYER, R., G. BOCH, and J. A. SHAPIRO. 1979. Transposition of DNA inserted into deletions of the Tn*5* kanamycin resistance element. *Mol. Gen. Genet.* **171:** 7.

MICKEL, S., E. OHTSUBO, and W. BAUER. 1977. Heteroduplex mapping of small plasmids derived from R-factor R12: *In vivo* recombination occurs at IS*1* insertion sequences. *Gene* **2:** 193.

NISEN, P. D., D. J. KOPECKO, J. CHOU, and S. N. COHEN. 1977. Site-specific DNA deletions occurring adjacent to the termini of a transposable ampicillin resistance element (Tn*3*). *J. Mol. Biol.* **117:** 975.

NOMURA, N., H. YAMAGISHI, and A. OKA. 1978. Isolation and characterization of transducing coliphage fd carrying a kanamycin resistance gene. *Gene* **3:** 39.

NYMAN, K., K. NAKAMURA, H. OHTSUBO, and E. OHTSUBO. 1981. Distribution of the insertion sequence IS*1* in gram-negative bacteria. *Nature* (in press).

OHSUMI, M., G. F. VOVIS, and N. D. ZINDER. 1978. The isolation and characterization of an *in vivo* recombinant between the filamentous bacteriophage f1 and the plasmid pSC101. *Virology* **89:** 438.

OHTSUBO, E. and M. T. HSU. 1978. Electron microscope heteroduplex studies of sequence relations among *Escherichia coli.* Isolation of a new F-prime factor, F80, and its implication for the mechanism of F integration into the chromosome. *J. Bacteriol.* **134:** 795.

OHTSUBO, E., M. ZENILMAN, and H. OHTSUBO. 1980. Plasmids containing insertion elements are potential transposons. *Proc. Natl. Acad. Sci.* **77:** 750.

OHTSUBO, E., R. C. DEONIER, H. J. LEE, and N. DAVIDSON. 1974. Electron microscope heteroduplex studies of sequence among plasmids of *Escherichia coli.* IV. The F sequences in F14. *J. Mol. Biol.* **89:** 565.

OHTSUBO, E., M. ZENILMAN, J. RIFKIN, and H. OHTSUBO. 1979. *In vivo* recombination between two plasmid genomes. In *Abstracts of the Annual Meeting of the American Society for Microbiology,* p. 124.

OHTSUBO, E., J. FEINGOLD, H. OHTSUBO, S. MICKEL, and W. BAUER. 1977. Unidirectional replication in *Escherichica coli* of three small plasmids derived from R factor R12. *Plasmid* **1:** 8.

OHTSUBO, H. and E. OHTSUBO. 1978. Nucleotide sequence of an insertion element, IS*1*. *Proc. Natl. Acad. Sci.* **75:** 615.

OHTSUBO, H., H. OHMORI, and E. OHTSUBO. 1979. Nucleotide sequence analysis of Tn*3* (Ap): Implications for insertion and deletion. *Cold Spring Harbor Symp. Quant. Biol.* **43:** 1269.

OHTSUBO, H., M. ZENILMAN, and E. OHTSUBO. 1980. An insertion element IS*102* resides in the plasmid pSC101. *J. Bacteriol.* **144:** 131.

OKA, A., N. NOMURA, K. SUGIMOTO, H. SUGISHAKI, and M. TAKANAMI. 1978. Nucleotide sequence at the insertion sites of a kanamycin transposon. *Nature* **276:** 845.

REIF, H. J. and H. SAEDLER. 1975. IS*1* is involved in deletion formation in the *gal* region of *E. coli* K12. *Mol. Gen. Genet.* **137:** 17.

ROSEN, J., H. OHTSUBO, and E. OHTSUBO. 1979. The nucleotide sequence of the region surrounding the replication origin of an R100 resistance factor derivative. *Mol. Gen. Genet.* **171:** 287.

SHAPIRO, J. A. 1979. Molecular model for the transposition and replication of bacteriophage Mu and other transposable elements. *Proc. Natl. Acad. Sci.* **76:** 1933.

SINSHEIMER, R. L. 1968. Bacteriophage ϕX174 and related viruses. *Prog. Nucleic Acid Res. Mol. Biol.* **8:** 115.

SO, M., F. HEFFRON, and B. McCARTHY. 1979. The *E. coli* gene encoding heat stable toxin is a bacterial transposon flanked by inverted repeats of IS*1*. *Nature* **277:** 453.

TOMIZAWA, J., Y. SAKAKIBARA, and T. KAKEFUDA. 1974. Replication of colicin E1 plasmid DNA in cell extracts. Origin and direction of replication. *Proc. Natl. Acad. Sci.* **71:** 2260.

Site-specific Recombination and Its Role in the Life Cycle of Bacteriophage P1

N. Sternberg, D. Hamilton, S. Austin, M. Yarmolinsky, and R. Hoess
Cancer Biology Program, Frederick Cancer Research Center, Frederick, Maryland 21701

A variety of bacteriophages, including λ, P2, P22, and Mu, encode site-specific recombination systems. The need for these systems is clear because the normal prophage DNA of these viruses is integrated into the bacterial chromosome. Integration occurs at specific sites within the phage DNA, and in the cases of λ, P2, and P22 also within the bacterial chromosome. In contrast, the prophage DNA of bacteriophage P1 is an autonomous plasmid. Consequently, the need for and the role of a P1 site-specific recombination system are not at all obvious. Despite this, the following observations suggest that P1 encodes such a system. First, the viral DNA of P1, like that of phages P22 and T4, is terminally redundant and cyclically permuted (Ikeda and Tomizawa 1969). Thus, one could expect that the genetic map of P1, like those of P22 and T4, would be circular; in fact, it is linear (Scott 1968; Walker and Walker 1975). The simplest explanation for this result is that P1 DNA contains a hotspot for genetic recombination, the location of which corresponds to the ends of the P1 genetic map. Also, P1 plasmid DNA can integrate into the bacterial chromosome of a *recA*⁻ strain with low efficiency; about 1 in every 10^5 P1-containing cells has an integrate P1 prophage (Chesney and Scott 1978; Chesney et al. 1979). The integrated P1 DNA is invariably arranged in the same way as is the DNA in the P1 genetic map, and a unique site on the bacterial chromosome is always used in the integration event in a *recA*⁻ strain. Thus, it would appear from these studies that the same recombination system that is responsible for the linearity of the P1 map is also responsible for the integration of P1 DNA into the bacterial chromosome. In this paper we describe the extent to which we have characterized this P1 site-specific recombination system and contrast its properties with those of the λ site-specific recombination system. We will also discuss the role of the recombination system in the P1 life cycle.

Demonstration That P1 *Eco*RI Fragment 7 Contains a Functional Site-specific (*lox*) Recombination System

To study P1 site-specific recombination, we inserted the P1 *Eco*RI fragment (*Eco*RI-7, 6.5 kb) that spans the ends of the P1 genetic map (Sternberg 1979b) at the single *Eco*RI site (*sr*Iλ3) in the λ*Dam**b*538*sr*Iλ3 vector (Sternberg et al. 1977). The resulting hybrid phage (λ-P1:7) was suitably marked on both sides of the fragment, and phage crosses were performed to measure recombination between the markers (Fig. 1). In these crosses a recombination frequency of 6–12% was detectable for λ markers flanking *Eco*RI-7. The following results indicate that this efficient recombination is mediated by elements encoded by the cloned *Eco*RI-7 fragment.

1. Recombination occurs normally in the absence of bacterial functions necessary for host general recombination (*recA* and *recBC* functions).
2. λ general and site-specific recombination systems are not responsible for the recombination because both systems are inactive in these phage crosses: Insertion of an *Eco*RI fragment at *sr*Iλ3 inactivates an essential component of the general recombination system, the λ*exo* gene (Murray and Murray 1974), and the *b*538 deletion in the λ vector removes essential components of the λ site-specific system. Moreover, bacterial mutations that affect λ site-specific recombination (*himA*, *himB* [Miller et al. 1979], and *hip* [Enquist et al. 1979]) do not affect λ-P1:7 recombination.
3. Substitution of other P1 *Eco*RI fragments for *Eco*RI-7 in one or both of the partners of a cross causes the recombination frequency to drop from 6–12% to 0.02–0.05% (Sternberg 1979a).
4. Our experiments indicate that the crossover event between two λ-P1:7 phages occurs within *Eco*RI-7 because the recombination frequency for markers that do not flank the cloned *Eco*RI-7, but are located on either side of the fragment, is 100 times lower (0.05%) than for markers that flank the fragment (N. Sternberg and D. Hamilton, in prep.).

The simplest interpretation of all these results is that *Eco*RI-7 contains a site, henceforth called *lox*P (for *lo*cus of crossing over [*x*] in P1), that must be present in both partners of a genetic cross for efficient recombination to be detected. Furthermore, it can be concluded that if a P1-encoded product is necessary for *Eco*RI-7-promoted recombination (that there is such a product is shown below), then the gene that codes for that product must also be on *Eco*RI-7.

Effect of the Orientation of *Eco*RI-7 on *lox*P Recombination

λ-P1:7 phage were isolated with the cloned fragment in both possible orientations. The results of crosses between these phages indicate that efficient recombination can occur between phages containing the fragment in either orientation, provided both phages in a cross have

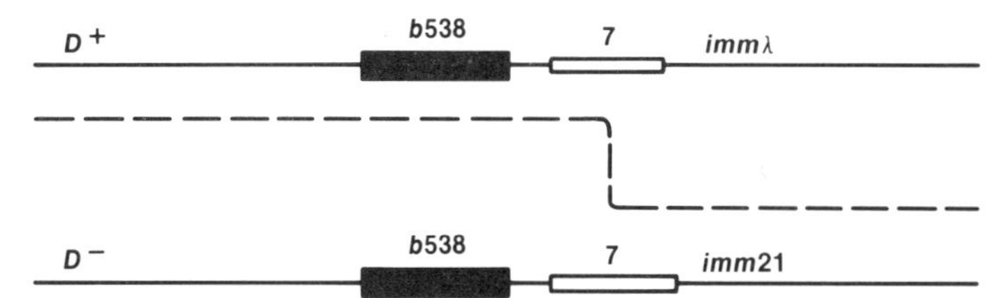

Figure 1. Recombination between λ-P1:7 phages. The figure illustrates the recombination event measured in phage crosses between two λ-P1:7 phages. The D^- mutation is *D*am15. The total phage yielded from the cross was assayed on a *supF* strain, and the yield of recombinants was assayed on a sup^+ strain containing a λ*D*am15*imm*λ prophage. (–) λ DNA; (▭) P1 DNA; (▬) deleted DNA.

the fragment in the same orientation. When the phages in a cross have the fragment in opposite orientations, then recombination for flanking markers cannot be detected above a background level of 0.06%. The first result implies that any function encoded by *Eco*RI-7 that is necessary for *lox*P recombination must be transcribed by a promoter present within that fragment. The second result implies an asymmetry in the *lox*P site, the significance of which will be discussed below.

Localization of *lox*P and a *lox* Recombinase Gene (*cre*) within *Eco*RI-7

A deletion analysis has allowed us to localize *lox*P and a *lox* recombinase gene, called *cre* (for *c*auses *re*combination), to specific segments of *Eco*RI-7 DNA (Fig. 2). The analysis involves crossing λ-P1:7 deletion mutants with each other and with normal λ-P1:7 and determining the efficiency of recombination for flanking λ markers. The only assumption made in the interpretation of the data is that when a phage contains a deletion that removes a P1 *lox* recombinase gene but leaves *lox*P intact, it should still be able to recombine efficiently with λ-P1:7, because the latter should be able to provide recombinase function in *trans*. However, when the deletion inactivates *lox*P, the mutant should not be able to recombine efficiently even with λ-P1:7. Thus, the observation that phage with deletions that enter *Bam*HI fragment 9, a subfragment of *Eco*RI-7, from opposite sides (7Δ201 and 7Δ103) can still recombine efficiently with λ-P1:7 indicates that both of these deletions leave *lox*P intact (Fig. 2, crosses 1 and 2). In contrast, phage with either a deletion of *Bam*HI-9 (7Δ*Bam*HI-9) or a deletion of most of *Bam*HI-9 (7Δ101) cannot recombine efficiently with λ-P1:7. These deletions presumably inactivate *lox*P. On the basis of these results, we can localize *lox*P to the 1.1-kb region of *Bam*HI-9 between 7Δ201 and 7Δ103 (Fig. 2). An extension of this analysis allows us to identify and localize the *cre* recombinase gene. Although we know from crosses 1 and 2 shown in Figure 2 that 7Δ201 and 7Δ103 are $loxP^+$, of these two mutants only 7Δ103 can recombine efficiently with itself (crosses 3 and 4 in Fig. 2). Presumably, the 7Δ201 deletion inactivates *cre*, and the 7Δ103 deletion does not. The results of cross 5 in Figure 2 support this contention, since the $loxP^-$ 7Δ101 deletion mutant can provide *cre* function in *trans* to promote recombination between two 7Δ201 $loxP^+$ DNA molecules. This result localizes the *cre* gene to the 2-kb region between the left end point of 7Δ101 and the left *Eco*RI site of fragment 7 (Fig. 2). As the 7Δ*Bam*HI-9 deletion mutant cannot provide *cre* function (cross 6 in Fig. 2), at least part, and conceivably all, of the *cre* gene must be within *Bam*HI-9. Additional support for these conclusions comes from our recent isolation of cre^- mutations, including several nonsense mutations, that map in the 2-kb region of P1 DNA to the left of deletion 7Δ101 (data not shown).

Recombination between *lox*P and the Bacterial Chromosome

In addition to recombination with itself, *lox*P can also recombine with a specific site in the bacterial chromosome, called *lox*B (Chesney et al. 1979; Sternberg 1979a). This reaction has been studied by measuring the frequency of lysogen formation by an infecting λ*b*538-P1:7 phage. Because this phage carries a deletion that removes the λ site-specific integration system, the only way it should be able to integrate into the bacterial chromosome of a $recA^-$ strain is by *lox*P recombination. The data presented in Table 1 show that whereas the lysogenization frequency is low, λ*imm*λ-P1:7 can lysogenize a *recA* strain 30 times more efficiently than can a λ-P1 phage that has a different P1 fragment. Once formed, λ-P1:7 lysogens are quite stable, with only a 0.2% loss of prophage in an overnight culture.

When a *recA* (λ*imm*λ-P1:7) lysogen is infected with a λimm^{21}-P1:7 phage and lysogens with imm^{21} are selected, then the frequency of lysogeny is 10^{-3}, and all of the resulting imm^{21} lysogens have lost the original *imm*λ prophage (Table 1, line 3). This lysogenization frequency is 100 times greater than the frequency observed when the λimm^{21}-P1:7 phage infects either a *recA* strain or a *recA* strain with a λ*imm*λ-P1 prophage containing a P1 fragment other than *Eco*RI-7 (Table 1, lines 1 and 4). When a *recA* (λ*imm*λ-P1:7) strain is infected with λimm^{21}-P1:7 and lysogens with both *imm*λ and imm^{21} are selected, then the lysogenization frequency is less than 1×10^{-6} (data not shown). Thus, efficient integration of λimm^{21}-P1:7 DNA requires the displacement of the resident prophage.

A model to explain these results is shown in Figure 3. In the reaction depicted in Figure 3a, we postulate that *lox*P recombines with *lox*B to integrate the λ*imm*λ-P1:7 DNA as prophage, creating in the process two hybrid *lox* sites, designated *lox*L and *lox*R. As both integration and excision reactions are inefficient, the model predicts that *lox*P × *lox*B and *lox*R × *lox*L recombination reactions also will be inefficient. When λimm^{21}-P1:7 infects a *recA* (λ*imm*λ-P1:7) lysogen (Fig. 3b), integration of the

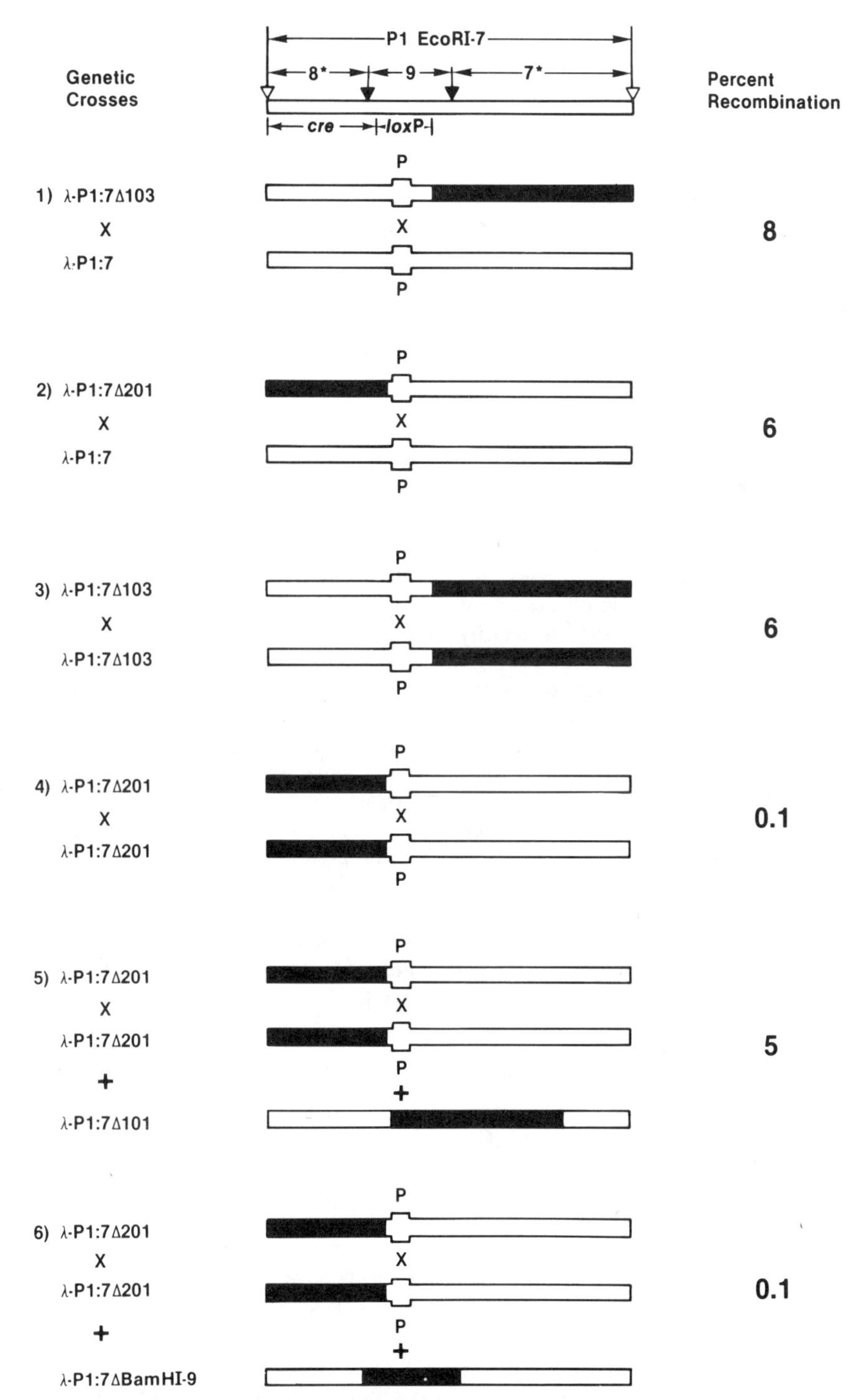

Figure 2. Recombination between λ-P1:7 deletion mutants. *Eco*RI-7 can be divided into three subfragments (7*, 9, and 8*) on the basis of two *Bam*HI sites (↓) in the fragment as shown at the top of the figure. Only P1 *Eco*RI-7 DNA is shown in these crosses; (▬) deleted DNA. The locations of deletion end points were determined by both heteroduplex and restriction enzyme analyses. The 7Δ101 and 7Δ*Bam*HI-9 deletion mutants were judged to be *lox*P$^-$ because they do not recombine with λ-P1:7. Consequently, they cannot physically take part in the recombination events occurring in crosses 5 and 6. One phage in each cross carries an amber mutation (*S*am7) at the right end of the λ chromosome and the other carries an amber mutation (*D*am15) at the left end of the λ chromosome. The yield of λD^+S^+ recombinants was measured on a *sup*$^+$ host. *loxP* and the *cre* gene are positioned on the basis of the phage crosses shown.

Table 1. The Frequency of Lysogeny by a λ-P1:7 Phage in a *recA* Bacterial Strain

Bacterial strain	Infecting phage	Lysogenization frequency by infecting phage $\times 10^{-5}$	Retention of resident prophage
A			
*rec*A	λ*imm*λ-P1:7	1	
*rec*A	λ*imm*λ-P1:2	0.03	
B			
*rec*A(λ*imm*λ-P1:7)	λ*imm*^{21}-P1:7	100	no
*rec*A(λ*imm*λ-P1:2)	λ*imm*^{21}-P1:7	1	yes

The frequency of λ lysogeny was measured as follows. *recA* strain N205 (Sternberg and Weisberg 1975) was infected with a λ*b*538-P1 phage containing either P1 *Eco*RI-7 or P1 *Eco*RI-2 (Sternberg 1979b), at a multiplicity of two phage per cell. The infected cells were then spread onto EMBO agar plates containing 10^8 selector phages: For λ*imm*λ lysogens the two *imm*λ-clear selector phages *W*248 and *W*30 (*WW;* Sternberg 1979b) were used, and for λ*imm*^{21} lysogens the two *imm*^{21}-clear selector phages *B*17 and *B*121 (*BB;* Sternberg 1979b) were used. Lysogens were scored as healthy pink colonies on the EMBO plate after incubation at 32°C for 48 hr (Gottesman and Yarmolinsky 1968). In the experiments described in *B,* λ*imm*^{21} lysogens were selected on EMBO plates with *BB* selector phage and then tested for retention of the resident *imm*λ prophage by transferring colonies to EMBO plates with *WW* selector phage. For the lysogen containing *Eco*RI-7, none of the 50 colonies tested retained the *imm*λ prophage, whereas for the lysogen containing *Eco*RI-2, all of the 50 colonies tested retained the *imm*λ prophage.

λ*imm*^{21} DNA can occur by either a *lox*P × *lox*R or a *lox*P × *lox*L reaction. The product of either of these reactions is a tandem prophage structure flanked by *lox*L, *lox*P, and *lox*R sites. This tandem prophage is unstable, with the *imm*λ prophage being lost when we select for *imm*^{21} prophage (Fig. 3c). Because stable integration of the λ-P1:7 DNA is at least 100 times more efficient by the reactions depicted in Figure 3 b and c than it is by the reaction depicted in Figure 3a, we postulate that the *lox*P × *lox*R and *lox*P × *lox*L reactions must be more efficient than the *lox*P × *lox*B reaction. Note that the reaction depicted in Figure 3c stabilizes the *imm*^{21} prophage because it causes prophage DNA to be flanked by two *lox* sites that recombine poorly with each other.

Isolation and Characterization of λ-P1 *lox*R, λ-P1 *lox*L, and λ-P1 *lox*B Phages

λ-P1 phage containing hybrid *lox* sites, *lox*L and *lox*R, were isolated as rare classes of phage produced by abnormal excision of the prophage DNA when a lysogen containing a λ-P1:7 prophage at *lox*B is induced (Fig. 4). The DNAs of these phages were analyzed by digestion with restriction endonucleases *Eco*RI, *Bam*HI, and *PvuI*, and the results are shown in Figure 4. λ-P1 *lox*L and λ-P1 *lox*R phages appear to carry reciprocal segments of *Eco*RI-7 DNA: The phage containing P1 fragment 8*, but not 9 or 7*, is arbitrarily designated λ-P1 *lox*L, whereas the phage containing P1 fragment 7*, but not 9 or 8*, is arbitrarily designated λ-P1 *lox*R. As both classes of phage are missing *Bam*HI fragment 9 (*Bam*HI-9), the crossover during the integration of λ-P1:7 DNA into *lox*B must have occurred within this fragment, thereby splitting the fragment so that part of it is on the *lox*R side and part is on the *lox*L side of the prophage. A surprising discovery from this analysis was that both λ-P1 *lox*R and λ-P1 *lox*L DNAs contain two different classes of bacterial DNA adjacent to the P1 sequences. This result means either the original λ-P1:7 DNA integrated in one orientation into at least two different bacterial sites, or it integrated in both possible orientations into one site (*lox*B). According to the first hypothesis, the two classes of bacterial DNA associated with *lox*L phage should be different from the two classes of bacterial DNA associated with *lox*R phage. According to the second hypothesis, they should be the same. Clearly, the data shown in Figure 4 support the second hypothesis, as class-1 *lox*L phage have the same bacterial DNA as class-2 *lox*R phage, and class-2 *lox*L phage have the same bacterial DNA as class-1 *lox*R phage. On the basis of an analysis of 42 λ-P1:7 lysogens isolated independently, there appears to be a 3:1 preference for orientation 1 (the orientation that gives rise to class-1 *lox*R and *lox*L phage) over orientation 2.

A λ *lox*B phage was constructed by recombination in a phage cross between a class-1 *lox*L phage and a class-1 *lox*R phage. Recombination between these phages at *lox* sites generates both *lox*P and *lox*B phages. A restriction map of the DNA of a λ *lox*B phage is shown in Figure 4.

Recombination between λ-P1 *lox* Phages: Verification of the Model for Prophage Integration and Displacement at *lox*B

The recombination frequencies in crosses between various pairs of λ-P1 *lox* phages are shown in Figure 5. The reactions *lox*L1 × *lox*L1, *lox*L1 or *lox*L2 × *lox*P, and *lox*P × *lox*P all occur with high frequency. The results obtained with *lox*L phage are particularly noteworthy because they indicate that the portion of *Eco*RI-7 missing from λ-P1 *lox*L DNA either contains no P1 sequences necessary for efficient *lox* recombination, or it contains such sequences and they can be substituted for by the bacterial DNA present in either λ-P1 *lox*L1 or λ-P1 *lox*L2 DNA. We show below that the second alternative is probably the correct one. Unlike results with DNAs that contain *lox*P and *lox*L sites, DNAs with *lox*R and *lox*B sites are poor substrates for recombination. The only recombination reaction with either *lox*R or *lox*B sites that occurs with any detectable efficiency is

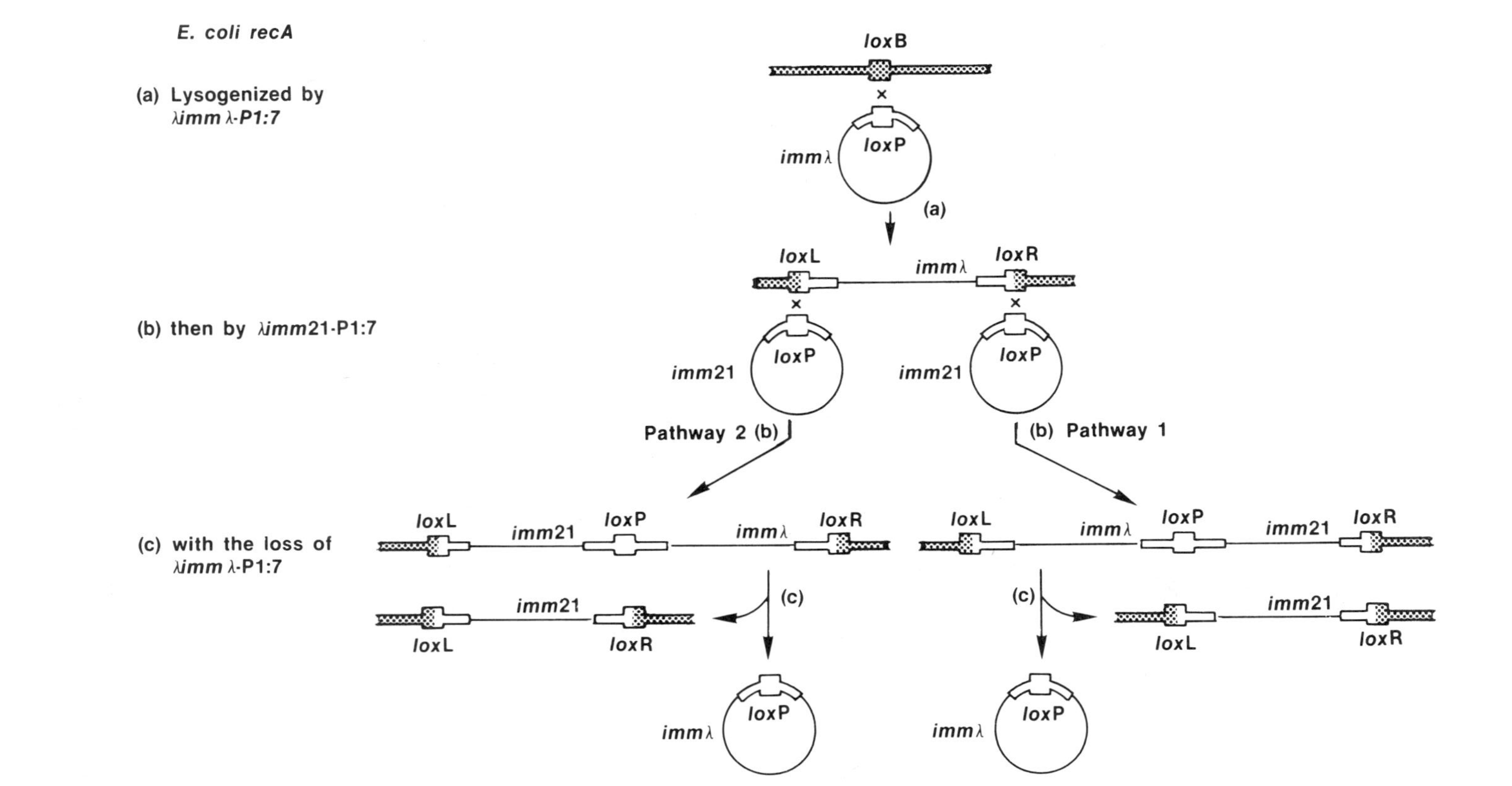

Figure 3. A model for prophage integration and displacement at *lox*B. Integration occurs by the single *lox*P × *lox*B recombination. In contrast, prophage displacement requires two recombination reactions, *lox*P × *lox*L and *lox*P × *lox*R. The two displacement pathways (reaction c) differ as to which of the two recombination reactions is involved in integration and which is involved in excision of λ-P1:7 DNA. (—) λ DNA; (□) P1 DNA; (▭) bacterial DNA.

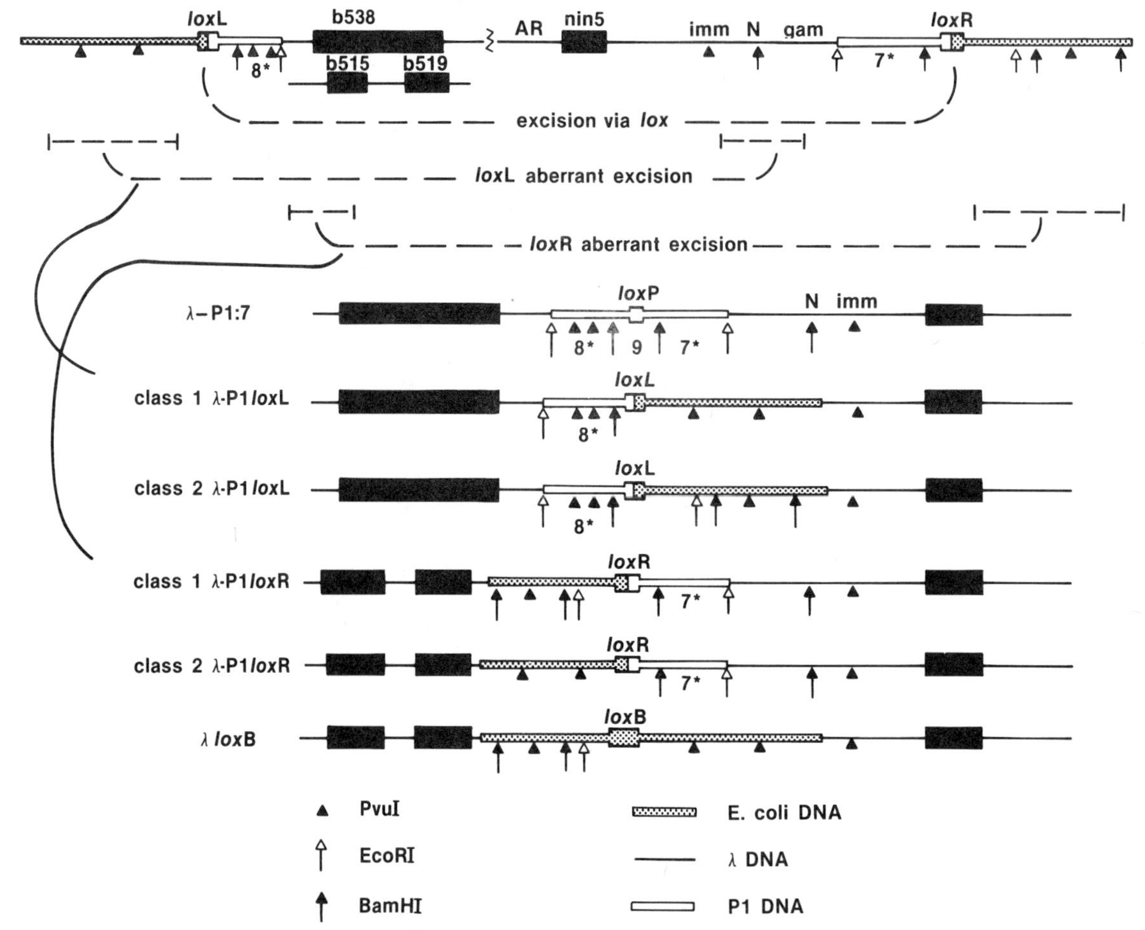

Figure 4. The formation of λ-P1 *lox*L, λ-P1 *lox*R, and λ *lox*B phages and a partial restriction map of their DNAs. Rare λ-P1 *lox*L phage were formed by the abnormal excision of a λ*b*538-P1:7 prophage at *lox*B, resulting in the loss of the λ*Gam* gene and, consequently, the ability to form plaques on a strain containing a P2 prophage (the Spi phenotype; Zissler et al. 1971). Rare λ-P1 *lox*R phages were formed by the abnormal excision of a λ*b*519, *b*515-P1:7 prophage at *lox*B, resulting in the loss of the λ *int* and *xis* genes. Phage lacking these genes can be selected because of their inability to make red plaques on a galactose-TB-TTC plate with an indicator strain containing a λ cryptic prophage in the *galT* gene (Enquist and Weisberg 1976). A λ *lox*B phage was isolated from the yield of a cross between a λ-P1 *lox*L1 and a λ-P1 *lox*R1 phage. Recombination at these *lox* sites should generate the desired *lox*B phage shown in the figure. The restriction maps of the various phages obtained using the enzymes *Eco*RI, *Bam*HI, and *Pvu*I are shown. *lox*L phage contain P1 fragment 8*, part of P1 fragment 9, and one of two different classes of adjacent bacterial sequences. *lox*R phage contain P1 fragment 7*, the part of fragment 9 not present in the *lox*L phage, and one of two different classes of adjacent bacterial sequences. On the basis of restriction patterns of these DNAs, we can conclude that the bacterial DNA adjacent to the *lox*L1 site is the same as the bacterial DNA adjacent to the *lox*R2 site, and the bacterial DNA adjacent to the *lox*L2 site is the same as the bacterial DNA adjacent to the *lox*R1 site. (▬) Deleted DNA; (▒▒▒) *E. coli* DNA; (—) λ DNA; (▭) P1 DNA; (▲) *Pvu*I; (↑) *Eco*RI; (↑) *Bam*HI.

that between *lox*R and *lox*P sites, and it occurs 20 times less efficiently than does *lox*L × *lox*P recombination. Note that the low efficiencies of the *lox*P × *lox*B integrative recombination reaction and the *lox*L × *lox*R excisive recombination reaction agree with the inefficiency of integration and excision of λ-P1:7 DNA.

The prophage displacement reaction shown in Figure 3 requires two recombination reactions—*lox*P × *lox*L and *lox*P × *lox*R. In pathway 1 (Fig. 3), the initial *lox*P × *lox*R integration reaction (0.4% recombination) is followed rapidly by efficient recombination between *lox*P × *lox*L sites (10% recombination), resulting in excision and subsequent loss of the original prophage DNA. In pathway 2 (Fig. 3), the initial integration reaction involving *lox*P × *lox*L recombination is followed most frequently by the more efficient of the two possible excision reactions (*lox*P × *lox*L), which has the effect of reversing the integration reaction. Only when the less-efficient *lox*P × *lox*R excision reaction occurs is the incoming prophage DNA stabilized within the chromosome. If the frequency of lysogen formation by a λ-P1:7 (*lox*P) phage, either by direct integration or by prophage displacement (Table 1), reflects the efficiency of the various recombination reactions (Fig. 4), then the product of the *lox*P × *lox*R and *lox*P × *lox*L recombination frequencies (0.04% = 10% × 0.4%) must be greater than the frequency of *lox*P × *lox*B recombination (0.002%, Fig. 4), and it is. Clearly, our data are con-

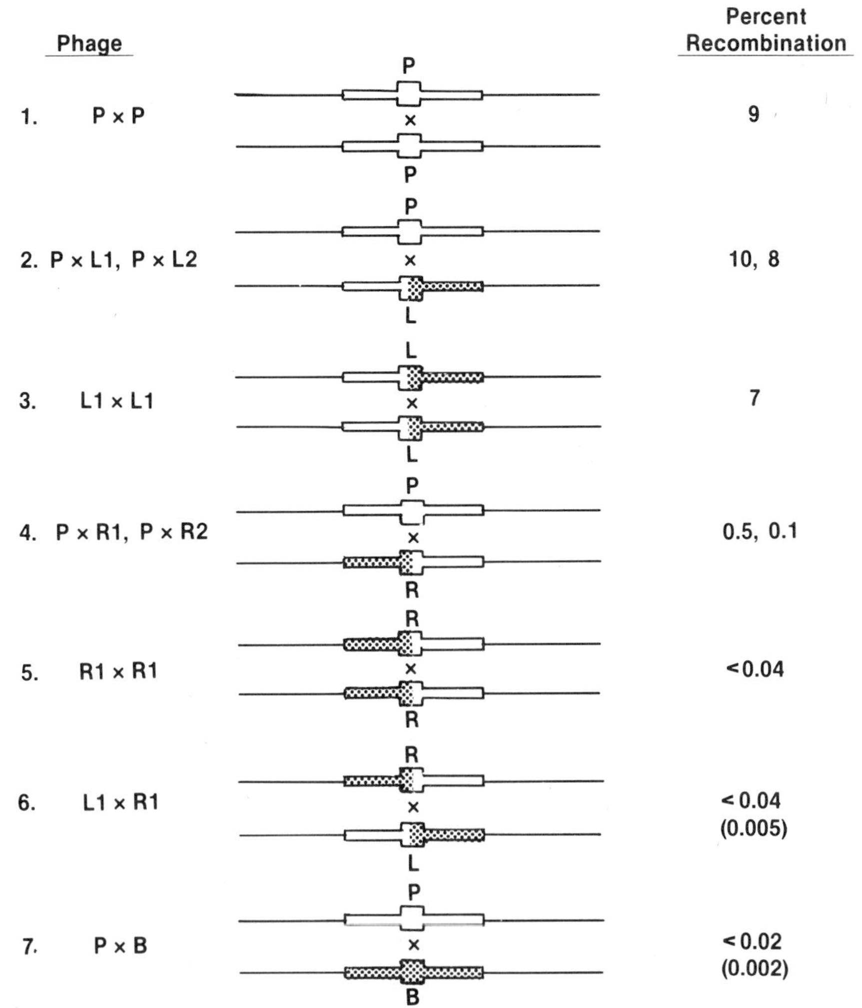

Figure 5. Recombination between λ-P1 *lox* phages. L1 and L2 and R1 and R2 refer to the two classes each of *lox*L and *lox*R phages, respectively (see Fig. 4). The *lox*R × *lox*R cross was carried out in the presence of λ-P1 *lox*L phage to provide *cre* function. In crosses 6 and 7, the higher recombination values represent the frequency of recombination for flanking λ markers, whereas the lower values (in parentheses) represent the percentage of recombinants that have crossed over at *lox* sites. (—) λ DNA; (□) P1 DNA: (▨) bacterial DNA.

sistent with the model for prophage integration and displacement proposed in Figure 3.

Nucleotide Sequence for DNA Regions Containing *lox*P, *lox*R1, and *lox*L1

The nucleotide sequence immediately surrounding the crossover region in *lox*P, *lox*R1, and *lox*L1 is shown in Figure 6. For *lox*P, this sequence is located within *Bam*HI-9 at about 650 bp from the 7* *Bam*HI site, and it is embedded in an AT-rich (~70%) region spanning at least 100 bp to either side. The crossover point in *lox*P × *lox*B recombination can be deduced by comparing the three *lox* sequences. It is illustrated as a flush cut on the 3′ side of the 5′ C-A 3′ sequence present in *lox*P. However, as the C-A sequence is present in all three *lox* sites, the crossover could just as well take place to the 5′ side of either A or C, or a staggered cut encompassing these sites could be made.

A striking feature of the *lox*P sequence is the presence of a 13-bp perfect inverted repeat hyphenated by an 8-bp region. The significance of this sequence is emphasized by the location of the crossover point in the 8-bp-hyphenated region between the inverted repeats. Interestingly, much of the inverted-repeat structure seen in the *lox*P sequence is also present in the *lox*L1 and *lox*R1 sequences. Thus, replacement of *lox*P DNA sequences by bacterial DNA in the λ-P1 *lox*L phage substitutes a critical portion of the P1 inverted-repeat sequence with an almost identical sequence from bacterial DNA. If we reconstruct *lox*B by using the bacterial sequences in *lox*R and *lox*L, then *lox*B contains a perfect inverted repeat of 10 bp hyphenated by a 5-bp region containing the C-A crossover point(s) (Fig. 6). A remarkable degree of homology exists between the *lox*P and *lox*B inverted-repeat sequences, suggesting that *lox*B may be a degenerate *lox*P left in the bacterial chromosome by a once-integrated P1.

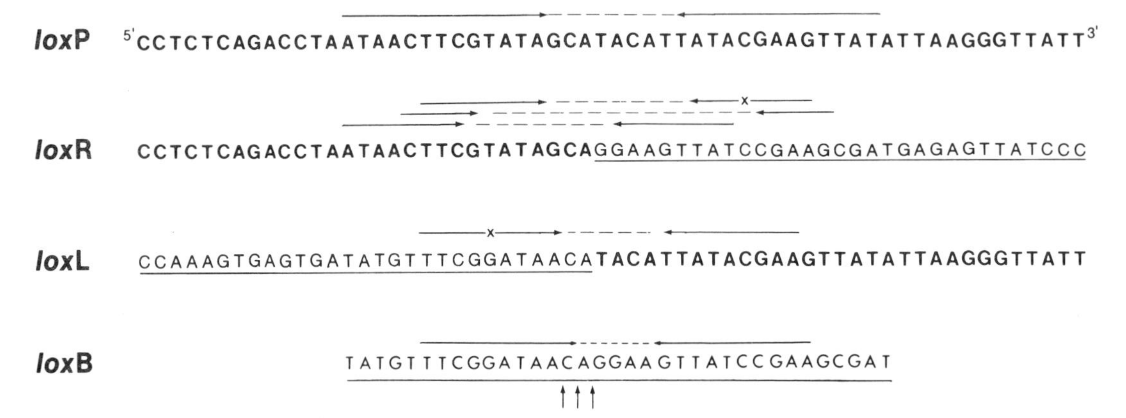

Figure 6. The nucleotide sequences of *lox*P, *lox*R1, *lox*L1, and reconstructed *lox*B sites. DNA was sequenced using the technique of Maxam and Gilbert (1977). For simplicity, only one strand of DNA is shown for each site. Bacterial sequences determined for *lox*R1 and *lox*L1 are combined to form the *lox*B sequence shown here. P1 DNA sequences are noted in bold type and bacterial sequences are underlined. (→) Inverted-repeat sequences with (--) lines connecting them; (x) mismatch in the inverted-repeat sequence.

A Model for P1 Site-specific Recombination

Any model used to explain how recombination occurs at *lox* sites must reflect the following facts. (1) All four *lox* sites shown in Figure 6 contain hyphenated inverted-repeat regions, which we have shown are important at least in the *lox*P × *lox*B recombination reaction. Despite the similarity of these regions, *lox*L and *lox*P sites are much better substrates for recombination than are *lox*B and *lox*R sites (Fig. 5). We believe that these results indicate that the P1 DNA present in λ-P1 *lox*L phage ("L" DNA) contains a sequence, other than the inverted-repeat region, that is essential for efficient *lox* recombination. (2) Both partners in a recombination reaction must carry this essential "L" DNA sequence if efficient recombination is to occur. Thus, of the reactions shown in Figure 5, only *lox*P × *lox*P, *lox*P × *lox*L, and *lox*L × *lox*L reactions occur with high efficiency. (3) The essential "L" DNA sequence must be on the same side of the inverted-repeat region in both partners of a recombination reaction for efficient recombination to occur between markers that flank the recombination site. Thus, if *Eco*RI-7 is oriented in opposite directions relative to flanking λ markers in λ-P1:7 phages, efficient recombination between these markers cannot be detected. (4) The crossover site in the *lox*P × *lox*B recombination reaction is in the hyphenated region between the inverted-repeat sequences (Fig. 6). We feel that the model described below (Fig. 7), although still highly speculative, can explain these results and provides a framework for further experiments.

The substrate for recombination between two *lox*P sites is the region of DNA containing the inverted-repeat segment. A P1 recombinase protein, presumably the product of the *cre* gene, is bound to specific "L" DNA sequences in this region. We postulate that two *lox*P sites are brought together when their bound protein molecules interact with each other (step 1). Conceivably, the recombinase monomer cannot dimerize until it first binds to "L" DNA. Moreover, we suggest that the formation of the resulting dimeric complex activates a single-strand nicking activity associated with the recombinase protein, with the consequence that this protein now nicks within the single-stranded loop of the stem-loop structure that can form at each *lox*P site (step 2). The stem-loop structures either could be forming and dissociating continuously or could be formed and stabilized through interactions between the inverted-repeat sequence and the recombinase. The requirements of the model are, up to this point, sufficient to explain the facts presented in the beginning of this section. To extend the model from step 2 to the formation of a recombinant molecule logically, one need only invoke the series of steps proposed by Sobell (1972) for recombination between complementary intrastrand stem-loop structures. Complementary loops nicked by the recombinase can come together by base-pairing (step 2). This would be followed by propagation of the hybrid DNA as shown in steps 3 and 4. In step 5, ligation of the nicked ends (J and N) produced in step 2 would give the last structure shown in the model, a Holliday hybrid DNA molecule (Holliday 1964). To simplify matters, we have not included the recombinase protein in the structures shown after step 2. Indeed, this protein might remain bound throughout steps 1–5, functioning to hold the nicked ends in close proximity to each other so that ligation can occur efficiently. As pointed out by Sobell (1972), the Holliday structure possesses twofold symmetry and can therefore be recognized by a nuclease(s) possessing twofold symmetry that would simultaneously act to nick strands of the same polarity at homologous sites. This would produce reciprocal recombination involving either single- or double-strand exchange, depending upon which strands are cut and joined (Holliday 1964).

It should be noted that in the model described in Figure 7, the importance of the inverted-repeat region for *lox*P × *lox*P recombination was emphasized because of its apparent role in the *lox*P × *lox*B reaction. However, as the former reaction occurs much more efficiently than does the latter, it is formally possible that the inverted-

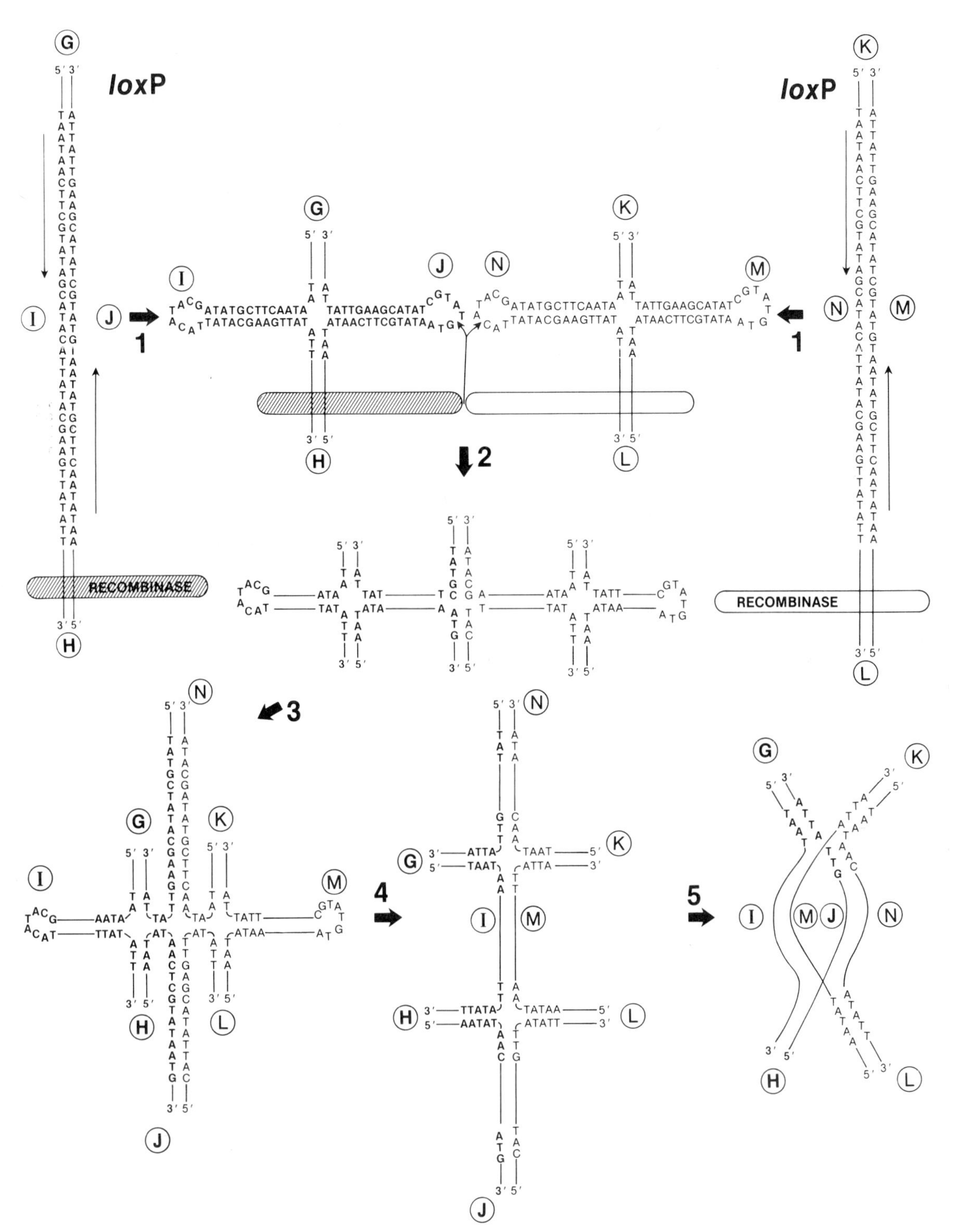

Figure 7. A model for *lox*P × *lox*P recombination. Homologous *lox*P sites, G-H and K-L, contain hyphenated inverted-repeat regions that can form stem-loop structures. These structures may become stabilized by interaction with recombinase proteins that bind to the *lox*L side of the sequences shown. *lox*P sites are brought together when two bound recombinase proteins interact to form a dimeric structure that can nick the DNA within the loop structure. When complementary loops J and N are nicked and opened, Watson-Crick base-pairing occurs, and this is followed by extensive propagation of the heteroduplex (steps 3 and 4). Ligation of the nicked ends produced in step 2 gives a Holliday structure, which in turn can be nicked in strands of the same polarity at homologous sites to produce recombinant molecules. Another feature of the Holliday structure is its ability to migrate along the parental DNA molecule, thereby allowing recombination to occur over a long stretch of DNA. Recent experiments (N. Sternberg and D. Hamilton, in prep.) indicate that *lox*P × *lox*P recombination can be associated with the migration of the crossover point for a distance of 1 kb from the inverted-repeat region.

repeat sequence does not play an important role in the *lox*P × *lox*P reaction. We hope to determine what sequences are necessary for *lox*P × *lox*P recombination by mutagenizing the *lox*P site, localizing the mutations, and then determining what effect the mutations have on the efficiency of recombination.

The Role of *lox* Recombination in the P1 Life Cycle

Efficient recombination between *lox*P sites appears to play an important role in at least three different aspects of the P1 life cycle: the faithful partitioning of P1 plasmid DNA at cell division, the cyclization of P1 DNA following its injection into a bacterial strain, and the production of P1 generalized transducing phage. Only a brief discussion of the involvement of *lox* recombination in each of these three processes is possible here.

The maintenance of P1 plasmid DNA. In Figure 8, we illustrate a problem confronted by most unit-copy plasmids, including P1, that have to partition faithfully the products of replication to daughter cells at cell division. It is proposed that the formation of plasmid dimers by *rec*-promoted recombination between sister copies of a plasmid prior to cell division would block the normal partition process and would give rise to a cell without a plasmid. Such recombination should occur during about 1% of the cell divisions and would, therefore, seriously compromise plasmid maintenance (Fig. 8A). Clearly, these events do not occur during the P1 plasmid life cycle, as this plasmid is lost with a frequency that is less than one cured cell in every 10^5 cell divisions. Because sister copies of the P1 plasmid contain *lox*P sites, the formation of stable dimers by *rec* is blocked, and thus the potential partition problem is avoided (Fig. 8B).

Two observations support this proposed role for *lox*P recombination. (1) When a λ-P1 chimeric phage that does not contain *Eco*RI-7 and that carries a deletion of the λ integration system infects a rec^+ (P1) strain, the infecting phage DNA can become established as a prophage by integration into the P1 plasmid DNA. This integration reaction involves homologous recombination between P1 sequences on the λ-P1 DNA and on P1 plasmid DNA and is dependent on the host general recombination (*rec*) system. The presence of a *lox*P site on the P1 fragment prevents the λ-P1 phage from stably integrating into the plasmid. Consequently, a λ-P1:7 phage lysogenizes a rec^+ (P1) strain 50 times less efficiently (at a frequency of 10^{-5} lysogens per infected cell) than does a λ-P1:7Δ*Bam*HI-9 phage. The latter contains a deletion of the *lox*P site (see discussion above). Moreover, we can show that the lysogens formed with λ-P1:7 carry the λ prophage in the bacterial chromosome, whereas the lysogens found with λ-P1:7Δ*Bam*HI-9 have the prophage integrated into the P1 plasmid (N. Sternberg and D. Hamilton, in prep.). (2) Miniplasmids of P1 that lack *lox*P are lost from a rec^+ strain at about 1% per generation but are far more stable in a *recA* host (<0.1% loss per generation). In contrast, a recombinant plasmid that contains both miniplasmid DNA and *Eco*RI-7 is far more stable than the miniplasmid alone in a rec^+

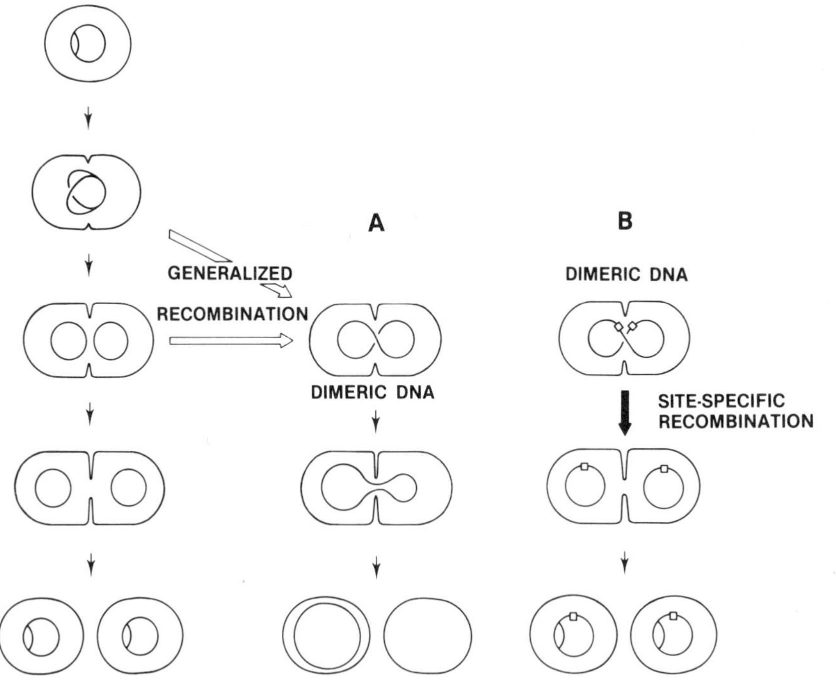

Figure 8. The role of P1 site-specific recombination in plasmid maintenance. Generalized recombination between replicating daughter DNA molecules will produce plasmid dimers that cannot be equipartitioned to daughter cells at cell division (pathway *A*). If the plasmid contains a site-specific recombination that can rapidly dissociate dimers, then the products of that dissociation can be equipartitioned (pathway *B*). The square in the plasmid DNA shown in pathway *B* represents *lox*P.

strain (=0.1% loss per cell division). Moreover, if the *Bam*HI-9 that contains *lox*P is removed from the recombinant plasmid, then the resulting plasmid is not more stable than the parental miniplasmid.

Cyclization of P1 DNA. As noted previously, the DNA of the P1 phage particle is cyclically permuted and terminally redundant (Ikeda and Tomizawa 1968). The terminal repeat regions encompass about 10% of the P1 genome and are important in the conversion of the infecting linear P1 DNA to a circular form. This reaction is probably catalyzed by the host *rec* system and presumably involves recombination between the terminal repeat regions. Cyclization of P1 DNA also occurs in a *recA* host, although with somewhat reduced efficiency (with about 10% the efficiency as in a *rec*$^+$ host) (Rosner 1972). How is P1 DNA cyclized in a *recA* host? With the discovery of *lox*P recombination, a simple explanation exists: Those linear P1 DNA molecules containing two copies of *lox*P are cyclized by the *lox* recombination system. As the experiments of Bächi and Arber (1977) indicate that only about one in every three or four viral DNA molecules contains two copies of *lox*P, one could account for the reduced efficiency of P1 cyclization in a *recA* host by arguing that only molecules with two *lox*P sites are ever cyclized in such a host. The evidence that *lox*P recombination is important in cyclization stems from a modification of an experiment originally performed by Hertman and Scott (1973). The original experiment showed that recombination between P1 markers in a *recA* strain was only 5–10 times lower than in a *rec*$^+$ strain when the recombinants were assayed on a *rec*$^+$ strain. This result led these authors to suggest that P1 encodes a general recombination system. The data shown in Table 2 confirm these results and extend them one step further. If the recombinants formed in a *recA* strain are cycled once through a *recA* strain at low multiplicity before they are assayed on a *rec*$^+$ strain, then these recombinants are largely eliminated. In contrast, recombinants formed in a *rec*$^+$ strain are not affected by cycling through a *recA* strain (Table 2). We suggest the following explanation for these results. P1 does not encode a general recombination system. Rather, the only pathway for recombination between P1 molecules in a *recA* strain is recombination between *lox*P sites. The product of this recombination will be a DNA multimer that can be a substrate for the P1 processive "headfull" packaging system. Some of the DNA "headfulls" packaged will be heterozygous for the markers being tested, and if that DNA is injected into a *rec*$^+$ host, the *rec* system, which acts at the terminal repeats to cyclize P1 DNA, will occasionally produce recombinant circular molecules. However, when the phage DNA is injected into a *recA* strain, we have postulated that the only means of cyclizing the P1 DNA is by recombination between terminal *lox*P sites. Thus, the cyclizing reaction will always reverse the recombination reaction that occurred in the *recA* strain in which the phage cross was originally performed. Consequently, heterozygous DNA molecules that might have been resolved into recombinant circles in a *rec*$^+$ host are eliminated by the requirement that they be cyclized by *lox*P recombination in a *recA* host. If this interpretation is correct, it suggests that *lox*P recombination is necessary for cyclizing P1 DNA in a *recA* host.

Table 2. P1 Recombination in a *recA* Host

1.4	+	*vir*^{s}	P1*vir*^{s} 1.4
			×
+	35	*vir*^{s}	P1*vir*^{s} 35

Host	Percentage recombination	Percentage recombinants after cycling phage yield at a low moi in	
		recA$^-$	*recA*$^+$
recA$^+$, *supF*	4.5	6.0	6.6
recA$^-$, *supF*	0.9	0.03	0.85

The P1 phage cross performed is illustrated at the top of the table. P1 markers 1.4 and 35 are amber mutations that are located approximately 2 kb and 20 kb, respectively, from the left end of the P1 genetic map (Yarmolinsky 1978). *vir*^{s} is a virulent mutation. The cross was performed as follows: Either strain YMC(*supF*) (Dennert and Henning 1968) or strain YMC *rec*A56(*supF*) was infected with the two P1 phages each at a multiplicity of five phage per cell. After 10 min adsorption, the infected cells were diluted 1000 times and incubated 90 min at 37 °C. At the end of this period, chloroform was added to complete cell lysis, and the total phage yield was assayed on strain YMC. The yield of wild-type recombinants was assayed on the *rec*$^+$ *sup*$^+$ strain W3350 (column 2). The phage from either the *rec*$^+$ or the *recA* cross were used also to infect either strain YMC or YMC *recA* at a multiplicity of 0.1 phage per cell. These infected cells were treated in the same way as were the cells in which the phage cross was performed, with the total phage yield being assayed in strain YMC and the yield of wild-type recombinants being assayed in strain W3350 (columns 3 and 4).

Production of P1 transducing phage. The *lox*B site characterized in this article is an inefficient substrate for recombination with *lox*P. This property might have been expected, as *lox*B was discovered by selecting for stable integration of a *lox*P-containing DNA molecule. Had *lox*B been an efficient site, it would have promoted not only efficient integration of *lox*P-containing DNA, but probably also efficient excision of that DNA. Consequently, it would not have been detected in an assay that depends on stability of the integrated prophage over a period of many cell generations. Thus, the possibility exists that there are efficient *lox* sites in the bacterial chromosome. These sites could be substrates for the transient integration of P1 DNA into the bacterial chromosome. Such an integration event during the lytic cycle of phage growth would position a normal P1 packaging site (*pac*) (Bächi and Arber 1977; N. Sternberg and R. Hoess, in prep.) adjacent to bacterial sequences. P1 processive "headfull" packaging from this *pac* site would result in the packaging of bacterial DNA into a phage head and the production of generalized transducing phage. Although we have not yet shown that efficient *lox* sites exist in the bacterial chromosome, we can simulate the effect of such sites on the yield of generalized transducing phage by preparing P1 lysates on a bacterial strain containing a λ-P1:7 prophage at *lox*B. As this prophage contains a flanking *lox*L site, a good substrate for *lox*P recombination, the infecting P1 DNA should integrate transiently into the chromsome at this site, thereby initiating the packaging of the adjacent bacterial DNA from the integrated *pac* site (Fig. 9). As the packaging of DNA from *pac* is unidirectional (Bächi and Arber 1977; N. Sternberg and R. Hoess, in prep.), only chromosomal markers on one side of the λ-P1:7 prophage will be packaged preferentially when P1 DNA integrates. Indeed, these properties have allowed us to map the λ-P1:7 prophage at *lox*B to a region of the *Escherichia coli* chromosome between *tolC* (66 min) and *dnaG* (67 min) (Bachmann and Low 1980). As illustrated in Figure 9, if P1 infects a *recA*⁻ strain with a λ-P1:7 prophage in orientation 1, we have found that the resulting P1 lysate transduces *tolC, metC,* and *serA*

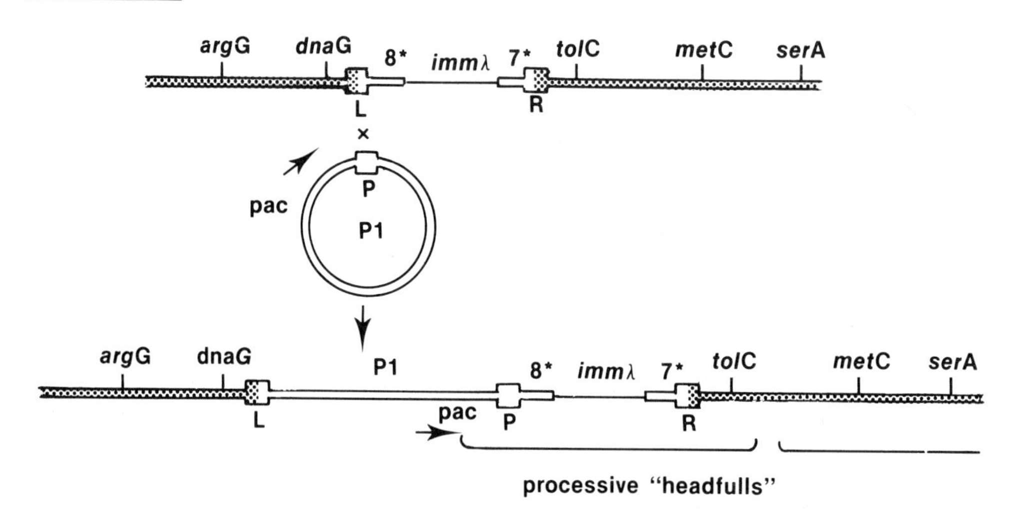

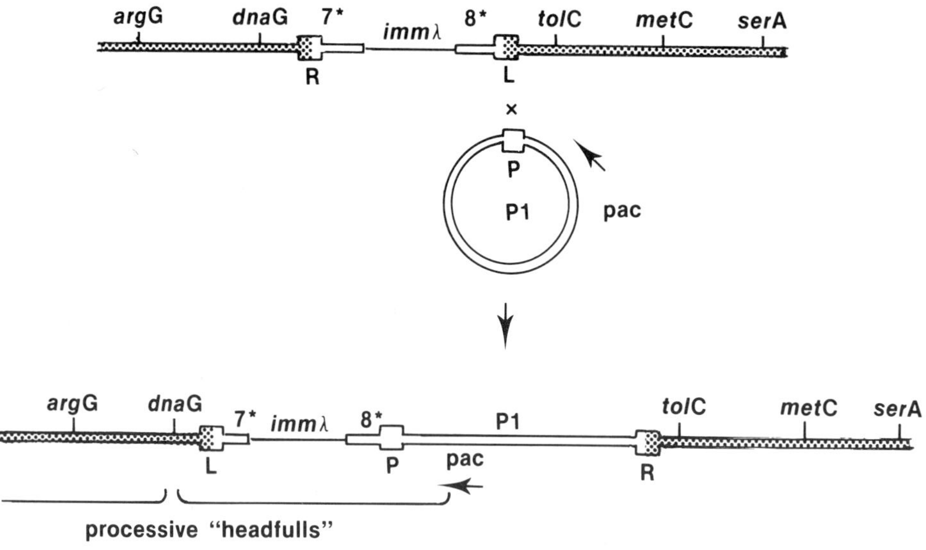

Figure 9. The enhanced production of P1 generalized transducing phage due to the presence of an efficient *lox* site in the bacterial chromosome. *recA* (λ-P1:7) lysogens with the prophage at *lox*B in either orientation 1 or 2 were infected with P1 to prepare lysates that were used to transduce a variety of bacterial markers, including *argG, dnaG, tolC, metC,* and *serA*. The figure illustrates the recombination between P1 DNA (*lox*P) and the *lox*L site of a resident λ prophage that results in the integration of the P1 DNA into the bacterial chromosome in either of two orientations. Processive "headfull" packaging of DNA from the P1 *pac* site into bacterial DNA increases the production of P1 transducing particles for markers on one side of the integrated λ prophage. This sort of experiment was used to map *lox*B between *tolC* and *dnaG*. (–) λ DNA; (▭) P1 DNA; (▬) bacterial DNA.

markers 50 times more efficiently than can a lysate made on a *recA*⁻ strain without a λ-P1:7 prophage. In contrast, if the P1 lysate is prepared on a strain with a λ-P1:7 prophage in orientation 2, then transduction for the *dnaG* and *argG* markers is enhanced. We hope to assess the role of *lox* recombination in the production of transducing particles by comparing the transduction frequencies of P1 wild-type and P1 *lox*⁻ lysates for a variety of chromosomal markers.

A Comparison between λ and P1 Site-specific Recombination Systems

These two recombination systems are more different than they are similar. This is not surprising in light of their different biological roles. The λ site-specific recombination system is designed primarily to integrate λ DNA into a site, *att*B, in the bacterial chromosome; *att*B × *att*P recombination is, consequently, very efficient. Once the prophage state is established, the prophage DNA is not excised from the chromosome because the synthesis of proteins necessary for excision is blocked by the repressor of λ lytic functions. In contrast, P1 is a plasmid prophage and thus need not integrate into the bacterial chromosome to be maintained in the cell population. It is therefore not surprising that the *lox*P × *lox*B integrative recombination reaction and the *lox*L × *lox*R excisive recombination reaction are inefficient. The more relevant recombination reaction in the P1 life cycle is the *lox*P × *lox*P reaction. Moreover, if this reaction is to play a role in the dissociation of P1 plasmid dimers, as we have suggested previously, *lox* recombination, unlike λ site-specific recombination, must be constitutive in the prophage state. We have confirmed this point by showing that a P1 plasmid prophage can recombine with a *lox*L site in the bacterial chromosome with the same efficiency as can an infecting λ-P1:7 phage (N. Sternberg et al., in prep.). Thus, the repressors that interfere with the expression of P1 lytic functions do not limit *lox* recombination.

The DNA sequences of the λ and P1 recombination sites reveal two obvious differences in the natures of the recombination processes themselves. First, there is no extensive common core structure maintained between the various *lox* sites, as is the case with λ*att*P, *att*B, *att*L, and *att*R (Landy and Ross 1977). Second, due to the lack of sequence homology, the cut made during crossing-over is probably not a staggered cut as in the λ site-specific recombination system (Landy et al., this volume). In addition to the differences in biological roles, the physical data underscore the considerable differences between these two recombination systems.

ACKNOWLEDGMENTS

We thank Drs. Richard Musso and Spenser Benson for their critical reading of this manuscript. This research was sponsored by the National Cancer Institute under contract NO1-CO-75380 with Litton Bionetics, Inc.

REFERENCES

Bächi, B. and W. Arber. 1977. Physical mapping of *Bgl*II, *Bam*HI, *Eco*RI, *Hin*dIII, and *Pst*I restriction fragments of bacteriophage P1 DNA. *Mol. Gen. Genet.* **153:** 311.

Bachmann, B. and K. B. Low. 1980. Linkage map of *Escherichia coli* K12, edition 6. *Microbiol. Rev.* **44:** 1.

Chesney, R. H. and J. R. Scott. 1978. Suppression of a thermosensitive *dnaA* mutation of *Escherichia coli* by bacteriophage P1 and P7. *Plasmid* **1:** 145.

Chesney, R. H., J. R. Scott, and D. Vapnek. 1979. Integration of the plasmid prophages P1 and P7 into the chromosome of *Escherichia coli. J. Mol. Biol.* **130:** 161.

Dennert, G. and V. Henning. 1968. Tyrosine-incorporating amber suppressors in *Escherichia coli. J. Mol. Biol.* **33:** 322.

Enquist, L. and R. A. Weisberg. 1976. The red plaque test: A rapid method for identification of excision defective variants of bacteriophage lambda. *Virology* **12:** 147.

Enquist, L., A. Kikuchi, and R. A. Weisberg. 1979. The role of λ integrase in integration and excision. *Cold Spring Harbor Symp. Quant. Biol.* **43:** 1150.

Gottesman, M. and M. B. Yarmolinsky. 1968. Integration-negative mutants of bacteriophage λ. *J. Mol. Biol.* **31:** 487.

Hertman, I. and J. R. Scott. 1973. Recombination of phage P1 in recombination deficient hosts. *Virology* **53:** 468.

Holliday, R. 1964. A mechanism for gene conversion in fungi. *Genet. Res.* **5:** 282.

Ikeda, H. and J. Tomizawa. 1969. Prophage P1, an extrachromosomal replication unit. *Cold Spring Harbor Symp. Quant. Biol.* **33:** 79.

Landy, A. and W. Ross. 1977. Viral integration and excision: Structure of the lambda *att* sites. *Science* **197:** 1147.

Maxam, A. and W. Gilbert. 1977. A new method for sequencing DNA. *Proc. Natl. Acad. Sci.* **74:** 560.

Miller, H. I., A. Kikuchi, H. A. Nash, R. A. Weisberg, and D. I. Friedman. 1979. Site-specific recombination of bacteriophage λ: The role of host gene products. *Cold Spring Harbor Symp. Quant. Biol.* **43:** 1121.

Murray, N. E. and K. Murray. 1974. Manipulation of restriction targets in phage λ to form receptor chromosomes for DNA fragments. *Nature* **251:** 446.

Rosner, J. L. 1972. Formation, induction, and curing of bacteriophage P1 lysogens. *Virology* **48:** 679.

Scott, J. R. 1968. Genetic studies on bacteriophage P1. *Viology* **41:** 66.

Sobell, H. M. 1972. Molecular mechanisms for genetic recombination. *Proc. Natl. Acad. Sci.* **69:** 2483.

Sternberg, N. 1979a. Demonstration and analysis of P1 site-specific recombination using λ-P1 hybrid phages constructed *in vitro. Cold Spring Harbor Symp. Quant. Biology* **43:** 1143.

———. 1979b. A characterization of bacteriophage P1 DNA fragments cloned in a λ vector. *Virology* **96:** 129.

Sternberg, N. and R. A. Weisberg. 1975. Packaging of prophage and host DNA by coliphage λ. *Nature* **71:** 568.

Sternberg, N., D. Tiemeier, and L. Enquist. 1977. *In vitro* packaging of a λ*Dam* vector containing *Eco*RI DNA fragments of *Escherichia coli* and phage P1. *Gene* **1:** 255.

Walker, D. and J. T. Walker. 1975. Genetic studies of coliphage P1. I. Mapping by use of prophage deletions. *J. Virol.* **16:** 525.

Yarmolinsky, M. B. 1978. Genetic and physical structure of bacteriophage P1 DNA. In *DNA insertion elements, plasmids, and episomes* (ed. A.I. Bukhari et al.), p. 721. Cold Spring Harbor Laboratory, Cold Spring Harbor, New York.

Zissler, J., E. Signer, and F. Schaefer. 1971. The role of recombination in the growth of bacteriophage λ. I. The gamma gene. In *The bacteriophage lambda* (ed. A. D. Hershey), p. 454. Cold Spring Harbor Laboratory, Cold Spring Harbor, New York.

Mechanism of Bacteriophage Mu DNA Transposition

G. CHACONAS, R. M. HARSHEY, N. SARVETNICK, AND A. I. BUKHARI
Cold Spring Harbor Laboratory, Cold Spring Harbor, New York 11724

The temperate bacteriophage Mu is a giant transposon under the cloak of a virus (see Bukhari 1976 for previous discussion). It has genes needed for its replication and transposition, genes for head and tail morphogenesis, and an intricate system to control the various functions. In addition, it contains genes for some highly interesting functions such as *gin,* which is required for the flip-flop of the G segment, and *mom,* which is involved in DNA modification, a function that is not yet clearly understood. The organization of the 37-kb-long Mu genome is shown in Figure 1.

Mu exhibits great recombinational versatility. It can carry out all the genetic rearrangements characteristic of movable genetic elements. The genetic rearrangements that have been well characterized are shown in Figure 2. These rearrangements cause fusion or dissociation of DNA molecules and rapidly rearrange host DNA when the Mu functions are fully active. It can be seen from the summary of genetic rearrangements in Figure 2 that the frequency of Mu transposition is extraordinarily high. During the normal Mu lytic cycle, the number of Mu DNA copies made is around 100 per cell. As each copy is integrated into the host chromosome, there are about 100 transposition events during the lytic cycle. This high frequency of transposition allows us to study the process of transposition not only genetically, but also biochemically, and to examine by electron microscopy structures of the intermediates involved.

It was suggested a few years ago that replication of Mu DNA is involved in its integration at a different site (Bukhari 1976; Bukhari et al. 1977; Ljungquist and Bukhari 1977; Ljungquist et al. 1979). We showed that upon induction, a prophage Mu is not excised, and yet copies of Mu DNA are found inserted at new places (Ljungquist and Bukhari 1977). Evidently, Mu DNA duplicates itself during the transposition process. Our efforts during the past two years have been to examine this idea critically and to develop a system for dissecting the transposition process in molecular terms. In this paper we discuss the current status of the Mu integration problem based on the experiments done in our laboratory. These experiments have sustained the replication-integration hypothesis and have shown that the donor molecules containing Mu DNA physically interact with the target DNA during the transposition process.

STUDIES WITH PLASMIDS CONTAINING MU INSERTIONS AND MINI-MU

One of the hallmarks of the lytic cycle of bacteriophage Mu is that Mu DNA is always found associated with the host DNA. No evidence has been obtained for the existence of a pool of free replicating copies of Mu (Ljungquist and Bukhari 1977, 1979; Ljungquist et al. 1979). It would then be a difficult proposition to find a replicative or integrative intermediate of Mu DNA. To solve this difficulty, we have isolated small plasmids containing either Mu or Mu derivatives (Chaconas et al. 1981). Some of these Mu derivatives are diagramed in Figure 3. From a prototype pSC101 plasmid carrying a whole Mu prophage, we removed (in vitro) most of the middle of the Mu DNA, generating an internally deleted Mu, referred to here as mini-Mu. In the mini-Mu, we further inserted selectable markers such as a gene for ampicillin resistance (Ap^r), a gene for kanamycin resistance (Km^r), and the entire *lac* operon of *Escherichia coli.* These plasmids offer three advantageous ways of examining the Mu transposition problem. (1) Their behavior in response to Mu induction can be examined easily, since all of their forms can be differentiated by agarose gel electrophoresis. (2) The selectable markers inserted within mini-Mu can be used to follow the transposition-replication of mini-Mu. (3) Various methods can be devised to isolate the plasmid molecules after Mu induction, and the structures can be examined by electron microscopy.

Our basic procedure for studying the behavior of these plasmids has been to induce the Mu functions (in the case of mini-Mu, a helper Mu prophage is needed), extract the total cellular DNA, and fractionate the DNA on dilute agarose gels by electrophoresis. The DNA fragments are then blotted onto nitrocellulose paper (Southern 1975) and probed with either ^{32}P-labeled pSC101 DNA or ^{32}P-labeled Mu DNA. The DNA is also examined by electron microscopy. We reported earlier that when a pSC101 Mu plasmid is induced, the plasmid begins to disappear from the cell as a separate entity and enters the chromosomal DNA (Chaconas et al. 1980). A similar result is obtained with mini-Mu plasmids, except that results are not as dramatic for various physiological reasons. For example, the induction process is slow presumably because of the extra copies of the Mu repressor gene. We inferred from studies using electron microscopy that the association of the mini-Mu plasmids with the chromosome requires the Mu *A* and the Mu *B* genes (Chaconas et al. 1980). Further evidence for this conclusion is shown in Figure 4. It can be seen in the figure that upon induction of the wild-type helper phage, the ^{32}P-labeled pSC101 plasmid hybridizes with the chromosomal DNA, but that in Mu*ctsA*am and Mu*ctsB*am mutants, such association of the plasmid with the chromosome does not occur.

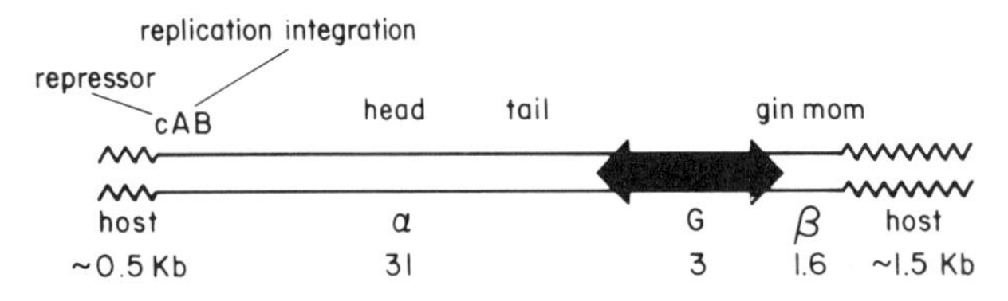

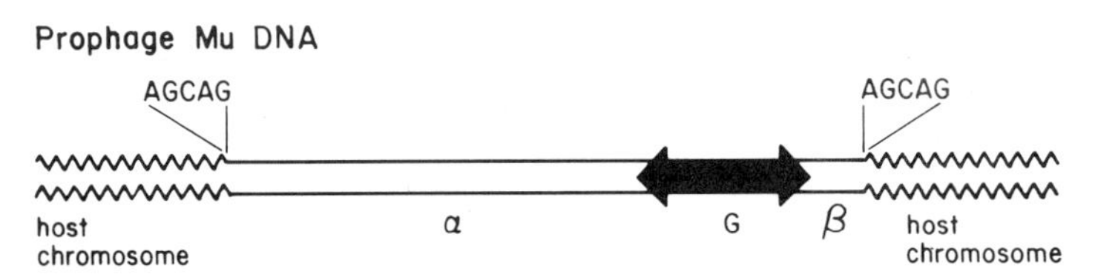

Figure 1. The genome of bacteriophage Mu. Mature Mu DNA is a linear molecule containing host DNA (heterogeneous in length and sequence) at both ends. The lengths of the different segments in Mu are given in kilobases. The G segment is an invertible region whose flip-flop depends upon the *gin* gene. The *mom* gene is involved in modification of DNA (for reviews, see Bukhari 1976 and O'Day et al. 1979). Prophage Mu DNA and mature Mu DNA are colinear. Like many other transposable elements, Mu generates the duplication of five host base pairs upon integration (Allet 1979; Kamp et al. 1979). A hypothetical duplication of the sequence AGCAG is indicated at the ends of the prophage. (The DNA segments are not drawn to scale.)

BIOCHEMICAL ANALYSIS OF PLASMID-HOST INTERACTION

To understand the structures of the Mu-containing plasmid associated with the host chromosome, we induced the Mu functions, allowing the plasmids to migrate with the host DNA, and then fractionated the DNA on agarose gels by electrophoresis. The host DNA bands were then eluted from the agarose gels. The eluted DNA thus contained no free plasmid DNA. The host DNA was then cut with restriction endonucleases *Kpn*I (which cuts only once in Mu) and *Sma*I (which does not cut Mu but makes two very closely spaced cuts in pSC101). In the case of pSC101::Mu plasmids, *Kpn*I digestion released a linear form of the original plasmid (G. Chaconas et al., in prep.). As shown in Figure 5, this linear form could arise only if the pSC101 Mu plasmid was integrated in such a manner that the pSC101 sequences were flanked by one copy of Mu on each side as direct repeats. As expected, the endonuclease *Sma*I did not release a linear form of the plasmid from the host DNA. Instead, it released

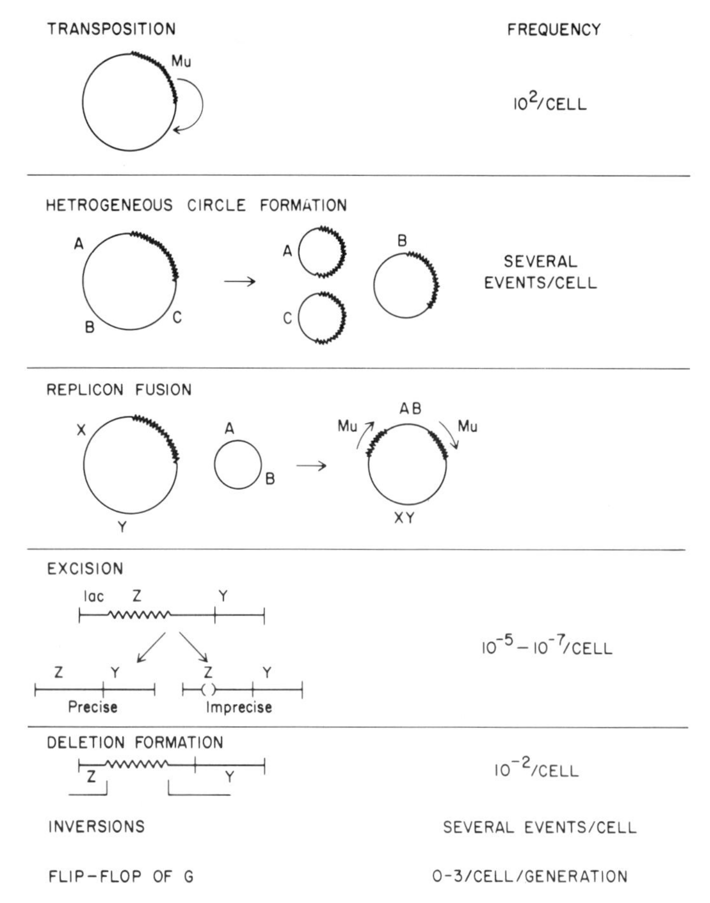

Figure 2. DNA rearrangements involving Mu DNA. Transposition refers to integration of Mu DNA copies at different sites during the lytic cycle. Heterogeneous circles arise during the Mu lytic cycle and contain at least one full-length copy of Mu DNA linked to different host sequences. Replicon fusion always results in the presence of one copy of Mu in direct orientation at each junction of the replicons. Excision of Mu DNA can be seen at a low frequency if prophages carry the *X* mutations (Bukhari 1975). For a discussion of the excision, deletion formation, and inversions, see Khatoon et al. (1979), Toussaint et al. (1977), and Faelen and Toussaint (1980). For a discussion of the G segment inversion, see Kamp et al. (1979) and van de Putte et al. (1980). All frequencies are approximate.

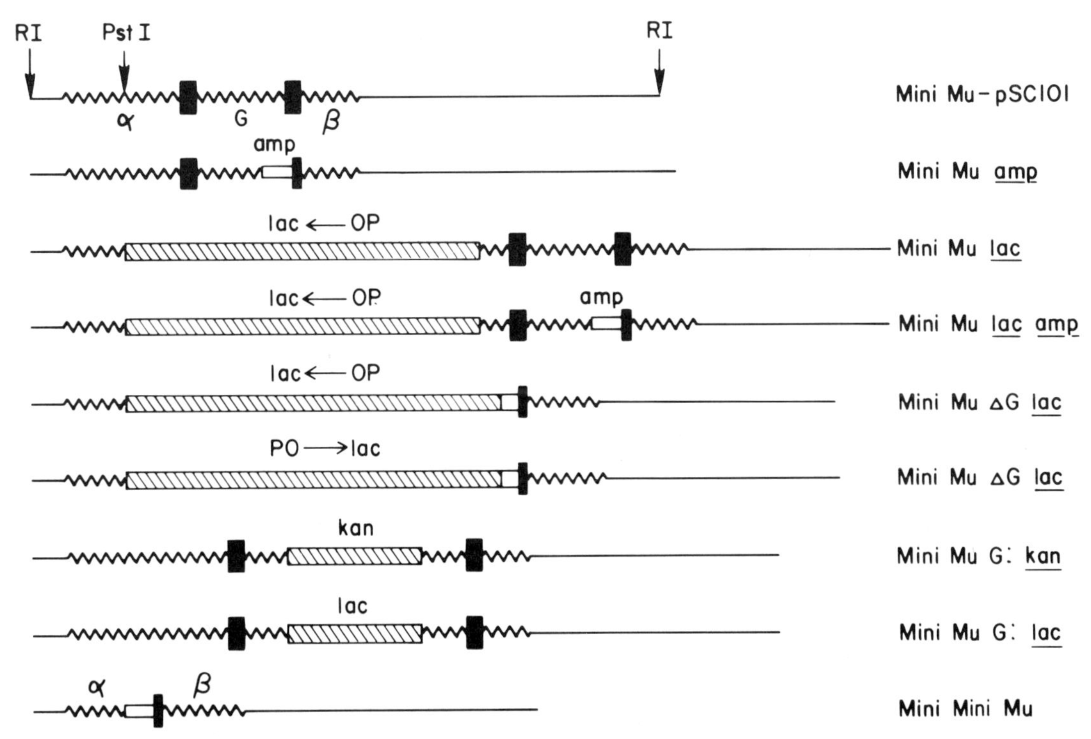

Figure 3. The pSC101 plasmids carrying deletions and insertions in Mu. A Mu*cts*62 (a temperature-inducible mutant) insertion in the pSC101 plasmid was isolated. The pSC101 plasmid is 9.09-kb long and is present in four to six copies per chromosome. (See Chaconas et al. [1981] for description of the methodology.) The 27-kb middle *Pst*I fragment of Mu was removed in vitro. This plasmid is the prototype pSC101::mini-Mu plasmid (mini-Mu121) (*top line*). A *Pst*I fragment carrying the *lac* operon (12.3 kb long) was inserted at the *Pst*I site in pSC101::mini-Mu (*line 3*). This is the mini-Mu*lac* plasmid (mini-Mu121*lac*). In mini-Mu121, a gene for ampicillin (Ap) resistance containing a *Pst*I cleavage site was recombined (*line 2*) and then the middle *Pst*I fragment removed to give mini-mini-Mu or mini-Mu222 (*line 9*). Mini-Mu*kan* contains a 6.6-kb insert containing a gene for kanamycin (Km) resistance at the *Kpn*I site. It should be noted that the gene for tetracycline (Tc) resistance is carried only on pSC101, and thus the presence of the Tc^r marker in genetic experiments indicates the presence of the plasmid sequences.

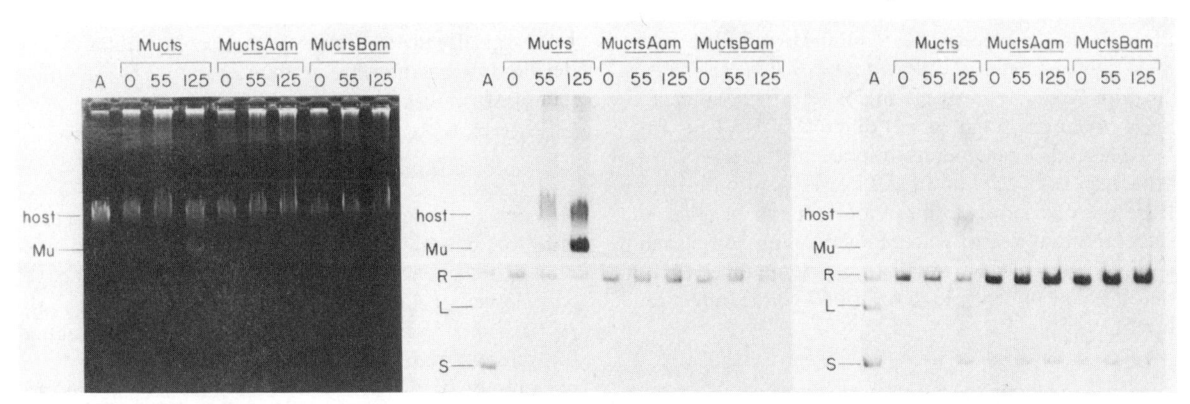

Figure 4. Association of mini-Mu plasmid pGC121 with host DNA following induction of a Mu*cts* prophage but not prophages carrying amber mutations in genes *A* or *B*. Total DNA was extracted from strains carrying the mini-Mu plasmid and a wild-type or defective Mu prophage before (0 time), at 55 minutes, and at 125 minutes after temperature shift up. The purified DNA was subjected to electrophoresis in 0.35% agarose gels and subsequently transferred to nitrocellulose paper by the method of Southern (1975). The blotted gels were hybridized to ^{32}P-labeled Mu DNA to monitor phage DNA replication (*middle panel*) and with ^{32}P-labeled pSC101 to follow association of the plasmid with the host DNA (*right panel*). The ethidium bromide stain is shown in the left panel and slot *A* consists of a plasmid marker containing all three forms of the plasmid (S, superhelical; L, linear; R, relaxed) mixed with *E. coli* DNA. There is no change in the plasmid bands in the presence of Mu*cts*62*A*am1093 and Mu*cts*62*B*am1066 (mutants described by O'Day et al. 1979), whereas with Mu*cts*62, the plasmid sequences begin to migrate with the host DNA at late times after induction.

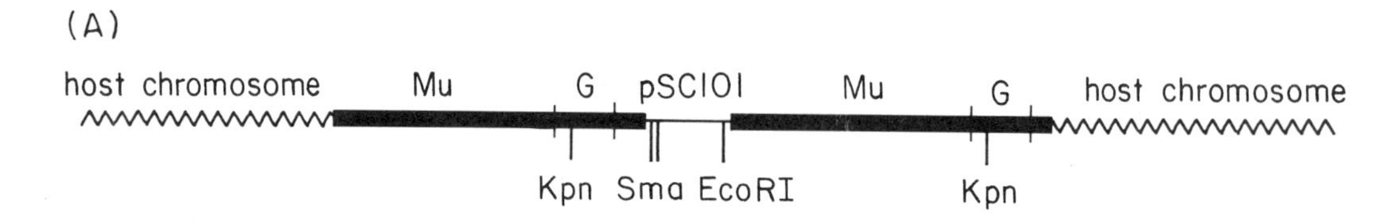

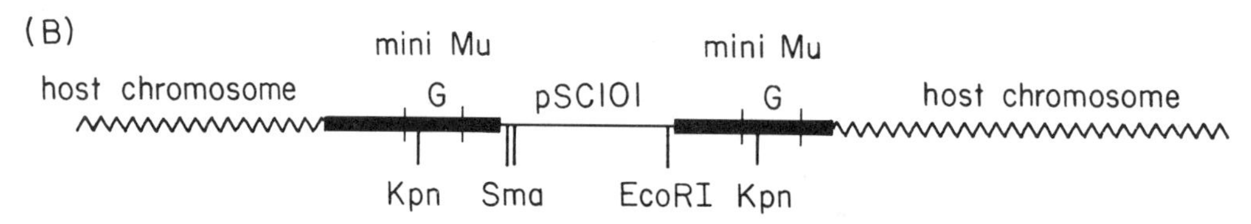

Figure 5. Structure of pSC101::Mu or pSC101::mini-Mu cointegrates. In response to Mu induction, the plasmids are inserted into random sites on the host chromosome, giving rise to the structures shown. (*A*) A cointegrate with full-length Mu; (*B*) a cointegrate with mini-Mu. *Kpn*I digestion would cut once in each copy of Mu, releasing a linear copy of the plasmid. *Sma*I, however, makes two very closely spaced cuts in pSC101. The *Sma*I cuts would be in random locations in the host chromosome. Thus, *Sma*I digestion would release variable fragments containing the plasmid and the Mu sequences. *Eco*RI makes only one cut in the plasmid and none in mini-Mu. (There are two *Eco*RI cuts in full-length Mu DNA that are not shown here.) *Eco*RI digestion of host DNA containing mini-Mu cointegrates would thus release variable fragments (see Fig. 6).

fragments of various sizes, but larger than the pSC101 Mu plasmid. Since *Sma*I makes two very closely spaced cuts in the pSC101 part of the Mu structure, the outside cuts would be in random places in host DNA.

When the restriction enzyme digestion experiment was repeated with the mini-Mu plasmids, a new form of association, in addition to the cointegrate formation, was seen. The experiment shown in Figure 6 provides evidence that only a small part of the mini-Mu plasmid is present in the host chromosome in the form of cointegrates, and the rest is attached to the chromosome. When the eluted host DNA containing the mini-Mu plasmid is cut with *Kpn*I, a linear mini-Mu plasmid would be expected to be released if the plasmid were in the mini-Mu–pSC101–mini-Mu form (see Fig. 5). However, in Figure 6 it can be seen that a linear form of the plasmid appears but that the radioactivity is distributed throughout the column, as if mini-Mu–pSC101 is attached to fragments of different sizes. When the DNA is cut with *Sma*I, no linear plasmid is seen. The radioactivity is present in fragments of different sizes. The mini-Mu plasmid is not merely trapped in the host DNA; otherwise, both *Kpn*I and *Sma*I would quantitatively release the plasmid in a linear form. It can be inferred, therefore, that a large fraction of the mini-Mu plasmid is physically attached to the chromosome of DNA but that it is not present in the form of a cointegrate structure.

ANALYSIS OF PLASMID-HOST INTERACTION BY ELECTRON MICROSCOPY

The mini-Mu plasmids are small enough in size to be easily visualized and measured by electron microscopy. When the total DNA is examined after Mu induction, up to 20% of the observable mini-Mu plasmids are found to be in association with the host DNA. Figure 7 shows two examples of such molecules. The circular plasmid appears to be attached at a point on the host DNA. It is not clear whether the attachment point on the plasmid is random or specific. We believe that these structures are the intermediates in the process of mini-Mu transposition. Structures in which a circular mini-Mu plasmid is attached to a long tail are also seen (Harshey and Bukhari 1981). These structures are called "key structures" (see Appendix) and presumably are also intermediates of the mini-Mu integration process.

ANALYSIS OF COINTEGRATES

As discussed above, the pSC101::Mu and pSC101::mini-Mu plasmids generate cointegrate structures in which the plasmid is sandwiched between two Mu or mini-Mu molecules. We have analyzed genetically and biochemically the formation of these cointegrates to understand the relationship of these structures to the problem of Mu transposition.

Transduction Experiments

When a Mu prophage is induced in the presence of a mini-Mu plasmid and a phage lysate is prepared, the mini-Mu sequences are found packaged in Mu phage particles as expected from previous results (Faelen et al. 1979). Table 1 shows the data on mini-Mu packaging for different mini-Mu's. It can be seen that mini-Mu is not present in overwhelming numbers as compared to the wild-type Mu. The prototype mini-Mu constitutes about 30% of the molecules extracted from the phage lysate. The proportion of mini-Mu's containing insertions is even less. Thus, the mini-Mu apparently has no particular advantage in replication (or may have a disadvantage in packaging). It should be noted that mini-Mu plasmids are too small to be packaged into Mu particles; thus, they must be attached to larger molecules as

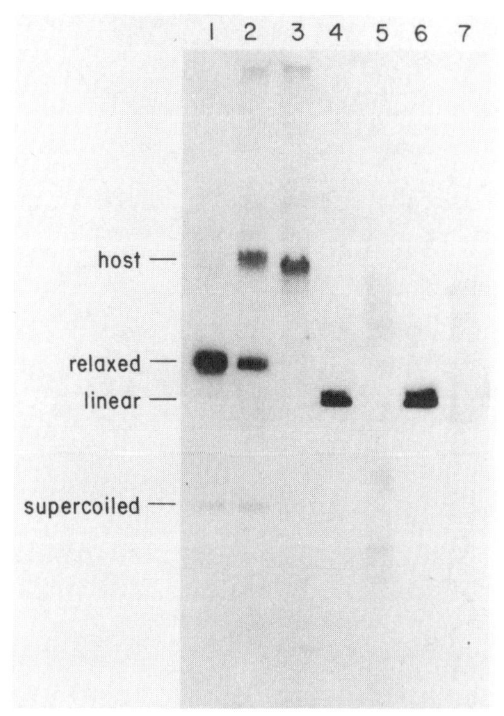

Figure 6. Evidence for the attachment of mini-Mu plasmid to the host chromosome after Mu induction. A strain containing mini-Mu121 (Fig. 3) and a Mu*cts*62 prophage was induced, and the DNA was treated and hybridized with pSC101 as outlined in Fig. 4. After electrophoresis, the agarose gel area containing the host DNA was cut and the DNA eluted. The eluted DNA was cut with *Kpn*I or *Sma*I. (*1*) 0-time DNA sample (before Mu induction). (*2*) Total cellular DNA extracted 75 min after induction; the host DNA at this point contains plasmid sequences. (*3*) The host DNA band shown in *2* was eluted and rerun on gel; the DNA has been purified away from the free plasmid. (*4*) The total DNA from a 0-time sample (before Mu induction) is cut with *Kpn*I; all of the plasmid is free and becomes linear. (*5*) The eluted host DNA shown in *3* is cut with *Kpn*I. pSC101 sequences appear all through the column; a faint linear plasmid band can be seen. (*6*) The 0-time total DNA is cut with *Eco*RI; again, the linear form of the plasmid is generated. (*7*) The eluted host DNA shown in *3* is cut with *Eco*RI; the pSC101 sequences are randomly distributed; no linear form of the plasmid can be seen. The fact that not all of the plasmid from the host DNA is released in a linear form after induction by *Kpn*I shows that some of the plasmid is merely attached to the chromosomal DNA and is not in the cointegrate form (see Fig. 5).

demanded by the headful packaging mode of Mu DNA (Bukhari and Taylor 1975). The mini-Mu-containing molecules evidently are injected into the Mu-sensitive cells upon infection, since transductants containing mini-Mu can be obtained.

When *E. coli* cells were transduced to Lac^+, Km^r, or Ap^r using lysates containing either mini-Mu*lac*, mini-Mu*kan*, or mini-Mu*amp*, it was found that a large number of transductants were also tetracycline-resistant (Tc^r). Since the Tc^r marker is only carried on the plasmid pSC101, it is clear that the plasmid sequences can be packaged along with mini-Mu. The results of transduction experiments are summarized in Tables 2 and 3. In general, the transduction frequency is relatively low in comparison with the number of mini-Mu particles present in the lysate. Presumably, only a fraction of the mini-Mu-containing molecules can give rise to stable transductants. When the recipient is rec^+, the number of transductants that are also Tc^r (that is, that carry the pSC101 sequences) is high. When the multiplicity of infection is increased, the number of plasmid-containing transductants goes down. In *recA* recipients, the transductants that carry the plasmid sequences are very infrequent. Two conclusions can be drawn from this result: (1) Retention of the plasmid sequences requires recombination mediated by *rec* functions of the host. (2) When Mu functions are adequately provided (high multiplicity of infection [moi]), the plasmid sequences can be lost. Apparently, in this situation, the mini-Mu has a better chance of transposition into the host chromosome. It can be seen in Table 2 that if the recipient is a Mu lysogen, then almost all of the transductants retain the plasmid sequences. This is consistent with the inference that transductants that contain the mini-Mu markers but are Tc-sensitive (Tc^s) arise from transposition of mini-Mu into the chromosome. Since this transposition requires Mu functions, it does not occur in a Mu lysogen where Mu functions are repressed. Similar results are obtained with all mini-Mu's whether or not they contain the G segment of Mu.

The transductants that retain the plasmid sequences were found, by agarose gel electrophoresis of DNA, to contain the original plasmid molecules. This result is shown in Figure 8. Whereas the Tc^r transductants contained the original plasmid, the Tc^s transductants did not. This result is obtained with all mini-Mu's. It is clear that these plasmids are generated by recombination between mini-Mu sequences that flank the plasmid sequences. Two control experiments were done to check this point: (1) The DNA from the transducing lysate was examined for circular plasmid DNA molecules by electrophoresis on agarose gels and by electron microscopy; none were found. (2) The DNA was used to transform the cells; transformants did not arise at a significant frequency as would be expected if mini-Mu's were not present in a circular plasmid. Thus, at least some mini-Mu's are packaged in the form of cointegrates. This was confirmed directly by digesting the DNA prepared from transducing lysates with *Kpn*I and *Sma*I. The DNA was digested, blotted onto nitrocellulose paper, and hybridized with ^{32}P-labeled pSC101 DNA. If cointegrate structures were packaged, then *Kpn*I would generate a linear form of the plasmid (see Fig. 5). As shown in Figure 9, *Kpn*I digestion of DNA did generate the linear form of the plasmid. The *Sma*I digestion did not. The *Sma*I cuts within pSC101 and would presumably also cut randomly within host sequences attached to the mini-Mu–pSC101–mini-Mu structures (see Fig. 5), giving rise to heterogeneous fragments that would not be seen as a discrete band.

The formation and packaging of cointegrates requires both ends of Mu because plasmids containing only the left end or the right end are not transduced significantly. This result is shown in Table 4. It should be noted that there is some measurable transduction of plasmids con-

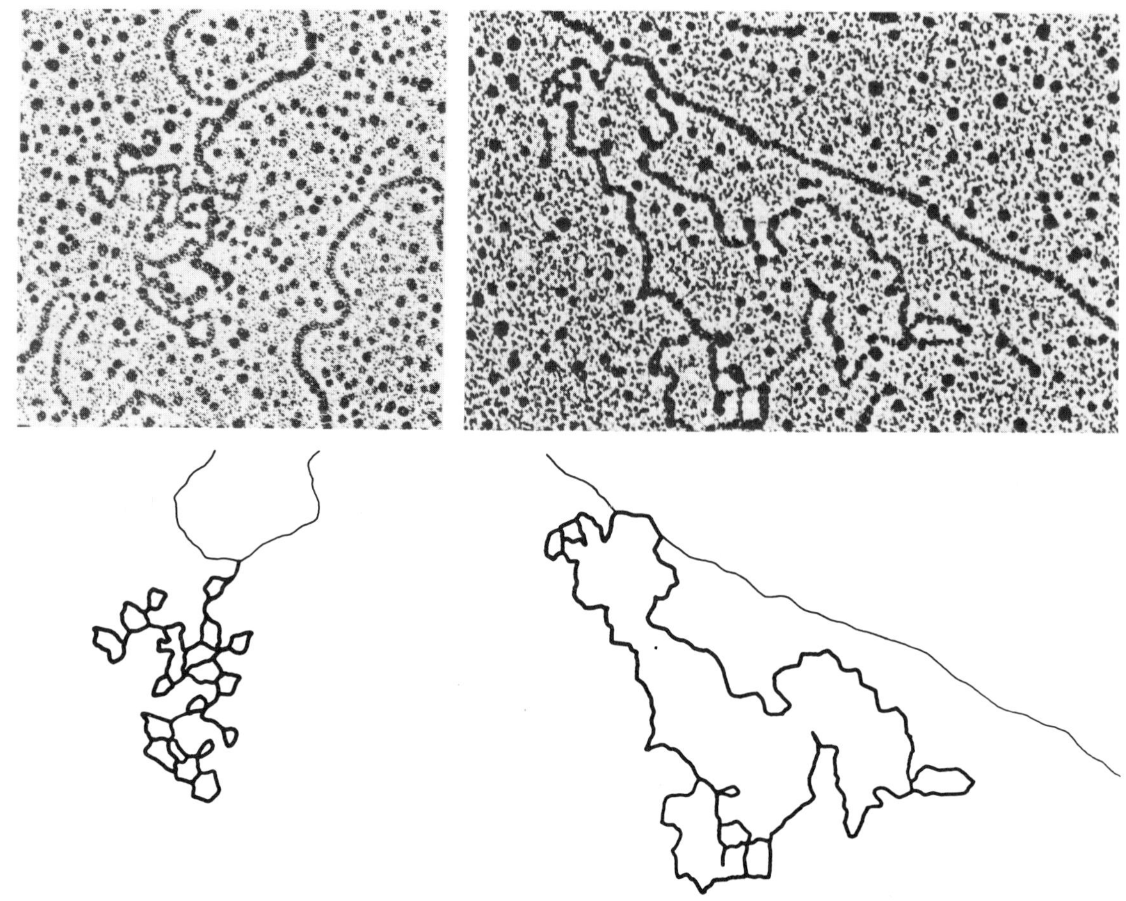

Figure 7. Association of mini-Mu plasmid molecules with the host chromosome. After Mu induction, the total DNA was extracted with phenol and spread for electron microscopy (see Chaconas et al. 1980). Either the supercoiled or open circles were seen to be associated with the chromosomal DNA. These circles were exactly the length of the mini-Mu plasmid.

taining only the left end of Mu in *rec*$^+$ recipients. This observation is being examined further in our laboratory.

Conjugation Experiments

Transposition of mini-Mu's and formation of cointegrates were also monitored by conjugation experiments.

Table 1. Packaging of Mini-Mu

Type of mini-Mu	Phage particles containing mini-Mu (%)
mini-Mu	30
mini-Mu*kan*	8
mini-Mu*amp*	10
mini-Mu*lac*	5

Strains containing mini-Mu's and a helper Mu*cts*62 prophage were induced and the phage particles were concentrated by the standard procedures (see Bukhari and Ljungquist 1977). The phage DNA molecules were then denatured and renatured, and the heteroduplexes were examined by electron microscopy. Full-length Mu DNA molecules and those showing mini-Mu's with large deletion loops were counted. Number of mini-Mu heteroduplexes among total number of molecules counted was used to estimate the percentage of phage particles containing Mu. For description of mini-Mu's, see Fig. 3.

The basic design of these experiments was to induce strains containing a mini-Mu plasmid, an F′ *pro*$^+$*lac*$^+$ episome, and a temperature-inducible prophage on the chromosome. At different times after induction, the strains were mated with F$^-$ Δ (*pro lac*) strains, and exconjugants were selected for either *pro*$^+$ marker or for the marker contained in the mini-Mu (e.g., Ap^r or Km^r). One out of ten episomes transferred carried the mini-Mu sequences. Most of the mini-Mu-carrying episomes also carried the plasmid sequences, as judged by the presence of the Tc^r marker. In a Mu$^-$ *recA* recipient, the frequency of mini-Mu and pSC101 cotransfer reached 98–100%. These episomes were shown to have standard cointegrate structures—mini-Mu–pSC101–mini-Mu inserted in the episome. Those transferred episomes that did not have the pSC101 sequences were found to have deletions of various sizes. It would seem therefore that the predominant event involving mini-Mu transposition after Mu induction is cointegrate formation. These cointegrates are stable and are not resolved at any appreciable frequency.

Conjugation experiments were also done with strains containing an F′ *pro*$^+$*lac*$^+$ episome and a Mu prophage Mu*cts*62*amp* (Leach and Symonds 1979) on the chromosome, but no mini-Mu plasmids. These experiments

Table 2. Transduction of Mini-Mu*lac* into *rec*$^+$ and *recA* Lysogenic and Nonlysogenic Recipients

Recipient strain	Relevant genotype	Mini-Mu plasmid	Mu G region	Selected marker	Helper prophage	moi	Transduction frequency[a]	Tc^r transductants (%)
610	*rec*$^+$	pCL198	yes	Lac$^+$	Mu*cts*	0.05	8.4×10^{-4}	76
610	*rec*$^+$	pCL198	yes	Lac$^+$	Mu*cts*	3.0	5.6×10^{-6}	33
BU5025	*recA*	pCL198	yes	Lac$^+$	Mu*cts*	0.05	6.8×10^{-5}	5
BU5025	*recA*	pCL198	yes	Lac$^+$	Mu*cts*	3.0	1.9×10^{-6}	3
610	*rec*$^+$	pGC404	no	Lac$^+$	Mu*cts amp*	0.05	2.0×10^{-5}	93
610	*rec*$^+$	pGC404	no	Lac$^+$	Mu*cts amp*	3.0	4.3×10^{-5}	86
BU5025	*recA*	pGC404	no	Lac$^+$	Mu*cts amp*	0.05	1.2×10^{-5}	2
BU5025	*recA*	pGC404	no	Lac$^+$	Mu*cts amp*	3.0	1.7×10^{-5}	2
BU5031	*rec*$^+$, Muc$^+$ lysogen	pGC404	no	Lac$^+$	Mu*cts amp*	0.05	1.0×10^{-4}	98
BU5031	*rec*$^+$, Muc$^+$ lysogen	pGC404	no	Lac$^+$	Mu*cts amp*	3.0	3.0×10^{-5}	99
BU5030	*recA*, Muc$^+$ lysogen	pGC404	no	Lac$^+$	Mu*cts amp*	0.05	5.2×10^{-6}	100
BU5030	*recA*, Muc$^+$ lysogen	pGC404	no	Lac$^+$	Mu*cts amp*	3.0	4.0×10^{-7}	97

Transducing lysates were prepared by inducing *recA*$^+$ strains containing mini-Mu and a helper temperature-inducible prophage. The lysates were mixed with the recipient strains at two different moi. The mini-Mu plasmids were pCL198 mini-Mu121*lac* (Fig. 3) and pGC404 mini-Mu222*lac* (Fig. 3). Mu G region refers to the presence or absence of the G segment in the mini-Mu. Relevant genotype refers to the genotype of the recipient strains. All recipient strains were isogenic derivatives of strain 610 with *lacZ*Δ and streptomycin sensitivity (Sm^s), The *recA* allele used was *recA*56.

[a] Transduction frequency is expressed as transductants recovered per input plaque forming unit.

were essentially a repetition of the earlier experiments that suggested that Mu integrates at different sites during its lytic cycle (Razzaki and Bukhari 1975). As expected, 1–3% of the transferred episomes had Mu*cts*62*amp* inserted in them. However, when these episomes were examined by gel electrophoresis, they were found to have insertions and deletions of various sizes. We infer from this result that Mu does not jump directly onto the episome from the chromosome but that the episome is first inserted in the host DNA as a cointegrate and comes out of the chromosome carrying various rearrangements.

INSERTION OF MU DURING LYSOGENIZATION

The result that insertion of a Mu prophage into F′ *pro*$^+$*lac*$^+$ episome after induction predominantly results in deletions and insertions is in complete contrast to the experiments in which insertion of Mu into F′ *pro*$^+$*lac*$^+$ was examined after lysogenization. When F′ *pro*$^+$*lac*$^+$-containing cells were infected with Mu and insertions of Mu in the *lac* operon were isolated, most of the episomes were found to carry faithful point insertions; that is, insertion of Mu occurred without causing a deletion. Many of the F′*pro*$^+$*lac* episomes carrying these insertions were analyzed on gels and were shown to be of the size expected for a normal F′*pro*$^+$*lac* containing a Mu insertion.

CONCLUSIONS

The experiments on the behavior of pSC101 Mu and pSC101 mini-Mu plasmids have provided strong evidence for the validity of the replication-integration model of Mu integration. No excision of Mu DNA is seen, and the association of the donor DNA containing Mu and the recipient DNA is obligatory in the twin processes of replication and transposition.

The predominant end products of the Mu transposition reaction during the lytic cycle appear to be the cointegrates and not simple transposition events. This stability of the cointegrates implies that there is no high-powered function that can resolve the cointegrates. On the other hand, Mu integration after infection leads to simple insertion in which the host DNA originally attached to the ends of mature Mu DNA is lost. This result implies that there may be an alternate mode of Mu integration in which cointegrate formation is not necessary. If during lysogenization Mu does go through a

Table 3. Transduction of mini-Mu*kan* and mini-Mu*amp* into *rec*$^+$ and *recA* Nonlysogenic Recipients

Recipient strain	Relevant genotype	Mini-Mu plasmid	Mu G region	Selectable marker	Helper prophage	moi	Transduction frequency[a]	Tc^r transductants (%)
40	*rec*$^+$	pGC102	yes	Km^r	Mu*cts amp*	0.05	5.0×10^{-4}	76
40	*rec*$^+$	pGC102	yes	Km^r	Mu*cts amp*	3.0	2.1×10^{-4}	67
BU8559	*recA*	pGC102	yes	Km^r	Mu*cts amp*	0.05	7.8×10^{-5}	3
BU8559	*recA*	pGC102	yes	Km^r	Mu*cts amp*	3.0	1.9×10^{-5}	1
40	*rec*$^+$	pGC501	no	Ap^r	Mu*cts*	0.05	2.6×10^{-4}	67
40	*rec*$^+$	pGC501	no	Ap^r	Mu*cts*	3.0	1.4×10^{-6}	12
BU8559	*recA*	pGC501	no	Ap^r	Mu*cts*	0.05	1.7×10^{-5}	2
BU8559	*recA*	pGC501	no	Ap^r	Mu*cts*	3.0	1.0×10^{-7}	5

The procedures were the same as described in the notes to Table 2. The mini-Mu plasmids were pGC102 mini-Mu121*kan* and pGC501 mini-Mu222*amp*. The recipient strains were 40 (Δ[*pro lac*], *trp-8*) and 8559 (Δ[*pro lac*], *met*, *recA*56).

[a] Transduction frequency is expressed as transductants recovered per input plaque forming unit.

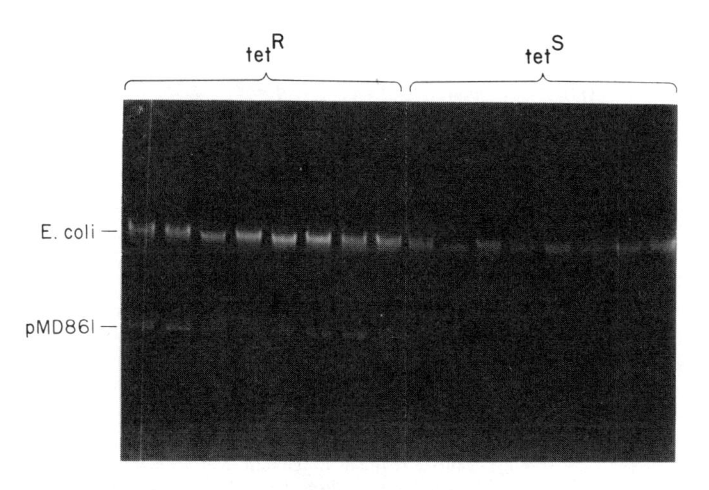

Figure 8. Transduction of a mini-Mu*amp* plasmid. Transductants obtained after mini-Mu*amp* transduction, as described in Table 3, were grown on plates. The cells were scraped off with toothpicks and lysed in 100 μl of 50 mM Tris-HCl, pH 8.0, containing 1% SDS, 2 mM EDTA, 0.4 M sucrose, and 0.01% bromophenol blue. The lysate was spun for 10 min in an Eppindorf Desk-Top centrifuge, and the supernatant was loaded directly onto a 0.35% agarose gel. The leftmost lane shows a control strain known to carry the mini-Mu*amp* plasmid (pMD861, line 2 in Fig. 3). It can be seen that all Tc^r transductants carry the original plasmid, but the Tc^s transductants do not. Since no circle plasmid is found in the phage particles, the circular plasmids in transductants must have arisen by recombination (by a *rec*-mediated resolution of the packaged cointegrate structures).

cointegrate mode of integration, then it would result in two copies of Mu in the chromosome, with the host sequences, originally attached to the ends of mature Mu DNA, in the middle. Since these sequences are not joined in the phage DNA (i.e., mature Mu DNA is linear), this would result in breakage of the chromosome. It can also be argued that some low-level recombination function recombines the two copies of Mu, regenerating a linear Mu DNA copy and a copy inserted into the chromosome as a prophage. In either case, some step in the transposition process in the lytic cycle must be different from the integration process during lysogenization.

We find that the mini-Mu sequences, although much smaller than the wild-type Mu, have no particular advantage in replication or integration. In fact, they seem to be at a disadvantage, since pSC101 containing the whole Mu enters the chromosome in toto, whereas only a portion of the mini-Mu plasmid interacts with the chromosome. Although some mini-Mu plasmids generate cointegrate structures, others apparently are merely attached to the chromosome, as shown by restriction endonuclease analysis. These mini-Mu plasmids are probably stuck at some intermediate step in transposition. This quantitative difference in the behavior of the mini-Mu plasmids and the plasmids containing the whole Mu prophage probably is a reflection of the relative amounts of the *A* and *B* proteins available. In the case of mini-Mu plasmids, the *A* and *B* proteins are supplied by a helper prophage. Thus, there are many more Mu ends than the number of copies of *A* and *B* genes. In the case of pSC101 Mu plasmids, however, each transposable copy of Mu carries the *A* and *B* genes. There is also the possibility that one or both of these proteins prefer to act in a *cis* manner, that is, they act preferentially on the DNA molecules from which their

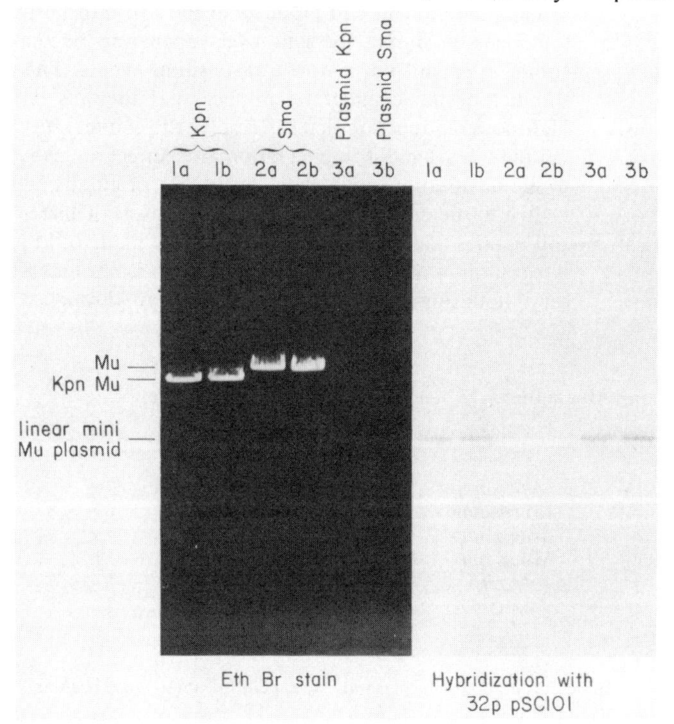

Figure 9. Packaging of mini-Mu plasmid cointegrates in mature virions. The phage particles prepared from strains containing mini-Mu and Mu*cts*62 were prepared, concentrated, and the DNA was extracted with phenol. The DNA samples were cut with either *Kpn*I or *Sma*I. Columns *3a* and *3b* show purified mini-Mu plasmid DNA cut with the enzymes (generating the linear form of the plasmid). *Kpn*I digestion of phage DNA gives rise to the linear form of the plasmid, as seen after hybridization with ^{32}P-labeled pSC101. *Kpn*I Mu refers to the large fragment released from wild-type Mu DNA by *Kpn*I. *Sma*I digestion does not show the linear form of the plasmid. These results are consistent with the structure shown in Fig. 5.

Table 4. Transduction of Tc^r by Lysates from Mu*cts* Lysogens Harboring Mu Left-end or Right-end Plasmids or pSC101

Recipient strain	Relevant genotype	Plasmid	Mu region	moi	Transduction frequency[a]
40	*rec*$^+$	pSC101	no	0.3	$<1 \times 10^{-7}$
BU8559	*recA*	pSC101	no	0.3	$<1 \times 10^{-7}$
40	*rec*$^+$	pGC302	c end	0.3	8.3×10^{-7}
BU8559	*recA*	pGC302	c end	0.3	$<1 \times 10^{-7}$
40	*rec*$^+$	pCL151	*S* end	0.3	$<1 \times 10^{-7}$
BU8559	*recA*	pCL151	*S* end	0.3	$<1 \times 10^{-7}$

The pGC302 plasmid carried the left-end *Pst*I fragment of Mu on pSC101; pCL151 carried the right-end *Eco*RI fragment on pSC101. (See Chaconas et al. [1981] for construction of strains.) For the genotype of the recipient strains, see Table 3.

[a] Transduction frequency expressed as transductants recovered per input plaque forming unit.

synthesis is initiated. The next step in understanding the Mu transposition problem would appear to be to study the synthesis, regulation, and the action of the *A* and *B* proteins.

Finally, it should be noted that the ends of Mu are the primary determinants of both replication and integration. This inference follows from studies on the mini-Mu's that retain only the sequences near the ends of Mu. Mu thus defines a new system of replication, which may be called integrative replication.

Mu Transposition in Perspective

It is now abundantly clear that insertion sequence (IS) elements, transposons (Tn), and bacteriophage Mu all exhibit parallel behavior. It follows, therefore, that the mechanisms by which these DNA sequences undergo transposition have important elements in common. We can compare Mu with the widely studied transposon for Ap^r, Tn*3*. Tn*3* contains a gene for transposase and a repressor gene that regulates the synthesis of the transposase. In Mu, the repressor controls the *A* and *B* genes that are required for the full-scale transposition of Mu during the lytic cycle. The *A* gene is absolutely required for Mu integration and transposition (Razzaki and Bukhari 1975; O'Day et al. 1978; Faelen et al. 1979). However, in the absence of the *B*-gene function, transposition occurs at a much lower frequency. An A^+ B^- mini-Mu cannot kill the host cells and undergoes transposition at a frequency of from 10^{-2} to 10^{-3} per cell (Faelen et al. 1979) and, in fact, is a genetic element almost exactly equivalent to a transposon.

In Tn*3*, the cointegrates (Tn*3*-plasmid-Tn*3* structures) are formed and are resolved at a high frequency. These structures may be intermediates in transposition, as originally pointed out by Gill et al. (1978). In Mu, the cointegrate structures are also formed at a high frequency, but we have found no evidence for the specific Mu function that would resolve these structures at a high frequency. Mu may be able to switch to an alternate mode of transposition, in which one copy of Mu is transposed to a new site and the other copy is released instantly. Thus, some transposable elements may exhibit alternate modes; they may form cointegrates or they may undergo simple transposition. The switching of modes may depend upon the availability of proteins that act at the ends of the elements. Thus, the elements that can form cointegrates but cannot switch the modes may then have evolved a specific recombination system to resolve the cointegrates. A protein-mediated switch in the mode of transposition of an element is a working hypothesis that can be put to a test in the Mu system.

ACKNOWLEDGMENTS

We are thankful to the National Science Foundation (grant no. PCM78-26710) and to the National Institutes of Health (grant no. GM-23566) for their support. We are also thankful to the Canadian Medical Research Council for a fellowship to G. C., and to the National Institutes of Health for a Career Development Award to A. I. B. Also appreciated is the hard work of Louisa Dalessandro and Mike Ockler in the preparation of this manuscript.

APPENDIX

A Model for Mu Transposition

R. M. HARSHEY AND A. I. BUKHARI

As discussed in the accompanying paper, electron microscopy studies of transposition using mini-Mu plasmids showed structures in which the plasmids seemed to be attached to the host DNA at a single point along their length. Structures in which plasmid-size circles were attached to tails of an indeterminate length were also seen. We will refer to the latter kind of a structure, in which a circle is attached to a tail, as a "key" structure. Such key structures have been reported earlier and are also seen when a Mu lysogen is induced (Shröder et al. 1974). They appear as early as ten minutes after induction and their number increases with time. Their occurrence is dependent on the presence of Mu genes *A* and *B*, and hence we assume that their appearance is related to the replication of Mu. The circular parts of the key structures vary greatly in length, as do the tails attached to them. Out of 50 key structures measured, the smallest circles were about 1 kb in length and the largest was about 200 kb long. The circular parts of most key structures examined were shorter than one Mu in length. An example of such a key structure is shown in Figure 10.

To explain the mini-Mu plasmid structures attached at a point on the host chromosome and the generation of the key structures, we have proposed a model for transposition (Fig. 11). In the figure, transposition of an element within a single replicon is depicted. We have divided the transposition process into four steps:

1. *Association.* The transposon is brought to the target site on the recipient molecule by proteins that recognize sequences at both of its ends.
2. *Attachment.* The target site undergoes a double-strand cut. Replication of the transposon is initiated at one of its ends by first nicking one strand and ligating it to the exposed 5′ phosphate of the target strand. A 3′-hydroxyl of the complementary target strand is used as a primer to extend into the transposon while the free target strand is protected and held in place by a replication complex, fixing the replication point.
3. *Roll-in replication.* DNA is replicated at this fixed complex as it reels through. DNA synthesis may be discontinuous on the opposite donor strand.
4. *Roll-in termination.* Replication is terminated when the other end of the element passes through the replication complex. Transposition within the same replicon in the manner shown in Figure 11 would lead to inversion of markers between the transposition sites (Fig. 11, 4b). If this process of transposition is disrupted during DNA extraction, it would result in the appearance of key structures (Fig. 11, 4a). These would also be generated during the formation of heterogeneous circles (see Harshey and Bukhari 1981).

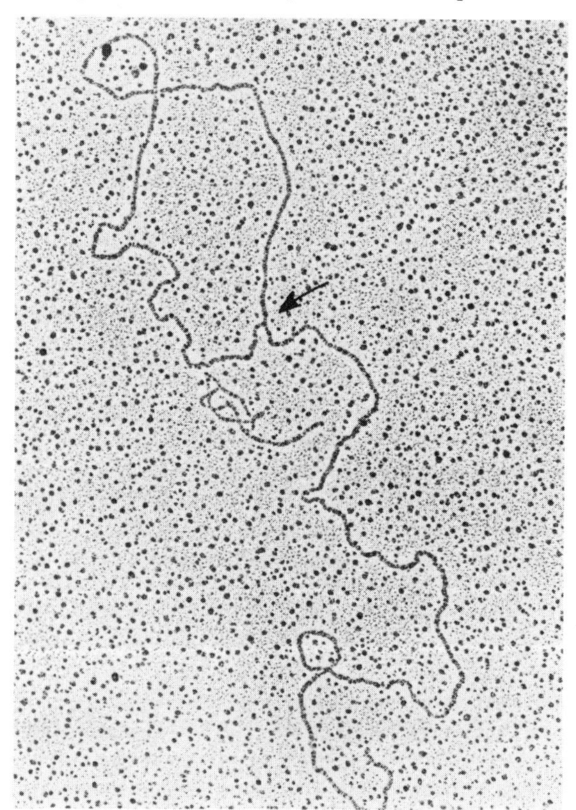

Figure 10. A "key" structure seen after Mu induction. The circular part is 17 kb long. (⟵) Indicates the beginning of the tail.

Cointegrate Formation versus Simple Transposition

If transposition is occurring from a plasmid onto the chromosome, one strand of the plasmid will attach to the cleaved target site, and the element will undergo roll-in replication. This would give the appearance of the plasmid being attached to the chromosome at a sin-

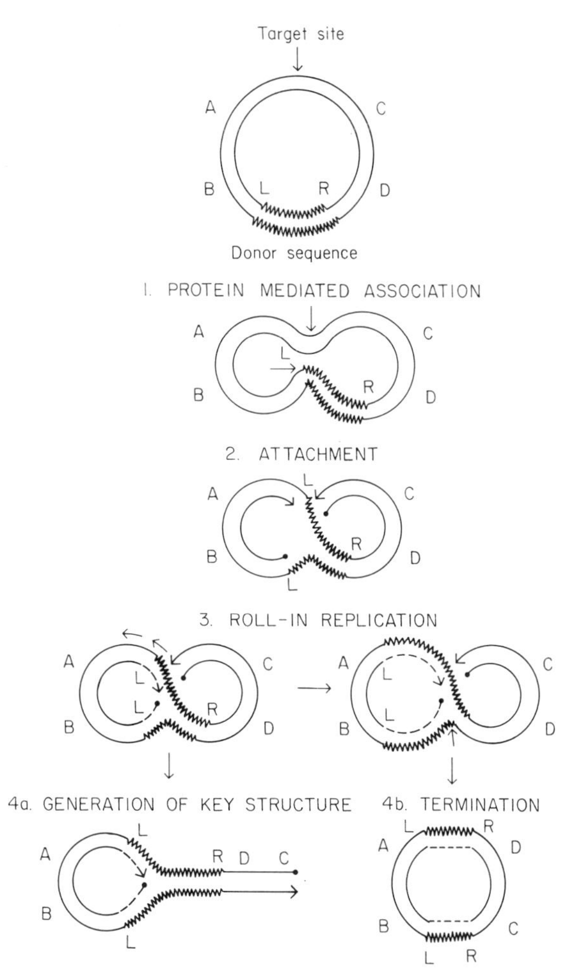

Figure 11. A model of transposition (within the same replicon). (〰〰) The transposable element; (→) free 3′-hydroxyl ends of DNA; (●) 5′ phosphate ends; (→) initial cleavage points; (---) newly replicated strands. (For a discussion of each step, see the text.) Polarities indicated are completely arbitrary and may well be reversed. The DNA ends are presumed to be held together by proteins. Upon disruption of the protein complex, a circular structure with a tail would be generated (*4a*). The length of the circle would depend upon the distance of the target site from the initial location of the element, and so also the length of the tail attached to the circle. Upon completion of replication, when the distal end of the element arrives at the replication fork, the 3′ end of the parental strand is nicked and ligated to the free 5′ phosphate of the target DNA. Upon sealing of all the relevant strands, a transposition accompanied by inversion (of *C* and *D* markers) would result (*4b*).

gle point. Whether or not the whole plasmid is inserted into the host DNA after replication is completed would depend upon how the termination structure is resolved, that is, how the strands are broken and rejoined. If the parental strand is nicked at the 3′ end after replication and ligated to the free 5′ phosphate of the host DNA, the whole plasmid will be inserted into the host DNA, giving rise to a cointegrate structure. If, however, the 3′ end of the newly synthesized strand is recognized for ligation, this would result in a simple transposition of the element into the target site, regenerating the original plasmid. The decision between the two modes of termination, resulting in either cointegrate formation or simple transposition, may be made by proteins.

Generation of Tandem Repeats

Any circular molecule that has the appropriate proteins can insert into the host DNA by the roll-in replication mechanism. If the site for termination is not fixed and the element can continue replicating until termination occurs randomly, this would result in the formation of tandem repeats.

Relationship to Previous Models

In the original model proposed by Faelen et al. (1975) (see also Toussaint et al. 1977), replication of the element occurs first, followed by recombination at the ends of the duplicated elements. The manner of replication was not specified. The mechanism of transposition proposed here defines exactly the mode of integrative replication; roll-in replication occurs through a fixed replication complex. The step for initiation of replication is the same as proposed by Grindley and Sherratt (1979). However, the subsequent steps are different. In the Grindley-Sherratt model, replication of the element when completed is essentially conservative, as opposed to a straightforward semiconservative replication described in the present proposal. The mechanism described here is different from the Shapiro (1979) model, which postulated simultaneous attachment at both ends of the element at the target site. Replication of the element then invariably leads to the cointegrate mode of integration; that is, if transposition were occurring within two replicons, they would always fuse. The cointegrates are then resolved by a discrete recombinational step. The proposal here provides alternate modes in which resolution of the replication structure occurs at the ends of the element after replication and is dependent upon the properties of specific proteins. A detailed version of the model proposed here will appear elsewhere (Harshey and Bukhari 1981).

REFERENCES

ALLET, B. 1979. Bacteriophage Mu integration creates five base pair duplications. *Cell* **16:** 123.

BUKHARI, A. I. 1975. Reversal of mutator phage Mu integration. *J. Mol. Biol.* **96:** 87.

———. 1976. Bacteriophage Mu as a transpositional element. *Annu. Rev. Genet.* **10:** 389.

BUKHARI, A. I. and E. LJUNGQUIST. 1977. Bacteriophage Mu: Methods for cultivation and use. In *DNA insertion elements, plasmids, and episomes* (ed. A. I. Bukhari et al.), p. 749. Cold Spring Harbor Laboratory, Cold Spring Harbor, New York.

BUKHARI, A. I. and A. L. TAYLOR. 1975. Influence of insertions on the packaging of host sequences covalently linked to mutator phage Mu DNA. *Proc. Natl. Acad. Sci.* **72:** 4399.

BUKHARI, A. I., E. LJUNGQUIST, F. DE BRUIJN, and H. KHATOON. 1977. The mechanism of bacteriophage Mu integration. In *DNA insertion elements, plasmids, and episomes* (ed. A. I. Bukhari et al.), p. 249. Cold Spring Harbor Laboratory, Cold Spring Harbor, New York.

CHACONAS, G., R. M. HARSHEY, and A. I. BUKHARI. 1980. Association of Mu-containing plasmids with the *Escherichia coli* chromosome upon prophage induction. *Proc. Natl. Acad. Sci.* **77:** 1778.

CHACONAS, G., F. J. DE BRUIJN, M. J. CASADABAN, J. R. LUPSKI, T. J. KWOH, R. M. HARSHEY, M. S. DUBOW, and A. I. BUKHARI. 1981. In vitro and in vivo manipulation of bacteriophage Mu DNA. *Gene* (in press).

FAELEN, M. and A. TOUSSAINT. 1980. Inversions induced by temperate bacteriophage Mu-1 in the chromosome of *Escherichia coli* K12. *J. Bacteriol.* **142:** 391.

FAELEN, M., A. RESIBOIS, and A. TOUSSAINT. 1979. Mini-Mu: An insertion element derived from temperate phage Mu-1. *Cold Spring Harbor Symp. Quant. Biol.* **43:** 1169.

FAELEN, M., A. TOUSSAINT, and J. DELAFONTEYNE. 1975. Model for the enhancement of λ-*gal* integration into partially induced Mu-1 lysogens. *J. Bacteriol.* **121:** 873.

GILL, R., F. HEFFRON, G. DOUGAN, and S. FALKOW. 1978. Analysis of sequences transposed by complementation of two classes of transposition-deficient mutant of Tn*3*. *J. Bacteriol.* **136:** 742.

GRINDLEY, N. D. F. and D. SHERRATT. 1979. Sequence analysis of IS*1* insertion sites: Models for transposition. *Cold Spring Harbor Symp. Quant. Biol.* **43:** 1257.

HARSHEY, R. M. and A. I. BUKHARI. 1981. A model of DNA transposition. *Proc. Natl. Acad. Sci.* (in press).

KAMP, D., L. T. CHOW, R. T. BROKER, D. KWOH, D. ZIPSER, and A. TOUSSAINT. 1979. Site-specific recombination in phage Mu. *Cold Spring Harbor Symp. Quant. Biol.* **43:** 1159.

KHATOON, H., G. CHACONAS, M. DUBOW, and A. I. BUKHARI. 1979. The Mu paradox: Excision versus replication. In *Extrachromosomal DNA* (ed. D. J. Cummings et al.), p. 143. Academic Press, New York.

LEACH, D. and W. SYMONDS. 1979. The isolation and characterization of a plaque-forming derivative of bacteriophage Mu carrying a fragment of Tn*3* conferring ampicillin resistance. *Mol. Gen. Genet.* **172:** 179.

LJUNGQUIST, E. and A. I. BUKHARI. 1977. The genomes of temperate viruses of bacteria. In *DNA insertion elements, plasmids, and episomes* (ed. A. I. Bukhari et al.), p. 705. Cold Spring Harbor Laboratory, Cold Spring Harbor, New York.

———. 1979. Behavior of bacteriophage Mu DNA upon infection of *Escherichia coli* cells. 1979. *J. Mol. Biol.* **133:** 339.

LJUNGQUIST, E., H. KHATOON, M. DUBOW, L. AMBROSIO, F. DE BRUIJN, and A. I. BUKHARI. 1979. Integration of bacteriophage Mu DNA. *Cold Spring Harbor Symp. Quant. Biol.* **43:** 1151.

O'DAY, J. D., D. W. SCHULTZ, and M. M. HOWE. 1978. Search for integration-deficient mutants of bacteriophage Mu. In *Microbiology—1978* (ed. D. Schlessinger), p. 48. American Society for Microbiology, Washington, D.C.

O'DAY, K., D. SCHULTZ, W. ERICSEN, L. RAWLUK, and M. HOWE. 1979. Correction and refinement of the genetic map of bacteriophage Mu. *Virology* **93:** 320.

RAZZAKI, T. and A. I. BUKHARI. 1975. Events following prophage Mu induction. *J. Bacteriol.* **122:** 437.

SCHRÖDER, W., E. B. BADE, and H. DELIUS. 1974. Participation of *E. coli* DNA in the replication of temperate bacteriophage Mu. *Virology* **60:** 534.

SHAPIRO, J. A. 1979. Molecular model for the transposition and replication of bacteriophage Mu and other transposable elements. *Proc. Natl. Acad. Sci.* **76:** 1933.

SOUTHERN, E. M. 1975. Deletion of specific sequences among DNA fragments. *J. Mol. Biol.* **98:** 503.

TOUSSAINT, A., M. FAELEN, and A. I. BUKHARI. 1977. In *DNA insertion elements, plasmids, and episomes* (ed. A. I. Bukhari et al.), p. 275. Cold Spring Harbor Laboratory, Cold Spring Harbor, New York.

VAN DE PUTTE, P., S. CRAMER, and M. GIPHART-GASSLER. 1980. Invertible DNA determines host specificity of bacteriophage Mu. *Nature* **286:** 218.

Transposition Studies Using a ColE1 Derivative Carrying Bacteriophage Mu

A. COELHO, D. LEACH, S. MAYNARD-SMITH, AND N. SYMONDS
School of Biological Sciences, University of Sussex, Brighton BN1 9QG, England

The Mu system has certain advantages for the study of transposition. (1) Induction of phage lysogens leads to a high frequency of Mu transposition, so that synchronized populations of cells undergoing transposition can be easily obtained. (2) Two Mu genes, namely, *A* and *B*, are known to affect the efficiency at which transposition occurs (Faelen et al. 1978; O'Day et al. 1978), and therefore both are likely candidates for genes coding for enzymes specifically mediating the tranposition reaction.

To capitalize on these attractive attributes of the system, we decided to introduce an antibiotic determinant into the Mu genome, so that Mu transposition could be easily followed by genetic methods, and then to construct a relatively small Mu-containing plasmid that was amenable to study by electrophoretic and electron microscopy techniques. These considerations led to the isolation of the Mu variant MupAp1 in which part of the G segment of Mu was replaced by a determinant conferring ampicillin (Ap) resistance, which was derived from Tn*3* and which can no longer transpose independently (Leach and Symonds 1979). By using straightforward genetic methods, this MupAp1 phage was inserted into the ColE1 derivative pML2 that contains a gene conferring kanamycin (Km) resistance (Hershfield et al. 1974). The structure of the resulting plasmid, pSU1, is shown in Figure 1. Two further plasmids were derived from pSU1, by genetic methods namely, pSU1A^- and pSU1B^-, which contain either the am1093 mutation in gene *A* or the am1066 mutation in gene *B* (Howe 1973)

Recently, the formation of cointegrates has become implicated in the transposition of Mu (Shapiro 1979). These cointegrate structures were first identified by Toussaint and Faelen (1973) and have been shown to occur frequently during Mu transposition between two plasmids (Maynard-Smith et al. 1980). Similar structures have also been identified as products of Tn*3* transposition (Gill et al. 1978) and, in this case, have been shown to be intermediates in the complete transposition reaction that results in the insertion of the transposon at a new location (Arthur and Sherratt 1979).

In this paper we report on two types of experiments with pSU1 plasmids. The first consisted of following the fate of both wild-type and mutant plasmid DNAs after induction to ascertain whether there was an intracellular phase during the life cycle when the physical integrity of the Mu plasmid was lost, as would be expected if cointegration occurred with the bacterial chromosome.

The second experiment was designed to investigate the relative frequencies of insertion and cointegration formation during transposition and the effect of the *A*- and *B*-gene products on these events. *recA* donor strains were constructed that contained one of the three pML2::Mu derivatives together with the conjugative trimethoprim-resistant (Tp^r) plasmid R388 (Datta and Hedges 1972) (Fig. 2) that mobilizes pML2 derivatives at only 10^{-5} to 10^{-6} of its own rate (Warren et al. 1979; Maynard-Smith et al. 1980). A culture of these donor strains was induced, and, after various lengths of time, the culture was mated with *recA* recipient cultures (which are lysogenic for Mu, to prevent any effects of zygotic induction, and are Mu^r, to obviate any secondary effects due to superinfection). Transposition of Mu from pML2::Mu into R388 was followed by selection of recipient cells resistant to both Tp and Ap. The plasmid content of these recipient cells was then analyzed to determine whether they contained cointegrates between pML2 and R388 or insertion of MupAp1 into R388.

Some earlier experiments of this latter type performed with pSU1 (Maynard-Smith et al. 1980) showed the presence in the Tp^rAp^r recipients of plasmids that satisfied all the criteria expected of cointegrates. They conferred kanamycin resistance to recipient cells, possessed the expected patterns upon restriction with *Eco*RI or *Hin*dIII endonuclease, and split into pSUI and R388::Mu upon transfer into Rec^+ recipients. A certain proportion of $Tp^rAp^rKm^s$ recipients was also observed in these experiments, but their analysis was complicated because Mu-sensitive recipients had been used, and superinfection with MupAp1 released from induced donor cells had clearly occurred. It was not possible, therefore, to compare the incidences of cointegration and insertion events but only to verify that cointegrates were formed at a measurable frequency.

RESULTS

The Fate of Mu Plasmids on Induction

Liquid cultures of strains carrying the plasmids pSU1, pSU1A^-, and pSU1B^- were induced for a series of time periods, and their DNA contents were examined by agarose gel electrophoresis of cleared lysates. These induction time courses can be seen in Figure 3, where a notable difference between the behaviors of wild-type and mutant plasmids is apparent. Both the supercoil

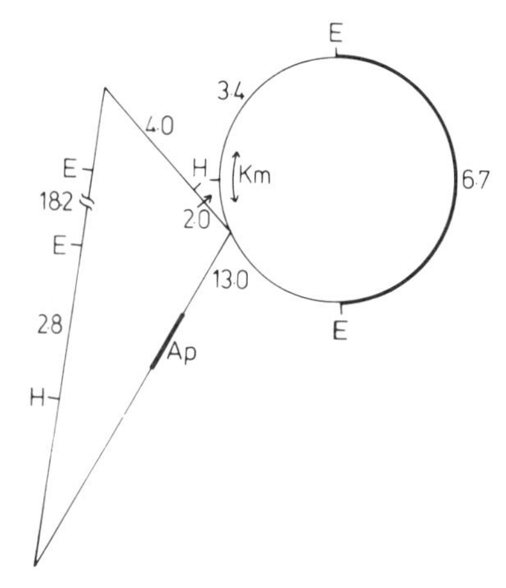

Figure 1. Diagrammatic representation of the plasmid pSU1, showing *Eco*RI (E) and *Hin*dIII (H) restriction sites. The figures represent distances in kilobases between adjacent restriction sites. The triangle represents the MupAp1 insertion into the circular pML2 molecule. (———) Mu DNA, Km^r fragment of pML2; (———) ColE1 DNA, Ap^r-determining DNA.

and open-circle bands of pSU1 rapidly disappear, leaving no trace of their presence after 20 minutes of induction, whereas in the cases of pSU1A^- and pSU1B^-, only a slow conversion of supercoils to open circles is observed. Figure 3a shows that the control plasmid R388 (which carries no part of the Mu genome) is unaffected by induction.

These results implicate both *A*- and *B*-gene functions in the reaction that leads to the disappearance of free pSU1 DNA.

A Time Course of MupAp1 Transposition from pSU1

A second induction time course was carried out for pSU1. A *recA* Mu^r strain carrying pSU1 and R388 was induced for a series of time periods and then mated with a *recA* Muc^+Mu^r recipient. Frequencies of exconjugants selected for Tp^rAp^r are shown in Table 1 and are expressed as a proportion of the transfer of Tp^r. Subsequently, 40 colonies derived from each mating were

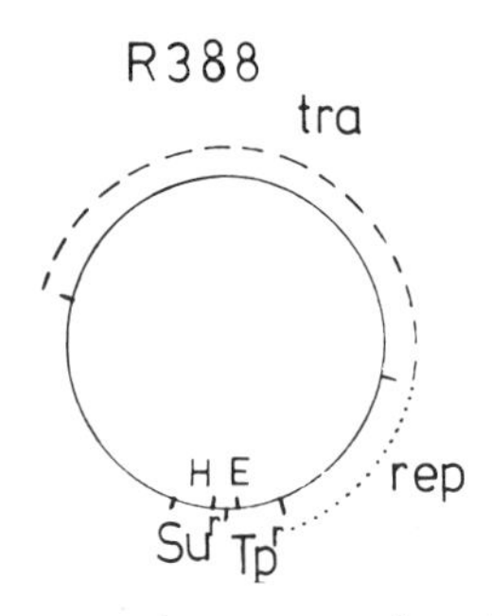

Figure 2. Diagrammatic representation of plasmid R388, showing the regions coding for drug resistances, replication, and transfer, as well as the *Eco*RI (E) and *Hin*dIII (H) restriction sites, which lie 0.6 kb apart. The total size of R388 is 32.2 kb (Ward and Grinsted 1978).

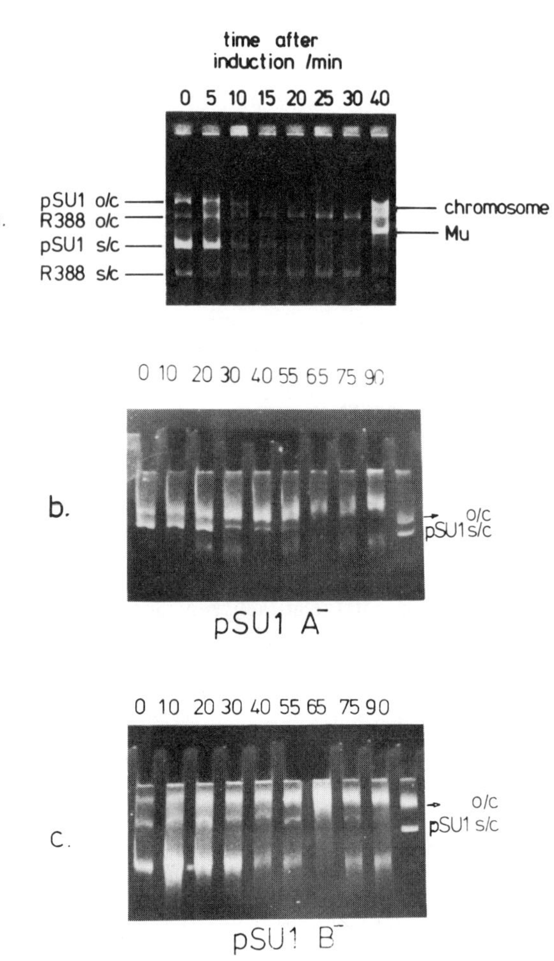

Figure 3. Induction time courses for pSU1 and its mutant derivatives. Liquid cultures of strains containing the plasmids were grown to 2×10^8 cells/ml, with aeration at 32 °C and induced by addition of an equal volume of warm broth. The cultures were then aerated at 43 °C for different time periods and then cooled by the addition of an equal volume of ice-cold buffer. Cleared lysates were prepared, and samples were run on agarose gels. (*a*) pSU1 (the strain induced also contained R388 as a control plasmid), 0.3% agarose; (*b*) pSU1A^-, 0.8% agarose; (*c*) pSU1B^-, 0.8% agarose. o/c and s/c indicate open circle and supercoil, respectively.

tested for the presence or absence of the Km^r marker, and these results are also shown in Table 1. These data show a gradual decrease with time during the first 20 minutes of induction in the proportion of Tp^rAp^r colonies that also carry Km^r. When the DNA contents of random samples of the exconjugants were analyzed by agarose gel electrophoresis, it was found that the $Tp^rAp^rKm^r$ colonies contained plasmids of a size compatible with a cointegrate structure (~118 kb) as well as others slightly larger or smaller. Because of their large size, these plasmids could not be classified unambiguously on the basis of the gel data.

The sizes of the $Tp^rAp^rKm^s$ plasmids (~50–100 kb) were more amenable to study, and their size distribution relative to R388::MupAp1 (68.2 kb) in six randomly chosen clones at each time point is shown in Table 2.

Table 1. Frequencies of Drug-resistance Transfer in Transposition Experiments

Time of induction (min)	pSU1 Tp^{r}Ap^r [a] / Tp^r	pSU1 proportion Km^r [b]	pSU1A^- Tp^{r}Ap^r / Tp^r	pSU1A^- proportion Km^r	pSU1B^- Tp^{r}Ap^r / Tp^r	pSU1B^- proportion Km^r
0	1.1×10^{-5}	45/78	1.4×10^{-5}	141/141	1.5×10^{-5}	210/215
5	1.6×10^{-3}	26/40				
10	2.3×10^{-3}	13/40				
15	2.9×10^{-3}	8/40				
20	1.4×10^{-3}	6/40				
25	2.5×10^{-3}	13/40				
30	2.8×10^{-3}	20/132	5.5×10^{-6}	162/162	4.3×10^{-6}	58/125
35	3.8×10^{-3}	13/40				

Donor strains were SU1702 (CSH26, *recA*$^-$, Mu^r, pSU1, R388), SU1703 (CSH26, *recA*$^-$, Mu^r, pSU1A^-am1093, R388), and SU1704 (CSH26, *recA*$^-$, Mu^r, pSU1B^-am1066, R388) (for CSH26, see Miller 1972). The recipient strain was SU1688 (CSH52, Mu^r, Muc$^+$). After induction of donor strains at 43 °C, the cultures were mixed with the recipient and incubated at 32 °C on Millipore filters for 2 hr. Minimal agar selection plates were supplemented with 0.1% casamino acids, 1 μg/ml of thiamine, 50 μg/ml of Tp, and 50 μg/ml of Ap. Km was used at 50 μg/ml.

[a] Frequency of Tp^{r}Ap^r transfer expressed as a proportion of the transfer of Tp^r.

[b] Proportion of Tp^{r}Ap^r colonies that are also Km^r.

From this table it can be seen that a variety of plasmid sizes were found at all time points except $t = 0$, where all were comparable to R388::MupAp1. At all other times, at least some plasmids of R388::MupAp1 size were found. It must be stressed here that these plasmids can only be considered precise insertions to the level of resolution of 0.3% agarose gel electrophoresis (about ±1 kb).

Effect of A^- and B^- Mutations on Transposition

The behavior of pSU1A^- and pSU1B^- has been compared with that of pSU1 in similar mating experiments without induction and with induction for 30 minutes.

Table 1 gives a comparison of the transfer frequencies of Tp^{r}Ap^r for the three plasmids under these conditions. With no induction prior to mating, Tp^{r}Ap^r is transferred at about 10^{-5} of the transfer of Tp^r in all three cases. The 30-minute induction of the donors increases the number of pSU1 exconjugants 100-fold but does not affect the numbers found with pSU1A^- or pSU1B^-, indicating that the *A*- and *B*-gene products are required for efficient transposition.

About 150 clones from each mating were tested for Km^r (see Table 1), and 20–80 clones from each mating were analyzed on agarose gels for their plasmid contents (see Table 3). The clones fall broadly into three classes: (1) Tp^{r}Ap^{r}Km^r clones containing pSU1 (or amber derivative) and R388, which probably arise from the low-level mobilization of pSU1; (2) those containing cointegrate-type plasmids, about half of which are Tp^{r}Ap^{r}Km^s; and (3) Tp^{r}Ap^{r}Km^s clones containing insertion-type plasmids.

With no induction, the pSU1 donor gave roughly equal proportions of the three classes: the pSU1A^- donor gave 100% mobilizations and the pSU1B^- donor gave 98% mobilizations and 2% insertion-type plasmids.

Table 2. Sizes of Tp^{r}Ap^{r}Km^s Plasmids, Formed after Different Times of Induction, Relative to R388::MupAp1

Time of induction (min)	Plasmid size: R388::MupAp1 size	smaller than R388::MupAp1	larger than R388::MupAp1
0	6	0	0
5	3	1	2
10	3	2	1
15	1	2	3
20	1	4	1
25	2	2	2
30	3	2	1
35	4	1	1

Induced pSU1 Donor

After the 30-minute induction of the pSU1 donor, the transfer of drug resistances is nearly always associated with the presence of either cointegrate- or insertion-type plasmids, with mobilization of pSU1 being rarely found. Thus, the raised level of drug-resistance transfer is due to the formation and transfer of new types of plasmids. Figure 4 shows some typical examples of plasmids found in exconjugants from this mating.

To examine the structure of the cointegrate plasmids, they were transferred to a Rec$^+$Mu^{r}Muc$^+$ background with selection for the drug resistances that they carried. In some cases no transfer was observed, which would be expected if MupAp1 had integrated into the R388 transfer region (Fig. 2). Two of the Tp^{r}Ap^{r}Km^r cointegrate plasmids break down in the Rec$^+$ background to R388::MupAp1 and pSU1, as was found in previous work (Maynard-Smith et al. 1980).

Of the three Tp^{r}Ap^{r}Km^s cointegrates tested, two were transferable, and, in the Rec$^+$ background, these gave only an R388::MupAp1-size plasmid. This observation suggests that these cointegrates contain R388 and two copies of MupAp1 but not the complete pML2 genome.

The pSU1 Km^r exconjugants formed both large and small colonies, 80% of them being small after induction; small colonies were not observed with the mutant plasmid donors. Upon purification, the small clones gave

Table 3. Comparison of the Plasmids Formed in Transposition Experiments with pSU1, pSU1A^-, and pSU1B^-

	Uninduced			Induced for 30 min		
Donor plasmids	mobilization (%)	cointegrate-type plasmid[a] (%)	insertion-type plasmid[b] (%)	mobilization (%)	cointegrate-type plasmid (%)	insertion-type plasmid (%)
R388 + pSU1	39.4	25.6	35.0	0.2	29.2	70.6
R388 + pSU1A^-	100			100		
R388 + pSU1B^-	98.0		2.0	42.7	11.0	46.3

[a] This class included plasmids of the expected size for cointegrates and plasmids larger or slightly smaller than this, whether or not they conferred Km^r.

[b] This class consisted of plasmids of the expected size for R388::MupAp1 as well as plasmids slightly larger or smaller than this, all of which were Km^s.

rise to some normal-size colonies whose DNA contents were examined on agarose gels. Two classes were found: those containing R388::MupAp1 and pSU1 (which have not been examined further) and those containing cointegrate plasmids together with a smaller plasmid that varied in size around that of pML2. In some cases, both classes were found in cells derived from a single colony. Transfer of these cointegrate plasmids into Rec^+ recipients gave the expected breakdown products, and transformation with DNA isolated from one clone, with selection for ColE1 immunity, indicated that the small plasmid is a ColE1 derivative. In this one case, at least, it carries no drug resistance.

Ten clones harboring $Tp^rAp^rKm^s$ insertion-type plasmids have been shown to contain MupAp1, and seven of them transferred Tp^r by conjugation. Those clones smaller than R388::MupAp1 presumably contain deletions of R388 DNA; however, the larger ones have not yet been characterized but may be deleted cointegrates.

Induced pSU1A^- and pSU1B^- Donors

The pSU1A^- experiments show no evidence for the insertion of MupAp1A^- into R388 or for the formation of pSU1A^- cointegrates. All exconjugants tested were $Tp^rAp^rKm^r$, and agarose gels showed only the mobilization of pSU1A^-.

The low level of Tp^rAp^r exconjugants recovered from the pSU1B^- mating meant that a substantial number of them represented mobilization of the mutant plasmid. However, after induction, 11% of the exconjugants contained either $Tp^rAp^rKm^r$ or $Tp^rAp^rKm^s$ cointegrate-type plasmids and nearly half harbored insertion-type plasmids. As with pSU1, some of the latter were smaller than R388::MupAp1.

The three $Tp^rAp^rKm^s$ cointegrate plasmids tested did not transfer Tp^r, but nine of ten $Tp^rAp^rKm^r$ cointegrates were transferred into the Rec^+ recipients and gave exconjugants containing R388::MupAp1 together with the original cointegrate. The expected breakdown product, pSU1B^-, was not observed. The retention of the cointegrate is presumably due to the selection for kanamycin resistance.

The B^- insertion-type plasmids were tested for the transfer of Tp^r, and six of ten gave positive results. All ten isolates showed the presence of MupAp1B^-, and therefore any deletions that they may contain lie in the R388 moiety.

Restriction Analysis of pSU1B^- Cointegrates

DNAs from 12 of these plasmids, 9 Km^r and 3 Km^s, were prepared and restricted with *Eco*RI and *Hin*dIII endonucleases. Some of the gel patterns are shown in

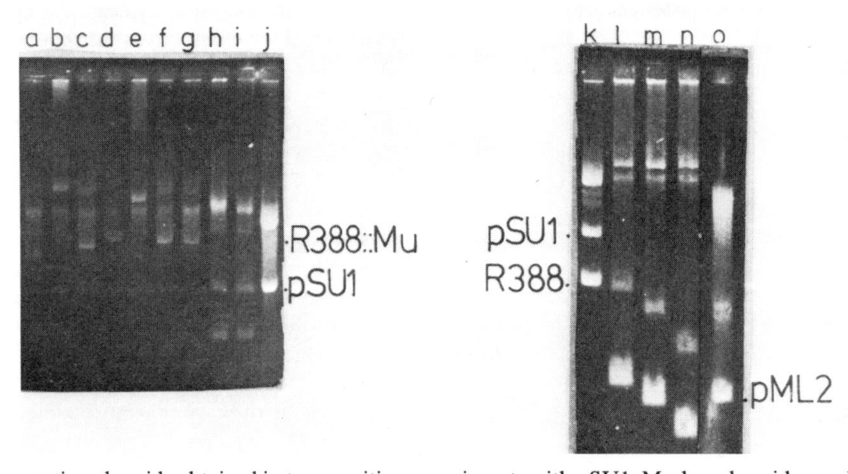

Figure 4. Representative plasmids obtained in transposition experiments with pSU1. Marker plasmids are shown in tracks *d*, *j*, *k*, and *o*. (*a,c,f,g*) Insertion-type plasmids (*a* and *c* carry deletions); (*b,e,l,m,n*) cointegrate-type plasmids (*b* and *e* are of different sizes, and *l,m,n* also contain small plasmids in the size range of pML2). Gels contained 0.3% agarose.

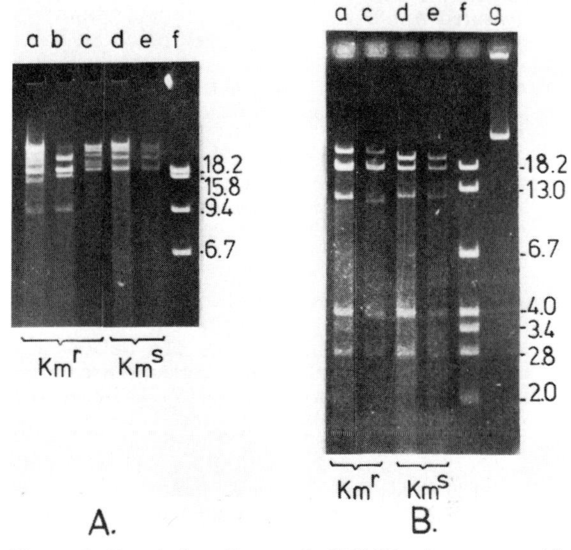

Figure 5. Restriction digests of pSU1B^- cointegrates with *Eco*RI (*A*) and *Eco*RI + *Hin*dIII (*B*). Tracks labeled with the same letter refer to digests of the same plasmid. (*f*) Restriction patterns obtained with pSU1; (*g*) unrestricted Mu DNA; (*a,b,c*) digests of Tp^{r}Ap^{r}Km^r cointegrates; (*d,e*) digests of Tp^{r}Ap^{r}Km^s cointegrates. The sizes of the fragments are given in kilobases. (*A*) 0.8% agarose; (*B*) 0.5% agarose.

Figure 5, and similar restriction patterns from pSU1 are shown for comparison. The *Eco*RI + *Hin*dIII internal Mu fragments (18.2 kb, 4.0 kb, and 2.8 kb; see Fig. 1) are present in all the digests, and their intensities suggest that they are present in two copies. The 6.7-kb *Eco*RI band formed by the ColE1 fragment of pML2 and which contains the replication functions of this plasmid is missing in all the digests. The 9.4-kb *Eco*RI Mu(*c*-end)–pML2 junction fragment containing the Km^r gene is also missing from some of them (Fig. 5A, tracks c, d, and e). Figure 5B shows that the 13.0-kb *Eco*RI + *Hin*dIII Mu(*S*-end)–pML2 junction fragment is also missing.

In each case studied, new bands appeared that can account for the R388–Mu junction fragments without detectable deletions and for the pML2–Mu junction fragments with extensive deletions into pML2.

The loss from these cointegrates of the fragment carrying the pML2 replication functions is consistent with their behavior when transferred to Rec$^+$.

DISCUSSION

Bacteriophage Mu has previously been demonstrated to perform both insertion and cointegration reactions (Toussaint and Faelen 1973; Faelen et al. 1975, 1977; Razzaki and Bukhari 1975; Parker and Bukhari 1976; Maynard-Smith et al. 1980). We have shown that these two reactions occur under *recA*$^-$ conditions in a system consisting of the MupAp1-containing plasmid pSU1 and the sex factor R388. The results here have implications for the understanding of the pathway(s) used by Mu in transposition.

The fact that the insertion of MupAp1 into R388 occurs efficiently in the absence of *rec*-mediated recombination indicates that a major insertion pathway exists for Mu. This pathway could proceed via a cointegrate intermediate and would require a *rec*-independent recombination system for resolution (Shapiro 1979), or a separate pathway to insertion could exist that does not implicate a cointegrate intermediate (Leach 1980).

Our results do not distinguish between these two possibilities; there is, however, no doubt that a cointegration pathway does exist that can account for the multiplication of Mu, the formation of heterogeneous circles, and deletions and inverted insertions (Shapiro 1979). If insertions occur by the resolution of cointegrates, the process cannot be highly efficient considering the large number of cointegrates detected. As insertion is not required for Mu multiplication, it is likely that it is essential for another feature of the life cycle, which may be lysogenization with linear Mu following infection, since, in this situation, unresolved cointegration would result in the linearization of the chromosome.

When the behavior of A^- or B^- mutant derivatives of pSU1 were compared with that of wild type in transposition experiments with R388, radical differences were observed. The A^- mutation abolished all detectable insertions and cointegrations, an observation consistent with previous reports that this mutation renders the phage unable to transpose (Faelen et al. 1978; O'Day et al. 1978). The B^- mutation, on the other hand, reduced the level of induced transposition to 10^{-2} of that observed with wild type. The transposed derivatives that were detected fell into two classes. The first class was composed of plasmids of a size consistent with insertions of MupAp1B^- into R388 (sometimes accompanied by deletions), and the second class was composed of plasmids of approximately cointegrate size, which, when analyzed, were shown to carry deletions of the pML2 replication genes. This may reflect an interaction between the plasmid replication and the transposition of Mu in the absence of the *B*-gene product, since the *B* gene has previously been implicated in Mu replication (Wijffelman and van de Putte 1974; Faelen et al. 1978). In the experiment with wild-type pSU1, plasmids were obtained that can be interpreted as insertions of MupAp1 into R388, insertions accompanied by deletions, cointegrates between pSU1 and R388, cointegrates accompanied by deletions, or cointegrates with increased size, all of which can be understood in terms of pathways that lead to insertion and cointegration. Some of the cointegrates appear to have deleted the replication functions of pML2, as in the B^- cointegrates, but this may not represent the same phenomenon, since the wild type is known to carry out multiple rounds of transposition during growth. In addition, some of the cells that contained pSU1 cointegrates and grew very poorly were able to overcome this defect via the generation of small plasmids of variable size approximating that of pML2 or by the formation of R388::MupAp1 and pSU1. At present, we do not understand the former reaction, but the latter shows that it is possible to resolve a cointegrate in *recA* cells when selection is applied.

When the natures of the plasmid DNAs present in induced cells containing pSU1, pSU1A^-, and pSU1B^- were compared at different times of induction, it was clear that only the wild-type-plasmid bands disappeared rapidly. Both mutant plasmids showed only a slow conversion from supercoils to open circles. This implicates both *A*-gene and *B*-gene functions in steps that lead up to the disappearance of the bands, a reaction that has previously been shown to be accompanied by association with the chromosome (Chaconas et al. 1980). If the *B*-gene function is primarily involved in the replication of Mu, this result suggests that Mu replication may precede association with the chromosome. This result is consistent with a model of Mu transposition in which newly replicated DNA strands are ligated to the target site (Leach 1980), as opposed to ligation of the parental DNA strands (Shapiro 1979).

ACKNOWLEDGMENTS

We would like to thank Dotti Dries and Joy Chessell, who provided media and clean laboratoryware, and Jo Harper, who typed the manuscript. A. C. is supported by a grant from C.A.P.E.S. (Brazil), and D. L. is supported by a Medical Research Council fellowship.

REFERENCES

Arthur, A. and D. J. Sherratt. 1979. Dissection of the transposition process: A transposon-encoded site-specific recombination system. *Mol. Gen. Genet.* **175:** 267.

Chaconas, G., R. Harshey, and A. I. Bukhari. 1980. Association of Mu-containing plasmids with the *E. coli* chromosome upon prophage induction. *Proc. Natl. Acad. Sci.* **77:** 1778.

Datta, N. and R. W. Hedges. 1972. Trimethoprim resistance conferred by W plasmids in Enterobacteriaceae. *J. Gen. Microbiol.* **72:** 349.

Faelen, M., O. Huisman, and A. Toussaint. 1978. Involvement of phage Mu-1 early functions in Mu-mediated chromosomal rearrangements. *Nature* **271:** 580.

Faelen, M., A. Toussaint and J. de Lafonteyne. 1975. Model for the enhancement of λ-*gal* integration into partially induced Mu-1 lysogens. *J. Bacteriol.* **121:** 873.

Faelen, M., A. Toussaint, M. Van Montagu, S. van den Elsacker, G. Engler, and J. Schell. 1977. *In vivo* genetic engineeering: The Mu-mediated transposition of chromosomal DNA segments onto transmissible plasmids. In *DNA insertion elements, plasmids, and episomes* (ed. A. I. Bukhari et al.) p. 521. Cold Spring Harbor Laboratory, Cold Spring Harbor, New York.

Gill, R., F. Heffron, G. Dougan, and S. Falkow. 1978. Analysis of sequences transposed by complementation of two classes of transposition-deficient mutants of Tn*3*. *J. Bacteriol.* **136:** 742.

Hershfield, V., H. W. Boyer, C. Yanofsky, M. A. Lovett, and D. R. Helinski. 1974. Plasmid ColE1 as a molecular vehicle for cloning and amplification of DNA. *Proc. Natl. Acad. Sci.* **71:** 3455.

Howe, M. 1973. Prophage deletion mapping of bacteriophage Mu-1. *Virology* **54:** 93.

Leach, D. 1980. "Bacteriophage Mu transposition." Ph.D. thesis, University of Sussex, Brighton, England.

Leach, D. and N. Symonds. 1979. The isolation and characterisation of a plaque-forming derivative of bacteriophage Mu carrying a fragment of Tn*3* conferring ampicillin resistance. *Mol. Gen. Genet.* **172:** 179.

Maynard-Smith, S., D. Leach, A. Coelho, J. Carey, and N. Symonds. 1980. The isolation and characteristics of plasmids derived from the insertion of MupAp1 into pML2: Their behaviour during transposition. *Plasmid* **4:** 34.

Miller, J. H. 1972. *Experiments in molecular genetics.* Cold Spring Harbor Laboratory, Cold Spring Harbor, New York.

O'Day, K. J., D. W. Schultz, and M. Howe. 1978. Search for integration-deficient mutants of bacteriophage Mu. In *Microbiology—1978* (ed. D. Schlessinger), p. 48. American Society for Microbiology, Washington, D.C.

Parker, V. and A. I. Bukhari. 1976. Genetic analysis of heterogeneous DNA circles formed after prophage Mu induction. *J. Virol.* **19:** 756.

Razzaki, T. and A. I. Bukhari. 1975. Events following prophage Mu induction. *J. Bacteriol.* **122:** 437.

Shapiro, J. A. 1979. Molecular model for the transposition and replication of bacteriophage Mu and other transposable elements. *Proc. Natl. Acad. Sci.* **76:** 1933.

Toussaint, A. and M. Faelen. 1973. Connecting two unrelated DNA sequences by a Mu dimer. *Nat. New Biol.* **242:** 1.

Ward, J. M. and J. Grinsted. 1978. Mapping of functions in the R-plasmid R388 by examination of deletion mutants generated *in vitro*. *Gene* **3:** 87.

Warren, G. J., M. W. Saul, and D. J. Sherratt. 1979. ColE1 plasmid mobility: Essential and conditional functions. *Mol. Gen. Genet.* **170:** 103.

Wijffelman, C. and P. van de Putte. 1974. Transcription of bacteriophage Mu. An analysis of the transcription pattern in the early phase of phage development. *Mol. Gen. Genet.* **135:** 327.

Two Pathways in Bacteriophage Mu Transposition?

D. KAMP AND R. KAHMANN
Max-Planck-Institut für Biochemie, 8033 Martinsried bei München, Federal Republic of Germany

The bacteriophage Mu can insert its DNA at a large number of sites in the *Escherichia coli* chromosome and mediates a variety of chromosomal rearrangements (Taylor 1963; Bukhari and Zipser 1972; Daniell et al. 1972; Faelen and Toussaint 1978). Since Mu DNA has never been found in its free form but always appears to be associated with bacterial DNA, the integration of Mu resembles the transposition of insertion elements (for review, see Bukhari 1976). However, the only direct proof that the mechanisms of transposition of Mu and insertion elements are related stems from the observation that upon integration of Mu, as upon transposition of insertion elements, a segment of host DNA is duplicated at the target site (for review, see Calos and Miller 1980). The main difference between Mu and insertion elements lies in the efficiency of the respective transposition systems. Furthermore, transposition of Mu in lytic development seems to be the main cause for the rapid amplification of Mu DNA. The linkage between transposition and replication is illustrated by the fact that the phage attachment sites are also required for replication, and the products of genes *A* and *B,* which are required for replication, are also involved in transposition (O'Day et al. 1978; van de Putte et al. 1978). However, the *A*-gene product is essential for integration, whereas the *B*-gene product has only a stimulatory effect (O'Day et al. 1978). Both precise and imprecise excision of Mu require the *A*-gene product and the absence of the *B*-gene product (Bukhari 1975).

Our studies have focused on the structural features of bacteriophage Mu integration, and we shall discuss possible implications of these results with respect to the existence of different pathways for transposition.

RESULTS AND DISCUSSION

Phage Attachment Sites of Mu

For all insertion elements and transposons that have been analyzed thus far, the terminal sequences are inverted repeats (for review, see Calos and Miller 1980). Presumably, these terminal repeat structures constitute the recognition sites for the element-specific transposase. The attachment sites of Mu are not homologous and therefore differ from the principal structure of the ends of transposable elements (Kahmann and Kamp 1979). To reflect this physical and probably functional difference of the two attachment sites, we have designated them as Mu *att*L and Mu *att*R, respectively. Although the Mu attachment sites are not simply inverted repeats, they are also not completely different; they contain two sets of nearly homologous sequences that are 12 and 15 nucleotides long and are arranged at different positions within the two attachment sites. Because these are the only common features of *att*L and *att*R and we think it more likely that one protein can recognize both ends of Mu instead of each being recognized by separate proteins, we propose that these two blocks of homologous sequences represent the recognition sites for the Mu-specific transposase. Since the two potential recognition sites share no homology, we would then expect the existence of different subunits in the Mu transposase complex, each of which binds specifically to one of these recognition units. The DNA-protein complex could have a structure as depicted in Figure 1. The asymmetry of this complex could conceivably affect the reactions taking place at the two attachment sites; for instance, it could cause a different rate of initiation of DNA replication. The two polypeptides involved in recognition could be the *A*- and *B*-gene products of Mu. Alternatively, the complex could be formed by the *A* protein of Mu and a host function. In favor of this hypothesis is that other transposable elements that generate a 5-bp duplication contain in their inverted repeats a 7-bp-long region of strong homology with Mu (Reed et al. 1979) that is part of one of the putative recognition blocks. Limited homology is also present at the ends of Mu and IS*5*, where 11 out of 16 nucleotides of Mu *att*L and 10 out of 16 nucleotides of Mu *att*R are homologous to the terminal sequences of IS*5* (M. Lusky, pers. comm.). Consistent with the role of the *B*-gene product, it is also conceivable that both types of transposase complexes exist and initiate different pathways of Mu transposition.

Bacterial Insertion Site of Mu

Different insertion elements display distinct insertion specificities; these range from one insertion site to regional specificity, hotspots, favored sites, or little if any specificity (for review, see Calos and Miller 1980). On the basis of genetic analysis of Mu insertions in *lacZ,* Mu has been thought to be a representative of the latter group (Bukhari and Zipser 1972; Daniell et al. 1972).

We have studied the problem of insertion specificity of Mu in a different way by analyzing the host DNA that is found at the ends of Mu DNA in mature phage particles. The host DNA in every phage particle is different (Daniell et al. 1973), and the accepted view of how this heterogeneity is generated is that during lytic development Mu DNA is transposed to many chromosomal sites and then packaged in situ. Thus, the host sequences that remain attached to the ends of the Mu DNA represent the various bacterial insertion sites of

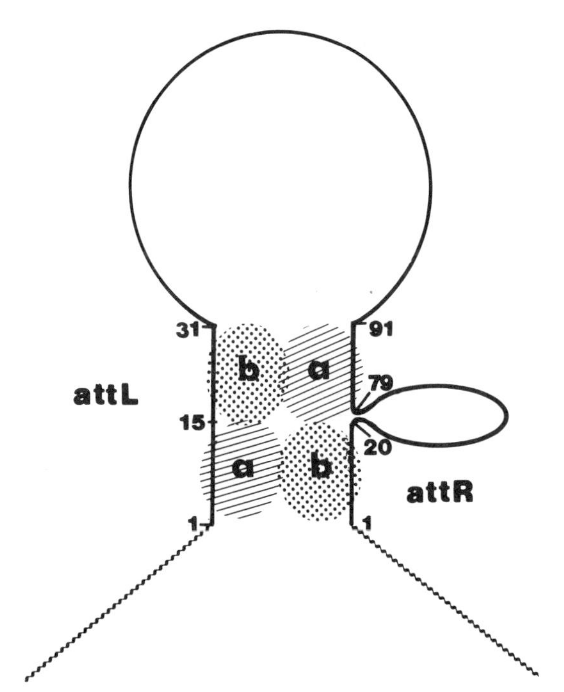

Figure 1. Proposed mechanism for the interaction between the two attachment sites of Mu. (~~~~) Bacterial DNA; (—) Mu DNA. *att*L and *att*R indicate the left-end and right-end attachment sites, respectively. The shaded circles represent protein subunits of the transposase complex. Numbers refer to distances (in base pairs) from the junctions.

Mu in the lytic cycle. We have found similarities in the host DNA sequences adjacent to the attachment sites through cloning and subsequent sequence analysis (Kahmann and Kamp 1979). However, there are too few sequenced junctions to assess the significance of these similarities. The chain-termination sequencing method of Sanger et al. (1977) allows a different approach to this question. Small primer fragments, originating from regions within Mu *att*L and Mu *att*R, respectively, can be extended, and the DNA sequence can be determined starting within the attachment site and reading over the junction into the bacterial DNA. In this way, we have analyzed the sequence of the Mu-host-DNA junctions directly in phage DNA. Since phage DNA represents a population with respect to the bacterial sequences, we look at the features of a large number of bacterial sequences in a single experiment. The typical result of such an experiment is shown in Figure 2. At both ends of Mu DNA, unique sequences are generated that perfectly correspond to the Mu attachment site sequences, followed by a heterogeneous sequence in which all four nucleotides occur with equal intensities. However, the heterogeneous bacterial DNA and the homogeneous Mu DNA sequences are separated by a stretch of five nucleotides, where at each position one or two nucleotides predominate. The same five predominant nucleotides are found when Mu DNA is derived from different lysogens and from phage propagated on *Enterobacter cloacae* and, therefore, do not represent a relic of the original insertion site. Instead, it indicates that during lytic development Mu DNA is transposed to preferred bacterial sites that are reflected in the consensus sequences seen next to Mu *att*L and Mu *att*R (Figs. 2 and 3). Since the sequences are complementary to each other, it would appear that the majority of Mu DNA molecules are inserted in the middle of a 10-bp-long, more or less palindromic bacterial sequence. In spite of the degeneracy of this sequence, this degree of specificity is not quite compatible with the virtually random insertion of Mu in *lacZ* (Bukhari and Zipser 1972; Daniell et al. 1972). The discrepancy could be explained in two ways. (1) *lacZ* does not contain a preferred insertion site, and, in its absence, insertion may then occur more or less at random, as it was found, for example, with Tn*10* (Botstein and Kleckner 1977; Foster 1977; Kleckner et al. 1979). This interpretation is supported by the finding of a hotspot for Mu insertion in the *malK-lamB* operon (Raibaud et al. 1979). (2) Differences in the transposition mechanisms operating in the lysogenic and lytic pathways could result in a different selection of bacterial insertion sites.

An unexpected feature of the consensus sequences is that they are complementary to each other and thus do not reflect a 5-bp duplication. This indicates that in contrast to prophage DNA (Allet 1979; Kahmann and Kamp 1979), the majority of Mu phage DNA molecules lack the host DNA duplication at the attachment sites.

Another attraction of this sequencing scheme is its potential application in the analysis of a large number of mutants at the attachment sites. We have used primer fragments isolated from Mu DNA to derive the junction sequences of a Mu-related phage D108 (Hull et al. 1978) (Fig. 2). D108 and Mu are completely homologous for nucleotides 1 to 31 at *att*L. In addition, nucleotides 32 to 54 at *att*L must share a high degree of homology since efficient pairing of the Mu primer fragment to the D108 template is necessary for the priming reaction. With respect to D108 *att*R, we were unable to achieve extension of a fragment that is located between positions 8 and 60 at Mu *att*R. We conclude that at least within the region from which the primer is derived, significant differences between the *att*R attachment sites of Mu and D108 must exist. The bacterial sequences next to D108 *att*L show the same consensus sequence as for Mu.

Arrangement of Mu DNA in the Bacterial Chromosome during Lytic Growth

To develop an assay system for the structural requirements of Mu DNA transposition, we have constructed a plasmid (pTM2) in vitro that contains both attachment sites of Mu. The Mu sequences of pTM2 are the leftmost 1000 bp and the rightmost 850 bp of Mu 445 (Chow et al. 1977) joined at a *Hin*dIII cleavage site. The Mu sequences of pTM2 are designated Tn*M1*. The plasmid does not encode any functions involved in transposition, and therefore the bacterial strain harboring pTM2 was superinfected with a Mu*cIts*62*c*II helper to test transposition of the Tn*M1* sequence. The assay system is based on the following rationale. If the structural requirements for transposition are met, the Tn*M1* sequence will be transposed to the chromosome. The headful packag-

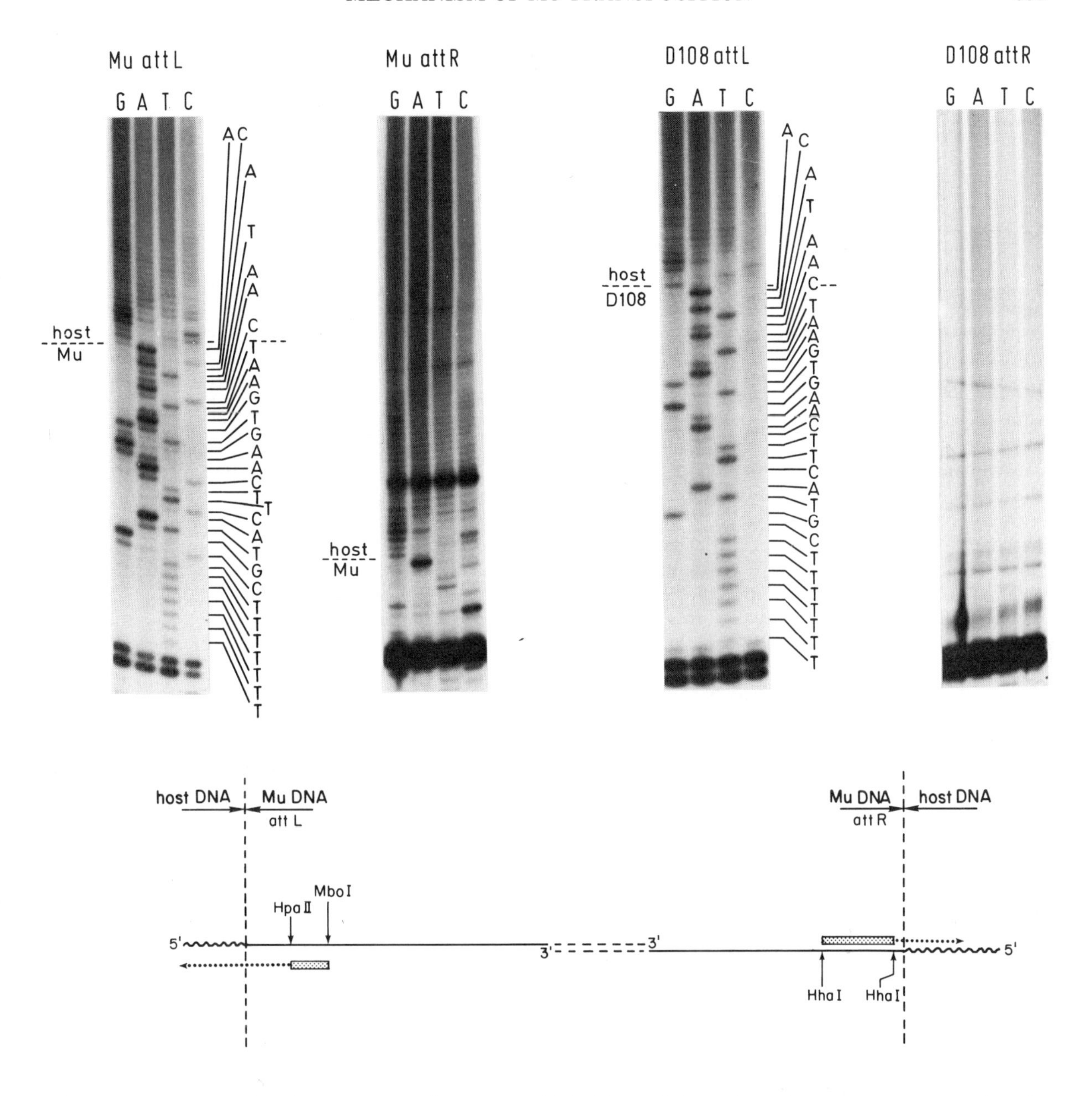

Figure 2. Autoradiographs of sequence gels of Mu–host and D108–host DNA junctions. Sequences were generated by the chain-termination sequencing technique of Sanger et al. (1977). The templates for the sequencing reactions were Mu and D108 DNA, respectively, which had been isolated from phage particles and had been made partially single-stranded by exonuclease-III treatment (Smith 1979). Primer fragments were isolated from Mu DNA and their positions on the Mu map are depicted schematically in the lower part of the figure. (•••••••) The direction of extension by DNA polymerase; (-----) the Mu–host and D108–host junctions.

ing system of Mu that initiates packaging at the left end of the Mu DNA molecule will selectively package those Tn*M1* sequences that have been transposed to the chromosome, since the original pTM2 plasmid is too small to fill a phage head. Thus, the presence of Tn*M1* sequences in the phage progeny demonstrates transposition. The phage progeny were purified by density gradient centrifugation, and the phage DNA was extracted and analyzed with the restriction endonuclease *Hin*dIII (Fig. 4). Two large restriction fragments are visible that correspond to the helper phage DNA. In addition, two fragments that comigrate with the *Hin*dIII cleavage products of pTM2 are present, plus one small fragment with a diffuse appearance that comigrates with the leftmost *Hin*dIII fragment of wild-type Mu DNA. This fragment cannot be generated from the helper phage DNA due to the absence of the leftmost *Hin*dIII cleavage site in this mutant phage and must originate from transposed and subsequently packaged Tn*M1* sequences. Since no new junction fragments of pTM2 are visible in this digestion, this must mean that the plasmid sequences are flanked by direct repeats of Tn*M1* and hence are present in the

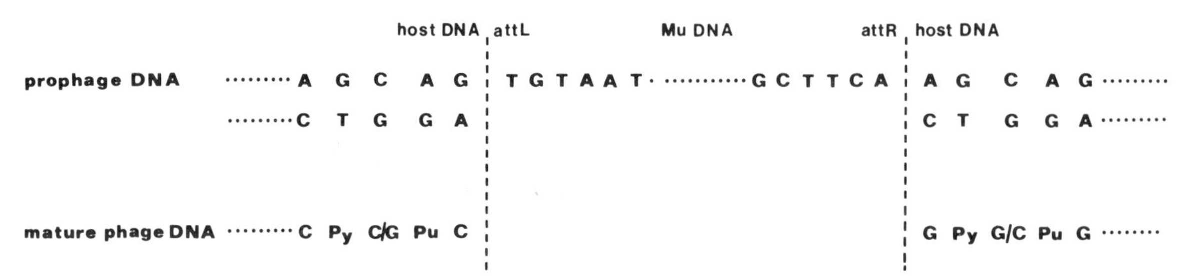

Figure 3. Sequence of bacterial insertion sites. The two Mu prophage sequences have been published (Allet 1979; Kahmann and Kamp 1979). The junction sequence of mature phage DNA is the consensus sequence derived by the chain-termination sequencing technique as described in the legend to Fig. 2. Sequences are written in the 5′ to 3′ direction.

chromosome as cointegrate structures. To examine the structure of the plasmid DNA in these packaged DNA molecules further, the biological properties of the phage progeny were examined. Upon infection of *E. coli* cells, the ampicillin-resistance (Ap^r) marker present on pTM2 is transduced with high efficiency (10^{-2}). The Ap^r cells harbor the original pTM2 plasmid. *rec*-mediated resolution of cointegrates is responsible for restoring the circular pTM2 DNA from a linear DNA molecule present in phage particles. This sequence of events is depicted schematically in Figure 5.

We have used this physical and biological assay for transposition to determine the effect of the host DNA duplication. For this purpose, the *att*R fragment of Tn*M1* was replaced by two different *Hin*dIII *att*R fragments of phage Mu 445-5 DNA (Chow et al. 1977) such that the Tn*M1* sequence is regenerated but the host sequences adjacent to *att*R are different from the one in pTM2. When exposed to a Mu helper, these plasmids (designated pTM3 and pTM5) behave like pTM2 (Fig. 4). A duplicated host sequence is therefore not a prerequisite of transposition.

The high percentage of cointegrates in Mu lysates grown lytically on strains containing pTM plasmids suggests that Mu does not resolve cointegrates very efficiently. It should also be pointed out that there is no need for resolution of cointegrates during lytic development, since the DNA rearrangements that are associated with cointegrate formation (Shapiro 1979) are of no consequence in a cell that is destined to perish in any case by virtue of the phage-induced lysis. Transposable elements have developed special systems that assure resolution of cointegrates (Gill et al. 1978; Heffron et al. 1979), and likewise, during lysogenization of a cell with Mu, a mechanism must be active that allows transposition without cointegrate formation or efficient resolution of cointegrate structures.

Integration of Mu DNA in Other Bacterial Species

The bacteriophage Mu can infect a variety of gram-negative bacteria. Most of these strains, like *Enterobacter cloacae, Serratia marcescens, Proteus vulgaris,* and *Agrobacterium tumefaciens,* are sensitive to Mu G(−) phage particles. Our preliminary studies show different behavior of Mu in some of these bacterial species. Growth of Mu G(−) phage in *E. cloacae* is normal compared to its growth in *E. coli* K12. *E. cloacae* can be efficiently lysogenized with Mu G(−) phage; 30% of the survivors of an infection become lysogenic, and 4% of

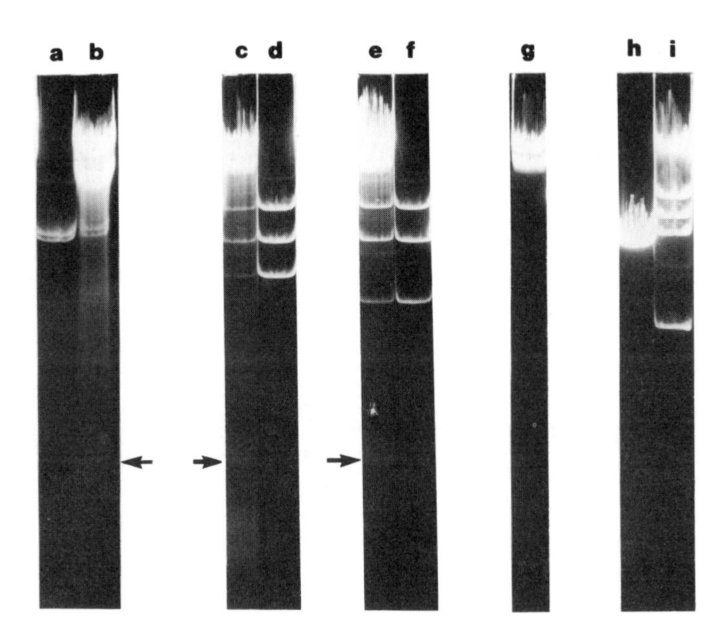

Figure 4. Restriction enzyme analysis of packaged cointegrate structures. Exponentially growing cells of *E. coli* C600, harboring plasmids pTM2, pTM3, or pTM5, were infected with Mu-c*I*ts62cII31 phage (D. Kamp, unpubl.) at a multiplicity of 5 pfu/bacterium. The phage lysate obtained was purified, and the DNA was extracted. Cleavage products of these DNAs obtained with the restriction endonuclease *Hin*dIII were run on a 1.2% agarose slab gel next to plasmid DNAs cleaved with *Hin*dIII. (*a*) pTM2; (*b*) pTM2 + Mu; (*c*) pTM3 + Mu; (*d*) pTM3; (*e*) pTM5 + Mu; (*f*) pTM5; (*g*) Muc*I*ts62cII31 cleaved with *Hin*dIII; (*h*) pBR322 cleaved with *Hin*dIII; (*i*) λ*b*522 cleaved with *Eco*RI as size marker. (→) The diffuse *Hin*dIII fragment originating from the left end of packaged Tn*M1* sequences.

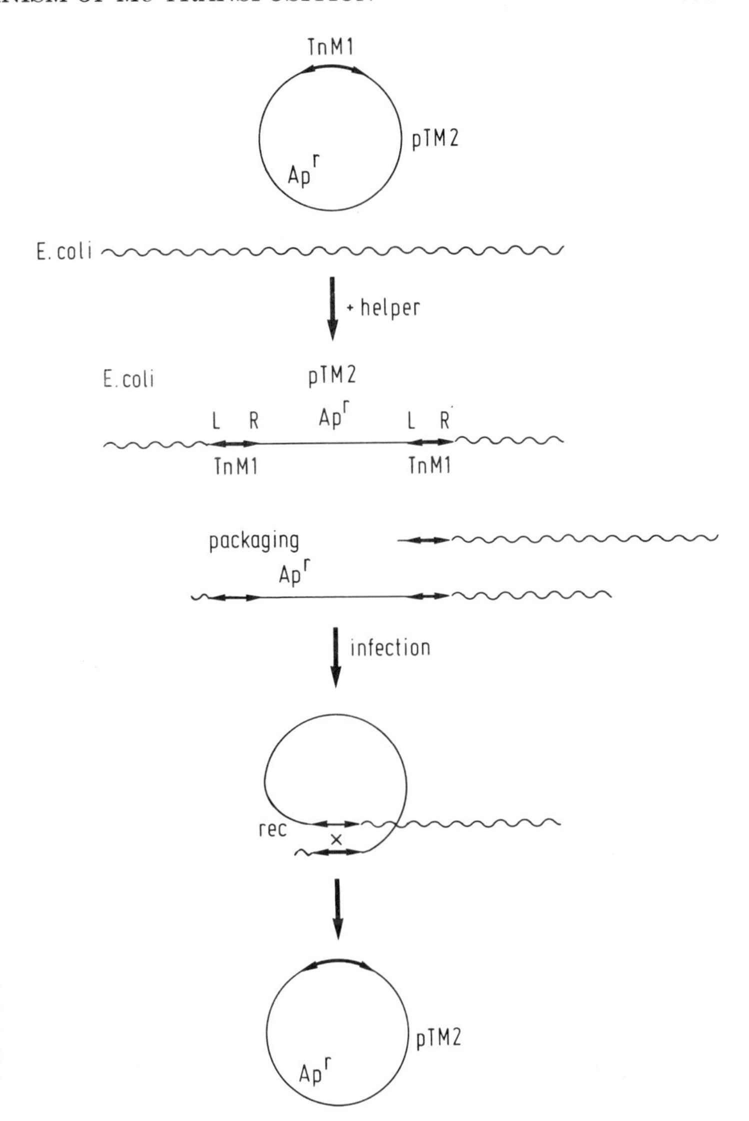

Figure 5. Schematic presentation of the transposition assay for Tn*M1*. (—) Tn*M1* sequences; (—) plasmid DNA; (∼∼) chromosomal DNA. The left and right attachment sites on Tn*M1* are indicated by L and R, respectively.

the lysogenic bacteria are auxotrophs. Twenty-four auxotrophs have been characterized further, and among these we found two *cys*⁻ mutants, two *his*⁻, two *thr*⁻, two *trp*⁻, one *leu*⁻, and one *ser*⁻ mutant. This indicates that insertion of Mu can occur at many sites on the *E. cloacae* chromosome. Bacteria that are more distantly related to *E. coli* K12 show a different response to infection with Mu. *S. marcescens* plates Mu*cts*62 G(−) phage at 40 °C; however, it cannot plate or be lysogenized by the same phage at 30 °C.

More interesting in this context of transposition of Mu is the phenomenon of abortive infection. Undiluted lysates of Mu G(−) phage clear a lawn of *A. tumefaciens*, but no single plaques are observed when appropriate dilutions are plated. Also, no lysogens have been found among 100 tested colonies of survivors. Abortive infection originally observed with phage P1 has been explained by the failure to express phage functions normally. The observed killing, on the other hand, suggests that some expression of early functions can take place (Amati 1962). In view of the key role of transposition for Mu growth, it is likely that abortive infection of Mu in *A. tumefaciens* is due to a defect in transposition and replication. The same effect of abortive infection is observed when Mu A^- or Mu B^- phage are plated on *E. coli* K12 (O'Day et al. 1978). This interpretation would imply that host factors, which could be missing in *A. tumefaciens*, for instance, play an essential role in Mu transposition. This is consistent with a previous report by van Vliet et al. (1978) that Mu does not produce phage in *A. tumefaciens* after having been introduced via an episome. It is interesting to note, however, that the same authors found normal transposition of Tn*7* in *A. tumefaciens*.

Model for the Transposition of Bacteriophage Mu

To accommodate some of our results and other facts about Mu transposition, we are proposing a model for the mechanism of replication and transposition of Mu. Some of the ideas originally proposed in general models for transposition (Grindley and Sherratt 1979; Shapiro 1979) have been incorporated. In addition, with respect to the initial events of transposition and replication of Mu, we have drawn on mechanisms of replication of single-stranded phage (Denhardt et al. 1978). We suggest that a Mu-specific transposase complex recognizes both attachment sites of the Mu genome and introduces nicks at the 5′ ends of Mu DNA (Fig. 6A). The transposase remains attached to and thereby protects the 5′ ends of the DNA from degradation. The free 3′-OH termini of the bacterial DNA prime synthesis of Mu DNA with the simultaneous displacement of the 5′ ends of Mu (Fig. 6B). The displaced 5′ ends, still held in a complex with the transposase, become transferred to a new insertion site (Fig. 6C). This could be the protruding 3′ ends of a target cut in a staggered fashion (as proposed by Grindley and Sherratt 1979), or, alternatively, the 5′ ends of Mu could be ligated to 3′-flush ends of a target, which could account for the absence of the host DNA duplication in lytically grown Mu phage. The DNA transfer could happen at any stage of one round of replication. Synthesis of Mu DNA will be terminated at the new insertion site (Fig. 6 D and E). The remaining nicks are then closed in a ligation step (Fig. 6F). The resulting molecule is a cointegrate in which Mu, in a step of integrative replication (Campbell, this volume), has become duplicated and fused to a new target. If a duplication of host DNA is generated during transposition, the two copies are not flanking the same molecule in a cointegrate.

On the other hand, one can envision how a modula-

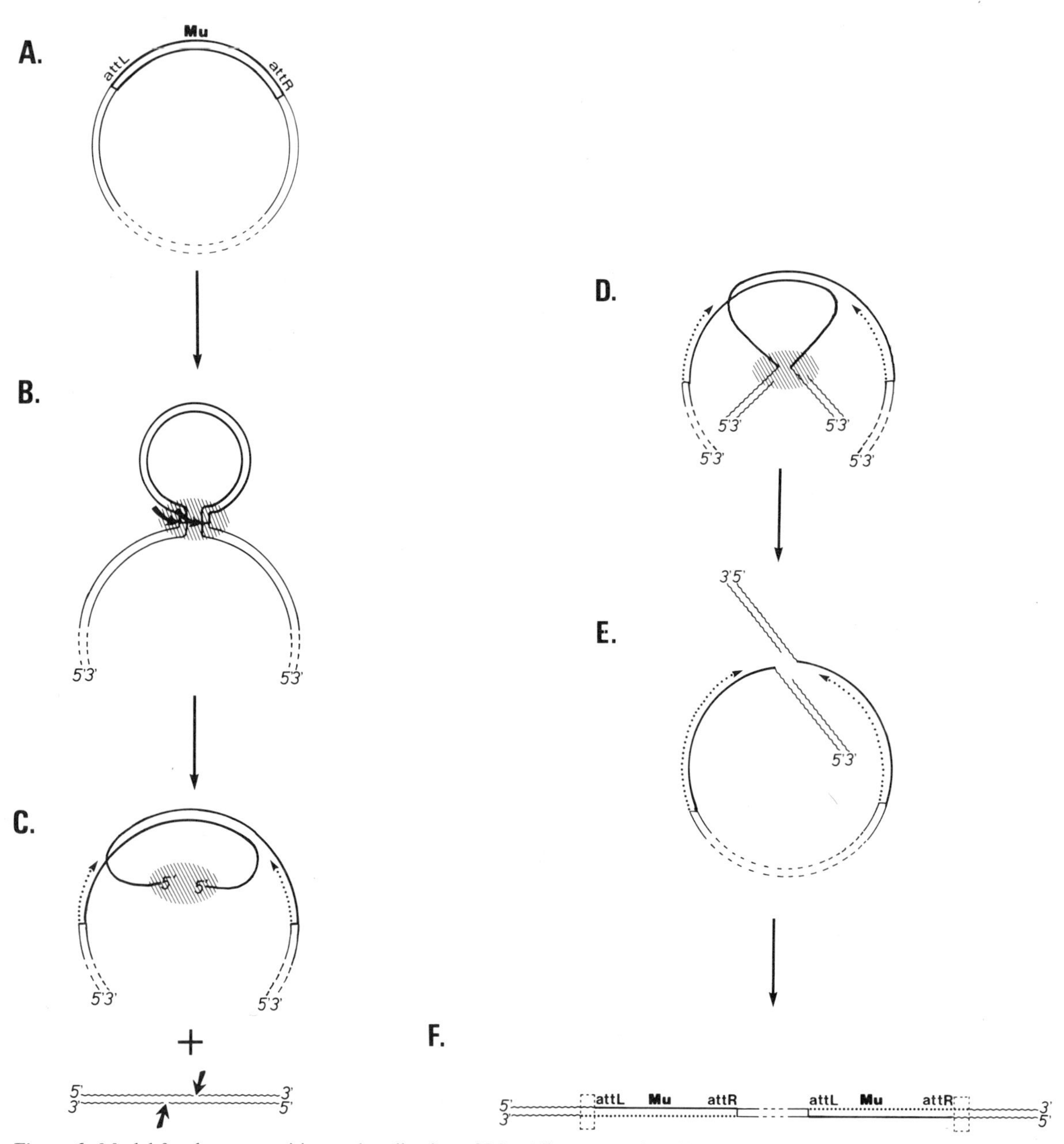

Figure 6. Model for the transposition and replication of Mu. All steps are described in the text. The shaded areas represent the Mu transposase. As an alternative to a staggered cut being introduced at the bacterial insertion site (as depicted in *C*), a flush-end cut could be made. This possibility is discussed in the text.

tion of this mechanism avoids the problem of cointegrate structures in the lysogenic pathway. If the transfer of the displaced 5′ ends to a new insertion site is delayed, a second round of replication can be initiated and proceed. Recombination, for instance by a template switch at a four-stranded DNA structure formed by the newly replicated Mu DNA, will liberate a newly synthesized Mu DNA molecule that can be transferred to a new site leading to single insertion.

CONCLUSIONS

On the basis of experiments presented here, we would suggest that the transposition mechanism of Mu is more complex than anticipated. In particular, the mechanism of integration during lysogenization and of transposition during lytic growth appear to differ at least in some aspects. This is not quite unexpected, considering that the physiologies of lysogenization and lytic growth are different. Upon lysogenization, the damage to the host cell and in particular to its chromosome must be kept to a minimum. Cointegrate structures would lead to deletions or inversions of large parts of the host chromosome unless they are resolved efficiently. No such restraints exist in the lytic pathway. In view of the imminent lysis of the cell, even severe scrambling of the host chromosome should be irrelevant.

The most intriguing result in this context is that the host DNA duplication is apparently absent in lytically grown Mu phage. Since the duplication does not have a function in transposition of Mu, it can only be speculated what the purpose of the duplication could be. Precise excision of Mu, which requires the bacterial *rec* function (Bukhari 1975), could be promoted by recombination between the duplicated host sequences and could therefore provide a safeguard in case the insertion of Mu in a particular site results in a growth disadvantage for the lysogenic cell. Again, this does not apply to lytic development of Mu and eliminates the need for duplicated DNA at the Mu ends. The question remains as to which mechanisms account for the presence or absence of duplicated host DNA. We consider three possibilities.

1. Different enzymes are involved in transposition to preferred and nonpreferred host sites. Preferred sites are cleaved to yield flush ends and nonpreferred sites are cleaved to yield staggered ends.
2. Different enzymes participate in lysogenization and lytic development. All insertions—in preferred or nonpreferred sites—are associated with a duplication in lysogenization but not in lytic development.
3. All transposition events are associated with a duplication, and the apparent absence of duplications in phage DNA is the consequence of cointegrate formation during lytic development. As already pointed out, the matching duplications are not flanking the same Mu DNA molecule but could be anywhere on the chromosome. Selective packaging could account for the disappearance of duplications in the DNA present in phage particles.

It is obvious that these possibilities can be distinguished by sequencing a large number of Mu insertion sites. Of particular interest are insertion sites of prophages that have not been selected to be in a particular gene, insertion hotspots, and cointegrate structures at various stages of Mu development.

ACKNOWLEDGMENTS

We thank Gaby Sowa for technical help. Part of this work was carried out at the Cold Spring Harbor Laboratory and was supported by National Institutes of Health grant GM-23996 awarded to Louise T. Chow and by National Science Foundation grant PCM76-82448 awarded to Richard J. Roberts.

REFERENCES

Allet, B. 1979. Mu insertion duplicates a five base pair sequence at the host inserted site. *Cell* **16:** 123.

Amati, P. 1962. Abortive infection of *Pseudomonas aeruginosa* and *Serratia marcescens* with coliphage P1. *J. Bacteriol.* **83:** 433.

Botstein, D. and N. Kleckner. 1977. Translocation and illegitimate recombination by the tetracycline resistance element Tn*10*. In *DNA insertion elements, plasmids, and episomes* (ed. A. I. Bukhari et al.), p. 185. Cold Spring Harbor Laboratory, Cold Spring Harbor, New York.

Bukhari, A. I. 1975. Reversal of mutator phage Mu integration. *J. Mol. Biol.* **96:** 87.

———. 1976. Bacteriophage Mu as a transposition element. *Annu. Rev. Genet.* **10:** 389.

Bukhari, A. I. and D. Zipser. 1972. Random insertion of Mu-1 DNA within a single gene. *Nat. New Biol.* **236:** 240.

Calos, M. P. and J. H. Miller. 1980. Transposable elements. *Cell* **20:** 579.

Chow, L. T., R. Kahmann, and D. Kamp. 1977. Electron microscopic characterization of DNAs of non-defective deletion mutants of bacteriophage Mu. *J. Mol. Biol.* **113:** 591.

Daniell, E., R. Roberts, and J. Abelson. 1972. Mutations in the lactose operon caused by bacteriophage Mu. *J. Mol. Biol.* **69:** 1.

Daniell, E., J. Abelson, J. S. Kim, and N. Davidson. 1973. Heteroduplex structures of bacteriophage Mu DNA. *Virology* **51:** 237.

Denhardt, D. T., D. Dressler, and D. S. Ray (eds.) 1978. *The single-stranded DNA phages.* Cold Spring Harbor Laboratory, Cold Spring Harbor, New York.

Faelen, M. and A. Toussaint. 1978. Mu mediated deletions in the chromosome of *E. coli. J. Bacteriol.* **136:** 477.

Foster, T. 1977. Insertion of the tetracycline resistance translocation unit Tn*10* into the *lac* operon of *Escherichia coli* K12. *Mol. Gen. Genet.* **154:** 305.

Gill, R., F. Heffron, G. Dougan, and S. Falkow. 1978. Analysis of sequences transposed by complementation of two classes of transposition-deficient mutants of Tn*3*. *J. Bacteriol.* **136:** 742.

Grindley, N. D. F., and D. Sherratt. 1979. Sequence analysis at IS*1* insertion sites: Models for transposition. *Cold Spring Harbor Symp. Quant. Biol.* **43:** 1257.

Heffron, F., B. McCarthy, H. Ohtsubo, and E. Ohtsubo. 1979. DNA sequence analysis of the transposon Tn*3*: Three genes and three sites involved in transposition of Tn*3*. *Cell* **18:** 1153.

Hull, R. A., G. S. Gill, and R. Curtiss, III. 1978. Genetic characterization of Mu-like bacteriophage D108. *J. Virol.* **27:** 513.

Kahmann, R. and D. Kamp. 1979. The nucleotide sequences of the attachment sites of bacteriophage Mu. *Nature* **280:** 247.

KLECKNER, N., D. STEELE, K. REICHARDT, and D. BOTSTEIN. 1979. Specificity of insertion by the translocatable tetracycline-resistance element Tn*10*. *Genetics* **92:** 1023.

O'DAY, K. J., D. W. SCHULTZ, and M. H. HOWE. 1978. Search for integration deficient mutants of bacteriophage Mu. In *Microbiology—1978* (ed. D. Schlessinger), p. 48. American Society for Microbiology, Washington, D.C.

RAIBAUD, O., M. ROA, C. BRAUN-BRETON, and M. SCHWARTZ. 1979. Structure of the *mal*B region in *Escherichia coli* K12. *Mol. Gen. Genet.* **174:** 241.

REED, R., R. YOUNG, J. A. STEITZ, N. GRINDLEY, and M. GUYER. 1979. Transposition of the *Escherichia coli* insertion element $\gamma\delta$ generates a five-base-pair repeat. *Proc. Natl. Acad. Sci.* **76:** 4882.

SANGER, F., S. NICKLEN, and A. R. COULSON. 1977. DNA sequencing with chain-terminating inhibitors. *Proc. Natl. Acad. Sci.* **74:** 5463.

SHAPIRO, J. A. 1979. Molecular model for the transposition of bacteriophage Mu and other transposable elements. *Proc. Natl. Acad. Sci.* **76:** 1933.

SMITH, A. J. H. 1979. The use of exonuclease III for preparing single stranded DNA for use as a template in the chain terminator sequencing method. *Nucleic Acids Res.* **6:** 831.

TAYLOR, A. L. 1963. Bacteriophage induced mutations in *Escherichia coli. Proc. Natl. Acad. Sci.* **50:** 1043.

VAN DE PUTTE, P., M. GIPHART-GASSLER, T. GOOSEN, A. VAN MEETEREN, and C. WIJFFELMAN. 1978. Is integration essential for Mu development? In *Integration and excision of DNA molecules* (ed. P. Hofschneider and P. Starlinger), p. 33. Springer-Verlag, Berlin.

VAN VLIET, F., B. SILVA, M. VAN MONTAGU, and J. SCHELL. 1978. Transfer of RP4: Mu plasmids to *Agrobacterium tumefaciens. Plasmid* **1:** 446.

Transposition of Bacteriophage Mu: Properties of λ Phages Containing Both Ends of Mu

M. M. Howe and J. W. Schumm
Department of Bacteriology, University of Wisconsin, Madison, Wisconsin 53706

Ever since Taylor (1963) discovered bacteriophage Mu and its ability to cause mutations by prophage integration into the DNA of its host, *Escherichia coli* K12, there has been great interest in understanding the molecular mechanism of the integration event. This interest has been enhanced by the realization that Mu is only one of a number of prokaryotic transposable elements able to integrate at many sites in DNA and by the belief, now being substantiated (see work presented in this volume), that elements with similar properties also exist in eukaryotic cells. Work in a number of laboratories, including our own, has been focused on elucidating the mechanism by which this integration occurs.

Integration of Mu into Host DNA

The pioneering work of Taylor and his collaborators (Taylor 1963; Martuscelli et al. 1971) demonstrated that Mu-induced mutations arise by linear integration of Mu DNA into host DNA during the process of lysogenization. A Mu prophage is tightly linked genetically to the new mutation (Taylor 1963), and circular F′ plasmid DNA containing a Mu-induced mutation is longer than the parent F′ DNA by a length approximately equal to the length of mature Mu DNA (Martuscelli et al. 1971). The fact that Mu-induced mutations arise during initial infection and establishment of lysogens, but only very rarely during subsequent growth or superinfection of a lysogen, indicates that the integration process is under control of the phage repressor (Taylor 1963). This control is extremely useful for analysis of the integration process because it allows the investigator to direct expression of the integration process using phages carrying temperature-sensitive mutations in the repressor gene (Howe 1973).

The generation of many different kinds of mutations by Mu (Taylor 1963) suggests that there are many sites at which Mu can integrate. The essentially random distribution of those sites is indicated by the large number of sites and the lack of "hotspots" for insertion of Mu within the *lacZ* gene (Bukhari and Zipser 1972; Daniell et al. 1972). These results demonstrate that Mu is the most random in integration ability of the prokaryotic transposable elements. The insertion sequence (IS) and transposon (Tn) drug-resistance elements all show some degree of integration specificity, which is characterized by repeated integration into a limited number of sites (site specificity) or preference for integration into different sites within a preferred region (regional specificity) (for review, see Kleckner 1977). The nature of the specificity in these latter cases is presumed to result from recognition of a primary sequence or secondary structure in the target DNA, but the precise mechanism of recognition is not yet clear (Kleckner 1977; Grindley 1979).

Most insertions of Mu appear to arise by simple point insertion of Mu DNA into the mutated host gene without loss of host DNA. This conclusion is supported by genetic mapping of Mu-induced mutations (Bukhari and Zipser 1972), by the ability of Mu-induced mutations to revert (reversion requires an *X* mutation in Mu) (Bukhari 1975), and by DNA sequence analysis of integrated Mu prophages (Allet 1979; Kahmann and Kamp 1979; Kamp et al. 1979). However, some Mu insertions are associated with deletion of extensive amounts of host DNA, and the frequency of such deletions among Mu-induced mutations varies depending on the genes examined. Up to 15% of *lac* mutations arising after Mu infection are deleted for the entire *lac* region and contain a Mu prophage substituted for the deleted DNA; yet only approximately 1% of *gal* mutations and *ara* mutations appear to be deletions (M. M. Howe et al., in prep.). The formation of deletions by transposable elements is quite common; one finds deletions extending into nearby host DNA from the ends of integrated IS or Tn elements (for reviews, see Starlinger and Saedler 1976; Kleckner 1977) or from the ends of integrated Mu after exposure to partially inducing conditions (Faelen and Toussaint 1978). The deletions in *lac* appear to differ from these deletions in that they arise during initial lysogenization and appear at a higher frequency. The mechanism of generation of these deletions is currently under investigation.

Attachment Sites on Mu DNA

Once it became clear that the attachment sites for Mu integration within host DNA were randomly distributed, a number of laboratories turned their attention toward determining the location on Mu DNA of the attachment site(s) where integrative recombination occurs. Their approach was to map the order of Mu genes in prophages integrated at a number of different sites in the host chromosome by isolating deletion mutations extending from nearby host genes into Mu and using marker rescue of Mu amber mutant phages to map the extent of the deletions into the prophage. The results demonstrated that the order of Mu genes is identical in all prophages and therefore suggested that Mu contains unique attachment site(s) where integration occurs (Abelson et al. 1973; Howe 1973).

Studies of the structure of mature Mu DNA and Mu prophage DNA revealed that there are two (rather than one) attachment sites and defined their precise locations in Mu DNA. Mu DNA is a linear double-stranded DNA of approximately 37 kb (Martuscelli et al. 1971; Daniell et al. 1973). When it is denatured and reannealed, the resulting DNA heteroduplexes reveal two regions of heterogeneity: single-stranded, nonhybridizing "split ends" of 1.0–2.0 kb at the right or variable end of the molecule and a 3-kb single-stranded G bubble located near the split ends and separated from them by the 1.7-kb double-stranded β DNA segment (Daniell et al. 1973). The G bubble arises by annealing of DNA from phages containing the G segment in opposite orientations (Hsu and Davidson 1972, 1974); it is the orientation of G that determines the host-range properties of the phage (for review, see Howe 1978, 1980; van de Putte et al. 1980). The split ends arise because the variable end of Mu is composed of host DNA, which differs in different phage particles and which contains in a population a random distribution of the entire host chromosome (Daniell et al. 1975). Hybridization analysis of Mu DNA restriction fragments revealed that the left end also contains a segment of host DNA, but one that is only 50–150 bp in length (Bukhari et al. 1976). These host DNA sequences at the left and right ends of Mu DNA are acquired by headful packaging of Mu DNA from a precursor containing Mu integrated in host DNA sequences (Bukhari and Taylor 1975). DNA heteroduplex (Hsu and Davidson 1974) and sequence analyses (Allet 1978, 1979; Kahmann and Kamp 1979; Kamp et al. 1979) have shown that the host sequences are lost during the integration process, which results in the generation of a Mu prophage that lacks the host DNA present on the infecting phage. Therefore, the junctions between the variable host DNA sequences and the conserved Mu DNA sequences define the attachment sites of Mu where integrative recombination occurs. These sites flank the conserved Mu DNA and separate it from the variable host DNA present at each end of Mu in the prophage and mature phage states.

It is believed that both attachment sites of Mu are required for the integration process. This belief is based on the observations that (1) both attachment sites are joined to the target DNA as a result of integration, (2) Mu prophages with large insertions that prevent incorporation into phage particles of the right attachment site produce phages that are defective in lysogenization (Bukhari and Froshauer 1978), and (3) deletions of the right attachment site are not found in plaque-forming phages deleted for nearby β and G-segment DNA (Chow et al. 1977). The belief is strengthened by the analogy of Mu to Tn elements, in which both ends are required for integration (Heffron et al. 1977; N. Kleckner, pers. comm.).

DNA sequence analyses of Mu attachment sites and integrated prophages have revealed two additional features common to Mu and other transposable elements (Allet 1979; Kahmann and Kamp 1979; Kamp et al. 1979). The first is that a 5-bp sequence originally present in the target DNA is duplicated and appears as a direct repeat immediately flanking each side of the integrated prophage. Such a duplication of 5 bp or 9 bp is characteristic of transposable elements and is presumed to arise by staggered single-strand cuts within the target DNA during the transposition process (Grindley and Sherratt 1979; Shapiro 1979). The second feature is the presence of a short segment of Mu DNA found as inverted repeats near the left and right attachment sites of Mu. IS and Tn elements contain sequences of varying lengths present as inverted repeats at the left and right ends of the element. The case with Mu is somewhat different. There are inverted repeats of 11 bp present near the two Mu attachment sites; however, only 2 of these base pairs are at the termini. The remaining 9 (with occasional mismatches and insertions) are located subterminally at positions 21–31 from the left end and positions 12–20 from the right end (Kahmann and Kamp 1979; Kamp et al. 1979). Whether these inverted repeats have a role in the transposition process is not yet clear.

Functions Involved in Integration of Mu

It is pertinent to ask what functions are involved in Mu integration. The host *recA* function is not required for integration, as the frequency of Mu lysogen formation and the frequency and distribution of Mu-induced *lac* mutations are similar in *rec*$^+$ and *recA* hosts (M. M. Howe, unpubl.). The lack of the requirement for *recA* is typical for integrative and chromosomal rearrangement events carried out by transposable elements (for review, see Kleckner 1977). To date, no host functions have been implicated in the Mu integration process; however, searches for such functions have not been extensive. Two early candidates for host mutants affecting integration were *E. coli himA* and *himB*. These mutants are defective for λ integration and Mu growth (Miller et al. 1979); however, they appear to allow normal integration of Mu, implying that it is some event other than integration that is blocked in these strains (R. Yoshida and M. M. Howe, in prep.).

It is clear that expression of Mu functions is required for Mu integration, since there is little or no integration in a lysogen (Taylor 1963). Early analysis of Mu amber mutant phages defective in essential genes revealed that mutations in the *A* and *B* genes reduce the frequency of lysogenization (Faelen and Toussaint 1973). A direct search for integration-deficient mutants of Mu was carried out by O'Day et al. (1978), who looked for both plaque-forming integration-deficient and conditional-lethal integration-deficient mutants. Among more than 10,000 heavily mutagenized plaque-forming phages assayed, none was found to be defective in integration, suggesting that integration is essential to Mu development. Among 1000 conditional-lethal (amber mutant) phages defective in essential genes, four new mutants defective in the *A* gene were found to be defective in integration. Quantitative analysis of integration ability revealed that *A* mutants are absolutely defective in integration, whereas *B* mutants integrate at a frequency approximately tenfold less than wild-type Mu (O'Day et al. 1978). These observations are corroborated by the demonstration that only the *A* gene is required for Mu*X* prophage excision (Bukhari 1975), Mu-promoted deletion formation (Faelen et al. 1978), and Mu-promoted integration of λ*gal*, an event that requires concomitant integration of Mu (Toussaint and Faelen 1973; Faelen et al. 1975). The determination of precise roles for the products of the *A* and *B* genes in integration has been made difficult by their additional roles in DNA and late RNA synthesis (Wijffelman et al. 1974); however, the current hypothesis is that the product of the *A* gene is the Mu-specific transposase and that of the *B* gene is involved primarily in replication (Faelen et al. 1978; O'Day et al. 1978).

Molecular Mechanism of Integration

The biochemical mechanism of Mu integration has been relatively refractory to elucidation. Ljungquist and Bukhari (1979) found no formation of supercoils by infecting Mu DNA, which suggests that the mechanism of Mu integration may differ significantly from that of λ. They also found no evidence for excision of a Mu prophage from the chromosome when phage development was induced (Ljungquist and Bukhari 1977). These observations have led to the suggestion that Mu may transpose by in situ replication followed by (Ljungquist and Bukhari 1977) or concurrent with (Shapiro 1979) the transposition process. Efforts to dissect the molecular events occurring during transposition are under way in a number of laboratories.

One source of difficulty in analyzing the fate of infecting Mu DNA is the heterogeneity in length and base sequence of the host DNA at the ends of Mu DNA. Therefore, our approach has been to construct a phage in which the ends of Mu are unique, defined DNA sequences whose fates can be followed in infected cells. This was done by cloning both Mu attachment sites and Mu *A* and *B* genes into a λ vector and using that vector to study genetically and biochemically the events occurring upon phage infection.

RESULTS

Design of Experimental System

The strategy for developing a Mu DNA substrate with unique ends was to put Mu into a vector DNA that could be manipulated easily to allow both genetic and biochemical analysis of its fate in vivo and in vitro. Both plasmid and phage vectors were considered. Bacteriophage λ was chosen as the vector because (1) it has great potential for genetic manipulation and analysis, (2) it allows initiation of Mu gene expression by infection rather than by heat induction, (3) it allows differential labeling of the host and Mu-containing DNAs by growth of the host cells and infecting lysate under appropriately different conditions, and (4) it provides for alternate experimental control of gene expression of Mu and vector sequences, for example, by alternate growth in a λ- or Mu-lysogenic strain.

Since Mu is only slightly smaller than λ, it was not possible to design a plaque-forming λ phage that could contain all of Mu. Therefore, we decided to leave out the middle portion of the Mu genome because the known genes in that region are not required for replication or integration. Since a Mu prophage with the desired internal deletion did not already exist, the approach taken to generate the desired structure was to clone the left and right ends of Mu on two separate DNA restriction fragments into the same λ cloning vehicle. This was convenient as λ phages containing each end of Mu were already available and could serve as sources of DNA for cloning. An important advantage to this approach was that the random joining of the two Mu fragments with each other and with λ would theoretically produce eight possible structures that differed in the relative orientations of the Mu DNA fragments to each other and to the λ DNA. Analysis of the different phages would then demonstrate whether the orientation of the fragments was important in the expression of Mu-specific properties.

The most useful system was deemed to be one in which the constructed λ-Mu phage would carry all the sites and functions known to be needed for integration and replication of Mu so that no helper phage would be required. This meant that the restriction fragments cloned should contain intact attachment sites and functional *A* and *B* genes. In addition, the presence of a heat-inducible *cts* allele of Mu repressor would enable analysis of events occurring at high temperature in the absence of lysogeny and would facilitate characterization of lysogenic forms generated at low temperature. Because the attachment sites and the replication and integration functions of Mu should not be required for growth of the λ-Mu phages, these phages should also be useful for the isolation and characterization of mutations in sites and functions required for Mu integration and replication. Although mutations in these functions can be characterized in Mu itself, mutations in the sites would be lethal and would therefore go undetected. Analysis of such mutant phages should better define the role of specific DNA sequences and proteins in the integration and replication processes.

Isolation of λ Phages Containing Both Ends of Mu

Previous work in our laboratory had resulted in the generation of a collection of λpMu phages containing varying amounts of Mu DNA from the left, right, or middle of the Mu genome (Moore et al. 1977; O'Day et al. 1979). Two of these phages required only minor modifications in order to serve as sources of Mu DNA ends for cloning into λ.

The primary source for the left end of Mu was λpMu508, a phage carrying approximately 6.3 kb of Mu DNA from the left end of a Mu prophage inserted in *lacZ*. This phage carried a heat-inducible Mu repressor and Mu *A* and *B* genes, whose ability to function was demonstrated by the ability of λpMu508 to complement Mu *A* and *B* amber mutants for growth (O'Day et al. 1979; M. M. Howe, unpubl.). A phage that would serve more suitably as the left-end donor due to the lack of some restriction sites was constructed by replacing the right arm of λpMu508 with the right arm of the λ cloning vector Charon 4 (containing *bio*256, KH54, BW1, *nin*5, and *QSR*-80 mutations and substitutions [Blattner et al. 1977]). The resulting left-end-donor phage, LED 508 (Fig. 1), contained the left end of Mu on a single *Eco*RI restriction fragment flanked by cleavage sites to the right of the Mu *kil* gene and approximately 0.1 kb to the left of the Mu attachment site within *lacZ* DNA. Because the location of the cleavage site within *lacZ* was as far from the left Mu attachment site as the left end of mature Mu DNA, this fragment should contain an intact left Mu attachment site. It should also contain functional genes for the heat-inducible Mu repressor, *A* and *B* genes, and probably the nonessential *kil* gene involved in host-cell killing.

The initial source of the right end of Mu was λpMu4M134, a Charon 4 clone carrying the right-end *Eco*RI restriction fragment from mature Mu DNA (Moore et al. 1977). Since the size of the *Eco*RI fragment was too large to fit stably into a λ phage also carrying the left end of Mu, a derivative of λpMu4M134 containing a deletion within the Mu DNA was isolated. First, an $imm^{434}nin^{+}$ recombinant of λpMu4M134 was constructed to provide a sufficient length of λ DNA to allow selection of deletions, then a deleted derivative was selected by growth on plates containing citrate (O'Day et al. 1979), and finally the right arm of Charon 4 was recombined back onto the deleted phage. The resulting right-end-donor phage, RED 25 (Fig. 1), contained an *Eco*RI restriction fragment bounded on one side by a cleavage site between the Mu *M* gene and vector DNA and on the other side by a site within variable-end DNA approximately 1.3 kb from the right-end attachment site of Mu. This length of variable-end DNA, which was comparable to that present in some plaque-forming, insertion-carrying Mu phages (Chow et al. 1977), ensured that the right-end attachment site on the *Eco*RI fragment would be intact. The Mu genes on this *Eco*RI fragment (and those deleted from it) were late genes believed to be involved in phage morphogenesis and presumed to be irrelevant to the integration process (O'Day et al. 1978).

To construct phages containing both ends of Mu, DNAs isolated from LED 508 and RED 25 were cleaved with *Eco*RI, mixed, ligated with T4 ligase, and packaged into λ phage particles in vitro (Blattner et al. 1978). The resulting phages were plated on the Mu-lysogenic strain MH131 (*araD*pΔ[*leu*::Mu*cts*Δ*L*-*U*-*S*---*ara*]; Howe 1973), and individual plaques were isolated and tested for the presence of *bio* DNA (Blattner et al. 1977), which is indicative of the presence of an unwanted *Eco*RI fragment from LED 508. bio^{-} phages were then tested for the presence of the two Mu *Eco*RI fragments by assaying the ability of each phage to donate, by recombination, the wild-type allele of amber mutant phages defective in Mu *A, B, M, Y,* and *N* genes (O'Day et al. 1979). Phages containing both ends of Mu were then saved for further analysis.

At this point, we found that the *QSR*-80 substitution in the right arm of the Charon 4 vector caused an inhibition of λ growth in a *himA* host. Since strains containing *himA* mutations also inhibited Mu growth, we wished to study Mu-specific processes in the *himA* host. Therefore, we replaced the *QSR*-80 region of each phage with a *QSR*-NRE region, which lacks restriction sites near the right end of λ (Blattner et al. 1977), by recombination in vivo with λ*lac*5 KH54, BW1, *nin*5, NRE.

Theoretically, the cloning process could result in eight possible configurations of restriction fragments: four types with different orientations of the two Mu fragments relative to each other and each type in either of

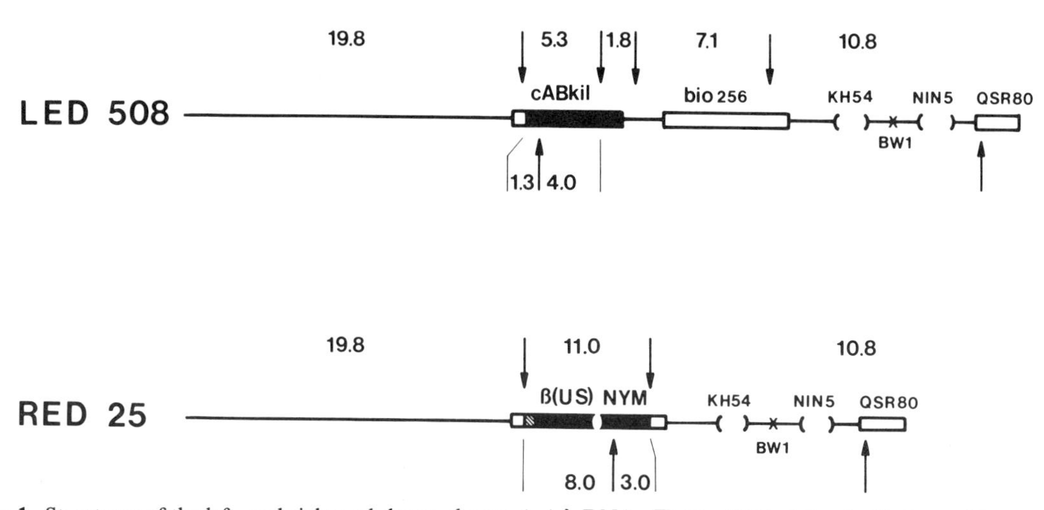

Figure 1. Structures of the left- and right-end-donor phages. (—) λ DNA; (■) Mu DNA; (□) substitutions of host or other phage DNA (Blattner et al. 1977); parentheses indicate deletions (Blattner et al. 1977); (▧) host variable-end DNA attached to the *β* end of Mu (Moore et al. 1977); (↓↑) locations of cleavage sites for *Eco*RI and *Hin*dIII, respectively. Numbers above and below the lines indicate the sizes (kbp) of the *Eco*RI restriction fragments (*above*) and of the relevant restriction fragments produced by digestion with both *Hin*dIII and *Eco*RI (*below*). Letters above the Mu DNA segments indicate the Mu genes encoded by the segment; the *U* and *S* genes are in parentheses because they are in the invertible G segment and may be in the order *U S* or *S U* (Howe et al. 1979).

two orientations within the λ DNA (Fig. 2). The type-1 phages have the normal relative orientation of Mu attachment sites; that is, the attachment sites are at the outside ends of the Mu DNA fragments. Phages designated type 2 have both attachment sites located close to each other in the middle of the Mu DNA fragments. The type-3 phages have only the right attachment site in the middle of the Mu DNA sequences, and the type-4 phages have only the left attachment site in the middle of the Mu DNA sequences. Within each type, the a and b notation designates the two alternative orientations of the Mu DNA with respect to λ DNA.

The orientation of Mu DNA fragments in each phage was determined by digesting phage DNA with the restriction endonuclease *Hin*dIII and separating the fragments by agarose gel electrophoresis (J. W. Schumm and M. M. Howe, in prep.). As shown in Figure 2, each phage should produce a unique set of three *Hin*dIII fragments due to the presence of an asymmetrically located *Hin*dIII cleavage site within each cloned Mu DNA *Eco*RI fragment. This analysis allowed the identification of at least two phages each of types 1a and 1b, 3a and 3b, and 4a and 4b. Among the 18 phages analyzed, none had the configuration of a type-2 phage; however, three phages were found that appeared upon further analysis to be deleted derivatives of type-2 phages. These will be discussed in more detail later.

The assignments of Mu fragment configurations derived by endonuclease analysis were confirmed by observation of DNA heteroduplexes for each of the following combinations of DNAs: (1) types 1a and 1b with Mu, (2) types 3a and 4a with type 1a, and (3) types 3b and 4b with type 1b. In each case, the heteroduplex structure observed was identical to that predicted from the fragment configurations shown in Figure 2 (J. W. Schumm and M. M. Howe, in prep.).

Growth Properties of λ Phages Containing Both Ends of Mu

The phage isolation and lysate preparation described previously were performed in a Mu lysogen or a *himA* host to prevent expression of Mu functions. The production of Mu repressor by a lysogen was expected to prevent Mu gene expression on the infecting λ-Mu phage unless replication of the infecting phage surpassed the repressor-producing capability of the Mu prophage. The precise mechanism of inhibition of Mu development in a *himA* host is not known; however, the lack of cell killing (R. Yoshida and M. M. Howe, in prep.) and the absence of Mu DNA synthesis (M. Pato, pers. comm.) suggest that Mu gene expression is blocked at a very early stage. These conditions of reduced Mu gene expression were chosen because we hypothesized that expression of Mu functions might result in Mu-specific excision, replication, or integration that might be lethal to further λ-specific development. This hypothesis was tested by measuring the efficiency of plaque formation of the different phage types in isogenic Mu-lysogenic and nonlysogenic strains. The results (Table 1) demonstrated that the type-1 phages showed reduced growth on a nonlysogenic strain; they plated approximately 1000-fold better on a Mu lysogenic strain than on a nonlysogenic strain. In contrast, type-3 and type-4 phages plated with equally high efficiency on both strains, as did the parental λ phages containing none or only one end of Mu. The plaques of type-1 phages growing on the nonlysogenic strain contained mutants that had lost the reduced growth characteristic. In subsequent tests, they plated equally well on both Mu-lysogenic and nonlysogenic strains. These mutants are currently being studied to determine the nature of the growth inhibition and the Mu sites and functions required for its expression.

To quantitate more precisely the degree of λ growth

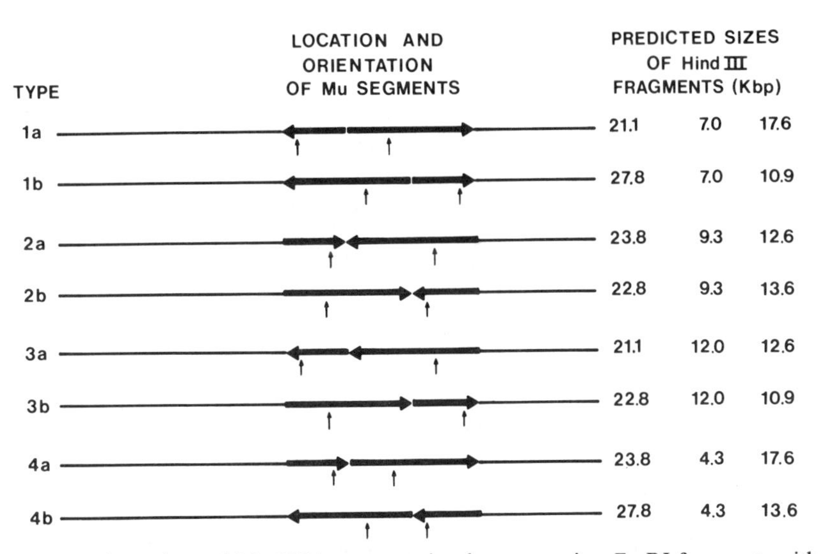

Figure 2. Eight possible configurations of Mu DNA segments in phages carrying *Eco*RI fragments with the left and right ends of Mu. (—) λ DNA; (▬) cloned Mu DNA fragments, with the shorter and longer fragments representing the left- and right-end fragments, respectively; (→←) attachment sites; (↑) locations of *Hin*dIII cleavage sites. The sizes (kbp) expected for the DNA restriction fragments arising by *Hin*dIII cleavage of the *QSR*-NRE derivative of each phage (*QSR*-NRE is 1.2 kbp smaller than *QSR*-80) are listed at the right.

Table 1. Growth of λ Phages Carrying Both Ends of Mu

Phage	Type	Relative efficiency of plating on RS54(Muc+)/RS54 cells
N33	1a	1860
N371	1a	930
N318	1b	2100
N323	1b	1100
N187	3a	0.5
N144	3b	0.8
N159	4a	0.6
N189	4b	0.9
NCh4	—	1.0
NLED 508	—	1.3
NRED 25	—	0.9

Lysates of each phage were grown by infecting strain K750 (SuIII+ *pro mel himA;* from H. I. Miller and D. Friedman) and titered by spotting 0.01 ml of tenfold serial dilutions in SM (0.2 M NaCl, 0.02 M Tris-HCl, 0.002 M $MgSO_4$ at pH 7.5) (Howe 1973) onto cell lawns of strains RS54 (*araD;* Howe 1973) and RS54 (Muc+) on TCMG plates (per liter: 10 g of BBL trypticase peptone, 5 g of NaCl, 8.5 g of agar, 1 mg of thiamine, and 0.01 M $MgSO_4$) (Weil et al. 1972) that were incubated at 42°C overnight. The relative efficiency of plating is the titer observed on strain RS54 (Muc+) divided by the titer found on strain RS54. The N prefix for each phage number indicates that the phage carries a *QSR*-NRE region.

inhibition, the burst sizes of λ-Mu phages produced during a single cycle of infection in several hosts were determined (J. W. Schumm and M. M. Howe, in prep.). All the phages grew normally in a *himA* host, producing bursts of 35–117 phage per cell. The type-3, type-4, and parental phages also grew equally well in the nonlysogenic, Mu-lysogenic, and *him*+ strains. In contrast, the type-1 phages showed a 5- to 20-fold decrease in growth in the nonlysogenic hosts and a 2- to 6-fold decrease in the Mu-lysogenic host compared with the *himA* strain. These results confirmed the inhibitory effect of Mu gene expression in the type-1 phages and pointed out that the presence of a Mu prophage in the host was not sufficient to block the inhibitory effect completely. The most notable feature of these results is that the relative orientation of Mu DNA fragments is critical to the expression of the λ growth inhibition effect. Only those phages with the Mu fragments in their normal relative orientation exhibited such inhibition.

Integration Properties of λ Phages Containing Both Ends of Mu

The λ-Mu phages were next assayed for their ability to induce mutations in the host (J. W. Schumm and M. M. Howe, in prep.), a property characteristic of transposition of Mu. In this experiment, a *rec*+ λ-lysogenic host was used to prevent killing of the infected cells by λ-specific growth. After infection, the cells were grown overnight to allow establishment of lysogeny and segregation of the mutant phenotypes; they were then plated to select for trimethoprim-resistant (Tp^r) *thyA* mutants (O'Day et al. 1978). The resulting mutants were tested for the presence of Mu DNA by assaying for immunity to superinfection by Mu and for cell death at high temperature due to release of repression by the temperature-sensitive Mu repressor. Only the type-1 phages, with Mu fragments in their normal relative orientation, produced a significant increase (3- to 13-fold) in the frequency of Tp^r mutants as compared with the uninfected control culture. The conclusion that many of the mutants were caused by the infecting type-1 phage was derived from the observation that the majority (65–85%) of mutants were now immune to superinfection by Mu and died at high temperature. In contrast, the type-3, type-4, and parental phages gave no significant increase in the frequency of Tp^r mutants, and none of the mutants tested had acquired Mu immunity or the property of death at high temperature. These results again demonstrated that the relative orientation of Mu fragments to each other was critical in the expression of a Mu-specific event, namely induction of host mutations. As expected, this event was dependent upon Mu gene expression; no increase in mutants was obtained upon infection of a Mu-lysogenic, λ-lysogenic host.

To determine the nature of the mutations induced in *thyA* by the type-1 phages, five Mu-immune mutants from each infected culture were analyzed in detail. The results indicated that the mutations were caused by integration of both Mu and λ sequences into the *thyA* gene rather than by integration of the Mu DNA segment alone. First, genetic recombination experiments demonstrated that 95% of the type-1-phage-induced mutants contained the wild-type alleles corresponding to amber mutations in the *A, B, O,* and *S* genes of λ and the *A, B, M, Y,* and *N* genes of Mu. In addition, these mutants spontaneously released plaque-forming phages that retained two properties characteristic of the infecting type-1 phage; namely, imm^{λ} and reduced growth in a nonlysogenic strain. These results demonstrated that the Mu and λ sequences in the host cells remained joined or were retained in a form that could be rejoined to regenerate the initial phage structure. A second experiment demonstrated that both the Mu and λ sequences were present within or very near the *thyA* gene; phage-P1-mediated generalized transduction of the *thyA* mutants to *thy*+ resulted in loss of both Mu and λ sequences in the majority of mutants. Tests of progeny cells derived by growth of the *thyA* mutants revealed that Mu immunity and the Thy− mutant phenotype were very stable; however, a small number of the progeny had lost the λ genes and the ability to release plaque-forming λ-Mu phages.

The presence of both Mu and λ sequences in the *thyA* region and the generation of λ− Mu-immune segregants from the mutant cells suggested that the mutation in *thyA* was caused by insertion of a Mu-λ-Mu cointegrate structure (Fig. 3) in which the λ sequences were flanked by two complete copies of the entire Mu DNA segment originally present within the type-1 phage. Such cointegrates have been found previously for Mu after the rare event of Mu-promoted integration of λ*gal* (Toussaint and Faelen 1973); however, the high frequency of generation of cointegrates by type-1 phages is more

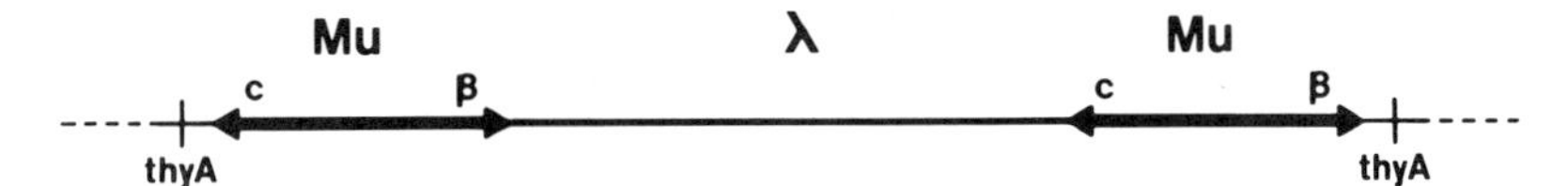

Figure 3. Proposed structure of *thyA* region of chromosome in cell containing *thyA* mutation induced by type-1 phage. (—) Host chromosomal DNA; (▬) Mu DNA; (→←) Mu attachment sites; (—) λ DNA. (*c*, *β*) Locations of the immunity (left) and variable (right) ends of Mu, respectively. The *thyA* gene is represented as being split by the insertion of the Mu-λ-Mu cointegrate.

comparable to the frequency of cointegrate formation observed after transposition of certain mutants of the ampicillin (Ap)-resistance element Tn*3* (Gill et al. 1978). One property of such cointegrates is the *recA*-dependent loss of the middle sequence. This is presumed to occur by homologous recombination between the directly repeated flanking sequences, thereby resulting in loss of one copy of the flanking sequences as well. Therefore, the effect of a *recA* mutation on the generation of type-1-induced *thyA* mutations and on the production of λ^- Mu-immune segregants was assayed. Infection of rec^+ and *recA* λ-lysogenic hosts with the type-1 phages resulted in similar increases in the frequency of Tp^r mutants and similar proportions of Mu-immune cells among those mutants for both the rec^+ and *recA* hosts. However, the properties of the mutants differed; there were no λ^- Mu-immune segregants produced or plaque-forming λ-Mu phages released from the *recA* host. This was exactly the result predicted from the cointegrate structure. To test whether the cointegrate structure was present in the *recA* host, several *thyA* Mu-immune mutants from each infection were made rec^+ by P1 transduction (cotransduction of rec^+ with tetracycline (Tc) resistance due to a closely linked Tn*10* in *srl*) and then tested for the same properties as the original rec^+ *thyA* mutants. In each case, the rec^+ transductants behaved exactly like the original rec^+ *thyA* mutants; that is, they retained both Mu and λ sequences, released plaque-forming λ-Mu phages, and generated λ^- Mu-immune segregants. Therefore, these genetic results support the hypothesis that the type-1-phage-induced mutations in *thyA* arose by insertion of a Mu-λ-Mu cointegrate. A physical test of this hypothesis, namely, restriction mapping of the integrated λ-Mu phage, is currently in progress.

DISCUSSION

λ phages carrying both ends of Mu were constructed by cloning two separate *Eco*RI restriction fragments, each carrying one end of Mu, into a λ Charon 4 cloning vehicle. Phages representing six of the eight possible configurations of fragments were isolated and analyzed for their abilities to carry out two reactions that depend on Mu gene expression: reduced growth of the λ-Mu phage in a nonlysogenic host and induction of mutations in a λ-lysogenic host. The relative orientation of the Mu fragments to each other determined the ability of the phage to carry out these reactions. Type-3 and type-4 phages, which each had one attachment site within the Mu sequences, did not show either effect. Only the type-1 phages, with Mu fragments in their normal relative orientation, showed reduced growth and induction of host mutations. Analysis of host mutations induced by type-1 phages suggested that the mutations were caused by integration of a Mu-λ-Mu cointegrate.

Orientation Effect

It is not yet clear why the orientation of Mu fragments determines the ability of the phage to carry out these Mu-specific reactions. A number of possible explanations can be considered. One is that the orientation affects the synthesis of Mu gene products that are required for the reactions. For example, if the type-3 and type-4 phages were unable to make the product of the *A* gene, they would probably be unable to integrate or express other Mu genes. The results of two experiments in which the expression of the *A* and/or *B* genes was tested suggested that the defect in type-3 and type-4 phages was not due to lack of the products of the *A* and *B* genes. All of the type-3 and type-4 phages and the parent LED 508 complemented Mu*A* and Mu*B* mutants for growth and also complemented *A*-defective, nonintegrating mutants of type-1 phages for integration (M. M. Howe and J. W. Schumm, in prep.). It is conceivable that orientation affects expression of Mu genes other than *A* and *B* that are required for these reactions. Late genes in the *MYNSU* region probably are not involved because deletion mutants of type-1 phages lacking various portions of that region still carried out both Mu-specific reactions (M. M. Howe and J. W. Schumm, in prep.). However, at this time it is not possible to rule out the participation of genes in other regions, for example, nonessential genes *ner* (negative regulation of early transcription) (Giphart-Gassler et al. 1979), *cim* (control of immunity) (Giphart-Gassler et al. 1979), *kil* (host-cell killing) (Giphart-Gassler et al. 1979), *gin* (G inversion) (Kamp et al. 1978), or *mom* (Mu DNA modification) (Toussaint 1976).

A second possible explanation for the different properties of the various λ-Mu phages is that the orientation of Mu fragments affects the synthesis of Mu gene products that prevent the reactions in type-3 and type-4 phages; however, there is no evidence to date for such inhibitors. A third possibility is that synthesis or action of λ functions differs in the various configurations and that these differences result in altered expression of the Mu-specific events. Again, there is no existing evidence for such differences, but further experiments investigating this possibility are needed. A fourth and attractive possibility is that the relative orientation of DNA sequences, for example, the inverted repeats near the Mu attachment sites, is the critical factor affecting expression of Mu-specific events. If an inverted-repeat structure were essential for the reactions, the orientation of the sequences as direct repeats in the type-3 and type-4 phages might prevent the reactions even in the presence of the necessary gene products. Analysis of mutants of type-1 phages unable to carry out one or both reactions should help to distinguish which of these possibilities is correct.

Lack of Recovery of Intact Type-2 Phages

Only six of the eight possible configurations of λ and Mu DNA fragments were found in the phages isolated; neither of the type-2 phages with both Mu attachment sites located close to each other within the Mu DNA sequences were recovered. Instead, three phages were found that had the type-2b configuration, except that each contained a deletion of at least one Mu attachment site. These phages carried out neither of the Mu-specific reactions (J. W. Schumm and M. M. Howe, in prep.). It is unlikely that intact type-2 phages were present but remained undetected due to a small sample size being studied, since large numbers of phages were examined to allow detection of multiple examples of each phage type. It seems more likely that there is a structural or functional reason why type-2 phages cannot be maintained under the conditions used for phage isolation.

One possible explanation is that type-2 phages might be like type-1 phages but might exhibit an even stronger growth inhibition effect. Circularization of infecting type-2 phages would result in a structure in which the attachment sites of Mu would be in their normal orientation but held close together (approximately 1.4 kb apart) and in which the λ sequences would be located within the middle of the Mu sequences, thereby substituting for the central Mu DNA originally eliminated from the phage. If these phages behaved like type-1 phages in their ability to inhibit their own growth and if the inhibition was more severe or faster due to the proximity of the attachment sites, the inhibition reaction might effectively prevent intact type-2 phages from growing; then only phages defective in the reaction, such as those deleted for the attachment sites, would be found. A second possibility is that the structure or gene expression of type-2 phages generates Mu-promoted deletions at a very high frequency, resulting in the loss of the intact type-2 phages and growth of the deleted derivatives. Although similar Mu-promoted deletions might be expected in the type-1 phages, they might occur at a low frequency or they might delete essential λ DNA, thereby producing a defective phage that would remain undetected. A third possibility is that the interaction of Mu- and λ-specific functions in the type-2 phages is different from that in the other phages and that this difference leads to an inability to grow that is controlled by some undefined mechanism different from the ones described previously. Further analysis will be needed to determine which of these possibilities is correct.

Integration by Cointegrates

Our original expectation was that the type-1 phages would cause host mutations by insertion of only the Mu DNA segment located between the Mu attachment sites. This expectation was based on the observation that when Mu infects a cell, it is only the Mu DNA between the attachment sites that is recovered in the integrated prophage (Allet 1979; Kamp et al. 1979; Kahmann and Kamp 1979). Therefore, it is necessary to consider possible reasons why type-1 phages produce primarily cointegrates. One possibility is that the DNA sequences near the Mu attachment sites are different from those in intact Mu and that these differences lead to the production of cointegrates rather than simple transpositions. Although the lengths of host DNA cloned with the Mu attachment sites were at least as long as those in viable Mu phages, they were derived from independent insertions (*lacZ*::Mu prophage for the left and mature Mu DNA for the right) and might possess different host sequences than those resulting from a single insertion. For example, it is likely that the 5 bp immediately flanking the Mu sequences in the type-1 phages are no longer direct repeats of one another; the absence of those direct repeats might lead to the production of cointegrates. In the case of Tn*10*, the presence of different 9-bp sequences flanking the element has no effect on Tn*10* transposition (Kleckner 1979), but the effect of these sequences is not yet known for Mu.

A second possibility is that cointegrates are formed due to the presence of λ-specific sites or to transient expression of λ-specific functions. This seems unlikely because cointegrates are also formed upon transposition of internally deleted Mu prophages located on plasmids (Chaconas et al. 1980; M. Faelen and A. Toussaint, pers. comm.; N. Symonds and D. Leach, pers. comm.). This does not rule out the possibility, however, that plasmids and λ share common features responsible for generating cointegrates.

A third possibility is that the removal of sequences from the middle of Mu (the entire central *Eco*RI fragment and genes *P–R*) has eliminated functions or sites critical to the "normal" transposition process. For example, Shapiro (1979) proposed a model in which cointegrates are an obligatory intermediate and are resolved by recombination to produce the final simple transposition product structure. Heffron et al. (1979) further proposed that Tn*3* contains a specific internal resolution site that is required for recombination to convert cointegrates to simple transposed products. If Mu integration did involve a cointegrate intermediate and if the regions deleted from Mu in the type-1 phages were required for resolution, cointegrates might be the resulting product. This explanation was made less likely by the observation that a plasmid carrying a MuAp prophage, which contains the regions of Mu deleted in the type-1 phages, also forms cointegrates upon transposition (D. Leach and N. Symonds, pers. comm.). Although this experiment is not totally conclusive due to the absence of a small amount of Mu G-segment DNA and the presence of Tn*3* DNA in the prophage (Leach and Symonds

1979), it certainly suggests that the absence of the deleted regions in the type-1 phages may not be entirely responsible for the production of cointegrates.

In fact, it may be that the generation of cointegrates is due to the structure of the DNA in which Mu is inserted. In all the cases described, cointegrates were formed by transposition of Mu or deleted Mu prophages from circular DNA forms, either plasmids or circular λ forms arising after infection. In contrast, normal infecting Mu DNA enters the cell in a nonsupercoiled, presumably linear form (Ljungquist and Bukhari 1979); this lack of supercoiling may affect the mechanism of transposition. If it is the donor DNA conformation that determines the mechanism of the transposition process, one possible result is that stable lysogen formation and lytic growth might occur by different processes; for example, lysogen formation might occur by integration from a nonsupercoiled form, and lytic growth might involve transposition of integrated Mu from supercoiled forms. The possibility that lysogen formation and lytic growth might involve different processes has also been suggested by D. Kamp and R. Kahmann (pers. comm.) on the basis of their observation that mature Mu DNA and Mu prophage DNA contain different configurations of flanking host DNA sequences. At present, the role of cointegrates in Mu transposition is not understood; however, further studies of their generation and possible resolution should help to elucidate the mechanism of the transposition process.

Future Uses of the Experimental System

These studies have provided a set of phages that have been useful in demonstrating the effect of orientation of Mu sequences on their ability to carry out Mu-specific reactions. The phages constructed have attributes that will make them suitable for analyzing a number of features of the Mu transposition process. For example, first, the isolation and characterization of mutants of type-1 phages defective in these reactions should allow a definition of the DNA sequences and proteins involved in the reactions. Second, the presence of unique λ DNA sequences at the ends of Mu should allow sensitive analysis of the fate of the attached DNA during the infection process, for example, during integration and replication. Third, the λ-Mu DNA can be isolated and used as a DNA substrate in an in vitro reaction to characterize the processes of transposition and replication. Such experiments directed toward elucidating mechanisms of replication and transposition are currently in progress in our laboratory.

ACKNOWLEDGMENTS

This work was supported by the College of Agricultural and Life Sciences, University of Wisconsin, Madison, Wisconsin; by National Science Foundation grant PCM 75-02465, Public Health Service grant AI-12731 from the National Institute of Allergy and Infectious Diseases, and American Cancer Society grant NP-264 to M. M. H.; and by National Institutes of Health Pre-doctoral Training grant GM-07133 to J. W. S. M. M. H. is the recipient of Research Career Development Award AI-00274 from the National Institute of Allergy and Infectious Diseases.

REFERENCES

Abelson, J., W. Boram, A. I. Bukhari, M. Faelen, M. Howe, M. Metlay, A. L. Taylor, A. Toussaint, P. van de Putte, G. C. Westmaas, and C. A. Wijffelman. 1973. Summary of the genetic mapping of prophage Mu. *Virology* **54:** 90.

Allet, B. 1978. Nucleotide sequences at the ends of bacteriophage Mu DNA. *Nature* **274:** 553.

———. 1979. Mu insertion duplicates a 5 bp sequence at the host inserted site. *Cell* **16:** 123.

Blattner, F. R., A. E. Blechl, K. Denniston-Thompson, H. E. Faber, J. E. Richards, J. L. Slightom, P. W. Tucker, and O. Smithies. 1978. Cloning human fetal γ globin and mouse α-type globin DNA: Preparation and screening of shotgun collections. *Science* **202:** 1279.

Blattner, F. R., B. G. Williams, A. E. Blechl, K. Denniston-Thompson, K. E. Faber, L.-A. Furlong, D. J. Grunwald, D. O. Kiefer, D. D. Moore, J. W. Schumm, E. L. Sheldon, and O. Smithies. 1977. Charon phages: Safer derivatives of bacteriophage lambda for DNA cloning. *Science* **196:** 161.

Bukhari, A. I. 1975. Reversal of mutator phage Mu integration. *J. Mol. Biol.* **96:** 87.

Bukhari, A. I. and S. Froshauer. 1978. Insertion of a transposon for chloramphenicol resistance into bacteriophage Mu. *Gene* **3:** 303.

Bukhari, A. I. and A. L. Taylor. 1975. Influence of insertions on packaging of host sequences covalently linked to bacteriophage Mu DNA. *Proc. Natl. Acad. Sci.* **72:** 4399.

Bukhari, A. I. and D. Zipser. 1972. Random insertion of Mu-1 DNA within a single gene. *Nat. New Biol.* **236:** 240.

Bukhari, A. I., S. Froshauer, and M. Botchan. 1976. Ends of bacteriophage Mu DNA. *Nature* **264:** 580.

Chaconas, G., R. M. Harshey, and A. I. Bukhari. 1980. Association of Mu-containing plasmids with the *Escherichia coli* chromosome upon prophage induction. *Proc. Natl. Acad. Sci.* **77:** 1778.

Chow, L. T., R. Kahmann, and D. Kamp. 1977. Electron microscopic characterization of DNAs of non-defective deletion mutants of bacteriophage Mu. *J. Mol. Biol.* **113:** 591.

Daniell, E., D. E. Kohne, and J. Abelson. 1975. Characterization of the inhomogeneous DNA in virions of bacteriophage Mu by DNA reannealing kinetics. *J. Virol.* **15:** 739.

Daniell, E., R. Roberts, and J. Abelson. 1972. Mutations in the lactose operon caused by bacteriophage Mu. *J. Mol. Biol.* **69:** 1.

Daniell, E., J. Abelson, J. S. Kim, and N. Davidson. 1973. Heteroduplex structures of bacteriophage Mu DNA. *Virology* **51:** 237.

Faelen, M. and A. Toussaint. 1973. Isolation of conditional defective mutants of temperate phage Mu-1 and deletion mapping of the Mu prophage. *Virology* **54:** 117.

———. 1978. Stimulation of deletions in the *Escherichia coli* chromosome by partially induced Mu*cts*62 prophages. *J. Bacteriol.* **136:** 477.

Faelen, M., O. Huisman, and A. Toussaint. 1978. Involvement of phage Mu-1 early functions in Mu-mediated chromosomal rearrangements. *Nature* **271:** 580.

Faelen, M., A. Toussaint, and J. de Lafonteyne. 1975. Model for the enhancement of λ-*gal* integration into partially induced Mu-1 lysogens. *J. Bacteriol.* **121:** 873.

Gill, R., F. Heffron, G. Dougan, and S. Falkow. 1978.

Analysis of sequences transposed by complementation of two classes of transposition-deficient mutants of Tn*3* *J. Bacteriol.* **136:** 742.

GIPHART-GASSLER, M., T. GOOSEN, A. VAN MEETEREN, C. WIJFFELMAN, and P. VAN DE PUTTE. 1979. Properties of the recombinant plasmid pGP1 containing part of the early region of bacteriophage Mu. *Cold Spring Harbor Symp. Quant. Biol.* **43:** 1179.

GRINDLEY, N. D. F. 1979. Integration of transposable DNA elements: Analysis by DNA sequencing. In *Proceedings of the 1979 ICN-UCLA Symposium: Extrachromosomal DNA* (ed. J. Cummings), p.155. Academic Press, New York.

GRINDLEY, N. D. F. and D. J. SHERRATT. 1979. Sequence analysis at IS*1* insertion sites: Models for transposition. *Cold Spring Harbor Symp. Quant. Biol.* **43:** 1257.

HEFFRON, F., P. BEDINGER, J. J. CHAMPOUX, and S. FALKOW. 1977. Deletions affecting the transposition of an antibiotic resistance gene. *Proc. Natl. Acad. Sci.* **74:** 702.

HEFFRON, F., B. J. MCCARTHY, H. OHTSUBO, and E. OHTSUBO. 1979. DNA sequence analysis of the transposon Tn*3*: Three genes and three sites involved in transposition of Tn*3*. *Cell* **18:** 1153.

HOWE, M. M. 1973. Prophage deletion mapping of bacteriophage Mu-1. *Virology* **54:** 93.

———. 1978. Invertible DNA in phage Mu. *Nature* **271:** 608.

———. 1980. The invertible G segment of phage Mu. *Cell* **21:** 605.

HOWE, M. M., J. W. SCHUMM, and A. L. TAYLOR. 1979. The *S* and *U* genes of bacteriophage Mu are located in the invertible G segment of Mu DNA. *Virology* **92:** 108.

HSU, M.-T. and N. DAVIDSON. 1972. Structure of inserted bacteriophage Mu-1 DNA and physical mapping of bacterial genes by Mu-1 DNA insertion. *Proc. Natl. Acad. Sci.* **69:** 2823.

———. 1974. Electron microscope heteroduplex study of the heterogeneity of Mu phage and prophage DNA. *Virology* **58:** 229.

KAHMANN, R. and D. KAMP. 1979. Nucleotide sequences of the attachment sites of bacteriophage Mu DNA. *Nature* **280:** 247.

KAMP, D., L. T. CHOW, T. R. BROKER, D. KWOH, D. ZIPSER, and R. KAHMANN. 1979. Site-specific recombination in phage Mu. *Cold Spring Harbor Symp. Quant. Biol.* **43:** 1159.

KAMP, D., R. KAHMANN, D. ZIPSER, T. R. BROKER, and L. T. CHOW. 1978. Inversion of the G DNA segment of phage Mu controls phage infectivity. *Nature* **271:** 577.

KLECKNER, N. 1977. Translocatable elements in procaryotes. *Cell* **11:** 11.

———. 1979. DNA sequence analysis of Tn*10* insertions: Origin and role of 9 base pair flanking repetitions during Tn*10* translocation. *Cell* **16:** 711.

LEACH, D. and N. SYMONDS. 1979. The isolation and characterization of a plaque-forming derivative of bacteriophage Mu carrying a fragment of Tn*3* conferring ampicillin resistance. *Mol. Gen. Genet.* **172:** 179.

LJUNGQUIST, E. and A. I. BUKHARI. 1977. State of prophage Mu upon induction. *Proc. Natl. Acad. Sci.* **74:** 3143.

———. 1979. Behavior of bacteriophage Mu DNA upon infection of *Escherichia coli* cells. *J. Mol. Biol.* **133:** 339.

MARTUSCELLI, J., A. L. TAYLOR, D. CUMMINGS, V. CHAPMAN, S. DELONG, and L. CANEDO. 1971. Electron microscopic evidence for linear insertion of bacteriophage Mu-1 in lysogenic bacteria. *J. Virol.* **8:** 551.

MILLER, H. I., A. KIKUCHI, H. A. NASH, R. A. WEISBERG, and D. I. FRIEDMAN. 1979. Site-specific recombination of bacteriophage λ: The role of host gene products. *Cold Spring Harbor Symp. Quant. Biol.* **43:** 1121.

MOORE, D. D., J. W. SCHUMM, M. M. HOWE, and F. R. BLATTNER. 1977. Insertion of Mu DNA fragments into phage λ *in vitro.* In *DNA insertion elements, plasmids, and episomes* (ed. A.I. Bukhari et al.), p. 567. Cold Spring Harbor Laboratory, Cold Spring Harbor, New York.

O'DAY, K. J., D. SCHULTZ, and M. M. HOWE. 1978. Search for integration-deficient mutants of bacteriophage Mu. In *Microbiology—1978* (ed. D. Schlessinger), p. 48. American Society for Microbiology, Washington, D.C.

O'DAY, K., D. SCHULTZ, W. ERICSEN, L. RAWLUK, and M. HOWE. 1979. Correction and refinement of the genetic map of bacteriophage Mu. *Virology* **93:** 320.

SHAPIRO, J. 1979. Molecular model for the transposition and replication of bacteriophage Mu and other transposable elements. *Proc. Natl. Acad. Sci.* **76:** 1933.

STARLINGER, P. and H. SAEDLER. 1976. IS elements in microorganisms. *Curr. Top. Microbiol. Immunol.* **75:** 111.

TAYLOR, A. L. 1963. Bacteriophage induced mutation in *Escherichia coli. Proc. Natl. Acad. Sci.* **50:** 1043.

TOUSSAINT, A. 1976. The DNA modification function of temperate phage Mu-1. *Virology* **70:** 17.

TOUSSAINT, A. and M. FAELEN. 1973. Connecting two unrelated DNA sequences with a Mu dimer. *Nat. New Biol.* **242:** 1.

VAN DE PUTTE, P., S. CRAMER, and M. GIPHART-GASSLER. 1980. Invertible DNA determines host specificity of bacteriophage Mu. *Nature* **286:** 218.

WEIL, J., R. CUNNINGHAM, R. MARTIN, III, E. MITCHELL, and B. BOLLING. 1972. Characteristics of λp4, a λ derivative containing 9% excess DNA. *Virology* **50:** 373.

WIJFFELMAN, C., M. GASSLER, W. F. STEVENS, P. VAN DE PUTTE. 1974. On the control of transcription of bacteriophage Mu. *Mol. Gen. Genet.* **131:** 85.

Regulation of Integration and Replication Functions of Bacteriophage Mu

P. VAN DE PUTTE, M. GIPHART-GASSLER, N. GOOSEN, T. GOOSEN, AND E. VAN LEERDAM
Department of Molecular Genetics, State University of Leiden, 2333 Al Leiden, The Netherlands

Upon infection of *Escherichia coli* cells, bacteriophage Mu, like bacteriophage λ, can either become stably integrated into a host chromosome or develop lytically, leading to a phage burst and death of the cell. The choice between these two alternatives is, for λ, dependent on the concentrations of repressor and *int*-gene proteins, which are influenced by a large number of phage and host factors.

The regulation of Mu development also seems to be complex and is in many aspects similar to that of λ (van de Putte et al. 1978). The relationship between integration and replication, however, is substantially different with λ and with Mu. With λ the choice between lysogeny or lytic development is, in fact, a choice between integration and replication, and these two processes need distinct functions. With Mu, however, integration seems to be an essential feature of its replication process, and both processes are dependent on the same proteins, the products of the *A* and *B* genes. This suggests that the regulation of the lytic development and lysogenization of Mu is different from that of λ and that different factors may play a role in establishing Mu lysogeny. In this paper the current state of our knowledge on the regulation of Mu integration and replication functions is summarized and discussed.

RESULTS

Early Mu Operon

The integration and replication genes *A* and *B* of Mu are part of the early region. All early genes are located in one large operon of approximately 8 kb, since an IS*1* insertion to the left of gene *A* blocks the expression of all distal genes in the same operon (see Fig. 1). Transcription of the early region occurs on the *r* strand of Mu (Wijffelman et al. 1974) and is initiated at a promoter that is presumably situated to the left of the *Hin*dIII site (Giphart-Gassler 1980).

Besides genes *A* and *B,* two regulatory functions, *ner* and *cim,* have been found. The *ner*-gene protein (Ner) negatively controls the transcription of the early operon (Wijffelman et al. 1974). The *cim*-gene protein (Cim) stimulates Mu immunity presumably by a positive control of repressor synthesis (van de Putte et al. 1978; Giphart-Gassler et al. 1980). Three more genes, *kil, gam,* and *sot,* whose functions in the Mu life cycle are not yet known, have been recognized. The *kil*-gene product (Kil) is responsible for the killing of the host cell in the absence of Mu replication, *gam* codes for an anti-ATP nuclease, and *sot* is a gene that must be expressed in the host to obtain optimal transfection of Mu DNA (van de Putte et al. 1977a,b). In addition to *cim, kil, gam,* and *sot,* the early region beyond gene *B* encodes at least seven more proteins (Giphart-Gassler et al. 1980).

The part of the early operon to the right of gene *B* is only semiessential for Mu development. This is concluded from the properties of a mutant, Mu*cts*62-13/4, that contains an IS*1* insertion in the *cim* gene. The replication of this mutant is strongly impaired (see Positive Control of Mu Replication and Integration), and, consequently, only pinpoint plaques are observed in a lawn of *E. coli* cells. Normal-size plaques of Mu*cts*62-13/4 are formed on an *E. coli suA* host in which the polarity of IS*1* is relieved. The region of Mu DNA beyond the *cim* gene should therefore contain a function that amplifies Mu replication. Additional proof for the existence of such a function has been obtained from experiments with mini-Mu (M. Pato, pers. comm.) (see Results).

The early Mu operon has been physically mapped, and the early Mu proteins have been characterized with the help of mutant phages and a large number of recombinant plasmids, as well as deletion and insertion mutants of these plasmids (Giphart-Gassler et al. 1980). The plasmid that has been used as a basis for these experiments is pGP1, which contains the *Eco*RI C fragment of Mu cloned on pMB9 (Fig. 2). Minicells containing pGP1 synthesize at least five Mu-specific polypeptides (Fig. 3). The *ner*-gene product could not be detected using this system. pGP1 contains at least three regulatory genes: the repressor *c, cim,* and *ner.* pGP1 and its derivatives were also used to study the regulation of Mu.

Mu Repressor and Its Regulation

The Mu repressor blocks the expression of the early Mu operon by binding to the promoter-operator region. Preliminary results indicate that the *Hin*dIII site (1050 bp from the left end of the Mu genome) is located in the repressor-binding site (van Meeteren 1980; Giphart-Gassler 1980). A second repressor-binding site is located in the leftmost *Hin*dIII Mu fragment (Zipser et al. 1977; Schumann et al. 1979). In contrast to the remainder of the Mu genome, the repressor gene is transcribed from the *l* strand (Giphart-Gassler et al. 1979; van Meeteren et al. 1980).

Synthesis of Mu repressor seems to be both positively and negatively regulated, as well as regulated by the repressor itself. This is deduced by measuring Mu immunity against superinfecting Mu in strains containing dif-

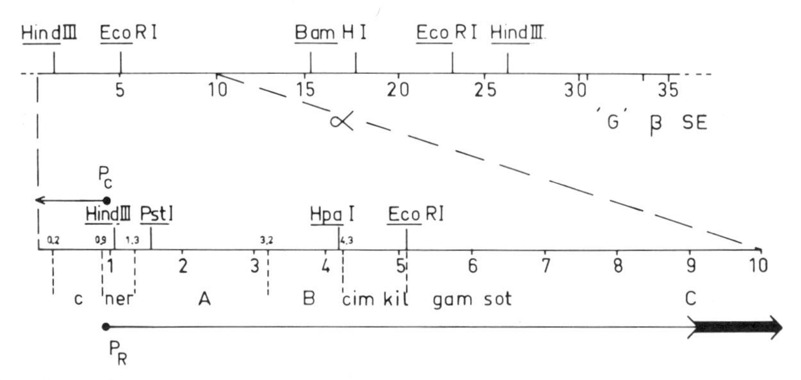

Figure 1. Physical and genetic maps of phage Mu. (*Top line*) The entire Mu DNA comprising the α, G, and β segments; (*bottom line*) the early Mu operon transcribed from p_R. Vertical dashed lines indicate the physical positions of the various early genes, described by Giphart-Gassler et al. (1980). Arrows indicate the direction of transcription (Wijffelman et al. 1974; van Meeteren et al. 1980). The approximate positions of the promoters (p_R and p_c) have been determined (Giphart-Gassler 1980; van Meeteren 1980).

ferent insertion mutants of pGP1 (Fig. 2; Table 1) or by measuring the amount of synthesis of repressor in minicells containing these plasmids (Giphart-Gassler et al. 1980). Strains containing pGP1 are Mu-immune at 32 °C and nonimmune at 42 °C because of the presence of the *cts* gene in the plasmid (Table 1). DNA insertions in *cim* and in *B* (pGP33 and pGP31) lead to a loss of immunity at 32 °C. In minicells, however, repressor synthesis can still be observed. Comparable amounts of repressor are synthesized when the DNA insertion is located in gene *A*. Nevertheless, an aberrant pattern of Mu immunity is obtained with such plasmids (pGP34; Table 1). This A^- immunity is discussed in the next section.

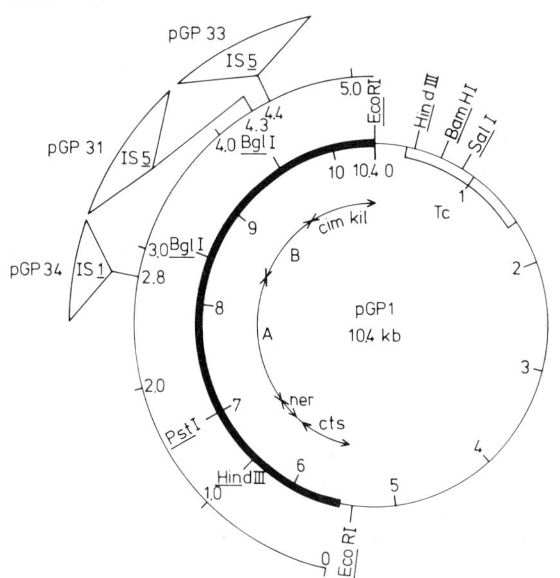

Figure 2. Circular restriction map of pGP1 and three derivatives of pGP1 that contain IS elements. The isolation and characterization of pGP1 have been described previously (Giphart-Gassler and van de Putte 1979). Insertion mutants of pGP1 were isolated from temperature-resistant survivors of bacteria containing pGP1 (Giphart-Gassler et al. 1979). The physical positions of the insertion sequences are as determined by Giphart-Gassler (1980). (▬) Mu DNA; (—) pMB9 DNA (Rodriguez et al. 1976).

pGP1-derived plasmids from which the Mu part to the right of the *Hin*dIII site is deleted (pGP7; Fig. 4a) show an enhanced repressor synthesis in minicells. Strains containing pGP7 are immune both at 32 °C and at 42 °C. This suggests that a function that maps to the right of the *Hin*dIII site and to the left of the IS*1* insertion in pGP34 (Fig. 2) negatively regulates the repressor synthesis. Moreover, the regulation of Mu repressor seems autogenous, as is observed in an experiment with minicells harboring pGP7. At 32 °C, these minicells synthesize a polypeptide with a molecular weight of 26,000, which is the Mu repressor (Fig. 4b); at 42 °C, the repressor is inactivated, but its synthesis is stimulated simultaneously. Also, an increase in the amount of pre-β-lactamase (m.w. 32,000) is observed at 42 °C. This is presumably due to an enhanced transcription of the ampicillin-resistance (*bla*) gene from the Mu repressor promoter at an elevated temperature.

Negative Control of the Early Mu Operon

The expression of the early Mu operon is negatively regulated not only by the repressor, but also by another Mu-specific protein. Evidence for the existence of this protein has been obtained from transcription studies with a Mu*cts*62 prophage (Wijffelman et al. 1974). These experiments showed that the early Mu RNA synthesis was reduced 4 minutes after induction (Fig. 5), and it was shown that this reduction was due to the synthesis of a Mu protein, which was designated Ner. Ner was also shown to regulate the early RNA synthesis throughout Mu development, resulting in a constant level of early RNA (Fig. 5). This implies that the amount of early RNA synthesis is independent of the number of Mu copies present. Negative regulation of early transcription has also been observed in bacteria lysogenic for Mu*cts*62-*X* or in bacteria containing pGP1 (Wijffelman and van de Putte 1974; Giphart-Gassler and van de Putte 1979). These results indicate that Ner is encoded by a gene proximal to the IS*1* insertion of Mu*cts*62-*X*. As *A* and *B* are adjacent genes (Giphart-Gassler et al. 1980), it can be concluded that *ner* is lo-

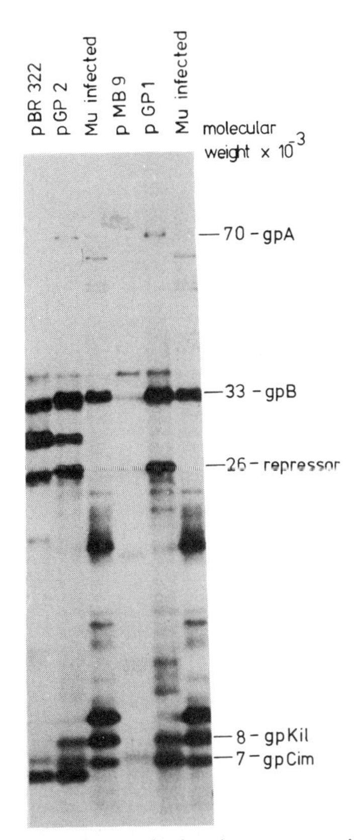

Figure 3. Autoradiograph showing a comparison of polypeptides synthesized in *E. coli* minicells containing pBR322 (Bolivar et al. 1977) and pMB9, the recombinant plasmids pGP2 and pGP1, and Mu-infected minicells. pGP2 is identical to pGP1 except that pBR322 has been used as a vector. Plasmid-containing minicells were incubated prior to and during labeling with L-[^{35}S]methionine at 42°C to induce expression of Mu genes. The characterization of the various Mu gene products (gp) and the details of the methods used are presented elsewhere (Giphart-Gassler et al. 1980).

cated in the leftmost 1300 bp, proximal to gene *A*, unless *ner* overlaps gene *A* or *B* or unless *ner* is not an exclusive gene.

New, although indirect, evidence for the existence and position of *ner* as a separate gene was obtained by studying Mu immunity of bacteria containing various recombinant plasmids that constitutively express some early Mu functions. With some of these plasmids, we observed immunity against superinfecting phage Mu, which is independent of the Mu repressor since such plasmids did not contain the Mu repressor gene. This pseudoimmunity is caused by the product of a gene that starts proximal to gene *A* and that may be identical to *ner*. The recombinant plasmids mentioned above were obtained from pLP103, a pACYC177 derivative carrying the *Hin*dIII fragment of pGP1 containing the Mu genes *A, B,* and *kil* in the *Hin*dIII site of the kanamycin-resistance (Km^r) gene (Fig. 6). The repressor gene and presumably the early Mu promoter (p_R) are not present in pLP103. Consequently, a constitutive expression of the *A, B,* and *kil* genes from p_{km} is obtained. Due to the expression of *kil*, pLP103 can be maintained only in a Hek$^-$ strain in which *kil* is not functional (Giphart-Gassler et al. 1979). Deletion derivatives of pLP103 were isolated by transformation of plasmid DNA linearized with *Eco*RI (Thompson and Achtman 1979). By selection for the Kil$^-$ phenotype, three classes of deletion mutants were obtained (E. van Leerdam and P. van de Putte, in prep.).

As shown in Table 2, strains carrying class-I or class-II plasmids are immune to superinfecting phage, in spite of the absence of Mu repressors. We conclude, therefore, that these plasmids constitutively express an early Mu function that blocks Mu development. This Mu function is expressed by a gene located to the right of the *Hin*dIII site and is different from the *A* gene, since class-I plasmids, which do not contain a complete *A* gene, also show this immunity. We propose that the function involved is *ner*. Constitutive expression of *ner* might almost completely block the early transcription of the superinfecting phage, resulting in a level of *A*-gene or *B*-gene protein too low to permit Mu development. Only when both the *A* and *B* genes are located in the plasmid (class III; Table 2) is the pseudoimmunity not observed.

In some of the deletion mutants of pLP103, the *Hin*dIII site close to the *Eco*RI site was also deleted, leaving a plasmid with a unique *Hin*dIII site in the *kan* gene. Deletions spanning this unique *Hin*dIII site were obtained after transformation of plasmid DNA linearized with *Hin*dIII. Several deletion mutants were found that

Table 1. Control of Mu Immunity

Plasmid	Genotype of cloned Mu fragment	Immunity 32°C	Immunity 42°C	Amount of repressor synthesized in minicells[a]
pGP1	*-cts ner A B cim kil-*	+[b]	–	+
pGP33	*-cts ner A B cim kil-* Δ IS*5*	–	–	±
pGP31	*-cts ner A B cim kil-* Δ IS*5*	–	–	±
pGP34	*-cts ner A B cim kil-* Δ IS*1*	+	+	±
pGP7	*-cts*	+	+	++

[a] Data from Giphart-Gassler et al. (1980).
[b] Determined by titration of Mu*cts*.

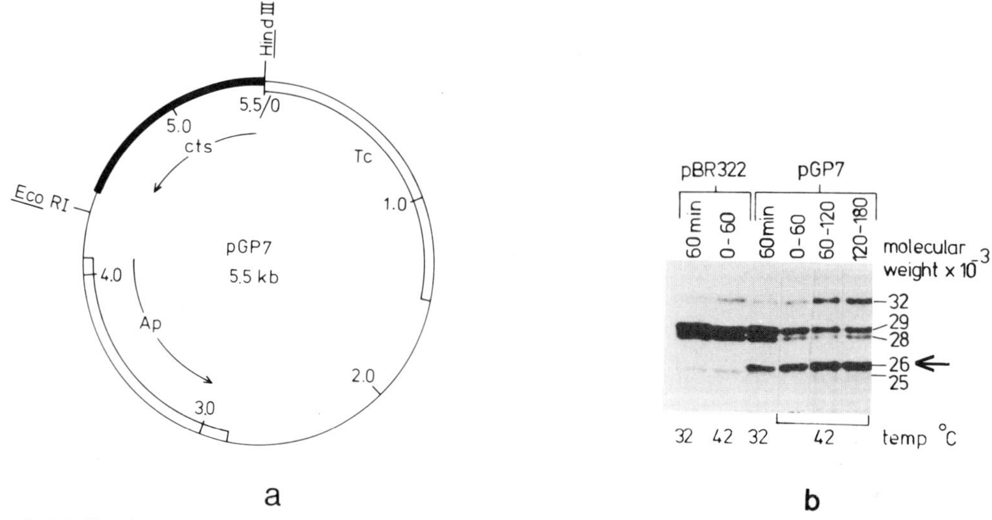

Figure 4. (*a*) Circular restriction map of pGP7. pGP7 has been constructed in vitro by cloning the 1150-bp *Eco*RI-*Hin*dIII fragment of pGP1 into the vector pBR322 (van Meeteren et al. 1980). Arrows indicate the direction of transcription of the *cts* and *amp* genes. (▬) Mu DNA; (—) pBR322 DNA. (*b*) Influence of a temperature shift to 42 °C on the synthesis of Mu repressor and the various molecular forms of β-lactamase in minicells containing pBR322 and pGP7. Minicells containing pBR322 or pGP7 were incubated at 32 °C for 60 min. They were labeled either for 60 min at 32 °C or at the various times indicated after a temperature shift to 42 °C. The time of the shift is indicated as zero. The polypeptide indicated by the arrow is the Mu repressor, whereas polypeptides indicated by 29 and 28 (M_r = 29,000 and 28,000) correspond to purified β-lactamase (Achtman et al. 1979). The polypeptide indicated by 32 (M_r = 32,000) is the presumed β-lactamase precursor (Dougan et al. 1979). The conditions for labeling and electrophoresis have been described previously (Giphart-Gassler et al. 1980).

left the *A* gene intact (as was shown by complementation) but that lost pseudoimmunity (class IV; Table 2). These results support our suggestion that the Mu function that confers pseudoimmunity to cells is identical to *ner* and that *ner* is located proximal to gene *A*.

As shown in Table 1, derivatives of pGP1 containing an IS*1* insertion in gene *A* (pGP34) also were Mu-immune both at 32 °C and at 42 °C. This A^- immunity also seems unrelated to repressor synthesis and is presumably identical with the pseudoimmunity caused by *ner*. Several observations support this idea: (1) Minicells containing pGP34 synthesize no more repressor than those containing plasmids with insertions in gene *B* or *cim* (Table 1). (2) A^- immunity is then observed only when the Mu fragment is located in a multicopy plasmid, since A^--defective Mu*cts* prophages do not show this immunity (N. Goosen and P. van de Putte, in prep.). The A^- immunity of multicopy plasmids can be explained by the negative regulation of *ner* as follows: Due to Ner, early Mu RNA synthesis is independent of the number of Mu copies present (Fig. 5). Consequently, in Mu-infected strains harboring pGP1 or derivatives, the early Mu RNA synthesis of the incoming phage is only a small fraction of the total early transcription. If bacteria contain plasmids with an IS element in the *A* gene, the level of *A*-gene protein (A) may become too low to permit Mu development, leading to pseudoimmunity. If, on the other hand, the insertion is located in gene *B* (Table 1), the strains are not Mu-im-

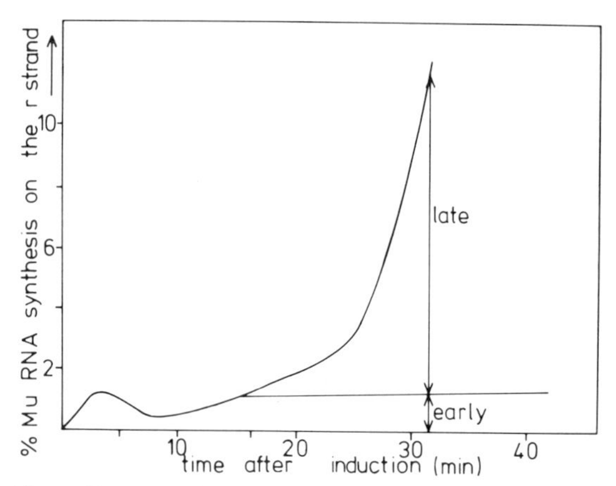

Figure 5. Transcription of the entire Mu genome and that of early genes during phage development. Bacteria lysogenic for Mu*cts*62 were labeled with [^{3}H]uridine for 1 min at various times after induction, followed by extraction of RNA. Total Mu RNA synthesis was determined by hybridization of the extracted RNA to the Mu *r* strand. Early RNA represents the fraction of the extracted RNA hybridizing to the *r* strand of λpMu phage, which contains the early region of Mu. The methods used were as described by Wijffelman and van de Putte (1974) and van Meeteren (1980).

Table 2. Mu Immunity of Deletion Mutants of pLP103

Plasmid class	Relevant genotype	Immunity 32 °C	Immunity 42 °C
I	p_{km}*ner*– –	+	+
II	p_{km}*ner A* –	+	+
III	p_{km}*ner A B*	–	–
IV	p_{km} – *A* –	–	–

Details will be presented elsewhere (E. van Leerdam and P. van de Putte, in prep.).

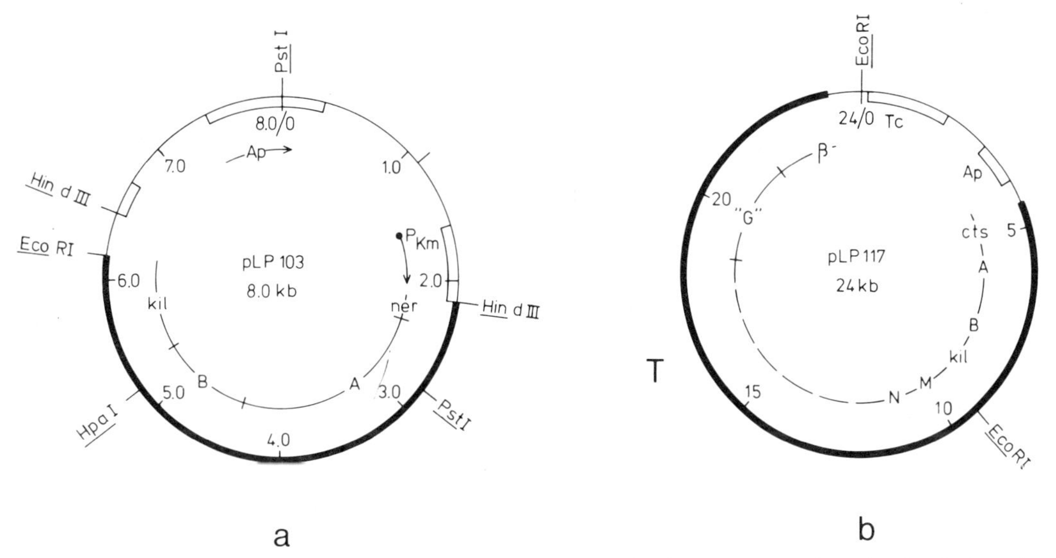

Figure 6. Circular restriction maps of pLP103 and pLP117. (*a*) pLP103 has been constructed in vitro by the insertion of the *Hin*dIII fragment of pGP1, which contains the Mu genes *ner, A, B,* and *kil,* into the *Hin*dIII site of pACYC177 (Chang and Cohen 1978). Arrows indicate the direction of transcription. Details of the construction, physical mapping, and genetic properties of pLP103 will be described elsewhere (E. van Leerdam and P. van de Putte, in prep.). (▬) Mu DNA; (—) pACYC DNA. (*b*) pLP117 contains the Mu fragment of pGP1 and the rightmost *Eco*RI fragment of Mu DNA cloned in pBR322 (E. van Leerdam and P. van de Putte, in prep). (▬) Mu DNA; (—) pBR322 DNA.

mune, although due to *ner* the level of *B*-gene protein (B) should be reduced also. Apparently, a reduction in the amount of B does not inhibit the development of the incoming phage Mu. However, when *ner* is constitutively expressed by transcription from the *kan* promoter, the amount of B produced by the superinfecting phage will be reduced further, which also results in immunity (class II; Table 2).

Positive Control of Mu Replication and Integration

A plasmid containing the *Eco*RI B and the C fragments of Mu was constructed in vitro and cloned in the vector pBR322 in the same orientation as in phage Mu (Fig. 6). Strains containing this plasmid, pLP117, do show Mu-specific replication after thermoinduction without the need for helper functions (Fig. 7). The amount of replication, however, is low as compared with that of a complete phage Mu, presumably because pLP117 lacks the part of the early region beyond gene *B*. The Mu-specific replication of pLP117 can be stimulated by induction of a defective prophage that contains the complete early region. The level of replication still remains lower than that of a complete phage Mu (Fig. 7). Similar results were obtained with Mu*cts*62-13/4, a Mu mutant that contains an IS*1* insertion in gene *cim* (van de Putte et al. 1978). Replication of this phage is markedly reduced (Fig. 8) as a result of the polar effect of the insertion on the expression of genes distal to gene *B*. Also, Mu*cts*62-13/4 replication can be complemented by defective prophages expressing all early functions (results not shown). The nature of the function(s) that amplifies replication and should be located beyond gene *B* is not yet known.

It also seems, however, that without expression of the early region beyond gene *B* a normal amount of Mu replication can be obtained. Normal-size plaques are formed both by Mu*cts*62 and by Mu*cts*62-13/4 on bacteria containing pLP103 (Fig. 6), on which the *ner, A,* and *B* genes are constitutively expressed. Although we presume that the transcription of the superinfecting phage Mu will be shut off by Ner (see previous section), constitutive expression of the *A* and *B* genes on pLP103 allows the superinfecting phage to replicate. These experiments suggest that if enough *A*- and *B*-gene products are provided, no additional replication functions are required.

DISCUSSION

The regulation of replication of Mu shows in many aspects similarities with that of λ. The replication functions are transcribed from a promoter situated on the *r* strand (p_R), whereas the repressor is transcribed from a promoter (p_c) on the *l* strand. In Mu, these two promoters are also located close to each other in a DNA region 900–1050 bp from the immunity end of the Mu genome. A similar regulatory organization has been observed for Tn*3* (Chou et al. 1979; Gill et al. 1979).

Mu immunity (e.g., presumably the repressor) is positively controlled by *cim*, a function comparable with *c*II or *c*III of λ. The positive effect, however, seems small, as mutations in *cim* only affect Mu immunity markedly if the repressor (*c*) contains a temperature-sensitive mutation. Moreover, loss of immunity at 32°C (Cim$^-$) can be observed only in defective prophages or plasmid clones lacking the *β* end of Mu. The *β* end provides for a second positive effect on immunity (P. van de Putte, unpubl.), which allows a phage Mu*cts cim* to establish immunity.

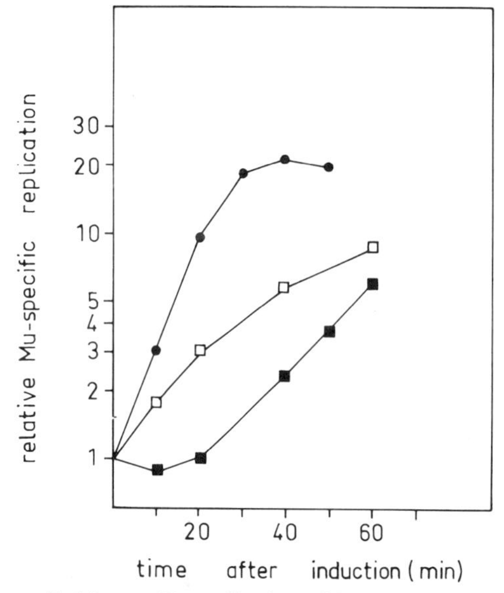

Figure 7. Mu-specific replication of Mu*cts*62 and pLP117 after thermoinduction. Plasmid- or prophage-containing strains were grown at 32 °C and induced at 42 °C. DNA was pulse-labeled with [^{3}H]thymidine and extracted (T. Goosen et al., in prep.), and the purified DNA was analyzed by filter hybridization to Mu DNA. Mu-specific replication is expressed relative to the amount of hybridization found with DNA of noninduced cultures. (■) KA52 (pLP117); (□) KA52, *trp*::Mu*cts*ΔD-*β* (pLP117); (●) KA52, *trp*::Mu*cts*62. KA52 is *thi galK*.

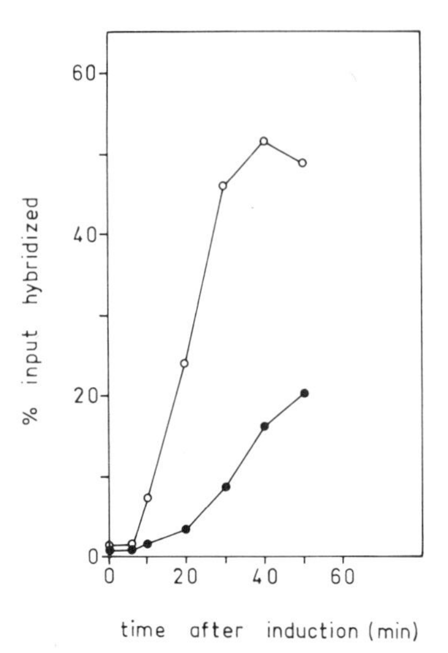

Figure 8. Replication of Mu*cts*62 and Mu*cts*62-13/4 after thermoinduction. Pulse labeling and hybridization were carried out as described in the legend to Fig. 7. Mu-specific replication is expressed as the fraction of [^{3}H]DNA input that hybridizes to Mu DNA. (○) KA52 (Mu*cts*62); (●) KA52 (Mu*cts*62-13/4).

We have presented evidence that the repressor regulates its own synthesis and also that repressor synthesis is negatively controlled by a function located to the left of gene *A*. The gene that negatively controls the transcription from p_R (*ner*) was also mapped proximal to gene *A*, and therefore we presume that these functions are supplied by one gene, as has been found for *cro* and *tof* of λ (Pero 1971).

We have been able to characterize and map the *ner* gene more accurately using the phenomenon of pseudoimmunity, which is observed when *ner* is constitutively expressed from a foreign promoter in a multicopy plasmid. This pseudoimmunity is presumably caused by an effective block of transcription of the superinfecting phage. Transcription studies are in progress to prove that pseudoimmunity is indeed caused by an overproduction of Ner. A *ner*-like function has not yet been reported to be essential for the regulation of transposable elements other than Mu. This type of negative regulation might only be important for controlling phage development or for the choice between lysogeny or lytic growth.

As a consequence of Ner activity, the amount of early RNA is kept constant during replication, and through Ner the amount of the early Mu gene products will be controlled. In this respect, when we consider the products of the important essential genes *A* and *B* that are needed for replication and integration, a reduction in the amount of A seems to have many more consequences for the phage than a similar reduction in the amount of B. Genetic data have shown that one copy of a defective Mu prophage does not provide enough A to complement superinfecting Mu*A*am, whereas the same defective prophage provides enough B to complement for at least 50–100 *B*am phage (Westmaas et al. 1976). Moreover, the pseudoimmunity that is observed with A$^-$ multicopy plasmids derived from pGP1 illustrates the effect of a reduction of A and not of B (see Results). An explanation for the observation that a reduction in the amount of A has more consequences than a similar reduction in the amount of B might be that, at least in minicells, the total amount of A synthesized is lower than that of B (see Fig. 3). As the *A* and *B* genes are located in the same operon, the low amount of A synthesized presumably reflects inefficient translation.

A low level of Mu-specific replication is observed when the early region between genes *B* and *C* is not expressed, as in the phage Mu*cts*-13/4, or when this region is absent, as in pLP117. It is tempting to suggest that this low level of replication is also due to a low amount of A. As a support for this notion, we have found that the expression of this second part of the early region is no longer needed when enough A is provided by a multicopy plasmid. The region between genes *A* and *B* might therefore be involved in the transcriptional or translational control of the *A* gene or might stabilize the *A*-gene protein. Whether the mode of Mu replication in the presence of a high level of A is identical to that in the presence of normal amounts of A plus the replication function between *B* and *C* remains to be seen.

Unlike the situation for λ, the replication and integration functions of Mu are not separated. According to a recent transposition model (Shapiro 1979), lysogenization of Mu involves at least an integration and a replication step. To obtain a stable Mu lysogen, further Mu replication must be blocked. This predicts that Mu replication has to be controlled in such a way that it is restricted to one round only, which may be achieved when the *A*-gene product is labile. Alternatively, upon establishing lysogeny, Mu is completely dependent on the replication of the host for the replication step. The observation that Mu seems to integrate near or in the replication forks could be related to this hypothesis (Paolozzi et al. 1978; Fitts and Taylor 1980). Nevertheless, for its integration step, Mu remains dependent on A. The level of A might therefore be of importance for the choice between lysogeny or lytic development. Experiments are in progress to determine the effect of a relatively high concentration of A on establishment of lysogeny.

REFERENCES

Achtman, M., P. A. Manning, C. Edelbluth, and P. Herrlich. 1979. Export without proteolytic processing of inner and outer membrane proteins encoded by F sex factor *tra* cistrons in *Escherichia coli* minicells. *Proc. Natl. Acad. Sci.* **76:** 4837.

Bolivar, F., R. L. Rodriguez, P. J. Greene, M. C. Betlach, H. L. Heijneker, H. W. Boyer, J. H. Crosa, and S. Falkow. 1977. Construction and characterization of new cloning vehicles. II. A multipurpose cloning system. *Gene* **2:** 95.

Chang, A. C. Y. and S. N. Cohen. 1978. Construction and characterization of amplifiable multicopy DNA cloning vehicles derived from the P15A cryptic miniplasmid. *J. Bacteriol.* **134:** 1141.

Chou, J., P. G. Lemaux, M. J. Casadaban, and S. N. Cohen. 1979. Transposition protein of Tn*3*: Identification and characterization of an essential repressor-controlled gene product. *Nature* **282:** 801.

Dougan, G., M. Saul, A. Twigg, R. Gill, and D. Sherratt. 1979. Polypeptides expressed *Escherichia coli* K-12 minicells by transposition elements Tn*1* and Tn*3*. *J. Bacteriol.* **138:** 48.

Fitts, A. and A. L. Taylor. 1980. Integration of bacteriophage Mu at host chromosomal replication forks during lytic development. *Proc. Natl. Acad. Sci.* **77:** 2801.

Gill, R. E., F. Heffron, and S. Falkow. 1979. Identification of the protein encoded by the transposable element Tn*3* which is required for its transposition. *Nature* **282:** 797.

Giphart-Gassler, M. 1980. "Gene products of bacteriophage Mu." Ph.D thesis, State University of Leiden, The Netherlands.

Giphart-Gassler, M. and P. van de Putte. 1979. Thermo-inducible expression of cloned early genes of bacteriophage Mu. *Gene* **7:** 33.

Giphart-Gassler, M., J. Reeve, and P. van de Putte. 1980. Polypeptides encoded by the early region of bacteriophage Mu synthesized in minicells of *Escherichia coli*. *J. Mol. Biol.* (in press).

Giphart-Gassler, M., T. Goosen, A. van Meeteren, C. Wijffelman, and P. van de Putte. 1979. Properties of the recombinant plasmid pGP1 containing part of the early region of bacteriophage Mu. *Cold Spring Harbor Symp. Quant. Biol.* **43:** 1179.

Paolozzi, L., R. Jucker, and E. Calef. 1978. Mechanism of phage Mu-1 integration: Nalidixic acid treatment causes clustering of Mu-1 induced mutations near replication origin. *Proc. Natl. Acad. Sci.* **75:** 4940.

Pero, J. 1971. Deletion mapping of the site of action of the *tof* gene product. In *The bacteriophage lambda* (ed. A. D. Hershey), p. 599. Cold Spring Harbor Laboratory, Cold Spring Harbor, New York.

Rodriguez, R. L., F. Bolivar, H. M. Goodman, H. W. Boyer, and M. Betlach. 1976. Construction and characterization of cloning vehicles. In *Molecular mechanisms in the control of gene expression* (ed. D. P. Nierlich et al.), p. 471. Academic Press, New York.

Schumann, W., C. Westphal, E. G. Bade, and L. Holzer. 1979. Origin and binding specificity of protein(s) coded for by Mu prophages. *Mol. Gen. Genet.* **173:** 189.

Shapiro, J. A. 1979. Molecular model for the transposition and replication of bacteriophage Mu and other transposable elements. *Proc. Natl. Acad. Sci.* **76:** 1933.

Thompson, R. and M. Achtman. 1979. The control region of the F sex factor DNA transfer cistrons: Physical mapping by deletion analysis. *Mol. Gen. Genet.* **169:** 49.

van Meeteren, A. 1980. "Transcription of bacteriophage Mu." Ph.D. thesis, State University of Leiden, The Netherlands.

van Meeteren, A., M. Giphart-Gassler, and P. van de Putte. 1980. Transcription of bacteriophage Mu. II. Transcription of the repressor gene. *Mol. Gen. Genet.* **179:** 185.

van de Putte, P., G. C. Westmaas, and C. Wijffelman. 1977a. Transfection with Mu DNA. *Virology* **81:** 152.

van de Putte, P., G. Westmaas, M. Giphart, and C. Wijffelman. 1977b. On the *kil* gene of bacteriophage Mu. In *DNA insertion elements, plasmids, and episomes* (ed. A. I. Bukhari et al.), p. 287. Cold Spring Harbor Laboratory, Cold Spring Harbor, New York.

van de Putte, P., M. Giphart-Gassler, T. Goosen, A. van Meeteren, and C. Wijffelman. 1978. Is integration essential for Mu development? In *Integration and excision of DNA molecules* (ed. P. Hofschneider and P. Starlinger), p. 33. Springer-Verlag, Berlin.

Westmaas, G. C., W. L. van der Maas, and P. van de Putte. 1976. Defective prophages of bacteriophage Mu. *Mol. Gen. Genet.* **145:** 81.

Wijffelman, C. A. and P. van de Putte. 1974. Transcription of bacteriophage Mu. *Mol. Gen. Genet.* **135:** 327.

Wijffelman, C. A., M. Gassler, W. F. Stevens, and P. van de Putte. 1974. On the control of transcription of bacteriophage Mu. *Mol. Gen. Genet.* **131:** 85.

Zipser, D., P. Moses, R. Kahmann, and D. Kamp. 1977. The molecular cloning of the immunity gene of phage Mu. *Gene* **2:** 263.

Genetic Study of Mu Transposition and Mu-mediated Chromosomal Rearrangements

RHODOMU: L. DESMET,* M. FAELEN,* N. LEFÈBVRE,† A. RÉSIBOIS,‡
A. TOUSSAINT,* AND F. VAN GIJSEGEM*§

*Laboratoire de Génétique, † Laboratoire de Microbiologie et d'Immunologie, ‡ Laboratoire de Microscopie Electronique, Université Libre de Bruxelles, Brussels, Belgium

Temperate phage Mu-1 (Taylor 1963) embodies the properties of a transposable element and of a bacteriophage. Its integrative properties are those of a transposable element; Mu DNA as extracted from viral particles is covalently linked at both ends to random host DNA sequences ±0.1 kb long at the left or *c* end (Allet and Bukhari 1975; Bukhari et al. 1976) and ±2 kb long at the right or *S* end (Daniell et al. 1973), and thus it is actually an integrated prophage. Upon lysogenization, Mu DNA is transposed from the linear viral DNA into random sites on the host genome (Bukhari and Zipser 1972; Daniell et al. 1972). To transpose, it uses its ends in a way similar to other transposable elements, and it generates a 5-bp duplication at the target site (Allet 1979; Kahman and Kamp 1979) as do IS*2* (Ghosal et al. 1979), Tn*3* (Ohtsubo et al. 1977), γδ (Reed et al. 1979), and some eukaryotic transposable sequences (this volume). The ends of Mu share more than fortuitous homology with the ends of Tn*3*, γδ (Reed et al. 1979), and IS*5* (Lusky et al., this volume).

Mu is a generalized transducing phage (Howe 1973a). Although it was first isolated in *Escherichia coli* K12, it grows in most, if not all, enterobacteria and in other gram-negative species (e.g., a strain of *Pseudomonas fluorescens* [M. Mergeay and A. Toussaint, unpubl.]). It can switch between two host-range specificities (van de Putte et al. 1980; D. Kamp, in prep.) that appear to be mutually exclusive. Some strains of *E. coli, Klebsiella, Salmonella,* and *Citrobacter* are sensitive to Mu phages with one or the other host range. Mu can also exchange host ranges with other temperate bacteriophages by recombination (Toussaint et al. 1978).

Infection is not the only way the phage can enter a new host; it can use a broad, transferrable host-range plasmid into which it integrates upon infection of a sensitive host to invade even Mu-resistant bacteria further (Dénarié et al. 1977; McNeil et al. 1978; van Vliet et al. 1978a; A. L. Taylor, unpubl.).

Mu *mom* gene codes for a nonspecific modification function that protects its DNA (and other DNA present in the same cell) against restriction when traveling among different bacterial species (Allet and Bukhari 1975; Toussaint 1976). The modification is not a methylation (Hattman 1979), although it requires a host methylation function to be fully expressed (Toussaint 1977; Khatoon and Bukhari 1978).

§ Authors have been listed in alphabetical order.

Aside from the fact that it can integrate into the host genome at random locations, Mu transposes very efficiently and provokes all kinds of chromosomal rearrangements during its lytic cycle, either upon infection or upon induction. Rearrangements include: deletions that are always linked to a Mu prophage, inversions where the inverted segment is surrounded by two copies of Mu in opposite orientation, replicon fusions in which the two replicons are fused by means of two Mu prophages in the same orientation, and transpositions of host DNA segments where the transposed DNA is surrounded by two Mu prophages in the same orientation (see Fig. 1 and Faelen et al. 1971, 1975; van de Putte and Gruijthuijsen 1972; Toussaint and Faelen 1973; Faelen and Toussaint 1976, 1978, 1980; Toussaint et al. 1977). The formation of these rearrangements could be explained by the model proposed by Shapiro (1979). One consequence of these properties is that the presence of Mu in a plasmid increases its potential to interact with other replicons. The RP4 plasmid, for instance, does not mobilize the chromosome of *E. coli* K12 at a detectable frequency, whereas RP4::Mu*cts*62 does (Table 1).

Only the ends of the Mu genome and the product of Mu gene *A* are essential for Mu transposition and Mu-mediated chromosomal rearrangements (Faelen et al. 1979a); this is also the case for lysogenization (O'Day et al. 1978). The Mu *B*-gene product increases the frequency of all of these events, although its exact function remains to be elucidated (see below). Since Mu gene *A* is located near the *c* end, it was easy to isolate mini-Mu derivatives that carry an internal deletion removing the whole central part of Mu DNA, and with it all the lytic functions, but leaving intact the ends and the *A* gene. These mini-MuA^+ derivatives retain the integrative abilities of the Mu, i.e., the ability to transpose or to induce rearrangements, although they do so at a lower frequency (±100 times) due to the lack of the *B* gene (Faelen et al. 1979a). These mini-Mu replicate and are maturated in the presence of a Mu*cts*62 helper. As schematically represented in Figure 2, lysates obtained upon induction of a strain lysogenic for Mu*cts*62 and mini-Mu consist of phage particles containing various types of DNA, all of which are ±38 kb long; ±90% contain normal Mu DNA (type I in Fig. 2), and the remaining ±10% contain either a mini-Mu linked to a long segment of random host DNA (type II in Fig. 2) or linked as follows: host DNA–mini-Mu–host DNA–mini-Mu–

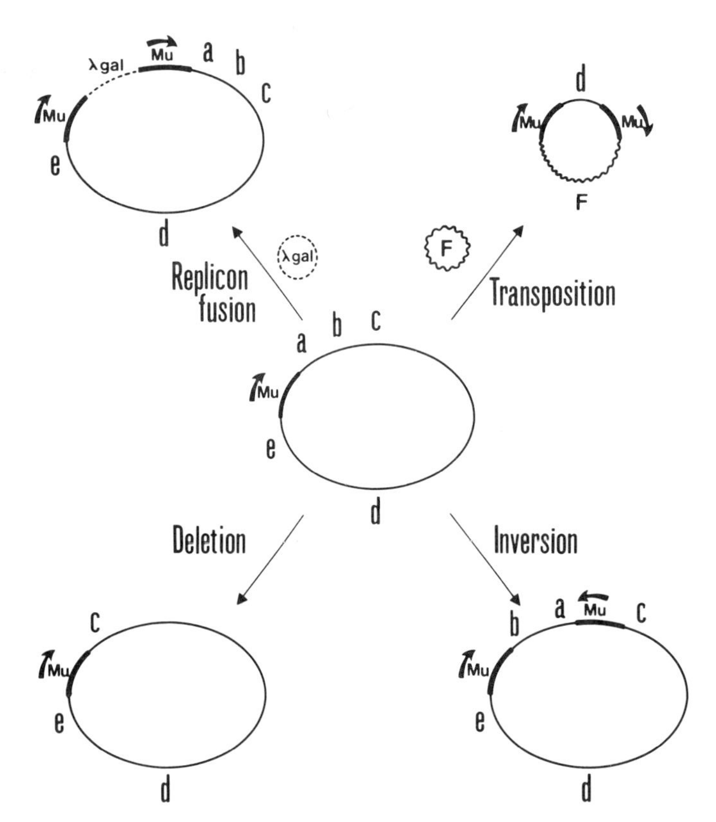

Figure 1. Chromosome rearrangements mediated by Mu include: (1) Replicon fusion. In a cell containing a λ-*gal* phage, the λ-*gal* becomes integrated into the chromosome between two copies of Mu in the same orientation. (2) Adjacent deletion. This occurs at either side of the prophage; here deletion of the host markers (*a* and *b*) is represented. (3) Adjacent inversion. The *a-b* host segment has been inverted and, as a result, is surrounded by two Mu prophages in opposite orientation. (4) Transposition of chromosome segments to a plasmid. Here the host DNA segment *d* has been transposed to an *F* episome and is surrounded by two Mu prophages in the same orientation. All of these rearrangements occur either after infection of a sensitive host or after induction of a Mu lysogen, in which case they can occur either adjacent to the original prophage or anywhere else in the chromosome, most probably after or while the Mu transposes to a new location. (———) Bacterial DNA; (———) Mu DNA; (~~~~) F DNA; (----) λ-*gal* DNA.

host DNA, where the mini-Mu's are in the same or in opposite orientation (types III and IV in Fig. 2). Upon reinfection of a sensitive host by such a mixed lysate, the frequency of generalized transduction is ±10 times higher than with a Mu*cts*62 pure lysate, most probably due to the presence of type-II DNA. In addition, a new type of transduction is observed, called mini-Muduction, where the transductants have acquired the transduced marker surrounded by two mini-Mu's in the same orientation at any site in their chromosomes. The mini-Mu–X^+–mini-Mu structures behave as transposons and transpose at 42 °C approximately 20 times less frequently than the mini-Mu itself. Mini-Muduction is *recA*-independent, requires Mu *A*-gene product, and represents 10% of the total transduction frequency obtained with Mu–mini-Mu mixed lysates (Faelen et al. 1979b).

Table 1. Chromosome Mobilization by Plasmid RP4 and RP4::Mu*cts*62 in *E. coli* K12

	Donor strain	
Selected marker	MXR/RP4	MXR/RP4::Mu*cts*62
Ilv	2×10^{-9}	2×10^{-4}
Thr Leu	2×10^{-9}	4.8×10^{-5}
Ura	2×10^{-9}	1.2×10^{-3}
His	10^{-9}	6×10^{-4}
Thy	4×10^{-9}	2×10^{-4}

Overnight cultures of MXR (*galE recA* [Faelen et al. 1979a]) carrying an RP4 or an RP4::Mu*cts*62 plasmid were grown at 30 °C and mated on plates at 42 °C (as will be described by F. van Gijsegem and A. Toussaint, in prep.) with the polyauxotrophic recipient KMBL146 Sm^r (*supE thr22, leu22 thi22, pyrF101, thyA22, arg118 ilvA108 his107 lac22 ton1 tsx*). The donors were counterselected with streptomycin (Sm). Exconjugants prototrophic for one marker were selected on appropriate minimal medium. The frequency of mobilization of a given marker is given by the ratio, number of prototrophic Sm^r: input R$^+$.

The ability to transduce host DNA, increase interactions between different replicons, connect unrelated DNA sequences, and generate transposons, combined with the ability to switch between two host ranges and to protect its DNA against restriction, make Mu a very powerful agent for exchanging genetic information between different bacterial species.

Aside from the fact that it can rearrange and transport DNA, Mu also interferes with host-gene expression. Below, we shall describe a mini-Mu that can turn on host-gene expression.

The molecular mechanisms of Mu transposition and Mu-mediated chromosomal rearrangements, including the exact function of the *A*- and *B*-gene products, remain to be determined. Since all prokaryotic transposable elements seem to share the same mechanism of transposition, it seems very likely that some host functions are involved in the transposition process. Therefore, we decided to test whether different mutations of *E. coli* K12 that alter either recombination or the spontaneous mutation frequency also affect in any way Mu and mini-Mu transposition.

To try to understand why the *B*-gene product increases the frequency of Mu transposition and Mu-induced chromosomal rearrangements, we also attempted to define the *B*-gene function further.

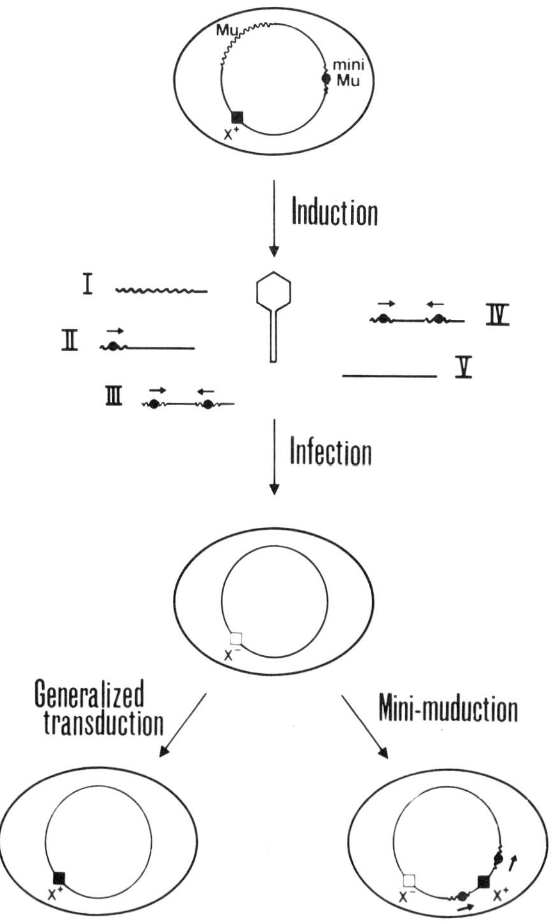

Figure 2. Induction of an X^+ strain lysogenic for Mu and mini-Mu provides lysates in which 90% of the viral particles contain normal Mu DNA (type I). The remaining 10% contain either a mini-Mu DNA covalently linked to ~0.1 kb of host DNA at one end and 38−L kb, where L is the length of the mini-Mu, of host DNA at the other end (type II) or two mini-Mu's in opposite orientation, covalently linked to host DNA at their ends (type III). DNAs of types I, II, and III are easily detected by electron microscopy (A. Résibois et al., in prep.). In addition, the lysates must contain particles with only host DNA (type V) and particles with DNA consisting of two mini-Mu's in parallel orientations covalently linked to host DNA at their ends (type IV). Due to the headful packaging of Mu DNA, DNAs of type I to type V are ±38 kb long. A fraction of types II, III, IV, and V DNAs should carry the X^+ host marker. DNAs of types I and V are thought to be intermediates in generalized transduction, which occurs by *rec*-dependent substitution of the X^- allele by the X^+ allele. Type-III DNA is supposed to be the intermediate in mini-Muduction, which occurs by the *rec*-independent, Mu*A*-dependent addition of a mini-Mu–X^+–mini-Mu transposable element.

RESULTS AND DISCUSSION

Mu-controlled Expression of Host Genes

In most cases when a change in host-gene expression occurs after Mu infection, it results from inactivation of a gene as a consequence of Mu insertion in that gene or in a proximal gene of the same operon. It can result in either constitutive expression or no expression of a set of genes if integration occurs in a negative or positive regulatory gene.

Precise excision of Mu can be detected by looking for reversion of mutations at 42 °C resulting from the integration of Mu*cts*62*X* or mini-Mu*cts*62A^+, which upon induction express the *A* but not the *B* and *kil* genes. Precise excision requires the *A*-gene product and is *recA*-dependent (Bukhari 1975; Faelen et al. 1978).

We recently isolated a mini-Mu (Mu3A-53) that carries the *cts*62 thermosensitive mutation in its repressor gene. When Mu3A-53 is integrated in the *lacZ* gene, it controls the expression of the distal *lacY* gene (see Fig. 3); *lacY* is expressed in the absence of Mu immunity (i.e., at 42 °C) but not in the presence of Mu immunity (i.e., at 30 °C or in the presence of an additional Muc$^+$ prophage). Mu3A-53 is ±2 kb long, has intact ends, and has lost at least part of the *A* gene. Expression of the host gene distal to the mini-Mu probably occurs because of the Mu early promoter located around 0.8 kb from the *c* end.

Since Mu3A-53 is A^-, this mini-Mu cannot induce rearrangements by itself, but when complemented by Mu*cts*62, it can mediate replicon fusion (Table 2) and most probably all other kinds of Mu-induced chromosomal rearrangements. Thus, Mu3A-53-dependent expression of the *lacY* gene requires the presence of the mini-Mu *cis* of that gene but can be controlled by a Muc$^+$ prophage located *trans* of *lacY*, and chromosomal rearrangements induced by Mu3A-53 are controlled by a Mu*cts*62 prophage located *trans* of that gene. This situation is somewhat reminiscent of certain controlling elements in maize (McClintock 1965).

Effect of *E. coli* Mutations on Mu and Mini-Mu Integrative Properties

We wondered whether *E. coli* K12 mutations that affect recombination (*recA, recB, sbcB, sbcA, recF*), excision repair (*uvrA*), error-prone repair (*spr lexB*), spontaneous-mutation frequency (*mutU, mutR, mutS*), and N6-methylation of adenine (*dam*), would alter (1) mini-Mu precise excision, (2) transposition of Mu or mini-Mu from the chromosome onto an episome, and (3) Mu-mediated transposition of host chromosomal DNA on an episome. The principal characteristics of the host mutants used are summarized in Table 3.

We measured the frequency of precise excision of mini-MuA^+ Mu3A (Faelen et al. 1979a) as the frequency of reversion to Lac$^+$ of strains carrying a Δ (*pro lac*) and an F *lacZ*::Mu3A *proAB*, after overnight growth at 42 °C. As shown in Table 3, precise excision is lower (±500 times) in *recA* (exoV overproducer) and *recA recB* (exoV$^-$) than in all the other hosts tested, confirming that it is highly *recA*-dependent and showing that the *rec* dependence is not due to the presence of altered levels of exoV. The frequency of transposition of Mu and mini-Mu from the chromosome onto an F *lac pro* episome was measured using spectinomycin-sensitive (Sp^s) strains carrying in the chromosome either a Mu*cts*62 pApl (Leach and Symonds 1979) or a mini-Mu*cts*62 18A-1 (Toussaint et al. 1980), both of which have a nontransposable ampicillin-resistance (Ap^r)

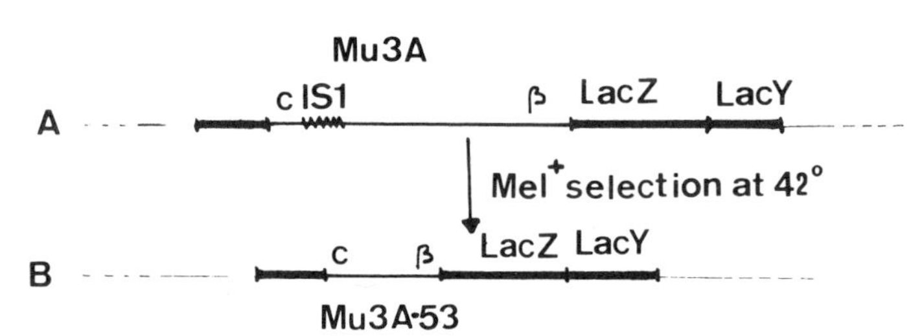

Figure 3. Mu3A (*A*) is ±8 kb long and carries an IS*1* at 4.4 kb from the *c* end, adjacent to a deletion of ~30 kb of Mu DNA. It is A^+. Mu3A-53 (*B*) derives from Mu3A. It is ~2 kb long and has lost the IS*1* and most of gene *A*. It was found in a strain that originally carried Mu3A in *lacZ*, was therefore Lac^-, and was selected for by its ability to grow on melibiose at 42 °C, which requires expression of the *lacY* gene. The Z^+, Z^- and Y^+, Y^- phenotypes correspond to the ability or inability to ferment lactose (at any temperature) and melibiose (at 42 °C), respectively. The strain with the structure shown on line *A* is Z^-Y^- at 30 °C and 42 °C whether or not it is also lysogenic for Muc^+. The strain with the structure shown on line *B* is Z^-Y^+ at 42 °C, Z^-Y^- at 30 ° C if not lysogenic for Muc^+, and Z^-Y^- at 30 °C and 42 °C if lysogenic for Muc^+.

marker and the episome F′ 128 (Low 1972). These strains were mated for 1 hour at 42 °C (to induce the prophage) with a $Sp^r(Muc^+)Mu^rRec^+$ recipient. Pro^+Sp^r and $Pro^+Sp^rAp^r$ exconjugants were selected on appropriate media. The frequency of transposition was calculated as the ratio of $Pro^+Sp^rAp^r$ to Pro^+Sp^r exconjugants (i.e., number of bacteria that received an F′ 128 [Mu] or F′ 128 [mini-Mu] to the number of bacteria that received an F′ 128).

As shown in Table 3, the frequency of transposition of Mu is almost the same (less than fivefold difference) in all strains tested, except in *recB sbcA* where it is ten times higher and in *mutR* where it is approximately sevenfold lower. The frequency of transposition of mini-Mu is not altered on *recB sbcA* and *mutR* but is approximately ten times higher on *mutS* (which does not affect Mu transposition) and approximately ten times lower on *recA* compared with other strains. The fact that various host mutations affect the frequencies of Mu and mini-Mu transposition differently suggests that transposition of Mu and mini-Mu might proceed through partially different pathways.

Mu-mediated transposition of the $thyA^+$ and $pyrF^+$ markers from the chromosome onto an F *lac pro* episome was measured by mating, at 42 °C, the same strains lysogenic for Mu*cts*62 pAp1 as above, with a *recA thyA pyrF* $Sp^r(Muc^+)Mu^r$ recipient, and selecting the Sp^r Lac^+, $Sp^rLac^+Thy^+$, and $Sp^rLac^+Pyr^+$ transconjugants. The Thy^+ and Pyr^+ exconjugants received an F *lac pro* with a Mu–$thyA^+$–Mu or Mu–$pyrF^+$–Mu DNA segment, respectively. The frequencies of Mu-mediated transposition are given by the ratio of the number of Sp^r Lac^+Thy^+ or Pyr^+ exconjugants to the number of Sp^r Lac^+ exconjugants.

As shown in Table 3, none of the mutant strains tested affect Mu-mediated transposition of either $thyA^+$ or $pyrF^+$. We do not as yet have any straightforward explanation for these results. It is obvious that a decrease in adenine methylation, a constitutive expression of the "SOS" repair system, or the absence of excision repair have no effect on precise excision of mini-Mu, on Mu and mini-Mu transposition, or on Mu-mediated transposition; this suggests that the enzymes involved in methylation, "SOS" repair, and excision repair are not involved in precise excision and transposition. However, some enzymes involved in recombination and some enzymes encoded or controlled by *mut* genes seem to play a role in these two processes.

The *B*-Gene Function Increases the Frequency of Resolution of Cointegrate Structures Resulting from Mini-Mu-mediated Replicon Fusion

Mini-MuA^+B^- transpose and mediate chromosomal rearrangements approximately 100 times less efficiently than Mu (see Tables 3 and 4). However, when mini-MuA^+ are complemented for the *B*-gene product, they transpose at about the same frequency as Mu (Table 4). We had proposed that *B* was a replication function and that the low frequency of transposition and rearrangements induced by Mu*X*, MuB^-, and mini-MuA^+B^- were due to the absence of replication (Faelen et al. 1978, 1979a,b). However, recent observations do not support such an hypothesis. Whether complemented

Table 2. Replicon Fusion Induced by Mu3A-53

	Frequency of F (mini-Mu [λ-*gal*] mini-Mu) formation
RH3936/F *lacZ*::Mu3A-53 *pro*	3×10^{-7}
RH3936 (Mu*cts*62*mom*3452)/F *lacZ*::Mu3A-53 *pro*	6×10^{-5}
RH3936 (Mu*cts*62*mom*3452)/F *lacZ*::Mu18A-1 *pro*	4×10^{-5}

The three strains shown in the first column were infected with λ*Nsus7Nsus53r14gal8* (Rambach and Brachet 1971) at a moi of 10 and mated at 42 °C for 1 hr with N100 (Gal^-Sm^r). Sm^rGal^+ exconjugants were selected on minimal medium supplemented with casamino acids and galactose. RH3936 is Δ(*lac pro*) *galE thi* and derives from CSH41 (Miller 1972). F *lacZ*::Mu3A-53 *pro* derives from F′ 128 (Low 1972). The strain carrying the F *lacZ*::Mu18A-1 *pro* episome was used as a positive control.

Table 3. Precise Excision of Mu3A, Transposition of Mu and Mu18A-1, and Mu-mediated Transposition of *thyA*$^+$ and *pyrF*$^+$ in Different Strains of *E. coli* K12

Parental Strain	Relevant host mutations and their main characteristics	Frequency of reversion to His$^+$		Frequency of reversion to Lac$^+$		Frequency of transposition				Frequency of Mu-mediated transposition of	
						Mu*cts*62	pAp1	Mu18A-1		*thyA*$^+$	*pyrF*$^+$
1. KL630	parent of 2–6	n.t.		2×10^{-6}	1	1.7×10^{-3}	1	4×10^{-5}	1	6.5×10^{-4}	1.1×10^{-4}
2. KL631	*recA1;* pleiotropic effect, protein X$^-$, UV^s	n.t.		$\pm 10^{-8}$	5×10^{-3}	4.5×10^{-3}	2.6	5.5×10^{-6}	0.13	2.4×10^{-4}	3.8×10^{-5}
3. KL632	*recB21;* exoV$^-$	n.t.		5×10^{-7}	0.25	3×10^{-3}	1.8	2.7×10^{-5}	0.7	1.3×10^{-3}	3.2×10^{-4}
4. KL639	*recA1 recB21;* Rec$^-$, UV^s	n.t.		$<1.7 \times 10^{-8}$	5×10^{-3}	2.7×10^{-3}	1.6	n.t.	n.t.	2.4×10^{-4}	8.7×10^{-5}
5. KL633	*recB21 sbcA8;* constitutive for exoVIII Rec$^+$	n.t.		5×10^{-7}	0.25	1.8×10^{-2}	10.5	3.7×10^{-5}	0.9	5.8×10^{-4}	3.2×10^{-4}
6. KL634	*recB21 sbcB15;* exoI$^-$, Rec$^+$	n.t.		2.7×10^{-6}	1.3	5.5×10^{-3}	3.2	5×10^{-5}	1.2	7.8×10^{-4}	2.6×10^{-4}
7. JC7623	*recB21, recC22, sbcB15;* taken as parent for 8 and 9	n.t.		n.t.		7×10^{-3}	1	n.t.		n.t.	n.t.
8. JC9239	*recF143;* Rec$^-$, UV^s	n.t.		n.t.		5×10^{-3}	0.7	n.t.		n.t.	n.t.
9. DM1137	*spr lexB sfiA;* constitutive expression of "SOS" repair	n.t.		2.5×10^{-6}		3×10^{-3}	0.4	$\pm 7 \times 10^{-6}$		n.t.	n.t.
10. KL684	parent of 11	n.t.		1.6×10^{-6}	1	4.4×10^{-3}	1	7.5×10^{-6}	1	7×10^{-4}	10^{-4}
11. KL685	*uvrA6;* no excision of pyrimidine dimers	n.t.		3.7×10^{-6}	2.3	4×10^{-3}	1	1.3×10^{-5}	1.7	2×10^{-4}	3×10^{-5}
12. NR3435	parent of 13	5×10^{-9}	1	9×10^{-6}	1	3×10^{-2}	1	4×10^{-5}	1	n.t.	n.t.
13. NR3437	*dam4;* reduced amount of N6-methyladenosine causes hyper Rec phenotype, increased spontaneous mutagenesis	5×10^{-7}	100	4×10^{-6}	0.5	4×10^{-2}	1	3×10^{-5}	1	n.t.	n.t.
14. KL686	parent of 15–17	4×10^{-9}	1	5.4×10^{-6}	1	2.6×10^{-3}	1	4.7×10^{-6}	1	2.5×10^{-4}	1.3×10^{-4}
15. KL687	*mutS;* high rate of transition	2.6×10^{-8}	5.3	5.3×10^{-6}	1	3.4×10^{-3}	1.3	6×10^{-5}	12.7	4.8×10^{-4}	2×10^{-4}
16. KL688	*mutR;* high rate of transition and frameshifts	9.9×10^{-8}	20	1.8×10^{-6}	0.3	4×10^{-4}	0.15	10^{-5}	2.1	5.6×10^{-4}	2.2×10^{-4}
17. KL689	*mutU;* same as *uvrE,* and probably *uvrD,* and *recL* increases transversions and frameshifts, UV^s	5.7×10^{-8}	12	7.7×10^{-6}	1.4	10^{-3}	0.4	3.5×10^{-6}	0.7	10^{-3}	10^{-4}

Numbers 1–17 each refer to a set of three strains derived from a single parent, the main properties of which are given in the second column. One strain carries an F *lacZ*::Mu3A *proAB* episome and was used to measure precise excision. The second strain carries a Mu*cts*62 pAp1 prophage in its chromosome and F′128 and was used to measure the frequencies of transposition of Mu and of Mu-mediated transposition of *thyA*$^+$ and *pyrF*$^+$. The third strain carries a Mu18A-1 prophage in the chromosome and F′128 and was used to measure transposition of mini-Mu. The frequencies shown are mean values of at least two independent measurements. All strains carry the Δ(*pro lac*) XIII deletion except 7, 8, and 9, which carry *lacY* and *pro* point mutations. Strains KL630 to 634, 639, and 684 to 689 were a gift from K. B. Low; NR3435 and NR3437 come from B. Glickman via M. Radman; JC7623 and JC9239 come from A. Clark; and DM1187 comes from D. Mount via O. Huisman. Lac$^+$ revertants were selected on minimal Lac medium at 42 °C and His$^+$ revertants on minimal Glu medium at 42 °C. CSH54 (Muc$^+$)Mu^{r}Sm^r, derives from CSH54 (Miller 1972) and was the recipient for measuring the frequencies of Mu and Mu18A-1 transposition. KMBL241 Sm^r (van de Putte et al. 1966) (same as KMBL146, see Table 1, except it is *recA*) was the recipient to measure the frequency of Mu-mediated transposition. n.t. means not tested.

Table 4. Frequency of Mu and Mini-Mu Transposition and Mu-mediated Transposition of Host DNA

Strain	Prophage present	Frequency of Mu18A-1 transposition[a]	Frequency of Mu18A-1-mediated transposition of Arg[+b]	Met[+b]
MF333	Mu18A-1	3×10^{-5}	7×10^{-7}	7×10^{-7}
MF393	Mu18A-1, Mu*cts*62A^+B^+	1.5×10^{-3}	10^{-4}	5×10^{-5}
MF394	Mu18A-1, Mu*cts*62	2×10^{-3}	8×10^{-5}	2×10^{-5}

MF333 is MXR(λ540::Mu18A-1)/F′ 128 (Toussaint et al. 1980); MF393 is *srl*::Tn*10 recA* (λ540::Mu18A-1)/F′ 128, carries the Mu*cts*62 Δ(*kil-β*) A^+B^+ prophage of MH133 (Howe 1973b), and derives from M5020G (Castellazzi et al. 1972). MF394 is similar to MF393 but lysogenic for Mu*cts*62. These three strains were mated for 1 hr at 42 °C with DS11 *recA* Mu^r $λ^r$, a Δ(*ppc-argE*) 100 (Elseviers et al. 1972) Sp^r strain. $Sp^rAp^rPro^+$, Sp^r $Ap^rPro^+Arg^+$, and $Sp^rAp^rPro^+Met^+$ exconjugants were selected on appropriate minimal media.

[a] Given by the ratio of number of $Sp^rAp^rPro^+$ exconjugants to number of Sp^rPro^+ exconjugants.

[b] Given by the ratio of number of $Sp^rAp^rPro^+Arg^+$ or Met^+ exconjugants to number of Sp^rPro^+ exconjugants.

only for *B* or for all the functions, Mu18A-1 transposes and mediates transposition of host DNA at about the same frequencies (Table 4); but it replicates very poorly, if at all, when provided with only the *B*-gene product, and replicates (although approximately ten times less than Mu) when complemented for all Mu early functions (B. Wagonner et al., in prep.). This suggested to us that no direct correlation exists between frequency of transposition and Mu replication and led us to reinvestigate the possible role of the *B* function.

The model proposed by Shapiro (1979) and the data that have been accumulated on the mechanism of Tn*3* transposition (Casadaban et al.; Heffron et al.; Ohtsubo et al.; Sherratt et al.; all this volume) suggest that coin-

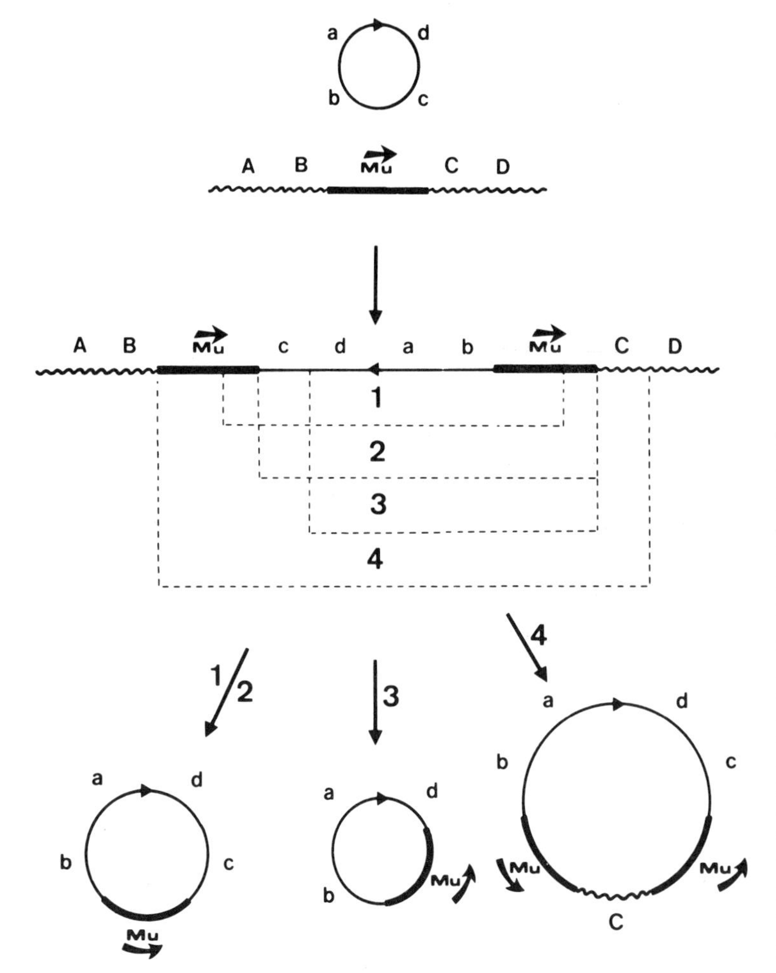

Figure 4. Upon transposition of Mu from the bacterial chromosome onto an F episome, Mu-mediated replicon fusion would result in the insertion of the F episome in the chromosome between two Mu prophages in the same orientation. In theory, different events could regenerate an F′ (Mu) from the resulting Hfr. Recombination could occur: (*1*) in Mu either by reciprocal recombination through the site-specific G inversion recombination system or through another hypothetical Mu-mediated reciprocal recombination system; (*2*) by recombination between the right (or left) extremities of the two Mu prophages surrounding the F episome; or (*3* and *4*) by Mu-mediated deletions generating two circular DNAs, one of which contains the F genome associated to a Mu prophage. Some of these episomes lack part of the F (*3*), and some carry a piece of host DNA flanked by two Mu prophages in the same orientation (*4*). (——) F DNA; (━━━) Mu DNA; (〰〰) bacterial DNA.

tegrate structures might be compulsory intermediates in transposition. If this is the case, in the system we use to measure transposition (i.e., detection of Mu or mini-Mu on an F′ episome after transfer of that episome in an appropriate background) the cointegrate intermediate would be an Hfr, where the integrated F is surrounded by two Mu (or mini-Mu) genomes in the same orientation (see Fig. 4). Since in most cases we used *recA*Δ(*pro lac*) bacteria as donors and recipients, the episome, in order to be transferred, should first be regenerated by resolution of the cointegrate structure. Resolution could occur in several ways, e.g., by reciprocal recombination between the two copies of Mu (or mini-Mu); such a recombination could occur either through the site-specific G inversion recombination system (for description see Kamp et al. 1978, 1979) or through a hypothetical Mu-mediated reciprocal recombination, which could be site-specific and occur in Mu or at the ends of Mu (see Fig. 4). In that case, resolution should regenerate an episome identical to the original one except that it should carry a Mu or a mini-Mu prophage. Resolution could also occur by Mu-mediated deletion (see Fig. 4) that would generate episomes of various sizes, some of which should lack part of the F and some of which should carry an additional segment of host DNA.

We wondered whether *B* were not involved in cointegrate resolution, thereby increasing the frequency of F′ regeneration and causing an increased frequency of mini-Mu transposition. Cointegrates represent an important fraction of the structures recovered after transposition of mini-Mu from one replicon to another (Schumm and Howe 1980; Toussaint and Faelen 1981). We constructed F *lac pro* (mini-Mu::λ*N*⁻*r*14*gal*8 *bio*1::mini-Mu) episomes with different mini-Mu A^- or A^+. Resolution of such a cointegrate structure should result in the formation of an F *lac pro* (mini-Mu) and a λ*gal* (mini-Mu) DNA that should be lost upon subsequent host divisions, generating Gal⁻ segregants. Note that the λ phage used does not carry the *red* gene. *recA* Gal⁻ strains containing those F *lac pro* (mini-Mu::λ*gal*::mini-Mu) episomes were grown overnight at 30 °C or 42 °C; the cultures were streaked for isolated colonies on MacConkey galactose agar plates at 30 °C to screen for Gal⁻ segregants. These were further tested for the presence or absence of Lac, Pro, Tra, and λ markers. As shown in Table 5, except for the F *lac pro* (Mu18A-1::λ*N*⁻*r*14*gal*8*bio*1::Mu18A-1) episome, which produced ±2% Gal⁻ segregants both at 30 °C and 42 °C, none of the episomes tested did lose the Gal⁺ marker unless they lost the whole episome.

The F *lac pro* (Mu18A-1::λ*N*⁻*r*14*gal*8 *bio*1::Mu18A-1) episome was transferred into a *recA* strain that carries a deleted Mu prophage retaining only genes *c* (*cts*62), *A*, and *B*. The F′ derivative was grown for different lengths of time at 42 °C and mated with a Gal⁻ *recA* recipient lysogenic for Muc^+, to look for Gal⁻ segregants (transfer was necessary because the F′ donor strain is killed at 42 °C due to the simultaneous presence of the *A* and *B* products and the two Mu ends, making the analysis of the segregants impossible). As shown in Table 6, the number of episomes that retained the Ap^r marker (thus most probably the mini-Mu), but lost the Gal⁺ marker, increases with the time of incubation at 42 °C, to reach approximately 70% after 90 minutes.

Most of the Gal⁻ episomes (18 of 20) recovered from the mating after a 30-minute induction period were still Lac⁺, Pro⁺, Tra⁺, carried a Mu18A-1 prophage, and had lost all the λ genes tested. However, only about 50% (13 of 24 tested) of the Gal⁻ episomes that were recovered from the mating after a 90-minute induction period had such properties. The remaining 50% had also lost Tra gene(s).

The same experiment was repeated using an Su⁻ *recA* donor carrying Mu*cts*62*A*am1093 on the chromosome and an F *lac pro* (Mu18-1::λ*N*⁻*r*14*gal*8*bio*1::Mu18-1) episome. Mu18-1 is A^- so that here only the *B* gene is expressed. No Gal⁻ exconjugants (<1%) were found at any time after induction.

These experiments show that in the presence of both the *A* and *B* products, the cointegrates are resolved at high frequency. However, the type of recombination involved remains to be characterized. Most of the "resolved" F′ episomes that were recovered after a 30-minute induction period had lost only the λ-*gal* markers, suggesting that resolution might occur by reciprocal re-

Table 5. Segregation of F (Mini-Mu) (λ*N*⁻*r*14*gal*8*bio*1) (Mini-Mu) Episomes

Mini-Mu	Gal⁻ segregants 30 °C no.	30 °C properties	42 °C no.	42 °C properties
Mu3A-53 (A^-,B^-)	0/418		2/495	lost the whole episome
Mu40 (A^-,B^-)	2/406	lost the whole episome	0/429	
Mu18-1 (A^-,B^- pAp1)	0/418		2/224	lost the whole episome
Mu3A (A^+,B^-)	3/433	lost the whole episome	4/454	lost the whole episome
Mu18A-1 (A^+,B^- pAp1)	4/210	one lost the episome, acquired Mu-18A-1 in chromosome; three contain F *lac pro* with Mu18A-1 but no λ markers	13/490	two lost the whole episome; one lost the episome, acquired Mu18A-1 on the chromosome; two retained an episome deleted of λ, Pro, and Lac but carrying Mu18A-1; eight retained an F *lac pro* (Mu18A-1) that lost all λ genes

The Gal⁻ segregants were tested for the presence or absence of Lac, Pro, Tra, Mu immunity, Mu Gov (van Vliet et al. 1978b), Ap^r, λ immunity, and λ genes *P, R, F, K,* and *J.*

Table 6. Resolution of F *lac pro* (Mu18A-1) (λN^-*r*14*gal*8*bio*1) (Mu18A-1) Cointegrates at 42 °C in the Presence of *A*- and *B*-Gene Products

Time spent at 42 °C before mating (min)	Temperature (°C) at mating	Frequency of Ap^rSp^r clones	No. of Gal^- clones
0	30	7.2×10^{-2}	0/170
0	42	7×10^{-2}	3/223
30	42	6.6×10^{-2}	17/247
60	42	7.2×10^{-2}	84/220
90	42	6×10^{-2}	132/192

Strain MF376 (M5020G *srl*::Tn*10, recA*, Δ(*pro lac*), *leu*::Mu*cts*62 Δ*kil-β*) containing the F *lac pro* (Mu18A-1) (λN^-*r*14*gal*8*bio*1) (Mu18A-1) episome was grown at 30 °C to a cell density of $\pm 2 \times 10^8$ bacteria/ml. The bacteria were incubated for different lengths of time at 42 °C and mated with MR96 (Muc^+) (a Gal^- *recA* Sp^r recipient) for 2 hr at the indicated temperature. Ap^rSp^r exconjugants were selected on L agar supplemented with ampicillin and spectinomycin. Gal^- and Gal^+ clones were detected by replica plating onto MacConkey galactose agar.

combination between the two copies of the mini-Mu. The fact that only 50% of the resolved F′ recovered after a 90-minute induction period have such properties is not necessarily contradictory with that hypothesis. Indeed, after a 60-minute induction period, approximately 40% of the cointegrates have been resolved, so that at 90 minutes some episomes might have undergone two Mu-mediated events, the first being reciprocal recombination between two mini-Mu segments, leaving behind a mini-Mu that could further mediate deletion of part of the F genome. We are currently trying to characterize further this new type of Mu-mediated recombination.

ACKNOWLEDGMENTS

We thank K. B. Low for providing us with many mutant strains of *E. coli* and O. Doubleday for his critical reading of this manuscript. This work was carried out under an agreement between the Université Libre de Bruxelles and the Belgian Government, concerning priority actions in collective basic research, and with the support of the Fonds National de la Recherche Scientifique.

REFERENCES

Allet, B. 1979. Mu insertion duplicates a 5 base pair sequence at the host insertion site. *Cell* **16:** 123.

Allet, B. and A. I. Bukhari. 1975. Analysis of Mu and λ-Mu hybrid DNAs by specific endonucleases. *J. Mol. Biol.* **92:** 529.

Bukhari, A. I. 1975. Reversal of mutator phage Mu integration. *J. Mol. Biol.* **96:** 87.

Bukhari, A. I. and D. Zipser. 1972. Random insertion of Mu-1 DNA within a single gene. *Nat. New Biol.* **236:** 240.

Bukhari, A. I., S. Froshauer, and M. Botchan. 1976. The ends of bacteriophage Mu DNA. *Nature* **264:** 580.

Castellazzi, M., P. Brachet, and H. Eisen. 1972. Isolation and characterisation of deletions in bacteriophage λ residing as prophage in *E. coli* K12. *Mol. Gen. Genet.* **117:** 211.

Daniell, E., R. Roberts, and J. Abelson. 1972. Mutations in the lactose operon caused by bacteriophage Mu. *J. Mol. Biol.* **69:** 1.

Daniell, E., J. Abelson, J. S. Kim, and N. Davidson. 1973. Heteroduplex structures of bacteriophage Mu DNA. *Virology* **51:** 237.

Dénarié, J., C. Rosenberg, B. Bergeron, C. Boucher, M. Michel, and B. de Bertalmio. 1977. Potential of RP4::Mu plasmids for in vivo genetic engineering of gram-negative bacteria. In *DNA insertion elements, plasmids, and episomes* (ed. A. I. Bukhari et al.), p. 507. Cold Spring Harbor Laboratory, Cold Spring Harbor, New York.

Elseviers, D., R. Cunin, N. Glansdorff, S. Baumberg, and E. Arschroft. 1972. Control regions in the *arg*-ECBH gene cluster of *Escherichia coli* K12. *Mol. Gen. Genet.* **117:** 349.

Faelen, M. and A. Toussaint. 1976. Bacteriophage Mu-1, a tool to transpose and to localize bacterial genes. *J. Mol. Biol.* **104:** 525.

———. 1978. Mu mediated deletions in the chromosome of *E. coli. J. Bacteriol.* **136:** 477.

———. 1980. Inversion induced by bacteriophage Mu-1 in the chromosome of *E. coli* K12. *J. Bacteriol.* **142:** 391.

Faelen, M., O. Huisman, and A. Toussaint. 1978. Involvement of phage Mu-1 early functions in Mu mediated chromosomal rearrangements. *Nature* **271:** 580.

Faelen, M., A. Resibois, and A. Toussaint. 1979a. Mini-Mu, an insertion element derived from temperature phage Mu-1. *Cold Spring Harbor Symp. Quant. Biol.* **43:** 1169.

Faelen, M., A. Toussaint, and M. Couturier. 1971. Mu-1 promoted integration of a λ-*gal* phage in the chromosome of *E. coli. Mol. Gen. Genet.* **113:** 367.

Faelen, M., A. Toussaint, and J. de Lafonteyne. 1975. A model for the enhancement of λ-*gal* integration into partially induced Mu lysogens. *J. Bacteriol.* **121:** 873.

Faelen, M., A. Toussaint, and A. Resibois. 1979b. Mini-Muduction: A new mode of gene transfer mediated by mini-Mu. *Mol. Gen. Genet.* **176:** 191.

Ghosal, D., H. Sommer, and H. Saedler. 1979. Nucleotide sequence of the transposable DNA-element IS*2*. *Nucleic Acids Res.* **6:** 1111.

Hattman, S. 1979. Unusual modification of bacteriophage Mu DNA. *J. Virol.* **32:** 468.

Howe, M. M. 1973a. Transduction by bacteriophage Mu-1. *Virology* **55:** 103.

———. 1973b. Prophage deletion mapping of bacteriophage Mu-1. *Virology* **54:** 93.

Kahmann, R. and D. Kamp. 1979. Nucleotide sequences of the attachment sites of bacteriophage Mu DNA. *Nature* **280:** 247.

Kamp, D., R. Kahmann, D. Zipser, T. R. Broker, and L. T. Chow. 1978. Inversion of the G segment of phage Mu controls phage infectivity. *Nature* **271:** 577.

Kamp, D., L. T. Chow, T. R. Broker, D. Kwoh, D. Zipser, and R. Kahmann.1979. Site-specific recombination in phage Mu. *Cold Spring Harbor Symp. Quant. Biol* **43:** 1159.

Khatoon, H. and A. I. Bukhari. 1978. Bacteriophage Mu-induced modification of DNA is dependent upon a host function. *J. Bacteriol.* **136:** 423.

Leach D. and N. Symonds. 1979. The isolation and characterisation of a plaque forming derivative of bacteriophage Mu carrying a fragment of Tn*3* conferring ampicillin resistance. *Mol. Gen. Genet.* **172:** 179.

LOW, K. B. 1972. *Escherichia coli* K12 F-prime factors, old and new. *Bacteriol. Rev.* **36:** 587.

MCCLINTOCK, B. 1965. The control of gene action in maize. *Brookhaven Symp. Biol.* **18:** 162.

MCNEIL, T. W., J. BRILL, and M. M. HOWE. 1978. Bacteriophage Mu-induced deletions in a plasmid containing the *nif* (N2 fixation) genes of *Klebsiella pneumoniae. J. Bacteriol.* **134:** 821.

MILLER, J. H. 1972. *Experiments in molecular genetics.* Cold Spring Harbor Laboratory, Cold Spring Harbor, New York.

O'DAY, K. J., D. W. SCHULTZ, and M. M. HOWE. 1978. A search for integration deficient mutants of bacteriophage Mu-1. *Microbiology—1978* (ed. D. Schlessinger), p. 48. American Society for Microbiology, Washington, D.C.

OHTSUBO, H., H. OHMORI, and E. OHTSUBO. 1979. Nucleotide-sequence analysis of Tn*3* (Ap): Implications for insertion and deletion. *Cold Spring Harbor Symp. Quant. Biol.* **43:** 1269.

RAMBACH, A. and P. BRACHET. 1971. Sélection de mutants du bactériophage lambda incapables de se répliquer. *C.R. Acad. Sci.* (D) **272:** 149.

REED, R., R. YOUNG, J. A. STEITZ, N. GRINDLEY, and M. GUYER. 1979. Transposition of the *Escherichia coli* insertion element γ generates a five-base-pair repeat. *Proc. Natl. Acad. Sci.* **76:** 4882.

SCHUMM, J. W. and M. M. HOWE. 1980. Analysis of integration and excision of bacteriophage Mu: A new approach. *Microbiology—1980* (ed. D. Schlessinger), p. 237. American Society for Microbiology, Washington, D.C.

SHAPIRO, J. A. 1979. Molecular model for the transposition and replication of bacteriophage Mu and other transposable elements. *Proc. Natl. Acad. Sci.* **76:** 1933.

TAYLOR, A. L. 1963. Bacteriophage-induced mutation in *Escherichia coli. Proc. Natl. Acad. Sci.* **50:** 1043.

TOUSSAINT, A. 1976. The DNA modification function of temperate phage Mu-1. *Virology* **70:** 17.

———. 1977. The modification function of bacteriophage Mu-1 requires both a bacterial and a phage function. *J. Virol.* **23:** 825.

TOUSSAINT, A. and M. FAELEN. 1973. Connecting two unrelated DNA sequences with a Mu dimer. *Nat. New Biol.* **242:** 1.

———. 1981. Formation and resolution of cointegrates upon transposition of mini-Mu. *Microbiology—1981* (ed. D. Schlessinger). American Society for Microbiology, Washington, D.C. (In press.)

TOUSSAINT, A., M. FAELEN, and A. I. BUKHARI. 1977. Mu-mediated illegitimate recombination as a part of the Mu life cycle. In *DNA insertion elements, plasmids, and episomes* (ed. A. I. Bukhari et al.), p. 275. Cold Spring Harbor Laboratory, Cold Spring Harbor, New York.

TOUSSAINT, A., L. DESMET, F. VAN GIJSEGEM, and M. FAELEN. 1980. Genetic analysis of the fate of F *pro lac* episomes carrying a Mu or a mini-Mu after prophage induction. *Mol. Gen. Genet.* (in press).

TOUSSAINT, A., N. LEFEBVRE, J. SCOTT, J. A. COWAN, F. DE BRUIJN, and A. I. BUKHARI. 1978. Relationships between temperate phages Mu and P1. *Virology* **89:** 146.

VAN DE PUTTE, P. and M. GRUIJTHUIJSEN. 1972. Chromosome mobilization and integration of F-factors in the chromosome of *recA* strains of *E. coli* under the influence of bacteriophage Mu-1. *Mol. Gen. Genet.* **118:** 173.

VAN DE PUTTE, P., S. CRAMER, M. GIPHART-GASSLER. 1980. Invertible DNA determines host specificity of bacteriophage Mu. *Nature* **286:** 218.

VAN DE PUTTE, P., H. ZWENK, and A. RORSH. 1966. Properties of four mutants of *Escherichia coli* defective in genetic recombination. *Mutat. Res.* **3:** 381.

VAN VLIET, F., B. SILVA, M. VAN MONTAGU, J. SCHELL. 1978a. Transfer of RP4::Mu plasmids to *Agrobacterium tumefaciens. Plasmid* **1:** 446.

VAN VLIET, F., M. COUTURIER, L. DESMET, M. FAELEN, and A. TOUSSAINT. 1978b. Virulent mutants of temperate phage Mu-1. *Mol. Gen. Genet.* **160:** 195.

Specificity of Bacteriophage Mu Integration into DNAs of Different Origins

E. Piruzian, V. Andrianov, M. Mogutov, E. Krivtsova, V. Yuzeeva, A. Vetoshkin, and N. Kobets

USSR Academy of Sciences, Institute of Molecular Genetics, Moscow, USSR 123182 D-182

Bacteriophage Mu along with transposons (Tn) and insertion sequence (IS) elements make up a class of so-called transposable elements. The most interesting thing about Mu is its extraordinary ability to become integrated at multiple sites in the host DNA and to cause various kinds of chromosomal rearrangements (Bukhari 1976). The ability of Mu to integrate into nonhomologous DNA regions allows it to be used for exchange of DNA segments both within a species as well as between phylogenetically unrelated organisms. This fact imparts utmost importance to studies on the integration of Mu into the cells of different organisms.

The range of hosts has been considerably widened owing to the introduction of the Mu genome within transmissible plasmids. The most convenient vector for interspecific transfer of Mu is the R factor RP4, which has an exceedingly wide range of hosts among various families of gram-negative bacteria (Datta and Hedges 1972; Olsen and Shipley 1973). In all cases examined, the genome of Mu was transcribed and translated in the new genetic environment.

The nonspecific integration of Mu into the DNAs of various bacteria raises the question, "Is Mu capable of integration into eukaryotic DNA?" To answer this question, we have elaborated a system that involves the construction of hybrid plasmids consisting of fragments of *Streptomyces* DNA and of DNA of plant origin and that determines Mu integration into alien DNA fragments cloned in *Escherichia coli* cells.

In this paper we summarize the results of our genetic research into the specificity of Mu insertion into DNAs of different origins.

RESULTS

Construction of Recombinant Plasmids

To assess the possibility of Mu integration into alien DNA cloned in *E. coli* cells, we used as targets fragments of plant DNA within hybrid pCPS plasmids and *Streptomyces coelicolor* SCP2 plasmid DNA within hybrid pPV plasmids.

Plasmid pRSF2124 was used for cloning *Eco*RI fragments of *Pisum sativum* chloroplast DNA. The cloning resulted in a series of recombinant plasmids that contained different *Eco*RI fragments of chloroplast DNA (Andrianov et al. 1978, 1979). Table 1 shows the main characteristics of the hybrid pCPS plasmids used.

Hybrid plasmids of the pPV series with fragments of plasmid SCP2 from *S. coelicolor* A3(2) were obtained by means of the pBR322 vector. SCP2 plasmid DNA was isolated from *S. coelicolor* A3(2) strain A585 (Troost et al. 1979) according to the method of Schrempf and Goebel (1977). Hybrid plasmids were obtained by treating pBR322 and SCP2 plasmid DNAs with restriction endonuclease *Bam*HI and a subsequent ligation with T4 ligase. After the transformation of *E. coli* strain 1100 (r^-) with the ligated mixture, ampicillin-resistant (Ap^r), tetracycline-sensitive (Tc^s) clones containing hybrid plasmids were selected. The hybrid plasmids were analyzed by electrophoresis on agarose gels (Fig. 1). The results show that pPV hybrid plasmids carry different *Bam*HI fragments of plasmid SCP2. Electrophoresis in a linear polyacrylamide gel (Kozlov et al. 1977) was used to construct physical maps of the original SCP2 plasmid from *S. coelicolor* A3(2) and of the hybrid pPV plasmids (Fig. 2). It should be noted that the locations of the restriction sites agree with data obtained independently by other authors (Kirby and Wotton 1979; Bibb et al. 1980).

The major difficulty facing any study of the integration and transposition of Mu is the lack of easily selectable markers in its genome. Therefore, to facilitate the selection of clones carrying the integrated phage, we used bacteriophage Mu-Cm^r (chloramphenicol-resistant), which was kindly provided by T. S. Ilyina (Institute of Epidemiology and Microbiology, N. F. Gamalei AMS, USSR). This bacteriophage was obtained from *E. coli* strain KS7201 (which contains the unstable transposon Tn*9* [Cm^r] in its chromosome) by selection for variants that showed resistance to Cm in a concentration of 25 μg/ml and displayed the thermoinducible phage Mu*cts*62.

Mu Integration into Recombinant Plasmids

To obtain plasmids with integrated Mu-Cm^r, *E. coli* HB101 (*recA* r^-m^-) cells lysogenic for Mu-Cm^r were transformed with hybrid pCPS or pPV plasmid DNA. The phage was partially induced in the transformants for 18 hours at 37 °C, which resulted in Mu integration into plasmid DNA. To identify such plasmids, plasmid DNA was isolated according to the method of Birnboim and Doly (1979) and used to transform the Mu-immune *E. coli* strain C600. The cells carrying plasmids with integrated Mu-Cm^r were selected on the basis of Ap^rCm^r.

Table 1. Hybrid pCPS Plasmids

Hybrid plasmids	Size of fragment of chloroplast + DNA (kD)	G + C (%)
pCPS117	4.2	38.5
pCPS121	4.1	40.0
pCPS152	2.3	33.5
pCPS120	1.9	34.5
pCPS164	1.2	53.5

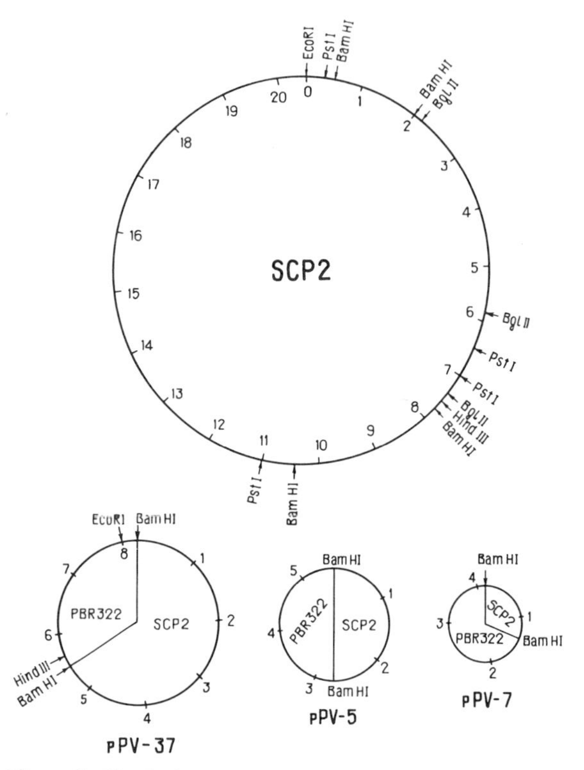

Figure 2. Physical maps of plasmid SCP2 from *S. coelicolor* A3(2) and of hybrid pPV plasmids. Arrows show the sites of digestions with restriction endonucleases. Fragment sizes are given in megadaltons.

Plasmid DNA was again isolated from these clones, treated with an appropriate restriction endonuclease, and analyzed by electrophoresis on an agarose gel.

The experimental results (Fig. 3) demonstrate the integration of Mu into the hybrid plasmids. A comparison of the cleavage patterns of the initial hybrid plasmids and the plasmids with integrated Mu-Cm^r shows that in some cases the phage was integrated into the cloned fragments of alien DNA (Fig. 3e,f).

This conclusion is confirmed by the fact that the band corresponding to the cloned fragment disappears after Mu integration and new bands appear corresponding to Mu-Cm^r DNA. Mu-Cm^r is integrated with equal probability into the vector parts of the hybrid plasmids or into the cloned fragments. Figure 3(c–f) shows the integration into plant DNA cloned in plasmid pCPS120.

A comparison of the electrophoretic patterns of Mu-Cm^r DNA before and after integration into plasmid DNA indicates the instability of Mu-Cm^r integration. Preliminary calculations reveal a discrepancy between the expected molecular weight of the plasmid with the whole Mu-Cm^r genome integrated and the sum of the molecular weights of the DNA fragments obtained. Nevertheless, when *E. coli* C600 cells and *E. coli* C600 cells lysogenic for defective Mu were transformed with the DNAs of plasmids pCPS120-3::Mu-Cm^r and pCPS120-25::Mu-Cm^r (Table 2), we observed a decrease in the frequency of *E. coli* C600 transformation with pCPS120-3::Mu-Cm^r DNA, which indicates a zygotic induction. No zygotic induction was observed in the case of transformation with pCPS120-25::Mu-Cm^r DNA. Presumably, different sections of the Mu-Cm^r genome were split out of the hybrid plasmid in each case. This supposition was confirmed by electrophoresis (Fig. 3). The same data were obtained with the recombinant plasmid pPV5.

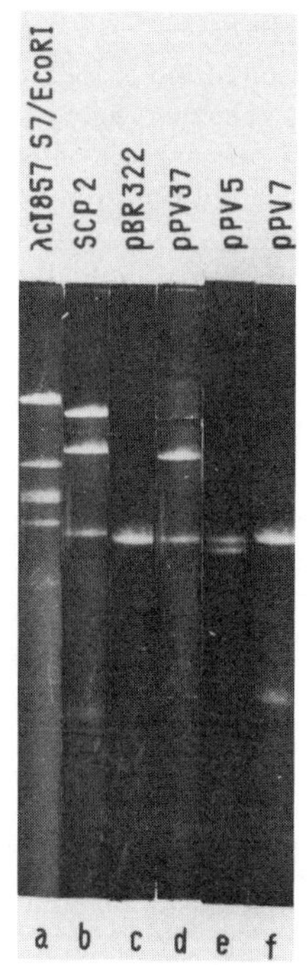

Figure 1. Electrophoresis of *Bam*HI digests of DNAs from SCP2 and hybrid pPV plasmids. The samples were electrophoresed in Tris-borate buffer in a 0.7% agarose gel at 1 mA per tube for 4 hr and photographed under UV irradiation.

The instability of whole-Mu integration in colicin plasmids is a well-known fact (Inselburg 1974). However, since in this study we were interested in the possibility of Mu integration into various DNAs and did not specifically attempt to achieve a stable integration of Mu into cloned DNA fragments, the system just described sufficed for our purposes.

At present, we are trying to obtain a deletion variant of Mu genetically marked with kanamycin resistance (Km^r). The mini-Mu variant will permit a more detailed study of Mu integration into eukaryotic DNAs.

Table 2. Frequency of Transformation by DNAs of Hybrid Plasmids

Plasmid DNA	*E. coli* C600			*E. coli* C600 (Mu*X*)[a]		
	Ap	Cm	Ap + Cm	Ap	Cm	Ap + Cm
pCPS120-3::Mu-Cm^r	<3 × 10^1	<3 × 10^1	<3 × 10^1	4.4 × 10^3	1.9 × 10^3	6.0 × 10^3
pCPS120-25::Mu-Cm^r	9.5 × 10^3	4.1 × 10^3	6.3 × 10^3	1.4 × 10^4	6.3 × 10^3	9.0 × 10^3

[a] *E. coli* C600 (Mu*X*) was obtained according to the method of Bukhari (1976) by plating Mu*cts* lysogens at 42 °C, at which temperature most of the lysogenic cells are killed because of Mu induction. Transformation was carried out in calcium-treated *E. coli* cells. The number of antibiotic-resistant transformants per 1 μg of plasmid PTK16 DNA (control) was 2.8 × 10^4 (Tc^r) and 5.2 × 10^4 (Km^r) in *E. coli* C600, and 4.2 × 10^{-4} (Tc^r) and 4.9 × 10^4 (Km^r) in *E. coli* C600 (Mu*X*).

DISCUSSION

We have recently witnessed an awakening of interest in the insertion elements and the beginning of intensive studies of Mu as a transposition element. The random integration of Mu at various sites in the host DNA followed by the induction of various chromosomal aberrations in the host genome make Mu a convenient means for genetic analysis of those microorganisms that defy study by the traditional techniques.

The range of hosts for Mu is confined to a few representatives of Enterobacteriaceae (Howe and Bade 1975). The resistance of many bacterial strains to Mu seems to be of an adsorptive nature, as has been proved for *Klebsiella pneumoniae* (Bachhuber et al. 1976). Mutant *E. coli* clones selected for Mu resistance often acquire resistance to phages P1 and P2 as well. However, the region of genes *nadA-gal* (16.5 min on the *E. coli* chromosome map) is believed to contain a gene that is specifically responsible for the Mu sensitivity of *E. coli* strains (Koretskaya and Piruzian 1978).

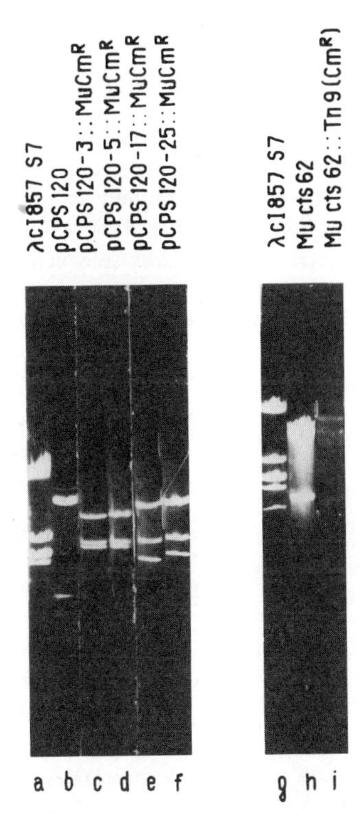

Figure 3. Electrophoresis of *Eco*RI-cleaved DNAs of recombinant plasmids carrying Mu DNA insertions. The samples were electrophoresed in Tris-borate buffer in a 0.7% agarose gel at 1 mA per tube for 4 hr and photographed under UV irradiation. (*b*) Original plasmid pCPS120: top line is vector part (pRSF2124) and bottom line is chloroplast *Eco*RI fragment; (*c,d,e,f*) plasmids after Mu-Cm^r integration.

The insertion of Mu into the cells of various bacterial genera and species (apart from the natural host *E. coli* K12) that possess a natural resistance may be achieved through selection of Mu-sensitive mutants (Rao 1976). A simpler and more promising way of transferring Mu into resistant strains involves the use of highly transmissible plasmids. At present, the list of bacteria into which Mu has been integrated includes (apart from the natural host *E. coli* K12 [Taylor 1963]) representatives of various species, genera, and families: from the family Enterobacteriaceae—*Shigella dysenteriae* (Taylor 1963), *Citrobacter freundii* (De Graaff et al. 1973), *Erwinia carotovora* (Perombelon and Boucher, as cited in Dénarié et al. 1977), *Erwinia stewartii* (Coplin 1979), non–nitrogen-fixing strains of *Klebsiella pneumoniae* (Rao and Pereira 1975), nitrogen-fixing strains of *K. pneumoniae* (Dénarié et al. 1977), and *Proteus* and *Serratia* (Faelen et al. 1977); from other bacterial groups—the plant pathogenic bacterium *Agrobacterium tumefaciens* (van Vliet et al. 1978), the root-nodule bacteria *Rhizobium* of the family Rhizobiaceae (Boucher et al. 1977; Dénarié et al. 1977), the phytopathogenic bacteria *Pseudomonas solanacearum* of the family Pseudomonadaceae (Boucher et al. 1977; Dénarié et al. 1977), and photosynthetic bacteria of the family Rhodospirillaceae (nonsulfur purple bacteria), *Rhodopseudomonas sphaeroides* (Tucker and Pemberton 1979; Piruzian et al., unpubl.).

An important outcome of all studies involving Mu transfer into microorganisms phylogenetically remote from *E. coli* is that it has been shown that Mu expresses its integrative functions in the new genetic environments.

In the present study we have looked into the possibility of integrating Mu into the DNAs of various organisms, both prokaryotic and eukaryotic, that have different levels of genome development. The Mu genome introduced into *R. sphaeroides* by means of the hybrid plasmid RP4 becomes integrated and expresses itself, yielding mature phage particles (data not shown). We have also demonstrated the integration of Mu into fragments of *Streptomyces* DNA cloned in *E. coli* cells, *Streptomyces* being a differentiated, gram-positive, mycelial bacterium that has a complex life cycle. Finally, the experimental data obtained in this study indicate that Mu can also be integrated into eukaryotic nucleotide sequences, namely, DNA fragments of *Pisum sativum* cloned in *E. coli* cells.

Since Mu is capable of joining unrelated DNA segments in new combinations without the necessity for large homologous regions, this bacteriophage offers a means for genetic engineering in vivo in bacterial systems and hopefully in eukaryotic systems as well. The idea that Mu might be useful for molecular genetic manipulations in eukaryotic cells is based on two major facts. First, we have obtained preliminary data on the insertion of Mu into plant DNA fragments cloned in vector plasmids, which demonstrates that, in principle, Mu DNA can be integrated into eukaryotic DNA sequences. Second, in spite of the powerful natural barriers hindering genetic exchange between remote taxonomic groups, information exchanges between prokaryotes and eukaryotes seem to be possible. For instance, one may recall the Ti plasmids of *A. tumefaciens*, some of whose segments can be effectively transferred into plant cells and then at least partially be transcribed in these eukaryotic cells, resulting in neoplastic changes in the form of crown-gall tumors (Chilton et al. 1977; Drummond et al. 1977). The Ti plasmids may be a good example of a "shuttle" vector whose genetic information is expressed partly in the plant cell and partly in the bacterial cell.

The transfer of RP4::Mu into *A. tumefaciens* has already been carried out (van Vliet et al. 1978), although far less effectively than in the case of the original RP4 plasmid. Presumably, one or several functions of Mu keep it from establishing itself normally in some gram-negative bacterial strains. Besides, van Vliet et al. (1978) believe that the cells of *A. tumefaciens* have a system of restriction and modification whose target is located within the Mu genome.

If these difficulties can be overcome, the Ti plasmids of *A. tumefaciens* might form the basis for work on the insertion of Mu into eukaryotic cells. The fragment of Ti plasmid that becomes integrated into plant DNA might be used as a vector for Mu.

The results of the study reported here have so far been promising and they suggest the possibility of Mu integration into eukaryotic DNA sequences. We are at present continuing with this research.

Bacteriophage Mu is unique in that as a prophage it is part of the host genome but in the active phase it becomes a transposable element subject to independent replication. Further studies aimed at widening the host range of Mu integration may bring us closer to understanding which of the known phenomena in eukaryotic organisms might be similar to the effects of prokaryotic transposable elements, and perhaps it will be possible to draw some interesting parallels.

REFERENCES

Andrianov, V. M., M. B. Amerkhanova, and Y. P. Vinetsky. 1979. Cloning of chloroplast DNA in *Escherichia coli*. II. Some properties of recombinant plasmids with *Eco*RI fragments of *Pisum sativum* chloroplast DNA and cloning of chloroplast DNA. *Genetika* **15:** 1918.

Andrianov, V. M., N. Y. Zemlyanaya, M. Karimov, M. B. Amerkhanova, and Y. P. Vinetsky. 1978. Cloning of chloroplast DNA in *Escherichia coli. Genetika* **14:** 1503.

Bachhuber, M., N. J. Brill, and M. Howe. 1976. Use of bacteriophage Mu to isolate deletions in the *his-nif* region of *Klebsiella pneumoniae. J. Bacteriol.* **128:** 749.

Bibb, M., J. L. Schottel, and S. N. Cohen. 1980. A DNA cloning system for interspecies gene transfer in antibiotic-producing *Streptomyces. Nature* **284:** 526.

Birnboim, H. C. and J. Doly. 1979. A rapid alkaline extraction procedure for screening recombinant plasmid DNA. *Nucleic Acids Res.* **7:** 1513.

Boucher, C., B. Bergeron, M. Barate de Bertalmio, and J. Dénarié. 1977. Introduction of bacteriophage Mu into *Pseudomonas salanacearum* and *Rhizobium meliloti* using the R factor RP4. *J. Gen. Microbiol.* **98:** 1253.

Bukhari, A. I. 1976. Bacteriophage Mu as a transposition element. *Annu. Rev. Genet.* **10:** 389.

Chilton, M. D., M. H. Drummond, M. H. Merlo, D. J. Sciaky, A. L. Montoya, M. P. Gordon, and E. W. Nester. 1977. Stable incorporation of plasmid DNA into higher plant cells: The molecular basis of crown-gall tumorigenesis. *Cell* **11:** 263.

Coplin, D. L. 1979. Introduction of bacteriophage Mu into *Erwinia stewartii* by use of a PK2::Mu hybrid plasmid. *J. Gen. Microbiol.* **113:** 181.

Datta, N. and R. W. Hedges. 1972. Host ranges of R factors. *J. Gen. Microbiol.* **70:** 453.

De Graaff, J., P. C. Kreuning, and P. van de Putte. 1973. Host controlled restriction and modification of bacteriophage Mu and Mu-promoted chromosome mobilization in *Citrobacter freundii. Mol. Gen. Genet.* **123:** 283.

Dénarié, J., C. Rosenberg, B. Bergeron, C. Boucher, M. Michel, and M. Barate de Bertalmio. 1977. Potential of RP4::Mu plasmids for in vivo genetic engineering of gram-negative bacteria. In *DNA insertion elements, plasmids, and episomes* (ed. A. I. Bukhari et al.), p. 507. Cold Spring Harbor Laboratory, Cold Spring Harbor, New York.

Drummond, M. H., M. P. Gordon, E. W. Nester, and M. D. Chilton. 1977. Foreign DNA of bacterial origin is transcribed in crown-gall tumors. *Nature* **269:** 535.

Faelen, M., A. Toussaint, M. Van Montagu, S. Van den Elsacker, G. Engler, and J. Schell. 1977. In vivo genetic engineering: The Mu-mediated transposition of chromosomal DNA segments onto transmissible plasmids. In *DNA insertion elements, plasmids, and episomes* (ed. A. I. Bukhari et al.), p. 521. Cold Spring Harbor Laboratory, Cold Spring Harbor, New York.

Howe, M. M. and E. G. Bade. 1975. Molecular biology of bacteriophage Mu. *Science* **190:** 624.

Inselburg, J. 1974. Isolation and characterization of mutants of colicin plasmids E1 and E2 after Mu bacteriophage infection. *J. Bacteriol.* **119:** 469.

Kirby, R. and S. Wotton. 1979. Restriction studies on the SCP2 plasmid of *Streptomyces coelicolor* A3(2). *FEMS Microbiol. Lett.* **6:** 321.

Koretskaya, N. G. and E. S. Piruzian. 1978. Investigation of the adsorption ability of bacteriophage Mu on *Escherichia coli* strains with deletions in *att*λ region. *Genetika* **14:** 1908.

Kozlov, J. I., N. A. Kalinina, L. V. Gening, B. A. Rebentish, A. Y. Strongin, V. G. Bogush, and V. G. Debabov. 1977. A suitable method for construction and cloning hybrid plasmids, containing *Eco*RI-fragments of *E. coli* genome. *Mol. Gen. Genet.* **150:** 211.

Olsen, R. and P. Shipley. 1973. Host range and properties of the *Pseudomonas aeruginosa* R factor R1822. *J. Bacteriol.* **113:** 772.

RAO, N. R. 1976. Mutational alteration of nitrogen-fixing bacterium to sensitivity to infection by bacteriophage Mu: Isolation of *nif* mutations of *Klebsiella pneumoniae* M5a1 induced by Mu. *J. Bacteriol.* **128:** 356.

RAO, N. and M. G. PEREIRA. 1975. Behavior of a hybrid F'*ts*114 *lac*$^+$*his*$^+$ factor (F42-400) in *Klebsiella pneumoniae* M5a1. *J. Bacteriol.* **123:** 792.

SCHREMPF, H. and W. GOEBEL. 1977. Characterization of a plasmid from *Streptomyces coelicolor* A3(2). *J. Bacteriol.* **131:** 251.

TAYLOR, A. L. 1963. Bacteriophage-induced mutations in *Escherichia coli. Proc. Natl. Acad. Sci.* **50:** 1043.

TROOST, T. R., V. N. DANILENKO, and N. D. LOMOVSKAYA. 1979. Fertility properties and regulation of antimicrobial substance production by plasmid SCP2 of *Streptomyces coelicolor. J. Bacteriol.* **140:** 359.

TUCKER, W. and J. M. PEMBERTON. 1979. The introduction of RP4::Mu*cts*62 in *Rhodopseudomonas sphaeroides. FEMS Microbiol. Lett.* **5:** 215.

VAN VLIET, F., B. SILVA, M. VAN MONTAGU, and J. SCHELL. 1978. Transfer of RP4::Mu plasmids to *Agrobacterium tumefaciens. Plasmid* **1:** 496.

Genome Fusion

H. Potter and D. Dressler
Department of Biochemistry and Molecular Biology, Harvard University, Cambridge, Massachusetts 02138

Two enzymes that carry out genome fusion reactions have recently been purified from *Escherichia coli.* In our laboratory, we have purified an enzyme, called DNA synaptase, that is capable of joining two DNA molecules together at a region of homology. In the laboratories of Shibata, DasGupta, Cunningham, and Radding, of Weinstock, McEntee, and Lehman, and of West, Cassuto, Mursalim, and Howard-Flanders, the *recA* protein (*recA*-gene product) has been purified and has been found also to have a genome fusion capability.

In this paper we will describe our current knowledge of DNA synaptase and discuss this enzyme with respect to the *recA* protein.

FUSING DNA MOLECULES IN VITRO

When an extract containing *E. coli* enzymes is incubated with exogenous monomer-size plasmid DNA rings, one detects a genome fusion reaction (Potter and Dressler 1978, 1979a). Over a period of hours, an increasing number of the input monomer rings are fused to form multimeric circles, as observed by electron microscopy.

In a typical experiment, the percentage of plasmids appearing as multimers rose from a background of less than 0.25% at 0 hours to about 15% at 20 hours (Fig. 1). Early during the incubation period, the multimer circles formed in vitro had the shape of simple figure eights (8). Examples are shown in Figure 2 A and B. Later on, more complex multimers arose (Fig. 2C); these involved several circular genomes and were interpreted as resulting from the interaction of monomers with previously formed multimers. Finally, if the reaction was allowed to proceed to completion, huge networks were formed, containing hundreds of fused monomer rings (Fig. 2D).

The figure-eight configuration of the product multimers could have two interpretations. This geometry could result from two monomer rings that had become interlocked like links in a chain or, alternatively, from two genomes that had become fused at a region of homology. We were able to distinguish between these two possibilities by opening the plasmid DNA molecules with the restriction enzyme *Eco*RI prior to electron microscopy. This enzyme cleaves the monomer-size plasmid DNA rings once, at a unique site, generating unit-length rods. Upon enzyme digestion, it is expected that a pair of interlocked monomer rings will be cleaved into two separable unit-length rods. On the other hand, if the figure-eight geometry represents two plasmid genomes fused at a region of DNA homology, then the restriction enzyme is expected to leave the fusion point intact and convert the figure eight into a bilaterally symmetric, dimer-size structure shaped like the Greek letter chi (χ). We were, in fact, able to observe such chi-shaped molecules following *Eco*RI digestion of the product multimers. Figure 3 shows two electron micrographs that are representative of more than 1500 chi forms derived from figure eights formed in vitro.

Could the crossed molecules observed in the electron microscope merely represent the accidental overlap of two monomer rods? This is unlikely for two reasons: First, the DNA is spread for the microscope at a low concentration, so that accidental overlaps are virtually impossible. Second, the crossed molecules display a special symmetry: The point of contact between the unit-size plasmid genomes always occurs so as to form a structure with two pairs of equal-length arms (as in Fig. 3). Since the contact point is doubly equidistant from a defined base sequence (the *Eco*RI-cut ends), it must almost certainly occur at a region of DNA homology. The fusion point is observed with equal probability at numerous and perhaps all locations along the plasmid DNA molecule, that is, at varying distances from the *Eco*RI cutting site (Fig. 4A).

In sum, the molecules fused in vitro by the unfractionated cell extract consist of two genomes stably held together at a region of DNA homology.

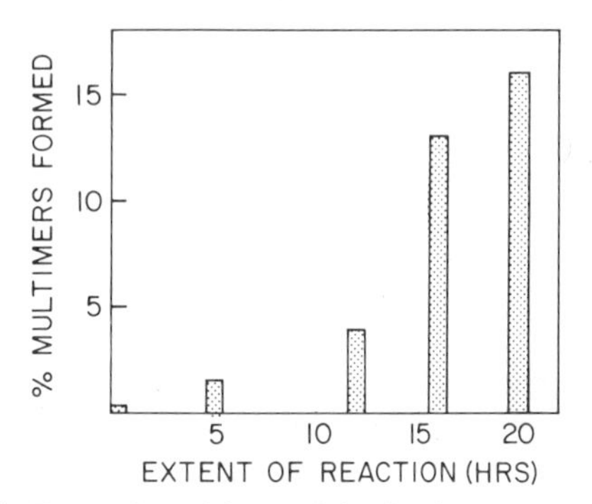

Figure 1. Formation of figure eights in vitro. A recombination-proficient *E. coli* strain (MM294) was grown in nutrient broth and made into a gentle cell lysate by a lysozyme freeze-thaw treatment (as described by Potter and Dressler 1978). The extract was supplemented with potentially useful metabolites (50 μM deoxyribonucleoside triphosphates; 2 mM ATP and 100 μM for the other ribonucleoside triphosphates; 1 mM each for NAD^+, $NADP^+$, and NADPH; and 2 mM phosphoenol pyruvate), 2.5 mM spermidine, 3 mM β-mercaptoethanol, 10 mM $MgCl_2$, and 50 mM potassium phosphate buffer (pH 7.5). Monomer-size plasmid DNA molecules (pMB-9) were then added to the crude extract at about 100 μg/ml, and the reaction mixture was incubated at 37 °C. At various times aliquots were withdrawn for analysis by electron microscopy. Over time, an increasing number of the input monomer-size plasmid rings were fused to form figure eights and higher multimers. Fig. 2 shows electron micrographs of the fused DNA circles.

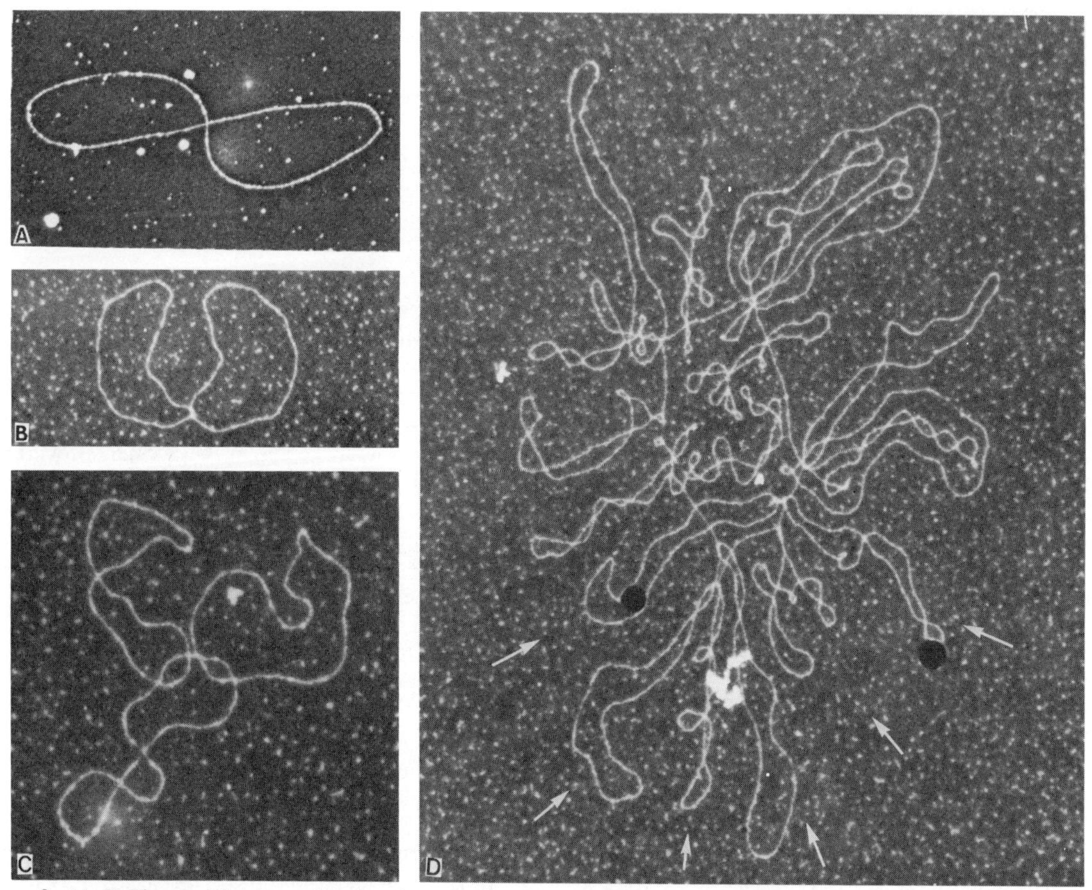

Figure 2. (*A, B*) Figure eights formed in vitro by an unfractionated extract containing *E. coli* enzymes. The extended region of pairing in the example in *B* may indicate the presence of two or more crossover connections close together. (*C*) A multimer consisting of three fused monomer rings. (*D*) A small network of fused DNA monomers, containing about 20 circles. In this network, almost every component circle can be traced; each appears to be interacting with a central monomer circle in a key-ring fashion. The six monomers marked with arrows at the bottom are examples of the key-ring geometry that appears to characterize the network. This geometry may indicate that the fusion enzyme remains bound to one DNA ring and processively adds new units to it. All of these structures can be cleaved with *Eco*RI to yield simple and complex chi forms with fused DNA connections (see Fig. 3). Electron microscopy was carried out as described by Wolfson et al. (1972). (*B* and *C* reprinted with permission, from Potter and Dressler 1978.)

Multiple activities. The above experimental results were obtained with an unfractionated cell extract that had been supplemented with supercoiled DNA rings and a full set of nucleotide precursors and cofactors (see the legend to Fig. 1). In most instances (then and now), the fusion reaction has been dependent on a functional *recA* gene (Potter and Dressler 1978). However, there is also a clear, and sometimes dominant, *recA*-independent fusion reaction that can be detected by using a mixture of supercoiled and nicked duplex rings and Mg^{++}, spermidine, and β-mercaptoethanol. It is this DNA fusion activity that we have first purified from the cell extract. As will be discussed, this protein appears to be representative of a class of *recA*-independent DNA fusion enzymes.

PURIFICATION OF DNA SYNAPTASE

Although the electron microscope provided an excellent method for the detection of genome fusion in vitro—because of the precision with which the products of an incubation could be analyzed—this technique would not be easy to use in enzyme purification. This is because the electron microscopic analysis of each reaction mixture requires about an hour. To process the many samples that would arise as the proteins of a cell extract are fractionated across chromatographic columns, we therefore developed an assay for detecting genome fusion that did not require the electron microscope (Potter and Dressler 1979b). The design of this radioisotope assay is shown in Figure 5. A cellular protein fraction is incubated with two different but closely related plasmid DNA molecules. The first plasmid is heavily labeled with radioactive thymidine. The second plasmid is essentially homologous to the first but is not radioactive and has been supplemented through recombinant DNA techniques to contain multiple copies of the *lac* operator. The presence of the operator DNA allows the second plasmid to be bound to a membrane filter in the presence of the *lac* repressor protein. To assay for genome fusion, the radioactive plasmid and the operator-containing plasmid are incubated together

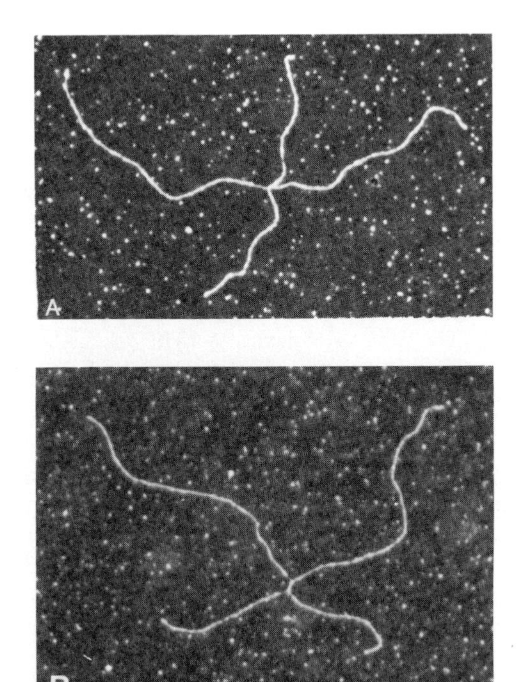

Figure 3. (*A* and *B*) Chi forms, resulting from the *Eco*RI cleavage of figure eights formed in vitro.

with a cellular protein fraction and then exposed to repressor and passed through a nitrocellulose filter. The radioactive plasmid will be retained on the filter only if it has become fused to an operator-containing plasmid. In a matter of hours, 100 or more column fractions can be assayed for their ability to carry out a DNA fusion reaction.

Thus far, the radio-biochemical assay has been used to purify one genome fusion activity from the crude cell extract (Potter and Dressler 1980). The data in Figures 6 through 9 show how this fusion enzyme, which we call DNA synaptase, is purified to apparent homogeneity by the application of the following methods: (1) ion-exchange chromatography on DEAE-Sephacel, (2) molecular sieve chromatography on Bio-Gel P-200, (3) affinity chromatography on DNA-cellulose, and (4) affinity chromatography on phenyl-Sepharose.

DEAE ion-exchange chromatography. Figure 6 shows the purification of the fusion activity resulting from ion-exchange chromatography on DEAE-Sephacel. Following an initial ammonium sulfate fractionation, the fusion activity was adsorbed to the DEAE resin in low salt and eluted with a KCl gradient. Each column fraction was then assayed for its ability to fuse radioactive and operator-containing plasmids. The fusion activity eluted in the range of 100 mM KCl and showed evidence for a distribution into a small number of discrete peaks (Fig. 6).

The activity profile shown in Figure 6 was typical of five independent preparations and suggests that *E. coli* may contain at least three distinct genome fusion enzymes. Another explanation for the multiple peaks would be that a single enzyme (a core fusion protein) occurs in multiple forms, associated with different auxiliary subunits or factors. A third possibility, which cannot be ruled out until all the activity peaks have been

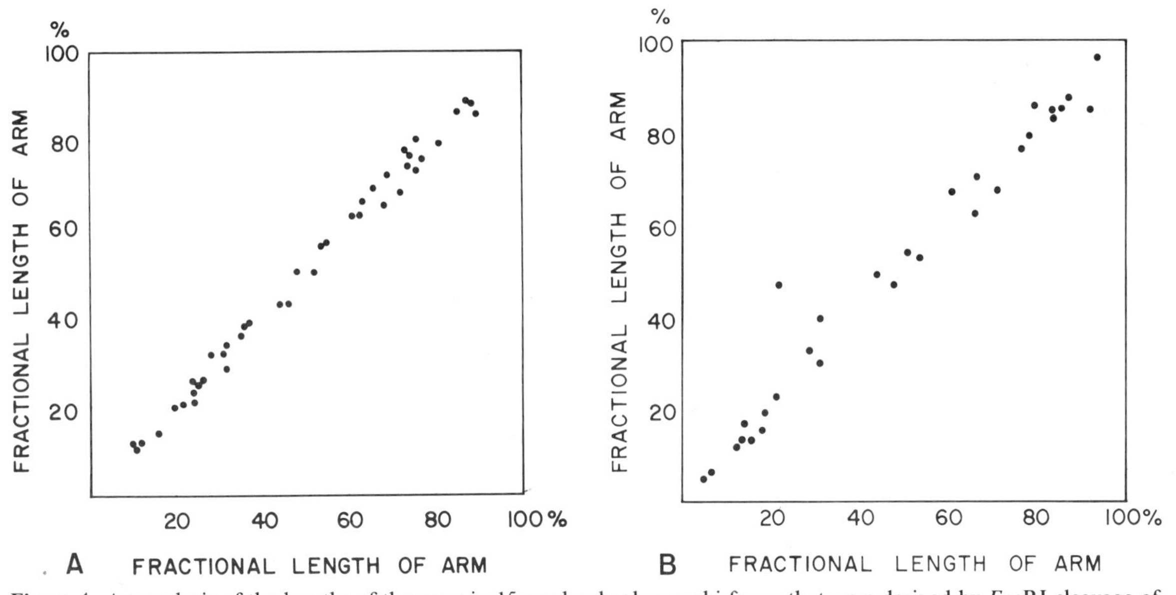

Figure 4. An analysis of the lengths of the arms in 15 randomly chosen chi forms that were derived by *Eco*RI cleavage of figure eights. For each molecule the lengths of the four arms were measured, totaled, and divided by two to obtain the unit genome length. The proportional lengths of the two shorter arms were then plotted, one as the abcissa and the other as the ordinate of a single point. Similarly, the two longer arms were used to produce a single point for the curve. The fact that the points generate essentially a straight line of slope 1 establishes that the chi forms contain pairs of equal-length arms and thus that fusion has occurred at a region of DNA homology. Furthermore, the finding that pairs of arms have different lengths indicates that the point of contact between the two genomes can occur at many locations, perhaps randomly. (*A*) Chi forms from figure eights made in vitro by the crude cell extract; (*B*) chi forms from figure eights made in vitro by the purified DNA synaptase.

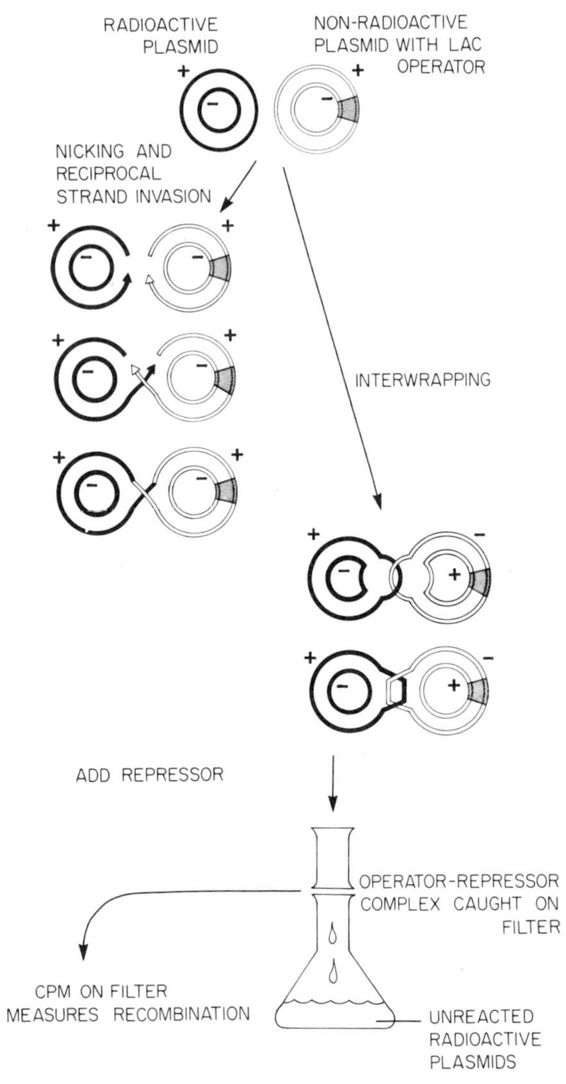

Figure 5. A radio-biochemical assay designed to detect genome fushion in vitro. Two plasmids that are almost entirely homologous are incubated with an *E. coli* cell extract or with a protein fraction derived by column chromatography. The first plasmid is radioactive (■) and the second is a nonradioactive form (□) that carries *lac* operator DNA (stippled area). If the plasmids become fused, for instance, by a nick and reciprocal strand exchange mechanism (discussed in connection with Fig. 10) or by an interwrapping mechanism (discussed in connection with Fig. 13), then both radioactive and *lac* operator DNA will be joined together in the same DNA molecule. Such composite structures can be bound to a nitrocellulose filter by *lac* repressor.

purified and comparatively analyzed, is that the three peaks represent domains of the same fusion enzyme, related to each other by proteolysis. As shown in the inset to Figure 6, it is clear that $RecA^-$ cells share with $RecA^+$ cells a family of genome fusion activities.

Because the DEAE-Sephacel column appeared to resolve the fusion activity into several peaks, it was decided to focus first on one well-defined peak (Fig. 6, fractions 137–139, marked with an arrow) for further purification. The activity in this peak is associated with about 3% of the protein originally loaded on the column.

Molecular sieve chromatography. The genome fusion activity recovered from the ion-exchange column (Fig. 6, arrow) was next purified on the basis of size using molecular sieve chromatography. The column resin chosen was Bio-Gel P-200, which, as shown in Figure 7A, was capable of resolving three marker proteins into distinct peaks: immunoglobulin G (150,000), hemoglobin (63,000), and cytochrome *c* (12,500).

An assay for fusion activity across the molecular sieve column gave the profile shown by the solid curve in Figure 7B. The peak of biological activity was associated with about 5% of the protein originally loaded on the column and eluted at a position corresponding to a molecular weight of less than 60,000.

Affinity chromatography on DNA-cellulose. The third column used for the purification of the fusion activity was DNA-cellulose. As shown in Figure 8, more than 95% of the protein loaded on the column (vertical bars) was recovered in the flow-through, whereas the fusion activity bound to the DNA-cellulose column and eluted at about 120 mM KCl.

At this stage of the purification, polyacrylamide gel analysis of the active fractions showed only two proteins, with molecular weights of 33,000 and 30,000 (Fig. 8, inset). The fusion activity quantitatively copurified with the 33,000-m.w. protein when chromatographed on the final column, phenyl-Sepharose.

Affinity chromatography on phenyl-Sepharose. The last step in the purification of the fusion activity involved chromatography on phenyl-Sepharose. This resin retains proteins on the basis of hydrophobic areas on their surfaces.

The material recovered from the DNA-cellulose column was applied to phenyl-Sepharose, and the bound proteins were eluted with a decreasing salt gradient. The fusion activity was recovered in a single peak (Fig. 9).

At this stage the enzyme had been purified to apparent homogeneity. This was determined by iodination and polyacrylamide gel analysis of the material in the phenyl-Sepharose column fractions. As shown in the inset of Figure 9, a single protein appears for which the amount rises and falls coincident with the peak of biological activity. The monomer molecular weight of the fusion activity is judged to be about 33,000, by comparison with the gel profile of four marker proteins run in parallel.

FUSION POINTS IN THE PRODUCT MULTIMERS

In Figures 6 through 9 the fusion enzyme has been purified on the basis of its ability to join radioactive and operator-containing plasmids. To demonstrate the existence of DNA fusion points in the product multimers and show that they occur at regions of homology, the most highly purified synaptase preparation (from the phenyl-Sepharose column) was assayed with the electron microscope for its ability to form figure eights that could be cleaved into chi forms. With the purified enzyme, the percentage of input monomer circles that

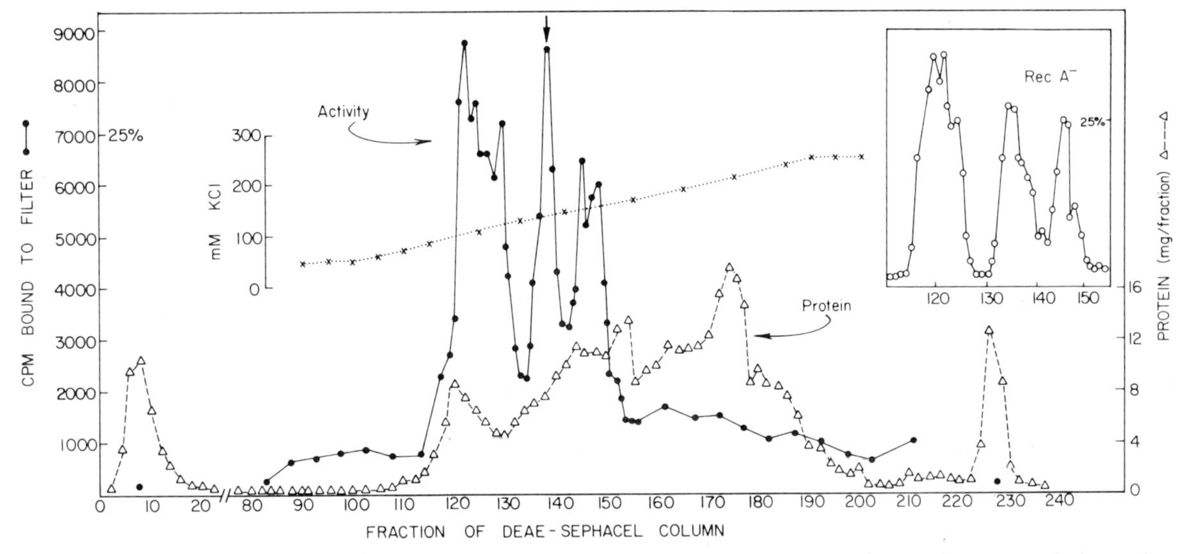

Figure 6. Purification of DNA synaptase by ion-exchange chromatography on DEAE-Sephacel. (●) Genome fusion activity; (△) protein content; (••••) KCl elution gradient; (↓) peak chosen for further purification. The inset shows the fusion activity recovered from an extract of RecA⁻ cells (MM-152) chromatographed on a similar DEAE column. (For experimental details of this and other column purifications, see Potter and Dressler 1980.)

were fused to form multimers rose to 16% over a period of 20 hours (as in Fig. 1), and when an aliquot of the product DNA was cleaved with *Eco*RI, a roughly equivalent number of chi forms was generated (as in Fig. 3).

By measuring the lengths of the arms in different chi forms (Fig. 4B), it was possible to demonstrate that each molecule contained two pairs of equal-length arms and, thus, that (1) fusion had occurred at a region of homology and (2) the fusion points were distributed at numerous sites along the length of the plasmid genome.

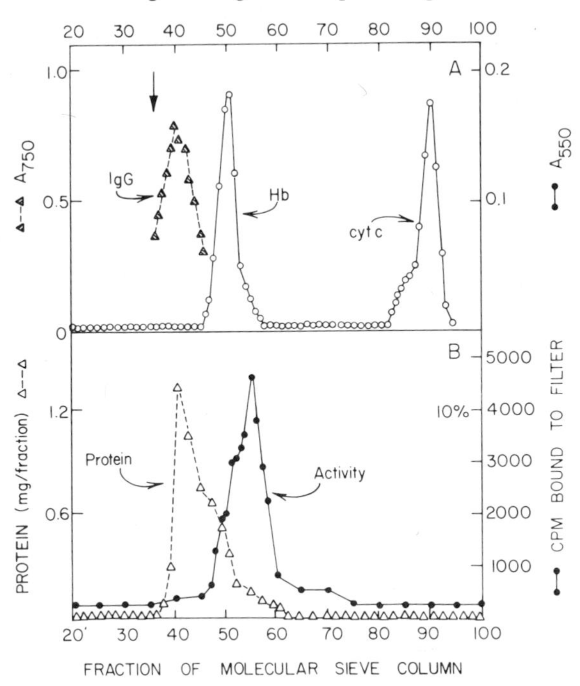

Figure 7. Purification of DNA synaptase by molecular sieve chromatography on Bio-Gel P-200. (*A*) Resolution of three marker proteins into distinct peaks; (*B*) the total protein content and genome fusion activity of each column fraction.

We conclude from the data shown in Figure 4B that the purified synaptase fuses DNA molecules at a region of homology, as was originally observed when the enzyme was present with numerous other proteins in the crude cell extract (Fig. 4A). Both in the crude extract and as a purified protein, this enzyme carries out its genome fusion reaction in the presence of Mg^{++}, spermidine, and β-mercaptoethanol. A high-energy cofactor is not required.

Because the protein that we have found fuses genomes at a region of homology, we have named the enzyme DNA synaptase (from the Greek word συναπτειν, meaning to join together).

GENOME FUSION REACTIONS AND GENETIC RECOMBINATION

Genome fusion reactions, in general, are most likely to be associated with the process of genetic recombina-

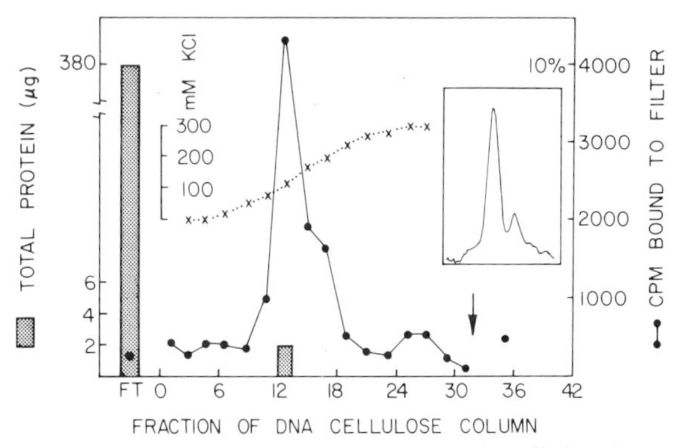

Figure 8. Purification of DNA synaptase by affinity chromatography on DNA cellulose. (▭) Protein content; (●) genome fusion activity. The inset shows a densitometer tracing of the middle area of a polyacrylamide gel of the peak fraction.

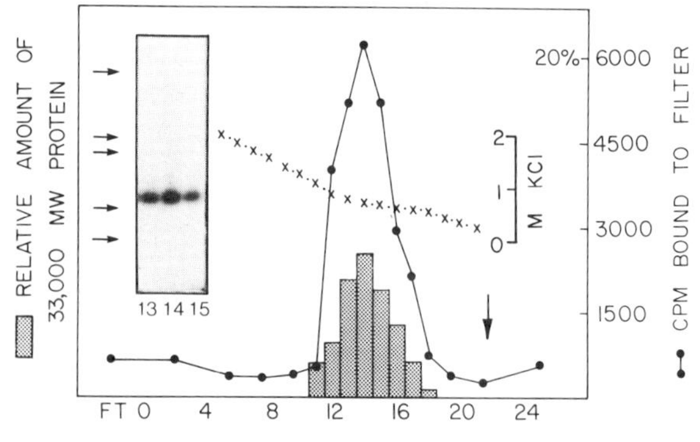

Figure 9. Purification of DNA synaptase by affinity chromatography on phenyl-Sepharose. (●) Genome fusion activity. (*Inset*) SDS-polyacrylamide gel analysis of the protein present in column fractions 13, 14, and 15. The five arrows show the positions of the gel origin and four marker proteins: bovine serum albumin (68,000), ϕX coat protein F (48,000), carbonic anhydrase (29,000), and chymotrypsin (25,000). (▭) Relative amount of the purified protein (33,000), which is seen to rise and fall coincident with the peak of fusion activity. (Reprinted from Potter and Dressler 1980.)

tion. Figures 10, 12, and 13 show three mechanisms for recombination that are representative of current thinking in this field. These models all use a recombination intermediate first proposed by Holliday in 1964 (Holliday 1964, 1968, 1974), but they differ markedly in the nature of the genome fusion reaction used to form the intermediate.

The Holliday Genome Fusion Reaction

Figure 10 shows the prototype genome fusion reaction proposed by Holliday and the recombination event that results. Two homologous double helixes are aligned, and the positive strands (or, alternatively, the negative strands) in each are broken open at a given region. The free ends thus created leave the complementary strands to which they had been hydrogen-bonded and become associated instead with the complementary strands in the homologous double helix (Fig. 10A–D). The result of this reciprocal strand invasion is to establish a tentative physical connection between the two DNA molecules. This linkage can be made stable through a process of DNA repair, which in this case can be as simple as the formation of two phosphodiester bonds by the enzyme ligase (Fig. 10E).

If homologous pairing of the two DNA molecules can occur prior to breakage and reunion, then an enzyme with nicking-closing activity could be used to carry out the genome fusion reaction (McGavin 1971; Kikuchi and Nash 1979; Wilson 1979). In this case, the breakage and reunion events would be accomplished in a concerted reaction, still leading to the structure shown in Figure 10E.

The structure shown in Figure 10E is the Holliday recombination intermediate. Although the intermediate is expected to be stable, it need not be static. A continuing strand transfer by the two polynucleotide chains involved in the crossover can occur, making and breaking an equivalent number of hydrogen bonds and allowing the point of linkage between the two molecules to move laterally along the DNA (Fig. 10E, F). This dynamic property of the Holliday crossover connection (Meselson 1972; Sigal and Alberts 1972; Warner et al. 1979) can lead to the development of regions of heteroduplex DNA during recombination and, when mismatched bases occur, to the formation of areas of double helix that are genetically heterozygous. Such physical and genetic heterozygosity in the immediate region of a crossover is known to be a property of chromosomes emerging from recombination (Kitani et al. 1962; Hurst et al. 1972; Russo 1973; Lam et al. 1974; White and Fox 1974; Enea and Zinder 1976; Fogel et al. 1979; Rossignol et al. 1979).

The remainder of Figure 10 shows the maturation mechanism proposed for the Holliday intermediate. Because of its structural symmetry, it is expected that the intermediate can be processed in either of two related ways to give rise to two different pairs of recombinant chromosomes (Holliday 1964, 1968, 1974; Emerson 1969; Sigal and Alberts 1972). This dual maturation potential is most easily appreciated if one rotates the intermediate into another planar configuration (as shown in Fig. 10I). Then, cleavage on an east-west axis, or on a north-south axis, leads to the release of unit-size DNA molecules in which potentially heterozygous regions exist and in which the flanking genes either remain in their parental linkage or, with equal probability, emerge in a recombinant linkage (Fig. 10L). Genetic distances are measured from the 50% of the cases in which the flanking genes are exchanged. The remaining maturation events are silent from the point of view of traditional recombination—except for the occurrence of the region of heterozygous DNA, which remains as a footprint of the former crossover connection.

Figure 11 shows fused genomes, recovered from intact cells, that have the structure of a Holliday intermediate (Potter and Dressler 1976, 1977a).

Alternative Genome Fusion Reactions

Several other genome fusion reactions, distinct from the one proposed by Holliday (Fig. 10A–E) but leading to the formation of the same intermediate, have also been proposed. Two of these alternatives are shown in Figures 12 and 13.

The Meselson-Radding mechanism. Because genetic studies in fungi have shown that the regions of heterozygous DNA in a finished pair of recombinant chromosomes are not perfectly reciprocal (as in Fig. 10L), Meselson and Radding (1975) were motivated to propose a two-stage initiation mechanism for the formation of the Holliday structure (Fig. 12). They sought to explain the dissimilarity in the heterozygous regions by suggesting that, initially, only one of the two participating double helixes is broken open. After nicking, DNA synthesis would occur on this double helix and lead to the displacement of a single-stranded DNA tail. The single-stranded tail would then invade the second double helix

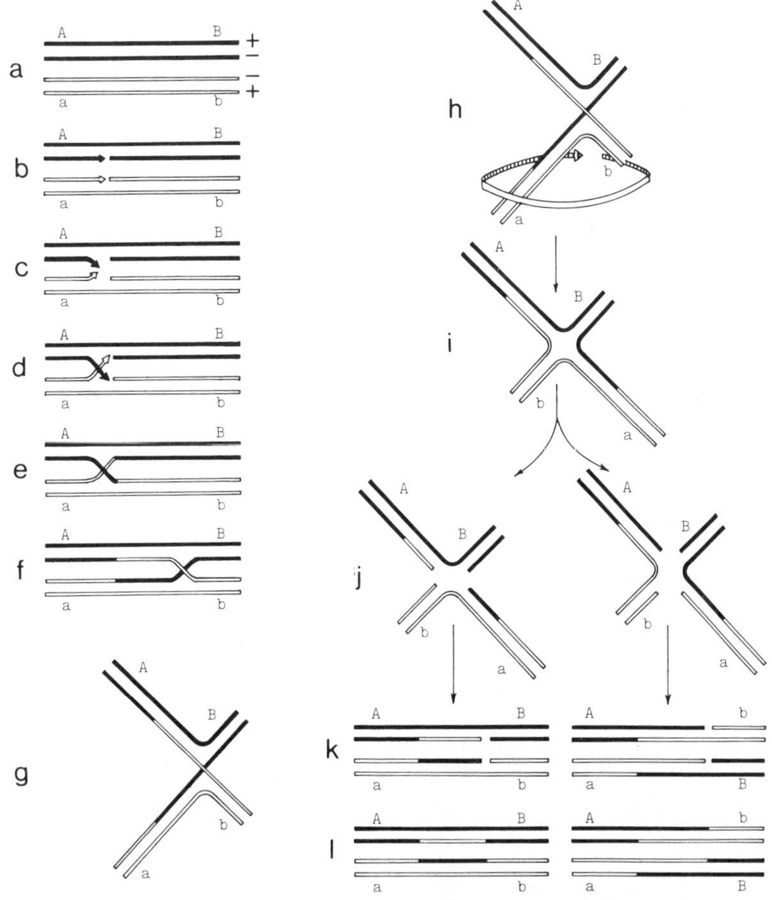

Figure 10. A prototype mechanism for genetic recombination based on a genome fusion reaction suggested by Holliday. See text for further explanation. (Reprinted, with permission, from Potter and Dressler 1976.)

at a region of homology, forming the first heteroduplex region. This process would eventually provoke a reciprocal nicking and strand invasion, which would occur at a slightly different place, giving rise to a second, shorter heteroduplex region (Fig. 12F). After the filling of any resultant gaps and the trimming of any remaining tails, this two-stage initiation mechanism would yield a Holliday intermediate with asymmetric heteroduplex regions at the outset of recombination (Fig. 12H).

The interwrapping mechanism. A final proposal for genome fusion leading to the formation of a Holliday-type structure is shown in Figure 13. In this case, as distinct from the Meselson-Radding mechanism, DNA-pairing precedes strand breakage. Genome fusion occurs when two DNA molecules undergo a localized denaturation at homologous places and the exposed pairs of positive and negative strands become interwrapped (Cross and Lieb 1967; Sobell 1975; Potter and Dressler 1978, 1979a). After the initial contact, strand interwrapping continues, with transient nicks being introduced by a nicking-closing enzyme (Champoux 1977; Kierkegaard and Wang 1978) so as to allow the rotation necessary for helix formation and extended interwrapping. The structure shown in Figure 13D results from and is characterized by two crossover connections of the type proposed by Holliday, one at either end of the wrapped segment (compare Fig. 13D with Fig. 10E).

If the two crossover connections are far enough apart, such a wrapped structure would appear to have an eye-loop in the center or an extended region of pairing. Indeed, when examined in the electron microscope, figure eights and chi forms sometimes have this appearance (see Figs. 2 and 3, and Potter and Dressler 1978, 1979a).

Although the wrapped structure is twofold more complex, its crossover connections are not fundamentally different from those of the simple Holliday structure shown in Figures 10E and 12H. Therefore, we imagine that the interwrapped structure may be matured by the same type of strand-nicking events that have been proposed for the Holliday intermediate. A pair of maturation nicks are introduced independently at each of the crossover connections (Fig. 13G–I). After one crossover is matured, a structure with a single crossover connection results (compare Fig. 13H with Fig. 10I). Maturation at the second crossover finally separates the two recombinant genomes. The finished molecules will each have a region of potentially heterozygous DNA in the area of the crossover, which, as discussed earlier, is known from genetic studies to be a property of recombinant chromosomes, and, depending on the axis of cutting at the two crossovers, there will be either a retention of the flanking genes in their parental linkage or, with equal probability, the production of a recombinant linkage for the flanking genes (again in accord with genetic findings about recombinant chromosomes). Asymmetry in the regions of heterozygous DNA would arise from differential nick translation at the time of maturation, as in Figure 13J (Potter and Dressler 1978, 1979a).

Recombinational Repair

Although genetic recombination is the most apparent cellular process that is expected to involve genome fusion, it is not the only such process. Another would be the recombinational repair of DNA (for reviews, see Rupp et al. 1971; Clark and Volkert 1978). Each of the fused structures we have discussed could also function as an intermediate in a special type of DNA repair particularly useful for correcting errors in regions of single-stranded DNA (such as occur in replication forks) or for correcting nearly opposite lesions in damaged double-helical DNA. In both of these situations, ordinary excision-repair would not be possible, whereas recombinational pairing would allow homologous DNA sequences to intertwine temporarily and serve as correcting templates for damage in complementary strands. In this context, one can speculate that the process of recombination as it is observed today grew out of a preexisting and more directly life sustaining DNA repair process. Thus, there may be a close relationship between the enzymology of the apparently distinct processes of DNA repair and DNA recombination.

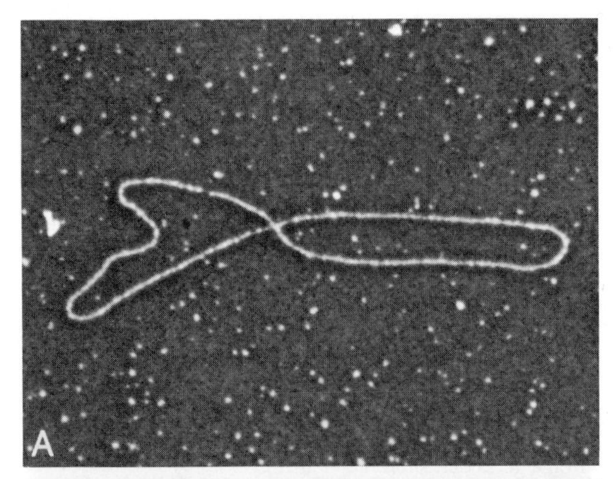

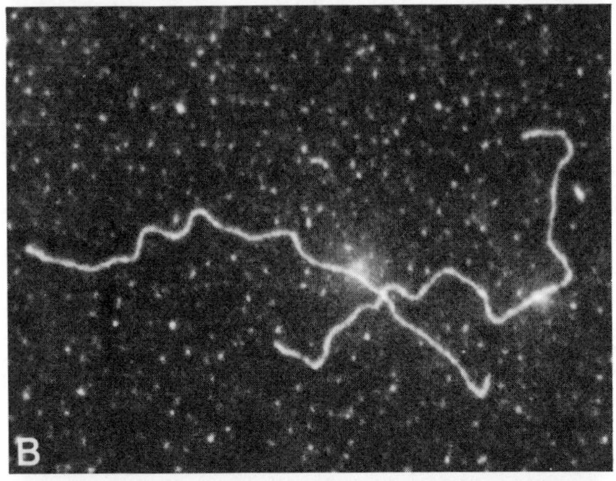

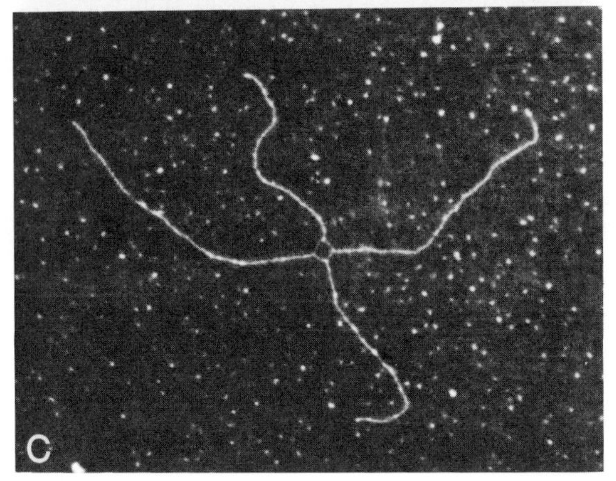

Figure 11. Electron micrographs of fused DNA molecules recovered from intact cells. (*A*) In a number of studies involving viruses and plasmids, dimer-size DNA forms shaped like figure eights have been observed among the DNA molecules recovered from intact cells (Doniger et al. 1973; Benbow et al. 1974, 1975; Thompson et al. 1975; Valenzuela and Inman 1975; Potter and Dressler 1976, 1977a,b); an in vivo plasmid figure-eight molecule is shown here. (*B*) Such molecules can be linearized with *Eco*RI to obtain chi forms, showing the existence of DNA fusion points in these molecules (Potter and Dressler 1976; Thompson et al. 1976). (*C*) With respect to the plasmid chi forms recovered from intact cells, it has proved possible in about 100 instances to observe the nature of the polynucleotide strand connections in the region of the crossover (Potter and Dressler 1976). In these instances, the crossover region became partially denatured during spreading for the electron microscope, and the individual strands connecting the two DNA molecules could be seen crossing over from one genome to the other. The connecting strands always form a characteristic quadriradial structure, with the homologous (equal-length) arms oriented in a *trans* configuration, as shown in the electron micrograph. The chi forms in *B* and *C* are interpreted as representing the two planar configurations of the Holliday intermediate (Fig. 10E,I). We have observed similar figure eights and chi forms (including open chi forms, as in *C*) among the SV40 DNA molecules recovered from monkey cells infected in culture. (Reprinted, with permission, from Potter and Dressler 1976, 1977a.)

THE REACTION CATALYZED BY DNA SYNAPTASE

What kind of a reaction does DNA synaptase catalyze, and how does this reaction relate to the proposed mechanisms for genome fusion just discussed?

One characteristic that distinguishes the interwrapping model in Figure 13 from the strand assimilation models in Figures 10 and 12 is the obligatory association of a nicking-closing activity with the genome fusion reaction. We have tested the first synaptase (which is purified) for its ability to function as a nicking-closing enzyme, or topoisomerase. We find that while the DNA synaptase is carrying out its genome fusion reaction, there is no evidence for a nicking-closing activity. If nicked and supercoiled DNA molecules are incubated

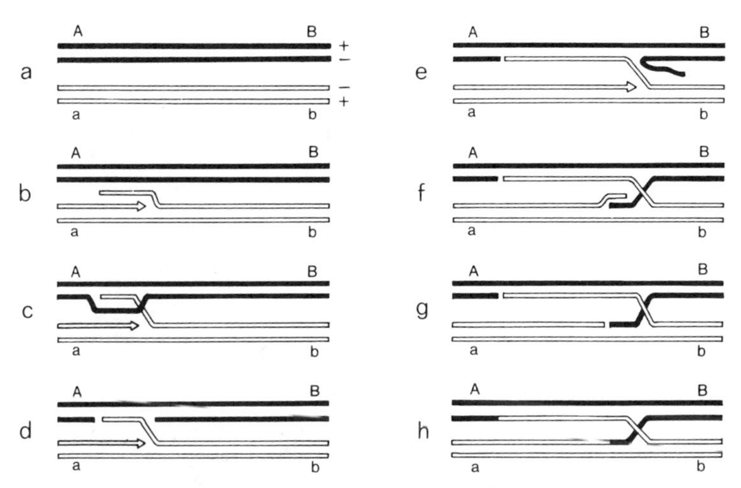

Figure 12. The Meselson-Radding initiation mechanism that couples DNA synthesis to the formation of the Holliday recombination intermediate. See text for further explanation. (Reprinted, with permission, from Potter and Dressler 1979a).

together in a reaction mixture, the enzyme fuses the nicked molecules into multimeric structures, but the supercoils, to a first approximation, are neither fused nor relaxed. It appears, therefore, that under the reaction conditions used, the synaptase does not display a nicking-closing activity, and that, instead, a nick or a free end in the reacting DNA molecule is required by the purified enzyme.

The data of Figure 14 demonstrate the stimulatory effect of nicks on the activity of synaptase. The upper curve shows the rate of fusion in a reaction mixture containing nicked duplex rings; the lower curve shows the nonreactivity of a parallel incubation mixture containing covalently closed DNA circles. Evidently, a nick in one, and perhaps both, participating DNA molecules is essential for genome fusion. It will be im-

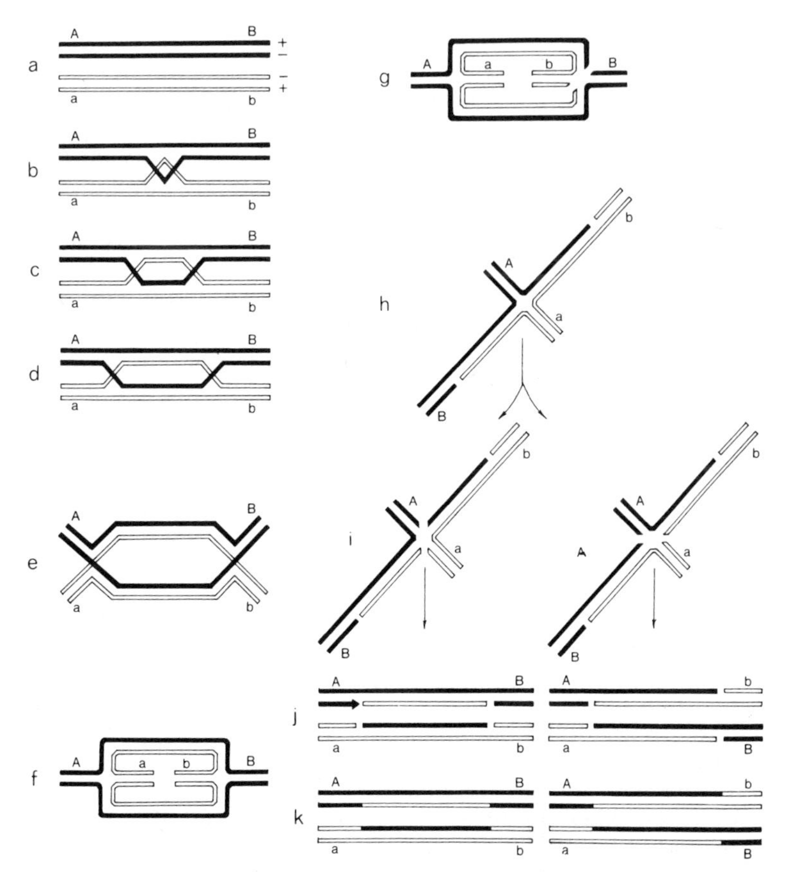

Figure 13. A prototype model for DNA recombination in which the recombination intermediate is formed by the interwrapping of complementary DNA strands after localized denaturation of two double helixes at a region of homology. DNA synthesis is not involved in this initiation. (Reprinted, with permission, from Potter and Dressler 1978, 1979a).

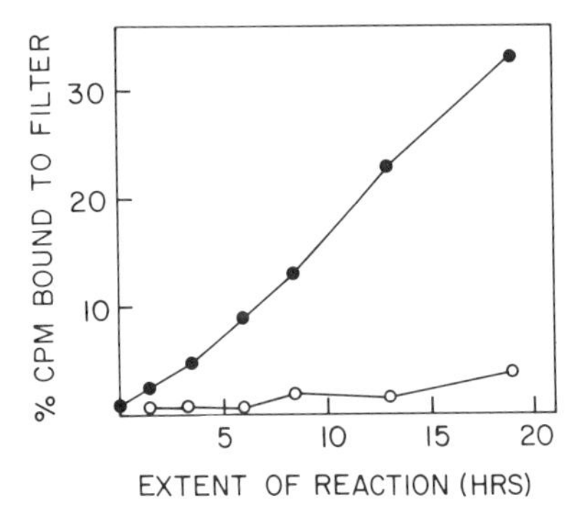

Figure 14. Covalently closed DNA circles are not efficient substrates for the purified DNA synaptase (○). In contrast, the introduction of a limited number of nicks into the same DNA preparation (as by X-irradiation or pancreatic DNase) converts the substrates to an active form (●). We do not yet know whether fusion occurs at the position of the nick or whether the nick serves solely as a swivel. Flush-ended linear plasmid DNA molecules, prepared by the cleavage of monomer plasmid rings with the restriction enzyme *Hpa*II, are also substrates for the fusion reaction. In this case, the reaction products appear directly as chi forms (Potter and Dressler 1979a). This result demonstrates that circularity and supercoiling are not essential for the fusion reaction.

portant to learn whether the genome fusion reaction catalyzed by the purified synaptase occurs at the site of the nick (implying that a free end is required) or at a distant location (indicating that the nick serves only in a swiveling capacity).

Strand assimilation. Given the apparent absence of a nicking-closing activity, the data so far suggest that DNA synaptase may operate in a manner that is more closely approximated by the genome fusion reactions shown in Figures 10 and 12. Both of these models propose a DNA fusion reaction that is characterized by the initial assimilation of a single strand into a recipient double helix. In fact, the synaptase is able to carry out just such a strand assimilation reaction (Potter and Dressler 1980).

In addition to fusing nicked duplex rings (as discussed above), the purified enzyme will also fuse a piece of single-stranded DNA to a homologous duplex molecule. Figure 15 shows the result of incubating the purified synaptase with (nicked) circular duplex plasmid molecules and linear positive strands (derived from the plasmid after *Eco*RI digestion). As seen in Figure 15A, during incubation with the synaptase, the single strands become partially assimilated into the duplex rings. Cleavage of the fusion products with *Eco*RI linearizes the duplex ring element and shows that the single strand has become associated at a region of DNA homology (Fig. 15B).

This reaction most clearly demonstrates the strand assimilation properties of DNA synaptase and suggests that during the fusion of two nicked duplex rings, a linear single-strand is withdrawn from one duplex molecule and inserted into the second duplex molecule to establish the homologous connection.

In future experiments, using the purified enzyme and

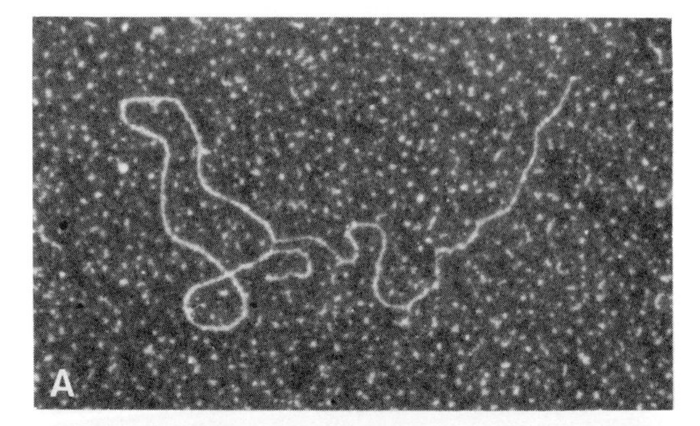

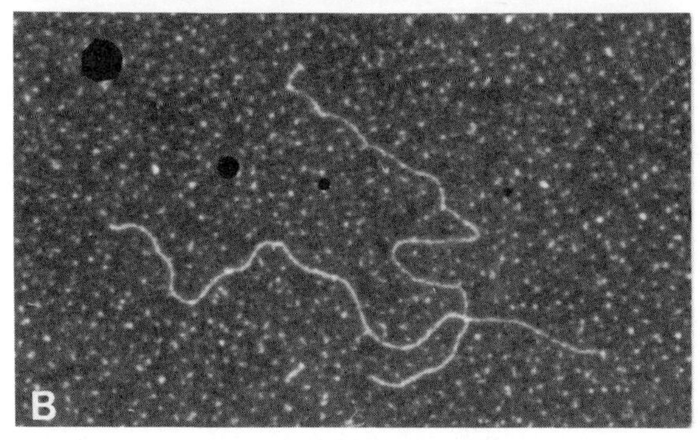

Figure 15. (*A*) The purified DNA synaptase has partially assimilated a unit-length positive strand into a plasmid duplex ring (the unit-length positive strand was derived from a plasmid after *Eco*RI digestion). (*B*) That the positive-strand rod is connected to the duplex ring at a region of homology is demonstrated by the fact that upon *Eco*RI digestion of the composite structure, the duplex ring is linearized and a chi-shaped structure results. (Reprinted from Potter and Dressler 1980.)

a variety of DNA substrates containing precisely located nicks, it should be possible to determine the exact biochemical mechanism by which the enzyme promotes genome fusion. Furthermore, it can be expected that purification of the other two genome fusion activities detected upon DEAE chromatography may reveal other synaptase enzymes that differ either with respect to the DNA substrate they use or in the nature of the fusion reaction they catalyze.

RELATION TO RECOMBINATION GENES

As yet, we do not know which *E. coli* gene is responsible for DNA synaptase. As will be discussed below, synaptase is not the product of the *recA* gene; nor is the synaptase the product of the *recB* and *recC* genes (Goldmark and Linn 1972): The protein does not have a molecular weight of 270,000 and does not function as a nuclease.

However, DNA synaptase could be the product of the *recF* gene, which has been cloned and, by indirect analysis, estimated to have a molecular weight of about 30,000 (Ream et al. 1980). Little is known about the biochemical function of *recF*. It is not essential for recombination in male-female matings (Clark 1973) but is stimulatory when recombination is assayed in another way—that is, by the formation of dimer DNA circles

(Potter and Dressler 1977a, 1979a). Potentially, *recF* acts at an early stage in recombination, when the process begins with intact duplex DNA molecules, but is not required in male-female matings where a linear single-strand enters the female cell to begin the recombination event.

THE *recA* PROTEIN

Aside from DNA synaptase, one other genome fusion enzyme has thus far been isolated from *E. coli;* this is the product of the *recA* gene (*recA* protein).

It is known from genetic studies that the *recA* gene is highly pleiotropic. Mutations in this locus affect (1) genetic recombination, (2) sensitivity to ultraviolet light, (3) inducibility of the prophage λ, and (4) ability to support error-prone DNA repair (Clark and Margulies 1965; Howard-Flanders and Theriot 1966; Witkin 1966; Brooks and Clark 1967).

Normally, the *recA* protein is present in small amounts in actively growing cells, but it becomes a major protein if the cells are exposed to DNA-damaging agents such as ultraviolet light or nalidixic acid (Gudas and Pardee 1975; Emerson and West 1977; Gudas and Mount 1977; McEntee 1977). This inducibility has been a great aid in the purification of the protein. Beginning with cells in which the *recA*-gene product constitutes 5–10% of the cellular protein, several effective purification protocols have been developed (Roberts et al. 1978; Weinstock et al. 1979; West et al. 1980). The *recA* protein has been obtained in milligram amounts and has been shown to be a monomer with a molecular weight of approximately 40,000 that can exist in multimeric forms.

The first enzymatic activity to be associated with the *recA* protein was discovered by Roberts and Roberts (1975). They demonstrated the ability of purified *recA* protein to inactivate the bacteriophage λ repressor by cleaving the repressor in half (Roberts et al. 1978). This result strengthened an earlier suggestion that the *recA* protein might exert its pleiotropic effects by functioning as a regulatory protein (Witkin 1966), in this case by using its protease activity to activate proenzymes or to inactivate the repressors for various genes, including those involved in recombination and repair.

More recent results, however, have suggested ways in which the *recA* protein could participate directly in such processes as recombination. For instance, Shibata et al. (1979), McEntee et al. (1979), and West et al. (1980) have shown that stoichiometric amounts of *recA* protein allow single-stranded DNA fragments to be taken up locally into homologous duplex DNA to form triple-stranded structures. The final structure formed is similar to that shown in Figure 15, and thus, the synaptase and the *recA* protein show a certain operational similarity.

A current model for the action of the *recA* protein assumes that the protein binds to single-stranded DNA and then denatures an area of target duplex DNA to form a nonspecific triple-stranded complex. This process would occur reiteratively, powered by the hydrolysis of ATP, until the single strand finds a region of homology in the duplex DNA and forms a stable triple-stranded structure (D loop). This could be the first step in a recombination process; but, given the pleiotropic nature of the *recA* gene, this reaction could also be a step in DNA repair, for instance, recombinational repair.

A striking property of the *recA*-mediated reaction is that the amount of purified *recA* protein needed to catalyze the strand-annealing reaction effectively is very high—1 monomer with a molecular weight of approximately 40,000 for every 100 bases of single-stranded DNA (Shibata et al. 1979; Weinstock et al. 1979). In more recent experiments, 1 monomer for every 5 bases of DNA has been found to be optimal (Cunningham et al. 1979).

COMPARISON OF DNA SYNAPTASE AND THE *recA* PROTEIN

In that they both can promote the assimilation of a single strand into a duplex DNA molecule, DNA synaptase and the *recA* protein appear to be somewhat similar in terms of function. However, the proteins can be distinguished by several criteria:

1. The two fusion proteins are structurally and genetically distinct. The synaptase can be recovered from $RecA^-$ cells (Fig. 6, inset) and has a monomer molecular weight of 33,000 (Fig. 10). The molecular weight of *recA* is approximately 40,000.
2. DNA synaptase can fuse fully double-stranded DNA, whereas the *recA* protein is inert on this substrate.
3. The synaptase appears to function catalytically, whereas the *recA* protein is needed in stoichiometric amounts.
4. Finally, the lack of an ATP requirement by the synaptase serves to distinguish it from the *recA* protein.

As a purified protein, neither the synaptase nor the *recA*-gene product functions in precisely the manner predicted by any of the prototype recombination models—nor will either protein carry out a full recombination event. But both proteins show activities compatible with an early step in recombination, and, of course, within the intact cell, they could be parts of larger and more complex holoenzymes.

Given the early stage of knowledge about enzymes that catalyze DNA fusion reactions, discussions about their exact role in recombination must necessarily be speculative. One may expect, however, that the continuing development of in vitro systems that carry out recombination-type reactions will open the way to a deeper understanding of the enzymology of DNA recombination.

ACKNOWLEDGMENTS

The work of our laboratory is made possible by research grants from the National Institutes of Health

(GM-17088) and the National Science Foundation (PCM-7912315) and by the Dreyfus Teacher-Scholar Fund. H. P. is a Fellow of the American Cancer Society.

REFERENCES

Benbow, R., A. Zuccarelli, A. Shafer, and P. Sinsheimer. 1974. Exchange of parental DNA during genetic recombination in bacteriophage φX174. In *Mechanisms in recombination* (ed. R. Grell), p. 3. Plenum Press, New York.

———. 1975. Recombinant DNA molecules of bacteriophage φX174. *Proc. Natl. Acad. Sci.* **72:** 235.

Brooks, K. and A. J. Clark. 1967. Behavior of λ bacteriophage in a recombination deficient strain of *Escherichia coli* K12. *J. Virol.* **1:** 283.

Champoux, J. 1977. Renaturation of complementary single-stranded DNA circles: Complete rewinding facilitated by the DNA untwisting enzyme. *Proc. Natl. Acad. Sci.* **74:** 5328.

Clark, A. J. 1973. Recombination deficient mutants of *E. coli* and other bacteria. *Annu. Rev. Genet.* **7:** 67.

Clark, A. J. and A. D. Margulies. 1965. Isolation and characterization of recombination-deficient mutants of *Escherichia coli* K12. *Proc. Natl. Acad. Sci.* **53:** 451.

Clark, A. J. and M. Volkert. 1978. A new classification of pathways repairing dimer damage in DNA. In *DNA repair mechanisms* (ed. P. Hanawalt et al.), p. 57. Academic Press, New York.

Cross, R. and M. Lieb. 1967. Heat-inducible lambda phage. V. Induction of prophages with mutations in genes O, P and R. *Genetics* **57:** 549.

Cunningham, R., T. Shibata, C. DasGupta, and C. Radding. 1979. Homologous pairing in genetic recombination: Single strands induce RecA protein to unwind duplex DNA. *Nature* **281:** 191.

Doniger, J., R. Warner, and I. Tessman. 1973. Role of circular dimer DNA in the primary recombination mechanism of bacteriophage S-13. *Nat. New Biol.* **242:** 9.

Emerson, P. T. and S. C. West. 1977. Identification of protein X of *Escherichia coli* as the recA$^+$/tif$^+$ gene product. *Mol. Gen. Genet.* **155:** 77.

Emerson, S. 1969. Linkage and recombination at the chromosome level. In *Genetic organization; a comprehensive treatise* (ed. E. Caspari and A. Ravin), vol. 1, p. 267. Academic Press, New York.

Enea, V. and N. Zinder. 1976. Heteroduplex DNA: A recombinational intermediate in bacteriophage f1. *J. Mol. Biol.* **101:** 25.

Fogel, S., R. Mortimer, K. Lusnak, and F. Tavares. 1979. Meiotic gene conversion: A signal of the basic recombination event in yeast. *Cold Spring Harbor Symp. Quant. Biol.* **43:** 1325.

Goldmark, P. J. and S. Linn. 1972. Purification and properties of the Rec BC DNase of *Escherichia coli* K12. *J. Biol. Chem.* **247:** 1849.

Gudas, L. and D. Mount. 1977. Identification of the Rec-A (tif) gene product of *Escherichia coli. Proc. Natl. Acad. Sci.* **74:** 5280.

Gudas, L. and A. B. Pardee. 1975. Model for regulation of *Escherichia coli* DNA repair functions. *Proc. Natl. Acad. Sci.* **72:** 2330.

Holliday, R. 1964. A mechanism for gene conversion in fungi. *Genet. Res.* **5:** 282.

———. 1968. Genetic recombination in fungi. In *Replication and recombination of genetic material* (ed. W. Peacock and R. Brock), p. 157. Australian Academy of Science, Canberra.

———. 1974. Molecular aspects of genetic exchange and gene conversion. *Genetics* **78:** 273.

Howard-Flanders, P. and L. Theriot. 1966. Mutants of *E. coli* defective in DNA repair and in genetic recombination. *Genetics* **53:** 1137.

Hurst, D., S. Fogel, and R. Mortimer. 1972. Conversion-associated recombination in yeast. *Proc. Natl. Acad. Sci.* **69:** 101.

Kierkegaard, K. and J. Wang. 1978. *Escherichia coli* DNA topoisomerase I catalyzed linking of single-stranded rings of complementary base sequences. *Nucleic Acids Res.* **5:** 3811.

Kikuchi, Y. and H. Nash. 1979. Nicking-closing activity associated with bacteriophage λ *int* gene product. *Proc. Natl. Acad. Sci.* **76:** 3760.

Kitani, Y., L. Olive, and A. El-ani. 1962. Genetics of *Sordaria fimicola.* V. Aberrant segregation at the G. locus. *Am. J. Bot.* **49:** 697.

Lam, S., M. Stahl, K. McMilin, and F. Stahl. 1974. Rec-mediated recombinational hot spot activity in bacteriophage lambda. *Genetics* **77:** 425.

McEntee, K. 1977. Protein X is the product of the *recA* gene of *Escherichia coli. Proc. Natl. Acad. Sci.* **74:** 5275.

McEntee, K., G. Weinstock, and R. Lehman. 1979. Initiation of general recombination catalyzed *in vitro* by the *recA* protein of *Escherichia coli. Proc. Natl. Acad. Sci.* **76:** 2615.

McGavin, S. 1971. Models of specifically paired like (homologous) nucleic acid structures. *J. Mol. Biol.* **55:** 293.

Meselson, M. 1972. Formation of hybrid DNA by rotary diffusion during genetic recombination. *J. Mol. Biol.* **71:** 795.

Meselson, M. and C. Radding. 1975. A general model for genetic recombination. *Proc. Natl. Acad. Sci.* **72:** 358.

Potter, H. and D. Dressler. 1976. On the mechanism of genetic recombination: Electron microscopic observation of recombination intermediates. *Proc. Natl. Acad. Sci.* **73:** 3000.

———. 1977a. On the mechanism of genetic recombination: The maturation of recombination intermediates. *Proc. Natl. Acad. Sci.* **74:** 4168.

———. 1977b. An electron microscope study of actively recombining plasmid DNA molecules. In *DNA insertion elements, plasmids, and episomes* (ed. A. I. Bukhari et al.), p. 409. Cold Spring Harbor Laboratory, Cold Spring Harbor, New York.

———. 1978. An in vitro system from *E. coli* that catalyzes generalized genetic recombination. *Proc. Natl. Acad. Sci.* **75:** 3698.

———. 1979a. DNA recombination: In vivo and in vitro studies. *Cold Spring Harbor Symp. Quant. Biol.* **43:** 969.

———. 1979b. A biochemical assay designed to detect recombination intermediates formed in vitro. *Proc. Natl. Acad. Sci.* **76:** 1084.

———. 1980. DNA synaptase: An enzyme that fuses DNA molecules at a region of homology. *Proc. Natl. Acad. Sci.* **77:** 2390.

Ream, L. W., L. Margossian, A. J. Clark, F. G. Hansen, and K. V. Meyenburg. 1980. Genetic and physical mapping of *recF* in *Escherichia coli* K12. *Mol. Gen. Genet.* **180:** 115.

Roberts, J. and C. Roberts. 1975. Proteolytic cleavage of bacteriophage lambda repressor in induction. *Proc. Natl. Acad. Sci.* **72:** 147.

Roberts, J. and C. Roberts, and N. Craig. 1978. *Escherichia coli recA* gene product inactivates phage λ repressor. *Proc. Natl. Acad. Sci.* **75:** 4714.

Rossignol, J. L., N. Paquette, and A. Nicolas. 1979. Aberrant 4:4 asci, disparity in the direction of conversion, and frequencies of conversion in *Ascobolus immersus. Cold Spring Harbor Symp. Quant Biol.* **43:** 1343.

Rupp, W. D., C. Wilde, D. Reno, and P. Howard-Flanders. 1971. Exchanges between DNA strands in ultraviolet-irradiated *Escherichia coli. J. Mol. Biol.* **61:** 24.

Russo, V. 1973. On the physical structure of lambda recombinant DNA. *Mol. Gen. Genet.* **122:** 353.

SHIBATA, T., C. DASGUPTA, R. CUNNINGHAM, and C. RADDING. 1979. Homologous pairing in genetic recombination: Complexes of RecA protein and DNA. *Proc. Natl. Acad. Sci.* **76:** 1638.

SIGAL, N. and B. ALBERTS. 1972. Genetic recombination: The nature of a crossed strand-exchange between two homologous DNA molecules. *J. Mol. Biol.* **71:** 789.

SOBELL, H. 1975. A mechanism to activate branch migration between homologous DNA molecules in genetic recombination. *Proc. Natl. Acad. Sci.* **72:** 279.

THOMPSON, B., M. CAMIEN, and R. WARNER. 1976. Kinetics of branch migration in double-stranded DNA. *Proc. Natl. Acad. Sci.* **73:** 2299.

THOMPSON, B., C. ESCARMIS, B. PARKER, W. SLATER, J. DONIGER, I. TESSMAN, and R. WARNER. 1975. Figure-8 configuration of dimers of S-13 and φX174 replicative form DNA. *J. Mol. Biol.* **91:** 409.

VALENZUELA, M. and R. INMAN. 1975. Visualization of a novel junction in bacteriophage lambda DNA. *Proc. Natl. Acad. Sci.* **72:** 3024.

WARNER, R., R. FISHEL, and F. WHEELER. 1979. Branch migration in recombination. *Cold Spring Harbor Symp. Quant. Biol.* **43:** 957.

WEINSTOCK, G. M., K. MCENTEE, and I. R. LEHMAN. 1979. ATP-dependent renaturation of DNA catalyzed by the recA protein of *Escherichia coli. Proc. Natl. Acad. Sci.* **76:** 126.

WEST, S., E. CASSUTO, J. MURSALIM, and P. HOWARD-FLANDERS. 1980. Recognition of duplex DNA containing single-stranded regions by RecA protein. *Proc. Natl. Acad. Sci.* **77:** 2569.

WILSON, J. 1979. Nick-free formation of reciprocal heteroduplexes: A simple solution to the topological problem. *Proc. Natl. Acad. Sci.* **76:** 3671.

WHITE, R. and M. FOX. 1974. On the molecular basis of high negative interference. *Proc. Natl. Acad. Sci.* **71:** 1544.

WITKIN, E. 1966. Ultraviolet mutagenesis and inducible DNA repair in *Escherichia coli. Bacteriol. Rev.* **40:** 869.

WOLFSON, J., D. DRESSLER, and M. MAGAZIN. 1972. Bacteriophage T7 DNA replication: A linear replicating intermediate. *Proc. Natl. Acad. Sci.* **69:** 499.

Kinetics and Topology of Homologous Pairing Promoted by *Escherichia coli recA*-gene Protein

C. M. RADDING, T. SHIBATA,* C. DASGUPTA, R. P. CUNNINGHAM, AND L. OSBER

Departments of Human Genetics and of Molecular Biophysics and Biochemistry, Yale University School of Medicine, New Haven, Connecticut 06510

The study of genetic recombination in vitro has lagged behind that of replication by many years, in large part because of a lack of suitable assays. In recent years, however, assays have been devised to study site-specific recombination (Gottesman and Gottesman 1975; Mizuuchi et al. 1978) and general or homologous recombination (Beattie et al. 1977; Potter and Dressler 1979; Roeder and Sadowski 1979; Kolodner 1980). To study the interaction of single-stranded DNA with homologous double-stranded DNA, a reaction that occurs nonenzymatically when the duplex DNA is superhelical (Holloman et al. 1975; Liu and Wang 1975) and that produces a structure called a D loop (Wiegand et al. 1977), we devised a rapid assay (Beattie et al. 1977). Based on the properties of nitrocellulose filters, this assay measures the attachment, by whatever means, of duplex DNA to single-stranded DNA. Later, using the same assay, we (Shibata et al. 1979b) and McEntee et al. (1979) found that purified *Escherichia coli recA* protein (*recA*-gene product) catalyzes the formation of D loops. This was a possibility that we had inferred from several lines of evidence, including: marker rescue promoted by the *recA* gene and superhelical DNA (Holloman and Radding 1976), implication of *recA* protein directly in recombination (Kobayashi and Ikeda 1978), and DNA-dependent ATPase activity of purified *recA* protein (Ogawa et al. 1979; Roberts et al. 1979). The D-loop assay has proved particularly suitable for studying homologous pairing promoted by *recA* protein, because single-stranded DNA plays a central role in this reaction and because the assay can detect a variety of joint molecules that are made by *recA* protein, as indicated in this paper.

HOMOLOGOUS PAIRING PROMOTED BY THE *RECA* PROTEIN

The *recA* protein catalyzes the rapid uptake of homologous single-stranded fragments by duplex DNA (Fig. 1) thereby forming a D loop, the properties of which are identical to those of D loops formed by the reaction of superhelical DNA with homologous fragments (Shibata et al. 1979b and unpubl.). In addition to homologous DNA and *recA* protein, D-loop formation requires ATP and Mg^{++} (Fig. 1). Spermidine reduces the concentration of Mg^{++} required but does not supplant it completely (T. Shibata et al., unpubl.). Unlike the formation of D loops by superhelical DNA and homologous single-stranded fragments that occurs at elevated temperatures (Beattie et al. 1977), the formation of D loops when catalyzed by *recA* protein does not require superhelical DNA (Fig. 1; McEntee et al. 1979; Cunningham et al. 1979).

Path of the Reaction

Observations on the formation of D loops from single-stranded fragments and circular duplex DNA have shown that single strands play a key role in initiating the pairing reaction (see Fig. 2). The amount of single-stranded DNA determines the stoichiometric requirement for *recA* protein (Shibata et al. 1979a; McEntee et al. 1980). Moreover, single-stranded DNA, whether homologous or not, as well as oligodeoxynucleotides, stimulates *recA* protein to bind to and partially unwind duplex DNA (Fig. 2; Cunningham et al. 1979) in the presence of adenosine 5′-0-(3-thiotriphosphate) ($ATP_{\gamma}S$), an analog of ATP (Shibata et al. 1979a; McEntee et al. 1980). These observations led us to postulate that *recA* protein first brings single-stranded and double-stranded DNA into proximity and then partially unwinds the duplex DNA in order to search for homology (Cunningham et al. 1979; Shibata et al. 1979a). This hypothesis is clarified by considering the alternative, namely, that the binding of *recA* protein to the DNA exposes the bases of both single-stranded and double-stranded DNA independently, following which homologous sequences pair in a second-order reaction. Our observations on the inhibition of the ATPase activity of *recA* protein by $ATP_{\gamma}S$, which indicated that the analog binds at the same site as ATP (Shibata et al. 1979a), made it reasonable to suppose that the reactions observed in the presence of $ATP_{\gamma}S$ represent partial steps in the overall reaction. Making this assumption, we postulated that ternary complexes consisting of *recA* protein, single-stranded DNA, and double-stranded DNA are intermediates in the reaction. Subsequent experiments have shown directly that ternary complexes are formed in the presence of ATP (Radding et al. 1980) and are intermediates in the formation of D loops (see below).

Kinetics of the Pairing Reaction

To determine whether the formation of D loops via the action of *recA* protein occurs by a process akin to renaturation or by a process more like a classic enzy-

* Visiting Fellow from The Institute of Physical and Chemical Research, Saitama, Japan.

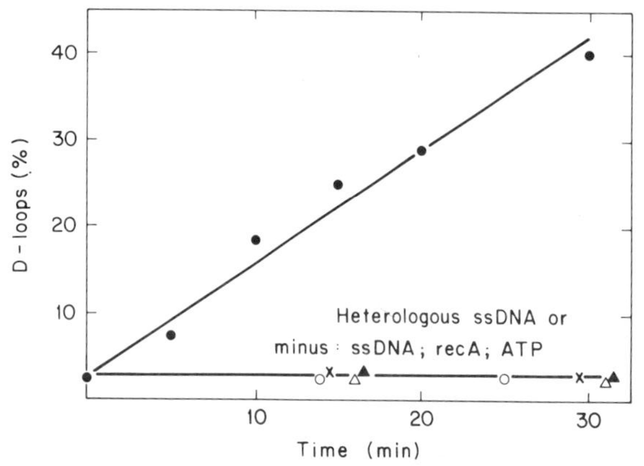

Figure 1. D loops were formed and assayed as described elsewhere (Shibata et al. 1980), except that ATP was the last addition to the reaction mixture. The DNA for these experiments was ϕX174 form-II DNA and fragments of single-stranded DNA from phage ϕX or fd. We use the following designations for the different forms of double-stranded DNA from the phages ϕX174 and fd: form I, superhelical DNA; form II, circular duplex DNA with one or more interruptions in either strand; form III, linear DNA.

matic reaction, we designed experiments to study the order of the reaction. The experiment cannot be done in precisely the classic fashion because stoichiometric amounts of *recA* protein are required, and there is an optimal ratio of *recA* protein to single-stranded DNA. These relationships are described in detail elsewhere (Shibata et al. 1979a, and in prep.). Holding the concentration of duplex form-II DNA constant at one of three different levels, we simultaneously increased the amount of single-stranded DNA and *recA* protein (s), at their optimal ratio, and measured the initial velocity (v_o) (Fig. 3). The relationship between v_o and $[s]$ closely resembled Michaelis-Menten kinetics; the rate approached a limit asymptotically as a function of the concentration of one of the reactants. The reaction remained first-order with regard to duplex DNA, which argues against an inhibitory effect of s, but became zero-order with regard to $[s]$. In other experiments, we observed similar kinetics when we held the concentration of *recA* protein constant, at a level that was in excess, and varied the amount of single-stranded DNA (T. Shibata et al., in prep.).

From the data in Figure 3 and from previous measurements of the spontaneous rate of formation of D loops at 37 °C (Beattie et al. 1977), we have estimated that *recA* protein accelerates the reaction more than 6000 times.

Renaturation, the pairing of complementary single strands, requires the unfolding of the single strands either by heat (Wetmur and Davidson 1968; Studier 1969) or by helix-destabilizing proteins (Alberts and Frey 1970; Christiansen and Baldwin 1977), both of which break intrastrand base pairs and thus allow the more stable interstrand pairing. Uptake of a third strand by duplex DNA would also seem to require the unfolding of a donor single strand (Beattie et al. 1977; McEntee et al. 1980; Shibata et al. 1980), as well as the partial unwinding of the recipient duplex molecule (Beattie et al. 1977). In addition, this reaction requires some energetic or kinetic barrier to keep the third strand in place (Radding et al. 1977). At an appropriate temperature

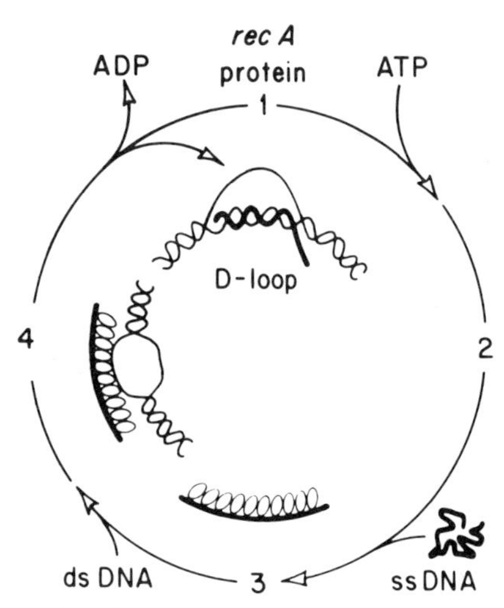

Figure 2. Model for the promotion of homologous pairing by *recA* protein (Cunningham et al. 1979; Shibata et al. 1979a). The binding of ATP promotes the binding of single strands by stoichiometric amounts of *recA* protein. Single strands stimulate *recA* protein to bind and partially unwind duplex DNA prior to homologous pairing. Alignment of homologous sequences and uptake of a third strand are accompanied by hydrolysis of ATP. The numbers represent states of *recA* protein, possibly different conformations; ellipses represent molecules of *recA* protein; (ss DNA and ds DNA) single- and double-stranded DNA, respectively.

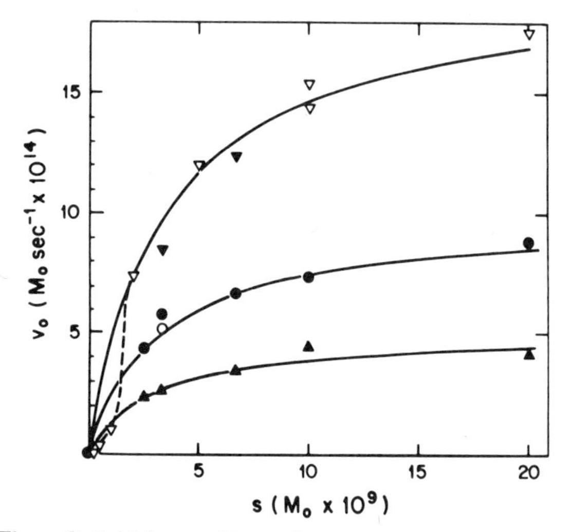

Figure 3. Initial rate of formation of D loops, v_o, versus the total concentration of homologous single-stranded fragments, s, expressed as the molar concentration of fragments, M_o, assuming a mean length of 600 nucleotide residues per fragment. The ratio of *recA* protein to single-stranded DNA was kept constant at 1 molecule of *recA* protein per 2.7 nucleotide residues of single-stranded DNA. Conditions of incubation and assay were similar to those described elsewhere (Shibata et al. 1980). The three curves represent experiments with three different concentrations of form-II fd [^{3}H]DNA: (▲) 1.7×10^{-10} M_o; (●, ○) 3.4×10^{-10} M_o; (▼,▽) 6.9×10^{-10} M_o.

and concentration of salt, superhelical DNA in the absence of any protein will spontaneously take up a homologous single-stranded fragment to form a D loop. In this uncatalyzed reaction, thermal energy appears both to unfold single-stranded DNA and to unwind duplex DNA sufficiently to permit nucleation (Beattie et al. 1977). The energy of superhelix formation enlarges the heteroduplex joint in the nascent D loop and stabilizes the product, thereby driving the reaction. Using the energy of ATP, *recA* protein by itself accomplishes all of these steps: *recA* protein coats single-stranded DNA (Shibata et al. 1979a; 1980; McEntee et al. 1980), unwinds duplex DNA while bound to a single strand (Cunningham et al. 1979), and, in some yet undefined way, stabilizes the product.

Helix-destabilizing proteins accelerate 10^3–10^4 times the renaturation of complementary single strands (Alberts and Frey 1970; Christiansen and Baldwin 1977). Although helix-destabilizing proteins can reduce the amount of *recA* protein required to form D loops (McEntee et al. 1980; Shibata et al. 1980), we have yet to observe the catalysis of D-loop formation by helix-destabilizing proteins alone (Shibata et al. 1980). *recA* protein, on the other hand, accelerates the formation of D loops by more than 6000 times. The renaturation of complementary single strands, whether catalyzed or not, is a second-order reaction (Wetmur and Davidson 1968; Studier 1969; Alberts and Frey 1970; Christiansen and Baldwin 1977). Whereas the uncatalyzed formation of D loops is also second order (Beattie et al. 1977), the reaction promoted by *recA* protein is not, but instead the kinetics more closely resemble those of an enzymatic reaction (Fig. 3) in which the rate is limited by the conversion of an enzyme-substrate complex to product plus free enzyme. The reaction catalyzed by *recA* protein is distinct in yet another way from renaturation promoted by helix-destabilizing proteins: Single strands act both as effector and as substrate in the formation of D loops by *recA* protein (Cunningham et al. 1979; Shibata et al. 1979a). In principle, this phenomenon is not different from an enzymatic reaction in which two substrates interact cooperatively, with the important exception that single strands also stimulate another activity of *recA* protein, the protease activity (Craig and Roberts 1980), in which case the single strand can hardly be considered a substrate.

TOPOLOGY OF HOMOLOGOUS PAIRING

Since *recA* protein does not require superhelical DNA as one of its substrates in the pairing reaction, we were able to examine the ability of *recA* protein to pair several topological variants of DNA (Fig. 4). Homologous pairing of the four combinations of duplex DNA with single-stranded or partially single-stranded DNA, illustrated in Figure 4, can be detected by the D-loop assay (Table 1; Cunningham et al. 1980).

Observations using the electron microscope confirmed the findings first indicated by the D-loop assay, namely, that circular single strands form joint molecules with linear or nicked circular duplex DNA at a high frequency. Three examples are illustrated in Figure 5.

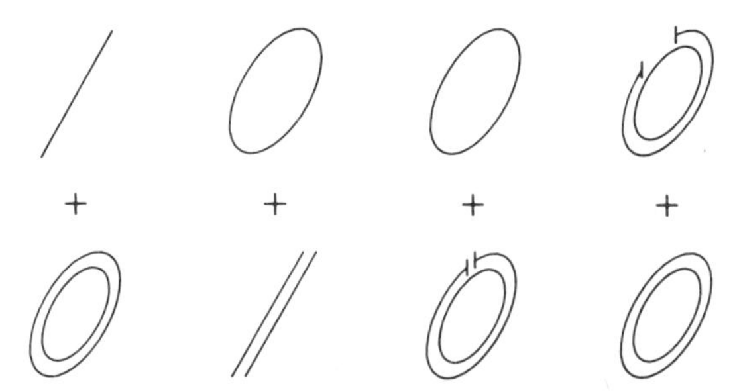

Figure 4. Topological variants of DNA from which *recA* protein will form joint molecules at a high frequency. These include: (*left to right*) linear single strands with superhelical DNA (McEntee et al. 1979; Shibata et al. 1979b); circular single strands with either linear or nicked, circular duplex DNA (DasGupta et al. 1980; T. Shibata et al., unpubl.); and gapped, circular duplex DNA with superhelical DNA (Cassuto et al. 1980; Cunningham et al. 1980).

The pairing by *recA* protein of circular single strands with linear duplex DNA is a particularly efficient reaction in which 80% or more of the duplex substrate is converted to joint molecules in 10 to 15 minutes. The kinetics of this reaction, like those of D-loop formation, resemble Michaelis-Menten kinetics (see Fig. 3); this resemblance suggests that the mechanisms of these reactions are similar (C. M. Radding et al., unpubl.).

Table 1. Formation of Stable Joint Molecules by *recA* Protein

Double-stranded DNA	Single-stranded DNA (%, measured by D-loop assay)		
		circular	
	fragments	homologous	heterologous
Form I	69	19	—
Form II	33	40	7
Form III	38	55	2
*Hae*III digest	18	28	1

Reaction mixtures contained either 4.4 μM duplex fd [^{3}H]DNA or 8.8 μM Endo•R•*Hae*III restriction fragments of duplex fd [^{3}H]DNA; 6 μM single-stranded DNA; 2.2 μM *recA* protein (Shibata et al. 1979b); plus the other components of the standard reaction mixture. Mixtures were incubated for 30 min (form-I DNA) or 60 min (other than form-I DNA) at 37°C. Complexes formed during incubation were treated with 0.5% Sarkosyl for 5 min at 17°C, followed by incubation for 4 min at 25°C in 1.5 M NaCl, 0.15 M sodium citrate, then measured by the D-loop assay. Examination of the preparation of circular single-stranded DNA by electron microscopy showed that some 90% of the molecules were circular. The preparation of form-I DNA contained 89% form I and 11% forms II and III; the preparation of form-II DNA contained 89% form II, 7% form I, and 4% form III; the form-IV DNA contained 79% form III, 1% form I, and 20% form II. The joint molecules made by form-I DNA and circular single-stranded DNA can be accounted for in large measure by the contamination of form-I DNA by form-II DNA and the contamination of circular single-stranded DNA by linear molecules. Electron microscopy reveals only 2% duplex DNA in joint molecules when superhelical DNA and circular single strands are the substrates (DasGupta et al. 1980).

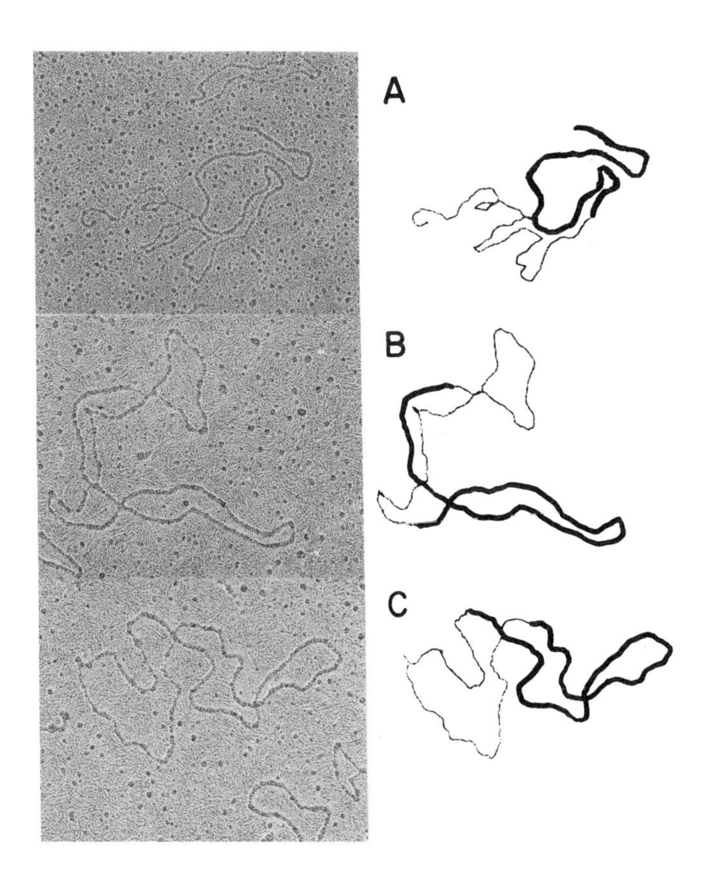

Figure 5. Electron micrographs of joint molecules formed by circular single-stranded fd DNA and (*A*) linear duplex fd DNA or (*B, C*) nicked, circular duplex fd DNA. In the tracings on the right, thick and thin lines represent our interpretation of which parts of the joint molecules are double-stranded and single-stranded, respectively. Interpretations of the molecular structures of these joint molecules are shown in Fig. 6, c and d. DNA was spread and examined microscopically as described elsewhere (Cunningham et al. 1980). In the present experiments, the products of the reaction promoted by *recA* protein were spread directly without removing the protein.

The reactions schematized in Figure 4 fit the rule that *recA* protein will efficiently produce stable joint molecules, i.e., molecules that survive the conditions of the D-loop assay, if one of a pair of homologous molecules is single-stranded or partially single-stranded and if either molecule has a free end. The need for single-stranded DNA reflects the special role that single strands play in stimulating *recA* protein both to bind and to unwind duplex DNA (Cunningham et al. 1979; Shibata et al. 1979a; McEntee et al. 1980). The need for a free end is thermodynamic. In the absence of topoisomerase activity (Cunningham et al. 1979, 1980, and unpubl.), a free end is necessary to form a stable heteroduplex joint with the normal Watson-Crick structure. The nonspecific nature of the requirement for a free end supports this view (Fig. 4): Single-stranded DNA and the free end can be part of the same molecule or different ones; the free end can be the terminus of a single-stranded molecule, the terminus of a duplex molecule, or it can be located at a nick or gap in circular duplex DNA. We can rationalize the array of reactions illustrated in Figure 4 by supposing the following: (1) Homologous pairing of a single strand and its complementary strand in a duplex molecule can occur initially at a site that is remote from a free end. This notion implies that recognition of complementarity requires neither a free end nor topological interwinding. (2) Once two molecules are aligned with homologous sequences in register, contact in the region of a free end can follow rapidly and lead to formation of a heteroduplex joint with the normal Watson-Crick structure. Recent experiments strongly support this view. By electron microscopy, we have observed, at a low frequency, the pairing of circular single strands and duplex DNA at homologous sites that are flanked by long heterologous regions (DasGupta et al. 1980). Moreover, we have recently observed that addition of a topoisomerase will stabilize such joint molecules and consequently increase their yields to levels comparable to those found when one molecule has a free end (R. P. Cunningham et al., unpubl.).

THE STRUCTURES OF JOINT MOLECULES

The joint molecule that *recA* protein makes from fragments of single-stranded DNA and homologous duplex DNA is a D loop (Fig. 6a, b), the properties of which are identical to those of D loops made by the spontaneous reaction of superhelical DNA with single strands at high temperature (Beattie et al. 1977; Cunningham et al. 1979; Shibata et al. 1979b). The structure of the joint molecule made from circular single-stranded DNA and linear duplex DNA can be inferred unambiguously from the micrographs (Fig. 5A). From a common point in these joint molecules, there emerges a single-stranded branch with a free end, a double-stranded branch with a free end, and one more each of single- and double-stranded arms that are part of a circle. Our interpretation of the structure is indicated in Figure 6c. A prominent and interesting feature of these joint molecules is the length of the putative heteroduplex regions, which can reach thousands of base pairs. Similar struc-

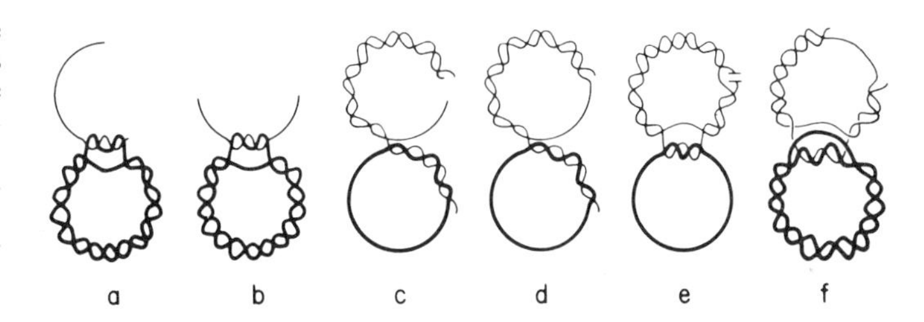

Figure 6. The structures of joint molecules made by *recA* protein. (*a, b*) The structures of D loops (Shibata et al. 1979a and unpubl.). (*c, d*) The structures of molecules shown in Fig. 5. (*e*) The possible structure of other joint molecules resulting from the pairing of circular single strands and nicked, circular duplex DNA (DasGupta et al. 1980). (*f*) Proposed structure of joint molecules made from gapped, circular DNA and superhelical DNA (Cunningham et al. 1980).

tures appear to result from the pairing of circular single strands with nicked duplex DNA (Fig. 5B, C and Fig. 6d). From this combination of substrates, there may also be joint molecules with the structure diagramed in Figure 6e, but here electron microscopy does permit an unambiguous interpretation. However, other experiments have indicated that *recA* protein may cause a strand that has paired with a new complement to return to its original partner (Fig. 6f). When we studied the pairing of gapped, circular duplex DNA with superhelical DNA, we found that (1) the two molecules were linked by noncovalent bonds, (2) the molecules were frequently joined at a site that was remote from the gap, and (3) many joint molecules survived heating until the temperature approached the T_m (Cunningham et al. 1980). These observations led us to postulate the structure illustrated in Figure 6f.

The substrate specificity of *recA* protein (Fig. 4) and the structures of certain products (Fig. 6) enlarge our view of the pairing reaction and reveal several features of particular biological interest: (1) *recA* protein plays a role not only in the receiving end of a strand transfer, as in the formation of D loops (McEntee et al. 1979; Shibata et al. 1979b), but also in the donating end, as in the formation of joint molecules from gapped DNA plus superhelical DNA (Fig. 6f; Cunningham et al. 1980), and from circular single-stranded DNA plus linear or nicked, circular duplex DNA (Figs. 4, 5, and 6). In the reactions of circular single-stranded DNA with duplex DNA (Fig. 5), *recA* protein, unaided by other proteins, caused the displacement of a strand from a duplex molecule and the pairing of that strand with a new complementary partner. Observations on the pairing of gapped duplex DNA and superhelical DNA led us to propose earlier that *recA* protein can cause a single strand to cross over from a gapped duplex molecule to pair with an intact circular strand of another duplex molecule (Fig. 6f; Cunningham et al. 1980). The action of single-stranded DNA in stimulating *recA* protein to unwind duplex DNA presumably plays a key role in the ability of *recA* protein to cause a strand to cross over. (2) *recA* protein not only initiates a strand transfer, but it can also extend or propagate the transfer. In the reaction of circular single-stranded DNA with linear duplex DNA, the action of *recA* protein produces heteroduplex joints that can exceed several thousand base pairs in length. The creation of such extensive heteroduplex joints is interesting genetically since recombination appears to involve such long joints (Wagner and Meselson 1976; Radding 1978; Fogel et al. 1979; Rossignol et al. 1979). (3) Several observations suggest that *recA* protein not only promotes the crossover of a strand, but also sometimes causes such a displaced strand to cross back and pair again with its original complement (Fig. 6f; Cunningham et al. 1980). Further work on the structure of the relevant products is required to establish that conclusion. The possible occurrence of multiple exchanges of a strand in vitro is interesting in relation to the repeated local exchanges (high negative interference) that often occur in genetic recombination.

Since *recA* protein promotes donation and receipt of a strand, initiation and extension of a strand transfer, the crossing over of a strand, and possibly the crossing back as well, we think of its activity as that of a strand transferase.

SUMMARY

recA protein, which is essential for recombination of *E. coli* (Clark 1973; Kobayashi and Ikeda 1978), promotes homologous pairing by a novel ATP-dependent mechanism in which single-stranded DNA plays a central role as effector as well as substrate and causes the binding and partial unwinding of duplex DNA. Unlike annealing reactions, which are second-order reactions, the kinetics of synthesis of joint molecules by *recA* protein resemble classic Michaelis-Menten kinetics, which confirms that a ternary complex consisting of *recA* protein and two DNA molecules is a precursor, the conversion of which to a joint molecule limits the rate of the reaction. Studies of purified *recA* protein have shown that it promotes the homologous pairing of a number of topological variants of DNA, including: (1) complementary linear single strands (Weinstock et al. 1979), (2) linear single strands and duplex DNA, (3) circular single strands and either linear duplex DNA or nicked, circular duplex DNA, and (4) duplex DNA with a gap in one strand and closed circular duplex DNA. According to these observations, *recA* protein will form stable joint molecules if one molecule is single-stranded or partially single-stranded and if either molecule has a free end. Circular single strands as well as linear ones stimulate *recA* protein to unwind duplex DNA. These observations, which revealed a lack of specificity with regard to ends, suggested that *recA* protein promotes homologous pairing of strands in a side-by-side fashion and that free ends subsequently serve to stabilize the product by permitting the formation of heteroduplex regions with the normal right-handed helical structure. This view has been confirmed by studies of the pairing of circular single strands with homologous duplex DNA that is flanked by long heteroduplex regions (DasGupta et al. 1980).

From the topological variations of the pairing reaction, we also infer that *recA* protein catalyzes a concerted strand transfer that, under some circumstances, can create very long heteroduplex regions and can possibly cause a strand to cross back and pair again with its original complement.

ACKNOWLEDGMENTS

This work was supported by grant NP90F from the American Cancer Society and grant USPHS CA-16038-06 from the National Cancer Institute.

REFERENCES

ALBERTS, B. M. and L. FREY. 1970. T4 Bacteriophage gene 32: A structural protein in the replication and recombination of DNA. *Nature* **227:** 1313.

BEATTIE, K. L., R. C. WIEGAND, and C. M. RADDING. 1977. Uptake of homologous single-stranded fragments by superhelical DNA. II. Characterization of the reaction. *J. Mol. Biol.* **116:** 783.

CASSUTO, E., S. C. WEST, J. MURSALIM, S. CONLON, and P. HOWARD-FLANDERS. 1980. Initiation of genetic recombination: Homologous pairing between duplex DNA molecules promoted by recA protein. *Proc. Natl. Acad. Sci.* **77:** 3962.

CHRISTIANSEN, C. and R. L. BALDWIN. 1977. Catalysis of DNA reassociation by the *Escherichia coli* DNA binding protein. *J. Mol. Biol.* **115:** 441.

CLARK, A. J. 1973. Recombination deficient mutants of *E. coli* and other bacteria. *Annu. Rev. Genet.* **7:** 67.

CRAIG, N. L. and J. W. ROBERTS. 1980. *E. coli* recA protein-directed cleavage of phage λ repressor requires polynucleotide. *Nature* **283:** 26.

CUNNINGHAM, R. P., C. DASGUPTA, T. SHIBATA, and C. M. RADDING. 1980. Homologous pairing in genetic recombination: RecA protein makes joint molecules of gapped circular DNA and closed circular DNA. *Cell* **20:** 223.

CUNNINGHAM, R. P., T. SHIBATA, C. DASGUPTA, and C. M. RADDING. 1979. Homologous pairing in genetic recombination: Single strands induce recA protein to unwind duplex DNA. *Nature* **281:** 191; **282:** 426 (erratum).

DASGUPTA, C., T. SHIBATA, R. P. CUNNINGHAM, and C. M. RADDING. 1980. The topology of homologous pairing promoted by recA protein. *Cell* **20:** 223.

FOGEL, S., R. MORTIMER, K. LUSNAK, and F. TAVARES. 1979. Meiotic gene conversion—A signal of the basic recombination event in yeast. *Cold Spring Harbor Symp. Quant. Biol.* **43:** 1325.

GOTTESMAN, S. and M. GOTTESMAN. 1975. Excision of prophage λ in a cell-free system. *Proc. Natl. Acad. Sci.* **72:** 2188.

HOLLOMAN, W. K. and C. M. RADDING. 1976. Recombination promoted by superhelical DNA and the *recA* gene of *Escherichia coli*. *Proc. Natl. Acad. Sci.* **73:** 3910.

HOLLOMAN, W. K., R. C. WIEGAND, C. HOESSLI, and C. M. RADDING. 1975. Uptake of homologous single-stranded fragments by superhelical DNA: A possible mechanism for initiation of genetic recombination. *Proc. Natl. Acad. Sci.* **72:** 2394.

KOBAYASHI, I. and H. IKEDA. 1978. On the role of *recA* gene product in genetic recombination: An analysis by in vitro packaging of recombinant DNA molecules formed in the absence of protein synthesis. *Mol. Gen. Genet.* **166:** 25.

KOLODNER, R. 1980. Genetic recombination of bacterial plasmid DNA: Electron microscopic analysis of *in vitro* intramolecular recombination. *Proc. Natl. Acad. Sci.* **77:** 4847.

LIU, L. F. and J. C. WANG. 1975. On the degree of unwinding of the DNA helix by ethidium. II. Studies by electron microscopy. *Biochim. Biophy. Acta* **395:** 405.

MCENTEE, K., G. M. WEINSTOCK, and I. R. LEHMAN. 1979. Initiation of general recombination catalyzed *in vitro* by the recA protein of *E. coli*. *Proc. Natl. Acad. Sci.* **76:** 2615.

———. 1980. RecA protein-catalyzed strand assimilation: Stimulation by *Escherichia coli* single-stranded DNA-binding protein. *Proc. Natl. Acad. Sci.* **77:** 857.

MIZUUCHI, K., M. GELLERT, and H. A. NASH. 1978. Involvement of supertwisted DNA in integrative recombination of bacteriophage lambda. *J. Mol. Biol.* **121:** 375.

OGAWA, T., H. WABIKO, T. TSURIMOTO, T. HORII, H. MASUKATA, and H. OGAWA. 1979. Characteristics of purified recA protein and the regulation of its synthesis *in vivo*. *Cold Spring Harbor Symp. Quant. Biol.* **43:** 909.

POTTER, H. and D. DRESSLER. 1979. Biochemical assay designed to detect formation of recombination intermediates *in vitro*. *Proc. Natl. Acad. Sci.* **76:** 1084.

RADDING, C. M. 1978. Kinetics and topology of genetic recombination: Strand transfer and mismatch repair. *Annu. Rev. Biochem.* **47:** 847.

RADDING, C. M., K. L. BEATTIE, W. K. HOLLOMAN, and R. C. WIEGAND. 1977. Uptake of homologous single-stranded fragments by superhelical DNA. IV. Branch migration. *J. Mol. Biol.* **116:** 825.

RADDING, C. M., T. SHIBATA, R. P. CUNNINGHAM, C. DASGUPTA, and L. OSBER. 1980. RecA protein of *E. coli* promotes homologous pairing of DNA molecules by a novel mechanism. *ICN-UCLA Symp. Mol. Cell Biol.* **19:** (in press).

ROBERTS, J. W., C. W. ROBERTS, N. L. CRAIG, and E. M. PHIZICKY. 1979. Activity of the *Escherichia coli recA*-gene product. *Cold Spring Harbor Symp. Quant. Biol.* **43:** 917.

ROEDER, G. S. and P. D. SADOWSKI. 1979. Pathways of recombination of bacteriophage T7 DNA *in vitro*. *Cold Spring Harbor Symp. Quant. Biol.* **43:** 1023.

ROSSIGNOL, J.-L., N. PAQUETTE, and A. NICHOLAS. 1979. Aberrant 4:4 asci, disparity in the direction of conversion and frequencies of conversion in *Ascobolus immersus*. *Cold Spring Harbor Symp. Quant. Biol.* **43:** 1343.

SHIBATA, T., R. P. CUNNINGHAM, C. DASGUPTA, and C. M. RADDING. 1979a. Homologous pairing in genetic recombination: Complexes of recA protein and DNA. *Proc. Natl. Acad. Sci.* **76:** 5100.

SHIBATA, T., C. DASGUPTA, R. P. CUNNINGHAM, and C. M. RADDING. 1979b. Purified *E. coli* recA protein catalyzes homologous pairing of superhelical DNA and single-stranded fragments. *Proc. Natl. Acad. Sci.* **76:** 1638.

———. 1980. Homologous pairing in genetic recombination: Formation of D-loops by combined action of recA protein and a helix-destabilizing protein. *Proc. Natl. Acad. Sci.* **77:** 2606.

STUDIER, F. W. 1969. Effects of the conformation of single-stranded DNA on renaturation and aggregation. *J. Mol. Biol.* **41:** 199.

WAGNER, R., JR, and M. MESELSON. 1976. Repair tracts in mismatched DNA heteroduplexes. *Proc. Natl. Acad. Sci.* **73:** 4135.

WEINSTOCK, G. M., K. MCENTEE, and I. R. LEHMAN. 1979. ATP-dependent renaturation of DNA catalyzed by the recA protein of *E. coli*. *Proc. Natl. Acad. Sci.* **76:** 126.

WETMUR, J. G. and N. DAVIDSON. 1968. Kinetics of renaturation of DNA. *J. Mol. Biol.* **31:** 349.

WIEGAND, R. C., K. L. BEATTIE, W. K. HOLLOMAN, and C. M. RADDING. 1977. Uptake of homologous single-stranded fragments by superhelical DNA. III. The product and its enzymic conversion to a recombinant molecule. *J. Mol. Biol.* **116:** 805.

DNA Gyrase: Site-specific Interactions and Transient Double-strand Breakage of DNA

M. Gellert, L. M. Fisher, H. Ohmori,* M. H. O'Dea, and K. Mizuuchi
*Laboratory of Molecular Biology, National Institute of Arthritis, Metabolism, and Digestive Diseases, National Institutes of Health, Bethesda, Maryland 20205; *Institute for Virus Research, Kyoto University, Kyoto, Japan*

Enzymes that catalyze the transient breakage and rejoining of DNA phosphodiester bonds occur widely in nature. In some cases, they have been initially identified as topoisomerases, so named because they are able to mediate the interconversion of topological isomers of DNA (e.g., by removing or introducing superhelical turns). With all topoisomerases, breakage of DNA appears to be coupled to the formation of a covalently linked enzyme-DNA intermediate that conserves the bond energy of the broken phosphodiester bond for later rejoining. Topoisomerases may be grouped into two classes. Some of these enzymes function by introducing a transient single-strand break into DNA. The archetype of this class of so-called nicking-closing enzymes is the ω protein first detected by Wang (1971) in extracts from *Escherichia coli.* Other enzymes belonging to this group are the eukaryotic nicking-closing enzymes (for review, see Champoux 1978) and the *int*-gene protein of phage λ (Nash et al., this volume).

In contrast, several recently discovered topoisomerases constitute a second mechanistic type in that they appear to act by passing a double-helical DNA segment through a transient double-strand break in DNA. These enzymes include the phage T4 ATP-dependent topoisomerase (Liu et al. 1979, 1980), DNA gyrase (Brown and Cozzarelli 1979; Mizuuchi et al 1980a), and the partially purified ATP-dependent eukaryotic topoisomerases isolated from HeLa cells (L. F. Liu, pers. comm.), *Drosophila melanogaster* embryos (Liu et al. 1980), and *Xenopus laevis* oocytes (Baldi et al. 1980).

An enzyme that breaks and rejoins DNA is an obvious candidate for a function that will transfer a broken-strand end from one DNA molecule to another, thus supplying one potential step in a recombination mechanism, as discussed by Champoux (1977). Indeed, there are some intriguing examples of the close connection between the two types of activity. The cistron-*A* protein of phage φX174, which produces a specific break in the replicative form of φX174 DNA (Henry and Knippers 1974) and can transfer the broken end to the homologous position on a newly replicated strand (Eisenberg et al. 1977), has also been reported to have a topoisomerase activity (Ikeda et al. 1976), as has the related gene-II protein of phage fd (Meyer and Geider 1979). Similarly, the *int*-gene protein of phage λ, which participates in strand transfer during λ integrative recombination, possesses a topoisomerase activity (Kikuchi and Nash 1979). Conversely, enzymes first studied as topoisomerases may well prove to participate in strand-transfer reactions. The work of Ikeda et al. (this volume) suggests that DNA gyrase is in some way involved in such a process, leading to illegitimate recombination events. From the viewpoint of this Symposium, the study of topoisomerases is thus a topic of obvious relevance.

DNA gyrase from *E. coli,* which catalyzes both the relaxation of supercoiled DNA and the ATP-dependent supercoiling of circular duplex DNA, is perhaps the best-characterized DNA topoisomerase. In this paper we describe several different experiments bearing on the mechanism of the enzyme. First, we describe how it has been established that DNA gyrase generates double-strand breaks at specific sites in DNA. Second, we examine the topography of complexes formed between gyrase and DNA. Finally, a mechanistic proposal is presented that accounts for many of the known catalytic features of DNA gyrase.

DNA Gyrase Interconverts DNA Topoisomers via Transient Double-strand Breaks

Following a line of investigation first opened by studies on T4 DNA topoisomerase, we and others have recently demonstrated that a salient feature of the reaction mechanism of DNA gyrase is the passage of a double-helical DNA segment through a transient double-strand break in DNA. Evidence for this conclusion has come from two types of experiments, one showing the formation and resolution of catenated and knotted duplex DNA structures (Liu et al. 1979, 1980; Kreuzer and Cozzarelli 1980; Mizuuchi et al. 1980a) and the other showing that the enzymes change the linking number of closed circular DNA in steps of two (Brown and Cozzarelli 1979; Liu et al. 1980; Mizuuchi et al. 1980a).

Catenanes and knotted molecules (Mizuuchi et al. 1980) were made for our studies by the in vitro phage λ integrative recombination reaction operating on plasmids that contained both the bacterial (*att*B) and phage (*att*P) attachment sites. It has been shown previously that recombination within a supercoiled DNA having the *att*B and *att*P sites in parallel orientation (i.e., with their identical core sequences arranged as a direct repeat) generates catenanes (Nash et al. 1977; Mizuuchi et al. 1980b). If, however, the substrate DNA contains the *att*B and *att*P sites in antiparallel orientation (i.e., with their core sequences arranged as an inverted repeat), then reciprocal recombination merely inverts the region between the sites, producing a knotted circular molecule. (It should be noted that this latter reaction corresponds exactly to the scheme proposed by Campbell [this volume] as a way of generating specific genomic inversions.) There are several advantages in producing catenanes and knots by this method. The intramolecular recombination reaction is very efficient: Typically, 75% or more of the input DNA is converted to product. The relative orientation of the *att* sites selectively controls whether knots or catenanes are made as products. Most important, because recombination occurs between sites within the same plasmid, a particularly simple product is obtained. Thus, the catenanes are made up of only two interlocked circular partners, and the knotted DNA consists predominantly of knotted single circles belonging to the torus knot group (i.e., trefoils, pentafoils, heptafoils, etc). These species are readily analyzed by electron microscopy or gel electrophoresis and therefore were useful substrates in testing the effects of DNA gyrase.

When the catenated or knotted DNA was incubated with DNA gyrase, both species were resolved to simple circular forms, as is shown in Figure 1 for catenanes. The reaction required ATP and Mg^{++} and was inhibited by novobiocin. The interconversion of catenated or knotted forms with simple closed circular DNA has the topological requirement that one double-stranded DNA segment must be passed through another. Therefore, a double-strand break (or, equivalently, two consecutive neighboring single-strand breaks) must be made and resealed in each event.

Topological interconversions between circular and knotted or catenated forms of duplex DNA show that DNA gyrase can make transient double-strand breaks through which a DNA segment is passed but they are not sufficient to prove that this step is an integral part of the supercoiling reaction. To show this, we made use of a different type of experiment. Passage of a double-stranded DNA segment through a double-strand break, followed by rejoining, will change the linking number of the DNA by 2 units at each event (Fuller 1978). In contrast, breakage of a single strand and rejoining implies that the linking number will change by 1 unit. With DNA samples in a suitable range of molecular weights and linking numbers, agarose gel electrophoresis is sensitive enough to register the change in mobility caused by unit changes in linking number. Normal closed circular DNA samples, however, start with a distribution over neighboring linking numbers and thus cannot be used to distinguish 1-unit changes from 2-unit changes in linking number. It is necessary to isolate a DNA with a single linking number by purifying a single band from an agarose gel electrophoresis of a circular DNA. When such a DNA sample was used as substrate for DNA gyrase, it was found that the enzyme does indeed change the linking number in steps of 2 units, during either supercoiling or relaxation (Brown and Cozzarelli 1979; Mizuuchi et al. 1980a). Figure 2 shows an example of such a supercoiling experiment, which was carried out at 0 °C to slow the reaction and enhance the proportion of intermediate supercoiled species. The linking-number results and the catenane and knotted DNA experiments together establish that DNA gyrase operates by the passing of a DNA segment through a transient double-strand break.

Site Specificity of DNA Gyrase

The next mechanistic problem that arises is whether DNA gyrase makes transient double-strand breaks at specific sites or at random on the DNA. Information about this question has come first of all from studies with the inhibitor oxolinic acid. This drug blocks the relaxation of DNA by DNA gyrase. It also blocks supercoiling; by inference, it does so by interfering with the breakage-rejoining reaction. Moreover, in the presence of oxolinic acid, DNA gyrase forms a complex with DNA, which upon treatment with anionic detergents results in double-strand breakage of the DNA (Gellert et al. 1977; Sugino et al. 1977). The breakage occurs at sites that are highly specific, though not equally favored. Typically, the strongest sites occur in DNA less often than once in a thousand base pairs.

The structure and sequences of a number of these sites have been analyzed by Morrison and Cozzarelli (1979) and ourselves (Gellert et al. 1979 and this paper). Our work has concentrated on cleavage sites in the plasmid pBR322, whose entire sequence is known (Sutcliffe 1979), and the colicin E1 derivative pVH51, for which a

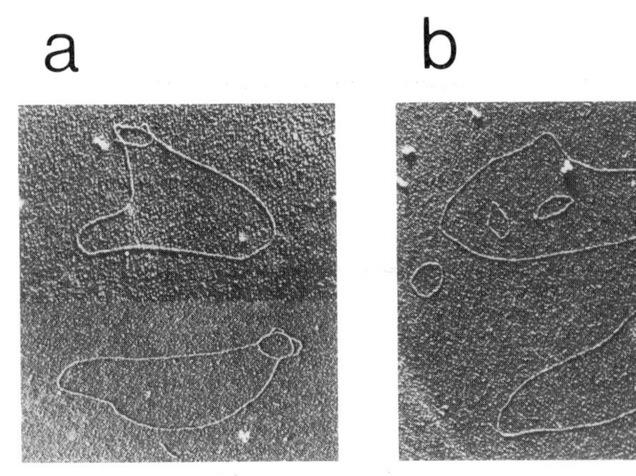

Figure 1. Electron micrographs of catenated DNA before (*a*) and after (*b*) incubation with DNA gyrase. Catenanes were made as described previously (Mizuuchi et al. 1980a) by the λ integrative recombination reaction in vitro, with pBP86 DNA, containing the *att*B and *att*P sites in parallel orientation, as substrate. The samples in *a* and *b* were then lightly nicked with pancreatic DNase, and the sample in *b* was incubated with DNA gyrase + ATP. (See Mizuuchi et al. 1980a for details.)

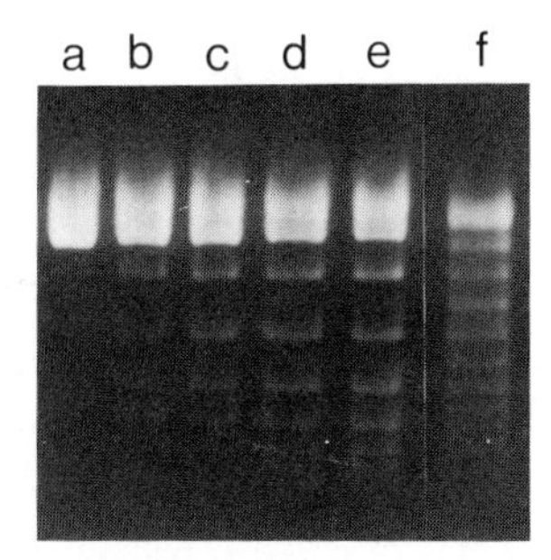

Figure 2. Supercoiling by DNA gyrase changes the linking number of DNA in steps of 2 units. Samples of relaxed pBR322 DNA with a unique linking number (*a–e*) and a polydisperse linking number (*f*) were first incubated with DNA gyrase (50 units) at 25 °C for 30 min and then chilled to 0 °C, and ATP was added. Incubation was continued at 0 °C for 0 min (*a*), 5 min (*b*), 10 min (*c*), 20 min (*d*), and 60 min (*e, f*). Reaction conditions were otherwise those of the standard assay (Mizuuchi et al. 1978).

large portion of the sequence has been determined (Tomizawa et al. 1977; H. Ohmori and J. Tomizawa, in prep.). As a preliminary to DNA sequencing, we ascertained the nature of the termini produced by the cleavage reaction. It was found that labeling of ends was possible using terminal nucleotidyl transferase but was not possible using polynucleotide kinase, suggesting the presence of free 3′-OH ends but blocked 5′-phosphate termini (Table 1). These findings parallel those of Morrison and Cozzarelli (1979). (Recently, Tse et al. [1980] have demonstrated that the 5′-phosphate termini generated by DNA gyrase cleavage are blocked by virtue of each being covalently linked to a protomer of the gyrase-A subunit.)

On the basis of this knowledge, our sequencing strategy was to isolate restriction fragments 100–250 bp long,

Table 1. Blockage of DNA 5′ Ends by Cleavage with DNA Gyrase

	Radioactivity incorporated at	
	3′ end	5′ end
DNA-gyrase-cleaved, − phosphatase	(1)	0.03
DNA-gyrase-cleaved, + phosphatase	0.94	0.03
*Eco*RI-cleaved, + phosphatase	(1)	0.41

Plasmid pNT1 DNA was cleaved by DNA gyrase in the presence of oxolinic acid (Gellert et al. 1977); purified by successive extraction with chloroform-isoamyl alcohol, phenol, and ether; and ethanol-precipitated. The redissolved sample was incubated with or without alkaline phosphatase at 37 °C for 2 hr; both samples were extracted with phenol and ether and ethanol-precipitated. The redissolved samples were divided into two portions: One portion was incubated with DNA terminal transferase and [α-^{32}P]ATP, and the other portion was incubated with polynucleotide kinase and [γ-^{32}P]ATP, both labeled compounds being used at a specific activity of 2500 Ci/mmole. Acid-precipitable radioactivity was determined. The control sample was digested with endonuclease *Eco*RI instead of DNA gyrase and then subjected to the same procedure. Radioactivity incorporated is normalized separately for gyrase-cleaved and *Eco*RI-cleaved samples.

which from previous rough mapping were known to contain a strong potential gyrase cleavage site. These fragments were labeled with ^{32}P at both 5′ ends and subjected to oxolinic-acid-induced gyrase cleavage followed by proteinase-K treatment. The two resulting double-stranded DNA fragments were then separated and analyzed in each of two ways:

1. They were electrophoresed in a denaturing gel alongside the Maxam-Gilbert (1977) sequencing products of the separated strands of the original uncleaved restriction fragment. This allowed the length of the cleaved fragments to be read directly. (Corrections have to be made for the facts that the Maxam-Gilbert method indicates bands from which a particular base has been removed and also that the sequencing reactions leave a 3′-phosphate group at the terminus. Consequently, gyrase-cleaved fragments [i.e., with 3′-OH termini] run 1.5-nucleotide positions slower than the corresponding sequencing product.)
2. The cleaved fragments were themselves subjected to the Maxam-Gilbert sequencing reactions, and the gel electrophoresis autoradiographs were read to identify exactly which base in the sequence was at the newly formed 3′-OH end.

These two independent methods gave concurrent results.

The sites we have examined include one in the tetracycline-resistance region of pBR322, one near the origin of replication of pVH51, and three within the *Alu*I D fragment of pVH51. Two of the latter three pVH51 sites are spaced 3 nucleotides apart from each other but can be distinguished by their differing extents of cleavage. The sequences around the cleavage sites are shown in Figure 3. At all the sites, the two strands are broken with a 4-bp stagger, yielding protruding 5′ ends (see also Morrison and Cozzarelli 1979). There is no readily apparent sequence rule that determines where cleavage occurs, even when the sequences are examined to greater distances from the cleavage sites. It has been proposed (Morrison and Cozzarelli 1979) that one of each pair of breaks generally occurs within the dinucleotide TpG, but this situation occurs only once in our series of sites.

Another possibility, base methylation, is not likely to be a factor in determining the sites at which cleavage takes place. In the first instance, the target sequences for the predominant methylating enzymes are known, and they do not always occur in the vicinity of gyrase-cleavage sites. Second, we isolated pBR322 DNA from an *E. coli dam dcm* strain (deficient in both adenine and cytosine methylation) and from an *E. coli hsd*M strain (lacking the K12 modification system). The DNA was linearized with restriction endonuclease *Eco*RI and subjected to gyrase cleavage. In each case, the pattern and intensity of bands observed in gel electrophoresis were essentially the same as those obtained by cleavage of linear pBR322 isolated from wild-type bacteria. The location of cleavage is also not perturbed by proximity to the end of a DNA fragment. We studied the cleavage

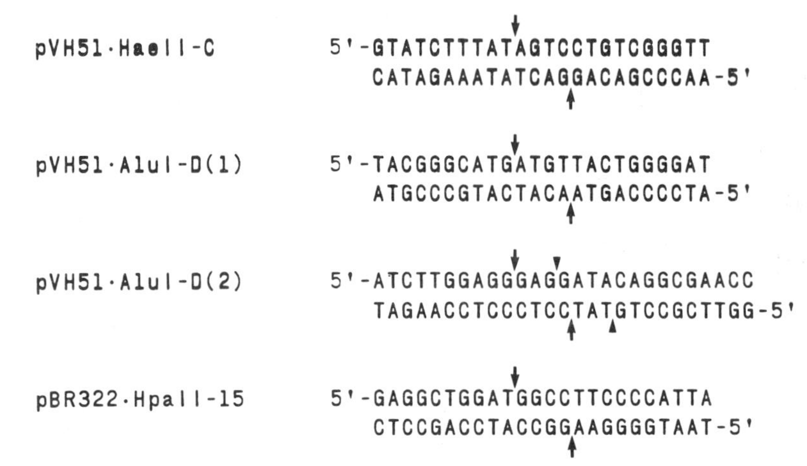

Figure 3. Specific DNA sites of oxolinic-acid-induced cleavage by DNA gyrase. Restriction fragments of pVH51 and pBR322 DNAs were labeled at their 5′ ends with ^{32}P and subjected to cleavage by DNA gyrase (reaction conditions were as described by Gellert et al. [1977]). Sites of breakage on both strands were determined by the use of sequencing gels (see text). The restriction fragments shown indicate the general locations of the sites. The actual restriction fragments used are listed below, as are the site locations (the nucleotide position halfway between the two staggered breaks is given). In the pVH51•*Hae*II C region, a *Hpa*II fragment of 206 bp was used. The cleavage site is at −77 relative to the origin of replication (Tomizawa et al. 1977). The sequence of pVH51 is identical to that of pNT1 in this region (H. Ohmori, unpubl.). In the pVH51•*Alu*I D region, two fragments were used: a *Hpa*II-*Tac*I fragment, designated *Alu*I-D(1), of 105 bp that overlaps the left end (oriented as in the restriction map of Tomizawa et al. [1977]) of the *Alu*I D region by 50 bp (cleavage site at +23 from left end of *Alu*I D) and a *Tac*I fragment, designated *Alu*I-D(2), of 91 bp immediately adjacent on the right (two cleavage sites at +86 and +89 from left end of the *Alu*I D region; cleavage is stronger at +86). The sequence of this region is known (H. Ohmori and J. Tomizawa, in prep.). In pBR322, the *Hpa*II 15 fragment of 90 bp was used. The cleavage site is at position 991 (Sutcliffe 1979).

of several restriction fragments containing the same pBR322 site at varying distances from the ends. Cleavage took place at the same nucleotide positions. The determinant governing the selection of sites in the gyrase-cleavage reaction remains a tantalizing enigma.

From the mechanistic standpoint, the existence of site specificity in the cleavage reaction raises the possibility that perhaps the normal binding and catalytic activities of DNA gyrase are also expressed at these sites. To examine this question, we studied the binding of DNA gyrase to DNA restriction fragments containing a potential cleavage site. The degree and extent of protection of the DNA against pancreatic DNase digestion was determined by the "footprinting" method of Galas and Schmitz (1978). Two different restriction fragments were used: a 203-bp pBR322 *Ava*II-*Alu*I fragment containing the pBR322 site (see above) centered 98 bp from the *Alu*I end and a 200-bp *Hae*II-*Hae*III fragment from pVH51 with the cleavage site centered 92 bp from the *Hae*III end. This latter restriction fragment (the *Hae*II-*Hae*III δ fragment of Tomizawa et al. [1977]) contained the cleavage site in pVH51 located near the origin of replication (see above). Footprinting experiments were carried out with the restriction fragments labeled with ^{32}P at one or the other 5′ end so that protection of each DNA strand could be followed. The resulting protection patterns are shown in Figure 4. For each restriction fragment, DNA gyrase protects both DNA strands in a region around the cleavage site, and this occurs whether or not oxolinic acid is included. In each case, the protected region is over 100 bp long.

Liu and Wang (1978b) have already shown that DNA gyrase can protect a stretch of 140 bp against micrococcal nuclease digestion. Our experiments demonstrate that these regions are preferentially located at cleavage sites. Protection in the absence of oxolinic acid implies that these sites are the locations of preferred binding under normal reaction conditions and, by extension, are the sites of transient double-strand breakage and catalytic activity of DNA gyrase.

Topography of DNA-Gyrase-DNA Complexes

The protection patterns contain detailed information about the structure of the predominant stable complex formed between DNA gyrase and DNA. A central region of 40 bp around the cleavage site is most strongly protected; beyond that, the protection is weaker, and bands appear corresponding to positions of enhanced DNase sensitivity (see Liu and Wang 1978b). These positions are spaced 10–11 bp apart. A similar pattern of DNase-sensitive sites has been found in nucleosomes (for review, see Felsenfeld 1978) and with DNA adsorbed to calcium phosphate precipitates (Liu and Wang 1978b). Such a pattern is generally thought to be characteristic of DNA adsorbed to a surface and suggests that at least a portion of the DNA is wrapped around the outside of DNA gyrase.

The protection patterns around the two sites are not entirely identical. In the pBR322 fragment, protection extends more or less symmetrically at least 50–60 bp on either side of the cleavage site; in the pVH51 fragment, protection is more asymmetric, extending at least 85 bp on one side of the site and stopping at 35 bp on the other side. It is not clear whether any functional significance is to be attributed to this difference.

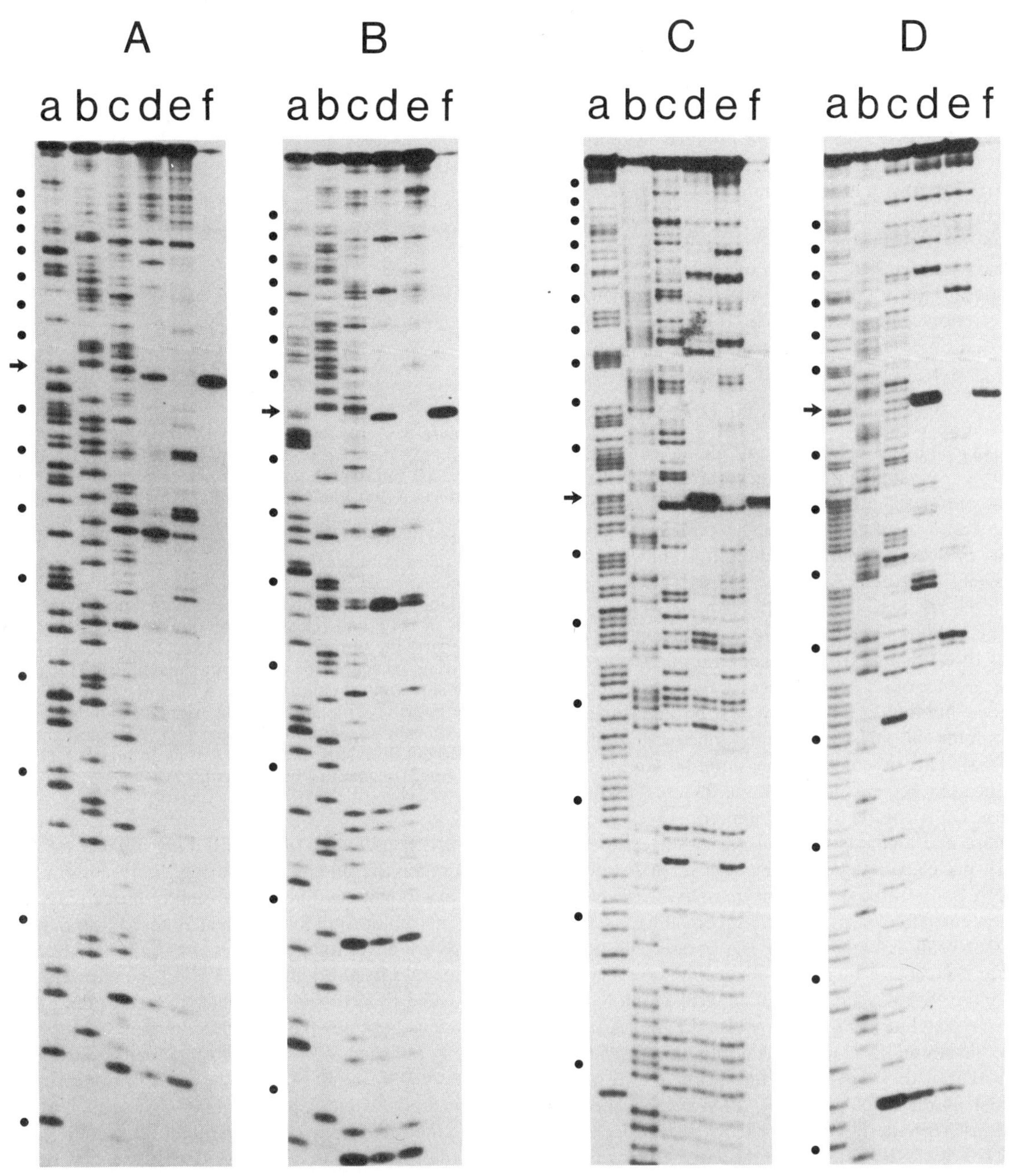

Figure 4. Footprinting analysis of DNA-gyrase binding to DNA restriction fragments containing a potential gyrase cleavage site. (*A, B*) Results of experiments on the 203-bp *Ava*II-*Alu*I fragment of pBR322 (spanning nucleotide positions 886–1089) (Sutcliffe 1979) ^{32}P-labeled at the *Ava*II 5′ end (*A*) or at the *Alu*I 5′ end (*B*). (*C, D*) Experiments on the 200-bp *Hae*II-*Hae*III δ fragment from pVH51 ^{32}P-labeled at the *Hae*III 5′ end (*C*) or at the *Hae*II 5′ end (*D*). Lanes *a* and *b* in *A* and *B* are the respective G + C Maxam-Gilbert sequencing reactions, and lanes *a* and *b* in *C* and *D* are G + A and C + T sequencing reactions. In all four panels, *c–f* refer to a particular experiment carried out on that end-labeled DNA fragment: (*c–e*) DNA samples treated with pancreatic DNase in the absence (*c*) or in the presence (*d, e*) of DNA gyrase; (*d, f*) oxolinic acid (73 μg/ml) included; (*f*) incubated with DNA gyrase under the footprinting conditions but subsequently treated with 5% SDS (4 μl) and proteinase K (8 μl of 0.2 mg/ml, 35 min at 37 °C) to induce DNA cleavage. Lanes *c–e* contained equal amounts of radioactivity. (→) The location of the center of the cleavage site, i.e., halfway between the two staggered breaks; (•) 10-base intervals from this position.

The footprinting assay buffer contained (in 48 μl) 46.4 mM Tris-HCl (pH 7.5), 6.9 mM $MgCl_2$, 75 mM KCl, 0.38 mM $CaCl_2$, 6 mM DTT, 0.19 mM Na_3EDTA, 4.1% (w/v) glycerol, 37.5 μg/ml of bovine serum albumin, and [^{32}P]DNA. For samples in which DNA gyrase was included, 500 units of gyrase-B protein and an excess of gyrase-A protein were used. After incubation at 25 °C for 75 min, pancreatic DNase (2 μl of 5 μg/ml) was added, and the solution was incubated at 25 °C for 1 min prior to the addition of stop solution (50 μl) containing 1.22 M ammonium acetate, 0.1 M Na_3EDTA, and 80 μg/ml of calf thymus DNA. The samples were heated to 75 °C for 10 min, and then the DNA was ethanol-precipitated and electrophoresed in a 10% polyacrylamide-urea gel.

Asymmetric binding of DNA gyrase at the pVH51 site was also indicated by experiments involving protection against exonuclease III. This enzyme processively degrades double-stranded DNA in the 3′→ 5′ direction beginning from the 3′ ends. The *Hae*II-*Hae*III δ fragment labeled with ^{32}P at one or the other 5′ end was treated with exonuclease III in the absence or presence of DNA gyrase. The digestion products were displayed by denaturing acrylamide gel electrophoresis and autoradiography. This procedure allowed determination of the positions at which exonuclease-III digestion had been arrested. In the presence of DNA gyrase, the nuclease was arrested 17 bp on one side and 90 bp on the opposite side of the cleavage site. The asymmetry of the protection against exonuclease III had the same polarity as that observed when pancreatic DNase was used.

Mechanistic Model for DNA Gyrase

Any plausible mechanistic proposal for DNA gyrase not only must take into account the aspects of site-specific binding and DNA passage through transient double-strand breaks, but also must explain the directionality of the supercoiling reaction. It is known that ATP-driven supercoiling caused by gyrase is always negative, i.e., the resulting DNA is always underwound (lacking in helical turns) relative to the relaxed form. We explain this directionality as a consequence of the fact that in the absence of ATP (and hence catalytic turnover), DNA gyrase binding to DNA induces a positive superhelical wrapping of DNA on the enzyme (Liu and Wang 1978a). This supercoiling, measured by binding the enzyme to nicked circular DNA and then sealing the DNA with DNA ligase, is stoichiometric in bound DNA gyrase and amounts to about one positive superhelical turn per enzyme tetramer. Passage of such a wrapped DNA loop through the enzyme-DNA complex will allow introduction of negative supercoiling.

Our proposed model for DNA gyrase then takes the form sketched in Figure 5 (Mizuuchi et al. 1980a). We suppose that the sites of oxolinic-acid-promoted cleavage (zigzag line in Fig. 5A) are also the sites of transient double-strand breakage in the normal reaction and that, together with the enzyme, they form a gate through which another DNA segment will be passed. We further suggest that the DNA segment to be translocated lies within or closely adjacent to the same protected region as the site and becomes wrapped over the enzyme (Fig. 5B) with the positive supercoiling indicated by the experiments of Liu and Wang (1978a). Binding of ATP then causes a conformational change in the enzyme that leads to a coupled opening of the gate and transport of the wrapped DNA segment through it, after which the gate closes again. This process decreases the linking number of the DNA by 2 units, introducing the observed negative supercoiling (Fig. 5C). Because the nonhydrolyzable ATP analog (β, γ-imido)ATP drives one cycle of supercoiling but then blocks further reaction (Sugino et al. 1978), one can plausibly assume that release of the translocated segment is coupled to hydrolysis of ATP (Fig. 5D). The system is then ready to undergo another cycle of reaction (Fig. 5E–G), resulting in

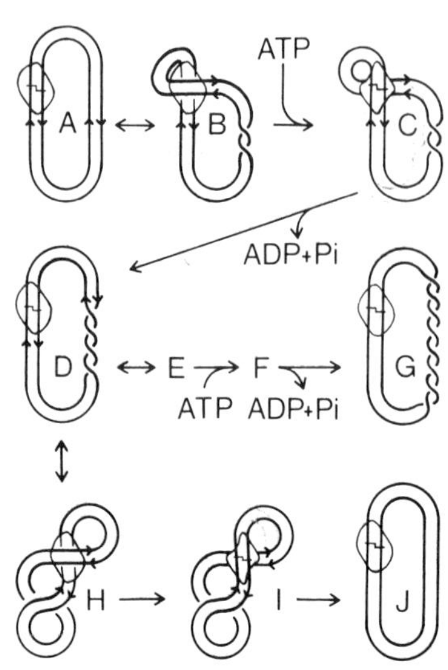

Figure 5. A model for DNA-gyrase-induced DNA supercoiling by means of transient double-strand breaks. The enzyme binds preferentially to certain sites on DNA and induces a left-hand (positive superhelical) wrapping of a local DNA region (*B*). ATP-binding then leads to transport of the upper double helix through the lower, via a transient double-strand break (*C*), with an accompanying conformational change in the enzyme. This reaction decreases the linking number of the DNA by 2 units. Subsequent hydrolysis of ATP and release of the transported DNA segment prepares the system for another cycle of supercoiling (*D–G*). During relaxation of negatively supercoiled DNA (*H–J*), the superhelical coiling causes a loop of DNA to fold over the enzyme with the opposite (right) handedness to that used in the supercoiling reaction (*H*). Transport through a transient double-strand break causes an increase of linking number (relaxation) by 2 units (*I, J*).

a further reduction of DNA linking number by 2 units. To retain the supercoiling during the reaction, it is necessary to assume that the transiently broken DNA ends do not rotate relative to each other. Presumably, the ends are never totally separated but are held by the enzyme in a fixed orientation. Two of the ends are probably held by reversible covalent linkage to the gyrase-A protein, because it is this protein that becomes covalently attached to both 5′-phosphate ends during cleavage by DNA gyrase (Tse et al. 1980). The enzyme form that binds to DNA is a tetrameric complex containing two molecules each of A protein and B protein (Liu and Wang 1978b); it is thus possible that the enzyme-DNA complex opens on one side to admit the transported DNA segment and then, in a coupled process, closes on that side and opens on the other to expel the segment.

Relaxation of negatively supercoiled DNA also changes the linking number in steps of 2 units and thus should also proceed by transient double-strand breakage. To relax negative supercoiling, the DNA segment to be translocated must be bound in the opposite sense from that used for supercoiling (Fig. 5H–J). Two kinds of experiments suggest that in this case the supercoiling energy itself is needed to drive the translocation. First, relaxation of negatively supercoiled DNA always stops short of fully relaxing the DNA, even when a large amount of enzyme is used, and second, the linking number of a nearly relaxed DNA with a unique linking number cannot be reequilibrated by DNA gyrase (unpubl.), contrary to the situation with a eukaryotic nicking-closing enzyme (Pulleyblank et al. 1975) and T4 DNA topoisomerase (Liu et al. 1980). Relaxation by DNA gyrase is always much less efficient than supercoiling, probably because it lacks the help of ATP in making the gating mechanism operate.

Two features of the mechanism proposed in Figure 5 were the subject of further investigation: (1) the ability of gyrase to bind DNA in both wrapped and unwrapped configurations and (2) the idea that nucleotide binding induces a conformation change in the enzyme. We consider the wrapping problem first.

Evidence that gyrase can also form an unwrapped complex has been obtained from temperature-shift experiments. We have followed the effect of a change in temperature from 25 °C to 0 °C on gyrase binding to DNA (as measured by a filter-binding assay) and at the same time we have measured the extent of positive superhelical wrapping of DNA on the enzyme. At 0 °C, gyrase binds very slowly (over a period of hours) to DNA. However, when the enzyme and DNA are mixed at 25 °C, binding is at least 2 orders of magnitude more rapid. Furthermore, the enzyme remains stably bound when the temperature is changed to 0 °C and maintained there for a prolonged time. The effect of temperature shift on DNA wrapping by gyrase was determined as follows. Nicked circular DNA was incubated with gyrase at 25 °C and then shifted to 0 °C. At successive time intervals, *E. coli* DNA ligase and NAD were added to seal the DNA rapidly and thus allow the resulting supercoiling to be determined (Liu and Wang 1978a). We observed that immediately following the shift to 0 °C, the extent of positive supercoiling (and thus wrapping) induced by gyrase was similar to that observed at 25 °C. However, on further incubation at 0 °C, the wrapping progressively decayed with a half-life of 20–30 minutes (Fig. 6). After prolonged incubation, the wrapping largely disappeared, although the enzyme was still bound to DNA, as indicated by filter-binding. When either form of complex was returned to 25 °C, full activity was present; therefore, the enzyme-DNA complex is not irreversibly inactivated at 0 °C.

These studies establish that (under the temperature-shift conditions) DNA gyrase can bind to DNA with or without positive wrapping of DNA on the enzyme. Although this observation is at least consistent with the scheme shown in Figure 5, further work will be required to determine whether the complexes detected here are intimately involved in catalysis.

It was suggested above that ATP binding to gyrase leads to an enzyme conformation change that is required for passage of one DNA segment through another during catalysis of DNA supercoiling. If this is correct, the conformational state of the enzyme should change in the presence of (β, γ-imido)ATP, which allows one cycle of negative supercoiling but then pre-

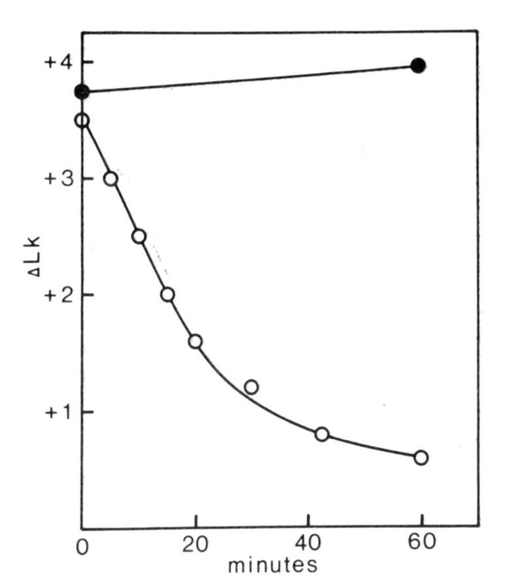

Figure 6. Loss of superhelical wrapping of DNA on DNA gyrase at low temperature. Nicked circular pBR322 DNA was incubated with DNA gyrase (100 units) under standard assay conditions (Mizuuchi et al. 1978) lacking ATP for 30 min at 25 °C. Incubation was then continued either at 25 °C (●) or at 0 °C (○), followed by sealing with *E. coli* DNA ligase (for 15 min at the temperature of the last incubation). Samples were analyzed by agarose gel electrophoresis. The change of the mean linking number (ΔLk) from that of a DNA sample sealed at the same temperature in the absence of DNA gyrase is plotted against the time of the second incubation.

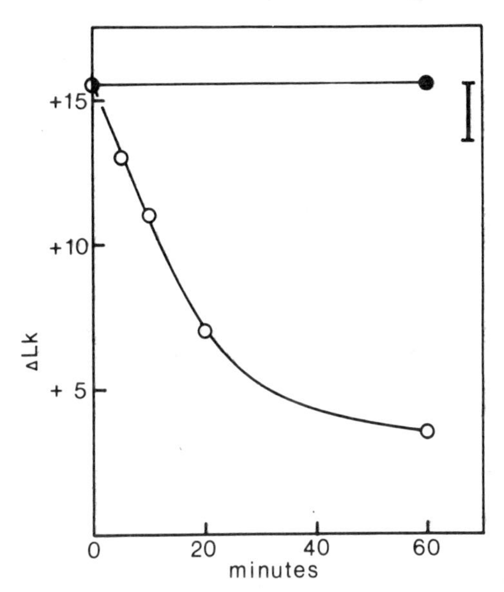

Figure 7. Relaxation of positive supercoiling by DNA gyrase in the presence of (β, γ-imido)ATP. The change of the mean linking number from that of relaxed DNA (ΔLk) is plotted against time of reaction. DNA gyrase (50 units) was incubated with positively supercoiled pBR322 DNA for various times under the conditions of the standard relaxation reaction (Gellert et al. 1977) with (○) or without (●) added (β, γ-imido)ATP (1 mM). Samples were analyzed by agarose gel electrophoresis both in Mg^{++}-containing gels run at 0 °C (Liu and Wang 1978a) and in Tris-borate-EDTA gels run at 20 °C, the former condition being used to shift the DNA band pattern toward negative supercoiling. Positively supercoiled DNA was made by sealing nicked pBR322 DNA with DNA ligase in the presence of a large amount of DNA gyrase (Liu and Wang 1978a), followed by deproteinization. The solid bar at top right shows the expected extent of single-cycle supercoiling induced by (β, γ-imido)ATP.

vents further reaction (Sugino et al. 1978). We have one piece of evidence that a functional (and therefore perhaps conformational) distinction does exist between these two situations. It has been shown that in the absence of nucleotides, DNA gyrase is unable to relax positively supercoiled DNA (Brown et al. 1979). However, we find that when (β, γ-imido)ATP is added, this relaxing activity is quite apparent (Fig. 7). Relaxation of positively supercoiled DNA by DNA gyrase appears to be catalytic, since a single cycle of negative supercoiling induced by (β, γ-imido)ATP could, under these conditions, account for at most only 20% of the observed change. The rate of relaxation of positively supercoiled DNA by gyrase was similar to that obtained in the absence of nucleotides for negatively supercoiled DNA of comparable superhelix density. The ability of (β, γ-imido)ATP to promote catalytic relaxation of positively supercoiled DNA may argue for conformational switching of DNA gyrase in the presence of nucleotides.

In this paper we have presented experiments dealing with several central aspects of DNA gyrase. These results should provide a useful framework in which to explore more detailed questions concerning the reaction mechanism of DNA gyrase.

ACKNOWLEDGMENT

L. M. F. was supported by a postdoctoral fellowship from the Damon Runyon–Walter Winchell Cancer Fund.

REFERENCES

BALDI, M. I., P. BENEDETTI, E. MATTOCCIA, and G. P. TOCCHINI-VALENTINI. 1980. *In vitro* catenation and decatenation of DNA and a novel ATP-dependent topoisomerase. *Cell* **20:** 461.

BROWN, P. O. and N. R. COZZARELLI. 1979. A sign inversion mechanism for enzymatic supercoiling of DNA. *Science* **206:** 1081.

BROWN, P. O., C. L. PEEBLES, and N. R. COZZARELLI. 1979. A topoisomerase from *Escherichia coli* related to DNA gyrase. *Proc. Natl. Acad. Sci.* **76:** 6110.

CHAMPOUX, J. J. 1977. Renaturation of complementary single-stranded DNA circles: Complete rewinding facilitated by the DNA untwisting enzyme. *Proc. Natl. Acad. Sci.* **74:** 5328.

———. 1978. Proteins that affect DNA conformation. *Annu. Rev. Biochem.* **47:** 449.

EISENBERG, S., J. GRIFFITH, and A. KORNBERG. 1977. φX174 cistron *A* protein is a multifunctional enzyme in DNA replication. *Proc. Natl. Acad. Sci.* **74:** 3198.

FELSENFELD, G. 1978. Chromatin. *Nature* **271:** 115.

FULLER, F. B. 1978. Decomposition of the linking number of a closed ribbon: A problem from molecular biology. *Proc. Natl. Acad. Sci.* **75:** 3557.

GALAS, D. J. and A. SCHMITZ. 1978. DNAase footprinting: A simple method for the detection of protein-DNA binding specificity. *Nucleic Acids Res.* **5:** 3157.

GELLERT, M., K. MIZUUCHI, M. H. O'DEA, T. ITOH, and J. TOMIZAWA. 1977. Nalidixic acid resistance: A second genetic character involved in DNA gyrase activity. *Proc. Natl. Acad. Sci.* **74:** 4772.

GELLERT, M., K. MIZUUCHI, M. H. O'DEA, H. OHMORI, and J. TOMIZAWA. 1979. DNA gyrase and DNA supercoiling. *Cold Spring Harbor Symp. Quant. Biol.* **43:** 35.

HENRY, T. J. and R. KNIPPERS. 1974. Isolation and function of the gene *A* initiator of bacteriophage φX174: A highly specific DNA endonuclease. *Proc. Natl. Acad. Sci.* **71:** 1549.

IKEDA, J., A. YUDELEVICH, and J. HURWITZ. 1976. Isolation and characterization of the protein coded by gene *A* of bacteriophage φX174. *Proc. Natl. Acad. Sci.* **73:** 2669.

KIKUCHI, Y. and H. A. NASH. 1979. Nicking-closing activity associated with bacteriophage λ *int* gene product. *Proc. Natl. Acad. Sci.* **76:** 3760.

KREUZER, K. N. and N. R. COZZARELLI. 1980. Formation and resolution of DNA catenanes by DNA gyrase. *Cell* **20:** 245.

LIU, L. F. and J. C. WANG. 1978a. *Micrococcus luteus* DNA gyrase: Active components and a model for its supercoiling of DNA. *Proc. Natl. Acad. Sci.* **75:** 2098.

———. 1978b DNA-DNA gyrase complex: The wrapping of the DNA duplex outside the enzyme. *Cell* **15:** 979.

LIU, L. F., C.-C. LIU, and B. M. ALBERTS. 1979. T4 DNA topoisomerase: A new ATP-dependent enzyme essential for initiation of T4 bacteriophage DNA replication. *Nature* **281:** 456.

———. 1980. Type II DNA topoisomerases: Enzymes that can unknot a topologically knotted DNA molecule via a reversible double-strand break. *Cell* **19:** 697.

MAXAM, A. M. and W. GILBERT. 1977. A new method for sequencing DNA. *Proc. Natl. Acad. Sci.* **74:** 560.

MEYER, T. F. and K. GEIDER. 1979. Bacteriophage fd gene II-protein. II. Specific cleavage and relaxation of supercoiled RF from filamentous phages. *J. Biol. Chem.* **254:** 12642.

MIZUUCHI, K., M. H. O'DEA, and M. GELLERT. 1978. DNA gyrase: Subunit structure and ATPase activity of the purified enzyme. *Proc. Natl. Acad. Sci.* **75:** 5960.

MIZUUCHI, K., L. M. FISHER, M. H. O'DEA, and M. GELLERT. 1980a. DNA gyrase action involves the introduction of transient double-strand breaks into DNA. *Proc. Natl. Acad. Sci.* **77:** 1847.

MIZUUCHI, K., M. GELLERT, R. A. WEISBERG, and H. A. NASH. 1980b. Catenation and supercoiling in the products of bacteriophage λ integrative recombination in vitro. *J. Mol. Biol.* **141:** 485.

MORRISON, A. and N. R. COZZARELLI. 1979. Site-specific cleavage of DNA by *E. coli* DNA gyrase. *Cell* **17:** 175.

NASH, H. A., K. MIZUUCHI, R. A. WEISBERG, Y. KIKUCHI, and M. GELLERT. 1977. Integrative recombination of bacteriophage lambda—The biochemical approach to DNA insertions. In *DNA insertion elements, plasmids, and episomes* (ed. A. I. Bukhari et al.), p. 363. Cold Spring Harbor Laboratory, Cold Spring Harbor, New York.

PULLEYBLANK, D. E., M. SHURE, D. TANG, J. VINOGRAD, and H.-P. VOSBERG. 1975. Action of nicking-closing enzyme on supercoiled and nonsupercoiled DNA: Formation of a Boltzmann distribution of topological isomers. *Proc. Natl. Acad. Sci.* **72:** 4280.

SUGINO, A., C. L. PEEBLES, K. N. KREUZER, and N. R. COZZARELLI. 1977. Mechanism of action of nalidixic acid: Purification of *Escherichia coli nal A* gene product and its relationship to DNA gyrase and a novel nicking-closing enzyme. *Proc. Natl. Acad. Sci.* **74:** 4767.

SUGINO, A., N. P. HIGGINS, P. O. BROWN, C. L. PEEBLES, and N. R. COZZARELLI. 1978. Energy coupling in DNA gyrase and the mechanism of action of novobiocin. *Proc. Natl. Acad. Sci.* **75:** 4838.

SUTCLIFFE, J. G. 1979. Complete nucleotide sequence of the *Escherichia coli* plasmid pBR322. *Cold Spring Harbor Symp. Quant. Biol.* **43:** 77.

TOMIZAWA, J., H. OHMORI, and R. E. BIRD. 1977. Origin of replication of colicin El plasmid DNA. *Proc. Natl. Acad. Sci.* **74:** 1865.

TSE, Y., K. KIRKEGAARD, and J. C. WANG. 1980. Covalent bonds between protein and DNA. *J. Biol. Chem.* **255:** 5560.

WANG, J. C. 1971. Interaction between DNA and an *Escherichia coli* protein ω. *J. Mol. Biol.* **55:** 523.

In Vitro Study of Illegitimate Recombination: Involvement of DNA Gyrase

H. Ikeda, K. Moriya, and T. Matsumoto
Institute of Medical Science, University of Tokyo, P.O. Takanawa, Tokyo 108, Japan

Illegitimate recombination is defined to be a recombination between nonhomologous regions or very short homologous regions of DNA (see Franklin 1971; Weisberg and Adhya 1977). In *Escherichia coli,* various kinds of illegitimate recombination have been known, including deletion, duplication, substitution, insertion, and transducing phage formation. Transposition of transposons (Tn) and insertion sequence (IS) elements may also be included in this classification. These illegitimate recombinations and transpositions take place independently of *E. coli recA* function, which mediates a general recombination between long homologous regions (Franklin 1967; Inselburg 1967; Gingery and Echols 1969; Gottesman and Yarmolinsky 1969; Emmons et al. 1975; Jaskunas et al. 1975; Kopecko and Cohen 1975). In eukaryotic systems, many examples of illegitimate recombination have been found. These have become an important subject of study in that they may relate to cellular regulation such as gene expression, differentiation, and evolution. However, molecular mechanisms of these recombinations have not been elucidated yet. Recent developments of gene technology have provided us an opportunity to study the mechanism of illegitimate recombination in prokaryotic as well as eukaryotic systems.

We have developed an in vitro system for studying the molecular mechanism of illegitimate recombination. The system uses an in vitro packaging mixture for phage λ DNA that consists of lysates of induced lysogens of *E. coli.* Incubation of a plasmid DNA with a packaging mixture resulted in the recombination with endogenous λ DNA, leading to the formation of λ-plasmid hybrid phages. The analysis of the resulting hybrid phages showed that they were formed by an insertion or a substitution of the plasmid into λ DNA. Although we used a plasmid containing a Tn*3* element as a substrate in most experiments, the recombination detected in our system is distinguishable from the in vivo translocation of transposons. This reaction might be related to the in vivo illegitimate recombinations such as insertion, deletion, and substitution.

We examined what function is involved in this recombination by using various bacterial and viral mutations. The results suggest that this recombination is promoted by *E. coli* functions but is not dependent on the *recA* function. In addition, we found that oxolinic acid, an inhibitor of DNA gyrase, stimulates the recombination. Moreover, coumermycin, another type of gyrase inhibitor, blocks the oxolinic-acid-induced recombination. Together with studies with $nalA^r$ and cou^r mutants, we concluded that the oxolinic-acid-induced illegitimate recombination is mediated by DNA gyrase. We propose a model of the illegitimate recombination that explains how DNA gyrase is involved in the recombination.

MATERIALS AND METHODS

In vitro recombination assay. The in vitro packaging system for phage λ DNA was used for the recombination assay (Kobayashi and Ikeda 1977). A standard packaging mixture was prepared from induced lysogens of *E. coli* K12: 594 (λ*c*I857*D*am15*F*Iam96B*S*am7) and 594 (λ*c*I857*E*am4*S*am7). Packaging mixtures were also prepared from lysogens containing various combinations of bacterial *rec* mutations and viral *int, red,* and deletion mutations. A plasmid DNA carrying the ampicillin resistance (Ap^r) determinant (about 1 μg/assay) was added to the packaging mixture and incubated for 90 minutes at 28 °C with gentle shaking. In addition to the DNA, the incubation mixture (35 μl) contained 1.5 × 10^9 induced cells, 50 mM Tris-HCl (pH 7.4), 7 mM $MgCl_2$, 7 mM KCl, 7 mM NaCl, 0.15 mM EDTA, 10% dimethyl sulfoxide, 7 mM spermidine, and 1 mM ATP. The reaction was stopped by adding 1 ml of SMC buffer (Kobayashi and Ikeda 1977) containing 10 μg/ml of pancreatic DNase and one drop of $CHCl_3$. The resulting phage suspension was purified by low-speed and high-speed centrifugation and assayed for plaque formation and Ap^r transduction capability. Ap^r transduction was carried out by infection to *E. coli* Ymel and spreading to a λ agar plate containing 20 μg/ml of ampicillin.

Heteroduplex analysis. Heteroduplex molecules were analyzed by the method of Davis et al. (1971) with some modifications. Phage particles were purified by CsCl density gradient centrifugation. An aliquot of phage suspension (5 × 10^9 particles) was mixed with 10 μl of 0.2 M EDTA, adjusted to 90 μl with water, and then mixed with 10 μl of 1 M NaOH for denaturation. After 10 minutes at room temperature, 100 μl of formamide and 10 μl of 2 M Tris-HCl (pH 6.7) were added to the mixture and incubated for 90 minutes at room temperature. A 3 μl portion of the mixture was mixed with 15 μl of formamide, 2.5 μl of 1 M Tris-HCl (pH 8.0), 1 μl of 0.25 M EDTA, 1.5 μl of 1 mg/ml of cytochrome *c* and adjusted to a volume of 25 μl with water. The mixture was spread onto a hypophase of water. The protein monolayer was picked up with a Parlodion-coated copper grid, dehydrated with 90% ethanol, stained with 5 × 10^{-5} M uranyl acetate for 30 seconds and destained with hexane. Grids were rotary-shadowed with a Pt-Pd wire

(80:20) wrapped around a 0.8-mm tungsten wire. Specimens were observed with a JEM 100Cx electron microscope, photographed at an instrument magnification of 2900, and optically enlarged 12-fold for length measurement by Numonic's Graphics Calculator Model 244-17. The data from 10 to 20 molecules were averaged for each DNA. Measured lengths were calibrated by the following intramolecular markers: *b*519 deletion (40.6–47.6), *b*515 deletion (49.6–53.4), and imm^{434} substitution (73.5–79.1) (Szybalski and Szybalski 1979). Other methods employed have been described previously (Ikeda and Kobayashi 1979).

RESULTS

In Vitro Recombination System

Illegitimate recombination usually occurs at a low frequency, so it is hard to distinguish the resulting recombinant DNA from the large amount of the remaining substrate DNA. To overcome this difficulty, we developed a recombination system in which phage λ DNA recombines with plasmid DNA in a packaging mixture, resulting in the formation of hybrid phage containing the plasmid. This packaging mixture contains a large amount of concatemeric λ DNA originating from prophage, and this λ DNA is a good substrate for packaging. When a plasmid DNA is incubated with the packaging mixture, the plasmid recombines with the λ DNA and is packaged into a λ head. Since the plasmid DNA alone is not able to be packaged, the remaining plasmid DNA can be easily inactivated by treatment with pancreatic deoxyribonuclease. Finally, transduction was carried out by selection with a genetic marker carried on the plasmid DNA, leading to the detection of the recombinant phages.

Recombination between λ and ColE1::Tn*3*

ColE1::Tn*3* (RSF2124) DNA was incubated with the standard packaging mixture, and the resulting phage particles were examined for plaque-forming activity and Ap^r transducing activity. The number of plaque formers was constant irrespective of the amounts of plasmid DNA added, whereas the frequencies of the Ap^r transducers relative to the plaque formers were proportional to the amounts of ColE1::Tn*3* DNA added to the packaging mixture (Table 1). When pancreatic deoxyribonuclease was mixed with ColE1::Tn*3* DNA and the packaging mixture, the formation of the Ap^r transducers was not detectable at all. These results indicate that ColE1::Tn*3* DNA, or at least the Ap^r gene, was packaged into phage particles during the packaging reaction.

To examine the properties of the resulting Ap^r transducers, the transductants were grown and induced by heating at 42°C after superinfection with λ*c*I857 helper phage. The lysates thus obtained were tested for frequencies of Ap^r phages. Among 20 Ap^r transductants tested, 15 produced high-frequency-transducing (HFT) lysates in which frequencies of Ap^r transducers were 10^{-2} to 10^{-1}. We further examined six Ap^r phage strains in detail. All six strains were not plaque-forming but carried *imm*λ, *imm*ColE1, and colicin genes. The contents of these phages, which were determined by CsCl density gradient centrifugation, were in the range of 92% to 104% of λ genome length. Since the genome length of ColE1::Tn*3* is 23.3% of the λ genome unit, the insertion of ColE1::Tn*3* into λ DNA should give rise to recombinant DNA containing 123% λ unit. We concluded, therefore, that most of the Ap^r particles formed in the packaging reaction are recombinant phages composed of λ and ColE1::Tn*3* genomes, though both or one of the genomes are deleted in part. We call them λ-ColE1::Tn*3* hybrid phages. In subsequent experiments, the frequency of Ap^r transducers per total plaque formers is taken as the frequency of λ-Ap^r recombinant DNA per total λ DNA.

Five Ap^r transductants failed to produce HFT lysates. To examine the properties of these transductants, cleared lysates were prepared from them by the lysozyme-brij method, as described previously (Ikeda and Kobayashi 1979), and then DNA of the lysates was examined by agarose gel electrophoresis. The results indicate that one Ap^r transductant contains a plasmid corresponding to the size of λ DNA, whereas another four

Table 1. Formation of λ-ColE1::Tn*3* Hybrid Phages in In Vitro Packaging Reaction

ColE1::Tn*3* DNA added per assay (μg)	Ap^r transducers per assay	pfu per assay	Frequency of Ap^r transducers per total pfu
0	0	1.4×10^9	$< 4 \times 10^{-10}$
0.45	25	2.3×10^9	1.1×10^{-8}
0.9	78	1.4×10^9	5.7×10^{-8}
1.8	97	0.8×10^9	1.3×10^{-7}
4.5	340	1.2×10^9	3.0×10^{-7}
4.5 + DNase (20 μg/ml)	0	2.5×10^6	n.d.

An aliquot of ColE1::Tn*3* DNA preparation (RSF2124) was incubated under a standard assay condition with a standard packaging mixture prepared from 594 (λcI857*D*am15*F*Iam96B*S*am7) and 594 (λ*c*I857*E*am4*S*am7). RSF2124 (So et al. 1975) has a molecular length of 11.6 kb and contains the ampicillin resistance (Ap^r) determinant. The resulting phage mixture was treated with pancreatic DNase (10 μg/ml) and one drop of $CHCl_3$ to inactivate the remaining DNA and surviving bacteria. After removing cell debris by low-speed centrifugation, the supernatant was sedimented by high-speed centrifugation to purify phage particles. The particles were then suspended in λ dilution buffer. An aliquot of phage suspension was mixed with *E. coli* Ymel (10^9 cells/ml), incubated at 32°C for 30 min, and spread on a λ agar plate containing ampicillin (20 μg/ml). The total plaque formers were also titrated on Ymel. n.d. indicates not determined.

contain plasmids corresponding to the size of ColE1::Tn*3*. We inferred that the former plasmid is λ-ColE1::Tn*3* recombinant DNA and the latter are free ColE1::Tn*3* that presumably resulted from unstable λ-ColE1::Tn*3* recombinant DNA.

Physical Mapping of the λ-ColE1::Tn*3* Hybrid Phages

DNA of the λ-ColE1::Tn*3* phages was characterized by mapping with *Eco*RI restriction enzyme. Figure 1 shows the electrophoretic pattern in an agarose gel of the *Eco*RI-digested DNA from these phages. Three of them (14, 74, and 77) contained all of the original *Eco*RI sites of λ DNA in addition to a new *Eco*RI site that is probably harbored in the inserted ColE1::Tn*3* region. Another three (11, 15, and 26) lost two of the original *Eco*RI sites of λ but acquired a new *Eco*RI site. In all of the DNA tested, the sum of the molecular weights of the fragments was smaller than the sum of those of λ and ColE1::Tn*3* DNAs.

To define the site of insertion of ColE1::Tn*3* on the λ genome, the hybrid phage DNA was heteroduplexed against λ*imm*^{434} DNA or λ*b*515*b*519 DNA. Figure 2, a and d, show examples of heteroduplex molecules between λ-ColE1::Tn*3*(11) and λ*imm*^{434} DNA and between λ-ColE1::Tn*3*(14) and λ*imm*^{434} DNA, in which substitution loops were observed at two different sites of λ DNA. One of the single-stranded parts in the loops had a length of 23.1 ± 0.4% λ unit, which corresponds to the size of ColE1::Tn*3*. Since the DNA content of the

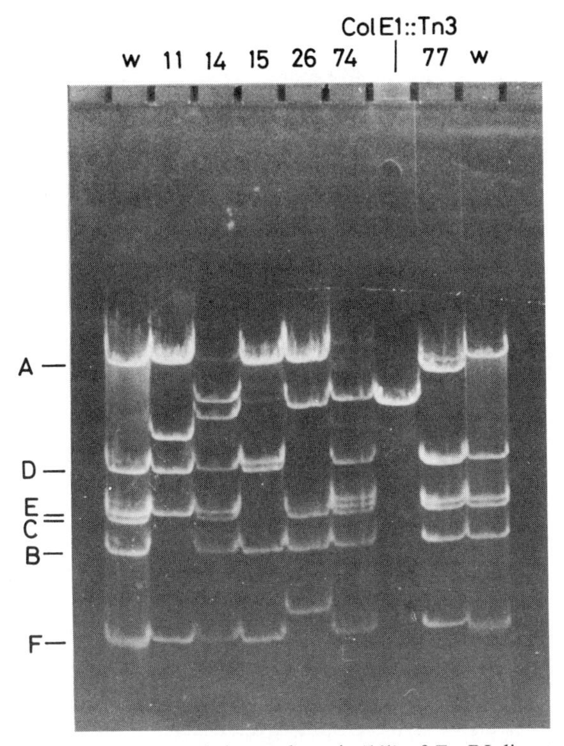

Figure 1. Agarose gel electrophoresis (1%) of *Eco*RI digests of the DNAs of wild-type λ (w), λ-ColE1::Tn*3* hybrid phages (11, 14, 15, 26, 74, 77), and ColE1::Tn*3*.

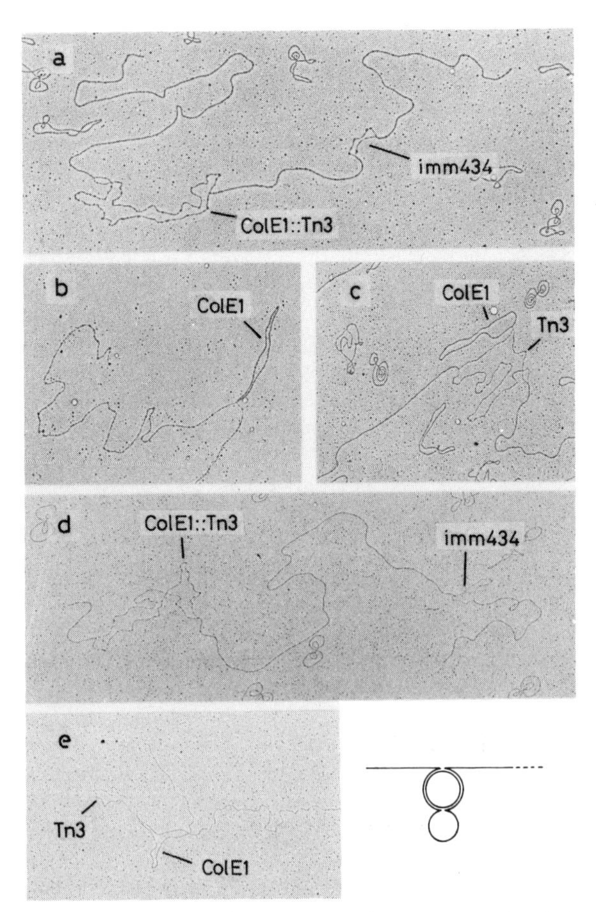

Figure 2. Heteroduplex molecules between λ-ColE1::Tn*3* hybrid phages and various DNAs. (*a*) Heteroduplex between λ-ColE1::Tn*3*(11) and λ*imm*^{434}. (*b*) Heteroduplex between λ-ColE1::Tn*3*(11) and ColE1. (*c*) Heterotriplex between λ-ColE1::Tn*3*(11), λ, and ColE1. (*d*) Heteroduplex between λ-ColE1::Tn*3*(14) and λ*imm*^{434}. (*e*) Heteroduplex between λ-ColE1::Tn*3*(14) and ColE1. Magnifications, 10,400× in *a–d*, and 9200× in *e*.

phage λ-ColE1::Tn*3*(11) is about equivalent to that of wild-type λ, based on the buoyant density of the phage, λ-ColE1::Tn*3*(11) seems to contain a whole ColE1::Tn*3* genome and to lack almost the same length of λ DNA as that of ColE1::Tn*3* at the same position as the insertion of the plasmid. λ-ColE1::Tn*3*(14) also contains a whole ColE1::Tn*3* genome and lacks 18.9 ± 0.5% of λ genome.

We have analyzed four other λ-ColE1::Tn*3* phages (15, 26, 74, and 77) by the same method and obtained essentially the same results, except that the sites of insertion of ColE1::Tn*3* are distributed over various portions of the λ genome and the extents of deletions of the λ DNA vary in each hybrid genome.

To determine the site of recombination within ColE1::Tn*3*, λ-ColE1::Tn*3* DNA was heteroduplexed against ColE1 DNA. Heteroduplexes between λ-ColE1::Tn*3*(11) and ColE1 consisted of a double-stranded circle and two large single-stranded branches (Fig. 2b). The length of the former was measured and calculated by using the length of double-stranded ColE1 DNA as a standard. It was 98 ± 2% of the length of the standard ColE1 DNA. Heteroduplexes between λ-ColE1::Tn*3*(14) and ColE1 consist of a double-stranded circle accompanied by a single-stranded circle and two large single-stranded branches (Fig. 2e). Lengths of the double-stranded and single-stranded circles were measured and calculated by the use of the lengths of double-stranded ColE1 DNA and single-stranded λ DNA, respectively, as standards. The length of the double-stranded circle was 102 ± 2% of that of the standard ColE1 DNA, and that of the single-stranded circle was 9.2 ± 0.3% λ unit, which corresponds to the length of the Tn*3* element. Hence the hybrid phages were divided into two classes. In the first class, which is represented by λ-ColE1::Tn*3*(11), the recombination occurred between λ and the Tn*3* element. In the second class, which is represented by λ-ColE1::Tn*3*(14), the recombination occurred between λ and the ColE1 element. Heteroduplex analyses of the other hybrid phages indicated that phages numbers 15 and 77 belong to the former class, whereas phages numbers 26 and 74 belong to the latter class.

λ-ColE1::Tn*3* DNAs of the former class (11, 15, and 77) were further analyzed by heterotriplexes between these DNAs, λ DNA, and ColE1 DNA. Figure 2c showed that ColE1 DNA forms a duplex circle inside a single-stranded substitution loop formed between λ and λ-ColE1::Tn*3*(11) DNAs. This result confirmed the conclusion that the insertion site of ColE1::Tn*3* is within the Tn*3* element in these phages and also enabled us to locate the precise insertion site.

To examine whether or not a part of the ColE1::Tn*3* genome is deleted in the hybrid phages, the phage DNAs were heteroduplexed against ColE1::Tn*3* DNA. In all six phages tested, heteroduplex molecules showed a structure that consists of a double-stranded circle, corresponding to the ColE1::Tn*3* element, and two single-stranded branches; however, no single-stranded deletion loop was found in them.

On the basis of the *Eco*RI and heteroduplex analyses, we constructed physical maps of six λ-ColE1::Tn*3* phages (Fig. 3). These maps show that the insertion of the ColE1::Tn*3* element occurs at various sites of the phage and accompanies a concomitant deletion of λ DNA at the point of insertion. The recombination within the plasmid is distributed on both the ColE1 and Tn*3* elements. No recombination site at the end of the Tn*3* element was found. It is therefore suggested that the recombination observed in our cell-free system is an illegitimate recombination that is distinguishable from the in vivo translocation of translocatable elements.

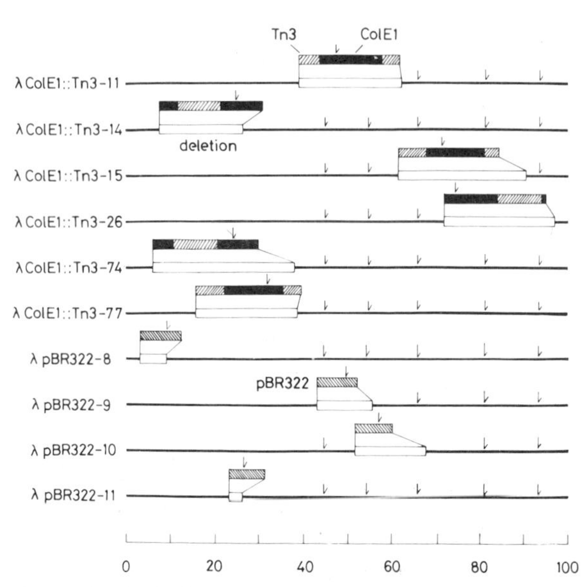

Figure 3. Physical maps of six λ-ColE1::Tn*3* and four λ-pBR322. Lengths of DNA segments were calculated on the basis of the following numbers: λ DNA, 49.5 kb (Daniels et al. 1979); Tn*3*, 4.96 kb (Heffron et al. 1979); ColE1, 6.6 kb (Ohmori and Tomizawa 1979); pBR322, 4.36 kb (Sutcliffe 1979). On the scale (percent), 0 and 100 represent the left and right cohesive ends of λ phage, respectively. (—) λ; (▭) λ deletion; (■) ColE1; (▨) Tn*3*; (▧) pBR322. Arrows show the cleavage sites of *Eco*RI enzyme.

Recombination between λ and pBR322

We examined the recombination between λ and pBR322, a plasmid containing a part of the Tn*3* element and lacking the ability to translocate itself in vivo. When pBR322 DNA (1 μg/assay) was incubated with a standard packaging mixture, Ap^r transducers were detected at a frequency of 1.3×10^{-7}. Heteroduplex analyses of Ap^r phage strains isolated from the resulting transductants showed that all four Ap^r phages tested were the substitution-type recombinant (Fig. 3). The insertion sites of pBR322 are located at various points on the phage genome. Hence, we confirmed that the recombination detected in this system is not related to the transposition mediated by the Tn*3* element.

To summarize the above results, we plotted the endpoints of the substitutions of ColE1::Tn*3* and pBR322 on the physical map of phage λ. Most of the points are distributed randomly over the λ genome, but some points seem to be clustered into several spots (Fig. 4).

Recombination between λ Deletion Mutants and Various Plasmids

In previous sections we showed that all of the hybrid phages tested are of the substitution-type. In those cases, λ DNA of full genome length was used as a substrate so

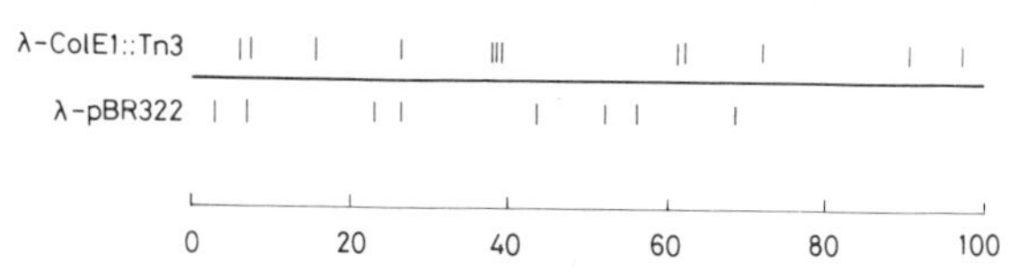

Figure 4. Distribution of endpoints of substitutions formed by recombination between λ and ColE1::Tn*3* and between λ and pBR322. The horizontal bar represents the map of λ phage. The vertical lines show endpoints of substitutions by ColE1::Tn*3* (above bar) and by pBR322 (below bar). On the scale (percent), 0 and 100 represent the left and right cohesive ends of λ phage, respectively.

Table 2. In Vitro Recombination between λ Deletion Mutants and Plasmids

λ Strain	Plasmid	Ap^r transducers per pfu	No. of insertion-type recombinants per no. of Ap^r phages tested
λ	ColE1::Tn*3*	1.4×10^{-7}	0/6
λ*b*515*b*519	ColE1::Tn*3*	1.8×10^{-7}	n.d.
λ	pBR322	1.3×10^{-7}	0/4
λ*b*515*b*519	pBR322	1.3×10^{-7}	2/2
λ*b*538	pYN30	2.6×10^{-8}	1/2

A standard packaging mixture containing wild-type λ was prepared as described in the note to Table 1. A mixture containing λ*b*515*b*519 was prepared by heat induction of 594 (λ*c*I857*A*am11*S*am7*xis*6*b*515*b*519) and 594 (λ*c*I857*E*am4*S*am7*xis*6*b*515*b*519) followed by superinfection with the same phages as prophages. A mixture containing λ*b*538 was prepared by infection of λ*c*I857*D*am15*F*Iam96B*S*am7*red*3*b*538 and λ*c*I857*E*am4*S*am7*red*3*b*538 into *E. coli* HI225. An aliquot of plasmid DNA (1 μg/assay) was incubated with these packaging mixtures and the resulting phage particles were titrated for Ap^r transducers and pfu as described in the note to Table 1. The type of recombinant DNA was determined by heteroduplex analysis. λ*b*515*b*519 and λ*b*538 contain deletions of 10.8% and 17.1% λ unit, respectively (Szybalski and Szybalski 1979). pBR322 (Bolivar et al. 1977) and pYN30 (Nakamura 1980) have molecular lengths of 4.36 kb and 7.5 kb, respectively, and contain the Ap^r determinant. n.d. indicates not determined.

that the combined length of λ and plasmid DNA exceeded the limitation of the packaging capacity of the λ head. Therefore, the above finding can be explained as a selective packaging of substitution-type recombinant DNA by the λ head. If λ deletion mutants and small plasmids are used as substrates of the recombination reaction, hybrid phages carrying insertion-type recombinant DNA might be picked up.

We constructed a packaging system using λ*b*515*b*519 derivatives and λ*b*538 derivatives. These phages contain deletions of 10.8% and 17.1% λ unit, respectively. Plasmids pBR322 and pYN30, which have respective molecular lengths of 8.8% and 15% λ unit, were incubated with these packaging mixtures. Ap^r transducers were found at the frequencies shown in Table 2. DNAs of the resulting Ap^r phages were analyzed by heteroduplex mapping. Figure 5 shows an example of a heteroduplex molecule between λ*b*515*b*519–pBR322(161) and λ DNA, in which an insertion loop was observed at a middle portion of λ DNA in addition to two loops corresponding to *b*515 and *b*519 deletions. The length of this single-stranded loop was calculated by using the length of the *b*519 loop as an internal standard and was 9.1 ± 0.3% of the λ unit, which corresponds to the length of pBR322. In sum, isolated recombinant phages between λ*b*515*b*519 and pBR322 were of insertion-type. One of two recombinant phages between λ*b*538 and pYN30 was an insertion-type, whereas another was a substitution-type. Physical maps of these insertion-type recombinant phages are summarized in Figure 6.

Effects of Bacterial and Viral Mutations on the Illegitimate Recombination

To examine the functions involved in illegitimate recombination, packaging mixtures were prepared from lysogens containing various combinations of bacterial *rec* mutations and viral *int* and *red* mutations as follows: $recA^-$ int^+ red^+, $recA^-$ int^+ red^-, $recA^-$ int^- red^+; $recA^-$ int^- red^-, and $recB^-$ $recC^-$ $recF^-$ int^+ red^-. ColE1::Tn*3* DNA was incubated with these mixtures and frequencies of Ap^r transducers were measured. The recombination took place in int^- and/or red^- mixtures at frequencies similar to that in a mixture prepared from int^+ red^+ lysogens (Table 3). A *recA, recB, recC,* or *recF* mutation does not affect the recombination (Table 3). These results suggest that this illegitimate recombination is promoted by *E. coli* functions but is independent of the *recA, recB, recC,* and *recF* functions.

The fact that recombination takes place in a packaging mixture prepared from λ*b*538 derivatives (Table 2) also indicates that the λ *att* site is not required for the recombination.

It should be noted that the functions encoded by the Tn*3* element are not present in these packaging mixtures. Furthermore, the expression of these functions is

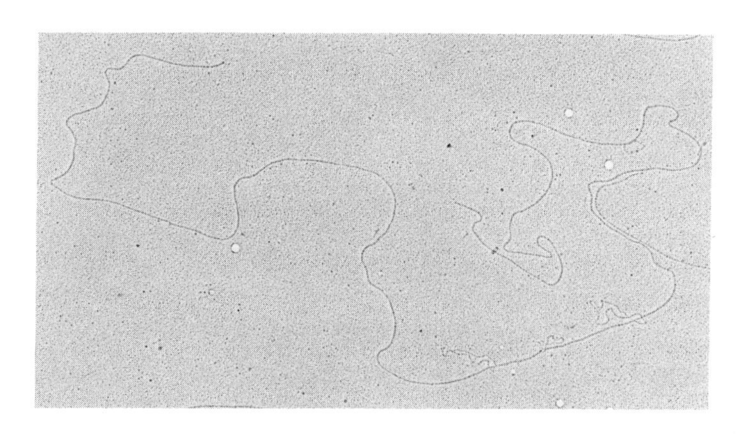

Figure 5. A heteroduplex molecule between λ and λ*b*515*b*519–pBR322(161). Deletion or insertion loops are *b*519, *b*515, and pBR322 from left to right. Magnification, 13,860×.

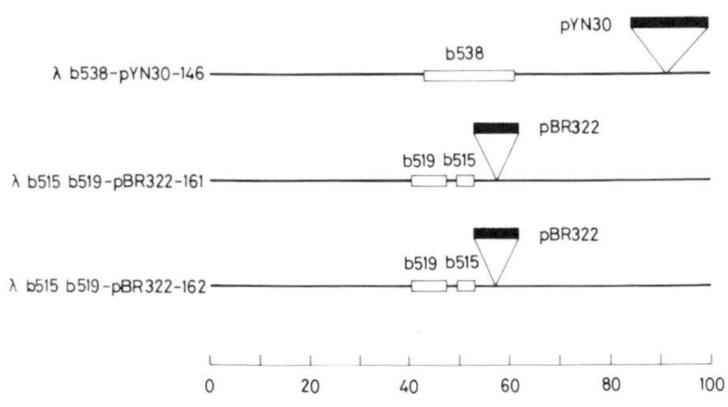

Figure 6. Physical maps of insertion-type hybrid phages. On the scale (percent), 0 and 100 represent the left and right cohesive ends of λ phage, respectively.

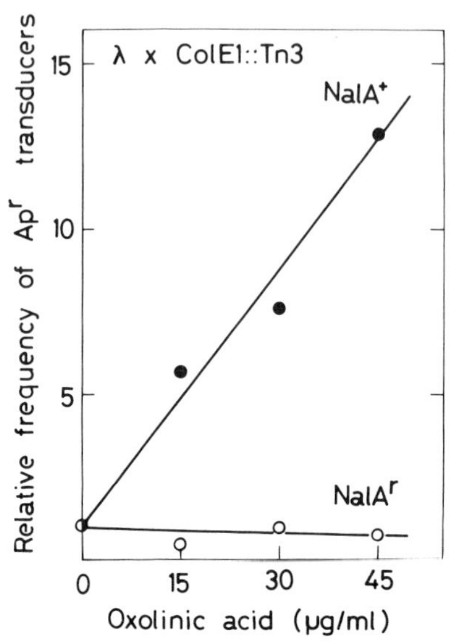

Figure 7. Effect of oxolinic acid on the recombination between λ and ColE1::Tn*3*. *nalA*$^+$ packaging mixture was a standard mixture as described in the note to Table 1. *nalA*^{r} packaging mixture was prepared from a pair of HI515 *nalA*26^r (λ*c*I857*D*am15*F*lam96*B**S*am7) and HI515 *nalA*26^r (λ*c*I857*E*am4*S*am7). HI515 *nalA*26^r was constructed by transduction of *nalA*26 from MO15124 *nalA*26 (Kubo et al. 1979) to 594 Su$^-$. ColE1::Tn*3* (1.8 µg/assay) was incubated with *nalA*$^+$ and *nalA*^{r} mixtures in the presence of increasing amounts of oxolinic acid. Frequencies of Ap^r transducers were determined as described in the note to Table 1. Relative frequency in each point was calculated from the frequency of Ap^r transducers in the absence of oxolinic acid, which were 2.1×10^{-7} and 2.7×10^{-7} in *nalA*$^+$ and *nalA*^{r} mixtures, respectively. (●) *nalA*$^+$ mixture; (○) *nalA*^{r} mixture. Oxolinic acid was generously provided by J. Tomizawa (National Institutes of Health).

not required for the recombination, because the addition of rifampin (3 µg/ml) or streptolydigin (200 µg/ml) into the reaction mixture does not affect the recombination.

Oxolinic Acid Stimulates the Illegitimate Recombination in Vitro

Since DNA gyrase is known to cause a transient breakage of the DNA double helix, it is interesting to know whether or not oxolinic acid, an inhibitor of DNA gyrase, affects the recombination between λ and plasmids. When ColE1::Tn*3* DNA was incubated in a packaging mixture in the presence of increasing amounts of oxolinic acid, frequencies of Ap^r transducers increased up to 13-fold at an inhibitor concentration of 45 µg/ml (Fig. 7). The inhibitor did not affect the packaging of bulk λ DNA. On the other hand, when a packaging mixture prepared from *nalA*^{r} lysogens was used for recombination, the frequency of Ap^r transducers was almost the same in the presence of oxolinic acid as that in the absence of oxolinic acid. Furthermore, oxolinic acid also stimulated the recombination between λ*b*515*b*519 and pBR322, which is expected to confer insertion-type recombinants. These results indicate that DNA gyrase, or at least gyrase-A protein, is involved in an illegitimate recombination induced by oxolinic acid.

Table 3. Effects of Bacterial and Viral Mutations on Recombination between λ and ColE1::Tn*3*

Source of mixture		Frequency of Ap^r transducers per plaque formers
host	phage	
rec$^+$	*int*$^+$ *red*$^+$	1.6×10^{-7}
recA	*int*$^+$ *red*$^+$	1.9×10^{-7}
recA	*int*$^+$*red*	2.3×10^{-7}
recA	*int red*$^+$	3.7×10^{-7}
recA	*int red*	4.5×10^{-7}
recB recC recF	*int*$^+$ *red*	2.3×10^{-7}

recA$^-$ *int*$^+$ *red*$^+$, *recA*$^-$ *int*$^+$ *red*$^-$, and *recB*$^-$ *recC*$^-$ *recF*$^-$ *int*$^+$ *red*$^-$ packaging mixtures were prepared from the following pairs of lysogens: HI225 *recA*1 (λ*c*I857*E*am4*S*am7) and HI225 *recA*1 (λ*c*I857*D*am15*F*lam96*B**S*am7); HI225 *recA*1 (λ*c*I857*E*am4*S*am7-*red*113) and HI225 *recA*1 (λ*c*I857*D*am15*F*lam96*B**S*am7*red*113); HI367 *recB*21 *recC*22 *recF*143 (λ*c*I857*E*am4*S*am7) and HI367 *recB*21 *recC*22 *recF*143 (λ*c*I857*D*am15*F*lam96*B**S*am7). *recA*$^-$ *int*$^-$ *red*$^+$ and *recA*$^-$ *int*$^-$ *red*$^-$ packaging mixtures were prepared by heat induction of the following pairs of lysogens followed by superinfection with the same phages as prophages: HI225 *recA*1 (λ*c*I857*E*am4*S*am7*int*6) and HI225 *recA*1 (λ*c*I857*D*am15*F*lam96*B**S*am7*int*6); HI225 *recA*1 (λ*c*I857-*E*am4*S*am7*int*6*red*3) and HI225 *recA*1 (λ*c*I857*D*am15-*F*lam96*B**S*am7*int*6*red*3). An aliquot of ColE1::Tn*3* DNA (4.5 µg/assay) was incubated with these packaging mixtures and frequencies of Ap^r transducers were determined as described in the note to Table 1.

Effect of Coumermycin on the Illegitimate Recombination Induced by Oxolinic Acid

Another question is whether gyrase-B protein participates in the illegitimate recombination or not. We examined the effect of coumermycin, an inhibitor whose target is gyrase-B protein, on the recombination induced by oxolinic acid. In an experiment using a *cou*$^+$ packaging mixture, the frequency of Ap^r recombinant phages formed in the presence of oxolinic acid and coumermycin (5 µg/ml) was about one fifth of that in the presence of oxolinic acid alone. Coumermycin did not affect the packaging of bulk λ DNA. In an experiment using a *cou*^{r} packaging mixture, on the other hand, the frequency of recombinant phages formed in the presence of both drugs was almost the same as that in the pres-

Table 4. Effect of Coumermycin on the Illegitimate Recombination Induced by Oxolinic Acid

Source of mixture	Oxolinic acid (μg/ml)	Coumermycin (μg/ml)	Ap^r transducers per pfu	Relative frequency of Ap^r transducers
cou^+	0	0	2.1×10^{-7}	1
cou^+	30	0	1.6×10^{-6}	7.6
cou^+	30	1	4.7×10^{-7}	2.2
cou^+	30	5	3.4×10^{-7}	1.6
cou^r	0	0	2.7×10^{-7}	1
cou^r	30	0	1.2×10^{-6}	6.7
cou^r	30	1	1.3×10^{-6}	6.9
cou^r	30	5	1.1×10^{-6}	6.1

cou^+ packaging mixture was a standard mixture as described in the note to Table 1. cou^r packaging mixture was prepared from a pair of HI520 *cou*51^r (λcI857*Dam*15*Fl*am96B*S*am7) and HI520 *cou*51^r (λcI857*E*am4*S*am7). HI520 *cou*51^r was isolated as a spontaneous coumermycin-resistant mutant of 594 and identified as a *gyrB* mutant by cotransduction with *ilv*. An aliquot of plasmid DNA (1.8 μg/assay) was incubated with these packaging mixtures with or without inhibitors and the resulting phage particles were examined for frequency of Ap^r recombinant phage as described in the note to Table 1. Coumermycin was generously provided by Dr. H. Tsukiura (Bristol-Banyu Research Institute, Ltd.).

ence of oxolinic acid alone (Table 4). The results showed that gyrase-B protein, in addition to gyrase-A protein, plays an essential role in the illegitimate recombination induced by oxolinic acid.

In the absence of oxolinic acid, coumermycin (5 μg/ml) reduced the frequency of recombinant phages only by 70–90%. We also examined the effect of coumermycin on the recombination in the absence of oxolinic acid in experiments using the *recA*$^-$ *int*$^-$ *red*$^-$ mixture or the *recA*$^-$ *b*538 *red*$^-$ mixture. Coumermycin did not affect the recombination in these mixtures in the absence of oxolinic acid. The recombination occurring in the absence of oxolinic acid may be mediated by a pathway different from that induced by oxolinic acid.

In Vivo Recombination between λ and pBR322

To compare the in vitro illegitimate recombination with the in vivo one, we examined the recombination between λ derivatives and pBR322 in *E. coli* cells. *E. coli* 594 (pBR322) cells were infected with wild-type λ, λ*b*515*b*519, or λ*b*538. The resulting phages were purified by low-speed and high-speed centrifugations and transduced into *E. coli* Ymel. Frequencies of Ap^r transducers in λ*b*515*b*519 and λ*b*538 preparations were 3.5×10^{-8} and 2.0×10^{-8}, respectively. These frequencies are comparable to those for the in vitro recombination of the same combinations. The frequency of Ap^r transducers in the wild-type-λ preparation, however, was 2.7×10^{-10}; this is lower than that of the in vitro recombination of the same combination.

DISCUSSION AND CONCLUSIONS

The hybrid phages between λ and plasmid DNAs were formed in vitro by incubating plasmid DNA with a packaging mixture. We concluded that these phages were produced by recombination between plasmid and endogenous λ DNAs existing in the packaging mixture. Depending on the phage strain used for the packaging system, two types of recombinant DNA were found: insertion-type and substitution-type recombinants. Experiments with the wild-type λ strain yielded the substitution-type recombinants, whereas those with deleted λ strains and small plasmids yielded both types of recombinants. We interpreted these results as follows: Both types of recombinant DNA were formed in this in vitro reaction irrespective of the length of λ DNA used, but only substitution-type recombinant DNA can be packaged in the experiments with the wild-type λ strain because of the capacity of the phage head.

One can suggest that the recombinant phages may be formed by packaging of free plasmid and recombination subsequent to transduction. However, this possibility is unlikely because most of the transductants produced HFT lysates (10^{-1} to 10^{-2} Ap^r transducers/pfu) upon induction, whereas infection of λ phage into cells carrying pBR322 yielded only low-frequency-transducing (LFT) lysates (10^{-8} to 10^{-10} Ap^r transducers/pfu).

Most of the endpoints of substitutions seem to be distributed randomly on λ DNA. Whether or not recombination sites on ColE1::Tn*3* or pBR322 DNA are distributed randomly is less clear. It is certain, at least, that both of the ColE1 and Tn*3* elements can recombine with the λ genome. Bacterial *recA, recB, recC,* and *recF* mutations and λ *int* and *red* mutations do not affect the recombination.

Several lines of evidence indicate that this recombination is distinguishable from the transposition of Tn*3*. First, the sites of insertion on ColE1::Tn*3* are not the ends of Tn*3* but are within the Tn*3* or ColE1 elements. Second, any function encoded by the Tn*3* element is not required for the recombination. Third, the recombination between λ and pBR322 (the latter element being incapable of carrying out transposition) takes place in this in vitro system. We concluded, therefore, that most, if not all, of the recombination detected in this system is due to an illegitimate recombination mediated by *E. coli* function(s). This recombination might be related to the illegitimate recombination observed in *E. coli* cells during the formation of deletions, substitutions, duplications, and specialized transducing phages.

How does the illegitimate recombination take place in our system? An intriguing clue to the answer to this question resulted from the finding that oxolinic acid, an inhibitor of DNA gyrase, stimulates the recombination. Studies with $nalA^+$ and $nalA^r$ packaging systems showed that DNA gyrase-A protein is involved in the recombination induced by oxolinic acid. Moreover, coumermycin (another inhibitor, whose target is gyrase-B protein) inhibited the induction of the recombination by oxolinic acid in the cou^+ packaging system but not in the cou^r packaging system. Hence, gyrase-B protein also participates in this recombination.

DNA gyrase is known to carry out supercoiling of relaxed circular DNA in the presence of ATP (Gellert et al. 1976a). Two types of gyrase inhibitors, oxolinic acid and coumermycin, inhibit this activity (Gellert et al. 1976b; Sugino et al. 1977). In addition, oxolinic acid inhibits gyrase-mediated conversion of supercoiled DNA into relaxed DNA. The most noticeable action of oxolinic acid is that it induces formation of a tight complex of DNA gyrase with DNA, which can then produce double-strand breaks in the DNA by treatment with a denaturing agent (Gellert et al. 1977; Sugino et al. 1977). Both 5′ termini of cleaved DNA are covalently bound with gyrase-A protein. This effect may be related to the ability of this drug to induce the recombination in our system.

On the basis of these biochemical characteristics of DNA gyrase, we constructed a model for the illegitimate recombination that explains how DNA gyrase is involved in the recombination (Fig. 8). It is most likely that the protomer structure of gyrase is $\alpha_2\beta_2$, in which α and β are encoded by *gyrA* and *gyrB*, respectively (Mizuuchi et al. 1978). Upon the binding of gyrase to DNA, gyrase transiently cleaves double strands of DNA, resulting in an intermediate structure in which each *gyrA* protomer covalently binds to each of the 5′ termini of the DNA (Fig. 8b, f). It is proposed that the structure $\alpha\beta$-$\alpha\beta$ bound to DNA undergoes subunit exchange with other $\alpha\beta$ structures bound to other 5′ termini of DNA as shown in Figure 8, c and g, leading to a concomitant exchange of DNA strands. Removal of gyrase by reverse reaction reseals the double-strand breaks in DNA, resulting in the formation of a recombinant DNA.

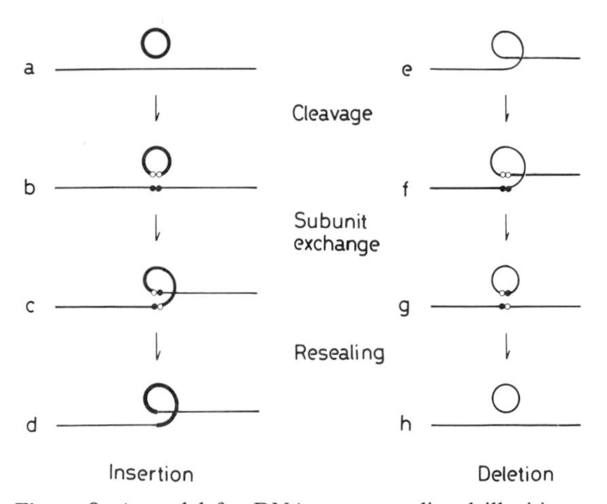

Figure 8. A model for DNA-gyrase-mediated illegitimate recombination. The circle and line in *a* and *e* represent plasmid and λ DNAs, respectively. Small circles attached to λ and plasmid DNAs represent the $\alpha\beta$ form of DNA gyrase, composed of one gyrase-A protomer and one gyrase-B protomer. Two adjacent circles constitute the $\alpha_2\beta_2$ form, which is a putative gyrase structure. DNA gyrase transiently cleaves DNA double helix, resulting in an intermediate structure (*b* or *f*). Subunit exchange of gyrase between two intermediate structures leads to the exchange of DNA strand (*c* and *g*). The resealing of the double-strand break in DNA results in the formation of a recombinant DNA (*d* and *h*). Intermolecular recombination between λ and plasmid gives rise to an insertion-type recombinant (*a–d*). Intramolecular recombination gives rise to a deletion-type recombinant (*e–h*).

The action of oxolinic acid on gyrase may be to make the gyrase-DNA complex, in which gyrase is covalently bound to DNA, more stable, leading to the accumulation of such intermediate structures. This effect caused by the inhibitor would facilitate the subunit exchange between two $\alpha\beta$ structures, resulting in the stimulation of exchange of DNA strands. According to this model, the subunit-exchange reaction can occur in the absence of oxolinic acid, even though the frequency of the exchange is low because of the small amount of the intermediate structures. Whether or not the recombination observed in the absence of the inhibitor is also mediated by DNA gyrase is an interesting question to be clarified.

Several groups recently found that DNA gyrase can both separate a catenated DNA circle and unknot knotted DNA in an ATP-dependent reaction (Brown and Cozzarelli 1979; Liu et al. 1980; Mizuuchi et al. 1980). It was also observed that the DNA-gyrase-catalyzed supercoiling involved changes in linking numbers of DNA in steps of 2 units. The interpretation of these findings was that these reactions can only occur by passage of a DNA segment through a transient double-strand break in DNA. These processes may be fundamentally the same process as the subunit exchange of gyrase that we propose. The role of gyrase-B protein in the recombination may be the promotion of the subunit exchange in this ATP-dependent process.

Figure 8a–d shows a model for the intermolecular recombination between λ and plasmid, leading to the formation of insertion-type recombinants. Figure 8e–h shows a model for the intramolecular recombination, leading to the formation of the deletion-type recombinants. An intermediate structure of an insertion-type recombinant, shown in Figure 8c, carries two covalently bound DNA-gyrase molecules. These enzymes still have an ability to induce the subunit exchange with other gyrase molecules bound to other sites of DNA. The formation of an additional deletion on this DNA would confer a substitution-type recombinant.

DNA gyrase also participates in integrative recombination of phage λ, because the λ *int* protein requires negatively supercoiled DNA to promote recombination in vitro (Mizuuchi and Nash 1976). The mode of action of DNA gyrase in the illegitimate recombination is not the same as that in the integrative recombination, because we used negatively supercoiled plasmid DNA as a substrate of the recombination and because oxolinic acid does not inhibit the recombination but stimulates it.

Models have been proposed for general genetic recombination (Holliday 1964; Meselson and Radding 1975). According to these models, the recombination is initiated by a single-strand transfer, which requires extensive homology. A model for RNA-polymerase-mediated recombination also requires a homology of DNA strands (Ikeda and Matsumoto 1979). In the integrative recombination system of phage λ, the viral and bacterial *att* sites share a common sequence of 15 bp, where crossover takes place (Landy and Ross 1977). The four-strand model proposed by Kikuchi and Nash (1979) assumes that an interaction between the homologous sequences has an essential role in the formation of recombination intermediates. A number of models for illegitimate recombination have been proposed, most of which also have been based on the concept that exchange of DNA takes place between short homologous regions (Ozeki and Ikeda 1968; see also Weisberg and Adhya 1977). Contrary to these models, our hypothesis proposes a novel mechanism in which the recombination is conducted by subunit exchange of DNA gyrase. It predicts that the site of recombination is the site of cleavage by gyrase. The double-strand breaks in DNA induced by DNA gyrase occur at specific sites (Gellert et al. 1977; Sugino et al. 1977). Sequences of the cleavage sites have already been determined (Morrison and Cozzarelli 1979; Gellert et al., this volume). The common sequence for several cleavage sites determined is only TG. Our model does not require homology of DNA sequence for crossing-over, providing us a novel concept for the mechanism of genetic recombination. A study to determine the sequence of the recombination site is in progress.

ACKNOWLEDGMENTS

We thank Drs. H. Uchida, Y, Kaziro, and K. Matsubara for valuable discussions; Dr. J. Inselberg for his careful reading of the manuscript; and Dr. F. Imamoto for providing the bacterial strain. This work was supported in part by a grant to H. I. from the Ministry of Education of the Japanese Government.

REFERENCES

Bolivar, F., R. L. Rodriguez, P. J. Greene, M. C. Betlack, H. L. Heyneker, and H. W. Boyer. 1977. Construction and characterization of new cloning vehicles. II. A multipurpose cloning system. *Gene* **2:** 95.

Brown, P. O. and N. R. Cozzarelli. 1979. A sign inversion mechanism for enzymatic supercoiling of DNA. *Science* **206:** 1081.

Daniels, D. L., J. R. deWet, and F. R. Blattner. 1979. A genetic and physical map of bacteriophage λ DNA. *J. Virol.* **33:** 390.

Davis, R. W., M. Simon, and N. Davidson. 1971. Electron microscope heteroduplex methods for mapping regions of base sequence homology in nucleic acids. *Methods Enzymol.* **21D:** 413.

Emmons, S. W., V. MacCosham, and R. L. Baldwin. 1975. On the mechanism of production of tandem genetic duplications in phage lambda. *J. Mol. Biol.* **95:** 83.

Franklin, N. 1967. Extraordinary recombinational events in *Escherichia coli.* Their independence of the rec^+ function. *Genetics* **55:** 699.

———. 1971. Illegitimate recombination. In *The bacteriophage lambda* (ed. A. D. Hershey), p. 175. Cold Spring Harbor Laboratory, Cold Spring Harbor, New York.

Gellert, M., K. Mizuuchi, M. H. O'Dea, and H. Nash. 1976a. DNA gyrase: An enzyme that introduces superhelical turns into DNA. *Proc. Natl. Acad. Sci.* **73:** 3872.

Gellert, M., M. H. O'Dea, T. Itoh, and J. Tomizawa. 1976b. Novobiocin and coumermycin inhibit DNA supercoiling catalysed by DNA gyrase. *Proc. Natl. Acad. Sci.* **73:** 4474.

Gellert, M., K. Mizuuchi, M. H. O'Dea. T. Itoh, and J. Tomizawa. 1977. Nalidixic acid resistance: A second genetic character involved in DNA gyrase activity. *Proc. Natl. Acad. Sci.* **74:** 4772.

Gingery, R. and H. Echols. 1969. Integration, excision, and transducing particle genesis by bacteriophage λ. *Cold Spring Harbor Symp. Quant. Biol.* **33:** 721.

Gottesman, M. E. and M. B. Yarmolinsky. 1969. The integration and excision of the bacteriophage lambda genome. *Cold Spring Harbor Symp. Quant. Biol.* **33:** 735.

Heffron, F., B. J. McCarthy, H. Ohtsubo, and E. Ohtsubo. 1979. DNA sequence analysis of the transposon Tn*3*: Three genes and three sites involved in transposition of Tn*3*. *Cell.* **18:** 1153.

Holliday, R. 1964. A mechanism for gene conversion in fungi. *Genet. Res.* **5:** 282.

Ikeda, H. and I. Kobayashi. 1979. *recA*-mediated recombination of bacteriophage λ: Structure of recombinant and intermediate DNA molecules and their packaging in vitro. *Cold Spring Harbor Symp. Quant. Biol.* **43:** 1009.

Ikeda, H. and T. Matsumoto. 1979. Transcription promotes *recA*-independent recombination mediated by DNA-dependent RNA polymerase in *Escherichia coli. Proc. Natl. Acad. Sci.* **76:** 4571.

Inselberg, J. 1967. Formation of deletion mutations in recombination-deficient mutants of *Escherichia coli. J. Bacteriol.* **94:** 1266.

Jaskunas, S. R., L. Lindahl, M. Nomura, and R. R. Burgess. 1975. Identification of two copies of the gene for the elongation factor EF-Tu in *E. coli. Nature* **257:** 458.

Kikuchi, Y. and H. A. Nash. 1979. Nicking-closing activity associated with bacteriophage λ *int* gene product. *Proc. Natl. Acad. Sci.* **76:** 3760.

Kobayashi, I. and H. Ikeda. 1977. Formation of recombinant DNA of bacteriophage lambda by *recA* function of *Escherichia coli* without duplication, transcription, translation, and maturation. *Mol. Gen. Genet.* **153:** 237.

Kopecko, D. J. and S. N. Cohen. 1975. Site-specific *recA*-independent recombination between bacterial plasmids: Involvement of palindromes at the recombinational loci. *Proc. Natl. Acad. Sci.* **72:** 1373.

Kubo, M., Y. Kano, H. Nakamura, A. Nagata, and F. Imamoto. 1979. *In vivo* enhancement of general and specific transcription in *Escherichia coli* by DNA gyrase activity. *Gene* **7:** 153.

Landy, A. and W. Ross. 1977. Viral integration and excision: Structure of the lambda *att* sites. *Science* **197:** 1147.

Liu, L. F., C. C. Liu, and B. M. Alberts. 1980. Type II DNA topoisomerase: Enzymes that can unknot a topologically knotted DNA molecule via a reversible double-strand break. *Cell* **19:** 697.

Meselson, M. S. and C. M. Radding. 1975. A general model for genetic recombination. *Proc. Natl. Acad. Sci.* **72:** 358.

Mizuuchi, K. and H. A. Nash. 1976. Restriction assay for integrative recombination of bacteriophage λ DNA *in vitro:* Requirement for closed circular DNA substrate. *Proc. Natl. Acad. Sci.* **73:** 3524.

Mizuuchi, K., M. H. O'Dea, and M. Gellert. 1978. DNA gyrase: Subunit structure and ATPase activity of the purified enzyme. *Proc. Natl. Acad. Sci.* **75:** 5960.

MIZUUCHI, K., L. M. FISHER, M. H. O'DEA, and M. GELLERT. 1980. DNA gyrase action involves the introduction of transient double-strand breaks into DNA. *Proc. Natl. Acad. Sci.* **77:** 1847.

MORRISON, A. and N. R. COZZARELLI. 1979. Site-specific cleavage of DNA by *E. coli* DNA gyrase. *Cell* **17:** 175.

NAKAMURA, Y. 1980. Hybrid plasmid carrying *Escherichia coli* genes for the primase (*dnaG*) and RNA polymerase sigma factor (*rpoD*); gene organization and control of their expression. *Mol. Gen. Genet.* **178:** 487.

OHMORI, H. and J. TOMIZAWA. 1979. Nucleotide sequence of the region required for maintenance of colicin E1 plasmid. *Mol. Gen. Genet.* **176:** 161.

OZEKI, H. and H. IKEDA. 1968. Transduction mechanisms. *Annu. Rev. Genet.* **2:** 245.

SO, M., R. GILL, and S. FALKOW. 1975. The generation of a ColE1-Ap^r cloning vehicle which allows detection of inserted DNA. *Mol. Gen. Genet.* **142:** 239.

SUGINO, A., C. L. PEEBLES, K. N. KREUZER, and N. R. COZZARELLI. 1977. Mechanism of action of nalidixic acid: Purification of *Escherichia coli nalA* gene product and its relationship to DNA gyrase and a novel nicking-closing enzyme. *Proc. Natl. Acad. Sci.* **74:** 4767.

SUTCLIFFE, J. G. 1979. Complete nucleotide sequence of the *Escherichia coli* plasmid pBR322. *Cold Spring Harbor Symp. Quant. Biol.* **43:** 77.

SZYBALSKI, E. and W. SZYBALSKI. 1979. A comprehensive molecular map of bacteriophage lambda. *Gene* **7:** 217.

WEISBERG, R. A. and S. ADHYA. 1977. Illegitimate recombination in bacteria and bacteriophage. *Annu. Rev. Genet.* **11:** 451.

Instability of Palindromic DNA in *Escherichia coli*

J. Collins
Gesellschaft für Biotechnolgische Forschung, Abt. Genetik, Mascheroder Weg 1, D-3300 Braunschweig, Federal Republic of Germany

The lack of palindrome formation during recombinant DNA cloning experiments is a paradox that has been largely overlooked. In the present context, palindromic molecules are defined as DNA molecules that have twofold rotational symmetry in which the symmetry is perfect right up to the axis of symmetry, as distinguished from inverted repeats, in which a nonsymmetrical region separates the inverted symmetries. Collins and Hohn (1978) demonstrated the absence of palindromes among recombinants during the first use of cosmid cloning vehicles in which polymers of the cloned fragment were preferentially produced. Earlier attempts to clone large palindromic DNAs, such as the ribosomal RNA genes of ciliates (J. Collins and J. Engberg, unpubl.; J. Gall, pers. comm.) or of slime molds (R. Firtel, pers. comm.), failed to give hybrids containing the axis of symmetry. Perricaudet et al. (1977) demonstrated that a large palindromic region in a λ clone was highly unstable and was excised exactly (it was speculated, depending on the *recA* system), whereas Bolivar et al. (1977) reported the presence in a plasmid vector of a small palindrome (58 bp) that appeared stable in spite of aberrations in DNA synthesis during replication (in the presence of chloramphenicol) caused by the presence of this palindromic structure.

It seems that no detailed study has been made up to now of the fate of palindromes in *E. coli.* Aside from the possible problem for those involved with cloning, it was considered that such a study was important to examine what happened to palindromic DNA in *E. coli* from the point of view of certain models (transposition, initiation of DNA replication, transcription control, phage packaging) in which hairpin or cruciform DNA structures are invoked (e.g., Sobell 1973). Since palindromes are the simplest structures that may migrate into the cruciform configuration, it was considered of interest to see whether this property was associated with any inherent instability.

In this paper I describe experiments in which palindromic regions are almost exclusively produced within the β-lactamase gene of pBR322 by means of the introduction of Tn*5* into the gene and the subsequent removal of the central region of the transposon by specific restriction endonuclease cleavage. A detailed analysis of the products obtained by transforming *E. coli* with this DNA leads to some surprising observations that relate to the serial manner in which subsequent deletions occur at high frequency, to the dependence of this phenomenon on temperature, and to the independence of deletion formation of host *recA* functions. It is concluded that this method or similar methods of introducing palindromic DNA within specific genes can be regarded as a novel method of mutagenesis, and a project designed to use this phenomenon for gene-gene fusion is described.

Test System

To test the stability of palindromic DNA in *E. coli*, pBR322::Tn*5* or pBR325::Tn*5* derivatives were cleaved with restriction enzymes that cut only within the inverted repeats of the transposon. Because of the preference for ring closure at low DNA concentration, dilution of the DNA before ligation (below 0.5 μg/ml) favored the production of palindromic structures, with the cleavage site at the axis of symmetry. Such structures are obtained, for example, by deletion of the small *Xho*I or *Bgl*II fragments of pBR322::Tn*5* derivatives (e.g., hatched regions in Fig. 1). The reinsertion of the small fragments can be screened for by testing for kanamycin resistance (Km^r), which is carried in the central region of Tn*5*, and by restriction mapping. Such molecules made up only a small percentage of the transformed tetracycline-resistant (Tc^r) population and were not examined further. Table 1 contains the frequencies of transformants obtained subsequent to transformation of *E. coli* 5K *thr leu thi ser* r^- m^+ with either pBR325::Tn*5* (Ap^s115) or pBR322::Tn*5* (Ap^s42) DNA subsequent to deletion of the central nonsymmetrical regions of Tn*5* with either *Xho*I or *Bgl*II as described above. The Ap^s115 derivative has Tn*5* inserted in pBR325 at almost the same position within the ampicillin-resistance (Ap^r) gene (± 30 bp).

Appearance of Ap^r Colonies

It was surprising to find that among Tc^r transformants a very large proportion (12–35%) were Ap^r. One possible explanation for this phenomenon could be the presence in the original pBR::Tn*5* plasmid DNA preparation of molecules in which exact excision of the transposon had already occurred or in which in vivo deletions prior to the isolation of the DNA had removed the restriction sites for *Bgl*II and/or *Xho*I (Berg 1977). Such molecules would have remained supercoiled throughout the indicated treatments and would have been transformed with high efficiency even if ligation had been omitted. However, this possibility is ruled out by the observation of a very low background of transformants in the nonligated control. It must be concluded, therefore, that the very high frequency of appearance of Ap^r clones is due to the complete deletion of the palindromic region (and one

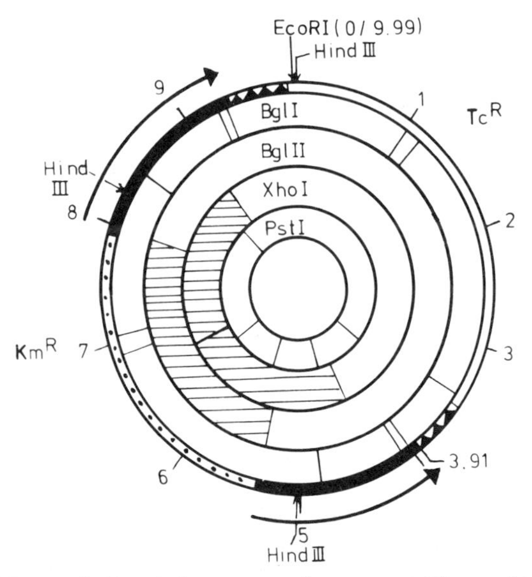

Figure 1. Restriction map of an ampicillin-sensitive (Ap^s42) Tn*5* (Km^r) derivative of pBR322. Molecular size is given in kilobase pairs (kb). (▬▶) Inverted repeats of Tn*5*; (▲▲) bisected β-lactamase gene; (▨) regions removed by *Xho*I or *Bgl*II digestion.

copy of the adjacent 9-bp direct repeat) at a correspondingly high frequency. This has been verified by the finding that the 258-bp *Sau*3A fragment of pBR322 (originally lost due to Tn*5* insertion) was restored (±5 bp) in all these Ap^r colonies.

Initial Characterization of Plasmid DNA Subsequent to Transformation

Thirteen of the Tc^rAp^s transformants from the *Xho*I-treated pBR322::Tn*5* (Ap^s42) transformation were examined in small, cleared lysates for the size of the plasmid DNA (Fig. 2). All isolates were found to be of different sizes, but all were larger than pBR322. Some isolates appeared to contain mixtures of plasmids (e.g., numbers 10, 11, and 17). Tc^rAp^r isolates from this same experiment (not shown) all contained at least a low percentage of a plasmid of about the same size as pBR322 and often one (or more) larger plasmids. All plasmids were smaller than the expected form that would have been obtained by straight ring closure. On subsequent restriction enzyme analysis, no isolate was found to contain an *Xho*I site.

A Class of Clones Segregating Ap^r with High Frequency

To avoid complications that might have arisen due to transposition proteins that might still have been expressed in the *Bgl*II deletions (Rothstein et al. 1980), further investigations concentrated on the pBR322::Tn*5* (Ap^s) x*Xho*I deletions (Δ*Xho*I). Samples of the primary cultures used to make the cleared lysates analyzed in Figure 2 were tested for the appearance of Ap^r segregants. Each clone gave rise to Ap^r colonies at a frequency of from 1×10^{-7} to 4×10^{-2} depending on the clone (Fig. 3). The frequency of segregation did not correlate with the length of the Tn*5* remnant in the Ap^r gene.

Is Normal β-Lactamase Responsible for the Ap^r Phenotype?

It was considered possible that new forms of β-lactamase might have been formed during the partial deletion process observed here. Using a Nitrocephin β-lactamase-specific color test (Matthew and Hedges 1976), the molecular weight (not shown) and the isoelectric point (Fig. 4) of the β-lactamase present at low levels in some Ap^s derivatives or in the Ap^r segregants from Δ*Xho*I-10, -13, -17, and -19 were compared with those of the parental pBR322. No differences were detected. The ratio of the enzyme level in the periplasmic space compared with the intracellular level was also the same in each case (93±4% in the periplasmic space). The possibility that the 9-bp direct repeat that borders the Tn*5* insertion is not removed in some cases could not be ruled out by these studies at the enzyme level. The reconstitution of the *Sau*3A 258-bp fragment of pBR322, however, has been seen in all Ap^r isolates examined (±5 bp), although usually a mixture of plasmids was present in the purified clones. This implies that the observed segregation of Ap^r colonies from these Tc^rAp^s transformants is due to exact excision of the entire symmetrical (palindromic) region and one copy of the adjacent 9-bp repeat.

Partially Deleted Molecules Still Contain Palindromes

Further analysis was concentrated on four Δ*Xho*I isolates that showed varying levels of instability. These were mapped in detail using a number of restriction endonucleases. It should be noted that by the time Δ*Xho*I-

Table 1. Transformation Frequencies of DNA Cleaved by Certain Restriction Endonucleases

Restriction Plasmid	Restriction endonuclease	Transformation per μg: ligated Tc^r	ligated Tc^rAp^r	nonligated Tc^r	nonligated Tc^rAp^r
pBR322::Tn*5* (Ap^s42)	*Xho*I	170	60	3	2
pBR322::Tn*5* (Ap^s42)	*Bgl*II	480	170	9	0
pBR325::Tn*5* (Ap^s115)	*Xho*I	270	33	2	1

Km^r colonies not included. Cleavage took place either in the absence of ligation or subsequent to ligation at 0.5 μg/ml.

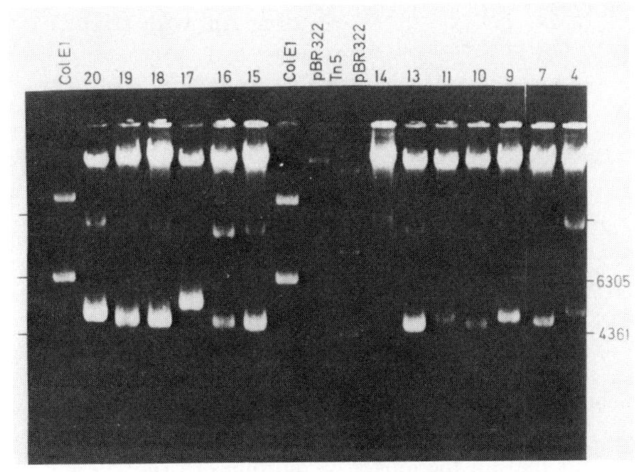

Figure 2. Gel electrophoresis of cleared lysates from *E. coli* 5K transformants carrying *Xho*I-deleted pBR322::Tn*5* (Ap^{s}42). Molecular-size markers pBR322 (4361 bp) and ColE1 supercoils (6305 bp) are indicated.

10 had been prepared in bulk for mapping, it had undergone a further spontaneous deletion to yield a more stable insert (Table 2) indistinguishable at the moment from Δ*Xho*I-13. The restriction map of these isolates (Fig. 5) clearly shows that they are derived from the strightforward ring-closure parent molecule (lower map) by symmetrical and congruent deletion from the axis of symmetry. Although these derivatives have not yet been sequenced, the restriction sites are so conveniently distributed that there are only very small regions in each molecule in which the extent of the symmetry right up to the axis is still in question (Δ*Xho*I-10 and Δ*Xho*I-13, 16 bp; Δ*Xho*I-17, 30 bp; Δ*Xho*I-19, 14 bp).

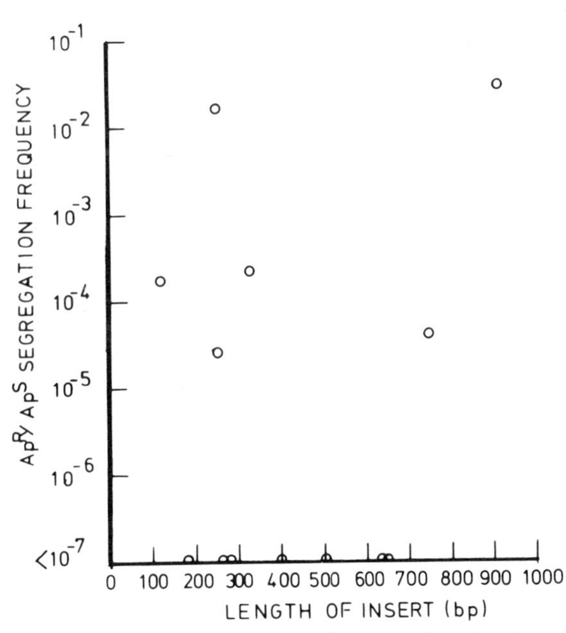

Figure 3. The frequency of Ap^r segregation is plotted against the length of the Tn*5* remnant in the *Xho*I-deleted pBR322::Tn*5* (Ap^{s}42) for the first 13 primary transformants obtained. Ap^r segregation in pBR322::Tn*5* (Ap^{s}42) was less than 10^{-7}.

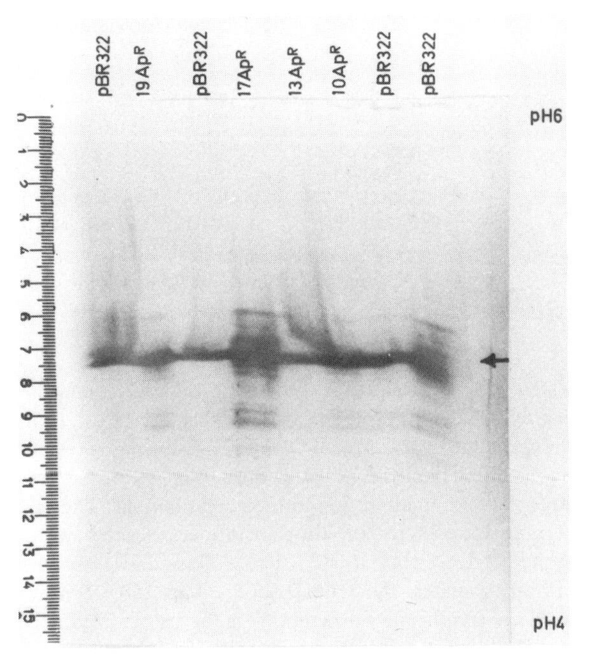

Figure 4. Isoelectric focusing of extracts from pBR322::Tn*5* (Ap^{s}42) *Xho*I Ap^r segregants in 2% ampholines, pH 4–pH 6. β-lactamase activity was visualized with a specific dye. The photograph shows the results after 30 min staining. After 2 min reaction, only the heavy band (→) was visible.

The Appearance of Plasmid Molecules Larger than the Original Isolates

In some examples of subcloning (e.g., during the selection of Ap^r colonies), and in other unrelated studies, two types of size increase of the plasmids were noted (estimated to be at least a few percent). First, many plasmids were found in the dimeric form. This was not considered an unusual feature for either pBR322 or its derivatives. More surprisingly, some plasmids were found that had increased their size by varying amounts, up to a few hundred base pairs. This phenomenon is being examined in more detail, but at present it is unclear whether these molecules were generated de novo from existing plasmids or were merely selected from a heterogeneous, but barely detectable, background of larger molecules present in the primary isolate. Local duplications and partial duplications have recently been observed in regions of palindromic symmetry in IS elements (Ghosal and Saedler 1978, 1979), and it is exciting to speculate that this is also occurring in these regions at a high frequency. If this were occurring, this would of course add a new dimension to the possible sequence diversity that could be generated at such a site.

The Independence of Exact Excision of the *recA* Genotype

Purified supercoiled DNA from Δ*Xho*I-10, -13, -17, and -19 was transformed into either *E. coli* HB101 *pro leu recA* r$^-$ m$^-$ or 5K at low DNA concentration (50 ng/ml) to ensure that each clone resulted from transformation by a single molecule. Isolated colonies were

Table 2. Frequency of Reversion to Ap^r in pBR322::Tn*5* (Ap^s42) *Xho*I Deletions

Isolate designation	Size of insert remaining	Frequency of reversion to Ap^r: original isolate 5K	transformed into HB101 *recA*[a]	transformed into 5K[a]
Δ*Xho*I-10	120	2.0×10^{-2}[b]	1.05×10^{-4}	4.6×10^{-5}
Δ*Xho*I-13	120	2.0×10^{-4}	3.3×10^{-4}	5.8×10^{-5}
Δ*Xho*I-17	910	4.0×10^{-2}	2.6×10^{-2}	1.3×10^{-3}
Δ*Xho*I-19	250	3.0×10^{-5}	4.2×10^{-5}	1.3×10^{-5}

[a] Averages from three single isolates (internal range 3- to 4-fold).
[b] DNA used in the Δ*Xho*I-10 transformations was 50 bp smaller than that observed in the original isolate.

tested for the frequency of Ap^r-segregant formation (Table 2) and for the size of the plasmid DNA (not shown). One of the three Δ*Xho*I-17 transformants tested was smaller than the parental molecule. All other isolates showed no alteration in molecular weight. The revertant frequencies of all transformed isolates were higher (2- to 20-fold) in HB101 *recA* than in 5K. For the primary isolates, the values obtained in HB101 *recA* were not significantly different from the values obtained in 5K (third column Table 2), with the exception of Δ*Xho*I-10. It has been pointed out, however, that the primary isolates often contained quite a mixture of plasmid molecules, and it is perhaps to be expected that the more unstable derivatives would be lost during the growing up of cultures for the bulk DNA preparation (e.g., Δ*Xho*I-10). The consistent trend that the reversion frequencies are higher in *recA* than in $RecA^+$ background approaches a significant level. The range from highest to lowest within each set of isolates was maximally 3- to 4-fold. The 20-fold difference observed with Δ*Xho*I-17 is therefore considered significant.

Temperature Dependence of the Deletion of Palindromic Regions

The frequency with which the exact excision of the palindromic regions occurred was measured as a function of temperature. Temperature was maintained as indicated in Figure 6. Each point on the graph is an average of three values obtained with separate colonies. It appears that for the longer palindromes (Δ*Xho*I-19 and -17), excision is very markedly dependent on temperature, 1–1½ orders of magnitude over the temperature range examined.

This temperature dependence of deletion formation is in strong contrast to that oberved with the exact excision of Tn*10* (Botstein and Kleckner 1977) and IS*1* (Reif and Saedler 1975), both of which showed no temperature dependence. In even more striking contrast is the observation of deletion formation caused by the presence of IS*1* on adjacent sequences. This process was found to be 10-fold higher at 32 °C than at 42 °C (Reif and Saedler 1975). In view of this difference, it would seem likely that the present phenomenon is unrelated to the deletion processes described previously.

CONCLUSION

On the basis of the negative results of earlier attempts to clone palindromic DNA sequences, it was expected that such sequences would be rapidly and completely deleted in a large proportion of transformants in which this type of symmetry was initially present. In the experiments described here, this was seen to occur with a high frequency, but, in addition, many different metastable derivatives were also found in which partial deletion of the palindromic DNA had occurred in such a

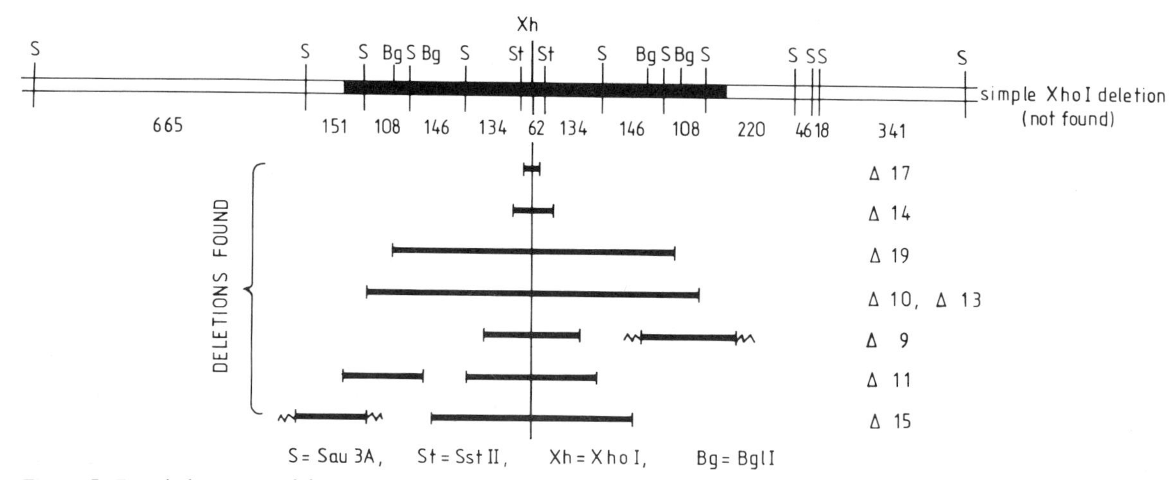

Figure 5. Restriction maps of the remnants left after partial deletion of the palindromic structures caused by deletion of the small *Xho*I fragments from pBR322::Tn*5* (Ap^s42) (see Fig. 1).

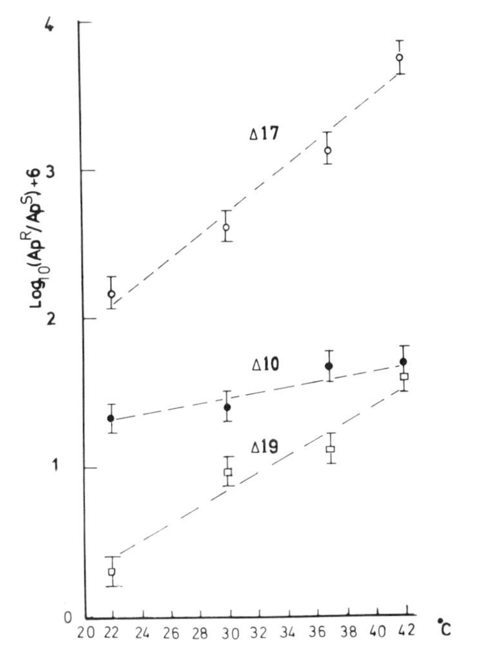

Figure 6. Single colonies from 5K pBR322::Tn*5* (Ap^{s}42) Δ*Xho*I derivatives Δ*Xho*I-10, -17, and -19 were streaked out on Luria-Bertani (LB)-broth plates at 22 °C, 30 °C, 37 °C, and 42 °C. Colonies were taken at a uniform size of about 3-mm diameter (i.e., 17 to 18 generations after plating) and tested for the ratio of Ap^r to Ap^s cells (37 °C, ±50 μg/ml of ampicillin).

fashion that new, shorter palindromes were created. Some of these were quite stable enough to allow accurate mapping of DNA amplified in cultures over 40 or so generations. However, complete excision, including excision of the 9-bp direct repeats bordering the palindromic region, was still occurring with very high frequency in many isolates. The stability of these structures appeared worse when introduced into a *recA* background. Previously, the deletion and/or further duplication of tandem direct repeats has been observed to be independent of the *recA* function of the host (Farabaugh et al. 1978), although at a level 5 orders of magnitude lower than that observed for these palindromic sequences. Egner and Berg (1980) have also demonstrated that exact excision of Tn*5* is independent of transposition functions of Tn*5* and can vary from 3×10^{-8} to 2×10^{-4} depending on the site of insertion. Some kind of DNA polymerase "slippage" mechanism (Streisinger et al. 1967) has been invoked to explain this type of deletion formation and its independence of host *recA* functions. This mechanism also requires that the ends of the deletion are close together during replication. The strong dependence of the frequency of exact excision of palindromes on the sequence at the axis of symmetry, as observed here, supports a model in which "snapback" of the palindromic region occurs during replication (Fig. 7), bringing the directly repeated sequences into juxtaposition for possible "slippage" deletion. That this should occur more frequently in a *recA* strain can be explained by the hypothesis that the editing repair system (step 4A, Fig. 7) competes with fixation of the deletion, which will only occur during a subsequent round of replication (step 5, Fig. 7). In other words, it is proposed in this model that errors of replication in which the polymerase has replicated across the stem of a foldback are preserved in a *recA* strain and later perpetuated.

Experiments are currently under way to use this property of the excision of palindromes in a scheme for gene-gene fusion. This system is depicted in Figure 8. Since Tn*5* (or other transposons) are used routinely in many laboratories for mapping the extent of a gene or operon on a physical map, such Tn*5* derivatives throughout the gene will in many cases be available. Cleavage of individual clones or of a mixture with a restriction endonuclease that cuts in the Tn*5* inverted repeats (*Bgl*II, *Hin*dIII, *Pst*I, *Xho*I, or *Sst*I) and cuts outside of the gene will yield (among others) a set of fragments containing one or the other end of the gene connected to a portion of the inverted repeat. By ligation of these fragments into a corresponding site in a vector carrying the same fragment of the inverted repeat at the cloning site, one should set in motion a train of events in which fragments that are correctly oriented will fuse with the adjacent gene fragment of the vector. In view of the observations presented above, it is expected that this will lead to many thousands of possible fusion products. Subsequent biological selection should lead to the isolation of many unusual fusion peptides. It is hoped that this method will provide a useful new weapon for the geneticist's armory. Using *Pst*I deletion of plasmids of the type shown in Figure 1 and *Hin*dIII deletions of pBR325::Tn*5* (Cm^s) derivatives, gene fusions are being attempted in which particular genes are fused either to the signal peptide for transport into the periplasmic space (N-terminal 40 amino acids of β-lactamase) or to the very strong promoter of the chloramphenicol transacetylase gene, respectively.

At present, there is no known bacterial host in which palindromic DNA can be cloned stably. In view of the multiple products that may arise from such attempts, care should be taken in the interpretation of experiments in which either purified fragments (the vector by itself may produce such artifacts!) or eukaryotic DNAs, which may contain extensive palindromes (Thomas et al. 1974), are being cloned.

Regions of twofold rotational symmetry may play important roles in DNA-protein interactions (Sobell 1973), as is seen, for example, in the *lac* operator and the CAP-protein-binding site (Dykes et al. 1975; Bertrand et al. 1975), in the attenuator region of the *trp* operon (Yanofsky 1976), in certain promoter regions (Walz and Pirrotta 1975), in the initiation of DNA replication (Hirota et al. 1979; Hobom et al. 1979), and in the recognition site for λ DNA packaging (Nichols and Donelson 1978). Apart from these examples, no long regions (longer than 8 bp) of perfect twofold symmetry have been found to occur naturally in the *E. coli* cell. In fact, only in the *trp* attenuator (10 bp), the *E. coli* origin (15 bp), and the λ *cos* site (10 bp and 15 bp) are the symmetry regions perfect palindromes. A mechanism that destroys longer palindromic regions could be imagined as being useful for correcting DNA replication errors

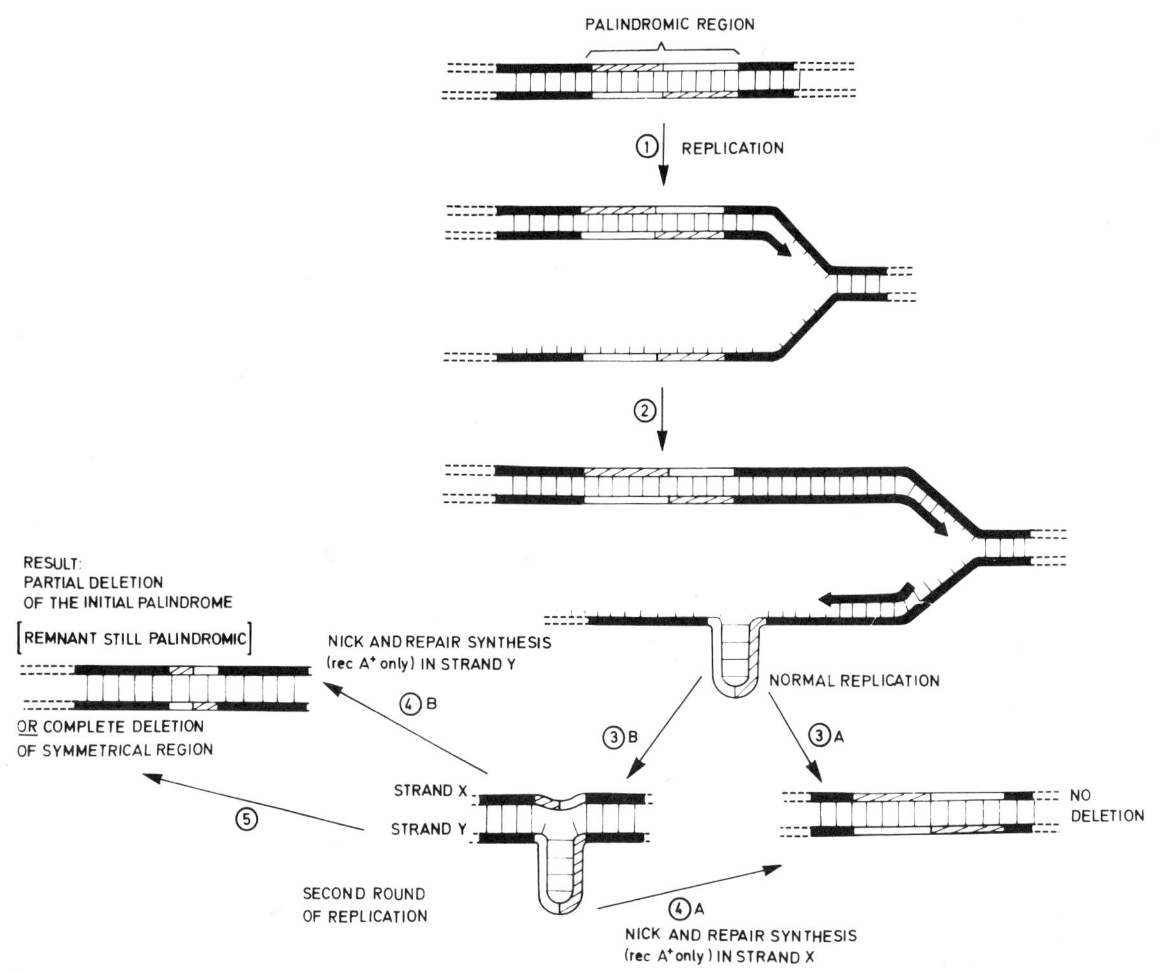

Figure 7. A model for the *recA*-independent excision of palindromic structures. Note in particular that in a *recA*$^+$ host, steps *4A* and *5* can be considered as competitive reactions in the establishment of deletions of the palindromic regions.

(destruction of hairpin molecules formed after polymerase had replicated back on the "lagging" strand).

It is also interesting to consider the possibility that this mechanism provides a strong impetus for the generation of natural diversity, i.e., for the basic raw material on which natural selection can operate. As demonstrated above, the deletion of the center of a transposon leads to a chain of site-specific mutagenic events at a very high frequency. This allows a very high rate of mutagenesis to occur in the cell over a considerable period without deleterious or mutagenic effects on other genes.

The possibility should also be considered that this mechanism may be initiated subsequent not only to partial deletion of transposons and insertion sequences, but also to defects in replication in which hairpin molecules are formed due to the DNA polymerase jumping to the opposite strand.

Questions Raised and Still Not Answered

The isolation of meta-stable palindromes within an otherwise intact structural gene will allow a number of the following problems to be studied in more detail.

Is exact and/or partial excision of palindromes

1. dependent on host functions other than *recA*, e.g., replication functions?
2. only apparent in ColE1-based replicons that use polymerase I for the replication of the first few hundred bases adjacent to the origin?
3. dependent on the adjacent direct repeats?
4. only a phenomenon associated with the ends of transposons?

Does the presence of a palindrome

5. generate larger structures by partial duplication (implied)?
6. lead to larger deletions extending beyond the ends of the symmetry regions (not yet observed)?
7. actually lead to hairpin or cruciform structures in vivo?

Finally, can one

8. define the level of stability in terms of the actual base sequence at the axis of symmetry?
9. find *E. coli* hosts in which palindromes are more stable?

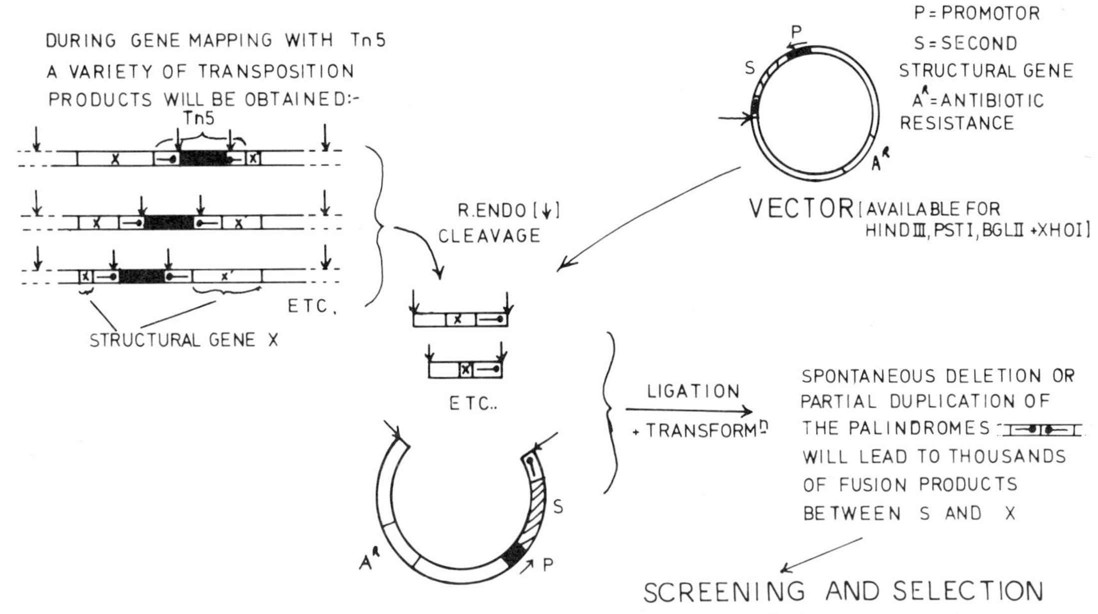

Figure 8. A scheme for creating gene-gene fusion using spontaneous deletion of palindromes.

Note Added in Proof

Plasmid DNA from clones giving very low segregation ($\leq 10^{-7}$) of Ap^r colonies was found to contain, in addition to symmetric deletions of the type described in Figure 5, asymmetric deletions and, in 1 case out of 20, more complicated rearrangements. These asymmetric deletions were also found to extend beyond the ends of the symmetry regions, indicating that the sequence diversity generated by the high-frequency deletion of palindromes may be several orders of magnitude higher than would have been expected on simple symmetric deletions.

ACKNOWLEDGMENTS

Nitrocephin for the β-lactamase staining was generously supplied by Dr. Muggleton of Glaxo Research, Ltd., England. I thank M. Van Montagu for stimulating discussions, D. E. Berg and C. Egner for sending a copy of their manuscript before publication, and E. A. Auerswald and H. Schaller for the DNA sequence of the Tn*5* inverted repeats.

REFERENCES

Berg, D. E. 1977. Insertion and exision of the transposable kanamycin resistance determinant Tn*5*. In *DNA insertion elements, plasmids, and episomes* (ed. A. I. Bukhari et al.), p. 205. Cold Spring Harbor Laboratory, Cold Spring Harbor, New York.

Bertrand, K., L. Korn, F. Lee, T. Platt, C. L. Squires, C. Squires, and C. Yanofsky. 1975. New features of the regulation of the tryptophan operon. *Science* **189:** 22.

Bolivar, F., M. C. Betlach, H. L. Heyneker, J. Shine, R. L. Rodriguez, and H. W. Boyer. 1977. Origin of replication of pBR345 plasmid DNA. *Proc. Natl. Acad. Sci.* **74:** 5265.

Botstein, D. and N. Kleckner. 1977. Translocation and illegitimate recombination by the tetracycline resistance element Tn*10*. In *DNA insertion elements, plasmids and episomes* (ed. A. I. Bukhari et al.), p. 185. Cold Spring Harbor Laboratory, Cold Spring Harbor, New York.

Collins, J. and B. Hohn. 1978. Cosmids: A type of plasmid gene-cloning vector that is packageable *in vitro* in bacteriophage lambda heads. *Proc. Natl. Acad. Sci.* **75:** 4242.

Dykes, G., R. Bambara, K. Marians, and R. Wu. 1975. On the statistical significance of primary structural features found in DNA-protein interaction sites. *Nucleic Acids Res.* **2:** 327.

Egner, C. and D. E. Berg. 1980. Excision of Tn*5* independent of Tn*5*-encoded transposition functions. *J. Bacteriol.* (in press).

Farabaugh, P. J., U. Schmeissner, M. Hofer, and J. Miller. 1978. Genetic studies of the *lac* repressor. VII. On the molecular nature of spontaneous hotspots in the *lac*I gene of *Escherichia coli*. *J. Mol. Biol.* **126:** 847.

Ghosal, D. and H. Saedler. 1978. DNA-sequence of the mini-insertion IS*2*-6 and its relation to the sequence of IS*2*. *Nature* **275:** 611.

———. 1979. IS*2*-61 and IS*2*-611 arise by illegitimate recombination from IS*2*-6. *Mol. Gen. Genet.* **176:** 233.

Hirota, Y., S. Yasuda, M. Yamada, A. Nishimura, K. Sugimoto, H. Sugisaki, A. Oka, and M. Takanami. 1979. Structural and functional properties of the *E. coli* origin of DNA replication. *Cold Spring Harbor Symp. Quant. Biol.* **43:** 129.

Hobom, G., R. Grosschedl, M. Lusky, G. Scherer, E. Schwarz, and H. Kössel. 1979. Functional analysis of the replicator structure of lambdoid bacteriophage DNAs. *Cold Spring Harbor Symp. Quant. Biol.* **43:** 165.

Matthew, M. and R. W. Hedges. 1976. Analytical isoelectric focusing of R-factor-determined β-lactamases: Correlation with plasmid compatibility. *J. Bacteriol.* **125:** 713.

Nichols, B. P. and J. E. Donelson. 1978. 178-Nucleotide sequence surrounding the *cos* site of bacteriophage lambda DNA. *J. Virol.* **26:** 429.

Perricaudet, M., A. Fritsch, U. Pettersson, L. Philipson, and P. Tiollais. 1977. Excision and recombination of adenovirus DNA fragments in *Escherichia coli*. *Science* **196:** 208.

REIF, H. J. and H. SAEDLER. 1975. IS*1* is involved in deletion formation in the *gal* region of *E. coli* K12. *Molec. Gen. Genet.* **137:** 17.

ROTHSTEIN, S. J., R. A. JORGENSEN, K. POSTLE, and W. S. REZNIKOFF. 1980. The inverted repeats of Tn*5* are functionally different. *Cell* **19:** 795.

SOBELL, H. M. 1973. The stereochemistry of actinomycin binding to DNA and its implications in molecular biology. *Prog. Nucleic Acid Res.* **13:** 153.

STREISINGER, G., Y. OKADA, J. EMRICH, J. NEWTON, A. TSUGITA, E. TERZAGHI, and M. INOUYE. 1967. Frameshift mutations and the genetic code. *Cold Spring Harbor Symp. Quant. Biol.* **31:** 77.

THOMAS, C. A., JR., R. E. PYERITZ, D. A. WILSON, B. M. DANCIS, C. S. LEE, M. D. BICK, H. L. HUANG, and B. H. ZIMM. 1974. Cyclodromes and palindromes in chromosomes. *Cold Spring Harbor Symp. Quant. Biol.* **38:** 353.

WALZ, A. and V. PIRROTTA. 1975. Sequence of the P_R promoter of phage lambda. *Nature* **254:** 118.

YANOFSKY, C. 1976. Control sites in the tryptophan operon. In *Alfred Benzon Symposium IX: Control of ribosome synthesis* (ed. O. Maaløe and N. O. Kjeldgaard), p. 149. Munksgaard, Copenhagen.

Strand Exchange in λ Integrative Recombination: Genetics, Biochemistry, and Models

H. A. NASH,* K. MIZUUCHI,† L. W. ENQUIST,‡ AND R. A. WEISBERG**

* *Laboratory of Neurochemistry, National Institute of Mental Health;* † *Laboratory of Molecular Biology, National Institute of Arthritis, Metabolism, and Digestive Diseases;* ‡ *Laboratory of Molecular Virology, National Cancer Institute;* ** *Laboratory of Molecular Genetics, National Institute of Child Health and Human Development; National Institutes of Health, Bethesda, Maryland 20205*

Phage λ integrates into its host by a single reciprocal recombination between the bacterial chromosome and a circular form of the phage chromosome. According to Campbell's original proposal, the integrative crossover was supposed to occur within a region of homology between phage and host (Campbell 1962). The estimated size of this homologous region, called the core, has fluctuated widely since it was originally proposed (Cowie and Hershey 1965; Dove 1970; Shimada et al. 1975). The question has recently been settled by DNA sequencing: the size of the core is 15 nucleotide pairs (Landy and Ross 1977).

In spite of the existence of the core, important differences between integrative and other homologous (or general) recombination are easy to find. We know that integrative crossing-over requires the phage-encoded *int*-gene product, which has only a small effect on crossing-over outside the core (Weil and Signer 1968; Echols et al. 1968; Echols and Green 1979; Enquist et al. 1979). Conversely, the absence of the *recA*$^+$ function of *Escherichia coli,* which markedly depresses all general recombination in the host, has little or no effect on integrative recombination. The frequency of *int*-dependent crossing-over, in striking contrast to that of *recA*$^+$-dependent crossing-over, is not increased by substituting large segments of fully homologous DNA for the heterologous regions immediately adjacent to the core; indeed, for certain substitutions, the frequency is markedly decreased (Guerrini 1969; Weisberg and Gottesman 1969). Because of these differences, a special term, site-specific recombination, is used to describe such processes as phage integration, prophage excision, and *int*-dependent recombination of chromosomes in a phage cross. The sites involved are called attachment (*att*) sites for historical reasons.

From what we have just said, it might appear that the mechanisms of general recombination and site-specific recombination have little in common. Nevertheless, the very existence of the core and the observation that the intensity of general recombination is not always uniform along the chromosome (Stahl 1979) have prompted us to consider an alternative view, namely, that site-specific recombination and general recombination proceed by similar steps and possibly share a common intermediate.

In this paper we address the question, "What are the details of the strand-exchange mechanism of λ site-specific recombination?" Before proceeding, we wish to point out that there are other genetic rearrangements that involve site-specific recombination that may be interesting variations of the λ system. For example, the *hin* system that promotes flagellar phase variation in *Salmonella* is site-specific, and, like λ, the DNA sites have a short homologous core (Silverman et al., this volume). In addition, the mechanism of inversion of the G and C segments of phages Mu and P1 appears to be related to that of the *Salmonella hin* system (Iino and Kutsukake, this volume). It seems likely that insights concerning the λ mechanism will be directly useful in understanding the mechanisms of these and other genetic rearrangements.

Characteristics of Recombinant Molecules

Plausible mechanisms for strand exchange abound. Several tens of mechanisms have been proposed that combine a choice of one, two, or four strands of parental DNA as the initial actors in provoking the exchange. DNA synthesis may be required before, during, or after the exchange (for review, see Signer 1971; Hotchkiss 1974). In contrast to the richness of proposed mechanisms, there are only a few attributes of completed recombinants that yield inferences about their history. In this section we briefly review those properties that are particularly applicable to λ integrative recombination.

The first attribute, reciprocality of the exchange, describes the degree of conservation of nucleic acid sequences. Reciprocal recombination preserves sequences; nonreciprocal recombination disrupts or alters them. Two types of nonreciprocality can be distinguished (see Fig 1A). In the first type, which we call here *macroscopic* nonreciprocality, some of the DNA strands broken during the recombination fail to be rejoined. The strand break can kill the affected genome. In the second type of nonreciprocal recombination, all the strand ends are reciprocally rejoined, but some sequence information at or near the recombination joint is altered, usually by replacement of sequences from one parent with sequences from the other parent. Because this replacement alters the number of copies of a particular allele entering and leaving a cross, this kind of *microscopic* nonreciprocality is frequently described as gene conversion. Although the underlying mechanisms are obviously distinct, the hallmark genetic consequences of both kinds of nonrecipro-

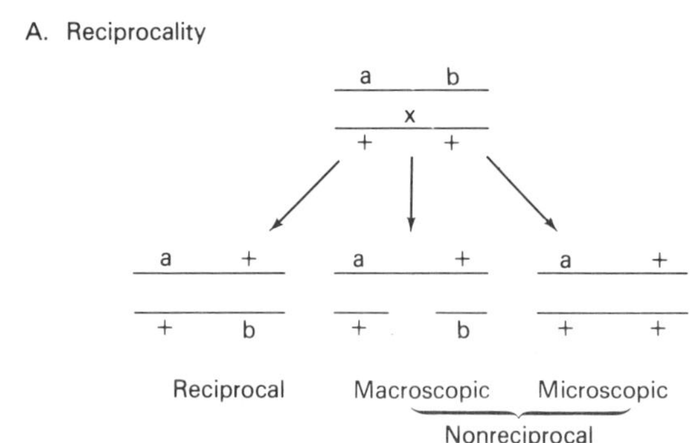

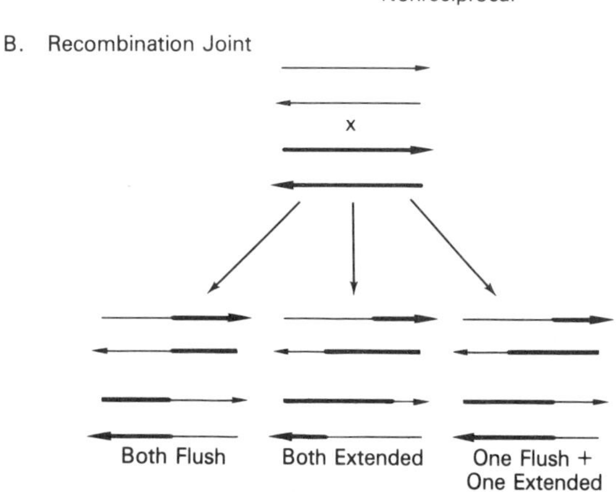

Figure 1. Schematic representation of some attributes of recombinants. (*A*) Genetic consequences of recombination. Each line represents a genome (double-stranded DNA) or a segment thereof containing marked alleles at two loci. (*B*) Alternatives at the molecular level for macroscopically reciprocal recombinants. Each line is a strand of DNA (5′→ 3′ polarity indicated by arrowheads).

cal recombination are the same: a single crossover between genotypes *ab* and ++ that produces *a*+ fails to produce the reciprocal +*b*.

The Campbell model for integration of a circular episome into a circular host chromosome postulates macroscopic reciprocality. However, this does not mean that site-specific recombination need always be macroscopically reciprocal. Under lytic conditions, although some phage crosses involving site-specific recombination yield equal numbers of reciprocal recombinant types (Weil 1969; Parkinson 1971), crosses involving certain pairs of attachment sites do not (Parkinson 1971). In the latter case, we do not know whether the yield of recombinants reflects the operation of a macroscopically nonreciprocal pathway for site-specific recombination or whether it is simply the consequence of a bias in the viability of the recombinants. Microscopic nonreciprocality is not forbidden by the Campbell model and, although not an adequate explanation for the results just described, remains an important consideration.

The second attribute that may be used to characterize a strand-exchange mechanism is the length of the recombination joint (Fig. 1B). On the one hand, a flush joint between two blocks of double-stranded DNA is formed when the exchange that produces one strand of the recombinant occurs at a phosphodiester link that is directly across the Watson-Crick helix from the site of the other exchange. On the other hand, an extended recombination joint means that the two strands of the recombinant have crossed over at different internucleotide levels. There is a direct genetic consequence of an extended recombination joint. Consider a single recombinant molecule whose extended joint includes a region of DNA that is genetically different in the two parents. Replication of the molecule containing such a heterozygous joint gives rise to progeny of both parental types. In contrast, recombinants with flush joints do not give rise to such "het" clones.

There is a connection between these two attributes in that microscopic nonreciprocality (gene conversion) can be a consequence of an extended heterozygous recombination joint. This can occur in two ways. First, at the site of heterozygosity in an extended joint, mismatched base pairs can be corrected to Watson-Crick pairs. Such mismatch repair could follow degradation of one polynucleotide strand and subsequent resynthesis using the remaining strand as template. Alternatively, only the mismatched base (and not the sugar or phosphate) may be removed and replaced through the action of the recently discovered base insertase enzymes (Deutsch and Linn 1979; Livneh et al. 1979), again using the remaining strand as template. In either case, mismatch repair allows for the possibility of nonreciprocal replacement of one parental allele by another. The second mechanism for conversion of parental alleles occurs without mismatch repair. This happens when a joint that is heterozygous for a particular allele is produced in only one of the two recombinant molecules emerging from an exchange of strands. In this case, it is the formation, rather than the processing, of the extended joint that alters the gene frequency.

What mechanisms of strand exchange are implied by the alternative attributes we have discussed? Of course, macroscopic reciprocality demands that a suitable mechanism ensure that all the strand ends broken during recombination are rejoined. Within this constraint, it is disappointing to note that many of the mechanisms that are envisioned to produce flush joints can, with slight alterations, also produce extended joints. For example, restriction endonuclease cleavage of two DNAs followed by reciprocal ligation of the broken ends can occur with projecting or blunt-ended cleavages. As a second example, consider the class of mechanisms in which recombination is initiated by the exchanging of a pair of strands, producing a crossed strand exchange or Holliday (1964) junction. According to such mechanisms, recombination is completed by the exchange of the remaining pair of strands. Depending on the distance between the first and second exchanges, a flush or extended joint may be produced. We can think of two kinds of recombinants with extended joints in which the nature of the recombinants favors a particular mechanism. First, a pair of recombinants in which only one molecule is heterozygous must have arisen, in the absence of mismatch repair, from replacement of one strand in the joint by a replica of another. This is a case for which DNA synthesis is an inescapable part of the mechanism of strand exchange. Moreover, because the synthesis occurs asymmetrically, the exchange can be initiated by a single strand (Meselson and Radding 1975), as opposed to initiation by two or more strands as required by the Holliday, restriction nuclease, and other mechanisms. Second, when the extended joint covers a large region (> 100 bp), it becomes hard to imagine that the projecting ends of the restriction-nuclease-like cuts can account for the overlap.

In summary, some basic information as to the nature of the recombination process can be revealed by a detailed examination of the genetic properties of recombinants. With this in mind, we review below genetic studies of λ site-specific recombination. However, the evaluation of a model that specifies molecular details of strand exchange requires input from biochemical experiments; these also will be explored later.

Recombination In Vivo

We begin by noting that site-specific recombination can occur with little or no concomitant synthesis of new DNA. Crosses between isotopically labeled parents show that parental DNA appears in recombinants without intervening degradation and resynthesis (Kellenberger-Gujer and Weisberg 1971; Nash and Robertson 1971; Stahl and Stahl 1971). Moreover, recombinants are still found when DNA synthesis is blocked either by mutations in essential genes (Stahl et al. 1974) or by chloramphenicol, rifampicin, or λ repressor (Nash 1975). The behavior of integrative recombination in vitro is also consistent with the conclusion that this reaction (and perhaps all λ site-specific recombination) involves only breakage, exchange, and rejoining of preexisting polynucleotide chains (see below). In the following, we attempt to describe site-specific recombination in terms of the location and timing of the four single-strand breaks and ligations that must occur.

Heterozygotes and reciprocality. As we have pointed out, the finding of heterozygosity and microscopic nonreciprocality in λ site-specific crosses would argue for an extended, rather than a flush, recombination joint (Fig. 1). Crosses between marked attachment sites have demonstrated the existence of heterozygosity but are inconclusive with respect to microscopic nonreciprocality. These experiments involved recombination of sites that differ from one another in the core. To discuss them, we must first introduce the nomenclature for attachment-site parts: The 15-nucleotide core is designated O, and the regions immediately to the left and to the right of the core are designated P and P′, respectively, in the phage and B and B′, respectively, in the host. Thus, a complete phage attachment site is POP′, a bacterial site is BOB′, and the recombinant sites lying to the left and to the right of an inserted prophage are BOP′ and POB′, respectively.

Shulman and Gottesman (1973) analyzed recombination between prophage attachment sites where one of the partners carried a mutation, which we designate 024, that we now know to be a single-base deletion in the core (A. Landy, pers. comm.). In crosses of the form B-O^+-P′ × P-024-B′, about 2% of the recombinant phage were P-O^+-P′/P-024-P′ hets, and this frequency increased to 7% when DNA replication was blocked during the cross. Similar crosses showed that 5–10% of individual infected cells contained both P-O^+-P′ and P-024-P′ recombinant phage particles. These results imply that the recombination joint produced by integrase encompasses more than 1 nucleotide pair at least some of the time. Such experiments may underestimate the frequency of extended recombination joints if DNA with mismatched base pairs frequently goes undetected as hets because of mismatch repair (Wildenberg and Meselson 1975) or frequent strand loss (Wagner and Meselson 1976). However, we must allow the caveat that a mutation in the attachment site may influence the mechanism of recombination.

Shulman and Gottesman (1973) also looked for nonreciprocality in the same kinds of crosses. They found that the recovery of mutant and wild-type recombinant progeny was sometimes unequal and that the direction and magnitude of the disparities depended on the regions immediately flanking the core. An example of this is shown in Table 1, where in cross 1, the relative recovery of wild-type *att*B is about 20 times that of mutant, whereas wild-type and mutant *att*P are recovered equally. Two explanations of the results have been considered:

1. The mutation causes macroscopic nonreciprocality; i.e., entire recombinant strands or chromosomes are lost. Although macroscopic nonreciprocality was not directly measured in these experiments, it should be recalled that site-specific recombination in phage crosses is sometimes grossly nonreciprocal in yield of recombinant types (Parkinson 1971). In this case, we would have to assume that, for example, strands carrying mutant BOB′ were preferentially lost in cross 1 but not in cross 2. This assumption may seem ad hoc, but it can be directly tested by performing the cross in such a way that individual macroscopically reciprocal pairs of recombinants can be selected and examined for microscopic reciprocality. Although this experiment has not yet been tried for the sites listed in Table 1, it has for other pairs of sites, namely, P-O^+-P′ × B-024-B′ and P-024-P′ × B-O^+-B′ (Table 6b in Shulman and Gottesman 1973). In these cases, recombination was microscopically reciprocal for the 21 pairs of recombinants examined.
2. Recombination of mutant with wild type frequently produces two extended joints, each containing mutant and wild-type information, and nonreciprocality results from mismatch repair of these heteroduplex joints. To explain the changes in parity in different crosses, we could assume either that the regions flanking the core direct mismatch repair or that these regions (perhaps together with the mutation) affect the size, location, or symmetry of the recombination joint. For example, if the recombination joint usually encompasses the mutational site in cross 1, but lies to its right in cross 2, this would account for the parity

Table 1. Segregation of the 024 Core Mutation

Cross	Parental genotypes	Recombinant genotypes			
		P-O$^+$-P′	P-024-P′	B-O$^+$-B′	B-024-B′
1	P-024-B′ × B-O$^+$-P′	28	36	19	1
2	P-O$^+$-B′ × B-024-P′	62	4	0	20

The results cited are taken from Table 6 in Shulman and Gottesman (1973) and the methods are described in full there. Briefly, phage carrying both of the attachment sites listed in column 2 above were allowed to grow for a single cycle in the presence of *int* and *xis* proteins, and the resulting recombinants were classified as mutant or nonmutant by further crosses. The numbers of recombinants scored of each type are noted. The frequencies of phage (POP′) and bacterial (BOB′) recombinants cannot be directly compared because they were isolated in different ways. If recombination were reciprocal, we expect that the following equality of ratios of numbers of recombinants would hold: P-O$^+$-P′/P-024-P′ = B-024-B′/B-O$^+$-B′.

of the latter cross as well as the linkage of the mutation to B, the left-hand flanking segment.

At this point, we have no strong reason to prefer either one of the above explanations. Therefore, the occurrence of microscopic nonreciprocality cannot be strongly argued from the results of these crosses, and, consequently, the information they might provide about the location and size of the recombination joint is uncertain.

A second type of cross between wild-type and mutant sites also allows us to ask about the reciprocality and the size of the recombination joint. In this case, the mutant sites are naturally occurring variants, otherwise known as secondary attachment sites (Shimada et al. 1972). More than ten secondary attachment sites have been sequenced to date; they show a fairly marked conservation of the wild-type sequence at several (but not all) positions within the core (Mizuuchi et al., this volume). The cross consists of inserting a wild-type genome into one of a number of different secondary sites, thereby assuring macroscopic reciprocality. The left and right (i.e., reciprocal) ends of insertions into five different sites have been sequenced, and in all cases the insertion occurred by a microscopically reciprocal recombination joint within the core. In the different insertions, the location of the joint varied over at least 4 bp within the core. Two extreme models are possible: (1) The joint is of size 0 (flush cut) and can occur at more than one position in the core. (2) The joint is 4 bp (or more) long, and mismatches within the joint are frequently not repaired (or are reciprocally repaired [Bidwell and Landy 1979]). Since crosses involving 024 (see above) and analysis of the recombination joint produced in vitro (Mizuuchi et al., this volume) are both indicative of an extended joint, model 2 is to be preferred. However, we must remain aware that the base changes in the cores of secondary attachment sites may change the nature of the recombination joint.

Synchrony of strand exchange. How are extended recombination joints produced? We can lump the many proposed mechanisms into two classes that differ in the degree of synchrony of the strand exchanges. On the one hand, there are models in which all strand exchange takes place simultaneously; in these, both strands of each parental helix are cut before pairing and/or rejoining occurs. On the other hand, there are models in which exchange of one parental strand is separated in time from exchange of the other.

According to the first kind of model, extended joints are produced when the two strands of a parent are cut at different points, producing an end with a protruding single strand. In the second class of model, extended joints result from the two exchanges being separated in space as well as in time. There are several ways in which this separation might be achieved. For example, in the Holliday model of recombination, after the first breakage and reunion creates a connection between the parental helices (Holliday 1964; Sigal and Alberts 1972), the rotation of the parental helices can move this connection to a new site (Meselson 1972). Breakage and reunion of the parental strands opposite this new site completes the crossover, creating recombinants with joints that extend over the region spanned by the migration. Later, we present a recombination model in which separate exchanges produce extended joints without branch migration.

The two classes of mechanisms can produce genetically distinguishable recombinant molecules. In the second mechanism, when exchange of one strand is followed not by exchange of the others, but instead by reexchange of the first strand at a distant point, an "insertion het" is formed. This is a molecule that contains an extended recombination joint (and is therefore potentially heterozygous) but has the parental configuration of markers outside the joint. In contrast, mechanisms of the first class cannot produce insertion hets (in a single step). One situation has been reported in which site-specific recombination of λ produces insertion hets. These crosses were designed to determine whether the recombination joint might, with some measurable frequency, encompass markers that were close to but outside of the attachment sites. The crosses were of the form a^+POP′b^+c × a POP′bc^+, where the location marker b was varied from 100 to 1200 nucleotide pairs to the right of the core, and a and c were located about 10 kb to the left and right of the core, respectively (Echols and Green 1979; Enquist et al. 1979). Note that the two attachment sites were identical; thus, the formation of recombinant joints extending outside the core could not be impeded by heterology. Even though the general recombination pathways of the phage and host were inactivated by mutations, recombination was detected in the bc interval. Such crossovers were rare: 1%

or less of all crossovers in the *a–c* interval. They were a product of site-specific recombination, since they required integrase and two functional attachment sites. Moreover, they were independent of several *E. coli* general recombination genes (*recA, recB, recC, recF,* and *recL*) but depended on an *E. coli* gene (*hip*) that is known to be required for site-specific recombination. Among the b^+c^+ recombinants, comparable numbers of $a^+b^+c^+$ and ab^+c^+ genotypes were found. The latter have retained the parental configuration of markers entering the cross; they could have arisen as insertion hets.

If *int*-gene-promoted recombination in the *bc* interval takes place by two sets of strand exchanges, it seems reasonable that one of these exchanges should be at the core. The recombinant joint would therefore be expected to extend from the core to the site of the second exchange, between markers *b* and *c*. Indeed, recombination in the *bc* interval requires homology between the core and *b* (Enquist et al. 1979). The existence of heteroduplex DNA extending from the attachment site through marker *b* was confirmed in the following way. Correction of mismatched DNA in the heteroduplex region will distort genetic linkage within the recombination joint. Such distortion—manifested as a correlation of multiple exchanges—was sought and found by analysis of crosses in which two markers were placed close to POP′ (Echols and Green 1979; Enquist et al. 1979).

What do these experiments tell us about the distribution of length of recombination joints? The frequency of b^+c^+ recombinants was 0.2–1.0% of the frequency of a^+c^+ recombinants when *b* was about 100 nucleotide pairs from the core. Therefore, more than 99% of the *a–c* crossovers occur to the left of 100 nucleotide pairs. We conclude that if extended joints are always formed during site-specific recombination, the great majority are shorter than 100 nucleotide pairs. However, when marker *b* was moved still farther from the core, the frequency of b^+c^+ recombinants fell gradually; in one study the frequency dropped tenfold in 1100 bp (Enquist et al. 1979), and in another it dropped less than twofold (Echols and Green 1979). (We have no explanation for the discrepancy between the two sets of results.) It therefore appears that the length distribution of recombination joints is not very uniform: Most are short and a small minority are long. Enquist et al. (1979) estimated the mean length of the second class at 600–800 bp, whereas the data of Echols and Green (1979) suggest a considerably larger value.

In summary, we conclude that the formation of long recombinant joints by the site-specific recombination pathway must be initiated by exchange of fewer than four strands. Initial exchange of two strands (Holliday 1964) or initial exchange of one strand (Meselson and Radding 1975) accounts for all of the results. What is the mechanism of formation of the short recombination joints that account for the vast majority of recombinants in these crosses? The most economical hypothesis is that the short joints are initiated in the same way as the long, but their extension is terminated earlier. Early termination could result from the action of the recombination enzymes on their normal site within the core. This would have the effect of producing a recombination joint that was heteroduplex only within the core, a structure consistent with the data on heterozygosity discussed above.

Biochemical Studies

Site-specific recombination of λ can be carried out readily in cell-free systems. Over the past five years, the study of integrative recombination in vitro has revealed much about the biochemistry of this reaction. The proteins that promote λ integration have been enumerated, purified, and in some cases identified as the products of known genes. The cofactor requirements have also been studied; the minimal set includes only buffer and monovalent ions, although even under the best conditions, the polyamine spermidine strongly stimulates the reaction. Recombination in vitro has also been invaluable in uncovering the requirement for a supercoiled DNA substrate in λ integration. We have not reviewed all these studies here; rather, we present in detail only those results that increase our understanding of the mechanism of strand exchange. Such results are of two kinds: (1) information on the structure of the completed products of in vitro recombination and (2) characterization of an activity of a recombination protein that may be used as the strand-exchange function in integrative recombination.

An informative way to study the products of recombination is to start with a supercoiled substrate that contains both *att*P and *att*B on the same circle of DNA. With such a substrate, the product is found to be a catenane of two supercoiled circles, each containing a single prophage attachment site (Mizuuchi et al. 1980b). Since many of our inferences about the nature of strand exchange derive from this finding, it is important to understand why catenanes are formed during recombination. As illustrated in Figure 2, we believe that catenation reflects no more than tangling of different parts of the supercoiled substrate circle prior to synapsis and strand exchange. Since tangling of the substrate is an expression of writhing of the DNA in response to its supertwisted state, nonsupercoiled circular DNA substrate should tend to yield recombinant products that are often unlinked to each other. Recently, it has been shown that at low ionic strength, nonsupercoiled DNA is a substrate, albeit a poor one, for integrative recombination (Pollock and Abremski 1979). Under these conditions, recombination of nicked circles containing both *att*B and *att*P again yields a pair of circular DNAs bearing prophage attachment sites, but some of these circles are unlinked (Pollock and Nash 1980). This means that catenation is not an intrinsic feature of the strand-exchange mechanism and supports our contention that catenanes are simply an accidental consequence of recombination with supertwisted substrate.

It seems reasonable to assume that the two supertwisted circles that are linked in a catenane are derived from a single recombination event from one substrate circle. If this is so, we must conclude that recombination in vitro, just as demanded by the Campbell model for

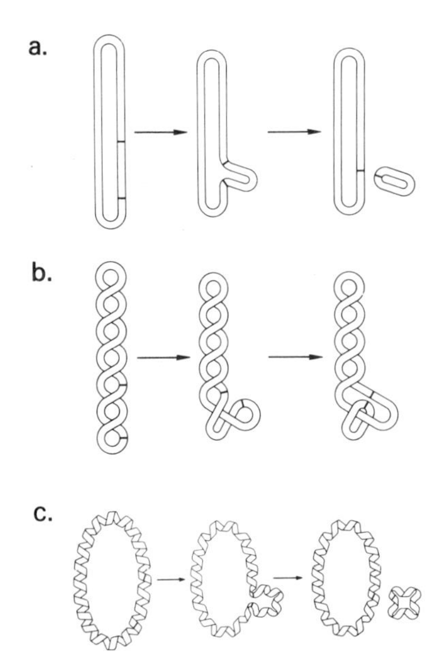

Figure 2. Formation of catenanes by intramolecular recombination. A circular substrate DNA carrying two attachment sites is pictured at the left of each panel. The cartoons to the right show the alignment of the attachment sites followed by their breakage and reunion to form (macroscopically) reciprocal recombinants. Catenanes are formed by the accidental interwinding of segments of the helix that are destined to reside in different recombinant products. This interwinding is much favored in supertwisted DNA (*b*), although it is possible to organize supertwists in a toroidal helix so as to avoid interwinding and consequent catenation (*c*). See Pollock and Nash (1980) for a discussion of how the frequency of catenation might be affected by alternative modes of alignment of attachment sites and by accidental interwinding of nonsupercoiled DNA.

integration in vivo, must be macroscopically reciprocal; i.e., all the strands broken during recombination are rejoined to produce continuous DNA helices. This conclusion is supported by other studies that fail to detect DNA broken at attachment sites after integrative recombination in vitro (Mizuuchi and Nash 1976). In addition, site-specific recombination between two DNA molecules, each containing an attachment site, produces recombinant structures that are consistent with a reciprocal mechanism; e.g., two circles yield a single dimer circle, circular and linear DNAs recombine to form a single linear DNA, etc. (Mizuuchi and Mizuuchi 1979).

Recombination in vitro is carried out with reaction mixtures that provide neither a substrate nor a catalyst for DNA synthesis (Kikuchi and Nash 1978). However, as mentioned above, the resulting recombinants are continuous circles of double-helical DNA. Moreover, the prophage sites produced in vitro contain complete core sequences, since they appear to be identical to the sites formed after integration in vivo in terms of their potential for further recombination. In addition, a recent experiment to follow radioactive label during recombination has shown that all of the phosphorous atoms in one parental *att* are recovered in the two recombinants (Mizuuchi et al., this volume). We must conclude that degradation and resynthesis of DNA are not part of the strand-exchange mechanism for λ integrative recombination. This rules out mechanisms of recombination that invoke formation of asymmetric heteroduplex joints, since such models require DNA synthesis as part of the recombination (see above). Indeed, one of us (K. M.) has recently obtained direct physical evidence indicating that λ integrative recombination in vitro forms heteroduplexes over a stretch of 5–7 bp, and, as expected, the heteroduplexes are symmetric (Mizuuchi et al., this volume).

Covalent closure of the catenated recombinant circles implies that none of the phosphodiester bonds of the substrate attachment sites are hydrolyzed during recombination. If they were, a high-energy cofactor would be required for ligation of the hydrolyzed bonds, but no such cofactor requirement is apparent. The attachment sites must therefore be broken in a way that stores the energy of the phosphodiester bond. This might be done in a manner similar to that described for several topoisomerases, namely, a phosphodiester bond is transiently formed between the phosphate at the site of cleavage and a nucleophilic residue on the protein (for review, see Wang and Liu 1979).

The retention of supercoiling in both circles in a catenated pair places limits on the dynamics of the strand exchange. In free solution, circular DNA containing a strand break is expected to lose its supercoils rapidly (Davison 1966). We find that the products of recombination have retained more than 70% of their superhelical density (Mizuuchi et al. 1980b). Therefore, between the breaking and reunion steps of recombination, free swiveling must not occur. There are several ways in which this might be arranged. One idea is presented below.

Integrative recombination in vitro absolutely requires two proteins, one encoded by the virus and the other by *E. coli.* (Kikuchi and Nash 1978). Of these two components, the viral protein, Int, has been the more extensively studied. It purifies as a single polypeptide (M_r~40,000) and has been shown to be encoded by the phage *int* gene (Kikuchi and Nash 1979a; Nash et al. 1977). A combination of filter-binding and sequence-protection studies has demonstrated that Int makes specific associations with both the phage and bacterial attachment sites (see Hsu et al. 1980, and references cited therein). This suggests that Int is a specificity element in recombination, perhaps playing an important role in synapsis of the attachment sites.

A recently discovered property of Int suggests an additional role in strand exchange. This property is the ability of Int to act as a topoisomerase (Kikuchi and Nash 1979b). Topoisomerases catalyze the conversion of supercoiled DNA to a form that is covalently closed but lacks supertwists. They do so by breaking and resealing the polynucleotide backbone. The relaxing activity of Int requires only buffer and monovalent salts, is rather sluggish at best, and displays no apparent preference for DNA carrying an attachment site. Recent experiments performed by two of us (K. M. and H. N.) clarify the mechanism of relaxation. Liu et al. (1980) have proposed that topoisomerases can be classified into two groups. Type-I enzymes relax DNA by transiently breaking a single strand to permit swiveling of the DNA around the polynucleotide backbone opposite the disruption; type-II enzymes transiently break both strands of the double helix and permit passage of a double-helical segment of DNA through the break. These mechanisms can be distinguished by determining the minimal enzyme-caused change in the topological linking number, a term defining the number of times the two strands of a double helix are intertwined. As shown by several groups, type-II enzymes change the topological linking number of DNA in steps of 2; type-I enzymes change the linking number in steps of 1, the smallest increment possible[1] (Brown and Cozzarelli 1979; Liu et al. 1980; Mizuuchi et al. 1980a). In Figure 3 we show the electrophoretic behavior of relaxed circular DNA. Initially, one sample contained a population of many different supercoiled species. The other two samples were unique isomers whose initial linking number differed by a single unit; these topoisomers were purified from the population by gel electrophoresis. After relaxation by Int, these samples display identical ladders of partially relaxed DNA (Fig. 3). This is the pattern one would expect if Int relaxed DNA by changing the linking number in steps of 1; i.e., if one sample is characterized by an initial linking number $Lk = n$ and the other by $Lk = n + 1$, then relaxation of each sample should produce ladders containing species corresponding to $Lk = n + 1$, $n + 2$, $n + 3$, etc. In contrast, relaxation of this pair of samples by DNA gyrase, a type-II topoisomerase, produced a pair of nonidentical but nested ladders corresponding to $Lk = n + 2, n + 4, n + 6$ for one sample and $Lk = n + 3, n + 5, n + 7$, etc., for the other (data not shown). We also found identical ladders of DNA when electrophoresis of the same Int-treated samples was carried out in the presence of chloroquine (Shure et al. 1977) to resolve circular species with linking numbers close to that of the initial sample (data not shown). Electrophoresis in the presence of chloroquine was also used to detect a small degree of relaxation by Int in the presence of inhibitory substances. In this way, we have found that the polyamine spermidine (at concentrations of 5–10 mM) does not completely inhibit the topoisomerase activity of Int (cf. Kikuchi and Nash 1979b). Like the uninhibited activity, the residual activity changes the linking number in steps of 1 and displays no obvious preference for *att* sites. The topoisomerase activity of Int is depressed only slightly when a highly purified preparation of the bacterial protein required for integration (H. A. Nash, in prep.) is added in quantities sufficient to stimulate recombination. Under these conditions, in the presence or absence of spermidine, this host protein changes neither the degree of specificity

[1] N. Cozzarelli (pers. comm.) has pointed out that current experimental evidence demands only that type-I topoisomerases change the linking number in steps of one some of the time; no unique restriction to steps of one needs to be invoked.

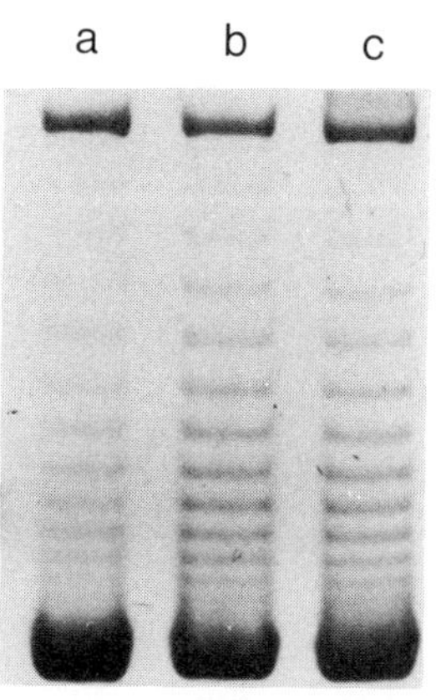

Figure 3. Relaxation of topoisomers by Int. Three samples of supercoiled DNA from plasmid pPA1, a derivative of pBR322 containing *att*P (Mizuuchi and Mizuuchi 1979), were incubated with purified Int. The reaction conditions used were essentially those described by Kikuchi and Nash (1979b). (*a*) An unselected mixture of topoisomers, purified from a cleared lysate by buoyant density centrifugation; (*b*) a unique topoisomer separated from this population by horizontal slab gel electrophoresis in 1% agarose containing chloroquine (6 μg/ml), followed by electroelution and buoyant density centrifugation; (*c*) a similarly purified topoisomer whose linkage number differed by 1 from that of the sample for *b*. After relaxation by Int, the samples were electrophoresed in 1% agarose gel, stained with ethidium bromide, and photographed as described previously (Kikuchi and Nash 1979b).

shown by Int topoisomerase nor its apparent type-I mechanism.

We have performed an additional experiment that confirms the assignment of Int as a nicking-closing type of topoisomerase. All type-II topoisomerases examined simplify catenanes to separate circles of DNA (Hsieh and Brutlag 1980; Kreuzer and Cozzarelli 1980; Mizuuchi et al. 1980a). In contrast, type-I enzymes carry out the reaction quite poorly (J. Wang; N. Cozzarelli; L. Liu; all pers. comm.), at least with catenanes composed of closed circles. We have asked whether Int, when acting as a topoisomerase, can unlink catenanes. The catenanes used for this test were produced by integrative recombination in vitro so that their component supertwisted circles each contained a prophage attachment site. This should increase the possibility for interaction of Int with the catenane, since one of the prophage sites, *att*L, is known to bind Int protein in a strong and specific manner (Kikuchi and Nash 1979a). It should be noted that recombination between the *att* sites in the catenane cannot occur, since both the product of the *xis* gene and the *E. coli* integration protein are absent. Decatenated forms of DNA are not found on agarose gel electrophoresis of material relaxed by Int (Fig. 4b,d). Control experiments show that, as previously reported by Mizuuchi et al. (1980a), treatment of these catenanes with DNA gyrase yields unlinked circles that are readily detectable on agarose gel electrophoresis (data not shown). The failure to find decatenation by Int argues against models that account for its ability to change linking number in steps of 1 by postulating a transient

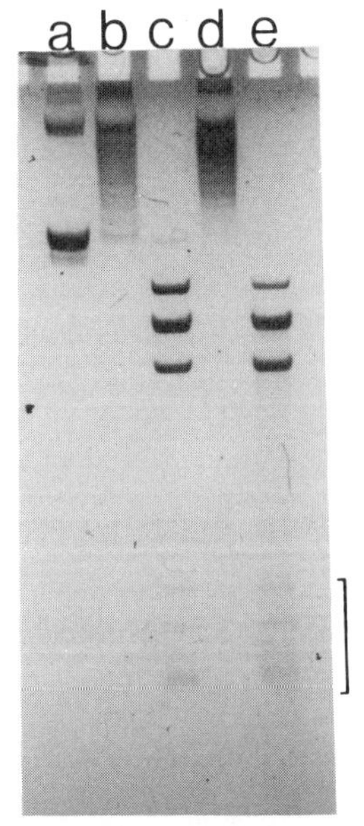

Figure 4. Relaxation of catenanes by Int. The catenanes for this experiment were prepared, as described previously (Mizuuchi et al. 1980a), by integrative recombination in vitro of the DNA from plasmid pBP86, a dimeric derivative of pBR322 in which *att*B and *att*P are located about 1.5 kb apart. (*a*) DNA from the recombination reaction mixture extracted with phenol and dialyzed. This mixture of DNA species was relaxed by incubation with Int (*b*) essentially as described in Fig. 3. Buoyant density centrifugation of a portion of the treated material confirmed that more than 90% of the DNA had been converted to relaxed closed circles. An additional sample was incubated with restriction endonuclease *Pst*I (New England BioLabs) for 1 hr at 37 °C in 50 mM Tris-HCl (pH 8.0) containing 50 mM NaCl and 10 mM $MgCl_2$ (*c*). This treatment linearizes unrecombined substrate DNA and one of the two recombinant circles; the remaining recombinant circle is freed from the catenane and migrates as a series of bands (indicated by bracket) that correspond to several topoisomers of this DNA. This ladder serves as a marker for the expected position of one of the circles that might be freed from the catenanes during relaxation by Int. The other circle in the catenane is much larger and would be expected to migrate with the ladder of relaxed substrate. An independently prepared sample of catenanes relaxed by Int and digested with *Pst*I is shown in *d* and *e*, respectively. Electrophoresis conditions were the same as described in the legend to Fig. 3.

double-strand break with additional features such as rotation of the broken end before rejoining. On the other hand, the data on decatenation are completely consistent with the proposition that Int is a nicking-closing (type-I) topoisomerase.

How can the topoisomerase activity of Int be used in site-specific recombination? One should recall the basic problem, i.e., that relaxation of DNA by a type-I enzyme does not exchange strands but simply breaks and reseals the same strand. However, since strand exchange in λ integrative recombination resembles relaxation by topoisomerases in that strands are broken without hydrolysis, we are tempted to consider mechanisms that circumvent this problem. One such scheme is presented below.

Proposed Mechanism for Strand Exchange in λ Integrative Recombination

In this section we review and expand on a previously proposed mechanism (Kikuchi and Nash 1979b) by which nicking-closing enzymes may catalyze strand exchange in a region of homology. The critical feature of this proposal is that prior to the cleavage of any strands, the two parental double helices are intimately juxtaposed. Other workers have postulated intimate contact between parental helices as the initial step in recombination (Cross and Lieb 1967; McGavin 1977; Wilson 1979). In the present model, the arrangement of the juxtaposed DNA molecules facilitates joining of broken strands from different parents and impedes simple rejoining.

McGavin (1971, 1979) has shown that it is stereochemically feasible to bring two double helices into close approximation by interwrapping them about their major-groove surfaces. The parental double helices are not greatly distorted in the resulting four-stranded DNA helix. Moreover, the parental double helices are connected to each other in the four-strand region by hydrogen bonds between identical base pairs; this feature indicates that only homologous stretches of DNA are likely to form McGavin structures. As shown in Figure 5, it is proposed that the core regions from a pair of attachment sites form a four-stranded helix. For simplicity, this helix is drawn with zero pitch. Even with a pitch of 10 bp/turn, this four-stranded helix should not comprise more than 1.5 turns, since the common core is only 15 bp long and the four-strand region may involve only a part of it. The resultant interwinding of the two parental DNAs can therefore be compensated by one to two left-handed interwinds at a position unrelated to the site of crossover (Wilson 1979).

Strand exchange is proposed to take place at the junction between this four-stranded DNA and the parental double helices. To begin, a pair of topoisomerases cleaves a pair of strands of like polarity. Note that because McGavin structures align homologous DNA, these cuts will be in register in the two parental DNAs. Each parental double helix is now free to rotate around its intact strand at a point opposite the disruption. In simple double-helical DNA, rotations of 360° and multiples thereof would bring the cleaved ends together again. They would then be rejoined, with concomitant release of the topoisomerase that is thought to be covalently attached to one end of the DNA (Wang and Liu 1979). However, in the structure shown in Figure 5, after rotation of 270° around each intact strand, the ends from two *different* parents are brought next to each other. Now if the topoisomerases promote rejoining, the result is a pair of strand exchanges. Indeed, Figure 5c shows a topological equivalent of a simple Holliday structure. This junction breaks no stereochemical rules; Wilson (1979) has built an acceptable model of just such a joint. Wilson's model (see Fig. 7 in Wilson 1979) was constructed to show the feasibility of joining two parental helices in a four-stranded DNA after an initial association across their minor-groove surfaces. After a re-

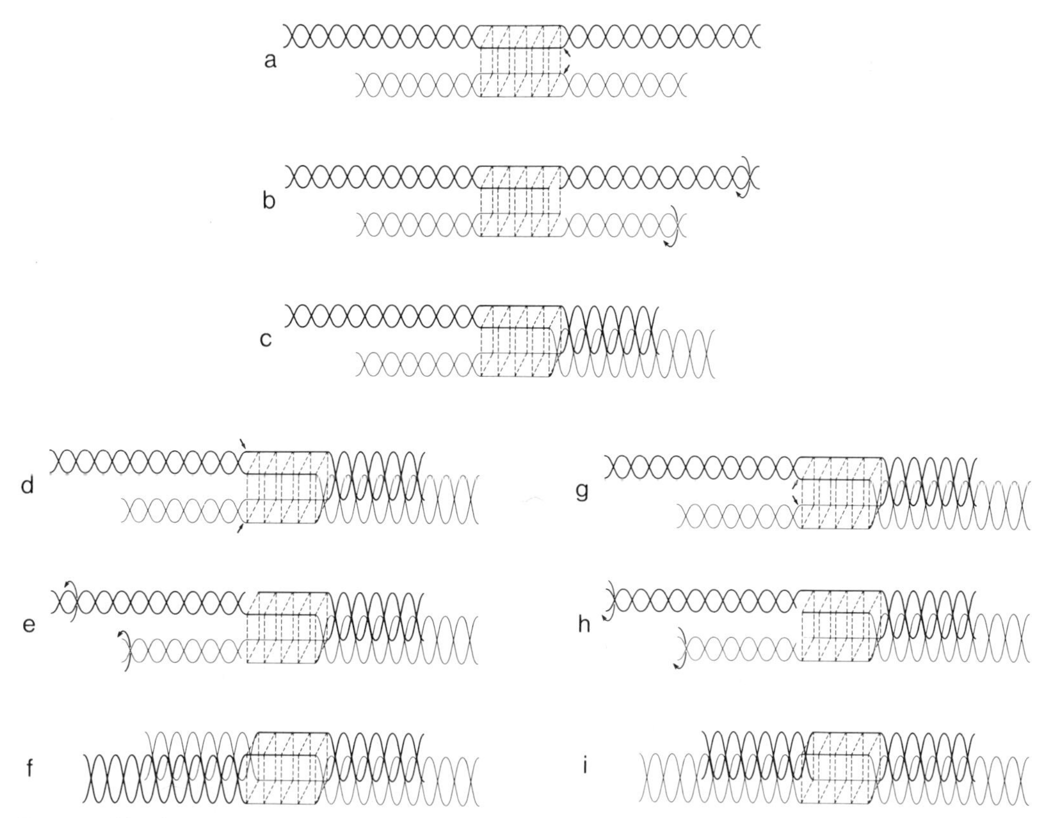

Figure 5. Recombination promoted by nicking-closing enzymes. Two parental DNA double helices are synapsed in a four-strand helix. One strand in each parent is incised by a type-I topoisomerase (*a*); after rotation (*b*), the strands are resealed in a new configuration to produce a Holliday structure (*c*). When these steps are repeated on the remaining strands (*d–f*), a pair of reciprocal recombinants is produced. Separation of the recombinant products from each other requires breakage of the parental hydrogen bonds in the four-strand region; new hydrogen bonds form to make a heteroduplex joint in each recipient. When the second exchange occurs on the same strand as the first (*g–i*), parental DNA is reformed, but it now contains a heteroduplex in the region of synapsis.

arrangement in base pairing (but no strand exchange) a stretch of four-stranded DNA is formed that is stereochemically identical to the McGavin structure, except at the junction between the parental helices and the four-strand region. J. H. Wilson has pointed out (pers. comm.) that the connections in his model are precisely those expected at the point of strand exchange in Figure 5c. A comparison of these two models for forming four-stranded DNA is presented later.

The crossover is completed by a second pair of exchanges carried out in a similar fashion to the first exchange. In principle, there are two choices for the site of the second exchange and two choices for the strands to be exchanged. If, as shown in Figure 5, the site of the second exchange is at the opposite border of the four-strand region as that for the first exchange, the recombinants are each heteroduplex for the region that formed the synapsis. On the other hand, if the site of the second exchange is at the same border as the first, no heteroduplex is formed. We favor the first choice for λ site-specific recombination because of the evidence cited above for the formation of heteroduplexes during this class of reactions in vivo and in vitro. It should be emphasized that although 15 bp is the maximum size of the overlap region, no intrinsic feature of four-stranded DNA suggests that it will involve all of this homology.

A macroscopically reciprocal recombination follows when the second pair of exchanges occurs on the strands that were not involved in the first exchanges. On the other hand, if the same strands are simply exchanged again, DNA flanking the site of the crossover is not recombined. In this case, as shown in Figure 5, g through i, the maximum possible genetic consequence is the formation of heteroduplex regions in the parental DNA, i.e., insertion hets. One of us (K. M.) is currently determining whether the λ integrative recombination system frequently involves transfer of a strand of DNA from within the core from one parent to another. At least one factor should work against this possibility: λ recombination involves supercoiled substrates; the limited swiveling of DNA during strand exchange can affect the topological linking number and thus will be biased in favor of rotation that relieves the strain of supercoiling. Stereochemical factors also affect the possibilities for rotation. Specifically, the free end from one parent must swivel in a way that avoids the intact strand of the other parent. It is for this reason that the ends from different parents are brought together by 270° rotations rather than by 90° rotations of the opposite sense. Taken together, these considerations should favor the assignment of strands to be cut and the direction of rotation that are shown in Figure 5. Starting from this point, the two alternatives for exchange of the second pair of strands lead to different energetic consequences. If the two parental attachment sites were located on separate circles of DNA, the reciprocal recombination (Fig. 5d–f) would increase the total topological linking number by 4,[2] whereas double exchange of the same strands would leave the linking number unperturbed. Depending on the initial superhelical density, a loss of four superhelical turns could provide a significant energetic bias in favor of crossing-over as opposed to formation of insertion hets. It can be estimated that at superhelical densities similar to that of intracellular DNA, each unit change in linking number is accompanied by a change in free energy of 4–9 kcal/mole (Hsieh and Wang 1975; Pulleyblank et al. 1975).

As mentioned above, two pathways for formation of four-stranded DNA have been proposed: major-groove pairing (McGavin 1971) and minor-groove association followed by shift in base pairing (Wilson 1979). Strand exchange by a pair of topoisomerases acting at the border of the four-strand region can convert either of these structures to Holliday junctions, and a second exchange produces reciprocal recombinants. Both modes of synapsis yield identical genetic results, but these alternatives can be distinguished by their effect on superhelical strain energy. Figure 6 illustrates the end-on views of the border of four-stranded DNA in the McGavin and Wilson schemes. Cleavage of a pair of strands and rotation around the remaining strands is shown. The choice of which pair of the four strands is cut is fixed by the recent demonstration (Mizuuchi et al., this volume) that λ integrative recombination produces extended recombination joints that are characterized by a 5′ overlap rather than a 3′ overlap. Note that the rotation sense is opposite in the two cases shown in Figure 6, and only in the McGavin scheme would rotation relieve superhelical strain. That is to say, strand exchange after major-groove pairing would remove four superhelical turns, whereas strand exchange after minor-groove pairing would add four superhelical turns. Using the estimates of the free energy of supercoiling mentioned above, one calculates that 16–36 kcal/mole would have to be supplied to achieve recombination after minor-groove pairing. No such difficulty is encountered by recombination after major-groove pairing.

The mechanism we have reviewed can account for essentially all of the major experimental observations concerning λ integrative recombination. The mechanism generates macroscopically reciprocal recombinants, produces symmetric heteroduplex, and carries out a cycle of breakage and reunion without an external energy source. Moreover, it can account for the occasional formation of long joints that extend into adjacent homologous DNA. This could happen by extensive migration of the Holliday junction formed at one edge of the four-strand region before a second exchange took place or by the initiation of recombination with an unusually long stretch of four-stranded DNA that extends beyond the core. Most importantly, this mechanism accounts for the completion of recombination without the loss of a major fraction of supercoiling from the substrate DNA. As more is learned about excisive recombination and site-specific recombination involving secondary or altered attachment sites, it will be of interest to determine whether the identical scheme or a related one will also accommodate experimental details of these processes. Finally, this type of mechanism should, with modifications, be able to account for general homologous recombination (Cassuto et al. 1980).

[2] Three of the four turns come from the rotations of the parental helices, and the fourth comes from the writhe introduced when the relative alignment of the *att* shifts from the configuration of synapsis to that of the completed recombinant.

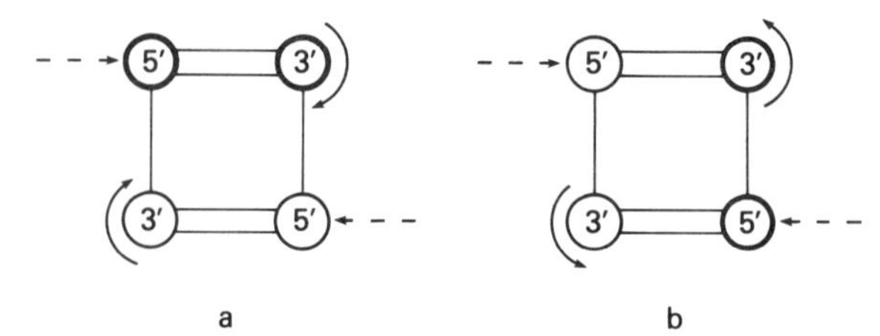

Figure 6. Details of strand exchange in four-stranded DNA. An end-on view of the border between four-stranded DNA and the parental helices is shown; the view is identical when looking in toward the core from either direction. Pairs of strands from one parental helix or the other are indicated by circles with bold and light outlines, respectively. Each strand is marked 5′ or 3′ to indicate the direction of the sugar that points toward the reader. Double lines connecting strands indicate Watson-Crick hydrogen bonding between bases in the four-stranded DNA. Single lines connecting strands indicate additional hydrogen bonds that connect the base pairs. (*a*) Four-stranded DNA formed by major-groove association (McGavin 1971); (*b*) four-stranded DNA formed by minor-groove association with a subsequent shift in base pairing (Wilson 1979). Dashed arrows point at the strands to be cut by a pair of topoisomerases, and curved arrows indicate the direction of rotation around the uncut strand required to produce strand exchange.

SUMMARY

We have asked, "What is the mechanism of strand exchange during site-specific recombination of phage λ?" Crosses carried out in vivo have shown that the recombination joint can be extended rather than flush and that the four-strand breaks and rejoinings needed to form a recombinant can occur asynchronously. Crosses carried out in vitro have shown that all the nucleotides at the site of crossover are conserved during recombination, as are most or all of the superhelical turns present in the substrate molecules. We have presented new data showing that topoisomerase activity of Int protein relaxes DNA by making transient single-strand, rather than double-strand, breaks in the phosphodiester backbone. These findings are incorporated into a model for strand exchange that has as its central intermediate a four-strand structure.

ACKNOWLEDGMENTS

We are especially indebted to Leroy Liu for suggesting the experiments on decatenation by Int, to Thomas Pollock for working out the change in linking number associated with strand exchange, and to John Wilson for clarifying the relationship between his model for synapsis and the model for synapsis and strand exchange presented herein. We also thank Martin Gellert and Thomas Pollock for their careful and thoughtful reading of this manuscript.

REFERENCES

Bidwell, K. and A. Landy. 1979. Structural features of λ site-specific recombination at a secondary *att* site in *galT*. *Cell* **16:** 397.

Brown, P. O. and N. R. Cozzarelli. 1979. A sign inversion mechanism for enzymatic supercoiling of DNA. *Science* **206:** 1081.

Campbell, A. M. 1962. Episomes. *Adv. Genet.* **11:** 101.

Cassuto, E., S. C. West, J. Mursalim, S. Conlon, and P. Howard-Flanders. 1980. Initiation of genetic recombination: Homologous pairing between duplex DNA molecules promoted by *recA* protein. *Proc. Natl. Acad. Sci.* **77:** 3962.

Cowie, D. B. and A. D. Hershey. 1965. Multiple sites of interaction with host-cell DNA in the DNA of phage λ. *Proc. Natl. Acad. Sci.* **53:** 57.

Cross, R. A. and M. Lieb. 1967. Heat-inducible λ phage. V. Induction of prophages with mutations in genes O, P, and R. *Genetics* **57:** 549.

Davison, P. F. 1966. The rate of strand separation in alkali-treated DNA. *J. Mol. Biol.* **22:** 97.

Deutsch, W. A. and S. Linn. 1979. Further characterization of a depurinated DNA-purine base insertion activity from cultured human fibroblasts. *J. Biol. Chem.* **254:** 12099.

Dove, W. F. 1970. An energy-level hypothesis for λ prophage insertion and excision. *J. Mol. Biol.* **47:** 585.

Echols, H. and L. Green. 1979. Some properties of site-specific and general recombination inferred from *int*-initiated exchanges by bacteriophage lambda. *Genetics* **93:** 297.

Echols, H., R. Gingery, and L. Moore. 1968. Integrative recombination function of bacteriophage λ: Evidence for a site-specific recombination enzyme. *J. Mol. Biol.* **34:** 251.

Enquist, L., H. Nash, and R. A. Weisberg. 1979. Strand exchange in site-specific recombination. *Proc. Natl. Acad. Sci.* **76:** 1363.

Guerrini, F. 1969. On the asymmetry of λ integration sites. *J. Mol. Biol.* **46:** 523.

Holliday, R. 1964. A mechanism for gene conversion in fungi. *Genet. Res.* **5:** 282.

Hotchkiss, R. D. 1974. Models of genetic recombination. *Annu. Rev. Microbiol.* **28:** 445.

Hsieh, T. and D. Brutlag. 1980. ATP-dependent DNA topoisomerase from *Drosophila melanogaster* reversibly catenates duplex DNA rings. *Cell* **21:** 115.

Hsieh, T.-S. and J. C. Wang. 1975. Thermodynamic properties of superhelical DNAs. *Biochemistry* **14:** 527.

Hsu, P., W. Ross, and A. Landy. 1980. The λ phage *att* site: Functional limits and interaction with Int protein. *Nature* **285:** 85.

Kellenberger-Gujer, G. and R. Weisberg. 1971. Recombination in bacteriophage lambda I. Exchange of DNA promoted by phage and bacterial recombination mechanism. In *The bacteriophage lambda* (ed. A. D. Hershey), p. 407. Cold Spring Harbor Laboratory, Cold Spring Harbor, New York.

Kikuchi, Y. and H. A. Nash. 1978. The bacteriophage λ *int* gene product. *J. Biol. Chem.* **253:** 7149.

———. 1979a. Integrative recombination of bacteriophage λ: Requirement for supertwisted DNA *in vivo* and characterization of Int. *Cold Spring Harbor Symp. Quant. Biol.* **43:** 1099.

———. 1979b. Nicking-closing activity associated with bacteriophage λ *int* gene product. *Proc. Natl. Acad. Sci.* **76:** 3760.

Kreuzer, K. N. and N. R. Cozzarelli. 1980. Formation and resolution of DNA catenanes by DNA gyrase. *Cell* **20:** 245.

Landy, A. and W. Ross. 1977. Viral integration and excision: Structure of the lambda *att* sites. *Science* **197:** 1147.

Liu, L. F., C.-C. Liu, and B. M. Alberts. 1980. Type II DNA topoisomerases: Enzymes that can unknot a topologically knotted DNA molecule via a reversible double-strand break. *Cell* **19:** 697.

Livneh, Z., D. Elad, and J. Sperling. 1979. Enzymatic insertion of purine bases into depurinated DNA *in vitro*. *Proc. Natl. Acad. Sci.* **76:** 1089.

McGavin, S. 1971. Models of specifically paired like (homologous) nucleic acid structures. *J. Mol. Biol.* **55:** 293.

———. 1977. A model for the specific pairing of homologous double-stranded nucleic acid molecules during genetic recombination. *Heredity* **39:** 15.

———. 1979. A reconsideration of the possibility of the specific pairing of base pairs. *J. Theor. Biol.* **77:** 83.

Meselson, M. 1972. Formation of hybrid DNA by rotary diffusion during genetic recombination. *J. Mol. Biol.* **71:** 895.

Meselson, M. and C. M. Radding. 1975. A general model for genetic recombination. *Proc. Natl. Acad. Sci.* **72:** 358.

Mizuuchi, K. and M. Mizuuchi. 1979. Integrative recombination of bacteriophage λ: *In vitro* study of the intermolecular reaction. *Cold Spring Harbor Symp. Quant. Biol.* **43:** 1111.

Mizuuchi, K. and H. A. Nash. 1976. Restriction assay for integrative recombination of bacteriophage λ DNA *in vitro:* Requirement for closed circular DNA substrate. *Proc. Natl. Acad. Sci.* **73:** 3524.

Mizuuchi, K., L. M. Fisher, M. H. O'Dea, and M. Gellert. 1980a. DNA gyrase action involves the introduction of transient double-strand breaks into DNA. *Proc. Natl. Acad. Sci.* **77:** 1847.

Mizuuchi, K., M. Gellert, R. A. Weisberg, and H. A. Nash. 1980b. Catenation and supercoiling in the products of *in vitro* λ integrative recombination *in vitro*. *J. Mol. Biol.* **141:** 485.

Nash, H. A. 1975. Integrative recombination in bacteriophage lambda: Analysis of recombinant DNA. *J. Mol. Biol.* **91:** 501.

Nash, H. A. and C. Robertson. 1971. On the mechanism of *int*-promoted recombination. *Virology* **44:** 446.

Nash, H. A., L. W. Enquist, and R. A. Weisberg. 1977. On the role of the bacteriophage λ *int* gene product in site specific recombination. *J. Mol. Biol.* **116:** 627.

Parkinson, J. S. 1971. Deletion mutants of bacteriophage lambda. II. Genetic properties of *att*-defective mutants. *J. Mol. Biol.* **56:** 385.

Pollock, T. J. and K. Abremski. 1979. DNA without supertwists can be an *in vitro* substrate for site-specific recombination of bacteriophage λ. *J. Mol. Biol.* **131:** 651.

Pollock, T. J. and H. A. Nash. 1980. Catenation of the products of integrative recombination: Comparison of supertwisted and nonsupertwisted substrates. In *Proceedings of the 1980 ICN-UCLA Symposium: DNA Replication* (ed. B. Alberts and C. F. Fox). Academic Press, New York. (In press.)

Pulleyblank, D. E., M Shure, D. Tang, J. Vinograd, and H.-P. Vosberg. 1975. Action of nicking-closing enzyme on supercoiled and nonsupercoiled closed circular

DNA: Formation of a Boltzmann distribution of topological isomers. *Proc. Natl. Acad. Sci.* **72:** 4280.

SHIMADA, K., R. A. WEISBERG, and M. E. GOTTESMAN. 1972. Prophage lambda at unusual chromosomal locations. I. Location of the secondary attachment sites and the properties of the lysogens. *J. Mol. Biol.* **63:** 483.

———. 1975. Prophage lambda at unusual chromosomal locations. III. The components of the secondary attachment sites. *J. Mol. Biol.* **93:** 415.

SHULMAN, M. and M. GOTTESMAN. 1973. Attachment site mutants of bacteriophage lambda. **81:** 461.

SHURE, M., D. E. PULLEYBLANK, and J. VINOGRAD. 1977. The problems of eukaryotic and prokaryotic DNA packaging and in vivo conformation posed by superhelix density heterogeneity. *Nucleic Acids Res.* **4:** 1183.

SIGAL, N. and B. ALBERTS. 1972. Genetic recombination: The nature of a crossed strand-exchange between two homologous DNA molecules. *J. Mol. Biol.* **71:** 789.

SIGNER, E. 1971. General recombination. In *The bacteriophage lambda* (ed. A. D. Hershey), p. 139. Cold Spring Harbor Laboratory, Cold Spring Harbor, New York.

STAHL, F. W. 1979. Special sites in generalized recombination. *Annu. Rev. Genet.* **13:** 7.

STAHL, F. W. and M. M. STAHL. 1971. DNA synthesis associated with recombination. II. Recombination between repressed chromosomes. In *The bacteriophage lambda.* (ed. A. D. Hershey), p. 443. Cold Spring Harbor Laboratory, Cold Spring Harbor, New York.

STAHL, F. W., K. D. MCMILIN, M. M. STAHL, J. E. CRASEMAN, and S. LAM. 1974. The distribution of crossovers along unreplicated lambda bacteriophage chromosomes. *Genetics* **77:** 395.

WAGNER, R., JR. and M. MESELSON. 1976. Repair tracts in mismatched DNA heteroduplexes. *Proc. Natl. Acad. Sci.* **73:** 4135.

WANG, J. C. and L. F. LIU. 1979. DNA topoisomerases: Enzymes which catalyze the concerted breaking and rejoining of DNA backbone bonds. In *Molecular genetics* (ed. J. H. Taylor), vol. 3, p. 65. Academic Press, New York.

WEIL, J. 1969. Reciprocal and non-reciprocal recombination in bacteriophage λ. *J. Mol. Biol.* **43:** 351.

WEIL, J. and E. R. SIGNER. 1968. Recombination in bacteriophage λ. II. Site-specific recombination promoted by the integration system. *J. Mol. Biol.* **34:** 273.

WEISBERG, R. A. and M. E. GOTTESMAN. 1969. The integration and excision defect of bacteriophage λdg. *J. Mol. Biol.* **46:** 565.

WILDENBERG, J. and M. MESELSON. 1975. Mismatch repair in heteroduplex DNA. *Proc. Natl. Acad. Sci.* **72:** 2202.

WILSON, J. H. 1979. Nick-free formation of reciprocal heteroduplexes: A simple solution to the topological problem. *Proc. Natl. Acad. Sci.* **76:** 3641.

Structure and Function of the Phage λ *att* Site: Size, Int-binding Sites, and Location of the Crossover Point

K. MIZUUCHI,* R. WEISBERG,† L. ENQUIST,‡ M. MIZUUCHI,* M. BURACZYNSKA,§ C. FOELLER,§ P.-L. HSU,§ W. ROSS,§ AND A. LANDY§

* *Laboratory of Molecular Biology, National Institute of Arthritis, Metabolism, and Digestive Diseases,* † *Laboratory of Molecular Genetics, National Institute of Child Health and Human Development,* ‡ *Laboratory of Molecular Virology, National Cancer Institute, National Institutes of Health, Bethesda, Maryland 20205;* § *Section of Microbiology and Molecular Biology, Division of Biology and Medicine, Brown University, Providence, Rhode Island 02912*

Reciprocal recombination between specific DNA sequences on the λ and *E. coli* chromosomes (*att*P and *att*B, respectively) gives rise to left (*att*L) and right (*att*R) prophage attachment (*att*) sites at the junctures of an integrated viral genome (for reviews, see Gottesman and Weisberg 1971; Nash 1977). All four of these *att* sites are genetically distinguishable from one another; however, the crossover occurs within (or at the boundaries of) a 15-bp "core region" that is common to all of them (Landy and Ross 1977). The crossover does not involve any degradation and resynthesis of the parental sequence nor a new round of DNA replication as a mandatory part of the mechanism. The most rigorous proof for this statement comes from recent in vitro studies of the reaction and is further supported by some of the experiments reported here (see also Nash et al., this volume). The reaction does not require any high-energy cofactor, and even the superhelicity of the substrate circular DNA does not undergo major alteration (Mizuuchi et al. 1980).

Both integrative as well as excisive (*att*L × *att*R) recombination require the phage-specific protein Int, as well as the products of several host *E. coli* genes (Miller et al. 1979). The phage protein Xis is only required in excisive recombination. Int, which has been purified to near homogeneity (Kikuchi and Nash 1978), has been shown to function as a type-I DNA topoisomerase (Kikuchi and Nash 1979; Nash et al., this volume). Since Int has this property of cutting and resealing phosphodiester bonds, it seems very likely to be directly responsible for executing the crossover event, for example, by a proposed mechanism involving two sets of specific nicking/closing reactions (Kikuchi and Nash 1979; Nash et al., this volume).

In this report we shall summarize experiments, in various stages of completion, that bear on two aspects of integrative recombination. We shall first discuss those features that make the phage and bacterial *att* sites different from one another in the reaction. We shall then consider that aspect of the reaction in which all the *att* sites participate in a similar manner, namely, the crossover event that takes place within the 15-bp common core region.

RESULTS AND DISCUSSION

Int-binding Sites

Since it is probably Int that promotes polynucleotide exchange within the core, we expected to find specific binding to this region. Using the footprint technique of Galas and Schmitz (1978), it was shown that purified Int protects a region of 30–35 bp, including the 15-bp core of the phage *att* site and extending symmetrically beyond it on either side (Ross et al. 1979). In contrast to this result, the other interactions of Int with the phage *att* site were not expected and are considerably more difficult to interpret. These interactions are summarized by the results shown in Figure 1. The 15-bp core region of the phage *att* site is located approximately in the middle of the 550-bp restriction fragment used for this experiment. Int protects four regions from digestion with the AT-specific antitumor agent neocarzinostatin. One of these, from −17 to +16, includes the 15-bp core region. To the right, in the P′ arm, there is another 30–35-bp protected region extending from approximately +50 to +85. As seen from the comparison between lanes 1 and 2 of Figure 1, the interaction between Int and this DNA site, in contrast to the other Int-binding sites, is resistant to challenge by the polyanion heparin (Ross et al. 1979). There are also two small Int-binding sites located in the P arm from −148 to −129 and from −116 to −98.

The P-arm Int-binding sites (15–20 bp) are each shorter in length than either the common core or the P′-arm sites but are comparable in size to the 15-bp protected region in the *att*B common core (Ross et al. 1979). However, as seen in the comparison in Figure 2, the P-arm sequences do not share homology with the common core sequence or with the core-arm juncture sequences.

The common-core and P′-arm Int-binding sites show very little sequence homology with each other (Fig. 2), possibly suggesting their interactions with different domains of Int. In contrast, however, each of the P-arm sites shares a considerable uninterrupted homology (11 bp in P-arm 1 and 8 bp in P-arm 2) with the left portion of the P′-arm site. The two P-arm sites also share this homologous sequence but with an inverted orientation, such that P-arm 1 is a direct repeat and P-arm 2 is an inverted repeat of the P′ homology. Included within the left half of the P′ site is the 7-bp sequence TCACTAT, which, in addition to being part of the homology with the two P-arm sites, is also found (with one mismatch) in the right half of the P′ site.

Since the unique heparin-resistant binding in the P′ arm is not dependent upon the presence of the other binding sites (W. Ross et al., unpubl.), the Int binding and resistance to heparin challenge are intrinsic features

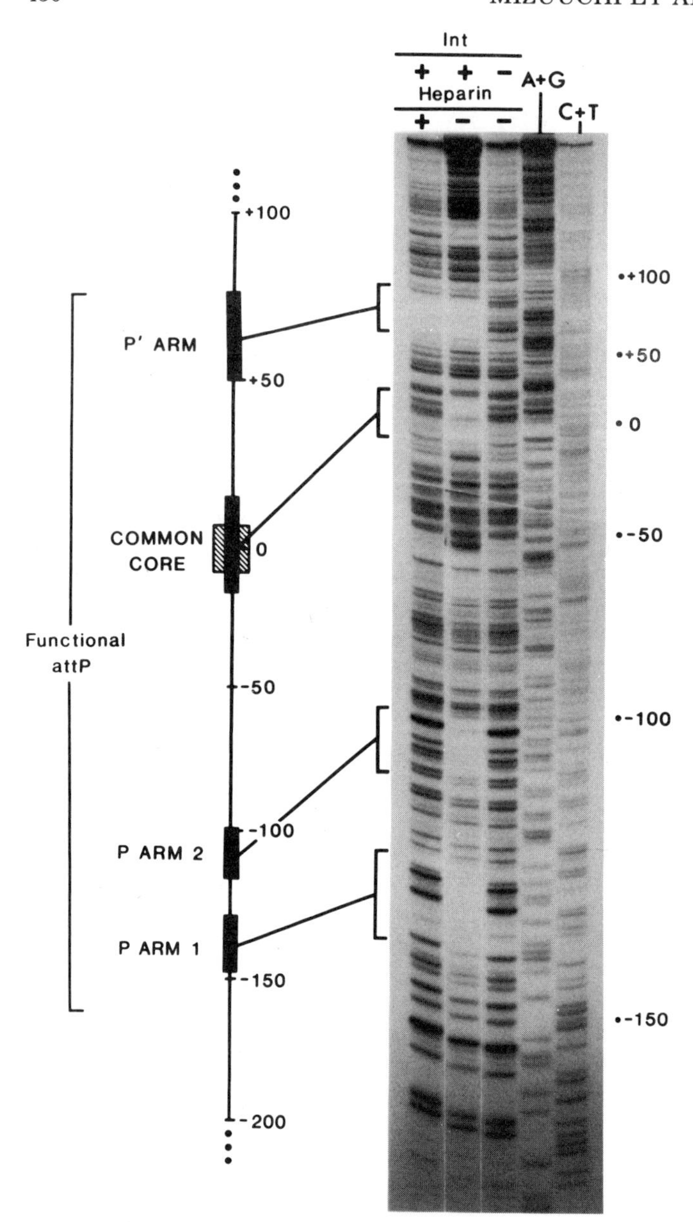

Figure 1. In lanes 1–3, the restriction fragment *Hin*dIII(−251)-*Hpa*II(+305), ^{32}P-labeled at the 5′ end of the top strand (*Hin*dIII), was partially digested with neocarzinostatin (D'Andrea and Haseltine 1978; Hatayama et al. 1978), in the absence (lane 3) or presence of purified Int at 10 μg/ml (lane 2) or 20 μg/ml (lane 1). The Int-DNA complex in lane 1 was challenged with heparin before neocarzinostatin digestion. Sequence markers for this restriction fragment, A+G (lane 4) and C+T (lane 5), were prepared according to the method of Maxam and Gilbert (1977). Details of these methods and electrophoresis conditions have been described previously (Ross et al. 1979). A linear map of the minimal phage *att*-site region depicts the relative sizes and positions of the four Int-protected sequences (■) seen in the neocarzinostatin footprint. Also indicated are the boundaries (−160 and +82) of the smallest functional *att*P region obtained in these experiments (see text) and the 15-bp common core sequence (▨). (Reprinted, with permission, from Hsu et al. 1980.)

of the P′-arm sequence. Int binding at sequences homologous to the left half of the P′-arm site (P-arm sites 1 and 2) did not show the property of heparin-resistance, and therefore the right half of the P′ site, which is essential for recombination, might confer heparin resistance.

Functional Limits of the *att* Sites

The significance of Int-binding sites outside of the common core region is not immediately obvious. Not only are these sites a considerable distance away from the region of the crossover event, but an analysis of their homologies indicates that the sequence relationships among them are not simple. Furthermore, Int, like many other nucleic-acid-binding proteins, also has a nonspecific interaction with DNA; at sufficiently high protein concentrations, all of the DNA in a restriction fragment can be protected by Int against nuclease digestion.

The importance of the P1 and P′ Int-binding sites for phage *att*-site function is strongly suggested by the results of exonuclease resection experiments. Cloned phage *att*-site fragments from which increasing amounts of DNA had been removed from one side or the other were tested for phage *att*-site function in an in vitro recombination reaction. It was shown that the presence of the P′ site and the P-arm 1 site are both required for the ability of an *att* site carried on supercoiled plasmid DNA to recombine with a linear form of *att*B DNA (Hsu et al. 1980; Mizuuchi and Mizuuchi 1980). Under the assay conditions used for those experiments, supercoiled plasmid DNAs carrying the bacterial *att* site or the left or right prophage *att* sites all failed to recombine efficiently with the linear bacterial *att* site, and therefore the competence for recombination must have reflected phage *att*-site function (and not prophage or bacterial *att*-site function). It is thus concluded that the sequences

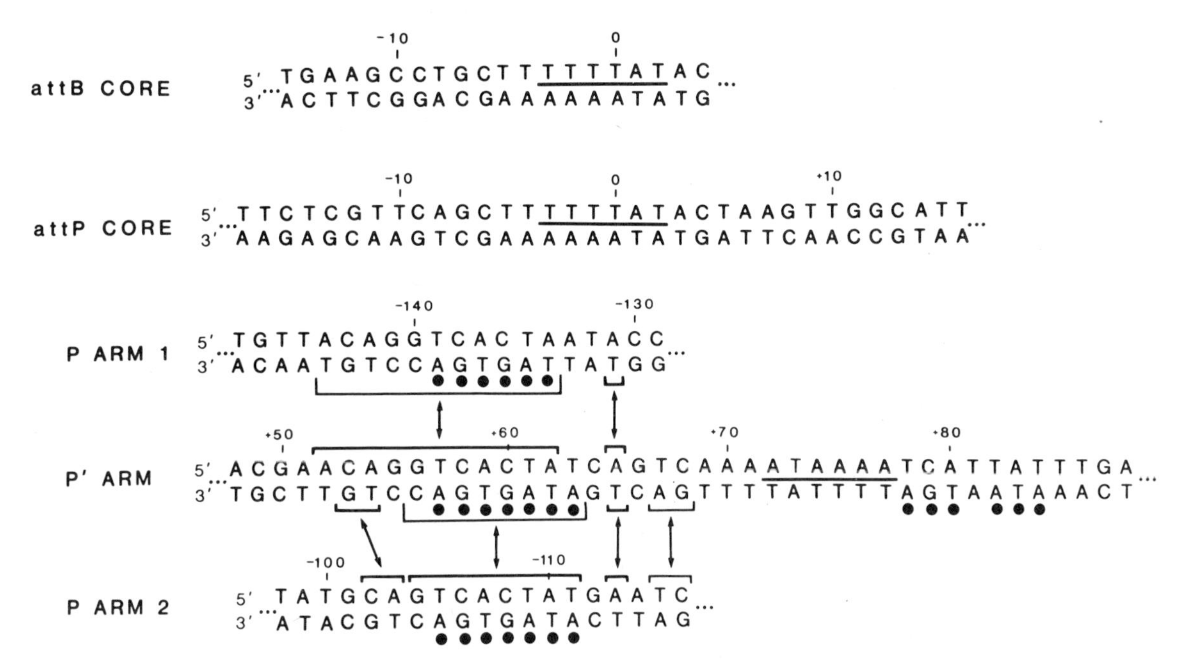

Figure 2. Comparison of Int-protected sequences in the bacterial and phage *att* sites. The sequence shown for each binding site is the maximum length of DNA protected by Int from either neocarzinostatin or DNase I digestion. (Ross et al. 1979; Hsu et al. 1980). The sequence of the P-arm 2 site has been inverted to facilitate comparison of sequence homologies. A 6-bp homology between the two common core sites and the P′-arm site is indicated by a line between the two strands of sequence (——). Homologies between the left portion of the P′-arm site and each of the two P-arm sites are indicated (⌐——¬) (see text). A 7-bp sequence in the left half of the P′-arm protected region, found within the sequence shared with the two P-arm sites, is also found, with one mismatch, in the right half of the P′-arm site (●). (Reprinted, with permission, from Hsu et al. 1980.)

delimited by the minimal P and P′ arms (−160 to +82) are both necessary and sufficient for phage *att*-site function.

A similar series of in vitro experiments was carried out to determine the minimal size of the bacterial *att* site (M. Mizuuchi, in prep.). In this case, each reaction contained the same phage *att* site on a supercoiled molecule. The linear recombination partner consisted of a cloned restriction fragment containing bacterial *att* sites resected to different extents from the right or from the left. The resected plasmids were also tested for in vivo bacterial *att*-site function by their ability to participate in Int-dependent recombination with an infecting phage genome and thereby produce recombinant plasmid-phage genomes that transduced the *amp*ʳ marker.

The first point to be made from Figure 3 is that the apparent boundaries of the bacterial *att* site are from −7 to +7 and from −11 to +11 in the in vivo and in vitro assays, respectively. Irrespective of the precise nucleotide boundaries of the bacterial *att* site, it is clear that this site does not extend much beyond the common core region and is therefore far less complex than its phage *att*-site partner. It is interesting that the in vivo test appears to be less demanding than the in vitro test. This might be expected if the in vivo reaction is so efficient as to be outside a linear range that is sensitive to small decreases in functionality.

One of the critical features distinguishing the phage and bacterial *att* sites from each other is the P′-arm Int-binding site. Not only is the presence of this arm a requirement for the phage *att* site, but its absence is a requirement for a bacterial *att* site (Hsu et al. 1980). Although it is clear from the results discussed above that Int-dependent recombination is not a mechanistically symmetrical reaction, i.e., *att*P is considerably more complex than *att*B, we still do not understand the role of these nonsymmetrical elements.

We shall now address that aspect of the recombination pathway that appears to be symmetrical, i.e., the crossover event that takes place within the 15-bp core region that is common to all four *att* sites. In particular, are there preferred loci within the common core for crossing over, and, if so, where are they located? To answer these questions we have used both genetic and chemical approaches. The former has the advantage of letting us look at events that occur in vivo. It suffers from the objection that the introduced markers may influence the location and frequency of the crossover. This objection does not apply to the chemical experiments in which isotopes rather than mutations are used as markers. To our gratification, the two approaches yield consistent conclusions.

Genetic Analysis of the Crossover Event

The core "mutants" we used in these experiments (R. Weisberg et al., in prep.) are naturally occurring variants of *att*B called secondary *att* sites (Shimada et al. 1972). Approximately ten have been sequenced to date. If we compare the regions around the crossover point, they

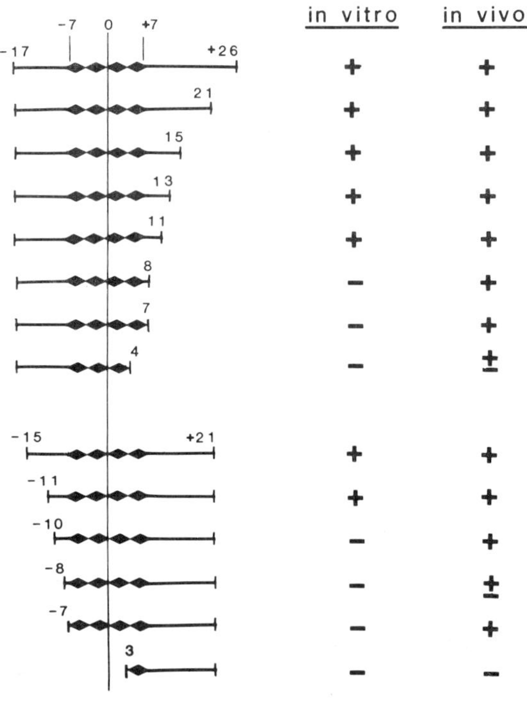

Figure 3. Integrative recombination of plasmids with resected B arms or resected B′ arms. Endpoints of resected *att*B fragments are shown in the figure. The ends of the B arm are joined to *Bam*HI linker molecules, and the ends of the B′ arm are joined to *Eco*RI linker molecules. Construction of the plasmid series with trimmed B′ arms started from a fragment extending from −17 to beyond +80, and the *Bam*HI-*Eco*RI fragment of pBR322 was used as the cloning vehicle. The series of plasmids with trimmed B arms was constructed starting from a plasmid that retained BOB′ sequence from −17 to +21. Circular DNA of the plasmid was cut with *Bam*HI, trimmed with nuclease S1 and recircularized using *Bam*HI linker. Thus, in this series, the sequences beyond the *Bam*HI linker in the B arm are different in each plasmid. It should be noted that in some cases the sequences brought in by the linkers correspond to the nucleotides originally found at those positions in the *att*B site. The possible significance of such partial homology will be discussed elsewhere (M. Mizuuchi, in prep.).

show high conservation of sequence at certain positions and considerable divergence at others (Bidwell and Landy 1979; Christie and Platt 1979; Csordás-Tóth et al. 1979; Landy et al. 1979; Pinkham et al. 1980; J. Chapman and J. Gardner, in prep.). We have determined the sequence of the ends of two λ insertions (called 614 and 791) into a secondary attachment site in the *E. coli galT* gene (Fig. 4). The same corelike sequence in *galT* was used for both insertions; it differs from the wild-type core at 7 out of 15 positions. Prophage 614 differs from 791 by a leftward shift of 4 bp to 7 bp in the prophage-host junction points. The shift has occurred in a coordinated way in *att*P and in *galT*. The two insertions also differ in the correlation of prophage excision with reversion of the host to $galT^+$: Upon excision of the prophage (Weisberg and Gallant 1967), 95% of 614 cells but only 5% of 791 cells reverted to $galT^+$. The correlation between excision and reversion was also determined for 13 other independently isolated λ insertions in *galT*: All resembled 614. It seems likely, therefore, that the 614 insertion is preferred to the 791 insertion.

The difference in sequence suggests an explanation of the difference in reversion. Let us assume that the 614 crossover point is preferred for excision, as it seems to be preferred for insertion. The consequences of this assumption are shown in Figure 4. When insertion and excision occur at different points, there is a reciprocal exchange of genetic information between virus and host. The information exchanged lies in the region between the crossover points (where the *galT* and wild-type cores differ in three positions). This 3-bp difference has been looked for and found in a $galT^-$ bacterial excisant and in five independent λ excisants derived from insertion 791. The excised phage are also defective in lysogenizing a wild-type host (Fig. 5, col. 1). This defect most probably reflects decreased *att* function (see below).

Some years ago, Erich Six (1963, 1966) described a similar phenomenon and proposed a similar model to account for it in coliphage P2. He called the mutants that arose from a cycle of insertion and excision "saf" for site affinity, and we have adopted this name for our mutants.

The cores of *att*P-*saf*, *galT-saf*, and $galT^+$ differ from the wild-type core by 3, 4, and 7 bp, respectively. How do these changes affect *att* function? If recombination frequency depends directly on the degree of match between the participating cores, then the crosses *att*P-*saf* × $galT^+$ and $attP^+$ × *galT-saf*, which match at 11/15 core positions, should produce recombinants with equal and relatively high efficiency. In contrast, the crosses $attP^+$ × $galT^+$ and *att*P-*saf* × *galT-saf*, which match at only 8/15 core positions, should produce recombinants with equal and relatively low efficiency. On the other hand, if recombination frequency is enhanced by an increase in the aggregate match of the two partners to the wild-type core sequence, then recombination frequency should increase as follows: *att*P-*saf* × $galT^+$ (20/30 matches) < *att*P-*saf* × *galT-saf* (23/30 matches) ≈ $attP^+$ × $galT^+$ (23/30 matches) < $attP^+$ × *galT-saf* (26/30 matches). The results (Fig. 5) confirm neither of these simple predictions. The first model is incorrect because the first of the predicted equalities was not found, and the second model is incorrect because *att*P-*saf* × $galT^+$ ranked second instead of fourth in recombination frequency. It appears, therefore, that the role of the 15-bp core region is not exclusively that of a homology element, nor is it exclusively a recognition element. A third model that combines features of the first two fits the data better: Increasing match between the recombining core sequences as well as increasing match to the wild-type core might both enhance the recombination efficiency.

Where are the crossover points located in other secondary attachment sites? A compilation of the available data is shown in Figure 6. What is the minimum number of crossover loci that will account for these results? The answer is two: one between +4 and +5, to account for *galT*791 and *b*522, and one between −3 and +1, to

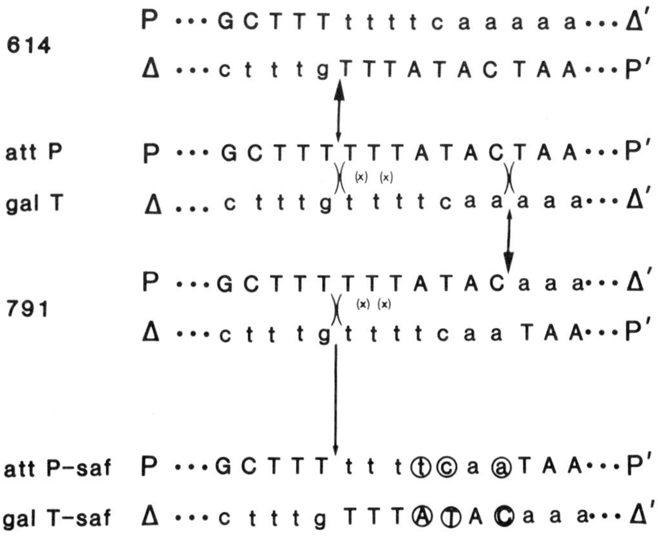

614	P •••G C T T T t t t t c a a a a a•••Δ′
	Δ •••c t t t g T T T A T A C T A A•••P′
att P	P ••• G C T T T T T T A T A C T A A•••P′
	(x) (x)
gal T	Δ ... c t t t g t t t t c a a a a a•••Δ′
791	P •••G C T T T T T T A T A C a a a•••Δ′
	(x) (x)
	Δ •••c t t t g t t t t c a a T A A•••P′
att P–saf	P •••G C T T T t t t t c a a T A A•••P′
gal T–saf	Δ •••c t t t g T T T A T A C a a a•••Δ′

Figure 4. Core sequences of the *att* sites derived by recombination with the secondary *att* site in *galT*. Rows three and four show the "starting sequences" for these experiments. The phage *att*-site core (*att*P) is shown in upper case (A, G, etc.) flanked by the P and P′ arm. The 15-bp sequence of wild-type *galT* that serves as the core of this secondary *att* site is shown in lower case (a, g, etc.), and the flanking sequences are designated Δ and Δ′ (Δ is in the direction of *galK*, and Δ′ is in the direction of *galE*). Integration of *att*P into the *galT* secondary *att* site gives rise to left (ΔP′) and right (PΔ′) prophage *att* sites in the 614 and 791 lysogens. The location of the crossover sites (X) for integrative (➜➜) or excisive (⇅) recombination determines the sequences of the product core regions, which are shown as hybrids of upper- and lower-case letters. The double-headed arrows indicate the reciprocity observed when excision and integration proceed via the same crossover site. The bottom two rows show the formation of *att*P-*saf* and *galT-saf* when excision from lysogen 791 occurs by means of the crossover site different from that used to form lysogen 791 (see text). Note that the leftmost crossover could have occurred between any of three adjacent Ts (x); however, the one indicated (X) fits best with the chemical results (see text and Figs. 7 and 8). Circled letters indicate those positions where *att*P-*saf* differs from *att*P and where *galT-saf* differs from *galT*. Details of the genetic and structural analyses of these *att* sites will be presented elsewhere (R. Weisberg et al., in prep.).

account for the seven others. (Clearly, *proA/B* and *b*511 could be in either group.) What must happen to DNA to produce a crossover point? If we assume that no new DNA synthesis occurs during recombination (see Nash 1977), one or both of the DNA strands must be broken and exchanged at that point. Can these two possibilities be distinguished? The only relevant genetic experiment is a cross of a phage with a wild-type core by a phage carrying the core mutation *att*24 (Shulman and Gottesman 1973). $att^{+}/att24$ heterozygotes were found among the recombinant progeny. This suggests that the recombinant joint can be the result of staggered cuts in the two DNA strands rather than flush cuts (see also Nash et al., this volume). In the next section we present evidence that confirms this conclusion.

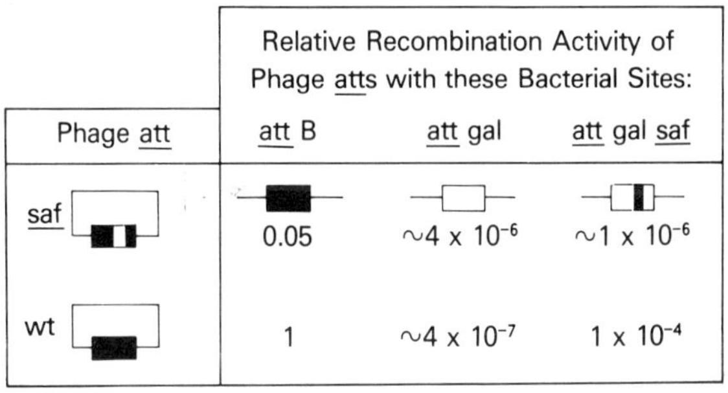

Phage att	Relative Recombination Activity of Phage atts with these Bacterial Sites: att B	att gal	att gal saf
saf	0.05	∼4 x 10^{-6}	∼1 x 10^{-6}
wt	1	∼4 x 10^{-7}	1 x 10^{-4}

Figure 5. Properties of site-affinity mutants. Recombination with *att*B was measured by determining the frequency of stable lysogens formed after infection of a wild-type host with the phage listed in column 1, as described by Shimada et al. (1972). The results were the same when an isogenic *recA* host was used. Recombination with *att-galT*$^{+}$ and *att-galT-saf* was measured after infection of mutant hosts that were also deleted for *att*B. Insertions into the *gal* operon were selected by resistance to galactose and immunity to λ*c*, as described by Shimada et al. (1973). The *att-galT* host carried a *galE* mutation, and the *att-galT-saf* host was *galT*$^{-}$ by virtue of the *saf* mutation. In both cases, the host phenotype before infection was galactose-sensitive. Five independently isolated λ*att*P-*saf* and two independently isolated *galT-saf* lines were used. There were no significant differences among them, and the results given are arithmetic means.

Chemical Analysis of the Crossover Event

Our chemical approach for studying the crossover event (K. Mizuuchi, in prep.) was made possible by the finding that a short DNA fragment carrying *att*B can be used efficiently as the substrate for recombination in vitro (M. Mizuuchi, in prep.). *att*B DNA was internally labeled with ^{32}P and recombined with unlabeled *att*P DNA. After the recombination reaction, distribution of the radioisotope in the reaction products was analyzed. In this experiment, the radioisotope not only played a role as the tracer, but also a second role—to generate cleavages by radioactive decay at the site where it was incorporated. This second role of the ^{32}P atoms facilitated precise localization of the isotope within the product molecules.

The design of the experiment is shown schematically in Figure 7. A 46-bp fragment of DNA carrying *att*B was prepared so that one of the strands carried ^{32}P at internal positions. Using one of the purified strands as the template, the ^{32}P atoms were incorporated from one of the four [α-^{32}P]deoxynucleoside triphosphates by the large fragment of DNA polymerase I. Thus, in total, a set of eight *att*B substrates was prepared (panel 1). Each one of the labeled *att*B fragments was recombined in vitro with circular DNA carrying *att*P (panel 2). The reaction produced a linear DNA that carried *att*L and *att*R at its ends (panel 3). This DNA was digested with

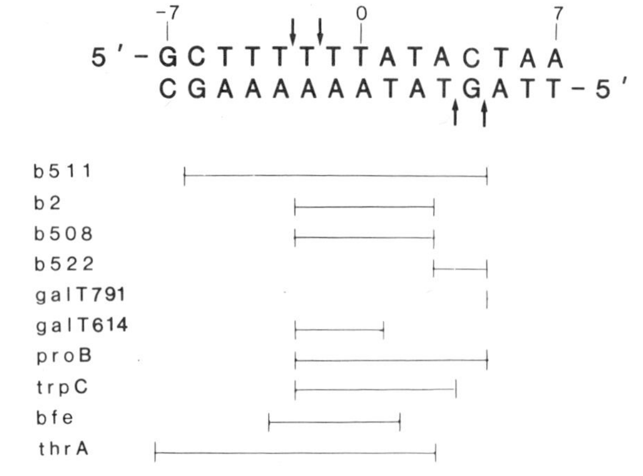

Figure 6. Potential crossover sites within the core regions of sequenced secondary *att* sites. The sequence of the 15-bp *att* core region (*top row*) was aligned with the analogous core region in each of the secondary *att* sites indicated. The vertical and horizontal bars (⊢⊣) indicate those positions where the crossover could have taken place when λ integrated into the respective secondary *att* site (as determined by sequence analysis of one or both recombination products). The locations of the cut sites for crossover in recombination between *att*P and *att*B (see text and Fig. 7) are indicated for the wild-type core sequence (⇅). The sources of the secondary *att*-site data are: *b*511, *b2*, *b*508, *b*522 (Landy et al. 1979); *galT*791 (Bidwell and Landy 1979), *galT*614 (R. Weisberg et al., in prep.); *proA/B* (Pinkham et al. 1980); *trpC* (Christie and Platt 1979); *bfe* (Csordás-Tóth et al. 1979); and *thrA* (J. Chapman and J. Gardner, in prep.).

restriction endonucleases (*Bam*HI and *Hin*fI), and the fragments carrying *att*L or *att*R were purified (panel 4). The purified fragments were stored two weeks to let half of the ^{32}P atoms disintegrate and generate cleavages in the DNA where they were located (panel 5). As is clear from the figure, each disintegration of ^{32}P generates two fragments, one containing the material derived exclusively from the *att*B parent and the other carrying material derived from both *att*B and *att*P. Thus, each sample contains two sets of fragments. The lengths of the P and P′ arms are longer than those of the B and B′ arms in the original restriction fragment, so the two sets of ^{32}P-generated fragments do not overlap in size. The resulting mixture of fragments was analyzed by electrophoresis on a sequencing-type polyacrylamide gel and an autoradiograph was prepared.

Figure 8 shows the larger series of fragments. In this series, the shortest fragment in each sample is generated by the disintegration of the innermost ^{32}P atom. Hence, it no longer contains ^{32}P atoms and is invisible on the autoradiograph. With this in mind, we can compare the pattern seen in Figure 8 with the sequence of the common core. Panel A shows the set of samples that contain BOP′ with ^{32}P atoms in the lower strand. The left end of the shortest radioactive fragment is at core position +1. Because decay of the rightmost ^{32}P atom in a recombinant yields a nonradioactive fragment, we infer the existence of an even shorter fragment whose left endpoint can be determined in the following way. The core sequence on the lower strand from the B end at −7 to +3 is: 3′-(C)(G)AAAAAAT(A)(T)-5′. The bases in parentheses are the rightmost C, G, A, and T that are adjacent to a 5′ ^{32}P atom, because radioactive fragments produced by decay of these atoms were *not* seen, whereas radioactive fragments produced by decay of ^{32}P atoms adjacent to the nearest leftward C, G, A, or T *were* seen. We conclude that the rightmost crossover point on the bottom strand is between the phosphate that is 5′ to the T at position +3 and the phosphate that is 5′ to the G at position +4. The cut could therefore occur on either the 5′ or the 3′ side of the G.

Is this the only crossover point on this strand, or can crossovers also occur to the left of this point? The distribution of ^{32}P in the lower strand of the POB′ product (panel B) shows that there is no ^{32}P to the left of core position +4 in this recombinant. This indicates that the crossover usually, or always, takes place at a unique point in the bottom strand. The same is true for the other strand (panels C and D), although the point is different; it is adjacent to the T at position −2. These results were further supported by also analyzing the families of shorter fragments that were produced by the ^{32}P decay (data not shown).

Now we can conclude that the recombination mechanism involves one of the following two possibilities as shown in Figure 7, panel 1. (1) The cut on each strand is staggered by 7 bp (position −2 to +4) and takes place between a 3′-phosphate and a 5′-hydroxyl. (2) The cut on each strand is staggered by 5 bp (position −1 to +3) and takes place between a 3′-hydroxyl and a 5′-phosphate.

The question as to which one of the two possibilities is correct remains to be tested directly. However, the results described in the previous section (see also Bidwell and Landy 1979) favor the first possibility since they place one of the cut points between positions +4 and +5.

How does this placement of cut sites fit with the observation that secondary *att* sites do not have a canonical core sequence and most of them cannot pair homologously with *att*P in the 7 (or 5)-bp region between the staggered cuts? First, it is interesting to note that from the sequence data presently available, the two most efficient secondary *att* sites, *galT-saf* (see above and R. Weisberg et al., in prep.) and *proA/B* (Shimada et al. 1972; J. L. Pinkham et al. 1980), both have perfect homology with *att*P in the 7-bp overlap region. The secondary *att* site in *trpC,* which only has a 6-bp homology in this region (Christie and Platt 1979), is a less efficient secondary *att* site (Shimada et al. 1972). For the majority of secondary *att* sites that lack perfect homology with *att*P in this region, we postulate the following. When λ inserts into secondary *att* sites, it uses the same 7-bp staggered cut as with normal *att*B, thus generating 7 bp of heterozygous region in its direct products. This heterozygote will segregate as the DNA replicates, generating two kinds of progeny. In the normal case, one might expect these two kinds of progeny to be recovered with equal probability. However, the only secondary *att* site for which several independent insertions have been ana-

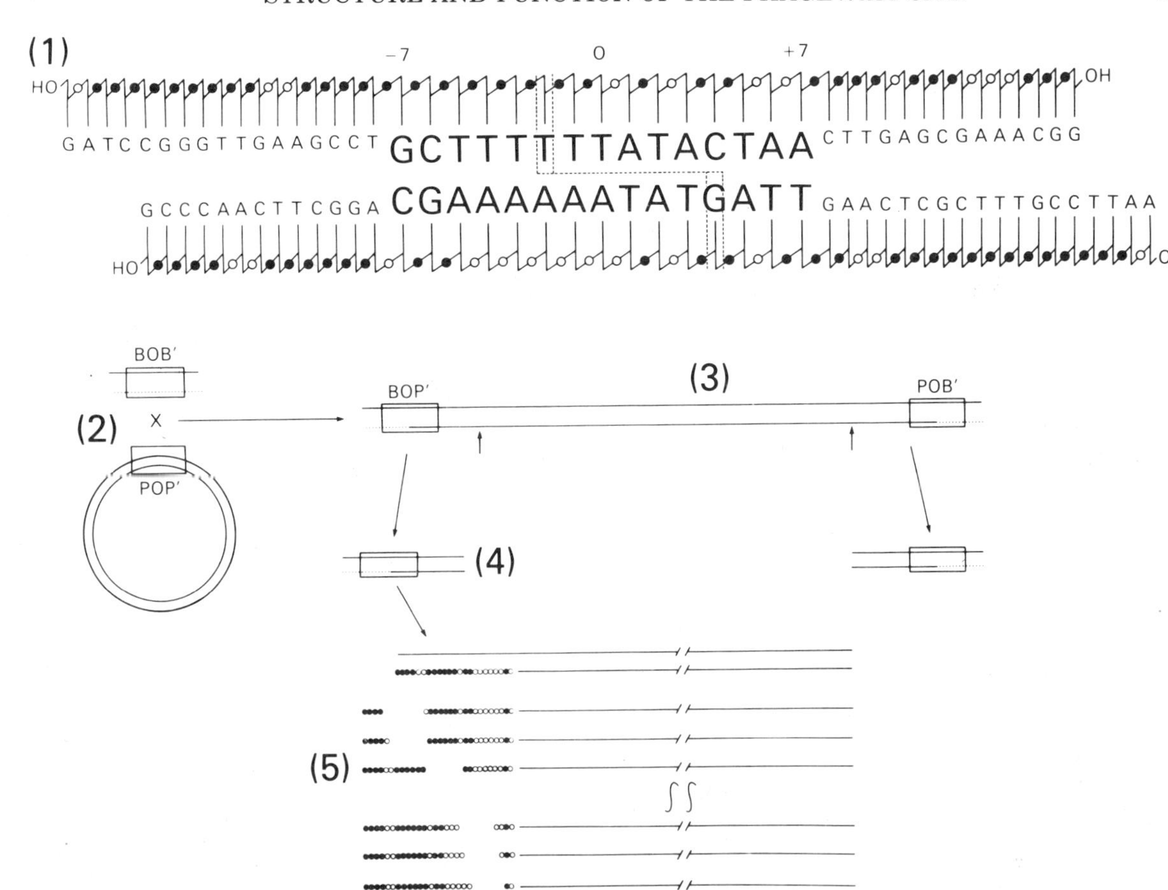

Figure 7. Scheme for the determination of the crossover point. A *Bam*HI-R1 fragment of pBA291 (M. Mizuuchi, in prep.) containing *att*B was prepared with one of the strands unlabeled and the other strand internally labeled with ^{32}P. The labeling was carried out using a purified strand of a larger fragment containing *att*B, an oligonucleotide primer, and the large fragment of DNA polymerase I. One of the four nucleoside triphosphates had ^{32}P in the α position; the fraction of ^{32}P atoms at the labeled position was about 0.1. This labeled fragment was treated with *Bam*HI and R1, and the *att*B fragment was purified. (*1*) (○) the phosphates on the 5′ side of deoxyadenosine. (*2*) The labeled *att*B fragment was recombined in vitro with the supercoiled form of *att*P plasmid DNA (pPA259 [Mizuuchi and Mizuuchi 1980]). The reaction conditions have been published (Mizuuchi and Mizuuchi 1980). (Purified Int and purified host integration factor used in this experiment were the generous gifts of H. Nash). (*3*) The reaction generated a linear DNA. (*4*) This was treated with *Bam*HI, *Hin*fI, and alkaline phosphatase, and the fragment carrying BOP′ (*att*L) or POB′ (*att*R) was purified by electrophoresis. (*5*) The fragment was stored for two weeks to allow the disintegration of ^{32}P atoms to generate cleavages in the DNA. The figure shows the example of the *att*L fragment derived from recombination, starting with the *att*B fragment labeled on the lower strand with [α-^{32}P]dATP. Each disintegration generates two fragments, of which the shorter carries the B arm and the longer carries the P′ arm. These series of fragments were separated on a denaturing 8% acrylamide gel, and an autoradiograph was prepared. A part of the results is shown in Fig. 8. The two possible crossover points on each strand that are equally consistent with the results are indicated with the dotted line in panel *1*.

lyzed is *galT*, and in this case there is approximately a tenfold difference in the frequency of the two progeny types (see results above). We assume that the existence of base changes in the substrate for the reaction biases the recovery of progeny inheriting one strand over those inheriting the other strand.

CONCLUSIONS

From both the structural and functional points of view, it is convenient to consider two aspects of integrative recombination. One of these, the crossover event, is symmetrical for the two recombining partners. It takes place within a 15-bp sequence that is identical in all four of the *att* sites. Determination of the fate of the core phosphorous atoms during recombination shows that it involves staggered cuts that produce 5′-ended overlaps of either 5 bp or 7 bp on the two DNA strands. Kikuchi and Nash (1979) and Nash et al. (this volume) have proposed a molecular model of the crossover event in which the two participating core regions are symmetrical. These chemical results are consistent with genetic analyses of the crossover, and the latter further suggest a spacing of 7 bp for the staggered cuts. The genetic analyses also indicate that the sequence of the core region may be important in two different ways. On the one hand, there is the degree of homology between the recombining partners; on the other hand, the degree of match to the wild-type core sequence may also be important.

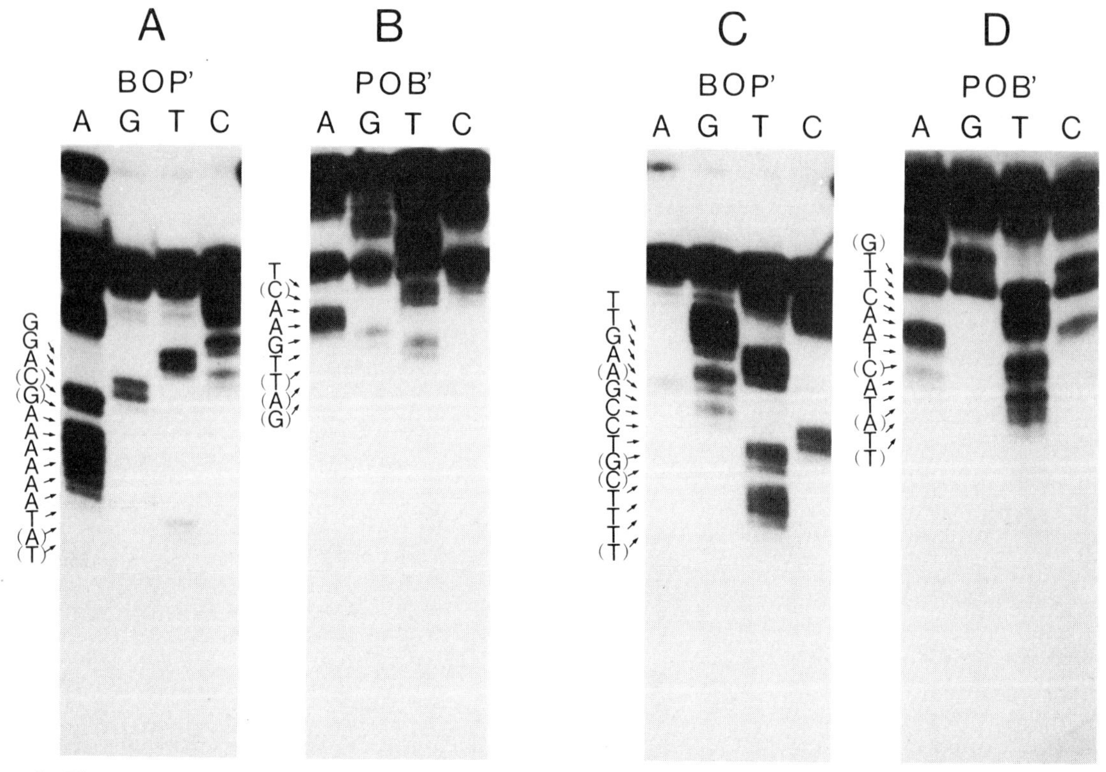

Figure 8. The crossover takes place with 5′ staggered cuts at unique points in the core. Some of the autoradiographs prepared as described in Fig. 7 are shown, with their interpretations. This part of the gel contained the series of ^{32}P-decay fragments that contained the P arm (POB′) or P′ arm (BOP′). Positions are indicated according to the base of the nucleoside whose 5′-phosphate disintegrated to generate the band in question. The position in parentheses is that of the hypothetical, longest-possible fragment in each sample that no longer carries any ^{32}P atom. (In other words, this is the shortest ^{32}P-decay fragment that carries the P arm or P′ arm in each sample.) Each band in the gel is accompanied by one or two weaker satellite bands with faster mobility. The interpretation of such bands will be discussed elsewhere (K. Mizuuchi, in prep.).

There is another aspect of integrative recombination in which the two recombining partners are not symmetrical. The minimal phage *att* site, which is 240 bp, is considerably larger than the minimal bacterial *att* site, which is 23 bp or less. The phage *att* site must be on a supercoiled molecule, whereas under certain in vitro conditions, the bacterial *att* site can be on a short, linear fragment. The phage *att* site is also more complex in its interaction with Int, having three binding sites in addition to that at the core region. The role of these Int-binding sites and the mechanistic basis for this asymmetry in the recombining partners has yet to be determined.

ACKNOWLEDGMENTS

We thank Howard Nash for purified Int protein and host factors, Sankar Adhya for help with mapping of *gal* mutants, Eric Johnson for technical assistance, and Susan Friedrich for preparation of the manuscript. A portion of this work was supported by grants AI-13544 from the National Institutes of Health and 1-543 from the National Foundation, March of Dimes. P-L., H. is the recipient of a National Institutes of Health Postdoctoral Research Fellowship (GM-07046) and W. R. is the recipient of a Brown University Graduate Fellowship. A. L. is a Faculty Research Associate of the American Cancer Society.

REFERENCES

Bidwell, K. and A. Landy. 1979. Structural features of λ site-specific recombination at a secondary *att* site in *gal*T. *Cell* **16:** 397.

Christie, G. E. and T. Platt. 1979. A secondary attachment site for bacteriophage lambda in *trp*C of *E. coli. Cell* **16:** 407.

Csordás-Tóth, E., I. Boros, and P. Venetianer. 1979. Nucleotide sequence of a secondary attachment site for bacteriophage lambda on the *Escherichia coli* chromosome. *Nucleic Acids Res.* **7:** 1335.

D'Andrea, A. and W. Haseltine. 1978. Sequence specific cleavage of DNA by the antitumor antibiotics neocarzinostatin and bleomycin. *Proc. Natl. Acad. Sci.* **75:** 3608.

Galas, D. J. and A. Schmitz. 1978. DNase footprinting: A simple method for the detection of protein-DNA binding specificity. *Nucleic Acids Res.* **5:** 3157.

Gottesman, M. E. and R. A. Weisberg. 1971. Prophage insertion and excision. In *The bacteriophage lambda* (ed. A. D. Hershey), p. 113. Cold Spring Harbor Laboratory, Cold Spring Harbor, New York.

Hatayama, T., I. Goldberg, M. Takeshita, and A. Grollman. 1978. Nucleotide specificity in DNA scission by neocarzinostatin. *Proc. Natl. Acad. Sci.* **75:** 3603.

Hsu, P-L., W. Ross, and A. Landy. 1980. The λ phage *att* site: Functional limits and interaction with Int protein. *Nature* **285:** 85.

Kikuchi, Y. and H. Nash. 1978. Purification and properties of a protein involved in genetic recombination: The bacteriophage λ *int* gene product. *J. Biol. Chem.* **253:** 7149.

———. 1979. Nicking-closing activity associated with bacteriophage λ *int* gene product. *Proc. Natl. Acad. Sci.* **76:** 3760.

LANDY, A. and W. ROSS. 1977. Viral integration and excision: Structure of the lambda *att* sites. *Science* **197:** 1147.

LANDY, A., R. H. HOESS, K. BIDWELL, and W. ROSS. 1979. Site-specific recombination in bacteriophage λ—Structural features of recombining sites. *Cold Spring Harbor Symp. Quant. Biol.* **43:** 1089.

MAXAM, A. and W. GILBERT. 1977. A new method for sequencing DNA. *Proc. Natl. Acad. Sci.* **74:** 560.

MILLER, H. I., A. KIKUCHI, H. NASH, R. A. WEISBERG, and D. I. FRIEDMAN. 1979. Site-specific recombination of bacteriophage λ: The role of host gene products. *Cold Spring Harbor Symp. Quant. Biol.* **43:** 1121.

MIZUUCHI, M., and K. MIZUUCHI. 1980. Integrative recombination of bacteriophage λ: Extent of the DNA sequence involved in attachment site function. *Proc. Natl. Acad. Sci.* **77:** 3220.

MIZUUCHI, K., M. GELLERT, and H. NASH. 1978. Involvement of supertwisted DNA in integrative recombination of bacteriophage lambda. *J. Mol. Biol.* **121:** 375.

MIZUUCHI, K., M. GELLERT, R. A. WEISBERG, and H. NASH. 1980. Catenation and supercoiling in the products of *in vitro* λ integrative recombination. *J. Mol. Biol.* **141:** 485.

NASH, H. 1977. Integration and excision of bacteriophage λ. *Curr. Top. Microbiol. Immunol.* **78:** 171.

PINKHAM, J. L., T. PLATT, L. W. ENQUIST, and R. WEISBERG. 1980. The secondary attachment site for bacteriophage λ in the *proA/B* gene of *E. coli. J. Mol. Biol.* (in press).

ROSS, W., A. LANDY, Y. KIKUCHI, and H. NASH. 1979. Interaction of Int protein with specific sites on λ *att* DNA. *Cell* **18:** 297.

SHIMADA, K., R. A. WEISBERG, and M. E. GOTTESMAN. 1972. Prophage lambda at unusual chromosomal locations. I. Location of the secondary attachment sites and the properties of the lysogens. *J. Mol. Biol.* **63:** 483.

———. 1973. Prophage lambda at unusual chromosomal locations. II. Mutations induced by bacteriophage lambda in *Escherichia coli* K12. *J. Mol. Biol.* **80:** 297.

SIX, E. W. 1963. Affinity of p2 *rd 1* for prophage sites on the chromosome of *E. coli* strain C. *Virology* **10:** 375.

———. 1966. Specificity of P2 for prophage site I on the chromosome of *E. coli* strain C. *Virology* **29:** 106.

SHULMAN, M. and M. GOTTESMAN. 1973. Attachment site mutants of bacteriophage lambda. *J. Mol. Biol.* **81:** 461.

WEISBERG, R. A. and J. A. GALLANT. 1967. Dual function of the λ prophage repressor. *J. Mol. Biol.* **25:** 537.

Regulation of the Integration-Excision Reaction by Bacteriophage λ

H. I. MILLER,* J. ABRAHAM,* M. BENEDIK,† A. CAMPBELL,† D. COURT,‡ H. ECHOLS,* R. FISCHER,* J. M. GALINDO,§ G. GUARNEROS,§ T. HERNANDEZ,§ D. MASCARENHAS,† C. MONTANEZ,§ D. SCHINDLER,* U. SCHMEISSNER,‡ AND L. SOSA§

** University of California, Berkeley, California 94720; † Stanford University, Stanford, California 94305; ‡ National Cancer Institute, Bethesda, Maryland 20205; § Centro de Investigacion, Mexico City, Mexico*

Bacteriophage λ regulates the integration-excision reaction as a crucial aspect of controlled development along the two viral pathways, lysogenic and productive. The events essential for the lysogenic response are complete repression of productive functions and integration of the viral DNA. The events essential for the productive response are an ordered expression of replication and maturation functions and, for an induced prophage, excision of the viral DNA. The choice between the two pathways available to the phage can be considered in terms of an initial regulatory partition toward one pathway or the other, stabilized in turn by subsequent regulatory events. The *c*II- and *c*III-gene products (cII and cIII proteins) of λ provide for the first stage of the lysogenic response, turning on production of the maintenance repressor cI and the integration protein Int and delaying the expression of late lytic functions. The cI protein stabilizes the lysogenic pathway by a switch-off of the productive pathway regulated by the λ *cro* and *Q* products (Cro and Q); an excess of Int over the excision protein Xis executes the recombinational switch toward insertion (reviewed by Herskowitz 1973; Weisberg et al. 1977; Echols 1979, 1980; Fig. 1).

The choice of lysogenic or productive growth involves the host as well as the phage. The cII protein is probably the primary partition function on which other regulatory influences act. These include the phage protein cIII and host proteins Hfl and HimA. Since HimA is also an accessory factor for the activity of Int in integrative recombination, there exists dual regulation of integrative recombination at the level of activity and synthesis of Int.

The basis for directional control of integration and excision is the different catalytic requirements for the forward and reverse reactions. As diagramed below, the only phage-specified protein required for integrative recombination is Int, whereas excisive recombination requires the viral proteins Int and Xis (Weisberg et al. 1977; Nash 1978; Echols 1980).

$$\underset{(attP)}{P{\bullet}P'} + \underset{(attB)}{B{\bullet}B'} \underset{\text{Int/Xis}}{\overset{\text{Int}}{\rightleftharpoons}} \underset{(attL)}{B{\bullet}P'} + \underset{(attR)}{P{\bullet}B'}$$

Thus, differential synthesis of Int with respect to Xis might be expected to favor integration. The phage cII protein provides this regulation by activating the promoter p_I (Fig. 1). Because cII turns on both the *int* and *c*I genes, the critical reactions of the lysogenic response, integration and repression, are coordinated.

In addition to positive regulation of *int* by cII, differential expression of the *int* and *xis* genes also occurs through an additional regulatory element, termed *sib*, on the opposite side of the *int* gene and phage attachment site from the p_I promoter. The presence of *sib* depresses *int*-gene expression from the N-controlled p_L promoter (Fig. 1). This remarkable *cis*-acting distal regulatory element presumably serves to prevent unwanted integration in a cell destined for the productive response. Because *sib* is separated from *int* in the integrated prophage, the induced prophage directs a different regulatory response in which Int as well as Xis can be synthesized from the p_L transcription unit; thus, prophage excision can proceed with maximal efficiency.

In this summary paper, we first review the evidence for control of the integration-excision reaction by differential expression of the *int* and *xis* genes from two promoters, p_I and p_L. We discuss the regulatory information that may be inferred from the DNA sequence of the p_I region and present evidence for the regulatory role of HimA. Finally, we consider some general principles that may be applicable to other developmental systems and transposition events.

Control of Integration-Excision by Differential Expression of the *int* and *xis* Genes

General principles inferred from genetics and protein synthesis in vivo. The *int-xis* region of λ DNA undergoes regulated transcription from two promoter sites, p_I and p_L (Fig. 1). Both the *int* and *xis* genes can be expressed fom the p_L promoter under positive regulation by the *N*-gene protein (N), which prevents termination of the p_L RNA (Adhya et al. 1974; Franklin 1974). The *int* gene can be expressed independently of *xis* from the p_I promoter under positive regulation by cII, which provides for initiation of new RNA chains (see below). In this section, we describe the control of the *int* and *xis* genes from these two promoters.

Two types of experiments have been used to study the regulation of Int and Xis synthesis: (1) biological activity by recombination frequency between different attachment sites (Chung and Echols 1977; Court et al. 1977; Enquist et al. 1979; Guarneros and Galindo 1979)

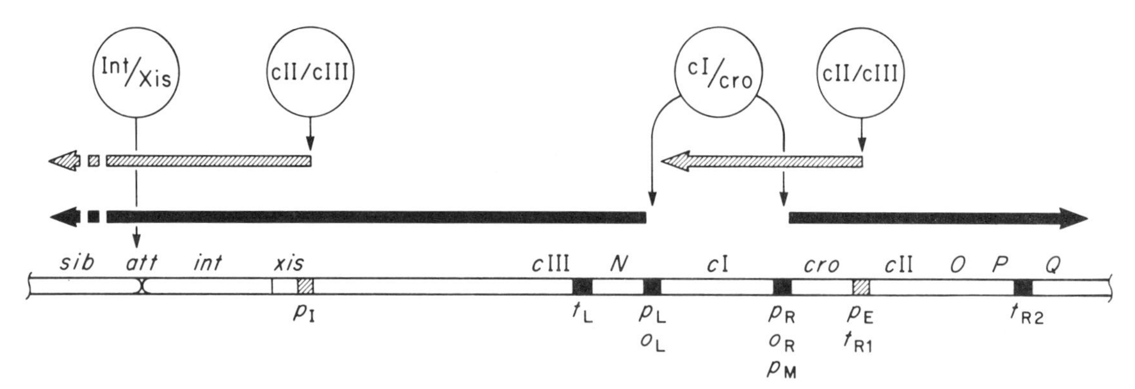

Figure 1. Regulation of transcription during the early stage of λ development. Immediately after infection, transcription from p_L and p_R is terminated mainly at t_L and t_{R1} with some RNA chains continuing to t_{R2}. The N protein eliminates these termination events and provides for transcription of the remainder of the early gene region (solid arrows). The cII and cIII proteins partition λ development toward the lysogenic response by stimulating RNA synthesis from p_E and p_I (hatched arrows) and by delaying the expression of late lytic functions. The cI protein binds to the operators o_L and o_R, shutting off the early λ transcripts from p_L and p_R; cI also maintains its own further synthesis by regulating the p_M promoter for the *c*I gene. The alternative late stage of productive development utilizes the Cro and Q proteins; Q turns on the genes for head, tail, and lysis proteins, and Cro turns off transcription of the *c*I gene from p_M and reduces the synthesis of the p_L and p_R transcripts for early proteins.

and (2) amounts of radioactive Int and Xis after gel electrophoresis (Katzir et al. 1976; Court et al. 1977; C. Epp et al., pers. comm.; L. Sosa and G. Guarneros; D. Schindler and H. Echols; both unpubl.). Neither experiment is ideal because the biological experiments are very indirect and the biochemical ones generally use a UV-irradiated host, which perturbs the physiology. However, both types of experiments have yielded similar conclusions, and therefore the general principles derived are likely to be sound. We will present one set of data from each type of experiment that is consistent with past observations and sets forth some new facts.

cII-dependent expression of the* int *gene. The p_L transcript extends through the *int-xis* region of λ DNA (Adhya et al. 1974) and was believed to be solely responsible for all *int*-gene expression (Signer 1970). However, if transcription of *int* and *xis* from p_L is eliminated, cII-dependent expression of *int* remains (Chung and Echols 1977; Court et al. 1977; Oppenheim and Oppenheim 1978) (Table 1, lines 1–4). Thus, protein cII stimulates Int synthesis from a promoter, p_I, independent of p_L (Shimada and Campbell 1974). Protein cII does not stimulate the activity of the *xis* gene (Chung and Echols 1977; Enquist et al. 1979). Moreover, *int*-constitutive (cII-independent) mutations have been isolated that are in the p_I promoter; these *int-c* mutations confer an Xis$^-$ phenotype (Shimada and Campbell 1974, see below). Therefore, cII is a positive regulator for Int but not Xis.

cII-independent expression of the* int *gene—the* sib *effect. The DNA of phage λ to the left of the attachment site and adjacent to the *int* gene is called the *b* region. Several deletions have been isolated that remove different parts of this 8-kb region; one extensively studied deletion, *b*2, has two distinct effects on site-specific recombination. The deletion alters part of the structural site (*att*P) required for the integrative recombination event and also eliminates an element that regulates the expression of the *int* gene (Lehman 1974; Roehrdanz and Dove 1977; Guarneros and Galindo 1979; C. Epp et al., pers. comm.).

The *b*2 deletion allows expression of Int activity under *c*II$^-$ conditions (Guarneros and Galindo 1979) (Table 1, lines 2 and 6). This implies the existence of an inhibitory function in the *b*2 region (termed *sib*). Since cII is essential for the expression of the p_I promoter, the *sib*-mediated inhibition of Int presumably acts on the expression of the distal promoter, p_L. The analysis of a set of deletions in *b* for their ability to cause inhibition of Int activity has revealed that *sib* is within the 250-bp DNA segment adjacent to the attachment site (J. M.

Table 1. Regulatory Elements for the *int-xis* Region

N site	cII protein	Sib site	Promoter for *int*	Int activity
+	+	+	$p_I + p_L$	1000
+	−	+	p_L	1
−	+	+	p_I	100
−	−	+	none	1
+	+	−	$p_I + p_L$	1000
+	−	−	p_L	1000
−	+	−	p_I	10

Int activity is expressed in relative numbers, based on the ability of infecting phages of differing genotypes to promote excision of a prophage flanked by the phage attachment site PP′ and the host attachment site BB′ (Guarneros and Galindo 1979). The moi was 10 in all cases. The promoter column lists the promoter(s) believed to be active for *int* expression under the conditions of the left three columns. The columns labeled N site, cII protein, and Sib site indicate the genotype of the helper phage: (+) wild-type allele; (−) mutant allele. An inactive N site is achieved by the mutation *nut*L, which prevents transcription of all λ p_L genes except *N* (Salstrom and Szybalski 1978); an inactive cII protein is achieved by the mutation *c*II41, which does not activate p_I expression of *int* (Katzir et al. 1976); and an inactive Sib site is achieved by the mutation *b*2, a 12% deletion that eliminates *sib* (Guarneros and Galindo 1979). Because the assay measures phage yield rather than excision directly, the numbers should be taken as qualitative measures of expression rather than quantitative measures of Int production. This table is taken from the unpublished work of G. Guarneros and J. M. Galindo.

Galindo and G. Guarneros, unpubl.). Recently, three recombinationally distinct point mutations (sib^-) have been isolated that eliminate the *b*-region inhibitory effect. These mutations map to the left of *att* and within this 250-bp segment (C. Montanez and G. Guarneros, unpubl.).

If *sib* is tested in *cis* or in *trans* to a normal *int* gene, inhibition of Int occurs only in the *cis* configuration (Guarneros and Galindo 1979). This result suggests that *sib* is a site rather than a diffusible substance. The possibility that *sib* is a *cis*-acting protein has not been ruled out, but examination of the entire nucleotide sequence for the *sib* and *att* regions reveals that the largest possible polypeptide coded by this region could contain only 27 amino acids (Hsu et al. 1980).

The *sib*-mediated inhibition is highly specific for the *int* gene; there appears to be little or no effect on the synthesis of Xis (Fig. 2). However, if the *int* and *xis* genes are deleted and *sib* is moved adjacent to the gene for the p_L protein Ea22, *sib* mediates inhibition of Ea22 production (D. Schindler and H. Echols, unpubl.).

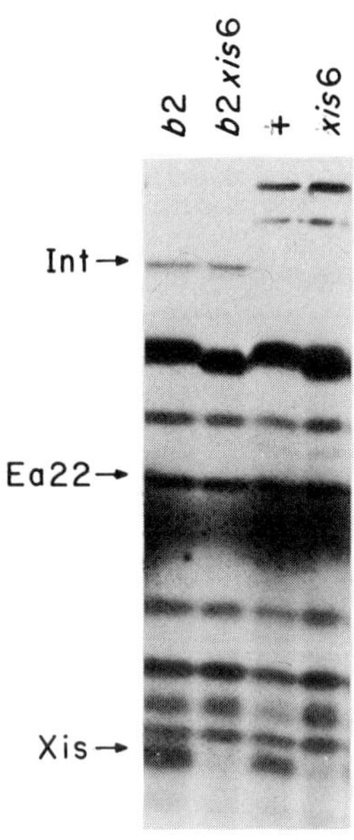

Figure 2. Effect of *b* region on Int and Xis production. *E. coli* strain 159 ($uvrA^-$) was irradiated with UV light (1200 J/m^2) to inactivate host-protein synthesis (Ptashne 1967; Hendrix 1971). The cells (in 200 μl) were infected with the designated phage at about 10 phage/cell; after a 15-min adsorption period, the cells were labeled with 5 μCi of ^{14}C-arginine (in 1.2 ml growth medium) for 15 min at 39°C. The cells were chilled, collected by centrifugation, lysed by sodium dodecyl sulfate in electrophoresis buffer, and the proteins were fractionated by electrophoresis in a 15% acrylamide gel (Laemmli 1970). The phage used for infection are from left to right, λ*b2*cI857*S7*, λ*b2xis*am6cI857*S7*, λcI857*S7*, and λ*xis*am6cI857*S7*. The migration positions are indicated for Int, Xis, and an additional early protein (Ea22) from the p_L transcript. The two high-molecular-weight proteins in the right two lanes are coded by the *b* region. The experiments were done under conditions in which cII-activated synthesis of Int from p_I does not occur. The *xis6* strains also carry a "silent mutation" in the β protein that migrates below Int (C. Epp and M. Pearson, pers. comm.).

The following posttranscriptional and transcriptional mechanisms have been suggested to explain the *sib* phenomenon: The ability of the p_L transcript of *int* to function in protein synthesis might be affected by *sib* nucleotides through alterations in degradation rate or in secondary structure; *sib* might also inhibit leftward transcription of the *int* gene by facilitating opposing rightward transcription from a promoter in the *b* region (Guarneros and Galindo 1979). (A rightward promoter in the proper location has been observed in transcription studies in vitro [Kravchenko et al. 1979].) If convergent transcription occurs, one would expect to see some *sib*-mediated inhibition of *int* expression from p_I as well as p_L. Contrary to this expectation, b^+ does not appear to inhibit Int production from p_I (Table 1, lines 3 and 7). Thus, unless some other specific transcriptional mechanism exists, the most likely basis for the *sib* effect appears to be translation or processing of *int* mRNA.

Other possible controls on Int activity. In addition to the *sib* inhibition of Int activity, two other forms of posttranscriptional control of Int may also exist. Both of these involve the expression of the *xis* gene. Int activity can be measured in vitro by recombination between DNA molecules that carry *att*P and *att*B. This activity is inhibited when Xis is added to the integrative recombination reaction either in crude extracts (Nash 1975a) or in partially purified form (K. Abremski, pers. comm.). A similar effect has been inferred from experiments in vivo (Nash 1975b; G. Guarneros and J. M. Galindo, unpubl.).

A second possible mode of posttranscriptional control has been suggested by the nucleotide sequence analysis of the *int* and *xis* genes. The end of the *xis*-coding region overlaps eight aminoterminal codons of the *int* gene (Davies 1980; Hoess et al. 1980). When transcription occurs from the p_L promoter, both the *int* and *xis* genes are part of the same polycistronic mRNA molecule. Thus, ribosomes translating *xis* sequences might inhibit the ability of other ribosomes to start translation at the initiator codon for the *int* gene. To examine this possibility, the biological activity of Int has been compared after infection by λcII^- phage carrying *xis*-missense or chain-termination mutations (nonsense and frameshift). All three chain-termination mutations tested increased Int-activity levels five to tenfold compared with the missense mutation, a result consistent with the idea that *xis* translation may inhibit *int* translation (D. Court et al., unpubl.).

Genetic Structure of the p_I Region: Inferences about the Mechanism of cII Activity

Three lines of genetic evidence have indicated that p_I is close to or overlaps the *xis* gene. First, *xis* is not activated by cII (Chung and Echols 1977; Enquist et al. 1979). Second, the *int-c* point mutations that allow a high rate of *int* expression in the absence of cII produce an Xis^- phenotype (Shimada and Campbell 1974); since *int-c* mutations generate an active promoter site in vitro, these mutations are likely candidates for base changes in the p_I sequence itself (Fischer et al. 1979). Third, among a set of λ*trp* fusion strains, one eliminates cII activation of p_I but leaves an intact *xis* gene (Heffernan et al. 1979).

The relationship between the *int* and *xis* genes and the p_I promoter has been clarified by a determination of the DNA sequence of the p_I region and the base changes of *int-c* and λ*trp* regulatory mutations (Abraham et al. 1980; Hoess et al. 1980). The results of this work are summarized interpretively in Figure 3. The left boxed sequence is a region of DNA with striking homology (11/15 bases) with a "consensus" (computer-generated) interaction site for RNA polymerase (Scherer et al. 1978); the "Pribnow box" subsequence is indicated by heavy lines (Pribnow 1975). The *int-c* mutations convert the interaction sequence into a "better" one for RNA polymerase; in particular, the mutationally produced A base in position 2 of the Pribnow box is one of the most conserved bases of promoter sites. Thus, we identify the left boxed region of Figure 3 as a likely candidate for a cII-activated interaction sequence for RNA polymerase.

Because cII acts at p_I and p_E, a common sequence essential for cII action should exist in both promoter regions. The right boxed sequence of Figure 3 exhibits notable homology (11/14 bases) between p_I and a region of the previously determined p_E sequence (Rosenberg et al. 1978; Schwarz et al. 1978; Schmeissner et al. 1980; Wulff et al. 1980). The deletion *trp*-λΔ303 removes most of the right boxed sequence and is cII-insensitive; the deletion *trp*-λΔ29, which leaves a few more λ nucleotides, is subject to cII activation. The homologous sequence in p_E has been identified also with cII-insensitive mutations (Wulff et al. 1980). Thus, we consider the right boxed region a likely candidate for a site essential for the activity of cII, perhaps a cII-binding site.

From the sequence analysis and properties of the regulatory mutations, we can conclude that cII probably acts to promote initiation of the p_I RNA by RNA polymerase, most likely by allowing the enzyme to use the interaction site noted in Figure 3. If this mechanism is correct, the RNA chain from p_I should begin some 5 bp downstream from the left boxed sequence (Fig. 3). To verify this prediction, we have analyzed the RNA products of transcription. For an *int-c* DNA template in vitro, two starts are found at adjacent bases, with the major start indicated in Figure 3 (J. Abraham and H. Echols, unpubl.). Data from transcription experiments in vivo indicate that the sequence of the cII-activated RNA is identical to the *int-c* RNA (U. Schmeissner, unpubl.).

From the sequence change associated with a nonsense mutation in *xis*, the initiation codon for Xis translation is within the boxed interaction sequence (Abraham et al. 1980; Hoess et al. 1980) (Fig. 3). Therefore, we conclude that cII provides for differential synthesis of Int with respect to Xis because the cII-activated RNA chain does not contain the entire coding sequence for Xis.

From the information described above, we can explain in a satisfactory manner the way in which cII provides for differential expression of the *int-xis* region to favor the lysogenic response. The "choice" for location of the p_I promoter is not so obvious; we will consider this point further in the final section.

Role of the *himA* Gene in Regulation of Integration-Excision

A successful lysogenic response by phage λ requires both virus- and host-encoded proteins. The virus provides the proteins cI and Int, and the bacterium supplies the proteins HimA and Hfl (and perhaps others as well). An efficient system for host-virus regulation of lysogeny should be able to couple the synthesis of Int and cI to intracellular sensors of host physiology; one such host regulatory element is probably HimA. As discussed above, the λ cII and cIII proteins control the viral contribution to the lysogenic response by activating coordinate transcription of the *int* and *cI* genes. The *Escherichia coli* HimA protein is essential to the lysogenic response at two levels: (1) as a host component of integrative recombination and (2) as a regulator for synthesis of both Int and cI.

The* himA *gene. Bacterial strains carrying mutations that define the *himA* gene were isolated by their inability to support λ integration (Miller and Friedman 1977). Experiments with strains carrying *himA* point mutations indicated that the integration defect was due to the absence of an active host protein required for integrative recombination both in vivo and in vitro (Miller et al.

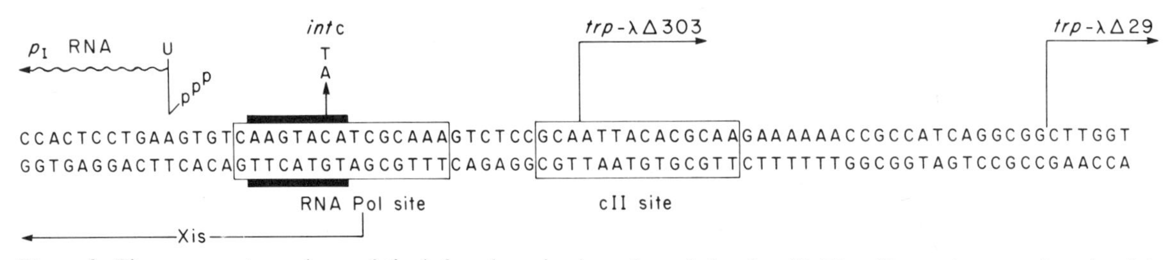

Figure 3. The p_I promoter region and the inferred mechanism of regulation by cII. The cII protein recognizes the right boxed sequence ("cII site") and promotes interaction of RNA polymerase with the left boxed sequence ("RNA Pol site"). The cII-activated RNA chain begins with UTP at adjacent bases, with the major start at the point noted above the sequence. Positive regulation of Int but not Xis ensues because the coding sequence for Xis begins with the ATG indicated on the figure, and the cII-stimulated RNA thus lacks the translation start and other aminoterminal codons for Xis. The base change of an *int-c* mutation is shown above the figure, as is the beginning point of two deletions, the cII-insensitive *trp*-λΔ303 and the cII-sensitive *trp*-λΔ29.

1979; Miller and Friedman 1980). The product of the *himA* gene has recently been identified by comparing the radioactive proteins synthesized from transducing λ phage carrying wild-type or mutant *himA* genes. The protein identified in this manner was found to be identical to a major component of the purified host factor that is active in integrative recombination in vitro (H. Miller and H. Nash, in prep.). Thus, the product of the *himA* gene directly participates in the integration reaction.

In addition to the inability to support phage λ integration and excision, *himA* mutants exhibit several other phenotypes. Bacteriophage Mu is incapable of lytic growth in *himA*⁻ strains, and the frequency of precise excision of transposable drug-resistance determinants is reduced. It is not known whether these phenomena are related to a defect in recombination. The *himA* mutations have no appreciable effect on bacterial growth, general recombination, or lytic growth of λ (Miller and Friedman 1980).

Recent experiments employing deletion mutations of the *himA* gene have focused attention on an additional characteristic of *himA*⁻ bacteria: λ exhibits a clear-plaque phenotype. This phenotype suggested that *himA* mutations affected either the establishment or maintenance of repression. However, no effect was found on *c*I-gene expression in lysogenic strains containing *himA* deletions (H. Miller, unpubl.). Thus, the effect appeared to be limited to the establishment of repression.

***Regulation of cI and Int synthesis by* himA.** To test the effects of a *himA* deletion on the synthesis of λ cI protein, we performed the experiments described in Figure 4. In these experiments, *himA*-deletion strains (otherwise isogenic) were infected with λ; at various times after infection, the cultures were pulse-labeled with [^{35}S]-methionine. The radioactive proteins were fractionated by two-dimensional gel electrophoresis (O'Farrell 1975), and the amount of radioactivity in the cI spot was determined. As shown in Figure 4, in wild-type strains cI synthesis begins between 5 and 10 minutes after infection and by 15 minutes has reached a very high rate. In the *himA*-deletion strain, cI is produced at a greatly reduced rate, five- to tenfold less than in the wild-type at 15 minutes. Thus, the clear-plaque phenotype can be explained by a reduced synthesis of cI from the establishment mode in *himA*⁻ strains, presumably reflecting lowered transcription from p_E.

Since the promoters for the *c*I gene and the *int* gene, p_E and p_I, exhibit similar regulatory controls, the synthesis of Int was also monitored. Figure 4 also shows the effect of a *himA* deletion on Int synthesis. Int synthesis exhibits kinetics identical to cI in wild-type cells. However, there is almost no detectable production of Int in the *himA*⁻ strain.

Similar experiments have been conducted with phage carrying mutationally induced constitutive promoters for both the *int* gene, *int-c* (Shimada and Campbell 1974), and the cI gene, *cin* (Wulff 1976). In these experiments, the constitutive promoters were expressed in both a wild-type and a *himA*-deleted host (data not shown); this indicates that the *himA* effect is specific for transcription from p_E and p_I.

The* himA *gene as a regulator of lysogeny. For the *himA* gene to function as a regulator of lysogeny under normal physiological conditions, the intracellular amount or activity of HimA must be limiting. Experiments to measure the amount of HimA in wild-type cells indicate that the level of HimA is low and dependent on growth conditions (H. Miller, unpubl.). Thus, physiological fluctuations in the concentration of HimA may be critical for the lysogenic response.

We have not determined the mechanism by which the *himA* gene regulates the transcription of the *int* and *c*I genes. HimA might control the synthesis or activity of cII or cIII or specify independently the capacity of RNA polymerase to use p_E and p_I.

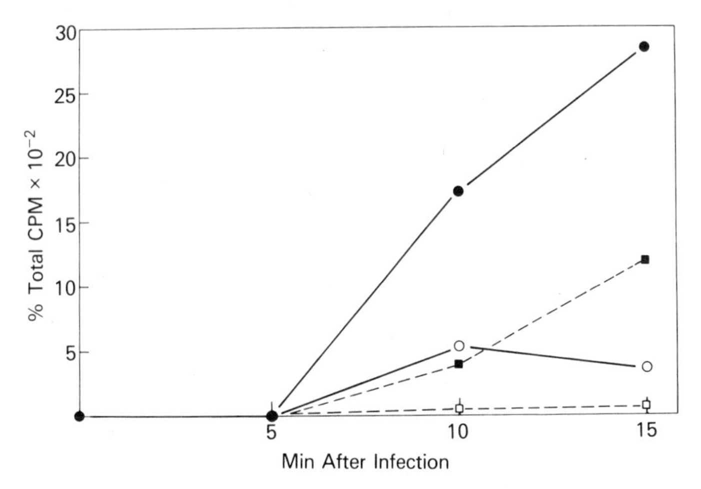

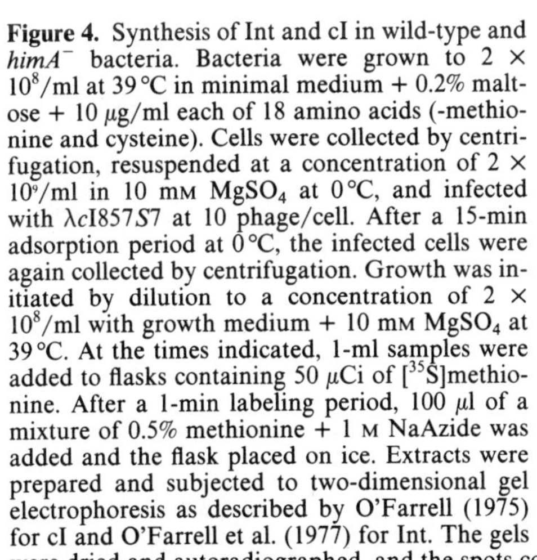
Figure 4. Synthesis of Int and cI in wild-type and *himA*⁻ bacteria. Bacteria were grown to 2 × 10^8/ml at 39 °C in minimal medium + 0.2% maltose + 10 μg/ml each of 18 amino acids (-methionine and cysteine). Cells were collected by centrifugation, resuspended at a concentration of 2 × 10^9/ml in 10 mM $MgSO_4$ at 0 °C, and infected with λ*c*I857*S*7 at 10 phage/cell. After a 15-min adsorption period at 0 °C, the infected cells were again collected by centrifugation. Growth was initiated by dilution to a concentration of 2 × 10^8/ml with growth medium + 10 mM $MgSO_4$ at 39 °C. At the times indicated, 1-ml samples were added to flasks containing 50 μCi of [^{35}S]methionine. After a 1-min labeling period, 100 μl of a mixture of 0.5% methionine + 1 M NaAzide was added and the flask placed on ice. Extracts were prepared and subjected to two-dimensional gel electrophoresis as described by O'Farrell (1975) for cI and O'Farrell et al. (1977) for Int. The gels were dried and autoradiographed, and the spots corresponding to Int or cI were excised from the gel and counted for radioactivity as described by Neidhardt et al. (1977). Total ^{35}S incorporation was determined by precipitation of a sample of the extract with trichloroacetic acid. The spot corresponding to Int was identified with purified Int (from H. Nash). The *c*I857 protein was identified by its synthesis after infection of UV-irradiated cells lysogenic for λ*c*I*ind*⁻ by λ*c*I857*cin*-1*S*7. (●) cI synthesis in wild-type cells; (○) cI in *himA*⁻; (■) Int in wild-type; (□) Int in *himA*⁻.

Some General Considerations for Developmental Pathways and Insertion Elements

Phage λ provides a simple and conveniently studied example of a developmental system in the sense that a single genome in a single cell executes a choice of temporal pathways. Phage λ is also an example of a self-directed transposable genetic element. As a model system for each biological phenomenon, λ will undoubtedly be the first example for which a molecular description will be possible. No one expects the λ model to provide a complete description of other systems; differences are already apparent. Nevertheless, many of the principles of λ biology are also likely to be widely applicable. In this section we consider the λ system in somewhat more general terms.

Structure and regulation of insertion elements. The genetic organization of the *int-xis* region is generally rationalizable with the role of integration and excision in λ biology. Being transcribable from two separate, differentially regulated promoters, the *int* gene effectively belongs to two separate operons. It is regulated coordinately with the *c*I gene during lysogenization and coordinately with *xis* following induction. The regulatory circuitry thus allows λ to exploit effectively the potential advantages of the differential catalytic requirements for insertion and excision.

The feature that was unanticipated on teleological grounds is the location of the p_I promoter, which might more logically be situated between *int* and *xis* rather than at the beginning of *xis*. Three possible (and not mutually exclusive) reasons for the actual location may be considered. The long leader sequence of the *int* message (comprising almost the entire *xis* gene) might play some role in the function or stabilization of that message. Or the overlap between the *int* and *xis* genes itself might serve some important function; because of that overlap, there is no available free space between *int* and *xis* in which a promoter could be located. Or the *att-int-xis* segment of λ DNA might be descended from a simple insertion sequence that became fused onto the rest of the viral genome during evolution; the p_I promoter might then be descended from this primordial promoter, so that its location reflects history rather than current function. At any rate, the end result is that the *att-int-xis* segment is so connected to the rest of λ that the recognition regions for λ-specific signals are located outside of the segment itself. This arrangement allows λ to control the insertion functions according to the dictates of its own life cycle, without necessitating evolutionary modification of the internal structure of the segment, which specifies interacting elements of the insertion system itself. If λ is not unique in this respect, we may anticipate that other site-specific recombination systems may likewise be controlled from outside, so that segments such as *att-int-xis* may be, in an evolutionary sense, modules that are capable of being plugged into different control systems (Campbell et al. 1979).

Developmental regulation. In general terms, one can think of two strategies for development: a master switch and a cascade. In the former, a single regulatory element determines the choice of a pathway; in the latter, a primary control signal provides for an initial partition between pathways, and later control functions stabilize the response along the favored pathway (e.g., Kauffman 1973; Echols 1980).

Phage λ development seems to be described most adequately by the cascade model. The cII protein appears to be the primary control element; a highly active cII function favors the lysogenic choice through efficient transcription of the *c*I and *int* genes. The activity of cII, the primary partition function, is dependent on the phage protein cIII and the bacterial proteins HimA and Hfl. Perhaps the viral cIII monitors the phage physiology (phage per cell or transcription rate per phage), and HimA and Hfl respond to a variety of physiological signals of the host environment.

The lysogenic pathway is stabilized by the λ cI and Int proteins. The competitive interaction between the maintenance repressor cI and the productive repressor Cro limits the eventual transcription pattern after infection to that characteristic of one pathway or the other; at sufficient levels of cI, the switch is thrown and maintained stably toward lysogeny (Echols 1980; Ptashne et al. 1980). Int turns the recombination switch to insertion.

For productive growth, cII is presumed to be less active than in the lysogenic response, and a deficiency of cI allows the transcription pattern to be stabilized toward lytic development by Cro. For a prophage, the recombination switch is thrown toward excision by Int and Xis production (from p_L); *sib* prevents unwanted Int production from p_L during productive growth after infection or excision of a prophage.

For λ, the basis for controlled choice of a pathway that becomes exclusive is achieving definition in biochemical terms. We hope that these studies will provide guidance for molecular hypotheses for developmentally controlled transcription and recombination switches in other creatures.

ACKNOWLEDGMENTS

We thank Mark Pearson and Arthur Landy for the communication of results prior to publication. This work was supported by National Institutes of Health grants GM-17078 and 5T32-GM0727 from the National Institute of General Medical Sciences and AI-08573 from the National Institute of Allergy and Infectious Diseases.

REFERENCES

Abraham, J., D. Mascarenhas, R. Fischer, M. Benedik, A. Campbell, and H. Echols. 1980. DNA sequence of regulatory region for integration gene of bacteriophage λ. *Proc. Natl. Acad. Sci.* **77:** 2477.

Adhya, S., M. Gottesman, and B. de Crombrugghe. 1974. Release of polarity in *Escherichia coli* by gene *N* of phage λ: Termination and antitermination of transcription. *Proc. Natl. Acad. Sci.* **77:** 2534.

Campbell, A., M. Benedik, and L. Heffernan. 1979. Viruses and inserting elements in chromosomal evolution. In *Concepts of the structure and function of DNA, chromatin and chromosomes* (ed. A. S. Dion), p. 51. Symposia Specialists, New York.

CHUNG, S. and H. ECHOLS. 1977. Positive regulation of integrative recombination by the *c*II and *c*III genes of bacteriophage λ. *Virology* **79:** 312.

COURT, D., S. ADHYA, H. NASH, and L. ENQUIST. 1977. The phage λ integration protein (Int) is subject to control by the *c*II and *c*III gene products. In *DNA insertion elements, plasmids and episomes* (ed. A. I. Bukhari et al.), p. 389. Cold Spring Harbor Laboratory, Cold Spring Harbor, New York.

DAVIES, R. W. 1980. DNA sequence of the *int-xis-*p_I region of the bacteriophage lambda; overlap of the *int* and *xis* gene. *Nucleic Acids Res.* **8:**1765.

ECHOLS, H. 1979. Bacteriophage and bacteria: Friend and foe. In *The bacteria* (ed. J. R. Sokatch and L. N. Ornston), vol. 7, p. 487. Academic Press, New York.

———. 1980. Bacteriophage λ development. In *Molecular genetics of development* (ed. T. Leighton and W. F. Loomis), p. 1. Academic Press, New York.

ENQUIST, L., A. HONIGMAN, S. L. HU, and W. SZYBALSKI. 1979. Expression of λ *int* gene function in ColE1 hybrid plasmid carrying the C fragment of bacteriophage λ. *Virology* **92:** 557.

FISCHER, R., Y. TAKEDA, and H. ECHOLS. 1979. Transcription of the *int* gene of bacteriophage λ. New RNA polymerase binding site and RNA start generated by *int*-constitutive mutations. *J. Mol. Biol.* **129:** 509.

FRANKLIN, N. C. 1974. Altered reading of genetic signals fused to the N operon of bacteriophage λ: Genetic evidence for modification of polymerase by the protein product of the N gene. *J. Mol. Biol.* **89:** 33.

GUARNEROS, G. and J. M. GALINDO. 1979. The regulation of integrative recombination by the *b*2 region and the *c*II gene of bacteriophage λ. *Virology* **95:** 119.

HEFFERNAN, L., M. BENEDIK, and A. CAMPBELL. 1979. Regulatory structure of the insertion region of bacteriophage λ. *Cold Spring Harbor Symp. Quant. Biol.* **43:** 1127.

HENDRIX, R. W. 1971. Identification of proteins coded in phage lambda. In *The bacteriophage lambda* (ed. A. D. Hershey), p. 355. Cold Spring Harbor Laboratory, Cold Spring Harbor, New York.

HERSKOWITZ, I. 1973. Control of gene expression in bacteriophage lambda. *Annu. Rev. Genet.* **7:** 289.

HOESS, R. H., C. FOELLER, K. BIDWELL, and A. LANDY. 1980. Site-specific recombination functions of bacteriophage λ: DNA sequence of regulatory regions and overlapping structural genes for Int and Xis. *Proc. Natl. Acad. Sci.* **77:** 2482.

HSU, P. L., W. ROSS, and A. LANDY. 1980. The λ phage *att* site: Functional limits and interaction with Int protein. *Nature* **285:** 85.

KATZIR, N., A. OPPENHEIM, M. BELFORT, and A. B. OPPENHEIM. 1976. Activation of the lambda *int* gene by the *c*II and *c*III gene products. *Virology* **74:** 324.

KAUFFMAN, S. A. 1973. Control circuits for determination and transdetermination. *Science* **181:** 310.

KRAVCHENKO, V. V., S. K. VASSILENKO, and M. A. GRACHEV. 1979. A rightward promoter to the left of the *att* site of λ phage DNA: Possible participant in site-specific recombination. *Gene* **7:** 181.

LAEMMLI, U. K. 1970. Cleavage of structural proteins during the assembly of the head of bacteriophage T4. *Nature* **227:** 680.

LEHMAN, J. F. 1974. λ Site-specific recombination: Local transcription and an inhibitor specified by the *b*2 region. *Mol. Gen. Genet.* **130:** 333.

MILLER, H. I. and D. I. FRIEDMAN. 1977. Isolation of *Escherichia coli* mutants unable to support lambda integrative recombination. In *DNA insertion elements, plasmids, and episomes* (ed. A. I. Bukhari et al.), p. 349. Cold Spring Harbor Laboratory, Cold Spring Harbor, New York.

———. 1980. An *Escherichia coli* gene product required for λ site-specific recombination. *Cell* **20:**711.

MILLER, H. I., A. KIKUCHI, H. A. NASH, R. A. WEISBERG, and D. I. FRIEDMAN. 1979. Site-specific recombination of bacteriophage lambda: The role of host gene products. *Cold Spring Harbor Symp. Quant. Biol.* **43:** 1121.

NASH, H. A. 1975a. Integrative recombination of bacteriophage lambda DNA in vitro. *Proc. Natl. Acad. Sci.* **72:** 1072.

———. 1975b. Integrative recombination in bacteriophage lambda: Analysis of recombinant DNA. *J. Mol. Biol.* **91:** 501.

———. 1978. Integration and excision of bacteriophage λ. *Curr. Top. Microbiol. Immunol.* **78:** 171.

NEIDHARDT, F. C., P. L. BLOCH, S. PEDERSON, and S. REEH. 1977. Chemical measurement of steady-state levels of ten aminoacyl-transfer ribonucleic acid synthetases in *Escherichia coli. J. Bacteriol.* **129:** 378.

O'FARRELL, P. H. 1975. High resolution two-dimensional electrophoresis of proteins. *J. Biol. Chem.* **250:** 4007.

O'FARRELL, P. Z., H. M. GOODMAN, and P. H. O'FARRELL. 1977. High resolution two-dimensional electrophoresis of basic as well as acidic proteins. *Cell* **12:** 1133.

OPPENHEIM, A. and A. B. OPPENHEIM. 1978. Regulation of the *int* gene of bacteriophage λ: Activation by the *c*II and *c*III gene products and the role of the p_I and p_L promoters. *Mol. Gen. Genet.* **165:** 39.

PRIBNOW, D. 1975. Nucleotide sequence of an RNA polymerase binding site at an early T7 promoter. *Proc. Natl. Acad. Sci.* **72:** 784.

PTASHNE, M. 1967. Isolation of the λ phage repressor. *Proc. Natl. Acad. Sci.* **57:** 306.

PTASHNE, M., A. JEFFREY, A. D. JOHNSON, R. MAURER, B. J. MEYER, C. D. PABO, T. M. ROBERTS, and R. T. SAUER. 1980. How the λ repressor and *cro* work. *Cell* **19:** 1.

ROEHRDANZ, R. L. and W. F. DOVE. 1977. A factor in the *b*2 region affecting site-specific recombinations in lambda. *Virology* **79:** 40.

ROSENBERG, M., D. COURT, H. SHIMATAKE, C. BRADY, and D. L. WULFF. 1978. The relationship between function and DNA sequence in an intercistronic regulatory region in phage λ. *Nature* **272:** 414.

SALSTROM, J. S. and W. SZYBALSKI. 1978. Coliphage λ *nut*L^-: A unique class of mutants defective in the site of gene *N* product utilization for antitermination of leftward transcription. *J. Mol. Biol.* **124:** 195.

SCHERER, G. E. F., M. D. WALKINSHAW, and S. ARNOTT. 1978. A computer aided oligonucleotide analysis provides a model sequence for RNA polymerase-promoter recognition in *E. coli. Nucleic Acids Res.* **5:** 3759.

SCHMEISSNER, V., D. COURT, H. SHIMATAKE, and M. ROSENBERG. 1980. Promoter for the establishment of repressor synthesis in bacteriophage λ. *Proc. Natl. Acad. Sci.* **77:**3191.

SCHWARZ, E., G. SCHERER, G. HOBOM, and H. KÖSSEL. 1978. Nucleotide sequence of *cro*, *c*II and part of the *O* gene in phage λ DNA. *Nature* **272:** 410.

SHIMADA, K. and A. CAMPBELL. 1974. Int-constitutive mutants of bacteriophage lambda. *Proc. Natl. Acad. Sci.* **71:** 237.

SIGNER, E. R. 1970. On the control of lysogeny in phage λ. *Virology* **40:** 624.

WEISBERG, R. A., S. GOTTESMAN, and M. E. GOTTESMAN. 1977. Bacteriophage λ: The lysogenic pathway. In *Comprehensive virology* (ed. H. Fraenkel-Conrat and R. R. Wagner), vol. 8, p. 197. Plenum Press, New York.

WULFF, D. L. 1976. Lambda *cin*-1, a new mutation which enhances lysogenization by bacteriophage lambda, and the genetic structure of the lambda cy region. *Genetics* **82:** 401.

WULFF, D. L., M. BEHER, S. IZUMI, J. BECK, M. MAHONEY, H. SHIMATAKE, C. BRADY, D. COURT, and M. ROSENBERG. 1980. Structure and function of the cy control region of bacteriophage lambda. *J. Mol. Biol.* **138:** 209.